Beginning Algebra

A TEXT/WORKBOOK

EIGHTH EDITION

Charles P. McKeague
CUESTA COLLEGE

BROOKS/COLE
CENGAGE Learning™

Australia • Brazil • Japan • Korea • Mexico • Singapore • Spain • United Kingdom • United States

***Beginning Algebra: A Text/Workbook*, Eighth Edition**
Charles P. McKeague

Mathematics Editor: Marc Bove
Publisher: Charlie Van Wagner
Consulting Editor: Richard T. Jones
Assistant Editor: Shaun Williams
Editorial Assistant: Mary De La Cruz
Media Editor: Maureen Ross
Marketing Manager: Joe Rogove
Marketing Assistant: Angela Kim
Marketing Communications Manager: Katy Malatesta
Project Manager, Editorial Production: Hal Humphrey
Art Director: Vernon Boes
Print Buyer: Judy Inouye
Permissions Editor: Bob Kauser
Production Service: XYZ Textbooks
Text Designer: Diane Beasley
Photo Researcher: Kathleen Olson
Cover Designer: Irene Morris
Cover Image: Pete McArthur
Compositor: Kelly Kaub/XYZ Textbooks
Illustrator: Kristina Chung/XYZ Textbooks

ISBN-13: 978-0-495-82681-1

ISBN-10: 0-495-82681-2

Brooks/Cole
10 Davis Drive
Belmont, CA 94002-3098
USA

Cengage Learning is a leading provider of customized learning solutions with office locations around the globe, including Singapore, the United Kingdom, Australia, Mexico, Brazil, and Japan. Locate your local office at: **www.cengage.com/international**

Cengage Learning products are represented in Canada by Nelson Education, Ltd.

For your course and learning solutions, visit **www.academic.cengage.com**

To learn more about Brooks/Cole, visit **www.cengage.com/brookscole**
Purchase any of our products at your local college store or at our preferred online store **www.ichapters.com**

Printed in the United States of America
1 2 3 4 5 6 7 11 10 09 08

Brief Contents

Contents

1 The Basics 1

2 Linear Equations and Inequalities 111

3 Linear Equations and Inequalities in Two Variables 213

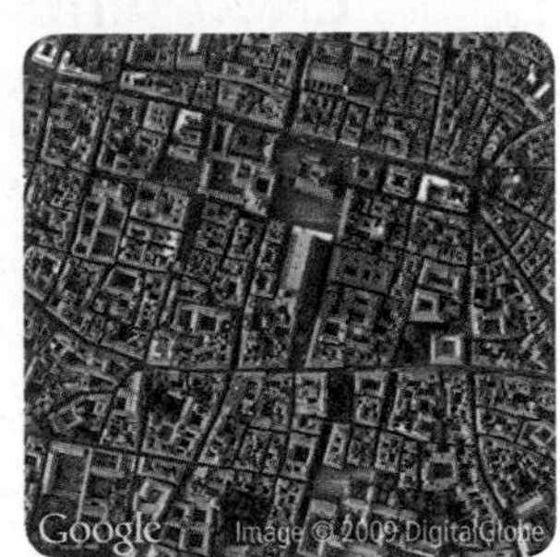

4 Systems of Linear Equations 293

5 Exponents and Polynomials 341

6 Factoring 423

7 Rational Expressions 489

8 Roots and Radicals 573

9 Quadratic Equations 633

Preface to the Instructor

I have a passion for teaching mathematics. That passion carries through to my textbooks. My goal is a textbook that is user-friendly for both students and instructors. For students, this book forms a bridge to intermediate algebra with clear, concise writing, continuous review, and interesting applications. For the instructor, I build features into the text that reinforce the habits and study skills we know will bring success to our students.

The eighth edition of *Beginning Algebra* builds upon these strengths. In this edition, renewal of the problem sets, new chapter openings, and a better connection of objectives to examples and problems were my priorities. This edition contains over 800 new problems. Most of these new problems have been written to shore up the midrange of exercises. The remaining new problems are new applications, most of which involve reading charts or graphs.

The Beginning Algebra Course as a Bridge to Further Success

Beginning algebra is a bridge course. The course and its syllabus bring the student to the level of ability required of college students, while getting them ready to make a successful start in intermediate algebra.

Our Proven Commitment to Student Success

After seven successful editions, we have developed several interlocking, proven features that will improve students' chances of success in the course. Here are some of the important success features of the book.

Chapter Pretest These are meant as a diagnostic test taken before the starting work in the chapter. Much of the material here is learned in the chapter so proficiency on the pretests is not necessary.

Getting Ready for Chapter X This is a set of problems from previous chapters that students need in order to be successful in the current chapter. These are review problems intended to reinforce the idea that all topics in the course are built on previous topics.

Practice Problems The Practice Problems, with their answers and solutions, are the key to moving students through the material in this book. We call it the EPAS system for Example, Practice Problem, Answer, Solution. Students begin by reading through the text, stopping after they have read through an example. Then they work the Practice Problem in the margin next to the example. When they are finished, they check the answers to the Practice Problem at the bottom of the page. If they have made a mistake, they try the problem again. If they still cannot get it right, they look at the Solutions to the Practice Problems in the back of the book. After working their way through the section in this manner, they are ready to start on the problems in the Problem Set. As an added bonus, there is a video lesson to accompany each section of the text.

Getting Ready for Class Just before each problem set is a list of four questions under the heading Getting Ready for Class. These problems require written responses from students and are to be done before students come to class. The answers can be found by reading the preceding section. These questions reinforce the importance of reading the section before coming to class.

Blueprint for Problem Solving Found in the main text, this feature is a detailed outline of steps required to successfully attempt application problems. Intended as a guide to problem solving in general, the blueprint takes the student through the solution process to various kinds of applications.

Organization of the Problem Sets

The problem sets begin with drill problems that are linked to the section objectives. This is where you will find most of the problems that are new to this edition. Following the drill problems we have the following categories of problems.

Applying the Concepts Students are always curious about how the mathematics they are learning can be applied, so we have included applied problems in most of the problem sets in the book and have labeled them to show students the array of uses of mathematics. These applied problems are written in an inviting way, many times accompanied by new interesting illustrations to help students overcome some of the apprehension associated with application problems.

Getting Ready for the Next Section Many students think of mathematics as a collection of discrete, unrelated topics. Their instructors know that this is not the case. The new Getting Ready for the Next Section problems reinforce the cumulative, connected nature of this course by showing how the concepts and techniques flow one from another throughout the course. These problems review all of the material that students will need in order to be successful in the next section, gently preparing students to move forward.

Maintaining Your Skills One of the major themes of our book is continuous review. We strive to continuously hone techniques learned earlier by keeping the important concepts in the forefront of the course. The Maintaining Your Skills problems review material from the previous chapter, or they review problems that form the foundation of the course—the problems that you expect students to be able to solve when they get to the next course.

End-of-Chapter Summary, Review, and Assessment

We have learned that students are more comfortable with a chapter that sums up what they have learned thoroughly and accessibly, and reinforces concepts and techniques well. To help students grasp concepts and get more practice, each chapter ends with the following features that together give a comprehensive reexamination of the chapter.

Chapter Summary The chapter summary recaps all main points from the chapter in a visually appealing grid. In the margin, next to each topic, is an example that illustrates the type of problem associated with the topic being reviewed. When students prepare for a test, they can use the chapter summary as a guide to the main concepts of the chapter.

Chapter Review Following the chapter summary in each chapter is the chapter review. It contains an extensive set of problems that review all the main topics in the chapter. This feature can be used flexibly, as assigned review, as a recommended self-test for students as they prepare for examinations, or as an in-class quiz or test.

Cumulative Review Starting in Chapter 2, following the chapter review in each chapter, is a set of problems that reviews material from all preceding chapters. This keeps students current with past topics and helps them retain the information they study.

Chapter Test A set of problems representative of all the main points of the chapter. These don't contain as many problems as the chapter review, and should be completed in 50 minutes.

Chapter Projects Each chapter closes with a pair of projects. One is a group project, suitable for students to work on in class. The second project is a research project for students to do outside of class and tends to be open ended.

Additional Features of the Book

Facts from Geometry Many of the important facts from geometry are listed under this heading. In most cases, an example or two accompanies each of the facts to give students a chance to see how topics from geometry are related to the algebra they are learning.

Chapter Openings Each chapter opens with an introduction in which a real-world application is used to stimulate interest in the chapter. We expand on these opening applications later in the chapter.

Supplements

If you are interested in any of the supplements below, please contact your sales representative.

For the Instructor

Annotated Instructor's Edition ISBN-10: 0495827258 | ISBN-13: 9780495827252
This special instructor's version of the text contains answers next to all exercises and instructor notes at the appropriate location.

Complete Solutions Manual ISBN-10: 0495827266 | ISBN-13: 9780495827269
This manual contains complete solutions for all problems in the text.

Enhanced WebAssign ISBN-10: 0495829145 | ISBN-13: 9780495829140

Enhanced WebAssign 1-Semester Printed Access Card for Lower Level Math
ISBN-10: 0495390801 | ISBN-13: 9780495390800
Enhanced WebAssign, used by over one million students at more than 1,100 institutions, allows you to assign, collect, grade, and record homework assignments via the web. This proven and reliable homework system includes

thousands of algorithmically generated homework problems, links to relevant textbook sections, video examples, problem-specific tutorials, and more.

ExamView®
ISBN-10: 0495829501 | ISBN-13: 9780495829508
Create, deliver, and customize tests and study guides (both print and online) in minutes with this easy-to-use assessment and tutorial system. ExamView offers both a Quick Test Wizard and an Online Test Wizard that guide you step-by-step through the process of creating tests—you can even see the test you are creating on the screen exactly as it will print or display online.

PowerLecture with JoinIn™ Student Response System, ExamView®, Lecture Video ISBN-10: 0495829528 | ISBN-13: 9780495829522

Text-Specific Videos ISBN-10: 0495829129 | ISBN-13: 9780495829126
This set of text-specific videos features segments taught by the author, worked-out solutions to many examples in the book. Available to instructors only.

For the Student

Enhanced WebAssign 1-Semester Printed Access Card for Lower Level Math
ISBN-10: 0495390801 | ISBN-13: 9780495390800
Enhanced WebAssign, used by over one million students at more than 1,100 institutions, allows you to assign, collect, grade, and record homework assignments via the web. This proven and reliable homework system includes thousands of algorithmically generated homework problems, links to relevant textbook sections, video examples, problem-specific tutorials, and more.

Acknowledgments

I would like to thank my editor at Cengage Learning, Marc Bove, for his help and encouragement with this project. Many thanks also to Rich Jones, my developmental editor, for his suggestions on content, and his availability for consulting. Devin Christ, the head of production at our office, was a tremendous help in organizing and planning the details of putting this book together. Mary Gentilucci, Michael Landrum and Tammy Fisher-Vasta assisted with error checking and proofreading. Special thanks to my other friends at Cengage Learning: Sam Subity and Shaun Williams for handling the media and ancillary packages on this project, and Hal Humphrey, my project manager, who did a great job of coordinating everyone and everything in order to publish this book.

Finally, I am grateful to the following instructors for their suggestions and comments: Patricia Clark, Sinclair CC; Matthew Hudock, St. Phillip's College; Bridget Young, Suffolk County CC; Bettie Truitt, Black Hawk College; Armando Perez, Laredo CC; Diane Allen, College of Technology Idaho State; Jignasa Rami, CCBC Catonsville; Yon Kim, Passaic Community College; Elizabeth Chu, Suffolk County CC, Ammerman; Marilyn Larsen, College of the Mainland; Sherri Ucravich, University of Wisconsin; Scott Beckett, Jacksonville State University; Nimisha Raval, Macon Technical Institute; Gary Franchy, Davenport University, Warren; Debbi Loeffler, CC of Baltimore County; Scott Boman, Wayne County CC; Dayna Coker, Southwestern Oklahoma State; Annette Wiesner, University of Wisconsin; Anne Kmet, Grossmont College; Mary Wagner-Krankel, St. Mary's University; Joseph Deguzman, Riverside CC, Norco; Deborah McKee, Weber State University; Gail Burkett, Palm Beach CC; Lee Ann Spahr, Durham Technical CC; Randall

Mills, KCTCS Big Sandy CC/Tech; Jana Bryant, Manatee CC; Fred Brown, University of Maine, Augusta; Jeff Waller, Grossmont College; Robert Fusco, Broward CC, FL; Larry Perez, Saddleback College, CA; Victoria Anemelu, San Bernardino Valley, CA; John Close, Salt Lake CC, UT; Randy Gallaher, Lewis and Clark CC; Julia Simms, Southern Illinois U; Julianne Labbiento, Lehigh Carbon CC; Joanne Kendall, Cy-Fair College; Ann Davis, Northeastern Tech.

Pat. McKeague
November 2008

Preface to the Student

I often find my students asking themselves the question "Why can't I understand this stuff the first time?" The answer is "You're not expected to." Learning a topic in mathematics isn't always accomplished the first time around. There are many instances when you will find yourself reading over new material a number of times before you can begin to work problems. That's just the way things are in mathematics. If you don't understand a topic the first time you see it, that doesn't mean there is something wrong with you. Understanding mathematics takes time. The process of understanding requires reading the book, studying the examples, working problems, and getting your questions answered.

How to Be Successful in Mathematics

1. If you are in a lecture class, be sure to attend all class sessions on time. You cannot know exactly what goes on in class unless you are there. Missing class and then expecting to find out what went on from someone else is not the same as being there yourself.

2. Read the book and work the Practice Problems. It is best to read the section that will be covered in class beforehand. Reading in advance, even if you do not understand everything you read, is still better than going to class with no idea of what will be discussed. As you read through each section, be sure to work the Practice Problems in the margin of the text. Each Practice Problem is similar to the example with the same number. Look over the example and then try the corresponding Practice Problem. The answers to the Practice Problems are at the bottom of the page. If you don't get the correct answer, see if you can rework the problem correcly. If you miss it a second time, check your solution with the solution to the Practice Problem in the back of the book.

3. Work problems every day and check your answers. The key to success in mathematics is working problems. The more problems you work, the better you will become at working them. The answers to the odd-numbered problems are given in the back of the book. When you have finished an assignment, be sure to compare your answers with those in the book. If you have made a mistake, find out what it is, and correct it.

4. Do it on your own. Don't be misled into thinking someone else's work is your own. Having someone else show you how to work a problem is not the same as working the same problem yourself. It is okay to get help when you are stuck. As a matter of fact, it is a good idea. Just be sure you do the work yourself.

5. Review every day. After you have finished the problems your instructor has assigned, take another 15 minutes and review a section you have already completed. The more you review, the longer you will retain the material you have learned.

6. Don't expect to understand every new topic the first time you see it. Sometimes you will understand everything you are doing, and sometimes you won't. That's just the way things are in mathematics. Expecting to understand each new topic the first time you see it can lead to disappointment and frustration. The process of understanding takes time. It requires that you read the book, work problems, and get your questions answered.

7. Spend as much time as it takes for you to master the material. No set formula exists for the exact amount of time you need to spend on mathematics to master it. You will find out as you go along what is or isn't enough time for you. If you end up spending 2 or more hours on each section in order to master the material there, then that's how much time it takes; trying to get by with less will not work.

8. Relax. It's probably not as difficult as you think.

The Basics

Introduction

Pisa, Italy, is most famous for its leaning tower. But for people involved in mathematics education, it is important because it is the birthplace of the Italian mathematician Leonardo Pisano, also known as Fibonacci. Fibonacci has a sequence of numbers named for him. It appears in a scrambled form as a clue to a secret in the beginning of the popular novel *The DaVinci Code*.

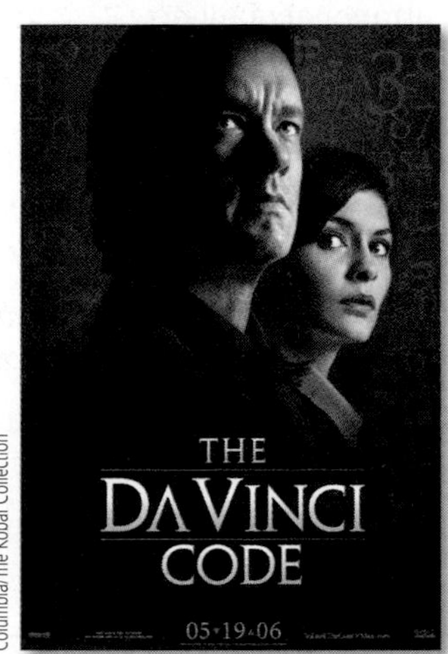

Columbia/The Kobal Collection

If you read the book and were familiar with the Fibonacci sequence, you would have solved part of the mystery yourself. Here are the scrambled clues from *The DaVinci Code* and the unscrambled Fibonacci sequence:

DaVinci Code clue: 13-3-2-21-1-1-8-5

Fibonacci sequence: 1, 1, 2, 3, 5, 8, 13, 21, . . .

In this chapter we will work with number sequences, including the Fibonacci sequence.

Chapter Pretest

Translate into symbols. [1.1]

1. The difference of 15 and x is 12.

2. The product of 6 and a is 30.

Simplify according to the rule for order of operations. [1.1]

3. $10 + 2(7 - 3) - 4^2$

4. $15 + 24 \div 6 - 3^2$

For each number, name the opposite, reciprocal, and absolute value. [1.2]

5. -6

6. $\frac{5}{3}$

Add. [1.3]

7. $5 + (-8)$

8. $|-2 + 7| + |-3 + (-4)|$

Subtract. [1.4]

9. $-7 - 4$

10. $5 - (6 + 3) - 3$

Multiply. [1.6]

11. $7(-4)$

12. $3(-4)(-2)$

13. $9\left(-\frac{1}{3}\right)$

14. $\left(-\frac{3}{2}\right)^3$

Simplify using the rule for order of operations. [1.1, 1.6, 1.7]

15. $-2(3) - 7$

16. $2(3)^3 - 4(-2)^4$

17. $\frac{-4(3) + 5(-2)}{-5 - 6}$

18. $\frac{4(3 - 5) - 2(-6 + 8)}{4(-2) + 10}$

Apply the associative property, and then simplify. [1.5, 1.6]

19. $5 + (7 + 3x)$

20. $3(-5y)$

Multiply by applying the distributive property. [1.5, 1.6]

21. $-5(2x - 3)$

22. $\frac{1}{3}(6x + 12)$

From the set of numbers $\{-3, -\frac{1}{2}, 2, \sqrt{5}, \pi\}$ list all the elements that are in the following sets. [1.8]

23. Integers

24. Rational numbers

25. Irrational numbers

26. Real numbers

Getting Ready for Chapter 1

To get started in this book, we assume that you can do simple addition and multiplication problems with whole numbers and decimals. To check to see that you are ready for this chapter, work each of the problems below.

1. $5 \cdot 5 \cdot 5$

2. $12 \div 4$

3. $1.7 - 1.2$

4. $10 - 9.5$

5. $7(0.2)$

6. $0.3(6)$

7. $14 - 9$

8. $39 \div 13$

9. $5 \cdot 9$

10. $2 \cdot 2 \cdot 2 \cdot 2 \cdot 2$

11. $15 + 14$

12. $28 \div 7$

13. $26 \div 2$

14. $125 - 81$

15. $630 \div 63$

16. $210 \div 10$

17. $2 \cdot 3 \cdot 5 \cdot 7$

18. $3 \cdot 7 \cdot 11$

19. $2 \cdot 3 \cdot 3 \cdot 5 \cdot 7$

20. $24 \div 8$

Notation and Symbols

1.1

Objectives

A Translate between phrases written in English and expressions written in symbols.

B Simplify expressions containing exponents.

C Simplify expressions using the rule for order of operations.

D Read tables and charts.

E Recognize the pattern in a sequence of numbers.

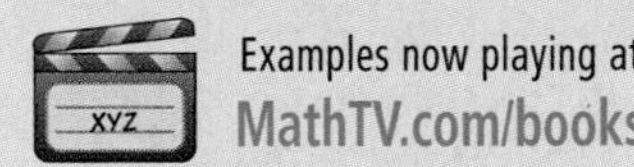

Suppose you have a checking account that costs you \$15 a month, plus \$0.05 for each check you write. If you write 10 checks in a month, then the monthly charge for your checking account will be

$$15 + 10(0.05)$$

Do you add 15 and 10 first and then multiply by 0.05? Or do you multiply 10 and 0.05 first and then add 15? If you don't know the answer to this question, you will after you have read through this section.

Because much of what we do in algebra involves comparison of quantities, we will begin by listing some symbols used to compare mathematical quantities. The comparison symbols fall into two major groups: equality symbols and inequality symbols.

We will let the letters a and b stand for (represent) any two mathematical quantities. When we use letters to represent numbers, as we are doing here, we call the letters *variables.*

A Variables: An Intuitive Look

When you filled out the application for the school you are attending, there was a space to fill in your first name. "First name" is a variable quantity because the value it takes depends on who is filling out the application. For example, if your first name is Manuel, then the value of "First Name" is Manuel. However, if your first name is Christa, then the value of "First Name" is Christa.

If we denote "First Name" as *FN,* "Last Name" as *LN*, and "Whole Name" as *WN*, then we take the concept of a variable further and write the relationship between the names this way:

$$FN + LN = WN$$

(We use the + symbol loosely here to represent writing the names together with a space between them.) This relationship we have written holds for all people who have only a first name and a last name. For those people who have a middle name, the relationship between the names is

$$FN + MN + LN = WN$$

A similar situation exists in algebra when we let a letter stand for a number or a group of numbers. For instance, if we say "let a and b represent numbers," then a and b are called *variables* because the values they take on vary. We use the variables a and b in the following lists so that the relationships shown there are true for all numbers that we will encounter in this book. By using variables, the following statements are general statements about all numbers, rather than specific statements about only a few numbers.

Comparison Symbols

Equality:	$a = b$	a is equal to b (a and b represent the same number)
	$a \neq b$	a is not equal to b
Inequality:	$a < b$	a is less than b
	$a \nless b$	a is not less than b
	$a > b$	a is greater than b
	$a \ngtr b$	a is not greater than b
	$a \geq b$	a is greater than or equal to b
	$a \leq b$	a is less than or equal to b

In the past you may have used the notation 3×5 to denote multiplication. In algebra it is best to avoid this notation if possible, because the multiplication symbol $\times$ can be confused with the variable x when written by hand.

For each example in the text there is a corresponding practice problem in the margin. After you read through an example in the text, work the practice problem with the same number in the margin. The answers to the practice problems are given on the same page as the practice problems. Be sure to check your answers as you work these problems. The worked-out solutions for the practice problems are given in the back of the book. So if you find a practice problem that you cannot work correctly, you can look up the correct solution to that problem in the back of the book.

The symbols for inequality, $<$ and $>$, always point to the smaller of the two quantities being compared. For example, $3 < x$ means 3 is smaller than x. In this case we can say "3 is less than x" or "x is greater than 3"; both statements are correct. Similarly, the expression $5 > y$ can be read as "5 is greater than y" or as "y is less than 5" because the inequality symbol is pointing to y, meaning y is the smaller of the two quantities.

Next, we consider the symbols used to represent the four basic operations: addition, subtraction, multiplication, and division.

Operation Symbols

Addition:	$a + b$	The *sum* of a and b
Subtraction:	$a - b$	The *difference* of a and b
Multiplication:	$a \cdot b$, $(a)(b)$, $a(b)$, $(a)b$, ab	The *product* of a and b
Division:	$a \div b$, a/b, $\frac{a}{b}$, $b\overline{)a}$	The *quotient* of a and b

When we encounter the word *sum,* the implied operation is addition. To find the sum of two numbers, we simply add them. *Difference* implies subtraction, *product* implies multiplication, and *quotient* implies division. Notice also that there is more than one way to write the product or quotient of two numbers.

Grouping Symbols

Parentheses ( ) and brackets [ ] are the symbols used for grouping numbers together. (Occasionally, braces { } are also used for grouping, although they are usually reserved for set notation, as we shall see.)

The following examples illustrate the relationship between the symbols for comparing, operating, and grouping and the English language.

EXAMPLES

Mathematical Expression	*English Equivalent*
1. $4 + 1 = 5$	The sum of 4 and 1 is 5.
2. $8 - 1 < 10$	The difference of 8 and 1 is less than 10.
3. $2(3 + 4) = 14$	Twice the sum of 3 and 4 is 14.
4. $3x \geq 15$	The product of 3 and x is greater than or equal to 15.
5. $\frac{y}{2} = y - 2$	The quotient of y and 2 is equal to the difference of y and 2.

PRACTICE PROBLEMS

Write an equivalent expression in English.

1. $3 + 7 = 10$

2. $9 - 6 < 4$

3. $4(2 + 3) \neq 6$

4. $6(8 - 1) = 42$

5. $\frac{4}{x} = 8 - x$

Answers

1. The sum of 3 and 7 is 10.
2. The difference of 9 and 6 is less than 4.
3. 4 times the sum of 2 and 3 is not equal to 6.
4. Six times the difference of 8 and 1 is 42.
5. The quotient of 4 and x is equal to the difference of 8 and x.

B Exponents

The last type of notation we need to discuss is the notation that allows us to write repeated multiplications in a more compact form—*exponents.* In the expression 2^3, the 2 is called the *base* and the 3 is called the *exponent.* The exponent 3 tells us the number of times the base appears in the product; that is,

$$2^3 = 2 \cdot 2 \cdot 2 = 8$$

The expression 2^3 is said to be in exponential form, whereas $2 \cdot 2 \cdot 2$ is said to be in expanded form. Here are some additional examples of expressions involving exponents.

EXAMPLES Expand and multiply.

6. $5^2 = 5 \cdot 5 = 25$ Base 5, exponent 2
7. $2^5 = 2 \cdot 2 \cdot 2 \cdot 2 \cdot 2 = 32$ Base 2, exponent 5
8. $10^3 = 10 \cdot 10 \cdot 10 = 1{,}000$ Base 10, exponent 3

Expand and multiply.
6. 7^2
7. 3^4
8. 10^5

Notation and Vocabulary Here is how we read expressions containing exponents.

Mathematical Expression	Written Equivalent
5^2	five to the second power
5^3	five to the third power
5^4	five to the fourth power
5^5	five to the fifth power
5^6	five to the sixth power

We have a shorthand vocabulary for second and third powers because the area of a square with a side of 5 is 5^2, and the volume of a cube with a side of 5 is 5^3.

5^2 can be read "five squared." 5^3 can be read "five cubed."

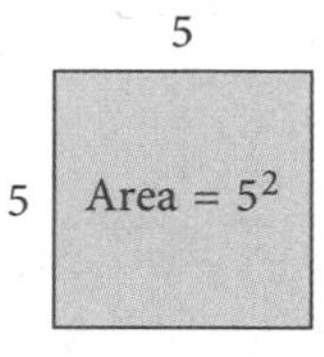

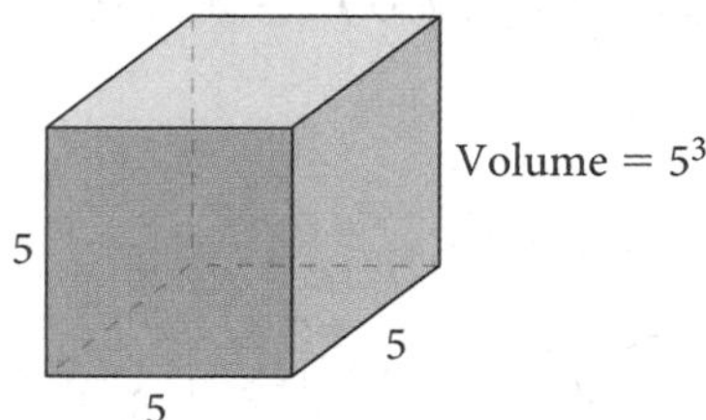

C Order of Operations

The symbols for comparing, operating, and grouping are to mathematics what punctuation marks are to English. These symbols are the punctuation marks for mathematics.

Consider the following sentence:

Paul said John is tall.

It can have two different meanings, depending on how it is punctuated.

1. "Paul," said John, "is tall."
2. Paul said, "John is tall."

Let's take a look at a similar situation in mathematics. Consider the following mathematical statement:

$5 + 2 \cdot 7$

If we add the 5 and 2 first and then multiply by 7, we get an answer of 49. However, if we multiply the 2 and the 7 first and then add 5, we are left with 19. We have a problem that seems to have two different answers, depending on whether we add first or multiply first. We would like to avoid this type of situation. Every problem like $5 + 2 \cdot 7$ should have only one answer. Therefore, we have developed the following rule for the order of operations.

Answers
6. 49 **7.** 81 **8.** 100,000

> **Rule (Order of Operations)**
> When evaluating a mathematical expression, we will perform the operations in the following order, beginning with the expression in the innermost parentheses or brackets first and working our way out.
>
> **1.** Simplify all numbers with exponents, working from left to right if more than one of these expressions is present.
>
> **2.** Then do all multiplications and divisions left to right.
>
> **3.** Perform all additions and subtractions left to right.

These next examples involve using the rule for order of operations.

Use the rule for order of operations to simplify each expression.

9. $4 + 6 \cdot 7$

10. $18 \div 6 \cdot 2$

11. $5[4 + 3(7 + 2 \cdot 4)]$

12. $12 + 8 \div 2 + 4 \cdot 5$

13. $3^4 + 2^5 \div 8 - 5^2$

EXAMPLE 9 Simplify $5 + 8 \cdot 2$.

SOLUTION

$$\begin{aligned} 5 + 8 \cdot 2 &= 5 + 16 && \text{Multiply } 8 \cdot 2 \text{ first} \\ &= 21 \end{aligned}$$

EXAMPLE 10 Simplify $12 \div 4 \cdot 2$.

SOLUTION

$$\begin{aligned} 12 \div 4 \cdot 2 &= 3 \cdot 2 && \text{Work left to right} \\ &= 6 \end{aligned}$$

EXAMPLE 11 Simplify $2[5 + 2(6 + 3 \cdot 4)]$.

SOLUTION

$$\begin{aligned} 2[5 + 2(6 + 3 \cdot 4)] &= 2[5 + 2(6 + 12)] && \text{Simplify within the innermost} \\ &= 2[5 + 2(18)] && \text{grouping symbols first} \\ &= 2[5 + 36] && \text{Next, simplify inside the brackets} \\ &= 2[41] \\ &= 82 && \text{Multiply} \end{aligned}$$

EXAMPLE 12 Simplify $10 + 12 \div 4 + 2 \cdot 3$.

SOLUTION

$$\begin{aligned} 10 + 12 \div 4 + 2 \cdot 3 &= 10 + 3 + 6 && \text{Multiply and divide left to right} \\ &= 19 && \text{Add left to right} \end{aligned}$$

EXAMPLE 13 Simplify $2^4 + 3^3 \div 9 - 4^2$.

SOLUTION

$$\begin{aligned} 2^4 + 3^3 \div 9 - 4^2 &= 16 + 27 \div 9 - 16 && \text{Simplify numbers with exponents} \\ &= 16 + 3 - 16 && \text{Then, divide} \\ &= 19 - 16 && \text{Finally, add and subtract} \\ &= 3 && \text{left to right} \end{aligned}$$

D Reading Tables and Bar Charts

The table below shows the average amount of caffeine in a number of beverages. The diagram in Figure 1 is a bar chart. It is a visual presentation of the information in Table 1. The table gives information in numerical form, whereas the chart gives the same information in a geometric way. In mathematics, it is important to be able to move back and forth between the two forms.

Answers
9. 46 **10.** 6 **11.** 245 **12.** 36
13. 60

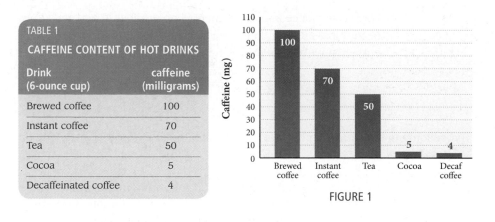

TABLE 1
CAFFEINE CONTENT OF HOT DRINKS

Drink (6-ounce cup)	caffeine (milligrams)
Brewed coffee	100
Instant coffee	70
Tea	50
Cocoa	5
Decaffeinated coffee	4

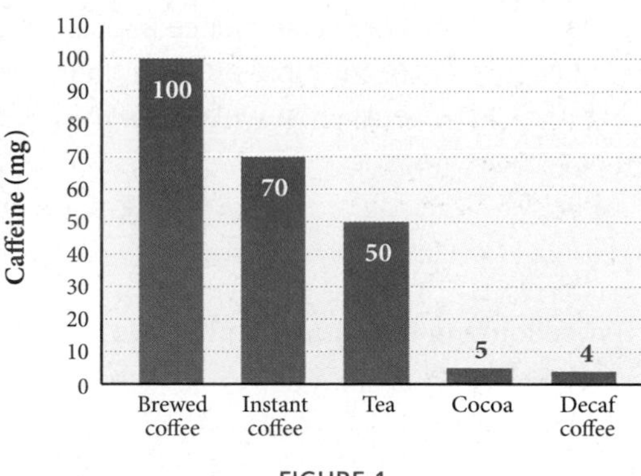

FIGURE 1

EXAMPLE 14 Referring to Table 1 and Figure 1, suppose you have 3 cups of brewed coffee, 1 cup of tea, and 2 cups of decaf in one day. Write an expression that will give the total amount of caffeine in these six drinks, then simplify the expression.

SOLUTION From the table or the bar chart, we find the number of milligrams of caffeine in each drink; then we write an expression for the total amount of caffeine:

$$3(100) + 50 + 2(4)$$

Using the rule for order of operations, we get 358 total milligrams of caffeine.

14. Suppose you drink 2 cups of instant coffee, 1 cup of cocoa, and 3 cups of tea. Write an expression that will give the total amount of caffeine in these six drinks, and then simplify the expression.

E Number Sequences and Inductive Reasoning

Suppose someone asks you to give the next number in the sequence of numbers below. (The dots mean that the sequence continues in the same pattern forever.)

$$2, 5, 8, 11, \ldots$$

If you notice that each number is 3 more than the number before it, you would say the next number in the sequence is 14 because $11 + 3 = 14$. When we reason in this way, we are using what is called *inductive reasoning*. In mathematics we use inductive reasoning when we notice a pattern to a sequence of numbers and then use the pattern to extend the sequence.

EXAMPLE 15 Find the next number in each sequence.

a. 3, 8, 13, 18, . . .

b. 2, 10, 50, 250, . . .

c. 2, 4, 7, 11, . . .

SOLUTION To find the next number in each sequence, we need to look for a pattern or relationship.

a. For the first sequence, each number is 5 more than the number before it; therefore, the next number will be $18 + 5 = 23$.

b. For the sequence in part (b), each number is 5 times the number before it; therefore, the next number in the sequence will be $5 \cdot 250 = 1{,}250$.

15. Find the next number in each sequence.

a. 3, 7, 11, 15, . . .

b. 1, 3, 9, 27, . . .

c. 2, 5, 9, 14, . . .

Answer

14. $2(70) + 5 + 3(50)$
$= 295$ milligrams

c. For the sequence in part (c), there is no number to add each time or multiply by each time. However, the pattern becomes apparent when we look at the differences between the numbers:

$$2 \quad 4 \quad 7 \quad 11 \quad \ldots$$
$$2 \quad 3 \quad 4$$

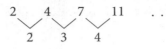

Proceeding in the same manner, we would add 5 to get the next term, giving us $11 + 5 = 16$.

In the introduction to this chapter we mentioned the mathematician known as Fibonacci. There is a special sequence in mathematics named for Fibonacci. Here it is.

Fibonacci sequence = 1, 1, 2, 3, 5, 8, . . .

Can you see the relationship among the numbers in this sequence? Start with two 1's, then add two consecutive members of the sequence to get the next number. Here is a diagram.

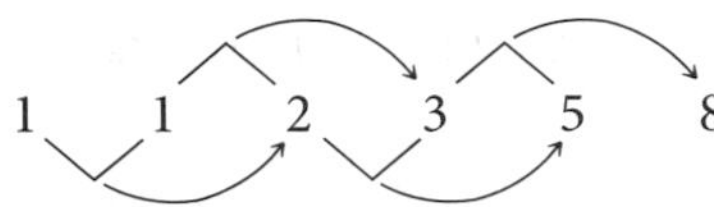

Sometimes we refer to the numbers in a sequence as *terms* of the sequence.

16. Write the first 10 terms of the following sequence.
2, 2, 4, 6, 10, 16, . . .

EXAMPLE 16 Write the first 10 terms of the Fibonacci sequence.

SOLUTION The first six terms are given above. We extend the sequence by adding 5 and 8 to obtain the seventh term, 13. Then we add 8 and 13 to obtain 21. Continuing in this manner, the first 10 terms in the Fibonacci sequence are

1, 1, 2, 3, 5, 8, 13, 21, 34, 55

Each section of the book will end with some problems and questions like the ones that follow. They are for you to answer after you have read through the section but before you go to class. All of them require that you give written responses in complete sentences. Writing about mathematics is a valuable exercise. If you write with the intention of explaining and communicating what you know to someone else, you will find that you understand the topic you are writing about even better than you did before you started writing. As with all problems in this course, you want to approach these writing exercises with a positive point of view. You will get better at giving written responses to questions as you progress through the course. Even if you never feel comfortable writing about mathematics, just the process of attempting to do so will increase your understanding and ability in mathematics.

Getting Ready for Class

After reading through the preceding section, respond in your own words and in complete sentences.

1. What is a variable?
2. Write the first step in the rule for order of operations.
3. What is inductive reasoning?
4. Explain the relationship between an exponent and its base.

Answers
15. a. 19 **b.** 81 **c.** 20
16. 2, 2, 4, 6, 10, 16, 26, 42, 68, 110

Problem Set 1.1

A For each sentence below, write an equivalent expression in symbols. **[Examples 1–5]**

1. The sum of x and 5 is 14.

2. The difference of x and 4 is 8.

3. The product of 5 and y is less than 30.

4. The product of 8 and y is greater than 16.

5. The product of 3 and y is less than or equal to the sum of y and 6.

6. The product of 5 and y is greater than or equal to the difference of y and 16.

7. The quotient of x and 3 is equal to the sum of x and 2.

8. The quotient of x and 2 is equal to the difference of x and 4.

B Expand and multiply. **[Examples 6–8]**

9. 3^2

10. 4^2

11. 7^2

12. 9^2

13. 2^3

14. 3^3

15. 4^3

16. 5^3

17. 2^4

18. 3^4

19. 10^2

20. 10^4

21. 11^2

22. 111^2

C Use the rule for order of operations to simplify each expression as much as possible. **[Examples 9–13]**

23. $2 \cdot 3 + 5$

24. $8 \cdot 7 + 1$

25. $2(3 + 5)$

26. $8(7 + 1)$

27. $5 + 2 \cdot 6$

28. $8 + 9 \cdot 4$

29. $(5 + 2) \cdot 6$

30. $(8 + 9) \cdot 4$

31. $5 \cdot 4 + 5 \cdot 2$

32. $6 \cdot 8 + 6 \cdot 3$

33. $5(4 + 2)$

34. $6(8 + 3)$

35. $8 + 2(5 + 3)$

36. $7 + 3(8 - 2)$

37. $(8 + 2)(5 + 3)$

38. $(7 + 3)(8 - 2)$

39. $20 + 2(8 - 5) + 1$

40. $10 + 3(7 + 1) + 2$

41. $5 + 2(3 \cdot 4 - 1) + 8$

42. $11 - 2(5 \cdot 3 - 10) + 2$

43. $8 + 10 \div 2$

44. $16 - 8 \div 4$

45. $4 + 8 \div 4 - 2$

46. $6 + 9 \div 3 + 2$

47. $3 + 12 \div 3 + 6 \cdot 5$

48. $18 + 6 \div 2 + 3 \cdot 4$

49. $3 \cdot 8 + 10 \div 2 + 4 \cdot 2$

50. $5 \cdot 9 + 10 \div 2 + 3 \cdot 3$

51. $(5 + 3)(5 - 3)$

52. $(7 + 2)(7 - 2)$

53. $5^2 - 3^2$

54. $7^2 - 2^2$

55. $(4 + 5)^2$

56. $(6 + 3)^2$

57. $4^2 + 5^2$

58. $6^2 + 3^2$

59. $3 \cdot 10^2 + 4 \cdot 10 + 5$

60. $6 \cdot 10^2 + 5 \cdot 10 + 4$

61. $2 \cdot 10^3 + 3 \cdot 10^2 + 4 \cdot 10 + 5$

62. $5 \cdot 10^3 + 6 \cdot 10^2 + 7 \cdot 10 + 8$

63. $10 - 2(4 \cdot 5 - 16)$

64. $15 - 5(3 \cdot 2 - 4)$

65. $4[7 + 3(2 \cdot 9 - 8)]$

66. $5[10 + 2(3 \cdot 6 - 10)]$

67. $5(7 - 3) + 8(6 - 4)$

68. $3(10 - 4) + 6(12 - 10)$

69. $3(4 \cdot 5 - 12) + 6(7 \cdot 6 - 40)$

70. $6(8 \cdot 3 - 4) + 5(7 \cdot 3 - 1)$

71. $3^4 + 4^2 \div 2^3 - 5^2$

72. $2^5 + 6^2 \div 2^2 - 3^2$

73. $5^2 + 3^4 \div 9^2 + 6^2$

74. $6^2 + 2^5 \div 4^2 + 7^2$

We are assuming that you know how to do arithmetic with decimals. Here are some problems to practice. Simplify each expression.

75. $0.08 + 0.09$ **76.** $0.06 + 0.04$ **77.** $0.10 + 0.12$ **78.** $0.08 + 0.06$

79. $4.8 - 2.5$ **80.** $6.3 - 4.8$ **81.** $2.07 + 3.48$ **82.** $4.89 + 2.31$

83. $0.12(2{,}000)$ **84.** $0.09(3{,}000)$ **85.** $0.25(40)$ **86.** $0.75(40)$

87. $510 \div 0.17$ **88.** $400 \div 0.1$ **89.** $240 \div 0.12$ **90.** $360 \div 0.12$

Use a calculator to find the following quotients. Round your answers to the nearest hundredth, if rounding is necessary.

91. $37.80 \div 1.07$ **92.** $85.46 \div 4.88$ **93.** $555 \div 740$ **94.** $740 \div 108$

95. $70 \div 210$ **96.** $15 \div 80$ **97.** $6{,}000 \div 22$ **98.** $51{,}000 \div 17$

Simplify each expression.

99. $20 \div 2 \cdot 10$ **100.** $40 \div 4 \cdot 5$ **101.** $24 \div 8 \cdot 3$ **102.** $24 \div 4 \cdot 6$

103. $36 \div 6 \cdot 3$ **104.** $36 \div 9 \cdot 2$ **105.** $48 \div 12 \cdot 2$ **106.** $48 \div 8 \cdot 3$

107. $16 - 8 + 4$ **108.** $16 - 4 + 8$ **109.** $24 - 14 + 8$ **110.** $24 - 16 + 6$

111. $36 - 6 + 12$ **112.** $36 - 9 + 20$ **113.** $48 - 12 + 17$ **114.** $48 - 13 + 15$

D Applying the Concepts [Example 14]

115. Babies According to the chart, in 2002 the number of babies born to mothers between the ages of 20 and 29 was 2,082,497. Write this number in words.

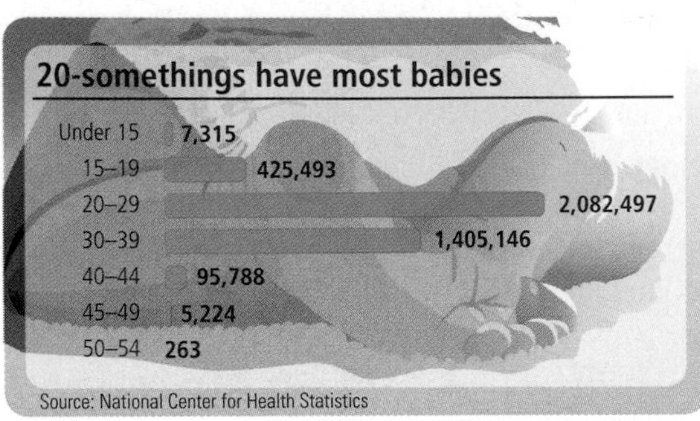

116. Cars The map shows the highest producers of cars in 2002. According to the map, Germany produced 6,213,000 cars. Write this number in words.

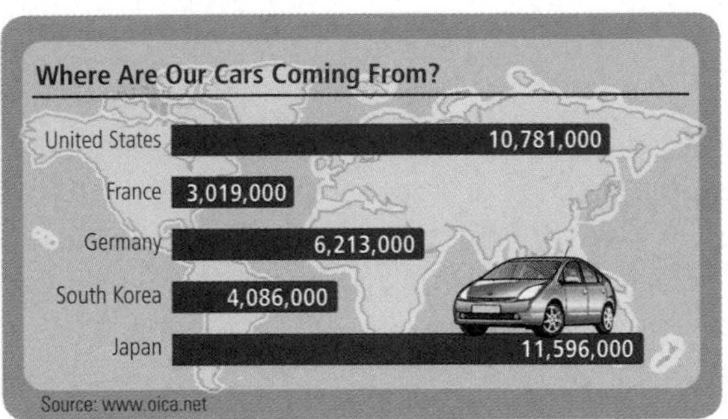

Food Labels In 1993 the government standardized the way in which nutrition information was presented on the labels of most packaged food products. Figure 2 shows a standardized food label from a package of cookies that I ate at lunch the day I was writing the problems for this problem set. Use the information in Figure 2 to answer the following questions.

117. How many cookies are in the package?

118. If I paid $0.50 for the package of cookies, how much did each cookie cost?

119. If the "calories" category stands for calories per serving, how many calories did I consume by eating the whole package of cookies?

120. Suppose that, while swimming, I burn 11 calories each minute. If I swim for 20 minutes, will I burn enough calories to cancel out the calories I added by eating 5 cookies?

Nutrition Facts
Serving Size 5 Cookies (about 43 g)
Servings Per Container 2
Calories 210
Fat Calories 90
* Percent Daily Values (DV) are based on a 2,000 calorie diet.

Amount/serving	%DV*	Amount/serving	%DV*
Total Fat 9 g	15%	**Total Carb.** 30 g	10%
Sat. Fat 2.5 g	12%	Fiber 1 g	2%
Cholest. less than 5 mg	2%	Sugars 14 g	
Sodium 110 mg	5%	**Protein** 3 g	

Vitamin A 0% • Vitamin C 0% • Calcium 2% • Iron 8%

Sandwich cremes

FIGURE 2

Education The chart shows the average income for people with different levels of education. This data is published recently by the U.S. Census. Use the information in the chart to answer the following questions.

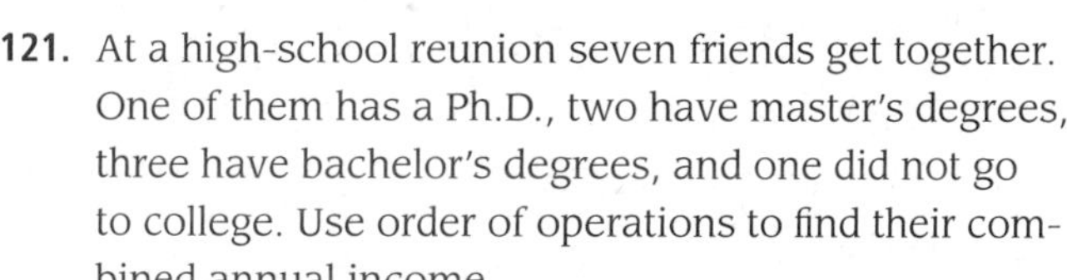

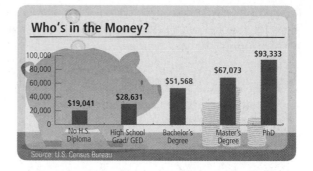

Who's in the Money?

100,000
80,000
60,000
40,000
20,000
0

$19,041 No H.S. Diploma
$28,631 High School Grad/ GED
$51,568 Bachelor's Degree
$67,073 Master's Degree
$93,333 PhD

Source: U.S. Census Bureau

121. At a high-school reunion seven friends get together. One of them has a Ph.D., two have master's degrees, three have bachelor's degrees, and one did not go to college. Use order of operations to find their combined annual income.

122. At a dinner party for six friends, one has a master's degree, two have bachelor's degrees, two have high school diplomas, and one did not finish high school. Use order of operations to find their combined annual income.

123. Reading Tables and Charts The following table and bar chart give the amount of caffeine in five different soft drinks. How much caffeine is in each of the following?

a. A 6-pack of Jolt

b. 2 Coca-Colas plus 3 Tabs

CAFFEINE CONTENT IN SOFT DRINKS

Drink	Caffeine (milligrams)
Jolt	100
Tab	47
Coca-Cola	45
Diet Pepsi	36
7 UP	0

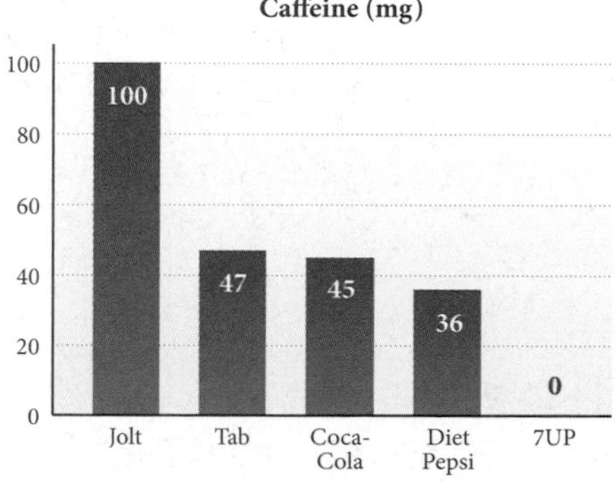

124. Reading Tables and Charts The following table and bar chart give the amount of caffeine in five different nonprescription drugs. How much caffeine is in each of the following?

a. A box of 12 Excedrin

b. 1 Dexatrim plus 4 Excedrin

CAFFEINE CONTENT IN NONPRESCRIPTION DRUGS

Nonprescription Drug	Caffeine (milligrams)
Dexatrim	200
NoDoz	100
Excedrin	65
Triaminicin tablets	30
Dristan tablets	16

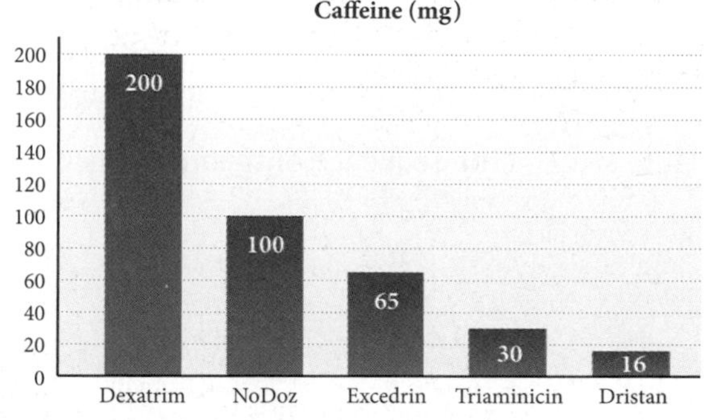

125. Reading Tables and Charts The following bar chart gives the number of calories burned by a 150-pound person during 1 hour of various exercises. The accompanying table should display the same information. Use the bar chart to complete the table.

CALORIES BURNED BY 150-POUND PERSON	
Activity	**Calories Burned in 1 Hour**
Bicycling	374
Bowling	
Handball	
Jogging	
Skiing	

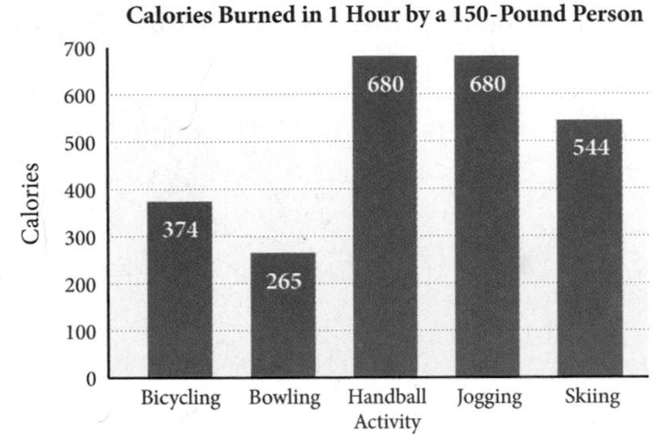

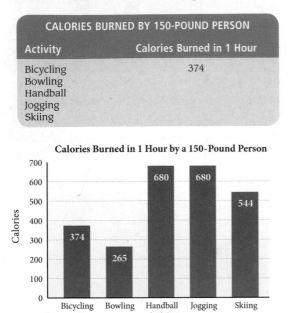

126. Reading Tables and Charts The following bar chart gives the number of calories consumed by eating some popular fast foods. The accompanying table should display the same information. Use the bar chart to complete the table.

CALORIES IN FAST FOOD	
Food	**Calories**
McDonald's Hamburger	270
Burger King Hamburger	
Jack in the Box Hamburger	
McDonald's Big Mac	
Burger King Whopper	

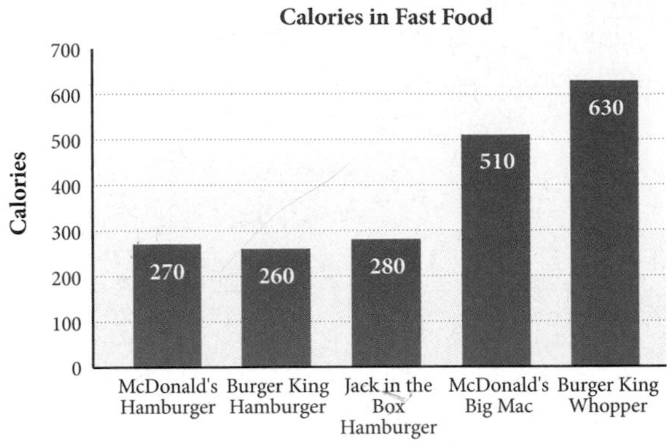

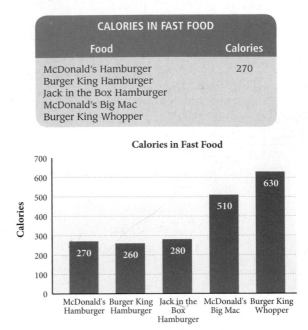

E Find the next number in each sequence. **[Examples 15–16]**

127. 1, 2, 3, 4, . . . (The sequence of counting numbers)

128. 0, 1, 2, 3, . . . (The sequence of whole numbers)

129. 2, 4, 6, 8, . . . (The sequence of even numbers)

130. 1, 3, 5, 7, . . . (The sequence of odd numbers)

131. 1, 4, 9, 16, . . . (The sequence of squares)

132. 1, 8, 27, 64, . . . (The sequence of cubes)

133. 2, 2, 4, 6, . . . (A Fibonacci-like sequence)

134. 5, 5, 10, 15, . . . (A Fibonacci-like sequence)

Real Numbers

1.2

Objectives

A Locate and label points on the number line.

B Change a fraction to an equivalent fraction with a new denominator.

C Simplify expressions containing absolute value.

D Identify the opposite of a number.

E Multiply fractions.

F Identify the reciprocal of a number.

G Find the perimeter and area of squares, rectangles, and triangles.

Table 1 and Figure 1 give the record low temperature, in degrees Fahrenheit, for each month of the year in the city of Jackson, Wyoming. Notice that some of these temperatures are represented by negative numbers.

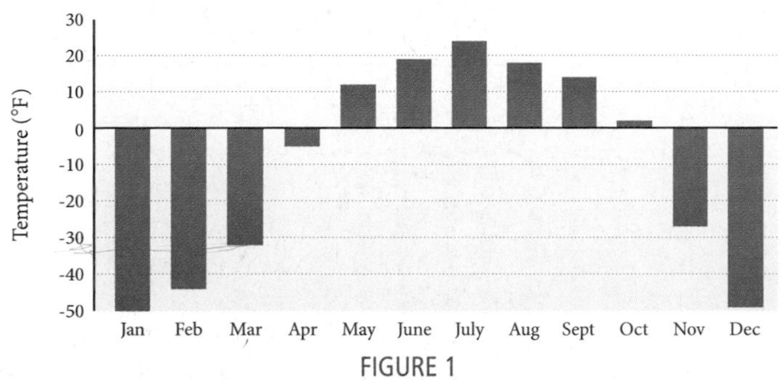

FIGURE 1

TABLE 1

RECORD LOW TEMPERATURES FOR JACKSON, WYOMING

Month	Temperature (Degrees Fahrenheit)	Month	Temperature (Degrees Fahrenheit)
January	−50	July	24
February	−44	August	18
March	−32	September	14
April	−5	October	2
May	12	November	−27
June	19	December	−49

A The Number Line

In this section we start our work with negative numbers. To represent negative numbers in algebra, we use what is called the *real number line*. Here is how we construct a real number line: We first draw a straight line and label a convenient point on the line with 0. Then we mark off equally spaced distances in both directions from 0. Label the points to the right of 0 with the numbers 1, 2, 3, . . . (the dots mean "and so on"). The points to the left of 0 we label in order, −1, −2, −3, ... Here is what it looks like.

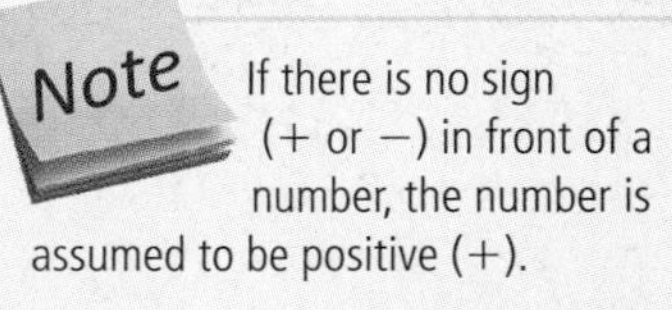

Note
If there is no sign (+ or −) in front of a number, the number is assumed to be positive (+).

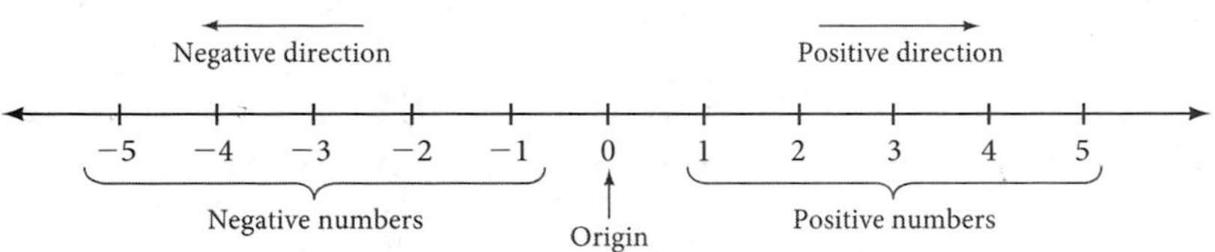

The numbers increase in value going from left to right. If we "move" to the right, we are moving in the positive direction. If we move to the left, we are moving in the negative direction. When we compare two numbers on the number line, the number on the left is always smaller than the number on the right. For instance, −3 is smaller than −1 because it is to the left of −1 on the number line.

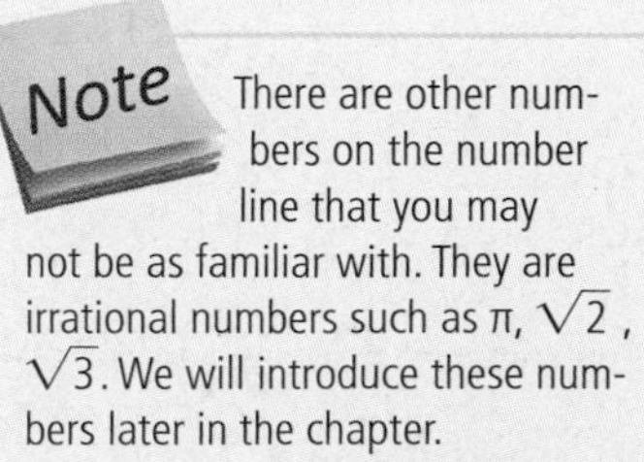

Note
There are other numbers on the number line that you may not be as familiar with. They are irrational numbers such as π, √2, √3. We will introduce these numbers later in the chapter.

PRACTICE PROBLEMS

1. Locate and label the points associated with $-2, -\frac{1}{2}, 0, 1.5, 2.75$.

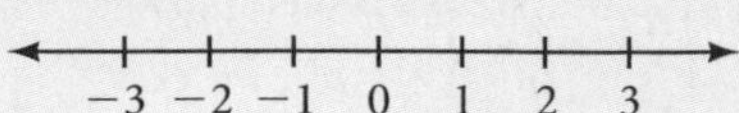

Answer

1. See graph in the appendix titled "Solutions to Selected Practice Problems."

EXAMPLE 1 Locate and label the points on the real number line associated with the numbers $-3.5, -1\frac{1}{4}, \frac{1}{2}, \frac{3}{4}, 2.5$.

SOLUTION We draw a real number line from -4 to 4 and label the points in question.

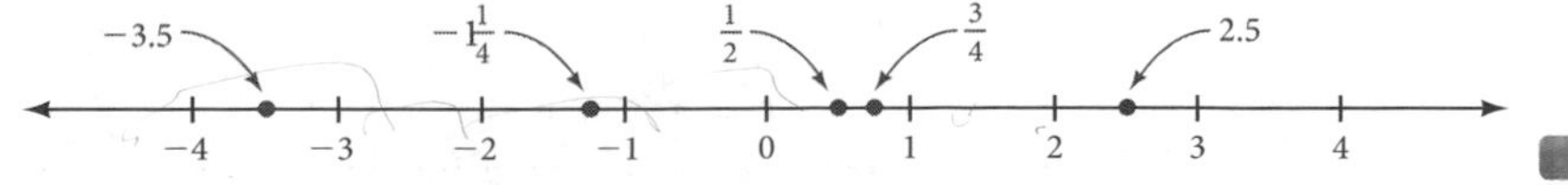

> **Definition**
> The number associated with a point on the real number line is called the **coordinate** of that point.

In the preceding example, the numbers $\frac{1}{2}, \frac{3}{4}, 2.5, -3.5$, and $-1\frac{1}{4}$ are the coordinates of the points they represent.

> **Definition**
> The numbers that can be represented with points on the real number line are called **real numbers**.

Real numbers include whole numbers, fractions, decimals, and other numbers that are not as familiar to us as these.

Fractions on the Number Line

As we proceed through Chapter 1, from time to time we will review some of the major concepts associated with fractions. To begin, here is the formal definition of a fraction.

> **Definition**
> If a and b are real numbers, then the expression
> $$\frac{a}{b} \qquad b \neq 0$$
> is called a **fraction.** The top number a is called the **numerator,** and the bottom number b is called the **denominator.** The restriction $b \neq 0$ keeps us from writing an expression that is undefined. (As you will see, division by zero is not allowed.)

The number line can be used to visualize fractions. Recall that for the fraction $\frac{a}{b}$, a is called the numerator and b is called the denominator. The denominator indicates the number of equal parts in the interval from 0 to 1 on the number line. The numerator indicates how many of those parts we have. If we take that part of the number line from 0 to 1 and divide it into *three equal parts,* we say that we have divided it into *thirds* (Figure 2). Each of the three segments is $\frac{1}{3}$ (one third) of the whole segment from 0 to 1.

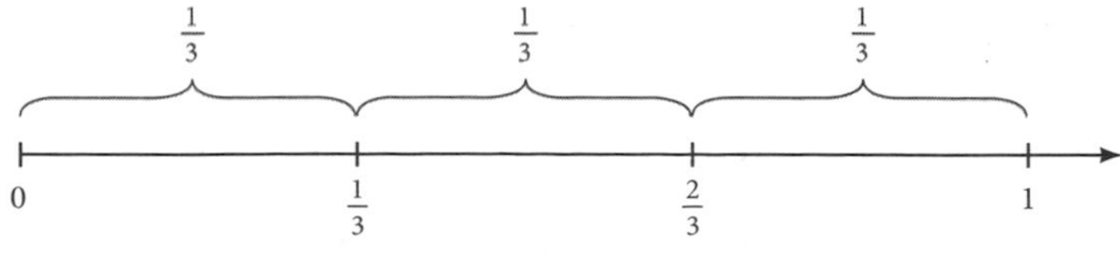

FIGURE 2

Two of these smaller segments together are $\frac{2}{3}$ (two thirds) of the whole segment. And three of them would be $\frac{3}{3}$ (three thirds), or the whole segment.

Let's do the same thing again with six equal divisions of the segment from 0 to 1 (Figure 3). In this case we say each of the smaller segments has a length of $\frac{1}{6}$ (one sixth).

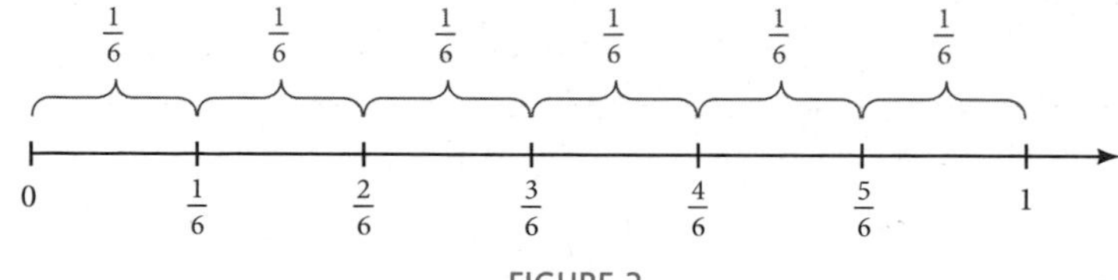

FIGURE 3

B Equivalent Fractions

The same point we labeled with $\frac{1}{3}$ in Figure 2 is labeled with $\frac{2}{6}$ in Figure 3. Likewise, the point we labeled earlier with $\frac{2}{3}$ is now labeled $\frac{4}{6}$. It must be true then that

$$\frac{2}{6} = \frac{1}{3} \quad \text{and} \quad \frac{4}{6} = \frac{2}{3}$$

Actually, there are many fractions that name the same point as $\frac{1}{3}$. If we were to divide the segment between 0 and 1 into 12 equal parts, 4 of these 12 equal parts $\left(\frac{4}{12}\right)$ would be the same as $\frac{2}{6}$ or $\frac{1}{3}$; that is,

$$\frac{4}{12} = \frac{2}{6} = \frac{1}{3}$$

Even though these three fractions look different, each names the same point on the number line, as shown in Figure 4. All three fractions have the same *value* because they all represent the same number.

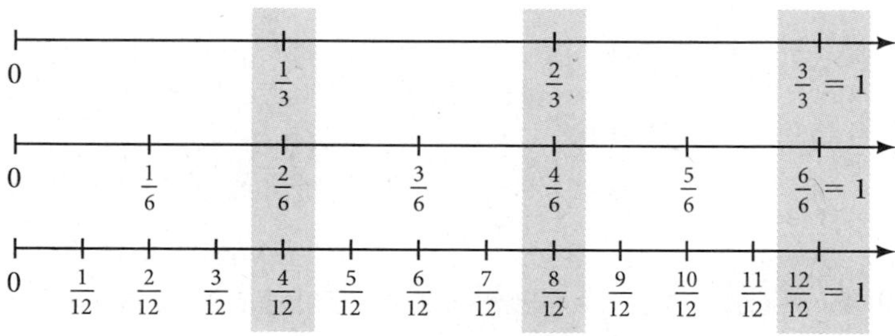

FIGURE 4

Definition

Fractions that represent the same number are said to be **equivalent**. Equivalent fractions may look different, but they must have the same value.

It is apparent that every fraction has many different representations, each of which is equivalent to the original fraction. The next two properties give us a way of changing the terms of a fraction without changing its value.

Property 1
Multiplying the numerator and denominator of a fraction by the same nonzero number never changes the value of the fraction.

Property 2
Dividing the numerator and denominator of a fraction by the same nonzero number never changes the value of the fraction.

2. Write $\frac{5}{8}$ as an equivalent fraction with denominator 48.

EXAMPLE 2 Write $\frac{3}{4}$ as an equivalent fraction with denominator 20.

SOLUTION The denominator of the original fraction is 4. The fraction we are trying to find must have a denominator of 20. We know that if we multiply 4 by 5, we get 20. Property 1 indicates that we are free to multiply the denominator by 5 as long as we do the same to the numerator.

$$\frac{3}{4} = \frac{3 \cdot 5}{4 \cdot 5} = \frac{15}{20}$$

The fraction $\frac{15}{20}$ is equivalent to the fraction $\frac{3}{4}$.

C Absolute Values

Representing numbers on the number line lets us give each number two important properties: a direction from zero and a distance from zero. The direction from zero is represented by the sign in front of the number. (A number without a sign is understood to be positive.) The distance from zero is called the absolute value of the number, as the following definition indicates.

Definition
The **absolute value** of a real number is its distance from zero on the number line. If x represents a real number, then the absolute value of x is written $|x|$.

Write each expression without absolute value bars.
3. $|7|$

EXAMPLE 3 Write the expression $|5|$ without absolute value bars.

SOLUTION $|5| = 5$ The number 5 is 5 units from zero

4. $|-7|$

EXAMPLE 4 Write the expression $|-5|$ without absolute value bars.

SOLUTION $|-5| = 5$ The number -5 is 5 units from zero

5. $\left|-\frac{3}{4}\right|$

EXAMPLE 5 Write the expression $\left|-\frac{1}{2}\right|$ without absolute value bars.

SOLUTION $\left|-\frac{1}{2}\right| = \frac{1}{2}$ The number $-\frac{1}{2}$ is $\frac{1}{2}$ unit from zero

The absolute value of a number is *never* negative. It is the distance the number is from zero without regard to which direction it is from zero. When working with the absolute value of sums and differences, we must simplify the expression

Answers
2. $\frac{30}{48}$ **3.** 7 **4.** 7 **5.** $\frac{3}{4}$

inside the absolute value symbols first and then find the absolute value of the simplified expression.

EXAMPLES Simplify each expression.

6. $|8 - 3| = |5| = 5$
7. $|3 \cdot 2^3 + 2 \cdot 3^2| = |3 \cdot 8 + 2 \cdot 9| = |24 + 18| = |42| = 42$
8. $|9 - 2| - |6 - 8| = |7| - |-2| = 7 - 2 = 5$

Simplify each expression.
6. $|7 - 2|$
7. $|2 \cdot 3^2 + 5 \cdot 2^2|$
8. $|10 - 4| - |9 - 11|$

D Opposites

Another important concept associated with numbers on the number line is that of opposites. Here is the definition.

> **Definition**
> Numbers the same distance from zero but in opposite directions from zero are called **opposites.**

EXAMPLE 9 Give the opposite of 5.

	Number	*Opposite*	
SOLUTION	5	-5	5 and -5 are opposites

Give the opposite of each number.
9. 8

EXAMPLE 10 Give the opposite of -3.

	Number	*Opposite*	
SOLUTION	-3	3	-3 and 3 are opposites

10. -5

EXAMPLE 11 Give the opposite of $\frac{1}{4}$.

	Number	*Opposite*	
SOLUTION	$\frac{1}{4}$	$-\frac{1}{4}$	$\frac{1}{4}$ and $-\frac{1}{4}$ are opposites

11. $-\frac{2}{3}$

EXAMPLE 12 Give the opposite of -2.3.

	Number	*Opposite*	
SOLUTION	-2.3	2.3	-2.3 and 2.3 are opposites

12. -4.2

Each negative number is the opposite of some positive number, and each positive number is the opposite of some negative number. The opposite of a negative number is a positive number. In symbols, if a represents a positive number, then

$$-(-a) = a$$

Opposites always have the same absolute value. And, when you add any two opposites, the result is always zero:

$$a + (-a) = 0$$

Answers
6. 5 **7.** 38 **8.** 4 **9.** -8 **10.** 5
11. $\frac{2}{3}$ **12.** 4.2

E Multiplication with Fractions

The last concept we want to cover in this section is the concept of reciprocals. Understanding reciprocals requires some knowledge of multiplication with fractions. To multiply two fractions, we simply multiply numerators and multiply denominators.

EXAMPLE 13 Multiply $\frac{3}{4} \cdot \frac{5}{7}$.

SOLUTION The product of the numerators is 15, and the product of the denominators is 28:

$$\frac{3}{4} \cdot \frac{5}{7} = \frac{3 \cdot 5}{4 \cdot 7} = \frac{15}{28}$$

EXAMPLE 14 Multiply $7\left(\frac{1}{3}\right)$.

SOLUTION The number 7 can be thought of as the fraction $\frac{7}{1}$:

$$7\left(\frac{1}{3}\right) = \frac{7}{1}\left(\frac{1}{3}\right) = \frac{7 \cdot 1}{1 \cdot 3} = \frac{7}{3}$$

EXAMPLE 15 Expand and multiply $\left(\frac{2}{3}\right)^3$.

SOLUTION Using the definition of exponents from the previous section, we have

$$\left(\frac{2}{3}\right)^3 = \frac{2}{3} \cdot \frac{2}{3} \cdot \frac{2}{3} = \frac{8}{27}$$

F Reciprocals

We are now ready for the definition of reciprocals.

> **Definition**
> Two numbers whose product is 1 are called **reciprocals.**

EXAMPLES Give the reciprocal of each number.

	Number	*Reciprocal*	
16.	5	$\frac{1}{5}$	Because $5\left(\frac{1}{5}\right) = \frac{5}{1}\left(\frac{1}{5}\right) = \frac{5}{5} = 1$
17.	2	$\frac{1}{2}$	Because $2\left(\frac{1}{2}\right) = \frac{2}{1}\left(\frac{1}{2}\right) = \frac{2}{2} = 1$
18.	$\frac{1}{3}$	3	Because $\frac{1}{3}(3) = \frac{1}{3}\left(\frac{3}{1}\right) = \frac{3}{3} = 1$
19.	$\frac{3}{4}$	$\frac{4}{3}$	Because $\frac{3}{4}\left(\frac{4}{3}\right) = \frac{12}{12} = 1$

13. Multiply $\frac{2}{3} \cdot \frac{7}{9}$

Note In past math classes you may have written fractions like $\frac{7}{3}$ (improper fractions) as mixed numbers, such as $2\frac{1}{3}$. In algebra it is usually better to write them as improper fractions rather than mixed numbers.

14. Multiply $8\left(\frac{1}{5}\right)$.

15. Expand and multiply $\left(\frac{11}{12}\right)^2$.

Give the reciprocal of each number.

16. 6

17. 3

18. $\frac{1}{2}$

19. $\frac{2}{3}$

Answers

13. $\frac{14}{27}$ **14.** $\frac{8}{5}$ **15.** $\frac{121}{144}$

16. $\frac{1}{6}$ **17.** $\frac{1}{3}$ **18.** 2 **19.** $\frac{3}{2}$

Although we will not develop multiplication with negative numbers until later in the chapter, you should know that the reciprocal of a negative number is also a negative number. For example, the reciprocal of -4 is $-\frac{1}{4}$.

Previously we mentioned that a variable is a letter used to represent a number or a group of numbers. An expression that contains any combination of numbers, variables, operation symbols, and grouping symbols is called an *algebraic expression* (sometimes referred to as just an expression). This definition includes the use of exponents and fractions. Each of the following is an algebraic expression.

$$3x + 5 \qquad 4t^2 - 9 \qquad x^2 - 6xy + y^2 \qquad -15x^2y^4z^5 \qquad \frac{a^2 - 9}{a - 3} \qquad \frac{(x - 3)(x + 2)}{4x}$$

In the last two expressions, the fraction bar separates the numerator from the denominator and is treated the same as a pair of grouping symbols; it groups the numerator and denominator separately.

The Value of an Algebraic Expression

An expression such as $3x + 5$ will take on different values depending on what x is. If we were to let x equal 2, the expression $3x + 5$ would become 11. On the other hand, if x is 10, the same expression has a value of 35:

When	$x = 2$	When	$x = 10$
the expression	$3x + 5$	the expression	$3x + 5$
becomes	$3(2) + 5$	becomes	$3(10) + 5$
	$= 6 + 5$		$= 30 + 5$
	$= 11$		$= 35$

Table 2 lists some other algebraic expressions, along with specific values for the variables and the corresponding value of the expression after the variable has been replaced with the given number.

TABLE 2

Original Expression	Value of the Variable	Value of the Expression
$5x + 2$	$x = 4$	$5(4) + 2 = 20 + 2$ $= 22$
$3x - 9$	$x = 2$	$3(2) - 9 = 6 - 9$ $= -3$
$4t^2 - 9$	$t = 5$	$4(5^2) - 9 = 4(25) - 9$ $= 100 - 9$ $= 91$
$\frac{a^2 - 9}{a - 3}$	$a = 8$	$\frac{8^2 - 9}{8 - 3} = \frac{64 - 9}{8 - 3}$ $= \frac{55}{5}$ $= 11$

G Formulas for Area and Perimeter

The vertical line labeled h in the triangle is its height, or altitude. It extends from the top of the triangle down to the base, meeting the base at an angle of 90°. The altitude of a triangle is always perpendicular to the base. The small square shown where the altitude meets the base is used to indicate that the angle formed is 90°.

FACTS FROM GEOMETRY: Formulas for Area and Perimeter

A square, rectangle, and triangle are shown in the following figures. Note that we have labeled the dimensions of each with variables. The formulas for the perimeter and area of each object are given in terms of its dimensions.

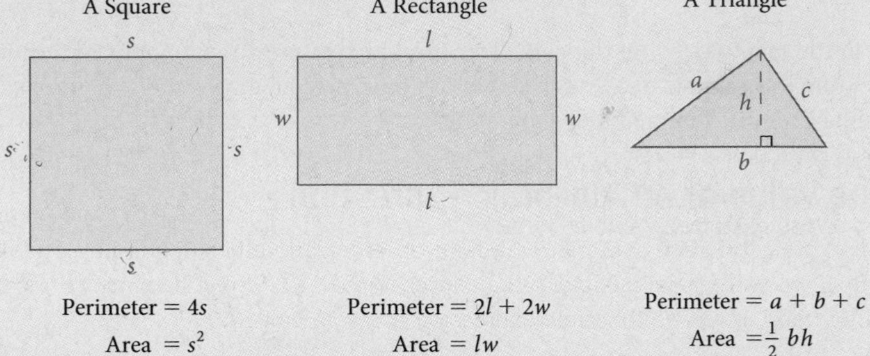

A Square	A Rectangle	A Triangle
Perimeter $= 4s$	Perimeter $= 2l + 2w$	Perimeter $= a + b + c$
Area $= s^2$	Area $= lw$	Area $= \frac{1}{2}bh$

The formula for perimeter gives us the distance around the outside of the object along its sides, whereas the formula for area gives us a measure of the amount of surface the object has.

EXAMPLE 20 Find the perimeter and area of each figure.

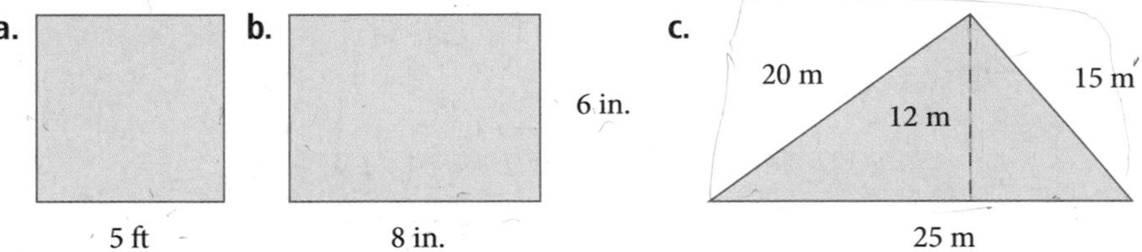

SOLUTION We use the preceding formulas to find the perimeter and the area. In each case, the units for perimeter are linear units, whereas the units for area are square units.

a. Perimeter $= 4s = 4 \cdot 5$ feet $= 20$ feet
Area $= s^2 = (5 \text{ feet})^2 = 25$ square feet

b. Perimeter $= 2l + 2w = 2(8 \text{ inches}) + 2(6 \text{ inches}) = 28$ inches
Area $= lw = (8 \text{ inches})(6 \text{ inches}) = 48$ square inches

c. Perimeter $= a + b + c$
$= (20 \text{ meters}) + (25 \text{ meters}) + (15 \text{ meters})$
$= 60$ meters

Area $= \dfrac{1}{2}bh = \dfrac{1}{2}(25 \text{ meters})(12 \text{ meters}) = 150$ square meters

20. Find the perimeter and area of the figures in Example 20, with the following changes:
a. $s = 4$ feet
b. $l = 7$ inches
$w = 5$ inches
c. $a = 40$ meters
$b = 50$ meters (base)
$c = 30$ meters
$h = 24$ meters

Answer
20. a. Perimeter = 16 feet
Area = 16 square feet
b. Perimeter = 24 inches
Area = 35 square inches
c. Perimeter = 120 meters
Area = 600 square meters

Getting Ready for Class

After reading through the preceding section, respond in your own words and in complete sentences.

1. What is a real number?
2. Explain multiplication with fractions.
3. How do you find the opposite of a number?

Bo

Please — P — simplify inside ( )

Excuse — E — remove exponents

My — M, D — Mult or Divide, go from left to right (D, M)

Dear

Aunt — Add

Sally — Sub.

$$-14 + 2(-3) + (-6)$$

$$-14 - 6 + (-6)$$

$$-20 + (-6)$$

$$-26$$

GENERAL INSTRUCTIONS

1. Obtain your textbook from the Bookstore.

2. Review the assignments. Begin reading and doing the homework.

3. Clearly and neatly do your homework on your paper. On the top of each assignment put the section, page and problem numbers.

4. Check all of your problems. The answers to the odd numbered problems can be found in the back of the textbook. If you have missed a problem and need help understanding your mistake, ask for help.

5. There are eight quizzes, four tests and a final exam (final exam is taken after you have completed all eight chapters).

ATTENDANCE REQUIREMENT

5 credits
If you are registered for 5 credits you MUST attend daily. You are allowed 5 unexcused absences. If you miss more than 5 days you will be penalized 2 points per day.

2 or 3 credits

Problem Set 1.2

A Draw a number line that extends from −5 to +5. Label the points with the following coordinates. [Example 1]

1. 5 **2.** −2 **3.** −4 **4.** −3

5. 1.5 **6.** −1.5 **7.** $\frac{9}{4}$ **8.** $\frac{8}{3}$

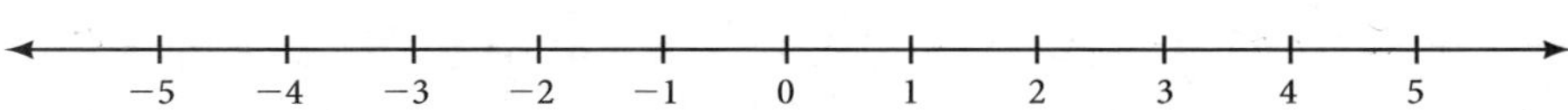

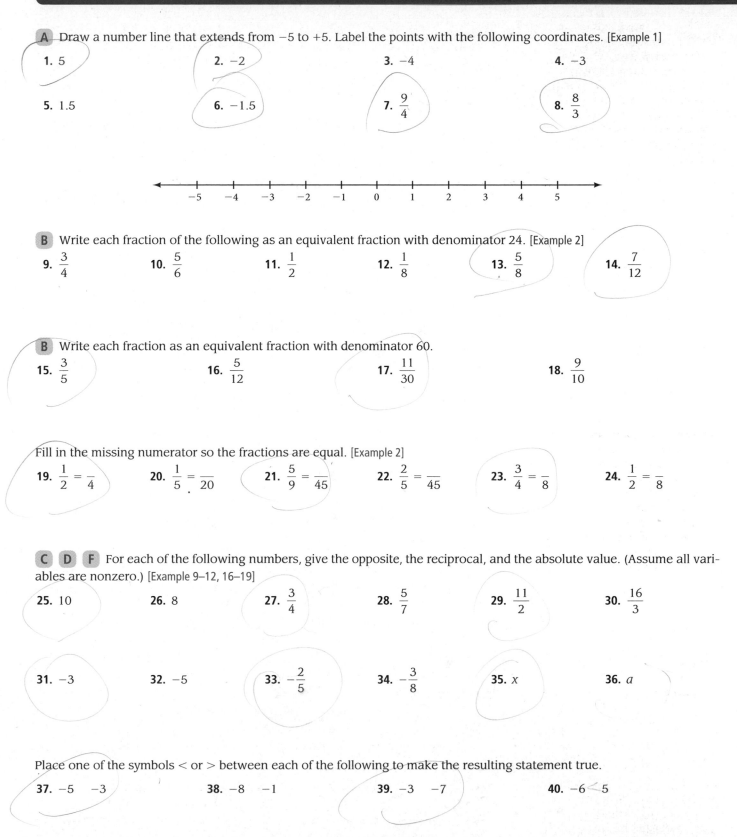

B Write each fraction of the following as an equivalent fraction with denominator 24. [Example 2]

9. $\frac{3}{4}$ **10.** $\frac{5}{6}$ **11.** $\frac{1}{2}$ **12.** $\frac{1}{8}$ **13.** $\frac{5}{8}$ **14.** $\frac{7}{12}$

B Write each fraction as an equivalent fraction with denominator 60.

15. $\frac{3}{5}$ **16.** $\frac{5}{12}$ **17.** $\frac{11}{30}$ **18.** $\frac{9}{10}$

Fill in the missing numerator so the fractions are equal. [Example 2]

19. $\frac{1}{2} = \frac{}{4}$ **20.** $\frac{1}{5} = \frac{}{20}$ **21.** $\frac{5}{9} = \frac{}{45}$ **22.** $\frac{2}{5} = \frac{}{45}$ **23.** $\frac{3}{4} = \frac{}{8}$ **24.** $\frac{1}{2} = \frac{}{8}$

C **D** **F** For each of the following numbers, give the opposite, the reciprocal, and the absolute value. (Assume all variables are nonzero.) [Example 9–12, 16–19]

25. 10 **26.** 8 **27.** $\frac{3}{4}$ **28.** $\frac{5}{7}$ **29.** $\frac{11}{2}$ **30.** $\frac{16}{3}$

31. −3 **32.** −5 **33.** $-\frac{2}{5}$ **34.** $-\frac{3}{8}$ **35.** x **36.** a

Place one of the symbols $<$ or $>$ between each of the following to make the resulting statement true.

37. −5 −3 **38.** −8 −1 **39.** −3 −7 **40.** −6 5

41. $|-4| \quad -|-4|$ **42.** $3 \quad -|-3|$ **43.** $7 \quad -|-7|$ **44.** $-7 \quad |-7|$

45. $-\frac{3}{4} \quad -\frac{1}{4}$ **46.** $-\frac{2}{3} \quad -\frac{1}{3}$ **47.** $-\frac{3}{2} \quad -\frac{3}{4}$ **48.** $-\frac{8}{3} \quad -\frac{17}{3}$

C Simplify each expression. [Example 3–8]

49. $|8 - 2|$ **50.** $|6 - 1|$ **51.** $|5 \cdot 2^3 - 2 \cdot 3^2|$ **52.** $|2 \cdot 10^2 + 3 \cdot 10|$

53. $|7 - 2| - |4 - 2|$ **54.** $|10 - 3| - |4 - 1|$ **55.** $10 - |7 - 2(5 - 3)|$ **56.** $12 - |9 - 3(7 - 5)|$

57. $15 - |8 - 2(3 \cdot 4 - 9)| - 10$ **58.** $25 - |9 - 3(4 \cdot 5 - 18)| - 20$

E Multiply the following. [Example 13–15]

59. $\frac{2}{3} \cdot \frac{4}{5}$ **60.** $\frac{1}{4} \cdot \frac{3}{5}$ **61.** $\frac{1}{2}(3)$ **62.** $\frac{1}{3}(2)$ **63.** $\frac{1}{4}(5)$ **64.** $\frac{1}{5}(4)$

65. $\frac{4}{3} \cdot \frac{3}{4}$ **66.** $\frac{5}{7} \cdot \frac{7}{5}$ **67.** $6\left(\frac{1}{6}\right)$ **68.** $8\left(\frac{1}{8}\right)$ **69.** $3 \cdot \frac{1}{3}$ **70.** $4 \cdot \frac{1}{4}$

Multiply.

71. a. $\frac{1}{2}(4)$ **b.** $\frac{1}{2}(8)$ **c.** $\frac{1}{2}(16)$ **d.** $\frac{1}{2}(0.06)$

72. a. $\frac{1}{4}(8)$ **b.** $\frac{1}{4}(24)$ **c.** $\frac{1}{4}(16)$ **d.** $\frac{1}{4}(0.20)$

73. a. $\frac{3}{2}(4)$ **b.** $\frac{3}{2}(8)$ **c.** $\frac{3}{2}(16)$ **d.** $\frac{3}{2}(0.06)$

74. a. $\frac{3}{4}(8)$ **b.** $\frac{3}{4}(24)$ **c.** $\frac{3}{4}(16)$ **d.** $\frac{3}{4}(0.20)$

Expand and multiply.

75. $\left(\frac{3}{4}\right)^2$ **76.** $\left(\frac{5}{6}\right)^2$ **77.** $\left(\frac{2}{3}\right)^3$ **78.** $\left(\frac{1}{2}\right)^3$ **79.** $\left(\frac{1}{10}\right)^4$ **80.** $\left(\frac{1}{10}\right)^5$

Problems here involve finding the value of an algrebraic expression.

81. Find the value of $2x - 6$ when

a. $x = 5$

b. $x = 10$

c. $x = 15$

d. $x = 20$

82. Find the value of $2(x - 3)$ when

a. $x = 5$

b. $x = 10$

c. $x = 15$

d. $x = 20$

83. Find the value of each expression when x is 10.

a. $x + 2$

b. $2x$

c. x^2

d. 2^x

84. Find the value of each expression when x is 3.

a. $x + 3$

b. $3x$

c. x^2

d. 3^x

85. Find the value of each expression when x is 4.

a. $x^2 + 1$

b. $(x + 1)^2$

c. $x^2 + 2x + 1$

86. Find the value of $b^2 - 4ac$ when

a. $a = 2, b = 6, c = 3$

b. $a = 1, b = 5, c = 6$

c. $a = 1, b = 2, c = 1$

Find the next number in each sequence.

87. $1, \frac{1}{3}, \frac{1}{5}, \frac{1}{7}, \ldots$ (Reciprocals of odd numbers)

88. $\frac{1}{2}, \frac{1}{4}, \frac{1}{6}, \frac{1}{8}, \ldots$ (Reciprocals of even numbers)

89. $1, \frac{1}{4}, \frac{1}{9}, \frac{1}{16}, \ldots$ (Reciprocals of squares)

90. $1, \frac{1}{8}, \frac{1}{27}, \frac{1}{64}, \ldots$ (Reciprocals of cubes)

G Find the perimeter and area of each figure. [Example 20]

91.

1 inch

1 inch

92.

15 millimeters

15 millimeters

93.

0.75 inches

1.5 inches

94.

1.5 centimeters

4.5 centimeters

95.

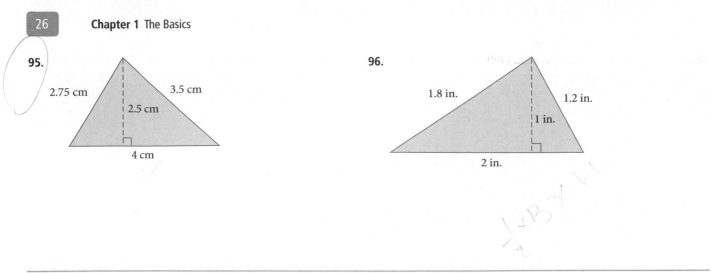

2.75 cm
3.5 cm
2.5 cm
4 cm

96.

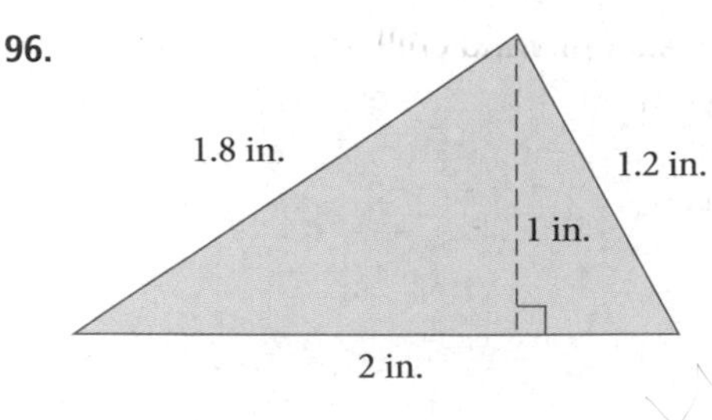

Applying the Concepts

97. Football Yardage A football team gains 6 yards on one play and then loses 8 yards on the next play. To what number on the number line does a loss of 8 yards correspond? The total yards gained or lost on the two plays corresponds to what negative number?

98. Checking Account Balance A woman has a balance of $20 in her checking account. If she writes a check for $30, what negative number can be used to represent the new balance in her checking account?

Temperature In the United States, temperature is measured on the Fahrenheit temperature scale. On this scale, water boils at 212 degrees and freezes at 32 degrees. To denote a temperature of 32 degrees on the Fahrenheit scale, we write

32°F, which is read "32 degrees Fahrenheit"

Use this information for Problems 99 and 100.

99. Temperature and Altitude Marilyn is flying from Seattle to San Francisco on a Boeing 737 jet. When the plane reaches an altitude of 35,000 feet, the temperature outside the plane is 64 degrees below zero Fahrenheit. Represent the temperature with a negative number. If the temperature outside the plane gets warmer by 10 degrees, what will the new temperature be?

100. Temperature Change At 10:00 in the morning in White Bear Lake, Minnesota, John notices the temperature outside is 10 degrees below zero Fahrenheit. Write the temperature as a negative number. An hour later it has warmed up by 6 degrees. What is the temperature at 11:00 that morning?

101. Google Earth The Google image shows the Stratosphere Tower in Las Vegas, Nevada. The Stratosphere Tower is $\frac{23}{12}$ the height of the Space Needle in Seattle, Washington. If the Space Needle is about 600 feet tall, how tall is the Stratosphere Tower?

102. Skyscrapers The chart shows the heights of the three tallest buildings in the world. The Transamerica Pyramid in San Francisco, California, is half the height of the Taipei 101 tower. How tall is the Transameric Pyramid?

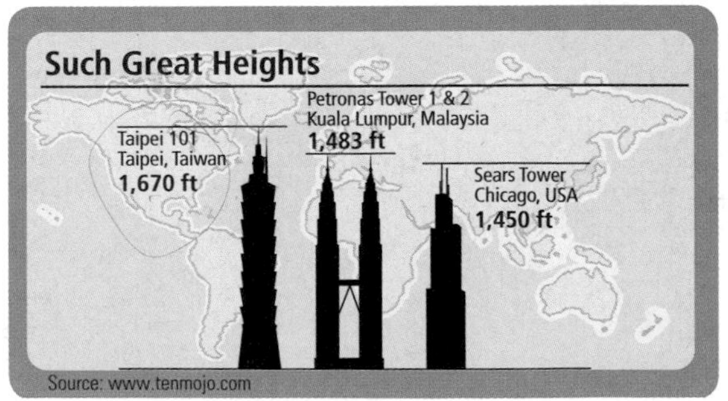

Wind Chill Table 2 is a table of wind chill temperatures. The top row gives the air temperature, and the first column is wind speed in miles per hour. The numbers within the table indicate how cold the weather will feel. For example, if the thermometer reads 30°F and the wind is blowing at 15 miles per hour, the wind chill temperature is 9°F. Use Table 2 to answer Questions 103 through 106.

TABLE 2

WIND CHILL TEMPERATURES

Wind Speed (mph)	Air Temperature (°F)							
	30°	25°	20°	15°	10°	5°	0°	−5°
10	16°	10°	3°	−3°	−9°	−15°	−22°	−27°
15	9°	2°	−5°	−11°	−18°	−25°	−31°	−38°
20	4°	−3°	−10°	−17°	−24°	−31°	−39°	−46°
25	1°	−7°	−15°	−22°	−29°	−36°	−44°	−51°
30	−2°	−10°	−18°	−25°	−33°	−41°	−49°	−56°

103. Reading Tables Find the wind chill temperature if the thermometer reads 20°F and the wind is blowing at 25 miles per hour.

104. Reading Tables Which will feel colder: a day with an air temperature of 10°F with a 25-mile-per-hour wind, or a day with an air temperature of −5° F and a 10-mile-per-hour wind?

105. Reading Tables Find the wind chill temperature if the thermometer reads 5°F and the wind is blowing at 15 miles per hour.

106. Reading Tables Which will feel colder: a day with an air temperature of 5°F with a 20-mile-per-hour wind, or a day with an air temperature of −5° F and a 10-mile-per-hour wind?

107. Scuba Diving Steve is scuba diving near his home in Maui. At one point he is 100 feet below the surface. Represent this number with a negative number. If he descends another 5 feet, what negative number will represent his new position?

108. Google Earth The Google Earth map shows Yellowstone National Park. If the boundary of the park is roughly a rectangle with a length of 63 miles and a width of 54 miles, find the perimeter and the area of the park.

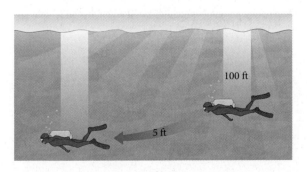

Calories and Exercise Table 3 gives the amount of energy expended per hour for various activities for a person weighing 120, 150, or 180 pounds. Use Table 3 to answer questions 109 – 112.

TABLE 3

ENERGY EXPENDED FROM EXERCISING

	Calories per Hour		
Activity	**120 lb**	**150 lb**	**180 lb**
Bicycling	299	374	449
Bowling	212	265	318
Handball	544	680	816
Horseback trotting	278	347	416
Jazzercise	272	340	408
Jogging	544	680	816
Skiing (downhill)	435	544	653

109. Suppose you weigh 120 pounds. How many calories will you burn if you play handball for 2 hours and then ride your bicycle for an hour?

110. How many calories are burned by a person weighing 150 pounds who jogs for $\frac{1}{2}$ hour and then goes bicycling for 2 hours?

111. Two people go skiing. One weighs 180 pounds and the other weighs 120 pounds. If they ski for 3 hours, how many more calories are burned by the person weighing 180 pounds?

112. Two people spend 3 hours bowling. If one weighs 120 pounds and the other weighs 150 pounds, how many more calories are burned during the evening by the person weighing 150 pounds?

Addition of Real Numbers

1.3

Objectives

A Add any combination of positive and negative numbers.

B Simplify expressions using the rule for order of operations.

C Extend an arithmetic sequence.

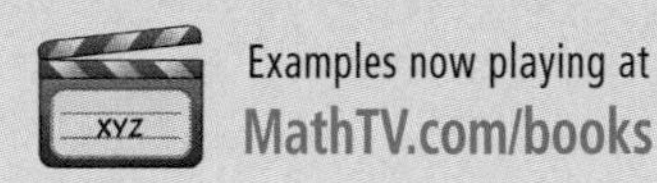

Suppose that you are playing a friendly game of poker with some friends, and you lose \$3 on the first hand and \$4 on the second hand. If you represent winning with positive numbers and losing with negative numbers, how can you translate this situation into symbols? Because you lost \$3 and \$4 for a total of \$7, one way to represent this situation is with addition of negative numbers:

$$(-\$3) + (-\$4) = -\$7$$

From this equation, we see that the sum of two negative numbers is a negative number. To generalize addition with positive and negative numbers, we use the number line.

Because real numbers have both a distance from zero (absolute value) and a direction from zero (sign), we can think of addition of two numbers in terms of distance and direction from zero.

A Adding Positive and Negative Numbers

Let's look at a problem for which we know the answer. Suppose we want to add the numbers 3 and 4. The problem is written 3 + 4. To put it on the number line, we read the problem as follows:

1. The 3 tells us to "start at the origin and move 3 units in the positive direction."

2. The + sign is read "and then move."

3. The 4 means "4 units in the positive direction."

To summarize, 3 + 4 means to start at the origin, move 3 units in the *positive* direction, and then move 4 units in the *positive* direction.

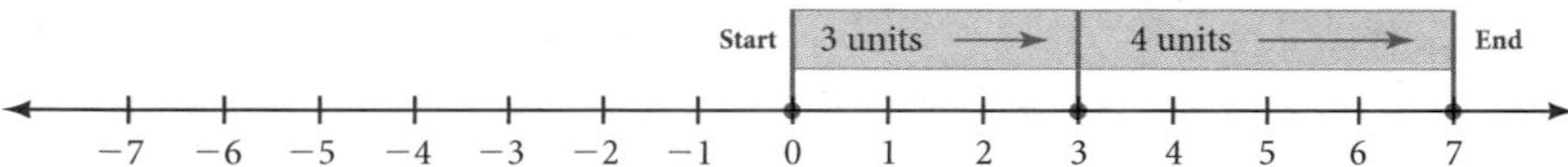

We end up at 7, which is the answer to our problem: 3 + 4 = 7.

Let's try adding other combinations of positive and negative 3 and 4 on the number line.

EXAMPLE 1 Add 3 + (−4).

SOLUTION Starting at the origin, move 3 units in the *positive* direction and then 4 units in the *negative* direction.

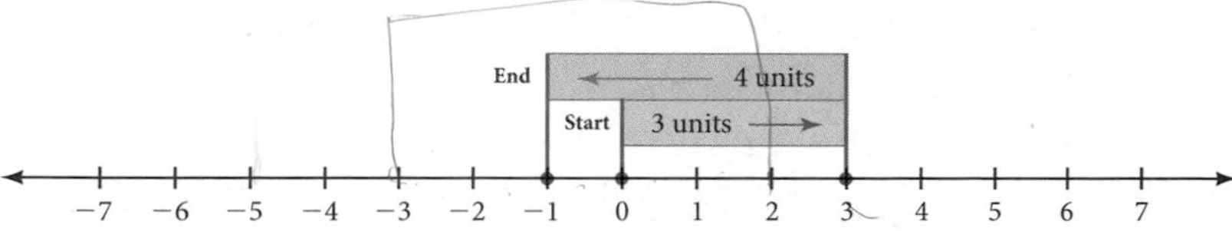

We end up at −1; therefore, 3 + (−4) = −1.

PRACTICE PROBLEMS

1. Add 2 + (−5). You can use the number line in Example 1 if you like.

Answer

1. −3

EXAMPLE 2 Add $-3 + 4$.

2. Add $-2 + 5$.

SOLUTION Starting at the origin, move 3 units in the *negative* direction and then 4 units in the *positive* direction.

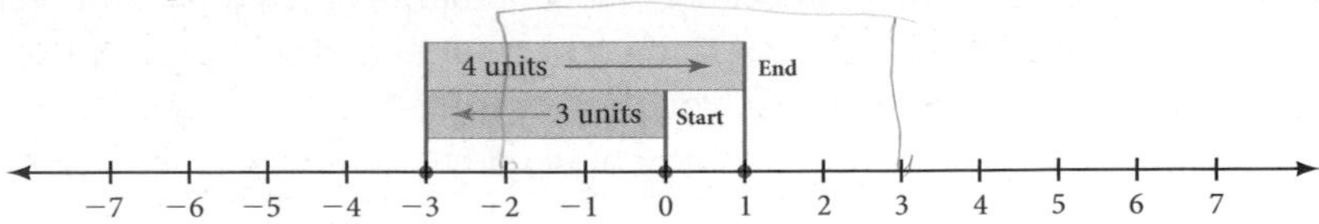

We end up at +1; therefore, $-3 + 4 = 1$.

EXAMPLE 3 Add $-3 + (-4)$.

3. Add $-2 + (-5)$.

SOLUTION Starting at the origin, move 3 units in the *negative* direction and then 4 units in the *negative* direction.

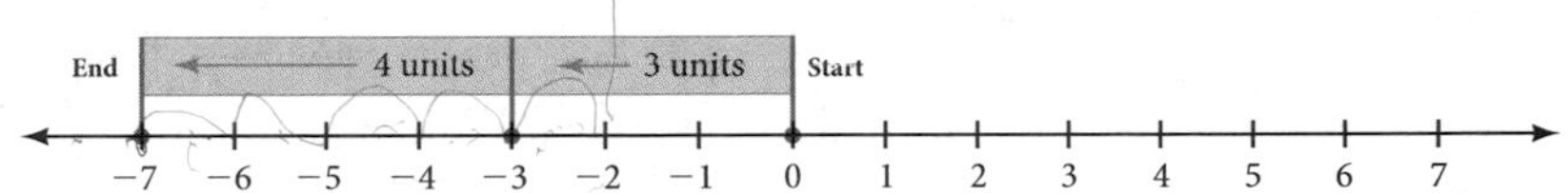

We end up at −7; therefore, $-3 + (-4) = -7$.

Here is a summary of what we have just completed:

$$3 + 4 = 7$$
$$3 + (-4) = -1$$
$$-3 + 4 = 1$$
$$-3 + (-4) = -7$$

Let's do four more problems on the number line and then summarize our results into a rule we can use to add any two real numbers.

EXAMPLE 4 Show that $5 + 7 = 12$.

4. Show $9 + (-4)$.

SOLUTION

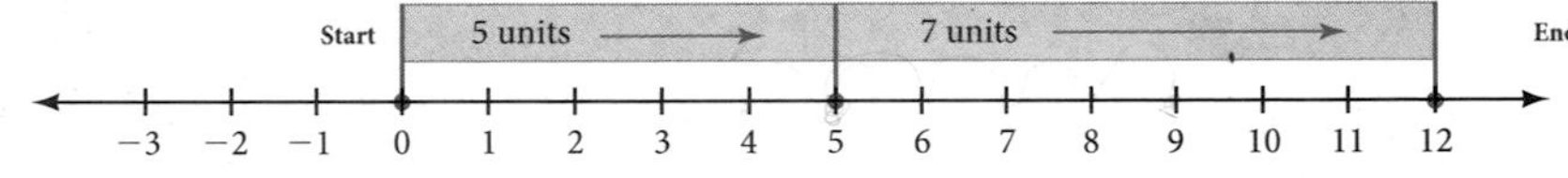

EXAMPLE 5 Show that $5 + (-7) = -2$.

5. Show $7 + (-3)$.

SOLUTION

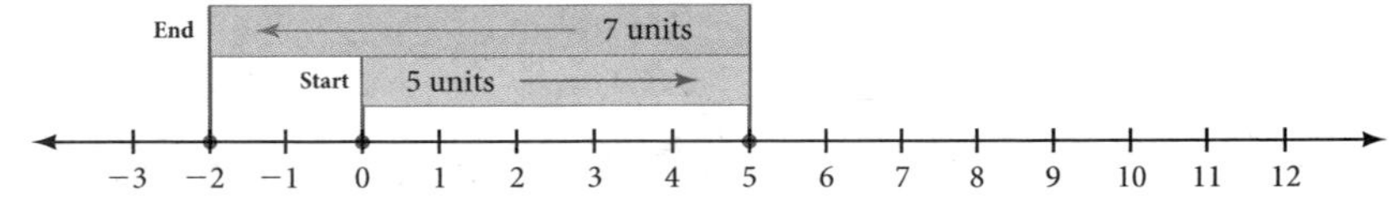

EXAMPLE 6 Show that $-5 + 7 = 2$.

6. Show $-10 + 12$.

SOLUTION

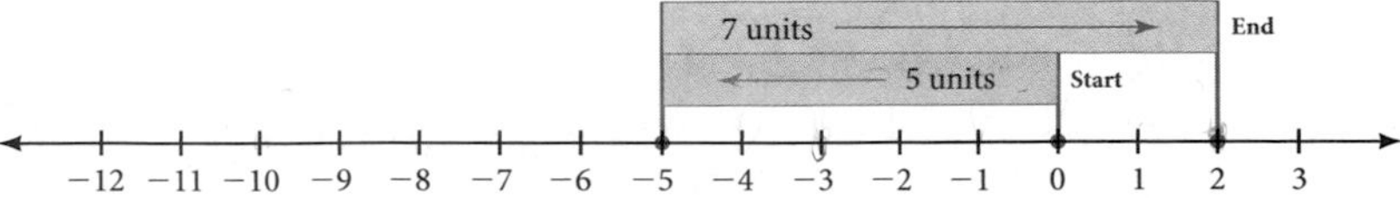

Answers
2. 3 **3.** −7 **4.** 5 **5.** 4 **6.** 2

EXAMPLE 7 Show that $-5 + (-7) = -12$.

SOLUTION

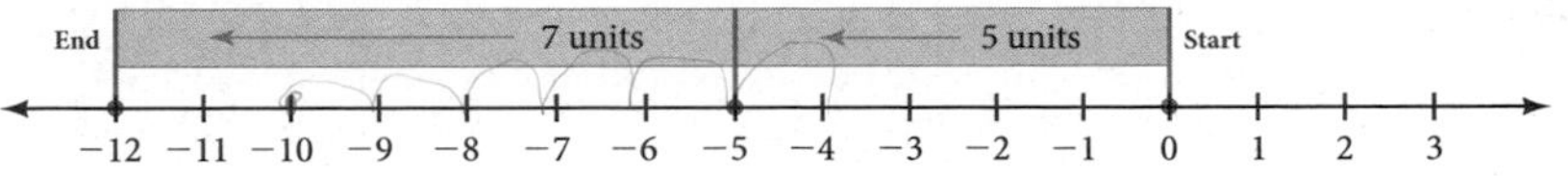

7. Show $-4 + (-6)$.

If we look closely at the results of the preceding addition problems, we can see that they support (or justify) the following rule.

Rule

To add two real numbers with

1. The *same* sign: Simply add their absolute values and use the common sign. (Both numbers are positive, the answer is positive. Both numbers are negative, the answer is negative.)
2. *Different* signs: Subtract the smaller absolute value from the larger. The answer will have the sign of the number with the larger absolute value.

Note This rule is what we have been working toward. The rule is very important. Be sure that you understand it and can use it. The problems we have done up to this point have been done simply to justify this rule. Now that we have the rule, we no longer need to do our addition problems on the number line.

This rule covers all possible combinations of addition with real numbers. You must memorize it. After you have worked a number of problems, it will seem almost automatic.

EXAMPLE 8 Add all combinations of positive and negative 10 and 13.

SOLUTION Rather than work these problems on the number line, we use the rule for adding positive and negative numbers to obtain our answers:

$$10 + 13 = 23$$
$$10 + (-13) = -3$$
$$-10 + 13 = 3$$
$$-10 + (-13) = -23$$

8. Add all combinations of positive and negative 9 and 12.

$9 + 12 =$

$9 + (-12) =$

$-9 + 12 =$

$-9 + (-12) =$

EXAMPLE 9 Add all combinations of positive and negative 12 and 17.

SOLUTION Applying the rule for adding positive and negative numbers, we have

$$12 + 17 = 29$$
$$12 + (-17) = -5$$
$$-12 + 17 = 5$$
$$-12 + (-17) = -29$$

9. Add all combinations of positive and negative 6 and 10.

$6 + 10 =$

$6 + (-10) =$

$-6 + 10 =$

$-6 + (-10) =$

B Order of Operations

EXAMPLE 10 Add $-3 + 2 + (-4)$.

SOLUTION Applying the rule for order of operations, we add left to right:

$$-3 + 2 + (-4) = -1 + (-4)$$
$$= -5$$

10. Add $-5 + 2 + (-7)$.

Answers

7. −10 8. 21, −3, 3, −21
9. 16, −4, 4, −16 10. −10

11. Add $-4 + [2 + (-3)] + (-1)$.

EXAMPLE 11 Add $-8 + [2 + (-5)] + (-1)$.

SOLUTION Adding inside the brackets first and then left to right, we have

$$\begin{aligned} -8 + [2 + (-5)] + (-1) &= -8 + (-3) + (-1) \\ &= -11 + (-1) \\ &= -12 \end{aligned}$$

12. Simplify $-5 + 3(-4 + 7) + (-10)$.

EXAMPLE 12 Simplify $-10 + 2(-8 + 11) + (-4)$.

SOLUTION First, we simplify inside the parentheses. Then, we multiply. Finally, we add left to right:

$$\begin{aligned} -10 + 2(-8 + 11) + (-4) &= -10 + 2(3) + (-4) \\ &= -10 + 6 + (-4) \\ &= -4 + (-4) \\ &= -8 \end{aligned}$$

C Arithmetic Sequences

The pattern in a sequence of numbers is easy to identify when each number in the sequence comes from the preceding number by adding the same amount.

Definition
An **arithmetic sequence** is a sequence of numbers in which each number (after the first number) comes from adding the same amount to the number before it.

13. Find the next two numbers in each of the following arithmetic sequences.
a. 5, 9, 13, . . .
b. 6, 6.5, 7, . . .
c. 4, 0, −4, . . .

EXAMPLE 13 Each sequence below is an arithmetic sequence. Find the next two numbers in each sequence.

a. 7, 10, 13, . . .

b. 9.5, 10, 10.5, . . .

c. 5, 0, −5, . . .

SOLUTION Because we know that each sequence is arithmetic, we know to look for the number that is added to each term to produce the next consecutive term.

a. 7, 10, 13, . . . : Each term is found by adding 3 to the term before it. Therefore, the next two terms will be 16 and 19.

b. 9.5, 10, 10.5, . . . : Each term comes from adding 0.5 to the term before it. Therefore, the next two terms will be 11 and 11.5.

c. 5, 0, −5, . . . : Each term comes from adding −5 to the term before it. Therefore, the next two terms will be $-5 + (-5) = -10$ and $-10 + (-5) = -15$.

Getting Ready for Class

After reading through the preceding section, respond in your own words and in complete sentences.

1. Explain how you would add 3 and −5 on the number line.
2. How do you add two negative numbers?
3. What is an arithmetic sequence?
4. Why is the sum of a number and its opposite always 0?

Answers
11. −6 **12.** −6
13. a. 17, 21 **b.** 7.5, 8 **c.** −8, −12

Problem Set 1.3

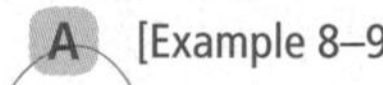

A [Example 8–9]

1. Add all combinations of positive and negative 3 and 5.

2. Add all combinations of positive and negative 6 and 4.

3. Add all combinations of positive and negative 15 and 20.

4. Add all combinations of positive and negative 18 and 12.

A Work the following problems. You may want to begin by doing a few on the number line. [Example 1–7]

5. $6 + (-3)$

6. $7 + (-8)$

7. $13 + (-20)$

8. $15 + (-25)$

9. $18 + (-32)$

10. $6 + (-9)$

11. $-6 + 3$

12. $-8 + 7$

13. $-30 + 5$

14. $-18 + 6$

15. $-6 + (-6)$

16. $-5 + (-5)$

17. $-9 + (-10)$

18. $-8 + (-6)$

19. $-10 + (-15)$

20. $-18 + (-30)$

B Work the following problems using the rule for addition of real numbers. You may want to refer back to the rule for order of operations. [Example 10–11]

21. $5 + (-6) + (-7)$

22. $6 + (-8) + (-10)$

23. $-7 + 8 + (-5)$

24. $-6 + 9 + (-3)$

25. $5 + [6 + (-2)] + (-3)$

26. $10 + [8 + (-5)] + (-20)$

27. $[6 + (-2)] + [3 + (-1)]$

28. $[18 + (-5)] + [9 + (-10)]$

29. $20 + (-6) + [3 + (-9)]$

30. $18 + (-2) + [9 + (-13)]$

31. $-3 + (-2) + [5 + (-4)]$

32. $-6 + (-5) + [-4 + (-1)]$

33. $(-9 + 2) + [5 + (-8)] + (-4)$

34. $(-7 + 3) + [9 + (-6)] + (-5)$

35. $[-6 + (-4)] + [7 + (-5)] + (-9)$

36. $[-8 + (-1)] + [8 + (-6)] + (-6)$

37. $(-6 + 9) + (-5) + (-4 + 3) + 7$

38. $(-10 + 4) + (-3) + (-3 + 8) + 6$

B The problems that follow involve some multiplication. Be sure that you work inside the parentheses first, then multiply, and finally, add left to right. [Example 12]

39. $-5 + 2(-3 + 7)$

40. $-3 + 4(-2 + 7)$

41. $9 + 3(-8 + 10)$

42. $4 + 5(-2 + 6)$

43. $-10 + 2(-6 + 8) + (-2)$

44. $-20 + 3(-7 + 10) + (-4)$

45. $2(-4 + 7) + 3(-6 + 8)$

46. $5(-2 + 5) + 7(-1 + 6)$

Add.

47. 3.9 + 7.1

48. 4.7 + 4.3

49. 8.1 + 2.7

50. 2.4 + 7.3

Solve.

51. Find the value of $0.06x + 0.07y$ when $x = 7{,}000$ and $y = 8{,}000$.

52. Find the value of $0.06x + 0.07y$ when $x = 7{,}000$ and $y = 3{,}000$.

53. Find the value of $0.05x + 0.10y$ when $x = 10$ and $y = 12$.

54. Find the value of $0.25x + 0.10y$ when $x = 3$ and $y = 11$.

C Each sequence below is an arithmetic sequence. In each case, find the next two numbers in the sequence. [Example 13]

55. 3, 8, 13, 18, . . .

56. 1, 5, 9, 13, . . .

57. 10, 15, 20, 25, . . .

58. 10, 16, 22, 28, . . .

59. 20, 15, 10, 5, . . .

60. 24, 20, 16, 12, . . .

61. 6, 0, −6, . . .

62 1, 0, −1, . . .

63. 8, 4, 0, . . .

64. 5, 2, −1, . . .

65. Is the sequence of odd numbers an arithmetic sequence?

66. Is the sequence of squares an arithmetic sequence?

Recall that the word *sum* indicates addition. Write the numerical expression that is equivalent to each of the following phrases and then simplify.

67. The sum of 5 and 9

68. The sum of 6 and −3

69. Four added to the sum of −7 and −5

70. Six added to the sum of −9 and 1

71. The sum of −2 and −3 increased by 10

72. The sum of −4 and −12 increased by 2

Answer the following questions.

73. What number do you add to −8 to get −5?

74. What number do you add to 10 to get 4?

75. The sum of what number and −6 is −9?

76. The sum of what number and −12 is 8?

Applying the Concepts

77. Music The chart shows the country music artists that earned the most in 2007. How much did Toby Keith and Tim McGraw make combined?

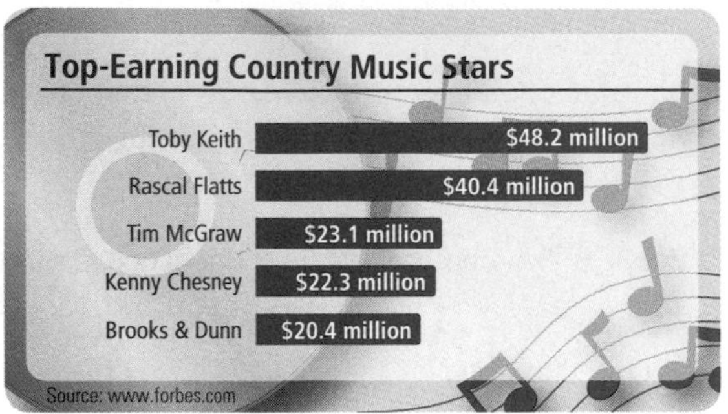

78. Computers The chart shows the three countries with the most computers. What is the total number of computers that can be found in the three countries?

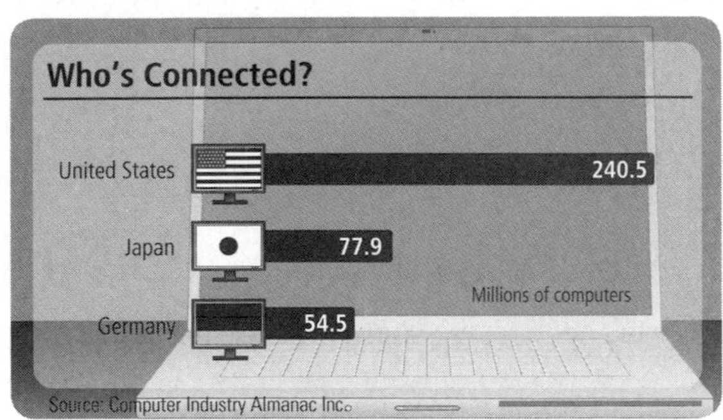

79. Temperature Change The temperature at noon is 12 degrees below 0 Fahrenheit. By 1:00 it has risen 4 degrees. Write an expression using the numbers -12 and 4 to describe this situation.

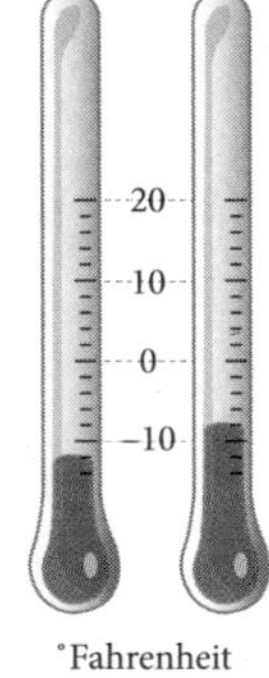

80. Stock Value On Monday a certain stock gains 2 points. On Tuesday it loses 3 points. Write an expression using positive and negative numbers with addition to describe this situation and then simplify.

81. Gambling On three consecutive hands of draw poker, a gambler wins \$10, loses \$6, and then loses another \$8. Write an expression using positive and negative numbers and addition to describe this situation and then simplify.

82. Number Problem You know from your past experience with numbers that subtracting 5 from 8 results in 3 $(8 - 5 = 3)$. What addition problem that starts with the number 8 gives the same result?

83. Checkbook Balance Suppose that you balance your checkbook and find that you are overdrawn by \$30; that is, your balance is $-\$30$. Then you go to the bank and deposit \$40. Translate this situation into an addition problem, the answer to which gives the new balance in your checkbook.

84. Checkbook Balance The balance in your checkbook is $-\$25$. If you make a deposit of \$75, and then write a check for \$18, what is the new balance?

1.4 Subtraction of Real Numbers

Objectives

A Subtract any combination of positive and negative numbers.

B Simplify expressions using the rule for order of operations.

C Translate sentences from English into symbols and then simplify.

D Find the complement and the supplement of an angle.

Suppose that the temperature at noon is 20° Fahrenheit and 12 hours later, at midnight, it has dropped to −15° Fahrenheit. What is the difference between the temperature at noon and the temperature at midnight? Intuitively, we know the difference in the two temperatures is 35°. We also know that the word difference indicates subtraction. The difference between 20 and −15 is written

$$20 - (-15)$$

It must be true that $20 - (-15) = 35$. In this section we will see how our definition for subtraction confirms that this last statement is in fact correct.

A Subtracting Positive and Negative Numbers

In the previous section we spent some time developing the rule for addition of real numbers. Because we want to make as few rules as possible, we can define subtraction in terms of addition. By doing so, we can then use the rule for addition to solve our subtraction problems.

Rule

To subtract one real number from another, simply add its opposite.

Algebraically, the rule is written like this: If a and b represent two real numbers, then it is always true that

$$\underbrace{a - b}_{\text{To subtract } b} = \underbrace{a + (-b)}_{\text{add the opposite of } b}$$

This is how subtraction is defined in algebra. This definition of subtraction will not conflict with what you already know about subtraction, but it will allow you to do subtraction using negative numbers.

EXAMPLE 1 Subtract all possible combinations of positive and negative 7 and 2.

SOLUTION

$$\left.\begin{aligned} 7 - 2 &= 7 + (-2) = 5 \\ -7 - 2 &= -7 + (-2) = -9 \end{aligned}\right\}$$ Subtracting 2 is the same as adding −2

$$\left.\begin{aligned} 7 - (-2) &= 7 + 2 = 9 \\ -7 - (-2) &= -7 + 2 = -5 \end{aligned}\right\}$$ Subtracting −2 is the same as adding 2

Notice that each subtraction problem is first changed to an addition problem. The rule for addition is then used to arrive at the answer.

We have defined subtraction in terms of addition, and we still obtain answers consistent with the answers we are used to getting with subtraction. Moreover, we now can do subtraction problems involving both positive and negative numbers.

As you proceed through the following examples and the problem set, you will begin to notice shortcuts you can use in working the problems. You will not always have to change subtraction to addition of the opposite to be able to get answers quickly. Use all the shortcuts you wish as long as you consistently get the correct answers.

PRACTICE PROBLEMS

1. Subtract
   $7 - 4 =$
   $-7 - 4 =$
   $7 - (-4) =$
   $-7 - (-4) =$

Answer

1. 3, −11, 11, −3

2. Subtract

$6 - 4 =$

$-6 - 4 =$

$6 - (-4) =$

$-6 - (-4) =$

EXAMPLE 2 Subtract all combinations of positive and negative 8 and 13.

SOLUTION

$$8 - 13 = 8 + (-13) = -5$$
$$-8 - 13 = -8 + (-13) = -21$$

Subtracting +13 is the same as adding −13

$$8 - (-13) = 8 + 13 = 21$$
$$-8 - (-13) = -8 + 13 = 5$$

Subtracting −13 is the same as adding +13

B Using Order of Operations

Simplify each expression as much as possible.

3. $4 + (-2) - 3$

EXAMPLE 3 Simplify the expression $7 + (-3) - 5$ as much as possible.

SOLUTION

$$\begin{aligned} 7 + (-3) - 5 &= 7 + (-3) + (-5) && \text{Begin by changing all subtractions to additions} \\ &= 4 + (-5) \\ &= -1 && \text{Then add left to right} \end{aligned}$$

4. $9 - 4 - 5$

EXAMPLE 4 Simplify the expression $8 - (-2) - 6$ as much as possible.

SOLUTION

$$\begin{aligned} 8 - (-2) - 6 &= 8 + 2 + (-6) && \text{Begin by changing all subtractions to additions} \\ &= 10 + (-6) \\ &= 4 && \text{Then add left to right} \end{aligned}$$

5. $-3 - (-5 + 1) - 6$

EXAMPLE 5 Simplify the each expression $-2 - (-3 + 1) - 5$ as much as possible.

SOLUTION

$$\begin{aligned} -2 - (-3 + 1) - 5 &= -2 - (-2) - 5 && \text{Do what is in the parentheses first} \\ &= -2 + 2 + (-5) \\ &= -5 \end{aligned}$$

The next two examples involve multiplication and exponents as well as subtraction. Remember, according to the rule for order of operations, we evaluate the numbers containing exponents and multiply before we subtract.

6. Simplify $3 \cdot 9 - 4 \cdot 10 - 5 \cdot 11$.

EXAMPLE 6 Simplify $2 \cdot 5 - 3 \cdot 8 - 4 \cdot 9$.

SOLUTION First, we multiply left to right, and then we subtract.

$$\begin{aligned} 2 \cdot 5 - 3 \cdot 8 - 4 \cdot 9 &= 10 - 24 - 36 \\ &= -14 - 36 \\ &= -50 \end{aligned}$$

7. Simplify $4 \cdot 2^5 - 3 \cdot 5^2$.

EXAMPLE 7 Simplify $3 \cdot 2^3 - 2 \cdot 4^2$.

SOLUTION We begin by evaluating each number that contains an exponent. Then we multiply before we subtract:

$$\begin{aligned} 3 \cdot 2^3 - 2 \cdot 4^2 &= 3 \cdot 8 - 2 \cdot 16 \\ &= 24 - 32 \\ &= -8 \end{aligned}$$

Answers

2. 2, −10, 10, −2 **3.** −1 **4.** 0
5. −5 **6.** −68 **7.** 53

C Writing Words in Symbols

EXAMPLE 8 Subtract 7 from −3.

SOLUTION First, we write the problem in terms of subtraction. We then change to addition of the opposite:

$$\begin{aligned} -3 - 7 &= -3 + (-7) \\ &= -10 \end{aligned}$$

EXAMPLE 9 Subtract −5 from 2.

SOLUTION Subtracting −5 is the same as adding +5:

$$\begin{aligned} 2 - (-5) &= 2 + 5 \\ &= 7 \end{aligned}$$

EXAMPLE 10 Find the difference of 9 and 2.

SOLUTION Written in symbols, the problem looks like this:

$$9 - 2 = 7$$

The difference of 9 and 2 is 7.

EXAMPLE 11 Find the difference of 3 and −5.

SOLUTION Subtracting −5 from 3 we have

$$\begin{aligned} 3 - (-5) &= 3 + 5 \\ &= 8 \end{aligned}$$

8. Subtract 8 from −5.

9. Subtract −3 from 8.

10. Find the difference of 8 and 6.

11. Find the difference of 4 and −7.

D Complementary and Supplementary Angles

We can apply our knowledge of algebra to help solve some simple geometry problems. Before we do, however, we need to review some of the vocabulary associated with angles.

Definition

In geometry, two angles that add to 90° are called **complementary angles.** In a similar manner, two angles that add to 180° are called **supplementary angles.** The diagrams at the left illustrate the relationships between angles that are complementary and between angles that are supplementary.

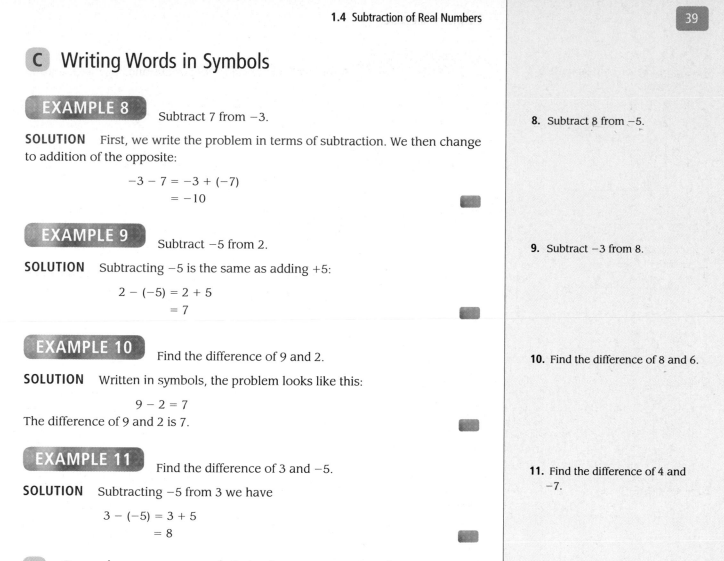

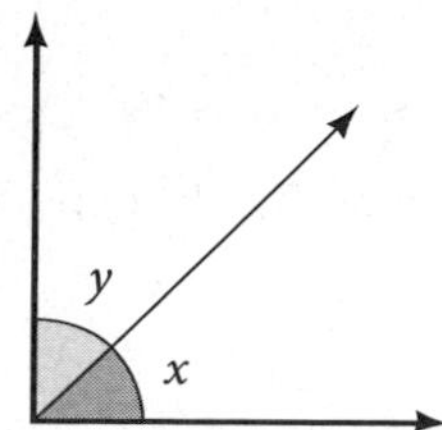

Complementary angles: $x + y = 90°$

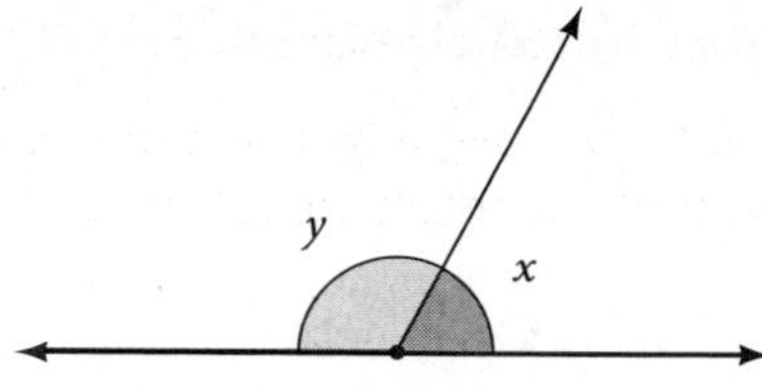

Supplementary angles: $x + y = 180°$

Answers
8. −13 **9.** 11 **10.** 2 **11.** 11

12. Find x in each of the following diagrams.

a.

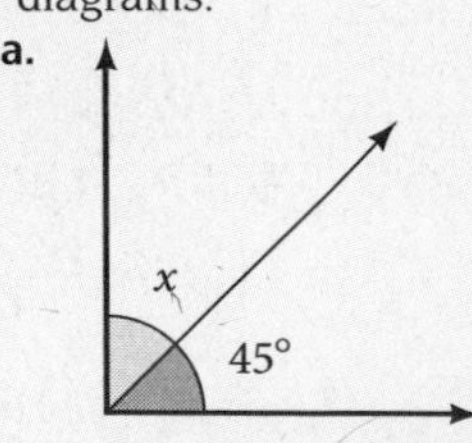

b.

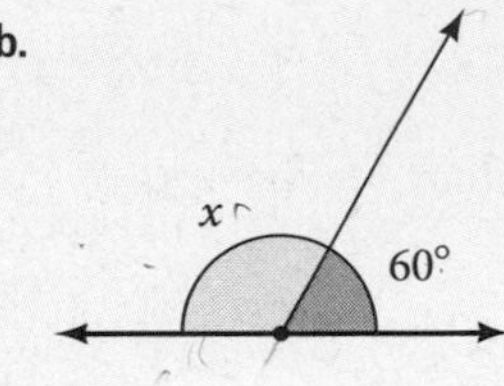

Answers
12. a. 45° b. 120°

EXAMPLE 12 Find x in each of the following diagrams.

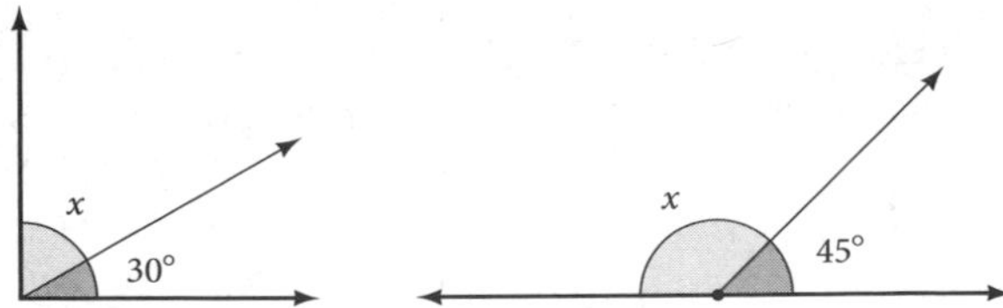

SOLUTION We use subtraction to find each angle.

a. Because the two angles are complementary, we can find x by subtracting 30° from 90°:

$$x = 90° - 30° = 60°$$

We say 30° and 60° are complementary angles. The complement of 30° is 60°.

b. The two angles in the diagram are supplementary. To find x, we subtract 45° from 180°:

$$x = 180° - 45° = 135°$$

We say 45° and 135° are supplementary angles. The supplement of 45° is 135°.

Subtraction and Taking Away

For some people taking algebra for the first time, subtraction of positive and negative numbers can be a problem. These people may believe that $-5 - 9$ should be -4 or 4, not -14. If this is happening to you, you probably are thinking of subtraction in terms of taking one number away from another. Thinking of subtraction in this way works well with positive numbers if you always subtract the smaller number from the larger. In algebra, however, we encounter many situations other than this. The definition of subtraction, that $a - b = a + (-b)$, clearly indicates the correct way to use subtraction; that is, when working subtraction problems, you should think "addition of the opposite," not "take one number away from another." To be successful in algebra, you need to apply properties and definitions exactly as they are presented here.

Getting Ready for Class

After reading through the preceding section, respond in your own words and in complete sentences.

1. Why do we define subtraction in terms of addition?
2. Write the definition for $a - b$.
3. Explain in words how you would subtract 3 from -7.
4. What are complementary angles?

Problem Set 1.4

The following problems are intended to give you practice with subtraction of positive and negative numbers. Remember in algebra subtraction is not taking one number away from another. Instead, subtracting a number is equivalent to adding its opposite.

A Subtract. [Examples 1–2}

1. $5 - 8$ **2.** $6 - 7$ **3.** $3 - 9$ **4.** $2 - 7$ **5.** $5 - 5$ **6.** $8 - 8$

7. $-8 - 2$ **8.** $-6 - 3$ **9.** $-4 - 12$ **10:** $-3 - 15$

11. $-6 - 6$ **12.** $-3 - 3$ **13.** $-8 - (-1)$ **14.** $-6 - (-2)$

15. $15 - (-20)$ **16.** $20 - (-5)$ **17.** $-4 - (-4)$ **18.** $-5 - (-5)$

B Simplify each expression by applying the rule for order of operations. [Examples 3–5]

19. $3 - 2 - 5$ **20.** $4 - 8 - 6$ **21.** $9 - 2 - 3$ **22.** $8 - 7 - 12$

23. $-6 - 8 - 10$ **24.** $-5 - 7 - 9$ **25.** $-22 + 4 - 10$ **26.** $-13 + 6 - 5$

27. $10 - (-20) - 5$ **28.** $15 - (-3) - 20$ **29.** $8 - (2 - 3) - 5$ **30.** $10 - (4 - 6) - 8$

31. $7 - (3 - 9) - 6$ **32.** $4 - (3 - 7) - 8$ **33.** $5 - (-8 - 6) - 2$ **34.** $4 - (-3 - 2) - 1$

35. $-(5 - 7) - (2 - 8)$

36. $-(4 - 8) - (2 - 5)$

37. $-(3 - 10) - (6 - 3)$

38. $-(3 - 7) - (1 - 2)$

39. $16 - [(4 - 5) - 1]$

40. $15 - [(4 - 2) - 3]$

41. $5 - [(2 - 3) - 4]$

42. $6 - [(4 - 1) - 9]$

43. $21 - [-(3 - 4) - 2] - 5$

44. $30 - [-(10 - 5) - 15] - 25$

B The following problems involve multiplication and exponents. Use the rule for order of operations to simplify each expression as much as possible. [Examples 6–7]

45. $2 \cdot 8 - 3 \cdot 5$

46. $3 \cdot 4 - 6 \cdot 7$

47. $3 \cdot 5 - 2 \cdot 7$

48. $6 \cdot 10 - 5 \cdot 20$

49. $5 \cdot 9 - 2 \cdot 3 - 6 \cdot 2$

50. $4 \cdot 3 - 7 \cdot 1 - 9 \cdot 4$

51. $3 \cdot 8 - 2 \cdot 4 - 6 \cdot 7$

52. $5 \cdot 9 - 3 \cdot 8 - 4 \cdot 5$

53. $2 \cdot 3^2 - 5 \cdot 2^2$

54. $3 \cdot 7^2 - 2 \cdot 8^2$

55. $4 \cdot 3^3 - 5 \cdot 2^3$

56. $3 \cdot 6^2 - 2 \cdot 3^2 - 8 \cdot 6^2$

Subtract.

57. $-3.4 - 7.9$

58. $-3.5 - 2.3$

59. $3.3 - 6.9$

60. $2.2 - 7.5$

Solve.

61. Find the value of $x + y - 4$ when

a. $x = -3$ and $y = -2$

b. $x = -9$ and $y = 3$

c. $x = -\frac{3}{5}$ and $y = \frac{8}{5}$

62. Find the value of $x - y - 3$ when

a. $x = -4$ and $y = 1$

b. $x = 2$ and $y = -1$

c. $x = -5$ and $y = -2$

C Rewrite each of the following phrases as an equivalent expression in symbols, and then simplify. [Examples 8–12]

63. Subtract 4 from -7.

64. Subtract 5 from -19.

65. Subtract -8 from 12.

66. Subtract -2 from 10.

67. Subtract -7 from -5.

68. Subtract -9 from -3.

69. Subtract 17 from the sum of 4 and -5.

70. Subtract -6 from the sum of 6 and -3.

C Recall that the word *difference* indicates subtraction. The difference of a and b is $a - b$, in that order. Write a numerical expression that is equivalent to each of the following phrases, and then simplify. [Example 12]

71. The difference of 8 and 5

72. The difference of 5 and 8

73. The difference of -8 and 5

74. The difference of -5 and 8

75. The difference of 8 and -5

76. The difference of 5 and -8

C Answer the following questions.

77. What number do you subtract from 8 to get -2?

78. What number do you subtract from 1 to get -5?

79. What number do you subtract from 8 to get 10?

80. What number do you subtract from 1 to get 5?

Applying the Concepts

81. Pitchers The chart shows the number of strikeouts for active starting pitchers with the most career strikeouts. How many more strikeouts does Randy Johnson have than Greg Maddux?

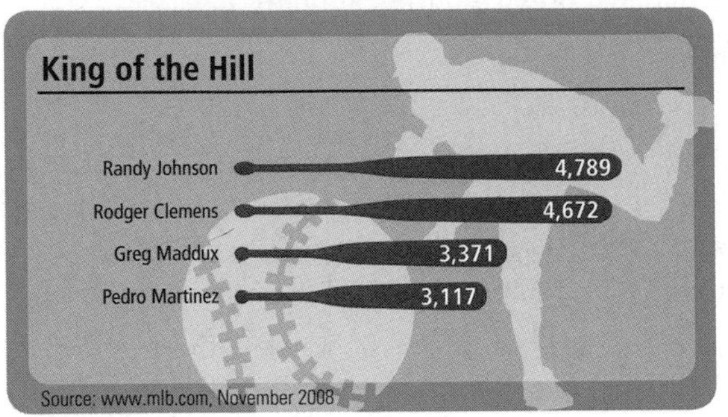

82. Sound The low-frequency pulses made by blue whales have been measured up to 188 decibels (dB) making it the loudest sound emitted by any living source. How much louder is a blue whale than normal conversation?

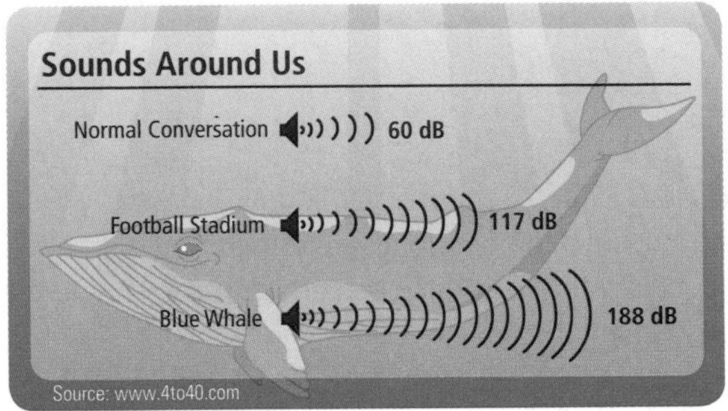

83. Savings Account Balance A man with $1,500 in a savings account makes a withdrawal of $730. Write an expression using subtraction that describes this situation.

First Bank — Account No. 12345

Date	Withdrawals	Deposits	Balance
1/1/08			1,500
2/2/08	730		

84. Temperature Change The temperature inside a space shuttle is 73°F before reentry. During reentry the temperature inside the craft increases 10°. On landing it drops 8°F. Write an expression using the numbers 73, 10, and 8 to describe this situation. What is the temperature inside the shuttle on landing?

85. Gambling A man who has lost $35 playing roulette in Las Vegas wins $15 playing blackjack. He then loses $20 playing the wheel of fortune. Write an expression using the numbers −35, 15, and 20 to describe this situation and then simplify it.

86. Altitude Change An airplane flying at 10,000 feet lowers its altitude by 1,500 feet to avoid other air traffic. Then it increases its altitude by 3,000 feet to clear a mountain range. Write an expression that describes this situation and then simplify it.

87. Checkbook Balance Bob has $98 in his checking account when he writes a check for $65 and then another check for $53. Write a subtraction problem that gives the new balance in Bob's checkbook. What is his new balance?

88. Temperature Change The temperature at noon is 23°F. Six hours later it has dropped 19°F, and by midnight it has dropped another 10°F. Write a subtraction problem that gives the temperature at midnight. What is the temperature at midnight?

89. Depreciation Stacey buys a used car for $4,500. With each year that passes, the car drops $550 in value. Write a sequence of numbers that gives the value of the car at the beginning of each of the first 5 years she owns it. Can this sequence be considered an arithmetic sequence?

90. Depreciation Wade buys a computer system for $6,575. Each year after that he finds that the system is worth $1,250 less than it was the year before. Write a sequence of numbers that gives the value of the computer system at the beginning of each of the first four years he owns it. Can this sequence be considered an arithmetic sequence?

Drag Racing In the sport of drag racing, two cars at the starting line race to the finish line $\frac{1}{4}$ mile away. The car that crosses the finish line first wins the race. Jim Rizzoli owns and races an alcohol dragster. On board the dragster is a computer that records data during each of Jim's races. Table 1 gives some of the data from a race Jim was in. In addition to showing the time and speed of Jim Rizzoli's dragster during a race, it also shows the distance past the starting line that his dragster has traveled. Use the information in the table shown here to answer the following questions.

SPEED AND DISTANCE FOR A RACE CAR

Time in Seconds	Speed in Miles/Hour	Distance Traveled in Feet
0	0	0
1	72.7	69
2	129.9	231
3	162.8	439
4	192.2	728
5	212.4	1,000
6	228.1	1,373

91. Find the difference in the distance traveled by the dragster after 5 seconds and after 2 seconds.

92. How much faster is he traveling after 4 seconds than he is after 2 seconds?

93. How far from the starting line is he after 3 seconds?

94. How far from the starting line is he when his speed is 192.2 miles per hour?

95. How many seconds have gone by between the time his speed is 162.8 miles per hour and the time at which he has traveled 1,000 feet?

96. How many seconds have gone by between the time at which he has traveled 231 feet and the time at which his speed is 228.1 miles per hour?

D Find x in each of the following diagrams. [Example 13]

97.

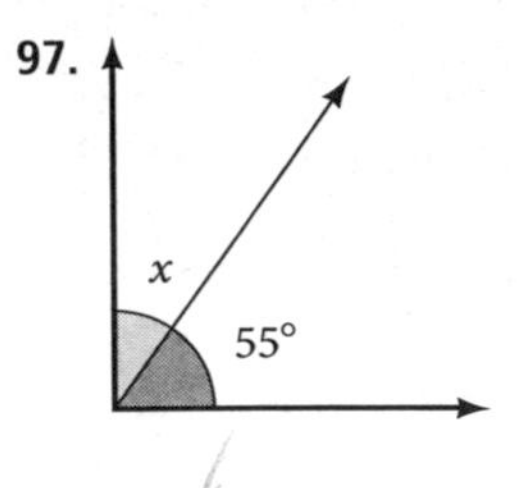

98.

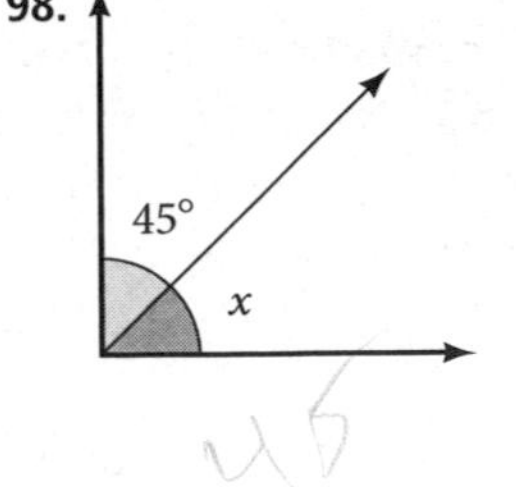

99.

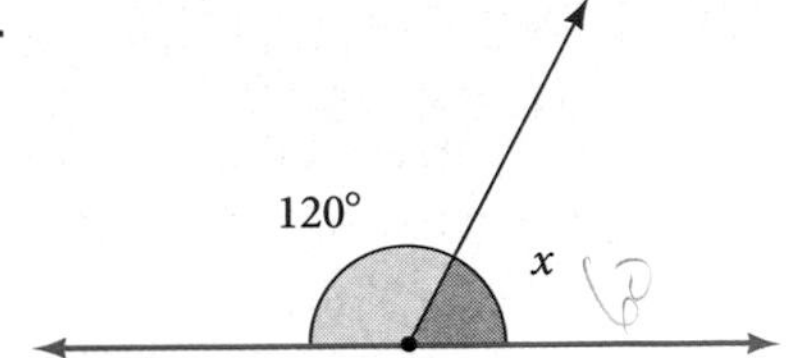

100.

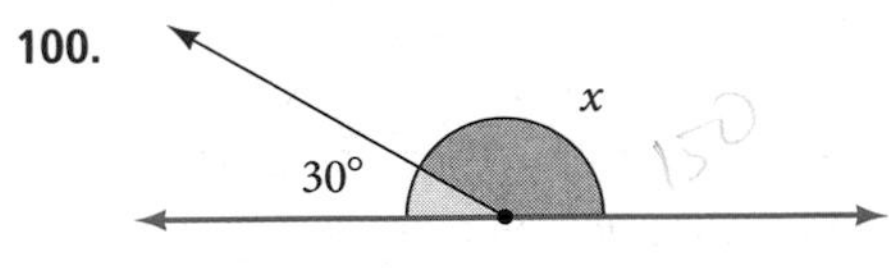

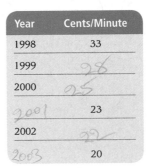

101. Grass Growth The bar chart below shows the growth of a certain species of grass over a period of 10 days.

a. Use the chart to fill in the missing entries in the table.

b. How much higher is the grass after 8 days than after 3 days?

Day	Plant Height (inches)	Day	Plant Height (inches)
0	0	6	
1	0.5	7	
2			13
3	1.5	9	18
4		10	
5	4		

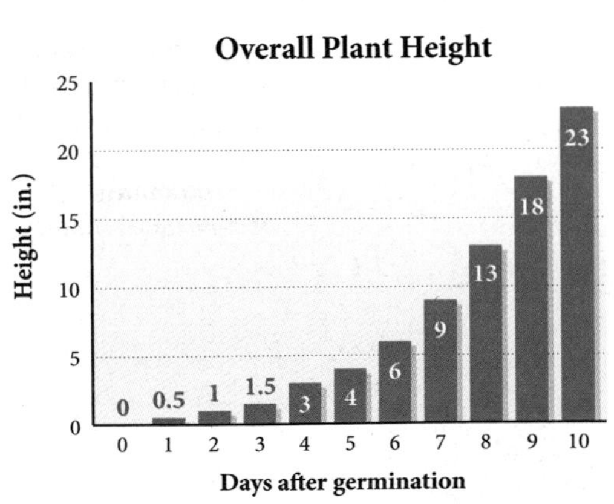

102. Wireless Phone Costs The bar chart below shows the projected cost of wireless phone use through 2003.

a. Use the chart to fill in the missing entries in the table.

b. What is the difference in cost between 1998 and 1999?

Year	Cents/Minute
1998	33
1999	
2000	
	23
2002	
	20

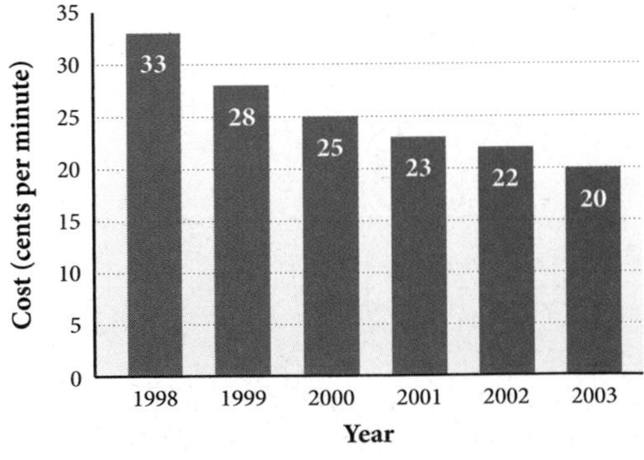

1.5 Properties of Real Numbers

Objectives

A Rewrite expressions using the commutative and associative properties.

B Multiply using the distributive property.

C Identify properties used to rewrite an expression.

In this section we will list all the facts (properties) that you know from past experience are true about numbers in general. We will give each property a name so we can refer to it later in this book. Mathematics is very much like a game. The game involves numbers. The rules of the game are the properties and rules we are developing in this chapter. The goal of the game is to extend the basic rules to as many situations as possible.

You know from past experience with numbers that it makes no difference in which order you add two numbers; that is, $3 + 5$ is the same as $5 + 3$. This fact about numbers is called the *commutative property of addition.* We say addition is a commutative operation. Changing the order of the numbers does not change the answer.

There is one other basic operation that is commutative. Because $3(5)$ is the same as $5(3)$, we say multiplication is a commutative operation. Changing the order of the two numbers you are multiplying does not change the answer.

A The Commutative and Associative Properties

For all properties listed in this section, a, b, and c represent real numbers.

Commutative Property of Addition

In symbols: $a + b = b + a$

In words: Changing the *order* of the numbers in a sum will not change the result.

Commutative Property of Multiplication

In symbols: $a \cdot b = b \cdot a$

In words: Changing the *order* of the numbers in a product will not change the result.

EXAMPLES

1. The statement $5 + 8 = 8 + 5$ is an example of the commutative property of addition.
2. The statement $2 \cdot y = y \cdot 2$ is an example of the commutative property of multiplication.
3. The expression $5 + x + 3$ can be simplified using the commutative property of addition:

$$5 + x + 3 = x + 5 + 3 \quad \text{Commutative property of addition}$$
$$= x + 8 \quad \text{Addition}$$

The other two basic operations, subtraction and division, are not commutative. The order in which we subtract or divide two numbers makes a difference in the answer.

Another property of numbers that you have used many times has to do with grouping. You know that when we add three numbers it makes no difference which two we add first. When adding $3 + 5 + 7$, we can add the 3 and 5 first and then the 7, or we can add the 5 and 7 first and then the 3. Mathematically, it looks like this: $(3 + 5) + 7 = 3 + (5 + 7)$. This property is true of multiplication as well. Operations that behave in this manner are called *associative* operations. The answer will not change when we change the association (or grouping) of the numbers.

PRACTICE PROBLEMS

1. Give an example of the commutative property of addition using 2 and y.
2. Give an example of the commutative property of multiplication using the numbers 5 and 8.
3. Use the commutative property of addition to simplify $4 + x + 7$.

Answers

1. $2 + y = y + 2$ 2. $5(8) = 8(5)$
3. $x + 11$

Associative Property of Addition

In symbols: $a + (b + c) = (a + b) + c$

In words: Changing the *grouping* of the numbers in a sum will not change the result.

Associative Property of Multiplication

In symbols: $a(bc) = (ab)c$

In words: Changing the *grouping* of the numbers in a product will not change the result.

The following examples illustrate how the associative properties can be used to simplify expressions that involve both numbers and variables.

Simplify.

4. $3 + (7 + x)$

5. $3(4x)$

6. $\frac{1}{3}(3x)$

7. $4\left(\frac{1}{4}x\right)$

8. $15\left(\frac{3}{5}x\right)$

Answers

4. $10 + x$ 5. $12x$ 6. x 7. x 8. $9x$

EXAMPLE 4 Simplify $4 + (5 + x)$.

SOLUTION

$4 + (5 + x) = (4 + 5) + x$ Associative property of addition

$= 9 + x$ Addition

EXAMPLE 5 Simplify $5(2x)$.

SOLUTION

$5(2x) = (5 \cdot 2)x$ Associative property of multiplication

$= 10x$ Multiplication

EXAMPLE 6 Simplify $\frac{1}{5}(5x)$.

SOLUTION

$\frac{1}{5}(5x) = \left(\frac{1}{5} \cdot 5\right)x$ Associative property of multiplication

$= 1x$ Multiplication

$= x$

EXAMPLE 7 Simplify $3\left(\frac{1}{3}x\right)$.

SOLUTION

$3\left(\frac{1}{3}x\right) = \left(3 \cdot \frac{1}{3}\right)x$ Associative property of multiplication

$= 1x$ Multiplication

$= x$

EXAMPLE 8 Simplify $12\left(\frac{2}{3}x\right)$.

SOLUTION

$12\left(\frac{2}{3}x\right) = \left(12 \cdot \frac{2}{3}\right)x$ Associative property of multiplication

$= 8x$ Multiplication

B The Distributive Property

The associative and commutative properties apply to problems that are either all multiplication or all addition. There is a third basic property that involves both addition and multiplication. It is called the *distributive property* and looks like this.

Distributive Property

In symbols: $a(b + c) = ab + ac$

In words: Multiplication *distributes* over addition.

You will see as we progress through the book that the distributive property is used very frequently in algebra. We can give a visual justification to the distributive property by finding the areas of rectangles. Figure 1 shows a large rectangle that is made up of two smaller rectangles. We can find the area of the large rectangle two different ways.

Note Because subtraction is defined in terms of addition, it is also true that the distributive property applies to subtraction as well as addition; that is, $a(b - c) = ab - ac$ for any three real numbers a, b, and c.

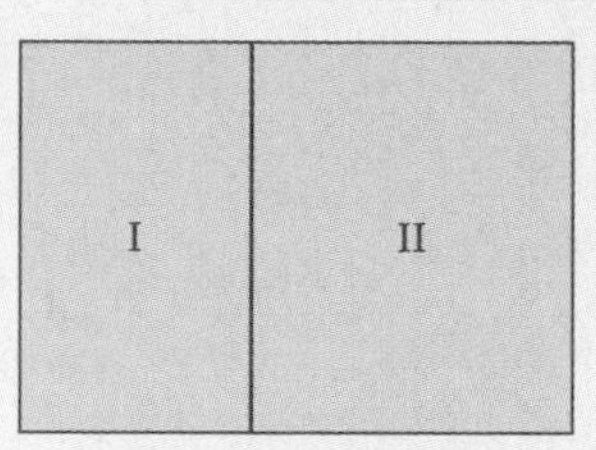

FIGURE 1

METHOD 1 We can calculate the area of the large rectangle directly by finding its length and width. The width is 5 inches, and the length is $(3 + 4)$ inches.

$$\begin{aligned}\text{Area of large rectangle} &= 5(3 + 4)\\ &= 5(7)\\ &= 35 \text{ square inches}\end{aligned}$$

METHOD 2 Because the area of the large rectangle is the sum of the areas of the two smaller rectangles, we find the area of each small rectangle and then add to find the area of the large rectangle.

$$\begin{aligned}\text{Area of large rectangle} &= \text{Area of rectangle I} + \text{Area of rectangle II}\\ &= 5(3) + 5(4)\\ &= 15 + 20\\ &= 35 \text{ square inches}\end{aligned}$$

In both cases the result is 35 square inches. Because the results are the same, the two original expressions must be equal. Stated mathematically, $5(3 + 4) = 5(3) + 5(4)$. We can either add the 3 and 4 first and then multiply that sum by 5, or we can multiply the 3 and the 4 separately by 5 and then add the products. In either case we get the same answer.

Here are some examples that illustrate how we use the distributive property.

EXAMPLE 9 Apply the distributive property to the expression $2(x + 3)$, and then simplify the result.

SOLUTION

$2(x + 3) = 2(x) + 2(3)$		Distributive property
$= 2x + 6$		Multiplication

EXAMPLE 10 Apply the distributive property to the expression $5(2x - 8)$, and then simplify the result.

SOLUTION

$5(2x - 8) = 5(2x) - 5(8)$		Distributive property
$= 10x - 40$		Multiplication

Notice in this example that multiplication distributes over subtraction as well as addition.

EXAMPLE 11 Apply the distributive property to the expression $4(x + y)$, and then simplify the result.

SOLUTION $4(x + y) = 4x + 4y$ Distributive property

Apply the distributive property to each expression, and then simplify the result.

9. $3(x + 2)$

10. $4(2x - 5)$

11. $5(x + y)$

Answers

9. $3x + 6$ **10.** $8x - 20$

11. $5x + 5y$

12. $3(7x - 6y)$

21x − 18Y

13. $\frac{1}{3}(3x + 6)$

14. $5(7a - 3) + 2$

15. $x\left(2 + \frac{2}{x}\right)$

16. $4\left(\frac{1}{4}x - 3\right)$

17. $8\left(\frac{3}{4}x + \frac{1}{2}y\right)$

EXAMPLE 12 Apply the distributive property to the expression $5(2x + 4y)$, and then simplify the result.

SOLUTION

$$\begin{aligned} 5(2x + 4y) &= 5(2x) + 5(4y) && \text{Distributive property} \\ &= 10x + 20y && \text{Multiplication} \end{aligned}$$

EXAMPLE 13 Apply the distributive property to the expression $\frac{1}{2}(3x + 6)$, and then simplify the result.

SOLUTION

$$\begin{aligned} \frac{1}{2}(3x + 6) &= \frac{1}{2}(3x) + \frac{1}{2}(6) && \text{Distributive property} \\ &= \frac{3}{2}x + 3 && \text{Multiplication} \end{aligned}$$

EXAMPLE 14 Apply the distributive property to the expression $4(2a + 3) + 8$, and then simplify the result.

SOLUTION

$$\begin{aligned} 4(2a + 3) + 8 &= 4(2a) + 4(3) + 8 && \text{Distributive property} \\ &= 8a + 12 + 8 && \text{Multiplication} \\ &= 8a + 20 && \text{Addition} \end{aligned}$$

EXAMPLE 15 Apply the distributive property to the expression $a\left(1 + \frac{1}{a}\right)$, and then simplify the result.

SOLUTION

$$\begin{aligned} a\left(1 + \frac{1}{a}\right) &= a \cdot 1 + a \cdot \frac{1}{a} && \text{Distributive property} \\ &= a + 1 && \text{Multiplication} \end{aligned}$$

EXAMPLE 16 Apply the distributive property to the expression $3\left(\frac{1}{3}x + 5\right)$, and then simplify the result.

SOLUTION

$$\begin{aligned} 3\left(\frac{1}{3}x + 5\right) &= 3 \cdot \frac{1}{3}x + 3 \cdot 5 && \text{Distributive property} \\ &= x + 15 && \text{Multiplication} \end{aligned}$$

EXAMPLE 17 Apply the distributive property to the expression $12\left(\frac{2}{3}x + \frac{1}{2}y\right)$, and then simplify the result.

SOLUTION

$$\begin{aligned} 12\left(\frac{2}{3}x + \frac{1}{2}y\right) &= 12 \cdot \frac{2}{3}x + 12 \cdot \frac{1}{2}y && \text{Distributive property} \\ &= 8x + 6y && \text{Multiplication} \end{aligned}$$

Special Numbers

In addition to the three properties mentioned so far, we want to include in our list two special numbers that have unique properties. They are the numbers zero and one.

Additive Identity Property

There exists a unique number 0 such that

In symbols: $a + 0 = a$ and $0 + a = a$

In words: Zero preserves identities under addition. (The identity of the number is unchanged after addition with 0.)

Answers

12. $21x - 18y$ 13. $x + 2$
14. $35a - 13$ 15. $2x + 2$
16. $x - 12$ 17. $6x + 4y$

Multiplicative Identity Property

There exists a unique number 1 such that

In symbols: $a(1) = a$ and $1(a) = a$

In words: The number 1 preserves identities under multiplication. (The identity of the number is unchanged after multiplication by 1.)

Additive Inverse Property

For each real number a, there exists a unique number $-a$ such that

In symbols: $a + (-a) = 0$

In words: Opposites add to 0.

Multiplicative Inverse Property

For every real number a, except 0, there exists a unique real number $\frac{1}{a}$ such that

In symbols: $a\left(\frac{1}{a}\right) = 1$

In words: Reciprocals multiply to 1.

C Identifying Properties

Of all the basic properties listed, the commutative, associative, and distributive properties are the ones we will use most often. They are important because they will be used as justifications or reasons for many of the things we will do.

The following examples illustrate how we use the preceding properties. Each one contains an algebraic expression that has been changed in some way. The property that justifies the change is written to the right.

EXAMPLE 18 State the property that justifies $x + 5 = 5 + x$.

SOLUTION $x + 5 = 5 + x$ Commutative property of addition

EXAMPLE 19 State the property that justifies $(2 + x) + y = 2 + (x + y)$.

SOLUTION $(2 + x) + y = 2 + (x + y)$ Associative property of addition

EXAMPLE 20 State the property that justifies $6(x + 3) = 6x + 18$.

SOLUTION $6(x + 3) = 6x + 18$ Distributive property

EXAMPLE 21 State the property that justifies $2 + (-2) = 0$.

SOLUTION $2 + (-2) = 0$ Additive inverse property

EXAMPLE 22 State the property that justifies $3\left(\frac{1}{3}\right) = 1$.

SOLUTION $3\left(\frac{1}{3}\right) = 1$ Multiplicative inverse property

EXAMPLE 23 State the property that justifies $(2 + 0) + 3 = 2 + 3$.

SOLUTION $(2 + 0) + 3 = 2 + 3$ Additive identity property

State the property that justifies the given statement.

18. $5 \cdot 4 = 4 \cdot 5$

19. $x + 7 = 7 + x$

20. $(3 + a) + b = 3 + (a + b)$

21. $6(x + 5) = 6x + 30$

22. $4\left(\frac{1}{4}\right) = 1$

23. $8 + (-8) = 0$

Answers

18. Commutative property of multiplication **19.** Commutative property of addition **20.** Associative property of addition **21.** Distributive property **22.** Multiplicative inverse **23.** Additive inverse

24. $(x + 2) + 3 = x + (2 + 3)$

EXAMPLE 24 State the property that justifies $(2 + 3) + 4 = 3 + (2 + 4)$.

SOLUTION $(2 + 3) + 4 = 3 + (2 + 4)$ Commutative and associative properties of addition

25. $5(1) = 5$

EXAMPLE 25 State the property that justifies $(x + 2) + y = (x + y) + 2$.

SOLUTION $(x + 2) + y = (x + y) + 2$ Commutative and associative properties of addition

As a final note on the properties of real numbers, we should mention that although some of the properties are stated for only two or three real numbers, they hold for as many numbers as needed. For example, the distributive property holds for expressions like $3(x + y + z + 5 + 2)$; that is,

$$3(x + y + z + 5 + 2) = 3x + 3y + 3z + 15 + 6$$

It is not important how many numbers are contained in the sum, only that it is a sum. Multiplication, you see, distributes over addition, whether there are two numbers in the sum or two hundred.

Getting Ready for Class

After reading through the preceding section, respond in your own words and in complete sentences.

1. What is the commutative property of addition?
2. Do you know from your experience with numbers that the commutative property of addition is true? Explain why.
3. Write the commutative property of multiplication in symbols and words.

Answers

24. Associative property of addtion
25. Multiplicative identity

Problem Set 1.5

C State the property or properties that justify the following. [Examples 18–25]

1. $3 + 2 = 2 + 3$
2. $5 + 0 = 5$
3. $4\left(\frac{1}{4}\right) = 1$
4. $10(0.1) = 1$
5. $4 + x = x + 4$
6. $3(x - 10) = 3x - 30$
7. $2(y + 8) = 2y + 16$
8. $3 + (4 + 5) = (3 + 4) + 5$
9. $(3 + 1) + 2 = 1 + (3 + 2)$
10. $(5 + 2) + 9 = (2 + 5) + 9$
11. $(8 + 9) + 10 = (8 + 10) + 9$
12. $(7 + 6) + 5 = (5 + 6) + 7$
13. $3(x + 2) = 3(2 + x)$
14. $2(7y) = (7 \cdot 2)y$
15. $x(3y) = 3(xy)$
16. $a(5b) = 5(ab)$
17. $4(xy) = 4(yx)$
18. $3[2 + (-2)] = 3(0)$
19. $8[7 + (-7)] = 8(0)$
20. $7(1) = 7$

Each of the following problems has a mistake in it. Correct the right-hand side.

21. $3(x + 2) = 3x + 2$
22. $5(4 + x) = 4 + 5x$
23. $9(a + b) = 9a + b$
24. $2(y + 1) = 2y + 1$
25. $3(0) = 3$
26. $5\left(\frac{1}{5}\right) = 5$
27. $3 + (-3) = 1$
28. $8(0) = 8$
29. $10(1) = 0$
30. $3 \cdot \frac{1}{3} = 0$

A Use the associative property to rewrite each of the following expressions, and then simplify the result. [Examples 4–6]

31. $4 + (2 + x)$

32. $5 + (6 + x)$

33. $(x + 2) + 7$

34. $(x + 8) + 2$

35. $3(5x)$

36. $5(3x)$

37. $9(6y)$

38. $6(9y)$

39. $\frac{1}{2}(3a)$

40. $\frac{1}{3}(2a)$

41. $\frac{1}{3}(3x)$

42. $\frac{1}{4}(4x)$

43. $\frac{1}{2}(2y)$

44. $\frac{1}{7}(7y)$

45. $\frac{3}{4}\left(\frac{4}{3}x\right)$

46. $\frac{3}{2}\left(\frac{2}{3}x\right)$

47. $\frac{6}{5}\left(\frac{5}{6}a\right)$

48. $\frac{2}{5}\left(\frac{5}{2}a\right)$

B Apply the distributive property to each of the following expressions. Simplify when possible. [Examples 9–12]

49. $8(x + 2)$

50. $5(x + 3)$

51. $8(x - 2)$

52. $5(x - 3)$

53. $4(y + 1)$

54. $4(y - 1)$

55. $3(6x + 5)$

56. $3(5x + 6)$

57. $2(3a + 7)$

58. $5(3a + 2)$

59. $9(6y - 8)$

60. $2(7y - 4)$

B Apply the distributive property to each of the following expressions. Simplify when possible. [Examples 13–14]

61. $\frac{1}{2}(3x - 6)$ **62.** $\frac{1}{3}(2x - 6)$ **63.** $\frac{1}{3}(3x + 6)$ **64.** $\frac{1}{2}(2x + 4)$

65. $3(x + y)$ **66.** $2(x - y)$ **67.** $8(a - b)$ **68.** $7(a + b)$

69. $6(2x + 3y)$ **70.** $8(3x + 2y)$ **71.** $4(3a - 2b)$ **72.** $5(4a - 8b)$

73. $\frac{1}{2}(6x + 4y)$ **74.** $\frac{1}{3}(6x + 9y)$ **75.** $4(a + 4) + 9$ **76.** $6(a + 2) + 8$

77. $2(3x + 5) + 2$ **78.** $7(2x + 1) + 3$ **79.** $7(2x + 4) + 10$ **80.** $3(5x + 6) + 20$

B Here are some problems you will see later in the book. Apply the distributive property and simplify, if possible. [Examples 15–17]

81. $0.09(x + 2{,}000)$ **82.** $0.04(x + 7{,}000)$ **83.** $0.05(3x + 1{,}500)$ **84.** $0.08(4x + 3{,}000)$

85. $6\left(\frac{1}{2}x - \frac{1}{3}y\right)$ **86.** $12\left(\frac{1}{4}x + \frac{2}{3}y\right)$ **87.** $12\left(\frac{1}{3}x + \frac{1}{4}y\right)$ **88.** $15\left(\frac{2}{5}x - \frac{1}{3}y\right)$

89. $\frac{1}{2}(4x + 2)$ **90.** $\frac{1}{3}(6x + 3)$ **91.** $\frac{3}{4}(8x - 4)$ **92.** $\frac{2}{5}(5x + 10)$

93. $\frac{5}{6}(6x + 12)$ **94.** $\frac{2}{3}(9x - 3)$ **95.** $10\left(\frac{3}{5}x + \frac{1}{2}\right)$ **96.** $8\left(\frac{1}{4}x - \frac{5}{8}\right)$

97. $15\left(\frac{1}{3}x + \frac{2}{5}\right)$ **98.** $12\left(\frac{1}{12}m + \frac{1}{6}\right)$ **99.** $12\left(\frac{1}{2}m - \frac{5}{12}\right)$ **100.** $8\left(\frac{1}{8} + \frac{1}{2}m\right)$

101. $21\left(\frac{1}{3} + \frac{1}{7}x\right)$ **102.** $6\left(\frac{3}{2}y + \frac{1}{3}\right)$ **103.** $6\left(\frac{1}{2}x - \frac{1}{3}y\right)$ **104.** $12\left(\frac{1}{4}x - \frac{2}{3}y\right)$

105. $0.09(x + 2{,}000)$ **106.** $0.04(x + 7{,}000)$ **107.** $0.12(x + 500)$ **108.** $0.06(x + 800)$

109. $a\left(1 + \frac{1}{a}\right)$ **110.** $a\left(1 - \frac{1}{a}\right)$ **111.** $a\left(\frac{1}{a} - 1\right)$ **112.** $a\left(\frac{1}{a} + 1\right)$

Applying the Concepts

113. Getting Dressed While getting dressed for work, a man puts on his socks and puts on his shoes. Are the two statements "put on your socks" and "put on your shoes" commutative?

114. Getting Dressed Are the statements "put on your left shoe" and "put on your right shoe" commutative?

115. Skydiving A skydiver flying over the jump area is about to do two things: jump out of the plane and pull the rip cord. Are the two events "jump out of the plane" and "pull the rip cord" commutative? That is, will changing the order of the events always produce the same result?

116. Commutative Property Give an example of two events in your daily life that are commutative.

117. Solar and Wind Energy The chart shows the cost to install either solar panels or a wind turbine. A homeowner buys 3 solar modules, and each module is $100 off the original price. Use the distributive property to calculate the total cost.

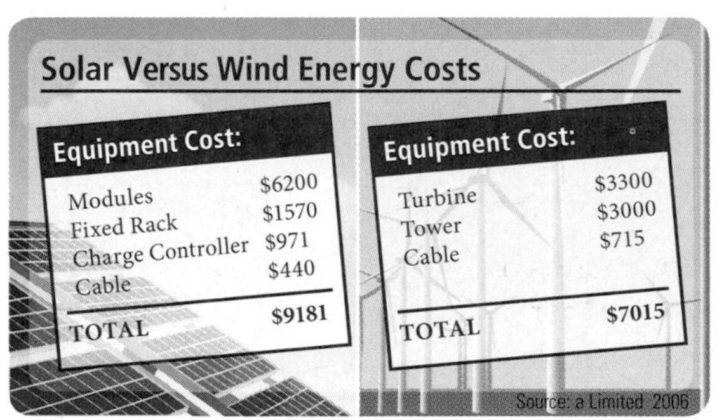

118. Pitchers The chart shows the number of saves for active relief pitchers. Which property of real numbers can be used to find the sum of the saves of Troy Percival and Billy Wagner in either order?

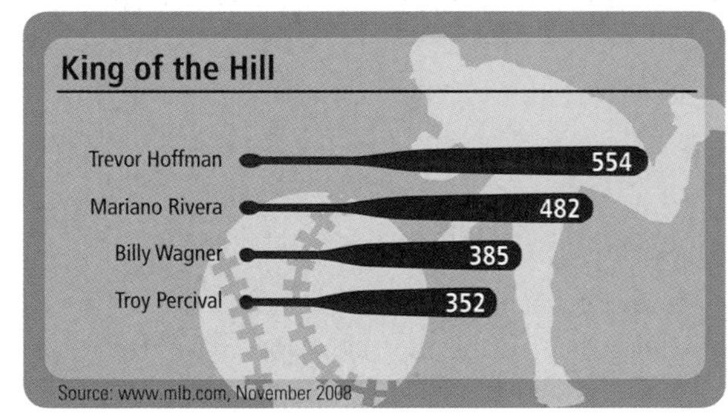

119. Division Give an example that shows that division is not a commutative operation; that is, find two numbers for which changing the order of division gives two different answers.

120. Subtraction Simplify the expression $10 - (5 - 2)$ and the expression $(10 - 5) - 2$ to show that subtraction is not an associative operation.

121. Take-Home Pay Jose works at a winery. His monthly salary is $2,400. To cover his taxes and retirement, the winery withholds $480 from each check. Calculate his yearly "take-home" pay using the numbers 2,400, 480, and 12. Do the calculation two different ways so that the results give further justification for the distributive property.

122. Hours Worked Carlo works as a waiter. He works double shifts 4 days a week. The lunch shift is 2 hours and the dinner shift is 3 hours. Find the total number of hours he works per week using the numbers 2, 3, and 4. Do the calculation two different ways so that the results give further justification for the distributive property.

Multiplication of Real Numbers

1.6

Objectives

A Multiply any combination of positive and negative numbers.

B Simplify expressions using the rule for order of operations.

C Multiply fractions.

D Multiply using the associative property.

E Multiply using the distributive property.

F Extend a geometric sequence.

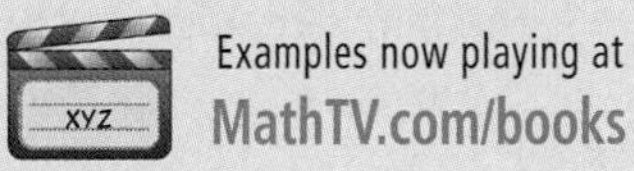

Suppose that you own 5 shares of a stock and the price per share drops \$3. How much money have you lost? Intuitively, we know the loss is \$15. Because it is a loss, we can express it as −\$15. To describe this situation with numbers, we would write

5 shares each lose \$3 for a total of \$15

$$5(-3) = -15$$

Reasoning in this manner, we conclude that the product of a positive number with a negative number is a negative number. Let's look at multiplication in more detail.

A Multiplication of Real Numbers

From our experience with counting numbers, we know that multiplication is simply repeated addition; that is, $3(5) = 5 + 5 + 5$. We will use this fact, along with our knowledge of negative numbers, to develop the rule for multiplication of any two real numbers. The following examples illustrate multiplication with all of the possible combinations of positive and negative numbers.

EXAMPLES Multiply.

1. Two positives: $3(5) = 5 + 5 + 5$
$= 15$ Positive answer

2. One positive: $3(-5) = -5 + (-5) + (-5)$
$= -15$ Negative answer

3. One negative: $-3(5) = 5(-3)$ Commutative property
$= -3 + (-3) + (-3) + (-3) + (-3)$
$= -15$ Negative answer

4. Two negatives: $-3(-5) = ?$

With two negatives, $-3(-5)$, it is not possible to work the problem in terms of repeated addition. (It doesn't "make sense" to write −5 down a −3 number of times.) The answer is probably +15 (that's just a guess), but we need some justification for saying so. We will solve a different problem and in so doing get the answer to the problem $(-3)(-5)$.

Here is a problem to which we know the answer. We will work it two different ways.

$$-3[5 + (-5)] = -3(0) = 0$$

The answer is zero. We also can work the problem using the distributive property.

$$-3[5 + (-5)] = -3(5) + (-3)(-5) \quad \text{Distributive property}$$
$$= -15 + ?$$

Because the answer to the problem is 0, our ? must be +15. (What else could we add to −15 to get 0? Only +15.)

Here is a summary of the results we have obtained from the first four examples.

Original Numbers Have		The Answer is
the same sign	$3(5) = 15$	positive
different signs	$3(-5) = -15$	negative
different signs	$-3(5) = -15$	negative
the same sign	$-3(-5) = 15$	positive

PRACTICE PROBLEMS

Multiply.

1. 4(2)

2. 4(−2)

3. −4(2)

4. −4(−2)

You may have to read the explanation for Example 4 several times before you understand it completely. The purpose of the explanation in Example 4 is simply to justify the fact that the product of two negative numbers is a positive number. If you have no trouble believing that, then it is not so important that you understand everything in the explanation.

Answers

1. 8 **2.** −8 **3.** −8 **4.** 8

Some students have trouble with the expression $-8(-3)$ because they want to subtract rather than multiply. Because we are very precise with the notation we use in algebra, the expression $-8(-3)$ has only one meaning—multiplication. A subtraction problem that uses the same numbers is $-8 - 3$. Compare the two following lists.

All Multiplication	No Multiplication
$5(4)$	$5 + 4$
$-5(4)$	$-5 + 4$
$5(-4)$	$5 - 4$
$-5(-4)$	$-5 - 4$

Multiply.

5. $-4(-9)$
6. $-11(-12)$
7. $-2(-7)$
8. $4(-3)$
9. $-2(10)$
10. $-3(13)$

Simplify as much as possible.

11. $-7(-6)(-1)$
12. $2(-8) + 3(-7) - 4$
13. $(-3)^5$
14. $-5(-2)^3 - 7(-3)^2$

Answers
5. 36 **6.** 132 **7.** 14 **8.** −12 **9.** −20 **10.** −39 **11.** −42 **12.** −41 **13.** −243 **14.** −23

By examining Examples 1 through 4 and the preceding table, we can use the information there to write the following rule. This rule tells us how to multiply any two real numbers.

Rule
To multiply any two real numbers, simply multiply their absolute values. The sign of the answer is

1. *Positive* if both numbers have the same sign (both + or both −).
2. *Negative* if the numbers have opposite signs (one +, the other −).

The following examples illustrate how we use the preceding rule to multiply real numbers.

EXAMPLES Multiply.

5. $-8(-3) = 24$
6. $-10(-5) = 50$
7. $-4(-7) = 28$

If the two numbers in the product have the same sign, the answer is positive

8. $5(-7) = -35$
9. $-4(8) = -32$
10. $-6(10) = -60$

If the two numbers in the product have different signs, the answer is negative

B Using Order of Operations

In the following examples, we combine the rule for order of operations with the rule for multiplication to simplify expressions. Remember, the rule for order of operations specifies that we are to work inside the parentheses first and then simplify numbers containing exponents. After this, we multiply and divide, left to right. The last step is to add and subtract, left to right.

EXAMPLE 11 Simplify $-5(-3)(-4)$ as much as possible.

SOLUTION
$$-5(-3)(-4) = 15(-4) = -60$$

EXAMPLE 12 Simplify $4(-3) + 6(-5) - 10$ as much as possible.

SOLUTION
$4(-3) + 6(-5) - 10 = -12 + (-30) - 10$ Multiply
$= -42 - 10$ Add
$= -52$ Subtract

EXAMPLE 13 Simplify $(-2)^3$ as much as possible.

SOLUTION
$(-2)^3 = (-2)(-2)(-2)$ Definition of exponents
$= -8$ Multiply, left to right

EXAMPLE 14 Simplify $-3(-2)^3 - 5(-4)^2$ as much as possible.

SOLUTION
$-3(-2)^3 - 5(-4)^2 = -3(-8) - 5(16)$ Exponents first
$= 24 - 80$ Multiply
$= -56$ Subtract

EXAMPLE 15 Simplify $6 - 4(7 - 2)$ as much as possible.

SOLUTION

$$\begin{aligned} 6 - 4(7 - 2) &= 6 - 4(5) && \text{Inside parentheses first} \\ &= 6 - 20 && \text{Multiply} \\ &= -14 && \text{Subtract} \end{aligned}$$

C Multiplying Fractions

Previously, we mentioned that to multiply two fractions we multiply numerators and multiply denominators. We can apply the rule for multiplication of positive and negative numbers to fractions in the same way we apply it to other numbers. We multiply absolute values: The product is positive if both fractions have the same sign and negative if they have different signs. Here are some examples.

EXAMPLE 16 Multiply $-\frac{3}{4}\left(\frac{5}{7}\right)$.

SOLUTION

$$\begin{aligned} -\frac{3}{4}\left(\frac{5}{7}\right) &= -\frac{3 \cdot 5}{4 \cdot 7} && \text{Different signs give a negative answer} \\ &= -\frac{15}{28} \end{aligned}$$

EXAMPLE 17 Multiply $-6\left(\frac{1}{2}\right)$.

SOLUTION

$$\begin{aligned} -6\left(\frac{1}{2}\right) &= -\frac{6}{1}\left(\frac{1}{2}\right) && \text{Different signs give a negative answer} \\ &= -\frac{6}{2} \\ &= -3 \end{aligned}$$

EXAMPLE 18 Multiply $-\frac{2}{3}\left(-\frac{3}{2}\right)$.

SOLUTION

$$\begin{aligned} -\frac{2}{3}\left(-\frac{3}{2}\right) &= \frac{2 \cdot 3}{3 \cdot 2} && \text{Same signs give a positive answer} \\ &= \frac{6}{6} \\ &= 1 \end{aligned}$$

EXAMPLE 19 Figure 1 gives the calories that are burned in 1 hour for a variety of forms of exercise by a person weighing 150 pounds. Figure 2 gives the calories that are consumed by eating some popular fast foods. Find the net change in calories for a 150-pound person playing handball for 2 hours and then eating a Whopper.

Calories Burned in 1 Hour by a 150-Pound Person

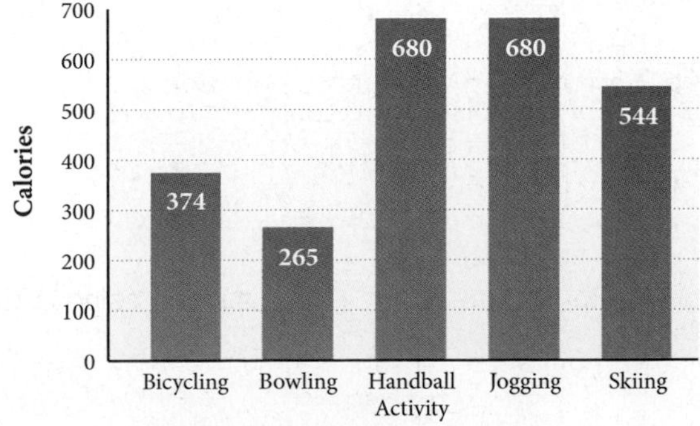

FIGURE 1

15. $7 - 3(4 - 9)$

Multiply.

16. $-\frac{2}{3}\left(\frac{5}{9}\right)$

17. $-8\left(\frac{1}{2}\right)$

18. $-\frac{3}{4}\left(-\frac{4}{3}\right)$

19. Find the net change in calories for a 150-pound person jogging for 2 hours then eating a Big Mac.

Answers

15. 22 **16.** $-\frac{10}{27}$ **17.** -4 **18.** 1

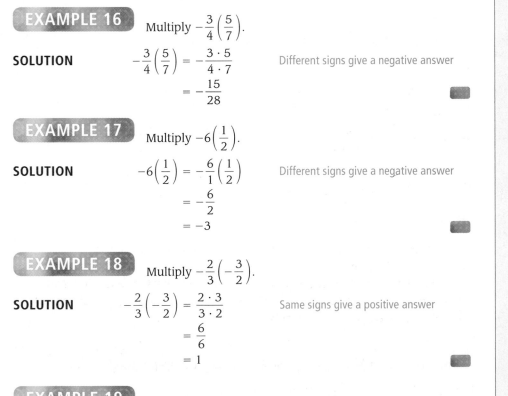

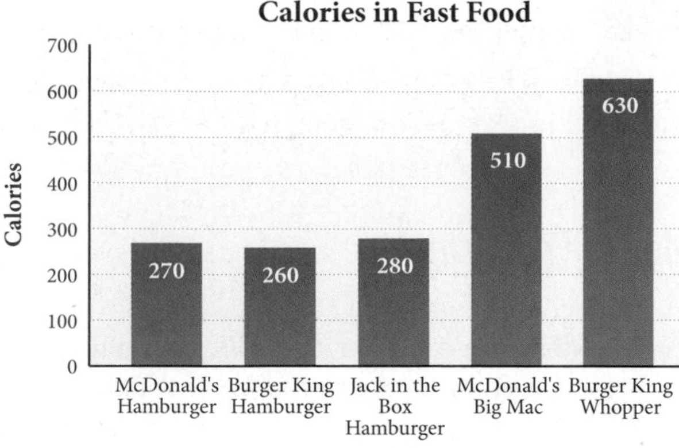

FIGURE 2

SOLUTION The net change in calories will be the difference of the calories gained from eating and the calories lost from exercise.

$$\text{Net change in calories} = 630 - 2(680) = -730 \text{ calories}$$

D Using the Associative Property

We can use the rule for multiplication of real numbers, along with the associative property, to multiply expressions that contain numbers and variables.

Apply the associative property, and then simplify.

20. $-5(3x)$

EXAMPLE 20 Apply the associative property, and then multiply.

SOLUTION

$$\begin{aligned} -3(2x) &= (-3 \cdot 2)x && \text{Associative property} \\ &= -6x && \text{Multiplication} \end{aligned}$$

21. $4(-8y)$

EXAMPLE 21 Apply the associative property, and then multiply.

SOLUTION

$$\begin{aligned} 6(-5y) &= [6(-5)]y && \text{Associative property} \\ &= -30y && \text{Multiplication} \end{aligned}$$

22. $-3\left(-\frac{1}{3}x\right)$

EXAMPLE 22 Apply the associative property, and then multiply.

SOLUTION

$$\begin{aligned} -2\left(-\frac{1}{2}x\right) &= \left[(-2)\left(-\frac{1}{2}\right)\right]x && \text{Associative property} \\ &= 1x && \text{Multiplication} \\ &= x && \text{Multiplication} \end{aligned}$$

E Using the Distributive Property

The following examples show how we can use both the distributive property and multiplication with real numbers.

Multiply by applying the distributive property.

23. $-3(a + 5)$

EXAMPLE 23 Apply the distributive property to $-2(a + 3)$.

SOLUTION

$$\begin{aligned} -2(a + 3) &= -2a + (-2)(3) && \text{Distributive property} \\ &= -2a + (-6) && \text{Multiplication} \\ &= -2a - 6 \end{aligned}$$

Answers

19. -850 calories **20.** $-15x$
21. $-32y$ **22.** x **23.** $-3a - 15$

EXAMPLE 24 Apply the distributive property to $-3(2x + 1)$.

SOLUTION

$$-3(2x + 1) = -3(2x) + (-3)(1) \quad \text{Distributive property}$$
$$= -6x + (-3) \quad \text{Multiplication}$$
$$= -6x - 3$$

EXAMPLE 25 Apply the distributive property to $-\frac{1}{3}(2x - 6)$.

SOLUTION

$$-\frac{1}{3}(2x - 6) = -\frac{1}{3}(2x) - \left(-\frac{1}{3}\right)(6) \quad \text{Distributive property}$$
$$= -\frac{2}{3}x - (-2) \quad \text{Multiplication}$$
$$= -\frac{2}{3}x + 2$$

EXAMPLE 26 Apply the distributive property to $-4(3x - 5) - 8$.

SOLUTION

$$-4(3x - 5) - 8 = -4(3x) - (-4)(5) - 8 \quad \text{Distributive property}$$
$$= -12x - (-20) - 8 \quad \text{Multiplication}$$
$$= -12x + 20 - 8 \quad \text{Definition of subtraction}$$
$$= -12x + 12 \quad \text{Subtraction}$$

The next examples continue the work we did previously with finding the value of an algebraic expression for given values of the variable or variables.

EXAMPLE 27 Find the value of $-\frac{2}{3}x - 4$ when

a. $x = 0$ **b.** $x = 3$ **c.** $x = -\frac{9}{2}$

SOLUTION Substituting the values of x into our expression one at a time we have

a. $-\frac{2}{3}(0) - 4 = 0 - 4 = -4$

b. $-\frac{2}{3}(3) - 4 = -2 - 4 = -6$

c. $-\frac{2}{3}\left(-\frac{9}{2}\right) - 4 = 3 - 4 = -1$

EXAMPLE 28 Find the value of $5x - 4y$ when

a. $x = 4$ and $y = 0$
b. $x = 0$ and $y = -8$
c. $x = -2$ and $y = 3$

SOLUTION Substituting the values of x into our expression one at a time we have

a. $5(4) - 4(0) = 20 - 0 = 20$
b. $5(0) - 4(-8) = 0 + 32 = 32$
c. $5(-2) - 4(3) = -10 - 12 = -22$

F Geometric Sequences

A *geometric sequence* is a sequence of numbers in which each number (after the first number) comes from the number before it by multiplying by the same amount each time. For example, the sequence

2, 6, 18, 54, . . .

24. $-5(2x + 1)$

25. $-\frac{1}{2}(4x - 12)$

26. $-3(2x - 6) - 7$

27. Find the value of $-\frac{3}{4}x + 2$ when
a. $x = 0$
b. $x = 4$
c. $x = -\frac{8}{3}$

28. Find the value of $3x - 5y$ when
a. $x = 4$ and $y = 0$
b. $x = 0$ and $y = -2$
c. $x = -2$ and $y = 3$

Answers
24. $-10x - 5$ **25.** $-2x + 6$
26. $-6x + 11$
27. a. 2 **b.** −1 **c.** 4
28. a. 12 **b.** 10 **c.** −21

is a geometric sequence because each number is obtained by multiplying the number before it by 3.

29. Find the next number in each of the following geometric sequences.

a. 4, 8, 16, . . .

b. 2, −6, 18, . . .

c. $\frac{1}{9}, \frac{1}{3}, 1, \ldots$

EXAMPLE 29 Each sequence below is a geometric sequence. Find the next number in each sequence.

a. 5, 10, 20, . . . **b.** 3, −15, 75, . . . **c.** $\frac{1}{8}, \frac{1}{4}, \frac{1}{2}, \ldots$

SOLUTION Because each sequence is a geometric sequence, we know that each term is obtained from the previous term by multiplying by the same number each time.

a. 5, 10, 20, . . . : The sequence starts with 5. After that, each number is obtained from the previous number by multiplying by 2 each time. The next number will be $20 \cdot 2 = 40$.

b. 3, −15, 75, . . . : The sequence starts with 3. After that, each number is obtained by multiplying by −5 each time. The next number will be $75(-5) = -375$.

c. $\frac{1}{8}, \frac{1}{4}, \frac{1}{2}, \ldots$: This sequence starts with $\frac{1}{8}$. Multiplying each number in the sequence by 2 produces the next number in the sequence. To extend the sequence, we multiply $\frac{1}{2}$ by 2: $\frac{1}{2} \cdot 2 = 1$.
The next number in the sequence is 1.

Getting Ready for Class

After reading through the preceding section, respond in your own words and in complete sentences.

1. How do you multiply two negative numbers?
2. How do you multiply two numbers with different signs?
3. Explain how some multiplication problems can be thought of as repeated addition.
4. What is a geometric sequence?

Answer
29. **a.** 32 **b.** −54 **c.** 3

Problem Set 1.6

A Use the rule for multiplying two real numbers to find each of the following products. [Examples 1–10]

1. $7(-6)$
2. $8(-4)$
3. $-8(2)$
4. $-16(3)$
5. $-3(-1)$
6. $-7(-1)$
7. $-11(-11)$
8. $-12(-12)$

B Use the rule for order of operations to simplify each expression as much as possible. [Examples 11–15]

9. $-3(2)(-1)$
10. $-2(3)(-4)$
11. $-3(-4)(-5)$
12. $-5(-6)(-7)$
13. $-2(-4)(-3)(-1)$
14. $-1(-3)(-2)(-1)$
15. $(-7)^2$
16. $(-8)^2$
17. $(-3)^3$
18. $(-2)^4$
19. $-2(2 - 5)$
20. $-3(3 - 7)$
21. $-5(8 - 10)$
22. $-4(6 - 12)$
23. $(4 - 7)(6 - 9)$
24. $(3 - 10)(2 - 6)$
25. $(-3 - 2)(-5 - 4)$
26. $(-3 - 6)(-2 - 8)$
27. $-3(-6) + 4(-1)$
28. $-4(-5) + 8(-2)$
29. $2(3) - 3(-4) + 4(-5)$
30. $5(4) - 2(-1) + 5(6)$
31. $4(-3)^2 + 5(-6)^2$
32. $2(-5)^2 + 4(-3)^2$

33. $7(-2)^3 - 2(-3)^3$

34. $10(-2)^3 - 5(-2)^4$

35. $6 - 4(8 - 2)$

36. $7 - 2(6 - 3)$

37. $9 - 4(3 - 8)$

38. $8 - 5(2 - 7)$

39. $-4(3 - 8) - 6(2 - 5)$

40. $-8(2 - 7) - 9(3 - 5)$

41. $7 - 2[-6 - 4(-3)]$

42. $6 - 3[-5 - 3(-1)]$

43. $7 - 3[2(-4 - 4) - 3(-1 - 1)]$

44. $5 - 3[7(-2 - 2) - 3(-3 + 1)]$

45. Simplify each expression.

a. $5(-4)(-3)$
b. $5(-4) - 3$
c. $5 - 4(-3)$
d. $5 - 4 - 3$

46. Simplify each expression.

a. $-2(-3)(-5)$
b. $-2(-3) - 5$
c. $-2 - 3(-5)$
d. $-2 - 3 - 5$

C Multiply the following fractions. [Examples 16–18]

47. $-\frac{2}{3} \cdot \frac{5}{7}$

48. $-\frac{6}{5} \cdot \frac{2}{7}$

49. $-8\left(\frac{1}{2}\right)$

50. $-12\left(\frac{1}{3}\right)$

51. $\left(-\frac{3}{4}\right)^2$

52. $\left(-\frac{2}{5}\right)^2$

53. Simplify each expression.

a. $\frac{5}{8}(24) + \frac{3}{7}(28)$
b. $\frac{5}{8}(24) - \frac{3}{7}(28)$
c. $\frac{5}{8}(-24) + \frac{3}{7}(-28)$
d. $-\frac{5}{8}(24) - \frac{3}{7}(28)$

54. Simplify each expression.

a. $\frac{5}{6}(18) + \frac{3}{5}(15)$
b. $\frac{5}{6}(18) - \frac{3}{5}(15)$
c. $\frac{5}{6}(-18) + \frac{3}{5}(-15)$
d. $-\frac{5}{6}(18) - \frac{3}{5}(15)$

Simplify.

55. $\left(\frac{1}{2} \cdot 6\right)^2$

56. $\left(\frac{1}{2} \cdot 10\right)^2$

57. $\left(\frac{1}{2} \cdot 5\right)^2$

58. $\left[\frac{1}{2}(0.8)\right]^2$

59. $\left[\frac{1}{2}(-4)\right]^2$

60. $\left[\frac{1}{2}(-12)\right]^2$

61. $\left[\frac{1}{2}(-3)\right]^2$

62. $\left[\frac{1}{2}(-0.8)\right]^2$

D Find the following products. [Examples 20–22]

63. $-2(4x)$ **64.** $-8(7x)$ **65.** $-7(-6x)$ **66.** $-8(-9x)$ **67.** $-\frac{1}{3}(-3x)$ **68.** $-\frac{1}{5}(-5x)$

E Apply the distributive property to each expression, and then simplify the result. [Examples 23–28]

69. $-4(a + 2)$ **70.** $-7(a + 6)$ **71.** $-\frac{1}{2}(3x - 6)$ **72.** $-\frac{1}{4}(2x - 4)$

73. $-3(2x - 5) - 7$ **74.** $-4(3x - 1) - 8$ **75.** $-5(3x + 4) - 10$ **76.** $-3(4x + 5) - 20$

77. $-4(3x + 5y)$ **78.** $5(5x + 4y)$ **79.** $-2(3x + 5y)$ **80.** $-2(2x - y)$

81. $\frac{1}{2}(-3x + 6)$ **82.** $\frac{1}{4}(5x - 20)$ **83.** $\frac{1}{3}(-2x + 6)$ **84.** $\frac{1}{5}(-4x + 20)$

85. $-\frac{1}{3}(-2x + 6)$ **86.** $-\frac{1}{2}(-2x + 6)$ **87.** $8\left(-\frac{1}{4}x + \frac{1}{8}y\right)$ **88.** $9\left(-\frac{1}{9}x + \frac{1}{3}y\right)$

89. Find the value of $-\frac{1}{3}x + 2$ when

a. $x = 0$

b. $x = 3$

c. $x = -3$

90. Find the value of $-\frac{2}{3}x + 1$ when

a. $x = 0$

b. $x = 3$

c. $x = -3$

91. Find the value of $2x + y$ when

a. $x = 2$ and $y = -1$

b. $x = 0$ and $y = 3$

c. $x = \frac{3}{2}$ and $y = -7$

92. Find the value of $2x - 5y$ when

a. $x = 2$ and $y = 3$

b. $x = 0$ and $y = -2$

c. $x = \frac{5}{2}$ and $y = 1$

93. Find the value of $2x^2 - 5x$ when

a. $x = 4$

b. $x = -\frac{3}{2}$

94. Find the value of $49a^2 - 16$ when

a. $a = \frac{4}{7}$

b. $a = -\frac{4}{7}$

95. Find the value of $y(2y + 3)$ when

a. $y = 4$

b. $y = -\frac{11}{2}$

96. Find the value of $x(13 - x)$ when

a. $x = 5$

b. $x = 8$

97. Five added to the product of 3 and -10 is what number?

98. If the product of -8 and -2 is decreased by 4, what number results?

99. Write an expression for twice the product of -4 and x, and then simplify it.

100. Write an expression for twice the product of -2 and $3x$, and then simplify it.

101. What number results if 8 is subtracted from the product of -9 and 2?

102. What number results if -8 is subtracted from the product of -9 and 2?

F Each of the following is a geometric sequence. In each case, find the next number in the sequence. [Example 29]

103. $1, 2, 4, \ldots$

104. $1, 5, 25, \ldots$

105. $10, -20, 40, \ldots$

106. $10, -30, 90, \ldots$

107. $1, \frac{1}{2}, \frac{1}{4}, \ldots$

108. $1, \frac{1}{3}, \frac{1}{9}, \ldots$

109. $3, -6, 12, \ldots$

110. $-3, 6, -12, \ldots$

Here are some problems you will see later in the book. Simplify.

111. $3(x - 5) + 4$

112. $5(x - 3) + 2$

113. $2(3) - 4 - 3(-4)$

114. $2(3) + 4(5) - 5(2)$

115. $\left(\frac{1}{2} \cdot 18\right)^2$

116. $\left[\frac{1}{2}(-10)\right]^2$

117. $\left(\frac{1}{2} \cdot 3\right)^2$

118. $\left(\frac{1}{2} \cdot 5\right)^2$

Applying the Concepts

119. Picture Messaging The graph shows the number of picture messages sent each of the first nine months of the year. Jan got a new phone in March that allowed picture messaging. If she gets charged $0.25 per message, how much did she get charged in April?

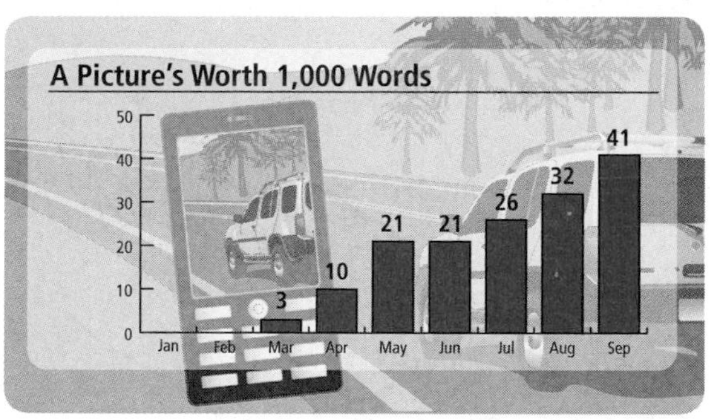

120. Google Earth The Google Earth map shows Yellowstone National Park. The park covers about 3,472 square miles. If there is an average of 2.3 moose per square mile, about how many moose live in Yellowstone? Round to the nearest moose.

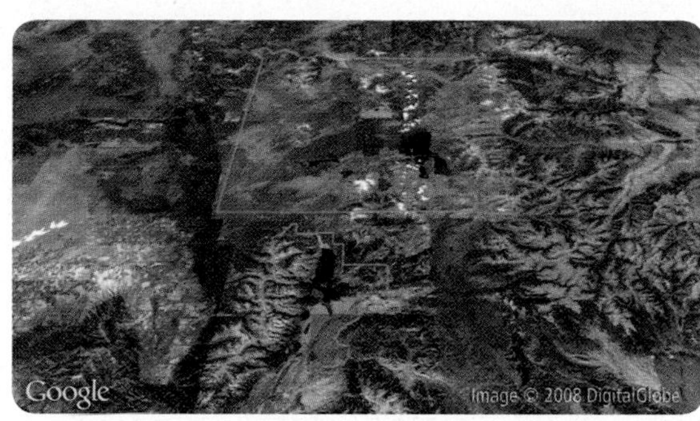

121. Stock Value Suppose you own 20 shares of a stock. If the price per share drops $3, how much money have you lost?

122. Stock Value Imagine that you purchase 50 shares of a stock at a price of $18 per share. If the stock is selling for $11 a share a week after you purchased it, how much money have you lost?

123. Temperature Change The temperature is 25°F at 5:00 in the afternoon. If the temperature drops 6°F every hour after that, what is the temperature at 9:00 in the evening?

124. Temperature Change The temperature is −5°F at 6:00 in the evening. If the temperature drops 3°F every hour after that, what is the temperature at midnight?

125. Nursing Problem A patient's prescription requires him to take two 25 mg tablets twice a day. What is the total dosage for the day?

126. Nursing Problem A patient takes three 50 mg capsules a day. How many milligrams is he taking daily?

127. Reading Charts Refer to Figures 3 and 4 to find the net change in calories for a 150-pound person who bowls for 3 hours and then eats 2 Whoppers.

128. Reading Charts Refer to Figures 3 and 4 to find the net change in calories for a 150-pound person who eats a Big Mac and then skis for 3 hours.

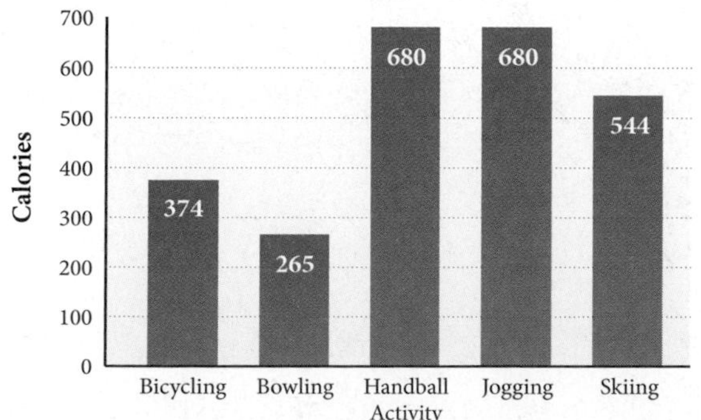

FIGURE 3

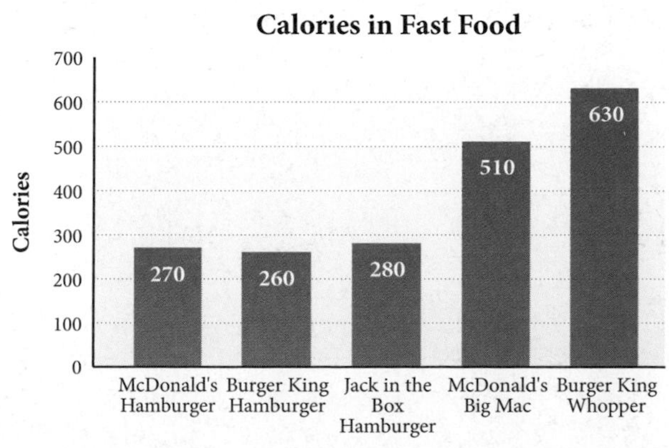

FIGURE 4

1.7 Division of Real Numbers

Suppose that you and four friends bought equal shares of an investment for a total of \$15,000 and then sold it later for only \$13,000. How much did each person lose? Because the total amount of money that was lost can be represented by −\$2,000, and there are 5 people with equal shares, we can represent each person's loss with division:

$$\frac{-\$2{,}000}{5} = -\$400$$

From this discussion it seems reasonable to say that a negative number divided by a positive number is a negative number. Here is a more detailed discussion of division with positive and negative numbers.

The last of the four basic operations is division. We will use the same approach to define division as we used for subtraction; that is, we will define division in terms of rules we already know.

Recall that we developed the rule for subtraction of real numbers by defining subtraction in terms of addition. We changed our subtraction problems to addition problems and then added to get our answers. Because we already have a rule for multiplication of real numbers, and division is the inverse operation of multiplication, we will simply define division in terms of multiplication.

We know that division by the number 2 is the same as multiplication by $\frac{1}{2}$; that is, 6 divided by 2 is 3, which is the same as 6 times $\frac{1}{2}$. Similarly, dividing a number by 5 gives the same result as multiplying by $\frac{1}{5}$. We can extend this idea to all real numbers with the following rule.

Objectives

A Divide any combination of positive and negative numbers.

B Divide fractions.

C Simplify expressions using the rule for order of operations.

Examples now playing at MathTV.com/books

Rule

If a and b represent any two real numbers (b cannot be 0), then it is always true that

$$a \div b = \frac{a}{b} = a\left(\frac{1}{b}\right)$$

A Dividing Positive and Negative Numbers

Division by a number is the same as multiplication by its reciprocal. Because every division problem can be written as a multiplication problem and because we already know the rule for multiplication of two real numbers, we do not have to write a new rule for division of real numbers. We will simply replace our division problem with multiplication and use the rule we already have.

Note We are defining division this way simply so that we can use what we already know about multiplication to do division problems. We actually want as few rules as possible. Defining division in terms of multiplication allows us to avoid writing a separate rule for division.

EXAMPLES Write each division problem as an equivalent multiplication problem, and then multiply.

1. $\frac{6}{2} = 6\left(\frac{1}{2}\right) = 3$ — The product of two positives is positive

2. $\frac{6}{-2} = 6\left(-\frac{1}{2}\right) = -3$

3. $\frac{-6}{2} = -6\left(\frac{1}{2}\right) = -3$

   (Examples 2 and 3:) The product of a positive and a negative is a negative

4. $\frac{-6}{-2} = -6\left(-\frac{1}{2}\right) = 3$ — The product of two negatives is positive

The second step in the previous examples is used only to show that we *can* write division in terms of multiplication. [In actual practice we wouldn't write $\frac{6}{2}$ as $6\left(\frac{1}{2}\right)$.]

PRACTICE PROBLEMS

Change each problem to an equivalent multiplication problem, and then multiply to get your answer.

1. $\frac{12}{4}$

2. $\frac{12}{-4}$

3. $\frac{-12}{4}$

4. $\frac{-12}{-4}$

Answers

1. 3 **2.** −3 **3.** −3 **4.** 3

The answers, therefore, follow from the rule for multiplication; that is, like signs produce a positive answer, and unlike signs produce a negative answer.

Here are some examples. This time we will not show division as multiplication by the reciprocal. We will simply divide. If the original numbers have the same signs, the answer will be positive. If the original numbers have different signs, the answer will be negative.

Note What we are saying here is that the work shown in Examples 1 through 4 is shown simply to justify the answers we obtain. In the future we won't show the middle step in these kinds of problems. Even so, we need to know that division is defined to be multiplication by the reciprocal.

EXAMPLE 5 Divide $\frac{12}{6}$.

SOLUTION $\frac{12}{6} = 2$ Like signs give a positive answer

EXAMPLE 6 Divide $\frac{12}{-6}$.

SOLUTION $\frac{12}{-6} = -2$ Unlike signs give a negative answer

EXAMPLE 7 Divide $\frac{-12}{6}$.

SOLUTION $\frac{-12}{6} = -2$ Unlike signs give a negative answer

EXAMPLE 8 Divide $\frac{-12}{-6}$.

SOLUTION $\frac{-12}{-6} = 2$ Like signs give a positive answer

EXAMPLE 9 Divide $\frac{15}{-3}$.

SOLUTION $\frac{15}{-3} = -5$ Unlike signs give a negative answer

EXAMPLE 10 Divide $\frac{-40}{-5}$.

SOLUTION $\frac{-40}{-5} = 8$ Like signs give a positive answer

EXAMPLE 11 Divide $\frac{-14}{2}$.

SOLUTION $\frac{-14}{2} = -7$ Unlike signs give a negative answer

Divide.

5. $\frac{18}{3}$

6. $\frac{18}{-3}$

7. $\frac{-18}{3}$

8. $\frac{-18}{-3}$

9. $\frac{30}{-10}$

10. $\frac{-50}{-25}$

11. $\frac{-21}{3}$

Answers

5. 6 **6.** −6 **7.** −6 **8.** 6 **9.** −3 **10.** 2 **11.** −7

B Division with Fractions

We can apply the definition of division to fractions. Because dividing by a fraction is equivalent to multiplying by its reciprocal, we can divide a number by the fraction $\frac{3}{4}$ by multiplying it by the reciprocal of $\frac{3}{4}$, which is $\frac{4}{3}$. For example,

$$\frac{2}{5} \div \frac{3}{4} = \frac{2}{5} \cdot \frac{4}{3} = \frac{8}{15}$$

You may have learned this rule in previous math classes. In some math classes, multiplication by the reciprocal is referred to as "inverting the divisor and multiplying." No matter how you say it, division by any number (except 0) is always equivalent to multiplication by its reciprocal. Here are additional examples that involve division by fractions.

EXAMPLE 12 Divide $\frac{2}{3} \div \frac{5}{7}$.

SOLUTION

$\frac{2}{3} \div \frac{5}{7} = \frac{2}{3} \cdot \frac{7}{5}$ Rewrite as multiplication by the reciprocal

$= \frac{14}{15}$ Multiply

EXAMPLE 13 Divide $-\frac{3}{4} \div \frac{7}{9}$.

SOLUTION

$-\frac{3}{4} \div \frac{7}{9} = -\frac{3}{4} \cdot \frac{9}{7}$ Rewrite as multiplication by the reciprocal

$= -\frac{27}{28}$ Multiply

EXAMPLE 14 Divide $8 \div \left(-\frac{4}{5}\right)$.

SOLUTION

$8 \div \left(-\frac{4}{5}\right) = \frac{8}{1}\left(-\frac{5}{4}\right)$ Rewrite as multiplication by the reciprocal

$= -\frac{40}{4}$ Multiply

$= -10$ Divide 40 by 4

Divide.

12. $\frac{3}{4} \div \frac{5}{7}$

13. $-\frac{7}{8} \div \frac{3}{5}$

14. $10 \div \left(-\frac{5}{6}\right)$

C Using Order of Operations

The last step in each of the following examples involves reducing a fraction to lowest terms. To reduce a fraction to lowest terms, we divide the numerator and denominator by the largest number that divides each of them exactly. For example, to reduce $\frac{15}{20}$ to lowest terms, we divide 15 and 20 by 5 to get $\frac{3}{4}$.

EXAMPLE 15 Simplify as much as possible: $\frac{-4(5)}{6}$

SOLUTION

$\frac{-4(5)}{6} = \frac{-20}{6}$ Simplify numerator

$= -\frac{10}{3}$ Reduce to lowest terms by dividing numerator and denominator by 2

EXAMPLE 16 Simplify as much as possible: $\frac{30}{-4-5}$

SOLUTION

$\frac{30}{-4-5} = \frac{30}{-9}$ Simplify denominator

$= -\frac{10}{3}$ Reduce to lowest terms by dividing numerator and denominator by 3

Simplify.

15. $\frac{-6(5)}{9}$

16. $\frac{10}{-7-1}$

In the examples that follow, the numerators and denominators contain expressions that are somewhat more complicated than those we have seen thus far. To apply the rule for order of operations to these examples, we treat fraction bars the same way we treat grouping symbols; that is, fraction bars separate numerators and denominators so that each will be simplified separately.

Answers

12. $\frac{21}{20}$ 13. $-\frac{35}{24}$ 14. -12

15. $-\frac{10}{3}$ 16. $-\frac{5}{4}$

Simplify.

17. $\frac{-6-6}{-2-4}$

18. $\frac{3(-4)+9}{6}$

19. $\frac{6(-2)+5(-3)}{5(4)-11}$

Simplify.

20. $\frac{4^2-2^2}{-4+2}$

21. $\frac{(4+3)^2}{-4^2-3^2}$

Answers

17. 2 **18.** $-\frac{1}{2}$ **19.** -3 **20.** -6
21. $-\frac{49}{25}$

EXAMPLE 17 Simplify: $\frac{-8-8}{-5-3}$

SOLUTION

$\frac{-8-8}{-5-3} = \frac{-16}{-8}$ Simplify numerator and denominator separately

$= 2$ Division

EXAMPLE 18 Simplify: $\frac{2(-3)+4}{12}$

SOLUTION

$\frac{2(-3)+4}{12} = \frac{-6+4}{12}$ In the numerator, we multiply before we add

$= \frac{-2}{12}$ Addition

$= -\frac{1}{6}$ Reduce to lowest terms by dividing numerator and denominator by 2

EXAMPLE 19 Simplify: $\frac{5(-4)+6(-1)}{2(3)-4(1)}$

SOLUTION

$\frac{5(-4)+6(-1)}{2(3)-4(1)} = \frac{-20+(-6)}{6-4}$ Multiplication before addition

$= \frac{-26}{2}$ Simplify numerator and denominator

$= -13$ Divide -26 by 2

We must be careful when we are working with expressions such as $(-5)^2$ and -5^2 that we include the negative sign with the base only when parentheses indicate we are to do so.

Unless there are parentheses to indicate otherwise, we consider the base to be only the number directly below and to the left of the exponent. If we want to include a negative sign with the base, we must use parentheses.

To simplify a more complicated expression, we follow the same rule. For example,

$7^2 - 3^2 = 49 - 9$ The bases are 7 and 3; the sign between the two terms is a subtraction sign

For another example,

$5^3 - 3^4 = 125 - 81$ We simplify exponents first, then subtract

EXAMPLE 20 Simplify: $\frac{5^2-3^2}{-5+3}$

SOLUTION

$\frac{5^2-3^2}{-5+3} = \frac{25-9}{-2}$ Simplify numerator and denominator separately

$= \frac{16}{-2}$

$= -8$

EXAMPLE 21 Simplify: $\frac{(3+2)^2}{-3^2-2^2}$

SOLUTION

$\frac{(3+2)^2}{-3^2-2^2} = \frac{5^2}{-9-4}$ Simplify numerator and denominator separately

$= \frac{25}{-13}$

$= -\frac{25}{13}$

We can combine our knowledge of the properties of multiplication with our definition of division to simplify more expressions involving fractions. Here are two examples:

EXAMPLE 22 Simplify $10\left(\frac{x}{2}\right)$.

SOLUTION

$$10\left(\frac{x}{2}\right) = 10\left(\frac{1}{2}x\right) \quad \text{Dividing by 2 is the same as multiplying by } \frac{1}{2}$$

$$= \left(10 \cdot \frac{1}{2}\right)x \quad \text{Associative property}$$

$$= 5x \quad \text{Multiplication}$$

EXAMPLE 23 Simplify $a\left(\frac{3}{a} - 4\right)$.

SOLUTION

$$a\left(\frac{3}{a} - 4\right) = a \cdot \frac{3}{a} - a \cdot 4 \quad \text{Distributive property}$$

$$= 3 - 4a \quad \text{Multiplication}$$

22. Simplify $8\left(\frac{x}{4}\right)$.

23. Simplify $x\left(\frac{4}{x} - 5\right)$.

Division with the Number 0

For every division problem there is an associated multiplication problem involving the same numbers. For example, the following two problems say the same thing about the numbers 2, 3, and 6:

Division	*Multiplication*
$\frac{6}{3} = 2$	$6 = 2(3)$

We can use this relationship between division and multiplication to clarify division involving the number 0.

First, dividing 0 by a number other than 0 is allowed and always results in 0. To see this, consider dividing 0 by 5. We know the answer is 0 because of the relationship between multiplication and division. This is how we write it:

$$\frac{0}{5} = 0 \quad \text{because} \quad 0 = 0(5)$$

However, dividing a nonzero number by 0 is not allowed in the real numbers. Suppose we were attempting to divide 5 by 0. We don't know if there is an answer to this problem, but if there is, let's say the answer is a number that we can represent with the letter n. If 5 divided by 0 is a number n, then

$$\frac{5}{0} = n \quad \text{and} \quad 5 = n(0)$$

This is impossible, however, because no matter what number n is, when we multiply it by 0 the answer must be 0. It can never be 5. In algebra, we say expressions like $\frac{5}{0}$ are undefined because there is no answer to them; that is, division by 0 is not allowed in the real numbers.

The only other possibility for division involving the number 0 is 0 divided by 0. We will treat problems like $\frac{0}{0}$ as if they were undefined also.

Getting Ready for Class

After reading through the preceding section, respond in your own words and in complete sentences.

1. Why do we define division in terms of multiplication?
2. What is the reciprocal of a number?
3. How do we divide fractions?
4. Why is division by 0 not allowed with real numbers?

Answers
22. $2x$ **23.** $4 - 5x$

Problem Set 1.7

A Find the following quotients (divide). [Examples 1–11]

1. $\frac{8}{-4}$

2. $\frac{10}{-5}$

3. $\frac{-48}{16}$

4. $\frac{-32}{4}$

5. $\frac{-7}{21}$

6. $\frac{-25}{100}$

7. $\frac{-39}{-13}$

8. $\frac{-18}{-6}$

9. $\frac{-6}{-42}$

10. $\frac{-4}{-28}$

11. $\frac{0}{-32}$

12. $\frac{0}{17}$

The following problems review all four operations with positive and negative numbers. Perform the indicated operations.

13. $-3 + 12$

14. $5 + (-10)$

15. $-3 - 12$

16. $5 - (-10)$

17. $-3(12)$

18. $5(-10)$

19. $-3 \div 12$

20. $5 \div (-10)$

B Divide and reduce all answers to lowest terms. [Examples 12–14]

21. $\frac{4}{5} \div \frac{3}{4}$

22. $\frac{6}{8} \div \frac{3}{4}$

23. $-\frac{5}{6} \div \left(-\frac{5}{8}\right)$

24. $-\frac{7}{9} \div \left(-\frac{1}{6}\right)$

25. $\frac{10}{13} \div \left(-\frac{5}{4}\right)$

26. $\frac{5}{12} \div \left(-\frac{10}{3}\right)$

27. $-\frac{5}{6} \div \frac{5}{6}$

28. $-\frac{8}{9} \div \frac{8}{9}$

29. $-\frac{3}{4} \div \left(-\frac{3}{4}\right)$

30. $-\frac{6}{7} \div \left(-\frac{6}{7}\right)$

C The following problems involve more than one operation. Simplify as much as possible. [Examples 15–23]

31. $\dfrac{3(-2)}{-10}$

32. $\dfrac{4(-3)}{24}$

33. $\dfrac{-5(-5)}{-15}$

34. $\dfrac{-7(-3)}{-35}$

35. $\dfrac{-8(-7)}{-28}$

36. $\dfrac{-3(-9)}{-6}$

37. $\dfrac{27}{4-13}$

38. $\dfrac{27}{13-4}$

39. $\dfrac{20-6}{5-5}$

40. $\dfrac{10-12}{3-3}$

41. $\dfrac{-3+9}{2\cdot 5-10}$

42. $\dfrac{-4+8}{2\cdot 4-8}$

43. $\dfrac{15(-5)-25}{2(-10)}$

44. $\dfrac{10(-3)-20}{5(-2)}$

45. $\dfrac{27-2(-4)}{-3(5)}$

46. $\dfrac{20-5(-3)}{10(-3)}$

47. $\dfrac{12-6(-2)}{12(-2)}$

48. $\dfrac{3(-4)+5(-6)}{10-6}$

49. $\dfrac{5^2-2^2}{-5+2}$

50. $\dfrac{7^2-4^2}{-7+4}$

51. $\dfrac{8^2-2^2}{8^2+2^2}$

52. $\dfrac{4^2-6^2}{4^2+6^2}$

53. $\dfrac{(5+3)^2}{-5^2-3^2}$

54. $\dfrac{(7+2)^2}{-7^2-2^2}$

55. $\dfrac{(8-4)^2}{8^2-4^2}$

56. $\dfrac{(6-2)^2}{6^2-2^2}$

57. $\dfrac{-4\cdot 3^2-5\cdot 2^2}{-8(7)}$

58. $\dfrac{-2\cdot 5^2+3\cdot 2^3}{-3(13)}$

59. $\dfrac{3 \cdot 10^2 + 4 \cdot 10 + 5}{345}$

60. $\dfrac{5 \cdot 10^2 + 6 \cdot 10 + 7}{567}$

61. $\dfrac{7 - [(2 - 3) - 4]}{-1 - 2 - 3}$

62. $\dfrac{2 - [(3 - 5) - 8]}{-3 - 4 - 5}$

63. $\dfrac{6(-4) - 2(5 - 8)}{-6 - 3 - 5}$

64. $\dfrac{3(-4) - 5(9 - 11)}{-9 - 2 - 3}$

65. $\dfrac{3(-5 - 3) + 4(7 - 9)}{5(-2) + 3(-4)}$

66. $\dfrac{-2(6 - 10) - 3(8 - 5)}{6(-3) - 6(-2)}$

67. $\dfrac{|3 - 9|}{3 - 9}$

68. $\dfrac{|4 - 7|}{4 - 7}$

69. $\dfrac{2 + 0.15(10)}{10}$

70. $\dfrac{5(5) + 250}{640(5)}$

71. $\dfrac{1 - 3}{3 - 1}$

72. $\dfrac{25 - 16}{16 - 25}$

73. Simplify.

a. $\dfrac{5 - 2}{3 - 1}$

b. $\dfrac{2 - 5}{1 - 3}$

74. Simplify.

a. $\dfrac{6 - 2}{3 - 5}$

b. $\dfrac{2 - 6}{5 - 3}$

75. Simplify.

a. $\dfrac{-4 - 1}{5 - (-2)}$

b. $\dfrac{1 - (-4)}{(-2) - 5}$

76. Simplify.

a. $\dfrac{-6 - 1}{4 - (-5)}$

b. $\dfrac{1 - (-6)}{-5 - 4}$

77. Simplify.

a. $\dfrac{3 + 2.236}{2}$

b. $\dfrac{3 - 2.236}{2}$

c. $\dfrac{3 + 2.236}{2} + \dfrac{3 - 2.236}{2}$

78. Simplify.

a. $\dfrac{1 + 1.732}{2}$

b. $\dfrac{1 - 1.732}{2}$

c. $\dfrac{1 + 1.732}{2} + \dfrac{1 - 1.732}{2}$

79. Simplify each expression.

a. $20 \div 4 \cdot 5$

b. $-20 \div 4 \cdot 5$

c. $20 \div (-4) \cdot 5$

d. $20 \div 4(-5)$

e. $-20 \div 4(-5)$

80. Simplify each expression.

a. $32 \div 8 \cdot 4$

b. $-32 \div 8 \cdot 4$

c. $32 \div (-8) \cdot 4$

d. $32 \div 8(-4)$

e. $-32 \div 8(-4)$

81. Simplify each expression.

a. $8 \div \frac{4}{5}$

b. $8 \div \frac{4}{5} - 10$

c. $8 \div \frac{4}{5}(-10)$

d. $8 \div \left(-\frac{4}{5}\right) - 10$

82. Simplify each expression.

a. $10 \div \frac{5}{6}$

b. $10 \div \frac{5}{6} - 12$

c. $10 \div \frac{5}{6}(-12)$

d. $10 \div \left(-\frac{5}{6}\right) - 12$

Apply the distributive property.

83. $10\left(\frac{x}{2} + \frac{3}{5}\right)$

84. $6\left(\frac{x}{3} + \frac{5}{2}\right)$

85. $15\left(\frac{x}{5} + \frac{4}{3}\right)$

86. $6\left(\frac{x}{3} + \frac{1}{2}\right)$

87. $x\left(\frac{3}{x} + 1\right)$

88. $x\left(\frac{4}{x} + 3\right)$

89. $21\left(\frac{x}{7} - \frac{y}{3}\right)$

90. $36\left(\frac{x}{4} - \frac{y}{9}\right)$

91. $a\left(\frac{3}{a} - \frac{2}{a}\right)$

92. $a\left(\frac{7}{a} + \frac{1}{a}\right)$

93. $2y\left(\frac{1}{y} - \frac{1}{2}\right)$

94. $5y\left(\frac{3}{y} - \frac{4}{5}\right)$

Answer the following questions.

95. What is the quotient of -12 and -4?

96. The quotient of -4 and -12 is what number?

97. What number do we divide by -5 to get 2?

98. What number do we divide by -3 to get 4?

99. Twenty-seven divided by what number is -9?

100. Fifteen divided by what number is -3?

101. If the quotient of -20 and 4 is decreased by 3, what number results?

102. If -4 is added to the quotient of 24 and -8, what number results?

Applying the Concepts

103. Golf The chart shows the lengths of the longest golf courses used for the U.S. open. If the length of the Bethpage course is 7,214 yards, what is the average length of each hole on the 18-hole course?

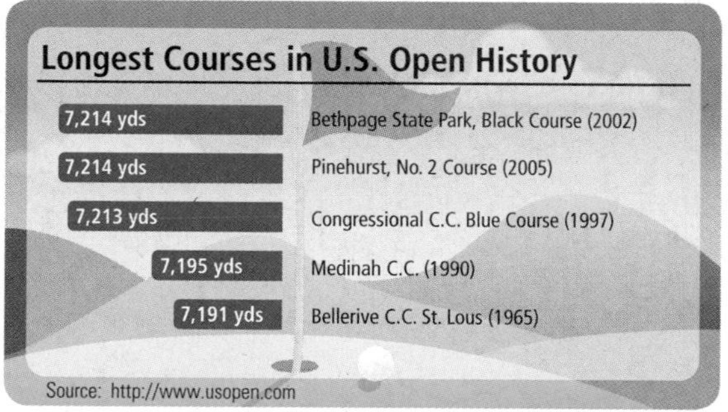

104. Broadway The chart shows the number of plays performed by the longest running Broadway musicals. If *Les Miserables* ran for 16 years, what was the mean number of shows per year? Round to the nearest show.

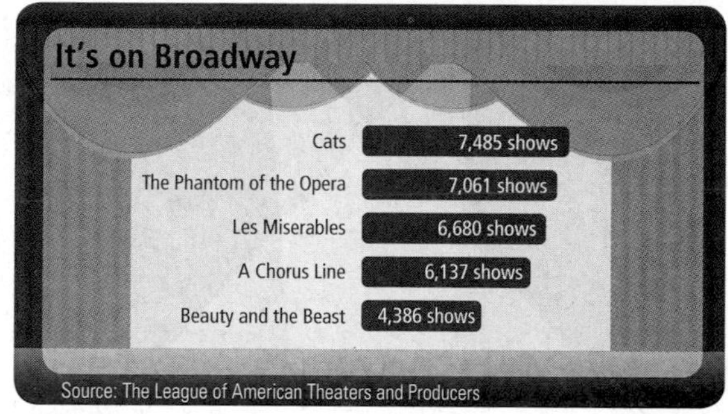

105. Investment Suppose that you and 3 friends bought equal shares of an investment for a total of $15,000 and then sold it later for only $13,600. How much did each person lose?

106. Investment If 8 people invest $500 each in a stamp collection and after a year the collection is worth $3,800, how much did each person lose?

107. Temperature Change Suppose that the temperature outside is dropping at a constant rate. If the temperature is 75°F at noon and drops to 61°F by 4:00 in the afternoon, by how much did the temperature change each hour?

108. Temperature Change In a chemistry class, a thermometer is placed in a beaker of hot water. The initial temperature of the water is 165°F. After 10 minutes the water has cooled to 72°F. If the water temperature drops at a constant rate, by how much does the water temperature change each minute?

109. Nursing Problem A patient is required to take 75 mg of a drug. If the capsule strength is 25 mg, how many capsules should she take?

110. Nursing Problem A patient is given prescribed a dosage of 25 mg, and the strength of each tablet is 50 mg. How many tablets should he take?

111. Nursing Problem A patient is required to take 1.2 mg of drug and is told to take 4 tablets. What is the strength of each tablet?

112. Nursing Problem A patient is required to take two tablets for a 3.6 mg dosage. What is the dosage strength of each tablet?

113. Internet Mailing Lists A company sells products on the Internet through an email list. They predict that they sell one $50 product for every 25 people on their mailing list.

- **a.** What is their projected revenue if their list contains 10,000 email addresses?
- **b.** What is their projected revenue if their list contains 25,000 email addresses?
- **c.** They can purchase a list of 5,000 email addresses for $5,000. Is this a wise purchase?

114. Internet Mailing Lists A new band has a following on the Internet. They sell their CDs through an email list. They predict that they sell one $15 CD for every 10 people on their mailing list.

- **a.** What is their projected revenue if their list contains 5,000 email addresses?
- **b.** What is their projected revenue if their list contains 20,000 email addresses?
- **c.** If they need to make $45,000, how many people do they need on their email list?

Subsets of the Real Numbers

1.8

Objectives

A Associate numbers with subsets of the real numbers.

B Factor whole numbers into the product of prime factors.

C Reduce fractions to lowest terms.

In Section 1.2 we introduced the real numbers and defined them as the numbers associated with points on the real number line. At that time, we said the real numbers include whole numbers, fractions, and decimals, as well as other numbers that are not as familiar to us as these numbers. In this section we take a more detailed look at the kinds of numbers that make up the set of real numbers.

The numbers that make up the set of real numbers can be classified as *counting numbers, whole numbers, integers, rational numbers,* and *irrational numbers;* each is said to be a *subset* of the real numbers.

> **Definition**
> Set A is called a **subset** of set B if set A is contained in set B, that is, if each and every element in set A is also a member of set B.

A Subsets of Real Numbers

Here is a detailed description of the major subsets of the real numbers.

The *counting numbers* are the numbers with which we count. They are the numbers 1, 2, 3, and so on. The notation we use to specify a group of numbers like this is *set notation.* We use the symbols { and } to enclose the members of the set.

$$\text{Counting numbers} = \{1, 2, 3, \ldots\}$$

EXAMPLE 1 Which of the numbers in the following set are not counting numbers?

$$\left\{-3, 0, \frac{1}{2}, 1, 1.5, 3\right\}$$

SOLUTION The numbers $-3, 0, \frac{1}{2}$, and 1.5 are not counting numbers.

The *whole numbers* include the counting numbers and the number 0.

$$\text{Whole numbers} = \{0, 1, 2, \ldots\}$$

The set of *integers* includes the whole numbers and the opposites of all the counting numbers.

$$\text{Integers} = \{\ldots, -3, -2, -1, 0, 1, 2, 3, \ldots\}$$

When we refer to *positive integers*, we are referring to the numbers 1, 2, 3, ... Likewise, the *negative integers* are $-1, -2, -3, \ldots$. The number 0 is neither positive nor negative.

EXAMPLE 2 Which of the numbers in the following set are not integers?

$$\left\{-5, -1.75, 0, \frac{2}{3}, 1, \pi, 3\right\}$$

SOLUTION The only numbers in the set that are not integers are -1.75, $\frac{2}{3}$, and π.

The set of *rational numbers* is the set of numbers commonly called "fractions" together with the integers. The set of rational numbers is difficult to list

PRACTICE PROBLEMS

1. Is 5 a counting number?

2. Is -7 a whole number?

Answers

1. Yes 2. No

3. a. Is $\frac{3}{4}$ a rational number?
b. Is 7 a rational number?

in the same way we have listed the other sets, so we will use a different kind of notation:

$$\text{Rational numbers} = \left\{ \frac{a}{b} \,\middle|\, a \text{ and } b \text{ are integers } (b \neq 0) \right\}$$

This notation is read "The set of elements $\frac{a}{b}$ such that a and b are integers (and b is not 0)." If a number can be put in the form $\frac{a}{b}$, where a and b are both from the set of integers, then it is called a rational number.

Rational numbers include any number that can be written as the *ratio* of two integers; that is, rational numbers are numbers that can be put in the form

$$\frac{\text{integer}}{\text{integer}}$$

EXAMPLE 3 Show why each of the numbers in the following set is a rational number.

$$\left\{-3, -\frac{2}{3}, 0, 0.333\ldots, 0.75\right\}$$

SOLUTION The number −3 is a rational number because it can be written as the ratio of −3 to 1; that is,

$$-3 = \frac{-3}{1}$$

Similarly, the number $-\frac{2}{3}$ can be thought of as the ratio of −2 to 3, whereas the number 0 can be thought of as the ratio of 0 to 1.

Any repeating decimal, such as 0.333 . . . (the dots indicate that the 3's repeat forever), can be written as the ratio of two integers. In this case 0.333 . . . is the same as the fraction $\frac{1}{3}$.

Finally, any decimal that terminates after a certain number of digits can be written as the ratio of two integers. The number 0.75 is equal to the fraction $\frac{3}{4}$ and is therefore a rational number.

Still other numbers exist, each of which is associated with a point on the real number line, that cannot be written as the ratio of two integers. In decimal form they never terminate and never repeat a sequence of digits indefinitely. They are called *irrational numbers* (because they are not rational):

Irrational numbers = {nonrational numbers}
= {nonrepeating, nonterminating decimals}

We cannot write any irrational number in a form that is familiar to us because they are all nonterminating, nonrepeating decimals. Because they are not rational, they cannot be written as the ratio of two integers. They have to be represented in other ways. One irrational number you have probably seen before is π. It is not 3.14. Rather, 3.14 is an approximation to π. It cannot be written as a terminating decimal number. Other representations for irrational numbers are $\sqrt{2}$, $\sqrt{3}$, $\sqrt{5}$, $\sqrt{6}$, and, in general, the square root of any number that is not itself a perfect square. (If you are not familiar with square roots, you will be after Chapter 8.) Right now it is enough to know that some numbers on the number line cannot be written as the ratio of two integers or in decimal form. We call them irrational numbers.

The set of *real numbers* is the set of numbers that are either rational or irrational; that is, a real number is either rational or irrational.

Real numbers = {all rational numbers and all irrational numbers}

Answer
3. a. Yes b. Yes

B Prime Numbers and Factoring

The following diagram shows the relationship between multiplication and factoring:

Multiplication

Factors $\longrightarrow 3 \cdot 4 = 12 \longleftarrow$ Product

Factoring

When we read the problem from left to right, we say the product of 3 and 4 is 12. Or we multiply 3 and 4 to get 12. When we read the problem in the other direction, from right to left, we say we have *factored* 12 into 3 times 4, or 3 and 4 are *factors* of 12.

The number 12 can be factored still further:

$$\begin{aligned} 12 &= 4 \cdot 3 \\ &= 2 \cdot 2 \cdot 3 \\ &= 2^2 \cdot 3 \end{aligned}$$

The numbers 2 and 3 are called *prime factors* of 12 because neither of them can be factored any further.

Definition

If a and b represent integers, then a is said to be a **factor** (or divisor) of b if a divides b evenly, that is, if a divides b with no remainder.

Definition

A **prime number** is any positive integer larger than 1 whose only positive factors (divisors) are itself and 1.

Here is a list of the first few prime numbers.

Prime numbers = {2, 3, 5, 7, 11, 13, 17, 19, 23, 29, 31, 37, 41, . . .}

When a number is not prime, we can factor it into the product of prime numbers. To factor a number into the product of primes, we simply factor it until it cannot be factored further.

Note The number 15 is not a prime number because it has factors of 3 and 5; that is, $15 = 3 \cdot 5$. When a whole number larger than 1 is not prime, it is said to be *composite.*

EXAMPLE 4 Factor the number 60 into the product of prime numbers.

SOLUTION We begin by writing 60 as the product of any two positive integers whose product is 60, like 6 and 10:

$$60 = 6 \cdot 10$$

We then factor these numbers:

$$\begin{aligned} 60 &= 6 \cdot 10 \\ &= (2 \cdot 3) \cdot (2 \cdot 5) \\ &= 2 \cdot 2 \cdot 3 \cdot 5 \\ &= 2^2 \cdot 3 \cdot 5 \end{aligned}$$

4. Factor 90 into the product of prime numbers.

Note It is customary to write prime factors in order from smallest to largest.

Answer
4. $2 \cdot 3^2 \cdot 5$

5. Factor 420 into the product of primes.

EXAMPLE 5 Factor the number 630 into the product of primes.

SOLUTION Let's begin by writing 630 as the product of 63 and 10:

$$\begin{aligned} 630 &= 63 \cdot 10 \\ &= (7 \cdot 9) \cdot (2 \cdot 5) \\ &= 7 \cdot 3 \cdot 3 \cdot 2 \cdot 5 \\ &= 2 \cdot 3^2 \cdot 5 \cdot 7 \end{aligned}$$

It makes no difference which two numbers we start with, as long as their product is 630. We always will get the same result because a number has only one set of prime factors.

$$\begin{aligned} 630 &= 18 \cdot 35 \\ &= 3 \cdot 6 \cdot 5 \cdot 7 \\ &= 3 \cdot 2 \cdot 3 \cdot 5 \cdot 7 \\ &= 2 \cdot 3^2 \cdot 5 \cdot 7 \end{aligned}$$

Note There are some "tricks" to finding the divisors of a number. For instance, if a number ends in 0 or 5, then it is divisible by 5. If a number ends in an even number (0, 2, 4, 6, or 8), then it is divisible by 2. A number is divisible by 3 if the sum of its digits is divisible by 3. For example, 921 is divisible by 3 because the sum of its digits is $9 + 2 + 1 = 12$, which is divisible by 3.

If we factor 210 into its prime factors, we have $210 = 2 \cdot 3 \cdot 5 \cdot 7$, which means that 2, 3, 5, and 7 divide 210, as well as any combination of products of 2, 3, 5, and 7; that is, because 3 and 7 divide 210, then so does their product 21. Because 3, 5, and 7 each divide 210, then so does their product 105:

21 divides 210

$$210 = 2 \cdot 3 \cdot 5 \cdot 7$$

105 divides 210

C Reducing to Lowest Terms

Recall that we reduce fractions to lowest terms by dividing the numerator and denominator by the same number. We can use the prime factorization of numbers to help us reduce fractions with large numerators and denominators.

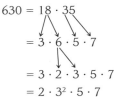

6. Reduce $\frac{154}{1,155}$ to lowest terms.

EXAMPLE 6 Reduce $\frac{210}{231}$ to lowest terms.

SOLUTION First we factor 210 and 231 into the product of prime factors. Then we reduce to lowest terms by dividing the numerator and denominator by any factors they have in common.

$$\begin{aligned} \frac{210}{231} &= \frac{2 \cdot 3 \cdot 5 \cdot 7}{3 \cdot 7 \cdot 11} && \text{Factor the numerator and denominator completely} \\ &= \frac{2 \cdot \cancel{3} \cdot 5 \cdot \cancel{7}}{\cancel{3} \cdot \cancel{7} \cdot 11} && \text{Divide the numerator and denominator by } 3 \cdot 7 \\ &= \frac{2 \cdot 5}{11} = \frac{10}{11} \end{aligned}$$

Note The small lines we have drawn through the factors that are common to the numerator and denominator are used to indicate that we have divided the numerator and denominator by those factors.

Getting Ready for Class

After reading through the preceding section, respond in your own words and in complete sentences.

1. What is a whole number?
2. How are factoring and multiplication related?
3. Is every integer also a rational number? Explain.
4. What is a prime number?

Answers

5. $2^2 \cdot 3 \cdot 5 \cdot 7$ **6.** $\frac{2}{15}$

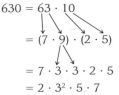

Problem Set 1.8

A Given the numbers in the set $\{-3, -2.5, 0, 1, \frac{3}{2}, \sqrt{15}\}$: [Examples 1–3]

1. List all the whole numbers.

2. List all the integers.

3. List all the rational numbers.

4. List all the irrational numbers.

5. List all the real numbers.

A Given the numbers in the set $\{-10, -8, -0.333\ldots, -2, 9, \frac{25}{3}, \pi\}$:

6. List all the whole numbers.

7. List all the integers.

8. List all the rational numbers.

9. List all the irrational numbers.

10. List all the real numbers.

Identify the following statements as either true or false.

11. Every whole number is also an integer.

12. The set of whole numbers is a subset of the set of integers.

13. A number can be both rational and irrational.

14. The set of rational numbers and the set of irrational numbers have some elements in common.

15. Some whole numbers are also negative integers.

16. Every rational number is also a real number.

17. All integers are also rational numbers.

18. The set of integers is a subset of the set of rational numbers.

B Label each of the following numbers as *prime* or *composite.* If a number is composite, then factor it completely.

19. 48

20. 72

21. 37

22. 23

23. 1,023

24. 543

B Factor the following into the product of primes. When the number has been factored completely, write its prime factors from smallest to largest. [Examples 4–5]

25. 144

26. 288

27. 38

28. 63

29. 105

30. 210

31. 180

32. 900

33. 385

34. 1,925

35. 121

36. 546

37. 420

38. 598

39. 620

40. 2,310

C Reduce each fraction to lowest terms by first factoring the numerator and denominator into the product of prime factors and then dividing out any factors they have in common. [Examples 6]

41. $\frac{105}{165}$

42. $\frac{165}{385}$

43. $\frac{525}{735}$

44. $\frac{550}{735}$

45. $\frac{385}{455}$

46. $\frac{385}{735}$

47. $\frac{322}{345}$

48. $\frac{266}{285}$

49. $\frac{205}{369}$

50. $\frac{111}{185}$

51. $\frac{215}{344}$

52. $\frac{279}{310}$

The next two problems are intended to give you practice reading, and paying attention to, the instructions that accompany the problems you are working. You will see a number of problems like this throughout the book. Working these problems is an excellent way to get ready for a test or a quiz.

53. Work each problem according to the instructions given. (Note that each of these instructions could be replaced with the instruction *Simplify*.)

a. Add: $50 + (-80)$

b. Subtract: $50 - (-80)$

c. Multiply: $50(-80)$

d. Divide: $\frac{50}{-80}$

54. Work each problem according to the instructions given. (Note that each of these instructions could be replaced with the instruction *Simplify*.)

a. Add: $-2.5 + 7.5$

b. Subtract: $-2.5 - 7.5$

c. Multiply: $-2.5(7.5)$

d. Divide: $\frac{-2.5}{7.5}$

Simplify each expression without using a calculator.

55. $\frac{6.28}{9(3.14)}$

56. $\frac{12.56}{4(3.14)}$

57. $\frac{9.42}{2(3.14)}$

58. $\frac{12.56}{2(3.14)}$

59. $\frac{32}{0.5}$

60. $\frac{16}{0.5}$

61. $\frac{5{,}599}{11}$

62. $\frac{840}{80}$

63. Find the value of $\frac{2 + 0.15x}{x}$ for each of the values of x given below. Write your answers as decimals, to the nearest hundredth.

a. $x = 10$

b. $x = 15$

c. $x = 20$

64. Find the value of $\frac{5x + 250}{640x}$ for each of the values of x given below. Write your answers as decimals, to the nearest thousandth.

a. $x = 10$

b. $x = 15$

c. $x = 20$

65. Factor 6^3 into the product of prime factors by first factoring 6 and then raising each of its factors to the third power.

66. Factor 12^2 into the product of prime factors by first factoring 12 and then raising each of its factors to the second power.

67. Factor $9^4 \cdot 16^2$ into the product of prime factors by first factoring 9 and 16 completely.

68. Factor $10^2 \cdot 12^3$ into the product of prime factors by first factoring 10 and 12 completely.

69. Simplify the expression $3 \cdot 8 + 3 \cdot 7 + 3 \cdot 5$, and then factor the result into the product of primes. (Notice one of the factors of the answer is 3.)

70. Simplify the expression $5 \cdot 4 + 5 \cdot 9 + 5 \cdot 3$, and then factor the result into the product of primes.

Applying the Concepts

71. **Cars** The chart shows the fastest cars in America. Factor the speed of the Saleen S7 Twin Turbo into the product of primes.

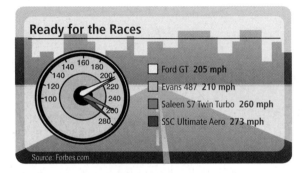

Ready for the Races

Ford GT 205 mph

Evans 487 210 mph

Saleen S7 Twin Turbo 260 mph

SSC Ultimate Aero 273 mph

Source: Forbes.com

72. **Classroom Energy** The chart shows how much energy is wasted in the classroom by leaving appliances on. Write the energy used by a ceiling fan over the energy used by the television and then reduce to lowest terms.

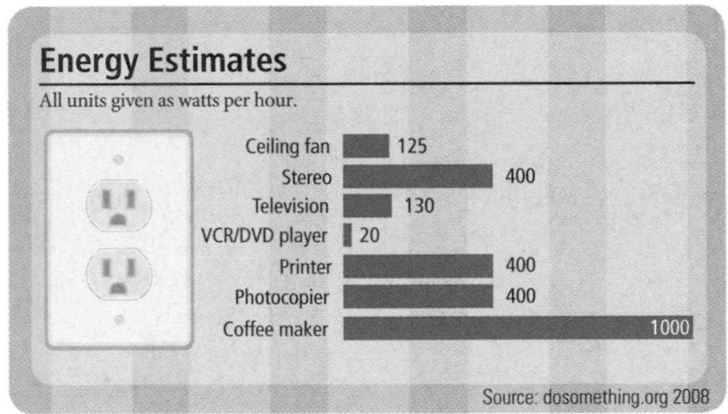

Recall the Fibonacci sequence we introduced earlier in this chapter.

$$\text{Fibonacci sequence} = 1, 1, 2, 3, 5, 8, \ldots$$

Any number in the Fibonacci sequence is a *Fibonacci number.*

73. The Fibonacci numbers are not a subset of which of the following sets: real numbers, rational numbers, irrational numbers, whole numbers?

74. Name three Fibonacci numbers that are prime numbers.

75. Name three Fibonacci numbers that are composite numbers.

76. Is the sequence of odd numbers a subset of the Fibonacci numbers?

Addition and Subtraction with Fractions

1.9

Objectives

A Add or subtract two or more fractions with the same denominator.

B Find the least common denominator for a set of fractions.

C Add or subtract fractions with different denominators.

D Extend a sequence of numbers containing fractions.

Examples now playing at **MathTV.com/books**

You may recall from previous math classes that to add two fractions with the same denominator, you simply add their numerators and put the result over the common denominator:

$$\frac{3}{4} + \frac{2}{4} = \frac{3+2}{4} = \frac{5}{4}$$

The reason we add numerators but do not add denominators is that we must follow the distributive property. To see this, you first have to recall that $\frac{3}{4}$ can be written as $3 \cdot \frac{1}{4}$, and $\frac{2}{4}$ can be written as $2 \cdot \frac{1}{4}$ (dividing by 4 is equivalent to multiplying by $\frac{1}{4}$). Here is the addition problem again, this time showing the use of the distributive property:

$$\begin{aligned}
\frac{3}{4} + \frac{2}{4} &= 3 \cdot \frac{1}{4} + 2 \cdot \frac{1}{4} \\
&= (3+2) \cdot \frac{1}{4} && \text{Distributive property} \\
&= 5 \cdot \frac{1}{4} \\
&= \frac{5}{4}
\end{aligned}$$

Note Most people who have done any work with adding fractions know that you add fractions that have the same denominator by adding their numerators but not their denominators. However, most people don't know why this works. The reason why we add numerators but not denominators is because of the distributive property. That is what the discussion at the left is all about. If you really want to understand addition of fractions, pay close attention to this discussion.

A Adding and Subtracting with the Same Denominator

What we have here is the sum of the numerators placed over the *common denominator.* In symbols we have the following.

> **Addition and Subtraction of Fractions**
>
> If a, b, and c are integers and c is not equal to 0, then
>
> $$\frac{a}{c} + \frac{b}{c} = \frac{a+b}{c}$$
>
> This rule holds for subtraction as well; that is,
>
> $$\frac{a}{c} - \frac{b}{c} = \frac{a-b}{c}$$

In Examples 1–4, find the sum or difference. (Add or subtract as indicated.) Reduce all answers to lowest terms. (Assume all variables represent nonzero numbers.)

EXAMPLE 1 Add $\frac{3}{8} + \frac{1}{8}$.

SOLUTION

$$\begin{aligned}
\frac{3}{8} + \frac{1}{8} &= \frac{3+1}{8} && \text{Add numerators; keep the same denominator} \\
&= \frac{4}{8} && \text{The sum of 3 and 1 is 4} \\
&= \frac{1}{2} && \text{Reduce to lowest terms}
\end{aligned}$$

EXAMPLE 2 Subtract: $\frac{a+5}{8} - \frac{3}{8}$.

SOLUTION

$$\begin{aligned}
\frac{a+5}{8} - \frac{3}{8} &= \frac{a+5-3}{8} && \text{Combine numerators; keep the same denominator} \\
&= \frac{a+2}{8}
\end{aligned}$$

PRACTICE PROBLEMS

Find the sum or difference. Reduce all answers to lowest terms. (Assume all variables represent nonzero numbers.)

1. $\frac{3}{10} + \frac{1}{10}$

2. $\frac{a-5}{12} + \frac{3}{12}$

Answers

1. $\frac{2}{5}$ **2.** $\frac{a-2}{12}$

3. $\frac{8}{x} - \frac{5}{x}$

4. $\frac{5}{9} - \frac{8}{9} + \frac{5}{9}$

5. Find the LCD for the fractions $\frac{5}{18}$ and $\frac{3}{14}$.

Note The ability to find least common denominators is very important in mathematics. The discussion here is a detailed explanation of how to do it.

Answers

3. $\frac{3}{x}$ 4. $\frac{2}{9}$ 5. 126

EXAMPLE 3 Subtract: $\frac{9}{x} - \frac{3}{x}$.

SOLUTION

$$\frac{9}{x} - \frac{3}{x} = \frac{9-3}{x}$$ Subtract numerators; keep the same denominator

$$= \frac{6}{x}$$ The difference of 9 and 3 is 6

EXAMPLE 4 Simplify: $\frac{3}{7} + \frac{2}{7} - \frac{9}{7}$.

SOLUTION

$$\frac{3}{7} + \frac{2}{7} - \frac{9}{7} = \frac{3+2-9}{7}$$

$$= \frac{-4}{7}$$

$$= -\frac{4}{7}$$ Unlike signs give a negative answer

B Finding the Least Common Denominator

We will now turn our attention to the process of adding fractions that have different denominators. To get started, we need the following definition.

Definition

The **least common denominator (LCD)** for a set of denominators is the smallest number that is exactly divisible by each denominator, also called the *least common multiple.*

In other words, all the denominators of the fractions in a problem must divide into the least common denominator without a remainder.

EXAMPLE 5 Find the LCD for the fractions $\frac{5}{12}$ and $\frac{7}{18}$.

SOLUTION The least common denominator for the denominators 12 and 18 must be the smallest number divisible by both 12 and 18. We can factor 12 and 18 completely and then build the LCD from these factors. Factoring 12 and 18 completely gives us

$$12 = 2 \cdot 2 \cdot 3 \qquad 18 = 2 \cdot 3 \cdot 3$$

Now, if 12 is going to divide the LCD exactly, then the LCD must have factors of $2 \cdot 2 \cdot 3$. If 18 is to divide it exactly, it must have factors of $2 \cdot 3 \cdot 3$. We don't need to repeat the factors that 12 and 18 have in common:

12 divides the LCD

$$\left.\begin{aligned} 12 &= 2 \cdot 2 \cdot 3 \\ 18 &= 2 \cdot 3 \cdot 3 \end{aligned}\right\} \quad \text{LCD} = 2 \cdot 2 \cdot 3 \cdot 3 = 36$$

18 divides the LCD

In other words, first we write down the factors of 12, then we attach the factors of 18 that do not already appear as factors of 12.

The LCD for 12 and 18 is 36. It is the smallest number that is divisible by both 12 and 18; 12 divides it exactly three times, and 18 divides it exactly two times.

C Adding and Subtracting with Different Denominators

We can use the results of Example 5 to find the sum of the fractions $\frac{5}{12}$ and $\frac{7}{18}$.

EXAMPLE 6 Add $\frac{5}{12} + \frac{7}{18}$.

SOLUTION We can add fractions only when they have the same denominators. In Example 5 we found the LCD for $\frac{5}{12}$ and $\frac{7}{18}$ to be 36. We change $\frac{5}{12}$ and $\frac{7}{18}$ to equivalent fractions that each have 36 for a denominator by applying Property 1, on page 18 for fractions:

$$\frac{5}{12} = \frac{5 \cdot \mathbf{3}}{12 \cdot \mathbf{3}} = \frac{15}{36} \qquad \frac{5}{12} \text{ is equivalent to } \frac{15}{36}$$

$$\frac{7}{18} = \frac{7 \cdot \mathbf{2}}{18 \cdot \mathbf{2}} = \frac{14}{36} \qquad \frac{7}{18} \text{ is equivalent to } \frac{14}{36}$$

The fraction $\frac{15}{36}$ is equivalent to $\frac{5}{12}$, because it was obtained by multiplying both the numerator and denominator by 3. Likewise, $\frac{14}{36}$ is equivalent to $\frac{7}{18}$ because it was obtained by multiplying the numerator and denominator by 2. All we have left to do is to add numerators:

$$\frac{15}{36} + \frac{14}{36} = \frac{29}{36}$$

The sum of $\frac{5}{12}$ and $\frac{7}{18}$ is the fraction $\frac{29}{36}$. Let's write the complete solution again step-by-step.

$$\frac{5}{12} + \frac{7}{18} = \frac{5 \cdot \mathbf{3}}{12 \cdot \mathbf{3}} + \frac{7 \cdot \mathbf{2}}{18 \cdot \mathbf{2}} \qquad \text{Rewrite each fraction as an equivalent fraction with denominator 36}$$

$$= \frac{15}{36} + \frac{14}{36}$$

$$= \frac{29}{36} \qquad \text{Add numerators; keep the common denominator}$$

6. Add $\frac{5}{18} + \frac{3}{14}$.

EXAMPLE 7 Find the LCD for $\frac{3}{4}$ and $\frac{1}{6}$.

SOLUTION We factor 4 and 6 into products of prime factors and build the LCD from these factors:

$$\left.\begin{aligned} 4 &= 2 \cdot 2 \\ 6 &= 2 \cdot 3 \end{aligned}\right\} \quad \text{LCD} = 2 \cdot 2 \cdot 3 = 12$$

The LCD is 12. Both denominators divide it exactly; 4 divides 12 exactly three times, and 6 divides 12 exactly two times.

7. Find the LCD for $\frac{2}{9}$ and $\frac{4}{15}$.

EXAMPLE 8 Add $\frac{3}{4} + \frac{1}{6}$.

SOLUTION In Example 7 we found that the LCD for these two fractions is 12. We begin by changing $\frac{3}{4}$ and $\frac{1}{6}$ to equivalent fractions with denominator 12:

$$\frac{3}{4} = \frac{3 \cdot \mathbf{3}}{4 \cdot \mathbf{3}} = \frac{9}{12} \qquad \frac{3}{4} \text{ is equivalent to } \frac{9}{12}$$

$$\frac{1}{6} = \frac{1 \cdot \mathbf{2}}{6 \cdot \mathbf{2}} = \frac{2}{12} \qquad \frac{1}{6} \text{ is equivalent to } \frac{2}{12}$$

To complete the problem, we add numerators:

$$\frac{9}{12} + \frac{2}{12} = \frac{11}{12}$$

The sum of $\frac{3}{4}$ and $\frac{1}{6}$ is $\frac{11}{12}$. Here is how the complete problem looks:

$$\frac{3}{4} + \frac{1}{6} = \frac{3 \cdot \mathbf{3}}{4 \cdot \mathbf{3}} + \frac{1 \cdot \mathbf{2}}{6 \cdot \mathbf{2}} \qquad \text{Rewrite each fraction as an equivalent fraction with denominator 12}$$

$$= \frac{9}{12} + \frac{2}{12} = \frac{11}{12} \qquad \text{Add numerators; keep the same denominator}$$

8. Add $\frac{2}{9} + \frac{4}{15}$.

Answers

6. $\frac{31}{63}$ 7. 45 8. $\frac{22}{45}$

9. Subtract $\frac{8}{25} - \frac{3}{20}$.

EXAMPLE 9 Subtract $\frac{7}{15} - \frac{3}{10}$.

SOLUTION Let's factor 15 and 10 completely and use these factors to build the LCD:

$$\left.\begin{aligned} 15 &= 3 \cdot 5 \\ 10 &= 2 \cdot 5 \end{aligned}\right\} \quad \text{LCD} = 2 \cdot 3 \cdot 5 = 30$$

15 divides the LCD (points to $3 \cdot 5$)

10 divides the LCD (points to $2 \cdot$ and 5)

Changing to equivalent fractions and subtracting, we have

$$\frac{7}{15} - \frac{3}{10} = \frac{7 \cdot \mathbf{2}}{15 \cdot \mathbf{2}} - \frac{3 \cdot \mathbf{3}}{10 \cdot \mathbf{3}}$$ Rewrite as equivalent fractions with the LCD for denominator

$$= \frac{14}{30} - \frac{9}{30}$$

$$= \frac{5}{30}$$ Subtract numerators; keep the LCD

$$= \frac{1}{6}$$ Reduce to lowest terms

As a summary of what we have done so far and as a guide to working other problems, we will now list the steps involved in adding and subtracting fractions with different denominators.

Strategy To Add or Subtract Any Two Fractions

Step 1: Factor each denominator completely and use the factors to build the LCD. (Remember, the LCD is the smallest number divisible by each of the denominators in the problem.)

Step 2: Rewrite each fraction as an equivalent fraction that has the LCD for its denominator. This is done by multiplying both the numerator and denominator of the fraction in question by the appropriate whole number.

Step 3: Add or subtract the numerators of the fractions produced in step 2. This is the numerator of the sum or difference. The denominator of the sum or difference is the LCD.

Step 4: Reduce the fraction produced in step 3 to lowest terms if it is not already in lowest terms.

The idea behind adding or subtracting fractions is really very simple. We can add or subtract only fractions that have the same denominators. If the fractions we are trying to add or subtract do not have the same denominators, we rewrite each of them as an equivalent fraction with the LCD for a denominator.

Here are some further examples of sums and differences of fractions.

10. Add $\frac{1}{9} + \frac{1}{4} + \frac{1}{6}$.

EXAMPLE 10 Add $\frac{1}{6} + \frac{1}{8} + \frac{1}{4}$.

SOLUTION We begin by factoring the denominators completely and building the LCD from the factors that result. We then change to equivalent fractions and add as usual:

Answers

9. $\frac{17}{100}$

$$\left.\begin{aligned}6 &= 2 \cdot 3\\ 8 &= 2 \cdot 2 \cdot 2\\ 4 &= 2 \cdot 2\end{aligned}\right\} \quad \text{LCD} = 2 \cdot 2 \cdot 2 \cdot 3 = 24$$

8 divides the LCD

4 divides the LCD

6 divides the LCD

$$\frac{1}{6} + \frac{1}{8} + \frac{1}{4} = \frac{1 \cdot \mathbf{4}}{6 \cdot \mathbf{4}} + \frac{1 \cdot \mathbf{3}}{8 \cdot \mathbf{3}} + \frac{1 \cdot \mathbf{6}}{4 \cdot \mathbf{6}}$$
$$= \frac{4}{24} + \frac{3}{24} + \frac{6}{24}$$
$$= \frac{13}{24}$$

EXAMPLE 11 Subtract $3 - \frac{5}{6}$.

SOLUTION The denominators are 1 $\left(\text{because } 3 = \frac{3}{1}\right)$ and 6. The smallest number divisible by both 1 and 6 is 6.

$$3 - \frac{5}{6} = \frac{3}{1} - \frac{5}{6}$$
$$= \frac{3 \cdot \mathbf{6}}{1 \cdot \mathbf{6}} - \frac{5}{6}$$
$$= \frac{18}{6} - \frac{5}{6}$$
$$= \frac{13}{6}$$

11. Subtract $2 - \frac{3}{4}$.

D Sequences Containing Fractions

EXAMPLE 12 Find the next number in each sequence.

a. $\frac{1}{2}, 0, -\frac{1}{2}, \ldots$ **b.** $\frac{1}{2}, 1, \frac{3}{2}, \ldots$ **c.** $\frac{1}{2}, \frac{1}{4}, \frac{1}{8}, \ldots$

SOLUTION

a. $\frac{1}{2}, 0, -\frac{1}{2}, \ldots$: Adding $-\frac{1}{2}$ to each term produces the next term. The fourth term will be $-\frac{1}{2} + \left(-\frac{1}{2}\right) = -1$. This is an arithmetic sequence.

b. $\frac{1}{2}, 1, \frac{3}{2}, \ldots$: Each term comes from the term before it by adding $\frac{1}{2}$. The fourth term will be $\frac{3}{2} + \frac{1}{2} = 2$. This sequence is also an arithmetic sequence.

c. $\frac{1}{2}, \frac{1}{4}, \frac{1}{8}, \ldots$: This is a geometric sequence in which each term comes from the term before it by multiplying by $\frac{1}{2}$ each time. The next term will be $\frac{1}{8} \cdot \frac{1}{2} = \frac{1}{16}$.

12. Find the next number in each sequence

a. $\frac{1}{3}, 0, -\frac{1}{3}, \ldots$

b. $\frac{1}{3}, \frac{2}{3}, 1, \ldots$

c. $1, \frac{1}{3}, \frac{1}{9}, \ldots$

EXAMPLE 13 Subtract: $\frac{x}{5} - \frac{1}{6}$

SOLUTION The LCD for 5 and 6 is their product, 30. We begin by rewriting each fraction with this common denominator:

$$\frac{x}{5} - \frac{1}{6} = \frac{x \cdot \mathbf{6}}{5 \cdot \mathbf{6}} - \frac{1 \cdot \mathbf{5}}{6 \cdot \mathbf{5}}$$
$$= \frac{6x}{30} - \frac{5}{30}$$
$$= \frac{6x - 5}{30}$$

13. Subtract: $\frac{x}{4} - \frac{1}{5}$

Answers

10. $\frac{19}{36}$ **11.** $\frac{5}{4}$

12. a. $-\frac{2}{3}$ **b.** $\frac{4}{3}$ **c.** $\frac{1}{27}$

13. $\frac{5x - 4}{20}$

14. Add: $\frac{5}{x} + \frac{2}{3}$

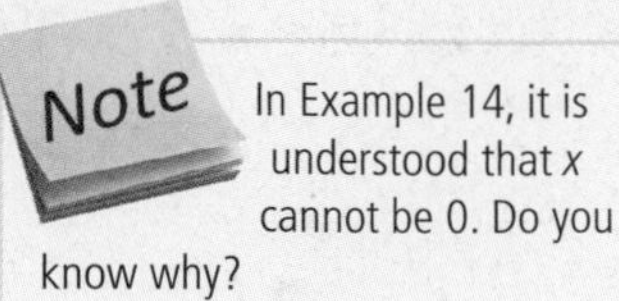

In Example 14, it is understood that x cannot be 0. Do you know why?

EXAMPLE 14 Add: $\frac{4}{x} + \frac{2}{3}$

SOLUTION The LCD for x and 3 is $3x$. We multiply the numerator and the denominator of the first fraction by 3 and the numerator and the denominator of the second fraction by x to get two fractions with the same denominator. We then add the numerators:

$$\frac{4}{x} + \frac{2}{3} = \frac{4 \cdot \mathbf{3}}{x \cdot \mathbf{3}} + \frac{2 \cdot \mathbf{x}}{3 \cdot \mathbf{x}} \quad \text{Change to equivalent fractions}$$
$$= \frac{12}{3x} + \frac{2x}{3x}$$
$$= \frac{12 + 2x}{3x} \quad \text{Add the numerators}$$

When we are working with fractions, we can change the form of a fraction without changing its value. There will be times when one form is easier to work with than another form. Look over the material below and be sure you see that the pairs of expressions are equal.

The expressions $\frac{x}{2}$ and $\frac{1}{2}x$ are equal.

The expressions $\frac{3a}{4}$ and $\frac{3}{4}a$ are equal.

The expressions $\frac{7y}{3}$ and $\frac{7}{3}y$ are equal.

15. Simplify: $\frac{1}{2}x + \frac{3}{4}x$

EXAMPLE 15 Add: $\frac{x}{3} + \frac{5x}{6}$.

SOLUTION We can do the problem two ways. One way probably seems easier, but both ways are valid methods of finding this sum. You should understand both of them.

METHOD 1:
$$\frac{x}{3} + \frac{5x}{6} = \frac{\mathbf{2} \cdot x}{\mathbf{2} \cdot 3} + \frac{5x}{6} \quad \text{LCD}$$
$$= \frac{2x + 5x}{6} \quad \text{Add numerators}$$
$$= \frac{7x}{6}$$

METHOD 2:
$$\frac{x}{3} + \frac{5x}{6} = \frac{1}{3}x + \frac{5}{6}x$$
$$= \left(\frac{1}{3} + \frac{5}{6}\right)x \quad \text{Distributive property}$$
$$= \left(\frac{\mathbf{2} \cdot 1}{\mathbf{2} \cdot 3} + \frac{5}{6}\right)x$$
$$= \left(\frac{2}{6} + \frac{5}{6}\right)x$$
$$= \frac{7}{6}x$$

Getting Ready for Class

After reading through the preceding section, respond in your own words and in complete sentences.

1. How do we add two fractions that have the same denominator?
2. What is a least common denominator?
3. What is the first step in adding two fractions that have different denominators?
4. What is the last thing you do when adding two fractions?

Answer

14. $\frac{15 + 2x}{3x}$ 15. $\frac{5}{4}x$

Problem Set 1.9

A Find the following sums and differences, and reduce to lowest terms. Add and subtract as indicated. Assume all variables represent nonzero numbers. [Examples 1–4]

1. $\frac{3}{6} + \frac{1}{6}$

2. $\frac{2}{5} + \frac{3}{5}$

3. $\frac{3}{8} - \frac{5}{8}$

4. $\frac{1}{7} - \frac{6}{7}$

5. $-\frac{1}{4} + \frac{3}{4}$

6. $-\frac{4}{9} + \frac{7}{9}$

7. $\frac{x}{3} - \frac{1}{3}$

8. $\frac{x}{8} - \frac{1}{8}$

9. $\frac{1}{4} + \frac{2}{4} + \frac{3}{4}$

10. $\frac{2}{5} + \frac{3}{5} + \frac{4}{5}$

11. $\frac{x+7}{2} - \frac{1}{2}$

12. $\frac{x+5}{4} - \frac{3}{4}$

13. $\frac{1}{10} - \frac{3}{10} - \frac{4}{10}$

14. $\frac{3}{20} - \frac{1}{20} - \frac{4}{20}$

15. $\frac{1}{a} + \frac{4}{a} + \frac{5}{a}$

16. $\frac{5}{a} + \frac{4}{a} + \frac{3}{a}$

B C Find the LCD for each of the following; then use the methods developed in this section to add and subtract as indicated. [Examples 5–11]

17. $\frac{1}{8} + \frac{3}{4}$

18. $\frac{1}{6} + \frac{2}{3}$

19. $\frac{3}{10} - \frac{1}{5}$

20. $\frac{5}{6} - \frac{1}{12}$

21. $\frac{4}{9} + \frac{1}{3}$

22. $\frac{1}{2} + \frac{1}{4}$

23. $2 + \frac{1}{3}$

24. $3 + \frac{1}{2}$

25. $-\frac{3}{4} + 1$

26. $-\frac{3}{4} + 2$

27. $\frac{1}{2} + \frac{2}{3}$

28. $\frac{2}{3} + \frac{1}{4}$

29. $\frac{5}{12} - \left(-\frac{3}{8}\right)$

30. $\frac{9}{16} - \left(-\frac{7}{12}\right)$

31. $-\frac{1}{20} + \frac{8}{30}$

32. $-\frac{1}{30} + \frac{9}{40}$

33. $\frac{17}{30} + \frac{11}{42}$

34. $\frac{19}{42} + \frac{13}{70}$

35. $\frac{25}{84} + \frac{41}{90}$

36. $\frac{23}{70} + \frac{29}{84}$

37. $\frac{13}{126} - \frac{13}{180}$

38. $\frac{17}{84} - \frac{17}{90}$

39. $\frac{3}{4} + \frac{1}{8} + \frac{5}{6}$

40. $\frac{3}{8} + \frac{2}{5} + \frac{1}{4}$

41. $\frac{1}{2} + \frac{1}{3} + \frac{1}{4} + \frac{1}{6}$

42. $\frac{1}{8} + \frac{1}{4} + \frac{1}{5} + \frac{1}{10}$

43. $1 - \frac{5}{2}$

44. $1 - \frac{5}{3}$

45. $1 + \frac{1}{2}$

46. $1 + \frac{2}{3}$

D Find the fourth term in each sequence. [Example 12]

47. $\frac{1}{3}, 0, -\frac{1}{3}, \ldots$

48. $\frac{2}{3}, 0, -\frac{2}{3}, \ldots$

49. $\frac{1}{3}, 1, \frac{5}{3}, \ldots$

50. $1, \frac{3}{2}, 2, \ldots$

51. $1, \frac{1}{5}, \frac{1}{25}, \ldots$

52. $1, -\frac{1}{2}, \frac{1}{4}, \ldots$

Use the rule for order of operations to simplify each expression.

53. $9 - 3\left(\frac{5}{3}\right)$

54. $6 - 4\left(\frac{7}{2}\right)$

55. $-\frac{1}{2} + 2\left(-\frac{3}{4}\right)$

56. $\frac{5}{4} - 3\left(\frac{7}{12}\right)$

57. $\frac{3}{5}(-10) + \frac{4}{7}(-21)$

58. $-\frac{3}{5}(10) - \frac{4}{7}(21)$

59. $16\left(-\frac{1}{2}\right)^2 - 125\left(-\frac{2}{5}\right)^2$

60. $16\left(-\frac{1}{2}\right)^3 - 125\left(-\frac{2}{5}\right)^3$

61. $-\frac{4}{3} \div 2 \cdot 3$

62. $-\frac{8}{7} \div 4 \cdot 2$

63. $-\frac{4}{3} \div 2(-3)$

64. $-\frac{8}{7} \div 4(-2)$

65. $-6 \div \frac{1}{2} \cdot 12$

66. $-6 \div \left(-\frac{1}{2}\right) \cdot 12$

67. $-15 \div \frac{5}{3} \cdot 18$

68. $-15 \div \left(-\frac{5}{3}\right) \cdot 18$

Add or subtract the following fractions. (Assume all variables represent nonzero numbers. [Examples 13–15]

69. $\frac{x}{4} + \frac{1}{5}$

70. $\frac{x}{3} + \frac{1}{5}$

71. $\frac{1}{3} + \frac{a}{12}$

72. $\frac{1}{8} + \frac{a}{32}$

73. $\frac{x}{2} + \frac{1}{3} + \frac{x}{4}$

74. $\frac{x}{3} + \frac{1}{4} + \frac{x}{5}$

75. $\frac{2}{x} + \frac{3}{5}$

76. $\frac{3}{x} - \frac{2}{5}$

77. $\frac{3}{7} + \frac{4}{y}$

78. $\frac{2}{9} + \frac{5}{y}$

79. $\frac{3}{a} + \frac{3}{4} + \frac{1}{5}$

80. $\frac{4}{a} + \frac{2}{3} + \frac{1}{2}$

81. $\frac{1}{2}x + \frac{1}{6}x$

82. $\frac{2}{3}x + \frac{5}{6}x$

83. $\frac{1}{2}x - \frac{3}{4}x$

84. $\frac{2}{3}x - \frac{5}{6}x$

85. $\frac{1}{3}x + \frac{3}{5}x$

86. $\frac{2}{3}x - \frac{3}{5}x$

87. $\frac{3x}{4} + \frac{x}{6}$

88. $\frac{3x}{4} - \frac{2x}{3}$

89. $\frac{2x}{5} + \frac{5x}{8}$

90. $\frac{3x}{5} - \frac{3x}{8}$

Use the rule for order of operations to simplify each expression.

91. $1 - \frac{1}{x}$

92. $1 + \frac{1}{x}$

The next two problems are intended to give you practice reading, and paying attention to, the instructions that accompany the problems you are working. As we mentioned previously, working these problems is an excellent way to get ready for a test or a quiz.

93. Work each problem according to the instructions given. (Note that each of these instructions could be replaced with the instruction *Simplify*.)

a. Add: $\frac{3}{4} + \left(-\frac{1}{2}\right)$

b. Subtract: $\frac{3}{4} - \left(-\frac{1}{2}\right)$

c. Multiply: $\frac{3}{4}\left(-\frac{1}{2}\right)$

d. Divide: $\frac{3}{4} \div \left(-\frac{1}{2}\right)$

94. Work each problem according to the instructions given.

a. Add: $-\frac{5}{8} + \left(-\frac{1}{2}\right)$

b. Subtract: $-\frac{5}{8} - \left(-\frac{1}{2}\right)$

c. Multiply: $-\frac{5}{8}\left(-\frac{1}{2}\right)$

d. Divide: $-\frac{5}{8} \div \left(-\frac{1}{2}\right)$

Simplify.

95. $\left(1 - \frac{1}{2}\right)\left(1 - \frac{1}{3}\right)$

96. $\left(1 + \frac{1}{2}\right)\left(1 + \frac{1}{3}\right)$

97. $\left(1 + \frac{1}{2}\right)\left(1 - \frac{1}{2}\right)$

98. $\left(1 + \frac{1}{3}\right)\left(1 - \frac{1}{3}\right)$

99. Find the value of $1 + \frac{1}{x}$ when x is

a. 2

b. 3

c. 4

100. Find the value of $1 - \frac{1}{x}$ when x is

a. 2

b. 3

c. 4

101. Find the value of $2x + \frac{6}{x}$ when x is

a. 1

b. 2

c. 3

102. Find the value of $x + \frac{4}{x}$ when x is

a. 1

b. 2

c. 3

Applying the Concepts

Rainfall The chart shows the average monthly rainfall for Death Valley. Use the chart to answer Problems 103 and 104.

103. What is the total rainfall for the months of December, October, and February?

104. How much more rain was there in August than in June?

Chapter 1 Summary

The number(s) in brackets next to each heading indicates the section(s) in which that topic is discussed.

We will use the margins in the chapter summaries to give examples that correspond to the topic being reviewed whenever it is appropriate.

Symbols [1.1]

$a = b$	*a* is equal to *b*
$a \neq b$	*a* is not equal to *b*
$a < b$	*a* is less than *b*
$a \nless b$	*a* is not less than *b*
$a > b$	*a* is greater than *b*
$a \ngtr b$	*a* is not greater than *b*
$a \geq b$	*a* is greater than or equal to *b*
$a \leq b$	*a* is less than or equal to *b*

Exponents [1.1]

Exponents are notation used to indicate repeated multiplication. In the expression 3^4, 3 is the *base* and 4 is the *exponent.*

$$3^4 = 3 \cdot 3 \cdot 3 \cdot 3 = 81$$

EXAMPLES

1. $2^5 = 2 \cdot 2 \cdot 2 \cdot 2 \cdot 2 = 32$
   $5^2 = 5 \cdot 5 = 25$
   $10^3 = 10 \cdot 10 \cdot 10 = 1{,}000$
   $1^4 = 1 \cdot 1 \cdot 1 \cdot 1 = 1$

Order of Operations [1.1]

When evaluating a mathematical expression, we will perform the operations in the following order, beginning with the expression in the innermost parentheses or brackets and working our way out.

1. Simplify all numbers with exponents, working from left to right if more than one of these numbers is present.
2. Then do all multiplications and divisions left to right.
3. Finally, perform all additions and subtractions left to right.

2. $10 + (2 \cdot 3^2 - 4 \cdot 2)$
   $= 10 + (2 \cdot 9 - 4 \cdot 2)$
   $= 10 + (18 - 8)$
   $= 10 + 10$
   $= 20$

Absolute Value [1.2]

The absolute value of a real number is its distance from zero on the real number line. Absolute value is never negative.

3. $|5| = 5$
   $|-5| = 5$

Opposites [1.2]

Any two real numbers the same distance from zero on the number line but in opposite directions from zero are called opposites. Opposites always add to zero.

4. The numbers 3 and −3 are opposites; their sum is 0:
   $3 + (-3) = 0$

Reciprocals [1.2]

5. The numbers 2 and $\frac{1}{2}$ are reciprocals; their product is 1:

$$2\left(\frac{1}{2}\right) = 1$$

Any two real numbers whose product is 1 are called reciprocals. Every real number has a reciprocal except 0.

Addition of Real Numbers [1.3]

6. Add all combinations of positive and negative 10 and 13.

$$10 + 13 = 23$$
$$10 + (-13) = -3$$
$$-10 + 13 = 3$$
$$-10 + (-13) = -23$$

To add two real numbers with

1. The same sign: Simply add their absolute values and use the common sign.
2. Different signs: Subtract the smaller absolute value from the larger absolute value. The answer has the same sign as the number with the larger absolute value.

Subtraction of Real Numbers [1.4]

7. Subtracting 2 is the same as adding -2:

$$7 - 2 = 7 + (-2) = 5$$

To subtract one number from another, simply add the opposite of the number you are subtracting; that is, if a and b represent real numbers, then

$$a - b = a + (-b)$$

Properties of Real Numbers [1.5]

	For Addition	*For Multiplication*
Commutative:	$a + b = b + a$	$a \cdot b = b \cdot a$
Associative:	$a + (b + c) = (a + b) + c$	$a \cdot (b \cdot c) = (a \cdot b) \cdot c$
Identity:	$a + 0 = a$	$a \cdot 1 = a$
Inverse:	$a + (-a) = 0$	$a\left(\frac{1}{a}\right) = 1$
Distributive:	$a(b + c) = ab + ac$	

Multiplication of Real Numbers [1.6]

8.

$$3(5) = 15$$
$$3(-5) = -15$$
$$-3(5) = -15$$
$$-3(-5) = 15$$

To multiply two real numbers, simply multiply their absolute values. Like signs give a positive answer. Unlike signs give a negative answer.

Division of Real Numbers [1.7]

9.

$$-\frac{6}{2} = -6\left(\frac{1}{2}\right) = -3$$
$$\frac{-6}{-2} = -6\left(-\frac{1}{2}\right) = 3$$

Division by a number is the same as multiplication by its reciprocal. Like signs give a positive answer. Unlike signs give a negative answer.

Subsets of the Real Numbers [1.8]

Counting numbers: $\{1, 2, 3, \ldots\}$
Whole numbers: $\{0, 1, 2, 3, \ldots\}$
Integers: $\{\ldots, -3, -2, -1, 0, 1, 2, 3, \ldots\}$
Rational numbers: {all numbers that can be expressed as the ratio of two integers}
Irrational numbers: {all numbers on the number line that cannot be expressed as the ratio of two integers}
Real numbers: {all numbers that are either rational or irrational}

10. a. 7 and 100 are counting numbers, but 0 and -2 are not.
b. 0 and 241 are whole numbers, but -4 and $\frac{1}{2}$ are not.
c. -15, 0, and 20 are integers.
d. -4, $-\frac{1}{2}$, 0.75, and 0.666 . . . are rational numbers.
e. $-\pi$, $\sqrt{3}$, and π are irrational numbers.
f. All the numbers listed above are real numbers.

Factoring [1.8]

Factoring is the reverse of multiplication.

Multiplication

Factors⟶ $3 \cdot 5 = 15$ ⟵ Product

Factoring

11. The number 150 can be factored into the product of prime numbers:

$$150 = 15 \cdot 10$$
$$= (3 \cdot 5)(2 \cdot 5)$$
$$= 2 \cdot 3 \cdot 5^2$$

Least Common Denominator (LCD) [1.9]

The *least common denominator* (LCD) for a set of denominators is the smallest number that is exactly divisible by each denominator.

12. The LCD for $\frac{5}{12}$ and $\frac{7}{18}$ is 36.

Addition and Subtraction of Fractions [1.9]

To add (or subtract) two fractions with a common denominator, add (or subtract) numerators and use the common denominator.

$$\frac{a}{c} + \frac{b}{c} = \frac{a+b}{c} \quad \text{and} \quad \frac{a}{c} - \frac{b}{c} = \frac{a-b}{c}$$

13. $\frac{5}{12} + \frac{7}{18} = \frac{5}{12} \cdot \frac{3}{3} + \frac{7}{18} \cdot \frac{2}{2}$

$$= \frac{15}{36} + \frac{14}{36}$$
$$= \frac{29}{36}$$

COMMON MISTAKES

1. Interpreting absolute value as changing the sign of the number inside the absolute value symbols. $|-5| = +5$, $|+5| = -5$. (The first expression is correct; the second one is not.) To avoid this mistake, remember: Absolute value is a distance and distance is always measured in positive units.
2. Using the phrase "two negatives make a positive." This works only with multiplication and division. With addition, two negative numbers produce a negative answer. It is best not to use the phrase "two negatives make a positive" at all.

Chapter 1 Review

Write the numerical expression that is equivalent to each phrase, and then simplify. [1.3, 1.4, 1.6, 1.7]

1. The sum of -7 and -10
2. Five added to the sum of -7 and 4
3. The sum of -3 and 12 increased by 5
4. The difference of 4 and 9
5. The difference of 9 and -3
6. The difference of -7 and -9
7. The product of -3 and -7 decreased by 6
8. Ten added to the product of 5 and -6
9. Twice the product of -8 and $3x$
10. The quotient of -25 and -5

Simplify. [1.2]

11. $|-1.8|$
12. $-|-10|$

For each number, give the opposite and the reciprocal. [1.2]

13. 6
14. $-\frac{12}{5}$

Multiply. [1.2, 1.6]

15. $\frac{1}{2}(-10)$
16. $\left(-\frac{4}{5}\right)\left(\frac{25}{16}\right)$

Add. [1.3]

17. $-9 + 12$
18. $-18 + (-20)$
19. $-2 + (-8) + [-9 + (-6)]$
20. $(-21) + 40 + (-23) + 5$

Subtract. [1.4]

21. $6 - 9$
22. $14 - (-8)$
23. $-12 - (-8)$
24. $4 - 9 - 15$

Find the products. [1.6]

25. $(-5)(6)$
26. $4(-3)$
27. $-2(3)(4)$
28. $(-1)(-3)(-1)(-4)$

Find the following quotients. [1.7]

29. $\frac{12}{-3}$
30. $-\frac{8}{9} \div \frac{4}{3}$

Simplify. [1.1, 1.6, 1.7]

31. $4 \cdot 5 + 3$

32. $9 \cdot 3 + 4 \cdot 5$

33. $2^3 - 4 \cdot 3^2 + 5^2$

34. $12 - 3(2 \cdot 5 + 7) + 4$

35. $20 + 8 \div 4 + 2 \cdot 5$

36. $2(3 - 5) - (2 - 8)$

37. $30 \div 3 \cdot 2$

38. $(-2)(3) - (4)(-3) - 9$

39. $3(4 - 7)^2 - 5(3 - 8)^2$

40. $(-5 - 2)(-3 - 7)$

41. $\dfrac{4(-3)}{-6}$

42. $\dfrac{3^2 + 5^2}{(3 - 5)^2}$

43. $\dfrac{15 - 10}{6 - 6}$

44. $\dfrac{2(-7) + (-11)(-4)}{7 - (-3)}$

State the property or properties that justify the following. [1.5]

45. $9(3y) = (9 \cdot 3)y$

46. $8(1) = 8$

47. $(4 + y) + 2 = (y + 4) + 2$

48. $5 + (-5) = 0$

49. $(4 + 2) + y = (4 + y) + 2$

50. $5(w - 6) = 5w - 30$

Use the associative property to rewrite each expression, and then simplify the result. [1.5]

51. $7 + (5 + x)$

52. $4(7a)$

53. $\dfrac{1}{9}(9x)$

54. $\dfrac{4}{5}\left(\dfrac{5}{4}y\right)$

Apply the distributive property to each of the following expressions. Simplify when possible. [1.5, 1.6]

55. $7(2x + 3)$

56. $3(2a - 4)$

57. $\dfrac{1}{2}(5x - 6)$

58. $-\dfrac{1}{2}(3x - 6)$

For the set $\{\sqrt{7}, -\frac{1}{3}, 0, 5, -4.5, \frac{2}{5}, \pi, -3\}$ list all the [1.8]

59. rational numbers

60. whole numbers

61. irrational numbers

62. integers

Factor into the product of primes. [1.8]

63. 90

64. 840

Combine. [1.9]

65. $\dfrac{18}{35} + \dfrac{13}{42}$

66. $\dfrac{x}{6} + \dfrac{7}{12}$

Find the next number in each sequence. [1.1, 1.2, 1.3, 1.6, 1.9]

67. 10, 7, 4, 1, . . .

68. 10, −30, 90, −270, . . .

69. 1, 1, 2, 3, 5, . . .

70. 4, 6, 8, 10, . . .

71. $1, \frac{1}{2}, 0, -\frac{1}{2}, \ldots$

72. $1, -\frac{1}{2}, \frac{1}{4}, -\frac{1}{8}, \ldots$

Chapter 1 Test

Translate into symbols. [1.1]

1. The sum of x and 3 is 8.

2. The product of 5 and y is 15.

Simplify according to the rule for order of operations. [1.1]

3. $5^2 + 3(9 - 7) + 3^2$ **4.** $10 - 6 \div 3 + 2^3$

For each number, name the opposite, reciprocal, and absolute value. [1.2]

5. -4 **6.** $\frac{3}{4}$

Add. [1.3]

7. $3 + (-7)$ **8.** $|-9 + (-6)| + |-3 + 5|$

Subtract. [1.4]

9. $-4 - 8$ **10.** $9 - (7 - 2) - 4$

Match each expression below with the letter of the property that justifies it. [1.5]

11. $(x + y) + z = x + (y + z)$ **12.** $3(x + 5) = 3x + 15$

13. $5(3x) = (5 \cdot 3)x$ **14.** $(x + 5) + 7 = 7 + (x + 5)$

a. Commutative property of addition
b. Commutative property of multiplication
c. Associative property of addition
d. Associative property of multiplication
e. Distributive property

Multiply. [1.6]

15. $-3(7)$ **16.** $-4(8)(-2)$

17. $8\left(-\frac{1}{4}\right)$ **18.** $\left(-\frac{2}{3}\right)^3$

Simplify using the rule for order of operations. [1.1, 1.6, 1.7]

19. $-3(-4) - 8$ **20.** $5(-6)^2 - 3(-2)^3$

21. $7 - 3(2 - 8)$

22. $4 - 2[-3(-1 + 5) + 4(-3)]$

23. $\dfrac{4(-5) - 2(7)}{-10 - 7}$ **24.** $\dfrac{2(-3 - 1) + 4(-5 + 2)}{-3(2) - 4}$

Apply the associative property, and then simplify. [1.5, 1.6]

25. $3 + (5 + 2x)$ **26.** $-2(-5x)$

Multiply by applying the distributive property. [1.5, 1.6]

27. $2(3x + 5)$ **28.** $-\frac{1}{2}(4x - 2)$

From the set of numbers $\{-8, \frac{3}{4}, 1, \sqrt{2}, 1.5\}$ list all the elements that are in the following sets. [1.8]

29. Integers **30.** Rational numbers

31. Irrational numbers **32.** Real numbers

Factor into the product of primes. [1.8]

33. 592 **34.** 1,340

Combine. [1.9]

35. $\frac{5}{15} + \frac{11}{42}$ **36.** $\frac{5}{x} + \frac{3}{x}$

Write an expression in symbols that is equivalent to each English phrase, and then simplify it.

37. The sum of 8 and -3 [1.1, 1.3]

38. The difference of -24 and 2 [1.1, 1.4]

39. The product of -5 and -4 [1.1, 1.6]

40. The quotient of -24 and -2 [1.1, 1.7]

Find the next number in each sequence. [1.1, 1.2, 1.3, 1.6, 1.9]

41. $-8, -3, 2, 7, \ldots$ **42.** $8, -4, 2, -1, \ldots$

Use the illustration below to answer the following questions. [1.1, 1.3, 1.4]

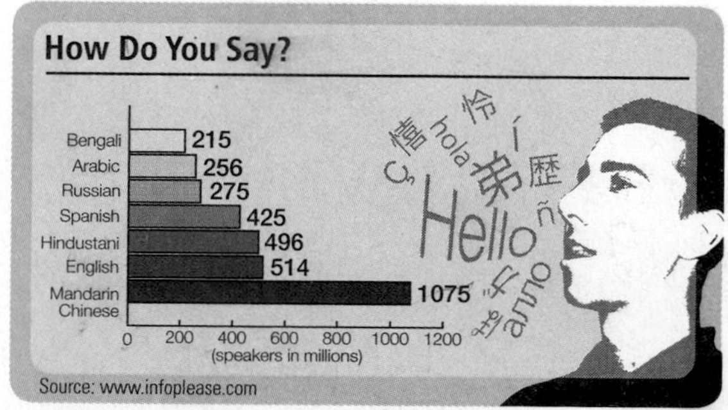

43. How many people speak Spanish?

44. What is the total number of people who speak Mandarin Chinese or Russian?

45. How many more people speak English than Hindustani?

Chapter 1 Projects

THE BASICS

GROUP PROJECT

Binary Numbers

Students and Instructors: The end of each chapter in this book will contain a section like this one containing two projects. The group project is intended to be done in class. The research projects are to be completed outside of class. They can be done in groups or individually. In my classes, I use the research projects for extra credit. I require all research projects to be done on a word processor and to be free of spelling errors.

Number of People 2 or 3

Time Needed 10 minutes

Equipment Paper and pencil

Background Our decimal number system is a base-10 number system. We have 10 digits—0, 1, 2, 3, 4, 5, 6, 7, 8, and 9—which we use to write all the numbers in our number system. The number 10 is the first number that is written with a combination of digits. Although our number system is very useful, there are other number systems that are more appropriate for some disciplines. For example, computers and programmers use both the binary number system, which is base 2, and the hexadecimal number system, which is base 16. The binary number system has only digits 0 and 1, which are used to write all the other numbers. Every number in our base 10 number system can be written in the base 2 number system as well.

Procedure To become familiar with the binary number system, we first learn to count in base 2. Imagine that the odometer on your car had only 0's and 1's. Here is what the odometer would look like for the first 6 miles the car was driven.

Odometer Reading	mileage
0 0 0 0 0 0	0
0 0 0 0 0 1	1
0 0 0 0 1 0	2
0 0 0 0 1 1	3
0 0 0 1 0 0	4
0 0 0 1 0 1	5
0 0 0 1 1 0	6

Continue the table at left to show the odometer reading for the first 32 miles the car is driven. At 32 miles, the odometer should read

1 0 0 0 0 0

RESEARCH PROJECT

Sophie Germain

The photograph at the right shows the street sign in Paris named for the French mathematician Sophie Germain (1776–1831). Among her contributions to mathematics is her work with prime numbers. In this chapter we had an introductory look at some of the classifications for numbers, including the prime numbers. Within the prime numbers themselves, there are still further classifications. In fact, a Sophie Germain prime is a prime number P, for which both P and $2P + 1$ are primes. For example, the prime number 2 is the first Sophie Germain prime because both 2 and $2 \cdot 2 + 1 = 5$ are prime numbers. The next Germain prime is 3 because both 3 and $2 \cdot 3 + 1 = 7$ are primes.

14e Arrt
RUE
SOPHIE
GERMAIN
1776 - 1831
MATHÉMATICIENNE

Cheryl Slaughter

Sophie Germain was born on April 1, 1776, in Paris, France. She taught herself mathematics by reading the books in her father's library at home. Today she is recognized most for her work in number theory, which includes her work with prime numbers. Research the life of Sophie Germain. Write a short essay that includes information on her work with prime numbers and how her results contributed to solving Fermat's Last Theorem almost 200 years later.

Linear Equations and Inequalities

2

Chapter Outline

Introduction

The Google Earth image is another view of the Leaning Tower of Pisa. As we mentioned in the previous chapter, Pisa is the birthplace of the Italian mathematician Fibonacci. It was in Pisa that Fibonacci wrote the first European algebra book *Liber Abaci*. It was published in 1202. Below is the cover of an English translation of that book, published in 2003, along with a quote from the Mathematics Association of America.

Permission granted by Judith Sigler Fell

MAA Online, March 2003:

> "The *Liber abaci* of Leonardo Pisano (today commonly called Fibonacci) is one of the fundamental works of European mathematics. No other book did more to establish the basic framework of arithmetic and algebra as they developed in the Western world."

In his book, Fibonacci shows how to solve linear equations in one variable, which is the main topic in this chapter.

Chapter Pretest

Simplify each of the following expressions. [2.1]

1. $5y - 3 - 6y + 4$
2. $3x - 4 + x + 3$
3. $4 - 2(y - 3) - 6$
4. $3(3x - 4) - 2(4x + 5)$
5. Find the value of $3x + 12 + 2x$ when $x = -3$. [2.1]
6. Find the value of $x^2 - 3xy + y^2$ when $x = -2$ and $y = -4$. [2.1]

Solve the following equations. [2.2, 2.3, 2.4]

7. $3x - 2 = 7$
8. $4y + 15 = y$
9. $\frac{1}{4}x - \frac{1}{12} = \frac{1}{3}x - \frac{1}{6}$
10. $-3(3 - 2x) - 7 = 8$
11. $3x - 9 = -6$
12. $0.05 + 0.07(100 - x) = 3.2$
13. $4(t - 3) + 2(t + 4) = 2t - 16$
14. $4x - 2(3x - 1) = 2x - 8$
15. What number is 40% of 56? [2.5]
16. 720 is 24% of what number? [2.5]
17. If $3x - 4y = 16$, find y when $x = 4$. [2.5]
18. If $3x - 4y = 16$, find x when $y = 2$. [2.5]
19. Solve $2x + 6y = 12$ for y. [2.5]
20. Solve $x^2 = v^2 + 2ad$ for a. [2.5]

Solve each inequality.

21. $\frac{1}{2}x - 2 > 3$
22. $-6y \leq 24$
23. $0.3 - 0.2x < 1.1$
24. $3 - 2(n - 1) \geq 9$

Write an inequality whose solution is the given graph. [2.8, 2.9]

25. 

26. 

Getting Ready for Chapter 2

Simplify each of the following.

1. $-3 + 7$
2. $-10 - 4$
3. $9 - (-24)$
4. $-6(5) - 5$
5. $-3(2y + 1)$
6. $8.1 + 2.7$
7. $-\frac{3}{4} + \left(-\frac{1}{2}\right)$
8. $-\frac{7}{10} + \left(-\frac{1}{2}\right)$
9. $\frac{1}{5}(5x)$
10. $4\left(\frac{1}{4}x\right)$
11. $\frac{3}{2}\left(\frac{2}{3}x\right)$
12. $\frac{5}{2}\left(\frac{2}{5}x\right)$
13. $-1(3x + 4)$
14. $-2.4 + (-7.3)$
15. $0.04(x + 7{,}000)$
16. $0.09(x + 2{,}000)$

Apply the distributive property and then simplify if possible.

17. $3(x - 5) + 4$
18. $5(x - 3) + 2$
19. $5(2x - 8) - 3$
20. $4(3x - 5) - 2$

2.1 Simplifying Expressions

Objectives

A Simplify expressions by combining similar terms.

B Simplify expressions by applying the distributive property and then combining similar terms.

C Calculate the value of an expression for a given value of the variable.

Examples now playing at MathTV.com/books

If a cellular phone company charges \$35 per month plus \$0.25 for each minute, or fraction of a minute, that you use one of their cellular phones, then the amount of your monthly bill is given by the expression $35 + 0.25t$. To find the amount you will pay for using that phone 30 minutes in one month, you substitute 30 for t and simplify the resulting expression. This process is one of the topics we will study in this section.

A Combining Similar Terms

As you will see in the next few sections, the first step in solving an equation is to simplify both sides as much as possible. In the first part of this section, we will practice simplifying expressions by combining what are called *similar* (or like) terms.

For our immediate purposes, a term is a number, or a number and one or more variables multiplied together. For example, the number 5 is a term, as are the expressions $3x$, $-7y$, and $15xy$.

> **Definition**
> Two or more terms with the same variable part are called **similar** (or **like**) terms.

The terms $3x$ and $4x$ are similar because their variable parts are identical. Likewise, the terms $18y$, $-10y$, and $6y$ are similar terms.

To simplify an algebraic expression, we simply reduce the number of terms in the expression. We accomplish this by applying the distributive property along with our knowledge of addition and subtraction of positive and negative real numbers. The following examples illustrate the procedure.

EXAMPLE 1 Simplify $3x + 4x$ by combining similar terms.

SOLUTION

$3x + 4x = (3 + 4)x$	Distributive property
$= 7x$	Addition of 3 and 4

EXAMPLE 2 Simplify $7a - 10a$ by combining similar terms.

SOLUTION

$7a - 10a = (7 - 10)a$	Distributive property
$= -3a$	Addition of 7 and -10

EXAMPLE 3 Simplify $18y - 10y + 6y$ by combining similar terms.

SOLUTION

$18y - 10y + 6y = (18 - 10 + 6)y$	Distributive property
$= 14y$	Addition of 18, -10, and 6

When the expression we intend to simplify is more complicated, we use the commutative and associative properties first.

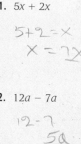

PRACTICE PROBLEMS

Simplify by combining similar terms.

1. $5x + 2x$

2. $12a - 7a$

3. $6y - 8y + 5y$

Answers

1. $7x$ 2. $5a$ 3. $3y$

Simplify each expression.

4. $2x + 6 + 3x - 5$

EXAMPLE 4 Simplify the expression $3x + 5 + 2x - 3$.

SOLUTION

$3x + 5 + 2x - 3 = 3x + 2x + 5 - 3$	Commutative property
$= (3x + 2x) + (5 - 3)$	Associative property
$= (3 + 2)x + (5 - 3)$	Distributive property
$= 5x + 2$	Addition

5. $8a - 2 - 4a + 5$

EXAMPLE 5 Simplify the expression $4a - 7 - 2a + 3$.

SOLUTION

$4a - 7 - 2a + 3 = (4a - 2a) + (-7 + 3)$	Commutative and associative properties
$= (4 - 2)a + (-7 + 3)$	Distributive property
$= 2a - 4$	Addition

6. $9x + 1 - x - 6$

EXAMPLE 6 Simplify the expression $5x + 8 - x - 6$.

SOLUTION

$5x + 8 - x - 6 = (5x - x) + (8 - 6)$	Commutative and associative properties
$= (5 - 1)x + (8 - 6)$	Distributive property
$= 4x + 2$	Addition

B Simplifying Expressions Containing Parentheses

If an expression contains parentheses, it is often necessary to apply the distributive property to remove the parentheses before combining similar terms.

7. Simplify the expression.

$4(3x - 5) - 2$

EXAMPLE 7 Simplify the expression $5(2x - 8) - 3$.

SOLUTION We begin by distributing the 5 across $2x - 8$. We then combine similar terms:

$5(2x - 8) - 3 = 10x - 40 - 3$	Distributive property
$= 10x - 43$	

8. Simplify $5 - 2(4y + 1)$.

EXAMPLE 8 Simplify $7 - 3(2y + 1)$.

SOLUTION By the rule for order of operations, we must multiply before we add or subtract. For that reason, it would be incorrect to subtract 3 from 7 first. Instead, we multiply -3 and $2y + 1$ to remove the parentheses and then combine similar terms:

$7 - 3(2y + 1) = 7 - 6y - 3$	Distributive property
$= -6y + 4$	

9. Simplify $6(x - 3) - (7x + 2)$.

EXAMPLE 9 Simplify $5(x - 2) - (3x + 4)$.

SOLUTION We begin by applying the distributive property to remove the parentheses. The expression $-(3x + 4)$ can be thought of as $-1(3x + 4)$. Thinking of it in this way allows us to apply the distributive property:

$$-1(3x + 4) = -1(3x) + (-1)(4)$$
$$= -3x - 4$$

The complete solution looks like this:

$5(x - 2) - (3x + 4) = 5x - 10 - 3x - 4$	Distributive property
$= 2x - 14$	Combine similar terms

Answers
4. $5x + 1$ **5.** $4a + 3$ **6.** $8x - 5$
7. $12x - 22$ **8.** $-8y + 3$
9. $-x - 20$

As you can see from the explanation in Example 9, we use the distributive property to simplify expressions in which parentheses are preceded by a negative sign. In general we can write

$$\begin{aligned}-(a + b) &= -1(a + b)\\ &= -a + (-b)\\ &= -a - b\end{aligned}$$

The negative sign outside the parentheses ends up changing the sign of each term within the parentheses. In words, we say "the opposite of a sum is the sum of the opposites."

C The Value of an Expression

Recall from Chapter 1, an expression like $3x + 2$ has a certain value depending on what number we assign to x. For instance, when x is 4, $3x + 2$ becomes $3(4) + 2$, or 14. When x is -8, $3x + 2$ becomes $3(-8) + 2$, or -22. The value of an expression is found by replacing the variable with a given number.

EXAMPLES Find the value of the following expressions by replacing the variable with the given number.

Expression	Value of the Variable	Value of the Expression
10. $3x - 1$	$x = 2$	$3(2) - 1 = 6 - 1 = 5$
11. $7a + 4$	$a = -3$	$7(-3) + 4 = -21 + 4 = -17$
12. $2x - 3 + 4x$	$x = -1$	$2(-1) - 3 + 4(-1)$ $= -2 - 3 + (-4) = -9$
13. $2x - 5 - 8x$	$x = 5$	$2(5) - 5 - 8(5)$ $= 10 - 5 - 40 = -35$
14. $y^2 - 6y + 9$	$y = 4$	$4^2 - 6(4) + 9 = 16 - 24 + 9 = 1$

Simplifying an expression should not change its value; that is, if an expression has a certain value when x is 5, then it will always have that value no matter how much it has been simplified, as long as x is 5. If we were to simplify the expression in Example 13 first, it would look like

$$2x - 5 - 8x = -6x - 5$$

When x is 5, the simplified expression $-6x - 5$ is

$$-6(5) - 5 = -30 - 5 = -35$$

It has the same value as the original expression when x is 5.

We also can find the value of an expression that contains two variables if we know the values for both variables.

10. Find the value of $4x - 7$ when $x = 3$.

11. Find the value of $2a + 4$ when $a = -5$.

12. Find the value of $2x - 5 + 6x$ when $x = -2$.

13. Find the value of $7x - 3 - 4x$ when $x = 10$.

14. Find the value of $y^2 - 10y + 25$ when $y = -2$.

Answers

10. 5 **11.** -6 **12.** -21 **13.** 27 **14.** 49

15. Find the value of $5a - 3b - 2$ when a is -3 and b is 4.

EXAMPLE 15 Find the value of the expression $2x - 3y + 4$ when x is -5 and y is 6.

SOLUTION Substituting -5 for x and 6 for y, the expression becomes

$$\begin{aligned} 2(-5) - 3(6) + 4 &= -10 - 18 + 4 \\ &= -28 + 4 \\ &= -24 \end{aligned}$$

16. Find the value of $x^2 - 4xy + 4y^2$ when x is 5 and y is 3.

EXAMPLE 16 Find the value of the expression $x^2 - 2xy + y^2$ when x is 3 and y is -4.

SOLUTION Replacing each x in the expression with the number 3 and each y in the expression with the number -4 gives us

$$\begin{aligned} 3^2 - 2(3)(-4) + (-4)^2 &= 9 - 2(3)(-4) + 16 \\ &= 9 - (-24) + 16 \\ &= 33 + 16 \\ &= 49 \end{aligned}$$

More about Sequences

As the next example indicates, when we substitute the counting numbers, in order, into algebraic expressions, we form some of the sequences of numbers that we studied in Chapter 1. To review, recall that the sequence of counting numbers (also called the sequence of positive integers) is

$$\text{Counting numbers} = 1, 2, 3, \ldots$$

17. Substitute 1, 2, 3, and 4 for n in the expression $2n + 1$.

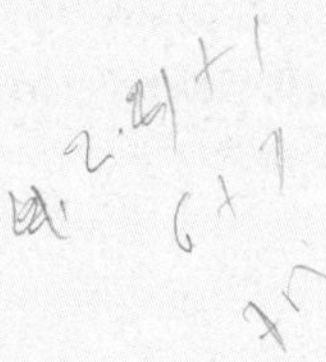

EXAMPLE 17 Substitute 1, 2, 3, and 4 for n in the expression $2n - 1$.

SOLUTION Substituting as indicated, we have

When $n = 1$, $2n - 1 = 2 \cdot 1 - 1 = 1$
When $n = 2$, $2n - 1 = 2 \cdot 2 - 1 = 3$
When $n = 3$, $2n - 1 = 2 \cdot 3 - 1 = 5$
When $n = 4$, $2n - 1 = 2 \cdot 4 - 1 = 7$

As you can see, substituting the first four counting numbers into the expression $2n - 1$ produces the first four numbers in the sequence of odd numbers.

Getting Ready for Class

After reading through the preceding section, respond in your own words and in complete sentences.

1. What are similar terms?
2. Explain how the distributive property is used to combine similar terms.
3. What is wrong with writing $3x + 4x = 7x^2$?
4. Explain how you would find the value of $5x + 3$ when x is 6.

Answers
15. -29 **16.** 1 **17.** 3, 5, 7, and 9

Problem Set 2.1

A Simplify the following expressions. [Examples 1–6]

1. $3x - 6x$

2. $7x - 5x$

3. $-2a + a$

4. $3a - a$

5. $7x + 3x + 2x$

6. $8x - 2x - x$

7. $3a - 2a + 5a$

8. $7a - a + 2a$

9. $4x - 3 + 2x$

10. $5x + 6 - 3x$

11. $3a + 4a + 5$

12. $6a + 7a + 8$

13. $2x - 3 + 3x - 2$

14. $6x + 5 - 2x + 3$

15. $3a - 1 + a + 3$

16. $-a + 2 + 8a - 7$

17. $-4x + 8 - 5x - 10$

18. $-9x - 1 + x - 4$

19. $7a + 3 + 2a + 3a$

20. $8a - 2 + a + 5a$

B Apply distributive property, then simplify. [Examples 7–9]

21. $5(2x - 1) + 4$

22. $2(4x - 3) + 2$

23. $7(3y + 2) - 8$

24. $6(4y + 2) - 7$

25. $-3(2x - 1) + 5$

26. $-4(3x - 2) - 6$

27. $5 - 2(a + 1)$

28. $7 - 8(2a + 3)$

B Simplify the following expressions. [Examples 7–9]

29. $6 - 4(x - 5)$

30. $12 - 3(4x - 2)$

31. $-9 - 4(2 - y) + 1$

32. $-10 - 3(2 - y) + 3$

33. $-6 + 2(2 - 3x) + 1$

34. $-7 - 4(3 - x) + 1$

35. $(4x - 7) - (2x + 5)$

36. $(7x - 3) - (4x + 2)$

37. $8(2a + 4) - (6a - 1)$

38. $9(3a + 5) - (8a - 7)$

39. $3(x - 2) + (x - 3)$

40. $2(2x + 1) - (x + 4)$

41. $4(2y - 8) - (y + 7)$

42. $5(y - 3) - (y - 4)$

43. $-9(2x + 1) - (x + 5)$

44. $-3(3x - 2) - (2x + 3)$

C Evaluate the following expressions when x is 2. [Examples 10–14]

45. $3x - 1$

46. $4x + 3$

47. $-2x - 5$

48. $-3x + 6$

49. $x^2 - 8x + 16$

50. $x^2 - 10x + 25$

51. $(x - 4)^2$

52. $(x - 5)^2$

C Find the value of each expression when x is -5. Then simplify the expression, and check to see that it has the same value for $x = -5$.

53. $7x - 4 - x - 3$

54. $3x + 4 + 7x - 6$

55. $5(2x + 1) + 4$

56. $2(3x - 10) + 5$

C Find the value of each expression when x is -3 and y is 5. [Examples 15–16]

57. $x^2 - 2xy + y^2$ **58.** $x^2 + 2xy + y^2$ **59.** $(x - y)^2$ **60.** $(x + y)^2$

61. $x^2 + 6xy + 9y^2$ **62.** $x^2 + 10xy + 25y^2$ **63.** $(x + 3y)^2$ **64.** $(x + 5y)^2$

C Find the value of $12x - 3$ for each of the following values of x.

65. $\frac{1}{2}$ **66.** $\frac{1}{3}$ **67.** $\frac{1}{4}$ **68.** $\frac{1}{6}$

69. $\frac{3}{2}$ **70.** $\frac{2}{3}$ **71.** $\frac{3}{4}$ **72.** $\frac{5}{6}$

73. Fill in the tables below to find the sequences formed by substituting the first four counting numbers into the expressions $3n$ and n^3.

a.

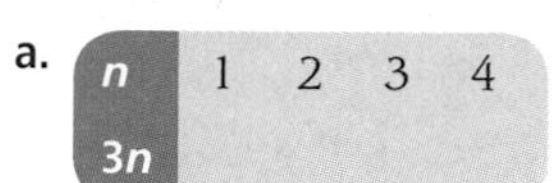

n	1	2	3	4
$3n$				

b.

n	1	2	3	4
n^3				

74. Fill in the tables below to find the sequences formed by substituting the first four counting numbers into the expressions $2n - 1$ and $2n + 1$.

a.

n	1	2	3	4
$2n - 1$				

b.

n	1	2	3	4
$2n + 1$				

C Find the sequences formed by substituting the counting numbers, in order, into the following expressions. [Examples 17]

75. $3n - 2$ **76.** $2n - 3$ **77.** $n^2 - 2n + 1$ **78.** $(n - 1)^2$

Here are some problems you will see later in the book.

Simplify.

79. $7 - 3(2y + 1)$ **80.** $4(3x - 2) - (6x - 5)$ **81.** $0.08x + 0.09x$ **82.** $0.04x + 0.05x$

83. $(x + y) + (x - y)$ **84.** $(-12x - 20y) + (25x + 20y)$ **85.** $3x + 2(x - 2)$ **86.** $2(x - 2) + 3(5x)$

87. $4(x + 1) + 3(x - 3)$ **88.** $5(x + 2) + 3(x - 1)$ **89.** $x + (x + 3)(-3)$ **90.** $x - 2(x + 2)$

91. $3(4x - 2) - (5x - 8)$

92. $2(5x - 3) - (2x - 4)$

93. $-(3x + 1) - (4x - 7)$

94. $-(6x + 2) - (8x - 3)$

95. $(x + 3y) + 3(2x - y)$

96. $(2x - y) - 2(x + 3y)$

97. $3(2x + 3y) - 2(3x + 5y)$

98. $5(2x + 3y) - 3(3x + 5y)$

99. $-6\left(\frac{1}{2}x - \frac{1}{3}y\right) + 12\left(\frac{1}{4}x + \frac{2}{3}y\right)$

100. $6\left(\frac{1}{3}x + \frac{1}{2}y\right) - 4\left(x + \frac{3}{4}y\right)$

101. $0.08x + 0.09(x + 2{,}000)$

102. $0.06x + 0.04(x + 7{,}000)$

103. $0.10x + 0.12(x + 500)$

104. $0.08x + 0.06(x + 800)$

105. Find a so the expression $(5x + 4y) + a(2x - y)$ simplifies to an expression that does not contain y. Using that value of a, simplify the expression.

106. Find a so the expression $(5x + 4y) - a(x - 2y)$ simplifies to an expression that does not contain x. Using that value of a, simplify the expression.

Find the value of $b^2 - 4ac$ for the given values of a, b, and c. (You will see these problems later in the book.)

107. $a = 1, b = -5, c = -6$

108. $a = 1, b = -6, c = 7$

109. $a = 2, b = 4, c = -3$

110. $a = 3, b = 4, c = -2$

Applying the Concepts

111. Temperature and Altitude If the temperature on the ground is 70°F, then the temperature at A feet above the ground can be found from the expression $-0.0035A + 70$. Find the temperature at the following altitudes.

a. 8,000 feet **b.** 12,000 feet **c.** 24,000 feet

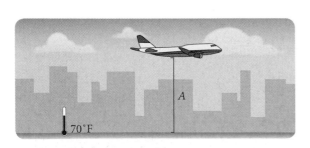

112. Perimeter of a Rectangle The expression $2l + 2w$ gives the perimeter of a rectangle with length l and width w. Find the perimeter of the rectangles with the following lengths and widths.

a. Length = 8 meters, width = 5 meters

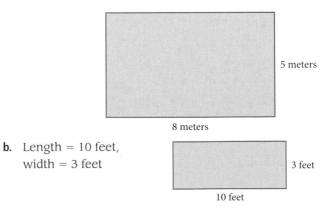

b. Length = 10 feet, width = 3 feet

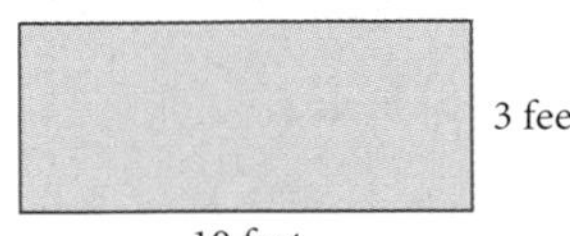

113. Cellular Phone Rates A cellular phone company charges \$35 per month plus \$0.25 for each minute, or fraction of a minute, that you use one of their cellular phones. The expression $35 + 0.25t$ gives the amount of money you will pay for using one of their phones for t minutes a month. Find the monthly bill for using one of their phones.

a. 10 minutes in a month
b. 20 minutes in a month
c. 30 minutes in a month

114. Cost of Bottled Water A water bottling company charges \$7.00 per month for their water dispenser and \$1.10 for each gallon of water delivered. If you have g gallons of water delivered in a month, then the expression $7 + 1.1g$ gives the amount of your bill for that month. Find the monthly bill for each of the following deliveries.

WBC WATER BOTTLE CO.

MONTHLY BILL

234 5th Street
Glendora, CA 91740
DUE 07/23/08

Water dispenser	1	\$7.00
Gallons of water	8	\$2.00
		\$23.00

a. 10 gallons
b. 20 gallons
c. 30 gallons

115. Taxes We all have to pay taxes. Suppose that 21% of your monthly pay is withheld for federal income taxes and another 8% is withheld for Social Security, state income tax, and other miscellaneous items. If G is your monthly pay before any money is deducted (your gross pay), then the amount of money that you take home each month is given by the expression $G - 0.21G - 0.08G$. Simplify this expression and then find your take-home pay if your gross pay is \$1,250 per month.

116. Taxes If you work H hours a day and you pay 21% of your income for federal income tax and another 8% for miscellaneous deductions, then the expression $0.21H + 0.08H$ tells you how many hours a day you are working to pay for those taxes and miscellaneous deductions. Simplify this expression. If you work 8 hours a day under these conditions, how many hours do you work to pay for your taxes and miscellaneous deductions?

117. Cars The chart shows the fastest cars in America. How fast is a car traveling if its speed is 56 miles per hour less than half the speed of the Saleen S7 Twin Turbo?

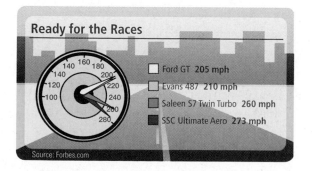

Ready for the Races

Ford GT 205 mph
Evans 487 210 mph
Saleen S7 Twin Turbo 260 mph
SSC Ultimate Aero 273 mph

Source: Forbes.com

118. Sound The chart shows the decibel level of various sounds. The loudness of a loud rock concert is 5 decibels less than twice the loudness of normal conversation. Find the sound level of a rock concert.

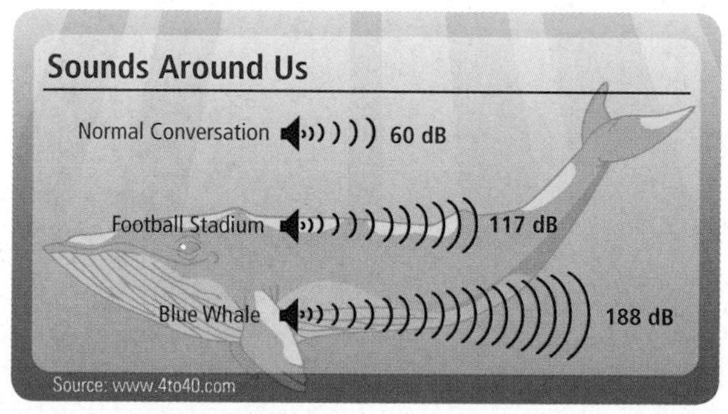

Getting Ready for the Next Section

Problems under this heading, *Getting Ready for the Next Section,* are problems that you must be able to work to understand the material in the next section. The problems below are exactly the types of problems you will see in the explanations and examples in the next section.

Simplify.

119. $17 - 5$

120. $12 + (-2)$

121. $2 - 5$

122. $25 - 20$

123. $-2.4 + (-7.3)$

124. $8.1 + 2.7$

125. $-\frac{1}{2} + \left(-\frac{3}{4}\right)$

126. $-\frac{1}{6} + \left(-\frac{2}{3}\right)$

127. $4(2 \cdot 9 - 3) - 7 \cdot 9$

128. $5(3 \cdot 45 - 4) - 14 \cdot 45$

129. $4(2a - 3) - 7a$

130. $5(3a - 4) - 14a$

131. $-3 - \frac{1}{2}$

132. $-5 - \frac{1}{3}$

133. $\frac{4}{5} + \frac{1}{10} + \frac{3}{8}$

134. $\frac{3}{10} + \frac{7}{25} + \frac{3}{4}$

135. Find the value of $2x - 3$ when x is 5.

136. Find the value of $3x + 4$ when x is -2.

Maintaining Your Skills

From this point on, each problem set will contain a number of problems under the heading *Maintaining Your Skills.* These problems cover the most important skills you have learned in previous sections and chapters. By working these problems regularly, you will keep yourself current on all the topics we have covered and possibly need less time to study for tests and quizzes.

137. $\frac{1}{8} - \frac{1}{6}$

138. $\frac{x}{8} - \frac{x}{6}$

139. $\frac{5}{9} - \frac{4}{3}$

140. $\frac{x}{9} - \frac{x}{3}$

141. $-\frac{7}{30} + \frac{5}{28}$

142. $-\frac{11}{105} + \frac{11}{30}$

2.2 Addition Property of Equality

Objectives

A Check the solution to an equation by substitution.

B Use the addition property of equality to solve an equation.

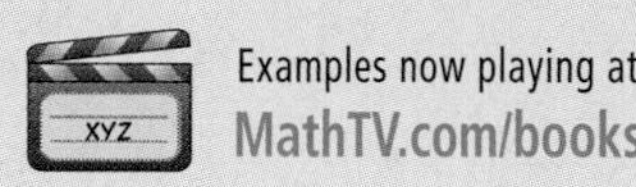

When light comes into contact with any object, it is reflected, absorbed, and transmitted, as shown here.

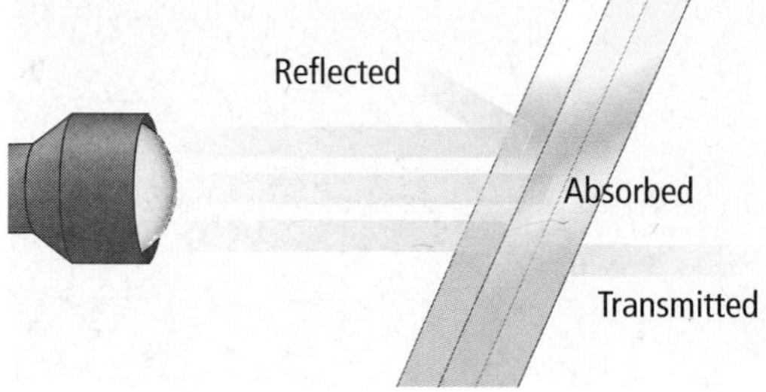

For a certain type of glass, 88% of the light hitting the glass is transmitted through to the other side, whereas 6% of the light is absorbed into the glass. To find the percent of light that is reflected by the glass, we can solve the equation

$$88 + R + 6 = 100$$

Solving equations of this type is what we study in this section. To solve an equation we must find all replacements for the variable that make the equation a true statement.

A Solutions to Equations

Definition
The **solution set** for an equation is the set of all numbers that when used in place of the variable make the equation a true statement.

For example, the equation $x + 2 = 5$ has solution set $\{3\}$ because when x is 3 the equation becomes the true statement $3 + 2 = 5$, or $5 = 5$.

EXAMPLE 1 Is 5 a solution to $2x - 3 = 7$?

SOLUTION We substitute 5 for x in the equation, and then simplify to see if a true statement results. A true statement means we have a solution; a false statement indicates the number we are using is not a solution.

When	$x = 5$	
the equation	$2x - 3 = 7$	
becomes	$2(5) - 3 \stackrel{?}{=} 7$	
	$10 - 3 \stackrel{?}{=} 7$	
	$7 = 7$	A true statement

Because $x = 5$ turns the equation into the true statement $7 = 7$, we know 5 is a solution to the equation.

EXAMPLE 2 Is -2 a solution to $8 = 3x + 4$?

SOLUTION Substituting -2 for x in the equation, we have

$8 \stackrel{?}{=} 3(-2) + 4$

$8 \stackrel{?}{=} -6 + 4$

$8 = -2$ A false statement

Substituting -2 for x in the equation produces a false statement. Therefore, $x = -2$ is not a solution to the equation.

PRACTICE PROBLEMS

1. Is 4 a solution to $2x + 3 = 7$?

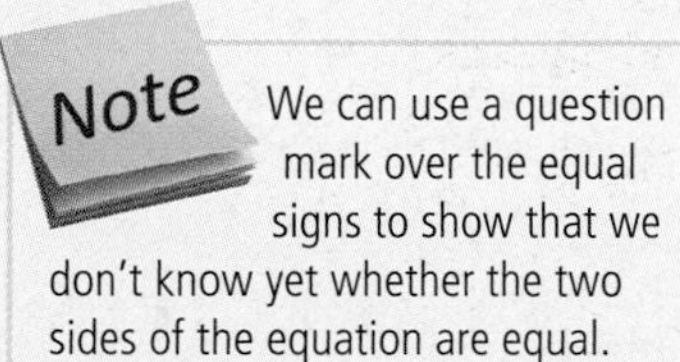

Note
We can use a question mark over the equal signs to show that we don't know yet whether the two sides of the equation are equal.

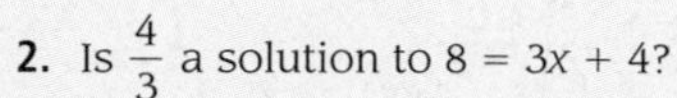

2. Is $\frac{4}{3}$ a solution to $8 = 3x + 4$?

Answers
1. No 2. Yes

The important thing about an equation is its solution set. We therefore make the following definition to classify together all equations with the same solution set.

Definition
Two or more equations with the same solution set are said to be **equivalent equations.**

Equivalent equations may look different but must have the same solution set.

3. a. Are the equations $x + 3 = 9$ and $x = 6$ equivalent?
b. Are the equations $a - 5 = 3$ and $a = 6$ equivalent?
c. Are $y + 2 = 5, y - 1 = 2$, and $y = 3$ equivalent equations?

EXAMPLE 3

a. $x + 2 = 5$ and $x = 3$ are equivalent equations because both have solution set {3}.
b. $a - 4 = 3$, $a - 2 = 5$, and $a = 7$ are equivalent equations because they all have solution set {7}.
c. $y + 3 = 4$, $y - 8 = -7$, and $y = 1$ are equivalent equations because they all have solution set {1}.

B Addition Property of Equality

If two numbers are equal and we increase (or decrease) both of them by the same amount, the resulting quantities are also equal. We can apply this concept to equations. Adding the same amount to both sides of an equation always produces an equivalent equation—one with the same solution set. This fact about equations is called the *addition property of equality* and can be stated more formally as follows.

Note We will use this property many times in the future. Be sure you understand it completely by the time you finish this section.

Addition Property of Equality
For any three algebraic expressions A, B, and C,

if $\quad A = B$

then $\quad A + C = B + C$

In words: Adding the same quantity to both sides of an equation will not change the solution set.

This property is just as simple as it seems. We can add any amount to both sides of an equation and always be sure we have not changed the solution set.

Consider the equation $x + 6 = 5$. We want to solve this equation for the value of x that makes it a true statement. We want to end up with x on one side of the equal sign and a number on the other side. Because we want x by itself, we will add -6 to both sides:

$$x + 6 + (\mathbf{-6}) = 5 + (\mathbf{-6}) \quad \text{Addition property of equality}$$
$$x + 0 = -1 \quad \text{Addition}$$
$$x = -1$$

All three equations say the same thing about x. They all say that x is -1. All three equations are equivalent. The last one is just easier to read.

Here are some further examples of how the addition property of equality can be used to solve equations.

Answer
3. a. Yes b. No c. Yes

EXAMPLE 4 Solve the equation $x - 5 = 12$ for x.

SOLUTION Because we want x alone on the left side, we choose to add +5 to both sides:

$$x - 5 + \mathbf{5} = 12 + \mathbf{5} \quad \text{Addition property of equality}$$
$$x + 0 = 17$$
$$x = 17$$

To check our solution, we substitute 17 for x in the original equation:

When	$x = 17$	
the equation	$x - 5 = 12$	
becomes	$17 - 5 \stackrel{?}{=} 12$	
	$12 = 12$	A true statement

As you can see, our solution checks. The purpose for checking a solution to an equation is to catch any mistakes we may have made in the process of solving the equation.

EXAMPLE 5 Solve for a: $a + \frac{3}{4} = -\frac{1}{2}$.

SOLUTION Because we want a by itself on the left side of the equal sign, we add the opposite of $\frac{3}{4}$ to each side of the equation.

$$a + \frac{3}{4} + \left(-\frac{\mathbf{3}}{\mathbf{4}}\right) = -\frac{1}{2} + \left(-\frac{\mathbf{3}}{\mathbf{4}}\right) \quad \text{Addition property of equality}$$
$$a + 0 = -\frac{1}{2} \cdot \frac{\mathbf{2}}{\mathbf{2}} + \left(-\frac{3}{4}\right) \quad \text{LCD on the right side is 4}$$
$$a = -\frac{2}{4} + \left(-\frac{3}{4}\right) \quad \frac{2}{4} \text{ is equivalent to } \frac{1}{2}$$
$$a = -\frac{5}{4} \quad \text{Add fractions}$$

The solution is $a = -\frac{5}{4}$. To check our result, we replace a with $-\frac{5}{4}$ in the original equation. The left side then becomes $-\frac{5}{4} + \frac{3}{4}$, which simplifies to $-\frac{1}{2}$, so our solution checks.

EXAMPLE 6 Solve for x: $7.3 + x = -2.4$.

SOLUTION Again, we want to isolate x, so we add the opposite of 7.3 to both sides:

$$7.3 + (\mathbf{-7.3}) + x = -2.4 + (\mathbf{-7.3}) \quad \text{Addition property of equality}$$
$$0 + x = -9.7$$
$$x = -9.7$$

The addition property of equality also allows us to add variable expressions to each side of an equation.

EXAMPLE 7 Solve for x: $3x - 5 = 4x$.

SOLUTION Adding $-3x$ to each side of the equation gives us our solution.

$$3x - 5 = 4x$$
$$3x + (\mathbf{-3x}) - 5 = 4x + (\mathbf{-3x}) \quad \text{Distributive property}$$
$$-5 = x$$

4. Solve $x - 3 = 10$ for x.

5. Solve for a: $a + \frac{2}{3} = -\frac{1}{6}$.

6. Solve $-2.7 + x = 8.1$.

7. Solve for x:

$$-4x + 2 = -3x$$

Answers

4. 13 5. $-\frac{5}{6}$ 6. 10.8 7. 2

Sometimes it is necessary to simplify each side of an equation before using the addition property of equality. The reason we simplify both sides first is that we want as few terms as possible on each side of the equation before we use the addition property of equality. The following examples illustrate this procedure.

8. Solve $5(3a - 4) - 14a = 25$.

EXAMPLE 8 Solve $4(2a - 3) - 7a = 2 - 5$.

SOLUTION We must begin by applying the distributive property to separate terms on the left side of the equation. Following that, we combine similar terms and then apply the addition property of equality.

$$4(2a - 3) - 7a = 2 - 5 \quad \text{Original equation}$$
$$8a - 12 - 7a = 2 - 5 \quad \text{Distributive property}$$
$$a - 12 = -3 \quad \text{Simplify each side}$$
$$a - 12 + \mathbf{12} = -3 + \mathbf{12} \quad \text{Add 12 to each side}$$
$$a = 9 \quad \text{Addition}$$

To check our solution, we replace a with 9 in the original equation.

$$4(2 \cdot 9 - 3) - 7 \cdot 9 \stackrel{?}{=} 2 - 5$$
$$4(15) - 63 \stackrel{?}{=} -3$$
$$60 - 63 \stackrel{?}{=} -3$$
$$-3 = -3 \quad \text{A true statement}$$

Note Again, we place a question mark over the equal sign because we don't know yet whether the expressions on the left and right side of the equal sign will be equal.

9. Solve $5x - 4 = 4x + 6$.

EXAMPLE 9 Solve $3x - 5 = 2x + 7$.

SOLUTION We can solve this equation in two steps. First, we add $-2x$ to both sides of the equation. When this has been done, x appears on the left side only. Second, we add 5 to both sides:

$$3x + (\mathbf{-2x}) - 5 = 2x + (\mathbf{-2x}) + 7 \quad \text{Add } -2x \text{ to both sides}$$
$$x - 5 = 7 \quad \text{Simplify each side}$$
$$x - 5 + \mathbf{5} = 7 + \mathbf{5} \quad \text{Add 5 to both sides}$$
$$x = 12 \quad \text{Simplify each side}$$

Note In my experience teaching algebra, I find that students make fewer mistakes if they think in terms of addition rather than subtraction. So, you are probably better off if you continue to use the addition property just the way we have used it in the examples in this section. But, if you are curious as to whether you can subtract the same number from both sides of an equation, the answer is yes.

A Note on Subtraction

Although the addition property of equality is stated for addition only, we can subtract the same number from both sides of an equation as well. Because subtraction is defined as addition of the opposite, subtracting the same quantity from both sides of an equation does not change the solution.

$$x + 2 = 12 \quad \text{Original equation}$$
$$x + 2 - \mathbf{2} = 12 - \mathbf{2} \quad \text{Subtract 2 from each side}$$
$$x = 10 \quad \text{Subtraction}$$

Getting Ready for Class

After reading through the preceding section, respond in your own words and in complete sentences.

1. What is a solution to an equation?
2. What are equivalent equations?
3. Explain in words the addition property of equality.
4. How do you check a solution to an equation?

Answers
8. 45 **9.** 10

Problem Set 2.2

B Find the solution for the following equations. Be sure to show when you have used the addition property of equality. [Examples 4–7]

1. $x - 3 = 8$
2. $x - 2 = 7$
3. $x + 2 = 6$
4. $x + 5 = 4$
5. $a + \frac{1}{2} = -\frac{1}{4}$
6. $a + \frac{1}{3} = -\frac{5}{6}$
7. $x + 2.3 = -3.5$
8. $x + 7.9 = -3.4$
9. $y + 11 = -6$
10. $y - 3 = -1$
11. $x - \frac{5}{8} = -\frac{3}{4}$
12. $x - \frac{2}{5} = -\frac{1}{10}$
13. $m - 6 = 2m$
14. $3m - 10 = 4m$
15. $6.9 + x = 3.3$
16. $7.5 + x = 2.2$
17. $5a = 4a - 7$
18. $12a = -3 + 11a$
19. $-\frac{5}{9} = x - \frac{2}{5}$
20. $-\frac{7}{8} = x - \frac{4}{5}$

B Simplify both sides of the following equations as much as possible, and then solve. [Example 7]

21. $4x + 2 - 3x = 4 + 1$
22. $5x + 2 - 4x = 7 - 3$
23. $8a - \frac{1}{2} - 7a = \frac{3}{4} + \frac{1}{8}$
24. $9a - \frac{4}{5} - 8a = \frac{3}{10} - \frac{1}{5}$
25. $-3 - 4x + 5x = 18$
26. $10 - 3x + 4x = 20$

B Simplify both sides of the following equations as much as possible, and then solve. **[Example 7]**

27. $-11x + 2 + 10x + 2x = 9$

28. $-10x + 5 - 4x + 15x = 0$

29. $-2.5 + 4.8 = 8x - 1.2 - 7x$

30. $-4.8 + 6.3 = 7x - 2.7 - 6x$

31. $2y - 10 + 3y - 4y = 18 - 6$

32. $15 - 21 = 8x + 3x - 10x$

B The following equations contain parentheses. Apply the distributive property to remove the parentheses, then simplify each side before using the addition property of equality. **[Example 8]**

33. $2(x + 3) - x = 4$

34. $5(x + 1) - 4x = 2$

35. $-3(x - 4) + 4x = 3 - 7$

36. $-2(x - 5) + 3x = 4 - 9$

37. $5(2a + 1) - 9a = 8 - 6$

38. $4(2a - 1) - 7a = 9 - 5$

39. $-(x + 3) + 2x - 1 = 6$

40. $-(x - 7) + 2x - 8 = 4$

41. $4y - 3(y - 6) + 2 = 8$

42. $7y - 6(y - 1) + 3 = 9$

43. $-3(2m - 9) + 7(m - 4) = 12 - 9$

44. $-5(m - 3) + 2(3m + 1) = 15 - 8$

B Solve the following equations by the method used in Example 9 in this section. Check each solution in the original equation.

45. $4x = 3x + 2$

46. $6x = 5x - 4$

47. $8a = 7a - 5$

48. $9a = 8a - 3$

49. $2x = 3x + 1$

50. $4x = 3x + 5$

51. $2y + 1 = 3y + 4$

52. $4y + 2 = 5y + 6$

53. $2m - 3 = m + 5$

54. $8m - 1 = 7m - 3$

55. $4x - 7 = 5x + 1$

56. $3x - 7 = 4x - 6$

57. $4x + \frac{4}{3} = 5x - \frac{2}{3}$

58. $2x + \frac{1}{4} = 3x - \frac{5}{4}$

59. $8a - 7.1 = 7a + 3.9$

60. $10a - 4.3 = 9a + 4.7$

61. Solve each equation.

a. $2x = 3$

b. $2 + x = 3$

c. $2x + 3 = 0$

d. $2x + 3 = -5$

e. $2x + 3 = 7x - 5$

62. Solve each equation.

a. $5t = 10$

b. $5 + t = 10$

c. $5t + 10 = 0$

d. $5t + 10 = 12$

e. $5t + 10 = 8t + 12$

Applying the Concepts

63. **Light** When light comes into contact with any object, it is reflected, absorbed, and transmitted, as shown in the following figure. If T represents the percent of light transmitted, R the percent of light reflected, and A the percent of light absorbed by a surface, then the equation $T + R + A = 100$ shows one way these quantities are related.

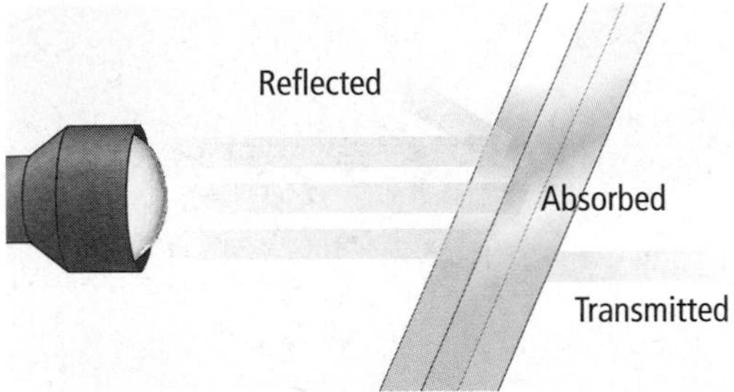

a. For glass, $T = 88$ and $A = 6$, meaning that 88% of the light hitting the glass is transmitted and 6% is absorbed. Substitute $T = 88$ and $A = 6$ into the equation $T + R + A = 100$ and solve for R to find the percent of light that is reflected.
b. For flat black paint, $A = 95$ and no light is transmitted, meaning that $T = 0$. What percent of light is reflected by flat black paint?
c. A pure white surface can reflect 98% of light, so $R = 98$. If no light is transmitted, what percent of light is absorbed by the pure white surface?
d. Typically, shiny gray metals reflect 70–80% of light. Suppose a thick sheet of aluminum absorbs 25% of light. What percent of light is reflected by this shiny gray metal? (Assume no light is transmitted.)

64. **Movie Tickets** A movie theater has a total of 300 seats. For a special Saturday night preview, they reserve 20 seats to give away to their VIP guests at no charge. If x represents the number of tickets they can sell for the preview, then x can be found by solving the equation $x + 20 = 300$.
a. Solve the equation for x.
b. If tickets for the preview are $7.50 each, what is the maximum amount of money they can make from ticket sales?

65. **Geometry** The three angles shown in the triangle at the front of the tent in the following figure add up to 180°. Use this fact to write an equation containing x, and then solve the equation to find the number of degrees in the angle at the top of the triangle.

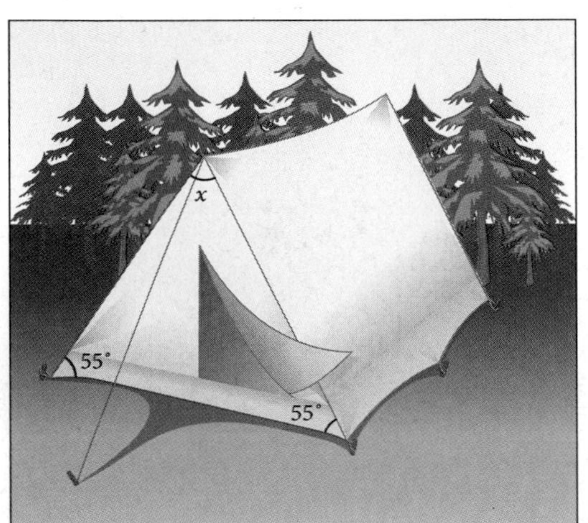

66. **Geometry** The figure shows part of a room. From a point on the floor, the angle of elevation to the top of the door is 47°, whereas the angle of elevation to the ceiling above the door is 59°. Use this diagram to write an equation involving x, and then solve the equation to find the number of degrees in the angle that extends from the top of the door to the ceiling.

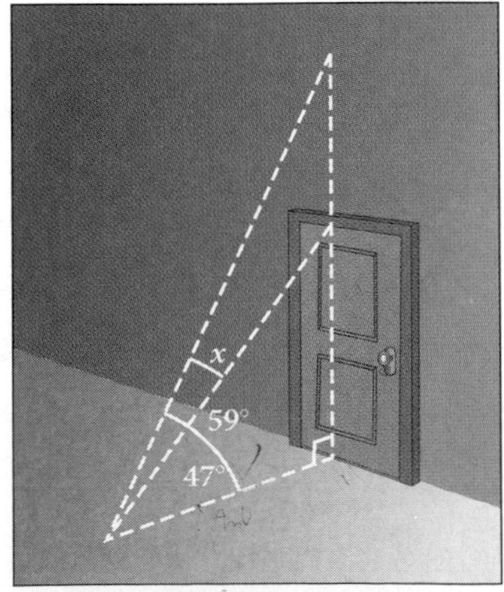

67. Sound The chart shows the decibel level of various sounds. The sum of the loudness of a motorcycle and 17 decibels is the loudness of a football stadium. What is the loudness of a motorcycle?

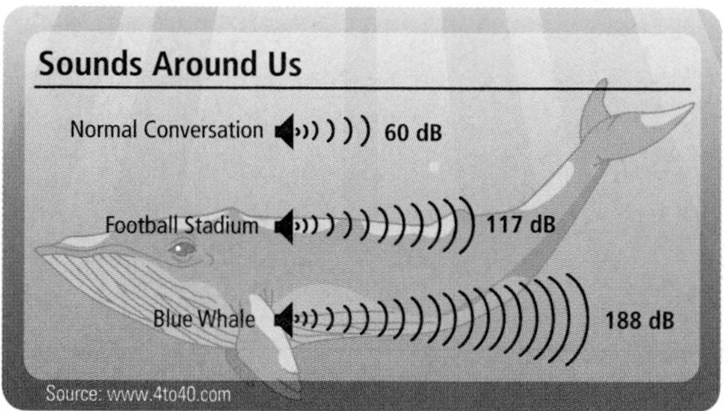

68. Skyscrapers The chart shows the heights of the three tallest buildings in the world. The height of the Sears Tower is 358 feet taller than the Tokyo Tower in Japan. How tall is the Tokyo Tower?

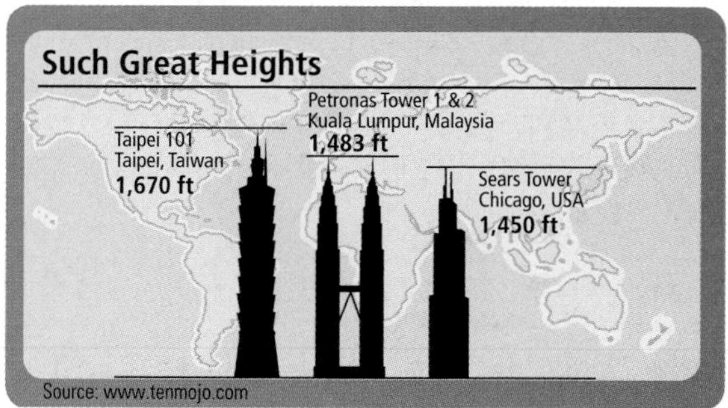

Getting Ready for the Next Section

To understand all of the explanations and examples in the next section you must be able to work the problems below.

Simplify.

69. $\frac{3}{2}\left(\frac{2}{3}y\right)$

70. $\frac{5}{2}\left(\frac{2}{5}y\right)$

71. $\frac{1}{5}(5x)$

72. $-\frac{1}{4}(-4a)$

73. $\frac{1}{5}(30)$

74. $-\frac{1}{4}(24)$

75. $\frac{3}{2}(4)$

76. $\frac{1}{26}(13)$

77. $12\left(-\frac{3}{4}\right)$

78. $12\left(\frac{1}{2}\right)$

79. $\frac{3}{2}\left(-\frac{5}{4}\right)$

80. $\frac{5}{3}\left(-\frac{6}{5}\right)$

81. $-13 + (-5)$

82. $-14 + (-3)$

83. $-\frac{3}{4} + \left(-\frac{1}{2}\right)$

84. $-\frac{7}{10} + \left(-\frac{1}{2}\right)$

85. $7x + (-4x)$

86. $5x + (-2x)$

Maintaining Your Skills

The problems that follow review some of the more important skills you have learned in previous sections and chapters. You can consider the time you spend working these problems as time spent studying for exams.

87. $3(6x)$

88. $5(4x)$

89. $\frac{1}{5}(5x)$

90. $\frac{1}{3}(3x)$

91. $8\left(\frac{1}{8}y\right)$

92. $6\left(\frac{1}{6}y\right)$

93. $-2\left(-\frac{1}{2}x\right)$

94. $-4\left(-\frac{1}{4}x\right)$

95. $-\frac{4}{3}\left(-\frac{3}{4}a\right)$

96. $-\frac{5}{2}\left(-\frac{2}{5}a\right)$

Multiplication Property of Equality

2.3

Objectives

A Use the multiplication property of equality to solve an equation.

B Use the addition and multiplication properties of equality together to solve an equation.

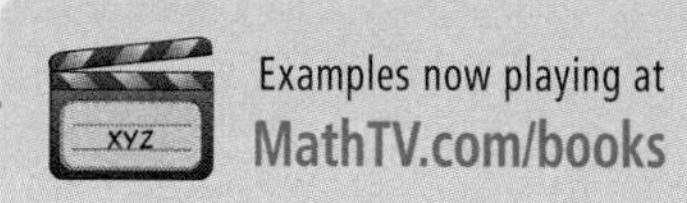

As we have mentioned before, we all have to pay taxes. According to Figure 1, people have been paying taxes for quite a long time.

FIGURE 1 Collection of taxes, ca. 3000 B.C. Clerks and scribes appear at the right, with pen and papyrus, and officials and taxpayers appear at the left.

If 21% of your monthly pay is withheld for federal income taxes and another 8% is withheld for Social Security, state income tax, and other miscellaneous items, leaving you with \$987.50 a month in take-home pay, then the amount you earned before the deductions were removed from your check is given by the equation

$$G - 0.21G - 0.08G = 987.5$$

In this section we will learn how to solve equations of this type.

A Multiplication Property of Equality

In the previous section, we found that adding the same number to both sides of an equation never changed the solution set. The same idea holds for multiplication by numbers other than zero. We can multiply both sides of an equation by the same nonzero number and always be sure we have not changed the solution set. (The reason we cannot multiply both sides by zero will become apparent later.) This fact about equations is called the *multiplication property of equality,* which can be stated formally as follows.

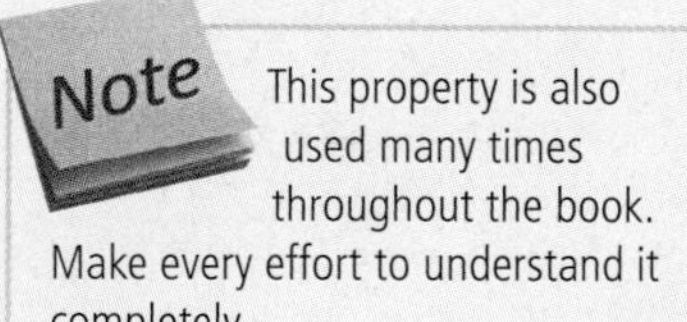

This property is also used many times throughout the book. Make every effort to understand it completely.

Multiplication Property of Equality

For any three algebraic expressions A, B, and C, where $C \neq 0$,

if $\quad A = B$

then $\quad AC = BC$

In words: Multiplying both sides of an equation by the same nonzero number will not change the solution set.

Suppose we want to solve the equation $5x = 30$. We have $5x$ on the left side but would like to have just x. We choose to multiply both sides by $\frac{1}{5}$ because $\left(\frac{1}{5}\right)(5) = 1$. Here is the solution:

$$5x = 30$$

$$\frac{\mathbf{1}}{\mathbf{5}}(5x) = \frac{\mathbf{1}}{\mathbf{5}}(30) \quad \text{Multiplication property of equality}$$

$$\left(\frac{1}{5} \cdot 5\right)x = \frac{1}{5}(30) \quad \text{Associative property of multiplication}$$

$$1x = 6$$

$$x = 6$$

We chose to multiply by $\frac{1}{5}$ because it is the reciprocal of 5. We can see that multiplication by any number except zero will not change the solution set. If, however, we were to multiply both sides by zero, the result would always be $0 = 0$ because multiplication by zero always results in zero. Although the statement $0 = 0$ is true, we have lost our variable and cannot solve the equation. This is the only restriction of the multiplication property of equality. We are free to multiply both sides of an equation by any number except zero.

Here are some more examples that use the multiplication property of equality.

PRACTICE PROBLEMS

1. Solve for a: $3a = -27$.

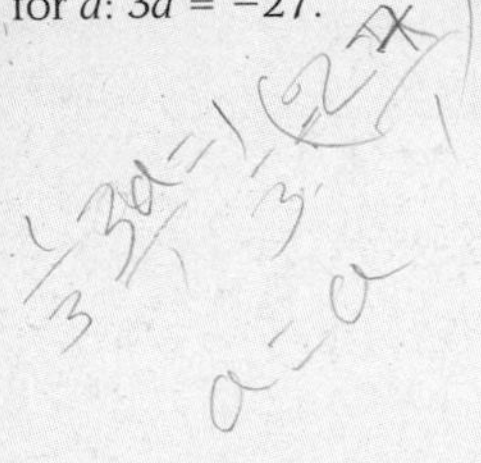

2. Solve for t: $\frac{t}{4} = 6$.

3. Solve $\frac{2}{5}y = 8$.

Note Notice in Examples 1 through 3 that if the variable is being multiplied by a number like -4 or $\frac{2}{3}$, we always multiply by the number's reciprocal, $-\frac{1}{4}$ or $\frac{3}{2}$, to end up with just the variable on one side of the equation.

4. Solve.
$3 + 7 = 6x + 8x - 9x$

Answers
1. -9 2. 24 3. 20 4. 2

EXAMPLE 1 Solve for a: $-4a = 24$.

SOLUTION Because we want a alone on the left side, we choose to multiply both sides by $-\frac{1}{4}$:

$$-\frac{\mathbf{1}}{\mathbf{4}}(-4a) = -\frac{\mathbf{1}}{\mathbf{4}}(24) \qquad \text{Multiplication property of equality}$$

$$\left[-\frac{1}{4}(-4)\right]a = -\frac{1}{4}(24) \qquad \text{Associative property}$$

$$a = -6$$

EXAMPLE 2 Solve for t: $-\frac{t}{3} = 5$.

SOLUTION Because division by 3 is the same as multiplication by $\frac{1}{3}$, we can write $-\frac{t}{3}$ as $-\frac{1}{3}t$. To solve the equation, we multiply each side by the reciprocal of $-\frac{1}{3}$, which is -3.

$$-\frac{t}{3} = 5 \qquad \text{Original equation}$$

$$-\frac{1}{3}t = 5 \qquad \text{Dividing by 3 is equivalent to multiplying by } \tfrac{1}{3}$$

$$\mathbf{-3}\left(-\frac{1}{3}t\right) = \mathbf{-3}(5) \qquad \text{Multiply each side by } -3$$

$$t = -15 \qquad \text{Multiplication}$$

EXAMPLE 3 Solve $\frac{2}{3}y = 4$.

SOLUTION We can multiply both sides by $\frac{3}{2}$ and have $1y$ on the left side:

$$\frac{\mathbf{3}}{\mathbf{2}}\left(\frac{2}{3}y\right) = \frac{\mathbf{3}}{\mathbf{2}}(4) \qquad \text{Multiplication property of equality}$$

$$\left(\frac{3}{2}\cdot\frac{2}{3}y\right) = \frac{3}{2}(4) \qquad \text{Associative property}$$

$$y = 6 \qquad \text{Simplify } \tfrac{3}{2}(4) = \tfrac{3}{2}\left(\tfrac{4}{1}\right) = \tfrac{12}{2} = 6$$

EXAMPLE 4 Solve $5 + 8 = 10x + 20x - 4x$.

SOLUTION Our first step will be to simplify each side of the equation:

$$13 = 26x \qquad \text{Simplify both sides first}$$

$$\frac{\mathbf{1}}{\mathbf{26}}(13) = \frac{\mathbf{1}}{\mathbf{26}}(26x) \qquad \text{Multiplication property of equality}$$

$$\frac{13}{26} = x \qquad \text{Multiplication}$$

$$\frac{1}{2} = x \qquad \text{Reduce to lowest terms}$$

B Solving Equations

In the next three examples, we will use both the addition property of equality and the multiplication property of equality.

EXAMPLE 5 Solve for x: $6x + 5 = -13$.

SOLUTION We begin by adding -5 to both sides of the equation:

$6x + 5 + (\mathbf{-5}) = -13 + (\mathbf{-5})$	Add -5 to both sides
$6x = -18$	Simplify
$\frac{\mathbf{1}}{\mathbf{6}}(6x) = \frac{\mathbf{1}}{\mathbf{6}}(-18)$	Multiply both sides by $\frac{1}{6}$
$x = -3$	

5. Solve for x: $4x + 3 = -13$.

EXAMPLE 6 Solve for x: $5x = 2x + 12$.

SOLUTION We begin by adding $-2x$ to both sides of the equation:

$5x + (\mathbf{-2x}) = 2x + (\mathbf{-2x}) + 12$	Add $-2x$ to both sides
$3x = 12$	Simplify
$\frac{\mathbf{1}}{\mathbf{3}}(3x) = \frac{\mathbf{1}}{\mathbf{3}}(12)$	Multiply both sides by $\frac{1}{3}$
$x = 4$	Simplify

6. Solve for x: $7x = 4x + 15$.

Note Notice that in Example 6 we used the addition property of equality first to combine all the terms containing x on the left side of the equation. Once this had been done, we used the multiplication property to isolate x on the left side.

EXAMPLE 7 Solve for x: $3x - 4 = -2x + 6$.

SOLUTION We begin by adding $2x$ to both sides:

$3x + \mathbf{2x} - 4 = -2x + \mathbf{2x} + 6$	Add $2x$ to both sides
$5x - 4 = 6$	Simplify

Now we add 4 to both sides:

$5x - 4 + \mathbf{4} = 6 + \mathbf{4}$	Add 4 to both sides
$5x = 10$	Simplify
$\frac{\mathbf{1}}{\mathbf{5}}(5x) = \frac{\mathbf{1}}{\mathbf{5}}(10)$	Multiply by $\frac{1}{5}$
$x = 2$	Simplify

7. Solve for x:

$$2x - 5 = -3x + 10$$

The next example involves fractions. You will see that the properties we use to solve equations containing fractions are the same as the properties we used to solve the previous equations. Also, the LCD that we used previously to add fractions can be used with the multiplication property of equality to simplify equations containing fractions.

EXAMPLE 8 Solve $\frac{2}{3}x + \frac{1}{2} = -\frac{3}{4}$.

SOLUTION We can solve this equation by applying our properties and working with the fractions, or we can begin by eliminating the fractions.

METHOD 1 Working with the fractions.

$\frac{2}{3}x + \frac{1}{2} + \left(-\frac{\mathbf{1}}{\mathbf{2}}\right) = -\frac{3}{4} + \left(-\frac{\mathbf{1}}{\mathbf{2}}\right)$	Add $-\frac{1}{2}$ to each side
$\frac{2}{3}x = -\frac{5}{4}$	Note that $-\frac{3}{4} + \left(-\frac{1}{2}\right) = -\frac{3}{4} + \left(-\frac{2}{4}\right)$
$\frac{\mathbf{3}}{\mathbf{2}}\left(\frac{2}{3}x\right) = \frac{\mathbf{3}}{\mathbf{2}}\left(-\frac{5}{4}\right)$	Multiply each side by $\frac{3}{2}$
$x = -\frac{15}{8}$	

8. Solve $\frac{3}{5}x + \frac{1}{2} = -\frac{7}{10}$.

Answers
5. -4 6. 5 7. 3

Note Our original equation has denominators of 3, 2, and 4. The LCD for these three denominators is 12, and it has the property that all three denominators will divide it evenly. Therefore, if we multiply both sides of our equation by 12, each denominator will divide into 12 and we will be left with an equation that does not contain any denominators other than 1.

METHOD 2 Eliminating the fractions in the beginning.

$\mathbf{12}\left(\frac{2}{3}x + \frac{1}{2}\right) = \mathbf{12}\left(-\frac{3}{4}\right)$	Multiply each side by the LCD 12
$\mathbf{12}\left(\frac{2}{3}x\right) + \mathbf{12}\left(\frac{1}{2}\right) = \mathbf{12}\left(-\frac{3}{4}\right)$	Distributive property on the left side
$8x + 6 = -9$	Multiply
$8x = -15$	Add -6 to each side
$x = -\frac{15}{8}$	Multiply each side by $\frac{1}{8}$

As the third line in Method 2 indicates, multiplying each side of the equation by the LCD eliminates all the fractions from the equation.

As you can see, both methods yield the same solution.

Property

Because division is defined as multiplication by the reciprocal, multiplying both sides of an equation by the same number is equivalent to dividing both sides of the equation by the reciprocal of that number; that is, multiplying each side of an equation by $\frac{1}{3}$ and dividing each side of the equation by 3 are equivalent operations. If we were to solve the equation $3x = 18$ using division instead of multiplication, the steps would look like this:

$3x = 18$	Original equation
$\frac{3x}{\mathbf{3}} = \frac{18}{\mathbf{3}}$	Divide each side by 3
$x = 6$	Division

Using division instead of multiplication on a problem like this may save you some writing. However, with multiplication, it is easier to explain "why" we end up with just one x on the left side of the equation. (The "why" has to do with the associative property of multiplication.) My suggestion is that you continue to use multiplication to solve equations like this one until you understand the process completely. Then, if you find it more convenient, you can use division instead of multiplication.

Getting Ready for Class

After reading through the preceding section, respond in your own words and in complete sentences.

1. Explain in words the multiplication property of equality.
2. If an equation contains fractions, how do you use the multiplication property of equality to clear the equation of fractions?
3. Why is it okay to divide both sides of an equation by the same nonzero number?
4. Explain in words how you would solve the equation $3x = 7$ using the multiplication property of equality.

Answer
8. -2

Problem Set 2.3

A Solve the following equations. Be sure to show your work. **[Examples 1–3]**

1. $5x = 10$

2. $6x = 12$

3. $7a = 28$

4. $4a = 36$

5. $-8x = 4$

6. $-6x = 2$

7. $8m = -16$

8. $5m = -25$

9. $-3x = -9$

10. $-9x = -36$

11. $-7y = -28$

12. $-15y = -30$

13. $2x = 0$

14. $7x = 0$

15. $-5x = 0$

16. $-3x = 0$

17. $\frac{x}{3} = 2$

18. $\frac{x}{4} = 3$

19. $-\frac{m}{5} = 10$

20. $-\frac{m}{7} = 1$

21. $-\frac{x}{2} = -\frac{3}{4}$

22. $-\frac{x}{3} = \frac{5}{6}$

23. $\frac{2}{3}a = 8$

24. $\frac{3}{4}a = 6$

25. $-\frac{3}{5}x = \frac{9}{5}$

26. $-\frac{2}{5}x = \frac{6}{15}$

27. $-\frac{5}{8}y = -20$

28. $-\frac{7}{2}y = -14$

A Simplify both sides as much as possible, and then solve. **[Example 4]**

29. $-4x - 2x + 3x = 24$

30. $7x - 5x + 8x = 20$

31. $4x + 8x - 2x = 15 - 10$

32. $5x + 4x + 3x = 4 + 8$

33. $-3 - 5 = 3x + 5x - 10x$

34. $10 - 16 = 12x - 6x - 3x$

35. $18 - 13 = \frac{1}{2}a + \frac{3}{4}a - \frac{5}{8}a$

36. $20 - 14 = \frac{1}{3}a + \frac{5}{6}a - \frac{2}{3}a$

Solve the following equations by multiplying both sides by -1.

37. $-x = 4$

38. $-x = -3$

39. $-x = -4$

40. $-x = 3$

41. $15 = -a$

42. $-15 = -a$

43. $-y = \frac{1}{2}$

44. $-y = -\frac{3}{4}$

B Solve each of the following equations. **[Examples 5–8]**

45. $3x - 2 = 7$

46. $2x - 3 = 9$

47. $2a + 1 = 3$

48. $5a - 3 = 7$

49. $\frac{1}{8} + \frac{1}{2}x = \frac{1}{4}$

50. $\frac{1}{3} + \frac{1}{7}x = -\frac{8}{21}$

51. $6x = 2x - 12$

52. $8x = 3x - 10$

53. $2y = -4y + 18$

54. $3y = -2y - 15$

55. $-7x = -3x - 8$

56. $-5x = -2x - 12$

57. $2x - 5 = 8x + 4$

58. $3x - 6 = 5x + 6$

59. $x + \frac{1}{2} = \frac{1}{4}x - \frac{5}{8}$

60. $\frac{1}{3}x + \frac{2}{5} = \frac{1}{5}x - \frac{2}{5}$

61. $m + 2 = 6m - 3$

62. $m + 5 = 6m - 5$

63. $\frac{1}{2}m - \frac{1}{4} = \frac{1}{12}m + \frac{1}{6}$

64. $\frac{1}{2}m - \frac{5}{12} = \frac{1}{12}m + \frac{5}{12}$

65. $6y - 4 = 9y + 2$

66. $2y - 2 = 6y + 14$

67. $\frac{3}{2}y + \frac{1}{3} = y - \frac{2}{3}$

68. $\frac{3}{2}y + \frac{7}{2} = \frac{1}{2}y - \frac{1}{2}$

69. $5x + 6 = 2$

70. $2x + 15 = 3$

71. $\frac{x}{2} = \frac{6}{12}$

72. $\frac{x}{4} = \frac{6}{8}$

73. $\frac{3}{x} = \frac{6}{7}$

74. $\frac{2}{9} = \frac{8}{x}$

75. $\frac{a}{3} = \frac{5}{12}$

76. $\frac{a}{2} = \frac{7}{20}$

77. $\frac{10}{20} = \frac{20}{x}$

78. $\frac{15}{60} = \frac{60}{x}$

79. $\frac{2}{x} = \frac{6}{7}$

80. $\frac{4}{x} = \frac{6}{7}$

Applying the Concepts

81. **Break-Even Point** Movie theaters pay a certain price for the movies that you and I see. Suppose a theater pays \$1,500 for each showing of a popular movie. If they charge \$7.50 for each ticket they sell, then the equation $7.5x = 1{,}500$ gives the number of tickets they must sell to equal the \$1,500 cost of showing the movie. This number is called the break-even point. Solve the equation for x to find the break-even point.

82. **Basketball** Laura plays basketball for her community college. In one game she scored 13 points total, with a combination of free throws, field goals, and three-pointers. Each free throw is worth 1 point, each field goal is 2 points, and each three-pointer is worth 3 points. If she made 1 free throw and 3 field goals, then solving the equation

$$1 + 3(2) + 3x = 13$$

will give us the number of three-pointers she made. Solve the equation to find the number of three-point shots Laura made.

83. **Taxes** If 21% of your monthly pay is withheld for federal income taxes and another 8% is withheld for Social Security, state income tax, and other miscellaneous items, leaving you with \$987.61 a month in take-home pay, then the amount you earned before the deductions were removed from your check is given by the equation

$$G - 0.21G - 0.08G = 987.61$$

Solve this equation to find your gross income.

84. **Rhind Papyrus** The *Rhind Papyrus* is an ancient document that contains mathematical riddles. One problem asks the reader to find a quantity such that when it is added to one-fourth of itself the sum is 15. The equation that describes this situation is

$$x + \frac{1}{4}x = 15$$

Solve this equation.

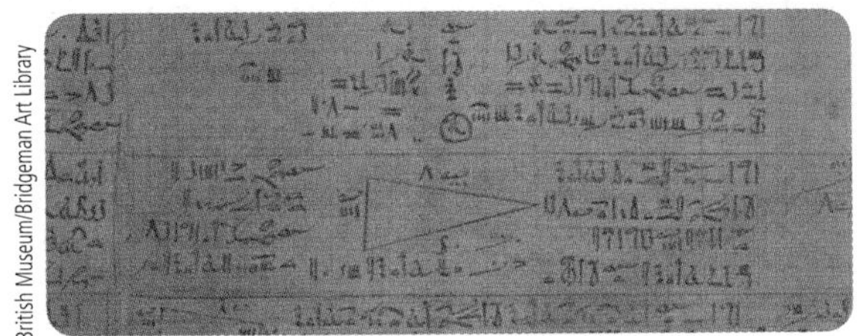

British Museum/Bridgeman Art Library

85. **Skyscrapers** The chart shows the heights of the three tallest buildings in the world. Two times the height of the Hoover Dam is the height of the Sears Tower. How tall is the Hoover Dam?

Such Great Heights

Taipei 101
Taipei, Taiwan
1,670 ft

Petronas Tower 1 & 2
Kuala Lumpur, Malaysia
1,483 ft

Sears Tower
Chicago, USA
1,450 ft

Source: www.tenmojo.com

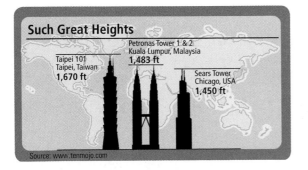

86. **Cars** The chart shows the fastest cars in America. The speed of the Ford GT in miles per hour is $\frac{15}{22}$ of its speed in feet per second. What is its speed in feet per second? Round to the nearest foot per second.

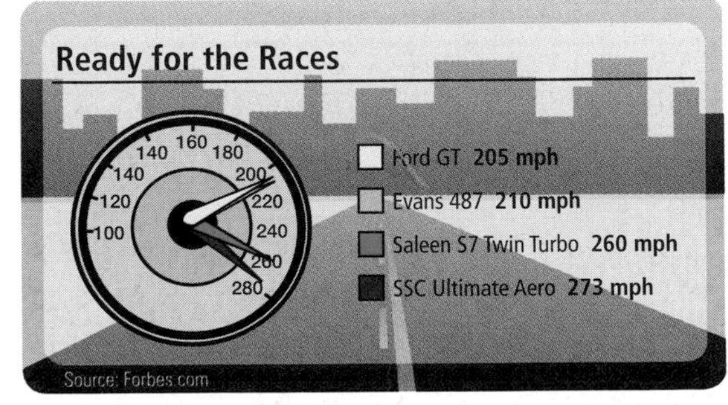

Getting Ready for the Next Section

To understand all of the explanations and examples in the next section you must be able to work the problems below. Solve each equation.

87. $2x = 4$ **88.** $3x = 24$ **89.** $30 = 5x$ **90.** $0 = 5x$ **91.** $0.17x = 510$ **92.** $0.1x = 400$

Apply the distributive property and then simplify if possible.

93. $3(x - 5) + 4$ **94.** $5(x - 3) + 2$ **95.** $0.09(x + 2{,}000)$ **96.** $0.04(x + 7{,}000)$

97. $7 - 3(2y + 1)$ **98.** $4 - 2(3y + 1)$ **99.** $3(2x - 5) - (2x - 4)$ **100.** $4(3x - 2) - (6x - 5)$

Simplify.

101. $10x + (-5x)$ **102.** $12x + (-7x)$ **103.** $0.08x + 0.09x$ **104.** $0.06x + 0.04x$

Maintaining Your Skills

The problems that follow review some of the more important skills you have learned in previous sections and chapters. You can consider the time you spend working these problems as time spent studying for exams.

Apply the distributive property, and then simplify each expression as much as possible.

105. $2(3x - 5)$ **106.** $4(2x - 6)$ **107.** $\frac{1}{2}(3x + 6)$

108. $\frac{1}{4}(2x + 8)$ **109.** $\frac{1}{3}(-3x + 6)$ **110.** $\frac{1}{2}(-2x + 6)$

Simplify each expression.

111. $5(2x - 8) - 3$ **112.** $4(3x - 1) + 7$ **113.** $-2(3x + 5) + 3(x - 1)$

114. $6(x + 3) - 2(2x + 4)$ **115.** $7 - 3(2y + 1)$ **116.** $8 - 5(3y - 4)$

Solving Linear Equations 2.4

Objectives

A Solve a linear equation in one variable.

Examples now playing at
MathTV.com/books

A Solving Linear Equations

We will now use the material we have developed in the first three sections of this chapter to build a method for solving any linear equation.

Definition

A **linear equation** in one variable is any equation that can be put in the form $ax + b = 0$, where a and b are real numbers and a is not zero.

Each of the equations we will solve in this section is a linear equation in one variable. The steps we use to solve a linear equation in one variable are listed here.

Strategy Solving Linear Equations in One Variable

Step 1a: Use the distributive property to separate terms, if necessary.

1b: If fractions are present, consider multiplying both sides by the LCD to eliminate the fractions. If decimals are present, consider multiplying both sides by a power of 10 to clear the equation of decimals.

1c: Combine similar terms on each side of the equation.

Step 2: Use the addition property of equality to get all variable terms on one side of the equation and all constant terms on the other side. A variable term is a term that contains the variable (for example, $5x$). A constant term is a term that does not contain the variable (the number 3, for example).

Step 3: Use the multiplication property of equality to get x (that is, $1x$) by itself on one side of the equation.

Step 4: Check your solution in the original equation to be sure that you have not made a mistake in the solution process.

You may have some previous experience solving equations. Even so, you should solve the equations in this section using the method developed here. Your work should look like the examples in the text. If you have learned shortcuts or a different method of solving equations somewhere else, you always can go back to them later. What is important now is that you are able to solve equations by the methods shown here.

As you will see as you work through the examples in this section, it is not always necessary to use all four steps when solving equations. The number of steps used depends on the equation. In Example 1 there are no fractions or decimals in the original equation, so step 1b will not be used. Likewise, after applying the distributive property to the left side of the equation in Example 1, there are no similar terms to combine on either side of the equation, making step 1c also unnecessary.

EXAMPLE 1 Solve $2(x + 3) = 10$.

SOLUTION To begin, we apply the distributive property to the left side of the equation to separate terms:

Step 1a: $2x + 6 = 10$ Distributive property

Step 2: $\begin{cases} 2x + 6 + (\mathbf{-6}) = 10 + (\mathbf{-6}) \\ 2x = 4 \end{cases}$ Addition property of equality

Step 3: $\begin{cases} \frac{\mathbf{1}}{\mathbf{2}}(2x) = \frac{\mathbf{1}}{\mathbf{2}}(4) \\ x = 2 \end{cases}$ Multiply each side by $\frac{1}{2}$; The solution is 2

PRACTICE PROBLEMS

1. Solve $3(x + 4) = 6$

The solution to our equation is 2. We check our work (to be sure we have not made either a mistake in applying the properties or an arithmetic mistake) by substituting 2 into our original equation and simplifying each side of the result separately.

CHECK:

Step 4: When $x = 2$, the equation $2(x + 3) = 10$ becomes

$$2(2 + 3) \stackrel{?}{=} 10$$
$$2(5) \stackrel{?}{=} 10$$
$$10 = 10 \quad \text{A true statement}$$

Our solution checks.

The general method of solving linear equations is actually very simple. It is based on the properties we developed in Chapter 1 and on two very simple new properties. We can add any number to both sides of the equation and multiply both sides by any nonzero number. The equation may change in form, but the solution set will not. If we look back to Example 1, each equation looks a little different from each preceding equation. What is interesting and useful is that each equation says the same thing about x. They all say x is 2. The last equation, of course, is the easiest to read, and that is why our goal is to end up with x by itself.

2. Solve for x: $2(x - 3) + 3 = 9$.

EXAMPLE 2 Solve for x : $3(x - 5) + 4 = 13$.

SOLUTION Our first step will be to apply the distributive property to the left side of the equation:

Step 1a: $3x - 15 + 4 = 13$ — Distributive property

Step 1c: $3x - 11 = 13$ — Simplify the left side

Step 2: $3x - 11 + \mathbf{11} = 13 + \mathbf{11}$ — Add 11 to both sides

$3x = 24$

Step 3: $\frac{\mathbf{1}}{\mathbf{3}}(3x) = \frac{\mathbf{1}}{\mathbf{3}}(24)$ — Multiply both sides by $\frac{1}{3}$

$x = 8$ — The solution is 8

CHECK:

Step 4: When $x = 8$, the equation $3(x - 5) + 4 = 13$ becomes

$$3(8 - 5) + 4 \stackrel{?}{=} 13$$
$$3(3) + 4 \stackrel{?}{=} 13$$
$$9 + 4 \stackrel{?}{=} 13$$
$$13 = 13 \quad \text{A true statement}$$

3. Solve.
$7(x - 3) + 5 = 4(3x - 2) - 8$

EXAMPLE 3 Solve $5(x - 3) + 2 = 5(2x - 8) - 3$.

SOLUTION In this case we apply the distributive property on each side of the equation:

Step 1a: $5x - 15 + 2 = 10x - 40 - 3$ — Distributive property

Step 1c: $5x - 13 = 10x - 43$ — Simplify each side

Step 2: $5x + (\mathbf{-5x}) - 13 = 10x + (\mathbf{-5x}) - 43$ — Add $-5x$ to both sides

$-13 = 5x - 43$

$-13 + \mathbf{43} = 5x - 43 + \mathbf{43}$ — Add 43 to both sides

$30 = 5x$

Answers
1. −2 2. 6

Step 3:
$$\frac{\mathbf{1}}{\mathbf{5}}(30) = \frac{\mathbf{1}}{\mathbf{5}}(5x) \qquad \text{Multiply both sides by } \tfrac{1}{5}$$
$$6 = x \qquad \text{The solution is 6}$$

CHECK: Replacing x with 6 in the original equation, we have

Step 4:
$$5(6 - 3) + 2 \stackrel{?}{=} 5(2 \cdot 6 - 8) - 3$$
$$5(3) + 2 \stackrel{?}{=} 5(12 - 8) - 3$$
$$5(3) + 2 \stackrel{?}{=} 5(4) - 3$$
$$15 + 2 \stackrel{?}{=} 20 - 3$$
$$17 = 17 \qquad \text{A true statement}$$

Note It makes no difference on which side of the equal sign x ends up. Most people prefer to have x on the left side because we read from left to right, and it seems to sound better to say x is 6 rather than 6 is x. Both expressions, however, have exactly the same meaning.

EXAMPLE 4 Solve the equation $0.08x + 0.09(x + 2{,}000) = 690$.

SOLUTION We can solve the equation in its original form by working with the decimals, or we can eliminate the decimals first by using the multiplication property of equality and solving the resulting equation. Both methods follow.

METHOD 1 Working with the decimals

$$0.08x + 0.09(x + 2{,}000) = 690 \qquad \text{Original equation}$$

Step 1a: $0.08x + 0.09x + 0.09(2{,}000) = 690$ — Distributive property

Step 1c: $0.17x + 180 = 690$ — Simplify the left side

Step 2:
$$0.17x + 180 + (\mathbf{-180}) = 690 + (\mathbf{-180}) \qquad \text{Add } -180 \text{ to each side}$$
$$0.17x = 510$$

Step 3:
$$\frac{0.17x}{\mathbf{0.17}} = \frac{510}{\mathbf{0.17}} \qquad \text{Divide each side by 0.17}$$
$$x = 3{,}000$$

Note that we divided each side of the equation by 0.17 to obtain the solution. This is still an application of the multiplication property of equality because dividing by 0.17 is equivalent to multiplying by $\frac{1}{0.17}$.

METHOD 2 Eliminating the decimals in the beginning

$$0.08x + 0.09(x + 2{,}000) = 690 \qquad \text{Original equation}$$

Step 1a: $0.08x + 0.09x + 180 = 690$ — Distributive property

Step 1b:
$$\mathbf{100}(0.08x + 0.09x + 180) = \mathbf{100}(690) \qquad \text{Multiply both sides by 100}$$
$$8x + 9x + 18{,}000 = 69{,}000$$

Step 1c: $17x + 18{,}000 = 69{,}000$ — Simplify the left side

Step 2: $17x = 51{,}000$ — Add $-18{,}000$ to each side

Step 3:
$$\frac{17x}{\mathbf{17}} = \frac{51{,}000}{\mathbf{17}} \qquad \text{Divide each side by 17}$$
$$x = 3{,}000$$

CHECK: Substituting 3,000 for x in the original equation, we have

Step 4:
$$0.08(3{,}000) + 0.09(3{,}000 + 2{,}000) \stackrel{?}{=} 690$$
$$0.08(3{,}000) + 0.09(5{,}000) \stackrel{?}{=} 690$$
$$240 + 450 \stackrel{?}{=} 690$$
$$690 = 690 \qquad \text{A true statement}$$

4. Solve the equation.
$0.06x + 0.04(x + 7{,}000) = 680$

Answers
3. 0 **4.** 4,000

5. Solve $4 - 2(3y + 1) = -16$.

EXAMPLE 5

Solve $7 - 3(2y + 1) = 16$.

SOLUTION We begin by multiplying -3 times the sum of $2y$ and 1:

Step 1a: $7 - 6y - 3 = 16$ — Distributive property

Step 1c: $-6y + 4 = 16$ — Simplify the left side

Step 2: $\begin{cases} -6y + 4 + (\mathbf{-4}) = 16 + (\mathbf{-4}) \\ -6y = 12 \end{cases}$ — Add -4 to both sides

Step 3: $\begin{cases} -\frac{\mathbf{1}}{\mathbf{6}}(-6y) = -\frac{\mathbf{1}}{\mathbf{6}}(12) \\ y = -2 \end{cases}$ — Multiply both sides by $-\frac{1}{6}$

Step 4: Replacing y with -2 in the original equation yields a true statement.

There are two things to notice about the example that follows: first, the distributive property is used to remove parentheses that are preceded by a negative sign, and, second, the addition property and the multiplication property are not shown in as much detail as in the previous examples.

6. Solve.
$4(3x - 2) - (6x - 5) = 6 - (3x + 1)$

EXAMPLE 6

Solve $3(2x - 5) - (2x - 4) = 6 - (4x + 5)$.

SOLUTION When we apply the distributive property to remove the grouping symbols and separate terms, we have to be careful with the signs. Remember, we can think of $-(2x - 4)$ as $-1(2x - 4)$, so that

$$-(2x - 4) = -1(2x - 4) = -2x + 4$$

It is not uncommon for students to make a mistake with this type of simplification and write the result as $-2x - 4$, which is incorrect. Here is the complete solution to our equation:

$3(2x - 5) - (2x - 4) = 6 - (4x + 5)$	Original equation
$6x - 15 - 2x + 4 = 6 - 4x - 5$	Distributive property
$4x - 11 = -4x + 1$	Simplify each side
$8x - 11 = 1$	Add $4x$ to each side
$8x = 12$	Add 11 to each side
$x = \frac{12}{8}$	Multiply each side by $\frac{1}{8}$
$x = \frac{3}{2}$	Reduce to lowest terms

The solution, $\frac{3}{2}$, checks when replacing x in the original equation.

Getting Ready for Class

After reading through the preceding section, respond in your own words and in complete sentences.

1. What is the first step in solving a linear equation containing parentheses?
2. What is the last step in solving a linear equation?
3. Explain in words how you would solve the equation $2x - 3 = 8$.
4. If an equation contains decimals, what can you do to eliminate the decimals?

Answers

5. 3 **6.** $\frac{8}{9}$

Problem Set 2.4

A Solve each of the following equations using the four steps shown in this section. [Examples 1–2]

1. $2(x + 3) = 12$

2. $3(x - 2) = 6$

3. $6(x - 1) = -18$

4. $4(x + 5) = 16$

5. $2(4a + 1) = -6$

6. $3(2a - 4) = 12$

7. $14 = 2(5x - 3)$

8. $-25 = 5(3x + 4)$

9. $-2(3y + 5) = 14$

10. $-3(2y - 4) = -6$

11. $-5(2a + 4) = 0$

12. $-3(3a - 6) = 0$

13. $1 = \frac{1}{2}(4x + 2)$

14. $1 = \frac{1}{3}(6x + 3)$

15. $3(t - 4) + 5 = -4$

16. $5(t - 1) + 6 = -9$

A Solve each equation. [Examples 2–6]

17. $4(2y + 1) - 7 = 1$

18. $6(3y + 2) - 8 = -2$

19. $\frac{1}{2}(x - 3) = \frac{1}{4}(x + 1)$

20. $\frac{1}{3}(x - 4) = \frac{1}{2}(x - 6)$

21. $-0.7(2x - 7) = 0.3(11 - 4x)$

22. $-0.3(2x - 5) = 0.7(3 - x)$

23. $-2(3y + 1) = 3(1 - 6y) - 9$

24. $-5(4y - 3) = 2(1 - 8y) + 11$

25. $\frac{3}{4}(8x - 4) + 3 = \frac{2}{5}(5x + 10) - 1$

26. $\frac{5}{6}(6x + 12) + 1 = \frac{2}{3}(9x - 3) + 5$

27. $0.06x + 0.08(100 - x) = 6.5$

28. $0.05x + 0.07(100 - x) = 6.2$

29. $6 - 5(2a - 3) = 1$

30. $-8 - 2(3 - a) = 0$

31. $0.2x - 0.5 = 0.5 - 0.2(2x - 13)$

32. $0.4x - 0.1 = 0.7 - 0.3(6 - 2x)$

33. $2(t - 3) + 3(t - 2) = 28$

34. $-3(t - 5) - 2(2t + 1) = -8$

35. $5(x - 2) - (3x + 4) = 3(6x - 8) + 10$

36. $3(x - 1) - (4x - 5) = 2(5x - 1) - 7$

37. $2(5x - 3) - (2x - 4) = 5 - (6x + 1)$

38. $3(4x - 2) - (5x - 8) = 8 - (2x + 3)$

39. $-(3x + 1) - (4x - 7) = 4 - (3x + 2)$

40. $-(6x + 2) - (8x - 3) = 8 - (5x + 1)$

41. $x + (2x - 1) = 2$

42. $x + (5x + 2) = 20$

43. $x - (3x + 5) = -3$

44. $x - (4x - 1) = 7$

45. $15 = 3(x - 1)$

46. $12 = 4(x - 5)$

47. $4x - (-4x + 1) = 5$

48. $-2x - (4x - 8) = -1$

49. $5x - 8(2x - 5) = 7$

50. $3x + 4(8x - 15) = 10$

51. $7(2y - 1) - 6y = -1$

52. $4(4y - 3) + 2y = 3$

53. $0.2x + 0.5(12 - x) = 3.6$

54. $0.3x + 0.6(25 - x) = 12$

55. $0.5x + 0.2(18 - x) = 5.4$

56. $0.1x + 0.5(40 - x) = 32$

57. $x + (x + 3)(-3) = x - 3$

58. $x - 2(x + 2) = x - 2$

59. $5(x + 2) + 3(x - 1) = -9$

60. $4(x + 1) + 3(x - 3) = 2$

61. $3(x - 3) + 2(2x) = 5$

62. $2(x - 2) + 3(5x) = 30$

63. $5(y + 2) = 4(y + 1)$

64. $3(y - 3) = 2(y - 2)$

65. $3x + 2(x - 2) = 6$

66. $5x - (x - 5) = 25$

67. $50(x - 5) = 30(x + 5)$

68. $34(x - 2) = 26(x + 2)$

69. $0.08x + 0.09(x + 2{,}000) = 860$

70. $0.11x + 0.12(x + 4{,}000) = 940$

71. $0.10x + 0.12(x + 500) = 214$

72. $0.08x + 0.06(x + 800) = 104$

73. $5x + 10(x + 8) = 245$

74. $5x + 10(x + 7) = 175$

75. $5x + 10(x + 3) + 25(x + 5) = 435$

76. $5(x + 3) + 10x + 25(x + 7) = 390$

The next two problems are intended to give you practice reading, and paying attention to, the instructions that accompany the problems you are working. Working these problems is a excellent way to get ready for a test or a quiz.

77. Work each problem according to the instructions.

a. Solve: $4x - 5 = 0$

b. Solve: $4x - 5 = 25$

c. Add: $(4x - 5) + (2x + 25)$

d. Solve: $4x - 5 = 2x + 25$

e. Multiply: $4(x - 5)$

f. Solve: $4(x - 5) = 2x + 25$

78. Work each problem according to the instructions.

a. Solve: $3x + 6 = 0$

b. Solve: $3x + 6 = 4$

c. Add: $(3x + 6) + (7x + 4)$

d. Solve: $3x + 6 = 7x + 4$

e. Multiply: $3(x + 6)$

f. Solve: $3(x + 6) = 7x + 4$

Applying the Concepts

79. Wind Energy The graph shows the world's capacity for generating power from wind energy. The year that produced 48,000 megawatts is given by the equation $48 = 7x - 13980$. What year is this?

World Wind Electricity-Generating Capacity

Megawatts: 70,000, 60,000, 50,000, 40,000, 30,000, 20,000, 10,000

1985 1990 1995 2000 2005 2010

Source: GWEC, WorldWatch 2006

80. Camera Phones The chart shows the estimated number of camera phones and non-camera phones sold from 2004 to 2010. The year in which 730 million phones were sold can be given by the equation $730 = 10(5x - 9{,}952)$. What year is this?

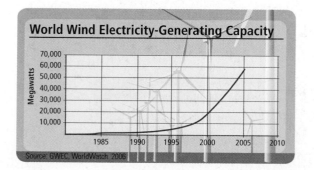

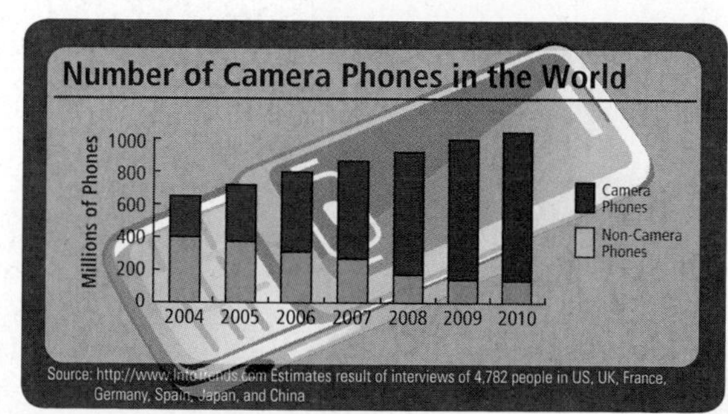

Getting Ready for the Next Section

To understand all of the explanations and examples in the next section you must be able to work the problems below. Solve each equation.

81. $40 = 2x + 12$

82. $80 = 2x + 12$

83. $12 + 2y = 6$

84. $3x + 18 = 6$

85. $24x = 6$

86. $45 = 0.75x$

87. $70 = x \cdot 210$

88. $15 = x \cdot 80$

Apply the distributive property.

89. $\frac{1}{2}(-3x + 6)$

90. $-\frac{1}{4}(-5x + 20)$

Maintaining Your Skills

The problems that follow review some of the more important skills you have learned in previous sections and chapters. You can consider the time you spend working these problems as time spent studying for exams.

Multiply.

91. $\frac{1}{2}(3)$

92. $\frac{1}{3}(2)$

93. $\frac{2}{3}(6)$

94. $\frac{3}{2}(4)$

95. $\frac{5}{9} \cdot \frac{9}{5}$

96. $\frac{3}{7} \cdot \frac{7}{3}$

Fill in the tables by finding the value of each expression for the given values of the variables.

97.

x	$3(x + 2)$	$3x + 2$	$3x + 6$
0			
1			
2			
3			

98.

x	$7(x - 5)$	$7x - 5$	$7x - 35$
−3			
−2			
−1			
0			

99.

a	$(2a + 1)^2$	$4a^2 + 4a + 1$
1		
2		
3		

100.

a	$(a + 1)^3$	$a^3 + 3a^2 + 3a + 1$
1		
2		
3		

2.5 Formulas

Objectives

A Find the value of a variable in a formula given replacements for the other variables.

B Solve a formula for one of its variables.

C Find the complement and supplement of an angle.

D Solve simple percent problems.

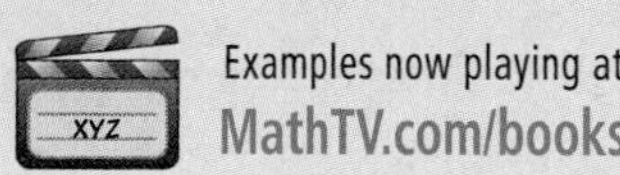

In this section we continue solving equations by working with formulas. To begin, here is the definition of a formula.

> **Definition**
> In mathematics, a **formula** is an equation that contains more than one variable.

The equation $P = 2l + 2w$, which tells us how to find the perimeter of a rectangle, is an example of a formula.

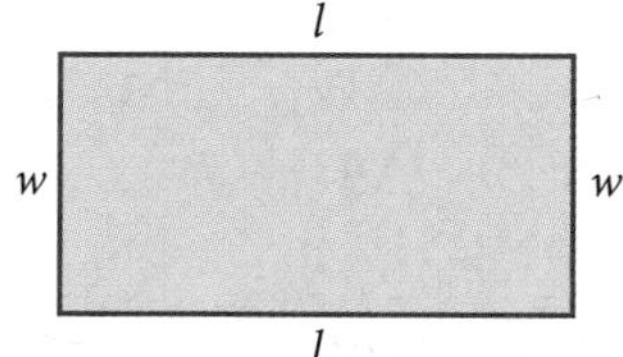

A Solving Formulas

To begin our work with formulas, we will consider some examples in which we are given numerical replacements for all but one of the variables.

EXAMPLE 1 The perimeter P of a rectangular livestock pen is 40 feet. If the width w is 6 feet, find the length.

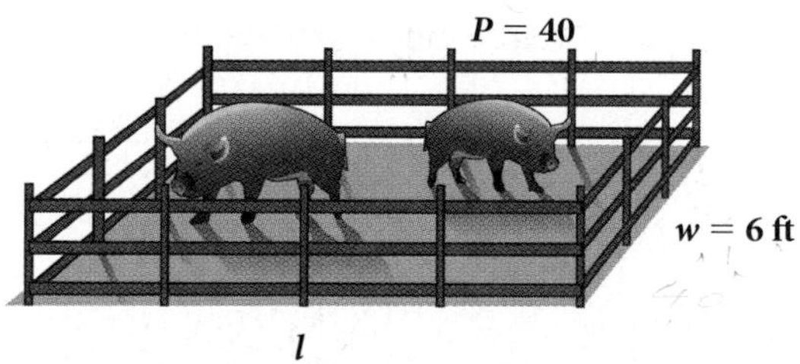

SOLUTION First we substitute 40 for P and 6 for w in the formula $P = 2l + 2w$. Then we solve for l:

When	$P = 40$ and $w = 6$	
the formula	$P = 2l + 2w$	
becomes	$40 = 2l + 2(6)$	
or	$40 = 2l + 12$	Multiply 2 and 6
	$28 = 2l$	Add -12 to each side
	$14 = l$	Multiply each side by $\frac{1}{2}$

To summarize our results, if a rectangular pen has a perimeter of 40 feet and a width of 6 feet, then the length must be 14 feet.

PRACTICE PROBLEMS

1. Suppose the livestock pen in Example 1 has a perimeter of 80 feet. If the width is still 6 feet, what is the new length?

Answer

1. 34 feet

2. Find x when y is 9 in the formula $3x + 2y = 6$.

EXAMPLE 2 Find y when $x = 4$ in the formula $3x + 2y = 6$.

SOLUTION We substitute 4 for x in the formula and then solve for y:

When	$x = 4$	
the formula	$3x + 2y = 6$	
becomes	$3(4) + 2y = 6$	
or	$12 + 2y = 6$	Multiply 3 and 4
	$2y = -6$	Add -12 to each side
	$y = -3$	Multiply each side by $\frac{1}{2}$

B Solving for a Variable

In the next examples we will solve a formula for one of its variables without being given numerical replacements for the other variables.

Consider the formula for the area of a triangle:

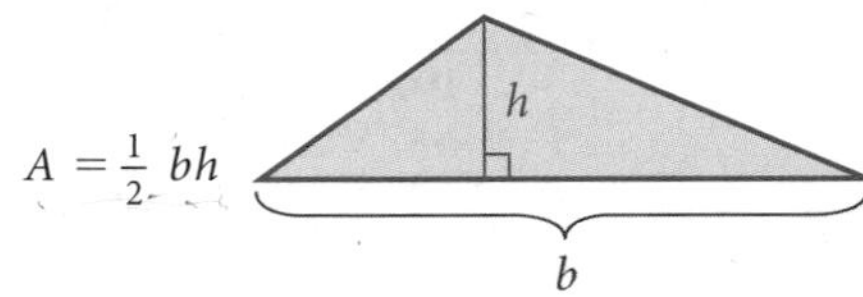

where A = area, b = length of the base, and h = height of the triangle.

Suppose we want to solve this formula for h. What we must do is isolate the variable h on one side of the equal sign. We begin by multiplying both sides by 2:

$$2 \cdot A = 2 \cdot \frac{1}{2} bh$$

$$2A = bh$$

Then we divide both sides by b:

$$\frac{2A}{b} = \frac{bh}{b}$$

$$h = \frac{2A}{b}$$

The original formula $A = \frac{1}{2}bh$ and the final formula $h = \frac{2A}{b}$ both give the same relationship among A, b, and h. The first one has been solved for A and the second one has been solved for h.

> **Rule**
> To solve a formula for one of its variables, we must isolate that variable on either side of the equal sign. All other variables and constants will appear on the other side.

Answer
2. -4

EXAMPLE 3

Solve $3x + 2y = 6$ for y.

SOLUTION To solve for y, we must isolate y on the left side of the equation. To begin, we use the addition property of equality to add $-3x$ to each side:

$$3x + 2y = 6 \quad \text{Original formula}$$
$$3x + (\mathbf{-3x}) + 2y = (\mathbf{-3x}) + 6 \quad \text{Add } -3x \text{ to each side}$$
$$2y = -3x + 6 \quad \text{Simplify the left side}$$
$$\frac{\mathbf{1}}{\mathbf{2}}(2y) = \frac{\mathbf{1}}{\mathbf{2}}(-3x + 6) \quad \text{Multiply each side by } \frac{1}{2}$$
$$y = -\frac{3}{2}x + 3 \quad \text{Multiplication}$$

3. Solve $5x - 4y = 20$ for y.

EXAMPLE 4

Solve $h = vt - 16t^2$ for v.

SOLUTION Let's begin by interchanging the left and right sides of the equation. That way, the variable we are solving for, v, will be on the left side.

$$vt - 16t^2 = h \quad \text{Exchange sides}$$
$$vt - 16t^2 + \mathbf{16t^2} = h + \mathbf{16t^2} \quad \text{Add } 16t^2 \text{ to each side}$$
$$vt = h + 16t^2$$
$$\frac{vt}{\mathbf{t}} = \frac{h + 16t^2}{\mathbf{t}} \quad \text{Divide each side by } t$$
$$v = \frac{h + 16t^2}{t}$$

We know we are finished because we have isolated the variable we are solving for on the left side of the equation and it does not appear on the other side.

4. Solve $P = 2w + 2l$ for l.

EXAMPLE 5

Solve for y: $\frac{y - 1}{x} = \frac{3}{2}$.

SOLUTION Although we will do more extensive work with formulas of this form later in the book, we need to know how to solve this particular formula for y in order to understand some things in the next chapter. We begin by multiplying each side of the formula by x. Doing so will simplify the left side of the equation and make the rest of the solution process simple.

$$\frac{y - 1}{x} = \frac{3}{2} \quad \text{Original formula}$$
$$x \cdot \frac{y - 1}{x} = \frac{3}{2} \cdot x \quad \text{Multiply each side by } x$$
$$y - 1 = \frac{3}{2}x \quad \text{Simplify each side}$$
$$y = \frac{3}{2}x + 1 \quad \text{Add 1 to each side}$$

This is our solution. If we look back to the first step, we can justify our result on the left side of the equation this way: Dividing by x is equivalent to multiplying by its reciprocal $\frac{1}{x}$. Here is what it looks like when written out completely:

$$x \cdot \frac{y - 1}{x} = x \cdot \frac{1}{x} \cdot (y - 1) = 1(y - 1) = y - 1$$

5. Solve for y: $\frac{y - 2}{x} = \frac{4}{3}$.

Answers

3. $y = \frac{5}{4}x - 5$ 4. $l = \frac{P - 2w}{2}$
5. $y = \frac{4}{3}x + 2$

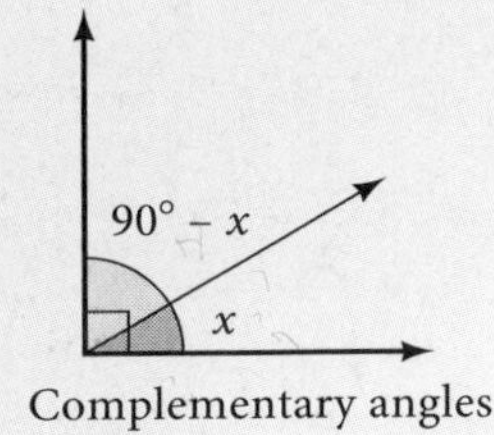

Complementary angles

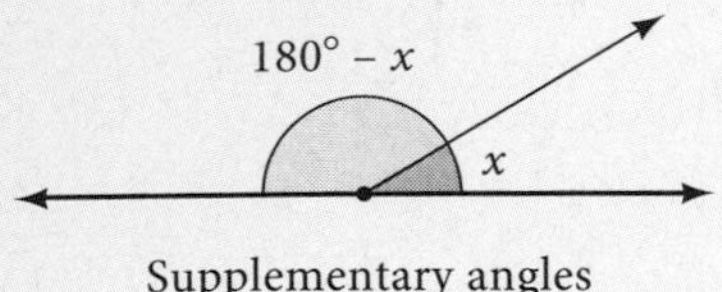

Supplementary angles

C Complementary and Supplementary Angles

FACTS FROM GEOMETRY: More on Complementary and Supplementary Angles
In Chapter 1 we defined complementary angles as angles that add to 90°; that is, if x and y are complementary angles, then

$$x + y = 90°$$

If we solve this formula for y, we obtain a formula equivalent to our original formula:

$$y = 90° - x$$

Because y is the complement of x, we can generalize by saying that the complement of angle x is the angle $90° - x$. By a similar reasoning process, we can say that the supplement of angle x is the angle $180° - x$. To summarize, if x is an angle, then

The complement of x is $90° - x$, and

The supplement of x is $180° - x$

If you go on to take a trigonometry class, you will see this formula again.

EXAMPLE 6 Find the complement and the supplement of 25°.

SOLUTION We can use the formulas above with $x = 25°$.
The complement of 25° is $90° - 25° = 65°$.
The supplement of 25° is $180° - 25° = 155°$.

6. Find the complement and the supplement of 35°.

D Basic Percent Problems

The last examples in this section show how basic percent problems can be translated directly into equations. To understand these examples, you must recall that *percent* means "per hundred." That is, 75% is the same as $\frac{75}{100}$, 0.75, and, in reduced fraction form, $\frac{3}{4}$. Likewise, the decimal 0.25 is equivalent to 25%. To change a decimal to a percent, we move the decimal point two places to the right and write the % symbol. To change from a percent to a decimal, we drop the % symbol and move the decimal point two places to the left. The table that follows gives some of the most commonly used fractions and decimals and their equivalent percents.

Fraction	Decimal	Percent
$\frac{1}{2}$	0.5	50%
$\frac{1}{4}$	0.25	25%
$\frac{3}{4}$	0.75	75%
$\frac{1}{3}$	$0.\overline{3}$	$33\frac{1}{3}\%$
$\frac{2}{3}$	$0.\overline{6}$	$66\frac{2}{3}\%$
$\frac{1}{5}$	0.2	20%
$\frac{2}{5}$	0.4	40%
$\frac{3}{5}$	0.6	60%
$\frac{4}{5}$	0.8	80%

Answer
6. 55°; 145°

EXAMPLE 7 What number is 25% of 60?

SOLUTION To solve a problem like this, we let $x =$ the number in question (that is, the number we are looking for). Then, we translate the sentence directly into an equation by using an equal sign for the word "is" and multiplication for the word "of." Here is how it is done:

$$\underbrace{\text{What number}}_{x}\ \underset{=}{\text{is}}\ \underset{0.25}{25\%}\ \underset{\cdot}{\text{of}}\ \underset{60}{60?}$$

$$x = 15$$

Notice that we must write 25% as a decimal to do the arithmetic in the problem.

The number 15 is 25% of 60.

7. What number is 25% of 74?

EXAMPLE 8 What percent of 24 is 6?

SOLUTION Translating this sentence into an equation, as we did in Example 7, we have:

$$\underbrace{\text{What percent}}_{x}\ \underset{\cdot}{\text{of}}\ \underset{24}{24}\ \underset{=}{\text{is}}\ \underset{6}{6?}$$

or $24x = 6$

Next, we multiply each side by $\frac{1}{24}$. (This is the same as dividing each side by 24.)

$$\frac{\mathbf{1}}{\mathbf{24}}(24x) = \frac{\mathbf{1}}{\mathbf{24}}(6)$$

$$x = \frac{6}{24}$$

$$= \frac{1}{4}$$

$$= 0.25, \text{ or } 25\%$$

The number 6 is 25% of 24.

8. What percent of 84 is 21?

EXAMPLE 9 45 is 75% of what number?

SOLUTION Again, we translate the sentence directly:

$$\underset{45}{45}\ \underset{=}{\text{is}}\ \underset{0.75}{75\%}\ \underset{\cdot}{\text{of}}\ \underbrace{\text{what number?}}_{x}$$

Next, we multiply each side by $\frac{1}{0.75}$ (which is the same as dividing each side by 0.75):

$$\frac{\mathbf{1}}{\mathbf{0.75}}(45) = \frac{\mathbf{1}}{\mathbf{0.75}}(0.75x)$$

$$\frac{45}{0.75} = x$$

$$60 = x$$

The number 45 is 75% of 60.

9. 35 is 40% of what number?

We can solve application problems involving percent by translating each problem into one of the three basic percent problems shown in Examples 7, 8, and 9.

Answers
7. 18.5 8. 25% 9. 87.5

10. In a third granola bar, 28% of the calories are from fat. If one serving of this bar contains 150 calories, how many of those calories are from fat?

EXAMPLE 10 The American Dietetic Association (ADA) recommends eating foods in which the calories from fat are less than 30% of the total calories. The nutrition labels from two kinds of granola bars are shown in Figure 1. For each bar, what percent of the total calories come from fat?

SOLUTION The information needed to solve this problem is located toward the top of each label. Each serving of Bar I contains 210 calories, of which 70 calories come from fat. To find the percent of total calories that come from fat, we must answer this question:

70 is what percent of 210?

Nutrition Facts
Serving Size 2 bars (47g)
Servings Per Container 6

Amount Per Serving	
Calories	210
Calories from Fat	**70**
	% Daily Value*
Total Fat 8g	**12%**
Saturated Fat 1g	**5%**
Cholesterol 0mg	**0%**
Sodium 150mg	**6%**
Total Carbohydrate 32g	**11%**
Dietary Fiber 2g	**10%**
Sugars 12g	
Protein 4g	

* Percent Daily Values are based on a 2,000 calorie diet. Your daily values may be higher or lower depending on your calorie needs.

Nutrition Facts
Serving Size 1 bar (21g)
Servings Per Container 8

Amount Per Serving	
Calories	80
Calories from Fat	**15**
	% Daily Value*
Total Fat 1.5g	**2%**
Saturated Fat 0g	**0%**
Cholesterol 0mg	**0%**
Sodium 60mg	**3%**
Total Carbohydrate 16g	**5%**
Dietary Fiber 1g	**4%**
Sugars 5g	
Protein 2g	

* Percent Daily Values are based on a 2,000 calorie diet. Your daily values may be higher or lower depending on your calorie needs.

FIGURE 1

For Bar II, one serving contains 80 calories, of which 15 calories come from fat. To find the percent of total calories that come from fat, we must answer this question:

15 is what percent of 80?

Translating each equation into symbols, we have

70 is what percent of 210	15 is what percent of 80
$70 = x \cdot 210$	$15 = x \cdot 80$
$x = \frac{70}{210}$	$x = \frac{15}{80}$
$x = 0.33$ to the nearest hundredth	$x = 0.19$ to the nearest hundredth
$x = 33\%$	$x = 19\%$

Comparing the two bars, 33% of the calories in Bar I are fat calories, whereas 19% of the calories in Bar II are fat calories. According to the ADA, Bar II is the healthier choice.

Getting Ready for Class

After reading through the preceding section, respond in your own words and in complete sentences.

1. What is a formula?
2. How do you solve a formula for one of its variables?
3. What are complementary angles?
4. What does percent mean?

Answer
10. 42 calories

Problem Set 2.5

A Use the formula $P = 2l + 2w$ to find the length l of a rectangular lot if **[Examples 1–2]**

1. The width w is 50 feet and the perimeter P is 300 feet

2. The width w is 75 feet and the perimeter P is 300 feet

3. For the equation $2x + 3y = 6$

a. Find y when x is 0.

b. Find x when y is 1.

c. Find y when x is 3.

4. For the equation $2x - 5y = 20$

a. Find y when x is 0.

b. Find x when y is 0.

c. Find x when y is 2.

5. For the equation $y = -\frac{1}{3}x + 2$

a. Find y when x is 0.

b. Find x when y is 3.

c. Find y when x is 3.

6. For the equation $y = -\frac{2}{3}x + 1$

a. Find y when x is 0.

b. Find x when y is -1.

c. Find y when x is -3.

A Use the formula $2x + 3y = 6$ to find y if

7. x is 3

8. x is -2

9. x is 0

10. x is -3

A Use the formula $2x - 5y = 20$ to find x if

11. y is 2

12. y is -4

13. y is 0

14. y is -6

A Use the equation $y = (x + 1)^2 - 3$ to find the value of y when

15. $x = -2$

16. $x = -1$

17. $x = 1$

18. $x = 2$

19. Use the formula $y = \frac{20}{x}$ to find y when

a. $x = 10$

b. $x = 5$

20. Use the formula $y = 2x^2$ to find y when

a. $x = 5$

b. $x = -6$

21. Use the formula $y = Kx$ to find K when

a. $y = 15$ and $x = 3$

b. $y = 72$ and $x = 4$

22. Use the formula $y = Kx^2$ to find K when

a. $y = 32$ and $x = 4$

b. $y = 45$ and $x = 3$

23. If $y = \frac{K}{x}$, find K if

a. x is 5 and y is 4.

b. x is 5 and y is 15.

24. If $I = \frac{K}{d^2}$, find K if

a. $I = 200$ and $d = 10$.

b. $I = 200$ and $d = 5$.

B Solve each of the following for the indicated variable. [Examples 3–4]

25. $A = lw$ for l

26. $d = rt$ for r

27. $V = lwh$ for h

28. $PV = nRT$ for P

29. $P = a + b + c$ for a

30. $P = a + b + c$ for b

31. $x - 3y = -1$ for x

32. $x + 3y = 2$ for x

33. $-3x + y = 6$ for y

34. $2x + y = -17$ for y

35. $2x + 3y = 6$ for y

36. $4x + 5y = 20$ for y

37. $y - 3 = -2(x + 4)$ for y

38. $y + 5 = 2(x + 2)$ for y

39. $y - 3 = -\frac{2}{3}(x + 3)$ for y

40. $y - 1 = -\frac{1}{2}(x + 4)$ for y

B Solve for the indicated variable. [Examples 3–5]

41. $P = 2l + 2w$ for w

42. $P = 2l + 2w$ for l

43. $h = vt + 16t^2$ for v

44. $h = vt - 16t^2$ for v

45. $A = \pi r^2 + 2\pi rh$ for h

46. $A = 2\pi r^2 + 2\pi rh$ for h

47. Solve for y.

a. $y - 3 = -2(x + 4)$

b. $y - 5 = 4(x - 3)$

48. Solve for y.

a. $y + 1 = -\frac{2}{3}(x - 3)$

b. $y - 3 = -\frac{2}{3}(x + 3)$

49. Solve for y.

a. $y - 1 = \frac{3}{4}(x - 1)$

b. $y + 2 = \frac{3}{4}(x - 4)$

50. Solve for y.

a. $y + 3 = \frac{3}{2}(x - 2)$

b. $y + 4 = \frac{4}{3}(x - 3)$

51. Solve for y.

a. $\frac{y - 1}{x} = \frac{3}{5}$

b. $\frac{y - 2}{x} = \frac{1}{2}$

52. Solve for y.

a. $\frac{y + 1}{x} = -\frac{3}{5}$

b. $\frac{y + 2}{x} = -\frac{1}{2}$

B Solve each formula for y.

53. $\frac{x}{7} - \frac{y}{3} = 1$

54. $\frac{x}{4} - \frac{y}{9} = 1$

55. $-\frac{1}{4}x + \frac{1}{8}y = 1$

56. $-\frac{1}{9}x + \frac{1}{3}y = 1$

The next two problems are intended to give you practice reading, and paying attention to, the instructions that accompany the problems you are working. As we have mentioned previously, working these problems is an excellent way to get ready for a test or a quiz.

57. Work each problem according to the instructions.

a. Solve: $4x + 5 = 20$

b. Find the value of $4x + 5$ when x is 3.

c. Solve for y: $4x + 5y = 20$

d. Solve for x: $4x + 5y = 20$

58. Work each problem according to the instructions.

a. Solve: $-2x + 1 = 4$

b. Find the value of $-2x + 1$ when x is 8.

c. Solve for y: $-2x + y = 20$

d. Solve for x: $-2x + y = 20$

C Find the complement and the supplement of each angle. [Example 6]

59. 30°

60. 60°

61. 45°

62. 15°

D Translate each of the following into an equation, and then solve that equation. **[Examples 7–9]**

63. What number is 25% of 40?

64. What number is 75% of 40?

65. What number is 12% of 2,000?

66. What number is 9% of 3,000?

67. What percent of 28 is 7?

68. What percent of 28 is 21?

69. What percent of 40 is 14?

70. What percent of 20 is 14?

71. 32 is 50% of what number?

72. 16 is 50% of what number?

73. 240 is 12% of what number?

74. 360 is 12% of what number?

Applying the Concepts

More About Temperatures As we mentioned in Chapter 1, in the U.S. system, temperature is measured on the Fahrenheit scale. In the metric system, temperature is measured on the Celsius scale. On the Celsius scale, water boils at 100 degrees and freezes at 0 degrees. To denote a temperature of 100 degrees on the Celsius scale, we write

100°C, which is read "100 degrees Celsius"

Table 1 is intended to give you an intuitive idea of the relationship between the two temperature scales. Table 2 gives the formulas, in both symbols and words, that are used to convert between the two scales.

TABLE 1

	Temperature	
Situation	Fahrenheit	Celsius
Water freezes	32°F	0°C
Room temperature	68°F	20°C
Normal body temperature	98.6°F	37°C
Water boils	212°F	100°C
Bake cookies	365°F	185°C

TABLE 2

To Convert from	Formula in Symbols	Formula in Words
Fahrenheit to Celsius	$C = \frac{5}{9}(F - 32)$	Subtract 32, multiply by 5, then divide by 9.
Celsius to Fahrenheit	$F = \frac{9}{5}C + 32$	Multiply by $\frac{9}{5}$, then add 32.

75. Let $F = 212$ in the formula $C = \frac{5}{9}(F - 32)$, and solve for C. Does the value of C agree with the information in Table 1?

76. Let $C = 100$ in the formula $F = \frac{9}{5}C + 32$, and solve for F. Does the value of F agree with the information in Table 1?

77. Let $F = 68$ in the formula $C = \frac{5}{9}(F - 32)$, and solve for C. Does the value of C agree with the information in Table 1?

78. Let $C = 37$ in the formula $F = \frac{9}{5}C + 32$, and solve for F. Does the value of F agree with the information in Table 1?

79. Solve the formula $F = \frac{9}{5}C + 32$ for C.

80. Solve the formula $C = \frac{5}{9}(F - 32)$ for F.

81. Population The map shows the most populated cities in the United States. If the population of California in 2006 was about 36.5 million, what percentage of the population are found in Los Angeles and San Diego? Round to the nearest tenth of a percent.

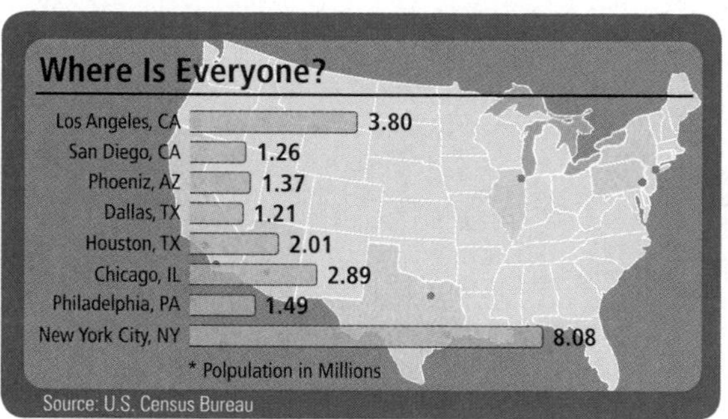

82. Winter Activities About 22.6 million Americans participate in winter activities each year. The pie graph shows which winter activities Americans enjoy. Use the graph to find how many more people enjoy snowboarding than ice-hockey. Round to the nearest tenth.

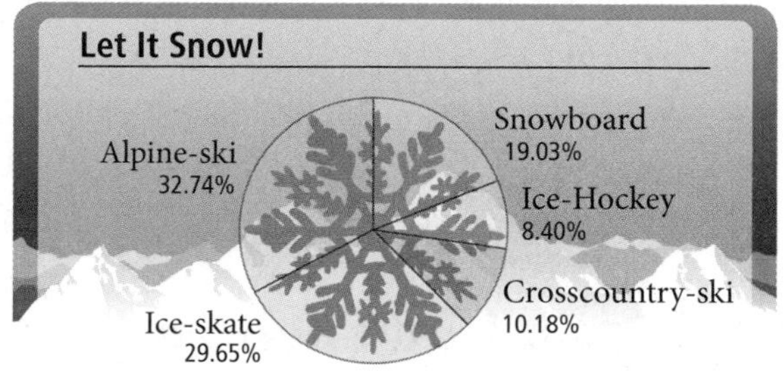

Nutrition Labels The nutrition label in Figure 2 is from a quart of vanilla ice cream. The label in Figure 3 is from a pint of vanilla frozen yogurt. Use the information on these labels for problems 83–86. Round your answers to the nearest tenth of a percent. **[Example 10]**

Nutrition Facts
Serving Size 1/2 cup (65g)
Servings 8

Amount/Serving	
Calories 150	Calories from Fat 90
	% Daily Value*
Total Fat 10g	**16%**
Saturated Fat 6g	**32%**
Cholesterol 35mg	**12%**
Sodium 30mg	**1%**
Total Carbohydrate 14g	**5%**
Dietary Fiber 0g	**0%**
Sugars 11g	
Protein 2g	
Vitamin A 6% •	Vitamin C 0%
Calcium 6% •	Iron 0%

* Percent Daily Values are based on a 2,000 calorie diet.

FIGURE 2 Vanilla ice cream

Nutrition Facts
Serving Size 1/2 cup (98g)
Servings Per Container 4

Amount Per Serving	
Calories 160	Calories from Fat 25
	% Daily Value*
Total Fat 2.5g	**4%**
Saturated Fat 1.5g	**7%**
Cholesterol 45mg	**15%**
Sodium 55mg	**2%**
Total Carbohydrate 26g	**9%**
Dietary Fiber 0g	**0%**
Sugars 19g	
Protein 8g	
Vitamin A 0% •	Vitamin C 0%
Calcium 25% •	Iron 0%

* Percent Daily Values are based on a 2,000 calorie diet.

FIGURE 3 Vanilla frozen yogurt

83. What percent of the calories in one serving of the vanilla ice cream are fat calories?

84. What percent of the calories in one serving of the frozen yogurt are fat calories?

85. One serving of frozen yogurt is 98 grams, of which 26 grams are carbohydrates. What percent of one serving are carbohydrates?

86. One serving of vanilla ice cream is 65 grams. What percent of one serving is sugar?

Circumference The circumference of a circle is given by the formula $C = 2\pi r$. Find r if

87. The circumference C is 44 meters and π is $\frac{22}{7}$

88. The circumference C is 176 meters and π is $\frac{22}{7}$

89. The circumference is 9.42 inches and π is 3.14

90. The circumference is 12.56 inches and π is 3.14

Volume The volume of a cylinder is given by the formula $V = \pi r^2 h$. Find the height h if

91. The volume V is 42 cubic feet, the radius is $\frac{7}{22}$ feet, and π is $\frac{22}{7}$

92. The volume V is 84 cubic inches, the radius is $\frac{7}{11}$ inches, and π is $\frac{22}{7}$

93. The volume is 6.28 cubic centimeters, the radius is 3 centimeters, and π is 3.14

94. The volume is 12.56 cubic centimeters, the radius is 2 centimeters, and π is 3.14

Getting Ready for the Next Section

To understand all of the explanations and examples in the next section you must be able to work the problems below. Write an equivalent expression in English. Include the words *sum* and *difference* when possible.

95. $4 + 1 = 5$

96. $7 + 3 = 10$

97. $6 - 2 = 4$

98. $8 - 1 = 7$

99. $x - 5 = -12$

100. $2x + 3 = 7$

101. $x + 3 = 4(x - 3)$

102. $2(2x - 5) = 2x - 34$

For each of the following expressions, write an equivalent equation.

103. Twice the sum of 6 and 3 is 18.

104. Four added to the product of 3 and -1 is 1.

105. The sum of twice 5 and 3 is 13.

106. Twice the difference of 8 and 2 is 12.

107. The sum of a number and five is thirteen.

108. The difference of ten and a number is negative eight.

109. Five times the sum of a number and seven is thirty.

110. Five times the difference of twice a number and six is negative twenty.

Maintaining Your Skills

The problems that follow review some of the more important skills you have learned in previous sections and chapters. You can consider the time you spend working these problems as time spent studying for exams.

111. **a.** $27 - (-68)$
b. $27 + (-68)$
c. $-27 - 68$
d. $-27 + 68$

112. **a.** $55 - (-29)$
b. $55 + (-29)$
c. $-55 - 29$
d. $-55 + 29$

113. **a.** $-32 - (-41)$
b. $-32 + (-41)$
c. $-32 + 41$
d. $-32 - 41$

114. **a.** $-56 - (-35)$
b. $-56 + (-35)$
c. $-56 + 35$
d. $-56 - 35$

Applications 2.6

The title of the section is "Applications," but many of the problems here don't seem to have much to do with "real life." Example 3 is what we refer to as an "age problem." But imagine a conversation in which you ask someone how old her children are and she replies, "Bill is 6 years older than Tom. Three years ago the sum of their ages was 21. You figure it out." Although many of the "application" problems in this section are contrived, they are also good for practicing the strategy we will use to solve all application problems.

Objectives

A Apply the Blueprint for Problem Solving to a variety of application problems.

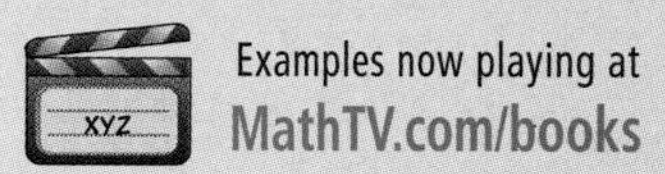

A Problem Solving

To begin this section, we list the steps used in solving application problems. We call this strategy the *Blueprint for Problem Solving.* It is an outline that will overlay the solution process we use on all application problems.

Blueprint for Problem Solving

Step 1: ***Read*** the problem, and then mentally ***list*** the items that are known and the items that are unknown.

Step 2: ***Assign a variable*** to one of the unknown items. (In most cases this will amount to letting x = the item that is asked for in the problem.) Then ***translate*** the other ***information*** in the problem to expressions involving the variable.

Step 3: ***Reread*** the problem, and then ***write an equation,*** using the items and variables listed in steps 1 and 2, that describes the situation.

Step 4: ***Solve the equation*** found in step 3.

Step 5: ***Write*** your ***answer*** using a complete sentence.

Step 6: ***Reread*** the problem, and ***check*** your solution with the original words in the problem.

There are a number of substeps within each of the steps in our blueprint. For instance, with steps 1 and 2 it is always a good idea to draw a diagram or picture if it helps visualize the relationship between the items in the problem. In other cases, a table helps organize the information. As you gain more experience using the blueprint to solve application problems, you will find additional techniques that expand the blueprint.

To help with problems of the type shown next in Example 1, here are some common English words and phrases and their mathematical translations.

English	Algebra
The sum of a and b	$a + b$
The difference of a and b	$a - b$
The product of a and b	$a \cdot b$
The quotient of a and b	$\frac{a}{b}$
of	$\cdot$ (multiply)
is	$=$ (equals)
A number	x
4 more than x	$x + 4$
4 times x	$4x$
4 less than x	$x - 4$

Number Problems

PRACTICE PROBLEMS

1. The sum of twice a number and five is eleven. Find the number.

EXAMPLE 1 The sum of twice a number and three is seven. Find the number.

SOLUTION Using the Blueprint for Problem Solving as an outline, we solve the problem as follows.

Step 1: ***Read*** the problem, and then mentally ***list*** the items that are known and the items that are unknown.

Known items: The numbers 3 and 7

Unknown items: The number in question

Step 2: ***Assign a variable*** to one of the unknown items. Then ***translate*** the other ***information*** in the problem to expressions involving the variable.

Let x = the number asked for in the problem,
then "The sum of twice a number and
three" translates to $2x + 3$.

Step 3: ***Reread*** the problem, and then ***write an equation,*** using the items and variables listed in steps 1 and 2, that describes the situation.

With all word problems, the word *is* translates to =.

$$\underbrace{\text{The sum of twice } x \text{ and } 3}_{2x+3}\ \underset{=}{\underset{\downarrow}{\text{is}}}\ \underset{7}{\underset{\downarrow}{7}}$$

Step 4: ***Solve the equation*** found in step 3.

$$2x + 3 = 7$$
$$2x + 3 + (\mathbf{-3}) = 7 + (\mathbf{-3})$$
$$2x = 4$$
$$\frac{\mathbf{1}}{\mathbf{2}}(2x) = \frac{\mathbf{1}}{\mathbf{2}}(4)$$
$$x = 2$$

Step 5: ***Write*** your ***answer*** using a complete sentence.

The number is 2.

Step 6: ***Reread*** the problem, and ***check*** your solution with the original words in the problem.

The sum of twice 2 and 3 is 7. A true statement.

You may find some examples and problems in this section that you can solve without using algebra or our blueprint. It is very important that you solve these problems using the methods we are showing here. The purpose behind these problems is to give you experience using the blueprint as a guide to solving problems written in words. Your answers are much less important than the work that you show to obtain your answer. You will be able to condense the steps in the blueprint later in the course. For now, though, you need to show your work in the same detail that we are showing in the examples in this section.

Answer
1. 3

EXAMPLE 2 One number is three more than twice another; their sum is eighteen. Find the numbers.

SOLUTION

Step 1: Read and list.

Known items: Two numbers that add to 18. One is 3 more than twice the other.

Unknown items: The numbers in question.

Step 2: Assign a variable, and translate information.
Let x = the first number. The other is $2x + 3$.

Step 3: Reread, and write an equation.

$$\underbrace{\text{Their sum}}\ \text{is } 18$$
$$x + (2x + 3) = 18$$

Step 4: Solve the equation.

$$\begin{aligned} x + (2x + 3) &= 18 \\ 3x + 3 &= 18 \\ 3x + 3 + (\mathbf{-3}) &= 18 + (\mathbf{-3}) \\ 3x &= 15 \\ x &= 5 \end{aligned}$$

Step 5: Write the answer.
The first number is 5. The other is $2 \cdot 5 + 3 = 13$.

Step 6: Reread, and check.
The sum of 5 and 13 is 18, and 13 is 3 more than twice 5.

2. If three times the difference of a number and two were decreased by six, the result would be three. Find the number.

Age Problem

Remember as you read through the steps in the solutions to the examples in this section that step 1 is done mentally. Read the problem, and then mentally list the items that you know and the items that you don't know. The purpose of step 1 is to give you direction as you begin to work application problems. Finding the solution to an application problem is a process; it doesn't happen all at once. The first step is to read the problem with a purpose in mind. That purpose is to mentally note the items that are known and the items that are unknown.

EXAMPLE 3 Bill is 6 years older than Tom. Three years ago Bill's age was four times Tom's age. Find the age of each boy now.

SOLUTION We apply the Blueprint for Problem Solving.

Step 1: Read and list.

Known items: Bill is 6 years older than Tom. Three years ago Bill's age was four times Tom's age.
Unknown items: Bill's age and Tom's age

Step 2: Assign a variable, and translate information.
Let x = Tom's age now. That makes Bill $x + 6$ years old now. A table like the one shown here can help organize the information in an age prob-

3. Pete is five years older than Mary. Five years ago Pete was twice as old as Mary. Find their ages now.

Answer
2. 5

lem. Notice how we placed the x in the box that corresponds to Tom's age now.

	Three Years Ago	Now
Bill		$x + 6$
Tom		x

If Tom is x years old now, 3 years ago he was $x - 3$ years old. If Bill is $x + 6$ years old now, 3 years ago he was $x + 6 - 3 = x + 3$ years old. We use this information to fill in the remaining entries in the table.

	Three Years Ago	Now
Bill	$x + 3$	$x + 6$
Tom	$x - 3$	x

Step 3: Reread, and write an equation.
Reading the problem again, we see that 3 years ago Bill's age was four times Tom's age. Writing this as an equation, we have Bill's age 3 years ago = 4 · (Tom's age 3 years ago):

$$x + 3 = 4(x - 3)$$

Step 4: Solve the equation.

$$x + 3 = 4(x - 3)$$
$$x + 3 = 4x - 12$$
$$x + (\mathbf{-x}) + 3 = 4x + (\mathbf{-x}) - 12$$
$$3 = 3x - 12$$
$$3 + \mathbf{12} = 3x - 12 + \mathbf{12}$$
$$15 = 3x$$
$$x = 5$$

Step 5: Write the answer.
Tom is 5 years old. Bill is 11 years old.

Step 6: Reread, and check.
If Tom is 5 and Bill is 11, then Bill is 6 years older than Tom. Three years ago Tom was 2 and Bill was 8. At that time, Bill's age was four times Tom's age. As you can see, the answers check with the original problem.

Geometry Problem

To understand Example 4 completely, you need to recall from Chapter 1 that the perimeter of a rectangle is the sum of the lengths of the sides. The formula for the perimeter is $P = 2l + 2w$.

EXAMPLE 4 The length of a rectangle is 5 inches more than twice the width. The perimeter is 34 inches. Find the length and width.

SOLUTION When working problems that involve geometric figures, a sketch of the figure helps organize and visualize the problem.

4. The length of a rectangle is three more than twice the width. The perimeter is 36 inches. Find the length and the width.

Answer
3. Mary is 10; Pete is 15

Step 1: *Read and list.*

Known items: The figure is a rectangle. The length is 5 inches more than twice the width. The perimeter is 34 inches.

Unknown items: The length and the width

Step 2: *Assign a variable, and translate information.*
Because the length is given in terms of the width (the length is 5 more than twice the width), we let x = the width of the rectangle. The length is 5 more than twice the width, so it must be $2x + 5$. The diagram below is a visual description of the relationships we have listed so far.

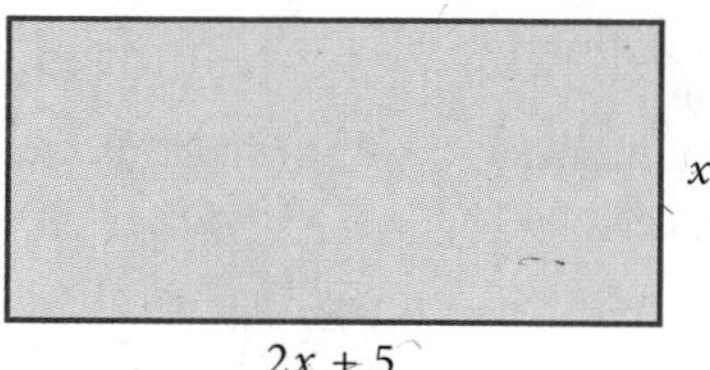

Step 3: *Reread, and write an equation.*
The equation that describes the situation is

Twice the length	+	twice the width	is	the perimeter
$2(2x + 5)$	$+$	$2x$	$=$	34

Step 4: *Solve the equation.*

$2(2x + 5) + 2x = 34$	Original equation
$4x + 10 + 2x = 34$	Distributive property
$6x + 10 = 34$	Add $4x$ and $2x$
$6x = 24$	Add -10 to each side
$x = 4$	Divide each side by 6

Step 5: *Write the answer.*
The width x is 4 inches. The length is $2x + 5 = 2(4) + 5 = 13$ inches.

Step 6: *Reread, and check.*
If the length is 13 and the width is 4, then the perimeter must be $2(13) + 2(4) = 26 + 8 = 34$, which checks with the original problem.

Coin Problem

EXAMPLE 5 Jennifer has \$2.45 in dimes and nickels. If she has 8 more dimes than nickels, how many of each coin does she have?

SOLUTION

Step 1: *Read and list.*

Known items: The type of coins, the total value of the coins, and that there are 8 more dimes than nickels

Unknown items: The number of nickels and the number of dimes

Step 2: *Assign a variable, and translate information.*
If we let x = the number of nickels, then $x + 8$ = the number of dimes. Because the value of each nickel is 5 cents, the amount of money in

5. Amy has \$1.75 in dimes and quarters. If she has 7 more dimes than quarters, how many of each coin does she have?

Answer
4. Width is 5; length is 13

nickels is $5x$. Similarly, because each dime is worth 10 cents, the amount of money in dimes is $10(x + 8)$. Here is a table that summarizes the information we have so far:

	Nickels	Dimes
Number	x	$x + 8$
Value (in cents)	$5x$	$10(x + 8)$

Step 3: ***Reread, and write an equation.***
Because the total value of all the coins is 245 cents, the equation that describes this situation is

Amount of money in nickels	+	Amount of money in dimes	=	Total amount of money
$5x$	+	$10(x + 8)$	=	245

Step 4: ***Solve the equation.***
To solve the equation, we apply the distributive property first.

$$\begin{aligned} 5x + 10x + 80 &= 245 && \text{Distributive property} \\ 15x + 80 &= 245 && \text{Add } 5x \text{ and } 10x \\ 15x &= 165 && \text{Add } -80 \text{ to each side} \\ x &= 11 && \text{Divide each side by 15} \end{aligned}$$

Step 5: ***Write the answer.***
The number of nickels is $x = 11$.
The number of dimes is $x + 8 = 11 + 8 = 19$.

Step 6: ***Reread, and check.***
To check our results

$$\begin{aligned} 11 \text{ nickels are worth } 5(11) &= 55 \text{ cents} \\ 19 \text{ dimes are worth } 10(19) &= 190 \text{ cents} \\ \hline \text{The total value is } 245 \text{ cents} &= \$2.45 \end{aligned}$$

When you begin working the problems in the problem set that follows, there are a couple of things to remember. The first is that you may have to read the problems a number of times before you begin to see how to solve them. The second thing to remember is that word problems are not always solved correctly the first time you try them. Sometimes it takes a couple of attempts and some wrong answers before you can set up and solve these problems correctly.

Getting Ready for Class

After reading through the preceding section, respond in your own words and in complete sentences.

1. What is the first step in the Blueprint for Problem Solving?
2. What is the last thing you do when solving an application problem?
3. What good does it do you to solve application problems even when they don't have much to do with real life?
4. Write an application problem whose solution depends on solving the equation $2x + 3 = 7$.

Answer
5. 3 quarters, 10 dimes

Problem Set 2.6

A Solve the following word problems. Follow the steps given in the Blueprint for Problem Solving.

Number Problems [Examples 1–2]

1. The sum of a number and five is thirteen. Find the number.

2. The difference of ten and a number is negative eight. Find the number.

3. The sum of twice a number and four is fourteen. Find the number.

4. The difference of four times a number and eight is sixteen. Find the number.

5. Five times the sum of a number and seven is thirty. Find the number.

6. Five times the difference of twice a number and six is negative twenty. Find the number.

7. One number is two more than another. Their sum is eight. Find both numbers.

8. One number is three less than another. Their sum is fifteen. Find the numbers.

9. One number is four less than three times another. If their sum is increased by five, the result is twenty-five. Find the numbers.

10. One number is five more than twice another. If their sum is decreased by ten, the result is twenty-two. Find the numbers.

Age Problems [Example 3]

11. Shelly is 3 years older than Michele. Four years ago the sum of their ages was 67. Find the age of each person now.

	four years ago	Now
Shelly	$x - 1$	$x + 3$
Michele	$x - 4$	x

12. Cary is 9 years older than Dan. In 7 years the sum of their ages will be 93. Find the age of each man now.

	now	in seven years
Cary	$x + 9$	
Dan	x	$x + 7$

13. Cody is twice as old as Evan. Three years ago the sum of their ages was 27. Find the age of each boy now.

	three years ago	Now
Cody		
Evan	$x - 3$	x

14. Justin is 2 years older than Ethan. In 9 years the sum of their ages will be 30. Find the age of each boy now.

	now	in nine years
Justin		
Ethan	x	

15. Fred is 4 years older than Barney. Five years ago the sum of their ages was 48. How old are they now? (Begin by filling in the table.)

	five years ago	Now
Fred		
Barney		x

16. Tim is 5 years older than JoAnn. Six years from now the sum of their ages will be 79. How old are they now?

	now	six years from now
Tim		
JoAnn	x	

17. Jack is twice as old as Lacy. In 3 years the sum of their ages will be 54. How old are they now?

18. John is 4 times as old as Martha. Five years ago the sum of their ages was 50. How old are they now?

19. Pat is 20 years older than his son Patrick. In 2 years Pat will be twice as old as Patrick. How old are they now?

20. Diane is 23 years older than her daughter Amy. In 6 years Diane will be twice as old as Amy. How old are they now?

Geometry Problems [Example 4]

21. The perimeter of a square is 36 inches. Find the length of one side.

22. The perimeter of a square is 44 centimeters. Find the length of one side.

23. The perimeter of a square is 60 feet. Find the length of one side.

24. The perimeter of a square is 84 meters. Find the length of one side.

25. One side of a triangle is three times the shortest side. The third side is 7 feet more than the shortest side. The perimeter is 62 feet. Find all three sides.

26. One side of a triangle is half the longest side. The third side is 10 meters less than the longest side. The perimeter is 45 meters. Find all three sides.

27. One side of a triangle is half the longest side. The third side is 12 feet less than the longest side. The perimeter is 53 feet. Find all three sides.

28. One side of a triangle is 6 meters more than twice the shortest side. The third side is 9 meters more than the shortest side. The perimeter is 75 meters. Find all three sides.

29. The length of a rectangle is 5 inches more than the width. The perimeter is 34 inches. Find the length and width.

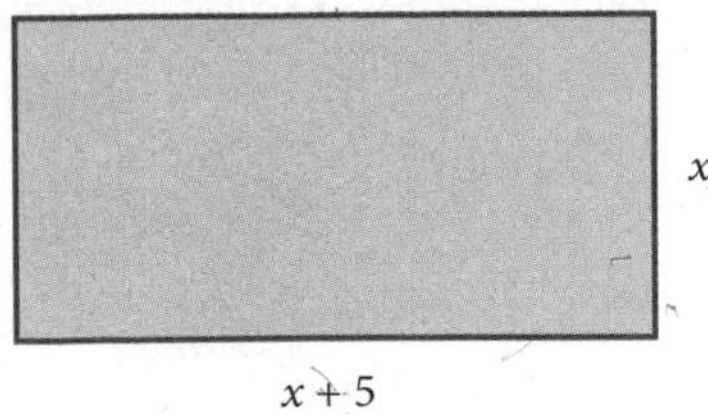

30. The width of a rectangle is 3 feet less than the length. The perimeter is 10 feet. Find the length and width.

31. The length of a rectangle is 7 inches more than twice the width. The perimeter is 68 inches. Find the length and width.

32. The length of a rectangle is 4 inches more than three times the width. The perimeter is 72 inches. Find the length and width.

33. The length of a rectangle is 6 feet more than three times the width. The perimeter is 36 feet. Find the length and width.

34. The length of a rectangle is 3 feet less than twice the width. The perimeter is 54 feet. Find the length and width.

Coin Problems [Example 5]

35. Marissa has $4.40 in quarters and dimes. If she has 5 more quarters than dimes, how many of each coin does she have?

	dimes	quarters
Number	x	$x + 5$
Value (cents)	$10(x)$	$25(x + 5)$

36. Kendra has $2.75 in dimes and nickels. If she has twice as many dimes as nickels, how many of each coin does she have?

	nickels	dimes
Number	x	$2x$
Value (cents)	$5(x)$	

37. Tanner has $4.35 in nickels and quarters. If he has 15 more nickels than quarters, how many of each coin does he have?

	nickels	quarters
Number	$x + 15$	x
Value (cents)		

38. Connor has $9.00 in dimes and quarters. If he has twice as many quarters as dimes, how many of each coin does he have?

	dimes	quarters
Number	x	$2x$
Value (cents)		

39. Sue has $2.10 in dimes and nickels. If she has 9 more dimes than nickels, how many of each coin does she have?

40. Mike has $1.55 in dimes and nickels. If he has 7 more nickels than dimes, how many of each coin does he have?

41. Katie has a collection of nickels, dimes, and quarters with a total value of $4.35. There are 3 more dimes than nickels and 5 more quarters than nickels. How many of each coin is in her collection?

	nickels	dimes	quarters
Number	x		
Value			

42. Mary Jo has $3.90 worth of nickels, dimes, and quarters. The number of nickels is 3 more than the number of dimes. The number of quarters is 7 more than the number of dimes. How many of each coin is in her collection?

	nickels	dimes	quarters
Number			
Value			

43. Cory has a collection of nickels, dimes, and quarters with a total value of $2.55. There are 6 more dimes than nickels and twice as many quarters as nickels. How many of each coin is in her collection?

	nickels	dimes	quarters
Number	x		
Value			

44. Kelly has a collection of nickels, dimes, and quarters with a total value of $7.40. There are four more nickels than dimes and twice as many quarters as nickels. How many of each coin is in her collection?

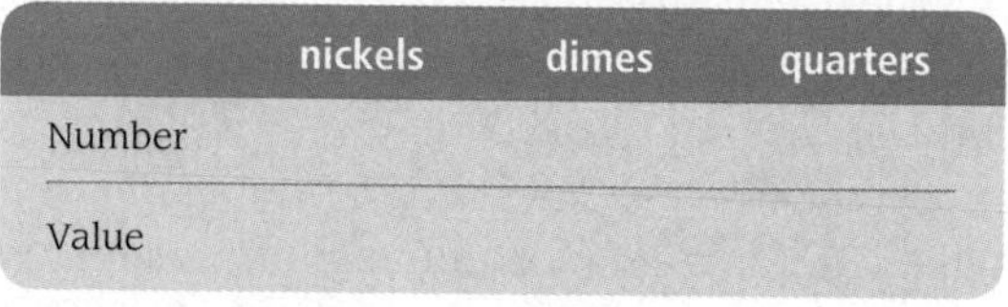

	nickels	dimes	quarters
Number			
Value			

Getting Ready for the Next Section

To understand all of the explanations and examples in the next section you must be able to work the problems below. Simplify the following expressions.

45. $x + 2x + 2x$

46. $x + 2x + 3x$

47. $x + 0.075x$

48. $x + 0.065x$

49. $0.09(x + 2{,}000)$

50. $0.06(x + 1{,}500)$

Solve each of the following equations.

51. $0.05x + 0.06(x - 1{,}500) = 570$

52. $0.08x + 0.09(x + 2{,}000) = 690$

53. $x + 2x + 3x = 180$

54. $2x + 3x + 5x = 180$

Maintaining Your Skills

Write an equivalent statement in English.

55. $4 < 10$

56. $4 \le 10$

57. $9 \ge -5$

58. $x - 2 > 4$

Place the symbol $<$ or the symbol $>$ between the quantities in each expression.

59. 12 20

60. −12 20

61. −8 −6

62. −10 −20

Simplify.

63. $|8 - 3| - |5 - 2|$

64. $|9 - 2| - |10 - 8|$

65. $15 - |9 - 3(7 - 5)|$

66. $10 - |7 - 2(5 - 3)|$

Extending the Concepts

67. The Alphabet and Area Have you ever wondered which of the letters in the alphabet is the most popular? No? The box shown here is called a *typecase*, or *printer's tray.* It was used in early typesetting to store the letters that would be used to lay out a page of type. Use the box to answer the questions below.

a. What letter is used most often in printed material?

b. Which letter is printed more often, the letter *i* or the letter *f*?

c. Regardless of what is shown in the figure, what letter is used most often in problems?

d. What is the relationship between area and how often a letter in the alphabet is printed?

More Applications 2.7

Objectives

A Apply the Blueprint for Problem Solving to a variety of application problems.

Now that you have worked through a number of application problems using our blueprint, you probably have noticed that step 3, in which we write an equation that describes the situation, is the key step. Anyone with experience solving application problems will tell you that there will be times when your first attempt at step 3 results in the wrong equation. Remember, mistakes are part of the process of learning to do things correctly. Many times the correct equation will become obvious after you have written an equation that is partially wrong. In any case it is better to write an equation that is partially wrong and be actively involved with the problem than to write nothing at all. Application problems, like other problems in algebra, are not always solved correctly the first time.

A Consecutive Integers

Our next example involves consecutive integers. When we ask for consecutive integers, we mean integers that are next to each other on the number line, like 5 and 6, or 13 and 14, or -4 and -3. In the dictionary, consecutive is defined as following one another in uninterrupted order. If we ask for consecutive *odd* integers, then we mean odd integers that follow one another on the number line. For example, 3 and 5, 11 and 13, and -9 and -7 are consecutive odd integers. As you can see, to get from one odd integer to the next consecutive odd integer we add 2.

If we are asked to find two consecutive integers and we let x equal the first integer, the next one must be $x + 1$, because consecutive integers always differ by 1. Likewise, if we are asked to find two consecutive odd or even integers, and we let x equal the first integer, then the next one will be $x + 2$, because consecutive even or odd integers always differ by 2. Here is a table that summarizes this information.

In Words	Using Algebra	Example
Two consecutive integers	$x, x + 1$	The sum of two consecutive integers is 15. $x + (x + 1) = 15$ or $7 + 8 = 15$
Three consecutive integers	$x, x + 1, x + 2$	The sum of three consecutive integers is 24. $x + (x + 1) + (x + 2) = 24$ or $7 + 8 + 9 = 24$
Two consecutive odd integers	$x, x + 2$	The sum of two consecutive odd integers is 16. $x + (x + 2) = 16$ or $7 + 9 = 16$
Two consecutive even integers	$x, x + 2$	The sum of two consecutive even integers is 18. $x + (x + 2) = 18$ or $8 + 10 = 18$

EXAMPLE 1 The sum of two consecutive odd integers is 28. Find the two integers.

SOLUTION

Step 1:* *Read and list.

Known items: Two consecutive odd integers. Their sum is equal to 28.

Unknown items: The numbers in question

PRACTICE PROBLEMS

1. The sum of two consecutive integers is 27. What are the two integers?

Step 2: ***Assign a variable, and translate information.***
If we let x = the first of the two consecutive odd integers, then $x + 2$ is the next consecutive one.

Step 3: ***Reread, and write an equation.***
Their sum is 28.

$$x + (x + 2) = 28$$

Step 4: ***Solve the equation.***

$2x + 2 = 28$	Simplify the left side
$2x = 26$	Add -2 to each side
$x = 13$	Multiply each side by $\frac{1}{2}$

Step 5: ***Write the answer.***
The first of the two integers is 13. The second of the two integers will be two more than the first, which is 15.

Step 6: ***Reread, and check.***
Suppose the first integer is 13. The next consecutive odd integer is 15. The sum of 15 and 13 is 28.

Interest

EXAMPLE 2 Suppose you invest a certain amount of money in an account that earns 8% in annual interest. At the same time, you invest $2,000 more than that in an account that pays 9% in annual interest. If the total interest from both accounts at the end of the year is $690, how much is invested in each account?

SOLUTION

Step 1: ***Read and list.***
Known items: The interest rates, the total interest earned, and how much more is invested at 9%

Unknown items: The amounts invested in each account

Step 2: ***Assign a variable, and translate information.***
Let x = the amount of money invested at 8%. From this, $x + 2{,}000$ = the amount of money invested at 9%. The interest earned on x dollars invested at 8% is $0.08x$. The interest earned on $x + 2{,}000$ dollars invested at 9% is $0.09(x + 2{,}000)$.

Here is a table that summarizes this information:

	Dollars Invested at 8%	Dollars Invested at 9%
Number of	x	$x + 2{,}000$
Interest on	$0.08x$	$0.09(x + 2{,}000)$

Step 3: ***Reread, and write an equation.***
Because the total amount of interest earned from both accounts is $690, the equation that describes the situation is

Interest earned at 8%	+	Interest earned at 9%	=	Total interest earned
$0.08x$	+	$0.09(x + 2{,}000)$	=	690

2. On July 1, 1992, Alfredo opened a savings account that earned 4% in annual interest. At the same time, he put $3,000 more than he had in his savings into a mutual fund that paid 5% annual interest. One year later, the total interest from both accounts was $510. How much money did Alfredo put in his savings account?

Answer
1. 13 and 14

Step 4: Solve the equation.

$$0.08x + 0.09(x + 2{,}000) = 690$$

$0.08x + 0.09x + 180 = 690$	Distributive property
$0.17x + 180 = 690$	Add $0.08x$ and $0.09x$
$0.17x = 510$	Add -180 to each side
$x = 3{,}000$	Divide each side by 0.17

Step 5: Write the answer:

The amount of money invested at 8% is \$3,000, whereas the amount of money invested at 9% is $x + 2{,}000 = 3{,}000 + 2{,}000 = \$5{,}000$.

Step 6: Reread, and check.

The interest at 8% is 8% of 3,000 = 0.08(3,000) = \$240
The interest at 9% is 9% of 5,000 = 0.09(5,000) = \$450
The total interest is \$690

FACTS FROM GEOMETRY: Labeling Triangles and the Sum of the Angles in a Triangle

One way to label the important parts of a triangle is to label the vertices with capital letters and the sides with small letters, as shown in Figure 1.

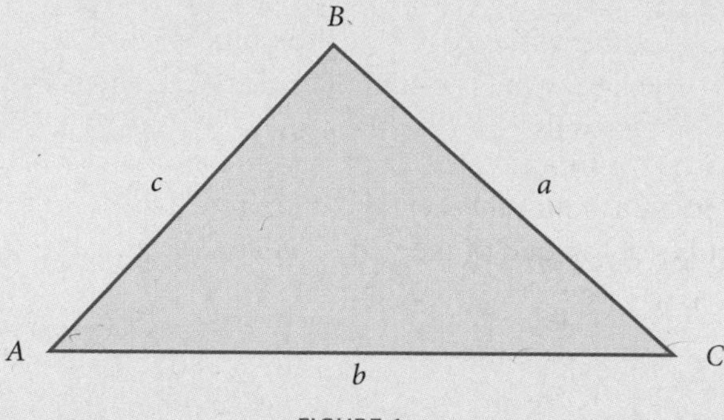

FIGURE 1

In Figure 1, notice that side a is opposite vertex A, side b is opposite vertex B, and side c is opposite vertex C. Also, because each vertex is the vertex of one of the angles of the triangle, we refer to the three interior angles as A, B, and C.

In any triangle, the sum of the interior angles is 180°. For the triangle shown in Figure 1, the relationship is written

$$A + B + C = 180°$$

Triangles

EXAMPLE 3 The angles in a triangle are such that one angle is twice the smallest angle, whereas the third angle is three times as large as the smallest angle. Find the measure of all three angles.

SOLUTION

Step 1: Read and list.

Known items: The sum of all three angles is 180°; one angle is twice the smallest angle; the largest angle is three times the smallest angle.

3. The angles in a triangle are such that one angle is three times the smallest angle, whereas the largest angle is five times the smallest angle. Find the measure of all three angles.

Answer

2. \$4,000

Unknown items: The measure of each angle

Step 2: ***Assign a variable, and translate information.***
Let x be the smallest angle, then $2x$ will be the measure of another angle and $3x$ will be the measure of the largest angle.

Step 3: ***Reread, and write an equation.***
When working with geometric objects, drawing a generic diagram sometimes will help us visualize what it is that we are asked to find. In Figure 2, we draw a triangle with angles A, B, and C.

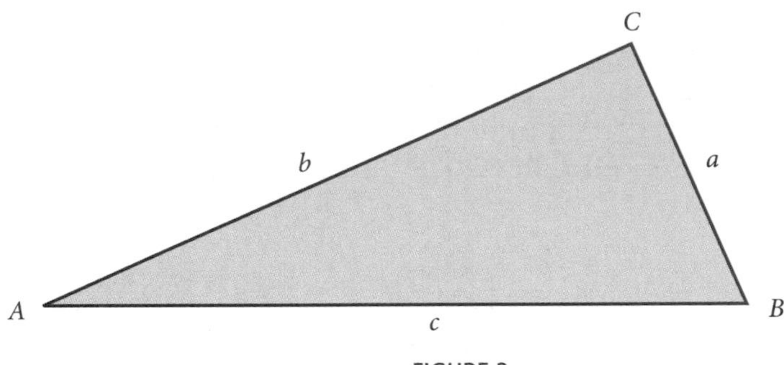

FIGURE 2

We can let the value of $A = x$, the value of $B = 2x$, and the value of $C = 3x$. We know that the sum of angles A, B, and C will be 180°, so our equation becomes

$$x + 2x + 3x = 180°$$

Step 4: ***Solve the equation.***

$$\begin{aligned} x + 2x + 3x &= 180° \\ 6x &= 180° \\ x &= 30° \end{aligned}$$

Step 5: ***Write the answer.***
The smallest angle A measures 30°
Angle B measures $2x$, or $2(30°) = 60°$
Angle C measures $3x$, or $3(30°) = 90°$

Step 6: ***Reread, and check.***
The angles must add to 180°:

$$\begin{aligned} A + B + C &= 180° \\ 30° + 60° + 90° &\stackrel{?}{=} 180° \\ 180° &= 180° \quad \text{Our answers check} \end{aligned}$$

Getting Ready for Class

After reading through the preceding section, respond in your own words and in complete sentences.

1. How do we label triangles?
2. What rule is always true about the three angles in a triangle?
3. Write an application problem whose solution depends on solving the equation $x + 0.075x = 500$.
4. Write an application problem whose solution depends on solving the equation $0.05x + 0.06(x + 200) = 67$.

Answer
3. 20°, 60°, and 100°

Problem Set 2.7

A Consecutive Integer Problems [Example 1]

1. The sum of two consecutive integers is 11. Find the numbers.

2. The sum of two consecutive integers is 15. Find the numbers.

3. The sum of two consecutive integers is -9. Find the numbers.

4. The sum of two consecutive integers is -21. Find the numbers.

5. The sum of two consecutive odd integers is 28. Find the numbers.

6. The sum of two consecutive odd integers is 44. Find the numbers.

7. The sum of two consecutive even integers is 106. Find the numbers.

8. The sum of two consecutive even integers is 66. Find the numbers.

9. The sum of two consecutive even integers is -30. Find the numbers.

10. The sum of two consecutive odd integers is -76. Find the numbers.

11. The sum of three consecutive odd integers is 57. Find the numbers.

12. The sum of three consecutive odd integers is -51. Find the numbers.

13. The sum of three consecutive even integers is 132. Find the numbers.

14. The sum of three consecutive even integers is -108. Find the numbers.

A Interest Problems [Example 2]

15. Suppose you invest money in two accounts. One of the accounts pays 8% annual interest, whereas the other pays 9% annual interest. If you have $2,000 more invested at 9% than you have invested at 8%, how much do you have invested in each account if the total amount of interest you earn in a year is $860? (Begin by completing the following table.)

	Dollars Invested at 8%	Dollars Invested at 9%
Number of	x	
Interest on		

16. Suppose you invest a certain amount of money in an account that pays 11% interest annually, and $4,000 more than that in an account that pays 12% annually. How much money do you have in each account if the total interest for a year is $940?

	Dollars Invested at 11%	Dollars Invested at 12%
Number of	x	
Interest on		

17. Tyler has two savings accounts that his grandparents opened for him. The two accounts pay 10% and 12% in annual interest; there is $500 more in the account that pays 12% than there is in the other account. If the total interest for a year is $214, how much money does he have in each account?

18. Travis has a savings account that his parents opened for him. It pays 6% annual interest. His uncle also opened an account for him, but it pays 8% annual interest. If there is $800 more in the account that pays 6%, and the total, interest from both accounts is $104, how much money is in each of the accounts?

19. A stockbroker has money in three accounts. The interest rates on the three accounts are 8%, 9%, and 10%. If she has twice as much money invested at 9% as she has invested at 8%, three times as much at 10% as she has at 8%, and the total interest for the year is $280, how much is invested at each rate? (*Hint:* Let $x =$ the amount invested at 8%.)

20. An accountant has money in three accounts that pay 9%, 10%, and 11% in annual interest. He has twice as much invested at 9% as he does at 10% and three times as much invested at 11% as he does at 10%. If the total interest from the three accounts is $610 for the year, how much is invested at each rate? (*Hint:* Let $x =$ the amount invested at 10%.)

A Triangle Problems [Example 3]

21. Two angles in a triangle are equal and their sum is equal to the third angle in the triangle. What are the measures of each of the three interior angles?

22. One angle in a triangle measures twice the smallest angle, whereas the largest angle is six times the smallest angle. Find the measures of all three angles.

23. The smallest angle in a triangle is $\frac{1}{5}$ as large as the largest angle. The third angle is twice the smallest angle. Find the three angles.

24. One angle in a triangle is half the largest angle but three times the smallest. Find all three angles.

25. A right triangle has one 37° angle. Find the other two angles.

26. In a right triangle, one of the acute angles is twice as large as the other acute angle. Find the measure of the two acute angles.

27. One angle of a triangle measures 20° more than the smallest, while a third angle is twice the smallest. Find the measure of each angle.

28. One angle of a triangle measures 50° more than the smallest, while a third angle is three times the smallest. Find the measure of each angle.

Miscellaneous Problems

29. **Ticket Prices** Miguel is selling tickets to a barbecue. Adult tickets cost $6.00 and children's tickets cost $4.00. He sells six more children's tickets than adult tickets. The total amount of money he collects is $184. How many adult tickets and how many children's tickets did he sell?

	Adult	Child
Number	x	
Income		

30. **Working Two Jobs** Maggie has a job working in an office for $10 an hour and another job driving a tractor for $12 an hour. One week she works in the office twice as long as she drives the tractor. Her total income for that week is $416. How many hours did she spend at each job?

Job	Office	Tractor
Hours Worked		x
Wages Earned		

31. Phone Bill The cost of a long-distance phone call is \$0.41 for the first minute and \$0.32 for each additional minute. If the total charge for a long-distance call is \$5.21, how many minutes was the call?

32. Phone Bill Danny, who is 1 year old, is playing with the telephone when he accidentally presses one of the buttons his mother has programmed to dial her friend Sue's number. Sue answers the phone and realizes Danny is on the other end. She talks to Danny, trying to get him to hang up. The cost for a call is \$0.23 for the first minute and \$0.14 for every minute after that. If the total charge for the call is \$3.73, how long did it take Sue to convince Danny to hang up the phone?

33. Hourly Wages JoAnn works in the publicity office at the state university. She is paid \$12 an hour for the first 35 hours she works each week and \$18 an hour for every hour after that. If she makes \$492 one week, how many hours did she work?

34. Hourly Wages Diane has a part-time job that pays her \$6.50 an hour. During one week she works 26 hours and is paid \$178.10. She realizes when she sees her check that she has been given a raise. How much per hour is that raise?

35. Office Numbers Professors Wong and Gil have offices in the mathematics building at Miami Dade College. Their office numbers are consecutive odd integers with a sum of 14,660. What are the office numbers of these two professors?

36. Cell Phone Numbers Diana and Tom buy two cell phones. The phone numbers assigned to each are consecutive integers with a sum of 11,109,295. If the smaller number is Diana's, what are their phone numbers?

37. Age Marissa and Kendra are 2 years apart in age. Their ages are two consecutive even integers. Kendra is the younger of the two. If Marissa's age is added to twice Kendra's age, the result is 26. How old is each girl?

38. Age Justin's and Ethan's ages form two consecutive odd integers. What is the difference of their ages?

39. Arrival Time Jeff and Carla Cole are driving separately from San Luis Obispo, California, to the north shore of Lake Tahoe, a distance of 425 miles. Jeff leaves San Luis Obispo at 11:00 am and averages 55 miles per hour on the drive, Carla leaves later, at 1:00 pm but averages 65 miles per hour. Which person arrives in Lake Tahoe first?

40. Piano Lessons Tyler is taking piano lessons. Because he doesn't practice as often as his parents would like him to, he has to pay for part of the lessons himself. His parents pay him \$0.50 to do the laundry and \$1.25 to mow the lawn. In one month, he does the laundry 6 more times than he mows the lawn. If his parents pay him \$13.50 that month, how many times did he mow the lawn?

At one time, the Texas Junior College Teachers Association annual conference was held in Austin. At that time a taxi ride in Austin was \$1.25 for the first $\frac{1}{5}$ of a mile and \$0.25 for each additional $\frac{1}{5}$ of a mile. Use this information for Problems 41 and 42.

41. Cost of a Taxi Ride If the distance from one of the convention hotels to the airport is 7.5 miles, how much will it cost to take at taxi from that hotel to the airport?

42. Cost of a Taxi Ride Suppose the distance from one of the hotels to one of the western dance clubs in Austin is 12.4 miles. If the fare meter in the taxi gives the charge for that trip as \$16.50, is the meter working correctly?

43. Geometry The length and width of a rectangle are consecutive even integers. The perimeter is 44 meters. Find the length and width.

44. Geometry The length and width of a rectangle are consecutive odd integers. The perimeter is 128 meters. Find the length and width.

45. Geometry The angles of a triangle are three consecutive integers. Find the measure of each angle.

46. Geometry The angles of a triangle are three consecutive even integers. Find the measure of each angle.

Ike and Nancy Lara give western dance lessons at the Elks Lodge on Sunday nights. The lessons cost \$3.00 for members of the lodge and \$5.00 for nonmembers. Half of the money collected for the lesson is paid to Ike and Nancy. The Elks Lodge keeps the other half. One Sunday night Ike counts 36 people in the dance lesson. Use this information to work Problems 47 through 50.

47. Dance Lessons What is the least amount of money Ike and Nancy will make?

48. Dance Lessons What is the largest amount of money Ike and Nancy will make?

49. Dance Lessons At the end of the evening, the Elks Lodge gives Ike and Nancy a check for \$80 to cover half of the receipts. Can this amount be correct?

50. Dance Lessons Besides the number of people in the dance lesson, what additional information does Ike need to know to always be sure he is being paid the correct amount?

Getting Ready for the Next Section

To understand all of the explanations and examples in the next section you must be able to work the problems below. Solve the following equations.

51. **a.** $x - 3 = 6$
b. $x + 3 = 6$
c. $-x - 3 = 6$
d. $-x + 3 = 6$

52. **a.** $x - 7 = 16$
b. $x + 7 = 16$
c. $-x - 7 = 16$
d. $-x + 7 = 16$

53. **a.** $\frac{x}{4} = -2$
b. $-\frac{x}{4} = -2$
c. $\frac{x}{4} = 2$
d. $-\frac{x}{4} = 2$

54. **a.** $3a = 15$
b. $3a = -15$
c. $-3a = 15$
d. $-3a = -15$

55. $2.5x - 3.48 = 4.9x + 2.07$

56. $2(1 - 3x) + 4 = 4x - 14$

57. $3(x - 4) = -2$

58. Solve $2x - 3y = 6$ for y.

Maintaining Your Skills

The problems that follow review some of the more important skills you have learned in previous sections and chapters. You can consider the time you spend working these problems as time spent studying for exams.

Simplify the expression $36x - 12$ for each of the following values of x.

59. $\frac{1}{4}$

60. $\frac{1}{6}$

61. $\frac{1}{9}$

62. $\frac{3}{2}$

63. $\frac{1}{3}$

64. $\frac{5}{12}$

65. $\frac{5}{9}$

66. $\frac{2}{3}$

Find the value of each expression when $x = -4$.

67. $3(x - 4)$

68. $-3(x - 4)$

69. $-5x + 8$

70. $5x + 8$

71. $\frac{x - 14}{36}$

72. $\frac{x - 12}{36}$

73. $\frac{16}{x} + 3x$

74. $\frac{16}{x} - 3x$

75. $7x - \frac{12}{x}$

76. $7x + \frac{12}{x}$

77. $8\left(\frac{x}{2} + 5\right)$

78. $-8\left(\frac{x}{2} + 5\right)$

Linear Inequalities

2.8

Linear inequalities are solved by a method similar to the one used in solving linear equations. The only real differences between the methods are in the multiplication property for inequalities and in graphing the solution set.

A Addition Property for Inequalities

An inequality differs from an equation only with respect to the comparison symbol between the two quantities being compared. In place of the equal sign, we use $<$ (less than), $\leq$ (less than or equal to), $>$ (greater than), or $\geq$ (greater than or equal to). The addition property for inequalities is almost identical to the addition property for equality.

Addition Property for Inequalities

For any three algebraic expressions A, B, and C,

if $\quad A < B$

then $\quad A + C < B + C$

In words: Adding the same quantity to both sides of an inequality will not change the solution set.

It makes no difference which inequality symbol we use to state the property. Adding the same amount to both sides always produces an inequality equivalent to the original inequality. Also, because subtraction can be thought of as addition of the opposite, this property holds for subtraction as well as addition.

EXAMPLE 1 Solve the inequality $x + 5 < 7$.

SOLUTION To isolate x, we add -5 to both sides of the inequality:

$$x + 5 < 7$$

$$x + 5 + (\mathbf{-5}) < 7 + (\mathbf{-5}) \qquad \text{Addition property for inequalities}$$

$$x < 2$$

We can go one step further here and graph the solution set. The solution set is all real numbers less than 2. To graph this set, we simply draw a straight line and label the center 0 (zero) for reference. Then we label the 2 on the right side of zero and extend an arrow beginning at 2 and pointing to the left. We use an open circle at 2 because it is not included in the solution set. Here is the graph.

EXAMPLE 2 Solve $x - 6 \leq -3$.

SOLUTION Adding 6 to each side will isolate x on the left side:

$$x - 6 \leq -3$$

$$x - 6 + \mathbf{6} \leq -3 + \mathbf{6} \qquad \text{Add 6 to both sides}$$

$$x \leq 3$$

The graph of the solution set is

Notice that the dot at the 3 is darkened because 3 is included in the solution set. We always will use open circles on the graphs of solution sets with $<$ or $>$ and closed (darkened) circles on the graphs of solution sets with $\leq$ or $\geq$.

Objectives

A Use the addition property for inequalities to solve an inequality.

B Use the multiplication property for inequalities to solve an inequality.

C Use both the addition and multiplication properties to solve an inequality.

D Graph the solution set for an inequality.

E Translate and solve application problems involving inequalities.

Examples now playing at MathTV.com/books

PRACTICE PROBLEMS

1. Solve the inequality $x + 3 < 5$, and then graph the solution.

2. Solve $x - 4 \leq 1$, and then graph the solution.

Answers

1. $x < 2$ 2. $x \leq 5$

B Multiplication Property for Inequalities

To see the idea behind the multiplication property for inequalities, we will consider three true inequality statements and explore what happens when we multiply both sides by a positive number and then what happens when we multiply by a negative number.

Consider the following three true statements:

$$3 < 5 \qquad -3 < 5 \qquad -5 < -3$$

Now multiply both sides by the positive number 4:

$$4(3) < 4(5) \qquad 4(-3) < 4(5) \qquad 4(-5) < 4(-3)$$
$$12 < 20 \qquad -12 < 20 \qquad -20 < -12$$

In each case, the inequality symbol in the result points in the same direction it did in the original inequality. We say the "sense" of the inequality doesn't change when we multiply both sides by a positive quantity.

Notice what happens when we go through the same process but multiply both sides by −4 instead of 4:

$$3 < 5 \qquad -3 < 5 \qquad -5 < -3$$
$$-4(3) > -4(5) \qquad -4(-3) > -4(5) \qquad -4(-5) > -4(-3)$$
$$-12 > -20 \qquad 12 > -20 \qquad 20 > 12$$

In each case, we have to change the direction in which the inequality symbol points to keep each statement true. Multiplying both sides of an inequality by a negative quantity *always* reverses the sense of the inequality. Our results are summarized in the multiplication property for inequalities.

Note This discussion is intended to show why the multiplication property for inequalities is written the way it is. You may want to look ahead to the property itself and then come back to this discussion if you are having trouble making sense out of it.

Multiplication Property for Inequalities

For any three algebraic expressions A, B, and C,

if	$A < B$	
then	$AC < BC$	when C is positive
and	$AC > BC$	when C is negative

In words: Multiplying both sides of an inequality by a positive number does not change the solution set. When multiplying both sides of an inequality by a negative number, it is necessary to reverse the inequality symbol to produce an equivalent inequality.

Note Because division is defined in terms of multiplication, this property is also true for division. We can divide both sides of an inequality by any nonzero number we choose. If that number happens to be negative, we must also reverse the direction of the inequality symbol.

We can multiply both sides of an inequality by any nonzero number we choose. If that number happens to be negative, we must also reverse the sense of the inequality.

EXAMPLE 3 Solve $3a < 15$ and graph the solution.

SOLUTION We begin by multiplying each side by $\frac{1}{3}$. Because $\frac{1}{3}$ is a positive number, we do not reverse the direction of the inequality symbol:

$$3a < 15$$
$$\frac{1}{3}(3a) < \frac{1}{3}(15) \quad \text{Multiply each side by } \frac{1}{3}$$
$$a < 5$$

0 5

3. Solve $4a < 12$, and graph the solution.

Answer

3. $a < 3$

EXAMPLE 4 Solve $-3a \le 18$, and graph the solution.

SOLUTION We begin by multiplying both sides by $-\frac{1}{3}$. Because $-\frac{1}{3}$ is a negative number, we must reverse the direction of the inequality symbol at the same time that we multiply by $-\frac{1}{3}$.

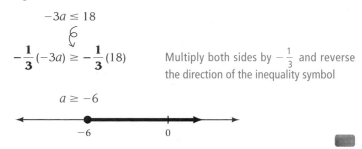

$$-3a \le 18$$

$$-\frac{1}{3}(-3a) \ge -\frac{1}{3}(18)$$ Multiply both sides by $-\frac{1}{3}$ and reverse the direction of the inequality symbol

$$a \ge -6$$

4. Solve $-2x \le 10$, and graph the solution.

EXAMPLE 5 Solve $-\frac{x}{4} > 2$ and graph the solution.

SOLUTION To isolate x, we multiply each side by -4. Because -4 is a negative number, we also must reverse the direction of the inequality symbol:

$$-\frac{x}{4} > 2$$

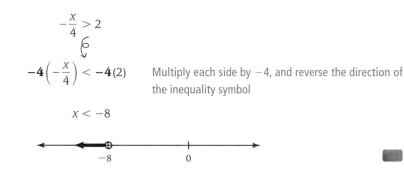

$$-4\left(-\frac{x}{4}\right) < -4(2)$$ Multiply each side by -4, and reverse the direction of the inequality symbol

$$x < -8$$

5. Solve $-\frac{x}{3} > 2$, and graph the solution.

C Solving Linear Inequalities in One Variable

To solve more complicated inequalities, we use the following steps.

Strategy Solving Linear Inequalities in One Variable

Step 1a: Use the distributive property to separate terms, if necessary.

1b: If fractions are present, consider multiplying both sides by the LCD to eliminate the fractions. If decimals are present, consider multiplying both sides by a power of 10 to clear the inequality of decimals.

1c: Combine similar terms on each side of the inequality.

Step 2: Use the addition property for inequalities to get all variable terms on one side of the inequality and all constant terms on the other side.

Step 3: Use the multiplication property for inequalities to get x by itself on one side of the inequality.

Step 4: Graph the solution set.

Answers
4. $x \ge -5$ 5. $x < -6$

6. Solve and graph.
$4.5x + 2.31 > 6.3x - 4.89$

EXAMPLE 6

Solve $2.5x - 3.48 < -4.9x + 2.07$.

SOLUTION We have two methods we can use to solve this inequality. We can simply apply our properties to the inequality the way it is currently written and work with the decimal numbers, or we can eliminate the decimals to begin with and solve the resulting inequality.

METHOD 1 Working with the decimals.

$2.5x - 3.48 < -4.9x + 2.07$	Original inequality
$2.5x + \mathbf{4.9x} - 3.48 < -4.9x + \mathbf{4.9x} + 2.07$	Add 4.9x to each side
$7.4x - 3.48 < 2.07$	
$7.4x - 3.48 + \mathbf{3.48} < 2.07 + \mathbf{3.48}$	Add 3.48 to each side
$7.4x < 5.55$	
$\frac{7.4x}{\mathbf{7.4}} < \frac{5.55}{\mathbf{7.4}}$	Divide each side by 7.4
$x < 0.75$	

METHOD 2 Eliminating the decimals in the beginning.

Because the greatest number of places to the right of the decimal point in any of the numbers is 2, we can multiply each side of the inequality by 100 and we will be left with an equivalent inequality that contains only whole numbers.

$2.5x - 3.48 < -4.9x + 2.07$	Original inequality
$\mathbf{100}(2.5x - 3.48) < \mathbf{100}(-4.9x + 2.07)$	Multiply each side by 100
$\mathbf{100}(2.5x) - \mathbf{100}(3.48) < \mathbf{100}(-4.9x) + \mathbf{100}(2.07)$	Distributive property
$250x - 348 < -490x + 207$	Multiplication
$740x - 348 < 207$	Add 490x to each side
$740x < 555$	Add 348 to each side
$\frac{740x}{\mathbf{740}} < \frac{555}{\mathbf{740}}$	Divide each side by 740
$x < 0.75$	

The solution by either method is $x < 0.75$. Here is the graph:

7. Solve and graph.
$2(x - 5) \ge -6$

EXAMPLE 7

Solve $3(x - 4) \ge -2$.

SOLUTION

$3x - 12 \ge -2$	Distributive property
$3x - 12 + \mathbf{12} \ge -2 + \mathbf{12}$	Add 12 to both sides
$3x \ge 10$	
$\frac{\mathbf{1}}{\mathbf{3}}(3x) \ge \frac{\mathbf{1}}{\mathbf{3}}(10)$	Multiply both sides by $\frac{1}{3}$
$x \ge \frac{10}{3}$	

0 $\frac{10}{3}$

Answers
6. $x < 4$ **7.** $x \ge 2$

D Graphing Inequalities

EXAMPLE 8 Solve and graph $2(1 - 3x) + 4 < 4x - 14$.

SOLUTION

$2 - 6x + 4 < 4x - 14$	Distributive property
$-6x + 6 < 4x - 14$	Simplify
$-6x + 6 + (\mathbf{-6}) < 4x - 14 + (\mathbf{-6})$	Add -6 to both sides
$-6x < 4x - 20$	
$-6x + (\mathbf{-4x}) < 4x + (\mathbf{-4x}) - 20$	Add $-4x$ to both sides
$-10x < -20$	
$\left(-\frac{1}{10}\right)(-10x) > \left(-\frac{1}{10}\right)(-20)$	Multiply by $-\frac{1}{10}$, reverse the sense of the inequality
$x > 2$	

8. Solve and graph.
$3(1 - 2x) + 5 < 3x - 1$

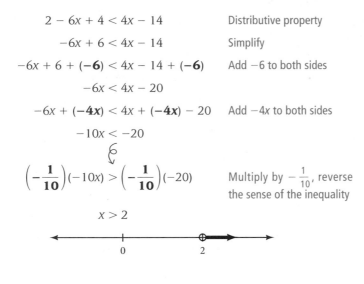

EXAMPLE 9 Solve $2x - 3y < 6$ for y.

SOLUTION We can solve this formula for y by first adding $-2x$ to each side and then multiplying each side by $-\frac{1}{3}$. When we multiply by $-\frac{1}{3}$ we must reverse the direction of the inequality symbol. Because this is an inequality in two variables, we will not graph the solution.

$2x - 3y < 6$	Original inequality
$2x + (\mathbf{-2x}) - 3y < (\mathbf{-2x}) + 6$	Add $-2x$ to each side
$-3y < -2x + 6$	
$-\frac{1}{3}(-3y) > -\frac{1}{3}(-2x + 6)$	Multiply each side by $-\frac{1}{3}$
$y > \frac{2}{3}x - 2$	Distributive property

9. Solve $2x - 3y < 6$ for x.

E Applications

When working application problems that involve inequalities, the phrases "at least" and "at most" translate as follows:

In Words	*In Symbols*
x is at least 30	$x \geq 30$
x is at most 20	$x \leq 20$

Our next example is similar to an example done earlier in this chapter. This time it involves an inequality instead of an equation.

We can modify our Blueprint for Problem Solving to solve application problems whose solutions depend on writing and then solving inequalities.

Answers

8. $x > 1$ **9.** $x < \frac{3}{2}y + 3$

10. The sum of two consecutive integers is at least 27. What are the possibilities for the first of the two integers?

EXAMPLE 10 The sum of two consecutive odd integers is at most 28. What are the possibilities for the first of the two integers?

SOLUTION When we use the phrase "their sum is at most 28," we mean that their sum is less than or equal to 28.

Step 1: Read and list.
Known items: Two consecutive odd integers. Their sum is less than or equal to 28.
Unknown items: The numbers in question

Step 2: Assign a variable, and translate information.
If we let x = the first of the two consecutive odd integers, then $x + 2$ is the next consecutive one.

Step 3: Reread, and write an inequality.
Their sum is at most 28.

$$x + (x + 2) \le 28$$

Step 4: Solve the inequality.

$2x + 2 \le 28$	Simplify the left side
$2x \le 26$	Add -2 to each side
$x \le 13$	Multiply each side by $\frac{1}{2}$

Step 5: Write the answer.
The first of the two integers must be an odd integer that is less than or equal to 13. The second of the two integers will be two more than whatever the first one is.

Step 6: Reread, and check.
Suppose the first integer is 13. The next consecutive odd integer is 15. The sum of 15 and 13 is 28. If the first odd integer is less than 13, the sum of it and the next consecutive odd integer will be less than 28.

Getting Ready for Class

After reading through the preceding section, respond in your own words and in complete sentences.

1. State the addition property for inequalities.
2. How is the multiplication property for inequalities different from the multiplication property of equality?
3. When do we reverse the direction of an inequality symbol?
4. Under what conditions do we not change the direction of the inequality symbol when we multiply both sides of an inequality by a number?

Answer

10. The first integer is at least 13; $x \ge 13$

Problem Set 2.8

A D Solve the following inequalities using the addition property of inequalities. Graph each solution set. [Examples 1–2]

1. $x - 5 < 7$

2. $x + 3 < -5$

3. $a - 4 \le 8$

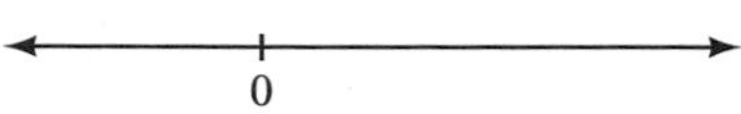

4. $a + 3 \le 10$

5. $x - 4.3 > 8.7$

6. $x - 2.6 > 10.4$

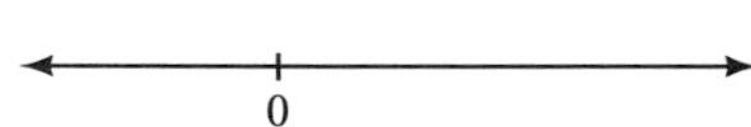

7. $y + 6 \ge 10$

8. $y + 3 \ge 12$

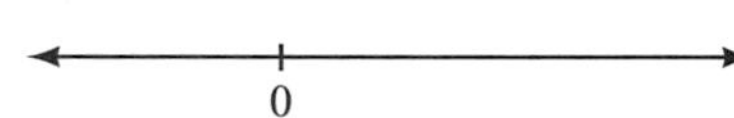

9. $2 < x - 7$

10. $3 < x + 8$

B D Solve the following inequalities using the multiplication property of inequalities. If you multiply both sides by a negative number, be sure to reverse the direction of the inequality symbol. Graph the solution set. [Examples 3–4]

11. $3x < 6$

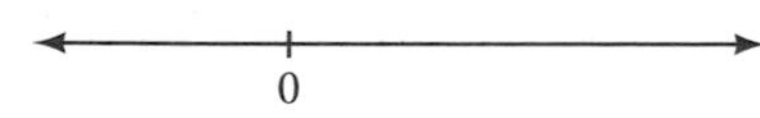

12. $2x < 14$

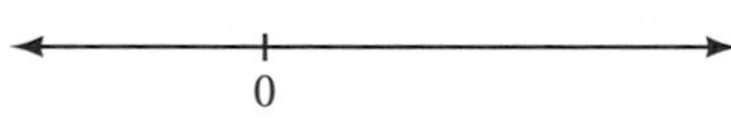

13. $5a \le 25$

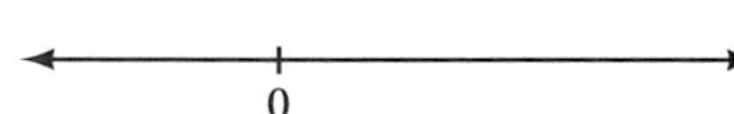

14. $4a \le 16$

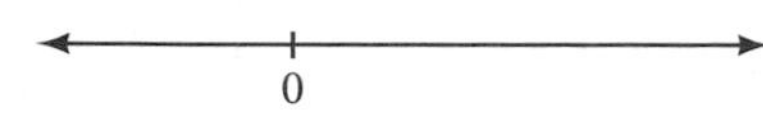

15. $\frac{x}{3} > 5$

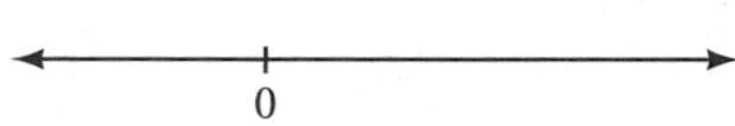

16. $\frac{x}{7} > 1$

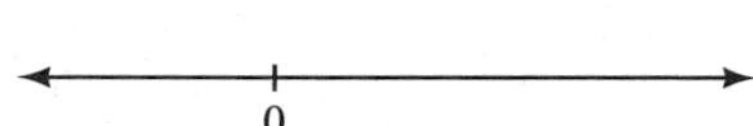

17. $-2x > 6$

18. $-3x \ge 9$

19. $-3x \ge -18$

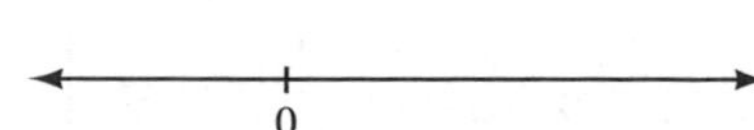

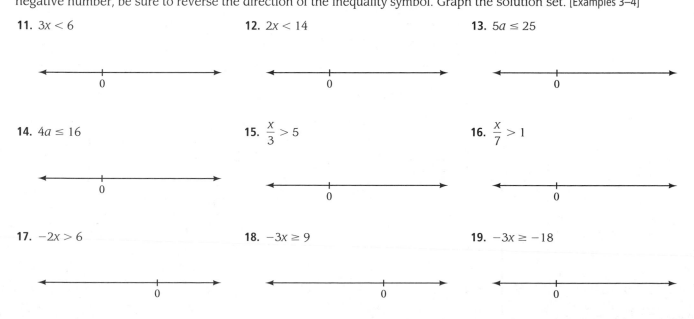

20. $-8x \geq -24$

21. $-\dfrac{x}{5} \leq 10$

22. $-\dfrac{x}{9} \geq -1$

23. $-\dfrac{2}{3}y > 4$

24. $-\dfrac{3}{4}y > 6$

C **D** Solve the following inequalities. Graph the solution set in each case. **[Examples 6–9]**

25. $2x - 3 < 9$

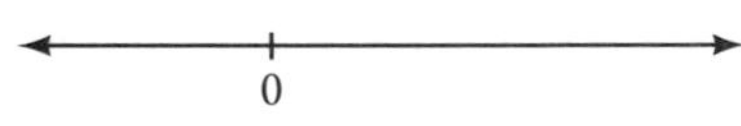

26. $3x - 4 < 17$

27. $-\dfrac{1}{5}y - \dfrac{1}{3} \leq \dfrac{2}{3}$

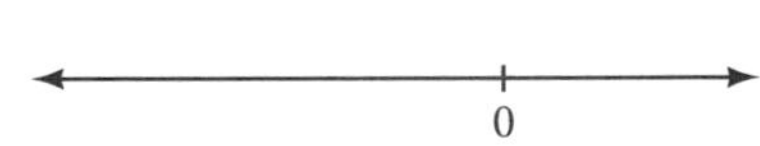

28. $-\dfrac{1}{6}y - \dfrac{1}{2} \leq \dfrac{2}{3}$

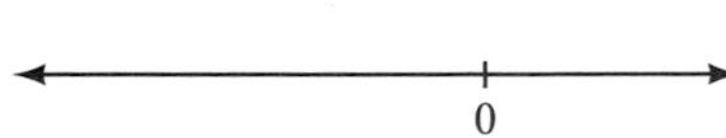

29. $-7.2x + 1.8 > -19.8$

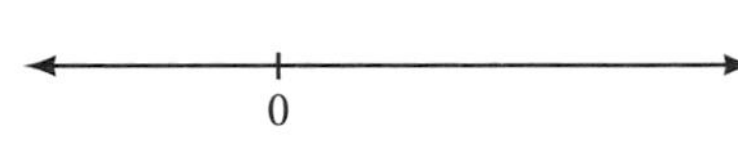

30. $-7.8x - 1.3 > 22.1$

31. $\dfrac{2}{3}x - 5 \leq 7$

32. $\dfrac{3}{4}x - 8 \leq 1$

33. $-\dfrac{2}{5}a - 3 > 5$

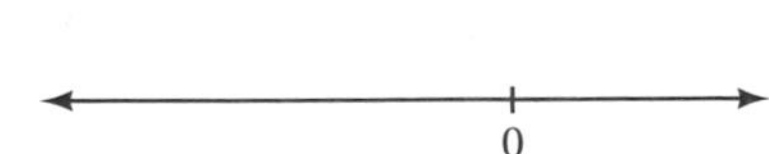

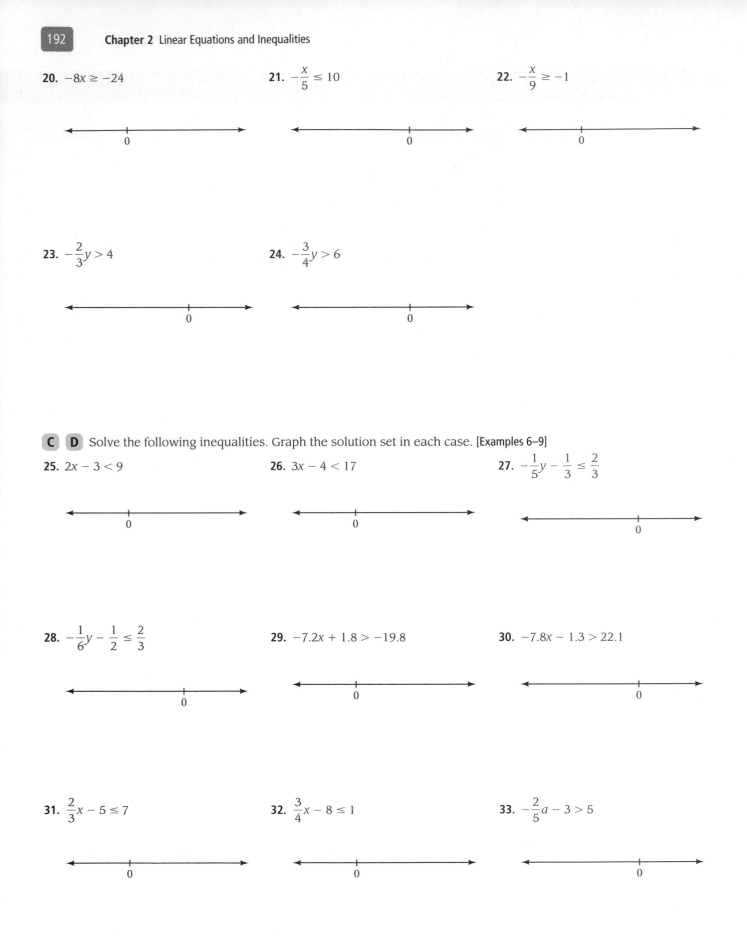

34. $-\frac{4}{5}a - 2 > 10$

35. $5 - \frac{3}{5}y > -10$

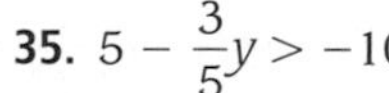

36. $4 - \frac{5}{6}y > -11$

37. $0.3(a + 1) \le 1.2$

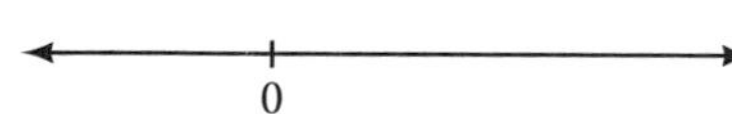

38. $0.4(a - 2) \le 0.4$

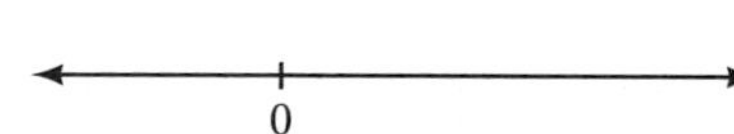

39. $2(5 - 2x) \le -20$

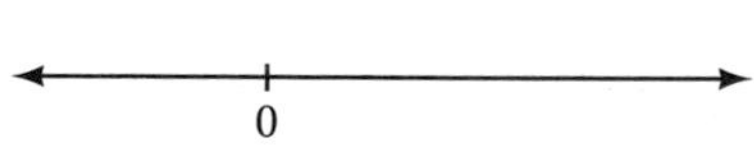

40. $7(8 - 2x) > 28$

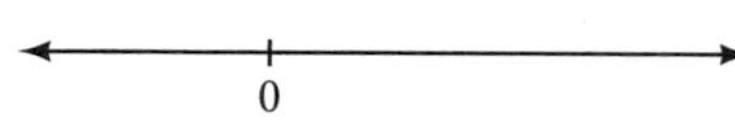

41. $3x - 5 > 8x$

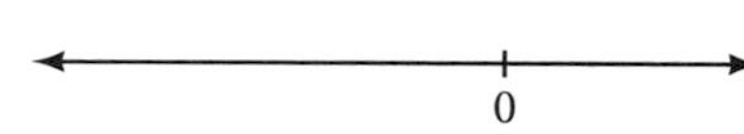

42. $8x - 4 > 6x$

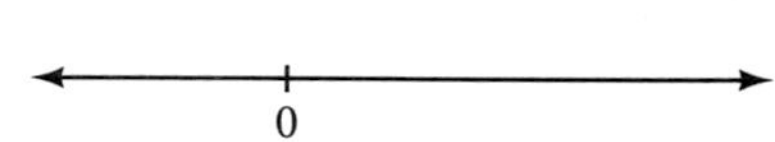

43. $\frac{1}{3}y - \frac{1}{2} \le \frac{5}{6}y + \frac{1}{2}$

44. $\frac{7}{6}y + \frac{4}{3} \le \frac{11}{6}y - \frac{7}{6}$

45. $-2.8x + 8.4 < -14x - 2.8$

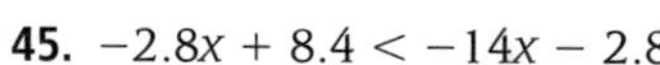

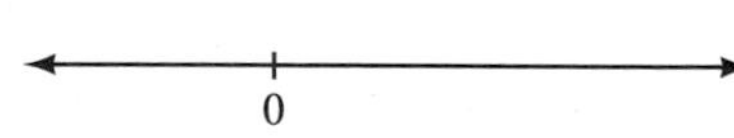

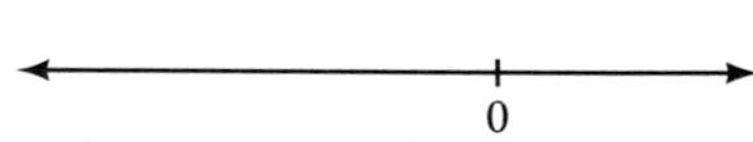

46. $-7.2x - 2.4 < -2.4x + 12$

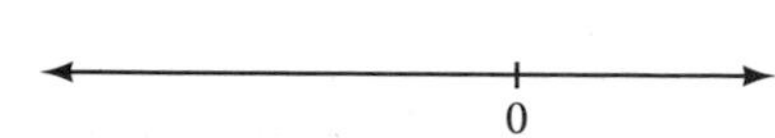

47. $3(m - 2) - 4 \ge 7m + 14$

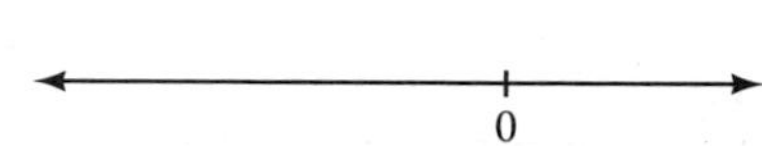

48. $2(3m - 1) + 5 \ge 8m - 7$

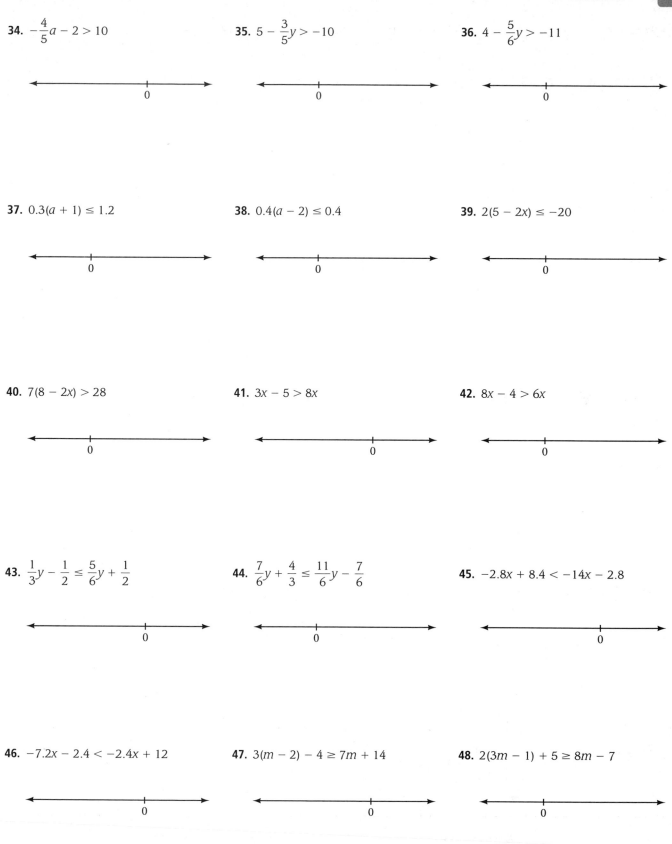

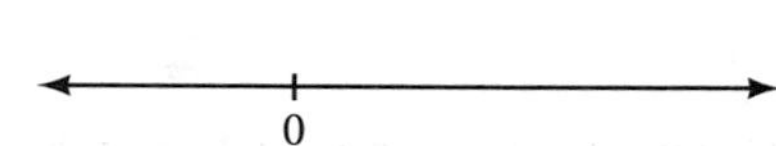

49. $3 - 4(x - 2) \leq -5x + 6$

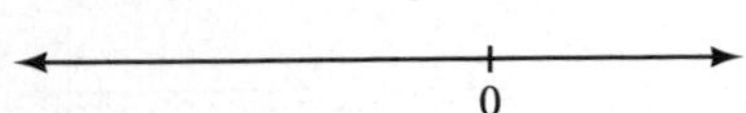

50. $8 - 6(x - 3) \leq -4x + 12$

Solve each of the following inequalities for y.

51. $3x + 2y < 6$

52. $-3x + 2y < 6$

53. $2x - 5y > 10$

54. $-2x - 5y > 5$

55. $-3x + 7y \leq 21$

56. $-7x + 3y \leq 21$

57. $2x - 4y \geq -4$

58. $4x - 2y \geq -8$

The next two problems are intended to give you practice reading, and paying attention to, the instructions that accompany the problems you are working.

59. Work each problem according to the instructions given.

- **a.** Evaluate when $x = 0$: $-5x + 3$
- **b.** Solve: $-5x + 3 = -7$
- **c.** Is 0 a solution to $-5x + 3 < -7$
- **d.** Solve: $-5x + 3 < -7$

60. Work each problem according to the instructions given.

- **a.** Evaluate when $x = 0$: $-2x - 5$
- **b.** Solve: $-2x - 5 = 1$
- **c.** Is 0 a solution to $-2x - 5 > 1$
- **d.** Solve: $-2x - 5 > 1$

For each graph below, write an inequality whose solution is the graph.

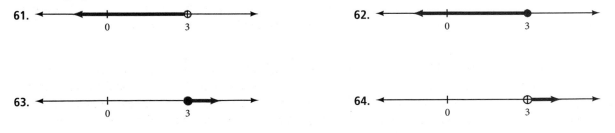

61. 

62. 

63. 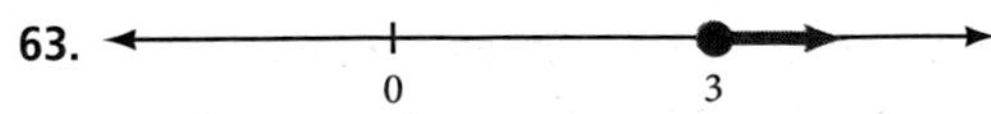

64. 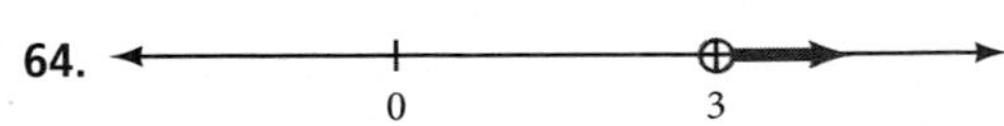

E Applying the Concepts [Example 10]

65. **Organic Groceries** The map shows a range of how many organic stores are found in each state. Write an inequality that describes the number of organic stores found in Florida.

66. **Population** The map shows the most populated cities in the United States. Write an inequality that describes the population of Chicago relative to San Diego.

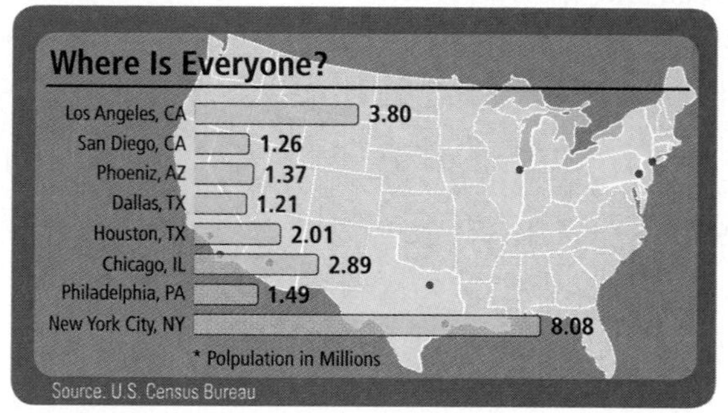

67. **Consecutive Integers** The sum of two consecutive integers is at least 583. What are the possibilities for the first of the two integers?

68. **Consecutive Integers** The sum of two consecutive integers is at most 583. What are the possibilities for the first of the two integers?

69. **Number Problems** The sum of twice a number and six is less than ten. Find all solutions.

70. **Number Problems** Twice the difference of a number and three is greater than or equal to the number increased by five. Find all solutions.

71. **Number Problems** The product of a number and four is greater than the number minus eight. Find the solution set.

72. **Number Problems** The quotient of a number and five is less than the sum of seven and two. Find the solution set.

73. **Geometry Problems** The length of a rectangle is 3 times the width. If the perimeter is to be at least 48 meters, what are the possible values for the width? (If the perimeter is at least 48 meters, then it is greater than or equal to 48 meters.)

74. **Geometry Problems** The length of a rectangle is 3 more than twice the width. If the perimeter is to be at least 51 meters, what are the possible values for the width? (If the perimeter is at least 51 meters, then it is greater than or equal to 51 meters.)

75. **Geometry Problems** The numerical values of the three sides of a triangle are given by three consecutive even integers. If the perimeter is greater than 24 inches, what are the possibilities for the shortest side?

76. **Geometry Problems** The numerical values of the three sides of a triangle are given by three consecutive odd integers. If the perimeter is greater than 27 inches, what are the possibilities for the shortest side?

77. **Car Heaters** If you have ever gotten in a cold car early in the morning you know that the heater does not work until the engine warms up. This is because the heater relies on the heat coming off the engine. Write an equation using an inequality sign to express when the heater will work if the heater works only after the engine is 100°F.

78. **Exercise** When Kate exercises, she either swims or runs. She wants to spend a minimum of 8 hours a week exercising, and she wants to swim 3 times the amount she runs. What is the minimum amount of time she must spend doing each exercise?

79. **Profit and Loss** Movie theaters pay a certain price for the movies that you and I see. Suppose a theater pays $1,500 for each showing of a popular movie. If they charge $7.50 for each ticket they sell, then they will lose money if ticket sales are less than $1,500. However, they will make a profit if ticket sales are greater than $1,500. What is the range of tickets they can sell and still lose money? What is the range of tickets they can sell and make a profit?

80. **Stock Sales** Suppose you purchase x shares of a stock at $12 per share. After 6 months you decide to sell all your shares at $20 per share. Your broker charges you $15 for the trade. If your profit is at least $3,985, how many shares did you purchase in the first place?

Getting Ready for the Next Section

Solve each inequality. Do not graph.

81. $2x - 1 \geq 3$

82. $3x + 1 \geq 7$

83. $-2x > -8$

84. $-3x > -12$

85. $-3 \leq 4x + 1$

86. $4x + 1 \leq 9$

Maintaining Your Skills

The problems that follow review some of the more important skills you have learned in previous sections and chapters. You can consider the time you spend working these problems as time spent studying for exams.

Apply the distributive property, then simplify.

87. $\frac{1}{6}(12x + 6)$

88. $\frac{3}{5}(15x - 10)$

89. $\frac{2}{3}(-3x - 6)$

90. $\frac{3}{4}(-4x - 12)$

91. $3\left(\frac{5}{6}a + \frac{4}{9}\right)$

92. $2\left(\frac{3}{4}a - \frac{5}{6}\right)$

93. $-3\left(\frac{2}{3}a + \frac{5}{6}\right)$

94. $-4\left(\frac{5}{6}a + \frac{4}{9}\right)$

Apply the distributive property, then find the LCD and simplify.

95. $\frac{1}{2}x + \frac{1}{6}x$

96. $\frac{1}{2}x - \frac{3}{4}x$

97. $\frac{2}{3}x - \frac{5}{6}x$

98. $\frac{1}{3}x + \frac{3}{5}x$

99. $\frac{3}{4}x + \frac{1}{6}x$

100. $\frac{3}{2}x - \frac{2}{3}x$

101. $\frac{2}{5}x + \frac{5}{8}x$

102. $\frac{3}{5}x - \frac{3}{8}x$

Compound Inequalities

2.9

Objectives

A Solve and graph compound inequalities.

A Solving Compound Inequalities

The *union* of two sets A and B is the set of all elements that are in A or in B. The word *or* is the key word in the definition. The *intersection* of two sets A and B is the set of elements contained in both A and B. The key word in this definition is *and*. We can put the words *and* and *or* together with our methods of graphing inequalities to find the solution sets for compound inequalities.

Definition

A **compound inequality** is two or more inequalities connected by the word *and* or *or*.

EXAMPLE 1 Graph the solution set for the compound inequality

$$x < -1 \quad \text{or} \quad x \geq 3$$

SOLUTION Graphing each inequality separately, we have

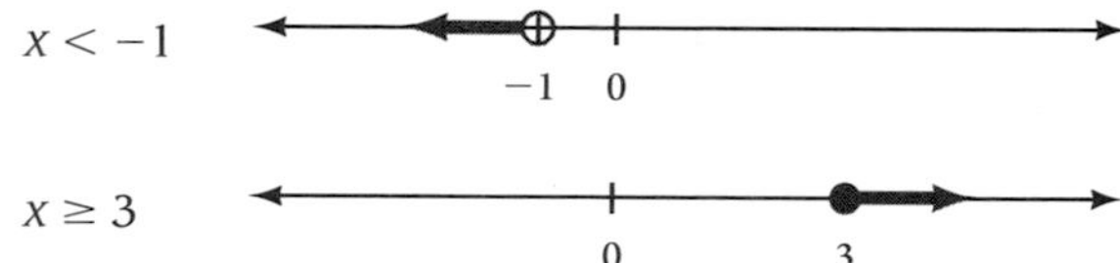

Because the two inequalities are connected by *or*, we want to graph their union; that is, we graph all points that are on either the first graph or the second graph. Essentially, we put the two graphs together on the same number line.

$$x < -1 \quad \text{or} \quad x \geq 3$$

EXAMPLE 2 Graph the solution set for the compound inequality

$$x > -2 \quad \text{and} \quad x < 3$$

SOLUTION Graphing each inequality separately, we have

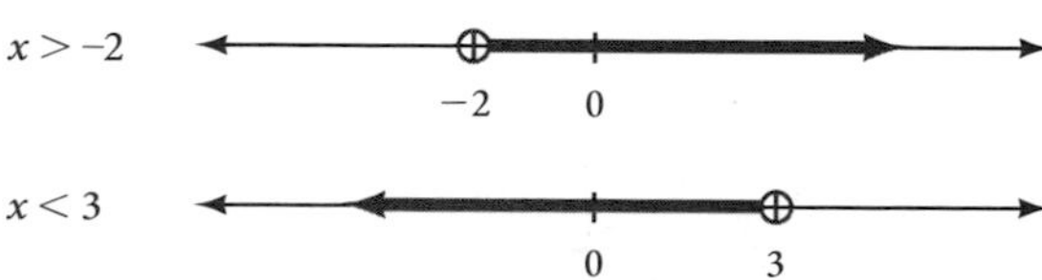

Because the two inequalities are connected by the word *and*, we will graph their intersection, which consists of all points that are common to both graphs; that is, we graph the region where the two graphs overlap.

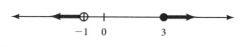

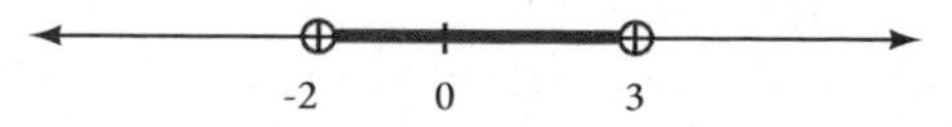

PRACTICE PROBLEMS

1. Graph $x < -3$ or $x \geq 2$.

2. Graph $x > -3$ and $x < 2$.

Answer

1. See Solutions to Selected Practice Problems.
2. See Solutions to Selected Practice Problems.

3. Solve and graph.
$3x + 1 \geq 7$ and $-2x > -8$

EXAMPLE 3

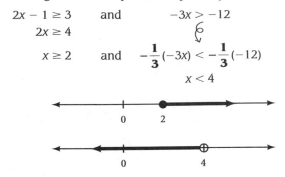

Solve and graph the solution set for

$$2x - 1 \geq 3 \quad \text{and} \quad -3x > -12$$

SOLUTION Solving the two inequalities separately, we have

$$\begin{aligned} 2x - 1 &\geq 3 & \text{and} \quad & -3x > -12 \\ 2x &\geq 4 & & \\ x &\geq 2 & \text{and} \quad & -\frac{1}{3}(-3x) < -\frac{1}{3}(-12) \\ & & & x < 4 \end{aligned}$$

Because the word *and* is used, their intersection is the points they have in common:

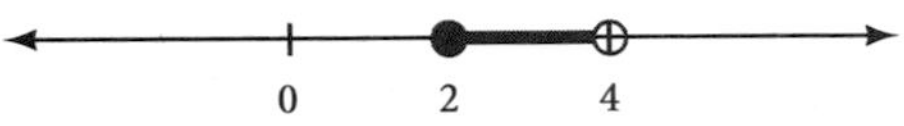

The compound inequality $-2 < x$ and $x < 3$ can be written as $-2 < x < 3$. The word *and* does not appear when an inequality is written in this form. It is implied. The solution set for $-2 < x$ and $x < 3$ is

It is all the numbers between -2 and 3 on the number line. It seems reasonable, then, that this graph should be the graph of $-2 < x < 3$. In both the graph and the inequality, x is said to be between -2 and 3.

4. Solve and graph.
$-3 \leq 4x + 1 \leq 9$

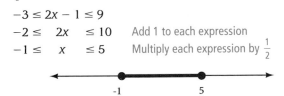

EXAMPLE 4

Solve and graph $-3 \leq 2x - 1 \leq 9$.

SOLUTION To solve for x, we must add 1 to the center expression and then divide the result by 2. Whatever we do to the center expression, we also must do to the two expressions on the ends. In this way we can be sure we are producing equivalent inequalities. The solution set will not be affected.

$$\begin{aligned} -3 \leq 2x - 1 &\leq 9 & & \\ -2 \leq \quad 2x \quad &\leq 10 & & \text{Add 1 to each expression} \\ -1 \leq \quad x \quad &\leq 5 & & \text{Multiply each expression by } \tfrac{1}{2} \end{aligned}$$

Getting Ready for Class

After reading through the preceding section, respond in your own words and in complete sentences.

1. What is a compound inequality?
2. Explain the shorthand notation that can be used to write two inequalities connected by the word *and*.
3. Write two inequalities connected by the word *and* that together are equivalent to $-1 < x < 2$.
4. Explain in words how you would graph the compound inequality

$$x < 2 \text{ or } x > -3$$

Answer
3. $x \geq 2$ and $x < 4$ **4.** $-1 \leq x \leq 2$

Problem Set 2.9

A Graph the following compound inequalities. [Examples 1–2]

1. $x < -1$ or $x > 5$

2. $x \le -2$ or $x \ge -1$

3. $x < -3$ or $x \ge 0$

4. $x < 5$ and $x > 1$

5. $x \le 6$ and $x > -1$

6. $x \le 7$ and $x > 0$

7. $x > 2$ and $x < 4$

8. $x < 2$ or $x > 4$

9. $x \ge -2$ and $x \le 4$

10. $x \le 2$ or $x \ge 4$

11. $x < 5$ and $x > -1$

12. $x > 5$ or $x < -1$

13. $-1 < x < 3$

14. $-1 \le x \le 3$

15. $-3 < x \le -2$

16. $-5 \le x \le 0$

A Solve the following compound inequalities. Graph the solution set in each case. [Examples 3–4]

17. $3x - 1 < 5$ or $5x - 5 > 10$

18. $x + 1 < -3$ or $x - 2 > 6$

19. $x - 2 > -5$ and $x + 7 < 13$

20. $3x + 2 \le 11$ and $2x + 2 \ge 0$

21. $11x < 22$ or $12x > 36$

22. $-5x < 25$ and $-2x \ge -12$

23. $3x - 5 < 10$ and $2x + 1 > -5$

24. $5x + 8 < -7$ or $3x - 8 > 10$

25. $2x - 3 < 8$ and $3x + 1 > -10$

26. $11x - 8 > 3$ or $12x + 7 < -5$

27. $2x - 1 < 3$ and $3x - 2 > 1$

28. $3x + 9 < 7$ or $2x - 7 > 11$

29. $-1 \le x - 5 \le 2$

30. $0 \le x + 2 \le 3$

31. $-4 \le 2x \le 6$

32. $-5 < 5x < 10$

33. $-3 < 2x + 1 < 5$

34. $-7 \le 2x - 3 \le 7$

35. $0 \le 3x + 2 \le 7$

36. $2 \le 5x - 3 \le 12$

37. $-7 < 2x + 3 < 11$

38. $-5 < 6x - 2 < 8$

39. $-1 \le 4x + 5 \le 9$

40. $-8 \le 7x - 1 \le 13$

For each graph below, write an inequality whose solution is the graph.

41. 

42. 

43. 

44. 

Applying the Concepts

45. Organic Groceries The map shows a range of how many organic stores are found in each state. If Oregon has 24 organic grocery stores, write an inequality that describes the difference, d, between the number of stores in Oregon and the number of stores found in Nevada.

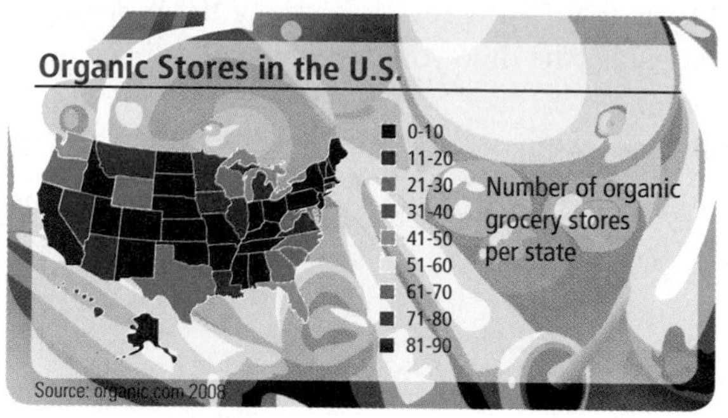

46. Organic Groceries The map shows a range of how many organic stores are found in each state. Write an inequality that describes the difference, d, in organic stores between California and the state of Washington.

Triangle Inequality The triangle inequality states that the sum of any two sides of a triangle must be greater than the third side.

47. The following triangle RST has sides of length x, $2x$, and 10 as shown.

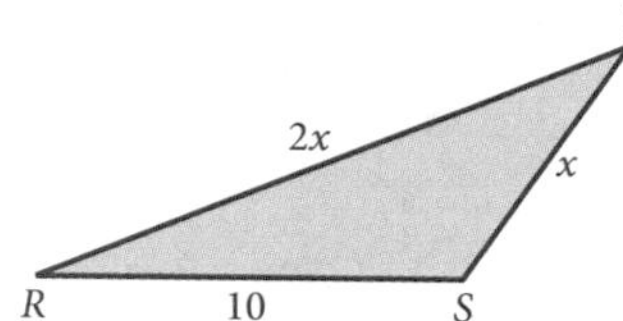

a. Find the three inequalities, which must be true based on the sides of the triangle.

b. Write a compound inequality based on your results above.

48. The following triangle ABC has sides of length x, $3x$, and 16 as shown.

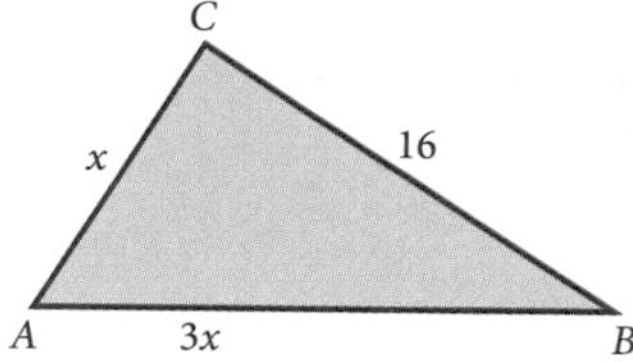

a. Find the three inequalities, which must be true based on the sides of the triangle.

b. Write a compound inequality based on your results above.

49. Engine Temperature The engine in a car gives off a lot of heat due to the combustion in the cylinders. The water used to cool the engine keeps the temperature within the range $50 \le F \le 266$ where F is in degrees Fahrenheit. Graph this inequality on the number line.

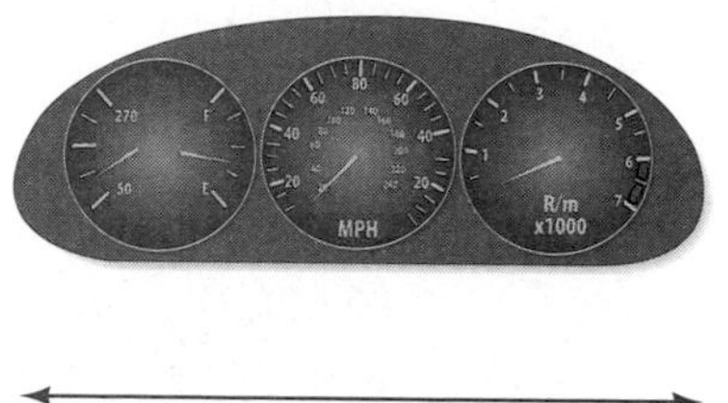

50. Engine Temperature To find the engine temperature range from Problem 47 in degrees Celsius, we use the fact that $F = \frac{9}{5}C + 32$ to rewrite the inequality as

$$50 \le \frac{9}{5}C + 32 \le 266$$

Solve this inequality and graph the solution set.

51. Number Problem The difference of twice a number and 3 is between 5 and 7. Find the number.

52. Number Problem The sum of twice a number and 5 is between 7 and 13. Find the number.

53. Perimeter The length of a rectangle is 4 inches longer than the width. If the perimeter is between 20 inches and 30 inches, find all possible values for the width.

54. Perimeter The length of a rectangle is 6 feet longer than the width. If the perimeter is between 24 feet and 36 feet, find all possible values for the width.

Maintaining Your Skills

The problems that follow review some of the more important skills you have learned in previous sections and chapters. You can consider the time you spend working these problems as time spent studying for exams.

Answer the following percent problems.

55. What number is 25% of 32?

56. What number is 15% of 75?

57. What number is 20% of 120?

58. What number is 125% of 300?

59. What percent of 36 is 9?

60. What percent of 16 is 9?

61. What percent of 50 is 5?

62. What percent of 140 is 35?

63. 16 is 20% of what number?

64. 6 is 3% of what number?

65. 8 is 2% of what number?

66. 70 is 175% of what number?

Simplify each expression.

67. $-|-5|$

68. $-\left(\frac{2}{3}\right)^3$

69. $-3 - 4(-2)$

70. $2^4 + 3^3 \div 9 - 4^2$

71. $5|3 - 8| - 6|2 - 5|$

72. $7 - 3(2 - 6)$

73. $5 - 2[-3(5 - 7) - 8]$

74. $\frac{5 + 3(7 - 2)}{2(-3) - 4}$

75. Find the difference of -3 and -9.

76. If you add -4 to the product of -3 and 5, what number results?

77. Apply the distributive property to $\frac{1}{2}(4x - 6)$.

78. Use the associative property to simplify $-6\left(\frac{1}{3}x\right)$.

For the set $\{-3, -\frac{4}{5}, 0, \frac{5}{8}, 2, \sqrt{5}\}$, which numbers are

79. Integers

80. Rational numbers

Chapter 2 Summary

Similar Terms [2.1]

A *term* is a number or a number and one or more variables multiplied together. *Similar terms* are terms with the same variable part.

EXAMPLES

1. The terms $2x$, $5x$, and $-7x$ are all similar because their variable parts are the same.

Simplifying Expressions [2.1]

In this chapter we simplified expressions that contained variables by using the distributive property to combine similar terms.

2. Simplify $3x + 4x$.

$$\begin{aligned} 3x + 4x &= (3 + 4)x \\ &= 7x \end{aligned}$$

Solution Set [2.2]

The *solution set* for an equation (or inequality) is all the numbers that, when used in place of the variable, make the equation (or inequality) a true statement.

3. The solution set for the equation $x + 2 = 5$ is $\{3\}$ because when x is 3 the equation is $3 + 2 = 5$, or $5 = 5$.

Equivalent Equations [2.2]

Two equations are called *equivalent* if they have the same solution set.

4. The equation $a - 4 = 3$ and $a - 2 = 5$ are equivalent because both have solution set $\{7\}$.

Addition Property of Equality [2.2]

When the same quantity is added to both sides of an equation, the solution set for the equation is unchanged. Adding the same amount to both sides of an equation produces an equivalent equation.

5. Solve $x - 5 = 12$.

$$\begin{aligned} x - 5 + 5 &= 12 + 5 \\ x + 0 &= 17 \\ x &= 17 \end{aligned}$$

Multiplication Property of Equality [2.3]

If both sides of an equation are multiplied by the same nonzero number, the solution set is unchanged. Multiplying both sides of an equation by a nonzero quantity produces an equivalent equation.

6. Solve $3x = 18$.

$$\begin{aligned} \mathbf{\frac{1}{3}}(3x) &= \mathbf{\frac{1}{3}}(18) \\ x &= 6 \end{aligned}$$

Strategy for Solving Linear Equations in One Variable [2.4]

Step 1a: Use the distributive property to separate terms, if necessary.

1b: If fractions are present, consider multiplying both sides by the LCD to eliminate the fractions. If decimals are present, consider multiplying both sides by a power of 10 to clear the equation of decimals.

1c: Combine similar terms on each side of the equation.

Step 2: Use the addition property of equality to get all variable terms on one side of the equation and all constant terms on the other side. A variable term is a term that contains the variable (for example, $5x$). A constant term is a term that does not contain the variable (the number 3, for example.)

7. Solve $2(x + 3) = 10$.

$$\begin{aligned} 2x + 6 &= 10 \\ 2x + 6 + (\mathbf{-6}) &= 10 + (\mathbf{-6}) \\ 2x &= 4 \\ \mathbf{\frac{1}{2}}(2x) &= \mathbf{\frac{1}{2}}(4) \\ x &= 2 \end{aligned}$$

Step 3: Use the multiplication property of equality to get x (that is, $1x$) by itself on one side of the equation.

Step 4: Check your solution in the original equation to be sure that you have not made a mistake in the solution process.

Formulas [2.5]

8. Solving $P = 2l + 2w$ for l, we have

$$P - 2w = 2l$$

$$\frac{P - 2w}{2} = l$$

A formula is an equation with more than one variable. To solve a formula for one of its variables, we use the addition and multiplication properties of equality to move everything except the variable in question to one side of the equal sign so the variable in question is alone on the other side.

Blueprint for Problem Solving [2.6, 2.7]

Step 1: ***Read*** the problem, and then mentally ***list*** the items that are known and the items that are unknown.

Step 2: ***Assign a variable*** to one of the unknown items. (In most cases this will amount to letting x = the item that is asked for in the problem.) Then ***translate*** the other ***information*** in the problem to expressions involving the variable.

Step 3: ***Reread*** the problem, and then ***write an equation,*** using the items and variables listed in steps 1 and 2, that describes the situation.

Step 4: ***Solve the equation*** found in step 3.

Step 5: ***Write*** your ***answer*** using a complete sentence.

Step 6: ***Reread*** the problem, and ***check*** your solution with the original words in the problem.

Addition Property for Inequalities [2.8]

9. Solve $x + 5 < 7$.

$$x + 5 + (\mathbf{-5}) < 7 + (\mathbf{-5})$$

$$x < 2$$

Adding the same quantity to both sides of an inequality produces an equivalent inequality, one with the same solution set.

Multiplication Property for Inequalities [2.8]

10. Solve $-3a \le 18$.

$$-\frac{\mathbf{1}}{\mathbf{3}}(-3a) \ge -\frac{\mathbf{1}}{\mathbf{3}}(18)$$

$$a \ge -6$$

Multiplying both sides of an inequality by a positive number never changes the solution set. If both sides are multiplied by a negative number, the sense of the inequality must be reversed to produce an equivalent inequality.

Strategy for Solving Linear Inequalities in One Variable [2.8]

11. Solve $3(x - 4) \ge -2$.

$$3x - 12 \ge -2$$

$$3x - 12 + \mathbf{12} \ge -2 + \mathbf{12}$$

$$3x \ge 10$$

$$\frac{\mathbf{1}}{\mathbf{3}}(3x) \ge \frac{\mathbf{1}}{\mathbf{3}}(10)$$

$$x \ge \frac{10}{3}$$

0 $\frac{10}{3}$

Step 1a: Use the distributive property to separate terms, if necessary.

1b: If fractions are present, consider multiplying both sides by the LCD to eliminate the fractions. If decimals are present, consider multiplying both sides by a power of 10 to clear the inequality of decimals.

1c: Combine similar terms on each side of the inequality.

Step 2: Use the addition property for inequalities to get all variable terms on one side of the inequality and all constant terms on the other side.

Step 3: Use the multiplication property for inequalities to get x by itself on one side of the inequality.

Step 4: Graph the solution set.

Compound Inequalities [2.9]

Two inequalities connected by the word *and* or *or* form a compound inequality. If the connecting word is *or,* we graph all points that are on either graph. If the connecting word is *and,* we graph only those points that are common to both graphs. The inequality $-2 \le x \le 3$ is equivalent to the compound inequality $-2 \le x$ and $x \le 3$.

12. $x < -3$ or $x > 1$

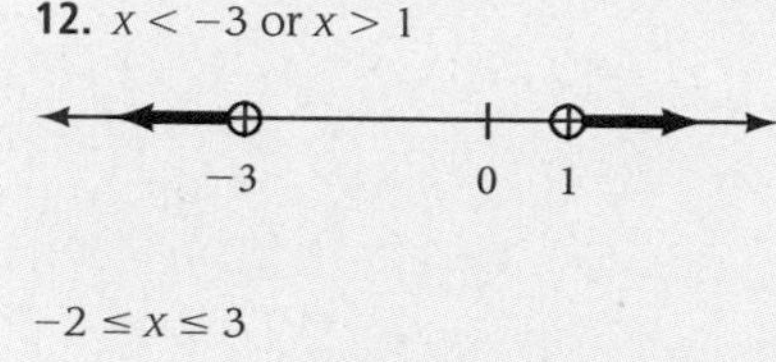

$-2 \le x \le 3$

COMMON MISTAKES

1. Trying to subtract away coefficients (the number in front of variables) when solving equations. For example:

$$4x = 12$$
$$4x \mathbf{- 4} = 12 \mathbf{- 4}$$
$$x = 8 \leftarrow \text{Mistake}$$

It is not incorrect to add (-4) to both sides; it's just that $4x - 4$ is not equal to x. Both sides should be multiplied by $\frac{1}{4}$ to solve for x.

2. Forgetting to reverse the direction of the inequality symbol when multiplying both sides of an inequality by a negative number. For instance:

$$-3x < 12$$
$$\mathbf{-\frac{1}{3}}(-3x) < \mathbf{-\frac{1}{3}}(12) \leftarrow \text{Mistake}$$
$$x < -4$$

It is not incorrect to multiply both sides by $-\frac{1}{3}$. But if we do, we must also reverse the sense of the inequality.

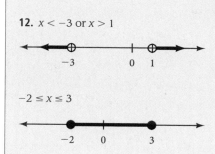

Chapter 2 Review

Simplify each expression as much as possible. [2.1]

1. $5x - 8x$ **2.** $6x - 3 - 8x$ **3.** $-a + 2 + 5a - 9$

4. $5(2a - 1) - 4(3a - 2)$ **5.** $6 - 2(3y + 1) - 4$ **6.** $4 - 2(3x - 1) - 5$

Find the value of each expression when x is 3. [2.1]

7. $7x - 2$ **8.** $-4x - 5 + 2x$ **9.** $-x - 2x - 3x$

Find the value of each expression when x is -2. [2.1]

10. $5x - 3$ **11.** $-3x + 2$ **12.** $7 - x - 3$

Solve each equation. [2.2, 2.3]

13. $x + 2 = -6$ **14.** $x - \frac{1}{2} = \frac{4}{7}$ **15.** $10 - 3y + 4y = 12$

16. $-3 - 4 = -y - 2 + 2y$ **17.** $2x = -10$ **18.** $3x = 0$

19. $\frac{x}{3} = 4$ **20.** $-\frac{x}{4} = 2$ **21.** $3a - 2 = 5a$

22. $\frac{7}{10}a = \frac{1}{5}a + \frac{1}{2}$ **23.** $3x + 2 = 5x - 8$ **24.** $6x - 3 = x + 7$

25. $0.7x - 0.1 = 0.5x - 0.1$ **26.** $0.2x - 0.3 = 0.8x - 0.3$

Solve each equation. Be sure to simplify each side first. [2.4]

27. $2(x - 5) = 10$ **28.** $12 = 2(5x - 4)$ **29.** $\frac{1}{2}(3t - 2) + \frac{1}{2} = \frac{5}{2}$

30. $\frac{3}{5}(5x - 10) = \frac{2}{3}(9x + 3)$ **31.** $2(3x + 7) = 4(5x - 1) + 18$ **32.** $7 - 3(y + 4) = 10$

Use the formula $4x - 5y = 20$ to find y if [2.5]

33. x is 5 **34.** x is 0 **35.** x is -5 **36.** x is 10

Solve each of the following formulas for the indicated variable. [2.5]

37. $2x - 5y = 10$ for y **38.** $5x - 2y = 10$ for y **39.** $V = \pi r^2 h$ for h **40.** $P = 2l + 2w$ for w

41. What number is 86% of 240? [2.5] **42.** What percent of 2,000 is 180? [2.5]

Solve each of the following word problems. In each case, be sure to show the equation that describes the situation. [2.6, 2.7]

43. Number Problem The sum of twice a number and 6 is 28. Find the number.

44. Geometry The length of a rectangle is 5 times as long as the width. If the perimeter is 60 meters, find the length and the width.

45. Investing A man invests a certain amount of money in an account that pays 9% annual interest. He invests $300 more than that in an account that pays 10% annual interest. If his total interest after a year is $125, how much does he have invested in each account?

46. Coin Problem A collection of 15 coins is worth $1.00. If the coins are dimes and nickels, how many of each coin are there?

Solve each inequality. [2.8]

47. $-2x < 4$

48. $-5x > -10$

49. $-\frac{a}{2} \le -3$

50. $-\frac{a}{3} > 5$

Solve each inequality, and graph the solution. [2.8]

51. $-4x + 5 > 37$

52. $2x + 10 < 5x - 11$

53. $2(3t + 1) + 6 \ge 5(2t + 4)$

Graph the solution to each of the following compound inequalities. [2.9]

54. $x < -2$ or $x > 5$

55. $-5x \ge 25$ or $2x - 3 \ge 9$

56. $-1 < 2x + 1 < 9$

Chapter 2 Cumulative Review

Simplify.

1. $7 - 9 - 12$
2. $-11 + 17 + (-13)$
3. $8 - 4 \cdot 5$
4. $30 \div 3 \cdot 10$
5. $6 + 3(6 + 2)$
6. $-6(5 - 11) - 4(13 - 6)$
7. $\left(-\frac{2}{3}\right)^3$
8. $\left[\frac{1}{2}(-8)\right]^2$
9. $\frac{-30}{120}$
10. $\frac{234}{312}$
11. $-\frac{3}{4} \div \frac{15}{16}$
12. $\frac{2}{5} \cdot \frac{4}{7}$
13. $\frac{-4(-6)}{-9}$
14. $\frac{2(-9) + 3(6)}{-7 - 8}$
15. $\frac{5}{9} + \frac{1}{3}$
16. $\frac{4}{21} - \frac{9}{35}$
17. $\frac{1}{5}(10x)$
18. $-8(9x)$
19. $\frac{1}{4}(8x - 4)$
20. $3(2x - 1) + 5(x + 2)$

Solve each equation.

21. $7x = 6x + 4$
22. $x + 12 = -4$
23. $-\frac{3}{5}x = 30$
24. $\frac{x}{5} = -\frac{3}{10}$
25. $3x - 4 = 11$
26. $5x - 7 = x - 1$
27. $4(3x - 8) + 5(2x + 7) = 25$
28. $15 - 3(2t + 4) = 1$
29. $3(2a - 7) - 4(a - 3) = 15$
30. $\frac{1}{3}(x - 6) = \frac{1}{4}(x + 8)$
31. Solve $P = a + b + c$ for c.
32. Solve $3x + 4y = 12$ for y.

Solve each inequality, and graph the solution.

33. $3(x - 4) \le 6$
34. $-5x + 9 < -6$
35. $x - 1 < 1$ or $2x + 1 > 7$
36. $-2 < x + 1 < 5$

Translate into symbols.

37. The difference of x and 5 is 12.
38. The sum of x and 7 is 4.

For the set $\{-5, -3, -1.7, 2.3, \frac{12}{7}, \pi\}$ list all the

39. Integers
40. Rational numbers
41. Evaluate $x^2 - 8x - 9$ when $x = -2$.
42. Evaluate $a^2 - 2ab + b^2$ when $a = 3, b = -2$.
43. What is 30% of 50?
44. What percent of 36 is 27?
45. **Number Problem** The sum of a number and 9 is 23. Find the number.
46. **Number Problem** Twice a number increased by 7 is 31. Find the number.
47. **Geometry** A right triangle has one 42° angle. Find the other two angles.
48. **Geometry** Two angles are complementary. If one is 25°, find the other one.
49. **Hourly Pay** Carol tutors in the math lab. She gets paid \$8 per hour for the first 15 hours and \$10 per hour for each hour after that. She made \$150 one week. How many hours did she work?
50. **Cost of a Letter** The cost of mailing a letter was 32¢ for the first ounce and 29¢ for each additional ounce. If the cost of mailing a letter was 90¢, how many ounces did the letter weigh?

The illustration shows the annual proceeds (in millions of dollars) from sales of the top five magazines. Use the illustration to answer the following questions.

Top Five Magazine Sales
1 TV Guide $548.9M
2 People $494.6M
3 Time The Weekly Magazine $313.8M
4 Sports Illustrated $306.9M
5 Reader's Digest $288.4M
Source: Magazine Publishers of America 2004

51. Find the total annual proceeds from sales of *TV Guide* and *Reader's Digest*.
52. How much more are the annual proceeds from sales of *People* than from sales of *Time*?

Chapter 2 Test

Simplify each of the following expressions. [2.1]

1. $3x + 2 - 7x + 3$

2. $4a - 5 - a + 1$

3. $7 - 3(y + 5) - 4$

4. $8(2x + 1) - 5(x - 4)$

5. Find the value of $2x - 3 - 7x$ when $x = -5$. [2.1]

6. Find the value of $x^2 + 2xy + y^2$ when $x = 2$ and $y = 3$. [2.1]

7. Fill in the tables below to find the sequences formed by substituting the first four counting numbers into the expressions $(n + 1)^2$ and $n^2 + 1$. [2.1]

a.

n	$(n + 1)^2$
1	
2	
3	
4	

b.

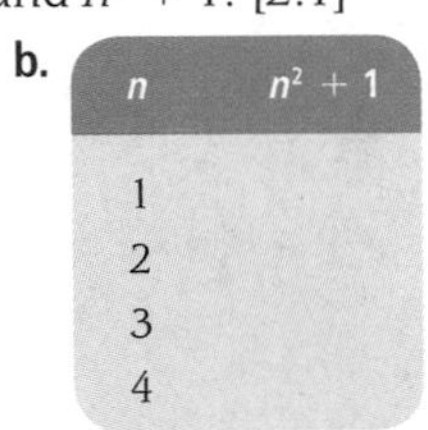

n	$n^2 + 1$
1	
2	
3	
4	

Solve the following equations. [2.2, 2.3, 2.4]

8. $2x - 5 = 7$

9. $2y + 4 = 5y$

10. $\frac{1}{2}x - \frac{1}{10} = \frac{1}{5}x + \frac{1}{2}$

11. $\frac{2}{5}(5x - 10) = -5$

12. $-5(2x + 1) - 6 = 19$

13. $0.04x + 0.06(100 - x) = 4.6$

14. $2(t - 4) + 3(t + 5) = 2t - 2$

15. $2x - 4(5x + 1) = 3x + 17$

16. What number is 15% of 38? [2.5]

17. 240 is 12% of what number? [2.5]

18. If $2x - 3y = 12$, find x when $y = -2$. [2.5]

19. If $2x - 3y = 12$, find y when $x = -6$. [2.5]

20. Solve $2x + 5y = 20$ for y. [2.5]

21. Solve $h = x + vt + 16t^2$ for v. [2.5]

Solve each word problem. [2.6, 2.7]

22. Age Problem Dave is twice as old as Rick. Ten years ago the sum of their ages was 40. How old are they now?

23. Geometry A rectangle is twice as long as it is wide. The perimeter is 60 inches. What are the length and width?

24. Coin Problem A man has a collection of dimes and quarters with a total value of \$3.50. If he has 7 more dimes than quarters, how many of each coin does he have?

25. Investing A woman has money in two accounts. One account pays 7% annual interest, whereas the other pays 9% annual interest. If she has \$600 more invested at 9% than she does at 7% and her total interest for a year is \$182, how much does she have in each account?

Solve each inequality, and graph the solution. [2.8]

26. $2x + 3 < 5$

27. $-5a > 20$

28. $0.4 - 0.2x \geq 1$

29. $4 - 5(m + 1) \leq 9$

Solve each inequality, and graph the solution. [2.9]

30. $3 - 4x \geq -5$ or $2x \geq 10$

31. $-7 < 2x - 1 < 9$

Write an inequality whose solution is the given graph. [2.8, 2.9]

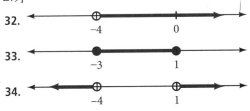

32. (number line: open circle at −4, shaded to the right; labels −4, 0)

33. (number line: closed circles at −3 and 1, shaded between; labels −3, 1)

34. (number line: open circles at −4 and 1, shaded left of −4 and right of 1; labels −4, 1)

Use the illustration below to find the percent of the hours in one week (168 hours) that a person from the given country spends watching television. Round your answers to the nearest tenth of a percent. [2.5]

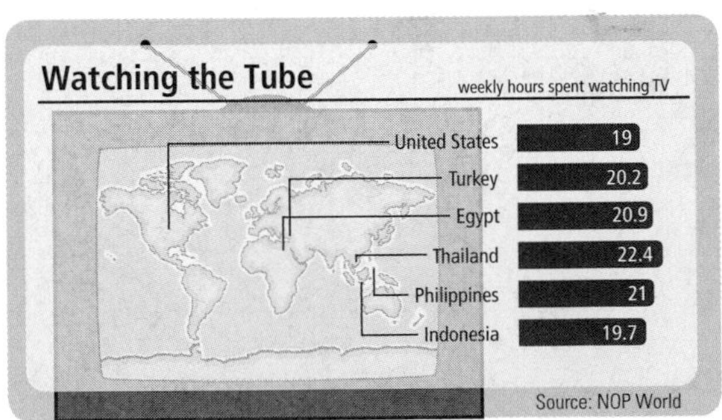

35. United States

36. Thailand

Chapter 2 Projects

LINEAR EQUATIONS AND INEQUALITIES

GROUP PROJECT

Tables and Graphs

Number of People 2-3

Time Needed 5–10 minutes

Equipment Pencil and graph paper

Background Building tables is a method of visualizing information. We can build a table from a situation (as below) or from an equation. In this project, we will first build a table and then write an equation from the information in the table.

Procedure A parking meter, which accepts only dimes and quarters, is emptied at the end of each day. The amount of money in the meter at the end of one particular day is \$3.15.

1. Complete the following table so that all possible combinations of dimes and quarters, along with the total number of coins, is shown. Remember, although the number of coins will vary, the value of the dimes and quarters must total \$3.15.

Number of Dimes	Number of Quarters	Total Coins	Value
29	1	30	\$3.15
24			\$3.15
			\$3.15
			\$3.15
			\$3.15
			\$3.15

2. From the information in the table, answer the following questions.
   a. What is the maximum possible number of coins taken from the meter?
   b. What is the minimum possible number of coins taken from the meter?
   c. When is the number of dimes equal to the number of quarters?

3. Let x = the number of dimes and y = the number of quarters. Write an equation in two variables such that the value of the dimes added to the value of the quarters is \$3.15.

RESEARCH PROJECT

Stand and Deliver

The 1988 film *Stand and Deliver* starring Edward James Olmos and Lou Diamond Phillips is based on a true story. Olmos, in his portrayal of high-school math teacher Jaime Escalante, earned an academy award nomination for best actor.

Watch the movie *Stand and Deliver.* After briefly describing the movie, explain how Escalante's students became successful in math. Make a list of specific things you observe that the students had to do to become successful. Indicate which items on this list you think will also help you become successful.

The Kobal Collection

Linear Equations and Inequalities in Two Variables

Chapter Outline

Introduction

The photo from Google Earth is of the University of Bologna where Maria Agnesi (1718-1799) was the first female professor of mathematics. She is also recognized as the first woman to publish a book of mathematics. The title of the book is *Instituzioni analitiche ad uso della gioventù italiana*. A diagram similar to the one below appears in an 1801 English translation of her book.

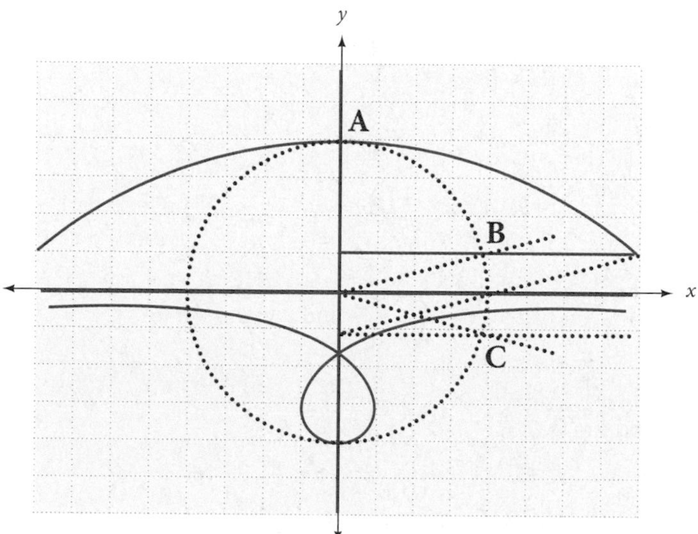

The diagram is drawn on a rectangular coordinate system. The rectangular coordinate system is the foundation upon which we build our study of graphing in this chapter.

Chapter Pretest

Graph the ordered pairs. [3.1]

1. (2, −1)

2. (−4, 3)

3. (−3, −2)

4. (0, −4)

5. Fill in the following ordered pairs for the equation $3x - 2y = 6$. [3.2]

(0,   ) (, 0) (4,   ) (, −6)

6. Which of the following ordered pairs are solutions to $y = -3x + 7$? [3.2]

(0, 7) (2, −1) (4, −5) (−5, −3)

Find the x- and y-intercepts. [3.4]

7. $8x - 4y = 16$

8. $y = \frac{3}{2}x + 6$

Find the slope of the line through each pair of points. [3.5]

9. (3, 2), (−5, 6)

10. (0, 9), (7, 1)

Find the slope of each line. [3.5]

11.

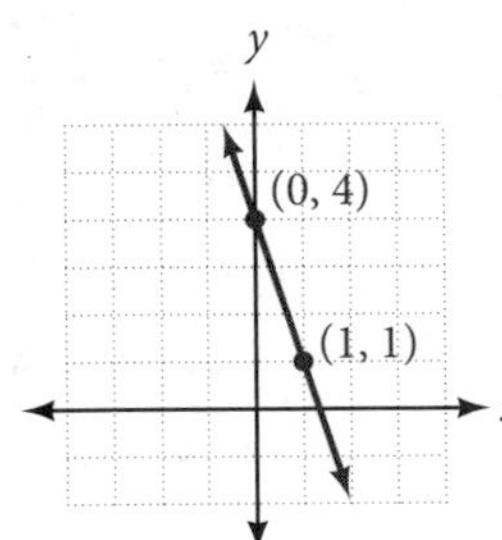

12.

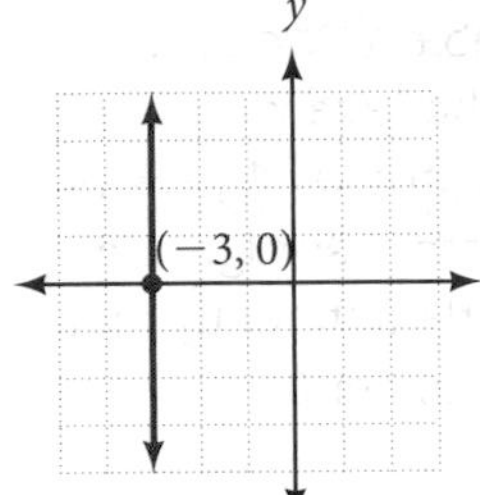

13. Find the equation of the line through (4, 1) with a slope of $-\frac{1}{2}$. [3.6]

14. Find the equation of the line with a slope of 3 and y-intercept −5. [3.6]

15. Find the equation of the line passing through the points (3, −4) and (−6, 2) [3.6]

16. A straight line has an x-intercept 3 and contains the point (−2, 6). Find its equation. [3.6]

Getting Ready for Chapter 3

Simplify

1. $2(2) + (-1)$

2. $2(5) - 1$

3. $-\frac{1}{3}(3) + 2$

4. $\frac{3}{2}(2) - 3$

5. $-\frac{1}{2}(-3x + 6)$

6. $-\frac{2}{3}(x - 3)$

Solve.

7. $6 + 3y = 6$

8. $2x - 10 = 20$

9. $7 = 2x - 1$

10. $3 = 2x - 1$

11. If $3x - 5y = 10$, find y when x is 0.

12. If $y = \frac{2}{3}x + 5$, find y when x is 6

Find the value of each expression when $x = -2$ and $y = 3$.

13. $\frac{y - 1}{x + 4}$

14. $\frac{y + 7}{x - 3}$

15. $\frac{4 - y}{-1 - x}$

16. $\frac{3 - y}{x + 8}$

Solve each of the following equations for y.

17. $7x - 2y = 14$

18. $-2x - 5y = 4$

19. $y - 3 = -2(x + 4)$

20. $y + 1 = -\frac{2}{3}x + 2$

Paired Data and Graphing Ordered Pairs

3.1

Objectives

A Create a bar chart, scatter diagram, or line graph from a table of data.

B Graph ordered pairs on a rectangular coordinate system.

In Chapter 1 we showed the relationship between the table of values for the speed of a racecar and the corresponding bar chart. Table 1 and Figure 1 from the introduction of Chapter 1 are reproduced here for reference. In Figure 1, the horizontal line that shows the elapsed time in seconds is called the *horizontal axis,* and the vertical line that shows the speed in miles per hour is called the *vertical axis.*

The data in Table 1 are called *paired data* because the information is organized so that each number in the first column is paired with a specific number in the second column. Each pair of numbers is associated with one of the solid bars in Figure 1. For example, the third bar in the bar chart is associated with the pair of numbers 3 seconds and 162.8 miles per hour. The first number, 3 seconds, is associated with the horizontal axis, and the second number, 162.8 miles per hour, is associated with the vertical axis.

A Scatter Diagrams and Line Graphs

The information in Table 1 can be visualized with a *scatter diagram* and *line graph* as well. Figure 2 is a scatter diagram of the information in Table 1. We use dots instead of the bars shown in Figure 1 to show the speed of the racecar at each second during the race. Figure 3 is called a *line graph.* It is constructed by taking the dots in Figure 2 and connecting each one to the next with a straight line. Notice that we have labeled the axes in these two figures a little differently than we did with the bar chart by making the axes intersect at the number 0.

TABLE 1

SPEED OF A RACECAR

Time in Seconds	Speed in Miles per Hour
0	0
1	72.7
2	129.9
3	162.8
4	192.2
5	212.4
6	228.1

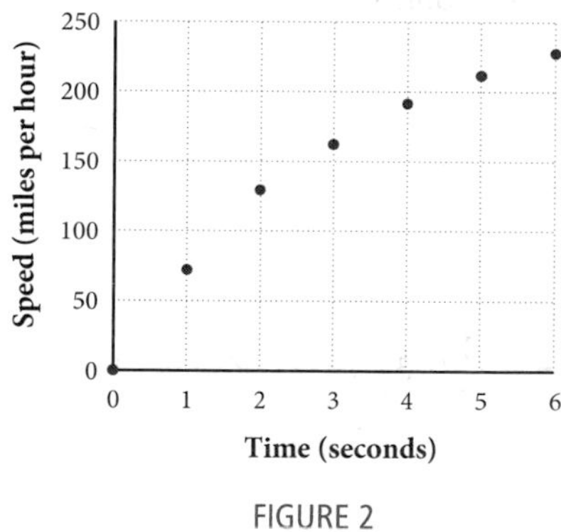

FIGURE 2

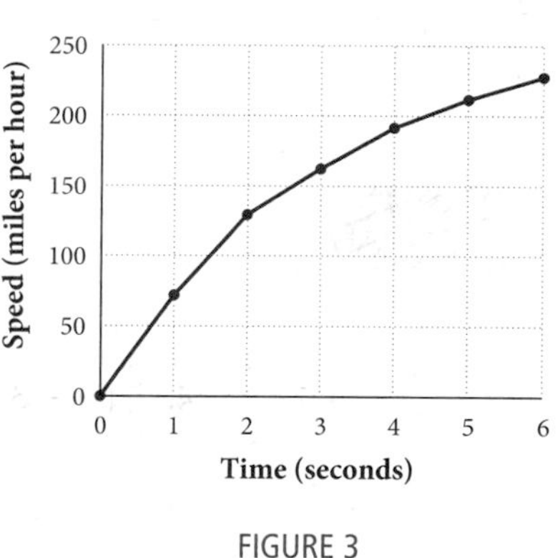

FIGURE 3

Visually

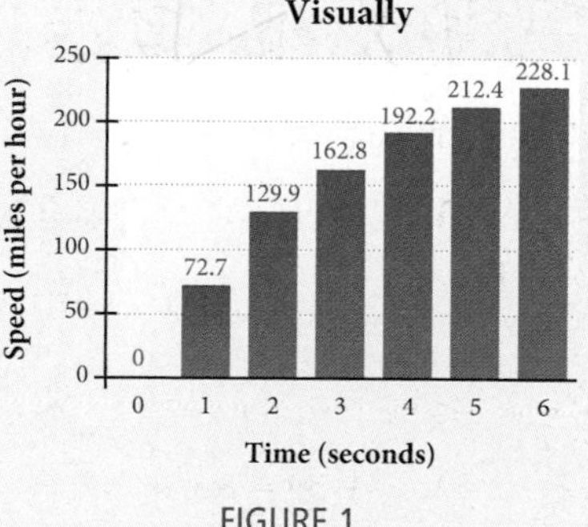

FIGURE 1

The number sequences we have worked with in the past can also be written as paired data by associating each number in the sequence with its position in the sequence. For instance, in the sequence of odd numbers

$$1, 3, 5, 7, 9, \ldots$$

the number 7 is the fourth number in the sequence. Its position is 4, and its value is 7. Here is the sequence of odd numbers written so that the position of each term is noted:

Position	1, 2, 3, 4, 5, . . .
Value	1, 3, 5, 7, 9, . . .

PRACTICE PROBLEMS

1. Table 4 gives the position and value of the first five terms of the sequence of cubes. Construct a scatter diagram using these data.

TABLE 4

CUBES	
Position	Value
1	1
2	8
3	27
4	64
5	125

EXAMPLE 1 Tables 2 and 3 give the first five terms of the sequence of odd numbers and the sequence of squares as paired data. In each case construct a scatter diagram.

TABLE 2

ODD NUMBERS	
Position	Value
1	1
2	3
3	5
4	7
5	9

TABLE 3

SQUARES	
Position	Value
1	1
2	4
3	9
4	16
5	25

SOLUTION The two scatter diagrams are based on the data from Tables 2 and 3 shown here. Notice how the dots in Figure 4 seem to line up in a straight line, whereas the dots in Figure 5 give the impression of a curve. We say the points in Figure 4 suggest a *linear* relationship between the two sets of data, whereas the points in Figure 5 suggest a *nonlinear* relationship.

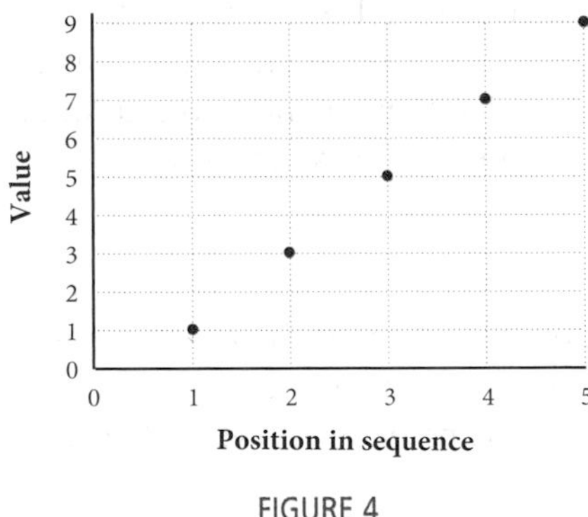

FIGURE 4

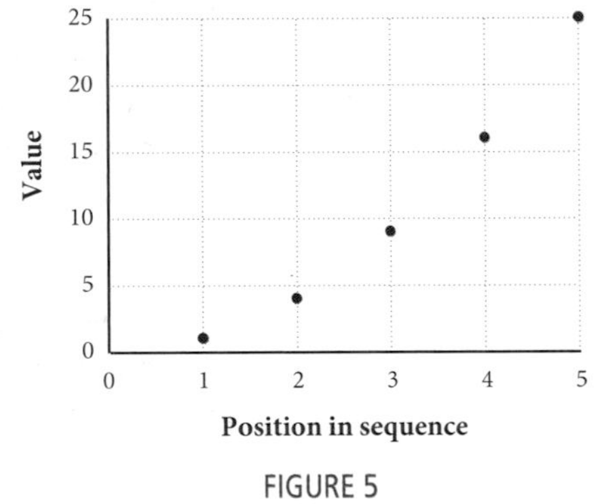

FIGURE 5

As you know, each dot in Figures 4 and 5 corresponds to a pair of numbers, one of which is associated with the horizontal axis and the other with the vertical axis. Paired data play a very important role in the equations we will solve in the next section. To prepare ourselves for those equations, we need to expand the concept of paired data to include negative numbers. At the same time, we want to standardize the position of the axes in the diagrams that we use to visualize paired data.

Definition

A pair of numbers enclosed in parentheses and separated by a comma, such as $(-2, 1)$, is called an **ordered pair** of numbers. The first number in the pair is called the ***x*-coordinate** of the ordered pair; the second number is called the ***y*-coordinate.** For the ordered pair $(-2, 1)$, the x-coordinate is -2 and the y-coordinate is 1.

Ordered pairs of numbers are important in the study of mathematics because they give us a way to visualize solutions to equations. To see the visual compo-

Answers

1–3. See Solutions to Selected Practice Problems.

nent of ordered pairs, we need the diagram shown in Figure 6. It is called the *rectangular coordinate system.*

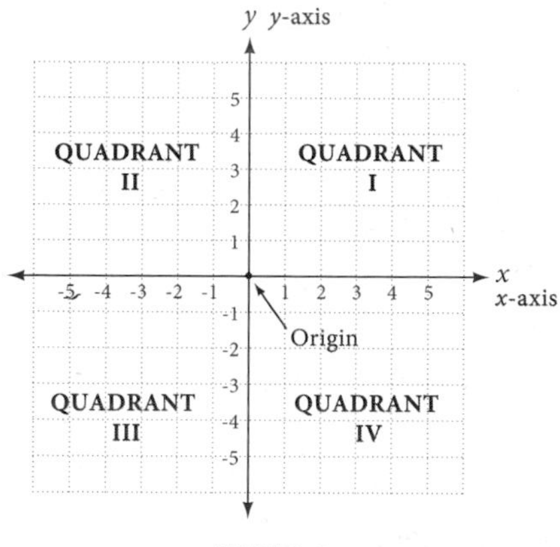

FIGURE 6

The rectangular coordinate system is built from two number lines oriented perpendicular to each other. The horizontal number line is exactly the same as our real number line and is called the *x-axis.* The vertical number line is also the same as our real number line with the positive direction up and the negative direction down. It is called the *y-axis.* The point where the two axes intersect is called the *origin.* As you can see from Figure 6, the axes divide the plane into four quadrants, which are numbered I through IV in a counterclockwise direction.

B Graphing Ordered Pairs

To graph the ordered pair (a, b), we start at the origin and move a units right or left (right if a is positive and left if a is negative). Then we move b units up or down (up if b is positive, down if b is negative). The point where we end up is the graph of the ordered pair (a, b).

EXAMPLE 2 Graph the ordered pairs (3, 4), (3, −4), (−3, 4), and (−3, −4).

SOLUTION

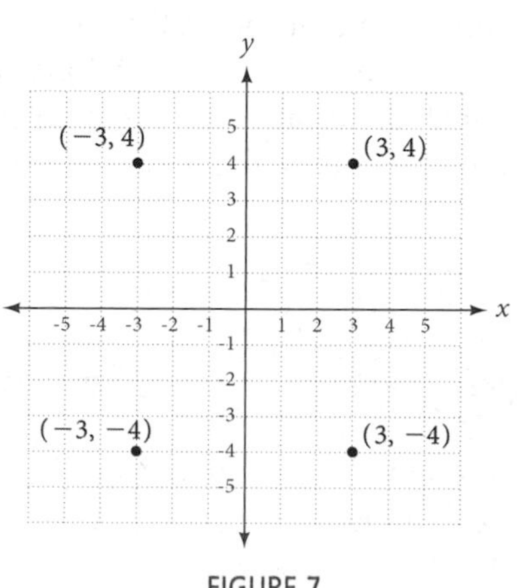

FIGURE 7

We can see in Figure 7 that when we graph ordered pairs, the x-coordinate corresponds to movement parallel to the x-axis (horizontal) and the y-coordinate corresponds to movement parallel to the y-axis (vertical).

2. Graph the ordered pairs (2, 3), (2, −3), (−2, 3), and (−2, −3).

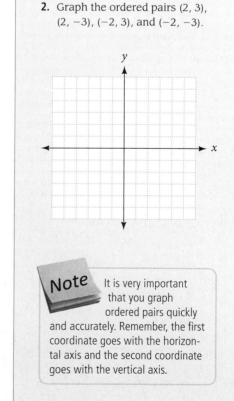

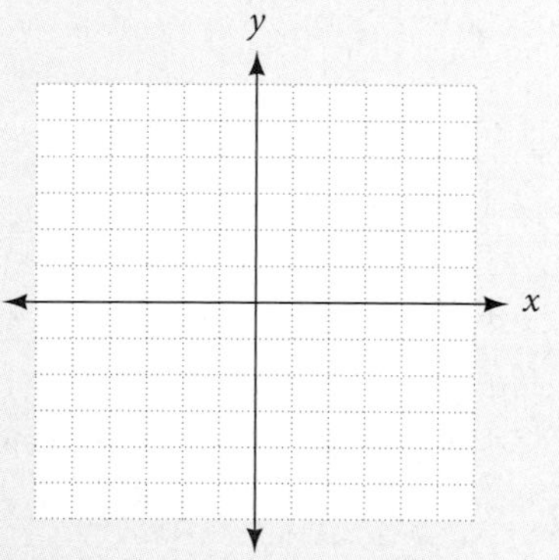

It is very important that you graph ordered pairs quickly and accurately. Remember, the first coordinate goes with the horizontal axis and the second coordinate goes with the vertical axis.

3. Graph the ordered pairs (−2, 1), (3, 5), (0, 2), (−5, 0), and (−3, −3).

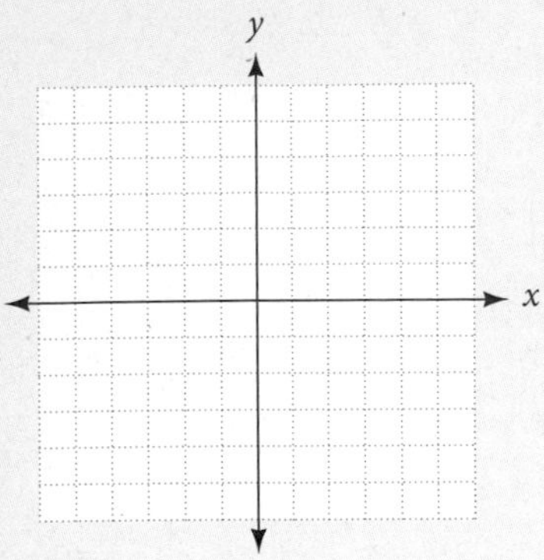

Note If we do not label the axes of a coordinate system, we assume that each square is one unit long and one unit wide.

EXAMPLE 3 Graph the ordered pairs (−1, 3), (2, 5), (0, 0), (0, −3), and (4, 0).

SOLUTION

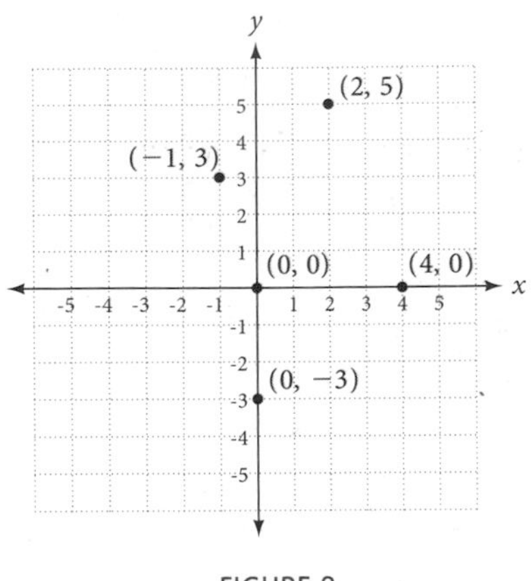

FIGURE 8

Getting Ready for Class

After reading through the preceding section, respond in your own words and in complete sentences.

1. What is an ordered pair of numbers?
2. Explain in words how you would graph the ordered pair (3, 4).
3. How do you construct a rectangular coordinate system?
4. Where is the origin on a rectangular coordinate system?

Problem Set 3.1

B Graph the following ordered pairs. [Examples 2–3]

1. (3, 2) **2.** (3, −2) **3.** (−3, 2) **4.** (−3, −2) **5.** (5, 1) **6.** (5, −1)

7. (1, 5) **8.** (1, −5) **9.** (−1, 5) **10.** (−1, −5) **11.** $\left(2, \frac{1}{2}\right)$ **12.** $\left(3, \frac{3}{2}\right)$

13. $\left(-4, -\frac{5}{2}\right)$ **14.** $\left(-5, -\frac{3}{2}\right)$ **15.** (3, 0) **16.** (−2, 0) **17.** (0, 5) **18.** (0, 0)

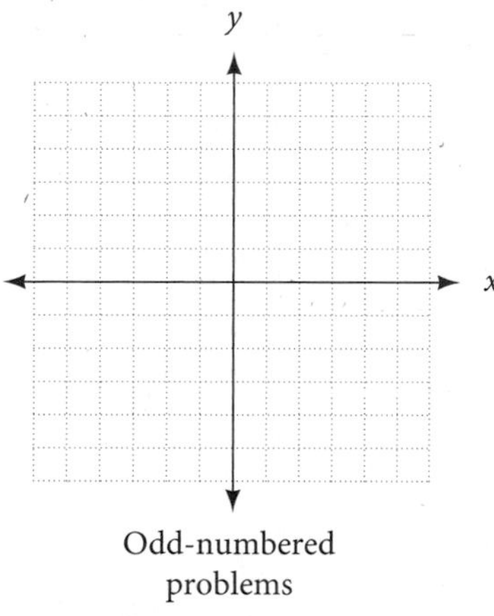

Odd-numbered problems

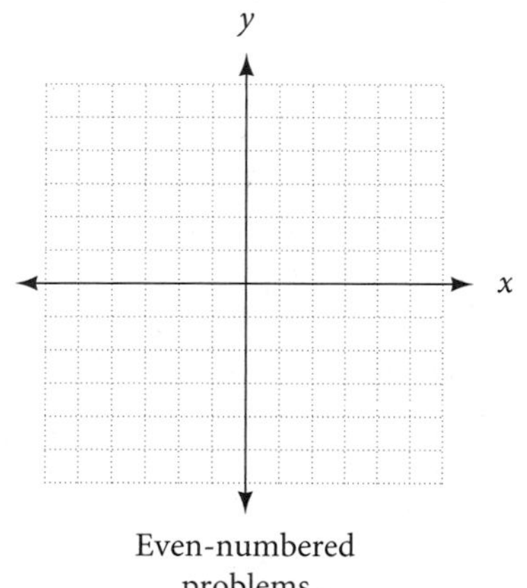

Even-numbered problems

Give the coordinates of each numbered point in the figure.

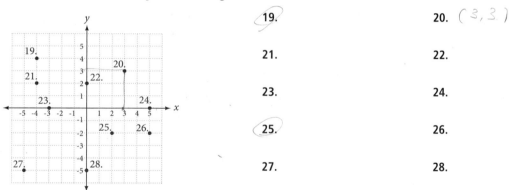

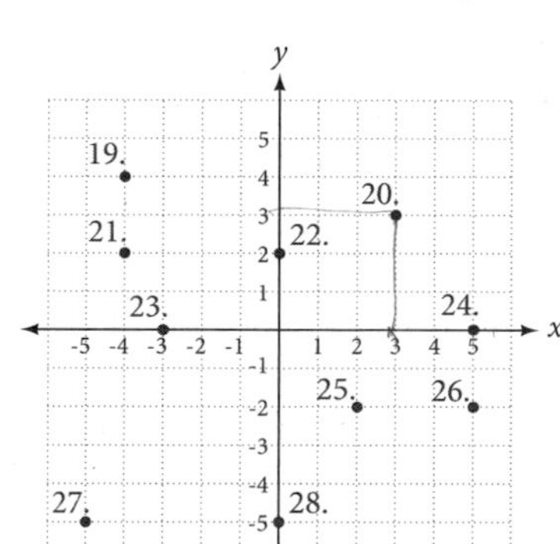

19. **20.**

21. **22.**

23. **24.**

25. **26.**

27. **28.**

Graph the points (4, 3) and (−4, −1), and draw a straight line that passes through both of them. Then answer the following questions.

29. Does the graph of (2, 2) lie on the line?

30. Does the graph of (−2, 0) lie on the line?

31. Does the graph of (0, −2) lie on the line?

32. Does the graph of (−6, 2) lie on the line?

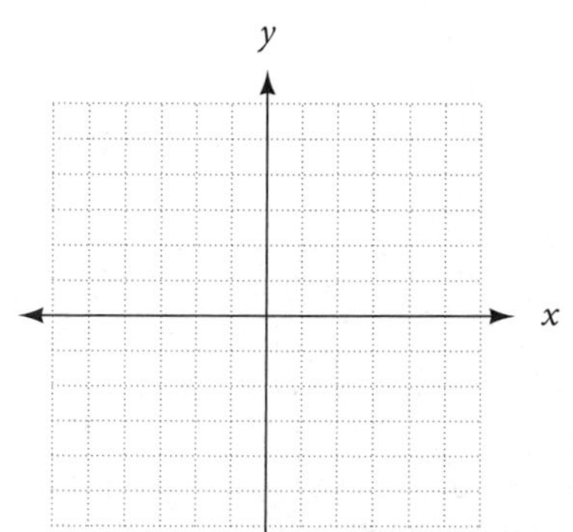

Graph the points $(-2, 4)$ and $(2, -4)$, and draw a straight line that passes through both of them. Then answer the following questions.

33. Does the graph of $(0, 0)$ lie on the line?

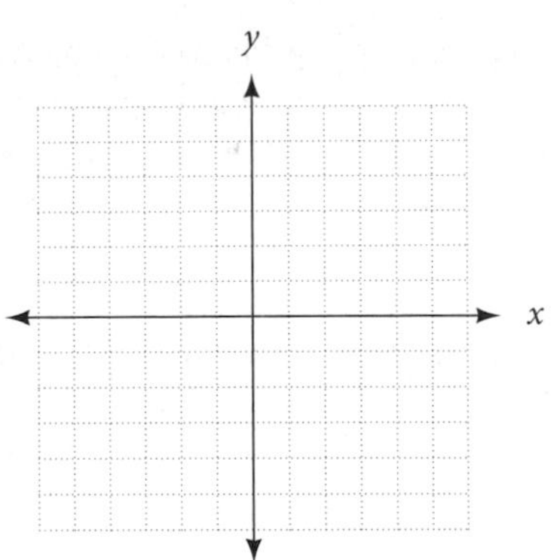

34. Does the graph of $(-1, 2)$ lie on the line?

35. Does the graph of $(2, -1)$ lie on the line?

36. Does the graph of $(1, -2)$ lie on the line?

Draw a straight line that passes through the points $(3, 4)$ and $(3, -4)$. Then answer the following questions.

37. Is the graph of $(3, 0)$ on this line?

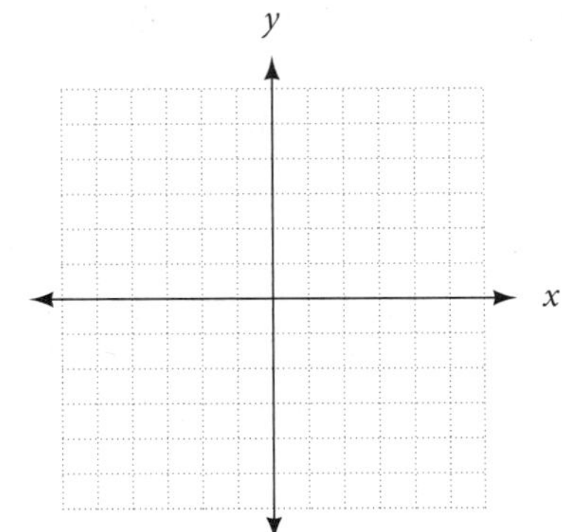

38. Is the graph of $(0, 3)$ on this line?

39. Is there any point on this line with an x-coordinate other than 3?

40. If you extended the line, would it pass through a point with a y-coordinate of 10

Draw a straight line that passes through the points $(3, 4)$ and $(-3, 4)$. Then answer the following questions.

41. Is the graph of $(4, 0)$ on this line?

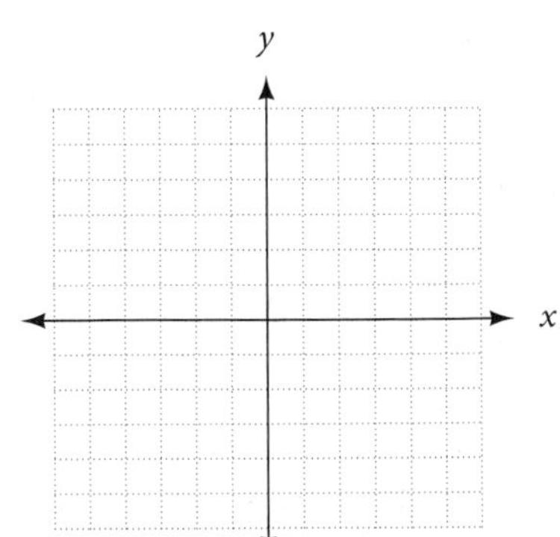

42. Is the graph of $(0, 4)$ on this line?

43. Is there any point on this line with a y-coordinate other than 4?

44. If you extended the line, would it pass through a point with an x-coordinate of 10?

A Applying the Concepts [Example 1]

45. Light Bulbs The chart shows a comparison of power usage between incandescent and energy efficient light bulbs. Using the graph, find the amount of watts used by the following bulbs.

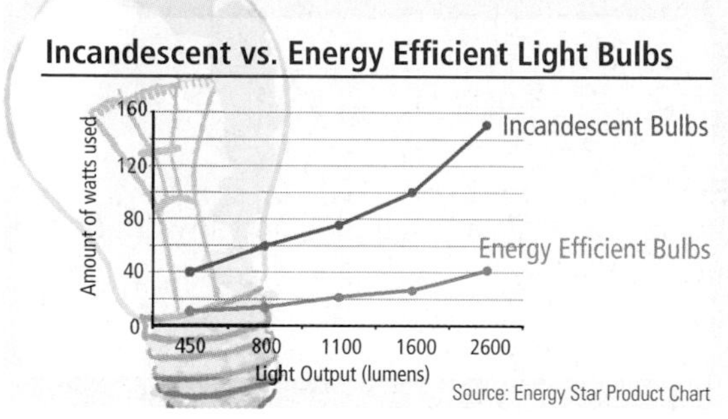

a. 800 lumen incandescent bulb

b. 1,100 lumen energy efficient bulb

c. 1,600 lumen incandescent bulb

46. Health Care The graph shows the rising cost of health care. Write the five ordered pairs that describe the information on the chart.

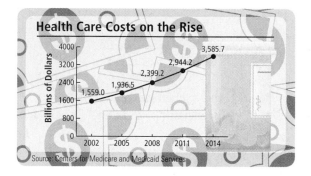

Health Care Costs on the Rise

Billions of Dollars: 0, 800, 1600, 2400, 3200, 4000

2002, 2005, 2008, 2011, 2014

1,559.0, 1,936.5, 2,399.2, 2,944.2, 3,585.7

Source: Centers for Medicare and Medicaid Services

47. Hourly Wages Jane takes a job at the local Marcy's department store. Her job pays $8.00 per hour. The graph shows how much Jane earns for working from 0 to 40 hours in a week.

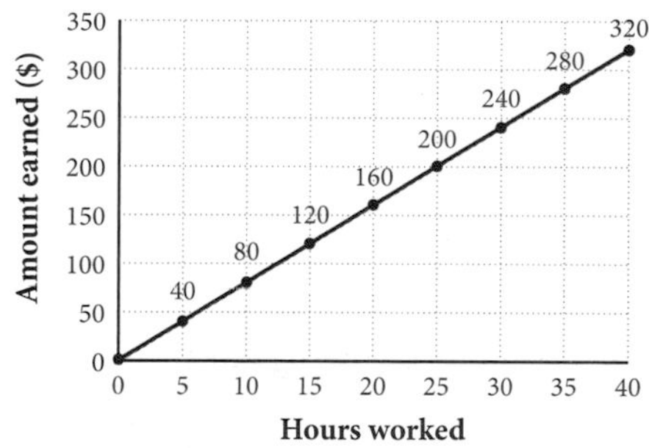

a. List three ordered pairs that lie on the line graph.

b. How much will she earn for working 40 hours?

c. If her check for one week is $240, how many hours did she work?

d. She works 35 hours one week, but her paycheck before deductions are subtracted is for $260. Is this correct? Explain.

48. Hourly Wages Judy takes a job at Gigi's boutique. Her job pays $6.00 per hour plus $50 per week in commission. The graph shows how much Judy earns for working from 0 to 40 hours in a week.

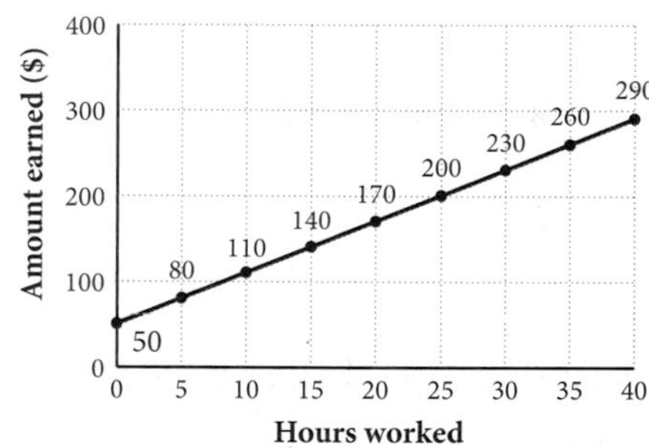

a. List three ordered pairs that lie on the line graph.

b. How much will she earn for working 40 hours?

c. If her check for one week is $230, how many hours did she work?

d. She works 35 hours one week, but her paycheck before deductions are subtracted is for $260. Is this correct? Explain.

49. Right triangle ABC has legs of length 5. Point C is the ordered pair (6, 2). Find the coordinates of A and B.

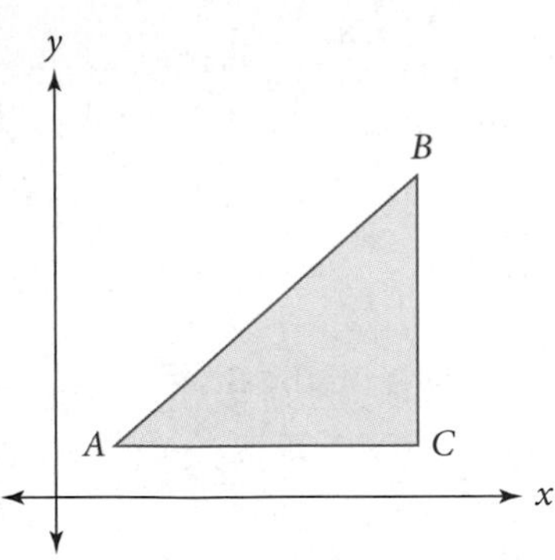

50. Right triangle ABC has legs of length 7. Point C is the ordered pair (−8, −3). Find the coordinates of A and B.

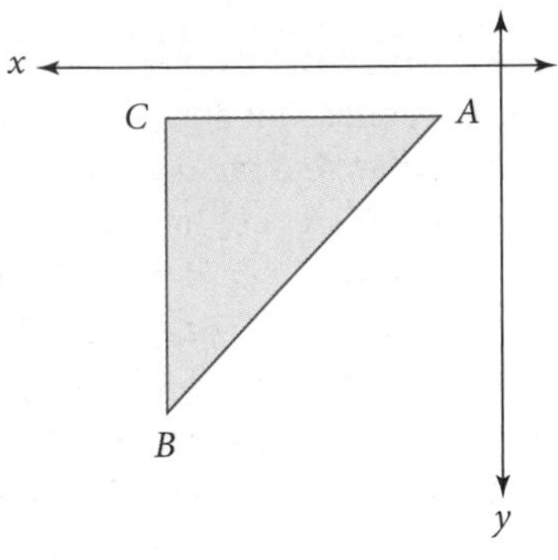

51. Rectangle $ABCD$ has a length of 5 and a width of 3. Point D is the ordered pair (7, 2). Find points A, B, and C.

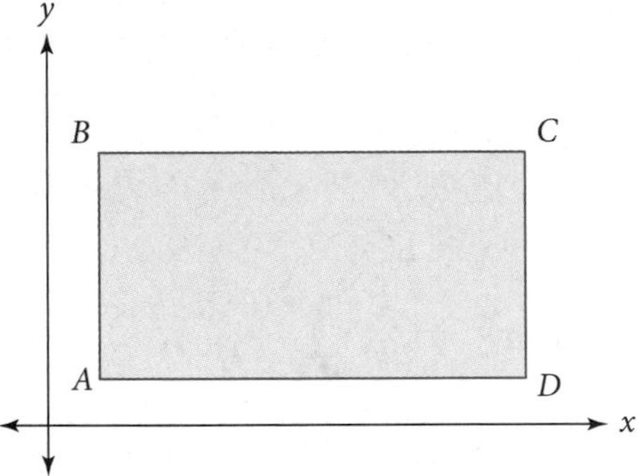

52. Rectangle $ABCD$ has a length of 5 and a width of 3. Point D is the ordered pair (−1, 1). Find points A, B, and C.

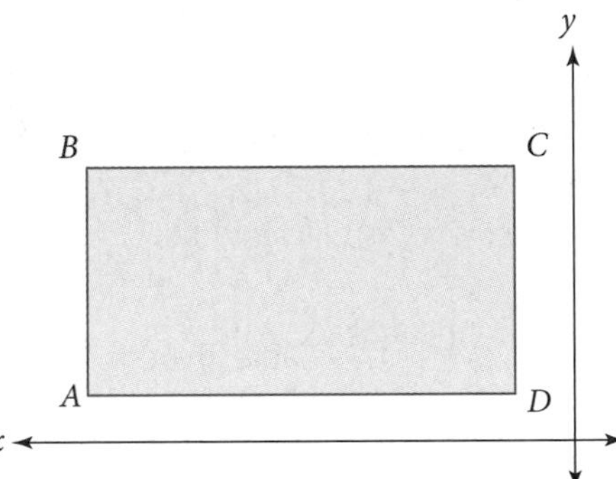

Getting Ready for the Next Section

53. Let $2x + 3y = 6$.

a. Find x if $y = 4$. **b.** Find x if $y = -2$.

c. Find y if $x = 3$. **d.** Find y if $x = 9$.

54. Let $2x - 5y = 20$.

a. Find x if $y = 0$. **b.** Find x if $y = -6$.

c. Find y if $x = 0$. **d.** Find y if $x = 5$.

55. Let $y = 2x - 1$.

a. Find x if $y = 7$. **b.** Find x if $y = 3$.

c. Find y if $x = 0$. **d.** Find y if $x = 5$.

56. Let $y = 3x - 2$.

a. Find x if $y = 4$. **b.** Find x if $y = 7$.

c. Find y if $x = 0$. **d.** Find y if $x = -3$.

Maintaining Your Skills

Add or subtract as indicated.

57. $\frac{x}{5} + \frac{3}{5}$ **58.** $\frac{x}{5} + \frac{3}{4}$ **59.** $\frac{2}{7} - \frac{a}{7}$ **60.** $\frac{2}{7} - \frac{a}{5}$ **61.** $\frac{1}{14} - \frac{y}{7}$ **62.** $\frac{3}{4} + \frac{x}{5}$

63. $\frac{1}{2} + \frac{3}{x}$ **64.** $\frac{2}{3} - \frac{6}{y}$ **65.** $\frac{5 + x}{6} - \frac{5}{6}$ **66.** $\frac{3 - x}{3} + \frac{2}{3}$ **67.** $\frac{4}{x} + \frac{1}{2}$ **68.** $\frac{3}{y} + \frac{2}{3}$

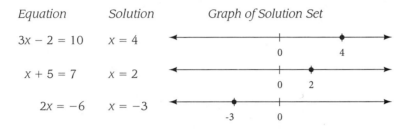

Solutions to Linear Equations in Two Variables

3.2

Objectives

A Find solutions to linear equations in two variables.

B Determine whether an ordered pair is a solution to a linear equation in two variables.

Examples now playing at
MathTV.com/books

A Solving Linear Equations

In this section we will begin to investigate equations in two variables. As you will see, equations in two variables have pairs of numbers for solutions. Because we know how to use paired data to construct tables, histograms, and other charts, we can take our work with paired data further by using equations in two variables to construct tables of paired data. Let's begin this section by reviewing the relationship between equations in one variable and their solutions.

If we solve the equation $3x - 2 = 10$, the solution is $x = 4$. If we graph this solution, we simply draw the real number line and place a dot at the point whose coordinate is 4. The relationship between linear equations in one variable, their solutions, and the graphs of those solutions look like this:

Equation	*Solution*	*Graph of Solution Set*
$3x - 2 = 10$	$x = 4$	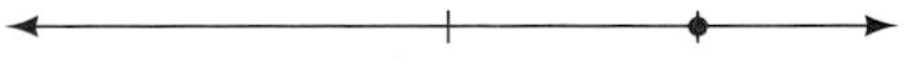
$x + 5 = 7$	$x = 2$	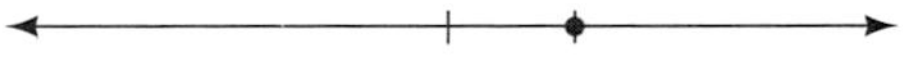
$2x = -6$	$x = -3$	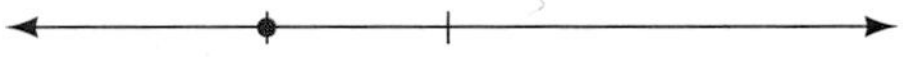

When the equation has one variable, the solution is a single number whose graph is a point on a line.

Now, consider the equation $2x + y = 3$. The first thing we notice is that there are two variables instead of one. Therefore, a solution to the equation $2x + y = 3$ will be not a single number but a pair of numbers, one for x and one for y, that makes the equation a true statement. One pair of numbers that works is $x = 2$, $y = -1$ because when we substitute them for x and y in the equation, we get a true statement.

$$2(2) + (-1) \stackrel{?}{=} 3$$
$$4 - 1 = 3$$
$$3 = 3 \quad \text{A true statement}$$

The pair of numbers $x = 2, y = -1$ is written as $(2, -1)$. As you know from Section 3.1, $(2, -1)$ is called an *ordered pair* because it is a pair of numbers written in a specific order. The first number is always associated with the variable x, and the second number is always associated with the variable y. We call the first number in the ordered pair the *x-coordinate* (or x component) and the second number the *y-coordinate* (or y component) of the ordered pair.

Let's look back to the equation $2x + y = 3$. The ordered pair $(2, -1)$ is not the only solution. Another solution is $(0, 3)$ because when we substitute 0 for x and 3 for y we get

$$2(0) + 3 \stackrel{?}{=} 3$$
$$0 + 3 = 3$$
$$3 = 3 \quad \text{A true statement}$$

Note If this discussion seems a little long and confusing, you may want to look over some of the examples first and then come back and read this. Remember, it isn't always easy to read material in mathematics. What is important is that you understand what you are doing when you work problems. The reading is intended to assist you in understanding what you are doing. It is important to read everything in the book, but you don't always have to read it in the order it is written.

Still another solution is the ordered pair $(5, -7)$ because

$$2(5) + (-7) \stackrel{?}{=} 3$$
$$10 - 7 = 3$$
$$3 = 3 \quad \text{A true statement}$$

As a matter of fact, for any number we want to use for x, there is another number we can use for y that will make the equation a true statement. There is an infinite number of ordered pairs that satisfy (are solutions to) the equation $2x + y = 3$; we have listed just a few of them.

PRACTICE PROBLEMS

1. For the equation $2x + 5y = 10$, complete the ordered pairs $(0,\ )$, $(\ , 1)$, and $(5,\ )$.

EXAMPLE 1 Given the equation $2x + 3y = 6$, complete the following ordered pairs so they will be solutions to the equation: $(0,\ )$, $(\ , 1)$, $(3,\ )$.

SOLUTION To complete the ordered pair $(0,\ )$, we substitute 0 for x in the equation and then solve for y:

$$2(0) + 3y = 6$$
$$3y = 6$$
$$y = 2$$

The ordered pair is $(0, 2)$.

To complete the ordered pair $(\ , 1)$, we substitute 1 for y in the equation and solve for x:

$$2x + 3(1) = 6$$
$$2x + 3 = 6$$
$$2x = 3$$
$$x = \frac{3}{2}$$

The ordered pair is $\left(\frac{3}{2}, 1\right)$.

To complete the ordered pair $(3,\ )$, we substitute 3 for x in the equation and solve for y:

$$2(3) + 3y = 6$$
$$6 + 3y = 6$$
$$3y = 0$$
$$y = 0$$

The ordered pair is $(3, 0)$.

Notice in each case in Example 1 that once we have used a number in place of one of the variables, the equation becomes a linear equation in one variable. We then use the method explained in Chapter 2 to solve for that variable.

2. Complete the table for $3x - 2y = 12$.

x	y
0	
	3
	0
−3	

EXAMPLE 2 Complete the following table for the equation $2x - 5y = 20$.

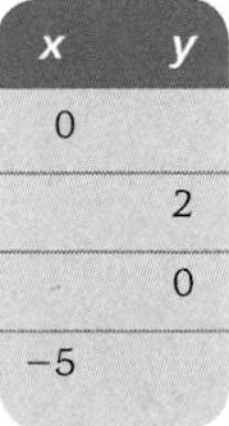

x	y
0	
	2
	0
−5	

Answer

1. $(0, 2)$, $\left(\frac{5}{2}, 1\right)$, $(5, 0)$

SOLUTION Filling in the table is equivalent to completing the following ordered pairs: (0,), (, 2), (, 0), (−5,). So we proceed as in Example 1.

When $x = 0$, we have

$$2(0) - 5y = 20$$
$$0 - 5y = 20$$
$$-5y = 20$$
$$y = -4$$

When $y = 2$, we have

$$2x - 5(2) = 20$$
$$2x - 10 = 20$$
$$2x = 30$$
$$x = 15$$

When $y = 0$, we have

$$2x - 5(0) = 20$$
$$2x - 0 = 20$$
$$2x = 20$$
$$x = 10$$

When $x = -5$, we have

$$2(-5) - 5y = 20$$
$$-10 - 5y = 20$$
$$-5y = 30$$
$$y = -6$$

The completed table looks like this:

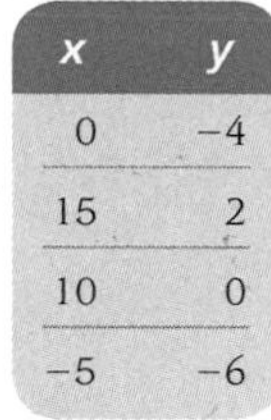

x	y
0	−4
15	2
10	0
−5	−6

which is equivalent to the ordered pairs (0, −4), (15, 2), (10, 0), and (−5, −6).

EXAMPLE 3 Complete the following table for the equation $y = 2x - 1$.

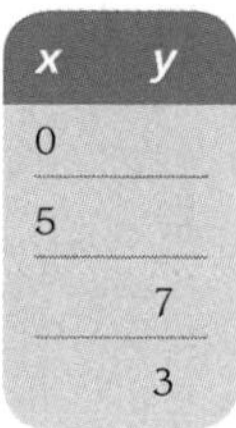

x	y
0	
5	
	7
	3

SOLUTION When $x = 0$, we have

$$y = 2(0) - 1$$
$$y = 0 - 1$$
$$y = -1$$

When $x = 5$, we have

$$y = 2(5) - 1$$
$$y = 10 - 1$$
$$y = 9$$

When $y = 7$, we have

$$7 = 2x - 1$$
$$8 = 2x$$
$$4 = x$$

When $y = 3$, we have

$$3 = 2x - 1$$
$$4 = 2x$$
$$2 = x$$

3. Complete the table for $y = 3x - 2$.

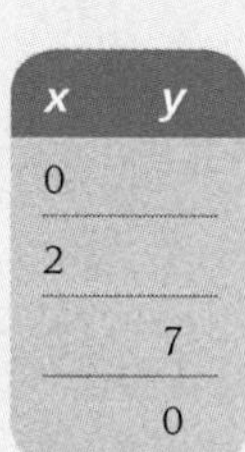

x	y
0	
2	
	7
	0

Answer

2. (0, −6), (6, 3), (4, 0), $\left(-3, -\frac{21}{2}\right)$

The completed table is

x	y
0	−1
5	9
4	7
2	3

which means the ordered pairs (0, −1), (5, 9), (4, 7), and (2, 3) are among the solutions to the equation $y = 2x - 1$.

B Testing Solutions

4. Which of the following ordered pairs is a solution to $y = 4x + 1$? (0, 1), (3, 11), (2, 9)

EXAMPLE 4 Which of the ordered pairs (2, 3), (1, 5), and (−2, −4) are solutions to the equation $y = 3x + 2$?

SOLUTION If an ordered pair is a solution to the equation, then it must satisfy the equation; that is, when the coordinates are used in place of the variables in the equation, the equation becomes a true statement.

Try (2, 3) in $y = 3x + 2$:

$$3 \stackrel{?}{=} 3(2) + 2$$
$$3 = 6 + 2$$
$$3 = 8 \qquad \text{A false statement}$$

Try (1, 5) in $y = 3x + 2$:

$$5 \stackrel{?}{=} 3(1) + 2$$
$$5 = 3 + 2$$
$$5 = 5 \qquad \text{A true statement}$$

Try (−2, −4) in $y = 3x + 2$:

$$-4 \stackrel{?}{=} 3(-2) + 2$$
$$-4 = -6 + 2$$
$$-4 = -4 \qquad \text{A true statement}$$

The ordered pairs (1, 5) and (−2, −4) are solutions to the equation $y = 3x + 2$, and (2, 3) is not.

Getting Ready for Class

After reading through the preceding section, respond in your own words and in complete sentences.

1. How can you tell if an ordered pair is a solution to an equation?
2. How would you find a solution to $y = 3x - 5$?
3. Why is (3, 2) not a solution to $y = 3x - 5$?
4. How many solutions are there to an equation that contains two variables?

Answers

3. (0, −2), (2, 4), (3, 7), $\left(\frac{2}{3}, 0\right)$
4. (0, 1) and (2, 9)

Problem Set 3.2

A For each equation, complete the given ordered pairs. [Examples 1–3]

1. $2x + y = 6$ $(0, \ ), (\ , 0), (\ , -6)$

2. $3x - y = 5$ $(0, \ ), (1, \ ), (\ , 5)$

3. $3x + 4y = 12$ $(0, \ ), (\ , 0), (-4, \ )$

4. $5x - 5y = 20$ $(0, \ ), (\ , -2), (1, \ )$

5. $y = 4x - 3$ $(1, \ ), (\ , 0), (5, \ )$

6. $y = 3x - 5$ $(\ , 13), (0, \ ), (-2, \ )$

7. $y = 7x - 1$ $(2, \ ), (\ , 6), (0, \ )$

8. $y = 8x + 2$ $(3, \ ), (\ , 0), (\ , -6)$

9. $x = -5$ $(\ , 4), (\ , -3), (\ , 0)$

10. $y = 2$ $(5, \ ), (-8, \ ), \left(\frac{1}{2}, \ \right)$

A For each of the following equations, complete the given table. **[Examples 2–3]**

11. $y = 3x$

x	y
1	3
−3	
	12
	18

12. $y = -2x$

x	y
−4	
0	
	10
	12

13. $y = 4x$

x	y
0	
	−2
−3	
	12

14. $y = -5x$

x	y
3	
	0
−2	
	−20

15. $x + y = 5$

x	y
2	
3	
	0
	−4

16. $x - y = 8$

x	y
0	
4	
	−3
	−2

17. $2x - y = 4$

x	y
	0
	2
1	
−3	

18. $3x - y = 9$

x	y
	0
	−9
5	
−4	

19. $y = 6x - 1$

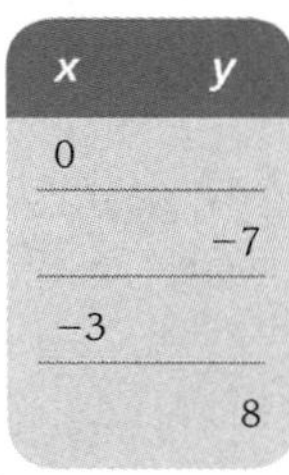

x	y
0	
	−7
−3	
	8

20. $y = 5x + 7$

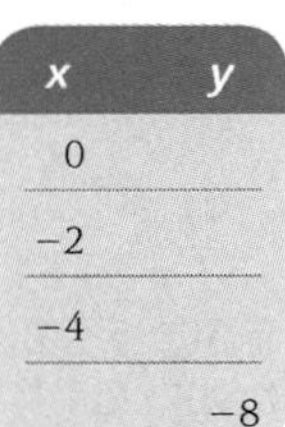

x	y
0	
−2	
−4	
	−8

B For the following equations, tell which of the given ordered pairs are solutions. **[Example 4]**

21. $2x - 5y = 10$ $(2, 3), (0, -2), \left(\frac{5}{2}, 1\right)$

22. $3x + 7y = 21$ $(0, 3), (7, 0), (1, 2)$

23. $y = 7x - 2$ $(1, 5), (0, -2), (-2, -16)$

24. $y = 8x - 3$ $(0, 3), (5, 16), (1, 5)$

25. $y = 6x$ $(1, 6), (-2, -12), (0, 0)$

26. $y = -4x$ $(0, 0), (2, 4), (-3, 12)$

27. $x + y = 0$ $(1, 1), (2, -2), (3, 3)$

28. $x - y = 1$ $(0, 1), (0, -1), (1, 2)$

29. $x = 3$ $(3, 0), (3, -3), (5, 3)$

30. $y = -4$ $(3, -4), (-4, 4), (0, -4)$

Applying the Concepts

31. **Wind Energy** The graph shows the world's capacity for generating power from wind energy. The line segment from 2000 to 2005 has the equation $y = 7{,}000x - 13{,}980{,}000$. Use it to complete the ordered pairs.
(, 20,000), (2003,), (, 48,000)

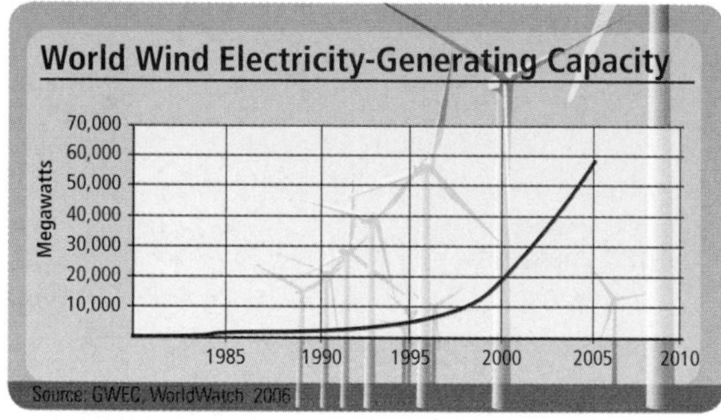

32. **Light Bulbs** The chart shows a comparison of power usage between incandescent and energy efficient light bulbs. The equation of the line segment from 1600 to 2600 lumens of an incandescent bulb is given by $y = \left(\frac{1}{20}\right)x + 20$. Use it to complete the ordered pairs.
(1800,), (, 120), (2400,)

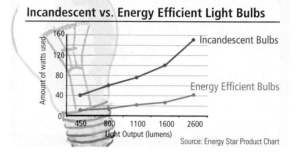

Incandescent vs. Energy Efficient Light Bulbs

Incandescent Bulbs
Energy Efficient Bulbs
Amount of watts used: 0, 40, 80, 120, 160
Light Output (lumens): 450, 800, 1100, 1600, 2600
Source: Energy Star Product Chart

33. **Perimeter** If the perimeter of a rectangle is 30 inches, then the relationship between the length l and the width w is given by the equation

$$2l + 2w = 30$$

What is the length when the width is 3 inches?

34. **Perimeter** The relationship between the perimeter P of a square and the length of its side s is given by the formula $P = 4s$. If each side of a square is 5 inches, what is the perimeter? If the perimeter of a square is 28 inches, how long is a side?

35. Janai earns \$12 per hour working as a math tutor. We can express the amount she earns each week, y, for working x hours with the equation $y = 12x$. Indicate with a *yes* or *no* which of the following could be one of Janai's paychecks. If you answer no, explain your answer.
   a. \$60 for working 5 hours
   b. \$100 for working nine hours
   c. \$80 for working 7 hours
   d. \$168 for working 14 hours

36. Erin earns \$15 per hour working as a graphic designer. We can express the amount she earns each week, y, for working x hours with the equation $y = 15x$. Indicate with a *yes* or *no* which of the following could be one of Erin's paychecks. If you answer no, explain your answer.
   a. \$75 for working 5 hours
   b. \$125 for working 9 hours.
   c. \$90 for working 6 hours
   d. \$500 for working 35 hours.

37. The equation $V = -45{,}000t + 600{,}000$ can be used to find the value, V, of a small crane at the end of t years.
   a. What is the value of the crane at the end of 5 years?
   b. When is the crane worth \$330,000?
   c. Is it true that the crane will be worth \$150,000 after 9 years?
   d. How much did the crane cost?

38. The equation $V = -400t + 2{,}500$, can be used to find the value, V, of a notebook computer at the end of t years.
   a. What is the price of the notebook computer at the end of 4 years?
   b. When is the notebook computer worth \$1,700?
   c. Is it true that the notebook computer will be worth \$100 after 5 years?
   d. How much did the notebook computer cost?

Getting Ready for the Next Section

39. Find y when x is 4 in the formula $3x + 2y = 6$.

40. Find y when x is 0 in the formula $3x + 2y = 6$.

41. Find y when x is 0 in $y = -\frac{1}{3}x + 2$.

42. Find y when x is 3 in $y = -\frac{1}{3}x + 2$.

43. Find y when x is 2 in $y = \frac{3}{2}x - 3$.

44. Find y when x is 4 in $y = \frac{3}{2}x - 3$.

45. Solve $5x + y = 4$ for y.

46. Solve $-3x + y = 5$ for y.

47. Solve $3x - 2y = 6$ for y.

48. Solve $2x - 3y = 6$ for y.

Maintaining Your Skills

49. $\frac{11(-5) - 17}{2(-6)}$

50. $\frac{12(-4) + 15}{3(-11)}$

51. $\frac{13(-6) + 18}{4(-5)}$

52. $\frac{9^2 - 6^2}{-9 - 6}$

53. $\frac{7^2 - 5^2}{(7 - 5)^2}$

54. $\frac{7^2 - 2^2}{-7 - 2}$

55. $\frac{-3 \cdot 4^2 - 3 \cdot 2^4}{-3(8)}$

56. $\frac{-4(8 - 13) - 2(6 - 11)}{-5(3) + 5}$

Graphing Linear Equations in Two Variables

3.3

Objectives

A Graph a linear equation in two variables.

B Graph horizontal lines, vertical lines, and lines through the origin.

Examples now playing at MathTV.com/books

A Graphing

In this section we will use the rectangular coordinate system introduced in Section 3.1 to obtain a visual picture of *all* solutions to a linear equation in two variables. The process we use to obtain a visual picture of all solutions to an equation is called *graphing*. The picture itself is called the *graph* of the equation.

EXAMPLE 1 Graph the solution set for $x + y = 5$.

SOLUTION We know from the previous section that an infinite number of ordered pairs are solutions to the equation $x + y = 5$. We can't possibly list them all. What we can do is list a few of them and see if there is any pattern to their graphs.

Some ordered pairs that are solutions to $x + y = 5$ are (0, 5), (2, 3), (3, 2), (5, 0). The graph of each is shown in Figure 1.

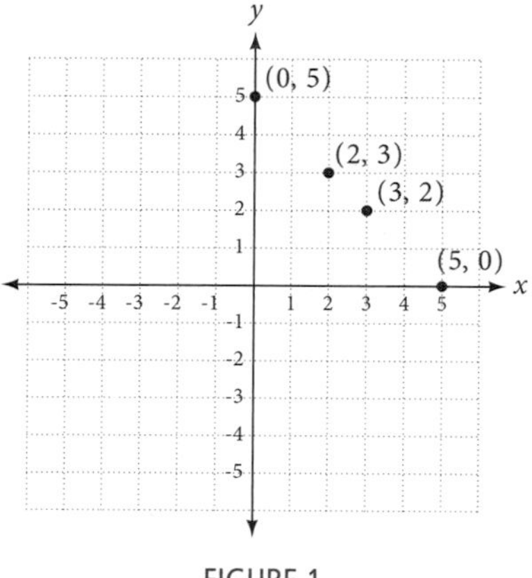

FIGURE 1

Now, by passing a straight line through these points, we can graph the solution set for the equation $x + y = 5$. Linear equations in two variables always have graphs that are straight lines. The graph of the solution set for $x + y = 5$ is shown in Figure 2.

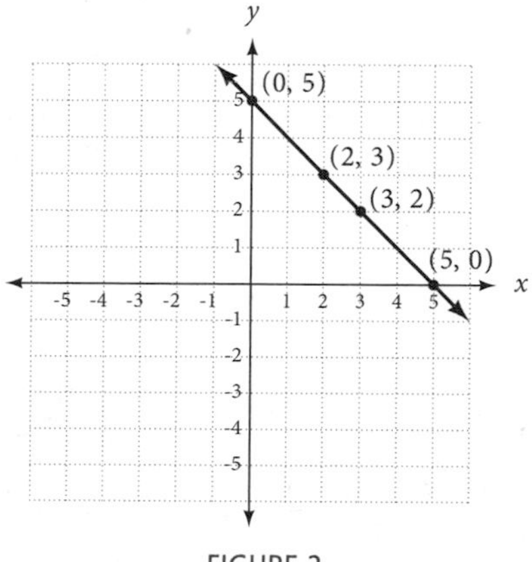

FIGURE 2

Every ordered pair that satisfies $x + y = 5$ has its graph on the line, and any point on the line has coordinates that satisfy the equation. So, there is a one-to-one correspondence between points on the line and solutions to the equation.

PRACTICE PROBLEMS

1. Graph the equation $x + y = 3$.

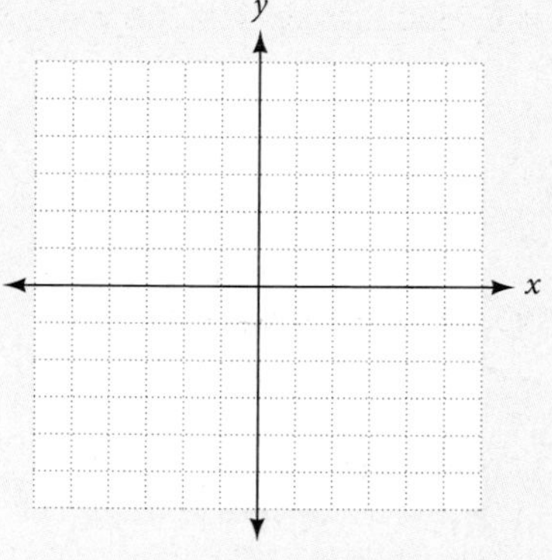

Answers

See Solutions to Selected Practice Problems for all answers in this section.

Our ability to graph an equation as we have done in Example 1 is due to the invention of the rectangular coordinate system. The French philosopher René Descartes (1596–1650) is the person usually credited with the invention of the rectangular coordinate system. As a philosopher, Descartes is responsible for the statement "I think, therefore I am." Until Descartes invented his coordinate system in 1637, algebra and geometry were treated as separate subjects. The rectangular coordinate system allows us to connect algebra and geometry by associating geometric shapes with algebraic equations.

Here is the precise definition for a linear equation in two variables.

Definition

Any equation that can be put in the form $ax + by = c$, where a, b, and c are real numbers and a and b are not both 0, is called a **linear equation in two variables.** The graph of any equation of this form is a straight line (that is why these equations are called "linear"). The form $ax + by = c$ is called **standard form.**

To graph a linear equation in two variables, we simply graph its solution set; that is, we draw a line through all the points whose coordinates satisfy the equation. Here are the steps to follow.

Note The meaning of the *convenient numbers* referred to in step 1 of the strategy for graphing a linear equation in two variables will become clear as you read the next two examples.

Strategy Graphing a Linear Equation in Two Variables

Step 1: Find any three ordered pairs that satisfy the equation. This can be done by using a convenient number for one variable and solving for the other variable.

Step 2: Graph the three ordered pairs found in step 1. Actually, we need only two points to graph a straight line. The third point serves as a check. If all three points do not line up, there is a mistake in our work.

Step 3: Draw a straight line through the three points graphed in step 2.

2. Graph the equation $y = 2x + 3$.

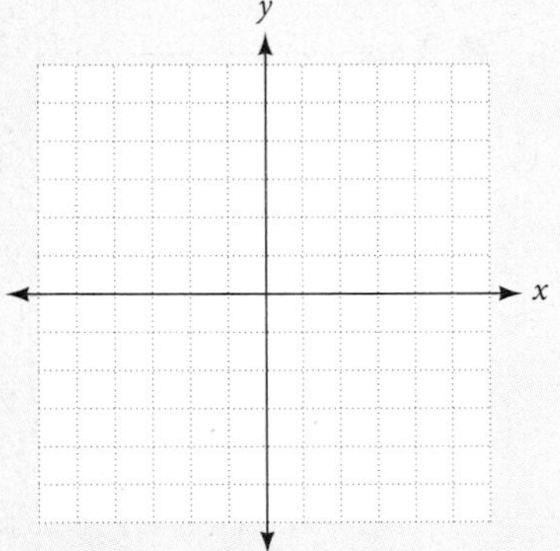

EXAMPLE 2 Graph the equation $y = 3x - 1$.

SOLUTION Because $y = 3x - 1$ can be put in the form $ax + by = c$, it is a linear equation in two variables. Hence, the graph of its solution set is a straight line. We can find some specific solutions by substituting numbers for x and then solving for the corresponding values of y. We are free to choose any numbers for x, so let's use 0, 2, and -1.

Let $x = 0$: $\quad y = 3(0) - 1$
$\quad y = 0 - 1$
$\quad y = -1$

The ordered pair $(0, -1)$ is one solution.

Let $x = 2$: $\quad y = 3(2) - 1$
$\quad y = 6 - 1$
$\quad y = 5$

The ordered pair $(2, 5)$ is a second solution.

Let $x = -1$: $\quad y = 3(-1) - 1$
$\quad y = -3 - 1$
$\quad y = -4$

The ordered pair $(-1, -4)$ is a third solution.

In table form

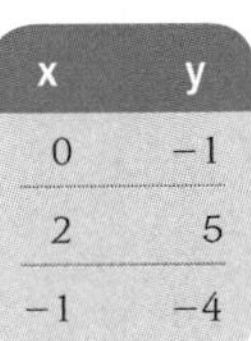

x	y
0	−1
2	5
−1	−4

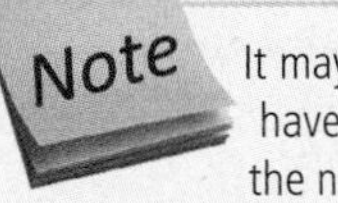

It may seem that we have simply picked the numbers 0, 2, and −1 out of the air and used them for x. In fact we have done just that. Could we have used numbers other than these? The answer is yes. We can substitute any number for x. There will always be a value of y to go with it.

Next, we graph the ordered pairs (0, −1), (2, 5), (−1, −4) and draw a straight line through them.

The line we have drawn in Figure 3 is the graph of $y = 3x - 1$.

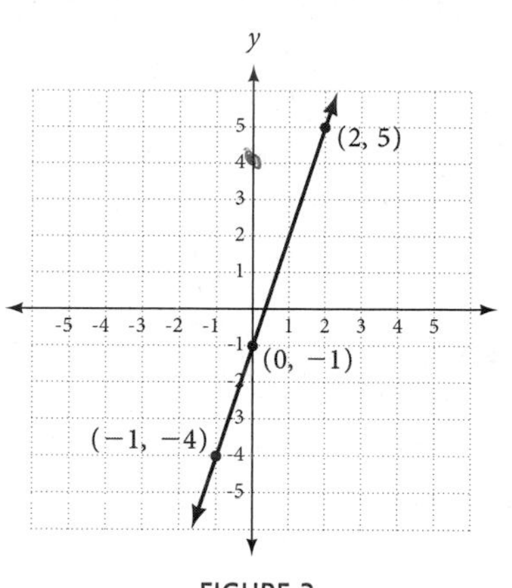

FIGURE 3

Example 2 again illustrates the connection between algebra and geometry that we mentioned previously. Descartes's rectangular coordinate system allows us to associate the equation $y = 3x - 1$ (an algebraic concept) with a specific straight line (a geometric concept). The study of the relationship between equations in algebra and their associated geometric figures is called *analytic geometry.* The rectangular coordinate system often is referred to as the *Cartesian coordinate system* in honor of Descartes.

EXAMPLE 3 Graph the equation $y = -\frac{1}{3}x + 2$.

SOLUTION We need to find three ordered pairs that satisfy the equation. To do so, we can let x equal any numbers we choose and find corresponding values of y. But, every value of x we substitute into the equation is going to be multiplied by $-\frac{1}{3}$. Let's use numbers for x that are divisible by 3, like −3, 0, and 3. That way, when we multiply them by $-\frac{1}{3}$, the result will be an integer.

Let $x = -3$: $y = -\frac{1}{3}(-3) + 2$

$y = 1 + 2$

$y = 3$

The ordered pair (−3, 3) is one solution.

Let $x = 0$: $y = -\frac{1}{3}(0) + 2$

$y = 0 + 2$

$y = 2$

The ordered pair (0, 2) is a second solution.

Let $x = 3$: $y = -\frac{1}{3}(3) + 2$

$y = -1 + 2$

$y = 1$

The ordered pair (3, 1) is a third solution.

In table form

x	y
-3	3
0	2
3	1

3. Graph $y = \frac{3}{2}x - 3$.

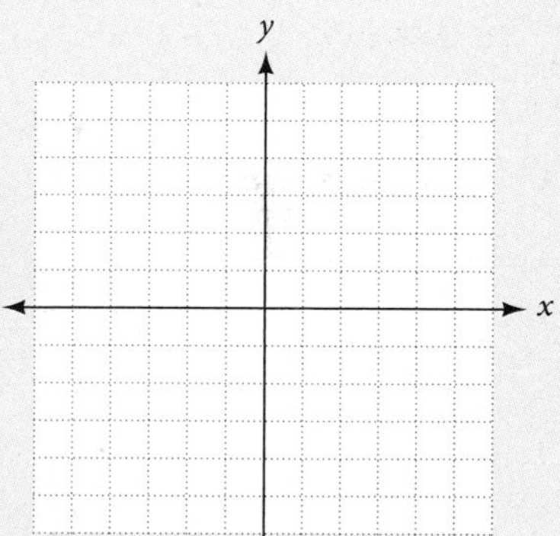

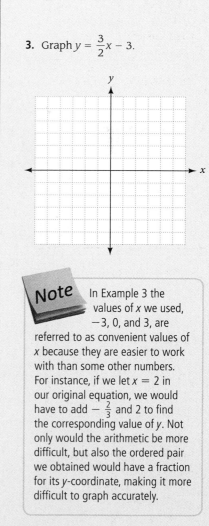

Note In Example 3 the values of x we used, −3, 0, and 3, are referred to as convenient values of x because they are easier to work with than some other numbers. For instance, if we let $x = 2$ in our original equation, we would have to add $-\frac{2}{3}$ and 2 to find the corresponding value of y. Not only would the arithmetic be more difficult, but also the ordered pair we obtained would have a fraction for its y-coordinate, making it more difficult to graph accurately.

Graphing the ordered pairs $(-3, 3)$, $(0, 2)$, and $(3, 1)$ and drawing a straight line through their graphs, we have the graph of the equation $y = -\frac{1}{3}x + 2$, as shown in Figure 4.

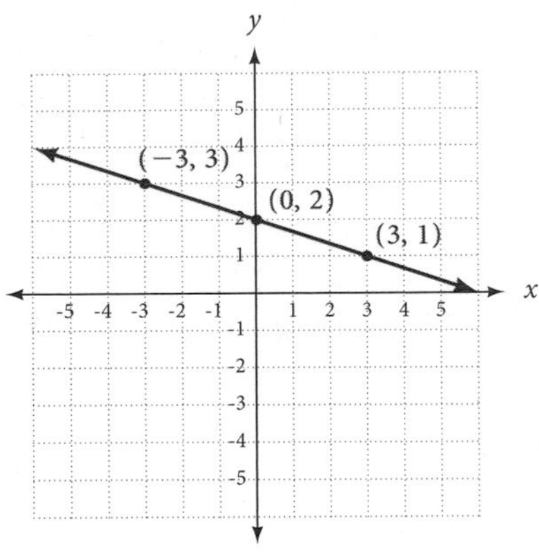

FIGURE 4

EXAMPLE 4 Graph the solution set for $3x - 2y = 6$.

SOLUTION It will be easier to find convenient values of x to use in the equation if we first solve the equation for y. To do so, we add $-3x$ to each side, and then we multiply each side by $-\frac{1}{2}$.

$3x - 2y = 6$	Original equation
$-2y = -3x + 6$	Add $-3x$ to each side
$-\frac{1}{2}(-2y) = -\frac{1}{2}(-3x + 6)$	Multiply each side by $-\frac{1}{2}$
$y = \frac{3}{2}x - 3$	Simplify each side

Now, because each value of x will be multiplied by $\frac{3}{2}$, it will be to our advantage to choose values of x that are divisible by 2. That way, we will obtain values of y that do not contain fractions. This time, let's use 0, 2, and 4 for x.

When $x = 0$:	$y = \frac{3}{2}(0) - 3$	
	$y = 0 - 3$	
	$y = -3$	$(0, -3)$ is one solution
When $x = 2$:	$y = \frac{3}{2}(2) - 3$	
	$y = 3 - 3$	
	$y = 0$	$(2, 0)$ is a second solution
When $x = 4$:	$y = \frac{3}{2}(4) - 3$	
	$y = 6 - 3$	
	$y = 3$	$(4, 3)$ is a third solution

4. Graph the solution set for $2x - 4y = 8$.

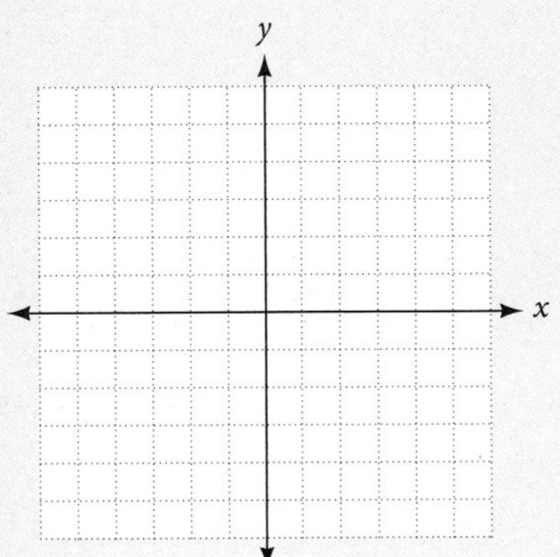

Graphing the ordered pairs (0, −3), (2, 0), and (4, 3) and drawing a line through them, we have the graph shown in Figure 5.

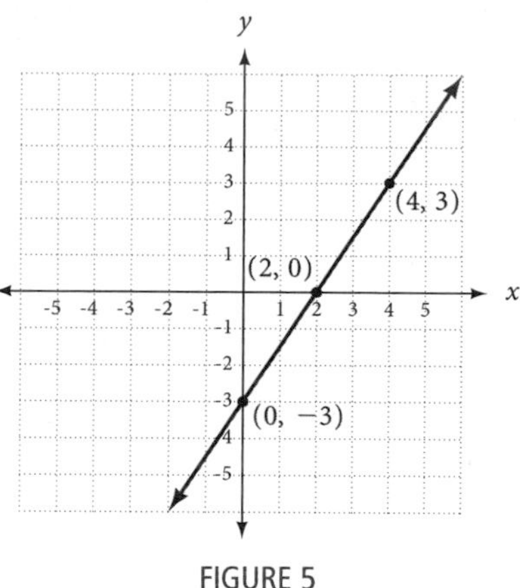

FIGURE 5

After reading through Example 4, many students ask why we didn't use −2 for x when we were finding ordered pairs that were solutions to the original equation. The answer is, we could have. If we were to let $x = -2$, the corresponding value of y would have been −6. As you can see by looking at the graph in Figure 5, the ordered pair (−2, −6) is on the graph.

B Special Equations

EXAMPLE 5 Graph each of the following lines.

a. $y = \frac{1}{2}x$ **b.** $x = 3$ **c.** $y = -2$

SOLUTION

a. The line $y = \frac{1}{2}x$ passes through the origin because (0, 0) satisfies the equation. To sketch the graph, we need at least one more point on the line. When x is 2, we obtain the point (2, 1), and when x is −4, we obtain the point (−4, −2). The graph of $y = \frac{1}{2}x$ is shown in Figure 6a.

b. The line $x = 3$ is the set of all points whose x-coordinate is 3. The variable y does not appear in the equation, so the y-coordinate can be any number. Note that we can write our equation as a linear equation in two variables by writing it as $x + 0y = 3$. Because the product of 0 and y will always be 0, y can be any number. The graph of $x = 3$ is the vertical line shown in Figure 6b.

c. The line $y = -2$ is the set of all points whose y-coordinate is −2. The variable x does not appear in the equation, so the x-coordinate can be any number. Again, we can write our equation as a linear equation in two variables by writing it as $0x + y = -2$. Because the product of 0 and x will always be 0, x can be any number. The graph of $y = -2$ is the horizontal line shown in Figure 6c.

5. Graph the line $x = -1$ and the line $y = 4$.

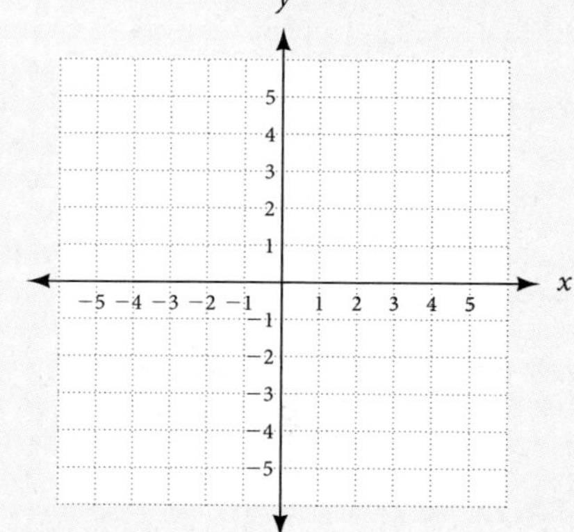

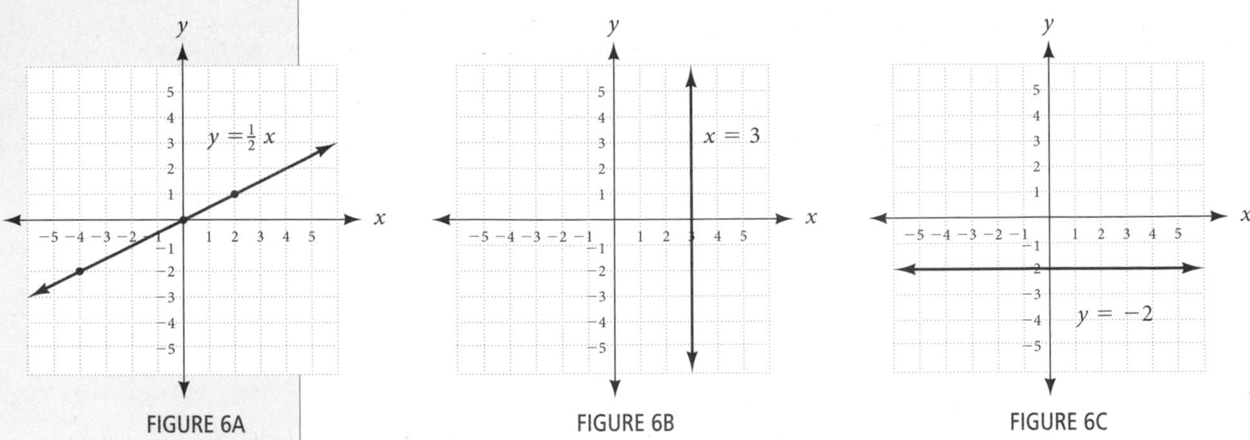

FIGURE 6A FIGURE 6B FIGURE 6C

FACTS FROM GEOMETRY: Special Equations and Their Graphs

For the equations below, m, a, and b are real numbers.

Through the Origin Vertical Line Horizontal Line

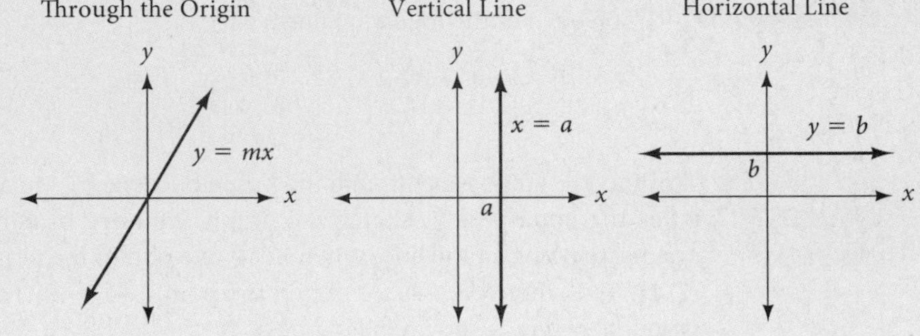

FIGURE 7A **Any equation of the form $y = mx$ has a graph that passes through the origin.**

FIGURE 7B **Any equation of the form $x = a$ has a vertical line for its graph.**

FIGURE 7C **Any equation of the form $y = b$ has a horizontal line for its graph.**

Getting Ready for Class

After reading through the preceding section, respond in your own words and in complete sentences.

1. Explain how you would go about graphing the line $x + y = 5$.
2. When graphing straight lines, why is it a good idea to find three points, when every straight line is determined by only two points?
3. What kind of equations have vertical lines for graphs?
4. What kind of equations have horizontal lines for graphs?

Problem Set 3.3

A B For the following equations, complete the given ordered pairs, and use the results to graph the solution set for the equation. **[Examples 1–3, 5]**

1. $x + y = 4$ (0,), (2,), (, 0)

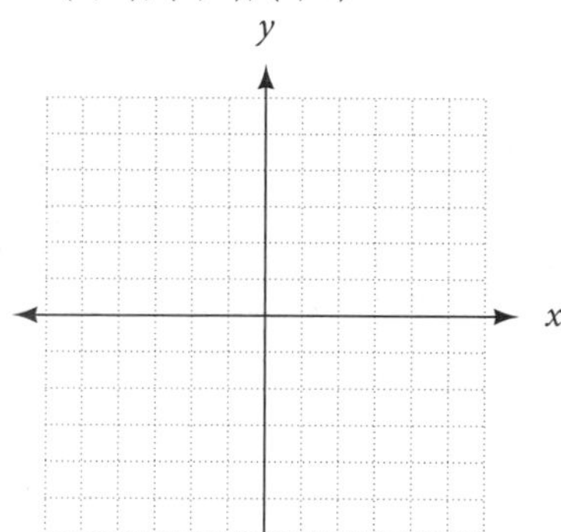

2. $x - y = 3$ (0,), (2,), (, 0)

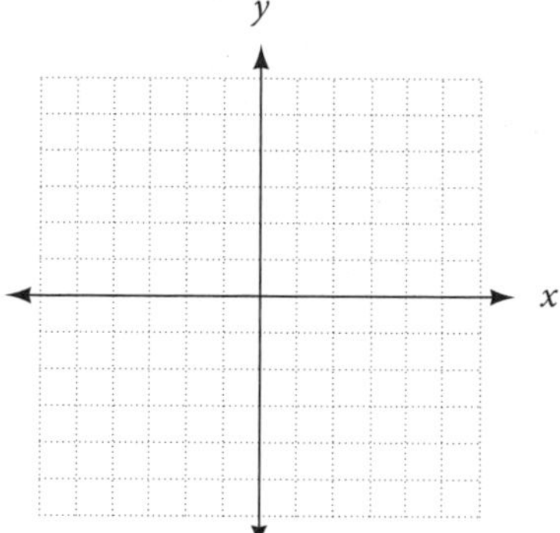

3. $x + y = 3$ (0,), (2,), (, −1)

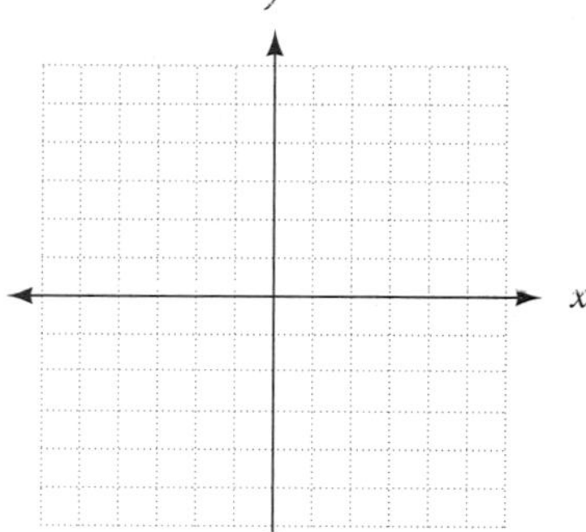

4. $x - y = 4$ (1,), (−1,), (, 0)

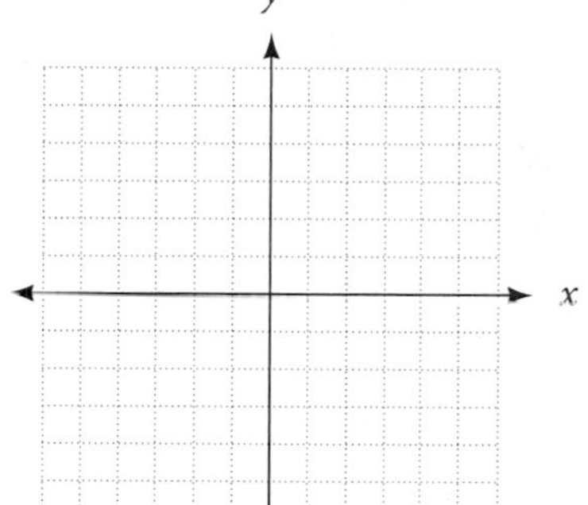

5. $y = 2x$ (0,), (−2,), (2,)

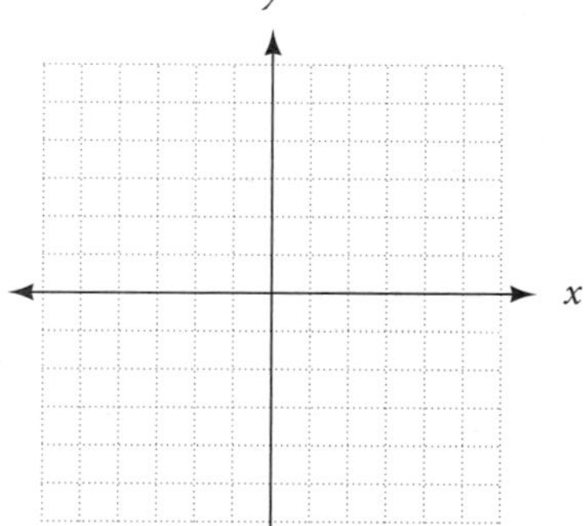

6. $y = \frac{1}{2}x$ (0,), (−2,), (2,)

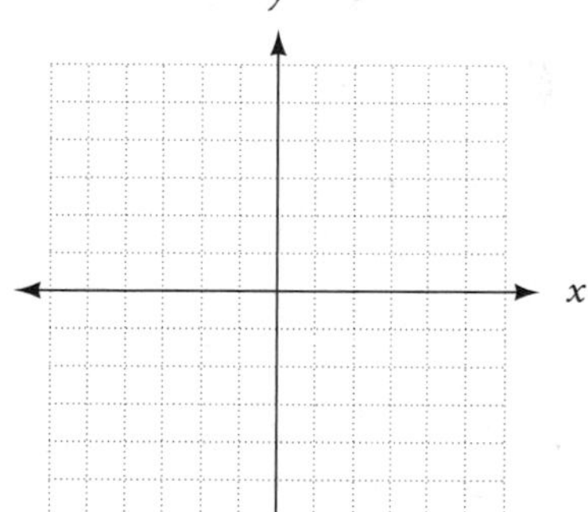

7. $y = \frac{1}{3}x$ (−3,), (0,), (3,)

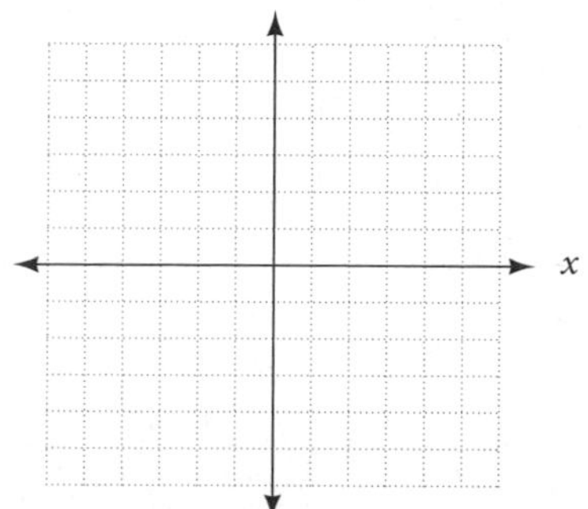

8. $y = 3x$ (−2,), (0,), (2,)

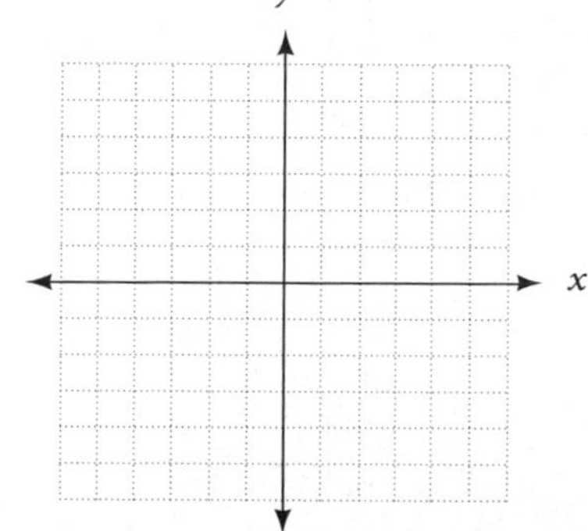

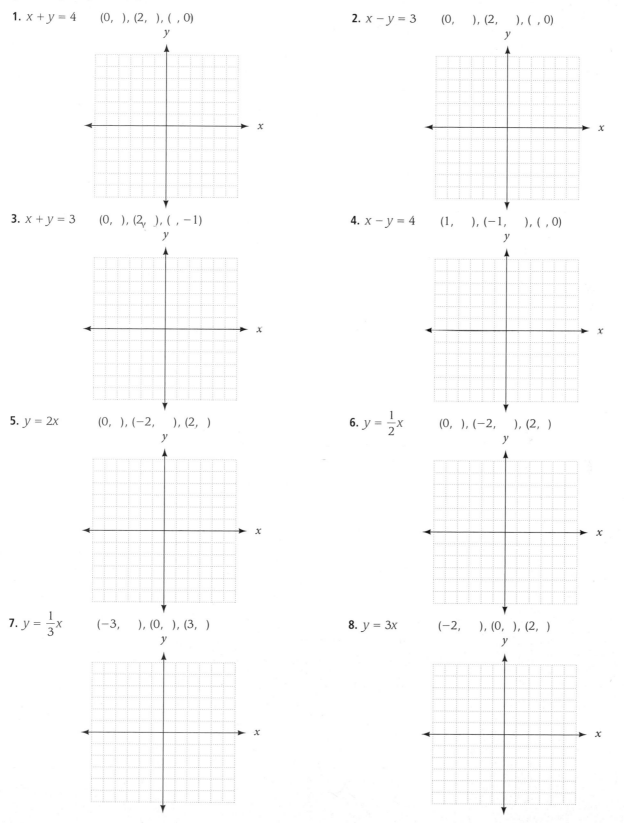

9. $y = 2x + 1$ (0,), (−1,), (1,)

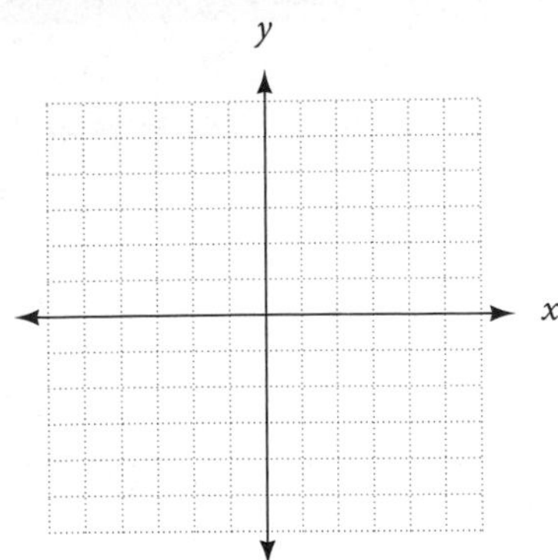

10. $y = -2x + 1$ (0,), (−1,), (1,)

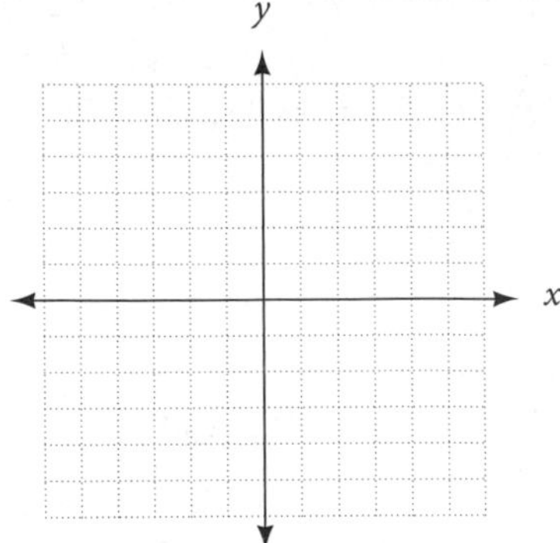

11. $y = 4$ (0,), (−1,), (2,)

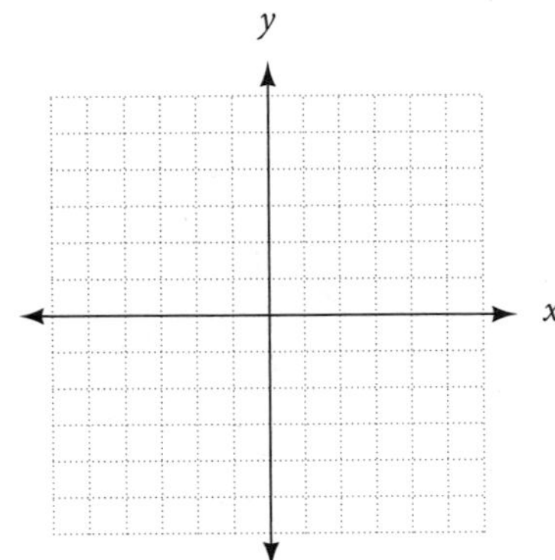

12. $x = 3$ (, −2), (, 0), (, 5)

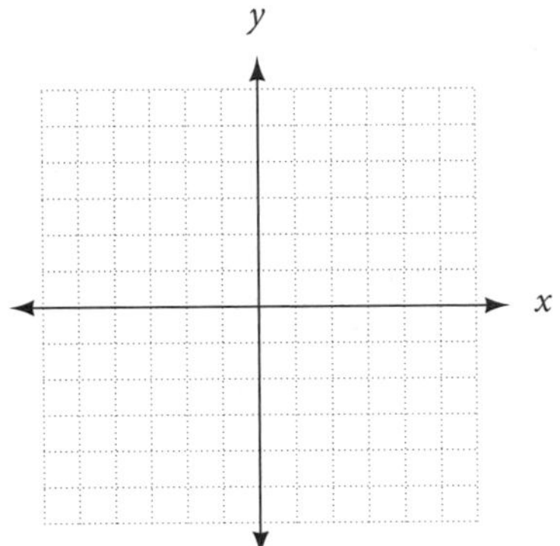

13. $y = \frac{1}{2}x + 3$ (−2,), (0,), (2,)

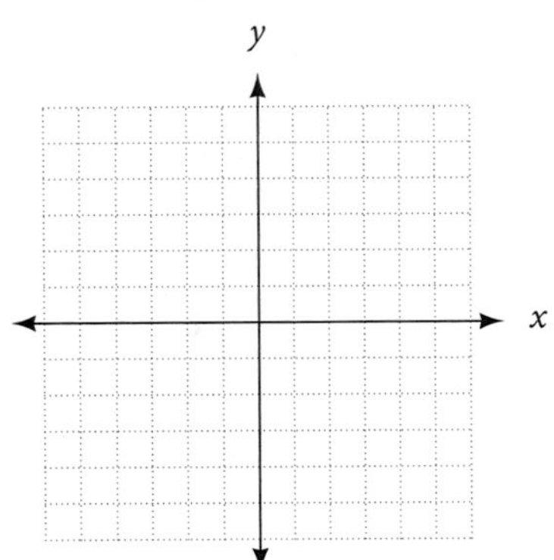

14. $y = \frac{1}{2}x - 3$ (−2,), (0, −), (2,)

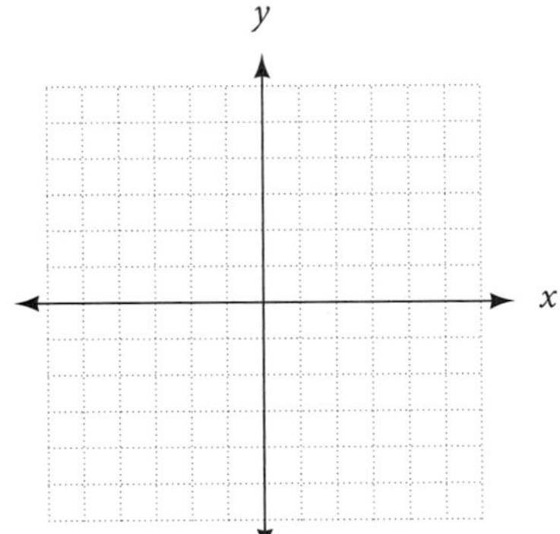

15. $y = -\frac{2}{3}x + 1$ (−3,), (0,), (3,)

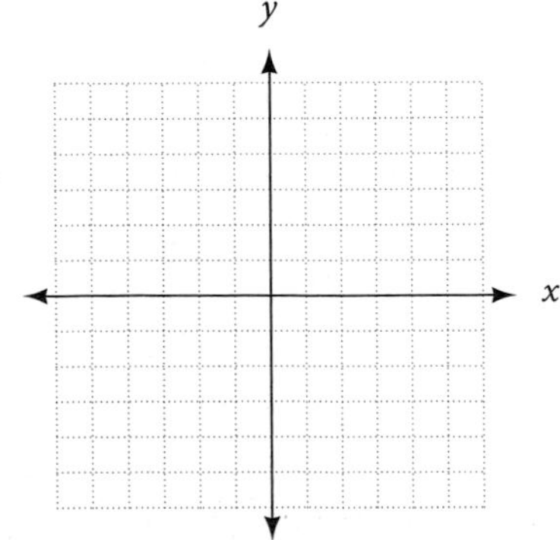

16. $y = -\frac{2}{3}x - 1$ (−3,), (0,), (3,)

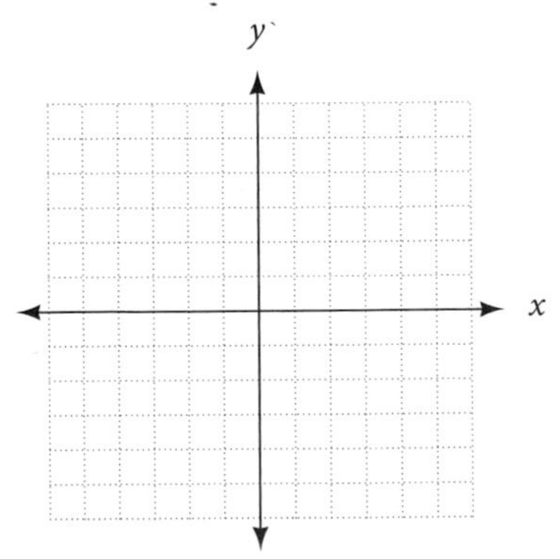

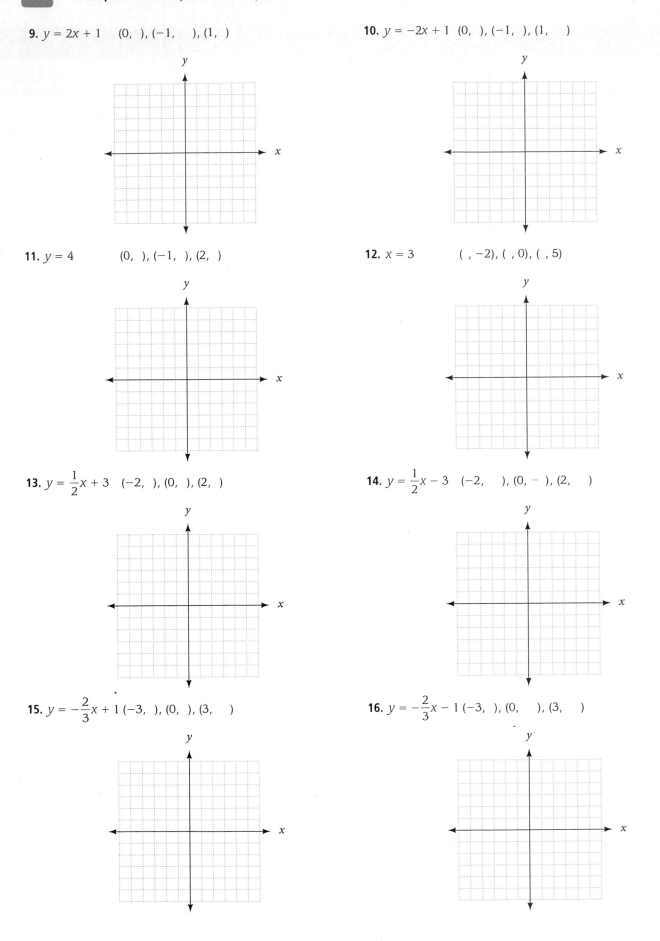

A Solve each equation for y. Then, complete the given ordered pairs, and use them to draw the graph. **[Example 4]**

17. $2x + y = 3$

$(-1, \ ), (0, \ ), (1, \ )$

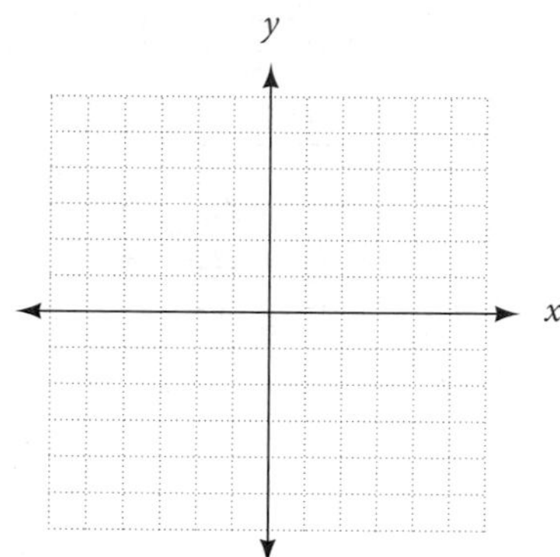

18. $3x + y = 2$

$(-1, \ ), (0, \ ), (1, \ )$

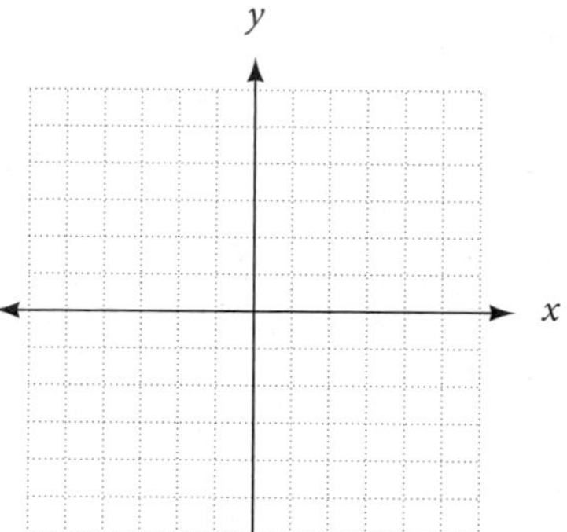

19. $3x + 2y = 6$

$(0, \ ), (2, \ ), (4, \ )$

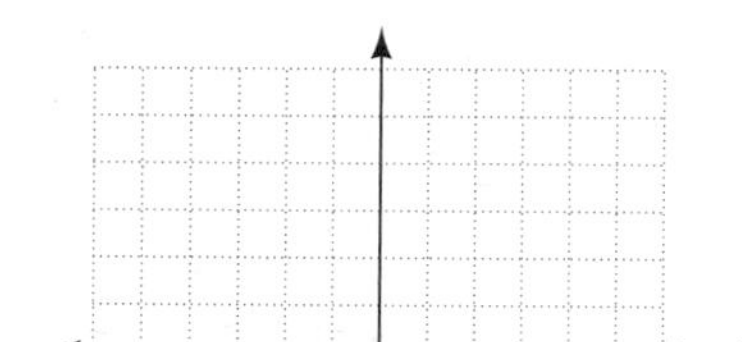

20. $2x + 3y = 6$

$(0, \ ), (3, \ ), (6, \ )$

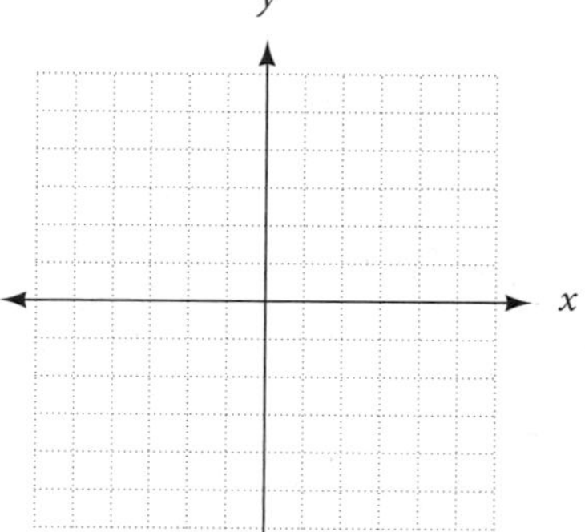

21. $-x + 2y = 6$

$(-2, \ ), (0, \ ), (2, \ )$

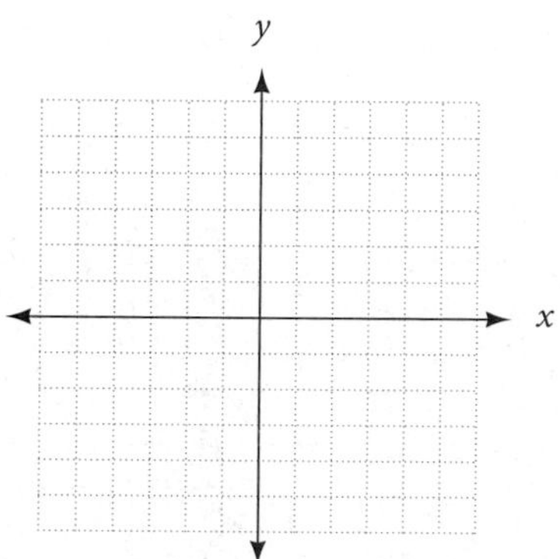

22. $-x + 3y = 6$

$(-3, \ ), (0, \ ), (3, \ )$

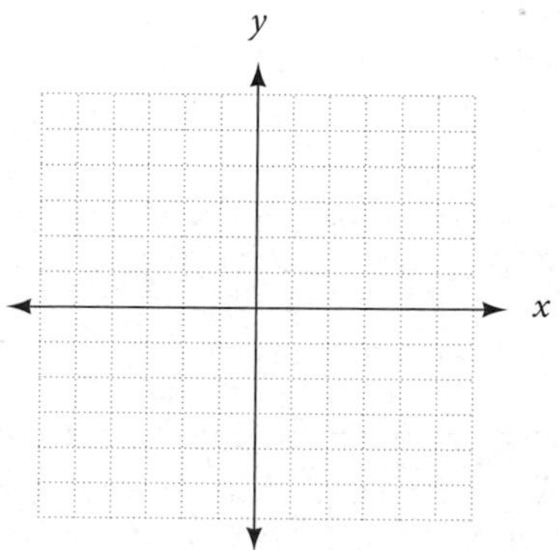

A B Find three solutions to each of the following equations, and then graph the solution set. [Examples 2–5]

23. $y = -\frac{1}{2}x$

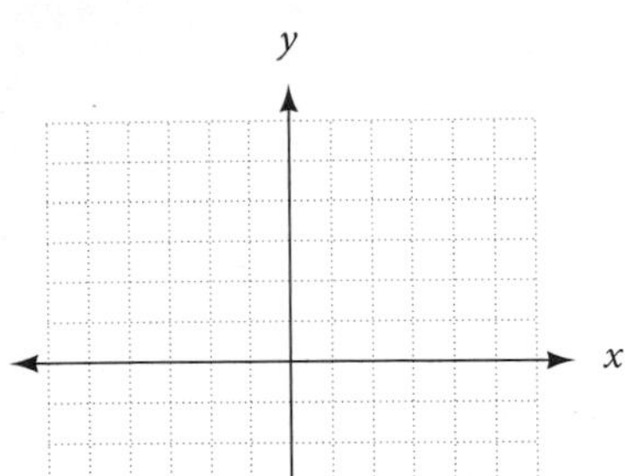

24. $y = -2x$

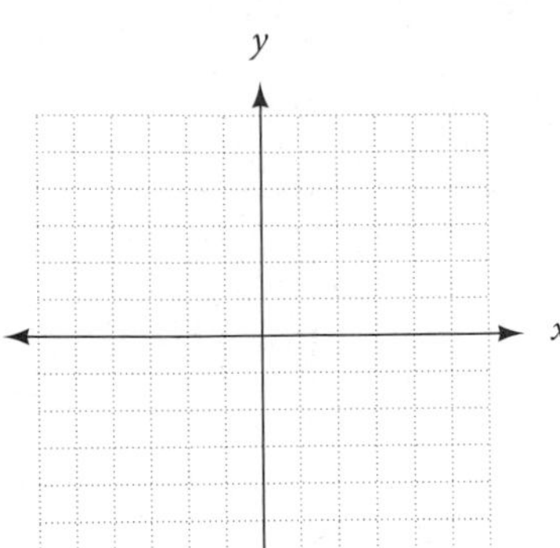

25. $y = 3x - 1$

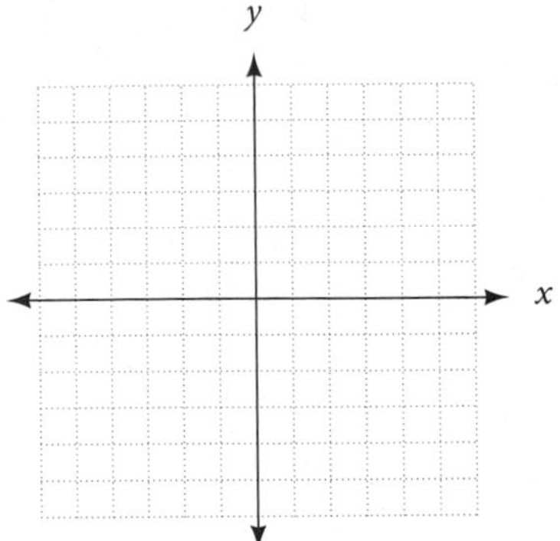

26. $y = -3x - 1$

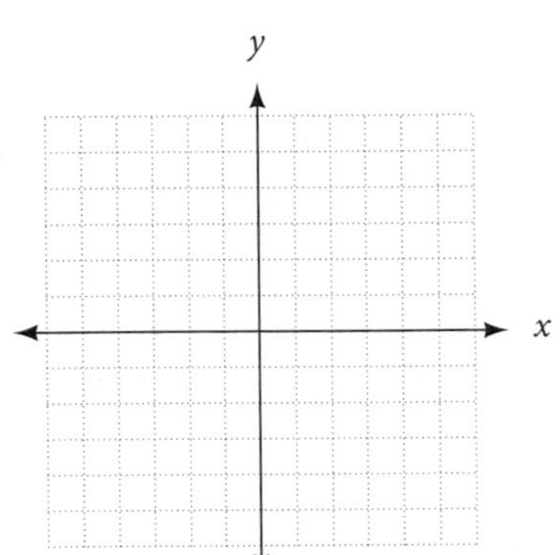

27. $-2x + y = 1$

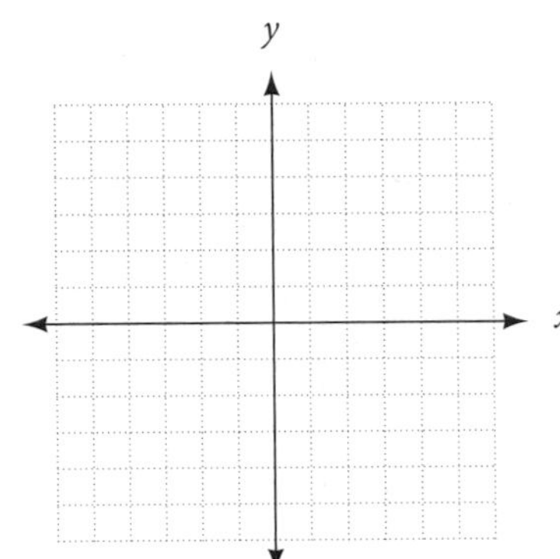

28. $-3x + y = 1$

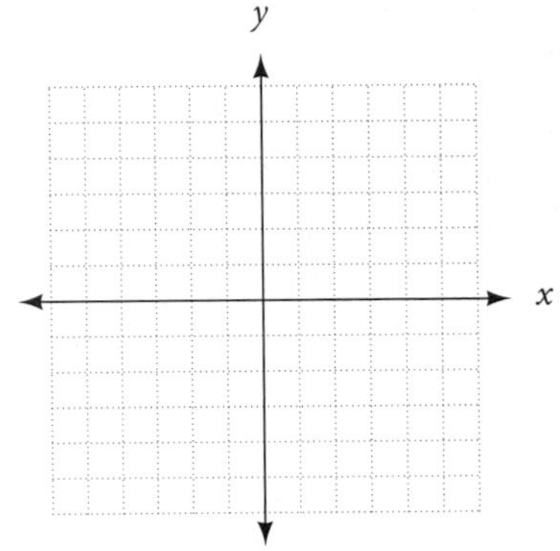

29. $3x + 4y = 8$

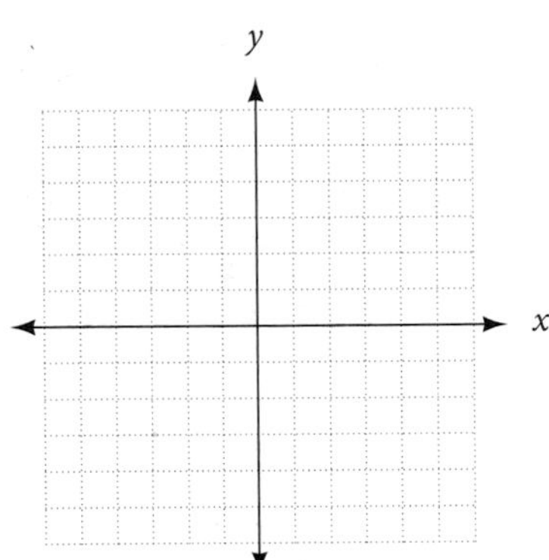

30. $3x - 4y = 8$

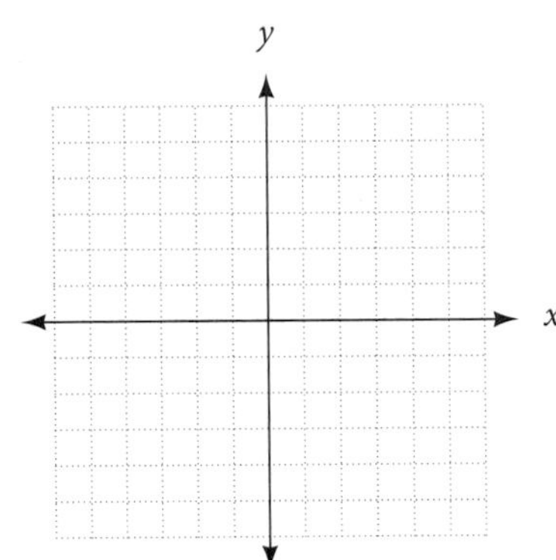

31. $x = -2$

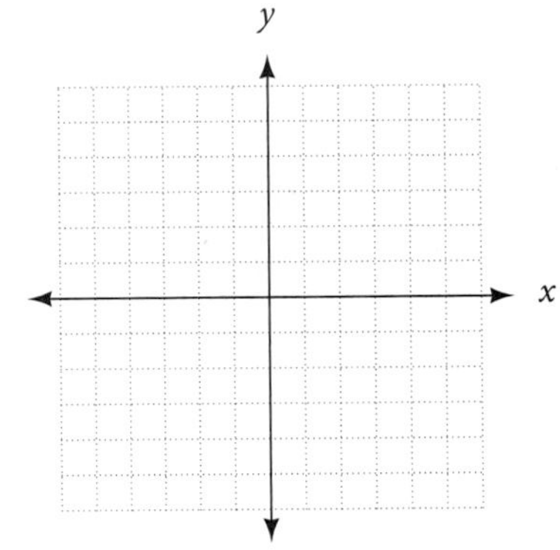

32. $y = 3$

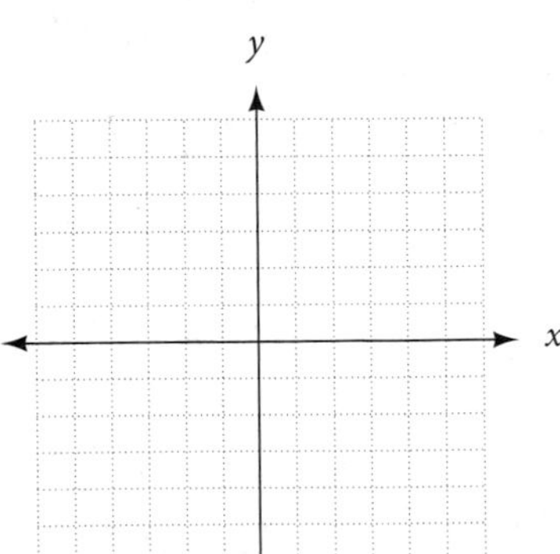

33. $y = 2$

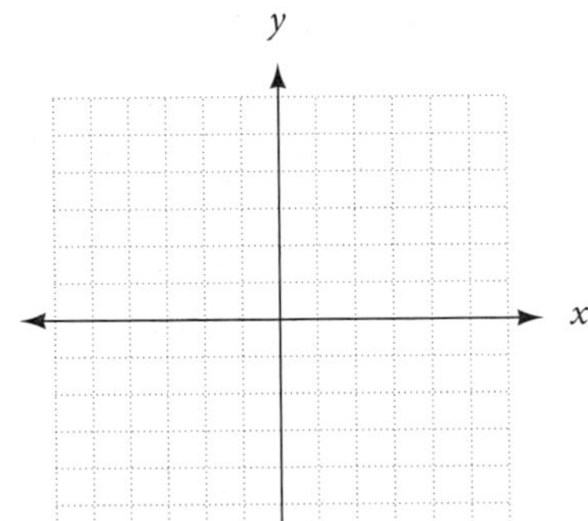

34. $x = -3$

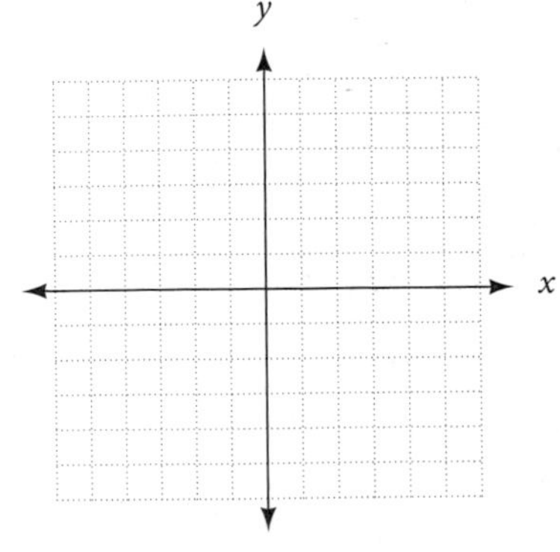

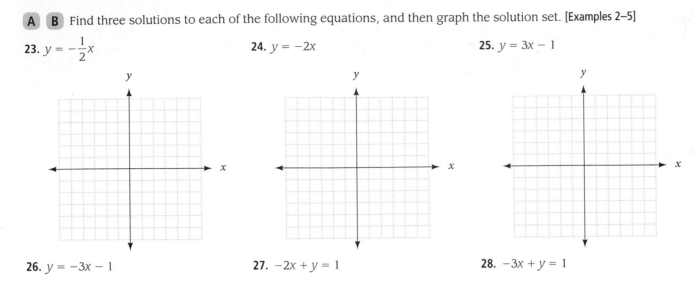

A Graph each equation. [Examples 2–3]

35. $y = \frac{3}{4}x + 1$

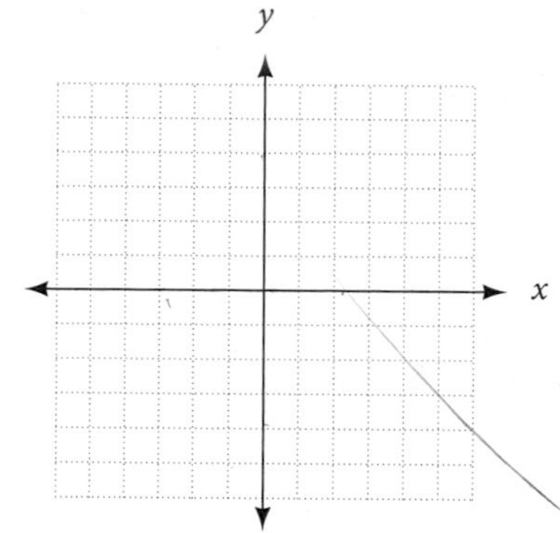

36. $y = \frac{2}{3}x + 1$

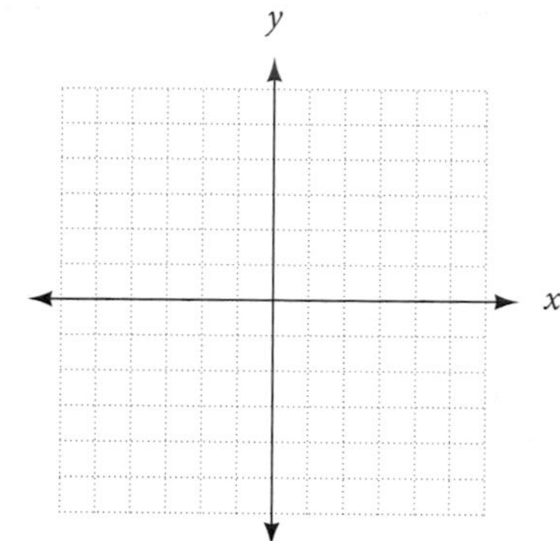

37. $y = \frac{1}{3}x + \frac{2}{3}$

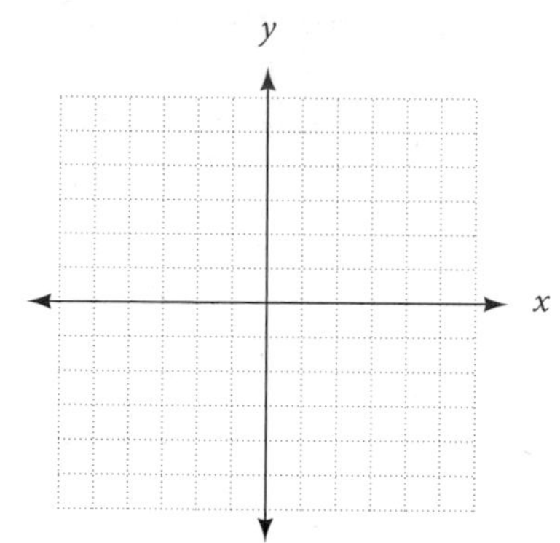

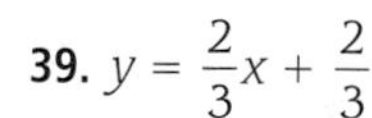

38. $y = \frac{1}{2}x + \frac{1}{2}$

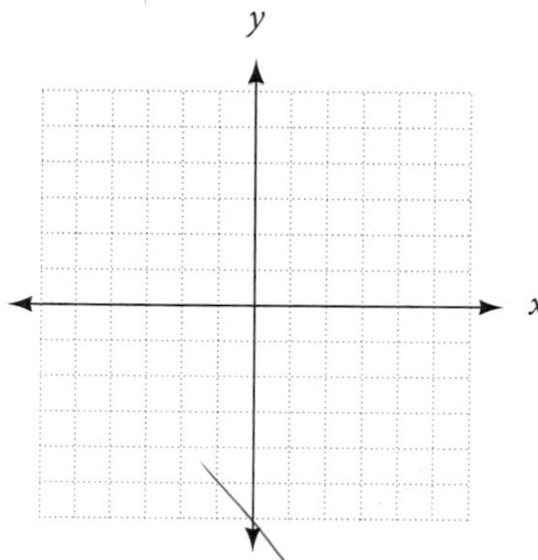

39. $y = \frac{2}{3}x + \frac{2}{3}$

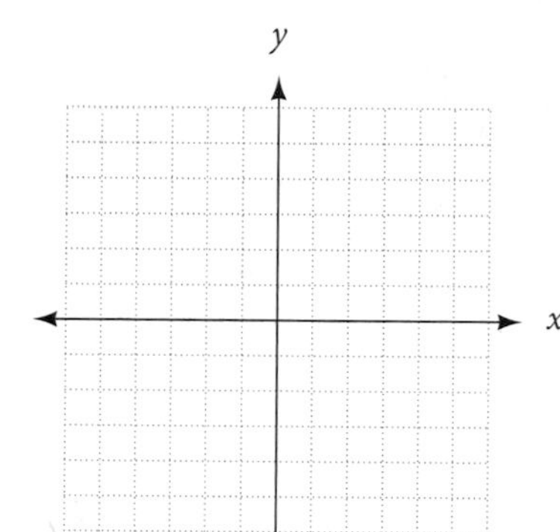

40. $y = -\frac{3}{4}x + \frac{3}{2}$

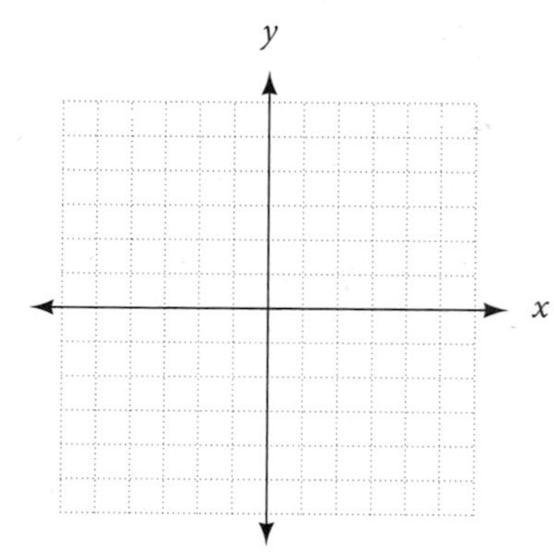

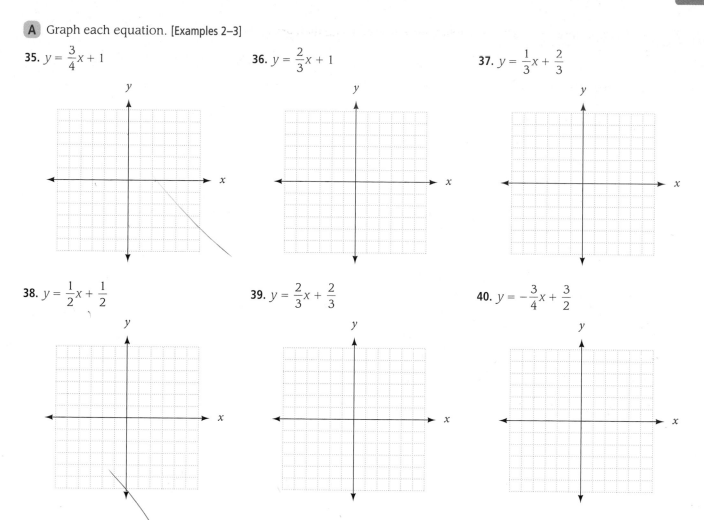

B For each equation in each table below, indicate whether the graph is horizontal (H), or vertical (V), or whether it passes through the origin (O). [Example 5]

41.

Equation	H, V, and/or O
$x = 3$	
$y = 3$	
$y = 3x$	
$y = 0$	

42.

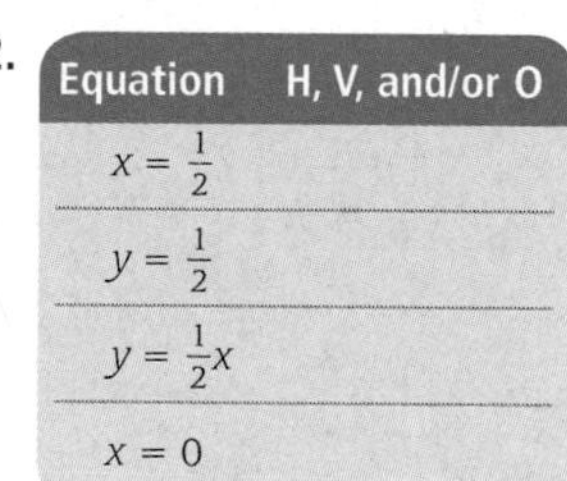

Equation	H, V, and/or O
$x = \frac{1}{2}$	
$y = \frac{1}{2}$	
$y = \frac{1}{2}x$	
$x = 0$	

43.

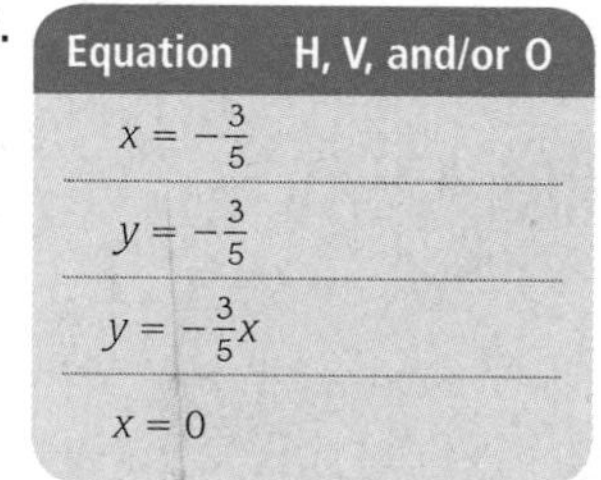

Equation	H, V, and/or O
$x = -\frac{3}{5}$	
$y = -\frac{3}{5}$	
$y = -\frac{3}{5}x$	
$x = 0$	

44.

Equation	H, V, and/or O
$x = -4$	
$y = -4$	
$y = -4x$	
$y = 0$	

The next two problems are intended to give you practice reading, and paying attention to, the instructions that accompany the problems you are working. Working these problems is an excellent way to get ready for a test or a quiz.

45. Work each problem according to the instructions given.

a. Solve: $2x + 5 = 10$

b. Find x when y is 0: $2x + 5y = 10$

c. Find y when x is 0: $2x + 5y = 10$

d. Graph: $2x + 5y = 10$

e. Solve for y: $2x + 5y = 10$

46. Work each problem according to the instructions given.

a. Solve: $x - 2 = 6$

b. Find x when y is 0: $x - 2y = 6$

c. Find y when x is 0: $x - 2y = 6$

d. Graph: $x - 2y = 6$

e. Solve for y: $x - 2y = 6$

Applying the Concepts

47. Picture Messaging The graph shows the number of text messages sent for the first nine months of the year. Make a line graph from the bar chart. What could be a reasonable equation of the line segment that connects May and June?

A Picture's Worth 1,000 Words

Month	Jan	Feb	Mar	Apr	May	Jun	Jul	Aug	Sep
Messages			3	10	21	21	26	32	41

48. Golf The chart shows the lengths of the longest golf courses used for the U.S. open. Use the bar chart to make a line graph with the name of the course on the x-axis and the length of the course on the y-axis. What is the equation of the line segment that connects the lengths of the Bethpage and the Pinehurst courses?

Longest Courses in U.S. Open History

Length	Course
7,214 yds	Bethpage State Park, Black Course (2002)
7,214 yds	Pinehurst, No. 2 Course (2005)
7,213 yds	Congressional C.C. Blue Course (1997)
7,195 yds	Medinah C.C. (1990)
7,191 yds	Bellerive C.C. St. Lous (1965)

Source: http://www.usopen.com

Getting Ready for the Next Section

49. Let $3x - 2y = 6$.
 a. Find x when $y = 0$.
 b. Find y when $x = 0$.

50. Let $2x - 5y = 10$.
 a. Find x when $y = 0$.
 b. Find y when $x = 0$.

51. Let $-x + 2y = 4$.
 a. Find x when $y = 0$.
 b. Find y when $x = 0$.

52. Let $3x - y = 6$.
 a. Find x when $y = 0$.
 b. Find y when $x = 0$.

53. Let $y = -\frac{1}{3}x + 2$.
 a. Find x when $y = 0$.
 b. Find y when $x = 0$.

54. Let $y = \frac{3}{2}x - 3$.
 a. Find x when $y = 0$.
 b. Find y when $x = 0$.

Maintaining Your Skills

Apply the distributive property.

55. $\frac{1}{2}(4x + 10)$

56. $\frac{1}{2}(6x - 12)$

57. $\frac{2}{3}(3x - 9)$

58. $\frac{1}{3}(2x + 12)$

59. $\frac{3}{4}(4x + 10)$

60. $\frac{3}{4}(8x - 6)$

61. $\frac{3}{5}(10x + 15)$

62. $\frac{2}{5}(5x - 10)$

63. $5\left(\frac{2}{5}x + 10\right)$

64. $3\left(\frac{2}{3}x + 5\right)$

65. $4\left(\frac{3}{2}x - 7\right)$

66. $4\left(\frac{3}{4}x + 5\right)$

67. $\frac{3}{4}(2x + 12y)$

68. $\frac{3}{4}(8x - 16y)$

69. $\frac{1}{2}(5x - 10y) + 6$

70. $\frac{1}{3}(5x - 15y) - 5$

More on Graphing: Intercepts

3.4

Objectives

A Find the intercepts of a line from the equation of the line.

B Use intercepts to graph a line.

A Intercepts

In this section we continue our work with graphing lines by finding the points where a line crosses the axes of our coordinate system. To do so, we use the fact that any point on the x-axis has a y-coordinate of 0 and any point on the y-axis has an x-coordinate of 0. We begin with the following definition.

> **Definition**
> The **x-intercept** of a straight line is the x-coordinate of the point where the graph crosses the x-axis. The **y-intercept** is defined similarly. It is the y-coordinate of the point where the graph crosses the y-axis.

B Using Intercepts to Graph Lines

If the x-intercept is a, then the point $(a, 0)$ lies on the graph. (This is true because any point on the x-axis has a y-coordinate of 0.)

If the y-intercept is b, then the point $(0, b)$ lies on the graph. (This is true because any point on the y-axis has an x-coordinate of 0.)

Graphically, the relationship is shown in Figure 1.

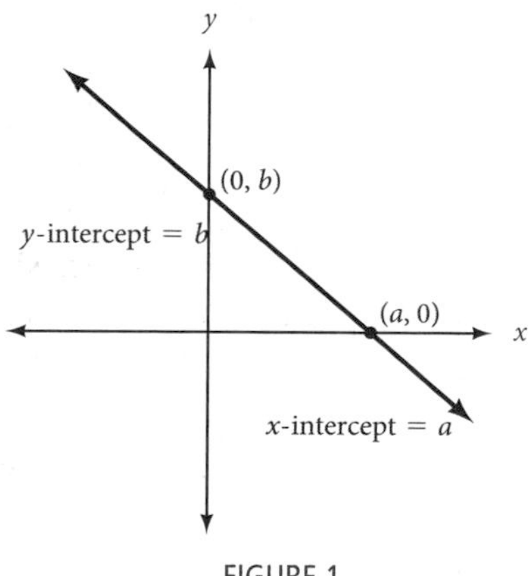

FIGURE 1

EXAMPLE 1 Find the x- and y-intercepts for $3x - 2y = 6$, and then use them to draw the graph.

SOLUTION To find where the graph crosses the x-axis, we let $y = 0$. (The y-coordinate of any point on the x-axis is 0.)

x-intercept:

When	$y = 0$
the equation	$3x - 2y = 6$
becomes	$3x - 2(0) = 6$
	$3x - 0 = 6$
	$x = 2$ Multiply each side by $\frac{1}{3}$

PRACTICE PROBLEMS

1. Find the x- and y-intercepts for $2x - 5y = 10$, and use them to draw the graph.

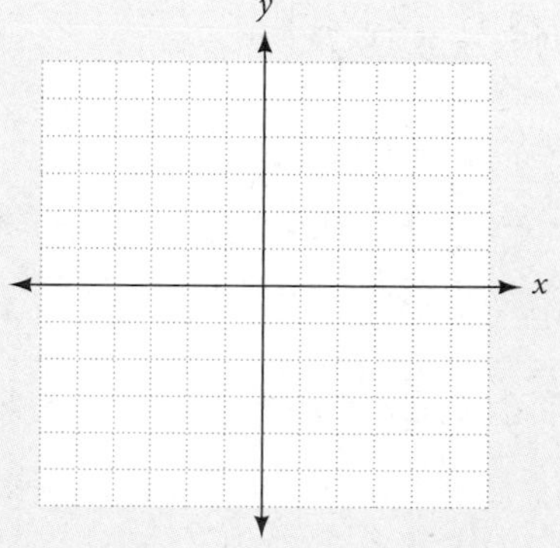

Answers
See Solutions to Selected Practice Problems for all answers in this section.

The graph crosses the x-axis at (2, 0), which means the x-intercept is 2.

***y*-intercept:**

$$\begin{aligned} \text{When} \quad && x &= 0 \\ \text{the equation} \quad && 3x - 2y &= 6 \\ \text{becomes} \quad && 3(0) - 2y &= 6 \\ && 0 - 2y &= 6 \\ && -2y &= 6 \\ && y &= -3 && \text{Multiply each side by } -\tfrac{1}{2} \end{aligned}$$

The graph crosses the y-axis at (0, −3), which means the y-intercept is −3.

Plotting the x- and y-intercepts and then drawing a line through them, we have the graph of $3x - 2y = 6$, as shown in Figure 2.

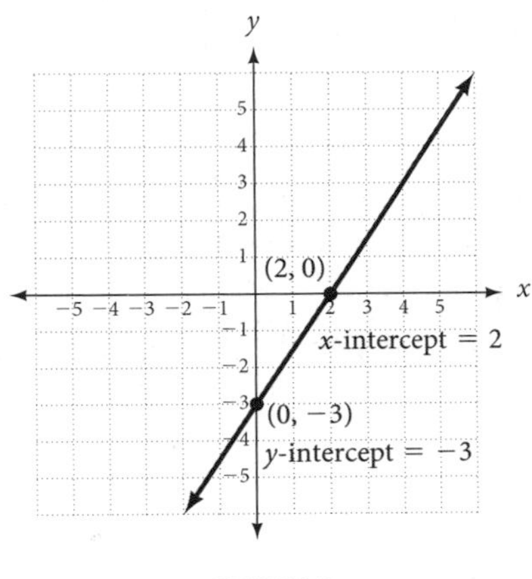

FIGURE 2

2. Graph $3x - y = 6$ by first finding the intercepts.

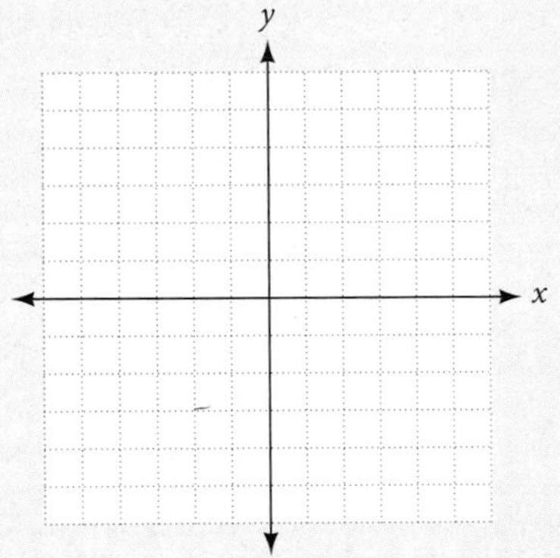

EXAMPLE 2 Graph $-x + 2y = 4$ by finding the intercepts and using them to draw the graph.

SOLUTION Again, we find the x-intercept by letting $y = 0$ in the equation and solving for x. Similarly, we find the y-intercept by letting $x = 0$ and solving for y.

***x*-intercept:**

$$\begin{aligned} \text{When} \quad && y &= 0 \\ \text{the equation} \quad && -x + 2y &= 4 \\ \text{becomes} \quad && -x + 2(0) &= 4 \\ && -x + 0 &= 4 \\ && -x &= 4 \\ && x &= -4 && \text{Multiply each side by } -1 \end{aligned}$$

The x-intercept is −4, indicating that the point (−4, 0), is on the graph of $-x + 2y = 4$.

***y*-intercept:**

$$\begin{aligned} \text{When} \quad && x &= 0 \\ \text{the equation} \quad && -x + 2y &= 4 \\ \text{becomes} \quad && -0 + 2y &= 4 \\ && 2y &= 4 \\ && y &= 2 && \text{Multiply each side by } \tfrac{1}{2} \end{aligned}$$

The y-intercept is 2, indicating that the point (0, 2) is on the graph of $-x + 2y = 4$.

Plotting the intercepts and drawing a line through them, we have the graph of $-x + 2y = 4$, as shown in Figure 3.

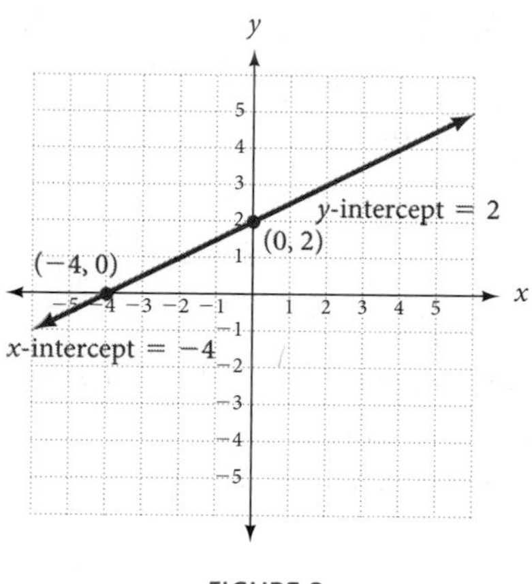

FIGURE 3

Graphing a line by finding the intercepts, as we have done in Examples 1 and 2, is an easy method of graphing if the equation has the form $ax + by = c$ and both the numbers a and b divide the number c evenly.

In our next example we use the intercepts to graph a line in which y is given in terms of x.

EXAMPLE 3 Use the intercepts for $y = -\frac{1}{3}x + 2$ to draw its graph.

SOLUTION We graphed this line previously in Example 3 of Section 3.3 by substituting three different values of x into the equation and solving for y. This time we will graph the line by finding the intercepts.

***x*-intercept:**

When	$y = 0$	
the equation	$y = -\frac{1}{3}x + 2$	
becomes	$0 = -\frac{1}{3}x + 2$	
	$-2 = -\frac{1}{3}x$	Add −2 to each side
	$6 = x$	Multiply each side by −3

The x-intercept is 6, which means the graph passes through the point (6, 0).

***y*-intercept:**

When	$x = 0$
the equation	$y = -\frac{1}{3}x + 2$
becomes	$y = -\frac{1}{3}(0) + 2$
	$y = 2$

The y-intercept is 2, which means the graph passes through the point (0, 2).

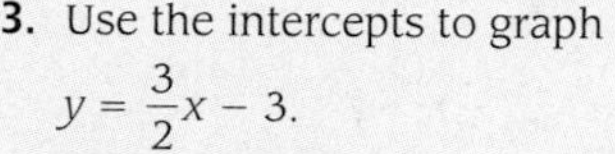

3. Use the intercepts to graph $y = \frac{3}{2}x - 3$.

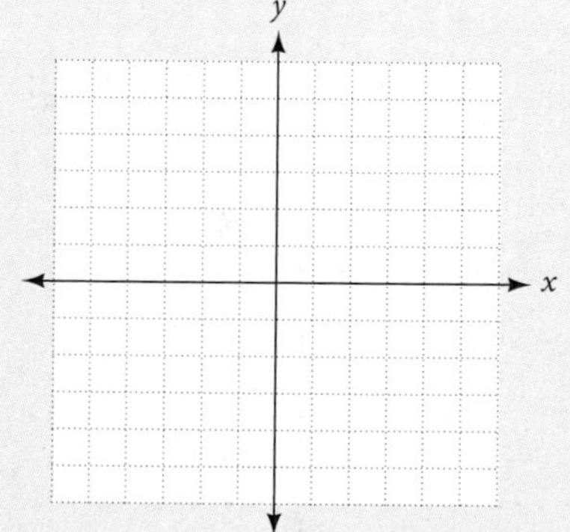

The graph of $y = -\frac{1}{3}x + 2$ is shown in Figure 4. Compare this graph, and the method used to obtain it, with Example 3 in Section 3.3.

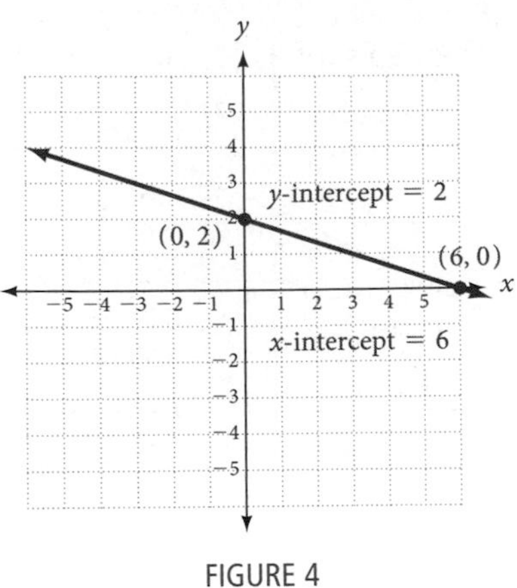

FIGURE 4

Getting Ready for Class

After reading through the preceding section, respond in your own words and in complete sentences.

1. What is the *x*-intercept for a graph?
2. What is the *y*-intercept for a graph?
3. How do we find the *y*-intercept for a line from the equation?
4. How do we graph a line using its intercepts?

Problem Set 3.4

A B Find the x- and y-intercepts for the following equations. Then use the intercepts to graph each equation. [Examples 1–3]

1. $2x + y = 4$

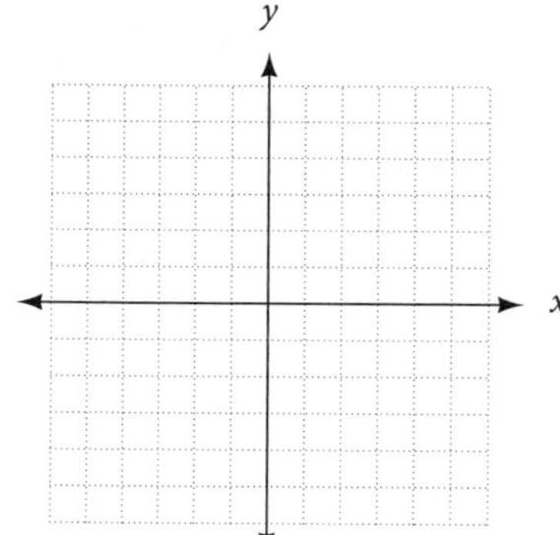

2. $2x + y = 2$

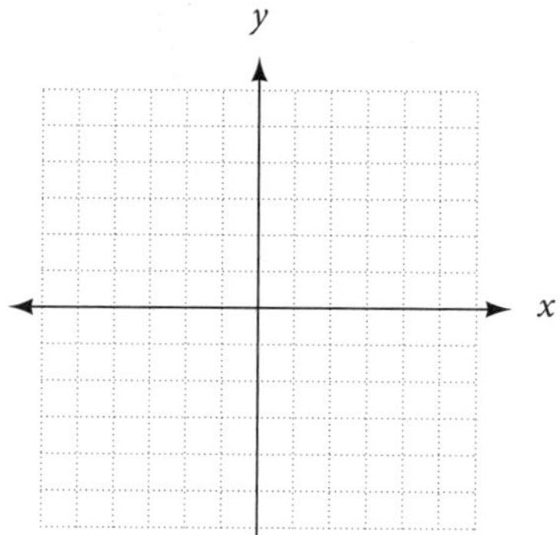

3. $-x + y = 3$

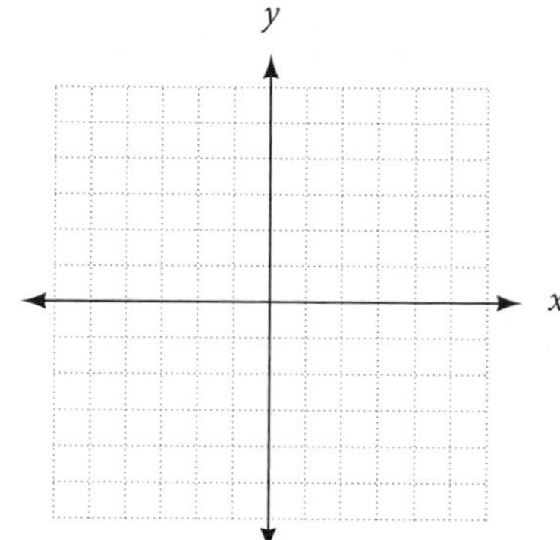

4. $-x + y = 4$

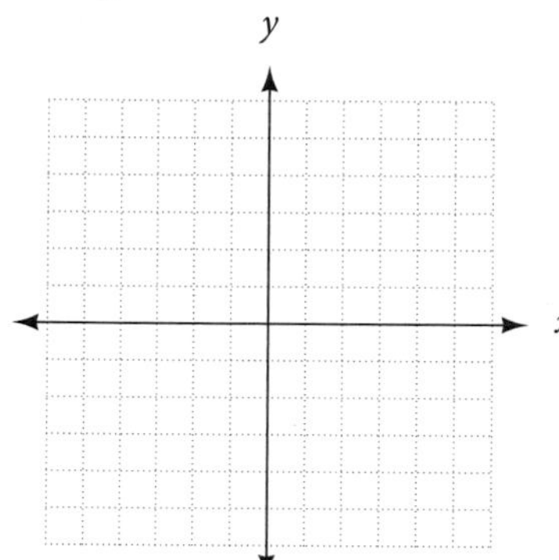

5. $-x + 2y = 2$

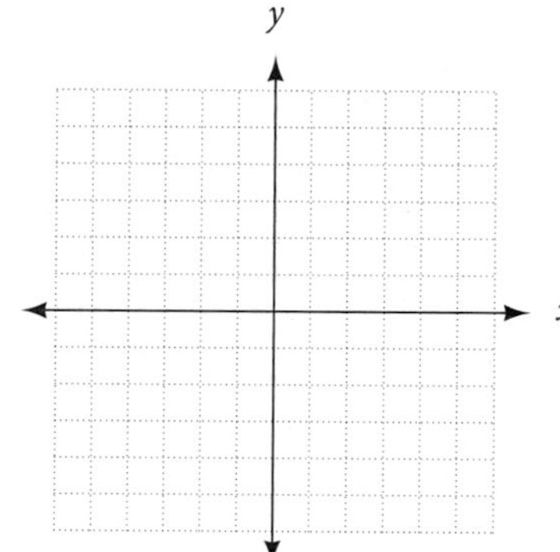

6. $-x + 2y = 4$

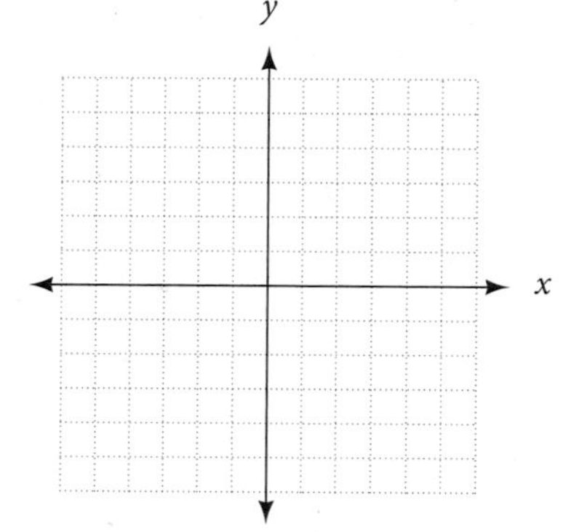

7. $5x + 2y = 10$

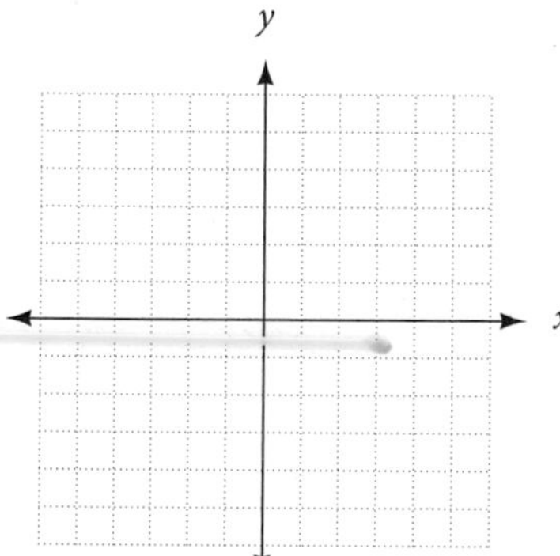

8. $2x + 5y = 10$

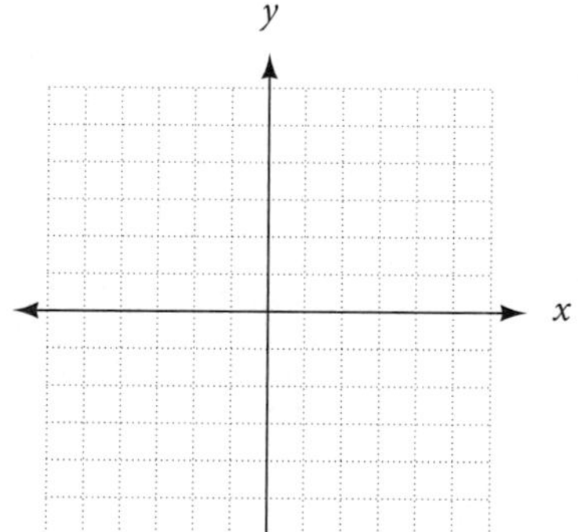

9. $-4x + 5y = 20$

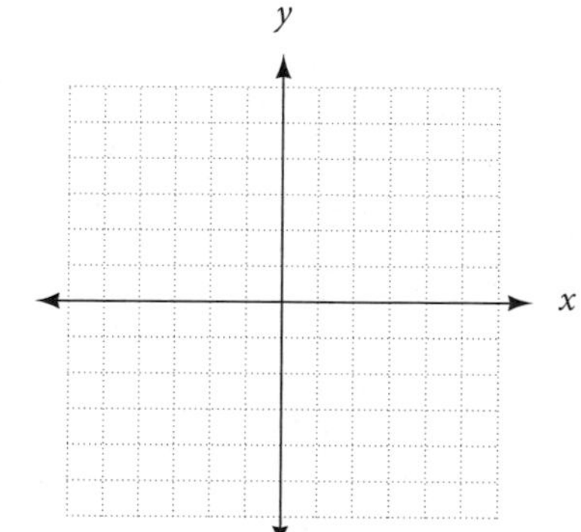

10. $-5x + 4y = 20$

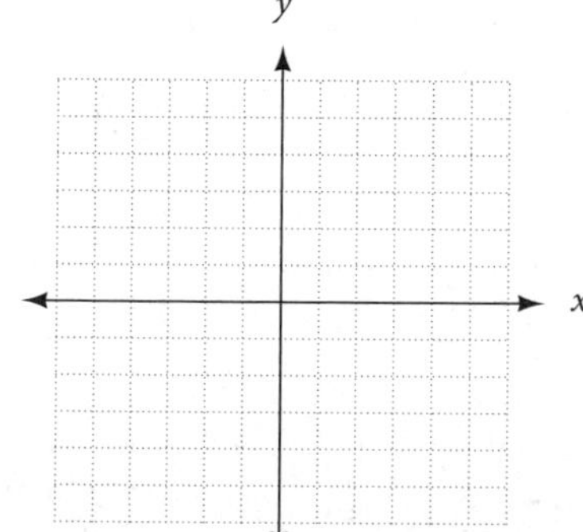

11. $3x - 4y = -4$

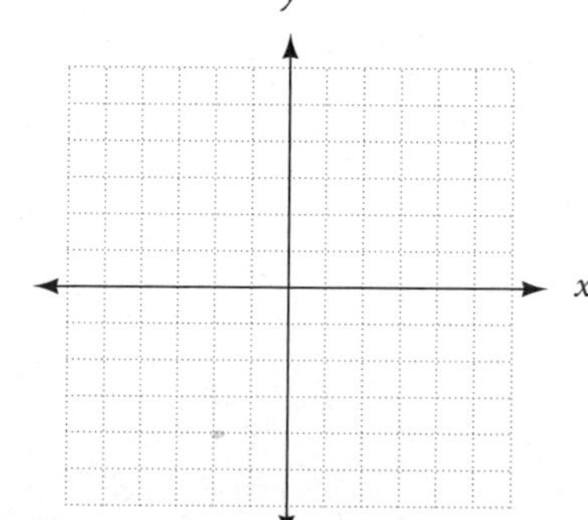

12. $-2x + 3y = 3$

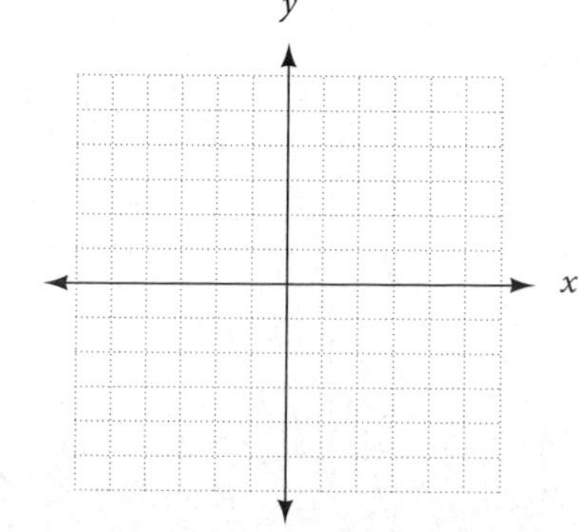

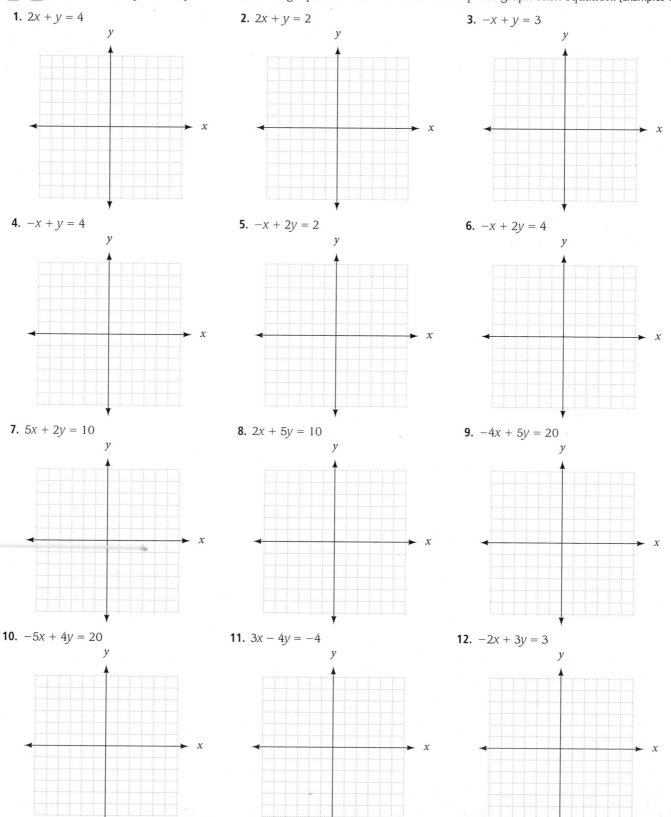

13. $x - 3y = 2$

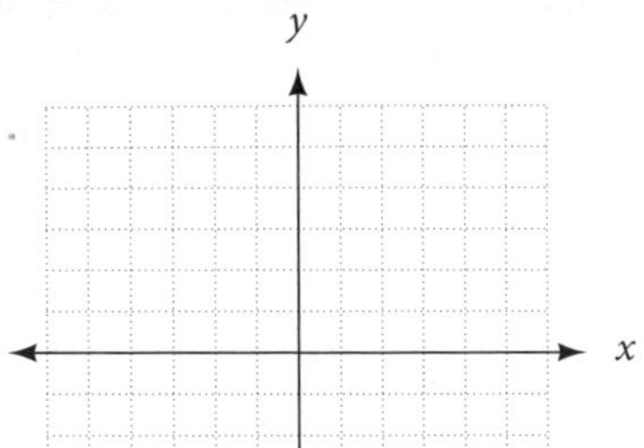

14. $x - 2y = 1$

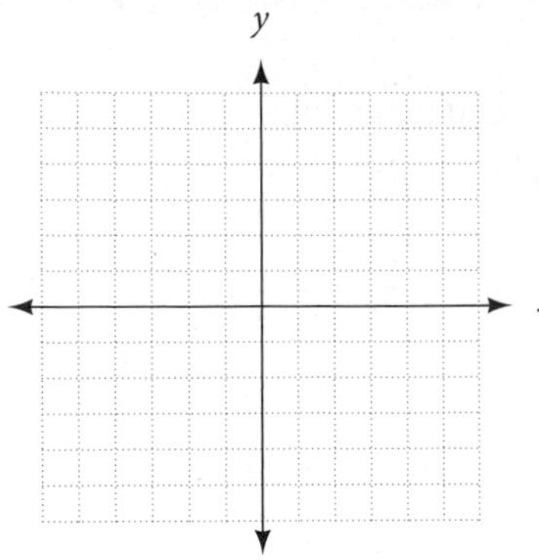

15. $2x - 3y = -2$

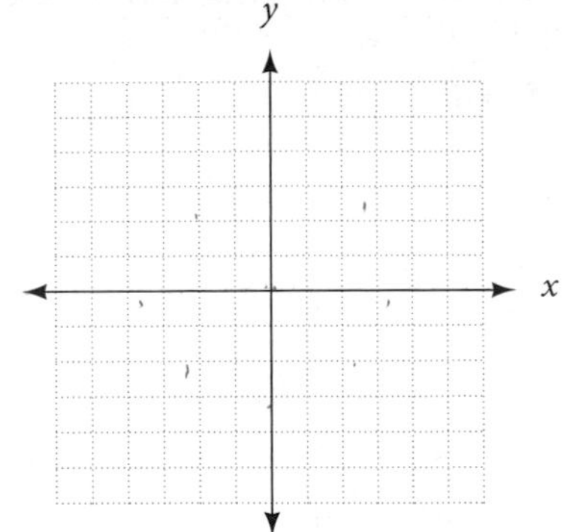

16. $3x + 4y = 6$

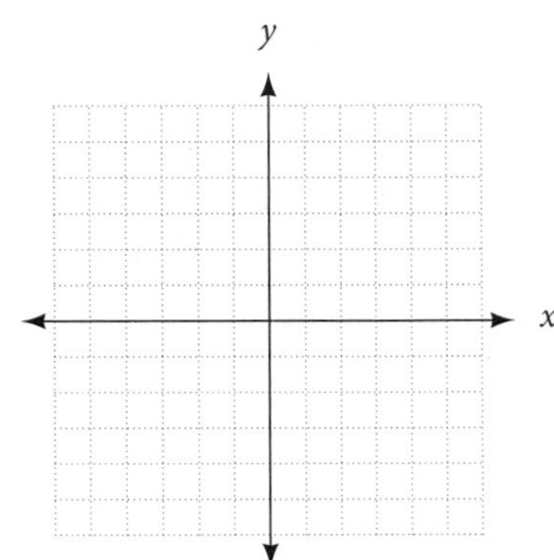

17. $y = 2x - 6$

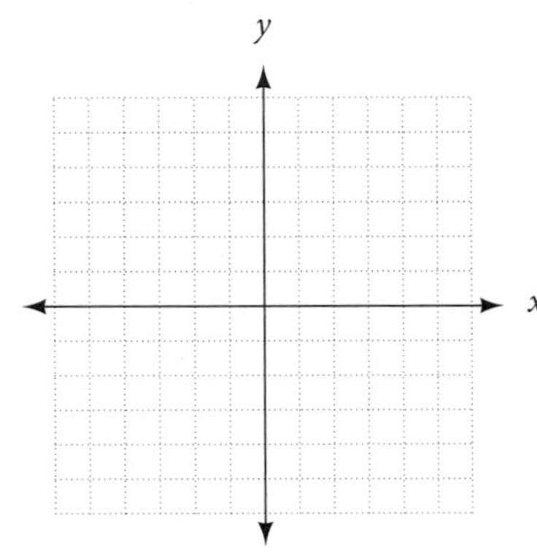

18. $y = 2x + 6$

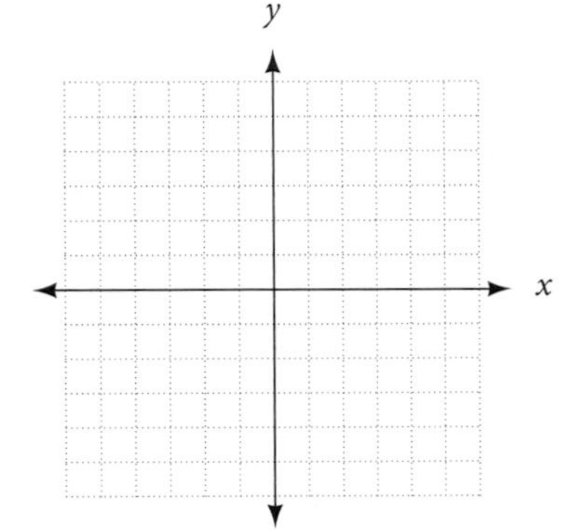

19. $y = 2x - 1$

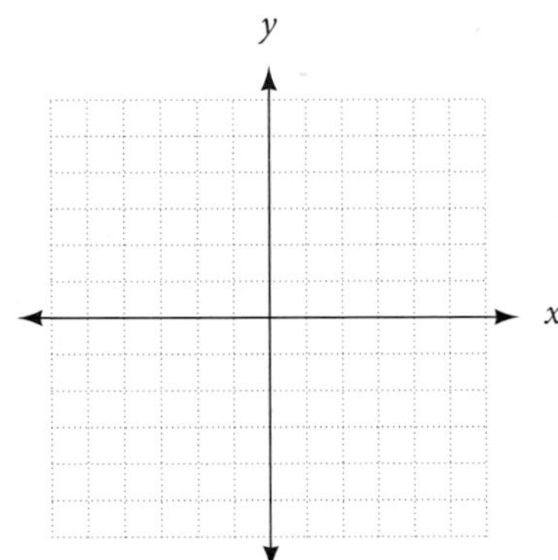

20. $y = -2x - 1$

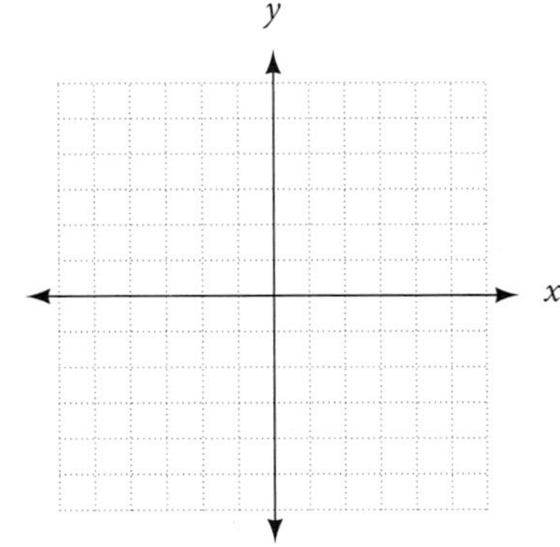

21. $y = \frac{1}{2}x + 3$

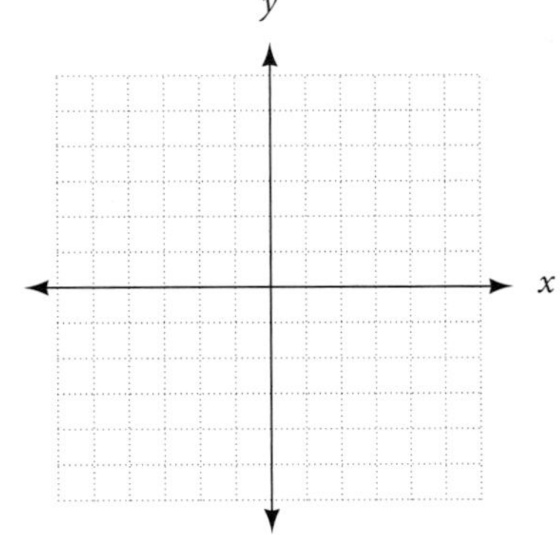

22. $y = \frac{1}{2}x - 3$

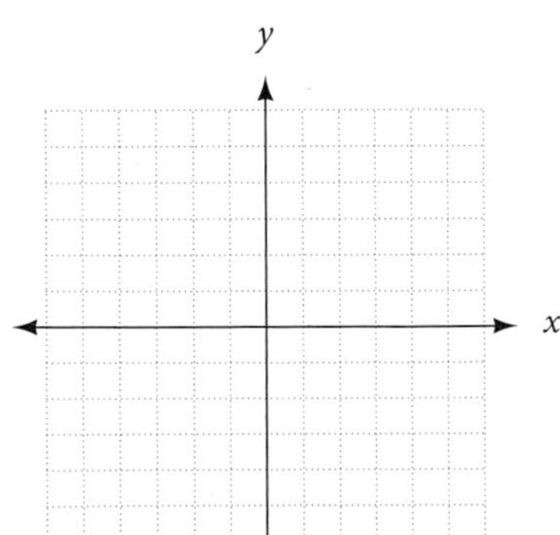

23. $y = -\frac{1}{3}x - 2$

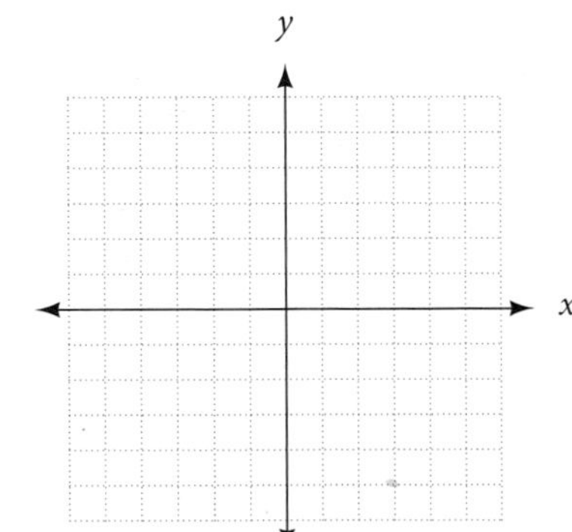

24. $y = -\frac{1}{3}x + 2$

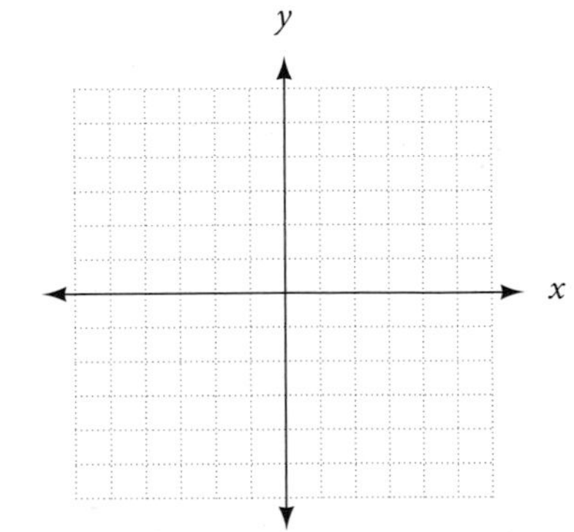

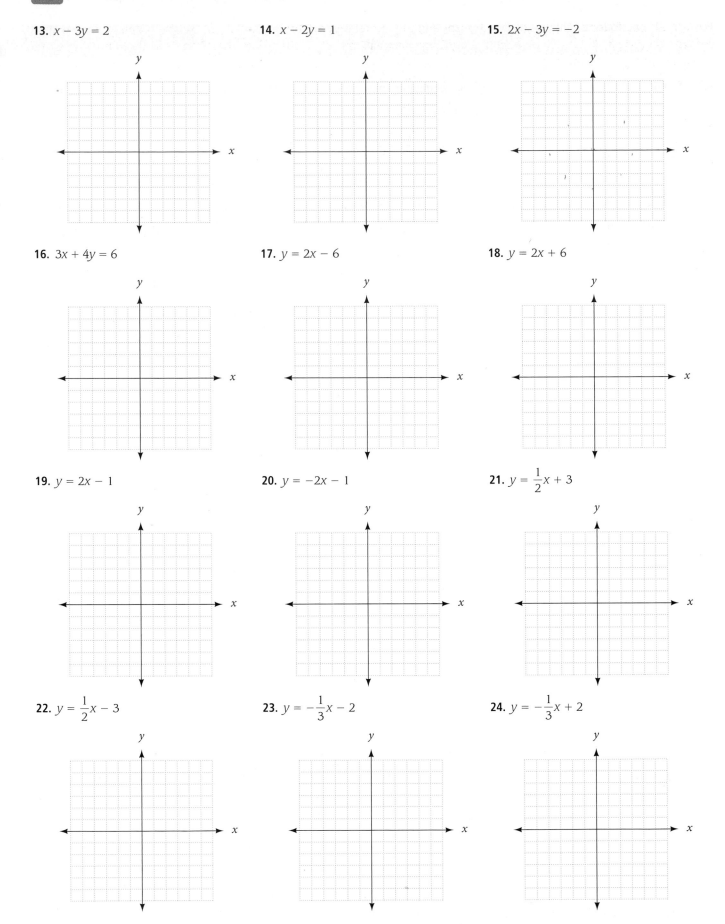

For each of the following lines, the x-intercept and the y-intercept are both 0, which means the graph of each will go through the origin, (0, 0). Graph each line by finding a point on each, other than the origin, and then drawing a line through that point and the origin.

25. $y = -2x$

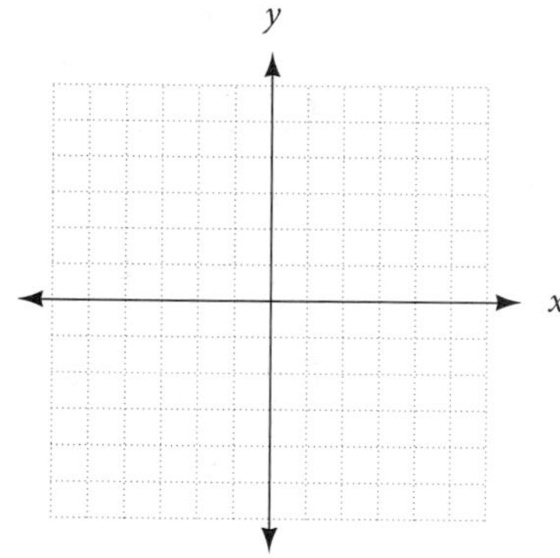

26. $y = \frac{1}{2}x$

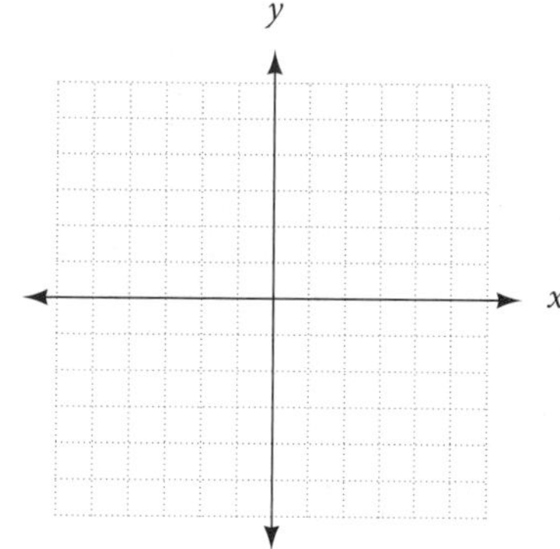

27. $y = -\frac{1}{3}x$

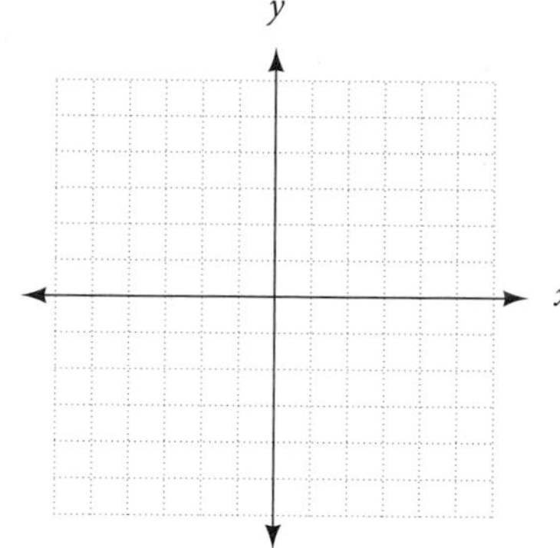

28. $y = -3x$

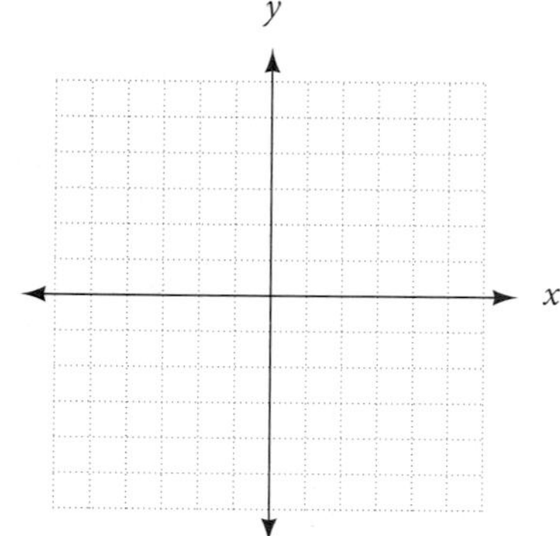

29. $y = \frac{2}{3}x$

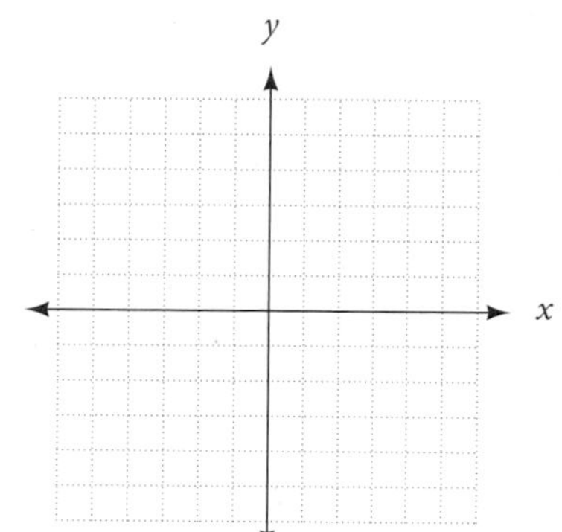

30. $y = \frac{3}{2}x$

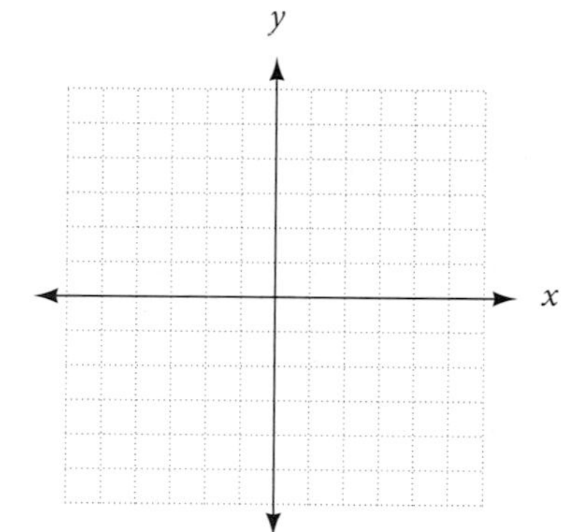

A Complete each table.

31.

Equation	x-intercept	y-intercept
$3x + 4y = 12$		
$3x + 4y = 4$		
$3x + 4y = 3$		
$3x + 4y = 2$		

32.

Equation	x-intercept	y-intercept
$-2x + 3y = 6$		
$-2x + 3y = 3$		
$-2x + 3y = 2$		
$-2x + 3y = 1$		

33.

Equation	x-intercept	y-intercept
$x - 3y = 2$		
$y = \frac{1}{3}x - \frac{2}{3}$		
$x - 3y = 0$		
$y = \frac{1}{3}x$		

34.

Equation	x-intercept	y-intercept
$x - 2y = 1$		
$y = \frac{1}{2}x - \frac{1}{2}$		
$x - 2y = 0$		
$y = \frac{1}{2}x$		

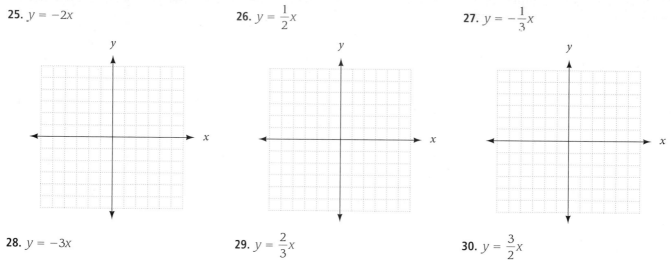

The next two problems are intended to give you practice reading, and paying attention to, the instructions that accompany the problems you are working. Working these problems is an excellent way to get ready for a test or a quiz.

35. Work each problem according to the instructions given.

a. Solve: $2x - 3 = -3$

b. Find the x-intercept: $2x - 3y = -3$

c. Find y when x is 0: $2x - 3y = -3$

d. Graph: $2x - 3y = -3$

e. Solve for y: $2x - 3y = -3$

36. Work each problem according to the instructions given.

a. Solve: $3x - 4 = -4$

b. Find the y-intercept: $3x - 4y = -4$

c. Find x when y is 0: $3x - 4y = -4$

d. Graph: $3x - 4y = -4$

e. Solve for y: $3x - 4y = -4$

37. Graph the line that passes through the points $(-2, 5)$ and $(5, -2)$. What are the x- and y-intercepts for this line?

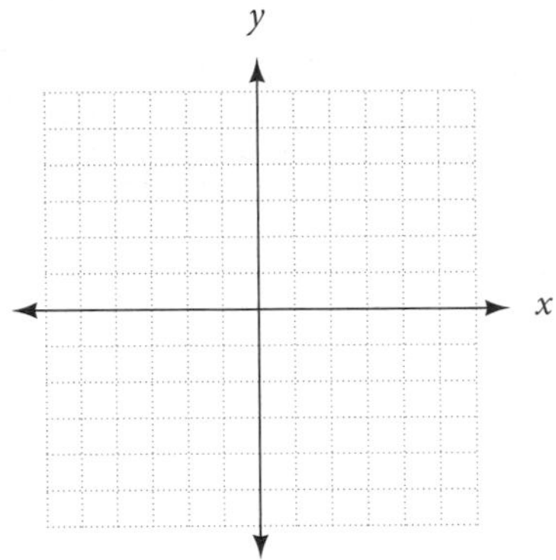

38. Graph the line that passes through the points $(5, 3)$ and $(-3, -5)$. What are the x- and y-intercepts for this line?

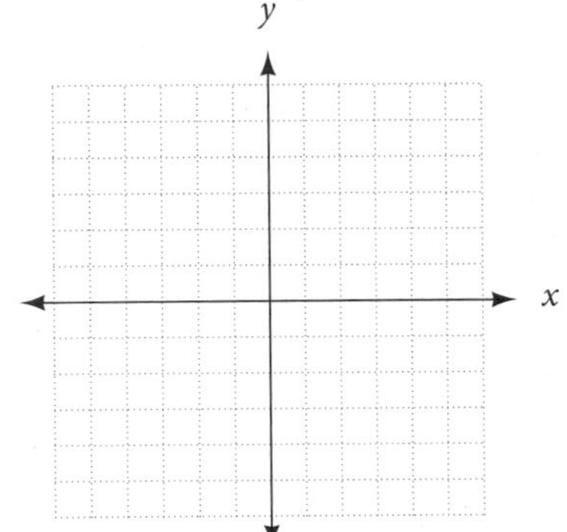

From the graphs below, find the x- and y-intercepts for each line.

39.

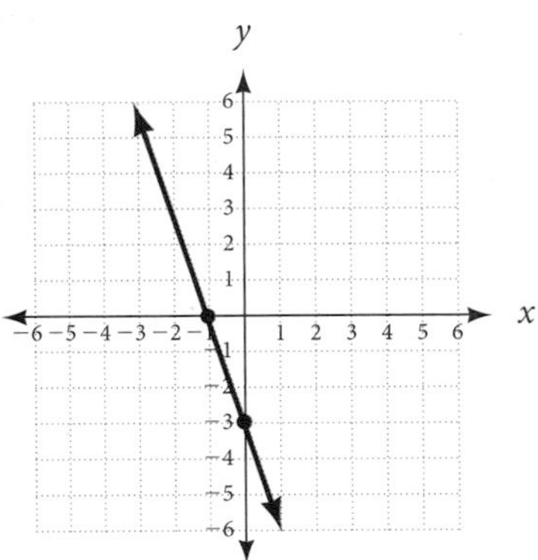

40.

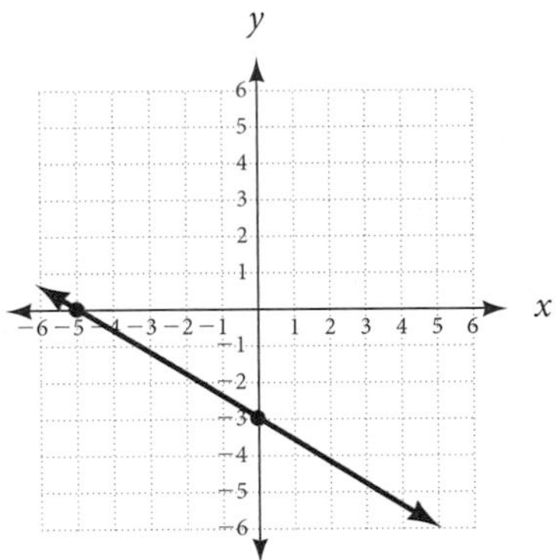

41. Use the graph at the right to complete the following table.

x	y
-2	
0	
	0
	-2

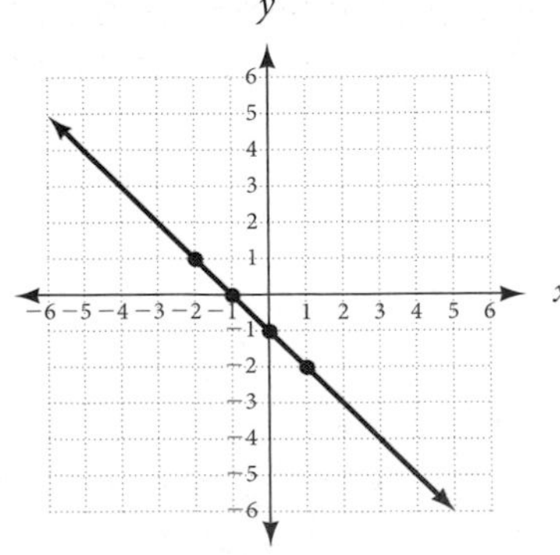

42. Use the graph at the right to complete the following table.

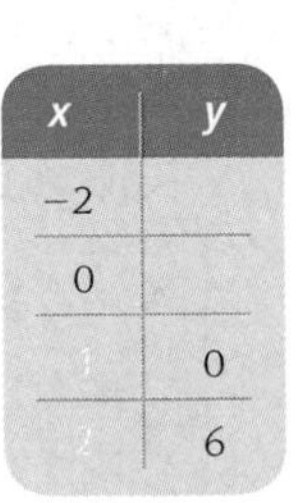

x	y
-2	
0	
	0
	6

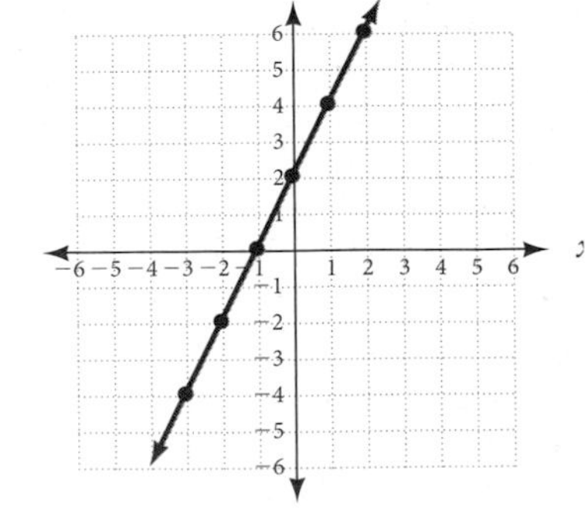

Applying the Concepts

43. **Light Bulbs** The chart shows a comparison of power usage between incandescent and energy efficient light bulbs. The equation of the line segment from 450 to 800 lumens of an energy efficient bulb can be given by, $y = \left(\frac{3}{175}\right)x + \frac{16}{7}$. Find the x- and y-intercepts of this equation.

Incandescent vs. Energy Efficient Light Bulbs

Amount of watts used: 160, 120, 80, 40, 0

Incandescent Bulbs

Energy Efficient Bulbs

450 800 1100 1600 2600

Light Output (lumens)

Source: Energy Star Product Chart

44. **Horse Racing** The graph shows the total amount of money wagered on the Kentucky Derby. The equation of the line segment from 1985 to 1990 is given by the equation $y = 2.84x - 5{,}617.2$. Find the x- and y-intercepts of this equation to the nearest tenth.

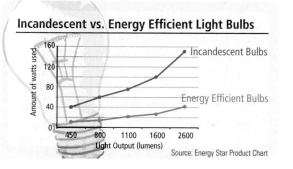

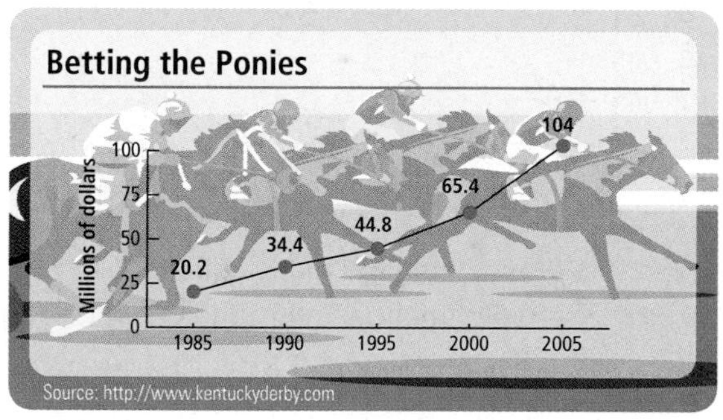

45. **Working Two Jobs** Maggie has a job working in an office for \$10 an hour and another job driving a tractor for \$12 an hour. Maggie works at both jobs and earns \$480 in one week.

   a. Write an equation that gives the number of hours x, she worked for \$10 per hour and the number of hours, y, she worked for \$12 per hour.

   b. Find the x- and y-intercepts for this equation.

   c. Graph this equation from the intercepts, using only the first quadrant.

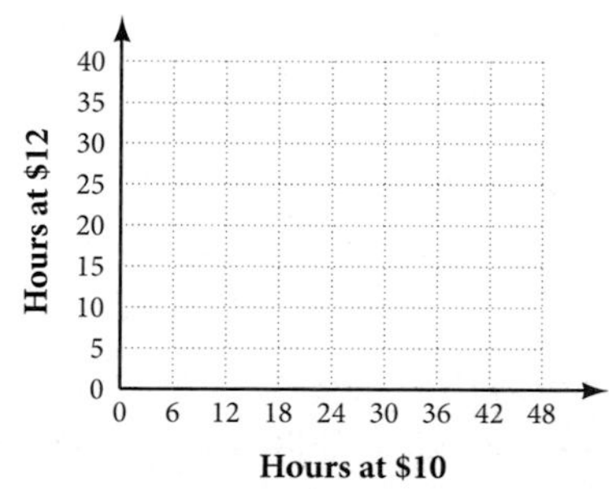

   d. From the graph, find how many hours she worked at \$12 if she worked 36 hours at \$10 per hour.

   e. From the graph, find how many hours she worked at \$10 if she worked 25 hours at \$12 per hour.

46. Ticket prices Devin is selling tickets to a barbecue. Adult tickets cost \$6.00 and children's tickets cost \$4.00. The total amount of money he collects is \$240.

a. Write an equation that gives the number of adult tickets, x, he sold for \$6 and the children's tickets, y, he sold for \$4.

b. Find the x- and y-intercepts for this equation.

c. Graph this equation from the intercepts, using only the first quadrant.

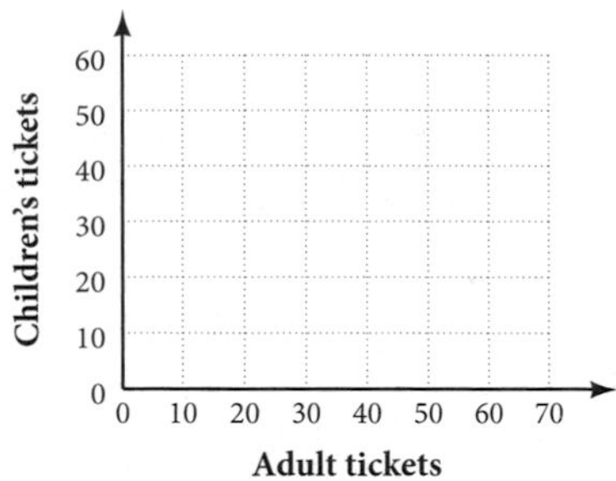

d. From the graph, find how many children's tickets were sold if 10 Adult tickets were sold.

e. From the graph, find how many adult tickets were sold if 30 children's tickets were sold.

Getting Ready for the Next Section

47. Evaluate:

a. $\dfrac{5-2}{3-1}$ b. $\dfrac{2-5}{1-3}$

48. Evaluate.

a. $\dfrac{-4-1}{5-(-2)}$ b. $\dfrac{1+4}{-2-5}$

49. Evaluate the following expressions when $x = 3$ and $y = 5$.

a. $\dfrac{y-2}{x-1}$ b. $\dfrac{2-y}{1-x}$

50. Evaluate the following expressions when $x = 3$ and $y = 2$.

a. $\dfrac{-4-y}{5-x}$ b. $\dfrac{y+4}{x-5}$

Maintaining Your Skills

Solve each equation.

51. $-12y - 4 = -148$

52. $-2x - 18 = 4$

53. $-5y - 4 = 51$

54. $-2y + 18 = -14$

55. $11x - 12 = -78$

56. $21 + 9y = -24$

57. $9x + 3 = 66$

58. $-11 - 15a = -71$

59. $-9c - 6 = 12$

60. $-7a + 28 = -84$

61. $4 + 13c = -9$

62. $-3x + 15 = -24$

63. $3y - 12 = 30$

64. $9x + 11 = -16$

65. $-11y + 9 = 75$

66. $9x - 18 = -72$

The Slope of a Line

3.5

Objectives

A Find the slope of a line from two points on the line.

B Graph a line given the slope and y-intercept.

Examples now playing at
MathTV.com/books

In defining the slope of a straight line, we are looking for a number to associate with a straight line that does two things. First of all, we want the slope of a line to measure the "steepness" of the line; that is, in comparing two lines, the slope of the steeper line should have the larger numerical value. Second, we want a line that *rises* going from left to right to have a *positive* slope. We want a line that *falls* going from left to right to have a *negative* slope. (A line that neither rises nor falls going from left to right must, therefore, have 0 slope.) These are illustrated in Figure 1.

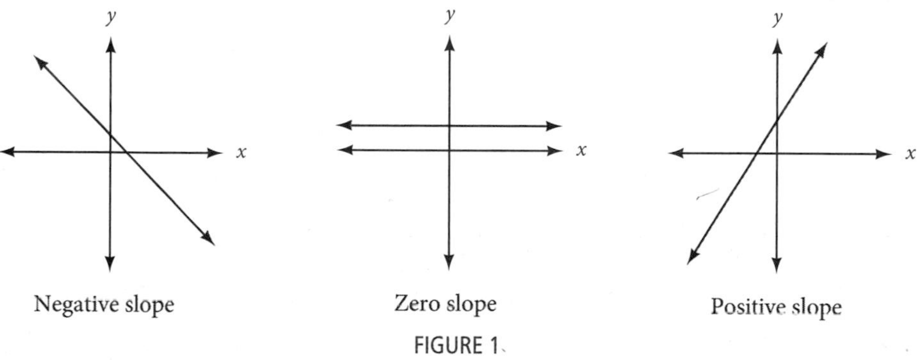

FIGURE 1

Suppose we know the coordinates of two points on a line. Because we are trying to develop a general formula for the slope of a line, we will use general points—call the two points $P_1(x_1, y_1)$ and $P_2(x_2, y_2)$. They represent the coordinates of any two different points on our line. We define the *slope* of our line to be the ratio of the vertical change to the horizontal change as we move from point (x_1, y_1) to point (x_2, y_2) on the line. (See Figure 2.)

The 2 in x_2 is called a subscript. It is notation that allows us to distinguish between the variables x_1 and x_2, while still showing that they are both x-coordinates.

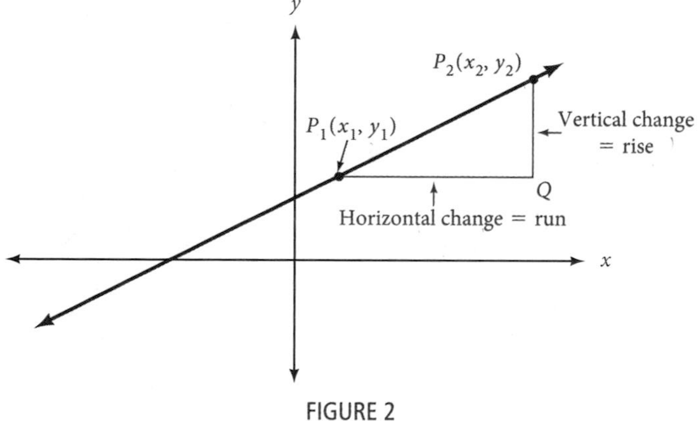

FIGURE 2

We call the vertical change the *rise* in the graph and the horizontal change the *run* in the graph. The slope, then, is

$$\text{Slope} = \frac{\text{vertical change}}{\text{horizontal change}} = \frac{\text{rise}}{\text{run}}$$

We would like to have a numerical value to associate with the rise in the graph and a numerical value to associate with the run in the graph. A quick study of Figure 2 shows that the coordinates of point Q must be (x_2, y_1), because Q is directly below point P_2 and right across from point P_1. We can draw our diagram again in the manner shown in Figure 3. It is apparent from this graph that the rise can be expressed as $(y_2 - y_1)$ and the run as $(x_2 - x_1)$. We usually denote the slope of a line by the letter m. The complete definition of slope follows along with a diagram (Figure 3) that illustrates the definition.

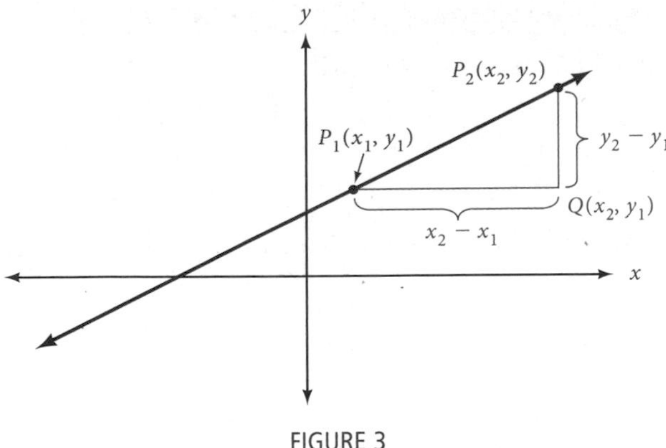

FIGURE 3

Definition

If points (x_1, y_1) and (x_2, y_2) are any two different points, then the **slope** of the line on which they lie is

$$\text{Slope} = m = \frac{\text{rise}}{\text{run}} = \frac{y_2 - y_1}{x_2 - x_1}$$

This definition of the slope of a line does just what we want it to do. If the line rises going from left to right, the slope will be positive. If the line falls from left to right, the slope will be negative. Also, the steeper the line, the larger numerical value the slope will have.

A Finding the Slope Given Two Points on a Line

PRACTICE PROBLEMS

1. Find the slope of the line through (5, 2) and (3, 6).

EXAMPLE 1 Find the slope of the line between the points (1, 2) and (3, 5).

SOLUTION We can let

$$(x_1, y_1) = (1, 2)$$

and

$$(x_2, y_2) = (3, 5)$$

then

$$m = \frac{y_2 - y_1}{x_2 - x_1} = \frac{5 - 2}{3 - 1} = \frac{3}{2}$$

The slope is $\frac{3}{2}$. For every vertical change of 3 units, there will be a corresponding horizontal change of 2 units. (See Figure 4.)

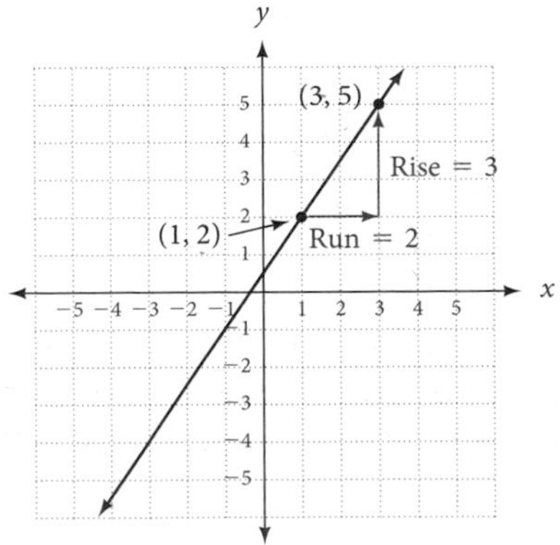

FIGURE 4

Answer

1. −2

EXAMPLE 2 Find the slope of the line through (−2, 1) and (5, −4).

SOLUTION It makes no difference which ordered pair we call (x_1, y_1) and which we call (x_2, y_2).

$$\text{Slope} = m = \frac{y_2 - y_1}{x_2 - x_1} = \frac{-4 - 1}{5 - (-2)} = \frac{-5}{7}$$

The slope is $-\frac{5}{7}$. Every vertical change of −5 units (down 5 units) is accompanied by a horizontal change of 7 units (to the right 7 units). (See Figure 5.)

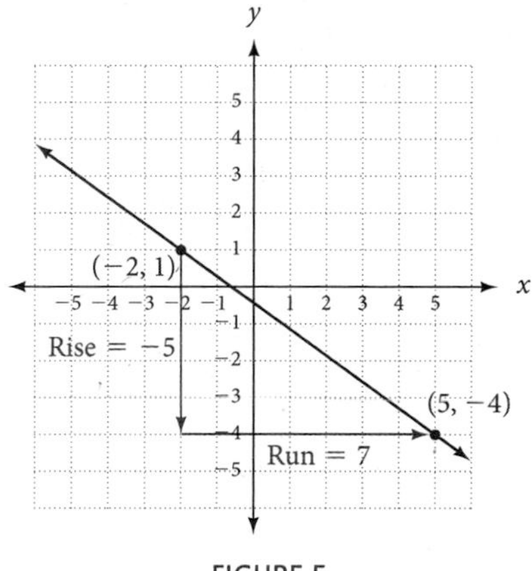

FIGURE 5

B Graphing a Line from Its Slope

EXAMPLE 3 Graph the line with slope $\frac{3}{2}$ and y-intercept 1.

SOLUTION Because the y-intercept is 1, we know that one point on the line is (0, 1). So, we begin by plotting the point (0, 1), as shown in Figure 6.

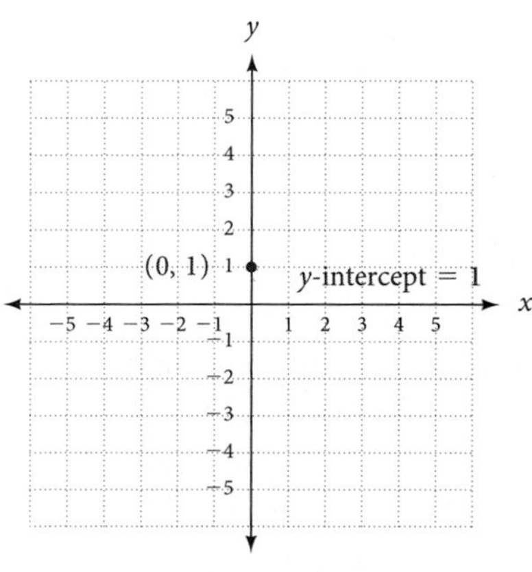

FIGURE 6

There are many lines that pass through the point shown in Figure 6, but only one of those lines has a slope of $\frac{3}{2}$. The slope, $\frac{3}{2}$, can be thought of as the rise in the graph divided by the run in the graph. Therefore, if we start at the point (0, 1) and move 3 units up (that's a rise of 3) and then 2 units to the right (a run of 2), we

2. Find the slope of the line through (−5, 1) and (4, −6).

3. Graph the line with slope $\frac{4}{3}$ and y-intercept 2.

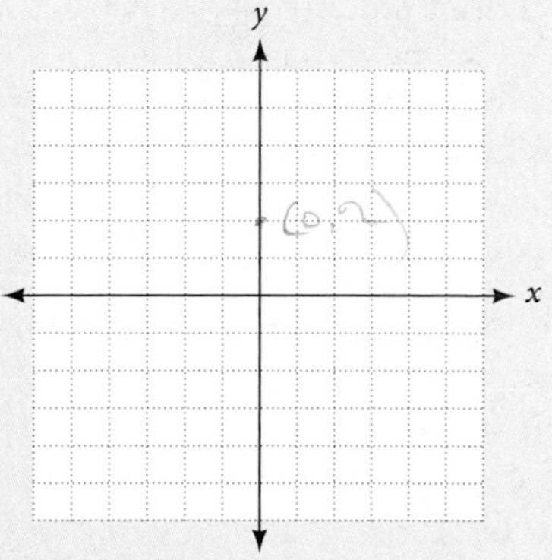

Answer

2. $-\frac{7}{9}$

will be at another point on the graph. Figure 7 shows that the point we reach by doing so is the point (2, 4).

$$\text{Slope} = m = \frac{\text{rise}}{\text{run}} = \frac{3}{2}$$

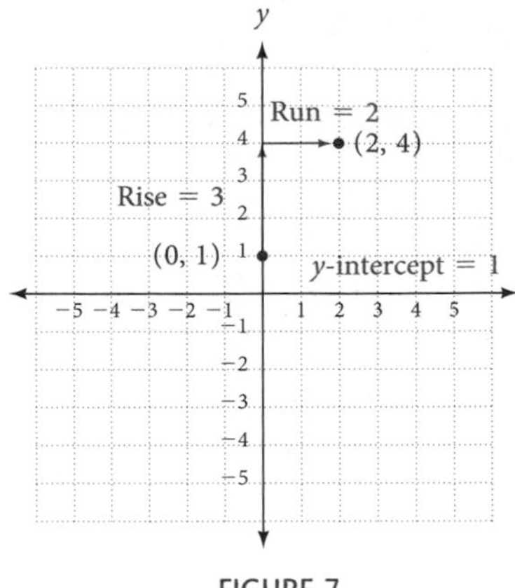

FIGURE 7

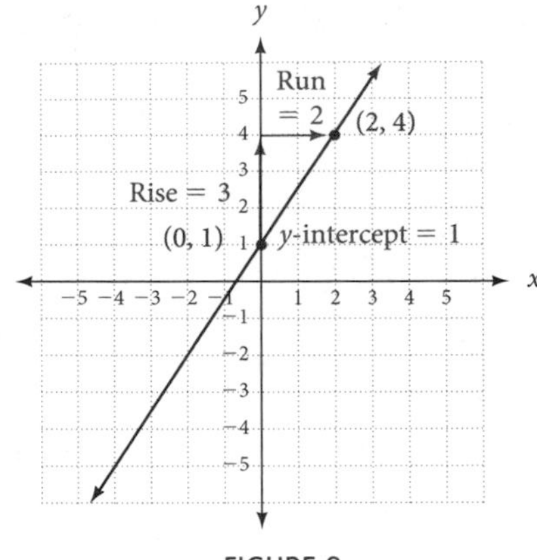

FIGURE 8

To graph the line with slope $\frac{3}{2}$ and y-intercept 1, we simply draw a line through the two points in Figure 7 to obtain the graph shown in Figure 8.

4. Find the slope of the line through (2, −3) and (−1, −3).

EXAMPLE 4 Find the slope of the line containing (3, −1) and (3, 4).

SOLUTION Using the definition for slope, we have

$$m = \frac{-1 - 4}{3 - 3} = \frac{-5}{0}$$

The expression $\frac{-5}{0}$ is undefined; that is, there is no real number to associate with it. In this case, we say the *slope is undefined.*

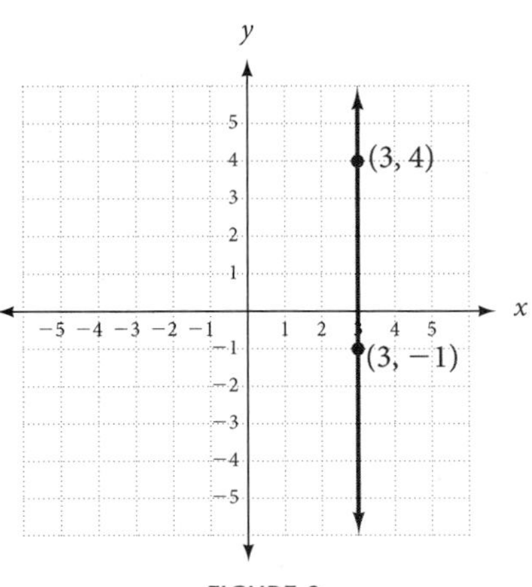

FIGURE 9

The graph of our line is shown in Figure 9. Our line is a vertical line. All vertical lines have an undefined slope. (And all horizontal lines, as we mentioned earlier, have 0 slope.)

Answers
3. See graph in the Solutions to Selected Practice Problems.
4. 0

As a final note, the summary reminds us that all horizontal lines have equations of the form $y = b$ and slopes of 0. Because they cross the y-axis at b, the y-intercept is b; there is no x-intercept. Vertical lines have an undefined slope and equations of the form $x = a$. Each will have an x-intercept at a and no y-intercept. Finally, equations of the form $y = mx$ have graphs that pass through the origin. The slope is always m and both the x-intercept and the y-intercept are 0.

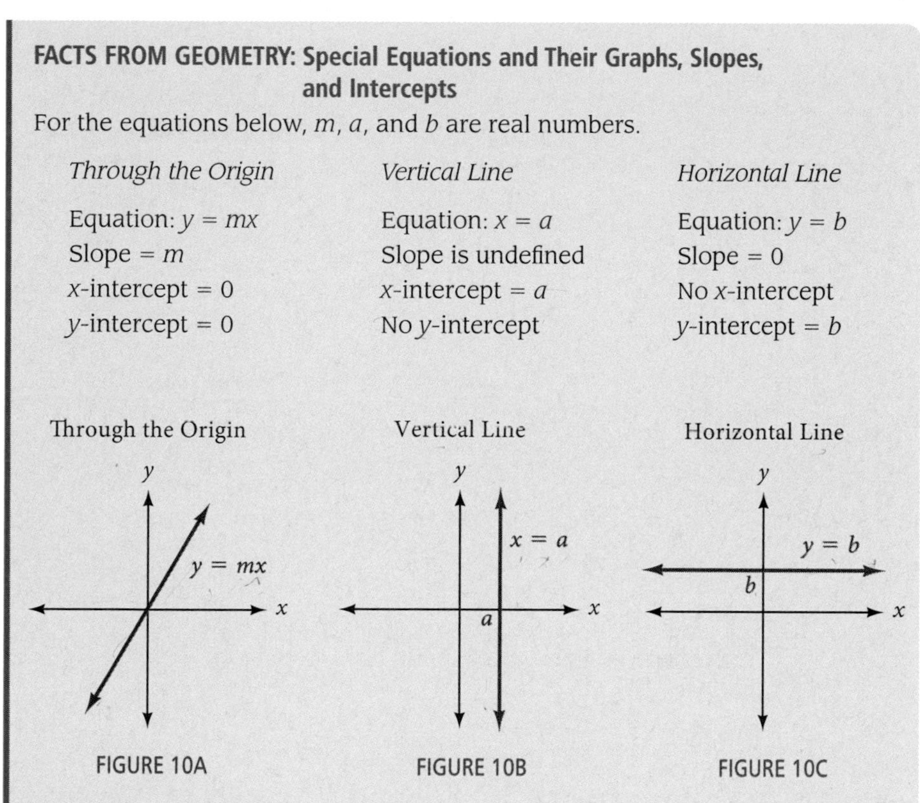

FACTS FROM GEOMETRY: Special Equations and Their Graphs, Slopes, and Intercepts

For the equations below, m, a, and b are real numbers.

Through the Origin	*Vertical Line*	*Horizontal Line*
Equation: $y = mx$	Equation: $x = a$	Equation: $y = b$
Slope $= m$	Slope is undefined	Slope $= 0$
x-intercept $= 0$	x-intercept $= a$	No x-intercept
y-intercept $= 0$	No y-intercept	y-intercept $= b$

FIGURE 10A FIGURE 10B FIGURE 10C

Getting Ready for Class

After reading through the preceding section, respond in your own words and in complete sentences.

1. What is the slope of a line?
2. Would you rather climb a hill with a slope of 1 or a slope of 3? Explain why.
3. Describe how to obtain the slope of a line if you know the coordinates of two points on the line.
4. Describe how you would graph a line from its slope and y-intercept.

Problem Set 3.5

A Find the slope of the line through the following pairs of points. Then plot each pair of points, draw a line through them, and indicate the rise and run in the graph. **[Examples 1–2]**

1. (2, 1), (4, 4)

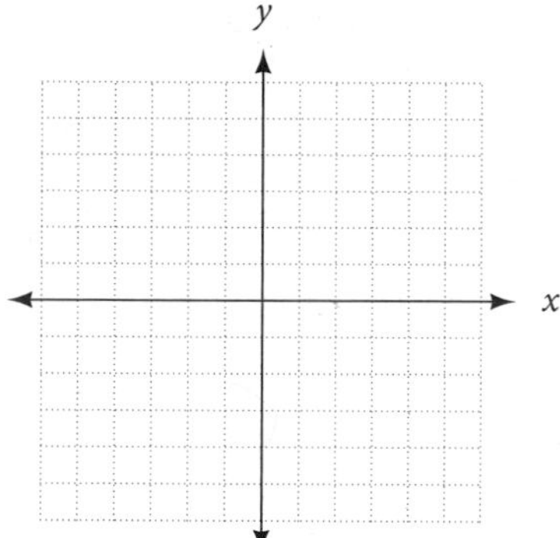

2. (3, 1), (5, 4)

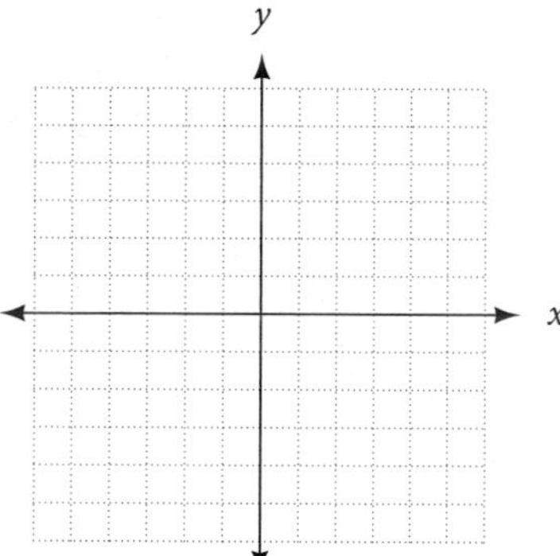

3. (1, 4), (5, 2)

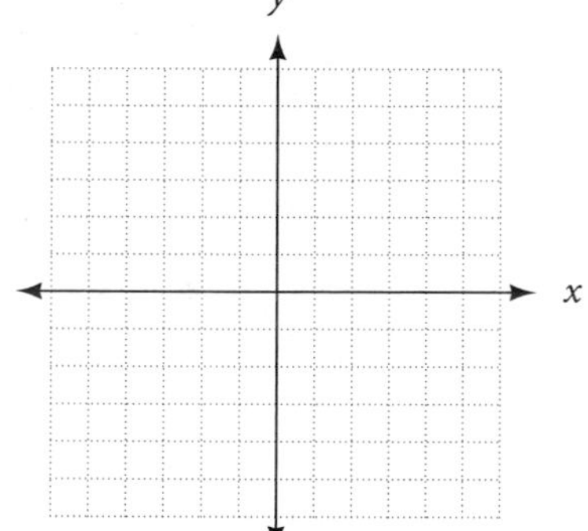

4. (1, 3), (5, 2)

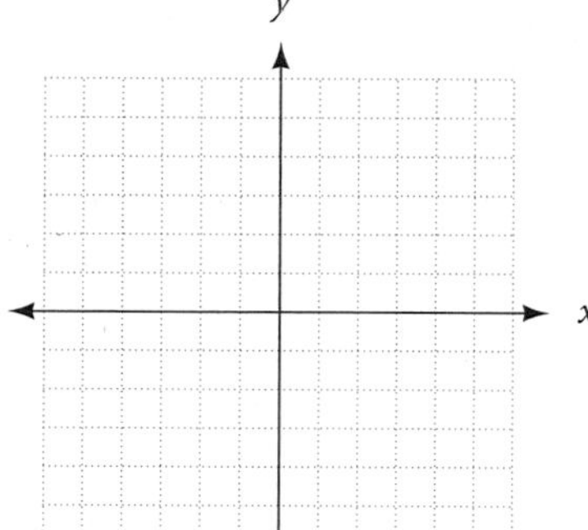

5. (1, −3), (4, 2)

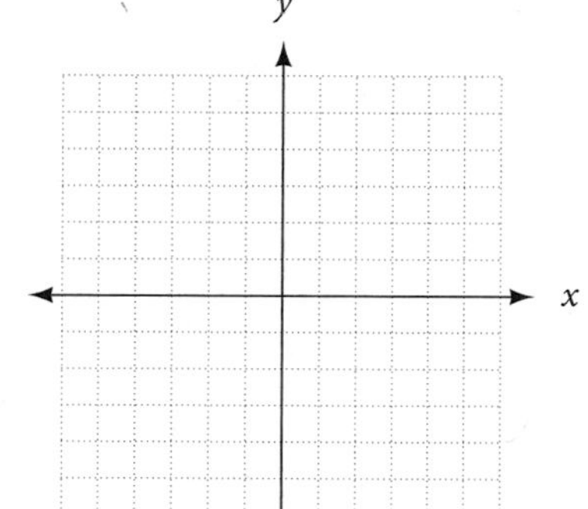

6. (2, −3), (5, 2)

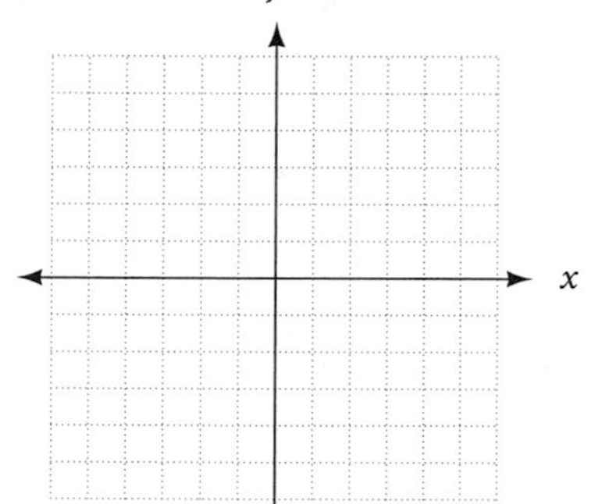

7. (−3, −2), (1, 3)

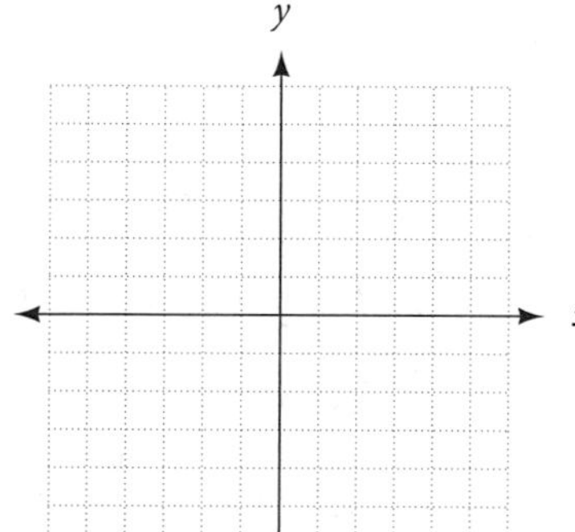

8. (−3, −1), (1, 4)

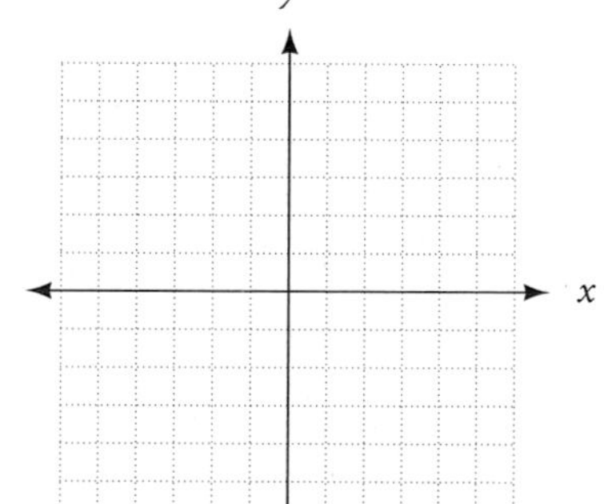

9. (−3, 2), (3, −2)

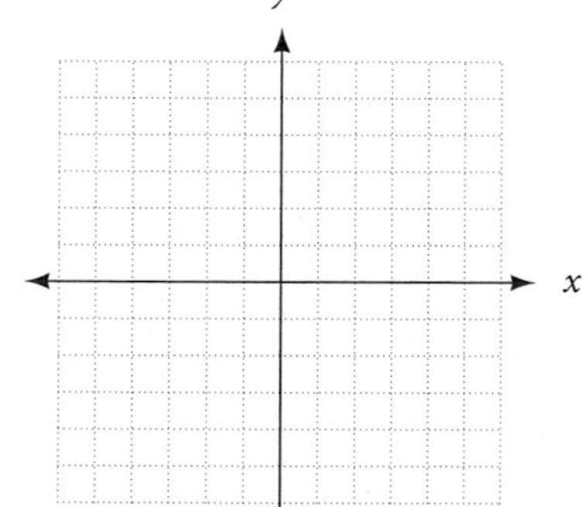

10. (−3, 3), (3, −1)

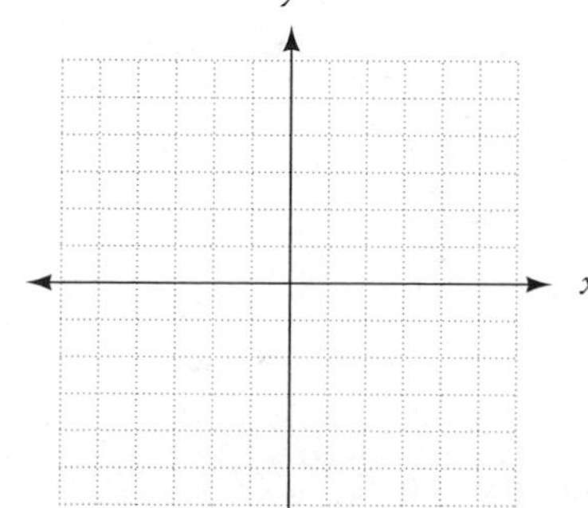

11. (2, −5), (3, −2)

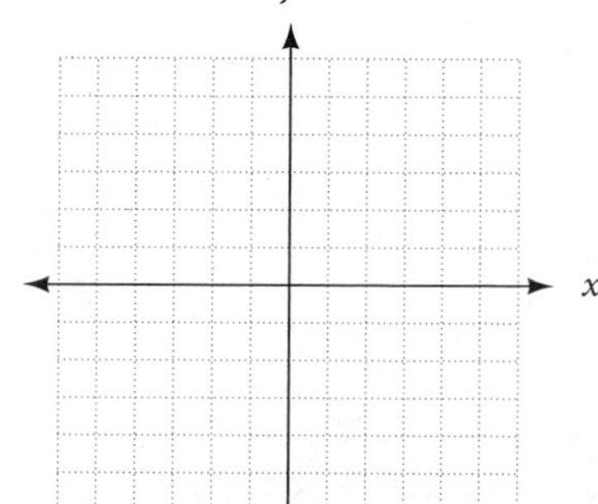

12. (2, −4), (3, −1)

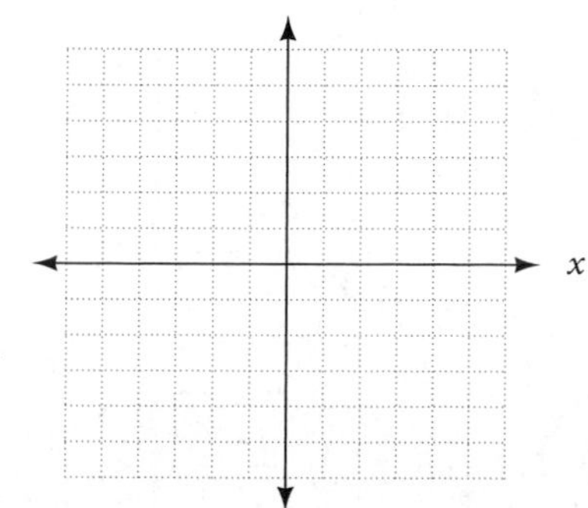

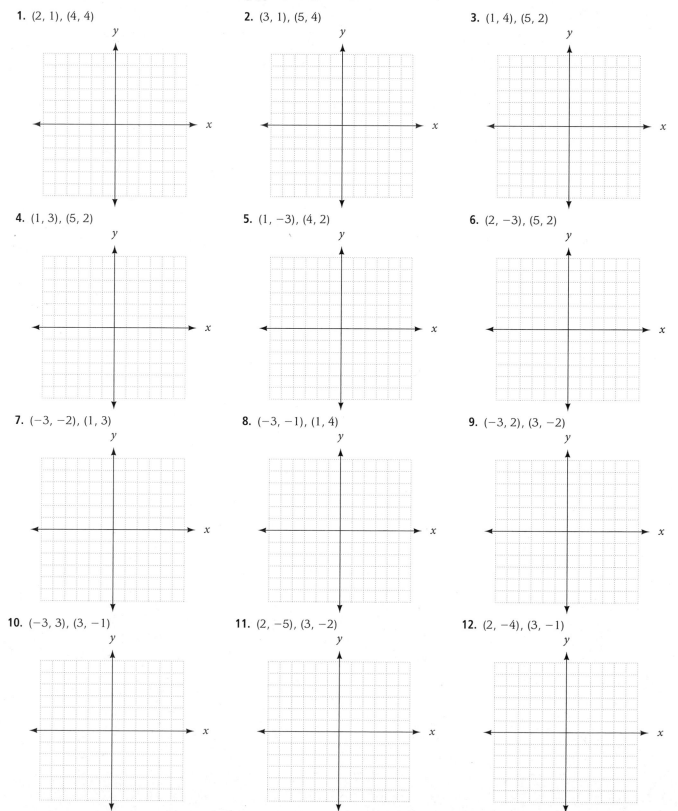

B In each of the following problems, graph the line with the given slope and y-intercept b. **[Example 3]**

13. $m = \frac{2}{3}, b = 1$

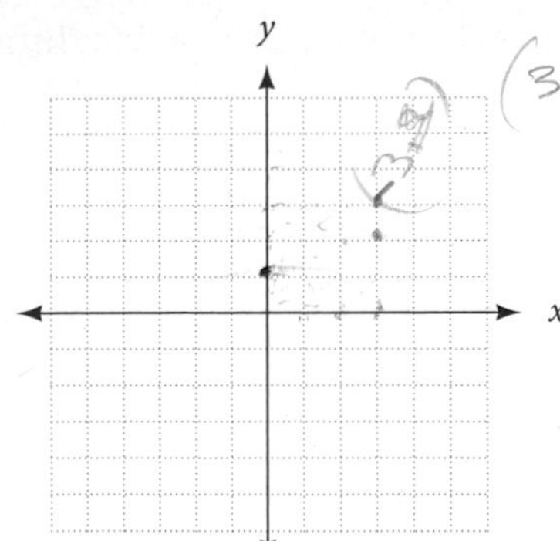

14. $m = \frac{3}{4}, b = -2$

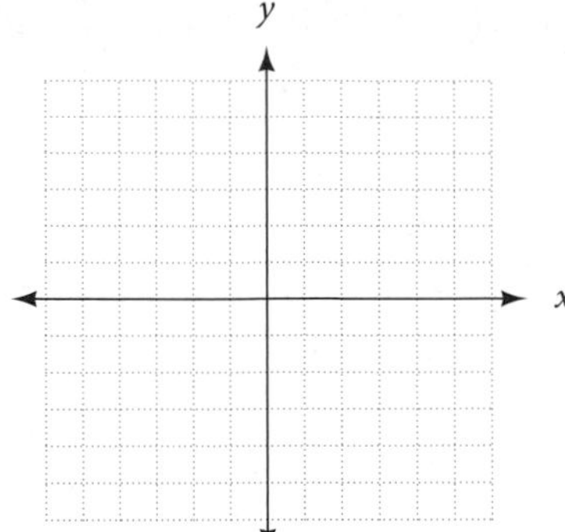

15. $m = \frac{3}{2}, b = -3$

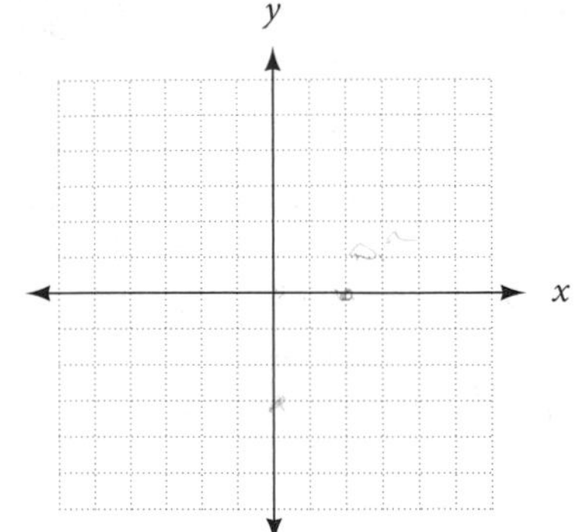

16. $m = \frac{4}{3}, b = 2$

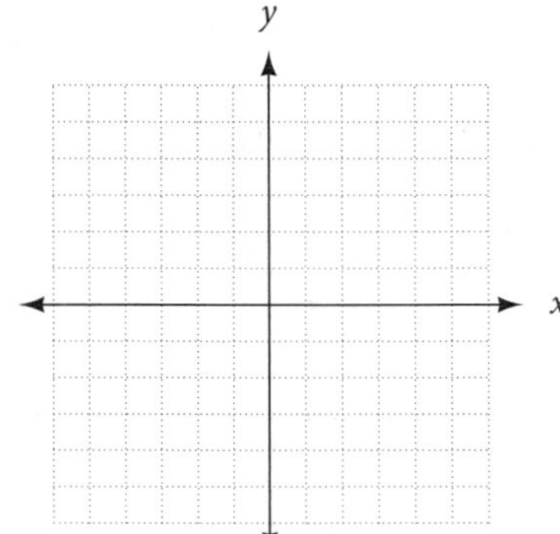

17. $m = -\frac{4}{3}, b = 5$

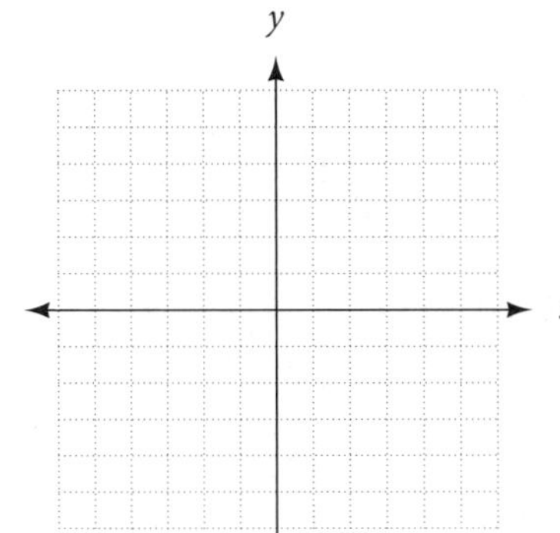

18. $m = -\frac{3}{5}, b = 4$

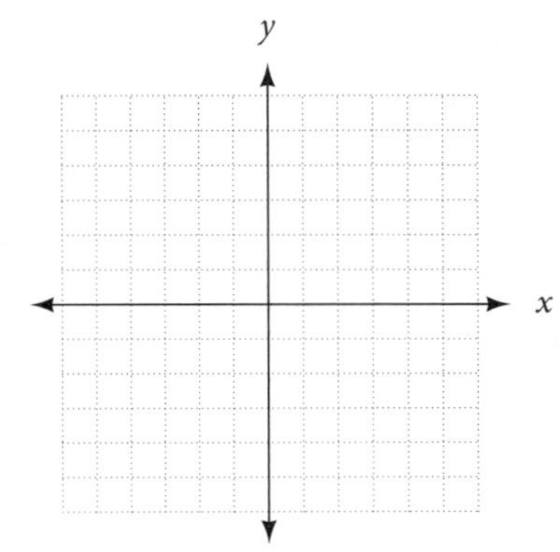

19. $m = 2, b = 1$

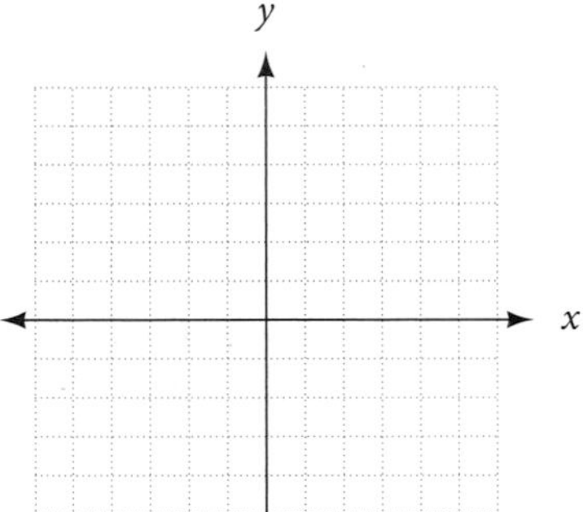

20. $m = -2, b = 4$

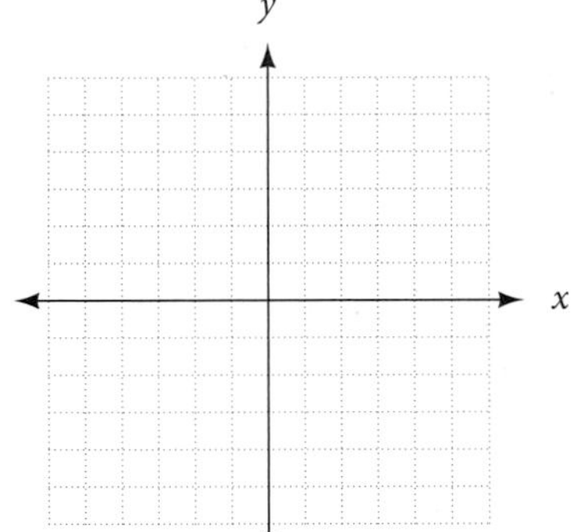

21. $m = 3, b = -1$

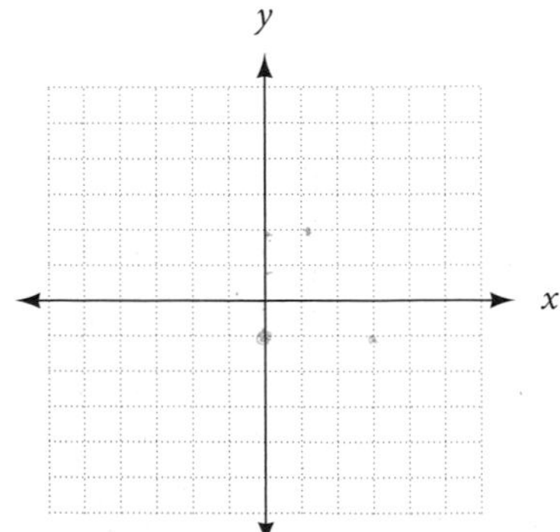

22. $m = 3, b = -2$

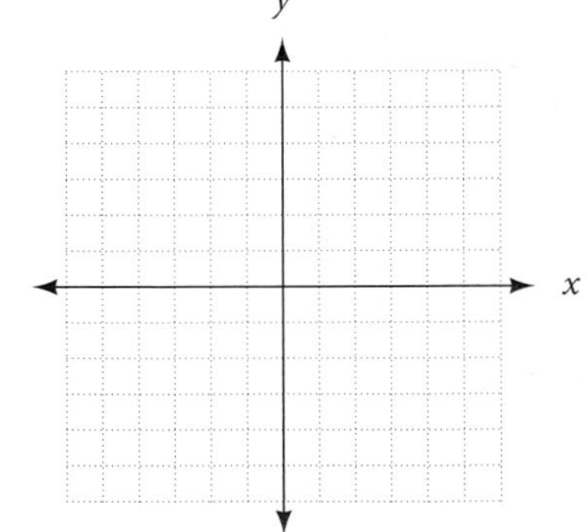

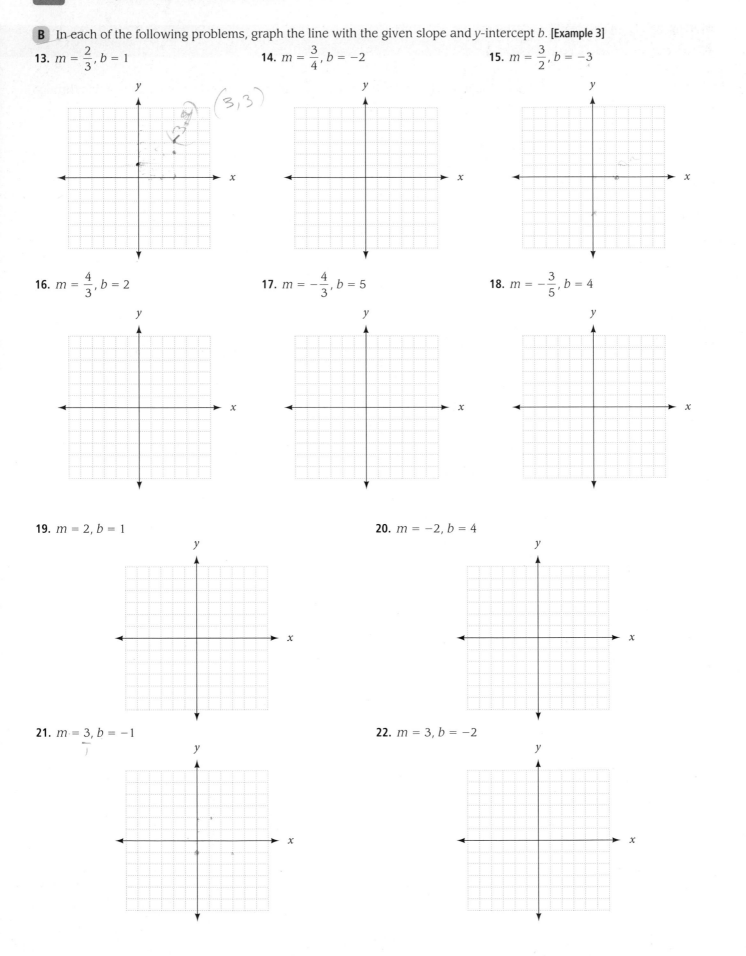

Find the slope and y-intercept for each line.

23.

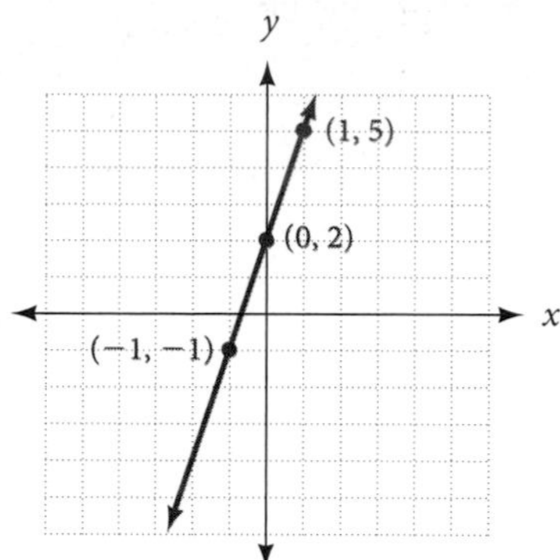

24.

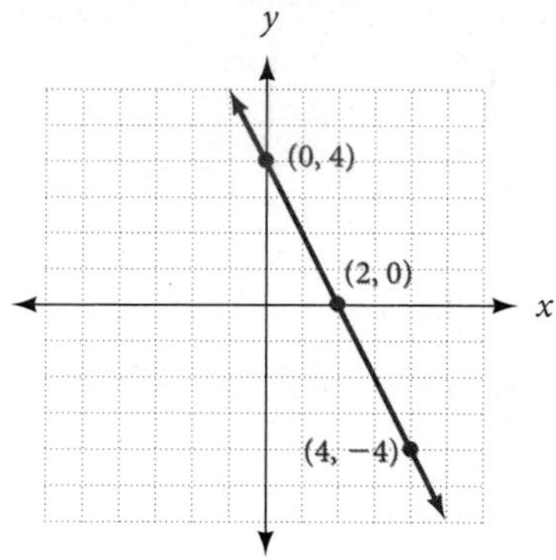

25.

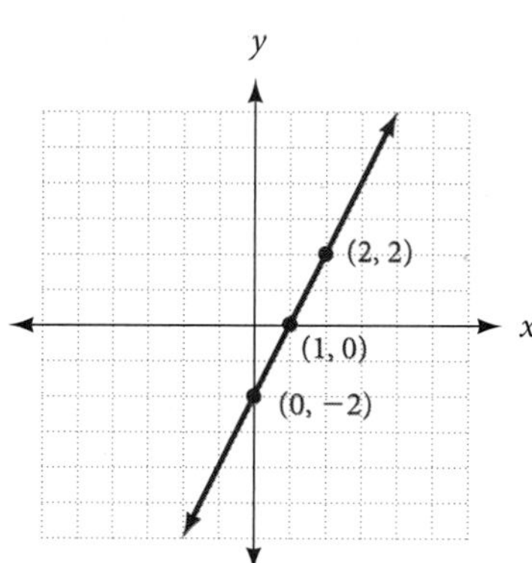

26.

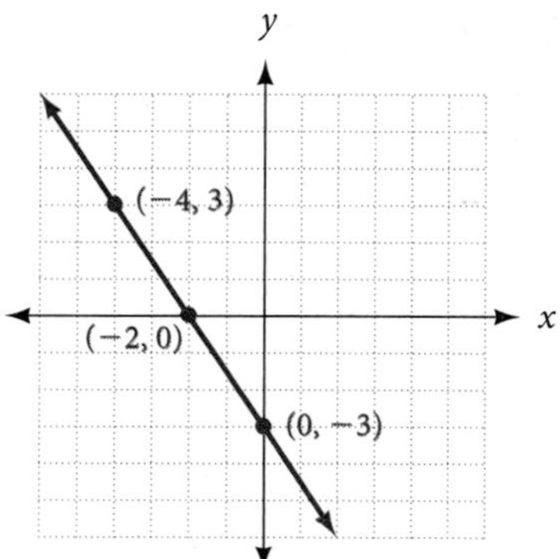

27. Graph the line that has an x-intercept of 3 and a y-intercept of -2. What is the slope of this line?

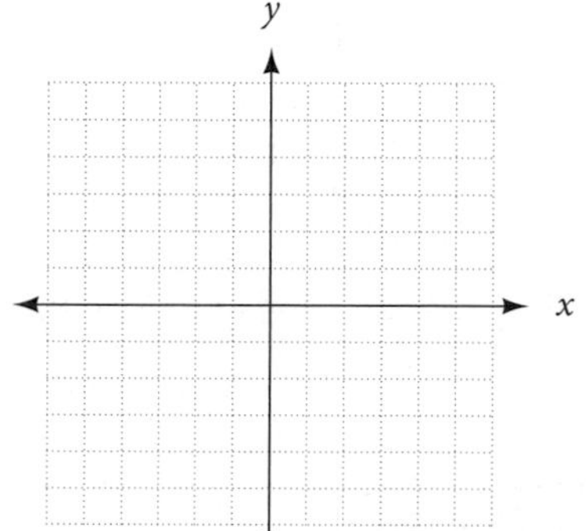

28. Graph the line that has an x-intercept of 2 and a y-intercept of -3. What is the slope of this line?

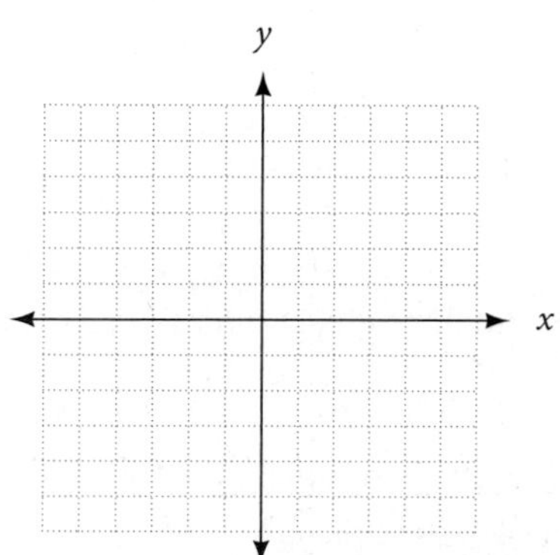

29. Graph the line with x-intercept 4 and y-intercept 2. What is the slope of this line?

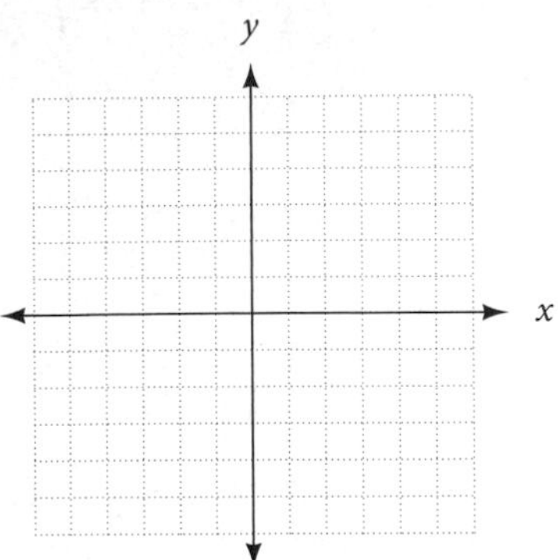

30. Graph the line with x-intercept -4 and y-intercept -2. What is the slope of this line?

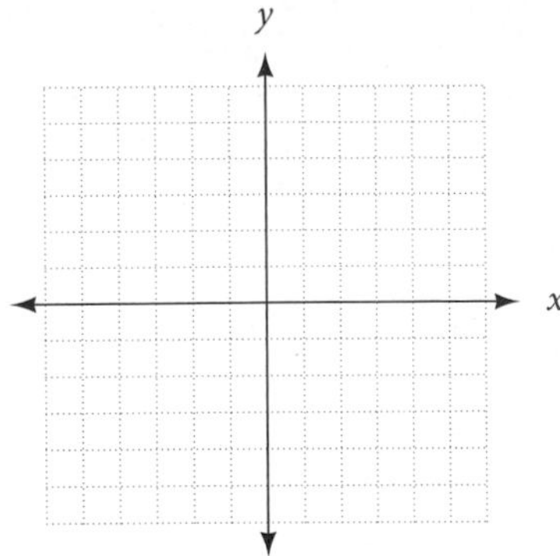

31. Graph the line $y = 2x - 3$, then name the slope and y-intercept by looking at the graph.

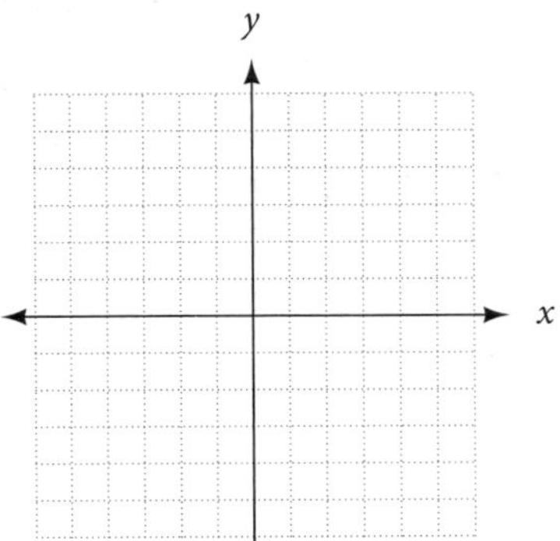

32. Graph the line $y = -2x + 3$, then name the slope and y-intercept by looking at the graph.

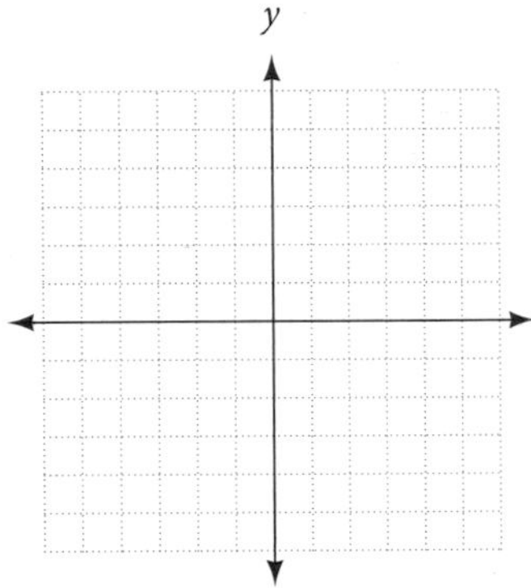

33. Graph the line $y = \frac{1}{2}x + 1$, then name the slope and y-intercept by looking at the graph.

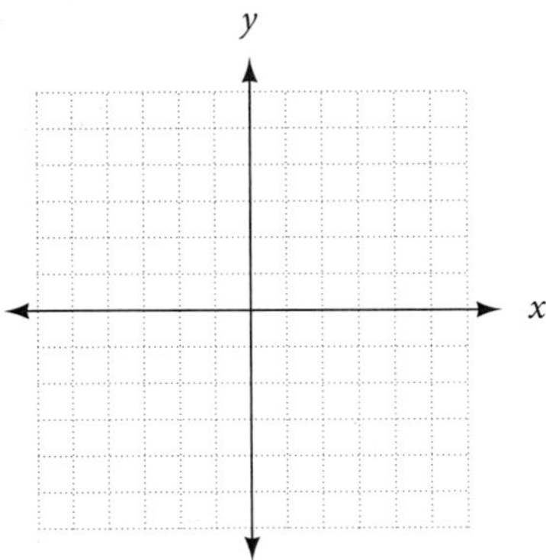

34. Graph the line $y = -\frac{1}{2}x - 2$, then name the slope and y-intercept by looking at the graph.

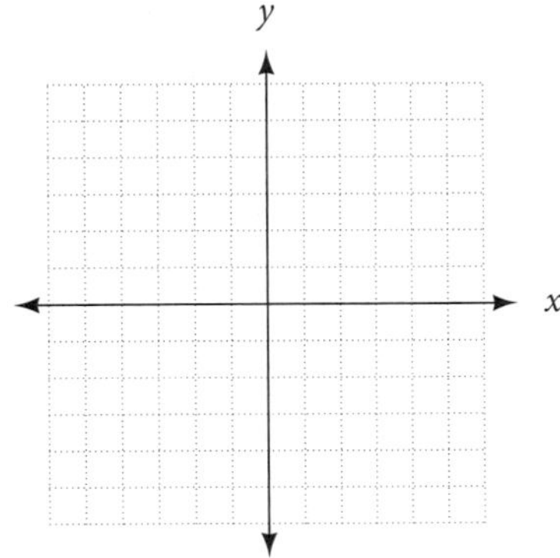

35. Find y if the line through $(4, 2)$ and $(6, y)$ has a slope of 2.

36. Find y if the line through $(1, y)$ and $(7, 3)$ has a slope of 6.

For each equation in each table, give the slope of the graph.

37.

Equation	slope
$x = 3$	
$y = 3$	
$y = 3x$	

38.

Equation	slope
$y = \frac{3}{2}$	
$x = \frac{3}{2}$	
$y = \frac{3}{2}x$	

39.

Equation	slope
$y = -\frac{2}{3}$	
$x = -\frac{2}{3}$	
$y = -\frac{2}{3}x$	

40.

Equation	slope
$x = -2$	
$y = -2$	
$y = -2x$	

Applying the Concepts

41. Health Care The graph shows the rising cost of health care. What is the slope between 2005 and 2008? Round to the nearest hundredth.

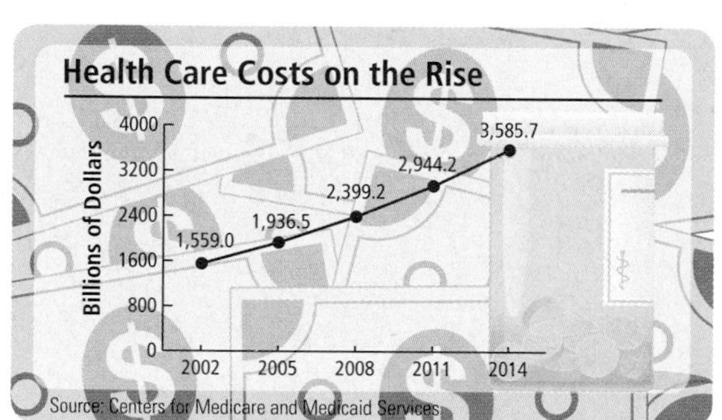

42. Wind Energy The graph shows the world's capacity for generating power from wind energy. Estimate the slope from 2000 to 2005.

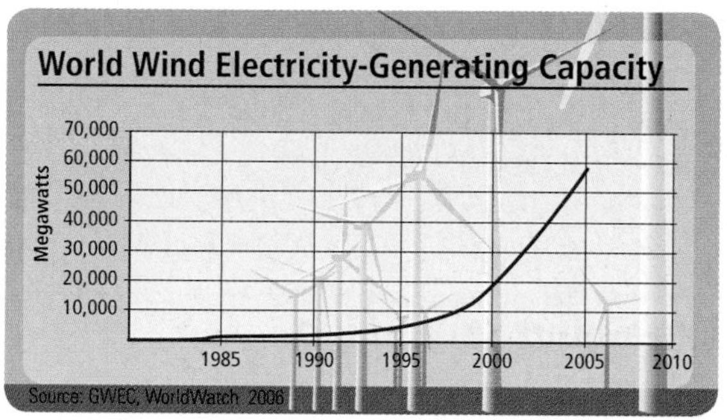

43. Garbage Production The table and completed line graph shown here give the annual production of garbage in the United States for some specific years. Find the slope of each of the four line segments, A, B, C, and D.

Year	Garbage (million of tons)
1960	88
1970	121
1980	152
1990	205
2000	224

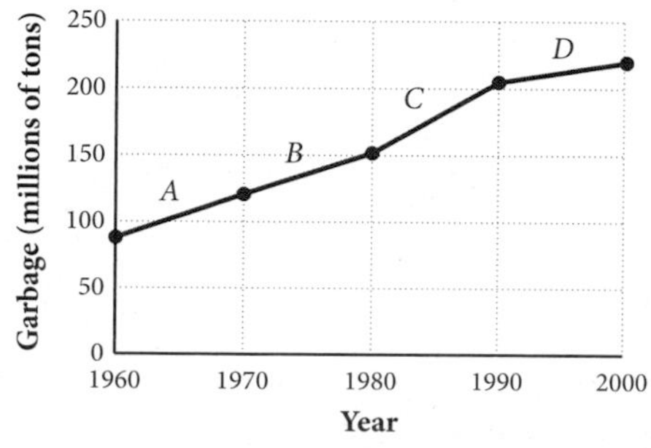

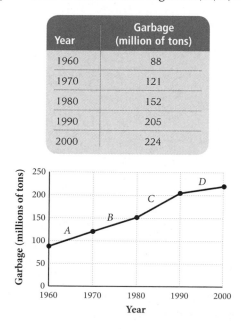

44. Grass Height The table and completed line graph shown here give the growth of a certain species over time. Find the slopes of the line segments labeled A, B, and C.

Day	Plant Height
0	0
1	0.5
2	1
3	1.5
4	3
5	4
6	6
7	9
8	13
9	18
10	23

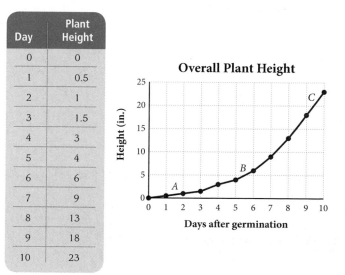

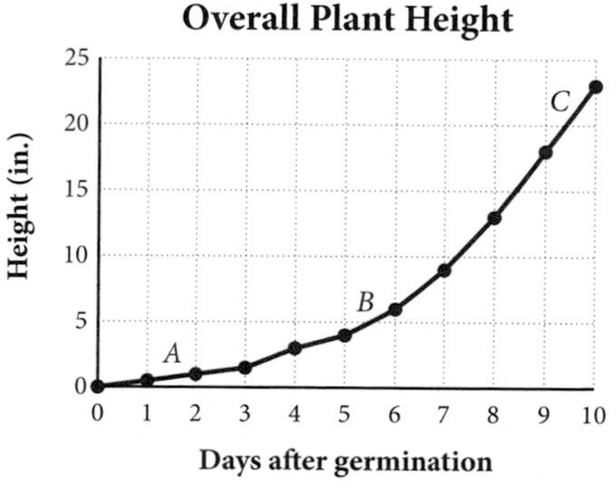

Getting Ready for Class

Solve each equation for y.

45. $-2x + y = 4$
46. $-4x + y = -2$
47. $3x + y = 3$
48. $3x + 2y = 6$
49. $4x - 5y = 20$
50. $-2x - 5y = 10$
51. $y - 3 = -2(x + 4)$
52. $y + 5 = 2(x + 2)$
53. $y - 3 = -\frac{2}{3}(x + 3)$
54. $y - 1 = -\frac{1}{2}(x + 4)$
55. $\frac{y - 1}{x} = \frac{3}{2}$
56. $\frac{y + 1}{x} = \frac{3}{2}$

Maintaining Your Skills

Solve each equation.

57. $\frac{1}{2}(4x + 10) = 11$
58. $\frac{1}{2}(6x - 12) = -18$
59. $\frac{2}{3}(3x - 9) = 24$
60. $\frac{1}{3}(2x + 15) = -29$
61. $\frac{3}{4}(4x + 8) = -12$
62. $\frac{3}{4}(8x - 16) = -36$
63. $\frac{3}{5}(10x + 15) = 45$
64. $\frac{2}{5}(5x - 10) = -10$
65. $5\left(\frac{2}{5}x + 10\right) = -28$
66. $3\left(\frac{2}{3}x + 5\right) = -13$
67. $4\left(\frac{3}{2}x - 7\right) = -4$
68. $4\left(\frac{3}{4}x + 5\right) = -40$
69. $\frac{3}{4}(2x + 12) = 24$
70. $-\frac{3}{4}(12x - 16) = -42$
71. $\frac{1}{2}(5x - 10) + 6 = -49$
72. $\frac{1}{3}(5x - 15) - 5 = 20$

Finding the Equation of a Line

3.6

Objectives

A Find the equation of a line given the slope and y-intercept of the line.

B Find the slope and y-intercept of a line given the equation of the line.

C Find the equation of a line given a point on the line and the slope of the line.

D Find the equation of a line given two points on the line.

To this point in the chapter, most of the problems we have worked have used the equation of a line to find different types of information about the line. For instance, given the equation of a line, we can find points on the line, the graph of the line, the intercepts, and the slope of the line. In this section we reverse things somewhat and move in the other direction. We will use information about a line, such as its slope and y-intercept, to find the equation of a line.

There are three main types of problems to solve in this section.

1. Find the equation of a line from the slope and y-intercept.
2. Find the equation of a line given one point on the line and the slope of the line.
3. Find the equation of a line given two points on the line.

Examples 1 and 2 illustrate the first type of problem. Example 5 solves the second type of problem. The third type of problem is solved in Example 6.

A The Slope-Intercept Form of an Equation of a Straight Line

EXAMPLE 1 Find the equation of the line with slope $\frac{3}{2}$ and y-intercept 1.

SOLUTION We graphed the line with slope $\frac{3}{2}$ and y-intercept 1 in Example 3 of the previous section. Figure 1 shows that graph.

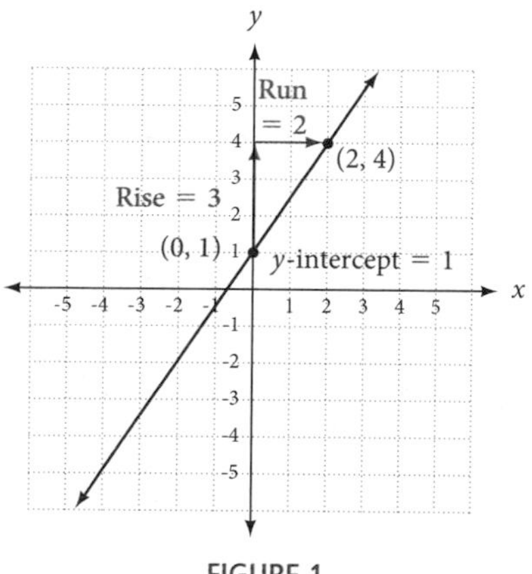

FIGURE 1

What we want to do now is find the equation of the line shown in Figure 1. To do so, we take any other point (x, y) on the line and apply our slope formula to that point and the point $(0, 1)$. We set that result equal to $\frac{3}{2}$, because $\frac{3}{2}$ is the slope of our line. The work is as follows, with a diagram of the situation following.

$$\frac{y-1}{x-0} = \frac{3}{2} \qquad \text{Slope} = \frac{\text{vertical change}}{\text{horizontal change}}$$

$$\frac{y-1}{x} = \frac{3}{2} \qquad x - 0 = x$$

$$y - 1 = \frac{3}{2}x \qquad \text{Multiply each side by } x$$

$$y = \frac{3}{2}x + 1 \qquad \text{Add 1 to each side}$$

PRACTICE PROBLEMS

1. Find the equation of the line with slope $\frac{4}{3}$ and y-intercept 2.

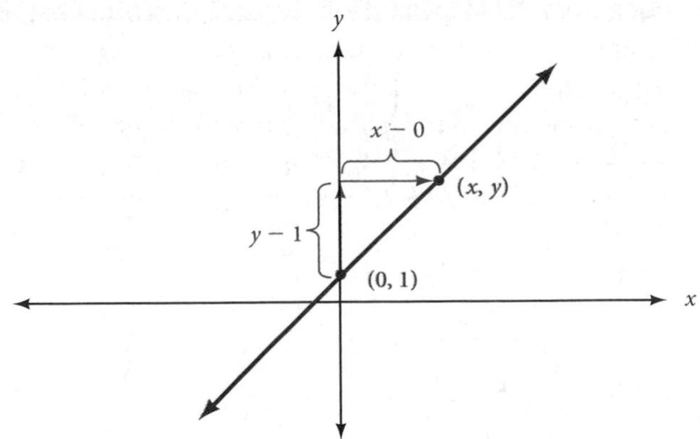

What is interesting and useful about the equation we found in Example 1 is that the number in front of x is the slope of the line and the constant term is the y-intercept. It is no coincidence that it turned out this way. Whenever an equation has the form $y = mx + b$, the graph is always a straight line with slope m and y-intercept b. To see that this is true in general, suppose we want the equation of a line with slope m and y-intercept b. Because the y-intercept is b,the point $(0, b)$ is on the line. If (x, y) is any other point on the line, then we apply our slope formula to get

$$\frac{y - b}{x - 0} = m \qquad \text{Slope} = \frac{\text{vertical change}}{\text{horizontal change}}$$

$$\frac{y - b}{x} = m \qquad x - 0 = x$$

$$y - b = mx \qquad \text{Multiply each side by } x$$

$$y = mx + b \qquad \text{Add } b \text{ to each side}$$

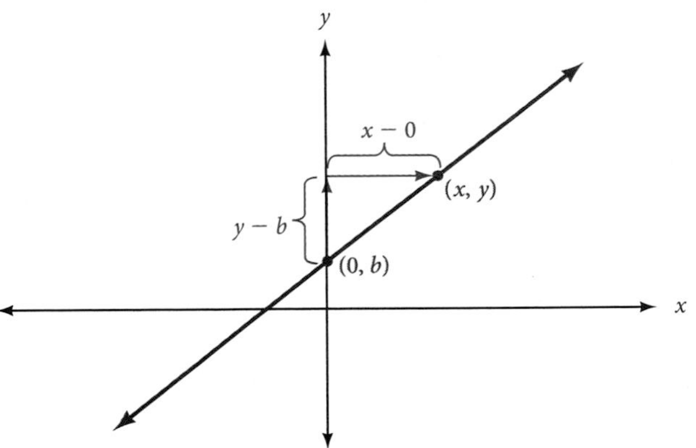

Here is a summary of what we have just found.

> **Property**
> The equation of the line with slope m and y-intercept b is always given by
>
> $$y = mx + b$$

EXAMPLE 2 Find the equation of the line with slope $-\frac{4}{3}$ and y-intercept 5. Then, graph the line.

SOLUTION Substituting $m = -\frac{4}{3}$ and $b = 5$ into the equation $y = mx + b$, we have

$$y = -\frac{4}{3}x + 5$$

2. Find the equation of the line with slope $-\frac{2}{3}$ and y-intercept 3.

Answer
1. $y = \frac{4}{3}x + 2$

Finding the equation from the slope and y-intercept is just that easy. If the slope is m and the y-intercept is b, then the equation is always $y = mx + b$.

Because the y-intercept is 5, the graph goes through the point (0, 5). To find a second point on the graph, we start at (0, 5) and move 4 units down (that's a rise of -4) and 3 units to the right (a run of 3). The point we reach is (3, 1). Drawing a line that passes through (0, 5) and (3, 1), we have the graph of our equation. (Note that we could also let the rise = 4 and the run = -3 and obtain the same graph.) The graph is shown in Figure 2.

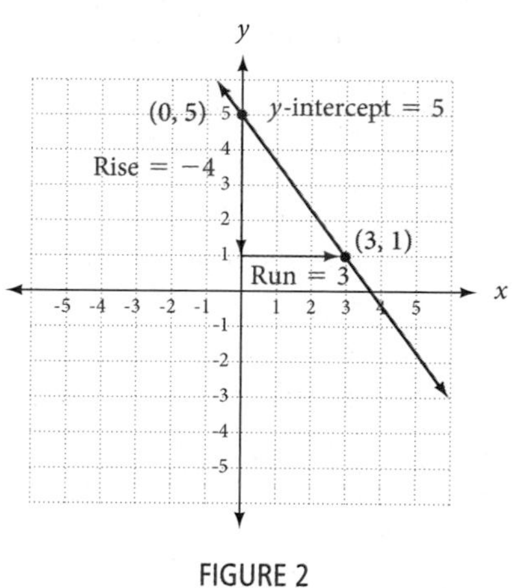

FIGURE 2

B Finding the Slope and y-Intercept

EXAMPLE 3 Find the slope and y-intercept for $-2x + y = -4$. Then, use them to draw the graph.

SOLUTION To identify the slope and y-intercept from the equation, the equation must be in the form $y = mx + b$ (slope-intercept form). To write our equation in this form, we must solve the equation for y. To do so, we simply add $2x$ to each side of the equation.

$$-2x + y = -4 \quad \text{Original equation}$$
$$y = 2x - 4 \quad \text{Add } 2x \text{ to each side}$$

The equation is now in slope-intercept form, so the slope must be 2 and the y-intercept must be -4. The graph, therefore, crosses the y-axis at (0, -4). Because the slope is 2, we can let the rise = 2 and the run = 1 and find a second point on the graph. The graph is shown in Figure 3.

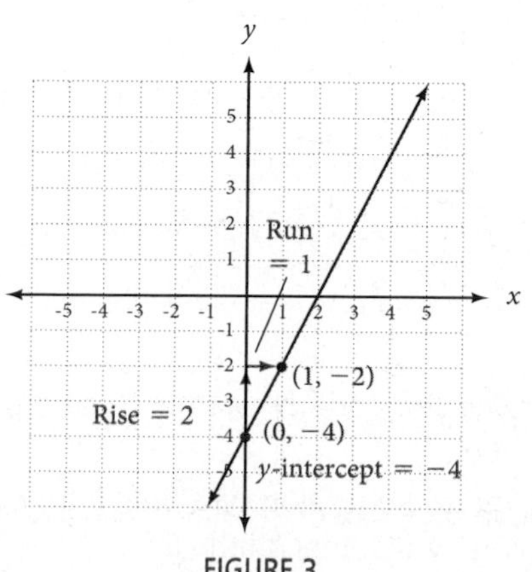

FIGURE 3

3. Find the slope and y-intercept for $-4x + y = -2$. Then, use them to draw the graph.

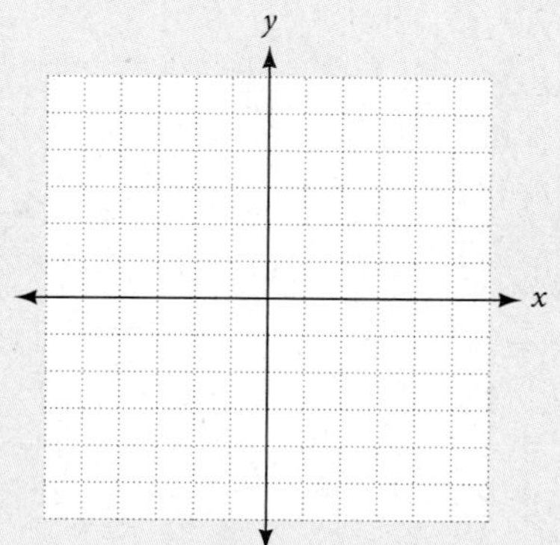

Answers

2. $y = -\frac{2}{3}x + 3$

3. Slope = 4; y-intercept = -2

4. Find the slope and y-intercept for $4x - 3y = 9$.

EXAMPLE 4 Find the slope and y-intercept for $3x - 2y = 6$.

SOLUTION To find the slope and y-intercept from the equation, we must write the equation in the form $y = mx + b$. This means we must solve the equation $3x - 2y = 6$ for y.

$$3x - 2y = 6 \quad \text{Original equation}$$

$$-2y = -3x + 6 \quad \text{Add } -3x \text{ to each side}$$

$$-\frac{1}{2}(-2y) = -\frac{1}{2}(-3x + 6) \quad \text{Multiply each side by } -\frac{1}{2}$$

$$y = \frac{3}{2}x - 3 \quad \text{Simplify each side}$$

Now that the equation is written in slope-intercept form, we can identify the slope as $\frac{3}{2}$ and the y-intercept as -3. The graph is shown in Figure 4.

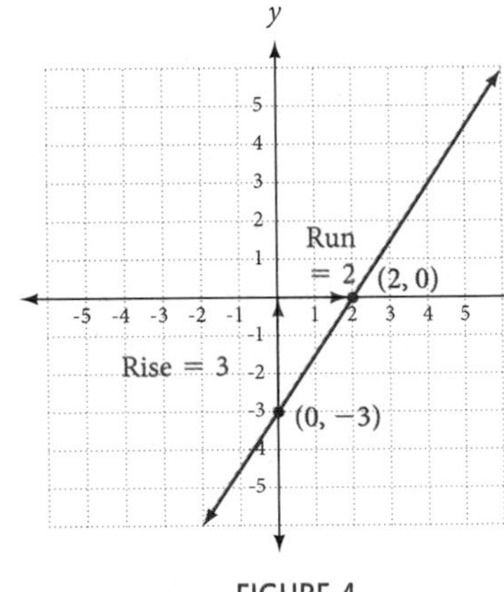

FIGURE 4

C The Point-Slope Form of an Equation of a Straight Line

A second useful form of the equation of a straight line is the point-slope form.

Let line l contain the point (x_1, y_1) and have slope m. If (x, y) is any other point on l, then by the definition of slope we have

$$\frac{y - y_1}{x - x_1} = m$$

Multiplying both sides by $(x - x_1)$ gives us

$$(x - x_1) \cdot \frac{y - y_1}{x - x_1} = m(x - x_1)$$

$$y - y_1 = m(x - x_1)$$

This last equation is known as the *point-slope form* of the equation of a straight line.

Property

The equation of the line through (x_1, y_1) with slope m is given by

$$y - y_1 = m(x - x_1)$$

This form is used to find the equation of a line, either given one point on the line and the slope, or given two points on the line.

Answer

4. Slope $= \frac{4}{3}$; y-intercept $= -3$

EXAMPLE 5 Find the equation of the line with slope −2 that contains the point (−4, 3). Write the answer in slope-intercept form.

SOLUTION

Using	$(x_1, y_1) = (-4, 3)$ and $m = -2$	
in	$y - y_1 = m(x - x_1)$	Point-slope form
gives us	$y - 3 = -2(x + 4)$	Note: $x - (-4) = x + 4$
	$y - 3 = -2x - 8$	Multiply out right side
	$y = -2x - 5$	Add 3 to each side

Figure 5 is the graph of the line that contains (−4, 3) and has a slope of −2. Notice that the y-intercept on the graph matches that of the equation we found.

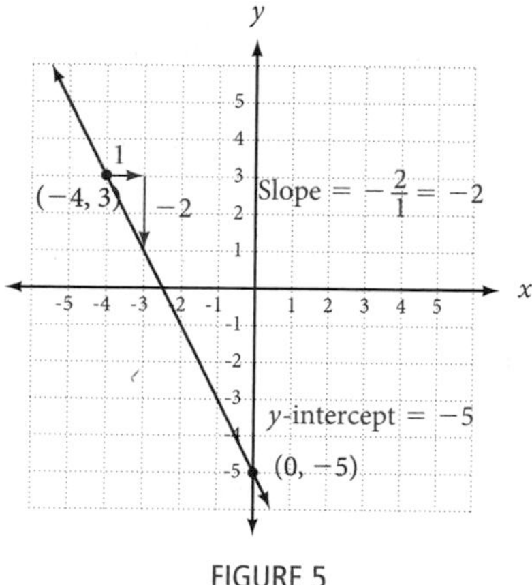

FIGURE 5

5. Find the equation of the line that contains (3, 5) and has a slope of 4.

D Finding an Equation Given Two Points

EXAMPLE 6 Find the equation of the line that passes through the points (−3, 3) and (3, −1).

SOLUTION We begin by finding the slope of the line:

$$m = \frac{3 - (-1)}{-3 - 3} = \frac{4}{-6} = -\frac{2}{3}$$

Using $(x_1, y_1) = (3, -1)$ and $m = -\frac{2}{3}$ in $y - y_1 = m(x - x_1)$ yields

$$y + 1 = -\frac{2}{3}(x - 3)$$

$$y + 1 = -\frac{2}{3}x + 2 \quad \text{Multiply out right side}$$

$$y = -\frac{2}{3}x + 1 \quad \text{Add } -1 \text{ to each side}$$

Figure 6 shows the graph of the line that passes through the points (−3, 3) and (3, −1). As you can see, the slope and y-intercept are $-\frac{2}{3}$ and 1, respectively.

6. Find the equation of the line that passes through the points (8, 1) and (4, −2). Write your answer in slope-intercept form.

Answer
5. $y = 4x - 7$

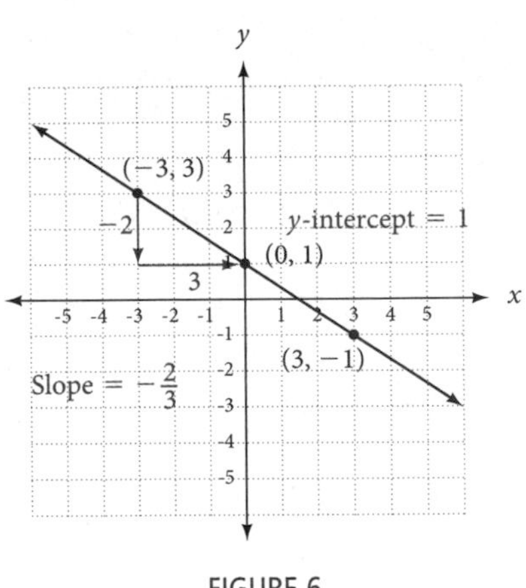

FIGURE 6

NOTE In Example 6 we could have used the point (−3, 3) instead of (3, −1) and obtained the same equation; that is, using $(x_1, y_1) = (-3, 3)$ and $m = -\frac{2}{3}$ in $y - y_1 = m(x - x_1)$ gives us

$$y - 3 = -\frac{2}{3}(x + 3)$$

$$y - 3 = -\frac{2}{3}x - 2$$

$$y = -\frac{2}{3}x + 1$$

which is the same result we obtained using (3, −1).

Methods of Graphing Lines

1. Substitute convenient values of x into the equation, and find the corresponding values of y. We used this method first for equations like $y = 2x - 3$. To use this method for equations that looked like $2x - 3y = 6$, we first solved them for y.
2. Find the x- and y-intercepts. This method works best for equations of the form $3x + 2y = 6$ where the numbers in front of x and y divide the constant term evenly.
3. Find the slope and y-intercept. This method works best when the equation has the form $y = mx + b$ and b is an integer.

Getting Ready for Class

After reading through the preceding section, respond in your own words and in complete sentences.

1. What are m and b in the equation $y = mx + b$?
2. How would you find the slope and y-intercept for the line $3x - 2y = 6$?
3. What is the point-slope form of the equation of a line?
4. How would you find the equation of a line from two points on the line?

Answer

6. $y = \frac{3}{4}x - 5$

Problem Set 3.6

A In each of the following problems, give the equation of the line with the given slope and y-intercept. [Examples 1–2]

1. $m = \frac{2}{3}, b = 1$

2. $m = \frac{3}{4}, b = -2$

3. $m = \frac{3}{2}, b = -1$

4. $m = \frac{4}{3}, b = 2$

5. $m = -\frac{2}{5}, b = 3$

6. $m = -\frac{3}{5}, b = 4$

7. $m = 2, b = -4$

8. $m = -2, b = 4$

B Find the slope and y-intercept for each of the following equations by writing them in the form $y = mx + b$. Then, graph each equation. [Examples 3–4]

9. $-2x + y = 4$

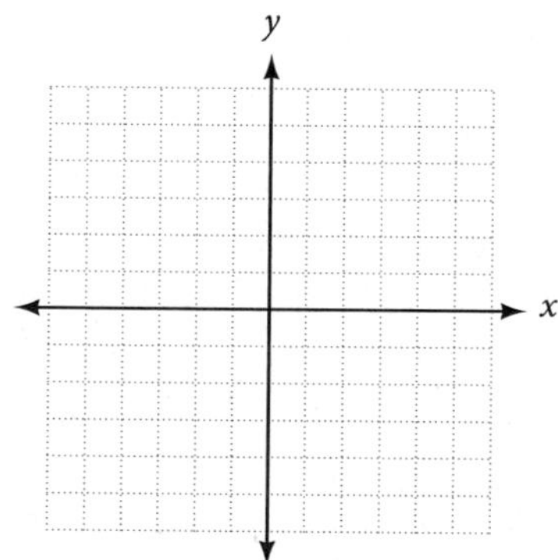

10. $-2x + y = 2$

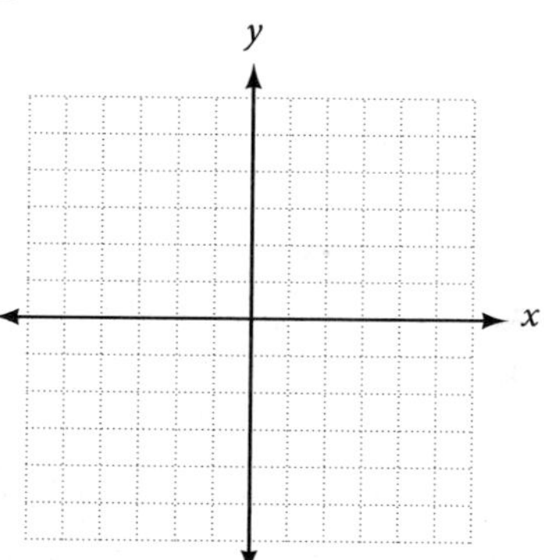

11. $3x + y = 3$

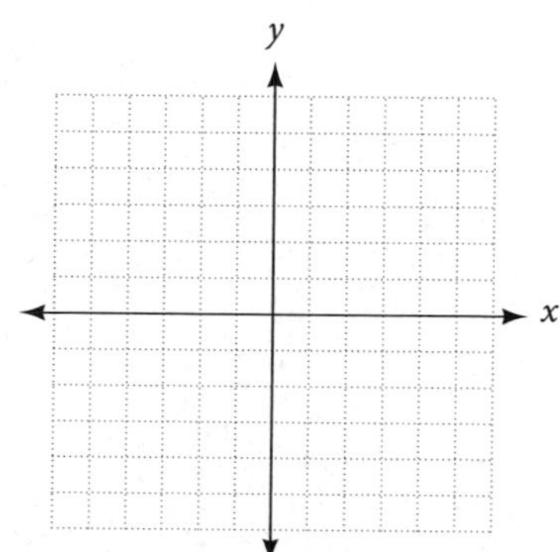

12. $3x + y = 6$

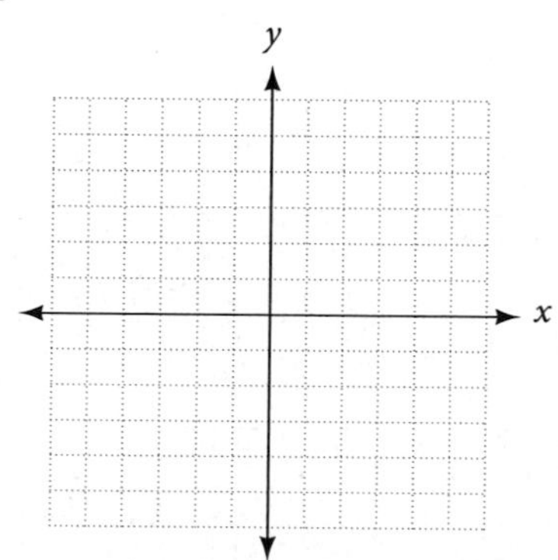

13. $3x + 2y = 6$

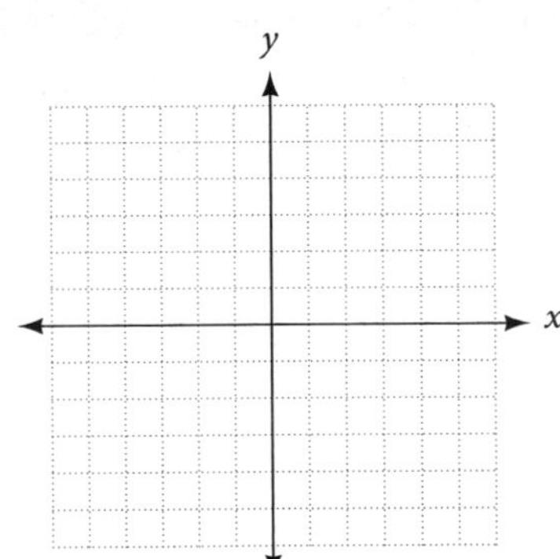

14. $2x + 3y = 6$

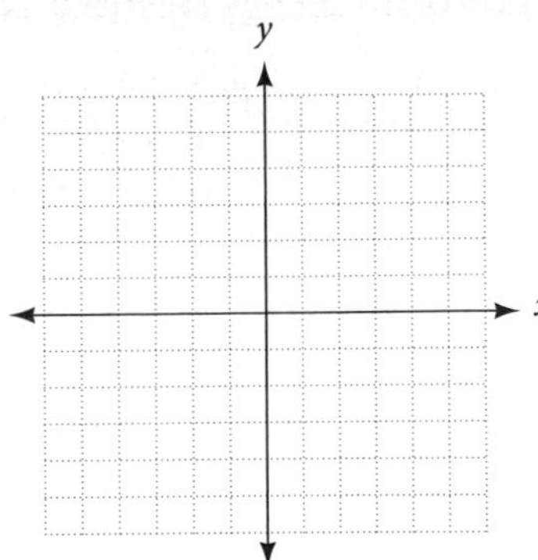

15. $4x - 5y = 20$

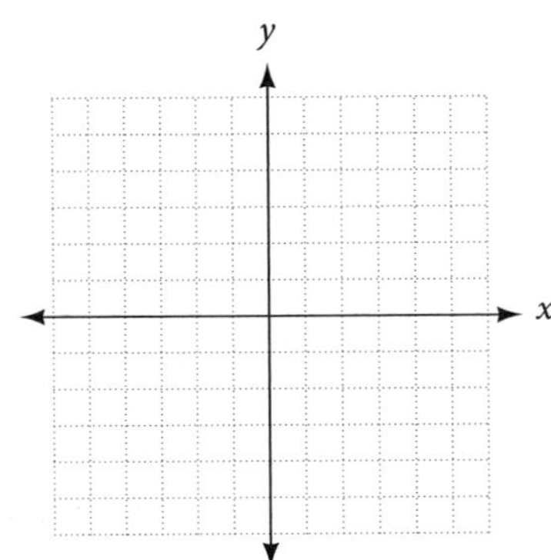

16. $2x - 5y = 10$

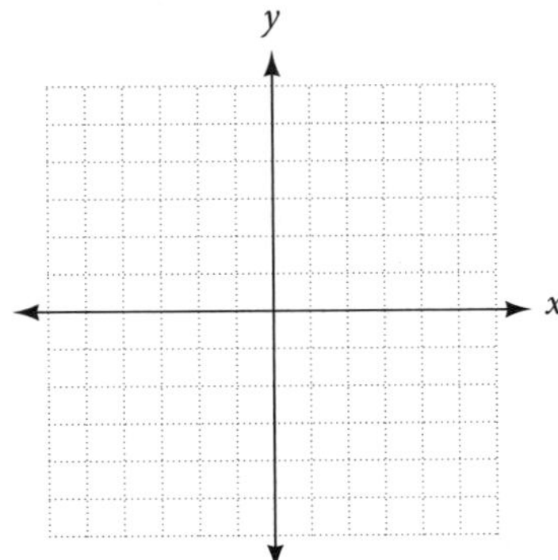

17. $-2x - 5y = 10$

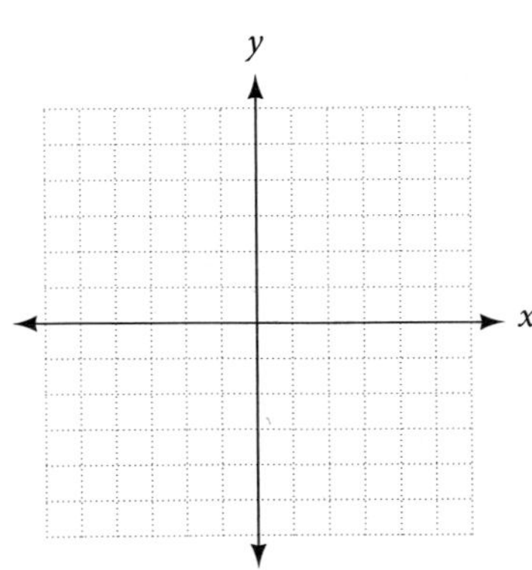

18. $-4x + 5y = 20$

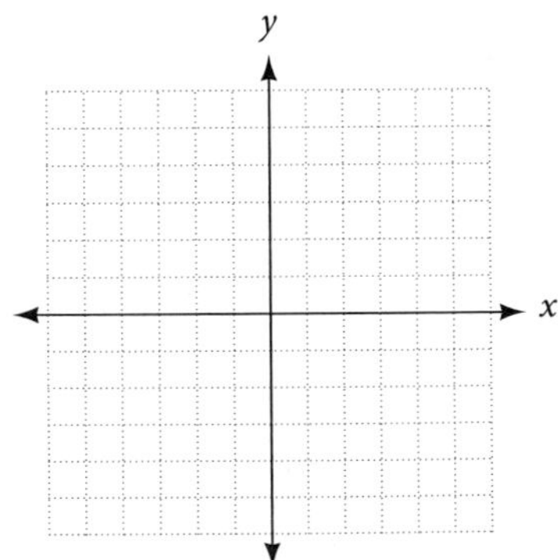

C For each of the following problems, the slope and one point on a line are given. In each case use the point-slope form to find the equation of that line. (Write your answers in slope-intercept form.) [Example 5]

19. $(-2, -5), m = 2$

20. $(-1, -5), m = 2$

21. $(-4, 1), m = -\frac{1}{2}$

22. $(-2, 1), m = -\frac{1}{2}$

23. $(2, -3), m = \frac{3}{2}$

24. $(3, -4), m = \frac{4}{3}$

25. $(-1, 4), m = -3$

26. $(-2, 5), m = -3$

D Find the equation of the line that passes through each pair of points. Write your answers in slope-intercept form. [Example 6]

27. $(-2, -4), (1, -1)$

28. $(2, 4), (-3, -1)$

29. $(-1, -5), (2, 1)$

30. $(-1, 6), (1, 2)$

31. $(-3, -2), (3, 6)$

32. $(-3, 6), (3, -2)$

33. $(-3, -1), (3, -5)$

34. $(-3, -5), (3, 1)$

Find the slope and y-intercept for each line. Then write the equation of each line in slope-intercept form.

35.

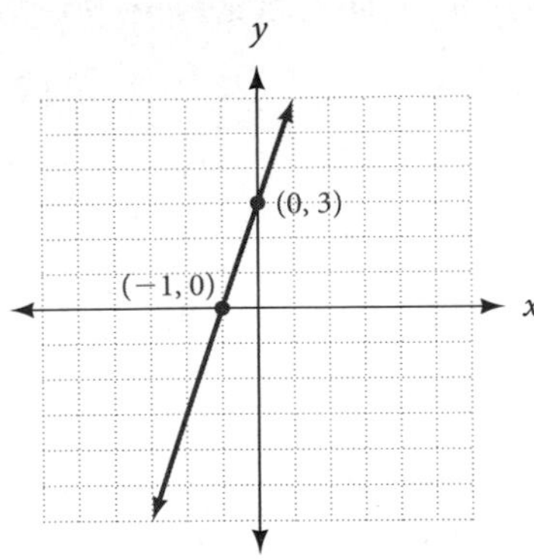

36.

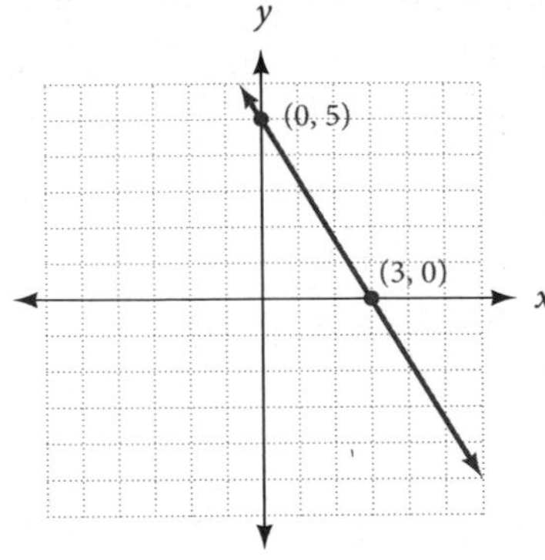

37.

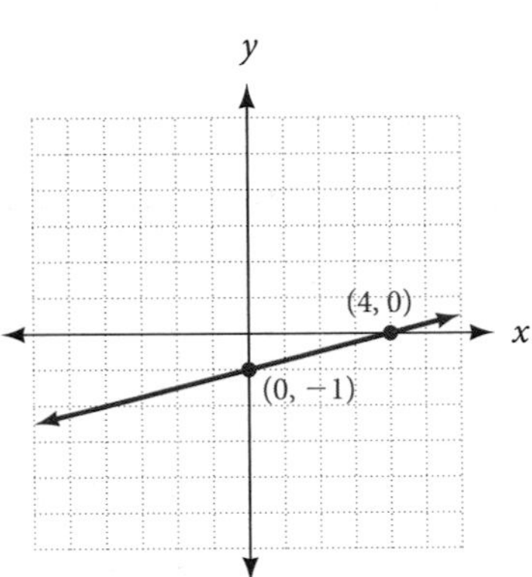

38.

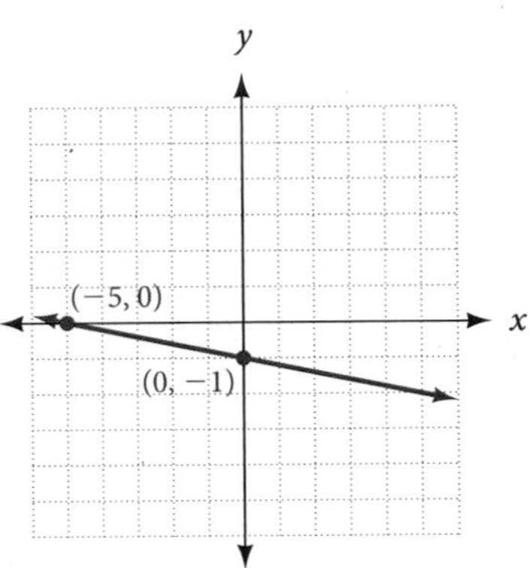

The next two problems are intended to give you practice reading, and paying attention to, the instructions that accompany the problems you are working. Working these problems is an excellent way to get ready for a test or a quiz.

39. Work each problem according to the instructions given.

a. Solve: $-2x + 1 = 6$

b. Write in slope-intercept form: $-2x + y = 6$

c. Find the y-intercept: $-2x + y = 6$

d. Find the slope: $-2x + y = 6$

e. Graph: $-2x + y = 6$

40. Work each problem according to the instructions given.

a. Solve: $x + 3 = -6$

b. Write in slope-intercept form: $x + 3y = -6$

c. Find the y-intercept: $x + 3y = -6$

d. Find the slope: $x + 3y = -6$

e. Graph: $x + 3y = -6$

41. Find the equation of the line with x-intercept 3 and y-intercept 2.

42. Find the equation of the line with x-intercept 2 and y-intercept 3.

43. Find the equation of the line with x-intercept -2 and y-intercept -5.

44. Find the equation of the line with x-intercept -3 and y-intercept -5.

45. The equation of the vertical line that passes through the points $(3, -2)$ and $(3, 4)$ is either $x = 3$ or $y = 3$. Which one is it?

46. The equation of the horizontal line that passes through the points $(2, 3)$ and $(-1, 3)$ is either $x = 3$ or $y = 3$. Which one is it?

Applying the Concepts

47. Horse Racing The graph shows the total amount of money wagered on the Kentucky Derby. Find an equation for the line segment between 2000 and 2005. Write the equation in slope-intercept form.

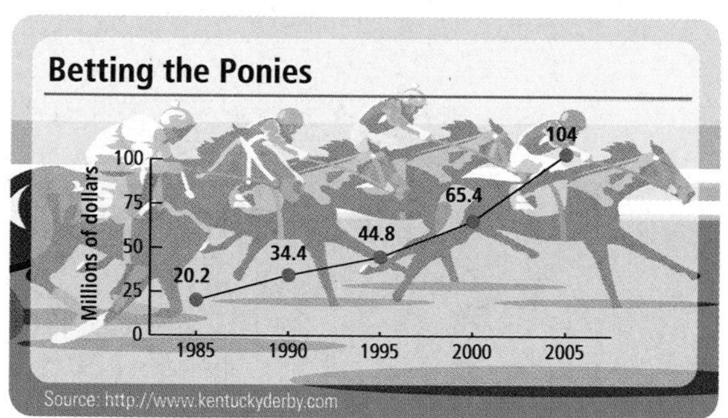

48. Solar Energy The graph shows the annual number of solar thermal collector shipments in the United States. Find an equation for the line segment from (2004, 13,750) to (2005, 15,000). Write your answer in slope-intercept form.

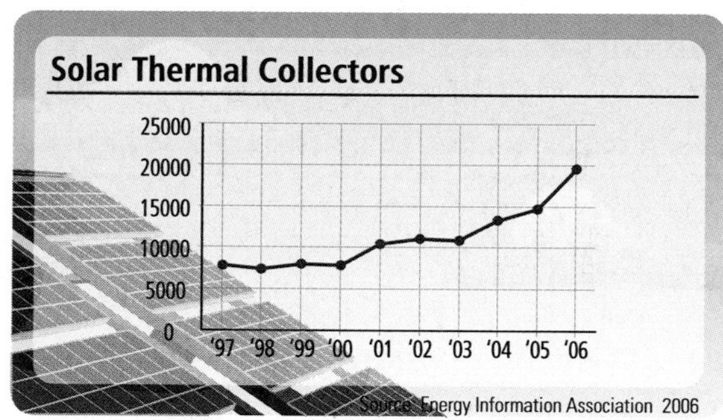

49. Value of a Copy Machine Cassandra buys a new color copier for her small business. It will cost \$21,000 and will decrease in value each year. The graph below shows the value of the copier after the first 5 years of ownership.

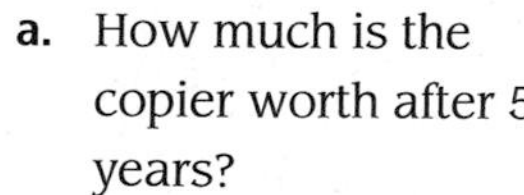

a. How much is the copier worth after 5 years?

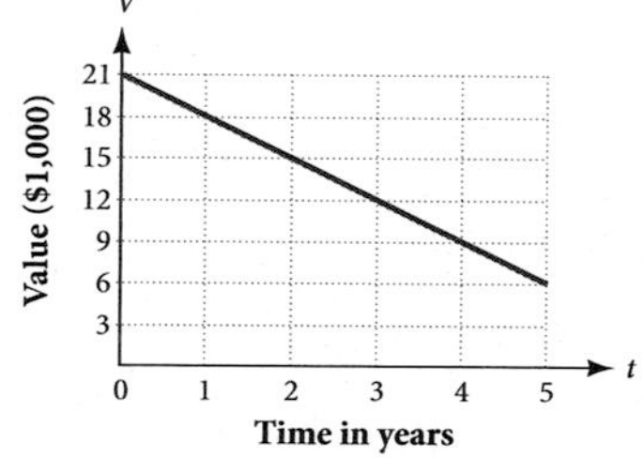

b. After how many years is the copier worth \$12,000?

c. Find the slope of this line.

d. By how many dollars per year is the copier decreasing in value?

e. Find the equation of this line where V is the value after t years.

50. Value of a Forklift Elliot buys a new forklift for his business. It will cost \$140,000 and will decrease in value each year. The graph below shows the value of the forklift after the first 6 years of ownership

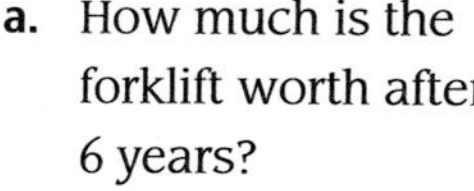

a. How much is the forklift worth after 6 years?

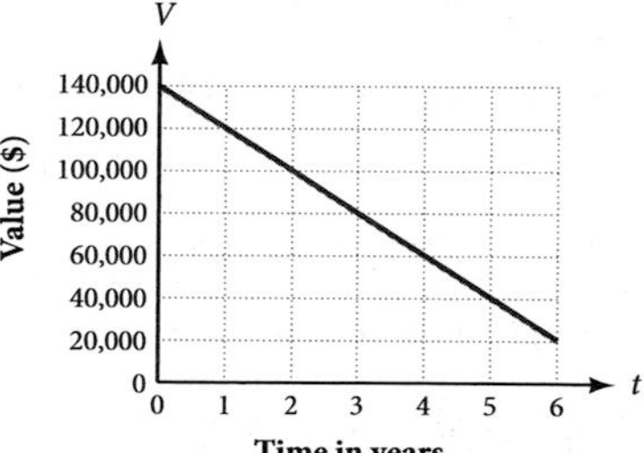

b. After how many years is the forklift worth \$80,000?

c. Find the slope of this line.

d. By how many dollars per year is the forklift decreasing in value?

e. Find the equation of this line where V is the value after t years.

Getting Ready for Class

Graph each of the following lines. [3.3, 3.4]

51. $x + y = 4$

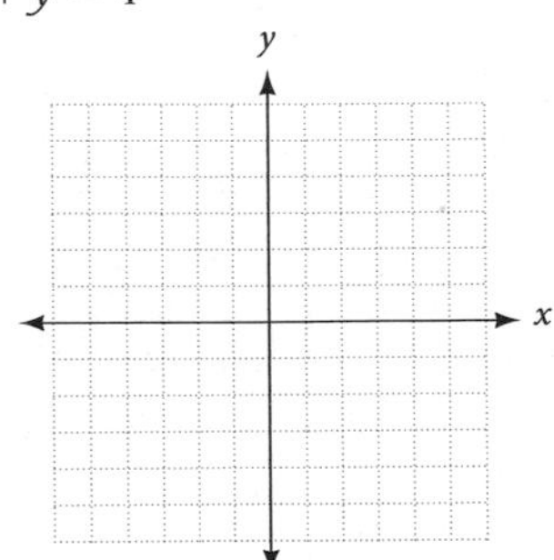

52. $x - y = -2$

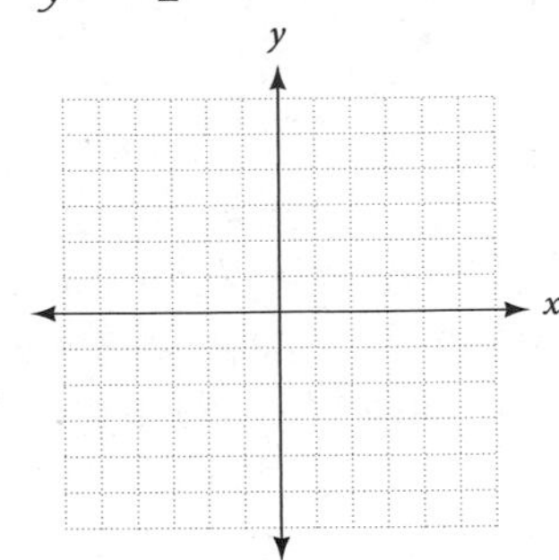

53. $y = 2x - 3$

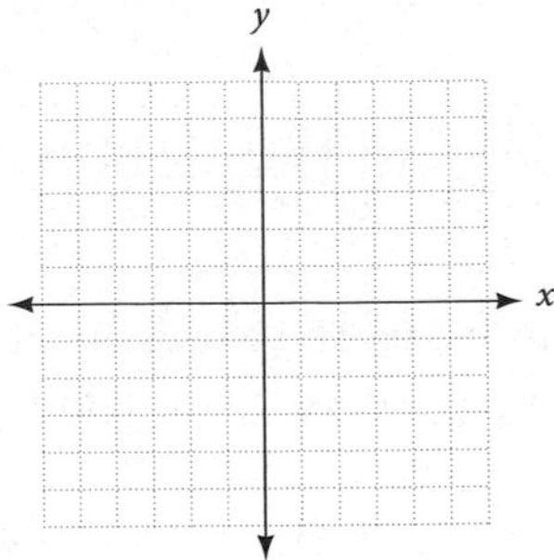

54. $y = 2x + 3$

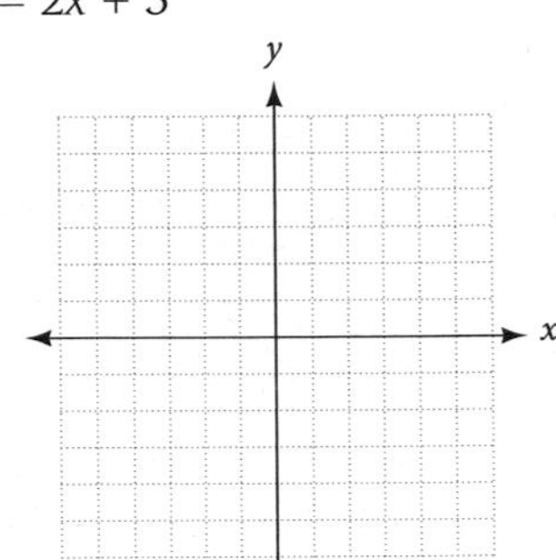

55. $y = 2x$

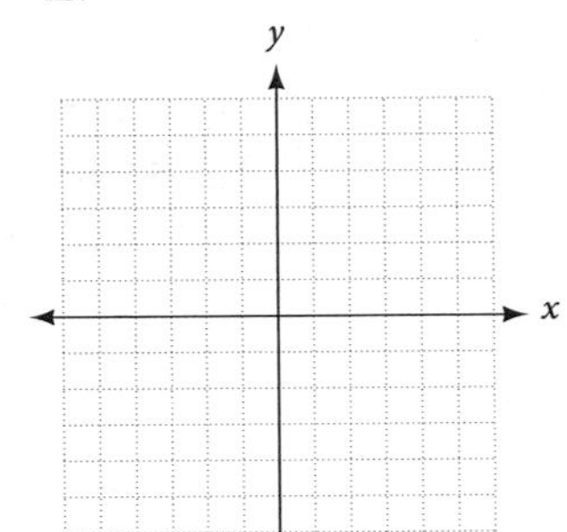

56. $y = -2x$

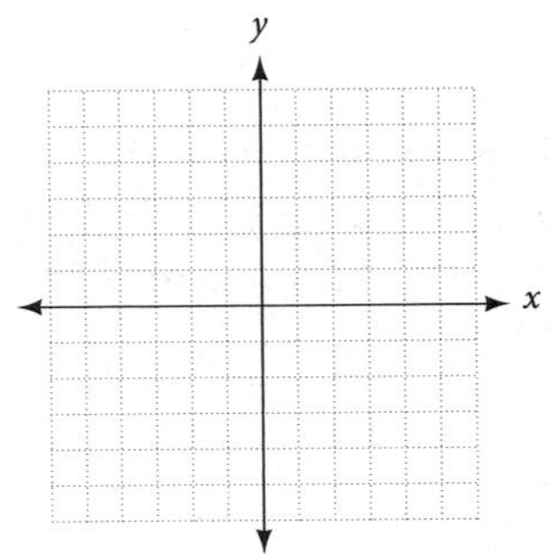

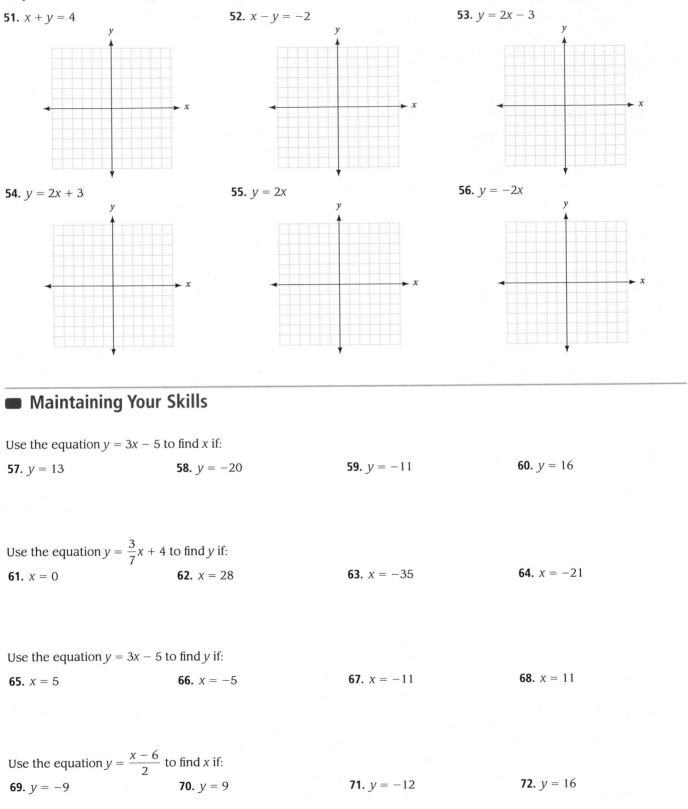

Maintaining Your Skills

Use the equation $y = 3x - 5$ to find x if:

57. $y = 13$ **58.** $y = -20$ **59.** $y = -11$ **60.** $y = 16$

Use the equation $y = \frac{3}{7}x + 4$ to find y if:

61. $x = 0$ **62.** $x = 28$ **63.** $x = -35$ **64.** $x = -21$

Use the equation $y = 3x - 5$ to find y if:

65. $x = 5$ **66.** $x = -5$ **67.** $x = -11$ **68.** $x = 11$

Use the equation $y = \frac{x - 6}{2}$ to find x if:

69. $y = -9$ **70.** $y = 9$ **71.** $y = -12$ **72.** $y = 16$

Linear Inequalities in Two Variables

3.7

Objectives

A Graph linear inequalities in two variables.

Examples now playing at MathTV.com/books

A Graphing Linear Inequalities in Two Variables

A linear inequality in two variables is any expression that can be put in the form

$$ax + by < c$$

where a, b, and c are real numbers (a and b not both 0). The inequality symbol can be any of the following four: $<, \le, >, \ge$.

Some examples of linear inequalities are

$$2x + 3y < 6 \qquad y \ge 2x + 1 \qquad x - y \le 0$$

Although not all of these inequalities have the form $ax + by < c$, each one can be put in that form.

The solution set for a linear inequality is a section of the coordinate plane. The boundary for the section is found by replacing the inequality symbol with an equal sign and graphing the resulting equation. The boundary is included in the solution set (and represented with a solid line) if the inequality symbol used originally is $\le$ or $\ge$. The boundary is not included (and is represented with a broken line) if the original symbol is $<$ or $>$.

Let's look at some examples.

EXAMPLE 1 Graph the solution set for $x + y \le 4$.

SOLUTION The boundary for the graph is the graph of $x + y = 4$. The boundary is included in the solution set because the inequality symbol is $\le$.

The graph of the boundary is shown in Figure 1.

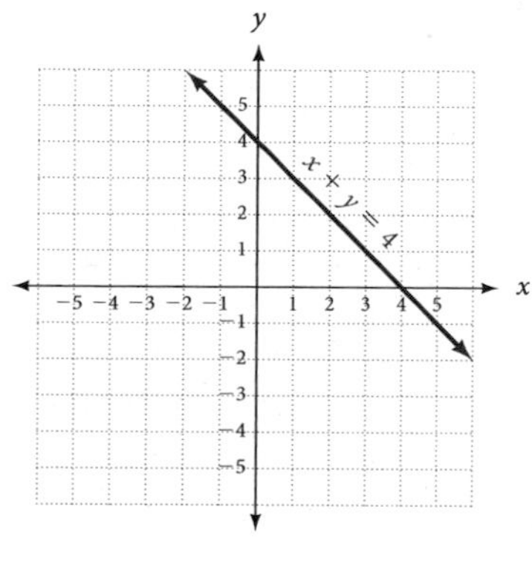

FIGURE 1

The boundary separates the coordinate plane into two sections, or regions: the region above the boundary and the region below the boundary. The solution set for $x + y \le 4$ is one of these two regions along with the boundary. To find the correct region, we simply choose any convenient point that is *not* on the boundary. We then substitute the coordinates of the point into the original inequality $x + y \le 4$. If the point we choose satisfies the inequality, then it is a member of the solution set, and we can assume that all points on the same side of the boundary as the chosen point are also in the solution set. If the coordinates of our point do not satisfy the original inequality, then the solution set lies on the other side of the boundary.

In this example a convenient point not on the boundary is the origin. Substituting (0, 0) into $x + y \le 4$ gives us

$$0 + 0 \overset{?}{\le} 4$$

$$0 \le 4 \qquad \text{A true statement}$$

PRACTICE PROBLEMS

1. Graph $x + y \ge 3$ by first graphing the boundary line $x + y = 3$ and then using a test point to decide which side of the boundary to shade.

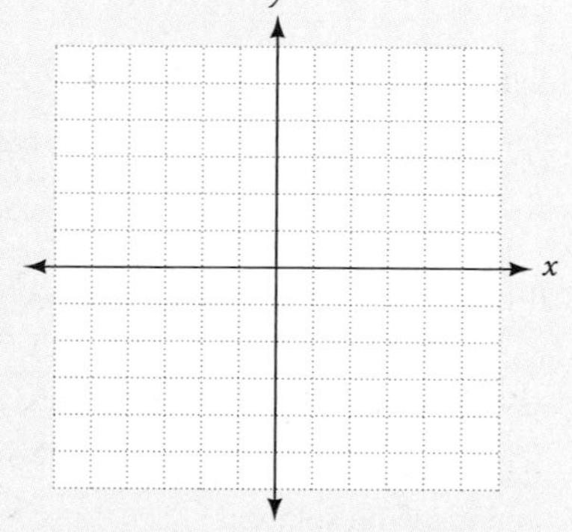

Answers

See Solutions to Selected Practice Problems.

Because the origin is a solution to the inequality $x + y \leq 4$, and the origin is below the boundary, all other points below the boundary are also solutions.

The graph of $x + y \leq 4$ is shown in Figure 2.

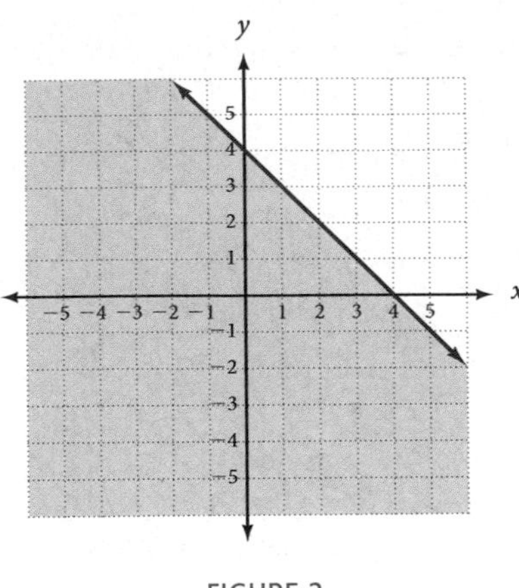

FIGURE 2

The region above the boundary is described by the inequality $x + y > 4$.

Here is a list of steps to follow when graphing the solution set for linear inequalities in two variables.

Strategy To Graph the Solution Set for Linear Inequalities in Two Variables

Step 1: Replace the inequality symbol with an equal sign. The resulting equation represents the boundary for the solution set.

Step 2: Graph the boundary found in step 1 using a *solid line* if the boundary is included in the solution set (that is, if the original inequality symbol was either $\leq$ or $\geq$). Use a *broken line* to graph the boundary if it is *not* included in the solution set. (It is not included if the original inequality was either $<$ or $>$.)

Step 3: Choose any convenient point not on the boundary and substitute the coordinates into the *original* inequality. If the resulting statement is *true,* the graph lies on the *same* side of the boundary as the chosen point. If the resulting statement is *false,* the solution set lies on the *opposite* side of the boundary.

2. Graph $y < \frac{1}{2}x - 3$. (Graph the boundary line $y = \frac{1}{2}x - 3$ first, but use a broken line because the inequality symbol is $<$, meaning the boundary is not part of the solution.)

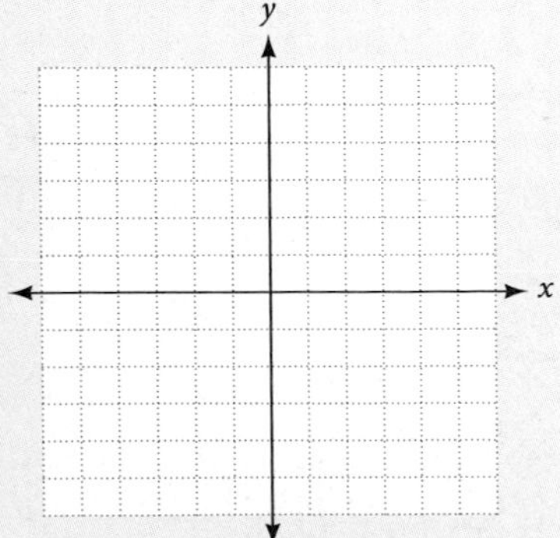

EXAMPLE 2 Graph the solution set for $y < 2x - 3$.

SOLUTION The boundary is the graph of $y = 2x - 3$. The boundary is not included because the original inequality symbol is $<$. We therefore use a broken line to represent the boundary, as shown in Figure 3.

A convenient test point is again the origin. Using (0, 0) in $y < 2x - 3$, we have

$$0 \overset{?}{<} 2(0) - 3$$

$$0 < -3 \qquad \text{A false statement}$$

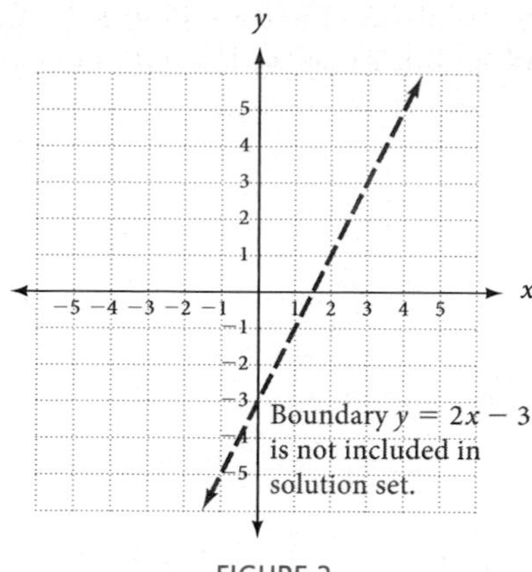

FIGURE 3

Because our test point gives us a false statement and it lies above the boundary, the solution set must lie on the other side of the boundary, as shown in Figure 4.

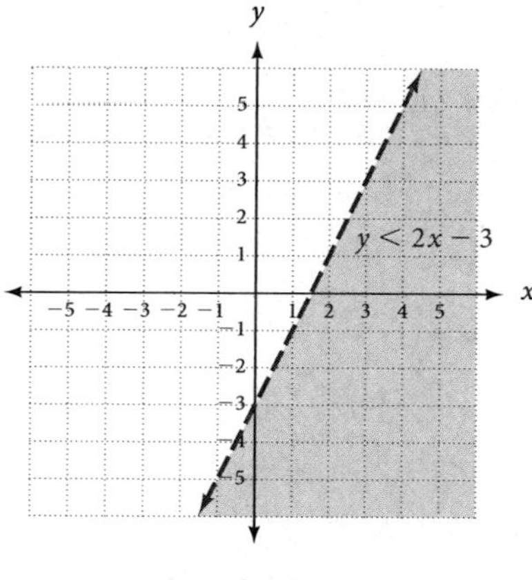

FIGURE 4

EXAMPLE 3 Graph the inequality $2x + 3y \le 6$.

SOLUTION We begin by graphing the boundary $2x + 3y = 6$. The boundary is included in the solution because the inequality symbol is $\le$.

If we use (0, 0) as our test point, we see that it yields a true statement when its coordinates are substituted into $2x + 3y \le 6$. The graph, therefore, lies below the boundary, as shown in Figure 5.

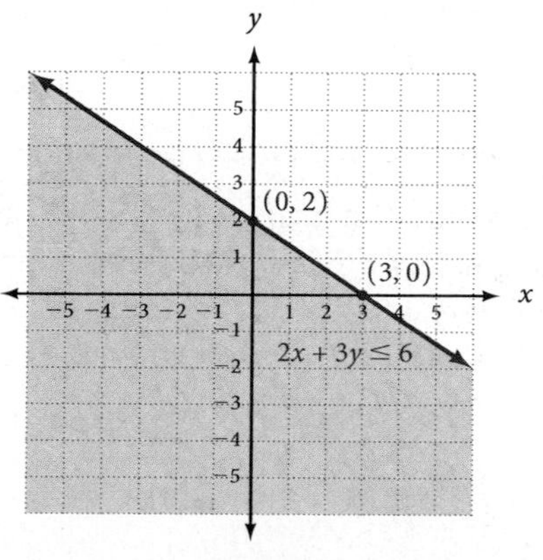

FIGURE 5

3. Graph $3x + 2y \le 6$.

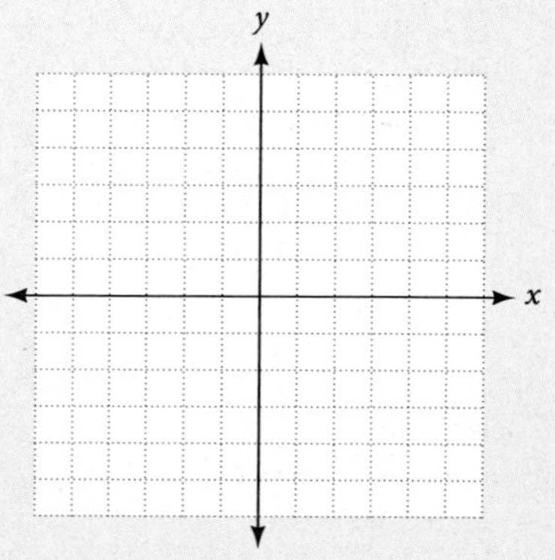

The ordered pair (0, 0) is a solution to $2x + 3y \le 6$; all points on the same side of the boundary as (0, 0) also must be solutions to the inequality $2x + 3y \le 6$.

4. Graph $y > -2$.

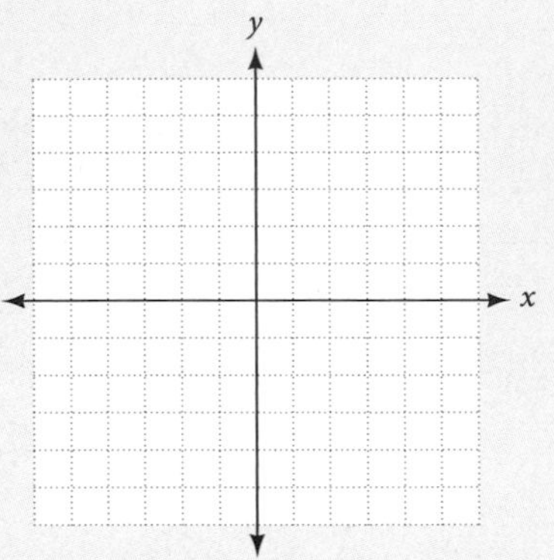

EXAMPLE 4

Graph the solution set for $x \le 5$.

SOLUTION The boundary is $x = 5$, which is a vertical line. All points to the left have x-coordinates less than 5, and all points to the right have x-coordinates greater than 5, as shown in Figure 6.

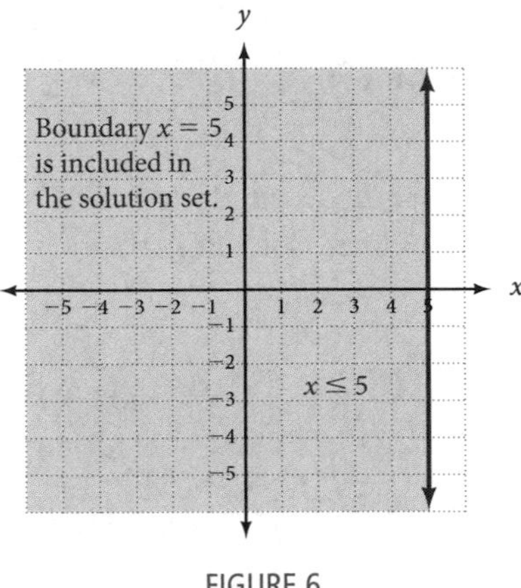

FIGURE 6

Getting Ready for Class

After reading through the preceding section, respond in your own words and in complete sentences.

1. When graphing a linear inequality in two variables, how do you find the equation of the boundary line?
2. What is the significance of a broken line in the graph of an inequality?
3. When graphing a linear inequality in two variables, how do you know which side of the boundary line to shade?
4. Describe the set of ordered pairs that are solutions to $x + y < 6$.

Problem Set 3.7

A Graph the following linear inequalities. [Examples 1–4]

1. $2x - 3y < 6$

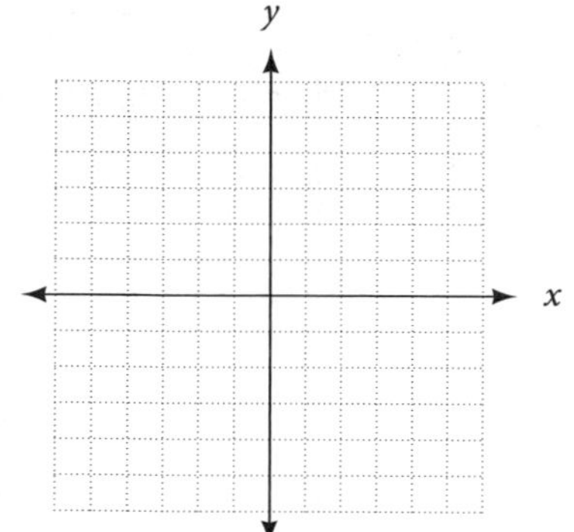

2. $3x + 2y \geq 6$

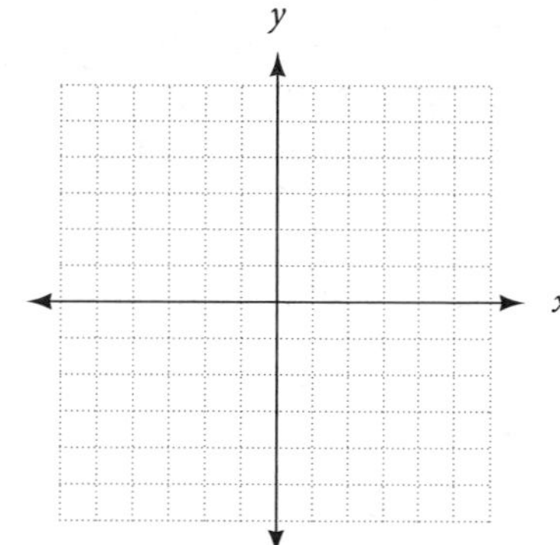

3. $x - 2y \leq 4$

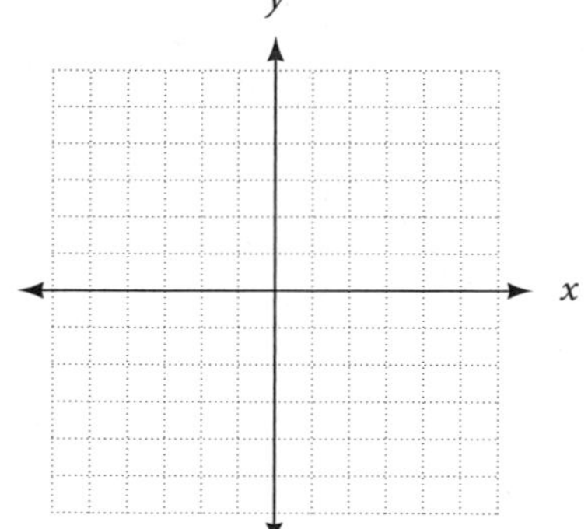

4. $2x + y > 4$

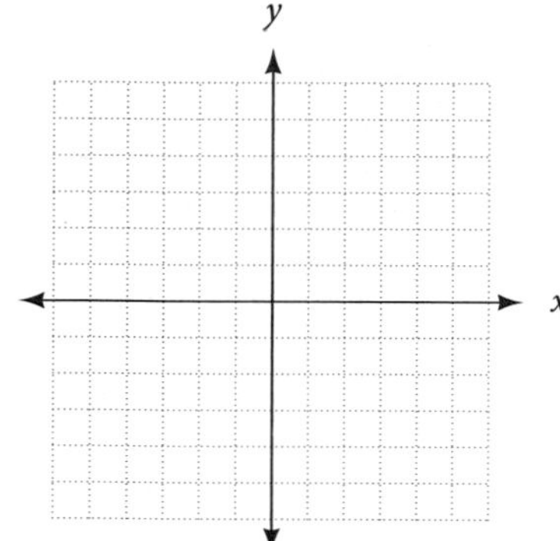

5. $x - y \leq 2$

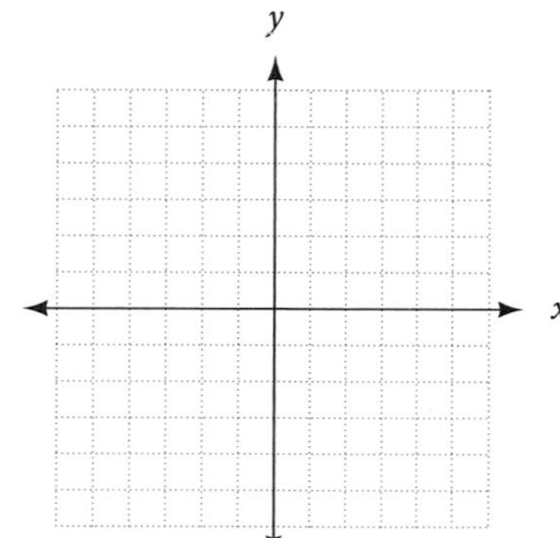

6. $x - y \leq 1$

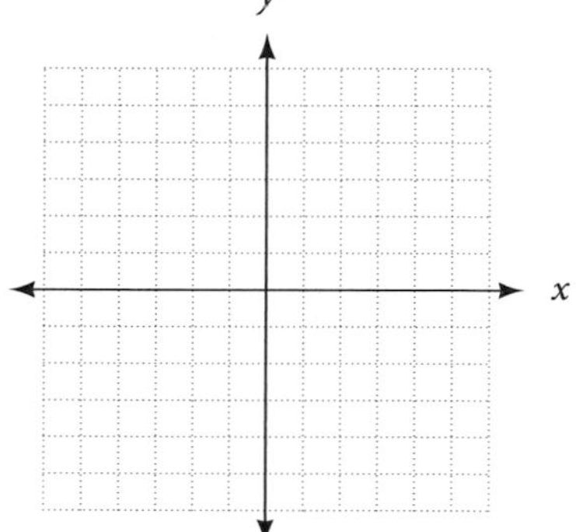

7. $3x - 4y \geq 12$

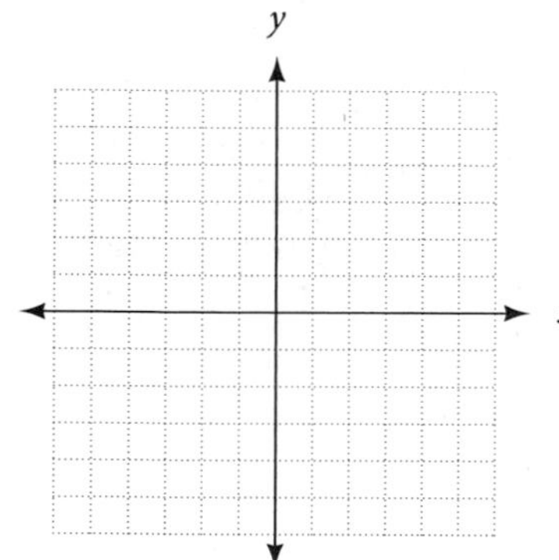

8. $4x + 3y < 12$

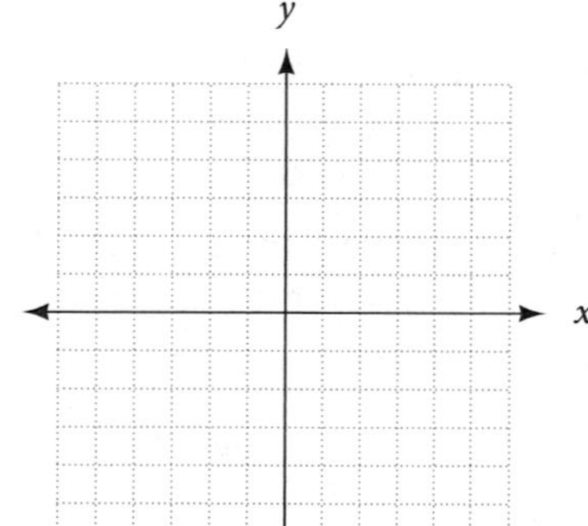

9. $5x - y \leq 5$

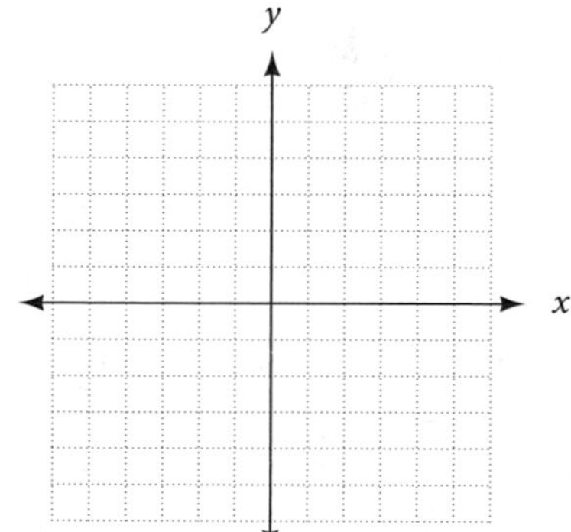

10. $4x + y > 4$

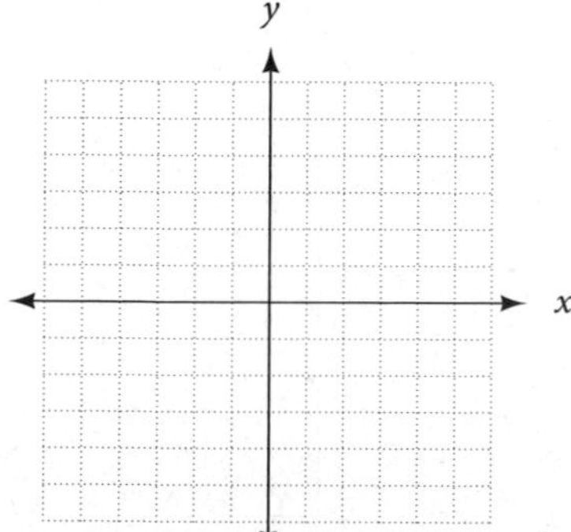

11. $2x + 6y \leq 12$

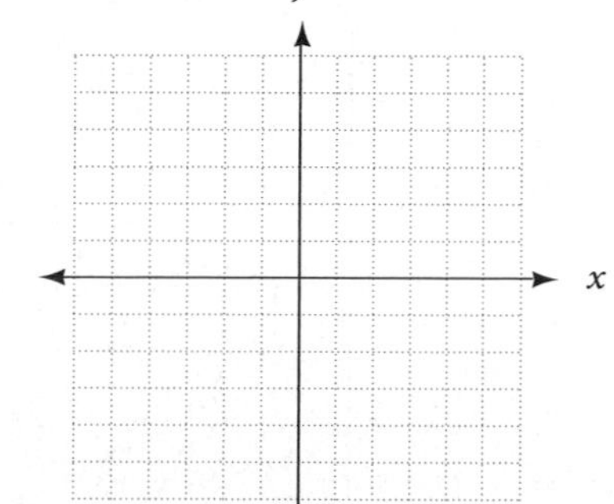

12. $x - 5y > 5$

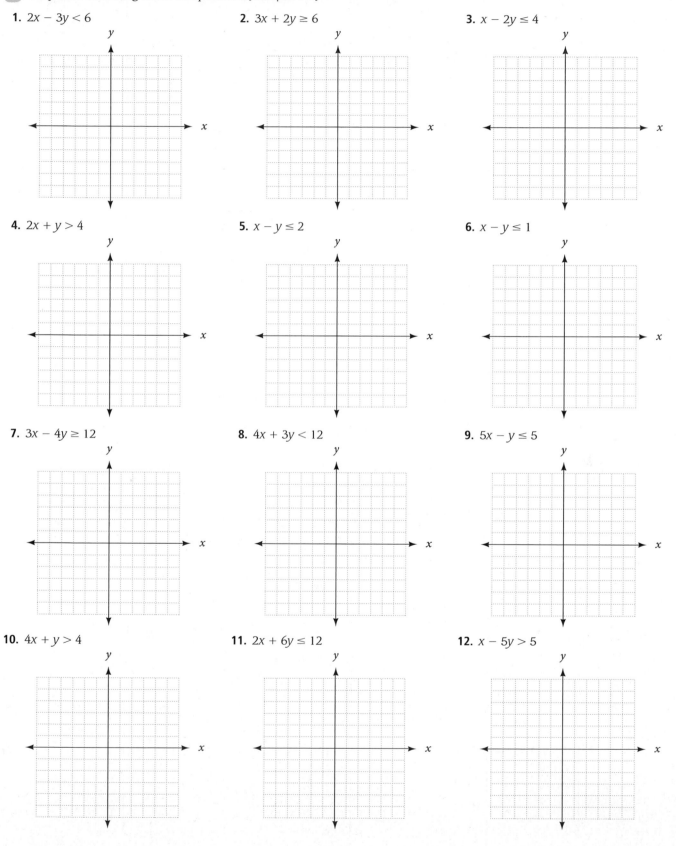

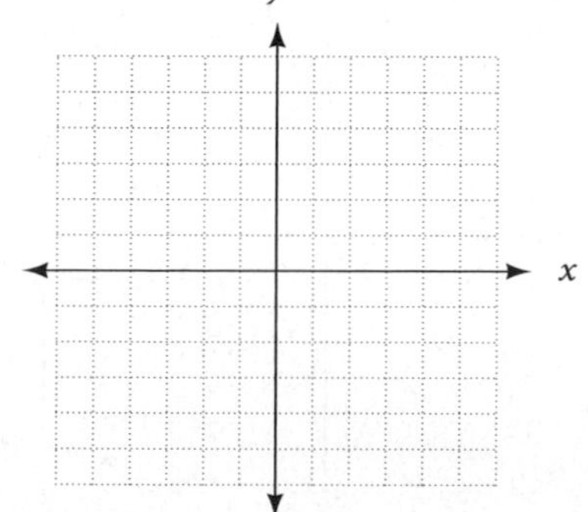

13. $x \geq 1$

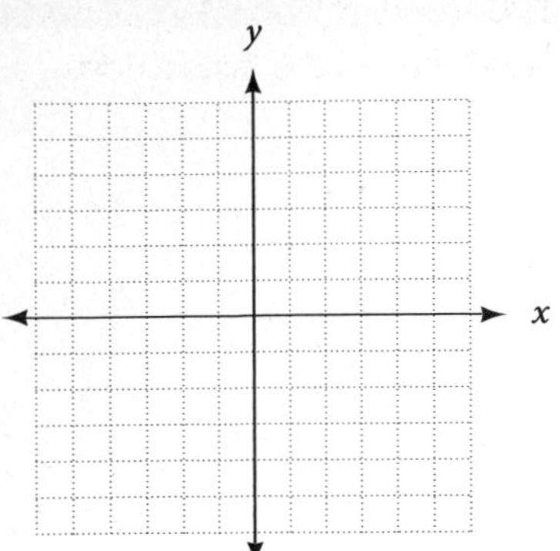

14. $x < 5$

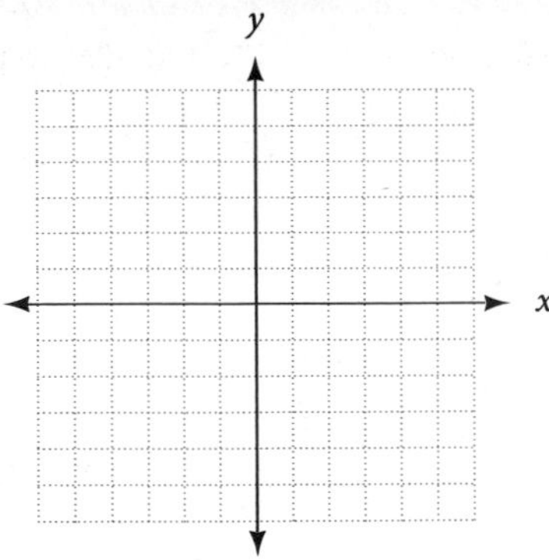

15. $x \geq -3$

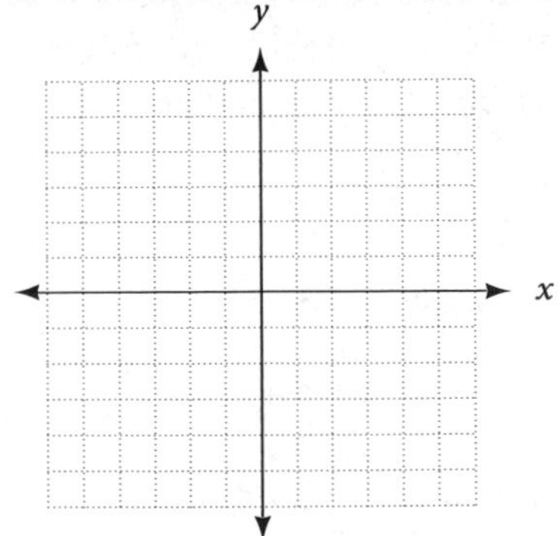

16. $y \leq -4$

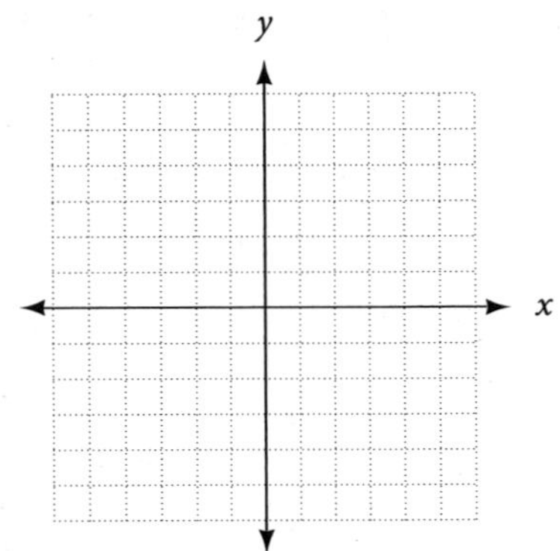

17. $y < 2$

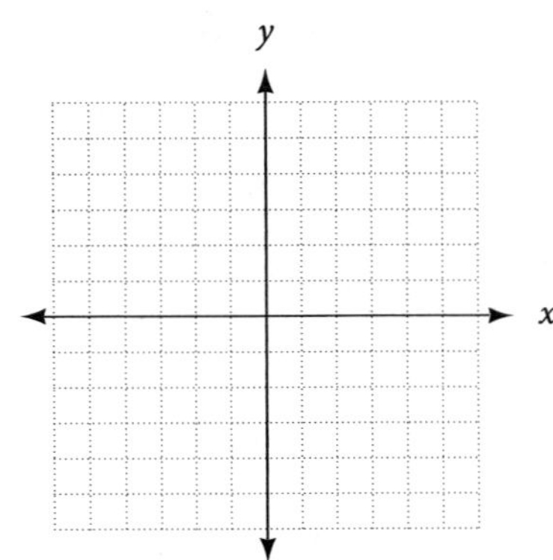

18. $3x - y > 1$

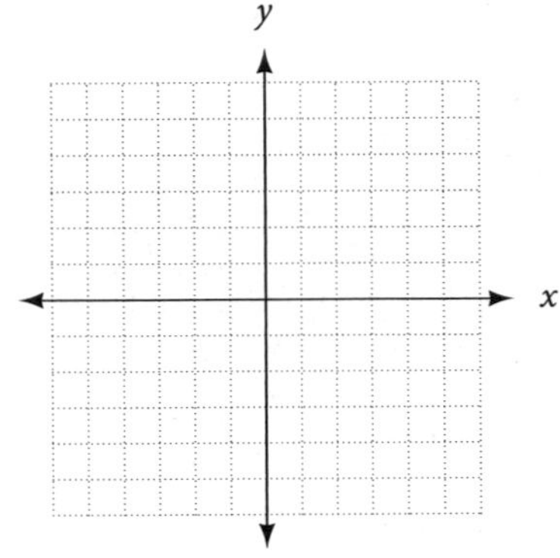

19. $2x + y > 3$

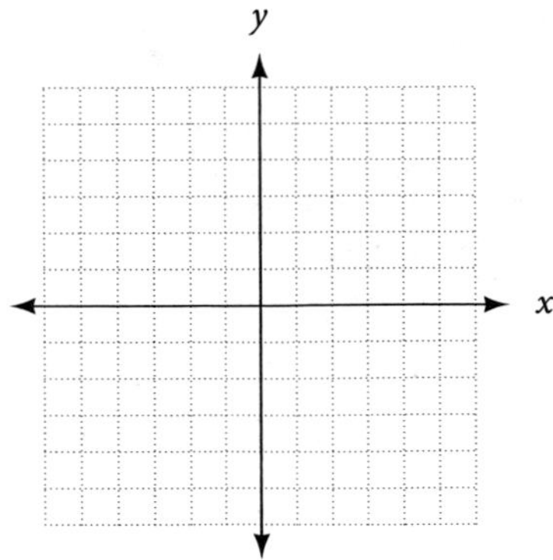

20. $5x + 2y < 2$

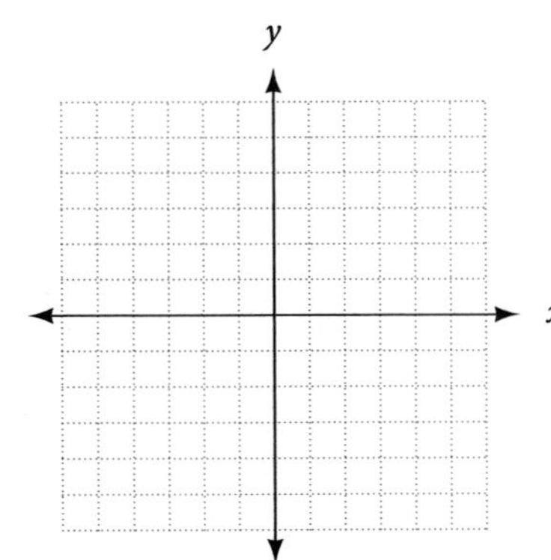

21. $y \leq 3x - 1$

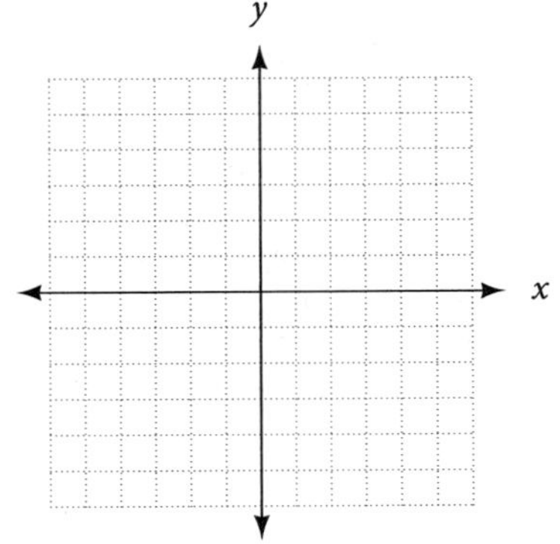

22. $y \geq 3x + 2$

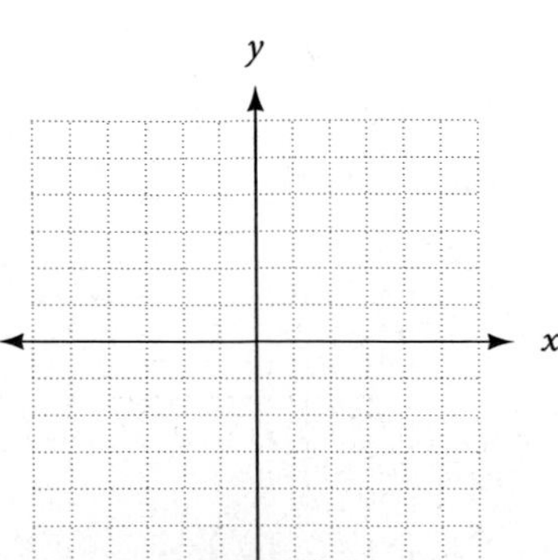

23. $y \leq -\frac{1}{2}x + 2$

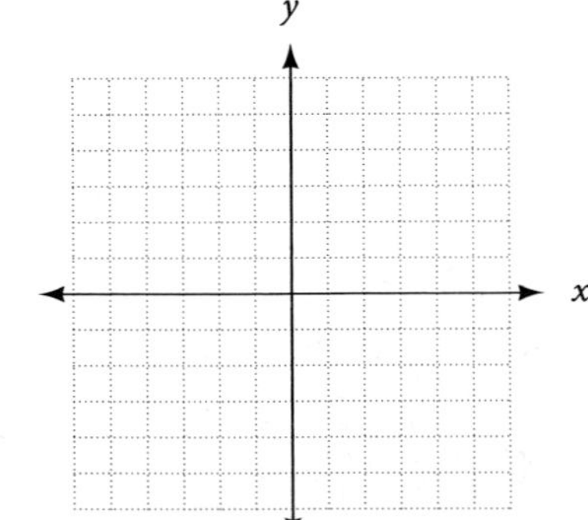

24. $y < \frac{1}{3}x + 3$

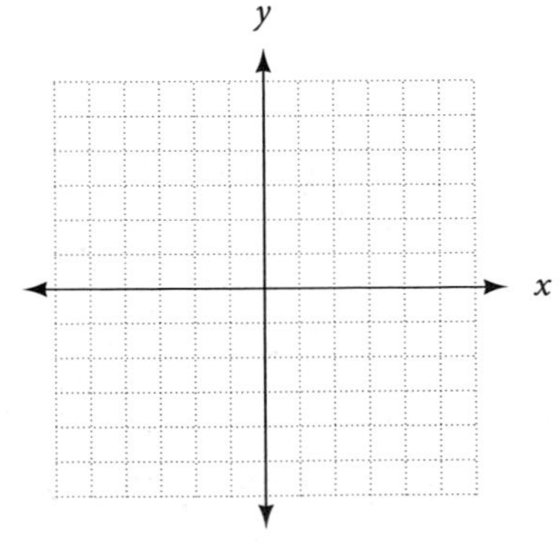

The next two problems are intended to give you practice reading, and paying attention to, the instructions that accompany the problems you are working.

25. Work each problem according to the instructions given.

a. Solve: $4 + 3y < 12$

b. Solve: $4 - 3y < 12$

c. Solve for y: $4x + 3y = 12$

d. Graph: $y < -\frac{4}{3}x + 4$

26. Work each problem according to the instructions given.

a. Solve: $3x + 2 \geq 6$

b. Solve: $-3x + 2 \geq 6$

c. Solve for y: $3x + 2y = 6$

d. Graph: $y \geq -\frac{3}{2}x + 3$

27. Find the equation of the line in part a, then use this information to find the inequalities for the graphs on parts b and c.

a.

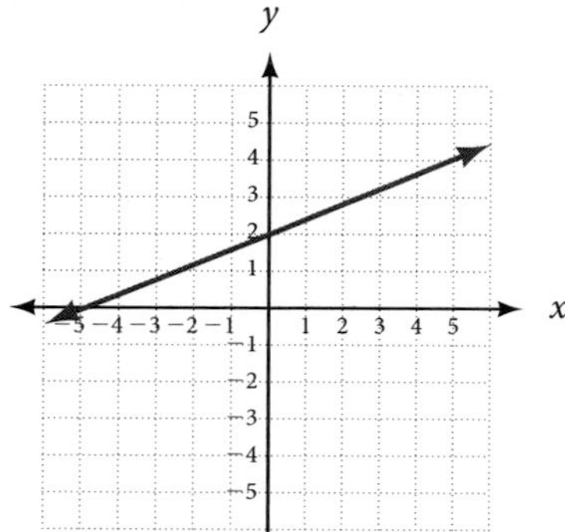

b.

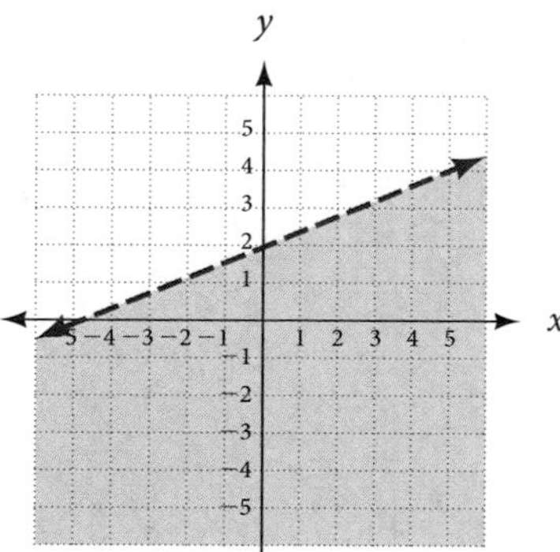

c.

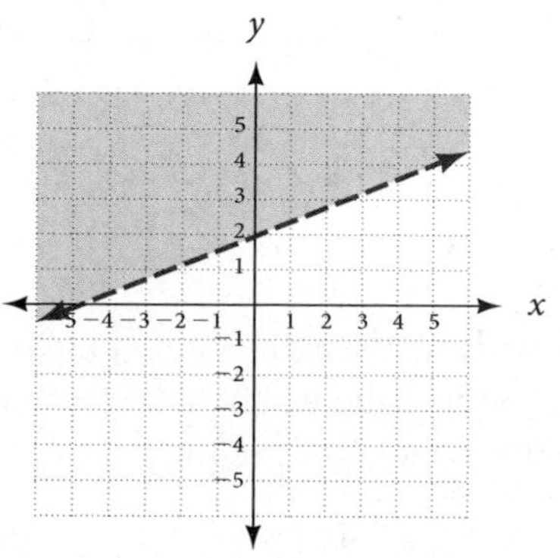

28. Find the equation of the line in part a, then use this information to find the inequalities for the graphs on parts b and c.

a.

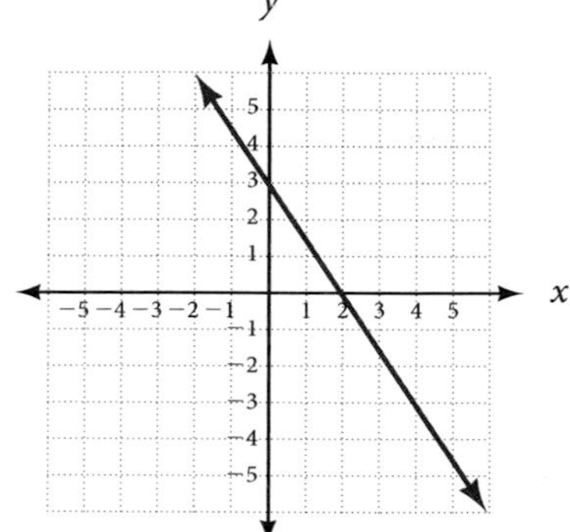

b.

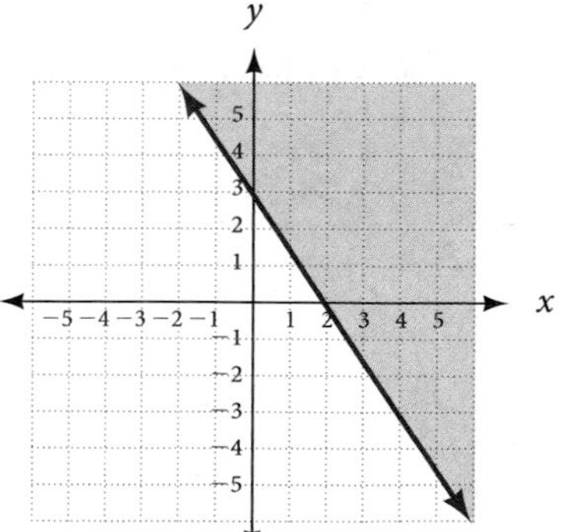

c.

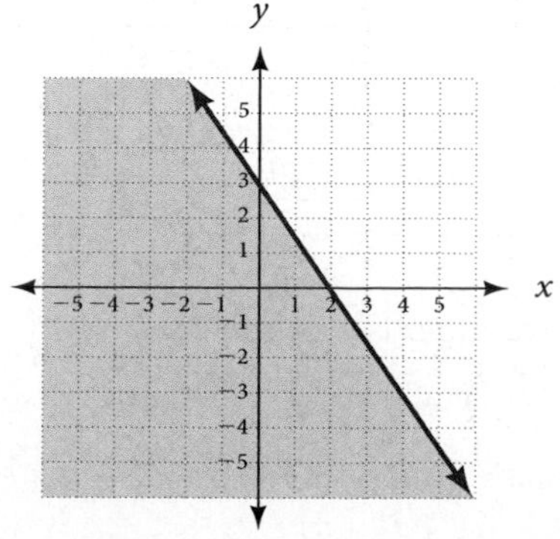

Applying the Concepts

29. **Horse Racing** The graph shows the total amount of money wagered on the Kentucky Derby. Suppose that the margin of error for this information was ±7 million. Redraw the graph to take into account the margin of error.

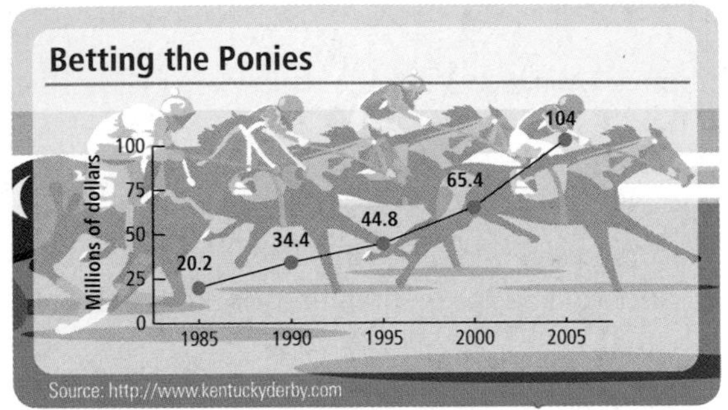

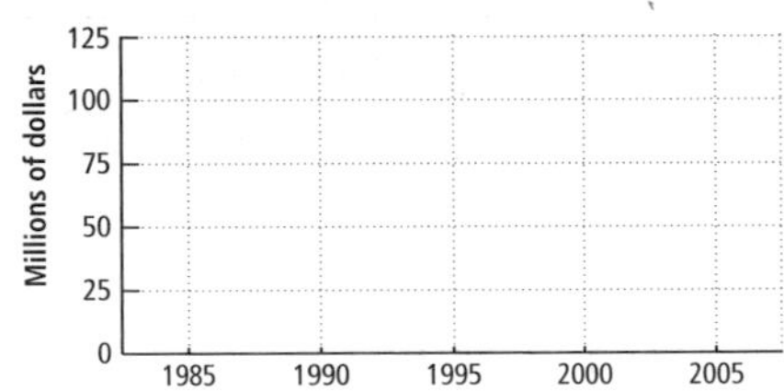

30. **Health Care** The graph shows the rising cost of health care. The information for 2008 to 2014 is projected costs. Suppose that there is a margin of error on this data of ±300 billion dollars. Redraw the graph from 2008 to 2014 so that it reflects the margin of error.

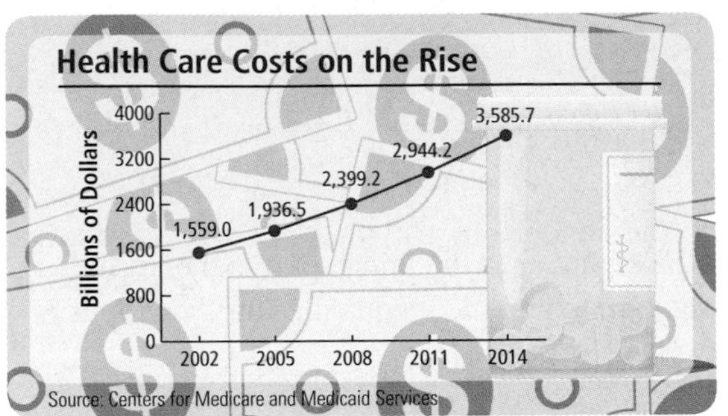

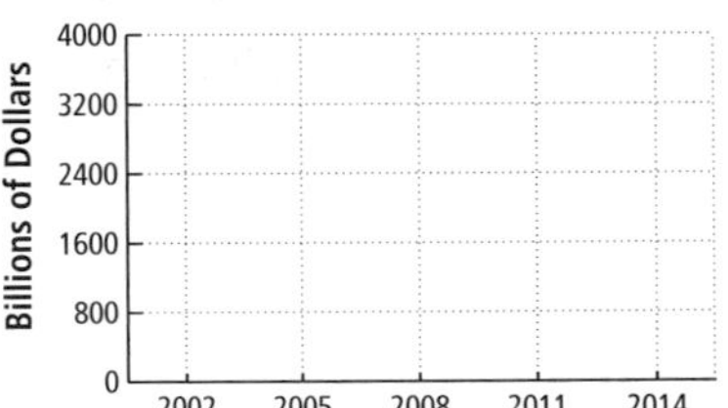

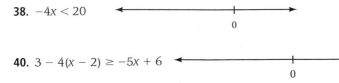

Maintaining Your Skills

31. Simplify the expression $7 - 3(2x - 4) - 8$.

32. Find the value of $x^2 - 2xy + y^2$ when $x = 3$ and $y = -4$.

Solve each equation.

33. $-\frac{3}{2}x = 12$

34. $2x - 4 = 5x + 2$

35. $8 - 2(x + 7) = 2$

36. $3(2x - 5) - (2x - 4) = 6 - (4x + 5)$

37. Solve the formula $P = 2l + 2w$ for w.

Solve each inequality, and graph the solution.

38. $-4x < 20$

0

39. $3 - 2x > 5$

40. $3 - 4(x - 2) \geq -5x + 6$

0

41. Solve the formula $3x - 2y \leq 12$ for y.

42. What number is 12% of 2,000?

43. **Geometry** The length of a rectangle is 5 inches more than 3 times the width. If the perimeter is 26 inches, find the length and width.

Chapter 3 Summary

Linear Equation in Two Variables [3.3]

A linear equation in two variables is any equation that can be put in the form $ax + by = c$. The graph of every linear equation is a straight line.

EXAMPLES

1. The equation $3x + 2y = 6$ is an example of a linear equation in two variables.

Strategy for Graphing Linear Equations in Two Variables [3.3]

Step 1 Find any three ordered pairs that satisfy the equation. This can be done by using a convenient number for one variable and solving for the other variable.

Step 2: Graph the three ordered pairs found in step 1. Actually, we need only two points to graph a straight line. The third point serves as a check. If all three points do not line up, there is a mistake in our work.

Step 3: Draw a straight line through the three points graphed in step 2.

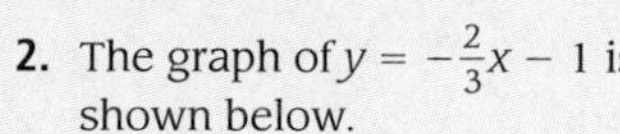

2. The graph of $y = -\frac{2}{3}x - 1$ is shown below.

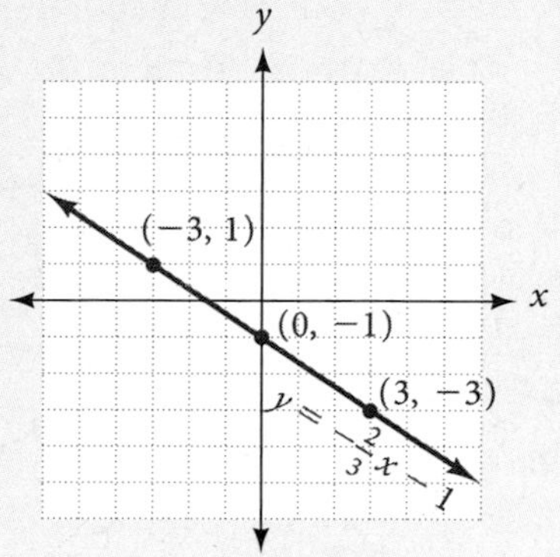

Intercepts [3.4]

The x-intercept of an equation is the x-coordinate of the point where the graph crosses the x-axis. The y-intercept is the y-coordinate of the point where the graph crosses the y-axis. We find the y-intercept by substituting $x = 0$ into the equation and solving for y. The x-intercept is found by letting $y = 0$ and solving for x.

3. To find the x-intercept for $3x + 2y = 6$, we let $y = 0$ and get

$$3x = 6$$
$$x = 2$$

In this case the x-intercept is 2, and the graph crosses the x-axis at (2, 0).

Slope of a Line [3.5]

The *slope* of the line containing the points (x_1, y_1) and $(x_2, y2)$ is given by

$$\text{Slope} = m = \frac{y_2 - y_1}{x_2 - x_1} = \frac{\text{rise}}{\text{run}}$$

4. The slope of the line through (3, −5) and (−2, 1) is

$$m = \frac{-5 - 1}{3 - (-2)} = \frac{-6}{5} = -\frac{6}{5}$$

Point-Slope Form of a Straight Line [3.6]

If a line has a slope of m and contains the point (x_1, y_1), the equation can be written as

$$y - y_1 = m(x - x_1)$$

5. The equation of the line through (1, 2) with a slope of 3 is

$$y - 2 = 3(x - 1)$$
$$y - 2 = 3x - 3$$
$$y = 3x - 1$$

Slope-Intercept Form of a Straight Line [3.6]

6. The equation of the line with a slope of 2 and a y-intercept of 5 is

$$y = 2x + 5$$

The equation of the line with a slope of m and a y-intercept of b is

$$y = mx + b$$

To Graph a Linear Inequality in Two Variables [3.7]

7. Graph $x - y \geq 3$.

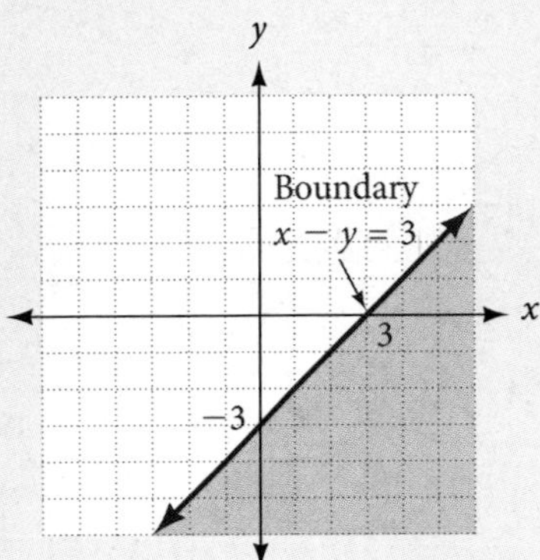

Step 1: Replace the inequality symbol with an equal sign. The resulting equation represents the boundary for the solution set.

Step 2: Graph the boundary found in step 1, using a *solid line* if the original inequality symbol was either $\leq$, or $\geq$. Use a *broken line* otherwise.

Step 3: Choose any convenient point not on the boundary and substitute the coordinates into the *original* inequality. If the resulting statement is *true,* the graph lies on the *same* side of the boundary as the chosen point. If the resulting statement is *false,* the solution set lies on the *opposite* side of the boundary.

Chapter 3 Review

For each equation, complete the given ordered pairs. [3.2]

1. $3x + y = 6$ (4,), (0,), (, 3), (, 0)
2. $2x - 5y = 20$ (5,), (0,), (, 2), (, 0)
3. $y = 2x - 6$ (4,), (, −2), (, 3)
4. $y = 5x + 3$ (2,), (, 0), (, −3)
5. $y = -3$ (2,), (−1,), (−3,)
6. $x = 6$ (, 5), (, 0), (, −1)

For the following equations, tell which of the given ordered pairs are solutions. [3.2]

7. $3x - 4y = 12$ $\left(-2, \frac{9}{2}\right)$, (0, 3), $\left(2, -\frac{3}{2}\right)$
8. $y = 3x + 7$ $\left(-\frac{8}{3}, -1\right)$, $\left(\frac{7}{3}, 0\right)$, (−3, −2)

Graph the following ordered pairs. [3.1]

9. (4, 2)
10. (−3, 1)
11. (0, 5)
12. (−2, −3)
13. (−3, 0)
14. $\left(5, -\frac{3}{2}\right)$

For the following equations, complete the given ordered pairs, and use the results to graph the solution set for the equations. [3.3]

15. $x + y = -2$
(, 0), (0,), (1,)
16. $y = 3x$
(−1,), (1,), (, 0)
17. $y = 2x - 1$
(1,), (0,), (, −3)
18. $x = -3$
(, 0), (, 5), (, −5)

Graph the following equations. [3.3]

19. $3x - y = 3$
20. $x - 2y = 2$
21. $y = -\frac{1}{3}x$
22. $y = \frac{3}{4}x$
23. $y = 2x + 1$
24. $y = -\frac{1}{2}x + 2$
25. $x = 5$
26. $y = -3$
27. $2x - 3y = 3$
28. $5x - 2y = 5$

Find the x- and y-intercepts for each equation. [3.4]

29. $3x - y = 6$ **30.** $2x - 6y = 24$ **31.** $y = x - 3$ **32.** $y = 3x - 6$ **33.** $y = -5$ **34.** $x = 4$

Find the slope of the line through the given pair of points. [3.5]

35. $(2, 3), (3, 5)$ **36.** $(-2, 3), (6, -5)$ **37.** $(-1, -4), (-3, -8)$ **38.** $\left(\frac{1}{2}, 4\right), \left(-\frac{1}{2}, 2\right)$

39. Find x if the line through $(3, 3)$ and $(x, 9)$ has slope 2.

40. Find y if the line through $(5, -5)$ and $(-5, y)$ has slope 2.

Find the equation of the line that contains the given point and has the given slope. Write answers in slope-intercept form. [3.6]

41. $(-1, 4)$; $m = -2$ **42.** $(4, 3)$; $m = \frac{1}{2}$ **43.** $(3, -2)$; $m = -\frac{3}{4}$ **44.** $(3, 5)$; $m = 0$

Find the equation of the line with the given slope and y-intercept. [3.6]

45. $m = 3, b = 2$ **46.** $m = -1, b = 6$ **47.** $m = -\frac{1}{3}, b = \frac{3}{4}$ **48.** $m = 0, b = 0$

For each of the following equations, determine the slope and y-intercept. [3.6]

49. $y = 4x - 1$ **50.** $-2x + y = -5$ **51.** $6x + 3y = 9$ **52.** $5x + 2y = 8$

Graph the following linear inequalities. [3.7]

53. $x - y < 3$ **54.** $x \geq -3$ **55.** $y \leq -4$ **56.** $y \leq -2x + 3$

Chapter 3 Cumulative Review

Simplify.

1. $7 - 2 \cdot 6$
2. $2 + 5(-4 + 8)$
3. $\frac{3}{8}(16) - \frac{2}{3}(9)$
4. $4 \cdot 6 + 12 \div 4 - 3^2$
5. $(4 - 9)(-3 - 8)$
6. $5[6 + (-12)] - 10(4)$
7. $\frac{18 + (-34)}{7 - 7}$
8. $\frac{(5 - 3)^2}{5^2 - 3^2}$
9. $\frac{75}{135}$
10. $\frac{0}{-25}$
11. $5(8x)$
12. $5a + 3 - 4a - 6$

Solve each equation.

13. $5x + 6 - 4x - 3 = 7$
14. $4x - 5 = 3$
15. $-4x + 7 = 5x - 11$
16. $6(t + 5) - 4 = 2$
17. $3 = -\frac{1}{4}(5x + 2)$
18. $0.05x + 0.07(200 - x) = 11$

Solve each inequality, and graph the solution.

19. $-4x > 28$
20. $5 - 7x \geq 19$

Graph on a rectangular coordinate system.

21. $y = 2x + 1$
22. $x + 2y = 3$
23. $y = x$
24. $y = -\frac{3}{2}x$
25. $x - 2y \leq 4$
26. $y \geq 3$
27. Graph the points $(-3, -2)$ and $(2, 3)$, and draw a line that passes through both of them. Does $(4, 7)$ lie on the line? Does $(1, 2)$?
28. Graph the line through $(-2, -2)$ with x-intercept -1. What is the y-intercept?
29. Find the x- and y-intercepts for $2x + 5y = 10$, and then graph the line.
30. Graph the line with slope $\frac{2}{3}$ and y-intercept -1.
31. Find the slope of the line through $(2, 3)$ and $(5, 7)$.
32. Find the equation of the line with slope 3 and y-intercept 1.
33. Find the slope of the line $2x + 3y = 6$.
34. Find the equation of the line with slope $\frac{1}{2}$ that passes through $(-4, 2)$. Write your answer in slope-intercept form.
35. Find the equation of the line through $(2, 3)$ and $(4, 7)$. Write your answer in slope-intercept form.
36. Find the equation of the line with x-intercept 3, y-intercept -4. Write your answer in slope-intercept form.
37. Complete the ordered pairs (, 3) and (0,) for $3x - 2y = 6$.
38. Which of the ordered pairs $(0, 2)$, $(-5, 0)$, and $\left(4, \frac{2}{5}\right)$ are solutions to $2x + 5y = 10$?
39. Write the expression, and then simplify: the difference of 15 and twice 11.
40. Five subtracted from the product of 6 and -2 is what number?

Give the opposite, reciprocal, and absolute value of the given number.

41. $-\frac{2}{3}$
42. 2

Find the next number in each sequence.

43. $\frac{1}{4}, \frac{1}{9}, \frac{1}{16}, \frac{1}{25}, \ldots$
44. $\frac{1}{5}, 1, \frac{9}{5}, \ldots$
45. Evaluate $x^2 + 6x - 7$ when $x = 2$.
46. Use the formula $2x - 3y = 7$ to find y when $x = -1$.
47. **Geometry** The length of a rectangle is 4 inches more than the width. The perimeter is 28 inches. Find the length and width.
48. **Coin Problem** Arbi has $2.20 in dimes and quarters. If he has 8 more dimes than quarters, how many of each coin does he have?

Use the line graph to answer the following questions.

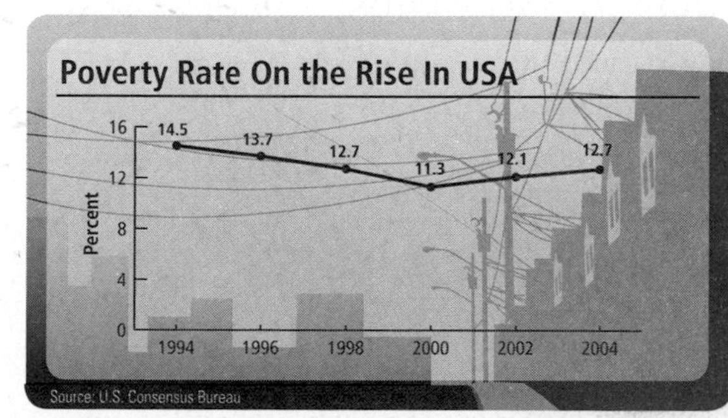

49. Find the slope of the line segment connecting the first two data points, and explain its significance.
50. Find the slope of the line segment connecting the last two data points, and explain its significance.

Chapter 3 Test

Graph the ordered pairs. [3.1]

1. (3, −4) **2.** (−1, −2)

3. (4, 0) **4.** (0, −1)

5. Fill in the following ordered pairs for the equation $2x - 5y = 10$. [3.2]

(0,) (, 0) (10,) (, −3)

6. Which of the following ordered pairs are solutions to $y = 4x - 3$? [3.2]

(2, 5) (0, −3) (3, 0) (−2, 11)

Graph each line. [3.3]

7. $y = 3x - 2$ **8.** $x = -2$

Find the *x*- and *y*-intercepts. [3.4]

9. $3x - 5y = 15$ **10.** $y = \frac{3}{2}x + 1$

11.

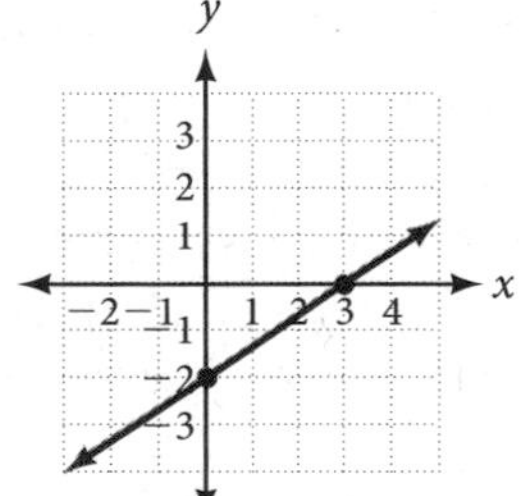

Find the slope of the line through each pair of points. [3.5]

12. (2, −3), (4, −7) **13.** (−3, 5), (2, −8)

Find the slope of each line. [3.5]

14.

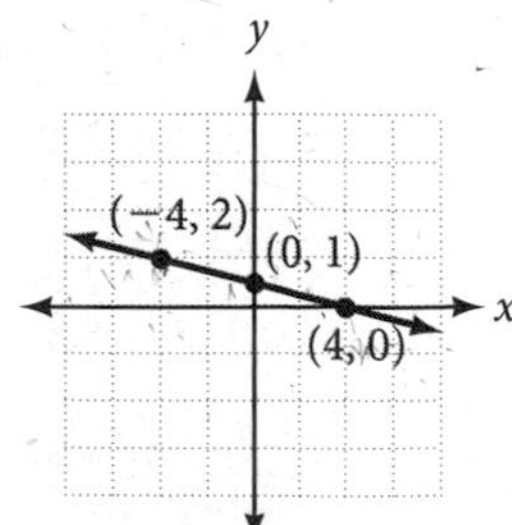

15.

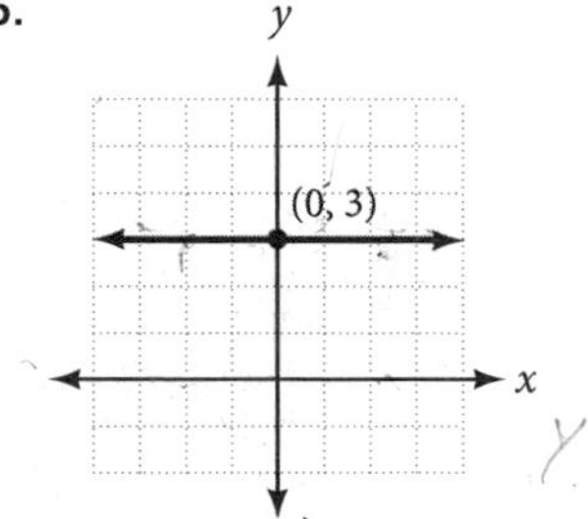

16.

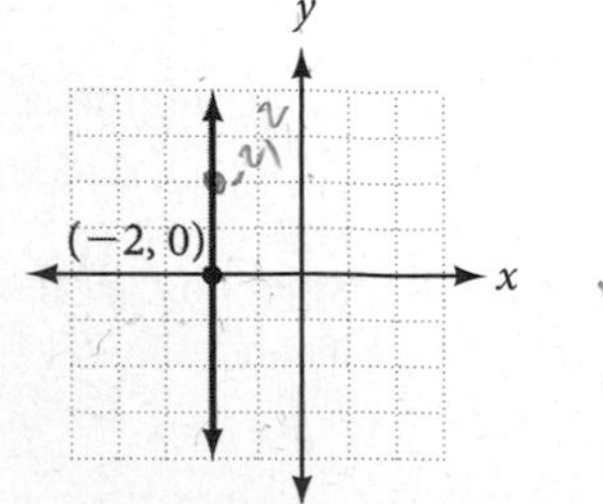

17. Find the equation of the line through (−5, 5) with a slope of 3. [3.6]

18. Find the equation of the line with a slope of 4 and *y*-intercept 8. [3.6]

19. Find the equation of the line passing through the points (−3, 1) and (−2, 4) [3.6]

20. A straight line has an *x*-intercept 1 and contains the point (3, 4). Find its equation. [3.6]

21. Write the equation, in slope-intercept form, of the line whose graph is shown below. [3.6]

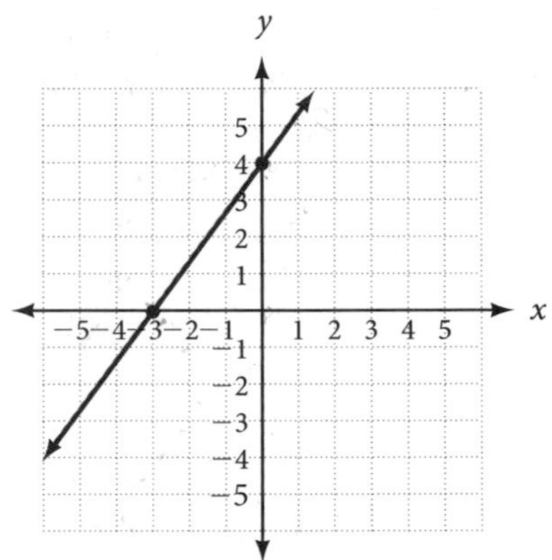

Graph each linear inequality in two variables. [3.7]

22. $y < -x + 4$ **23.** $3x - 4y \geq 12$

24. Write each data point on the graph below as an ordered pair. [3.1]

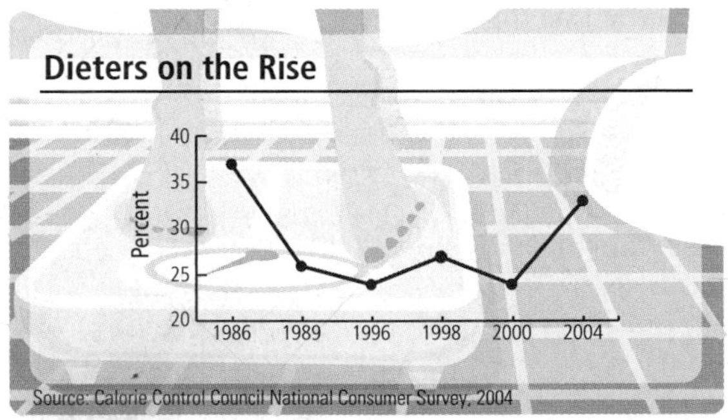

Chapter 3 Projects

LINEAR EQUATIONS AND INEQUALITIES IN TWO VARIABLES

GROUP PROJECT

Reading Graphs

Number of People 2-3

Time Needed 5–10 minutes

Equipment Pencil and paper

Background Although most of the graphs we have encountered in this chapter have been straight lines, many of the graphs that describe the world around us are not straight lines. In this group project we gain experience working with graphs that are not straight lines.

Procedure Read the introduction to each problem below. Then use the graphs to answer the questions.

1. A patient is taking a prescribed dose of a medication every 4 hours during the day to relieve the symptoms of a cold. Figure 1 shows how the concentration of that medication in the patient's system changes over time. The 0 on the horizontal axis corresponds to the time the patient takes the first dose of medication. (The units of concentration on the vertical axis are nanograms per milliliter.)
   - **a.** Explain what the steep vertical line segments show with regard to the patient and his medication.
   - **b.** What has happened to make the graph fall off on the right?
   - **c.** What is the maximum concentration of the medication in the patient's system during the time period shown in Figure 1?
   - **d.** Find the values of A, B, and C.

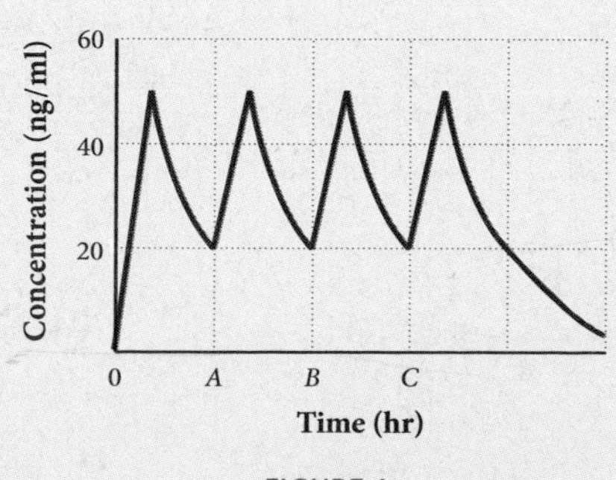

FIGURE 1

2. **Reading Graphs.** Figure 2 shows the number of people in line at a theater box office to buy tickets for a movie that starts at 7:30. The box office opens at 6:45.
   - **a.** How many people are in line at 6:30?
   - **b.** How many people are in line when the box office opens?
   - **c.** How many people are in line when the show starts?
   - **d.** At what times are there 60 people in line?
   - **e.** How long after the show starts is there no one left in line?

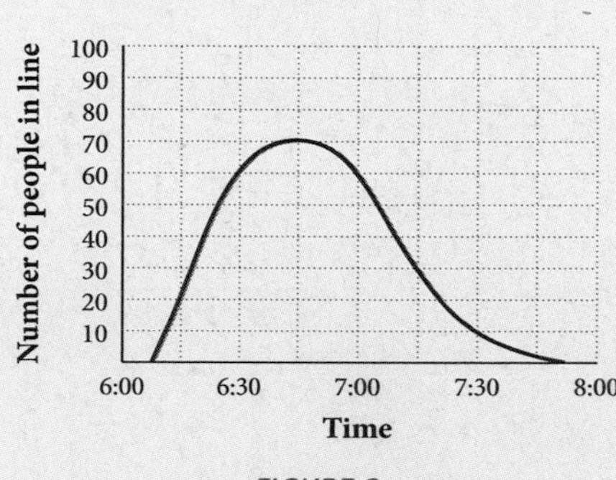

FIGURE 2

RESEARCH PROJECT

Least Squares Curve Fitting

In 1929, the astronomer Edwin Hubble (shown in Figure 1) announced his discovery that the other galaxies in the universe are moving away from us at velocities that increase with distance. The relationship between velocity and distance is described by the linear equation

$$v = Hr$$

where r is the distance of the galaxy from us, v is its velocity away from us, and H is "Hubble's constant." Figure 2 shows a plot of velocity versus distance, where each point represents a galaxy. The fact that the dots all lie approximately on a straight line is the basis of "Hubble's law."

Bettmann/Corbis

FIGURE 1

As you can imagine, there are many lines that could be drawn through the dots in Figure 2. The line shown in Figure 2 is called the *line of best fit* for the points shown in the figure. The method used most often in mathematics to find the line of best fit is called the *least squares method.* Research the least squares method of finding the line of best fit and write an essay that describes the method. Your essay should answer the question: "Why is this method of curve fitting called the *least squares method*?"

Hubble's Law:
velocity = Hubble's constant × distance

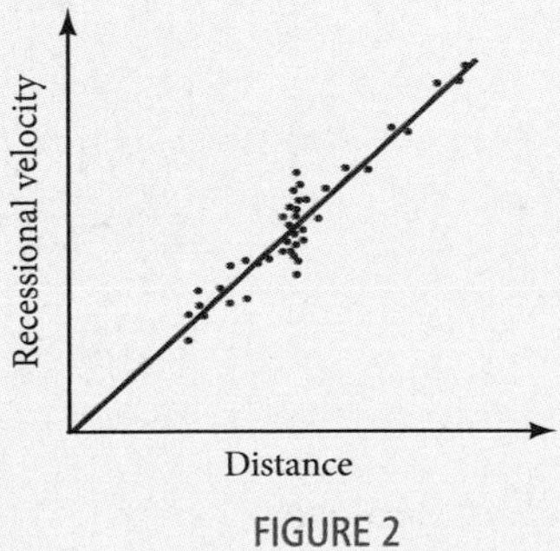

FIGURE 2

Systems of Linear Equations

4

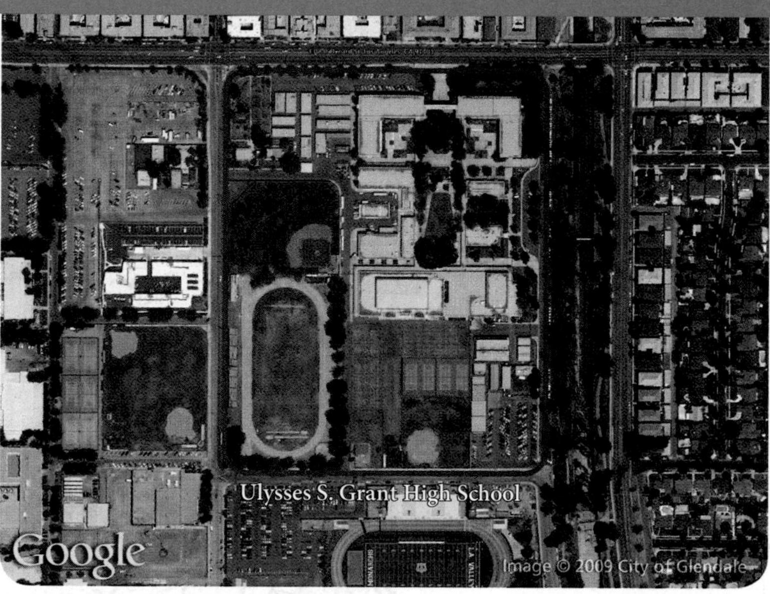

Chapter Outline

Introduction

One of the most innovative and successful textbook authors of our time is Harold Jacobs. Mr. Jacobs taught mathematics, physics, and chemistry at Ulysses S. Grant High School in Los Angeles for 35 years. The page below is from the transparency masters that accompany his elementary algebra textbook. It is a page from a French mathematics book showing an example of how to solve a system of linear equations in two variables, which is one of the main topics of this chapter.

SYSTÈME DE DEUX ÉQUATIONS À DEUX INCONNUES

1° **Elimination d'une inconnue par addition.**

Exemple $\begin{cases} 2x + y = 1 & (1) \\ 3x - 4y = 29 & (2) \end{cases}$

Multiplions par 4 les deux membres de (1) :

$$8x + 4y = 4 \qquad (3)$$

Ajoutons (2) et (3) :

$$11x = 33$$

$$x = \frac{33}{11} = \mathbf{3.}$$

Portons x = 3 dans (1) :

$$6 + y = 1$$

$$y = 1 - 6 = \mathbf{-5.}$$

Règle pratique : par le jeu des multiplications, faire apparaître dans les équations deux coefficients opposés.

After you have finished this chapter, come back to this introduction and see if you can understand the example written in French. I think you will be surprised how easy it is to understand the French words, when you understand the mathematics they are describing.

Chapter Pretest

1. Write the solution to the system graphed below. [4.1]

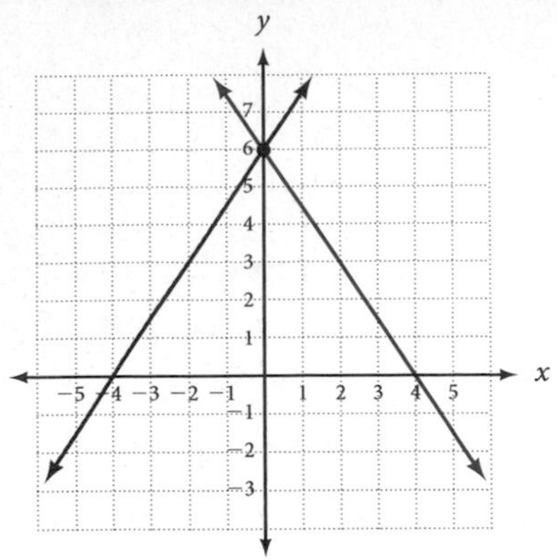

Solve each system by graphing. [4.1]

2. $4x - 2y = 8$
   $y = \frac{2}{3}x$

3. $3x - 2y = 13$
   $y = 4$

4. $2x - 2y = -12$
   $-3x - y = 2$

Solve each system by the elimination method. [4.2]

5. $x - y = -9$
   $2x + 3y = 7$

6. $3x - y = 1$
   $5x - y = 3$

7. $2x + 3y = -3$
   $x + 6y = 12$

8. $2x + 3y = 4$
   $4x + 6y = 8$

Solve each system by the substitution method. [4.3]

9. $3x - y = 12$
   $y = 2x - 8$

10. $3x + 6y = 3$
    $x = 4y - 17$

11. $2x - 3y = -18$
    $3x + y = -5$

12. $2x - 3y = 13$
    $x - 4y = -1$

Getting Ready for Chapter 4

Simplify each expression.

1. $(3x + 2y) + (-3x + y)$

2. $(6x - 7y) + (x + 7y)$

3. $(5x + 9y) + (-5x - 12y)$

4. $-2(3x + 4y)$

5. $4(3x + 4y)$

6. $9\left(\frac{2}{3}x - \frac{1}{4}y\right)$

Solve each equation.

7. $4x - (2x - 1) = 5$

8. $5x - 3(4x - 7) = -7$

9. $3(3y - 1) - 3y = -27$

10. $2x + 9\left(-\frac{5}{3}x - \frac{8}{3}\right) = 2$

11. Let $x + 2y = 7$. If $x = 3$, find y.

12. Let $2x - 3y = 4$. If $x = -4$, find y.

13. Let $x + 3y = 3$. If $x = 3$, find y.

14. Let $-2x + 2y = 4$. If $x = 3$, find y.

Solving Linear Systems by Graphing

4.1

Objectives

A Solve a system of linear equations in two variables by graphing.

A Solving by Graphing

Two linear equations considered at the same time make up what is called a *system of linear equations*. Both equations contain two variables and, of course, have graphs that are straight lines. The following are systems of linear equations:

$$\begin{aligned} x + y &= 3 \\ 3x + 4y &= 2 \end{aligned} \qquad \begin{aligned} y &= 2x + 1 \\ y &= 3x + 2 \end{aligned} \qquad \begin{aligned} 2x - y &= 1 \\ 3x - 2y &= 6 \end{aligned}$$

The solution set for a system of linear equations is all ordered pairs that are solutions to both equations. Because each linear equation has a graph that is a straight line, we can expect the intersection of the graphs to be a point whose coordinates are solutions to the system; that is, if we graph both equations on the same coordinate system, we can read the coordinates of the point of intersection and have the solution to our system. Here is an example.

EXAMPLE 1 Solve the following system by graphing.

$$\begin{aligned} x + y &= 4 \\ x - y &= -2 \end{aligned}$$

SOLUTION On the same set of coordinate axes we graph each equation separately. Figure 1 shows both graphs, without showing the work necessary to get them. We can see from the graphs that they intersect at the point (1, 3). The point (1, 3) therefore must be the solution to our system because it is the only ordered pair whose graph lies on both lines. Its coordinates satisfy both equations.

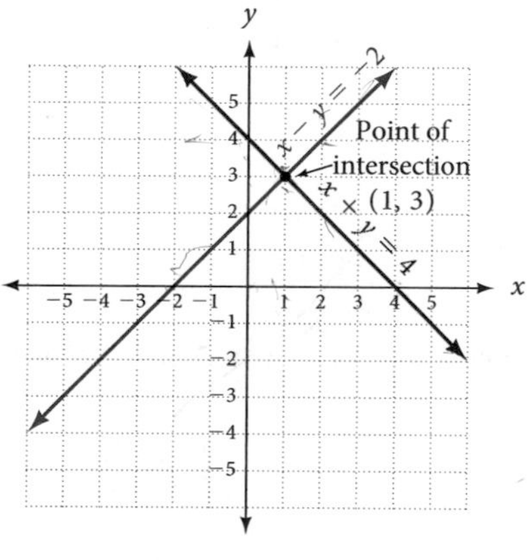

FIGURE 1

We can check our results by substituting the coordinates $x = 1, y = 3$ into both equations to see if they work.

When	$x = 1$	When	$x = 1$
and	$y = 3$	and	$y = 3$
the equation	$x + y = 4$	the equation	$x - y = -2$
becomes	$1 + 3 \stackrel{?}{=} 4$	becomes	$1 - 3 \stackrel{?}{=} -2$
or	$4 = 4$	or	$-2 = -2$

The point (1, 3) satisfies both equations.

PRACTICE PROBLEMS

1. Solve this system by graphing.

$$\begin{aligned} x + y &= 3 \\ x - y &= 5 \end{aligned}$$

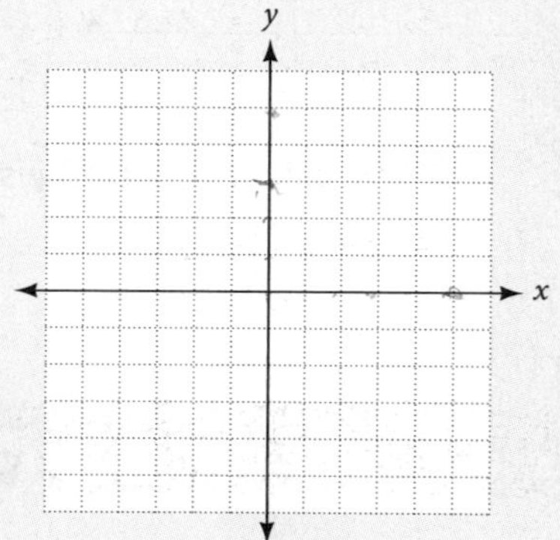

Answer

1. (4, −1)

Here are some steps to follow in solving linear systems by graphing.

Strategy **Solving a Linear System by Graphing**

Step 1: Graph the first equation by the methods described in Section 3.3 or 3.4.

Step 2: Graph the second equation on the same set of axes used for the first equation.

Step 3: Read the coordinates of the point of intersection of the two graphs.

Step 4: Check the solution in both equations.

2. Solve this system by graphing.

$$x + 2y = 6$$
$$3x - y = 4.$$

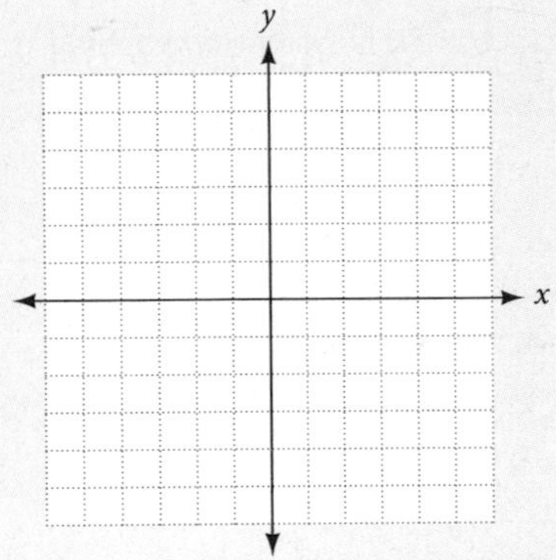

EXAMPLE 2 Solve the following system by graphing.

$$x + 2y = 8$$
$$2x - 3y = 2$$

SOLUTION Graphing each equation on the same coordinate system, we have the lines shown in Figure 2.

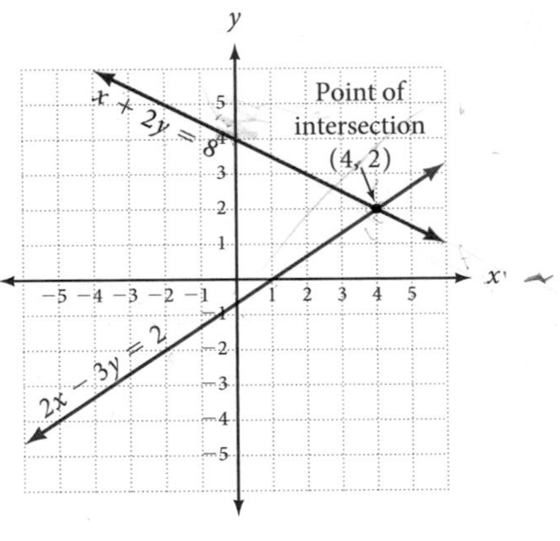

FIGURE 2

From Figure 2, we can see the solution for our system is (4, 2). We check this solution as follows.

When	$x = 4$	When	$x = 4$
and	$y = 2$	and	$y = 2$
the equation	$x + 2y = 8$	the equation	$2x - 3y = 2$
becomes	$4 + 2(2) \stackrel{?}{=} 8$	becomes	$2(4) - 3(2) \stackrel{?}{=} 2$
	$4 + 4 = 8$		$8 - 6 = 2$
	$8 = 8$		$2 = 2$

The point (4, 2) satisfies both equations and, therefore, must be the solution to our system.

Answer
2. (2, 2)

EXAMPLE 3 Solve this system by graphing.

$$y = 2x - 3$$
$$x = 3$$

SOLUTION Graphing both equations on the same set of axes, we have Figure 3.

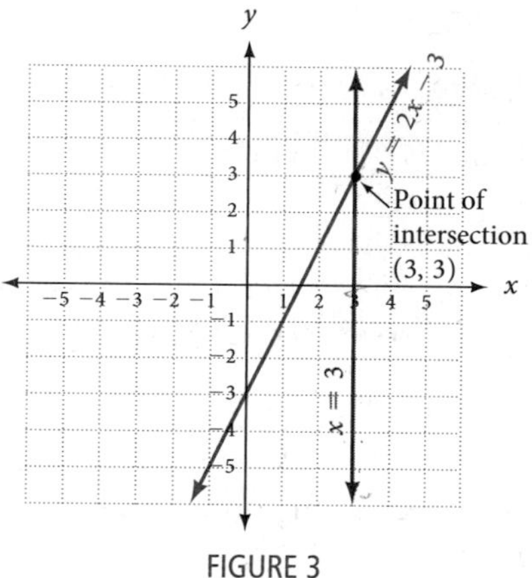

FIGURE 3

The solution to the system is the point (3, 3).

EXAMPLE 4 Solve by graphing.

$$y = x - 2$$
$$y = x + 1$$

SOLUTION Graphing both equations produces the lines shown in Figure 4. We can see in Figure 4 that the lines are parallel and therefore do not intersect. Our system has no ordered pair as a solution because there is no ordered pair that satisfies both equations. We say the solution set is the empty set and write $\varnothing$.

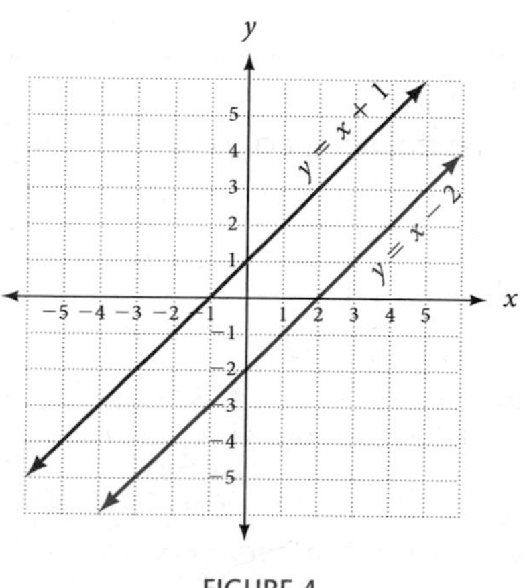

FIGURE 4

Example 4 is one example of two special cases associated with linear systems. The other special case happens when the two graphs coincide. Here is an example.

3. Solve by graphing.
$y = 3x - 1$
$x = 2$

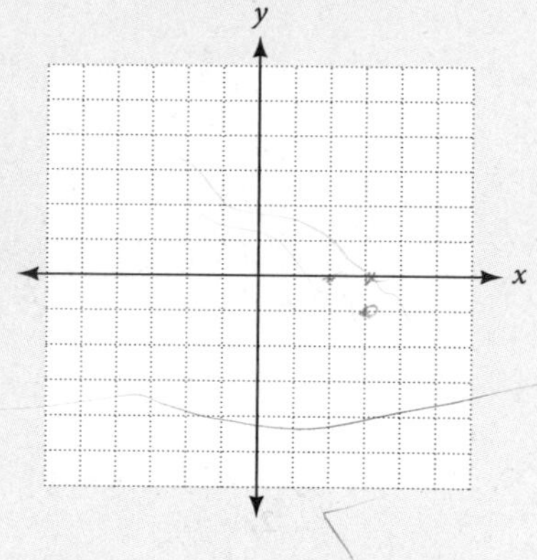

4. Solve by graphing.
$y = x + 3$
$y = x - 2$

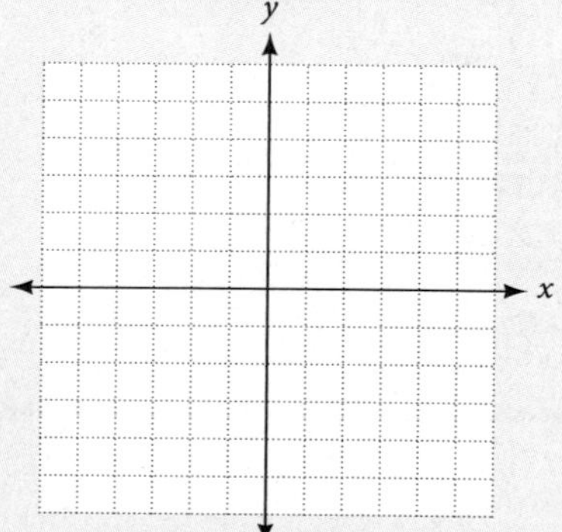

Answers
3. (2, 5)
4. No solution. The lines are parallel.

5. Graph the system.

$$6x + 2y = 12$$
$$9x + 3y = 18$$

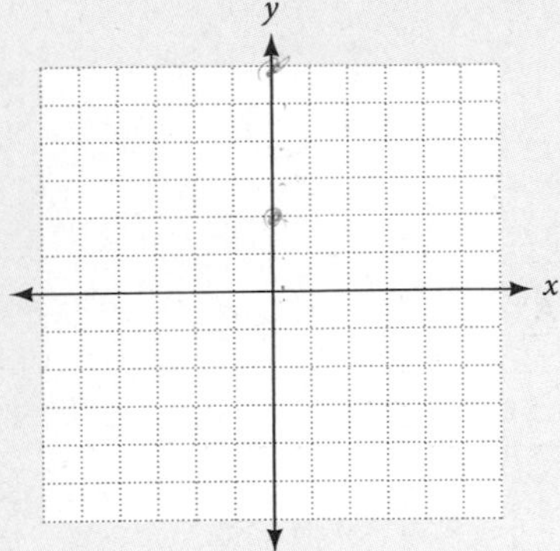

Note We sometimes use special vocabulary to describe the special cases shown in Examples 4 and 5. When a system of equations has no solution because the lines are parallel (as in Example 4), we say the system is *inconsistent*. When the lines coincide (as in Example 5), we say the system is *dependent*.

EXAMPLE 5 Graph the system.

$$2x + y = 4$$
$$4x + 2y = 8$$

SOLUTION Both graphs are shown in Figure 5. The two graphs coincide. The reason becomes apparent when we multiply both sides of the first equation by 2:

$$2x + y = 4$$
$$\mathbf{2}(2x + y) = \mathbf{2}(4) \quad \text{Multiply both sides by 2}$$
$$4x + 2y = 8$$

The equations have the same solution set. Any ordered pair that is a solution to one is a solution to the system. The system has an infinite number of solutions. (Any point on the line is a solution to the system.)

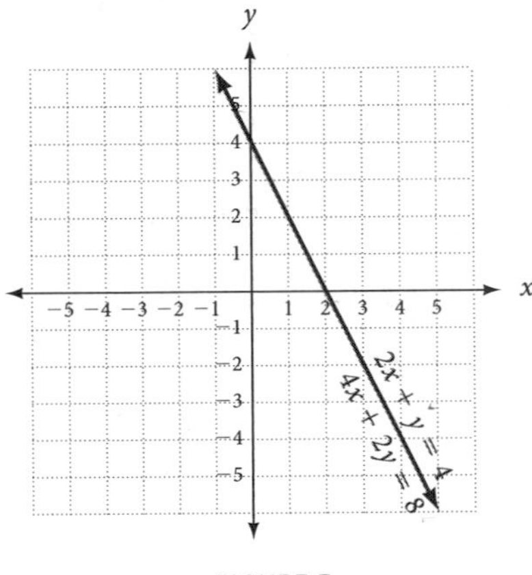

FIGURE 5

The two special cases illustrated in the previous two examples do not happen often. Usually, a system has a single ordered pair as a solution. Solving a system of linear equations by graphing is useful only when the ordered pair in the solution set has integers for coordinates. Two other solution methods work well in all cases. We will develop the other two methods in the next two sections.

Getting Ready for Class

After reading through the preceding section, respond in your own words and in complete sentences.

1. What is a system of two linear equations in two variables?
2. What is a solution to a system of linear equations?
3. How do we solve a system of linear equations by graphing?
4. Under what conditions will a system of linear equations not have a solution?

Answer

5. The lines coincide. Any solution to one is a solution to the other.

Problem Set 4.1

A Solve the following systems of linear equations by graphing. [Examples 1–5]

1. $x + y = 3$
$x - y = 1$

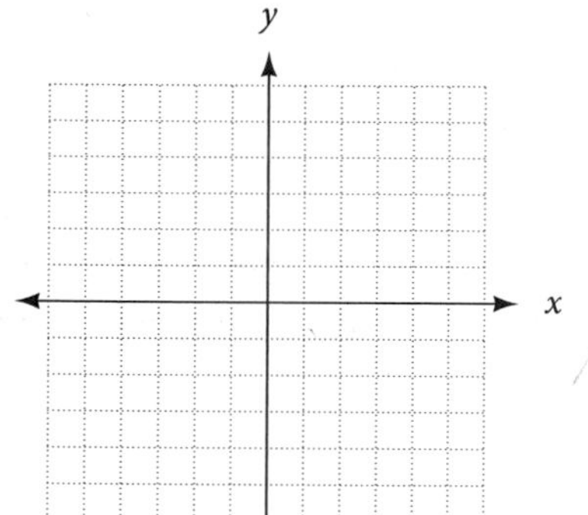

2. $x + y = 2$
$x - y = 4$

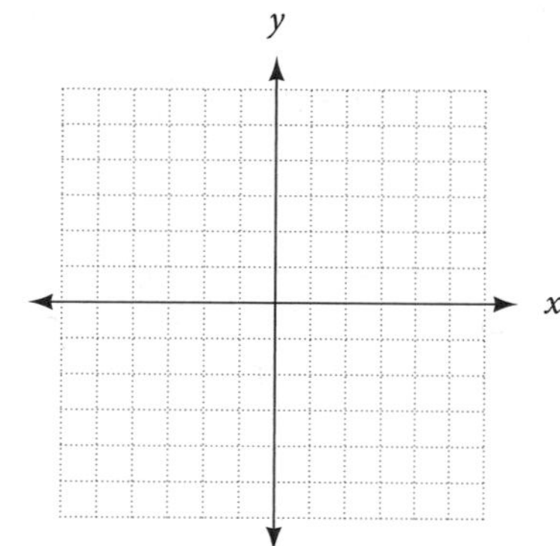

3. $x + y = 1$
$-x + y = 3$

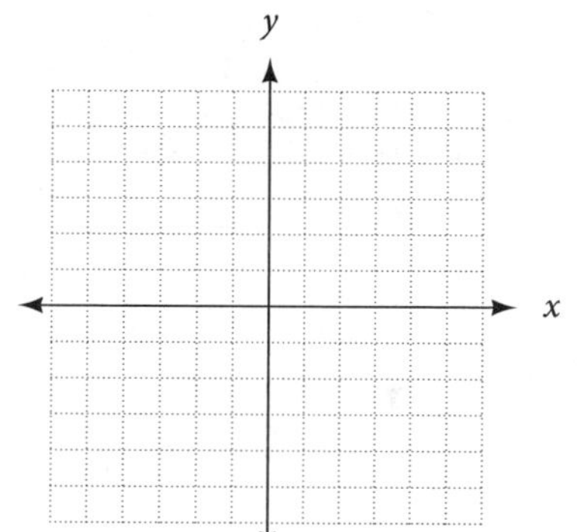

4. $x + y = 1$
$x - y = -5$

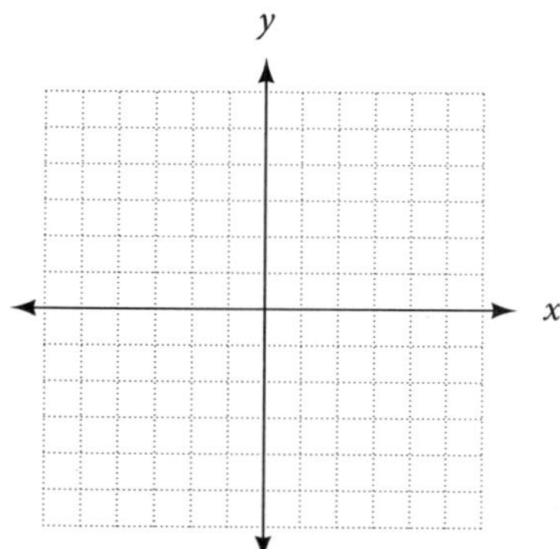

5. $x + y = 8$
$-x + y = 2$

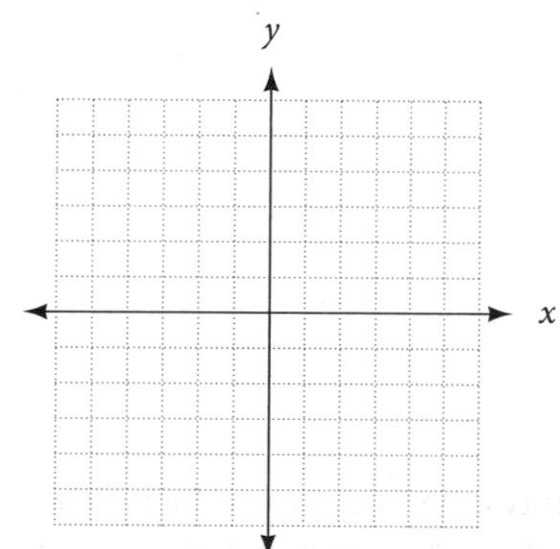

6. $x + y = 6$
$-x + y = -2$

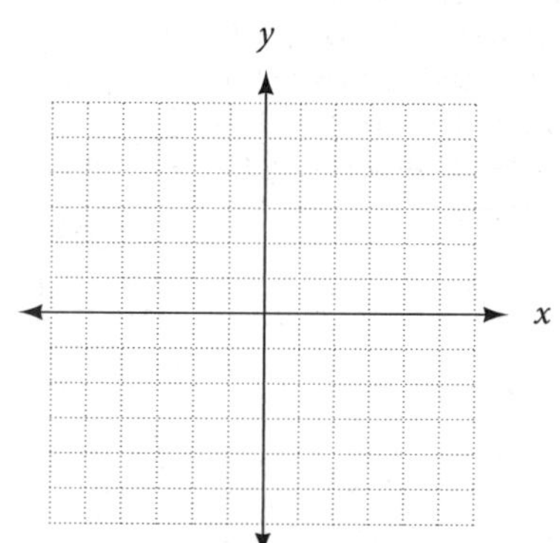

7. $3x - 2y = 6$
$x - y = 1$

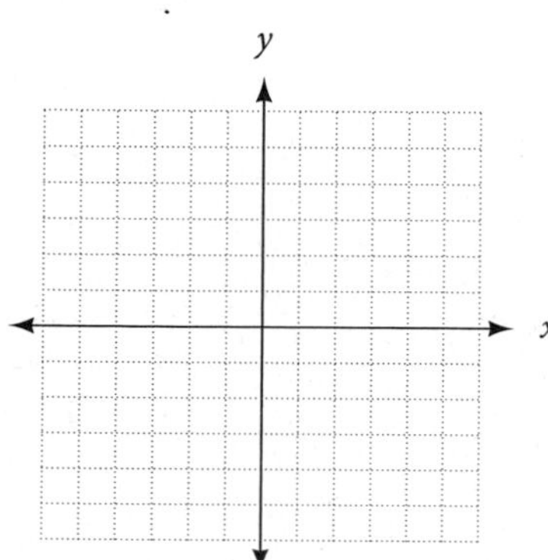

8. $5x - 2y = 10$
$x - y = -1$

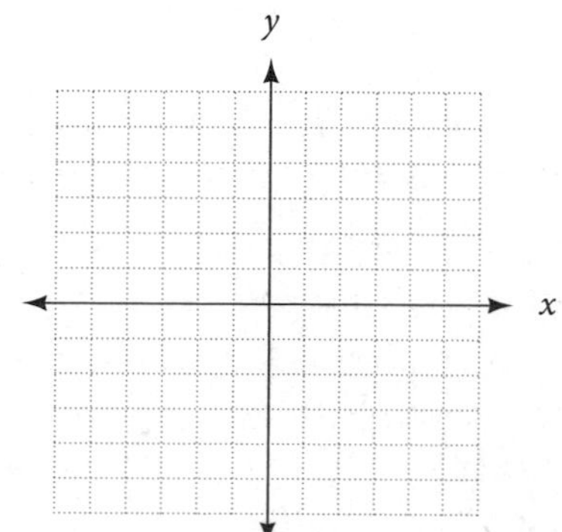

9. $6x - 2y = 12$
$3x + y = -6$

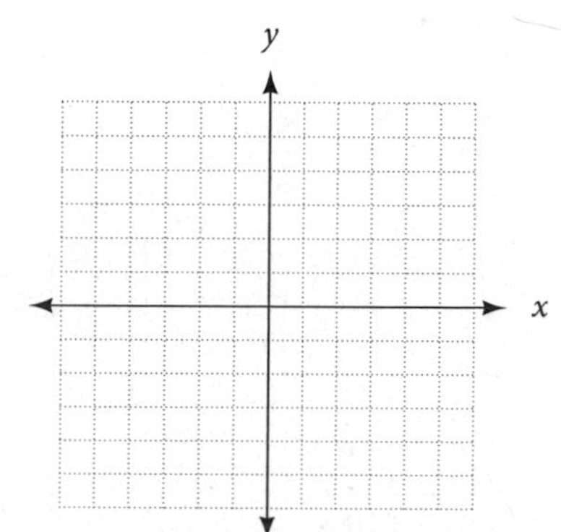

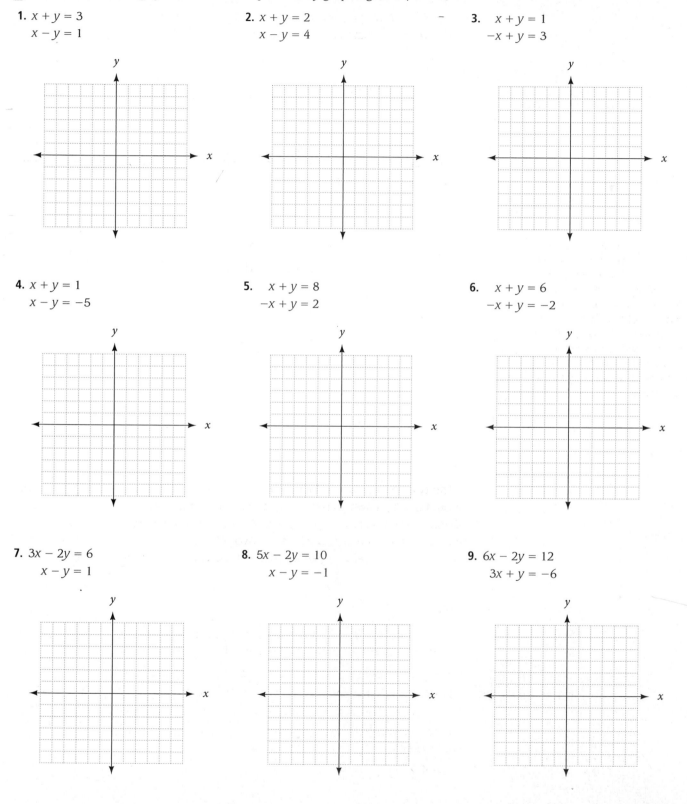

10. $4x - 2y = 8$
$2x + y = -4$

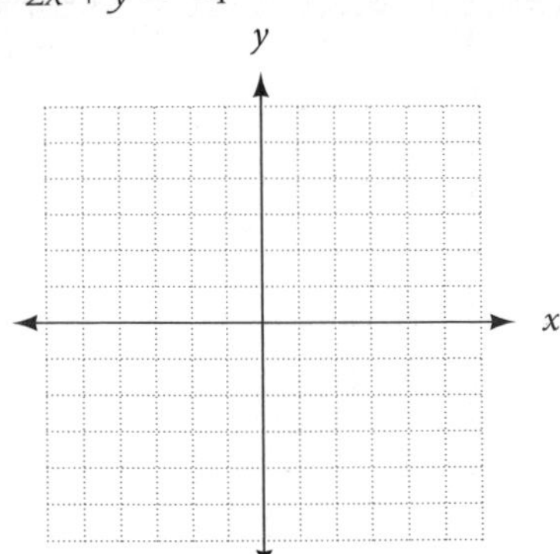

11. $4x + y = 4$
$3x - y = 3$

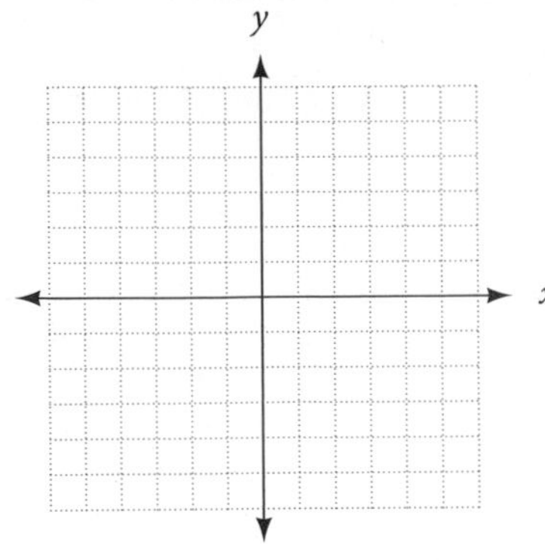

12. $5x - y = 10$
$2x + y = 4$

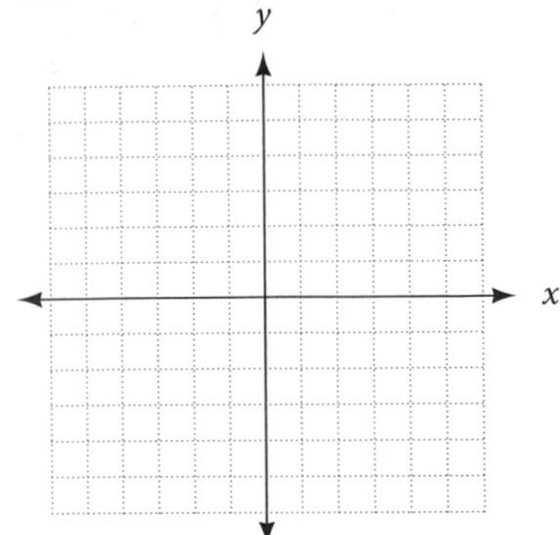

13. $x + 2y = 0$
$2x - y = 0$

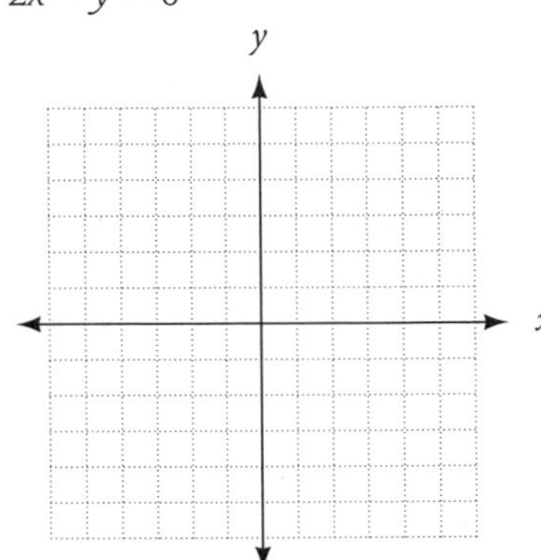

14. $3x + y = 0$
$5x - y = 0$

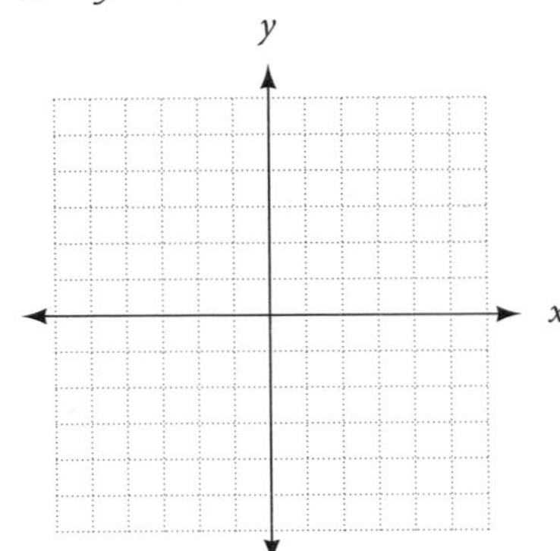

15. $3x - 5y = 15$
$-2x + y = 4$

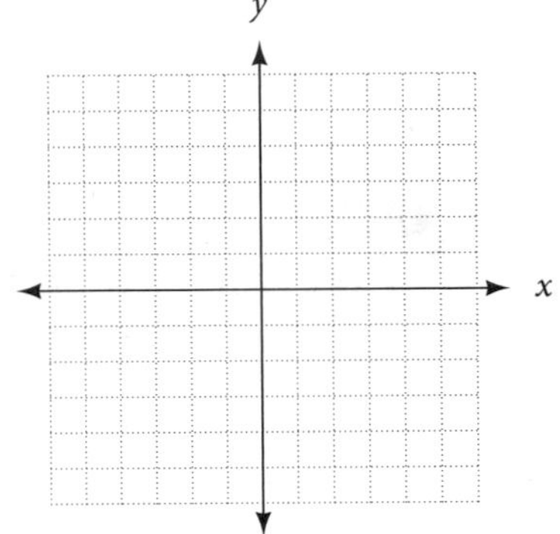

16. $2x - 4y = 8$
$2x - y = -1$

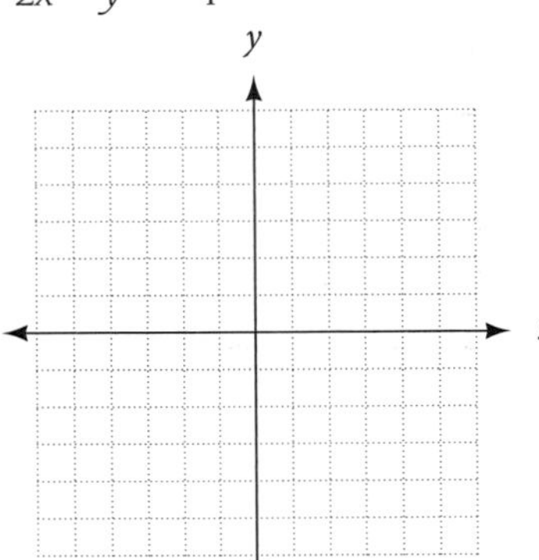

17. $y = 2x + 1$
$y = -2x - 3$

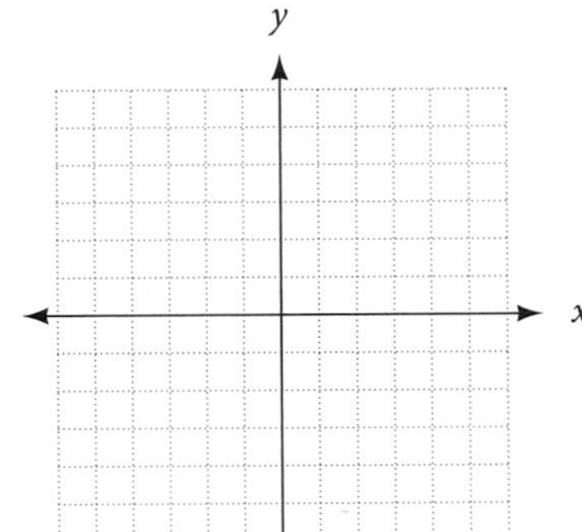

18. $y = 3x - 4$
$y = -2x + 1$

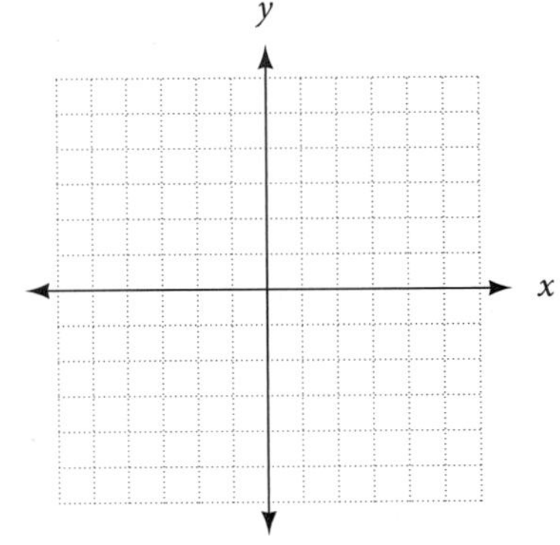

19. $x + 3y = 3$
$y = x + 5$

20. $2x + y = -2$
$y = x + 4$

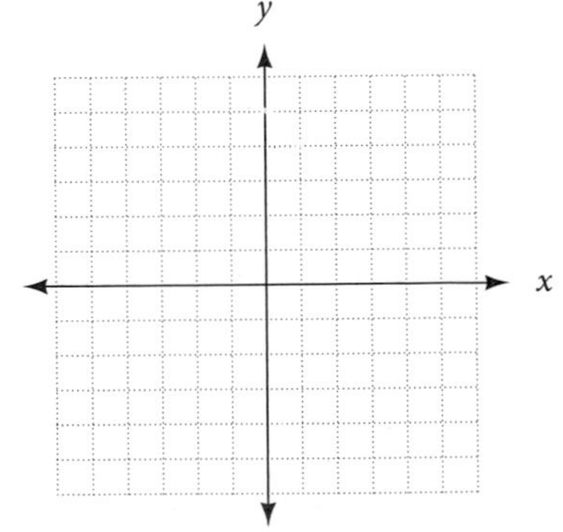

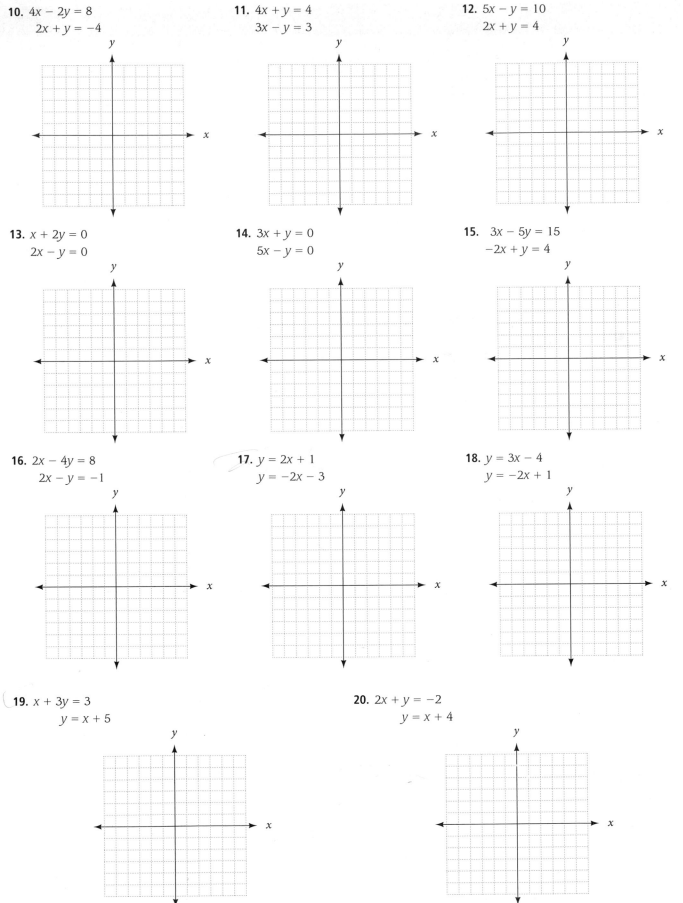

21. $x + y = 2$
$x = -3$

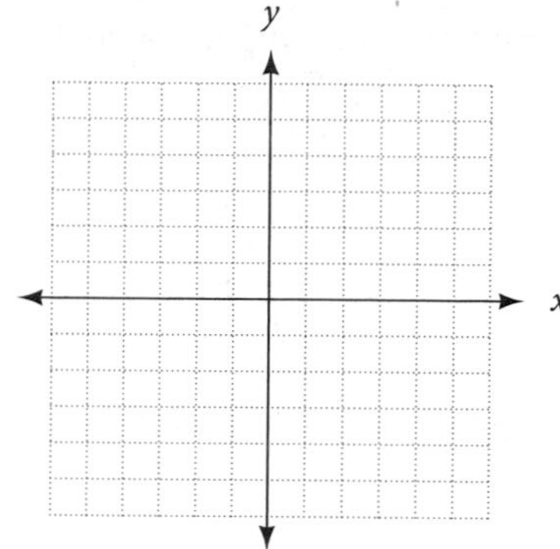

22. $x + y = 6$
$y = 2$

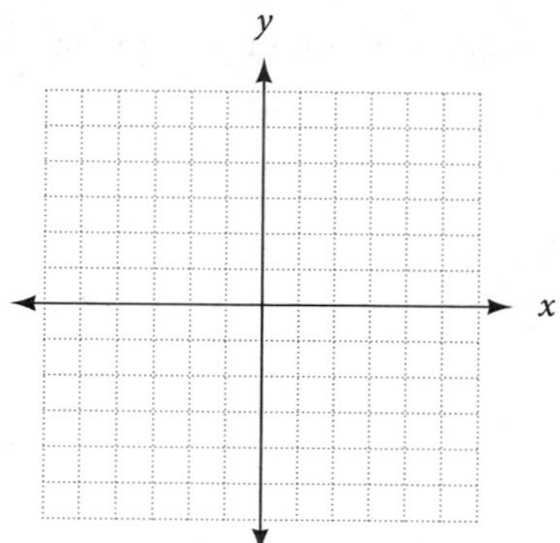

23. $x = -4$
$y = 6$

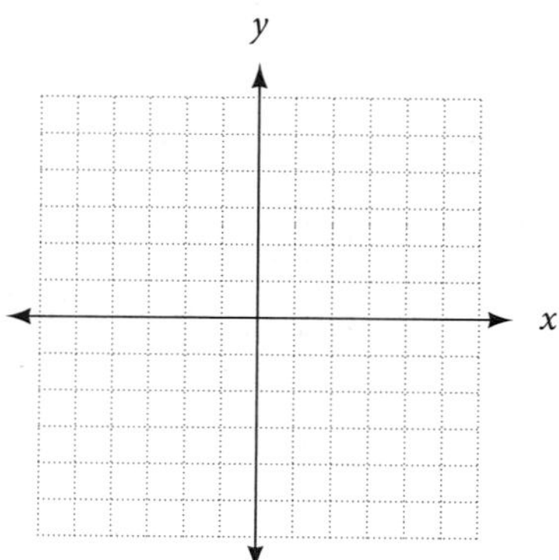

24. $x = 5$
$y = -1$

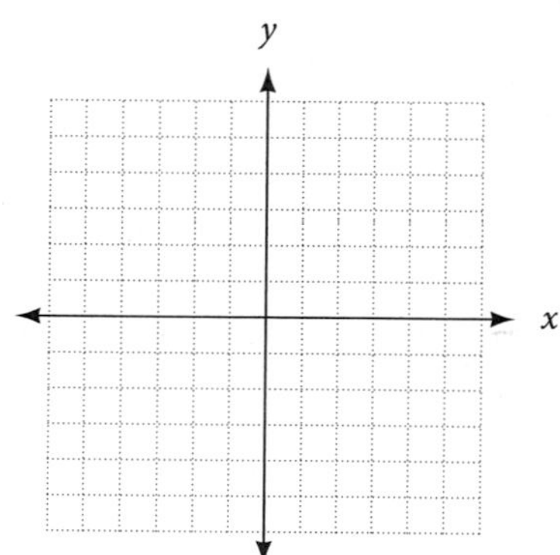

25. $x + y = 4$
$2x + 2y = -6$

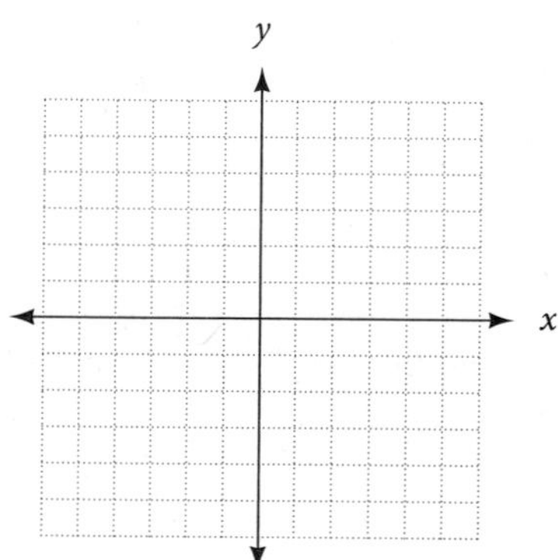

26. $x - y = 3$
$2x - 2y = 6$

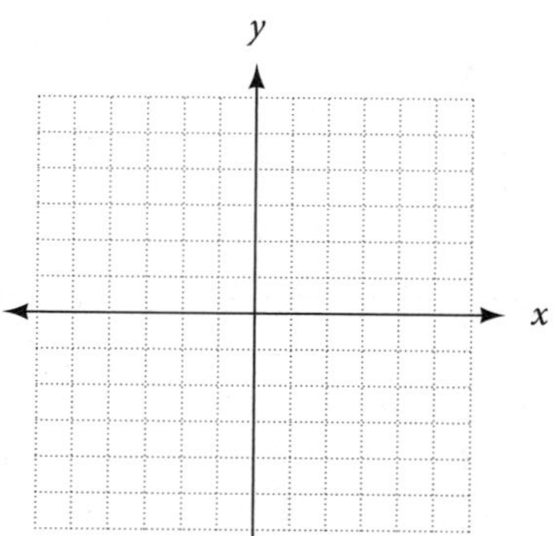

27. $4x - 2y = 8$
$2x - y = 4$

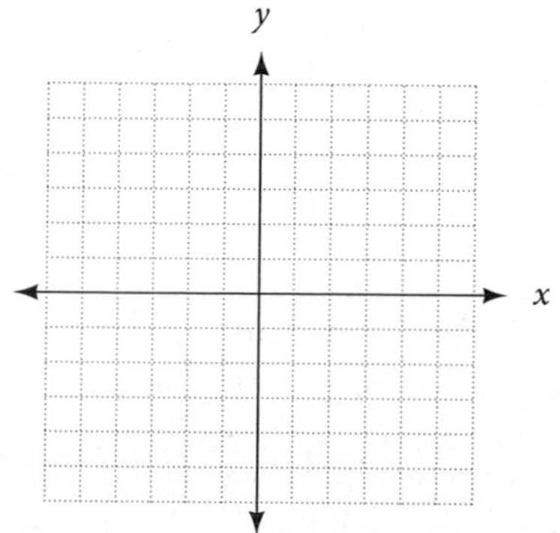

28. $3x - 6y = 6$
$x - 2y = 4$

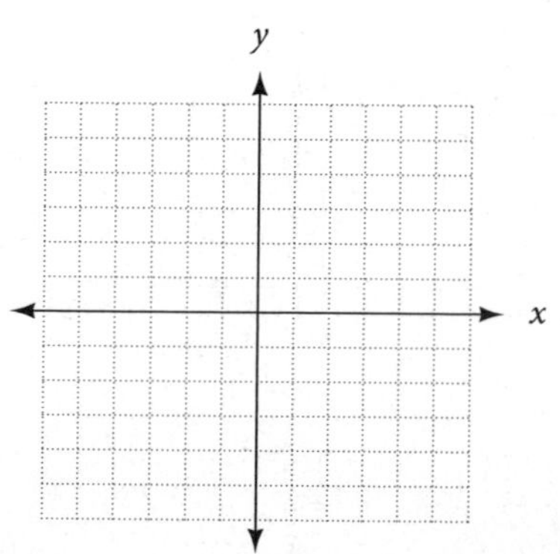

The next two problems are intended to give you practice reading, and paying attention to, the instructions that accompany the problems you are working.

29. Work each problem according to the instructions given.

a. Simplify: $(3x - 4y) + (x - y)$

b. Find y when x is 4 in $3x - 4y = 8$

c. Find the y-intercept: $3x - 4y = 8$

d. Graph: $3x - 4y = 8$

e. Find the point where the graphs of $3x - 4y = 8$ and $x - y = 2$ cross.

30. Work each problem according to the instructions given.

a. Simplify: $(x + 4y) + (-2x + 3y)$

b. Find y when x is 3 in $-2x + 3y = 3$

c. Find the y-intercept: $-2x + 3y = 3$

d. Graph: $-2x + 3y = 3$

e. Find the point where the graphs of $-2x + 3y = 3$ and $x + 4y = 4$ cross.

31. As you probably have guessed by now, it can be difficult to solve a system of equations by graphing if the solution to the system contains a fraction. The solution to the following system is $\left(\frac{1}{2}, 1\right)$. Solve the system by graphing.

$$y = -2x + 2$$
$$y = 4x - 1$$

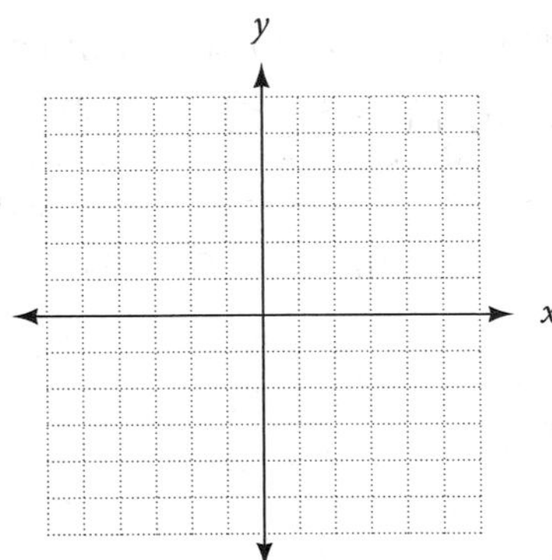

32. The solution to the following system is $\left(\frac{1}{3}, -2\right)$. Solve the system by graphing.

$$y = 3x - 3$$
$$y = -3x - 1$$

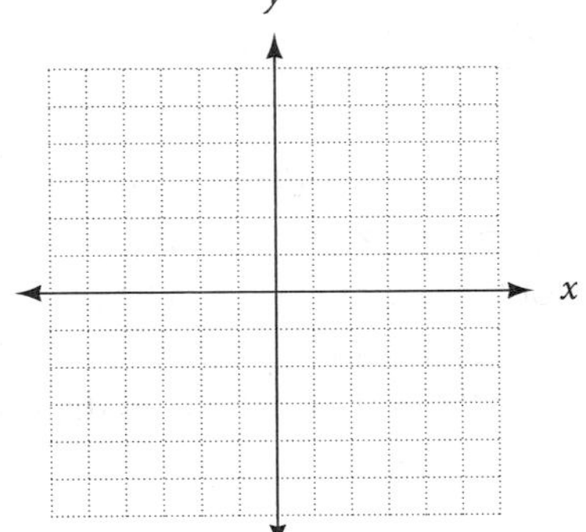

33. A second difficulty can arise in solving a system of equations by graphing if one or both of the equations is difficult to graph. The solution to the following system is (2, 1). Solve the system by graphing.

$$3x - 8y = -2$$
$$x - y = 1$$

34. The solution to the following system is $(-3, 2)$. Solve the system by graphing.

$$2x + 5y = 4$$
$$x - y = -5$$

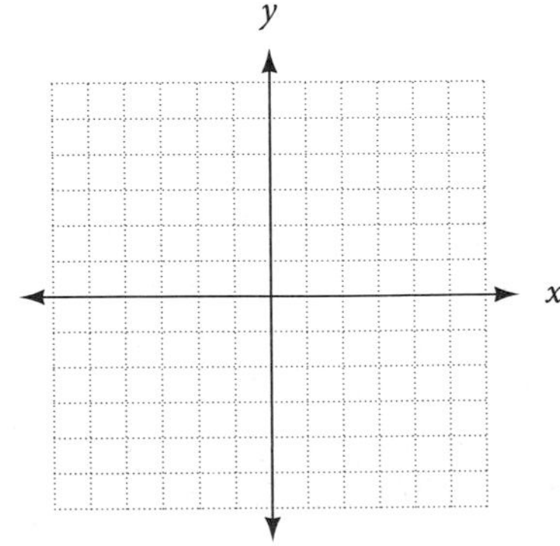

Applying the Concepts

35. **Light Bulbs** The chart shows a comparison of power usage between incandescent and energy efficient light bulbs. Would the line segments from 800 to 1100 lumens intersect in the first quadrant, if you were to extend the lines?

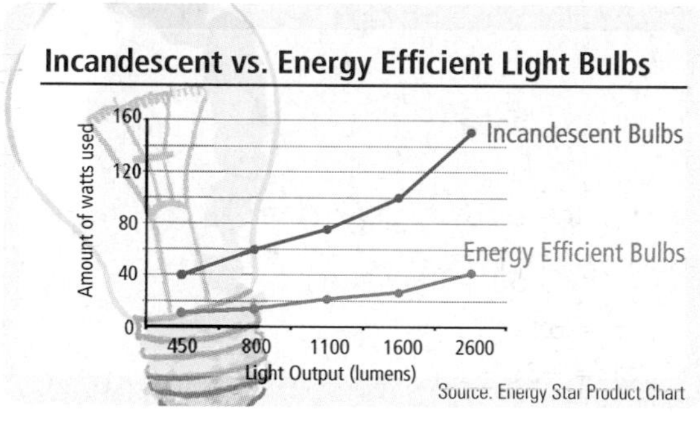

36. **Television** The graph shows how much television certain groups of people watch. In which years do the amounts of time that teens and children watch TV intersect?

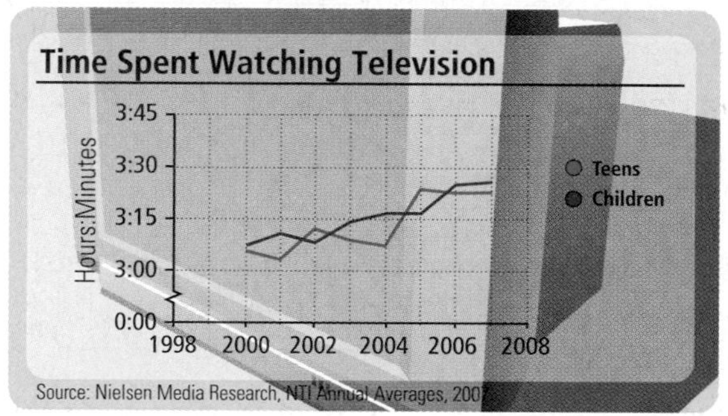

37. **Job Comparison** Jane is deciding between two sales positions. She can work for Marcy's and receive $8.00 per hour or for Gigi's, where she earns $6.00 per hour but also receives a $50 commission per week. The two lines in the following figure represent the money Jane will make for working at each of the jobs.

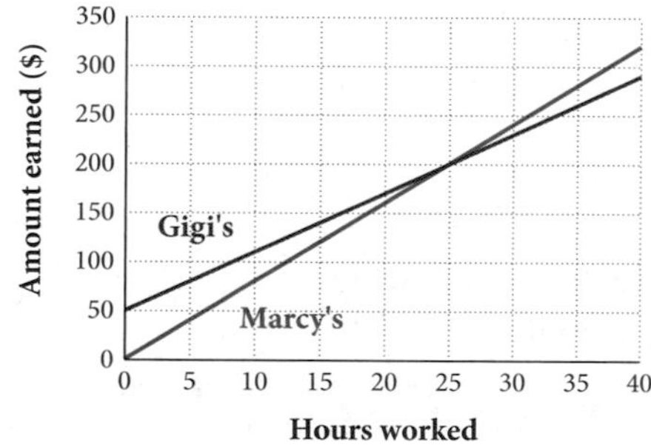

a. From the figure, how many hours would Jane have to work to earn the same amount at each of the positions?

b. If Jane expects to work less than 20 hours a week, which job should she choose?

c. If Jane expects to work more than 30 hours a week, which job should she choose?

38. **Truck Rental** You need to rent a moving truck for two days. Rider Moving Trucks charges $50 per day and $0.50 per mile. UMove Trucks charges $45 per day and $0.75 per mile. The following figure represents the cost of renting each of the trucks for two days.

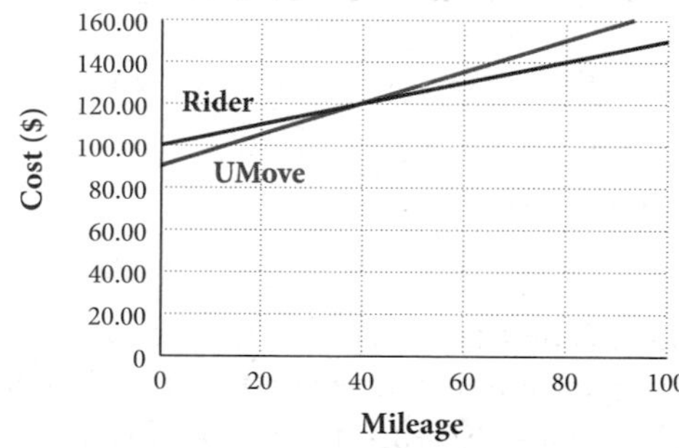

a. From the figure, after how many miles would the trucks cost the same?

b. Which company will give you a better deal if you drive less than 30 miles?

c. Which company will give you a better deal if you drive more than 60 miles?

Getting Ready for the Next Section

Simplify each of the following.

39. $(x + y) + (x - y)$

40. $(x + 2y) + (-x + y)$

41. $(6x - 3y) + (x + 3y)$

42. $(6x + 9y) + (-6x - 10y)$

43. $(-12x - 20y) + (25x + 20y)$

44. $(-3x + 2y) + (3x + 8y)$

45. $-4(3x + 5y)$

46. $6\left(\frac{1}{2}x - \frac{1}{3}y\right)$

47. $12\left(\frac{1}{4}x + \frac{2}{3}y\right)$

48. $5(5x + 4y)$

49. $-2(2x - y)$

50. $-2(4x - 3y)$

51. Let $x + y = 4$. If $x = 3$, find y.

52. Let $x + 2y = 4$. If $x = 3$, find y.

53. Let $x + 3y = 3$. If $x = 3$, find y.

54. Let $2x + 3y = -1$. If $y = -1$, find x.

55. Let $3x + 5y = -7$. If $x = 6$, find y.

56. Let $3x - 2y = 12$. If $y = 6$, find x.

Maintaining Your Skills

Simplify each expression.

57. $6x + 100(0.04x + 0.75)$

58. $5x + 100(0.03x + 0.65)$

59. $13x - 1{,}000(0.002x + 0.035)$

60. $9x - 1{,}000(0.023x + 0.015)$

61. $16x - 10(1.7x - 5.8)$

62. $43x - 10(3.1x - 2.7)$

63. $0.04x + 0.06(100 - x)$

64. $0.07x + 0.03(100 - x)$

65. $0.025x - 0.028(1{,}000 + x)$

66. $0.065x - 0.037(1{,}000 + x)$

67. $2.56x - 1.25(100 + x)$

68. $8.42x - 6.68(100 + x)$

The Elimination Method

4.2

Objectives

A Use the elimination method to solve a system of linear equations in two variables.

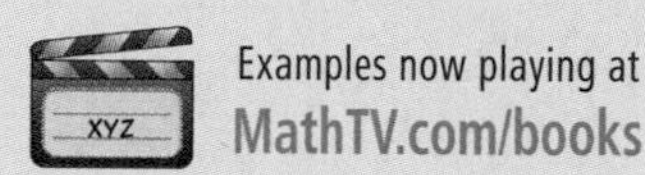

The addition property states that if equal quantities are added to both sides of an equation, the solution set is unchanged. In the past we have used this property to help solve equations in one variable. We will now use it to solve systems of linear equations. Here is another way to state the addition property of equality.

Let A, B, C, and D represent algebraic expressions.

$$\begin{array}{ll} \text{If} & A = B \\ \text{and} & \underline{C = D} \\ \text{then} & A + C = B + D \end{array}$$

Because C and D are equal (that is, they represent the same number), what we have done is added the same amount to both sides of the equation $A = B$. Let's see how we can use this form of the addition property of equality to solve a system of linear equations.

A The Elimination Method

EXAMPLE 1 Solve the following system.

$$x + y = 4$$
$$x - y = 2$$

SOLUTION The system is written in the form of the addition property of equality as written in this section. It looks like this:

$$A = B$$
$$C = D$$

where A is $x + y$, B is 4, C is $x - y$, and D is 2.

We use the addition property of equality to add the left sides together and the right sides together.

$$\begin{array}{r} x + y = 4 \\ \underline{x - y = 2} \\ 2x + 0 = 6 \end{array}$$

We now solve the resulting equation for x.

$$\begin{aligned} 2x + 0 &= 6 \\ 2x &= 6 \\ x &= 3 \end{aligned}$$

The value we get for x is the value of the x-coordinate of the point of intersection of the two lines $x + y = 4$ and $x - y = 2$. To find the y-coordinate, we simply substitute $x = 3$ into either of the two original equations. Using the first equation, we get

$$\begin{aligned} 3 + y &= 4 \\ y &= 1 \end{aligned}$$

The solution to our system is the ordered pair (3, 1). It satisfies both equations.

When	$x = 3$	When	$x = 3$
and	$y = 1$	and	$y = 1$
the equation	$x + y = 4$	the equation	$x - y = 2$
becomes	$3 + 1 \stackrel{?}{=} 4$	becomes	$3 - 1 \stackrel{?}{=} 2$
or	$4 = 4$	or	$2 = 2$

Figure 1 is visual evidence that the solution to our system is (3, 1).

PRACTICE PROBLEMS

1. Solve the following system.

$$x + y = 3$$
$$x - y = 5$$

Note The graphs shown in the margin next to our first three examples are not part of the solution shown in each example. The graphs are there simply to show you that the results we obtain by the elimination method are consistent with the results we would obtain by graphing.

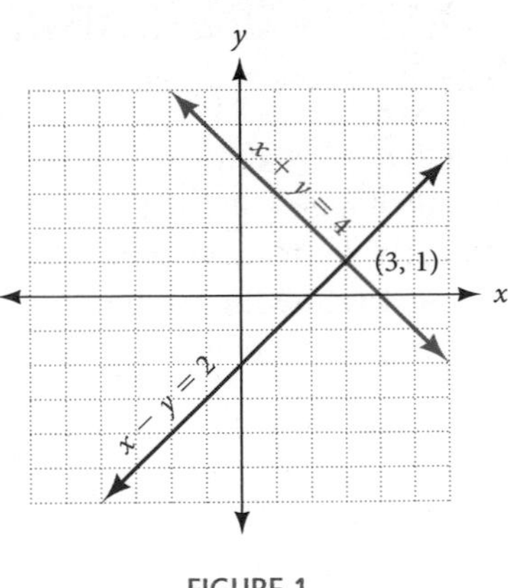

FIGURE 1

The most important part of this method of solving linear systems is eliminating one of the variables when we add the left and right sides together. In our first example, the equations were written so that the y variable was eliminated when we added the left and right sides together. If the equations are not set up this way to begin with, we have to work on one or both of them separately before we can add them together to eliminate one variable.

EXAMPLE 2 Solve the following system.

$$x + 2y = 4$$
$$x - y = -5$$

SOLUTION Notice that if we were to add the equations together as they are, the resulting equation would have terms in both x and y. Let's eliminate the variable x by multiplying both sides of the second equation by -1 before we add the equations together. (As you will see, we can choose to eliminate either the x or the y variable.) Multiplying both sides of the second equation by -1 will not change its solution, so we do not need to be concerned that we have altered the system.

$$x + 2y = 4 \xrightarrow{\text{No change}} x + 2y = 4$$
$$x - y = -5 \xrightarrow[\text{Multiply by } -1]{} \underline{-x + y = 5}$$
$$0 + 3y = 9 \quad \text{Add left and right sides to get}$$
$$3y = 9$$
$$y = 3 \quad \left\{ \text{y-coordinate of the point of intersection} \right.$$

Substituting $y = 3$ into either of the two original equations, we get $x = -2$. The solution to the system is $(-2, 3)$. It satisfies both equations. Figure 2 shows the solution to the system as the point where the two lines cross.

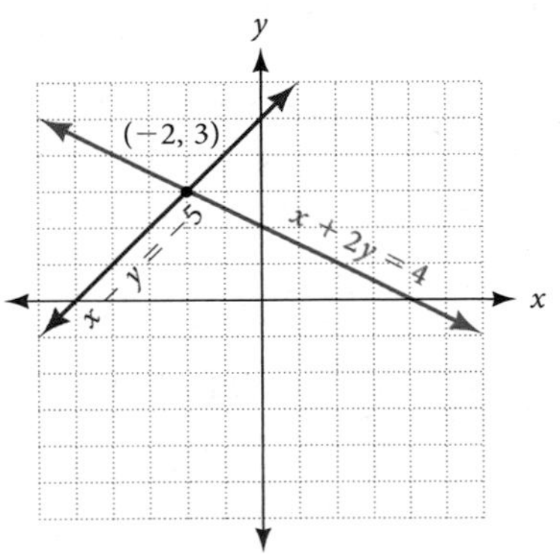

FIGURE 2

2. Solve the following system.
$x + 3y = 5$
$x - y = 1$

Answers
1. $(4, -1)$ 2. $(2, 1)$

EXAMPLE 3 Solve the following system.

$$2x - y = 6$$
$$x + 3y = 3$$

SOLUTION Let's eliminate the y variable from the two equations. We can do this by multiplying the first equation by 3 and leaving the second equation unchanged.

$$2x - y = 6 \xrightarrow{\text{3 times 3both sides}} 6x - 3y = 18$$

$$x + 3y = 3 \xrightarrow[\text{No change}]{} x + 3y = 3$$

The important thing about our system now is that the coefficients (the numbers in front) of the y variables are opposites. When we add the terms on each side of the equal sign, then the terms in y will add to zero and be eliminated.

$$\begin{array}{rl} 6x - 3y = 18 & \\ x + 3y = 3 & \\ \hline 7x \quad\quad = 21 & \text{Add corresponding terms} \end{array}$$

This gives us $x = 3$. Using this value of x in the second equation of our original system, we have

$$3 + 3y = 3$$
$$3y = 0$$
$$y = 0$$

We could substitute $x = 3$ into any of the equations with both x and y variables and also get $y = 0$. The solution to our system is the ordered pair (3, 0). Figure 3 is a picture of the system of equations showing the solution (3, 0).

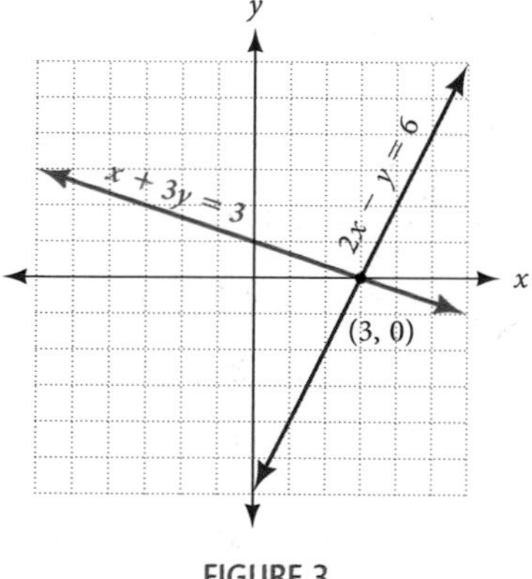

FIGURE 3

3. Solve the following system.

$$3x - y = 7$$
$$x + 2y = 7$$

EXAMPLE 4 Solve the system.

$$2x + 3y = -1$$
$$3x + 5y = -2$$

SOLUTION Let's eliminate x from the two equations. If we multiply the first equation by 3 and the second by -2, the coefficients of x will be 6 and -6, respectively. The x terms in the two equations will then add to zero.

$$2x + 3y = -1 \xrightarrow{\text{Multiply by 3}} 6x + 9y = -3$$

$$3x + 5y = -2 \xrightarrow[\text{Multiply by } -2]{} -6x - 10y = 4$$

4. Solve the system.

$$3x + 2y = 3$$
$$2x + 5y = 13$$

Answer

3. (3, 2)

If you are having trouble understanding this method of solution, it is probably because you can't see why we chose to multiply by 3 and −2 in the first step of Example 4. Look at the result of doing so: the 6*x* and −6*x* will add to 0. We chose to multiply by 3 and −2 because they produce 6*x* and −6*x*, which will add to 0.

We now add the left and right sides of our new system together.

$$\begin{aligned} 6x + 9y &= -3 \\ -6x - 10y &= 4 \\ \hline -y &= 1 \\ y &= -1 \end{aligned}$$

Substituting $y = -1$ into the first equation in our original system, we have

$$\begin{aligned} 2x + 3(-1) &= -1 \\ 2x - 3 &= -1 \\ 2x &= 2 \\ x &= 1 \end{aligned}$$

The solution to our system is $(1, -1)$. It is the only ordered pair that satisfies both equations.

5. Solve the system.

$$\begin{aligned} 5x + 4y &= -6 \\ 2x + 3y &= -8 \end{aligned}$$

EXAMPLE 5

Solve the system.

$$\begin{aligned} 3x + 5y &= -73 \\ 5x + 4y &= 10 \end{aligned}$$

SOLUTION Let's eliminate *y* by multiplying the first equation by −4 and the second equation by 5.

$$\begin{aligned} 3x + 5y = -7 &\xrightarrow{\text{Multiply by } -4} -12x - 20y = 28 \\ 5x + 4y = 10 &\xrightarrow[\text{Multiply by 5}]{} \underline{25x + 20y = 50} \\ &\qquad 13x \qquad\quad = 78 \\ &\qquad\qquad\qquad x = 6 \end{aligned}$$

Substitute $x = 6$ into either equation in our original system, and the result will be $y = -5$. The solution is therefore $(6, -5)$.

6. Solve the system.

$$\begin{aligned} \frac{1}{3}x + \frac{1}{2}y &= 1 \\ x + \frac{3}{4}y &= 0 \end{aligned}$$

EXAMPLE 6

Solve the system.

$$\begin{aligned} \frac{1}{2}x - \frac{1}{3}y &= 2 \\ \frac{1}{4}x + \frac{2}{3}y &= 6 \end{aligned}$$

SOLUTION Although we could solve this system without clearing the equations of fractions, there is probably less chance for error if we have only integer coefficients to work with. So let's begin by multiplying both sides of the top equation by 6 and both sides of the bottom equation by 12, to clear each equation of fractions.

$$\frac{1}{2}x - \frac{1}{3}y = 2 \xrightarrow{\text{Multiply by 6}} 3x - 2y = 12$$

$$\frac{1}{4}x + \frac{2}{3}y = 6 \xrightarrow[\text{Multiply by 12}]{} 3x + 8y = 72$$

Now we can eliminate *x* by multiplying the top equation by −1 and leaving the bottom equation unchanged.

$$\begin{aligned} 3x - 2y = 12 &\xrightarrow{\text{Multiply by } -1} -3x + 2y = -12 \\ 3x + 8y = 72 &\xrightarrow[\text{No change}]{} \underline{\;3x + 8y = \;\;72} \\ &\qquad\qquad\quad 10y = 60 \\ &\qquad\qquad\qquad y = 6 \end{aligned}$$

Answers
4. $(-1, 3)$ **5.** $(2, -4)$ **6.** $(-3, 4)$

We can substitute $y = 6$ into any equation that contains both x and y. Let's use $3x - 2y = 12$.

$$\begin{aligned} 3x - 2(6) &= 12 \\ 3x - 12 &= 12 \\ 3x &= 24 \\ x &= 8 \end{aligned}$$

The solution to the system is (8, 6).

Our next two examples will show what happens when we apply the elimination method to a system of equations consisting of parallel lines and to a system in which the lines coincide.

EXAMPLE 7 Solve the system.

$$\begin{aligned} 2x - y &= 2 \\ 4x - 2y &= 12 \end{aligned}$$

SOLUTION Let's choose to eliminate y from the system. We can do this by multiplying the first equation by -2 and leaving the second equation unchanged.

$$2x - y = 2 \xrightarrow{\text{Multiply by } -2} -4x + 2y = -4$$

$$4x - 2y = 12 \xrightarrow[\text{No change}]{} 4x - 2y = 12$$

If we add both sides of the resulting system, we have

$$\begin{aligned} -4x + 2y &= -4 \\ 4x - 2y &= 12 \\ \hline 0 + 0 &= 8 \\ \text{or} \quad 0 &= 8 \quad \text{A false statement} \end{aligned}$$

Both variables have been eliminated and we end up with the false statement $0 = 8$. We have tried to solve a system that consists of two parallel lines. There is no solution, and that is the reason we end up with a false statement. Figure 4 is a visual representation of the situation and is conclusive evidence that there is no solution to our system.

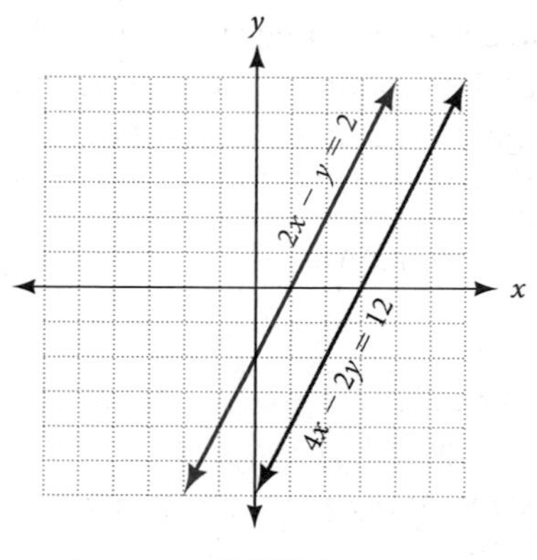

FIGURE 4

EXAMPLE 8 Solve the system.

$$\begin{aligned} 4x - 3y &= 2 \\ 8x - 6y &= 4 \end{aligned}$$

SOLUTION Multiplying the top equation by -2 and adding, we can eliminate the variable x.

7. Solve the system.
$$\begin{aligned} x - 3y &= 2 \\ -3x + 9y &= 2 \end{aligned}$$

8. Solve the system.
$$\begin{aligned} 5x - y &= 1 \\ 10x - 2y &= 2 \end{aligned}$$

Answers

7. Lines are parallel.

8. Lines coincide.

$$\begin{aligned} 4x - 3y &= 2 \xrightarrow{\text{Multiply by } -2} & -8x + 6y &= -4 \\ 8x - 6y &= 4 \xrightarrow[\text{No change}]{} & 8x - 6y &= 4 \\ \hline & & 0 &= 0 \end{aligned}$$

Both variables have been eliminated, and the resulting statement $0 = 0$ is true. In this case the lines coincide because the equations are equivalent. The solution set consists of all ordered pairs that satisfy either equation.

The preceding two examples illustrate the two special cases in which the graphs of the equations in the system either coincide or are parallel.

Here is a summary of our results from these two examples:

Both variables are eliminated and the resulting statement is false.	$\leftrightarrow$	The lines are parallel and there is no solution to the system.
Both variables are eliminated and the resulting statement is true.	$\leftrightarrow$	The lines coincide and there is an infinite number of solutions to the system.

The main idea in solving a system of linear equations by the elimination method is to use the multiplication property of equality on one or both of the original equations, if necessary, to make the coefficients of either variable opposites. The following box shows some steps to follow when solving a system of linear equations by the elimination method.

Strategy **Solving a System of Linear Equations by the Elimination Method**

Step 1: Decide which variable to eliminate. (In some cases one variable will be easier to eliminate than the other. With some practice you will notice which one it is.)

Step 2: Use the multiplication property of equality on each equation separately to make the coefficients of the variable that is to be eliminated opposites.

Step 3: Add the respective left and right sides of the system together.

Step 4: Solve for the variable remaining.

Step 5: Substitute the value of the variable from step 4 into an equation containing both variables and solve for the other variable.

Step 6: Check your solution in both equations, if necessary.

Getting Ready for Class

After reading through the preceding section, respond in your own words and in complete sentences.

1. How is the addition property of equality used in the elimination method of solving a system of linear equations?
2. What happens when we use the elimination method to solve a system of linear equations consisting of two parallel lines?
3. What does it mean when we solve a system of linear equations by the elimination method and we end up with the statement $0 = 8$?
4. What is the first step in solving a system of linear equations that contains fractions?

Problem Set 4.2

A Solve the following systems of linear equations by elimination. **[Examples 1–3, 7–8]**

1. $x + y = 3$
$x - y = 1$

2. $x + y = -2$
$x - y = 6$

3. $x + y = 10$
$-x + y = 4$

4. $x - y = 1$
$-x - y = -7$

5. $x - y = 7$
$-x - y = 3$

6. $x - y = 4$
$2x + y = 8$

7. $x + y = -1$
$3x - y = -3$

8. $2x - y = -2$
$-2x - y = 2$

9. $3x + 2y = 1$
$-3x - 2y = -1$

10. $-2x - 4y = 1$
$2x + 4y = -1$

A Solve each of the following systems by eliminating the y variable. **[Examples 3, 5, 7]**

11. $3x - y = 4$
$2x + 2y = 24$

12. $2x + y = 3$
$3x + 2y = 1$

13. $5x - 3y = -2$
$10x - y = 1$

14. $4x - y = -1$
$2x + 4y = 13$

15. $11x - 4y = 11$
$5x + y = 5$

16. $3x - y = 7$
$10x - 5y = 25$

A Solve each of the following systems by eliminating the x variable. [Examples 4, 6, 8]

17. $3x - 5y = 7$
$-x + y = -1$

18. $4x + 2y = 32$
$x + y = -2$

19. $-x - 8y = -1$
$-2x + 4y = 13$

20. $-x + 10y = 1$
$-5x + 15y = -9$

21. $-3x - y = 7$
$6x + 7y = 11$

22. $-5x + 2y = -6$
$10x + 7y = 34$

A Solve each of the following systems of linear equations by the elimination method. [Examples 3–8]

23. $6x - y = -8$
$2x + y = -16$

24. $5x - 3y = -3$
$3x + 3y = -21$

25. $x + 3y = 9$
$2x - y = 4$

26. $x + 2y = 0$
$2x - y = 0$

27. $x - 6y = 3$
$4x + 3y = 21$

28. $8x + y = -1$
$4x - 5y = 16$

29. $2x + 9y = 2$
$5x + 3y = -8$

30. $5x + 2y = 11$
$7x + 8y = 7$

31. $\frac{1}{3}x + \frac{1}{4}y = \frac{7}{6}$
$\frac{3}{2}x - \frac{1}{3}y = \frac{7}{3}$

32. $\frac{7}{12}x - \frac{1}{2}y = \frac{1}{6}$
$\frac{2}{5}x - \frac{1}{3}y = \frac{11}{15}$

33. $3x + 2y = -1$
$6x + 4y = 0$

34. $8x - 2y = 2$
$4x - y = 2$

35. $11x + 6y = 17$
$5x - 4y = 1$

36. $3x - 8y = 7$
$10x - 5y = 45$

37. $\frac{1}{2}x + \frac{1}{6}y = \frac{1}{3}$
$-x - \frac{1}{3}y = -\frac{1}{6}$

38. $-\frac{1}{3}x - \frac{1}{2}y = -\frac{2}{3}$
$-\frac{2}{3}x - y = -\frac{4}{3}$

Solve each system.

39. $x + y = 22$
$5x + 10y = 170$

40. $x + y = 14$
$10x + 25y = 185$

41. $x + y = 14$
$5x + 25y = 230$

42. $x + y = 11$
$5x + 10y = 95$

43. $x + y = 15{,}000$
$6x + 7y = 98{,}000$

44. $x + y = 10{,}000$
$6x + 7y = 63{,}000$

45. $x + y = 11{,}000$
$4x + 7y = 68{,}000$

46. $x + y = 20{,}000$
$8x + 6y = 138{,}000$

47. $x + y = 23$
$5x + 10y = 175$

48. $x + y = 45$
$25x + 5y = 465$

49. Multiply both sides of the second equation in the following system by 100, and then solve as usual.

$$x + y = 22$$
$$0.05x + 0.10y = 1.70$$

50. Multiply both sides of the second equation in the following system by 100, and then solve as usual.

$$x + y = 15{,}000$$
$$0.06x + 0.07y = 980$$

Applying the Concepts

51. **Camera Phones** The chart shows the estimated number of camera phones and non-camera phones sold from 2004 to 2010. Suppose the equation that describes the total sales in phones is $60x - y = 119{,}560$, and the equation that describes the sale of non-camera phones is $40x + y = 80{,}560$. What was the last year that the sale of non-camera phones accounted for the sale of all phones?

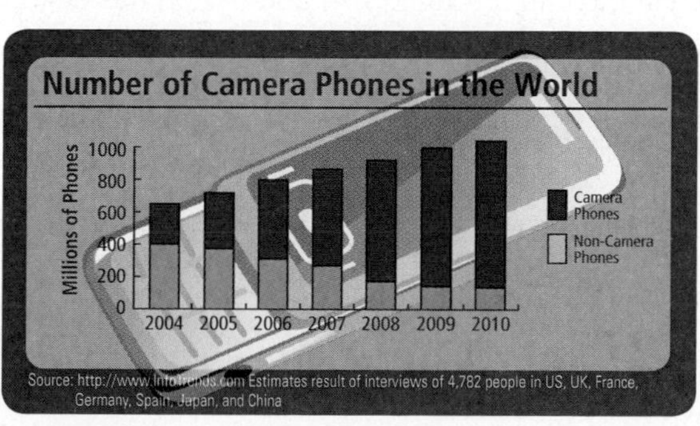

52. **Light Bulbs** The chart shows a comparison of power usage between incandescent and energy efficient light bulbs. Suppose that any output less than 450 lumens is just an extension of the line segment from 450 to 800 lumens. If the equations are $35y - 2x = 500$ for the incandescent bulb and $70y - x = 250$ for the energy efficient bulb, where do the lines intersect?

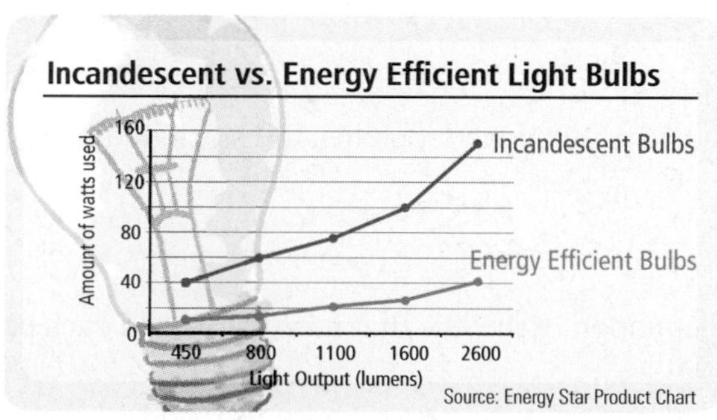

Getting Ready for the Next Section

Solve.

53. $x + (2x - 1) = 2$

54. $2x - 3(2x - 8) = 12$

55. $2(3y - 1) - 3y = 4$

56. $-2x + 4(3x + 6) = 14$

57. $-2x + 3(5x - 1) = 10$

58. $1.5x + 15 = 0.75x + 24.95$

Solve each equation for the indicated variable.

59. $x - 3y = -1$ for x

60. $-3x + y = 6$ for y

61. Let $y = 2x - 1$. If $x = 1$, find y.

62. Let $y = 2x - 8$. If $x = 5$, find y.

63. Let $x = 3y - 1$. If $y = 2$, find x.

64. Let $x = 4y - 5$. If $y = 2$, find x.

Let $y = 1.5x + 15$.

65. If $x = 13$, find y.

66. If $x = 14$, find y.

Let $y = 0.75x + 24.95$.

67. If $x = 12$, find y.

68. If $x = 16$, find y.

Maintaining Your Skills

For each of the equations, determine the slope and y-intercept.

69. $3x - y = 3$

70. $2x + y = -2$

71. $2x - 5y = 25$

72. $-3x + 4y = -12$

Find the slope of the line through the given points.

73. $(-2, 3)$ and $(6, -5)$

74. $(2, -4)$ and $(8, -2)$

75. $(5, 3)$ and $(2, -3)$

76. $(-1, -4)$ and $(-4, -1)$

77. Find y if the line through $(-2, 5)$ and $(-4, y)$ has a slope of -3.

78. Find y if the line through $(-2, 4)$ and $(6, y)$ has a slope of -2.

79. Find y if the line through $(3, -6)$ and $(6, y)$ has a slope of 5.

80. Find y if the line through $(3, 4)$ and $(-2, y)$ has a slope of -4.

For each of the following problems, the slope and one point on a line are given. Find the equation.

81. $(-2, -6)$, $m = 3$

82. $(4, 2)$, $m = \frac{1}{2}$

Find the equation of the line that passes through each pair of points.

83. $(-3, -5)$, $(3, 1)$

84. $(-1, -5)$, $(2, 1)$

The Substitution Method

4.3

Objectives

A Use the substitution method to solve a system of linear equations in two variables.

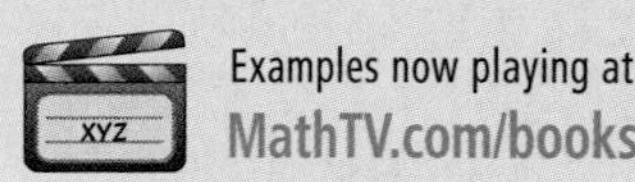

There is a third method of solving systems of equations. It is the substitution method, and like the elimination method, it can be used on any system of linear equations. Some systems, however, lend themselves more to the substitution method than others do.

A Substitution Method

EXAMPLE 1 Solve the following system.

$$x + y = 2$$
$$y = 2x - 1$$

SOLUTION If we were to solve this system by the methods used in the previous section, we would have to rearrange the terms of the second equation so that similar terms would be in the same column. There is no need to do this, however, because the second equation tells us that y is $2x - 1$. We can replace the y variable in the first equation with the expression $2x - 1$ from the second equation; that is, we *substitute* $2x - 1$ from the second equation for y in the first equation. Here is what it looks like:

$$x + (2x - 1) = 2$$

The equation we end up with contains only the variable x. The y variable has been eliminated by substitution.

Solving the resulting equation, we have

$$x + (2x - 1) = 2$$
$$3x - 1 = 2$$
$$3x = 3$$
$$x = 1$$

This is the x-coordinate of the solution to our system. To find the y-coordinate, we substitute $x = 1$ into the second equation of our system. (We could substitute $x = 1$ into the first equation also and have the same result.)

$$y = 2(1) - 1$$
$$y = 2 - 1$$
$$y = 1$$

The solution to our system is the ordered pair (1, 1). It satisfies both of the original equations. Figure 1 provides visual evidence that the substitution method yields the correct solution.

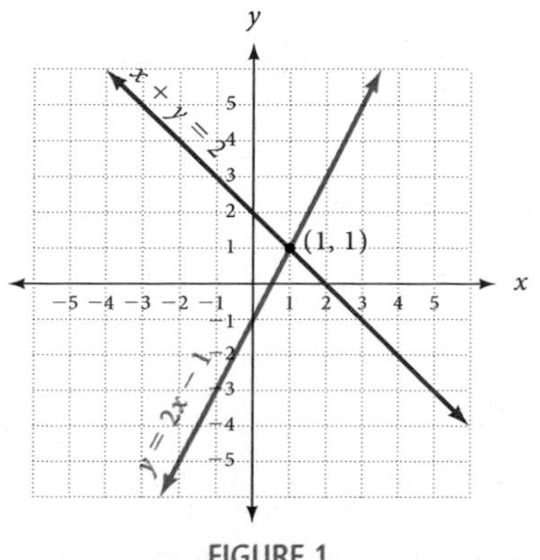

FIGURE 1

PRACTICE PROBLEMS

1. Solve by substitution.

$$x + y = 3$$
$$y = x + 5$$

Note Sometimes this method of solving systems of equations is confusing the first time you see it. If you are confused, you may want to read through this first example more than once and try it on your own.

Answer

1. (−1, 4)

2. Solve by substitution.

$$5x - 4y = -2$$
$$y = 2x + 2$$

EXAMPLE 2 Solve the following system by the substitution method.

$$2x - 3y = 12$$
$$y = 2x - 8$$

SOLUTION Again, the second equation says y is $2x - 8$. Because we are looking for the ordered pair that satisfies both equations, the y in the first equation must also be $2x - 8$. Substituting $2x - 8$ from the second equation for y in the first equation, we have

$$2x - 3(2x - 8) = 12$$

This equation can still be read as $2x - 3y = 12$ because $2x - 8$ is the same as y. Solving the equation, we have

$$\begin{aligned} 2x - 3(2x - 8) &= 12 \\ 2x - 6x + 24 &= 12 \\ -4x + 24 &= 12 \\ -4x &= -12 \\ x &= 3 \end{aligned}$$

To find the y-coordinate of our solution, we substitute $x = 3$ into the second equation in the original system.

When	$x = 3$
the equation	$y = 2x - 8$
becomes	$y = 2(3) - 8$
	$y = 6 - 8 = -2$

The solution to our system is $(3, -2)$.

3. Solve the first equation for x, and then substitute the result into the second equation to solve the system by substitution.

$$x - 4y = -5$$
$$3x - 2y = 5$$

EXAMPLE 3 Solve the following system by solving the first equation for x and then using the substitution method:

$$x - 3y = -1$$
$$2x - 3y = 4$$

SOLUTION We solve the first equation for x by adding $3y$ to both sides to get

$$x = 3y - 1$$

Using this value of x in the second equation, we have

$$\begin{aligned} 2(3y - 1) - 3y &= 4 \\ 6y - 2 - 3y &= 4 \\ 3y - 2 &= 4 \\ 3y &= 6 \\ y &= 2 \end{aligned}$$

Next, we find x.

When	$y = 2$
the equation	$x = 3y - 1$
becomes	$x = 3(2) - 1$
	$x = 6 - 1$
	$x = 5$

The solution to our system is $(5, 2)$.

Here are the steps to use in solving a system of equations by the substitution method.

Answers
2. $(-2, -2)$ 3. $(3, 2)$

Strategy Solving a System of Equations by the Substitution Method

Step 1: Solve either one of the equations for x or y. (This step is not necessary if one of the equations is already in the correct form, as in Examples 1 and 2.)

Step 2: Substitute the expression for the variable obtained in step 1 into the other equation and solve it.

Step 3: Substitute the solution from step 2 into any equation in the system that contains both variables and solve it.

Step 4: Check your results, if necessary.

EXAMPLE 4 Solve by substitution.

$$-2x + 4y = 14$$
$$-3x + y = 6$$

SOLUTION We can solve either equation for either variable. If we look at the system closely, it becomes apparent that solving the second equation for y is the easiest way to go. If we add $3x$ to both sides of the second equation, we have

$$y = 3x + 6$$

Substituting the expression $3x + 6$ back into the first equation in place of y yields the following result.

$$-2x + 4(3x + 6) = 14$$
$$-2x + 12x + 24 = 14$$
$$10x + 24 = 14$$
$$10x = -10$$
$$x = -1$$

Substituting $x = -1$ into the equation $y = 3x + 6$ leaves us with

$$y = 3(-1) + 6$$
$$y = -3 + 6$$
$$y = 3$$

The solution to our system is $(-1, 3)$.

EXAMPLE 5 Solve by substitution.

$$4x + 2y = 8$$
$$y = -2x + 4$$

SOLUTION Substituting the expression $-2x + 4$ for y from the second equation into the first equation, we have

$$4x + 2(-2x + 4) = 8$$
$$4x - 4x + 8 = 8$$
$$8 = 8 \quad \text{A true statement}$$

Both variables have been eliminated, and we are left with a true statement. Recall from the last section that a true statement in this situation tells us the lines coincide; that is, the equations $4x + 2y = 8$ and $y = -2x + 4$ have exactly the same graph. Any point on that graph has coordinates that satisfy both equations and is a solution to the system.

4. Solve by substitution.
$$5x - y = 1$$
$$-2x + 3y = 10$$

5. Solve by substitution.
$$6x + 3y = 1$$
$$y = -2x - 5$$

Answers

4. $(1, 4)$

5. No solution. The lines are parallel.

6. The rates for two garbage companies are given in the following table. How many bags of trash must be picked up in one month for the two companies to charge the same amount?

	Flat Rate	plus	Per Bag Charge
Company 1	\$13.00		\$1.50
Company 2	\$21.85		\$1.00

EXAMPLE 6 The following table shows two contract rates charged by GTE Wireless for cellular phone use. At how many minutes will the two rates cost the same amount?

	Flat Rate	plus	Per Minute Charge
Plan 1	\$15		\$1.50
Plan 2	\$24.95		\$0.75

SOLUTION If we let $y =$ the monthly charge for x minutes of phone use, then the equations for each plan are

$$\text{Plan 1:} \quad y = 1.5x + 15$$
$$\text{Plan 2:} \quad y = 0.75x + 24.95$$

We can solve this system by substitution by replacing the variable y in Plan 2 with the expression $1.5x + 15$ from Plan 1. If we do so, we have

$$1.5x + 15 = 0.75x + 24.95$$
$$0.75x + 15 = 24.95$$
$$0.75x = 9.95$$
$$x = 13.27 \quad \text{to the nearest hundredth}$$

The monthly bill is based on the number of minutes you use the phone, with any fraction of a minute moving you up to the next minute. If you talk for a total of 13 minutes, you are billed for 13 minutes. If you talk for 13 minutes, 10 seconds, you are billed for 14 minutes. The number of minutes on your bill always will be a whole number. So, to calculate the cost for talking 13.27 minutes, we would replace x with 14 and find y. Let's compare the two plans at $x = 13$ minutes and at $x = 14$ minutes.

$$\text{Plan 1:} \quad y = 1.5x + 15$$
$$\text{When} \quad x = 13, y = \$34.50$$
$$\text{When} \quad x = 14, y = \$36.00$$
$$\text{Plan 2:} \quad y = 0.75x + 24.95$$
$$\text{When} \quad x = 13, y = \$34.70$$
$$\text{When} \quad x = 14, y = \$35.45$$

The two plans will never give the same cost for talking x minutes. If you talk 13 or less minutes, Plan 1 will cost less. If you talk for more than 13 minutes, you will be billed for 14 minutes, and Plan 2 will cost less than Plan 1.

Getting Ready for Class

After reading through the preceding section, respond in your own words and in complete sentences.

1. What is the first step in solving a system of linear equations by substitution?
2. When would substitution be more efficient than the elimination method in solving two linear equations?
3. What does it mean when we solve a system of linear equations by the substitution method and we end up with the statement $8 = 8$?
4. How would you begin solving the following system using the substitution method?

$$x + y = 2$$
$$y = 2x - 1$$

Answer

6. See Solutions to Selected Practice Problems.

Problem Set 4.3

A Solve the following systems by substitution. Substitute the expression in the second equation into the first equation and solve. [Examples 1–2]

1. $x + y = 11$
   $y = 2x - 1$

2. $x - y = -3$
   $y = 3x + 5$

3. $x + y = 20$
   $y = 5x + 2$

4. $3x - y = -1$
   $x = 2y - 7$

5. $-2x + y = -1$
   $y = -4x + 8$

6. $4x - y = 5$
   $y = -4x + 1$

7. $3x - 2y = -2$
   $x = -y + 6$

8. $2x - 3y = 17$
   $x = -y + 6$

9. $5x - 4y = -16$
   $y = 4$

10. $6x + 2y = 18$
    $x = 3$

11. $5x + 4y = 7$
    $y = -3x$

12. $10x + 2y = -6$
    $y = -5x$

A Solve the following systems by solving one of the equations for x or y and then using the substitution method. [Examples 3–4]

13. $x + 3y = 4$
    $x - 2y = -1$

14. $x - y = 5$
    $x + 2y = -1$

15. $2x + y = 1$
    $x - 5y = 17$

16. $2x - 2y = 2$
    $x - 3y = -7$

17. $3x + 5y = -3$
    $x - 5y = -5$

18. $2x - 4y = -4$
    $x + 2y = 8$

19. $5x + 3y = 0$
    $x - 3y = -18$

20. $x - 3y = -5$
    $x - 2y = 0$

21. $-3x - 9y = 7$
    $x + 3y = 12$

22. $2x + 6y = -18$
    $x + 3y = -9$

A Solve the following systems using the substitution method. [Examples 1–2, 5]

23. $5x - 8y = 7$
$y = 2x - 5$

24. $3x + 4y = 10$
$y = 8x - 15$

25. $7x - 6y = -1$
$x = 2y - 1$

26. $4x + 2y = 3$
$x = 4y - 3$

27. $-3x + 2y = 6$
$y = 3x$

28. $-2x - y = -3$
$y = -3x$

29. $3x - 5y = -8$
$y = x$

30. $2x - 4y = 0$
$y = x$

31. $3x + 3y = 9$
$y = 2x - 12$

32. $7x + 6y = -9$
$y = -2x + 1$

33. $7x - 11y = 16$
$y = 10$

34. $9x - 7y = -14$
$x = 7$

35. $-4x + 4y = -8$
$y = x - 2$

36. $-4x + 2y = -10$
$y = 2x - 5$

Solve each system.

37. $2x + 5y = 36$
$y = 12 - x$

38. $3x + 6y = 120$
$y = 25 - x$

39. $5x + 2y = 54$
$y = 18 - x$

40. $10x + 5y = 320$
$y = 40 - x$

41. $2x + 2y = 96$
$y = 2x$

42. $x + y = 22$
$y = x + 9$

Solve each system by substitution. You can eliminate the decimals if you like, but you don't have to. The solution will be the same in either case.

43. $0.05x + 0.10y = 1.70$
$y = 22 - x$

44. $0.20x + 0.50y = 3.60$
$y = 12 - x$

The next two problems are intended to give you practice reading, and paying attention to, the instructions that accompany the problems you are working. Working these problems is an excellent way to get ready for a test or quiz.

45. Work each problem according to the instructions given.

a. Solve: $4y - 5 = 20$

b. Solve for y: $4x - 5y = 20$

c. Solve for x: $x - y = 5$

d. Solve the system: $4x - 5y = 20$
$x - y = 5$

46. Work each problem according to the instructions given.

a. Solve: $2x - 1 = 4$

b. Solve for y: $2x - y = 4$

c. Solve for x: $x + 3y = 9$

d. Solve the system: $2x - y = 4$
$x + 3y = 9$

Applying the Concepts [Example 6]

47. Camera Phones The chart shows the estimated number of camera phones and non-camera phones sold from 2004 to 2010. In the last section we solved this problem using the method of elimination. Now use substitution to solve for the last year that the sale of non-camera phones accounted for the sale of all phones, supposing that the equation that describes the total sales in phones is $60x - y = 119{,}560$ and the equation that describes the sale of non-camera phones is $40x + y = 80{,}560$.

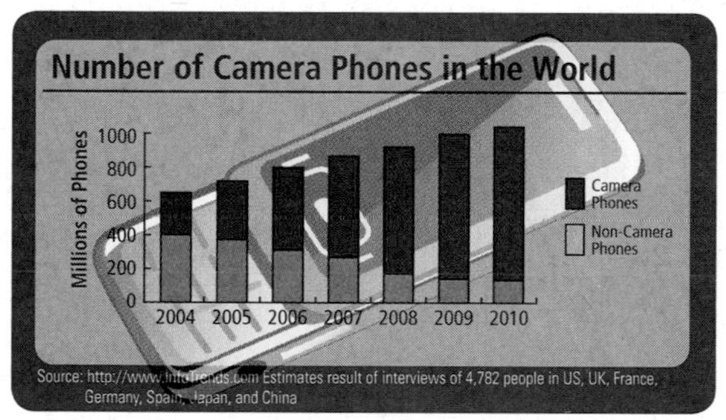

48. Light Bulbs The chart shows a comparison of power usage between incandescent and energy efficient light bulbs. The line segments from 1,600 to 2,600 lumens can be given by the equations $y = \left(\frac{1}{20}\right)x + 20$ and $y = \left(\frac{3}{200}\right)x + 1$. Find where these two equations intersect.

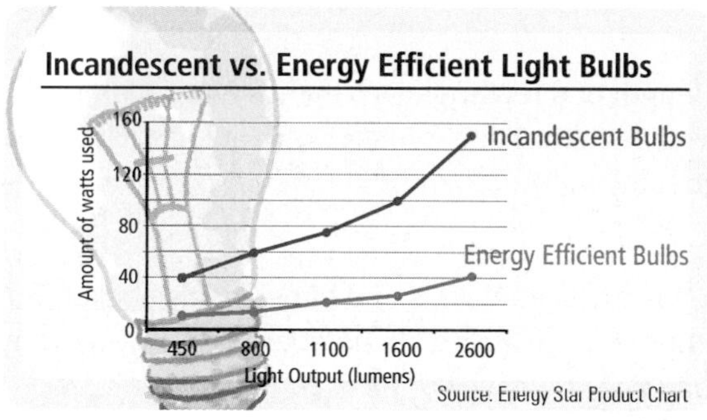

49. Gas Mileage Daniel is trying to decide whether to buy a car or a truck. The truck he is considering will cost him \$150 a month in loan payments, and it gets 20 miles per gallon in gas mileage. The car will cost \$180 a month in loan payments, but it gets 35 miles per gallon in gas mileage. Daniel estimates that he will pay \$1.40 per gallon for gas. This means that the monthly cost to drive the truck x miles will be $y = \frac{1.40}{20}x + 150$. The total monthly cost to drive the car x miles will be $y = \frac{1.40}{35}x + 180$. The following figure shows the graph of each equation.

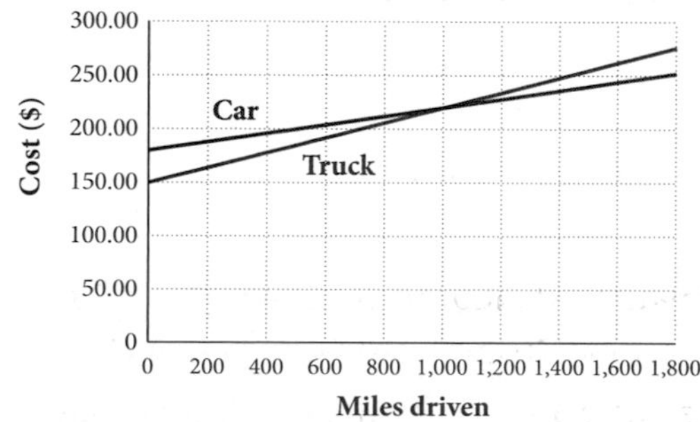

a. At how many miles do the car and the truck cost the same to operate?

b. If Daniel drives more than 1,200 miles, which will be cheaper?

c. If Daniel drives fewer than 800 miles, which will be cheaper?

d. Why do the graphs appear in the first quadrant only?

50. Video Production Pat runs a small company that duplicates DVD's. The daily cost and daily revenue for a company duplicating DVD's are shown in the following figure. The daily cost for duplicating x DVD's is $y = \frac{6}{5}x + 20$; the daily revenue (the amount of money he brings in each day) for duplicating x DVD's is $y = 1.7x$. The graphs of the two lines are shown in the following figure.

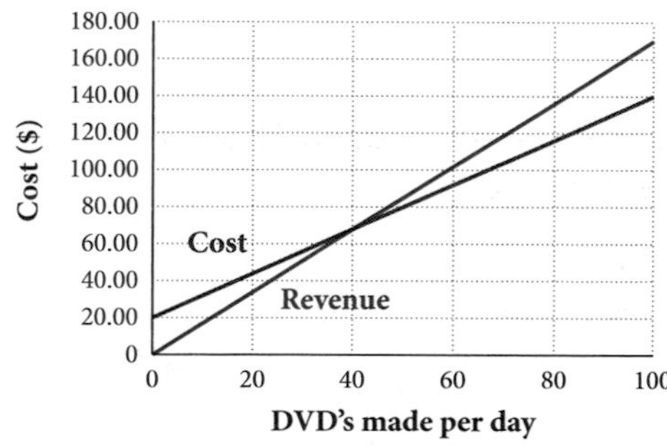

a. Pat will "break even" when his cost and his revenue are equal. How many DVD's does he need to duplicate to break even?

b. Pat will incur a loss when his revenue is less than his cost. If he duplicates 30 DVD's in one day, will he incur a loss?

c. Pat will make a profit when his revenue is larger than his costs. For what values of x will Pat make a profit?

d. Why does the graph appear in the first quadrant only?

Getting Ready for the Next Section

51. One number is eight more than five times another, their sum is 26. Find the numbers.

52. One number is three less than four times another; their sum is 27. Find the numbers.

53. The difference of two positive numbers is nine. The larger number is six less than twice the smaller number. Find the numbers.

54. The difference of two positive numbers is 17. The larger number is one more than twice the smaller number. Find the numbers.

55. The length of a rectangle is 5 inches more than three times the width. The perimeter is 58 inches. Find the length and width.

56. The length of a rectangle is 3 inches less than twice the width. The perimeter is 36 inches. Find the length and width.

57. John has $1.70 in nickels and dimes in his pocket. He has four more nickels than he does dimes. How many of each does he have?

58. Jamie has $2.65 in dimes and quarters in her pocket. She has two more dimes than she does quarters. How many of each does she have?

Solve the systems by any method.

59. $y = 5x + 2$
$x + y = 20$

60. $y = 2x + 3$
$x + y = 9$

61. $x + y = 15{,}000$
$0.06x + 0.07y = 980$

62. $x + y = 22$
$0.05x + 0.1y = 1.70$

Maintaining Your Skills

63. $6(3 + 4) + 5$

64. $[(1 + 2)(2 + 3)] + (4 \div 2)$

65. $1^2 + 2^2 + 3^2$

66. $(1 + 2 + 3)^2$

67. $5(6 + 3 \cdot 2) + 4 + 3 \cdot 2$

68. $(1 + 2)^3 + [(2 \cdot 3) + (4 \cdot 5)]$

69. $(1^3 + 2^3) + [(2 \cdot 3) + (4 \cdot 5)]$

70. $[2(3 + 4 + 5)] \div 3$

71. $(2 \cdot 3 + 4 + 5) \div 3$

72. $10^4 + 10^3 + 10^2 + 10^1$

73. $6 \cdot 10^3 + 5 \cdot 10^2 + 4 \cdot 10^1$

74. $5 \cdot 10^3 + 2 \cdot 10^2 + 8 \cdot 10^1$

75. $1 \cdot 10^3 + 7 \cdot 10^2 + 6 \cdot 10^1 + 0$

76. $4(2 - 1) + 5(3 - 2)$

77. $4 \cdot 2 - 1 + 5 \cdot 3 - 2$

78. $2^3 + 3^2 \cdot 4 - 5$

79. $(2^3 + 3^2) \cdot 4 - 5$

80. $4^2 - 2^4 + (2 \cdot 2)^2$

81. $2(2^2 + 3^2) + 3(3^2)$

82. $2 \cdot 2^2 + 3^2 + 3 \cdot 3^2$

Applications

4.4

Objectives

A Apply the Blueprint for Problem Solving to a variety of application problems involving systems of equations.

I often have heard students remark about the word problems in beginning algebra: "What does this have to do with real life?" Most of the word problems we will encounter don't have much to do with "real life." We are actually just practicing. Ultimately, all problems requiring the use of algebra are word problems; that is, they are stated in words first, then translated to symbols. The problem then is solved by some system of mathematics, like algebra. Most real applications involve calculus or higher levels of mathematics. So, if the problems we solve are upsetting or frustrating to you, then you probably are taking them too seriously.

The word problems in this section have two unknown quantities. We will write two equations in two variables (each of which represents one of the unknown quantities), which of course is a system of equations. We then solve the system by one of the methods developed in the previous sections of this chapter. Here are the steps to follow in solving these word problems.

A The Blueprint for Problem Solving

Blueprint: Problem Solving Using a System of Equations

Step 1: ***Read*** the problem, and then mentally ***list*** the items that are known and the items that are unknown.

Step 2: ***Assign variables*** to each of the unknown items; that is, let x = one of the unknown items and y = the other unknown item. Then ***translate*** the other ***information*** in the problem to expressions involving the two variables.

Step 3: ***Reread*** the problem, and then ***write a system of equations***, using the items and variables listed in steps 1 and 2, that describes the situation.

Step 4: ***Solve the system*** found in step 3.

Step 5: ***Write*** your ***answers*** using complete sentences.

Step 6: ***Reread*** the problem, and ***check*** your solution with the original words in the problem.

Remember, the more problems you work, the more problems you will be able to work. If you have trouble getting started on the problem set, come back to the examples and work through them yourself. The examples are similar to the problems found in the problem set.

Number Problem

EXAMPLE 1 One number is 2 more than 5 times another number. Their sum is 20. Find the two numbers.

SOLUTION We apply the steps in our blueprint.

Step 1: We know that the two numbers have a sum of 20 and that one of them is 2 more than 5 times the other. We don't know what the numbers themselves are.

PRACTICE PROBLEMS

1. One number is 3 more than twice another. Their sum is 9. Find the numbers.

Step 2: Let x represent one of the numbers and y represent the other. "One number is 2 more than 5 times another" translates to

$$y = 5x + 2$$

"Their sum is 20" translates to

$$x + y = 20$$

Step 3: The system that describes the situation must be

$$x + y = 20$$
$$y = 5x + 2$$

Step 4: We can solve this system by substituting the expression $5x + 2$ in the second equation for y in the first equation:

$$x + (5x + 2) = 20$$
$$6x + 2 = 20$$
$$6x = 18$$
$$x = 3$$

Using $x = 3$ in either of the first two equations and then solving for y, we get $y = 17$.

Step 5: So 17 and 3 are the numbers we are looking for.

Step 6: The number 17 is 2 more than 5 times 3, and the sum of 17 and 3 is 20.

Note We are using the substitution method here because the system we are solving is one in which the substitution method is the more convenient method.

Interest Problem

EXAMPLE 2 Mr. Hicks had \$15,000 to invest. He invested part at 6% and the rest at 7%. If he earns \$980 in interest, how much did he invest at each rate?

SOLUTION Remember, step 1 is done mentally.

Step 1: We do not know the specific amounts invested in the two accounts. We do know that their sum is \$15,000 and that the interest rates on the two accounts are 6% and 7%.

Step 2: Let x = the amount invested at 6% and y = the amount invested at 7%. Because Mr. Hicks invested a total of \$15,000, we have

$$x + y = 15{,}000$$

The interest he earns comes from 6% of the amount invested at 6% and 7% of the amount invested at 7%. To find 6% of x, we multiply x by 0.06, which gives us $0.06x$. To find 7% of y, we multiply 0.07 times y and get $0.07y$.

Interest at 6%	+	interest at 7%	=	total interest
$0.06x$	+	$0.07y$	=	980

Step 3: The system is

$$x + y = 15{,}000$$
$$0.06x + 0.07y = 980$$

2. Amy has \$10,000 to invest. She invests part at 6% and the rest at 7%. If she earns \$630 in interest for the year, how much does she have invested at each rate?

Answer
1. 2 and 7

Step 4: We multiply the first equation by −6 and the second by 100 to eliminate x:

$$\begin{aligned} x + \quad y &= 15{,}000 \xrightarrow{\text{Multiply by } -6} -6x - 6y = -90{,}000 \\ 0.06x + 0.07y &= 980 \xrightarrow[\text{Multiply by 100}]{} \underline{6x + 7y = 98{,}000} \\ & \qquad\qquad\qquad\qquad y = 8{,}000 \end{aligned}$$

Substituting $y = 8{,}000$ into the first equation and solving for x, we get $x = 7{,}000$.

Step 5: He invested \$7,000 at 6% and \$8,000 at 7%.

Step 6: Checking our solutions in the original problem, we have: The sum of \$7,000 and \$8,000 is \$15,000, the total amount he invested. To complete our check, we find the total interest earned from the two accounts:

The interest on \$7,000 at 6% is 0.06(7,000) = 420
The interest on \$8,000 at 7% is 0.07(8,000) = 560
The total interest is \$980

> **Note** In this case we are using the elimination method. Notice also that multiplying the second equation by 100 clears it of decimals.

Coin Problem

EXAMPLE 3 John has \$1.70 all in dimes and nickels. He has a total of 22 coins. How many of each kind does he have?

SOLUTION

Step 1: We know that John has 22 coins that are dimes and nickels. We know that a dime is worth 10 cents and a nickel is worth 5 cents. We do not know the specific number of dimes and nickels he has.

Step 2: Let x = the number of nickels and y = the number of dimes. The total number of coins is 22, so

$$x + y = 22$$

The total amount of money he has is \$1.70, which comes from nickels and dimes:

$$\begin{matrix} \text{Amount of money in nickels} & + & \text{amount of money in dimes} & = & \text{total amount of money} \\ 0.05x & + & 0.10y & = & 1.70 \end{matrix}$$

Step 3: The system that represents the situation is

$$\begin{aligned} x + y &= 22 && \text{The number of coins} \\ 0.05x + 0.10y &= 1.70 && \text{The value of the coins} \end{aligned}$$

Step 4: We multiply the first equation by −5 and the second by 100 to eliminate the variable x:

$$\begin{aligned} x + y &= \ 22 \xrightarrow{\text{Multiply by } -5} -5x - 5y = -110 \\ 0.05x + 0.10y &= 1.70 \xrightarrow[\text{Multiply by 100}]{} \underline{5x + 10y = 170} \\ & \qquad\qquad\qquad\qquad 5y = 60 \\ & \qquad\qquad\qquad\qquad y = 12 \end{aligned}$$

Substituting $y = 12$ into our first equation, we get $x = 10$.

3. Patrick has \$1.85 all in dimes and quarters. He has a total of 14 coins. How many of each kind does he have?

Answer
2. \$3,000 at 7%, \$7,000 at 6%

4. How much 30% alcohol solution and 60% alcohol solution must be mixed to get 25 gallons of 48% solution?

Step 5: John has 12 dimes and 10 nickels.

Step 6: Twelve dimes and 10 nickels total 22 coins.

$$\begin{aligned} \text{12 dimes are worth } 12(0.10) &= 1.20 \\ \underline{\text{10 nickels are worth } 10(0.05)} &\underline{= 0.50} \\ \text{The total value is } \$1.70 \end{aligned}$$

Mixture Problem

EXAMPLE 4 How much 20% alcohol solution and 50% alcohol solution must be mixed to get 12 gallons of 30% alcohol solution?

SOLUTION To solve this problem we must first understand that a 20% alcohol solution is 20% alcohol and 80% water.

Step 1: We know there are two solutions that together must total 12 gallons. 20% of one of the solutions is alcohol and the rest is water, whereas the other solution is 50% alcohol and 50% water. We do not know how many gallons of each individual solution we need.

Step 2: Let x = the number of gallons of 20% alcohol solution needed and y = the number of gallons of 50% alcohol solution needed. Because the total number of gallons we will end up with is 12, and this 12 gallons must come from the two solutions we are mixing, our first equation is

$$x + y = 12$$

To obtain our second equation, we look at the amount of alcohol in our two original solutions and our final solution. The amount of alcohol in the x gallons of 20% solution is $0.20x$, and the amount of alcohol in y gallons of 50% solution is $0.50y$. The amount of alcohol in the 12 gallons of 30% solution is $0.30(12)$. Because the amount of alcohol we start with must equal the amount of alcohol we end up with, our second equation is

$$0.20x + 0.50y = 0.30(12)$$

The information we have so far can also be summarized with a table. Sometimes by looking at a table like the one that follows it is easier to see where the equations come from.

	20% Solution	50% Solution	Final Solution
Number of Gallons	x	y	12
Gallons of Alcohol	$0.20x$	$0.50y$	$0.30(12)$

Step 3: Our system of equations is

$$x + y = 12$$
$$0.20x + 0.50y = 0.30(12)$$

Answer
3. 3 quarters, 11 dimes

Step 4: We can solve this system by substitution. Solving the first equation for y and substituting the result into the second equation, we have

$$0.20x + 0.50(12 - x) = 0.30(12)$$

Multiplying each side by 10 gives us an equivalent equation that is a little easier to work with.

$$2x + 5(12 - x) = 3(12)$$
$$2x + 60 - 5x = 36$$
$$-3x + 60 = 36$$
$$-3x = -24$$
$$x = 8$$

If x is 8, then y must be 4 because $x + y = 12$.

Step 5: It takes 8 gallons of 20% alcohol solution and 4 gallons of 50% alcohol solution to produce 12 gallons of 30% alcohol solution.

Step 6: Try it and see.

Getting Ready for Class

After reading through the preceding section, respond in your own words and in complete sentences.

1. If you were to apply the Blueprint for Problem Solving from Section 2.6 to the examples in this section, what would be the first step?
2. If you were to apply the Blueprint for Problem Solving from Section 2.6 to the examples in this section, what would be the last step?
3. Which method of solving these systems do you prefer? Why?
4. Write an application problem for which the solution depends on solving a system of equations.

Answer

4. 10 gallons of 30%, 15 gallons of 60%

Problem Set 4.4

A Solve the following word problems. Be sure to show the equations used.

Number Problems [Example 1]

1. Two numbers have a sum of 25. One number is 5 more than the other. Find the numbers.

2. The difference of two numbers is 6. Their sum is 30. Find the two numbers.

3. The sum of two numbers is 15. One number is 4 times the other. Find the numbers.

4. The difference of two positive numbers is 28. One number is 3 times the other. Find the two numbers.

5. Two positive numbers have a difference of 5. The larger number is one more than twice the smaller. Find the two numbers.

6. One number is 2 more than 3 times another. Their sum is 26. Find the two numbers.

7. One number is 5 more than 4 times another. Their sum is 35. Find the two numbers.

8. The difference of two positive numbers is 8. The larger is twice the smaller decreased by 7. Find the two numbers.

Interest Problems [Example 2]

9. Mr. Wilson invested money in two accounts. His total investment was $20,000. If one account pays 6% in interest and the other pays 8% in interest, how much does he have in each account if he earned a total of $1,380 in interest in 1 year?

10. A total of $11,000 was invested. Part of the $11,000 was invested at 4%, and the rest was invested at 7%. If the investments earn $680 per year, how much was invested at each rate?

11. A woman invested 4 times as much at 5% as she did at 6%. The total amount of interest she earns in 1 year from both accounts is $520. How much did she invest at each rate?

12. Ms. Hagan invested twice as much money in an account that pays 7% interest as she did in an account that pays 6% in interest. Her total investment pays her $1,000 a year in interest. How much did she invest at each rate?

Coin Problems [Example 3]

13. Ron has 14 coins with a total value of \$2.30. The coins are nickels and quarters. How many of each coin does he have?

14. Diane has \$0.95 in dimes and nickels. She has a total of 11 coins. How many of each kind does she have?

15. Suppose Tom has 21 coins totaling \$3.45. If he has only dimes and quarters, how many of each type does he have?

16. A coin collector has 31 dimes and nickels with a total face value of \$2.40. (They are actually worth a lot more.) How many of each coin does she have?

Mixture Problems [Example 4]

17. How many liters of 50% alcohol solution and 20% alcohol solution must be mixed to obtain 18 liters of 30% alcohol solution?

	50% Solution	20% Solution	Final Solution
Number of Liters	x	y	
Liters of Alcohol			

18. How many liters of 10% alcohol solution and 5% alcohol solution must be mixed to obtain 40 liters of 8% alcohol solution?

	10% Solution	5% Solution	Final Solution
Number of Liters	x	y	
Liters of Alcohol			

19. A mixture of 8% disinfectant solution is to be made from 10% and 7% disinfectant solutions. How much of each solution should be used if 30 gallons of 8% solution are needed?

20. How much 50% antifreeze solution and 40% antifreeze solution should be combined to give 50 gallons of 46% antifreeze solution?

Miscellaneous Problems

21. For a Saturday matinee, adult tickets cost \$5.50 and kids under 12 pay only \$4.00. If 70 tickets are sold for a total of \$310, how many of the tickets were adult tickets and how many were sold to kids under 12?

22. The Bishop's Peak 4-H club is having its annual fundraising dinner. Adults pay \$15 apiece and children pay \$10 apiece. If the number of adult tickets sold is twice the number of children's tickets sold, and the total income for the dinner is \$1,600, how many of each kind of ticket did the 4-H club sell?

23. A farmer has 96 feet of fence with which to make a corral. If he arranges it into a rectangle that is twice as long as it is wide, what are the dimensions?

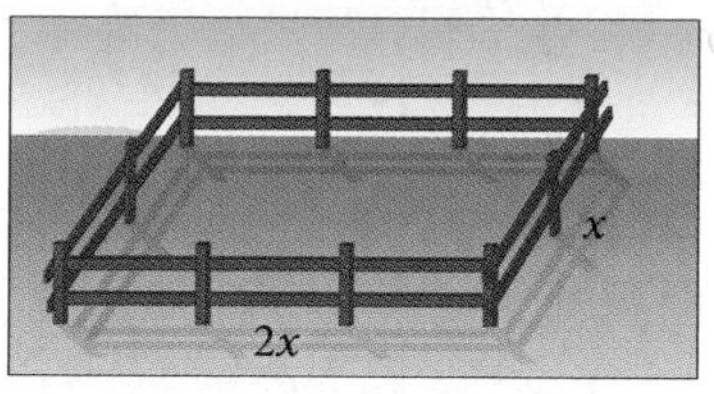

24. If a 22-inch rope is to be cut into two pieces so that one piece is 3 inches longer than twice the other, how long is each piece?

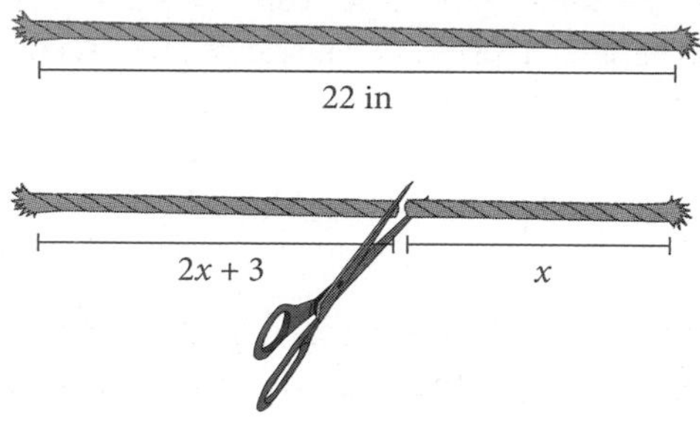

25. A gambler finishes a session of blackjack with \$5 chips and \$25 chips. If he has 45 chips in all, with a total value of \$465, how many of each kind of chip does the gambler have?

26. Tyler has been saving his winning lottery tickets. He has 23 tickets that are worth a total of \$175. If each ticket is worth either \$5 or \$10, how many of each does he have?

27. Mary Jo spends \$2,550 to buy stock in two companies. She pays \$11 a share to one of the companies and \$20 a share to the other. If she ends up with a total of 150 shares, how many shares did she buy at \$11 a share and how many did she buy at \$20 a share?

28. Kelly sells 62 shares of stock she owns for a total of \$433. If the stock was in two different companies, one selling at \$6.50 a share and the other at \$7.25 a share, how many of each did she sell?

Applying the Concepts

29. **Solar and Wind Energy** The chart shows the cost to install either solar panels or a wind turbine. A company that specializes in installing solar panels bought a number of modules and fixed racks to upgrade some their costumers. The company bought 8 items for a total of \$21,820. How many modules and how many fixed racks did they buy?

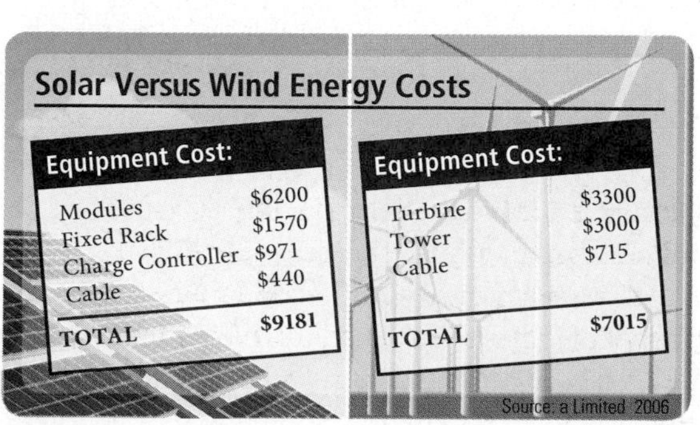

30. **Classroom Energy** The chart shows how much energy is wasted in the classroom by leaving appliances on. A certain number of printers and ceiling fans were left on in the computer lab. If the number of appliances left on was 7 and the total amount of energy wasted per hour was 2,250 watts, how many of each appliance was left on?

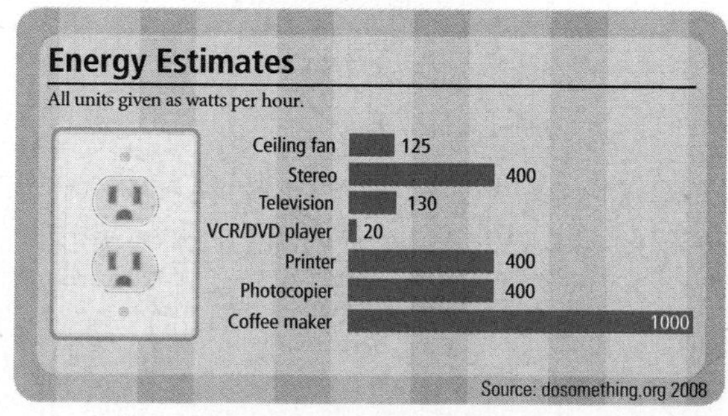

Maintaining Your Skills

31. Fill in each ordered pair so that it is a solution to $y = \frac{1}{2}x + 3$. $(-2,\ ),\ (0,\ ),\ (2,\ )$

32. Graph the line $y = \frac{1}{2}x + 3$.

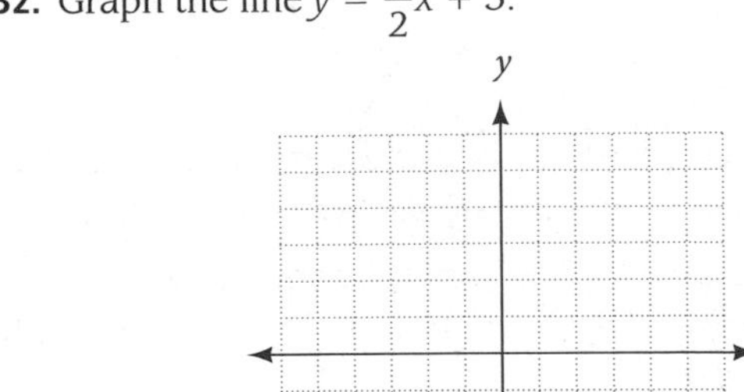

33. Graph the line $x = -2$.

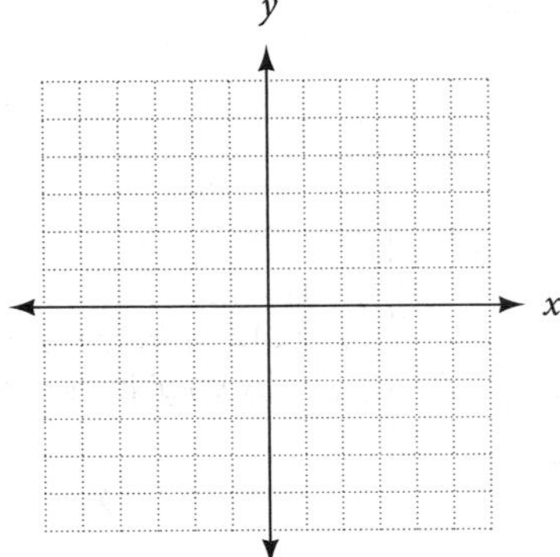

34. Graph $3x - 2y = 6$.

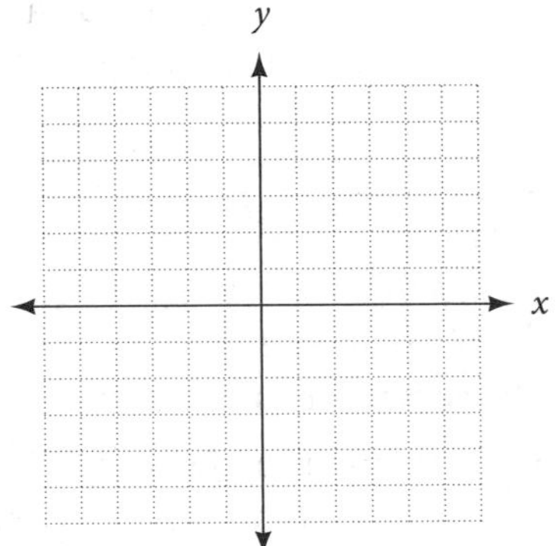

35. Find the slope of the line through (2, 5) and (0, 1).

36. Find the slope and y-intercept for the line $2x - 5y = 10$.

37. Find the equation of the line through $(-2, 1)$ with slope $\frac{1}{2}$.

38. Write the equation of the line with slope -2 and y-intercept $\frac{3}{2}$.

39. Find the equation of the line through (2, 5) and (0, 1).

40. Graph the solution set for $2x - y < 4$.

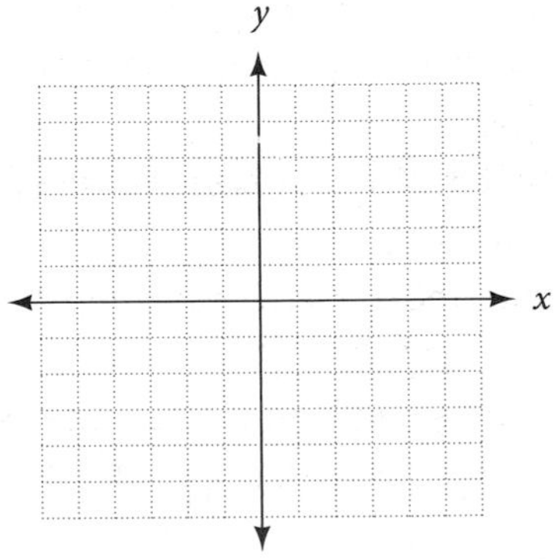

Chapter 4 Summary

Definitions [4.1]

1. A *system of linear equations,* as the term is used in this book, is two linear equations that each contain the same two variables.
2. The *solution set* for a system of equations is the set of all ordered pairs that satisfy *both* equations. The solution set to a system of linear equations will contain:

   Case I One ordered pair when the graphs of the two equations intersect at only one point (this is the most common situation)

   Case II No ordered pairs when the graphs of the two equations are parallel lines

   Case III An infinite number of ordered pairs when the graphs of the two equations coincide (are the same line)

EXAMPLES

1. The solution to the system

$$x + 2y = 4$$
$$x - y = 1$$

is the ordered pair (2, 1). It is the only ordered pair that satisfies both equations.

Strategy for Solving a System by Graphing [4.1]

Step 1: Graph the first equation.

Step 2: Graph the second equation on the same set of axes.

Step 3: Read the coordinates of the point where the graphs cross each other (the coordinates of the point of intersection).

Step 4: Check the solution to see that it satisfies *both* equations.

2. Solving the system in Example 1 by graphing looks like

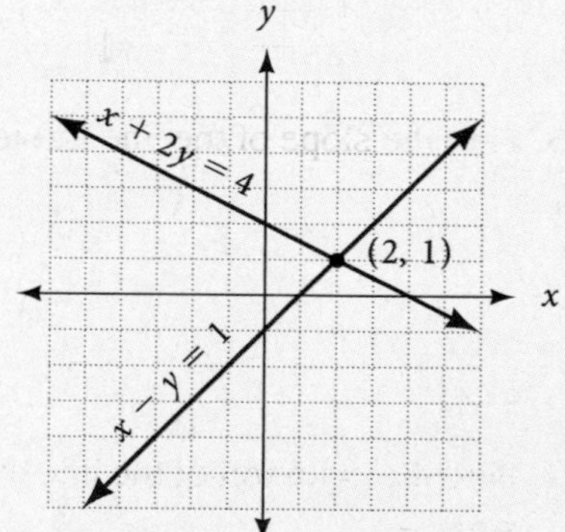

Strategy for Solving a System by the Elimination Method [4.2]

Step 1: Look the system over to decide which variable will be easier to eliminate.

Step 2: Use the multiplication property of equality on each equation separately to ensure that the coefficients of the variable to be eliminated are opposites.

Step 3: Add the left and right sides of the system produced in step 2, and solve the resulting equation.

Step 4: Substitute the solution from step 3 back into any equation with both x and y variables, and solve.

Step 5: Check your solution in both equations, if necessary.

3. We can eliminate the y variable from the system in Example 1 by multiplying both sides of the second equation by 2 and adding the result to the first equation

$$\begin{array}{lcr} x + 2y = 4 & & x + 2y = 4 \\ x - y = 1 & \xrightarrow{\text{Multiply by 2}} & 2x - 2y = 2 \\ \hline & & 3x \quad = 6 \\ & & x \quad = 2 \end{array}$$

Substituting $x = 2$ into either of the original two equations gives $y = 1$. The solution is (2, 1).

Strategy for Solving a System by the Substitution Method [4.3]

4. We can apply the substitution method to the system in Example 1 by first solving the second equation for x to get $x = y + 1$. Substituting this expression for x into the first equation, we have

$$\begin{aligned}(y + 1) + 2y &= 4\\ 3y + 1 &= 4\\ 3y &= 3\\ y &= 1\end{aligned}$$

Using $y = 1$ in either of the original equations gives $x = 2$.

Step 1: Solve either of the equations for one of the variables (this step is not necessary if one of the equations has the correct form already).

Step 2: Substitute the results of step 1 into the other equation, and solve.

Step 3: Substitute the results of step 2 into an equation with both x and y variables, and solve. (The equation produced in step 1 is usually a good one to use.)

Step 4: Check your solution, if necessary.

Special Cases [4.1, 4.2, 4.3]

In some cases, using the elimination or substitution method eliminates both variables. The situation is interpreted as follows.

1. If the resulting statement is *false,* then the lines are parallel and there is no solution to the system.
2. If the resulting statement is *true,* then the equations represent the same line (the lines coincide). In this case any ordered pair that satisfies either equation is a solution to the system.

COMMON MISTAKES

The most common mistake encountered in solving linear systems is the failure to complete the problem. Here is an example.

$$\begin{aligned}x + y &= 8\\ x - y &= 4\\ 2x &= 12\\ x &= 6\end{aligned}$$

This is only half the solution. To find the other half, we must substitute the 6 back into one of the original equations and then solve for y.

Remember, solutions to systems of linear equations always consist of ordered pairs. We need an x-coordinate and a y-coordinate; $x = 6$ can never be a solution to a system of linear equations.

Chapter 4 Review

Solve the following systems by graphing. [4.1]

1. $x + y = 2$
   $x - y = 6$

2. $x + y = -1$
   $-x + y = 5$

3. $2x - 3y = 12$
   $-2x + y = -8$

4. $4x - 2y = 8$
   $3x + y = 6$

5. $y = 2x - 3$
   $y = -2x + 5$

6. $y = -x - 3$
   $y = 3x + 1$

Solve the following systems by the elimination method. [4.2]

7. $x - y = 4$
   $x + y = -2$

8. $-x - y = -3$
   $2x + y = 1$

9. $5x - 3y = 2$
   $-10x + 6y = -4$

10. $2x + 3y = -2$
    $3x - 2y = 10$

11. $-3x + 4y = 1$
    $-4x + y = -3$

12. $-4x - 2y = 3$
    $2x + y = 1$

13. $2x + 5y = -11$
    $7x - 3y = -5$

14. $-2x + 5y = -15$
    $3x - 4y = 19$

Solve the following systems by substitution. [4.3]

15. $x + y = 5$
    $y = -3x + 1$

16. $x - y = -2$
    $y = -2x - 10$

17. $4x - 3y = -16$
    $y = 3x + 7$

18. $5x + 2y = -2$
    $y = -8x + 10$

19. $x - 4y = 2$
    $-3x + 12y = -8$

20. $4x - 2y = 8$
    $3x + y = -19$

21. $10x - 5y = 20$
    $x + 6y = -11$

22. $3x - y = 2$
    $-6x + 2y = -4$

Solve the following word problems. Be sure to show the equations used. [4.4]

23. **Number Problem** The sum of two numbers is 18. If twice the smaller number is 6 more than the larger, find the two numbers.

24. **Number Problem** The difference of two positive numbers is 16. One number is 3 times the other. Find the two numbers.

25. **Investing** A total of $12,000 was invested. Part of the $12,000 was invested at 4%, and the rest was invested at 5%. If the interest for one year is $560, how much was invested at each rate?

26. **Investing** A total of $14,000 was invested. Part of the $14,000 was invested at 6%, and the rest was invested at 8%. If the interest for one year is $1,060, how much was invested at each rate?

27. **Coin Problem** Barbara has $1.35 in dimes and nickels. She has a total of 17 coins. How many of each does she have?

28. **Coin Problem** Tom has $2.40 in dimes and quarters. He has a total of 15 coins. How many of each does he have?

29. **Mixture Problem** How many liters of 20% alcohol solution and 10% alcohol solution must be mixed to obtain 50 liters of a 12% alcohol solution?

30. **Mixture Problem** How many liters of 25% alcohol solution and 15% alcohol solution must be mixed to obtain 40 liters of a 20% alcohol solution?

Chapter 4 Cumulative Review

Simplify.

1. $3 \cdot 4 + 5$
2. $8 - 6(5 - 9)$
3. $4 \cdot 3^2 + 4(6 - 3)$
4. $7[8 + (-5)] + 3(-7 + 12)$
5. $\dfrac{12 - 3}{8 - 8}$
6. $\dfrac{5(4 - 12) - 8(14 - 3)}{3 - 5 - 6}$
7. $\dfrac{11}{60} - \dfrac{13}{84}$
8. $\dfrac{2}{3} + \dfrac{3}{4} - \dfrac{1}{6}$
9. $2(x - 5) + 8$
10. $7 - 5(2a - 3) + 7$

Solve each equation.

11. $-5 - 6 = -y - 3 + 2y$
12. $-2x = 0$
13. $3(x - 4) = 9$
14. $8 - 2(y + 4) = 12$

Solve each inequality, and graph the solution.

15. $0.3x + 0.7 \le -2$
16. $5x + 10 \le 7x - 14$

Graph on a rectangular coordinate system.

17. $y = -2x + 1$
18. $y = -\dfrac{2}{3}x$

Solve each system by graphing.

19. $2x + 3y = 3$
    $4x + 6y = -4$
20. $3x - 3y = 3$
    $2x + 3y = 2$

Solve each system.

21. $x + y = 7$
    $2x + 2y = 14$
22. $2x + y = -1$
    $2x - 3y = 11$
23. $2x + 3y = 13$
    $x - y = -1$
24. $x + y = 13$
    $0.05x + 0.10y = 1$
25. $2x + 5y = 33$
    $x - 3y = 0$
26. $3x + 4y = 8$
    $x - y = 5$
27. $3x - 7y = 12$
    $2x + y = 8$
28. $5x + 6y = 9$
    $x - 2y = 5$
29. $2x - 3y = 7$
    $y = 5x + 2$
30. $3x - 6y = 9$
    $x = 2y + 3$

Find the next number in each sequence.

31. 4, 1, −2, −5, . . .
32. 2, 4, 8, 16, . . .
33. What is the quotient of −30 and 6?
34. Subtract 8 from −9.

Factor into primes.

35. 180
36. 300

Find the value of each expression when x is 3.

37. $-3x + 7 + 5x$
38. $-x - 4x - 2x$
39. Given $4x - 5y = 12$, complete the ordered pair $(-2,\ \ )$.
40. Given $y = \frac{3}{5}x - 2$, complete the ordered pairs $(2,\ \ )$ and $(\ \ , -1)$.
41. Find the x- and y-intercepts for the line $3x - 4y = 12$.
42. Find the slope of the line $x = 3y - 6$.

Find the slope of the line through the given pair of points.

43. $(-1, 1), (-5, -4)$
44. $\left(6, \dfrac{1}{2}\right), \left(-3, \dfrac{5}{4}\right)$
45. Find the equation of the line with slope $\frac{2}{3}$ and y-intercept 3.
46. Find the equation of the line with slope 2 if it passes through (4, 1).
47. Find the equation of the line through (4, 3) and (6, 6).
48. Find the x- and y-intercepts for the line that passes through the points (−2, 3) and (6, −1).
49. **Coin Problem** Joy has 15 coins with a value of $1.10. The coins are nickels and dimes. How many of each does she have?
50. **Investing** I have invested money in two accounts. One account pays 5% annual interest, and the other pays 6%. I have $200 more in the 6% account than I have in the 5% account. If the total amount of interest was $56, how much do I have in each account?

Use the pie chart to answer the following questions.

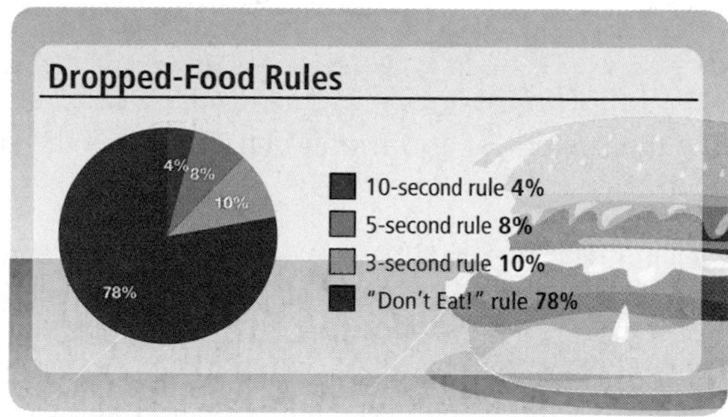

51. If 200 people were surveyed, how many said they don't eat food dropped on the floor?
52. If 20 people said they would eat food that stays on the floor for five seconds, how many people were surveyed?

Chapter 4 Test

1. Write the solution to the system which is graphed below. [4.1]

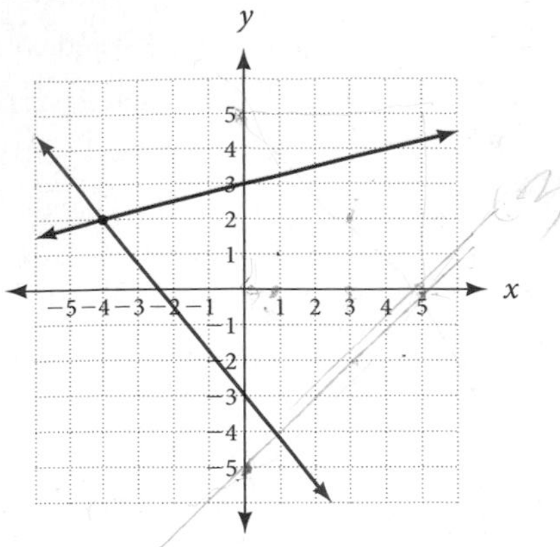

Solve each system by graphing. [4.1]

2. $x + 2y = 5$
   $y = 2x$

3. $x - y = 5$
   $x = 3$

4. $2x - y = 1$
   $4x - 2y = -2$

Solve each system by the elimination method. [4.2]

5. $x - y = 1$
   $2x + y = -10$

6. $2x + y = 7$
   $3x + y = 12$

7. $7x + 8y = -2$
   $3x - 2y = 10$

8. $6x - 10y = 6$
   $9x - 15y = 9$

Solve each system by the substitution method. [4.3]

9. $3x + 2y = 20$
   $y = 2x + 3$

10. $3x - 6y = -6$
    $x = y + 1$

11. $7x - 2y = -4$
    $-3x + y = 3$

12. $2x - 3y = -7$
    $x + 3y = -8$

Solve the following word problems. In each case, be sure to show the system of equations that describes the situation. [4.4]

13. **Number Problem** The sum of two numbers is 12. Their difference is 2. Find the numbers.

14. **Number Problem** The sum of two numbers is 15. One number is 6 more than twice the other. Find the two numbers.

15. **Investing** Dr. Stork has \$10,000 to invest. He would like to earn \$980 per year in interest. How much should he invest at 9% if the rest is to be invested at 11%?

16. **Coin Problem** Diane has 12 coins that total \$1.60. If the coins are all nickels and quarters, how many of each type does she have?

17. **Perimeter Problem** A rancher wants to build a rectangular corral using 170 feet of fence. If the length of the corral is to be 10 feet longer than twice the width, find the dimensions of the corral.

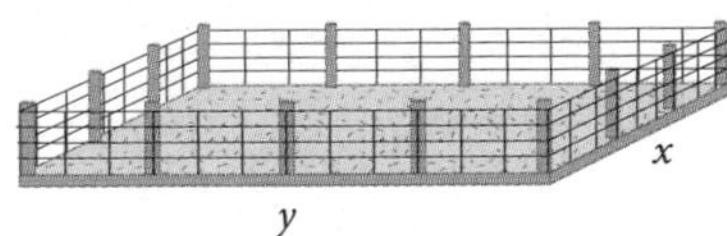

18. **Moving Truck Rental** One moving truck company charges \$40 per day plus \$0.80 per mile to rent a 26-foot truck for local use. Another company charges \$70 per day plus \$0.60 per mile. The total costs to use the trucks for a day and drive x miles are

$$\text{Company 1: } y = 0.80x + 40$$

$$\text{Company 2: } y = 0.60x + 70$$

The graphs of these equations are shown below. [4.3]

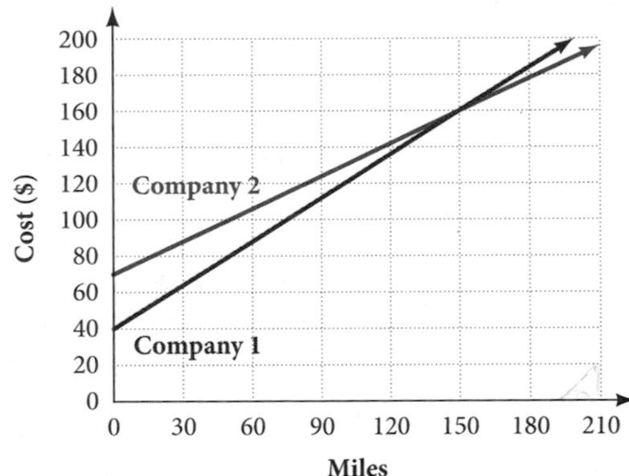

a. At how many miles do the two companies charge the same amount for the truck rental?

b. If the truck is driven less than 100 miles, which company would charge less?

c. For which values of x will Company 2 charge less than Company 1?

Chapter 4 Projects

SYSTEMS OF LINEAR EQUATIONS

GROUP PROJECT

Tables and Graphs

Number of People 2–3

Time Needed 8–12 minutes

Equipment Pencil, ruler, and graph paper

Background In the Chapter 2 project, we built the table below. As you may recall, we built the table based on the different combinations of coins from a parking meter that accepts only dimes and quarters. The amount of money in the meter at the end of one particular day was $3.15.

Number of Dimes	Number of Quarters	Total Coins	Value
29	1	30	$3.15
24	3	27	$3.15
19	5	24	$3.15
14	7	21	$3.15
9	9	18	$3.15
4	11	15	$3.15
x	y	$x + y$	

Procedure

1. From each row in the table, plot the points (x, y) where x is the number of dimes and y is the number of quarters. For example, the first point to plot is (29, 1). Then connect the points by drawing a straight line through them. The line should pass through all the points.

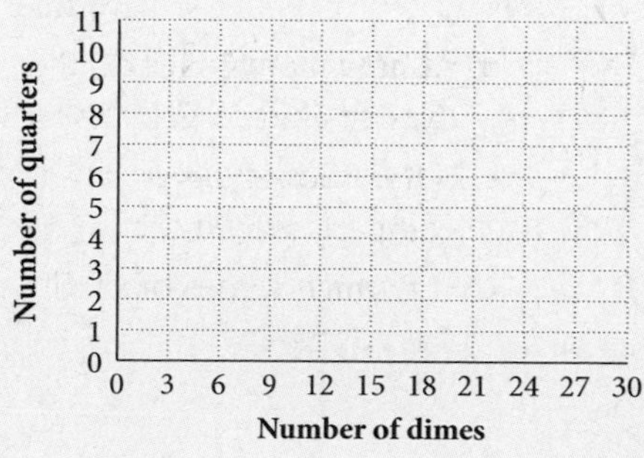

2. What is the equation of the line that passes through all the points?
3. Draw six additional lines on the same grid using the following equations:

   $x + y = 30$, $x + y = 27$, $x + y = 24$, $x + y = 21$, $x + y = 18$, and $x + y = 15$.

4. Write a parking meter application problem, the solution to which is the pair of lines $10x + 25y = 315$ and $x + y = 24$.

RESEARCH PROJECT

Cartesian Coordinate System

The stamp shown here was issued by Albania in 1966. It shows the French philosopher and mathematician Rene Descartes. As we mentioned earlier in this chapter, Descartes is credited with the discovery of the rectangular coordinate system. (Notice the coordinate system in the background of the stamp.)

Descartes insisted that his best theories came to him while resting in bed. He once said, "the only way to do good work in mathematics and preserve one's health is never to get up in the morning before one feels inclined to do so." One story of how he came to develop the Cartesian (from Des*cartes*) coordinate system is as follows: One morning, while lying in bed, he noticed a fly crawling on the ceiling. After studying it, he realized he could state the fly's position on the ceiling by giving its distance from each of the edges of the ceiling. Research this story and then put your results into an essay that shows the connection between the position of the fly on the ceiling and the coordinates of points in our rectangular coordinate system.

Exponents and Polynomials

Chapter Outline

Introduction

The French mathematician and philosopher, Blaise Pascal, was born in France in 1623. Both a scientist mathematician, he is credited for defining the scientific method. In the image below, Pascal carries a barometer to the top of the bell tower at the church of Saint-Jacques-de-la-Boucherie, overlooking Paris, to test a scientific theory.

Mary Evans Picture Library/Alamy

1
1 1
1 2 1
1 3 3 1
1 4 6 4 1
1 5 10 10 5 1
1 6 15 20 15 6 1

The triangular array of numbers to the right of the painting is called Pascal's triangle. Pascal's triangle is connected to the work we will do in this chapter when we find increasing powers of binomials.

Chapter Pretest

Simplify each of the following expressions. [5.1]

1. $(-2)^5$

2. $\left(\frac{2}{3}\right)^3$

3. $(4x^2)^2(2x^3)^3$

Simplify each expression. Write all answers with positive exponents only. [5.2]

4. 4^{-2}

5. $(4a^5b^3)^0$

6. $\frac{x^{-4}}{x^{-7}}$

7. $\frac{(x^{-3})^2(x^{-5})^{-3}}{(x^{-3})^{-4}}$

8. Write 0.04307 in scientific notation. [5.2]

9. Write 7.63×10^6 in expanded form. [5.1]

Simplify. Write all answers with positive exponents only. [5.3]

10. $\frac{17x^2y^5z^3}{51x^4y^2z}$

11. $\frac{(3a^3b)(4a^2b^5)}{24a^2b^4}$

12. $\frac{28x^4}{4x} + \frac{30x^7}{6x^4}$

13. $\frac{(1.1 \times 10^5)(3 \times 10^{-2})}{4.4 \times 10^{-5}}$

Add or subtract as indicated. [5.4]

14. $9x^2 - 2x + 7x + 4$

15. $(4x^2 + 5x - 6) - (2x^2 - x - 4)$

16. Subtract $2x + 7$ from $7x + 3$. [5.4]

17. Find the value of $3a^2 + 4a + 6$ when a is -3. [5.4]

Multiply. [5.5]

18. $3x^2(5x^2 - 2x + 4)$

19. $\left(x + \frac{1}{4}\right)\left(x - \frac{1}{3}\right)$

20. $(2x - 3)(5x + 6)$

21. $(x + 4)(x^2 - 4x + 16)$

Multiply. [5.6]

22. $(x - 6)^2$

23. $(2a + 4b)^2$

24. $(3x - 6)(3x + 6)$

25. $(x^2 - 4)(x^2 + 4)$

26. Divide $18x^3 - 36x^2 + 6x$ by$6x$. [5.7]

Divide. [5.8]

27. $\frac{9x^2 - 6x - 4}{3x - 1}$

28. $\frac{4x^2 - 5x + 6}{x - 4}$

Getting Ready for Chapter 5

Simplify.

1. $\frac{1}{2} \cdot \frac{5}{7}$

2. $\frac{9.6}{3}$

3. $9 - 20$

4. $1 - 8$

5. $-8 - 2(3)$

6. $2(3) - 4 - 3(-4)$

7. $2(3) + 4(5) - 5(2)$

8. $3(-2)^2 - 5(-2) + 4$

9. $-(4x^2 - 2x - 6)$

10. $3x \cdot 3x$

11. $2(2x)(-3)$

12. $(3x)(2x)$

13. $-3x - 10x$

14. $x - 2x$

15. $(2x^3 + 0x^2) - (2x^3 - 10x^2)$

16. $(4x - 14) - (4x - 10)$

17. $-4x(x + 5)$

18. $-2x(2x + 7)$

19. $2x^2(x - 5)$

20. $10x(x - 5)$

Multiplication with Exponents

5.1

Objectives

A Use the definition of integer exponents to evaluate expressions containing exponents.
B Use Property 1 for exponents.
C Use Property 2 for exponents.
D Use Property 3 for exponents.
E Write numbers in scientific notation and expanded form.

XYZ
Examples now playing at
MathTV.com/books

A Evaluating Exponents

Recall that an *exponent* is a number written just above and to the right of another number, which is called the *base*. In the expression 5^2, for example, the exponent is 2 and the base is 5. The expression 5^2 is read "5 to the second power" or "5 squared." The meaning of the expression is

$$5^2 = 5 \cdot 5 = 25$$

In the expression 5^3, the exponent is 3 and the base is 5. The expression 5^3 is read "5 to the third power" or "5 cubed." The meaning of the expression is

$$5^3 = 5 \cdot 5 \cdot 5 = 125$$

Here are some further examples.

EXAMPLE 1

$$4^3 = 4 \cdot 4 \cdot 4 = 16 \cdot 4 = 64$$ Exponent 3, base 4

EXAMPLE 2

$$-3^4 = -3 \cdot 3 \cdot 3 \cdot 3 = -81$$ Exponent 4, base 3

EXAMPLE 3

$$(-2)^5 = (-2)(-2)(-2)(-2)(-2) = -32$$ Exponent 5, base -2

EXAMPLE 4

$$\left(-\frac{3}{4}\right)^2 = \left(-\frac{3}{4}\right)\left(-\frac{3}{4}\right) = \frac{9}{16}$$ Exponent 2, base $-\frac{3}{4}$

PRACTICE PROBLEMS

Simplify the following:

1. 5^3
2. -2^4
3. $(-3)^4$
4. $\left(-\frac{2}{3}\right)^2$

QUESTION: In what way are $(-5)^2$ and -5^2 different?
ANSWER: In the first case, the base is -5. In the second case, the base is 5. The answer to the first is 25. The answer to the second is -25. Can you tell why? Would there be a difference in the answers if the exponent in each case were changed to 3?

We can simplify our work with exponents by developing some properties of exponents. We want to list the things we know are true about exponents and then use these properties to simplify expressions that contain exponents.

B Property 1 and the Properties of Exponents

The first property of exponents applies to products with the same base. We can use the definition of exponents, as indicating repeated multiplication, to simplify expressions like $7^4 \cdot 7^2$.

$$7^4 \cdot 7^2 = (7 \cdot 7 \cdot 7 \cdot 7)(7 \cdot 7)$$
$$= (7 \cdot 7 \cdot 7 \cdot 7 \cdot 7 \cdot 7)$$
$$= 7^6$$ *Notice:* $4 + 2 = 6$

Answers

1. 125 2. -16 3. 81 4. $\frac{4}{9}$

As you can see, multiplication with the same base resulted in addition of exponents. We can summarize this result with the following property.

Property 1 for Exponents
If a is any real number and r and s are integers, then

$$a^r \cdot a^s = a^{r+s}$$

In words: To multiply two expressions with the same base, add exponents and use the common base.

Here are some examples using Property 1.

Simplify, leaving your answers in terms of exponents:

5. $5^4 \cdot 5^5$
6. $x^3 \cdot x^7$
7. $4^5 \cdot 4^2 \cdot 4^6$

EXAMPLES Use Property 1 to simplify the following expressions. Leave your answers in terms of exponents:

5. $5^3 \cdot 5^6 = 5^{3+6} = 5^9$
6. $x^7 \cdot x^8 = x^{7+8} = x^{15}$
7. $3^4 \cdot 3^8 \cdot 3^5 = 3^{4+8+5} = 3^{17}$

Note In Examples 5, 6, and 7, notice that in each case the base in the original problem is the same base that appears in the answer and that it is written only once in the answer. A very common mistake that people make when they first begin to use Property 1 is to write a 2 in front of the base in the answer. For example, people making this mistake would get $2x^{15}$ or $(2x)^{15}$ as the result in Example 6. To avoid this mistake, you must be sure you understand the meaning of Property 1 exactly as it is written.

C Property 2 and the Properties of Exponents

Another common type of expression involving exponents is one in which an expression containing an exponent is raised to another power. The expression $(5^3)^2$ is an example:

$$\begin{aligned}(5^3)^2 &= (5^3)(5^3)\\ &= 5^{3+3}\\ &= 5^6 \qquad \textit{Notice: } 3 \cdot 2 = 6\end{aligned}$$

This result offers justification for the second property of exponents.

Property 2 for Exponents
If a is any real number and r and s are integers, then

$$(a^r)^s = a^{r \cdot s}$$

In words: A power raised to another power is the base raised to the product of the powers.

Simplify:

8. $(3^4)^5$
9. $(x^7)^2$

EXAMPLES Simplify the following expressions:

8. $(4^5)^6 = 4^{5 \cdot 6} = 4^{30}$
9. $(x^3)^5 = x^{3 \cdot 5} = x^{15}$

D Property 3 and the Properties of Exponents

The third property of exponents applies to expressions in which the product of two or more numbers or variables is raised to a power. Let's look at how the expression $(2x)^3$ can be simplified:

$$\begin{aligned}(2x)^3 &= (2x)(2x)(2x)\\ &= (2 \cdot 2 \cdot 2)(x \cdot x \cdot x)\\ &= 2^3 \cdot x^3 \qquad \textit{Notice: } \text{The exponent 3 distributes over the product } 2x\\ &= 8x^3\end{aligned}$$

Answers
5. 5^9 6. x^{10} 7. 4^{13} 8. 3^{20} 9. x^{14}

We can generalize this result into a third property of exponents.

Property 3 for Exponents
If a and b are any two real numbers and r is an integer, then

$$(ab)^r = a^r b^r$$

In words: The power of a product is the product of the powers.

Here are some examples using Property 3 to simplify expressions.

Simplify:

EXAMPLE 10 Simplify the following expression: $(5x)^2$

SOLUTION $(5x)^2 = 5^2 \cdot x^2$ Property 3
$= 25x^2$

EXAMPLE 11 Simplify the following expression: $(2xy)^3$

SOLUTION $(2xy)^3 = 2^3 \cdot x^3 \cdot y^3$ Property 3
$= 8x^3y^3$

EXAMPLE 12 Simplify the following expression: $(3x^2)^3$

SOLUTION $(3x^2)^3 = 3^3(x^2)^3$ Property 3
$= 27x^6$ Property 2

EXAMPLE 13 Simplify the following expression: $\left(-\frac{1}{4}x^2y^3\right)^2$

SOLUTION $\left(-\frac{1}{4}x^2y^3\right)^2 = \left(-\frac{1}{4}\right)^2(x^2)^2(y^3)^2$ Property 3
$= \frac{1}{16}x^4y^6$ Property 2

EXAMPLE 14 Simplify the following expression: $(x^4)^3(x^2)^5$

SOLUTION $(x^4)^3(x^2)^5 = x^{12} \cdot x^{10}$ Property 2
$= x^{22}$ Property 1

EXAMPLE 15 Simplify the following expression: $(2y)^3(3y^2)$

SOLUTION $(2y)^3(3y^2) = 2^3y^3(3y^2)$ Property 3
$= 8 \cdot 3(y^3 \cdot y^2)$ Commutative and associative properties
$= 24y^5$ Property 1

EXAMPLE 16 Simplify the following expression: $(2x^2y^5)^3(3x^4y)^2$

SOLUTION $(2x^2y^5)^3(3x^4y)^2 = 2^3(x^2)^3(y^5)^3 \cdot 3^2(x^4)^2y^2$ Property 3
$= 8x^6y^{15} \cdot 9x^8y^2$ Property 2
$= (8 \cdot 9)(x^6x^8)(y^{15}y^2)$ Commutative and associative properties
$= 72x^{14}y^{17}$ Property 1

10. $(7x)^2$

11. $(3xy)^4$

12. $(4x^3)^2$

13. $\left(-\frac{1}{3}x^3y^2\right)^2$

14. $(x^5)^2(x^4)^3$

15. $(3y)^2(2y^3)$

16. $(3x^3y^2)^2(2xy^4)^3$

Answers
10. $49x^2$ **11.** $81x^4y^4$ **12.** $16x^6$
13. $\frac{1}{9}x^6y^4$ **14.** x^{22} **15.** $18y^5$
16. $72x^9y^{16}$

If we include units with the dimensions of the diagrams, then the units for the area will be square units and the units for volume will be cubic units. More specifically:

If a square has a side 5 inches long, then its area will be

$A = (5 \text{ inches})^2 = 25 \text{ inches}^2$

where the unit inches^2 stands for square inches.

If a cube has a single side 5 inches long, then its volume will be

$V = (5 \text{ inches})^3 = 125 \text{ inches}^3$

where the unit inches^3 stands for cubic inches.

If a rectangular solid has a length of 5 inches, a width of 4 inches, and a height of 3 inches, then its volume is

$V = (5 \text{ inches})(4 \text{ inches})(3 \text{ inches})$
$= 60 \text{ inches}^3$

FACTS FROM GEOMETRY: Volume of a Rectangular Solid

It is easy to see why the phrase "five squared" is associated with the expression 5^2. Simply find the area of the square shown in Figure 1 with a side of 5.

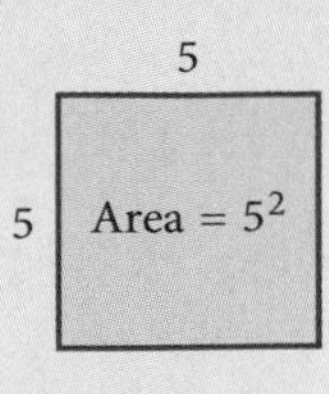

FIGURE 1

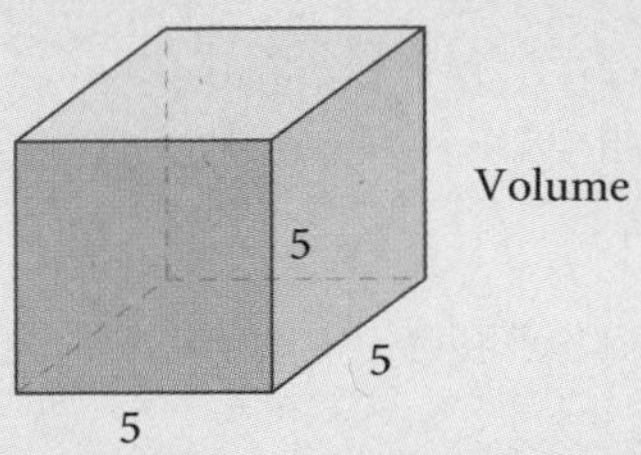

FIGURE 2

To see why the phrase "five cubed" is associated with the expression 5^3, we have to find the *volume* of a cube for which all three dimensions are 5 units long. The volume of a cube is a measure of the space occupied by the cube. To calculate the volume of the cube shown in Figure 2, we multiply the three dimensions together to get $5 \cdot 5 \cdot 5 = 5^3$.

The cube shown in Figure 2 is a special case of a general category of three-dimensional geometric figures called *rectangular solids.* Rectangular solids have rectangles for sides, and all connecting sides meet at right angles. The three dimensions are length, width, and height. To find the volume of a rectangular solid, we find the product of the three dimensions, as shown in Figure 3.

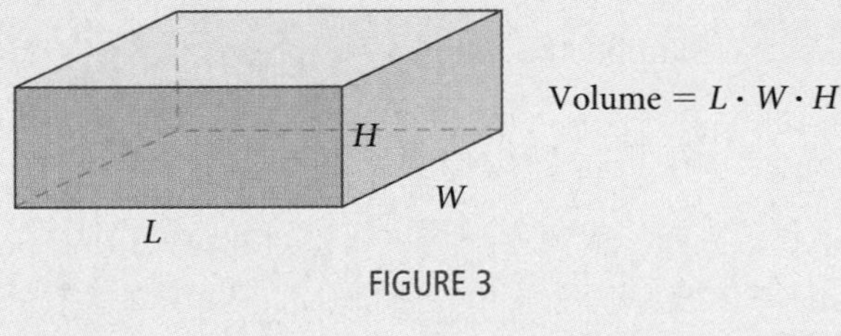

FIGURE 3

E Scientific Notation

Many branches of science require working with very large numbers. In astronomy, for example, distances commonly are given in light-years. A light-year is the distance light travels in a year. It is approximately

5,880,000,000,000 miles

This number is difficult to use in calculations because of the number of zeros it contains. Scientific notation provides a way of writing very large numbers in a more manageable form.

Definition

A number is in **scientific notation** when it is written as the product of a number between 1 and 10 and an integer power of 10. A number written in scientific notation has the form

$$n \times 10^r$$

where $1 \le n < 10$ and r = an integer.

EXAMPLE 17 Write 376,000 in scientific notation.

SOLUTION We must rewrite 376,000 as the product of a number between 1 and 10 and a power of 10. To do so, we move the decimal point 5 places to the left so that it appears between the 3 and the 7. Then we multiply this number by 10^5. The number that results has the same value as our original number and is written in scientific notation:

$$376{,}000 = 3.76 \times 10^5$$

Moved 5 places. Decimal point originally here. Keeps track of the 5 places we moved the decimal point.

17. Write 4,810,000 in scientific notation.

EXAMPLE 18 Write 4.52×10^3 in expanded form.

SOLUTION Since 10^3 is 1,000, we can think of this as simply a multiplication problem; that is,

$$4.52 \times 10^3 = 4.52 \times 1{,}000 = 4{,}520$$

On the other hand, we can think of the exponent 3 as indicating the number of places we need to move the decimal point to write our number in expanded form. Since our exponent is positive 3, we move the decimal point three places to the right:

$$4.52 \times 10^3 = 4{,}520$$

18. Write 3.05×10^5 in expanded form.

Getting Ready for Class

After reading through the preceding section, respond in your own words and in complete sentences.

1. Explain the difference between -5^2 and $(-5)^2$.
2. How do you multiply two expressions containing exponents when they each have the same base?
3. What is Property 2 for exponents?
4. When is a number written in scientific notation?

Answers
17. 4.81×10^6 **18.** 305,000

Problem Set 5.1

A Name the base and exponent in each of the following expressions. Then use the definition of exponents as repeated multiplication to simplify. [Examples 1–4]

1. 4^2 **2.** 6^2 **3.** $(0.3)^2$ **4.** $(0.03)^2$ **5.** 4^3 **6.** 10^3

7. $(-5)^2$ **8.** -5^2 **9.** -2^3 **10.** $(-2)^3$ **11.** 3^4 **12.** $(-3)^4$

13. $\left(\frac{2}{3}\right)^2$ **14.** $\left(\frac{2}{3}\right)^3$ **15.** $\left(\frac{1}{2}\right)^4$ **16.** $\left(\frac{4}{5}\right)^2$

17. a. Complete the following table.

Number x	Square x^2
1	
2	
3	
4	
5	
6	
7	

b. Using the results of part a, fill in the blank in the following statement: For numbers larger than 1, the square of the number is _____ than the number.

18. a. Complete the following table.

Number x	Square x^2
$\frac{1}{2}$	
$\frac{1}{3}$	
$\frac{1}{4}$	
$\frac{1}{5}$	
$\frac{1}{6}$	
$\frac{1}{7}$	
$\frac{1}{8}$	

b. Using the results of part a, fill in the blank in the following statement: For numbers between 0 and 1, the square of the number is _______ than the number.

B Use Property 1 for exponents to simplify each expression. Leave all answers in terms of exponents. [Examples 5–7]

19. $x^4 \cdot x^5$ **20.** $x^7 \cdot x^3$ **21.** $y^{10} \cdot y^{20}$ **22.** $y^{30} \cdot y^{30}$

23. $2^5 \cdot 2^4 \cdot 2^3$ **24.** $4^2 \cdot 4^3 \cdot 4^4$ **25.** $x^4 \cdot x^6 \cdot x^8 \cdot x^{10}$ **26.** $x^{20} \cdot x^{18} \cdot x^{16} \cdot x^{14}$

C Use Property 2 for exponents to write each of the following problems with a single exponent. (Assume all variables are positive numbers.) [Examples 8–9]

27. $(x^2)^5$ **28.** $(x^5)^2$ **29.** $(5^4)^3$ **30.** $(5^3)^4$ **31.** $(y^3)^3$ **32.** $(y^2)^2$

33. $(2^5)^{10}$ **34.** $(10^5)^2$ **35.** $(a^3)^x$ **36.** $(a^5)^x$ **37.** $(b^x)^y$ **38.** $(b^r)^s$

D Use Property 3 for exponents to simplify each of the following expressions. [Examples 10–16]

39. $(4x)^2$ **40.** $(2x)^4$ **41.** $(2y)^5$ **42.** $(5y)^2$ **43.** $(-3x)^4$ **44.** $(-3x)^3$

45. $(0.5ab)^2$ **46.** $(0.4ab)^2$ **47.** $(4xyz)^3$ **48.** $(5xyz)^3$

Simplify the following expressions by using the properties of exponents.

49. $(2x^4)^3$ **50.** $(3x^5)^2$ **51.** $(4a^3)^2$ **52.** $(5a^2)^2$

53. $(x^2)^3(x^4)^2$ **54.** $(x^5)^2(x^3)^5$ **55.** $(a^3)^1(a^2)^4$ **56.** $(a^4)^1(a^1)^3$

57. $(2x)^3(2x)^4$ **58.** $(3x)^2(3x)^3$ **59.** $(3x^2)^3(2x)^4$ **60.** $(3x)^3(2x^3)^2$

61. $(4x^2y^3)^2$ **62.** $(9x^3y^5)^2$ **63.** $\left(\frac{2}{3}a^4b^5\right)^3$ **64.** $\left(\frac{3}{4}ab^7\right)^3$

Write each expression as a perfect square.

65. $x^4 = (\quad)^2$ **66.** $x^6 = (\quad)^2$ **67.** $16x^2 = (\quad)^2$ **68.** $256x^4 = (\quad)^2$

Write each expression as a perfect cube.

69. $8 = (\quad)^3$ **70.** $27 = (\quad)^3$ **71.** $64x^3 = (\quad)^3$ **72.** $27x^6 = (\quad)^3$

73. Let $x = 2$ in each of the following expressions and simplify.

a. x^3x^2

b. $(x^3)^2$

c. x^5

d. x^6

74. Let $x = -1$ in each of the following expressions and simplify.

a. x^3x^4

b. $(x^3)^4$

c. x^7

d. x^{12}

75. Complete the following table, and then use the template to construct a line graph of the information in the table.

Number x	Square x^2
−3	
−2	
−1	
0	
1	
2	
3	

76. Complete the table, and then use the template to construct a line graph of the information in the table.

Number x	Cube x^3
−3	
−2	
−1	
0	
1	
2	
3	

77. Complete the table. When you are finished, notice how the points in this table could be used to refine the line graph you created in Problem 75.

Number x	Square x^2
−2.5	
−1.5	
−0.5	
0	
0.5	
1.5	
2.5	

78. Complete the following table. When you are finished, notice that this table contains exactly the same entries as the table from Problem 77. This table uses fractions, whereas the table from Problem 77 uses decimals.

Number x	Square x^2
$-\frac{5}{2}$	
$-\frac{3}{2}$	
$-\frac{1}{2}$	
0	
$\frac{1}{2}$	
$\frac{3}{2}$	
$\frac{5}{2}$	

E Write each number in scientific notation. [Example 17]

79. 43,200

80. 432,000

81. 570

82. 5,700

83. 238,000

84. 2,380,000

E Write each number in expanded form. [Example 18]

85. 2.49×10^3

86. 2.49×10^4

87. 3.52×10^2

88. 3.52×10^5

89. 2.8×10^4

90. 2.8×10^3

Applying the Concepts

91. **Google Earth** This Google Earth image is of the Luxor Hotel in Las Vegas, Nevada. The casino has a square base with sides of 525 feet. What is the area of the casino floor?

92. **Google Earth** This is a three dimensional model created by Google Earth of the Louvre Museum in Paris, France. The pyramid that dominates the Napoleon Courtyard has a square base with sides of 35.50 meters. What is the area of the base of the pyramid?

93. **Volume of a Cube** Find the volume of a cube if each side is 3 inches long.

94. **Volume of a Cube** Find the volume of a cube if each side is 0.3 feet long.

95. **Volume of a Cube** A bottle of perfume is packaged in a box that is in the shape of a cube. Find the volume of the box if each side is 2.5 inches long. Round to the nearest tenth.

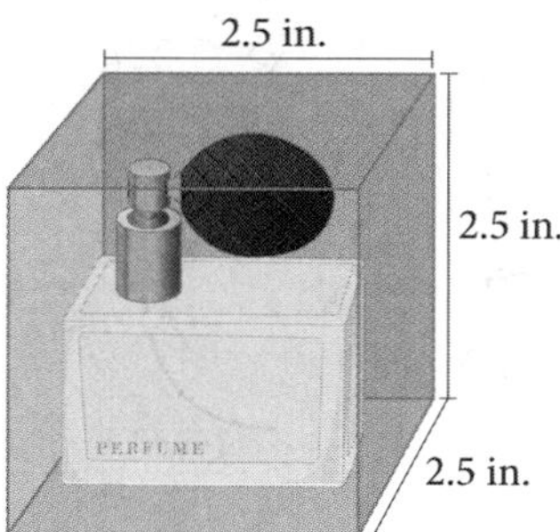

96. **Volume of a Cube** A television set is packaged in a box that is in the shape of a cube. Find the volume of the box if each side is 18 inches long.

97. **Volume of a Box** A rented videotape is in a plastic container that has the shape of a rectangular solid. Find the volume of the container if the length is 8 inches, the width is 4.5 inches, and the height is 1 inch.

98. **Volume of a Box** Your textbook is in the shape of a rectangular solid. Find the volume in cubic inches.

99. **Age in seconds** If you are 21 years old, you have been alive for more than 650,000,000 seconds. Write this last number in scientific notation.

100. **Distance Around the Earth** The distance around the Earth at the equator is more than 130,000,000 feet. Write this number in scientific notation.

101. **Lifetime Earnings** If you earn at least $12 an hour and work full-time for 30 years, you will make at least 7.4×10^5 dollars. Write this last number in expanded form.

102. **Heart Beats per Year** If your pulse is 72, then in one year your heart will beat at least 3.78×10^7 times. Write this last number in expanded form.

103. **Investing** If you put $1,000 into a savings account every year from the time you are 25 years old until you are 55 years old, you will have more than 1.8×10^5 dollars in the account when you reach 55 years of age (assuming 10% annual interest). Write 1.8×10^5 in expanded form.

104. **Investing** If you put $20 into a savings account every month from the time you are 20 years old until you are 30 years old, you will have more than 3.27×10^3 dollars in the account when you reach 30 years of age (assuming 6% annual interest compounded monthly). Write 3.27×10^3 in expanded form.

Displacement The displacement, in cubic inches, of a car engine is given by the formula

$$d = \pi \cdot s \cdot c \cdot \left(\frac{1}{2} \cdot b\right)^2$$

where s is the stroke and b is the bore, as shown in the figure, and c is the number of cylinders.

Calculate the engine displacement for each of the following cars. Use 3.14 to approximate π. Round your answers to the nearest cubic inch.

105. **Ferrari Modena** 8 cylinders, 3.35 inches of bore, 3.11 inches of stroke

106. **Audi A8** 8 cylinders, 3.32 inches of bore, 3.66 inches of stroke

107. **Mitsubishi Eclipse** 6 cylinders, 3.59 inches of bore, 2.99 inches of stroke

108. **Porsche 911 GT3** 6 cylinders, 3.94 inches of bore, 3.01 inches of stroke

Getting Ready for the Next Section

Subtract.

109. $4 - 7$

110. $-4 - 7$

111. $4 - (-7)$

112. $-4 - (-7)$

113. $15 - 20$

114. $15 - (-20)$

115. $-15 - (-20)$

116. $-15 - 20$

Simplify.

117. $2(3) - 4$

118. $5(3) - 10$

119. $4(3) - 3(2)$

120. $-8 - 2(3)$

121. $2(5 - 3)$

122. $2(3) - 4 - 3(-4)$

123. $5 + 4(-2) - 2(-3)$

124. $2(3) + 4(5) - 5(2)$

Maintaining Your Skills

Factor each of the following into its product of prime factors.

125. 128

126. 200

127. 250

128. 512

129. 720

130. 555

131. 820

132. 1,024

Factor the following by first factoring the base and then raising each of its factors to the third power.

133. 6^3

134. 10^3

135. 30^3

136. 42^3

137. 25^3

138. 8^3

139. 12^3

140. 36^3

Division with Exponents

5.2

Objectives

A Apply the definition for negative exponents.

B Use Property 4 for exponents.

C Use Property 5 for exponents.

D Simplify expressions involving exponents of 0 and 1.

E Simplify expressions using combinations of the properties of exponents.

F Write numbers in scientific notation and expanded form.

XYZ
Examples now playing at
MathTV.com/books

In Section 5.1 we found that multiplication with the same base results in addition of exponents; that is, $a^r \cdot a^s = a^{r+s}$. Since division is the inverse operation of multiplication, we can expect division with the same base to result in subtraction of exponents.

A Negative Exponents

To develop the properties for exponents under division, we again apply the definition of exponents:

$$\begin{aligned} \frac{x^5}{x^3} &= \frac{x \cdot x \cdot x \cdot x \cdot x}{x \cdot x \cdot x} \\ &= \frac{x \cdot x \cdot x}{x \cdot x \cdot x}(x \cdot x) \\ &= 1(x \cdot x) \\ &= x^2 \qquad \textit{Notice: } 5 - 3 = 2 \end{aligned}$$

$$\begin{aligned} \frac{2^4}{2^7} &= \frac{2 \cdot 2 \cdot 2 \cdot 2}{2 \cdot 2 \cdot 2 \cdot 2 \cdot 2 \cdot 2 \cdot 2} \\ &= \frac{2 \cdot 2 \cdot 2 \cdot 2}{2 \cdot 2 \cdot 2 \cdot 2} \cdot \frac{1}{2 \cdot 2 \cdot 2} \\ &= \frac{1}{2 \cdot 2 \cdot 2} \\ &= \frac{1}{2^3} \qquad \textit{Notice: } 7 - 4 = 3 \end{aligned}$$

In both cases division with the same base resulted in subtraction of the smaller exponent from the larger. The problem is deciding whether the answer is a fraction. The problem is resolved easily by the following definition.

Definition

If r is a positive integer, then $a^{-r} = \frac{1}{a^r} = \left(\frac{1}{a}\right)^r \quad (a \neq 0)$

The following examples illustrate how we use this definition to simplify expressions that contain negative exponents.

EXAMPLES Write each expression with a positive exponent and then simplify:

1. $2^{-3} = \frac{1}{2^3} = \frac{1}{8}$

Notice: Negative exponents do not indicate negative numbers. They indicate reciprocals

2. $5^{-2} = \frac{1}{5^2} = \frac{1}{25}$

3. $3x^{-6} = 3 \cdot \frac{1}{x^6} = \frac{3}{x^6}$

PRACTICE PROBLEMS

Write each expression with a positive exponent and then simplify.

1. 3^{-2}

2. 4^{-3}

3. $5x^{-4}$

Answers

1. $\frac{1}{9}$ **2.** $\frac{1}{64}$ **3.** $\frac{5}{x^4}$

B Property 4 and the Properties of Exponents

Now let us look back to one of our original problems and try to work it again with the help of a negative exponent. We know that $\frac{2^4}{2^7} = \frac{1}{2^3}$. Let's decide now that with division of the same base, we will always subtract the exponent in the denominator from the exponent in the numerator and see if this conflicts with what we know is true.

$$\begin{aligned} \frac{2^4}{2^7} &= 2^{4-7} && \text{Subtracting the bottom exponent from the top exponent} \\ &= 2^{-3} && \text{Subtraction} \\ &= \frac{1}{2^3} && \text{Definition of negative exponents} \end{aligned}$$

Subtracting the exponent in the denominator from the exponent in the numerator and then using the definition of negative exponents gives us the same result we obtained previously. We can now continue the list of properties of exponents we started in Section 5.1.

Property 4 for Exponents

If a is any real number and r and s are integers, then

$$\frac{a^r}{a^s} = a^{r-s} \qquad (a \neq 0)$$

In words: To divide with the same base, subtract the exponent in the denominator from the exponent in the numerator and raise the base to the exponent that results.

The following examples show how we use Property 4 and the definition for negative exponents to simplify expressions involving division.

Simplify each expression:

4. $\frac{x^{10}}{x^4}$

5. $\frac{x^5}{x^7}$

6. $\frac{2^{21}}{2^{25}}$

EXAMPLES Simplify the following expressions:

4. $\frac{x^9}{x^6} = x^{9-6} = x^3$

5. $\frac{x^4}{x^{10}} = x^{4-10} = x^{-6} = \frac{1}{x^6}$

6. $\frac{2^{15}}{2^{20}} = 2^{15-20} = 2^{-5} = \frac{1}{2^5} = \frac{1}{32}$

C Property 5 and the Properties of Exponents

Our final property of exponents is similar to Property 3 from Section 5.1, but it involves division instead of multiplication. After we have stated the property, we will give a proof of it. The proof shows why this property is true.

Property 5 for Exponents

If a and b are any two real numbers ($b \neq 0$) and r is an integer, then

$$\left(\frac{a}{b}\right)^r = \frac{a^r}{b^r}$$

In words: A quotient raised to a power is the quotient of the powers.

Proof

$$\left(\frac{a}{b}\right)^r = \left(a \cdot \frac{1}{b}\right)^r \qquad \text{By the definition of division}$$

$$= a^r \cdot \left(\frac{1}{b}\right)^r \qquad \text{By Property 3}$$

$$= a^r \cdot b^{-r} \qquad \text{By the definition of negative exponents}$$

$$= a^r \cdot \frac{1}{b^r} \qquad \text{By the definition of negative exponents}$$

$$= \frac{a^r}{b^r} \qquad \text{By the definition of division}$$

Simplify each expression.

7. $\left(\frac{x}{5}\right)^2$

8. $\left(\frac{2}{a}\right)^3$

9. $\left(\frac{3}{4}\right)^3$

EXAMPLES Simplify the following expressions.

7. $\left(\frac{x}{2}\right)^3 = \frac{x^3}{2^3} = \frac{x^3}{8}$

8. $\left(\frac{5}{y}\right)^2 = \frac{5^2}{y^2} = \frac{25}{y^2}$

9. $\left(\frac{2}{3}\right)^4 = \frac{2^4}{3^4} = \frac{16}{81}$

Answers

4. x^6 5. $\frac{1}{x^2}$ 6. $\frac{1}{16}$ 7. $\frac{x^2}{25}$ 8. $\frac{8}{a^3}$ 9. $\frac{27}{64}$

D Zero and One as Exponents

We have two special exponents left to deal with before our rules for exponents are complete: 0 and 1. To obtain an expression for x^1, we will solve a problem two different ways:

$$\left.\begin{aligned}\frac{x^3}{x^2} &= \frac{x \cdot x \cdot x}{x \cdot x} = x \\ \frac{x^3}{x^2} &= x^{3-2} = x^1\end{aligned}\right\} \quad \text{Hence } x^1 = x$$

Stated generally, this rule says that $a^1 = a$. This seems reasonable and we will use it since it is consistent with our property of division using the same base.

We use the same procedure to obtain an expression for x^0:

$$\left.\begin{aligned}\frac{5^2}{5^2} &= \frac{25}{25} = 1 \\ \frac{5^2}{5^2} &= 5^{2-2} = 5^0\end{aligned}\right\} \quad \text{Hence } 5^0 = 1$$

It seems, therefore, that the best definition of x^0 is 1 for all x except $x = 0$. In the case of $x = 0$, we have 0^0, which we will not define. This definition will probably seem awkward at first. Most people would like to define x^0 as 0 when they first encounter it. Remember, the zero in this expression is an exponent, so x^0 does not mean to multiply by zero. Thus, we can make the general statement that $a^0 = 1$ for all real numbers except $a = 0$.

Here are some examples involving the exponents 0 and 1.

EXAMPLES Simplify the following expressions:

10. $8^0 = 1$

11. $8^1 = 8$

12. $4^0 + 4^1 = 1 + 4 = 5$

13. $(2x^2y)^0 = 1$

Simplify each expression:

10. 10^0

11. 10^1

12. $6^1 + 6^0$

13. $(3x^5y^2)^0$

E Combinations of Properties

Here is a summary of the definitions and properties of exponents we have developed so far. For each definition or property in the list, a and b are real numbers, and r and s are integers.

Definitions	Properties
$a^{-r} = \frac{1}{a^r} = \left(\frac{1}{a}\right)^r \quad a \neq 0$	**1.** $a^r \cdot a^s = a^{r+s}$
$a^1 = a$	**2.** $(a^r)^s = a^{rs}$
$a^0 = 1 \quad a \neq 0$	**3.** $(ab)^r = a^r b^r$
	4. $\frac{a^r}{a^s} = a^{r-s} \quad a \neq 0$
	5. $\left(\frac{a}{b}\right)^r = \frac{a^r}{b^r} \quad b \neq 0$

Here are some additional examples. These examples use a combination of the preceding properties and definitions.

Answers
10. 1 **11.** 10 **12.** 7 **13.** 1

Simplify, and write your answers with positive exponents only:

14. $\dfrac{(2x^3)^2}{x^4}$

15. $\dfrac{x^{-6}}{(x^3)^4}$

16. $\left(\dfrac{y^8}{y^3}\right)^2$

17. $(2x^4)^{-2}$

18. $x^{-6} \cdot x^2$

19. $\dfrac{a^5(a^{-2})^4}{(a^{-3})^2}$

Answers

14. $4x^2$ **15.** $\frac{1}{x^{18}}$ **16.** y^{10} **17.** $\frac{1}{4x^8}$

18. $\frac{1}{x^4}$ **19.** a^3

EXAMPLE 14 Simplify the expression. Write the answer with a positive exponent: $\frac{(5x^3)^2}{x^4}$

SOLUTION

$$\begin{aligned} \frac{(5x^3)^2}{x^4} &= \frac{25x^6}{x^4} && \text{Properties 2 and 3} \\ &= 25x^2 && \text{Property 4} \end{aligned}$$

EXAMPLE 15 Simplify the expression. Write the answer with a positive exponent: $\frac{x^{-8}}{(x^2)^3}$

SOLUTION

$$\begin{aligned} \frac{x^{-8}}{(x^2)^3} &= \frac{x^{-8}}{x^6} && \text{Property 2} \\ &= x^{-8-6} && \text{Property 4} \\ &= x^{-14} && \text{Subtraction} \\ &= \frac{1}{x^{14}} && \text{Definition of negative exponents} \end{aligned}$$

EXAMPLE 16 Simplify the expression. Write the answer with a positive exponent: $\left(\frac{y^5}{y^3}\right)^2$

SOLUTION

$$\begin{aligned} \left(\frac{y^5}{y^3}\right)^2 &= \frac{(y^5)^2}{(y^3)^2} && \text{Property 5} \\ &= \frac{y^{10}}{y^6} && \text{Property 2} \\ &= y^4 && \text{Property 4} \end{aligned}$$

Notice that we could have simplified inside the parentheses first and then raised the result to the second power:

$$\left(\frac{y^5}{y^3}\right)^2 = (y^2)^2 = y^4$$

EXAMPLE 17 Simplify the expression. Write the answer with a positive exponent: $(3x^5)^{-2}$

SOLUTION

$$\begin{aligned} (3x^5)^{-2} &= \frac{1}{(3x^5)^2} && \text{Definition of negative exponents} \\ &= \frac{1}{9x^{10}} && \text{Properties 2 and 3} \end{aligned}$$

EXAMPLE 18 Simplify the expression. Write the answer with a positive exponent: $x^{-8} \cdot x^5$

SOLUTION

$$\begin{aligned} x^{-8} \cdot x^5 &= x^{-8+5} && \text{Property 1} \\ &= x^{-3} && \text{Addition} \\ &= \frac{1}{x^3} && \text{Definition of negative exponents} \end{aligned}$$

EXAMPLE 19 Simplify the expression. Write the answer with a positive exponent: $\frac{(a^3)^2a^{-4}}{(a^{-4})^3}$

SOLUTION

$$\begin{aligned} \frac{(a^3)^2a^{-4}}{(a^{-4})^3} &= \frac{a^6a^{-4}}{a^{-12}} && \text{Property 2} \\ &= \frac{a^2}{a^{-12}} && \text{Property 1} \\ &= a^{14} && \text{Property 4} \end{aligned}$$

In the next two examples we use division to compare the area and volume of geometric figures.

EXAMPLE 20 Suppose you have two squares, one of which is larger than the other. If the length of a side of the larger square is 3 times as long as the length of a side of the smaller square, how many of the smaller squares will it take to cover up the larger square?

SOLUTION If we let x represent the length of a side of the smaller square, then the length of a side of the larger square is $3x$. The area of each square, along with a diagram of the situation, is given in Figure 1.

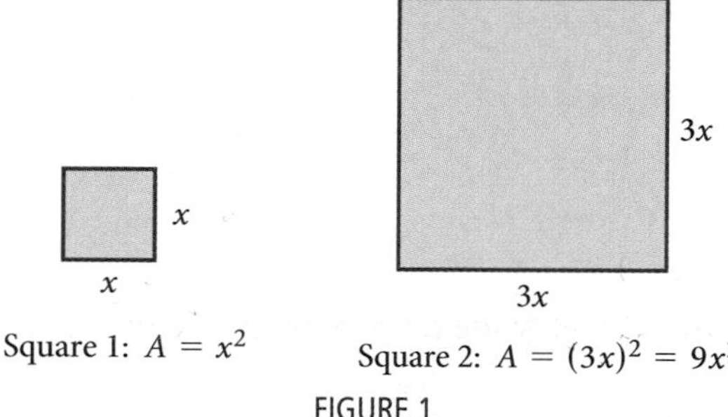

FIGURE 1

To find out how many smaller squares it will take to cover up the larger square, we divide the area of the larger square by the area of the smaller square.

$$\frac{\text{Area of square 2}}{\text{Area of square 1}} = \frac{9x^2}{x^2} = 9$$

It will take 9 of the smaller squares to cover the larger square.

EXAMPLE 21 Suppose you have two boxes, each of which is a cube. If the length of a side in the second box is 3 times as long as the length of a side of the first box, how many of the smaller boxes will fit inside the larger box?

SOLUTION If we let x represent the length of a side of the smaller box, then the length of a side of the larger box is $3x$. The volume of each box, along with a diagram of the situation, is given in Figure 2.

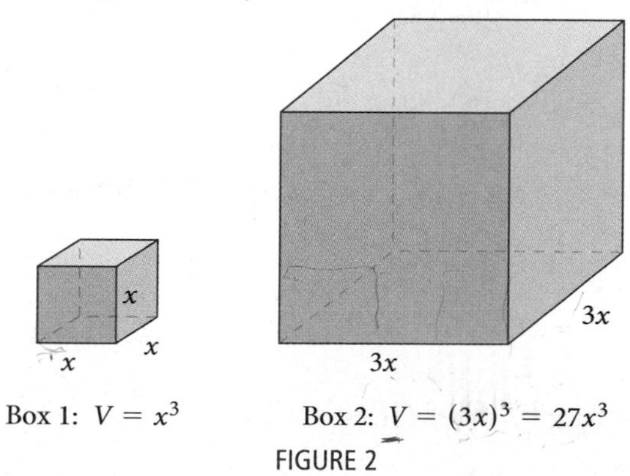

FIGURE 2

To find out how many smaller boxes will fit inside the larger box, we divide the volume of the larger box by the volume of the smaller box.

$$\frac{\text{Volume of box 2}}{\text{Volume of box 1}} = \frac{27x^3}{x^3} = 27$$

We can fit 27 of the smaller boxes inside the larger box.

20. Suppose the side of the larger square in Example 20 is 4 times as long as the side of the smaller square. How many smaller squares will it take to cover the larger square?

21. Suppose the side of the larger box in Example 21 is 4 times as long as the side of the smaller box. How many of the smaller boxes will fit into the larger box?

Answers

20. It will take 16 smaller squares to cover the larger square.
21. The larger box will hold 64 smaller boxes.

F More on Scientific Notation

Now that we have completed our list of definitions and properties of exponents, we can expand the work we did previously with scientific notation.

Recall that a number is in scientific notation when it is written in the form

$$n \times 10^r$$

where $1 \le n < 10$ and r is an integer.

Since negative exponents give us reciprocals, we can use negative exponents to write very small numbers in scientific notation. For example, the number 0.00057, when written in scientific notation, is equivalent to 5.7×10^{-4}. Here's why:

$$5.7 \times 10^{-4} = 5.7 \times \frac{1}{10^4} = 5.7 \times \frac{1}{10{,}000} = \frac{5.7}{10{,}000} = 0.00057$$

The table below lists some other numbers in both scientific notation and expanded form.

EXAMPLE 22

Number Written the Long Way		Number Written Again in Scientific Notation
376,000	=	3.76×10^5
49,500	=	4.95×10^4
3,200	=	3.2×10^3
591	=	5.91×10^2
46	=	4.6×10^1
8	=	8×10^0
0.47	=	4.7×10^{-1}
0.093	=	9.3×10^{-2}
0.00688	=	6.88×10^{-3}
0.0002	=	2×10^{-4}
0.000098	=	9.8×10^{-5}

22. Complete the following table:

	Expanded Form	Scientific Notation
a.	4,730	
b.		4.73×10^1
c.	0.473	
d.		4.73×10^{-3}

Notice that in each case in Example 22, when the number is written in scientific notation, the decimal point in the first number is placed so that the number is between 1 and 10. The exponent on 10 in the second number keeps track of the number of places we moved the decimal point in the original number to get a number between 1 and 10:

$$376{,}000 = 3.76 \times 10^5$$

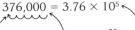

Moved 5 places.

Decimal point originally here.

Keeps track of the 5 places we moved the decimal point.

$$0.00688 = 6.88 \times 10^{-3}$$

Moved 3 places.

Keeps track of the 3 places we moved the decimal point.

Getting Ready for Class

After reading through the preceding section, respond in your own words and in complete sentences.

1. How do you divide two expressions containing exponents when they each have the same base?
2. Explain the difference between 3^2 and 3^{-2}.
3. If a positive base is raised to a negative exponent, can the result be a negative number?
4. Explain what happens when we use 0 as an exponent.

Answer

22. a. 4.73×10^3 b. 47.3 c. 4.73×10^{-1} d. 0.00473

Problem Set 5.2

A Write each of the following with positive exponents, and then simplify, when possible. [Examples 1–3]

1. 3^{-2} **2.** 3^{-3} **3.** 6^{-2} **4.** 2^{-6} **5.** 8^{-2} **6.** 3^{-4}

7. 5^{-3} **8.** 9^{-2} **9.** $2x^{-3}$ **10.** $5x^{-1}$ **11.** $(2x)^{-3}$ **12.** $(5x)^{-1}$

13. $(5y)^{-2}$ **14.** $5y^{-2}$ **15.** 10^{-2} **16.** 10^{-3}

17. Complete the following table.

Number x	Square x^2	Power of 2 2^x
−3		
−2		
−1		
0		
1		
2		
3		

18. Complete the following table.

Number x	Cube x^3	Power of 3 3^x
−3		
−2		
−1		
0		
1		
2		
3		

B Use Property 4 to simplify each of the following expressions. Write all answers that contain exponents with positive exponents only. [Examples 4–6]

19. $\frac{5^1}{5^3}$ **20.** $\frac{7^6}{7^8}$ **21.** $\frac{x^{10}}{x^4}$ **22.** $\frac{x^4}{x^{10}}$ **23.** $\frac{4^3}{4^0}$ **24.** $\frac{4^0}{4^3}$

25. $\frac{(2x)^7}{(2x)^4}$ **26.** $\frac{(2x)^4}{(2x)^7}$ **27.** $\frac{6^{11}}{6}$ **28.** $\frac{8^7}{8}$ **29.** $\frac{6}{6^{11}}$ **30.** $\frac{8}{8^7}$

31. $\frac{2^{-5}}{2^3}$ **32.** $\frac{2^{-5}}{2^{-3}}$ **33.** $\frac{2^5}{2^{-3}}$ **34.** $\frac{2^{-3}}{2^{-5}}$ **35.** $\frac{(3x)^{-5}}{(3x)^{-8}}$ **36.** $\frac{(2x)^{-10}}{(2x)^{-15}}$

E Simplify the following expressions. Any answers that contain exponents should contain positive exponents only. [Examples 7–19]

37. $(3xy)^4$ **38.** $(4xy)^3$ **39.** 10^0 **40.** 10^1

41. $(2a^2b)^1$ **42.** $(2a^2b)^0$ **43.** $(7y^3)^{-2}$ **44.** $(5y^4)^{-2}$

45. $x^{-3}x^{-5}$ **46.** $x^{-6} \cdot x^8$ **47.** $y^7 \cdot y^{-10}$ **48.** $y^{-4} \cdot y^{-6}$

49. $\frac{(x^2)^3}{x^4}$ **50.** $\frac{(x^5)^3}{x^{10}}$ **51.** $\frac{(a^4)^3}{(a^3)^2}$ **52.** $\frac{(a^5)^3}{(a^5)^2}$

53. $\frac{y^7}{(y^2)^8}$ **54.** $\frac{y^2}{(y^3)^4}$ **55.** $\left(\frac{y^7}{y^2}\right)^8$ **56.** $\left(\frac{y^2}{y^3}\right)^4$

57. $\frac{(x^{-2})^3}{x^{-5}}$ **58.** $\frac{(x^2)^{-3}}{x^{-5}}$ **59.** $\left(\frac{x^{-2}}{x^{-5}}\right)^3$ **60.** $\left(\frac{x^2}{x^{-5}}\right)^{-3}$

61. $\frac{(a^3)^2(a^4)^5}{(a^5)^2}$ **62.** $\frac{(a^4)^8(a^2)^5}{(a^3)^4}$ **63.** $\frac{(a^{-2})^3(a^4)^2}{(a^{-3})^{-2}}$ **64.** $\frac{(a^{-5})^{-3}(a^7)^{-1}}{(a^{-3})^5}$

65. Let $x = 2$ in each of the following expressions and simplify.

a. $\frac{x^7}{x^2}$

b. x^5

c. $\frac{x^2}{x^7}$

d. x^{-5}

66. Let $x = -1$ in each of the following expressions and simplify.

a. $\frac{x^{12}}{x^9}$

b. x^3

c. $\frac{x^{11}}{x^9}$

d. x^2

67. Write each expression as a perfect square.

a. $\frac{1}{25} = \left(\frac{}{}\right)^2$

b. $\frac{1}{64} = \left(\frac{}{}\right)^2$

c. $\frac{1}{x^2} = \left(\frac{}{}\right)^2$

d. $\frac{1}{x^4} = \left(\frac{}{}\right)^2$

68. Write each expression as a perfect cube.

a. $\frac{1}{125} = \left(\frac{}{}\right)^3$

b. $\frac{1}{27} = \left(\frac{}{}\right)^3$

c. $\frac{x^6}{125} = \left(\frac{}{}\right)^3$

d. $\frac{x^3}{27} = \left(\frac{}{}\right)^3$

69. Complete the following table, and then use the template to construct a line graph of the information in the table.

Number x	Power of 2 2^x
−3	
−2	
−1	
0	
1	
2	
3	

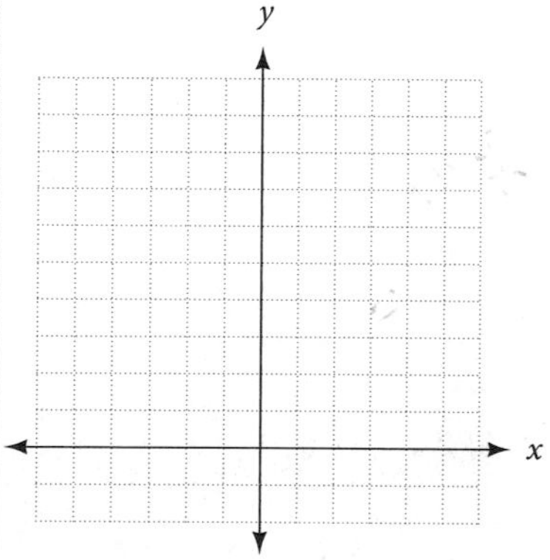

y

x

70. Complete the following table, and then use the template to construct a line graph of the information in the table.

Number x	Power of 3 3^x
−3	
−2	
−1	
0	
1	
2	
3	

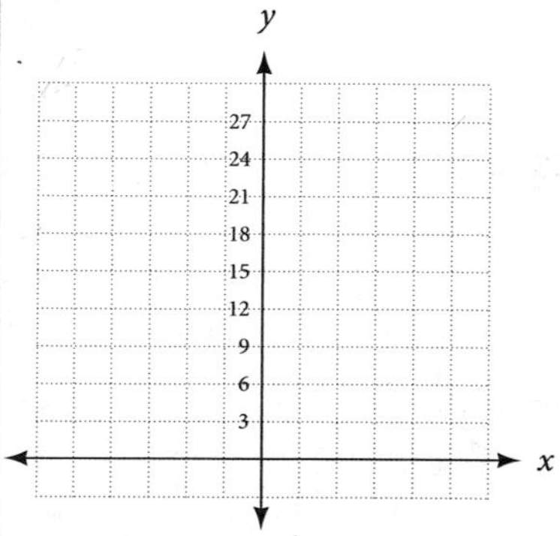

y

27 24 21 18 15 12 9 6 3

x

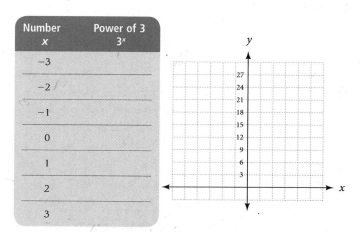

F Write each of the following numbers in scientific notation. [Example 22]

71. 0.0048

72. 0.000048

73. 25

74. 35

75. 0.000009

76. 0.0009

77. Complete the following table.

Expanded Form	Scientific Notation $n \times 10^r$
0.000357	3.57×10^{-4}
0.00357	
0.0357	
0.357	
3.57	
35.7	
357	
3,570	
35,700	

78. Complete the following table.

Expanded Form	Scientific Notation $n \times 10^r$
0.000123	1.23×10^{-4}
	1.23×10^{-3}
	1.23×10^{-2}
	1.23×10^{-1}
	1.23×10^{0}
	1.23×10^{1}
	1.23×10^{2}
	1.23×10^{3}
	1.23×10^{4}

F Write each of the following numbers in expanded form.

79. 4.23×10^{-3} **80.** 4.23×10^{3} **81.** 8×10^{-5} **82.** 8×10^{5} **83.** 4.2×10^{0} **84.** 4.2×10^{1}

Applying the Concepts [Examples 20–21]

85. Music The chart shows the country music singers that earned the most in 2007. Use the chart and write the earnings of Toby Keith in scientific notation. Remember the sales numbers are in millions.

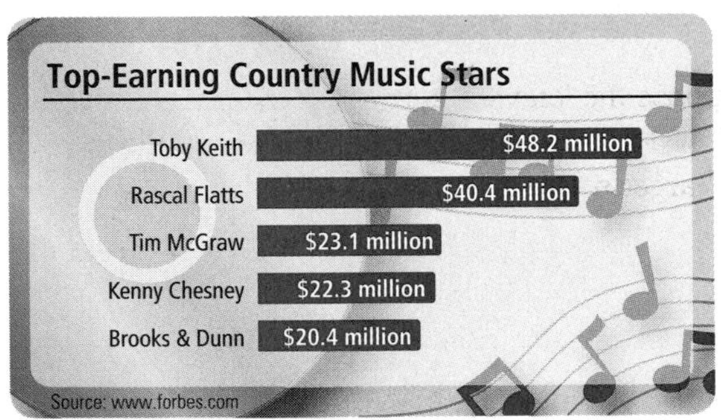

86. Cars The map shows the highest producers of cars for 2006. Find, to the nearest thousand, how many more cars Germany produced than the United States. Then write the answer in scientific notation.

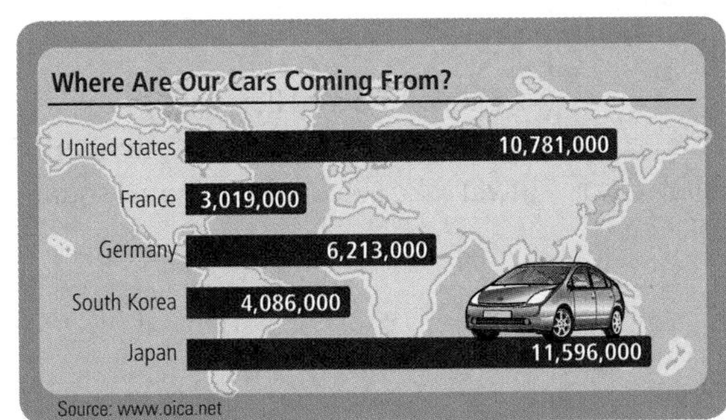

87. Some home computers can do a calculation in 2×10^{-3} seconds. Write this number in expanded form.

88. Some of the cells in the human body have a radius of 3×10^{-5} inches. Write this number in expanded form.

89. The number 25×10^3 is not in scientific notation because 25 is larger than 10. Write 25×10^3 in scientific notation.

90. The number 0.25×10^3 is not in scientific notation because 0.25 is less than 1. Write 0.25×10^3 in scientific notation.

91. The number 23.5×10^4 is not in scientific notation because 23.5 is not between 1 and 10. Rewrite 23.5×10^4 in scientific notation.

92. The number 375×10^3 is not in scientific notation because 375 is not between 1 and 10. Rewrite 375×10^3 in scientific notation.

93. The number 0.82×10^{-3} is not in scientific notation because 0.82 is not between 1 and 10. Rewrite 0.82×10^{-3} in scientific notation.

94. The number 0.93×10^{-2} is not in scientific notation because 0.93 is not between 1 and 10. Rewrite 0.93×10^{-2} in scientific notation.

Comparing Areas Suppose you have two squares, one of which is larger than the other. Suppose further that the side of the larger square is twice as long as the side of the smaller square.

95. If the length of the side of the smaller square is 10 inches, give the area of each square. Then find the number of smaller squares it will take to cover the larger square.

96. How many smaller squares will it take to cover the larger square if the length of the side of the smaller square is 1 foot?

97. If the length of the side of the smaller square is x, find the area of each square. Then find the number of smaller squares it will take to cover the larger square.

98. Suppose the length of the side of the larger square is 1 foot. How many smaller squares will it take to cover the larger square?

Comparing Volumes Suppose you have two boxes, each of which is a cube. Suppose further that the length of a side of the second box is twice as long as the length of a side of the first box.

99. If the length of a side of the first box is 6 inches, give the volume of each box. Then find the number of smaller boxes that will fit inside the larger box.

100. How many smaller boxes can be placed inside the larger box if the length of a side of the second box is 1 foot?

101. If the length of a side of the first box is x, find the volume of each box. Then find the number of smaller boxes that will fit inside the larger box.

102. Suppose the length of a side of the larger box is 12 inches. How many smaller boxes will fit inside the larger box?

Getting Ready for the Next Section

Simplify.

103. $3(4.5)$

104. $\frac{1}{2} \cdot \frac{5}{7}$

105. $\frac{4}{5}(10)$

106. $\frac{9.6}{3}$

107. $6.8(3.9)$

108. $9 - 20$

109. $-3 + 15$

110. $2x \cdot x \cdot \frac{1}{2}x$

111. $x^5 \cdot x^3$

112. $y^2 \cdot y$

113. $\frac{x^3}{x^2}$

114. $\frac{x^2}{x}$

115. $\frac{y^3}{y^5}$

116. $\frac{x^2}{x^5}$

Write in expanded form.

117. 3.4×10^2

118. 6.0×10^{-4}

Maintaining Your Skills

Simplify the following expressions.

119. $4x + 3x$

120. $9x + 7x$

121. $5a - 3a$

122. $10a - 2a$

123. $4y + 5y + y$

124. $6y - y + 2y$

Operations with Monomials

5.3

Objectives

A Multiply monomials.
B Divide monomials.
C Multiply and divide numbers written in scientific notation.
D Add and subtract monomials.

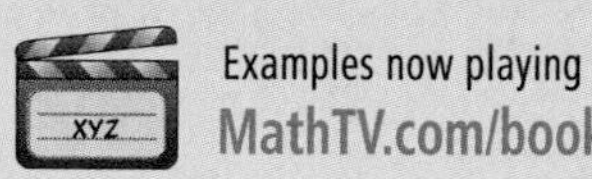
Examples now playing at
MathTV.com/books

We have developed all the tools necessary to perform the four basic operations on the simplest of polynomials: monomials.

> **Definition**
> A **monomial** is a one-term expression that is either a constant (number) or the product of a constant and one or more variables raised to whole number exponents.

The following are examples of monomials:

$$-3 \qquad 15x \qquad -23x^2y \qquad 49x^4y^2z^4 \qquad \frac{3}{4}a^2b^3$$

The numerical part of each monomial is called the *numerical coefficient,* or just *coefficient.* Monomials are also called *terms.*

A B Multiplication and Division of Monomials

There are two basic steps involved in the multiplication of monomials. First, we rewrite the products using the commutative and associative properties. Then, we simplify by multiplying coefficients and adding exponents of like bases.

EXAMPLE 1 Multiply: $(-3x^2)(4x^3)$

SOLUTION

$(-3x^2)(4x^3) = (-3 \cdot 4)(x^2 \cdot x^3)$ Commutative and associative properties

$= -12x^5$ Multiply coefficients, add exponents

EXAMPLE 2 Multiply: $\left(\frac{4}{5}x^5 \cdot y^2\right)(10x^3 \cdot y)$

SOLUTION

$\left(\frac{4}{5}x^5 \cdot y^2\right)(10x^3 \cdot y) = \left(\frac{4}{5} \cdot 10\right)(x^5 \cdot x^3)(y^2 \cdot y)$ Commutative and associative properties

$= 8x^8y^3$ Multiply coefficients, add exponents

You can see that in each example above the work was the same—multiply coefficients and add exponents of the same base. We can expect division of monomials to proceed in a similar way. Since our properties are consistent, division of monomials will result in division of coefficients and subtraction of exponents of like bases.

EXAMPLE 3 Divide: $\frac{15x^3}{3x^2}$

SOLUTION

$\frac{15x^3}{3x^2} = \frac{15}{3} \cdot \frac{x^3}{x^2}$ Write as separate fractions

$= 5x$ Divide coefficients, subtract exponents

EXAMPLE 4 Divide: $\frac{39x^2y^3}{3xy^5}$

SOLUTION

$\frac{39x^2y^3}{3xy^5} = \frac{39}{3} \cdot \frac{x^2}{x} \cdot \frac{y^3}{y^5}$ Write as separate fractions

$= 13x \cdot \frac{1}{y^2}$ Divide coefficients, subtract exponents

$= \frac{13x}{y^2}$ Write answer as a single fraction

PRACTICE PROBLEMS

Multiply:

1. $(-2x^3)(5x^2)$

2. $(4x^6y^3)(2xy^2)$

Divide:

3. $\frac{25x^4}{5x}$

4. $\frac{27x^4y^3}{9xy^2}$

Answers
1. $-10x^5$ **2.** $8x^7y^5$ **3.** $5x^3$ **4.** $3x^3y$

In Example 4, the expression $\frac{y^3}{y^5}$ simplifies to $\frac{1}{y^2}$ because of Property 4 for exponents and the definition of negative exponents. If we were to show all the work in this simplification process, it would look like this:

$$\frac{y^3}{y^5} = y^{3-5} \quad \text{Property 4 for exponents}$$
$$= y^{-2} \quad \text{Subtraction}$$
$$= \frac{1}{y^2} \quad \text{Definition of negative exponents}$$

The point of this explanation is this: Even though we may not show all the steps when simplifying an expression involving exponents, the result we obtain still can be justified using the properties of exponents. We have not introduced any new properties in Example 4. We have just not shown the details of each simplification.

5. Divide $\frac{13a^6b^2}{39a^4b^7}$.

EXAMPLE 5 Divide: $\frac{25a^5b^3}{50a^2b^7}$

SOLUTION
$$\frac{25a^5b^3}{50a^2b^7} = \frac{25}{50} \cdot \frac{a^5}{a^2} \cdot \frac{b^3}{b^7} \quad \text{Write as separate fractions}$$
$$= \frac{1}{2} \cdot a^3 \cdot \frac{1}{b^4} \quad \text{Divide coefficients, subtract exponents}$$
$$= \frac{a^3}{2b^4} \quad \text{Write answer as a single fraction}$$

Notice in Example 5 that dividing 25 by 50 results in $\frac{1}{2}$. This is the same result we would obtain if we reduced the fraction $\frac{25}{50}$ to lowest terms, and there is no harm in thinking of it that way. Also, notice that the expression $\frac{b^3}{b^7}$ simplifies to $\frac{1}{b^4}$ by Property 4 for exponents and the definition of negative exponents, even though we have not shown the steps involved in doing so.

C Multiplication and Division of Numbers Written in Scientific Notation

We multiply and divide numbers written in scientific notation using the same steps we used to multiply and divide monomials.

6. Multiply $(3 \times 10^6)(3 \times 10^{-8})$.

EXAMPLE 6 Multiply $(4 \times 10^7)(2 \times 10^{-4})$.

SOLUTION Because multiplication is commutative and associative, we can rearrange the order of these numbers and group them as follows:

$$(4 \times 10^7)(2 \times 10^{-4}) = (4 \times 2)(10^7 \times 10^{-4})$$
$$= 8 \times 10^3$$

Notice that we add exponents, $7 + (-4) = 3$, when we multiply with the same base.

7. Divide $\frac{4.8 \times 10^{20}}{2 \times 10^{12}}$.

EXAMPLE 7 Divide $\frac{9.6 \times 10^{12}}{3 \times 10^4}$.

SOLUTION We group the numbers between 1 and 10 separately from the powers of 10 and proceed as we did in Example 6:

$$\frac{9.6 \times 10^{12}}{3 \times 10^4} = \frac{9.6}{3} \times \frac{10^{12}}{10^4}$$
$$= 3.2 \times 10^8$$

Notice that the procedure we used in both of the examples above is very similar to multiplication and division of monomials, for which we multiplied or divided coefficients and added or subtracted exponents.

Answers

5. $\frac{a^2}{3b^5}$ 6. 9×10^{-2} 7. 2.4×10^8

D Addition and Subtraction of Monomials

Addition and subtraction of monomials will be almost identical since subtraction is defined as addition of the opposite. With multiplication and division of monomials, the key was rearranging the numbers and variables using the commutative and associative properties. With addition, the key is application of the distributive property. We sometimes use the phrase *combine monomials* to describe addition and subtraction of monomials.

> **Definition**
> Two terms (monomials) with the same variable part (same variables raised to the same powers) are called **similar** (or *like*) **terms.**

You can add only similar terms. This is because the distributive property (which is the key to addition of monomials) cannot be applied to terms that are not similar.

EXAMPLE 8 Combine the monomial: $-3x^2 + 15x^2$

SOLUTION $-3x^2 + 15x^2 = (-3 + 15)x^2$ Distributive property

$= 12x^2$ Add coefficients

EXAMPLE 9 Subtract: $9x^2y - 20x^2y$

SOLUTION $9x^2y - 20x^2y = (9 - 20)x^2y$ Distributive property

$= -11x^2y$ Add coefficients

EXAMPLE 10 Add: $5x^2 + 8y^2$

SOLUTION $5x^2 + 8y^2$ In this case we cannot apply the distributive property, so we cannot add the monomials

The next examples show how we simplify expressions containing monomials when more than one operation is involved.

EXAMPLE 11 Simplify $\dfrac{(6x^4y)(3x^7y^5)}{9x^5y^2}$.

SOLUTION We begin by multiplying the two monomials in the numerator:

$$\frac{(6x^4y)(3x^7y^5)}{9x^5y^2} = \frac{18x^{11}y^6}{9x^5y^2} \quad \text{Simplify numerator}$$

$$= 2x^6y^4 \quad \text{Divide}$$

EXAMPLE 12 Simplify $\dfrac{(6.8 \times 10^5)(3.9 \times 10^{-7})}{7.8 \times 10^{-4}}$.

SOLUTION We group the numbers between 1 and 10 separately from the powers of 10:

$$\frac{(6.8)(3.9)}{7.8} \times \frac{(10^5)(10^{-7})}{10^{-4}} = 3.4 \times 10^{5+(-7)-(-4)}$$

$$= 3.4 \times 10^2$$

Combine if possible.

8. $-4x^2 + 9x^2$

9. $12x^2y - 15x^2y$

10. $6x^2 + 9x^3$

11. Simplify $\dfrac{(10x^3y^2)(6xy^4)}{12x^2y^3}$.

12. Simplify

$$\frac{(1.2 \times 10^6)(6.3 \times 10^{-5})}{6 \times 10^{-10}}.$$

Answers
8. $5x^2$ **9.** $-3x^2y$
10. Cannot be combined
11. $5x^2y^3$ **12.** 1.26×10^{11}

13. Simplify $\frac{24x^7}{3x^2} + \frac{14x^9}{7x^4}$.

14. Simplify: $x^2\left(1 - \frac{4}{x}\right)$

15. Simplify: $3a\left(\frac{1}{a} - \frac{1}{3}\right)$

16. The width of a rectangular solid is half the length, whereas the height is 4 times the length. Find an expression for the volume in terms of the length, x.

Answers
13. $10x^5$ **14.** $x^2 - 4x$
15. $3 - a$ **16.** $2x^3$

EXAMPLE 13 Simplify $\frac{14x^5}{2x^2} + \frac{15x^8}{3x^5}$.

SOLUTION Simplifying each expression separately and then combining similar terms gives

$$\frac{14x^5}{2x^2} + \frac{15x^8}{3x^5} = 7x^3 + 5x^3 \quad \text{Divide}$$
$$= 12x^3 \quad \text{Add}$$

EXAMPLES Apply the distributive property, then simplify, if possible.

14. $x^2\left(1 - \frac{6}{x}\right) = x^2 \cdot 1 - x^2 \cdot \frac{6}{x} = x^2 - \frac{6x^2}{x} = x^2 - 6x$

15. $ab\left(\frac{1}{b} - \frac{1}{a}\right) = ab \cdot \frac{1}{b} - ab \cdot \frac{1}{a} = \frac{ab}{b} - \frac{ab}{a} = a - b$

EXAMPLE 16 A rectangular solid is twice as long as it is wide and one-half as high as it is wide. Write an expression for the volume in terms of the width, x.

SOLUTION We begin by making a diagram of the object (Figure 1) with the dimensions labeled as given in the problem.

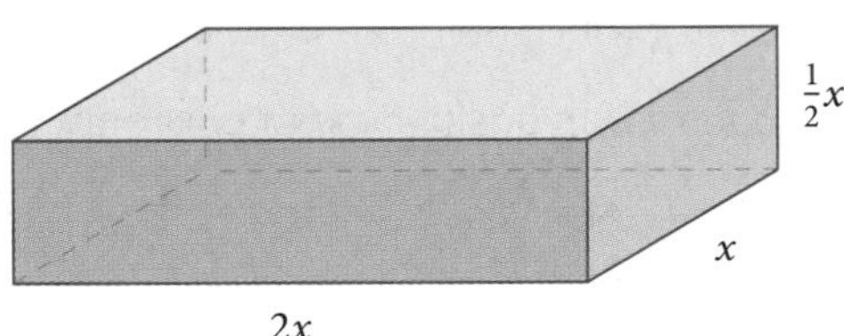

FIGURE 1

The volume is the product of the three dimensions:

$$V = 2x \cdot x \cdot \frac{1}{2}x = x^3$$

The box has the same volume as a cube with side x, as shown in Figure 2.

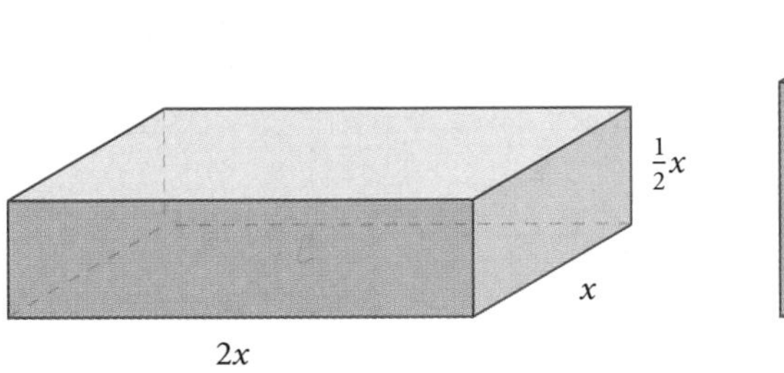

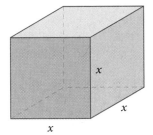

Equal Volumes

FIGURE 2

Getting Ready for Class

After reading through the preceding section, respond in your own words and in complete sentences.

1. What is a monomial?
2. Describe how you would multiply $3x^2$ and $5x^2$.
3. Describe how you would add $3x^2$ and $5x^2$.
4. Describe how you would multiply two numbers written in scientific notation.

Problem Set 5.3

A Multiply. [Examples 1–2]

1. $(3x^4)(4x^3)$

2. $(6x^5)(-2x^2)$

3. $(-2y^4)(8y^7)$

4. $(5y^{10})(2y^5)$

5. $(8x)(4x)$

6. $(7x)(5x)$

7. $(10a^3)(10a)(2a^2)$

8. $(5a^4)(10a)(10a^4)$

9. $(6ab^2)(-4a^2b)$

10. $(-5a^3b)(4ab^4)$

11. $(4x^2y)(3x^3y^3)(2xy^4)$

12. $(5x^6)(-10xy^4)(-2x^2y^6)$

B Divide. Write all answers with positive exponents only. [Examples 3–5]

13. $\frac{15x^3}{5x^2}$

14. $\frac{25x^5}{5x^4}$

15. $\frac{18y^9}{3y^{12}}$

16. $\frac{24y^4}{-8y^7}$

17. $\frac{32a^3}{64a^4}$

18. $\frac{25a^5}{75a^6}$

19. $\frac{21a^2b^3}{-7ab^5}$

20. $\frac{32a^5b^6}{8ab^5}$

21. $\frac{3x^3y^2z}{27xy^2z^3}$

22. $\frac{5x^5y^4z}{30x^3yz^2}$

23. Fill in the table.

a	b	ab	$\frac{a}{b}$	$\frac{b}{a}$
10	$5x$			
$20x^3$	$6x^2$			
$25x^5$	$5x^4$			
$3x^{-2}$	$3x^2$			
$-2y^4$	$8y^7$			

24. Fill in the table.

a	b	ab	$\frac{a}{b}$	$\frac{b}{a}$
$10y$	$2y^2$			
$10y^2$	$2y$			
$5y^3$	15			
5	$15y^3$			
$4y^{-3}$	$4y^3$			

C Find each product. Write all answers in scientific notation. [Example 6]

25. $(3 \times 10^3)(2 \times 10^5)$

26. $(4 \times 10^8)(1 \times 10^6)$

27. $(3.5 \times 10^4)(5 \times 10^{-6})$

28. $(7.1 \times 10^5)(2 \times 10^{-8})$

29. $(5.5 \times 10^{-3})(2.2 \times 10^{-4})$

30. $(3.4 \times 10^{-2})(4.5 \times 10^{-6})$

C Find each quotient. Write all answers in scientific notation. [Example 7]

31. $\dfrac{8.4 \times 10^5}{2 \times 10^2}$

32. $\dfrac{9.6 \times 10^{20}}{3 \times 10^6}$

33. $\dfrac{6 \times 10^8}{2 \times 10^{-2}}$

34. $\dfrac{8 \times 10^{12}}{4 \times 10^{-3}}$

35. $\dfrac{2.5 \times 10^{-6}}{5 \times 10^{-4}}$

36. $\dfrac{4.5 \times 10^{-8}}{9 \times 10^{-4}}$

D Combine by adding or subtracting as indicated. [Examples 8–10]

37. $3x^2 + 5x^2$

38. $4x^3 + 8x^3$

39. $8x^5 - 19x^5$

40. $75x^6 - 50x^6$

41. $2a + a - 3a$

42. $5a + a - 6a$

43. $10x^3 - 8x^3 + 2x^3$

44. $7x^5 + 8x^5 - 12x^5$

45. $20ab^2 - 19ab^2 + 30ab^2$

46. $18a^3b^2 - 20a^3b^2 + 10a^3b^2$

47. Fill in the table.

a	b	ab	a + b
$5x$	$3x$		
$4x^2$	$2x^2$		
$3x^3$	$6x^3$		
$2x^4$	$-3x^4$		
x^5	$7x^5$		

48. Fill in the table.

a	b	ab	a − b
$2y$	$3y$		
$-2y$	$3y$		
$4y^2$	$5y^2$		
y^3	$-3y^3$		
$5y^4$	$7y^4$		

A **B** Simplify. Write all answers with positive exponents only. [Example 11]

49. $\dfrac{(3x^2)(8x^5)}{6x^4}$

50. $\dfrac{(7x^3)(6x^8)}{14x^5}$

51. $\dfrac{(9a^2b)(2a^3b^4)}{18a^5b^7}$

52. $\dfrac{(21a^5b)(2a^8b^4)}{14ab}$

53. $\dfrac{(4x^3y^2)(9x^4y^{10})}{(3x^5y)(2x^6y)}$

54. $\dfrac{(5x^4y^4)(10x^3y^3)}{(25xy^5)(2xy^7)}$

C Simplify each expression, and write all answers in scientific notation. [Example 12]

55. $\dfrac{(6 \times 10^8)(3 \times 10^5)}{9 \times 10^7}$

56. $\dfrac{(8 \times 10^4)(5 \times 10^{10})}{2 \times 10^7}$

57. $\dfrac{(5 \times 10^3)(4 \times 10^{-5})}{2 \times 10^{-2}}$

58. $\dfrac{(7 \times 10^6)(4 \times 10^{-4})}{1.4 \times 10^{-3}}$

59. $\dfrac{(2.8 \times 10^{-7})(3.6 \times 10^4)}{2.4 \times 10^3}$

60. $\dfrac{(5.4 \times 10^2)(3.5 \times 10^{-9})}{4.5 \times 10^6}$

B **D** Simplify. [Example 13]

61. $\dfrac{18x^4}{3x} + \dfrac{21x^7}{7x^4}$

62. $\dfrac{24x^{10}}{6x^4} + \dfrac{32x^7}{8x}$

63. $\dfrac{45a^6}{9a^4} - \dfrac{50a^8}{2a^6}$

64. $\dfrac{16a^9}{4a} - \dfrac{28a^{12}}{4a^4}$

65. $\dfrac{6x^7y^4}{3x^2y^2} + \dfrac{8x^5y^8}{2y^6}$

66. $\dfrac{40x^{10}y^{10}}{8x^2y^5} + \dfrac{10x^8y^8}{5y^3}$

Apply the distributive property. [Examples 14–15]

67. $xy\left(x + \dfrac{1}{y}\right)$

68. $xy\left(y + \dfrac{1}{x}\right)$

69. $xy\left(\dfrac{1}{y} + \dfrac{1}{x}\right)$

70. $xy\left(\dfrac{1}{x} - \dfrac{1}{y}\right)$

71. $x^2\left(1 - \dfrac{4}{x^2}\right)$

72. $x^2\left(1 - \dfrac{9}{x^2}\right)$

73. $x^2\left(1 - \dfrac{1}{x} - \dfrac{6}{x^2}\right)$

74. $x^2\left(1 - \dfrac{5}{x} + \dfrac{6}{x^2}\right)$

75. $x^2\left(1 - \dfrac{5}{x}\right)$

76. $x^2\left(1 - \dfrac{3}{x}\right)$

77. $x^2\left(1 - \dfrac{8}{x}\right)$

78. $x^2\left(1 - \dfrac{6}{x}\right)$

79. Divide each monomial by $5a^2$.
   a. $10a^2$
   b. $-15a^2b$
   c. $25a^2b^2$

80. Divide each monomial by $36x^2$.
   a. $6x^2a$
   b. $12x^2a$
   c. $-6x^2a$

81. Divide each monomial by $8x^2y$.
   a. $24x^3y^2$
   b. $16x^2y^2$
   c. $-4x^2y^3$

82. Divide each monomial by $7x^2y$.
   a. $21x^3y^2$
   b. $14x^2y^2$
   c. $-7x^2y^3$

Getting Ready for the Next Section

Simplify.

83. $3 - 8$
84. $-5 + 7$
85. $-1 + 7$
86. $1 - 8$
87. $3(5)^2 + 1$
88. $3(-2)^2 - 5(-2) + 4$
89. $2x^2 + 4x^2$
90. $3x^2 - x^2$
91. $-5x + 7x$
92. $x - 2x$
93. $-(2x + 9)$
94. $-(4x^2 - 2x - 6)$
95. Find the value of $2x + 3$ when $x = 4$.
96. Find the value of $(3x)^2$ when $x = 3$.

Maintaining Your Skills

Find the value of each expression when $x = -2$.

97. $4x$
98. $-3x$
99. $-2x + 5$
100. $-4x - 1$
101. $x^2 + 5x + 6$
102. $x^2 - 5x + 6$

For each of the following equations complete the given ordered pairs so each is a solution to the equation, and then use the ordered pairs to graph the equation.

103. $y = 2x + 2$
$(-2,\quad), (0,\quad), (2,\quad)$

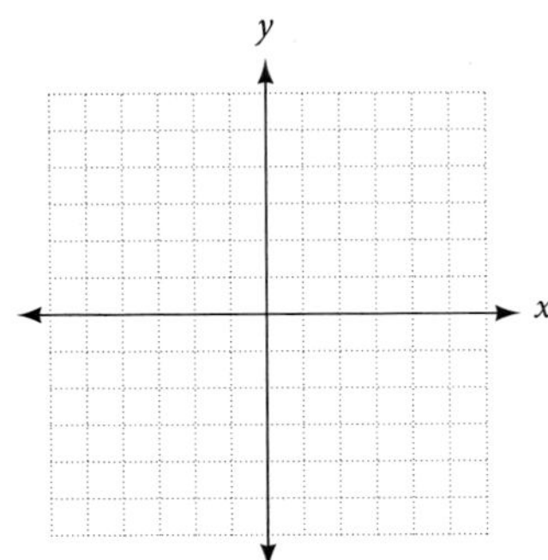

104. $y = 2x - 3$
$(-1,\quad), (0,\quad), (2,\quad)$

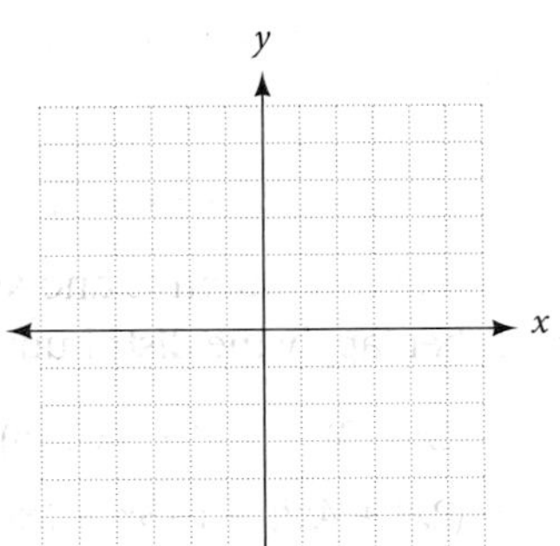

105. $y = \frac{1}{3}x + 1$
$(-3,\quad), (0,\quad), (3,\quad)$

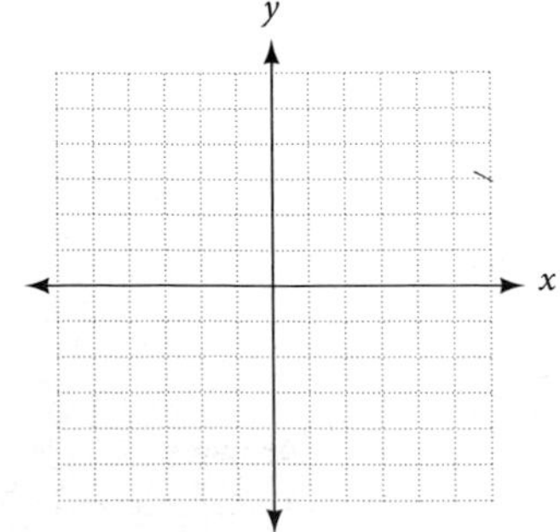

106. $y = \frac{1}{2}x - 2$
$(-2,\quad), (0,\quad), (2,\quad)$

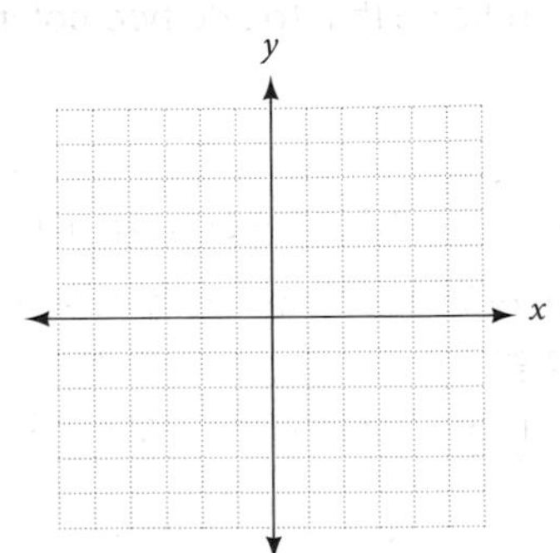

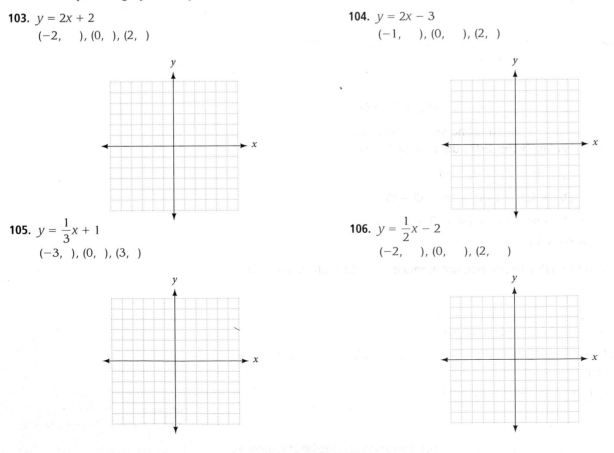

Addition and Subtraction of Polynomials

5.4

Objectives

A Add and subtract polynomials.

B Find the value of a polynomial for a given value of the variable.

Examples now playing at MathTV.com/books

In this section we will extend what we learned in Section 5.3 to expressions called polynomials. We begin this section with the definition of a polynomial.

Definition
A **polynomial** is a finite sum of monomials (terms).

Examples The following are polynomials:

$$3x^2 + 2x + 1 \qquad 15x^2y + 21xy^2 - y^2 \qquad 3a - 2b + 4c - 5d$$

Polynomials can be further classified by the number of terms they contain. A polynomial with two terms is called a binomial. If it has three terms, it is a trinomial. As stated before, a monomial has only one term.

Definition
The **degree** of a polynomial in one variable is the highest power to which the variable is raised.

Examples

$3x^5 + 2x^3 + 1$	A trinomial of degree 5
$2x + 1$	A binomial of degree 1
$3x^2 + 2x + 1$	A trinomial of degree 2
$3x^5$	A monomial of degree 5
-9	A monomial of degree 0

A Adding and Subtracting Polynomials

There are no new rules for adding one or more polynomials. We rely only on our previous knowledge. Here are some examples.

EXAMPLE 1 Add $(2x^2 - 5x + 3) + (4x^2 + 7x - 8)$.

SOLUTION We use the commutative and associative properties to group similar terms together and then apply the distributive property to add

$$(2x^2 - 5x + 3) + (4x^2 + 7x - 8)$$
$$= (2x^2 + 4x^2) + (-5x + 7x) + (3 - 8) \quad \text{Commutative and associative properties}$$
$$= (2 + 4)x^2 + (-5 + 7)x + (3 - 8) \quad \text{Distributive property}$$
$$= 6x^2 + 2x - 5 \quad \text{Addition}$$

The results here indicate that to add two polynomials, we add coefficients of similar terms.

EXAMPLE 2 Add $x^2 + 3x + 2x + 6$.

SOLUTION The only similar terms here are the two middle terms. We combine them as usual to get

$$x^2 + 3x + 2x + 6 = x^2 + 5x + 6$$

PRACTICE PROBLEMS

1. Add:
$(2x^2 + 3x + 4) + (5x^2 - 6x - 3)$

2. Add $x^2 + 7x + 4x + 28$.

Answers
1. $7x^2 - 3x + 1$ 2. $x^2 + 11x + 28$

You will recall from Chapter 1 the definition of subtraction: $a - b = a + (-b)$. To subtract one expression from another, we simply add its opposite. The letters a and b in the definition can each represent polynomials. The opposite of a polynomial is the opposite of each of its terms. When you subtract one polynomial from another you subtract each of its terms.

EXAMPLE 3 Subtract $(3x^2 + x + 4) - (x^2 + 2x + 3)$.

SOLUTION To subtract $x^2 + 2x + 3$, we change the sign of each of its terms and add. If you are having trouble remembering why we do this, remember that we can think of $-(x^2 + 2x + 3)$ as $-1(x^2 + 2x + 3)$. If we distribute the -1 across $x^2 + 2x + 3$, we get $-x^2 - 2x - 3$:

$$\begin{aligned}&(3x^2 + x + 4) - (x^2 + 2x + 3)\\ &\quad = 3x^2 + x + 4 - x^2 - 2x - 3 \qquad \text{Take the opposite of each term in the second polynomial}\\ &\quad = (3x^2 - x^2) + (x - 2x) + (4 - 3)\\ &\quad = 2x^2 - x + 1\end{aligned}$$

EXAMPLE 4 Subtract $-4x^2 + 5x - 7$ from $x^2 - x - 1$.

SOLUTION The polynomial $x^2 - x - 1$ comes first, then the subtraction sign, and finally the polynomial $-4x^2 + 5x - 7$ in parentheses.

$$\begin{aligned}&(x^2 - x - 1) - (-4x^2 + 5x - 7)\\ &\quad = x^2 - x - 1 + 4x^2 - 5x + 7 \qquad \text{Take the opposite of each term in the second polynomial}\\ &\quad = (x^2 + 4x^2) + (-x - 5x) + (-1 + 7)\\ &\quad = 5x^2 - 6x + 6\end{aligned}$$

B Finding the Value of a Polynomial

The last topic we want to consider in this section is finding the value of a polynomial for a given value of the variable.

To find the value of the polynomial $3x^2 + 1$ when x is 5, we replace x with 5 and simplify the result:

When	$x = 5$
the polynomial	$3x^2 + 1$
becomes	$3(5)^2 + 1 = 3(25) + 1$
	$= 75 + 1 = 76$

EXAMPLE 5 Find the value of $3x^2 - 5x + 4$ when $x = -2$.

SOLUTION

When	$x = -2$
the polynomial	$3x^2 - 5x + 4$
becomes	$3(-2)^2 - 5(-2) + 4 = 3(4) + 10 + 4$
	$= 12 + 10 + 4 = 26$

Getting Ready for Class

After reading through the preceding section, respond in your own words and in complete sentences.

1. What are similar terms?
2. What is the degree of a polynomial?
3. Describe how you would subtract one polynomial from another.
4. How you would find the value of $3x^2 - 5x + 4$ when x is -2?

---

3. Subtract: $(5x^2 + x + 2) - (x^2 + 3x + 7)$

4. Subtract $-4x^2 - 3x + 8$ from $x^2 + x + 1$.

Note There are two important points to remember when adding or subtracting polynomials. First, to add or subtract two polynomials, you always add or subtract *coefficients* of similar terms. Second, the exponents never increase in value when you are adding or subtracting similar terms.

5. Find the value of $2x^2 - x + 3$ when $x = -3$.

Answers
3. $4x^2 - 2x - 5$ 4. $5x^2 + 4x - 7$
5. 24

Problem Set 5.4

Identify each of the following polynomials as a trinomial, binomial, or monomial, and give the degree in each case. [Examples]

1. $2x^3 - 3x^2 + 1$
2. $4x^2 - 4x + 1$
3. $5 + 8a - 9a^3$
4. $6 + 12x^3 + x^4$
5. $2x - 1$
6. $4 + 7x$
7. $45x^2 - 1$
8. $3a^3 + 8$
9. $7a^2$
10. $90x$
11. -4
12. 56

A Perform the following additions and subtractions. [Examples 1–4]

13. $(2x^2 + 3x + 4) + (3x^2 + 2x + 5)$
14. $(x^2 + 5x + 6) + (x^2 + 3x + 4)$
15. $(3a^2 - 4a + 1) + (2a^2 - 5a + 6)$
16. $(5a^2 - 2a + 7) + (4a^2 - 3a + 2)$
17. $x^2 + 4x + 2x + 8$
18. $x^2 + 5x - 3x - 15$
19. $6x^2 - 3x - 10x + 5$
20. $10x^2 + 30x - 2x - 6$
21. $x^2 - 3x + 3x - 9$
22. $x^2 - 5x + 5x - 25$
23. $3y^2 - 5y - 6y + 10$
24. $y^2 - 18y + 2y - 12$
25. $(6x^3 - 4x^2 + 2x) + (9x^2 - 6x + 3)$
26. $(5x^3 + 2x^2 + 3x) + (2x^2 + 5x + 1)$
27. $\left(\frac{2}{3}x^2 - \frac{1}{5}x - \frac{3}{4}\right) + \left(\frac{4}{3}x^2 - \frac{4}{5}x + \frac{7}{4}\right)$
28. $\left(\frac{3}{8}x^3 - \frac{5}{7}x^2 - \frac{2}{5}\right) + \left(\frac{5}{8}x^3 - \frac{2}{7}x^2 + \frac{7}{5}\right)$
29. $(a^2 - a - 1) - (-a^2 + a + 1)$
30. $(5a^2 - a - 6) - (-3a^2 - 2a + 4)$
31. $\left(\frac{5}{9}x^3 + \frac{1}{3}x^2 - 2x + 1\right) - \left(\frac{2}{3}x^3 + x^2 + \frac{1}{2}x - \frac{3}{4}\right)$
32. $\left(4x^3 - \frac{2}{5}x^2 + \frac{3}{8}x - 1\right) - \left(\frac{9}{2}x^3 + \frac{1}{4}x^2 - x + \frac{5}{6}\right)$
33. $(4y^2 - 3y + 2) + (5y^2 + 12y - 4) - (13y^2 - 6y + 20)$
34. $(2y^2 - 7y - 8) - (6y^2 + 6y - 8) + (4y^2 - 2y + 3)$

Simplify.

35. $(x^2 - 5x) - (x^2 - 3x)$

36. $(-2x + 8) - (-2x + 6)$

37. $(6x^2 - 11x) - (6x^2 - 15x)$

38. $(10x^2 - 3x) - (10x^2 - 50x)$

39. $(x^3 + 3x^2 + 9x) - (3x^2 + 9x + 27)$

40. $(x^3 + 2x^2 + 4x) - (2x^2 + 4x - 8)$

41. $(x^3 + 4x^2 + 4x) + (2x^2 + 8x + 8)$

42. $(x^3 + 2x^2 + x) + (2x^2 + 2x + 1)$

43. $(x^2 - 4) - (x^2 - 4x + 4)$

44. $(4x^2 - 9) - (4x^2 - 12x + 9)$

45. Subtract $10x^2 + 23x - 50$ from $11x^2 - 10x + 13$.

46. Subtract $2x^2 - 3x + 5$ from $4x^2 - 5x + 10$.

47. Subtract $3y^2 + 7y - 15$ from $11y^2 + 11y + 11$.

48. Subtract $15y^2 - 8y - 2$ from $3y^2 - 3y + 2$.

49. Add $50x^2 - 100x - 150$ to $25x^2 - 50x + 75$.

50. Add $7x^2 - 8x + 10$ to $-8x^2 + 2x - 12$.

51. Subtract $2x + 1$ from the sum of $3x - 2$ and $11x + 5$.

52. Subtract $3x - 5$ from the sum of $5x + 2$ and $9x - 1$.

B [Example 5]

53. Find the value of the polynomial $x^2 - 2x + 1$ when x is 3.

54. Find the value of the polynomial $(x - 1)^2$ when x is 3.

55. Find the value of $100p^2 - 1{,}300p + 4{,}000$ when
a. $p = 5$
b. $p = 8$

56. Find the value $100p^2 - 800p + 1{,}200$ when
a. $p = 2$
b. $p = 6$

57. Find the value of $600 + 1{,}000x - 100x^2$ when
a. $x = 8$
b. $x = -2$

58. Find the value of $500 + 800x - 100x^2$ when
a. $x = 6$
b. $x = -1$

Applying the Concepts

59. **Google Earth** This Google Earth image is of Mount Rainier National Park in Washington. Suppose the park is roughly a square with sides of 1.2×10^6 inches. What is the area of the park?

60. **Google Earth** The Google Earth map shows a picture of Death Valley National Park in California. The park is 2.2×10^7 centimeters long and 6.9×10^6 centimeters wide. What is the area of the park in square centimeters?

61. **Packaging** A crystal ball with a diameter of 6 inches is being packaged for shipment. If the crystal ball is placed inside a circular cylinder with radius 3 inches and height 6 inches, how much volume will need to be filled with padding? (The volume of a sphere with radius r is $\frac{4}{3}\pi r^3$, and the volume of a right circular cylinder with radius r and height h is $\pi r^2 h$.)

62. **Packaging** Suppose the circular cylinder of Problem 45 has a radius of 4 inches and a height of 7 inches. How much volume will need to be filled with padding?

63. **Education** The chart shows the average income for people with different levels of education. In a high school's graduating class, x students plan to get their bachelor's degree and y students plan to go on and get their master's degree. The next year, twice as many students plan to get their bachelor's degree as the year before, but the number of students who plan to go on and get their master's degree stays the same as the year before. Write an expression that describes the total future annual income for each year's graduating class, and then find the total future yearly income for both graduating classes in terms of x and y.

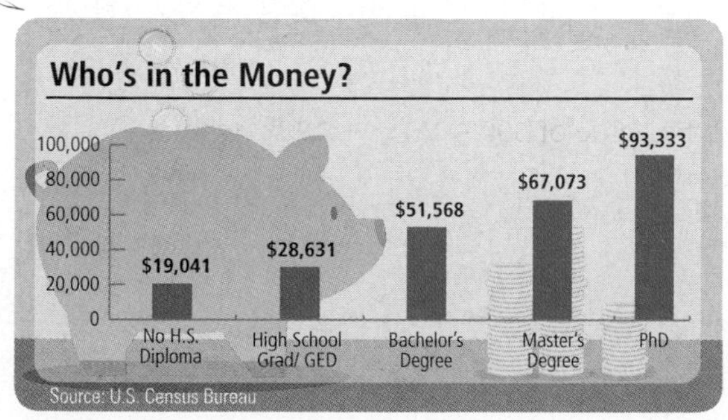

64. **Classroom Energy** The chart shows how much energy is wasted in the classroom by leaving appliances on. Suppose twice as many ceiling fans were left on as printers, and the DVD player was left on. Write an expression that describes the energy used in this situation if x represents the number of printers left on. Then find the value of the expression if 2 printers were left on.

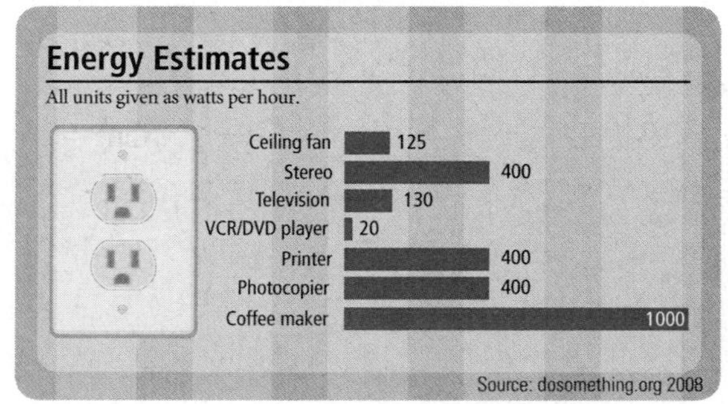

Getting Ready for the Next Section

Simplify.

65. $(-5)(-1)$

66. $3(-4)$

67. $(-1)(6)$

68. $(-7)8$

69. $(5x)(-4x)$

70. $(3x)(2x)$

71. $3x(-7)$

72. $3x(-1)$

73. $5x + (-3x)$

74. $-3x - 10x$

Multiply.

75. $3(2x - 6)$

76. $-4x(x + 5)$

Maintaining Your Skills

77. $3x\,(-5x)$

78. $-3x\,(-7x)$

79. $2x\,(3x^2)$

80. $x^2(3x)$

81. $3x^2(2x^2)$

82. $4x^2(2x^2)$

Multiplication with Polynomials

5.5

Objectives

A Multiply a monomial with a polynomial.
B Multiply two binomials.
C Multiply two polynomials.

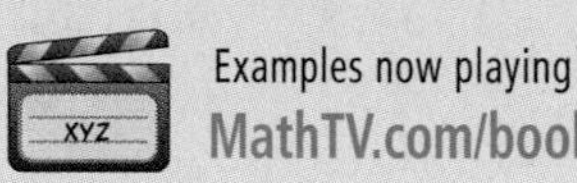
Examples now playing at
MathTV.com/books

A Multiplying Monomials with Polynomials

We begin our discussion of multiplication of polynomials by finding the product of a monomial and a trinomial.

EXAMPLE 1 Multiply $3x^2(2x^2 + 4x + 5)$.

SOLUTION Applying the distributive property gives us

$$\begin{aligned} 3x^2(2x^2 + 4x + 5) &= 3x^2(2x^2) + 3x^2(4x) + 3x^2(5) && \text{Distributive property} \\ &= 6x^4 + 12x^3 + 15x^2 && \text{Multiplication} \end{aligned}$$

B Multiplying Binomials

The distributive property is the key to multiplication of polynomials. We can use it to find the product of any two polynomials. There are some shortcuts we can use in certain situations, however. Let's look at an example that involves the product of two binomials.

EXAMPLE 2 Multiply $(3x - 5)(2x - 1)$.

SOLUTION

$$\begin{aligned} (3x - 5)(2x - 1) &= 3x(2x - 1) - 5(2x - 1) \\ &= 3x(2x) + 3x(-1) + (-5)(2x) + (-5)(-1) \\ &= 6x^2 - 3x - 10x + 5 \\ &= 6x^2 - 13x + 5 \end{aligned}$$

If we look closely at the second and third lines of work in the previous example, we can see that the terms in the answer come from all possible products of terms in the first binomial with terms in the second binomial. This result is generalized as follows.

Rule
To multiply any two polynomials, multiply each term in the first with each term in the second.

There are two ways we can put this rule to work.

FOIL Method

If we look at the original problem in Example 2 and then to the answer, we see that the first term in the answer came from multiplying the first terms in each binomial:

$$3x \cdot 2x = 6x^2 \qquad \text{FIRST}$$

The middle term in the answer came from adding the product of the two outside terms in each binomial and the product of the two inside terms in each binomial:

$$\begin{aligned} 3x(-1) &= -3x && \text{OUTSIDE} \\ -5(2x) &= \underline{-10x} && \text{INSIDE} \\ & \quad -13x \end{aligned}$$

PRACTICE PROBLEMS

1. Multiply $2x^3(4x^2 - 5x + 3)$.

2. Multiply $(2x - 4)(3x + 2)$.

Answers
1. $8x^5 - 10x^4 + 6x^3$
2. $6x^2 - 8x - 8$

The last term in the answer came from multiplying the two last terms:

$-5(-1) = 5$ LAST

To summarize the FOIL method, we will multiply another two binomials.

3. Multiply $(3x + 5)(2x - 1)$.

EXAMPLE 3 Multiply $(2x + 3)(5x - 4)$.

SOLUTION $(2x + 3)(5x - 4) = \underbrace{2x(5x)}_{\text{First}} + \underbrace{2x(-4)}_{\text{Outside}} + \underbrace{3(5x)}_{\text{Inside}} + \underbrace{3(-4)}_{\text{Last}}$

$= 10x^2 - 8x + 15x - 12$

$= 10x^2 + 7x - 12$

With practice $-8x + 15x = 7x$ can be done mentally.

C COLUMN Method

The FOIL method can be applied only when multiplying two binomials. To find products of polynomials with more than two terms, we use what is called the COLUMN method.

The COLUMN method of multiplying two polynomials is very similar to long multiplication with whole numbers. It is just another way of finding all possible products of terms in one polynomial with terms in another polynomial.

4. Multiply $(3x - 2)(2x^2 - 3x + 4)$.

EXAMPLE 4 Multiply $(2x + 3)(3x^2 - 2x + 1)$.

SOLUTION

$$\begin{array}{rl} 3x^2 - 2x + 1 & \\ 2x + 3 & \\ \hline 6x^3 - 4x^2 + 2x \quad\quad & \leftarrow 2x(3x^2 - 2x + 1) \\ 9x^2 - 6x + 3 & \leftarrow 3(3x^2 - 2x + 1) \\ \hline 6x^3 + 5x^2 - 4x + 3 & \leftarrow \text{Add similar terms} \end{array}$$

It will be to your advantage to become very fast and accurate at multiplying polynomials. You should be comfortable using either method. The following examples illustrate the three types of multiplication.

Multiply:

5. $5a^2(3a^2 - 2a + 4)$
6. $(x + 3)(y - 5)$
7. $(x + 2y)(a + b)$
8. $(4x + 1)(3x + 5)$

EXAMPLES Multiply:

5. $4a^2(2a^2 - 3a + 5) = 4a^2(2a^2) + 4a^2(-3a) + 4a^2(5)$
$= 8a^4 - 12a^3 + 20a^2$

6. $(x - 2)(y + 3) = \underset{F}{x(y)} + \underset{O}{x(3)} + \underset{I}{(-2)(y)} + \underset{L}{(-2)(3)}$
$= xy + 3x - 2y - 6$

7. $(x + y)(a - b) = \underset{F}{x(a)} + \underset{O}{x(-b)} + \underset{I}{y(a)} + \underset{L}{y(-b)}$
$= xa - xb + ya - yb$

8. $(5x - 1)(2x + 6) = \underset{F}{5x(2x)} + \underset{O}{5x(6)} + \underset{I}{(-1)(2x)} + \underset{L}{(-1)(6)}$
$= 10x^2 + 30x + (-2x) + (-6)$
$= 10x^2 + 28x - 6$

Answers

3. $6x^2 + 7x - 5$
4. $6x^3 - 13x^2 + 18x - 8$
5. $15a^4 - 10a^3 + 20a^2$
6. $xy - 5x + 3y - 15$
7. $ax + bx + 2ay + 2by$
8. $12x^2 + 23x + 5$

EXAMPLE 9 The length of a rectangle is 3 more than twice the width. Write an expression for the area of the rectangle.

SOLUTION We begin by drawing a rectangle and labeling the width with x. Since the length is 3 more than twice the width, we label the length with $2x + 3$.

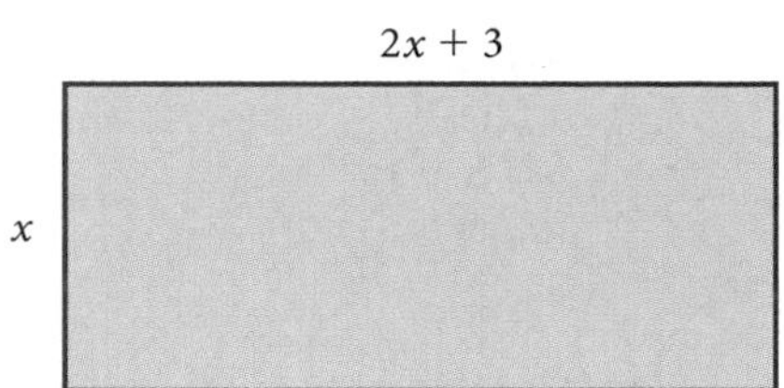

Since the area A of a rectangle is the product of the length and width, we write our formula for the area of this rectangle as

$$A = x(2x + 3)$$
$$A = 2x^2 + 3x \quad \text{Multiply}$$

Revenue

Suppose that a store sells x items at p dollars per item. The total amount of money obtained by selling the items is called the *revenue.* It can be found by multiplying the number of items sold, x, by the price per item, p. For example, if 100 items are sold for \$6 each, the revenue is $100(6) = \$600$. Similarly, if 500 items are sold for \$8 each, the total revenue is $500(8) = \$4,000$. If we denote the revenue with the letter R, then the formula that relates R, x, and p is

$$\text{Revenue} = (\text{number of items sold})(\text{price of each item})$$

In symbols: $R = xp$.

EXAMPLE 10 A store selling diskettes for home computers knows from past experience that it can sell x diskettes each day at a price of p dollars per diskette, according to the equation $x = 800 - 100p$. Write a formula for the daily revenue that involves only the variables R and p.

SOLUTION From our previous discussion we know that the revenue R is given by the formula

$$R = xp$$

But, since $x = 800 - 100p$, we can substitute $800 - 100p$ for x in the revenue equation to obtain

$$R = (800 - 100p)p$$
$$R = 800p - 100p^2$$

This last formula gives the revenue, R, in terms of the price, p.

Getting Ready for Class

After reading through the preceding section, respond in your own words and in complete sentences.

1. How do we multiply two polynomials?
2. Describe how the distributive property is used to multiply a monomial and a polynomial.
3. Describe how you would use the FOIL method to multiply two binomials.
4. Show how the product of two binomials can be a trinomial.

9. The length of a rectangle is 5 more than three times the width. Write an expression for the area.

10. Suppose the store in Example 10 can sell x diskettes for p dollars according to the equation $x = 700 - 110p$. Find the revenue.

Answers

9. $3x^2 + 5x$

10. $R = 700p - 110p^2$

Problem Set 5.5

A Multiply the following by applying the distributive property. **[Example 1]**

1. $2x(3x + 1)$
2. $4x(2x - 3)$
3. $2x^2(3x^2 - 2x + 1)$
4. $5x(4x^3 - 5x^2 + x)$
5. $2ab(a^2 - ab + 1)$
6. $3a^2b(a^3 + a^2b^2 + b^3)$
7. $y^2(3y^2 + 9y + 12)$
8. $5y(2y^2 - 3y + 5)$
9. $4x^2y(2x^3y + 3x^2y^2 + 8y^3)$
10. $6xy^3(2x^2 + 5xy + 12y^2)$

B C Multiply the following binomials. You should do about half the problems using the FOIL method and the other half using the COLUMN method. Remember, you want to be comfortable using both methods. **[Examples 2–8]**

11. $(x + 3)(x + 4)$
12. $(x + 2)(x + 5)$
13. $(x + 6)(x + 1)$
14. $(x + 1)(x + 4)$
15. $\left(x + \frac{1}{2}\right)\left(x + \frac{3}{2}\right)$
16. $\left(x + \frac{3}{5}\right)\left(x + \frac{2}{5}\right)$
17. $(a + 5)(a - 3)$
18. $(a - 8)(a + 2)$
19. $(x - a)(y + b)$
20. $(x + a)(y - b)$
21. $(x + 6)(x - 6)$
22. $(x + 3)(x - 3)$
23. $\left(y + \frac{5}{6}\right)\left(y - \frac{5}{6}\right)$
24. $\left(y - \frac{4}{7}\right)\left(y + \frac{4}{7}\right)$
25. $(2x - 3)(x - 4)$
26. $(3x - 5)(x - 2)$
27. $(a + 2)(2a - 1)$
28. $(a - 6)(3a + 2)$
29. $(2x - 5)(3x - 2)$
30. $(3x + 6)(2x - 1)$
31. $(2x + 3)(a + 4)$
32. $(2x - 3)(a - 4)$
33. $(5x - 4)(5x + 4)$
34. $(6x + 5)(6x - 5)$
35. $\left(2x - \frac{1}{2}\right)\left(x + \frac{3}{2}\right)$
36. $\left(4x - \frac{3}{2}\right)\left(x + \frac{1}{2}\right)$
37. $(1 - 2a)(3 - 4a)$
38. $(1 - 3a)(3 + 2a)$

B For each of the following problems, fill in the area of each small rectangle and square, and then add the results together to find the indicated product. **[Example 9]**

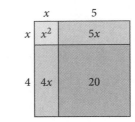

39. $(x + 2)(x + 3)$

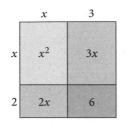

	x	3
x	x^2	$3x$
2	$2x$	6

40. $(x + 4)(x + 5)$

	x	5
x	x^2	$5x$
4	$4x$	20

41. $(x + 1)(2x + 2)$

	x	x	2
x	x^2	x^2	$2x$
1	x	x	2

42. $(2x + 1)(2x + 2)$

	x	x	2
x	x^2	x^2	$2x$
x	x^2	x^2	$2x$
1	x	x	2

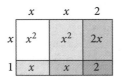

B C Multiply the following. **[Examples 2–8]**

43. $(a - 3)(a^2 - 3a + 2)$

44. $(a + 5)(a^2 + 2a + 3)$

45. $(x + 2)(x^2 - 2x + 4)$

46. $(x + 3)(x^2 - 3x + 9)$

47. $(2x + 1)(x^2 + 8x + 9)$

48. $(3x - 2)(x^2 - 7x + 8)$

49. $(5x^2 + 2x + 1)(x^2 - 3x + 5)$

50. $(2x^2 + x + 1)(x^2 - 4x + 3)$

Multiply.

51. $(x^2 + 3)(2x^2 - 5)$

52. $(4x^3 - 8)(5x^3 + 4)$

53. $(3a^4 + 2)(2a^2 + 5)$

54. $(7a^4 - 8)(4a^3 - 6)$

55. $(x + 3)(x + 4)(x + 5)$

56. $(x - 3)(x - 4)(x - 5)$

Simplify.

57. $(x - 3)(x - 2) + 2$

58. $(2x - 5)(3x + 2) - 4$

59. $(2x - 3)(4x + 3) + 4$

60. $(3x + 8)(5x - 7) + 52$

61. $(x + 4)(x - 5) + (-5)(2)$

62. $(x + 3)(x - 4) + (-4)(2)$

63. $2(x - 3) + x(x + 2)$

64. $5(x + 3) + 1(x + 4)$

65. $3x(x + 1) - 2x(x - 5)$

66. $4x(x - 2) - 3x(x - 4)$

67. $x(x + 2) - 3$

68. $2x(x - 4) + 6$

69. $a(a - 3) + 6$

70. $a(a - 4) + 8$

71. Find each product.

a. $(x + 1)(x - 1)$
b. $(x + 1)(x + 1)$
c. $(x + 1)(x^2 + 2x + 1)$
d. $(x + 1)(x^3 + 3x^2 + 3x + 1)$

72. Find each product.

a. $(x + 1)(x^2 - x + 1)$
b. $(x + 2)(x^2 - 2x + 4)$
c. $(x + 3)(x^2 - 3x + 9)$
d. $(x + 4)(x^2 - 4x + 16)$

73. Find each product.

a. $(x + 1)(x - 1)$
b. $(x + 1)(x - 2)$
c. $(x + 1)(x - 3)$
d. $(x + 1)(x - 4)$

74. Find each product.

a. $(x + 2)(x - 2)$
b. $(x - 2)(x^2 + 2x + 4)$
c. $(x^2 + 4)(x^2 - 4)$
d. $(x^3 + 8)(x^3 - 8)$

If the product of two expressions is 0, then one or both of the expressions must be zero. That is, the only way to multiply and get 0, is to multiply by 0. For each expression below, find all values of x that make the expression 0. (If the expression cannot be 0, say so.)

75. $5x$

76. $3x^2$

77. $x + 5$

78. $x^2 + 5$

79. $(x - 3)(x + 2)$

80. $x(x - 5)$

81. $x^2 + 16$

82. $x^2 - 100$

Applying the Concepts [Example 10]

Solar and Wind Energy The chart shows the cost to install either solar panels or a wind turbine. Use the chart to work Problems 83 and 84.

83. A homeowner is buying a certain number of solar panel modules. He is going to get a discount on each module that is equal to 25 times the number of modules he buys. Write an equation that describes this situation and simplify. Then find the cost if he buys 3 modules.

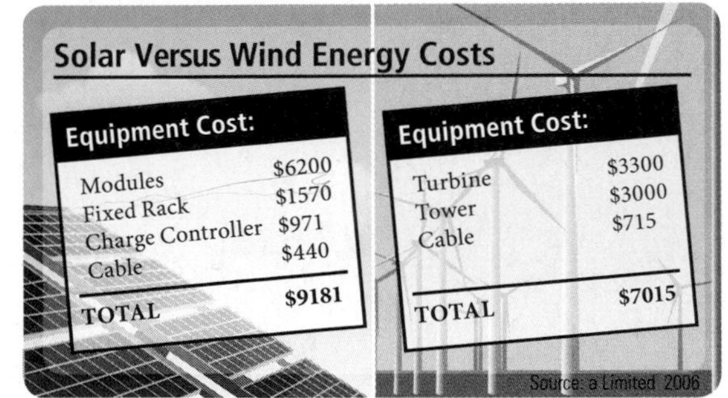

84. A farmer is replacing several turbines in his field. He is going to get a discount on each turbine that is equal to 50 times the number of turbines he buys. Write an expression that describes this situation and simplify. Then find the cost if he replaces 5 turbines.

85. Area The length of a rectangle is 5 units more than twice the width. Write an expression for the area of the rectangle.

$2x + 5$

x

86. Area The length of a rectangle is 2 more than three times the width. Write an expression for the area of the rectangle.

87. Area The width and length of a rectangle are given by two consecutive integers. Write an expression for the area of the rectangle.

88. Area The width and length of a rectangle are given by two consecutive even integers. Write an expression for the area of the rectangle.

89. Revenue A store selling typewriter ribbons knows that the number of ribbons it can sell each week, x, is related to the price per ribbon, p, by the equation $x = 1{,}200 - 100p$. Write an expression for the weekly revenue that involves only the variables R and p. (*Remember:* The equation for revenue is $R = xp$.)

90. Revenue A store selling small portable radios knows from past experience that the number of radios it can sell each week, x, is related to the price per radio, p, by the equation $x = 1{,}300 - 100p$. Write an expression for the weekly revenue that involves only the variables R and p.

Getting Ready for the Next Section

Simplify.

91. $13 \cdot 13$

92. $3x \cdot 3x$

93. $2(x)(-5)$

94. $2(2x)(-3)$

95. $6x + (-6x)$

96. $3x + (-3x)$

97. $(2x)(-3) + (2x)(3)$

98. $(2x)(-5y) + (2x)(5y)$

Multiply.

99. $-4(3x - 4)$

100. $-2x(2x + 7)$

101. $(x - 1)(x + 2)$

102. $(x + 5)(x - 6)$

103. $(x + 3)(x + 3)$

104. $(3x - 2)(3x - 2)$

Maintaining Your Skills

105. Solve this system by graphing:
$$x + y = 4$$
$$x - y = 2$$

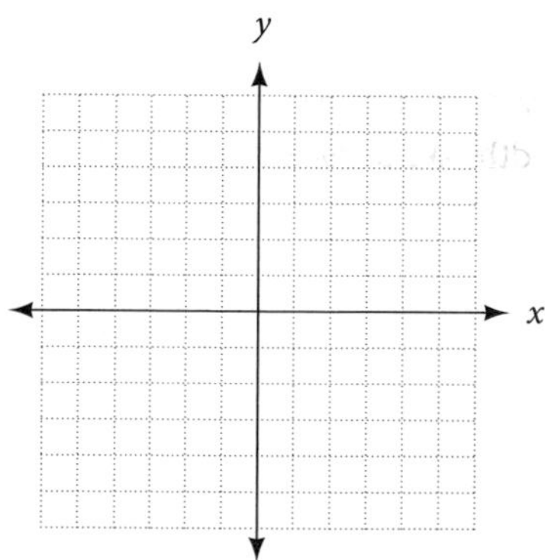

Solve each system by the elimination method.

106. $3x + 2y = 1$
$2x + y = 3$

107. $2x + 3y = -1$
$3x + 5y = -2$

Solve each system by the substitution method.

108. $x + y = 20$
$y = 5x + 2$

109. $2x - 6y = 2$
$y = 3x + 1$

110. Investing A total of $1,200 is invested in two accounts. One of the accounts pays 8% interest annually and the other pays 10% interest annually. If the total amount of interest earned from both accounts for the year is $104, how much is invested in each account?

111. Coin Problem Amy has $1.85 in dimes and quarters. If she has a total of 11 coins, how many of each coin does she have?

Binomial Squares and Other Special Products

5.6

Objectives

A Find the square of a binomial.

B Multiply expressions of the form $(a + b)(a - b)$.

XYZ
Examples now playing at
MathTV.com/books

A Squaring Binomials

In this section we will combine the results of the last section with our definition of exponents to find some special products.

EXAMPLE 1 Find the square of $(3x - 2)$.

SOLUTION To square $(3x - 2)$, we multiply it by itself:

$(3x - 2)^2 = (3x - 2)(3x - 2)$	Definition of exponents
$= 9x^2 - 6x - 6x + 4$	FOIL method
$= 9x^2 - 12x + 4$	Combine similar terms

PRACTICE PROBLEMS

1. Find the square of $(4x - 5)$.

Notice in Example 1 that the first and last terms in the answer are the squares of the first and last terms in the original problem and that the middle term is twice the product of the two terms in the original binomial.

EXAMPLES

2. $(a + b)^2 = (a + b)(a + b)$
   $= a^2 + 2ab + b^2$
3. $(a - b)^2 = (a - b)(a - b)$
   $= a^2 - 2ab + b^2$

Multiply:

2. $(x + y)^2$

3. $(x - y)^2$

Binomial squares having the form of Examples 2 and 3 occur very frequently in algebra. It will be to your advantage to memorize the following rule for squaring a binomial.

Note A very common mistake when squaring binomials is to write

$$(a + b)^2 = a^2 + b^2$$

which just isn't true. The mistake becomes obvious when we substitute 2 for a and 3 for b:

$$(2 + 3)^2 \neq 2^2 + 3^2$$
$$25 \neq 13$$

Exponents do not distribute over addition or subtraction.

Rule

The square of a binomial is the sum of the square of the first term, the square of the last term, and twice the product of the two original terms. In symbols this rule is written as follows:

$(x + y)^2$	$=$	x^2	$+$	$2xy$	$+$	y^2
		Square of first term		Twice product of the two terms		Square of last term

EXAMPLES Multiply using the preceding rule:

		First term squared		*Twice their product*		*Last term squared*		*Answer*
4. $(x - 5)^2$	$=$	x^2	$+$	$2(x)(-5)$	$+$	25	$=$	$x^2 - 10x + 25$
5. $(x + 2)^2$	$=$	x^2	$+$	$2(x)(2)$	$+$	4	$=$	$x^2 + 4x + 4$
6. $(2x - 3)^2$	$=$	$4x^2$	$+$	$2(2x)(-3)$	$+$	9	$=$	$4x^2 - 12x + 9$
7. $(5x - 4)^2$	$=$	$25x^2$	$+$	$2(5x)(-4)$	$+$	16	$=$	$25x^2 - 40x + 16$

Multiply using the method shown in Examples 4 through 7:

4. $(x - 3)^2$

5. $(x + 5)^2$

6. $(3x - 2)^2$

7. $(4x + 5)^2$

B More Special Products

Another special product that occurs frequently is $(a + b)(a - b)$. The only difference in the two binomials is the sign between the two terms. The interesting thing about this type of product is that the middle term is always zero. Here are some examples.

Answers

1. $16x^2 - 40x + 25$
2. $x^2 + 2xy + y^2$ 3. $x^2 - 2xy + y^2$
4. $x^2 - 6x + 9$ 5. $x^2 + 10x + 25$
6. $9x^2 - 12x + 4$
7. $16x^2 + 40x + 25$

Multiply using the FOIL method:

8. $(3x - 5)(3x + 5)$

9. $(x - 4)(x + 4)$

10. $(4x - 1)(4x + 1)$

EXAMPLES Multiply using the FOIL method:

8. $(2x - 3)(2x + 3) = 4x^2 + 6x - 6x - 9$ FOIL method
$= 4x^2 - 9$

9. $(x - 5)(x + 5) = x^2 + 5x - 5x - 25$ FOIL method
$= x^2 - 25$

10. $(3x - 1)(3x + 1) = 9x^2 + 3x - 3x - 1$ FOIL method
$= 9x^2 - 1$

Notice that in each case in the examples above, the middle term is zero and therefore doesn't appear in the answer. The answers all turn out to be the difference of two squares. Here is a rule to help you memorize the result.

Rule
When multiplying two binomials that differ only in the sign between their terms, subtract the square of the last term from the square of the first term.

$$(a - b)(a + b) = a^2 - b^2$$

Here are some problems that result in the difference of two squares.

Multiply using the formula $(a + b)(a - b) = a^2 - b^2$:

11. $(x + 2)(x - 2)$
12. $(a + 7)(a - 7)$
13. $(6a + 1)(6a - 1)$
14. $(8x - 2y)(8x + 2y)$
15. $(4a - 3b)(4a + 3b)$

EXAMPLES Multiply using the preceding rule:

11. $(x + 3)(x - 3) = x^2 - 9$
12. $(a + 2)(a - 2) = a^2 - 4$
13. $(9a + 1)(9a - 1) = 81a^2 - 1$
14. $(2x - 5y)(2x + 5y) = 4x^2 - 25y^2$
15. $(3a - 7b)(3a + 7b) = 9a^2 - 49b^2$

Although all the problems in this section can be worked correctly using the methods in the previous section, they can be done much faster if the two rules are *memorized*. Here is a summary of the two rules:

$$(a + b)^2 = (a + b)(a + b) = a^2 + 2ab + b^2$$
$$(a - b)^2 = (a - b)(a - b) = a^2 - 2ab + b^2$$
$$(a - b)(a + b) = a^2 - b^2$$

16. Expand and simplify:
$x^2 + (x - 1)^2 + (x + 1)^2$

EXAMPLE 16 Write an expression in symbols for the sum of the squares of three consecutive even integers. Then, simplify that expression.

SOLUTION If we let $x =$ the first of the even integers, then $x + 2$ is the next consecutive even integer, and $x + 4$ is the one after that. An expression for the sum of their squares is

$x^2 + (x + 2)^2 + (x + 4)^2$ Sum of squares
$= x^2 + (x^2 + 4x + 4) + (x^2 + 8x + 16)$ Expand squares
$= 3x^2 + 12x + 20$ Add similar terms

Getting Ready for Class

After reading through the preceding section, respond in your own words and in complete sentences.

1. Describe how you would square the binomial $a + b$.
2. Explain why $(x + 3)^2$ cannot be $x^2 + 9$.
3. What kind of products result in the difference of two squares?
4. When multiplied out, how will $(x + 3)^2$ and $(x - 3)^2$ differ?

Answers
8. $9x^2 - 25$ **9.** $x^2 - 16$
10. $16x^2 - 1$ **11.** $x^2 - 4$
12. $a^2 - 49$ **13.** $36a^2 - 1$
14. $64x^2 - 4y^2$ **15.** $16a^2 - 9b^2$
16. $3x^2 + 2$

Problem Set 5.6

A Perform the indicated operations. [Examples 1–7]

1. $(x - 2)^2$

2. $(x + 2)^2$

3. $(a + 3)^2$

4. $(a - 3)^2$

5. $(x - 5)^2$

6. $(x - 4)^2$

7. $\left(a - \frac{1}{2}\right)^2$

8. $\left(a + \frac{1}{2}\right)^2$

9. $(x + 10)^2$

10. $(x - 10)^2$

11. $(a + 0.8)^2$

12. $(a - 0.4)^2$

13. $(2x - 1)^2$

14. $(3x + 2)^2$

15. $(4a + 5)^2$

16. $(4a - 5)^2$

17. $(3x - 2)^2$

18. $(2x - 3)^2$

19. $(3a + 5b)^2$

20. $(5a - 3b)^2$

21. $(4x - 5y)^2$

22. $(5x + 4y)^2$

23. $(7m + 2n)^2$

24. $(2m - 7n)^2$

25. $(6x - 10y)^2$

26. $(10x + 6y)^2$

27. $(x^2 + 5)^2$

28. $(x^2 + 3)^2$

29. $(a^2 + 1)^2$

30. $(a^2 - 2)^2$

31. $\left(y + \frac{3}{2}\right)^2$

32. $\left(y - \frac{3}{2}\right)^2$

33. $\left(a + \frac{1}{2}\right)^2$

34. $\left(a - \frac{5}{2}\right)^2$

35. $\left(x + \frac{3}{4}\right)^2$

36. $\left(x - \frac{3}{8}\right)^2$

37. $\left(t + \frac{1}{5}\right)^2$

38. $\left(t - \frac{3}{5}\right)^2$

Comparing Expressions Fill in each table.

39.

x	$(x + 3)^2$	$x^2 + 9$	$x^2 + 6x + 9$
1			
2			
3			
4			

40.

x	$(x - 5)^2$	$x^2 + 25$	$x^2 - 10x + 25$
1			
2			
3			
4			

41.

a	b	$(a + b)^2$	$a^2 + b^2$	$a^2 + ab + b^2$	$a^2 + 2ab + b^2$
1	1				
3	5				
3	4				
4	5				

42.

a	b	$(a - b)^2$	$a^2 - b^2$	$a^2 - 2ab + b^2$
2	1			
5	2			
2	5			
4	3			

B Multiply. [Examples 8–15]

43. $(a + 5)(a - 5)$

44. $(a - 6)(a + 6)$

45. $(y - 1)(y + 1)$

46. $(y - 2)(y + 2)$

47. $(9 + x)(9 - x)$

48. $(10 - x)(10 + x)$

49. $(2x + 5)(2x - 5)$

50. $(3x + 5)(3x - 5)$

51. $\left(4x + \frac{1}{3}\right)\left(4x - \frac{1}{3}\right)$

52. $\left(6x + \frac{1}{4}\right)\left(6x - \frac{1}{4}\right)$

53. $(2a + 7)(2a - 7)$

54. $(3a + 10)(3a - 10)$

55. $(6 - 7x)(6 + 7x)$

56. $(7 - 6x)(7 + 6x)$

57. $(x^2 + 3)(x^2 - 3)$

58. $(x^2 + 2)(x^2 - 2)$

59. $(a^2 + 4)(a^2 - 4)$

60. $(a^2 + 9)(a^2 - 9)$

61. $(5y^4 - 8)(5y^4 + 8)$

62. $(7y^5 + 6)(7y^5 - 6)$

Multiply and simplify.

63. $(x + 3)(x - 3) + (x - 5)(x + 5)$

64. $(x - 7)(x + 7) + (x - 4)(x + 4)$

65. $(2x + 3)^2 - (4x - 1)^2$

66. $(3x - 5)^2 - (2x + 3)^2$

67. $(a + 1)^2 - (a + 2)^2 + (a + 3)^2$

68. $(a - 1)^2 + (a - 2)^2 - (a - 3)^2$

69. $(2x + 3)^3$

70. $(3x - 2)^3$

71. Find the value of each expression when x is 6.

a. $x^2 - 25$

b. $(x - 5)^2$

c. $(x + 5)(x - 5)$

72. Find the value of each expression when x is 5.

a. $x^2 - 9$

b. $(x - 3)^2$

c. $(x + 3)(x - 3)$

73. Evaluate each expression when x is -2.

a. $(x + 3)^2$

b. $x^2 + 9$

c. $x^2 + 6x + 9$

74. Evaluate each expression when x is -3.

a. $(x + 2)^2$

b. $x^2 + 4$

c. $x^2 + 4x + 4$

Applying the Concepts [Example 16]

75. Comparing Expressions Evaluate the expression $(x + 3)^2$ and the expression $x^2 + 6x + 9$ for $x = 2$.

76. Comparing Expressions Evaluate the expression $x^2 - 25$ and the expression $(x - 5)(x + 5)$ for $x = 6$.

77. Number Problem Write an expression for the sum of the squares of two consecutive integers. Then, simplify that expression.

78. Number Problem Write an expression for the sum of the squares of two consecutive odd integers. Then, simplify that expression.

79. Number Problem Write an expression for the sum of the squares of three consecutive integers. Then, simplify that expression.

80. Number Problem Write an expression for the sum of the squares of three consecutive odd integers. Then, simplify that expression.

81. Area We can use the concept of area to further justify our rule for squaring a binomial. The length of each side of the square shown in the figure is $a + b$. (The longer line segment has length a and the shorter line segment has length b.) The area of the whole square is $(a + b)^2$. However, the whole area is the sum of the areas of the two smaller squares and the two smaller rectangles that make it up. Write the area of the two smaller squares and the two smaller rectangles and then add them together to verify the formula $(a + b)^2 = a^2 + 2ab + b^2$.

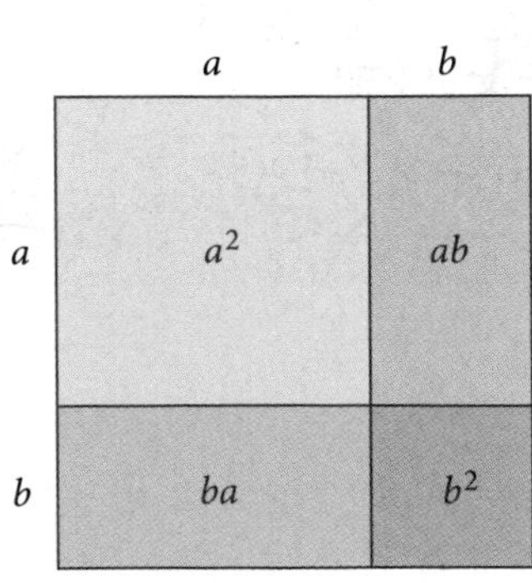

82. Area The length of each side of the large square shown in the figure is $x + 5$. Therefore, its area is $(x + 5)^2$. Find the area of the two smaller squares and the two smaller rectangles that make up the large square, then add them together to verify the formula $(x + 5)^2 = x^2 + 10x + 25$.

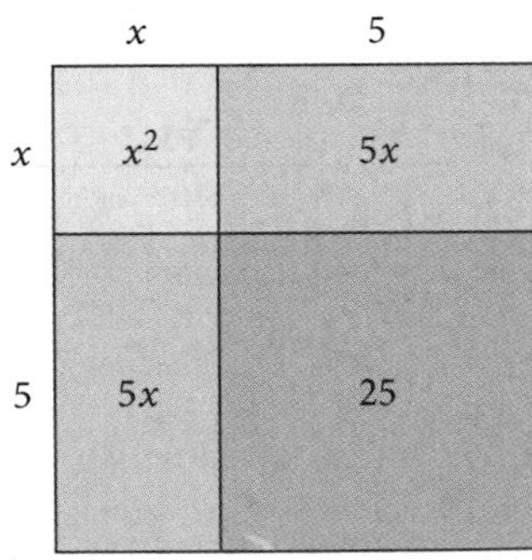

Getting Ready for the Next Section

Simplify each expression (divide).

83. $\dfrac{10x^3}{5x}$

84. $\dfrac{15x^2}{5x}$

85. $\dfrac{3x^2}{3}$

86. $\dfrac{4x^2}{2}$

87. $\dfrac{9x^2}{3x}$

88. $\dfrac{3x^4}{9x^2}$

89. $\dfrac{24x^3y^2}{8x^2y}$

90. $-\dfrac{4x^2y^3}{8x^2y}$

91. $\dfrac{15x^2y}{3xy}$

92. $\dfrac{21xy^2}{3xy}$

93. $\dfrac{35a^6b^8}{70a^2b^{10}}$

94. $\dfrac{75a^2b^6}{25a^4b^3}$

95. Biggest Hits The chart shows the number of hits for the three best charting artists in the United States. Write the number of hits the Beach Boys had over the numbers of hits the Beatles had, and then reduce to lowest terms.

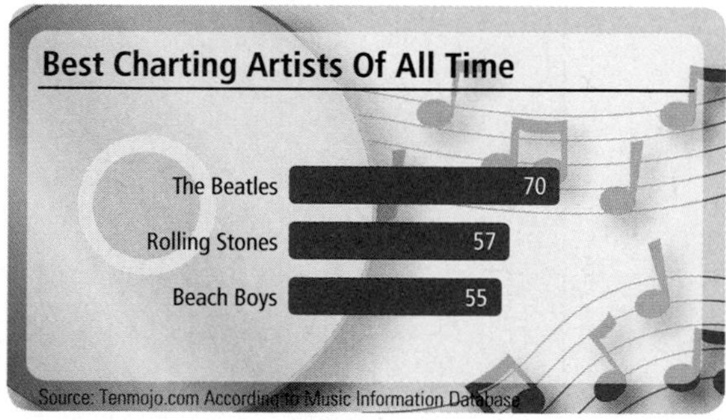

96. Picture Messaging The graph shows the number of text messages sent each month in 2008. Write the number of text messages during August over the number of text messages sent during April, and then reduce to lowest terms.

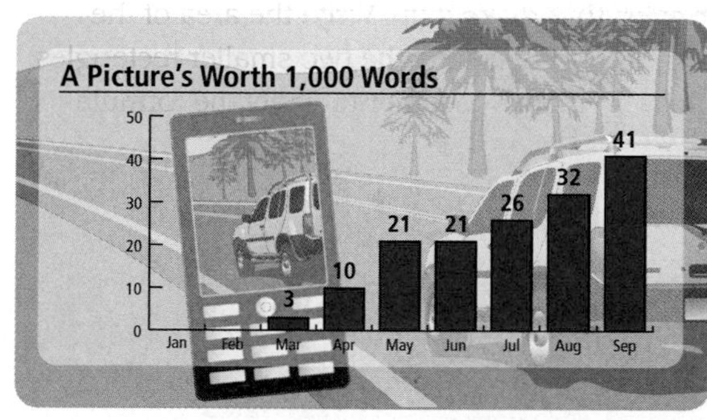

Maintaining Your Skills

Solve each system by graphing.

97. $x + y = 2$
$x - y = 4$

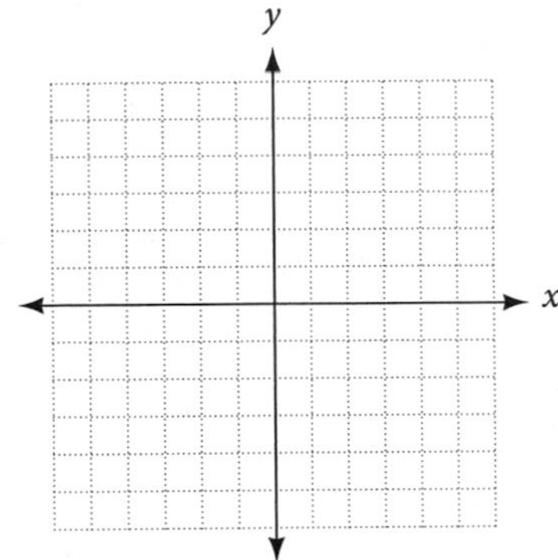

98. $x + y = 1$
$x - y = -3$

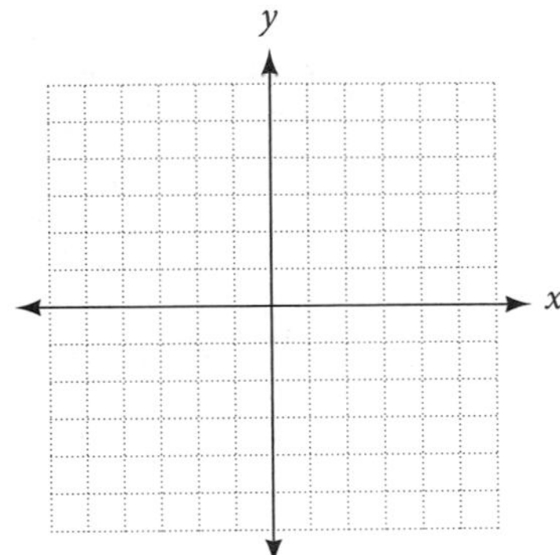

99. $y = 2x + 3$
$y = -2x - 1$

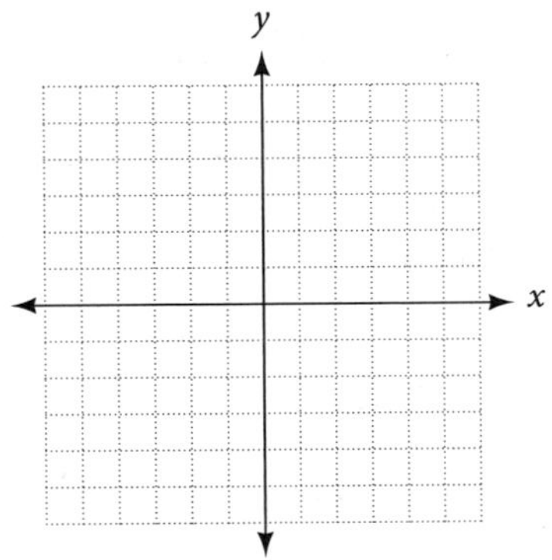

100. $y = 2x - 1$
$y = -2x + 3$

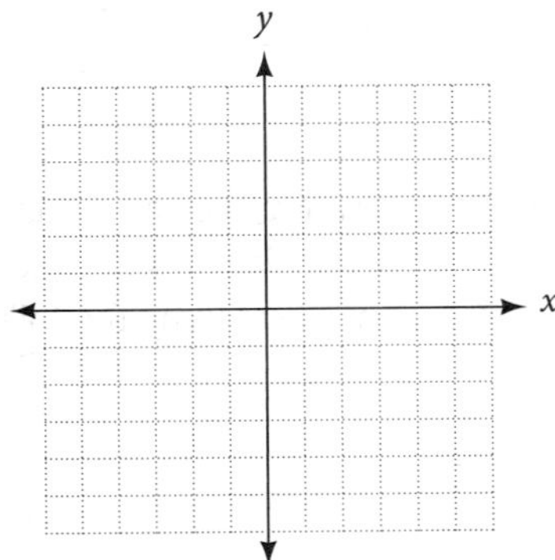

Dividing a Polynomial by a Monomial

5.7

Objectives

A Divide a polynomial by a monomial.

XYZ
Examples now playing at
MathTV.com/books

A Divide a Polynomial by a Monomial

To divide a polynomial by a monomial, we will use the definition of division and apply the distributive property. Follow the steps in this example closely.

EXAMPLE 1 Divide $10x^3 - 15x^2$ by $5x$.

SOLUTION

$$\frac{10x^3 - 15x^2}{5x} = (10x^3 - 15x^2)\frac{1}{5x}$$ Division by 5x is the same as multiplication by $\frac{1}{5x}$

$$= 10x^3\left(\frac{1}{5x}\right) - 15x^2\left(\frac{1}{5x}\right)$$ Distribute $\frac{1}{5x}$ to both terms

$$= \frac{10x^3}{5x} - \frac{15x^2}{5x}$$ Multiplication by $\frac{1}{5x}$ is the same as division by 5x

$$= 2x^2 - 3x$$ Division of monomials as done in Section 5.3

If we were to leave out the first steps in Examples, the problem would look like this:

$$\frac{10x^3 - 15x^2}{5x} = \frac{10x^3}{5x} - \frac{15x^2}{5x}$$
$$= 2x^2 - 3x$$

The problem is much shorter and clearer this way. You may leave out the first two steps from Example 1 when working problems in this section. They are part of Example 1 only to help show you why the following rule is true.

Rule

To divide a polynomial by a monomial, simply divide each term in the polynomial by the monomial.

Here are some further examples using our rule for division of a polynomial by a monomial.

EXAMPLE 2 Divide $\frac{3x^2 - 6}{3}$.

SOLUTION We begin by writing the 3 in the denominator under each term in the numerator. Then we simplify the result:

$$\frac{3x^2 - 6}{3} = \frac{3x^2}{3} - \frac{6}{3}$$ Divide each term in the numerator by 3

$$= x^2 - 2$$ Simplify

EXAMPLE 3 Divide $\frac{4x^2 - 2}{2}$.

SOLUTION Dividing each term in the numerator by 2, we have

$$\frac{4x^2 - 2}{2} = \frac{4x^2}{2} - \frac{2}{2}$$ Divide each term in the numerator by 2

$$= 2x^2 - 1$$ Simplify

PRACTICE PROBLEMS

1. Divide $8x^3 - 12x^2$ by $4x$.

Divide:

2. $\frac{5x^2 - 10}{5}$

3. $\frac{8x^2 - 2}{2}$

Answers

1. $2x^2 - 3x$ 2. $x^2 - 2$ 3. $4x^2 - 1$

4. Find the quotient of $15x^4 - 10x^3$ and $5x$.

5. Divide: $(9x^3y^2 - 18x^2y^3) \div (-9xy)$

6. Divide $\dfrac{21x^3y^2 + 14x^2y^2 - 7x^2y^3}{7x^2y}$.

EXAMPLE 4 Find the quotient of $27x^3 - 9x^2$ and $3x$.

SOLUTION We again are asked to divide the first polynomial by the second one:

$$\frac{27x^3 - 9x^2}{3x} = \frac{27x^3}{3x} - \frac{9x^2}{3x} \quad \text{Divide each term by } 3x$$

$$= 9x^2 - 3x \quad \text{Simplify}$$

EXAMPLE 5 Divide $(15x^2y - 21xy^2) \div (-3xy)$.

SOLUTION This is the same type of problem we have shown in the first four examples; it is just worded a little differently. Note that when we divide each term in the first polynomial by $-3xy$, the negative sign must be taken into account:

$$\frac{15x^2y - 21xy^2}{-3xy} = \frac{15x^2y}{-3xy} - \frac{21xy^2}{-3xy} \quad \text{Divide each term by } -3xy$$

$$= -5x - (-7y) \quad \text{Simplify}$$

$$= -5x + 7y \quad \text{Simplify}$$

EXAMPLE 6 Divide $\dfrac{24x^3y^2 + 16x^2y^2 - 4x^2y^3}{8x^2y}$.

SOLUTION Writing $8x^2y$ under each term in the numerator and then simplifying, we have

$$\frac{24x^3y^2 + 16x^2y^2 - 4x^2y^3}{8x^2y} = \frac{24x^3y^2}{8x^2y} + \frac{16x^2y^2}{8x^2y} - \frac{4x^2y^3}{8x^2y}$$

$$= 3xy + 2y - \frac{y^2}{2}$$

From the examples in this section, it is clear that to divide a polynomial by a monomial, we must divide each term in the polynomial by the monomial. Often, students taking algebra for the first time will make the following mistake:

$$\frac{x + \not{2}}{\not{2}} = x + 1 \quad \text{Mistake}$$

The mistake here is in not dividing both terms in the numerator by 2. The correct way to divide $x + 2$ by 2 looks like this:

$$\frac{x + 2}{2} = \frac{x}{2} + \frac{2}{2} = \frac{x}{2} + 1 \quad \text{Correct}$$

Getting Ready for Class

After reading through the preceding section, respond in your own words and in complete sentences.

1. What property of real numbers is the key to dividing a polynomial by a monomial?
2. Describe how you would divide a polynomial by $5x$.
3. Is our answer to Example 6 a polynomial?
4. Explain the mistake in the problem $\dfrac{x + 2}{2} = x + 1$.

Answers

4. $3x^3 - 2x^2$ 5. $-x^2y + 2xy^2$
6. $3xy + 2y - y^2$

Problem Set 5.7

A Divide the following polynomials by $5x$. [Example 1]

1. $5x^2 - 10x$
2. $10x^3 - 15x$
3. $15x - 10x^3$
4. $50x^3 - 20x^2$

5. $25x^2y - 10xy$
6. $15xy^2 + 20x^2y$
7. $35x^5 - 30x^4 + 25x^3$
8. $40x^4 - 30x^3 + 20x^2$

9. $50x^5 - 25x^3 + 5x$
10. $75x^6 + 50x^3 - 25x$

A Divide the following by $-2a$. [Example 1]

11. $8a^2 - 4a$
12. $a^3 - 6a^2$
13. $16a^5 + 24a^4$
14. $30a^6 + 20a^3$

15. $8ab + 10a^2$
16. $6a^2b - 10ab^2$
17. $12a^3b - 6a^2b^2 + 14ab^3$
18. $4ab^3 - 16a^2b^2 - 22a^3b$

19. $a^2 + 2ab + b^2$
20. $a^2b - 2ab^2 + b^3$

A Perform the following divisions (find the following quotients). [Examples 1–6]

21. $\frac{6x + 8y}{2}$
22. $\frac{9x - 3y}{3}$
23. $\frac{7y - 21}{-7}$
24. $\frac{14y - 12}{2}$

25. $\frac{2x^2 + 16x - 18}{2}$
26. $\frac{3x^2 - 3x - 18}{3}$
27. $\frac{3y^2 - 9y + 3}{3}$
28. $\frac{2y^2 - 8y + 2}{2}$

29. $\dfrac{10xy - 8x}{2x}$

30. $\dfrac{12xy^2 - 18x}{-6x}$

31. $\dfrac{x^2y - x^3y^2}{x}$

32. $\dfrac{x^2y - x^3y^2}{x^2}$

33. $\dfrac{x^2y - x^3y^2}{-x^2y}$

34. $\dfrac{ab + a^2b^2}{ab}$

35. $\dfrac{a^2b^2 - ab^2}{-ab^2}$

36. $\dfrac{a^2b^2c + ab^2c^2}{abc}$

37. $\dfrac{x^3 - 3x^2y + xy^2}{x}$

38. $\dfrac{x^2 - 3xy^2 + xy^3}{x}$

39. $\dfrac{10a^2 - 15a^2b + 25a^2b^2}{5a^2}$

40. $\dfrac{11a^2b^2 - 33ab}{-11ab}$

41. $\dfrac{26x^2y^2 - 13xy}{-13xy}$

42. $\dfrac{6x^2y^2 - 3xy}{6xy}$

43. $\dfrac{4x^2y^2 - 2xy}{4xy}$

44. $\dfrac{6x^2a + 12x^2b - 6x^2c}{36x^2}$

45. $\dfrac{5a^2x - 10ax^2 + 15a^2x^2}{20a^2x^2}$

46. $\dfrac{12ax - 9bx + 18cx}{6x^2}$

47. $\dfrac{16x^5 + 8x^2 + 12x}{12x^3}$

48. $\dfrac{27x^2 - 9x^3 - 18x^4}{-18x^3}$

A Divide. Assume all variables represent positive numbers. **[Example 6]**

49. $\dfrac{9a^{5m} - 27a^{3m}}{3a^{2m}}$

50. $\dfrac{26a^{3m} - 39a^{5m}}{13a^{3m}}$

51. $\dfrac{10x^{5m} - 25x^{3m} + 35x^{m}}{5x^{m}}$

52. $\dfrac{18x^{2m} + 24x^{4m} - 30x^{6m}}{6x^{2m}}$

A Simplify each numerator, and then divide.

53. $\dfrac{2x^3(3x + 2) - 3x^2(2x - 4)}{2x^2}$

54. $\dfrac{5x^2(6x - 3) + 6x^3(3x - 1)}{3x}$

55. $\dfrac{(x + 2)^2 - (x - 2)^2}{2x}$

56. $\dfrac{(x - 3)^2 - (x + 3)^2}{3x}$

57. $\dfrac{(x + 5)^2 + (x + 5)(x - 5)}{2x}$

58. $\dfrac{(x - 4)^2 + (x + 4)(x - 4)}{2x}$

59. Find the value of each expression when x is 2.

a. $2x + 3$

b. $\dfrac{10x + 15}{5}$

c. $10x + 3$

60. Find the value of each expression when x is 5.

a. $3x + 2$

b. $\dfrac{6x^2 + 4x}{2x}$

c. $6x^2 + 2$

61. Evaluate each expression for $x = 10$.

a. $\dfrac{3x + 8}{2}$

b. $3x + 4$

c. $\dfrac{3}{2}x + 4$

62. Evaluate each expression for $x = 10$.

a. $\dfrac{5x - 6}{2}$

b. $5x - 3$

c. $\dfrac{5}{2}x - 3$

63. Find the value of each expression when x is 2.

a. $2x^2 - 3x$

b. $\dfrac{10x^3 - 15x^2}{5x}$

64. Find the value of each expression when a is 3.

a. $-4a + 2$

b. $\dfrac{8a^2 - 4a}{-2a}$

65. Comparing Expressions Evaluate the expression $\dfrac{10x + 15}{5}$ and the expression $2x + 3$ when $x = 2$.

66. Comparing Expressions Evaluate the expression $\dfrac{6x^2 + 4x}{2x}$ and the expression $3x + 2$ when $x = 5$.

67. Comparing Expressions Show that the expression $\dfrac{3x + 8}{2}$ is not the same as the expression $3x + 4$ by replacing x with 10 in both expressions and simplifying the results.

68. Comparing Expressions Show that the expression $\dfrac{x + 10}{x}$ is not equal to 10 by replacing x with 5 and simplifying.

Getting Ready for the Next Section

Divide.

69. $27\overline{)3,962}$

70. $13\overline{)18,780}$

71. $\dfrac{2x^2 + 5x}{x}$

72. $\dfrac{7x^2 + 9x^3 + 3x^7}{x^3}$

Multiply

73. $(x - 3)x$

74. $(x - 3)(-2)$

75. $2x^2(x - 5)$

76. $10x(x - 5)$

Subtract.

77. $(x^2 - 5x) - (x^2 - 3x)$

78. $(2x^3 + 0x^2) - (2x^3 - 10x^2)$

79. $(-2x + 8) - (-2x + 6)$

80. $(4x - 14) - (4x - 10)$

Maintaining Your Skills

81. Sound The low-frequency pulses made by blue whales have been measured up to 188 decibels (dB) making it the loudest sound emitted by any living source. The chart shows the decibel level of various sounds. Half the sum of the loudness of a lawn mower from 3 feet away and 13 decibels is the loudness of normal conversation. What is the loudness of a lawn mower?

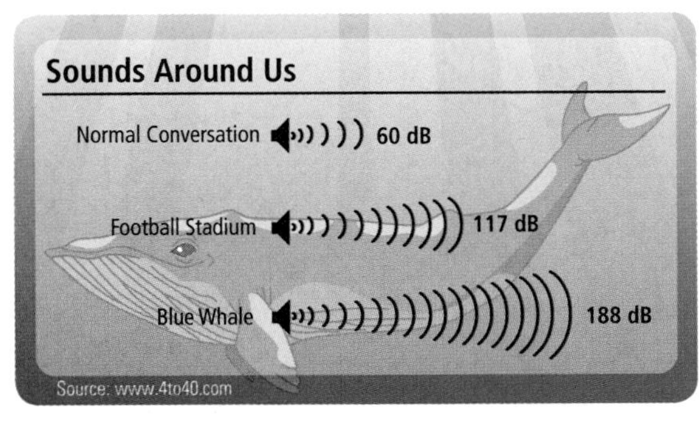

82. Skyscrapers The chart shows the heights of the three tallest buildings in the world. Five times the height of Big Ben in London, England, and 70 feet is the height of Taipei 101 tower. How tall is Big Ben?

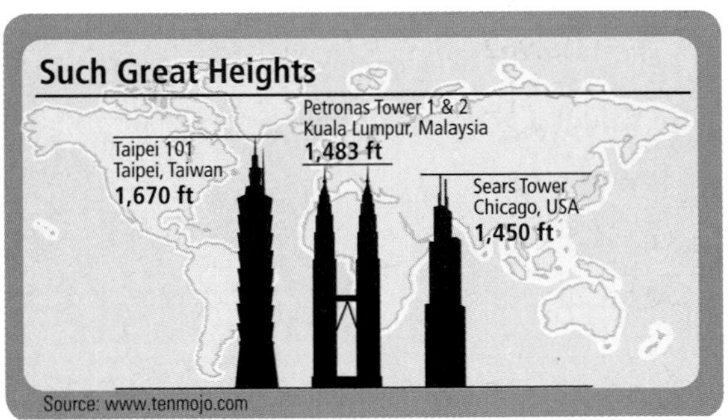

Solve each system of equations by the elimination method.

83. $x + y = 6$
$x - y = 8$

84. $2x + y = 5$
$-x + y = -4$

85. $2x - 3y = -5$
$x + y = 5$

86. $2x - 4y = 10$
$3x - 2y = -1$

Solve each system by the substitution method.

87. $x + y = 2$
$y = 2x - 1$

88. $2x - 3y = 4$
$x = 3y - 1$

89. $4x + 2y = 8$
$y = -2x + 4$

90. $4x + 2y = 8$
$y = -2x + 5$

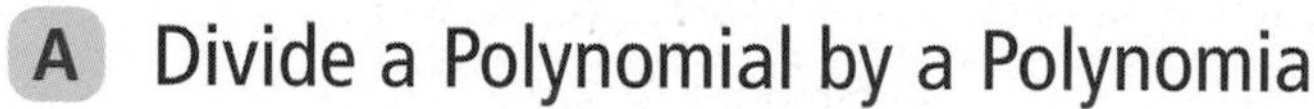

Dividing a Polynomial by a Polynomial

A Divide a Polynomial by a Polynomial

Since long division for polynomials is very similar to long division with whole numbers, we will begin by reviewing a division problem with whole numbers. You may realize when looking at Example 1 that you don't have a very good idea why you proceed as you do with long division. What you do know is that the process always works. We are going to approach the explanations in this section in much the same manner; that is, we won't always be sure why the steps we will use are important, only that they always produce the correct result.

EXAMPLE 1 Divide $27\overline{)3,962}$.

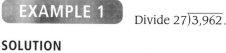

SOLUTION

```
     1        ← Estimate 27 into 39
27)3,962
   2 7        ← Multiply 1 × 27 = 27
   1 2        ← Subtract 39 − 27 = 12
```

```
     1
27)3,962
   2 7↓
   1 26       ← Bring down the 6
```

These are the four basic steps in long division. Estimate, multiply, subtract, and bring down the next term. To finish the problem, we simply perform the same four steps again:

```
     14       ← 4 is the estimate
27)3,962
   2 7↓
   1 26
   1 08       ← Multiply to get 108
     182      ← Subtract to get 18, then bring down the 2
```

One more time.

```
     146      ← 6 is the estimate
27)3,962
   2 7
   1 26
   1 08
     182
     162      ← Multiply to get 162
      20      ← Subtract to get 20
```

Since there is nothing left to bring down, we have our answer.

$$\frac{3,962}{27} = 146 + \frac{20}{27} \quad \text{or} \quad 146\frac{20}{27}$$

Here is how it works with polynomials.

Objectives

A Divide a polynomial by a polynomial.

Examples now playing at MathTV.com/books

PRACTICE PROBLEMS

1. Divide $35\overline{)4,281}$.

Answer

1. $122 + \frac{11}{35}$

2. Divide $\frac{x^2 - 5x + 8}{x - 2}$.

EXAMPLE 2 Divide $\frac{x^2 - 5x + 8}{x - 3}$.

SOLUTION

$$\begin{array}{r} x \\ x - 3\overline{)\ x^2 - 5x + 8} \\ \underline{\not- x^2 \not+ 3x} \\ -2x \end{array}$$

← Estimate $x^2 \div x = x$

← Multiply $x(x - 3) = x^2 - 3x$

← Subtract $(x^2 - 5x) - (x^2 - 3x) = -2x$

$$\begin{array}{r} x \\ x - 3\overline{)\ x^2 - 5x + 8} \\ \underline{\not- x^2 \not+ 3x \quad \downarrow} \\ -2x + 8 \end{array}$$

← Bring down the 8

Notice that to subtract one polynomial from another, we add its opposite. That is why we change the signs on $x^2 - 3x$ and add what we get to $x^2 - 5x$. (To subtract the second polynomial, simply change the signs and add.)

We perform the same four steps again:

$$\begin{array}{r} x - 2 \\ x - 3\overline{)\ x^2 - 5x + 8} \\ \underline{\not- x^2 \not+ 3x \quad \downarrow} \\ -2x + 8 \\ \underline{\not+ 2x \not- 6} \\ 2 \end{array}$$

← −2 is the estimate $(-2x \div x = -2)$

← Multiply $-2(x - 3) = -2x + 6$

← Subtract $(-2x + 8) - (-2x + 6) = 2$

Since there is nothing left to bring down, we have our answer:

$$\frac{x^2 - 5x + 8}{x - 3} = x - 2 + \frac{2}{x - 3}$$

To check our answer, we multiply $(x - 3)(x - 2)$ to get $x^2 - 5x + 6$. Then, adding on the remainder, 2, we have $x^2 - 5x + 8$.

3. Divide $\frac{8x^2 - 6x - 5}{2x - 3}$.

EXAMPLE 3 Divide $\frac{6x^2 - 11x - 14}{2x - 5}$.

SOLUTION

$$\begin{array}{r} 3x + 2 \\ 2x - 5\overline{)\ 6x^2 - 11x - 14} \\ \underline{\not- 6x^2 \not+ 15x \quad \downarrow} \\ +4x - 14 \\ \underline{\not- 4x \not+ 10} \\ -4 \end{array}$$

$$\frac{6x^2 - 11x - 14}{2x - 5} = 3x + 2 + \frac{-4}{2x - 5}$$

One last step is sometimes necessary. The two polynomials in a division problem must both be in descending powers of the variable and cannot skip any powers from the highest power down to the constant term.

Answers

2. $x - 3 + \frac{2}{x - 2}$ 3. $4x + 3 + \frac{4}{2x - 3}$

EXAMPLE 4 Divide $\frac{2x^3 - 3x + 2}{x - 5}$.

SOLUTION The problem will be much less confusing if we write $2x^3 - 3x + 2$ as $2x^3 + 0x^2 - 3x + 2$. Adding $0x^2$ does not change our original problem.

$$\begin{array}{rl}
 & 2x^2 \quad \leftarrow \text{Estimate } 2x^3 \div x = 2x^2 \\
x - 5 \,\big)\!\! & \overline{2x^3 + 0x^2 - 3x + 2} \\
 & \underline{\not{-}\,2x^3 \not{+}\, 10x^2} \downarrow \quad \leftarrow \text{Multiply } 2x^2(x - 5) = 2x^3 - 10x^2 \\
 & \quad + 10x^2 - 3x \quad \leftarrow \text{Subtract: } (2x^3 + 0x^2) - (2x^3 - 10x^2) = 10x^2 \\
 & \qquad \nwarrow \text{Bring down the next term}
\end{array}$$

Adding the term $0x^2$ gives us a column in which to write $10x^2$. (Remember, you can add and subtract only similar terms.)

Here is the completed problem:

$$\begin{array}{rl}
 & 2x^2 + 10x + 47 \\
x - 5 \,\big)\!\! & \overline{2x^3 + 0x^2 - 3x + 2} \\
 & \underline{\not{-}\,2x^3 \not{+}\, 10x^2} \downarrow \\
 & \quad + 10x^2 - 3x \\
 & \quad \underline{\not{-}\,10x^2 \not{+}\, 50x} \downarrow \\
 & \qquad + 47x + 2 \\
 & \qquad \underline{\not{-}\,47x \not{+}\, 235} \\
 & \qquad \qquad 237
\end{array}$$

Our answer is $\frac{2x^3 - 3x + 2}{x - 5} = 2x^2 + 10x + 47 + \frac{237}{x - 5}$.

As you can see, long division with polynomials is a mechanical process. Once you have done it correctly a couple of times, it becomes very easy to produce the correct answer.

Getting Ready for Class

After reading through the preceding section, respond in your own words and in complete sentences.

1. What are the four steps used in long division with whole numbers?
2. How is division of two polynomials similar to long division with whole numbers?
3. What are the four steps used in long division with polynomials?
4. How do we use 0 when dividing the polynomial $2x^3 - 3x + 2$ by $x - 5$?

4. Divide $\frac{3x^3 - 2x + 1}{x - 3}$.

Answer

4. $3x^2 + 9x + 25 + \frac{76}{x - 3}$

Problem Set 5.8

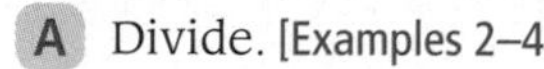

A Divide. [Examples 2–4]

1. $\dfrac{x^2 - 5x + 6}{x - 3}$

2. $\dfrac{x^2 - 5x + 6}{x - 2}$

3. $\dfrac{a^2 + 9a + 20}{a + 5}$

4. $\dfrac{a^2 + 9a + 20}{a + 4}$

5. $\dfrac{x^2 - 6x + 9}{x - 3}$

6. $\dfrac{x^2 + 10x + 25}{x + 5}$

7. $\dfrac{2x^2 + 5x - 3}{2x - 1}$

8. $\dfrac{4x^2 + 4x - 3}{2x - 1}$

9. $\dfrac{2a^2 - 9a - 5}{2a + 1}$

10. $\dfrac{4a^2 - 8a - 5}{2a + 1}$

11. $\dfrac{x^2 + 5x + 8}{x + 3}$

12. $\dfrac{x^2 + 5x + 4}{x + 3}$

13. $\dfrac{a^2 + 3a + 2}{a + 5}$

14. $\dfrac{a^2 + 4a + 3}{a + 5}$

15. $\dfrac{x^2 + 2x + 1}{x - 2}$

16. $\dfrac{x^2 + 6x + 9}{x - 3}$

17. $\dfrac{x^2 + 5x - 6}{x + 1}$

18. $\dfrac{x^2 - x - 6}{x + 1}$

19. $\dfrac{a^2 + 3a + 1}{a + 2}$

20. $\dfrac{a^2 - a + 3}{a + 1}$

21. $\dfrac{2x^2 - 2x + 5}{2x + 4}$

22. $\dfrac{15x^2 + 19x - 4}{3x + 8}$

23. $\dfrac{6a^2 + 5a + 1}{2a + 3}$

24. $\dfrac{4a^2 + 4a + 3}{2a + 1}$

25. $\dfrac{6a^3 - 13a^2 - 4a + 15}{3a - 5}$

26. $\dfrac{2a^3 - a^2 + 3a + 2}{2a + 1}$

Fill in the missing terms in the numerator, and then use long division to find the quotients (see Example 4).

27. $\dfrac{x^3 + 4x + 5}{x + 1}$

28. $\dfrac{x^3 + 4x^2 - 8}{x + 2}$

29. $\dfrac{x^3 - 1}{x - 1}$

30. $\dfrac{x^3 + 1}{x + 1}$

31. $\dfrac{x^3 - 8}{x - 2}$

32. $\dfrac{x^3 + 27}{x + 3}$

33. Find the value of each expression when x is 3.

a. $x^2 + 2x + 4$

b. $\dfrac{x^3 - 8}{x - 2}$

c. $x^2 - 4$

34. Find the value of each expression when x is 2.

a. $x^2 - 3x + 9$

b. $\dfrac{x^3 + 27}{x + 3}$

c. $x^2 + 9$

35. Find the value of each expression when x is 4.

a. $x + 3$

b. $\dfrac{x^2 + 9}{x + 3}$

c. $x - 3 + \dfrac{18}{x + 3}$

36. Find the value of each expression when x is 2.

a. $x + 1$

b. $\dfrac{x^2 + 1}{x + 1}$

c. $x - 1 + \dfrac{2}{x + 1}$

Applying the Concepts

37. Golf The chart shows the lengths of the longest golf courses used for the U.S. Open. If the length of the Bellerive course is 7,214 yards, what is the average length of each hole on the 18-hole course? Round to the nearest yard.

38. Cost of Living The charts shows how much it costs to live in some the most expensive cities in the United States. What is the monthly cost to live in Washington D.C.? Round to the nearest cent.

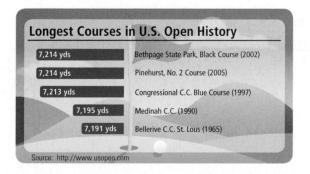

Longest Courses in U.S. Open History

Length	Course
7,214 yds	Bethpage State Park, Black Course (2002)
7,214 yds	Pinehurst, No. 2 Course (2005)
7,213 yds	Congressional C.C. Blue Course (1997)
7,195 yds	Medinah C.C. (1990)
7,191 yds	Bellerive C.C. St. Lous (1965)

Source: http://www.usopen.com

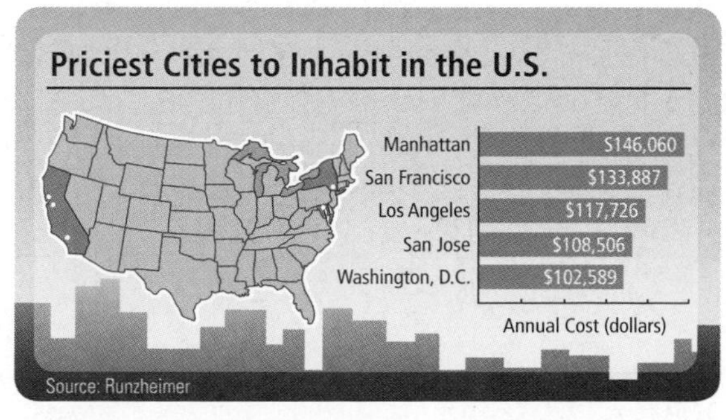

Maintaining Your Skills

Use systems of equations to solve the following word problems.

39. **Number Problem** The sum of two numbers is 25. One of the numbers is 4 times the other. Find the numbers.

40. **Number Problem** The sum of two numbers is 24. One of the numbers is 3 more than twice the other. Find the numbers.

41. **Investing** Suppose you have a total of $1,200 invested in two accounts. One of the accounts pays 8% annual interest and the other pays 9% annual interest. If your total interest for the year is $100, how much money did you invest in each of the accounts?

42. **Investing** If you invest twice as much money in an account that pays 12% annual interest as you do in an account that pays 11% annual interest, how much do you have in each account if your total interest for a year is $210?

43. **Money Problem** If you have a total of $160 in $5 bills and $10 bills, how many of each type of bill do you have if you have 4 more $10 bills than $5 bills?

44. **Coin Problem** Suppose you have 20 coins worth a total of $2.80. If the coins are all nickels and quarters, how many of each type do you have?

45. **Mixture Problem** How many gallons of 20% antifreeze solution and 60% antifreeze solution must be mixed to get 16 gallons of 35% antifreeze solution?

46. **Mixture Problem** A chemist wants to obtain 80 liters of a solution that is 12% hydrochloric acid. How many liters of 10% hydrochloric acid solution and 20% hydrochloric acid solution should he mix to do so?

Chapter 5 Summary

Exponents: Definition and Properties [5.1, 5.2]

Integer exponents indicate repeated multiplications.

$a^r \cdot a^s = a^{r+s}$ To multiply with the same base, you add exponents

$\dfrac{a^r}{a^s} = a^{r-s}$ To divide with the same base, you subtract exponents

$(ab)^r = a^r \cdot b^r$ Exponents distribute over multiplication

$\left(\dfrac{a}{b}\right)^r = \dfrac{a^r}{b^r}$ Exponents distribute over division

$(a^r)^s = a^{r \cdot s}$ A power of a power is the product of the powers

$a^{-r} = \dfrac{1}{a^r}$ Negative exponents imply reciprocals

EXAMPLES

1. a. $2^3 = 2 \cdot 2 \cdot 2 = 8$

b. $x^5 \cdot x^3 = x^{5+3} = x^8$

c. $\dfrac{x^5}{x^3} = x^{5-3} = x^2$

d. $(3x)^2 = 3^2 \cdot x^2 = 9x^2$

e. $\left(\dfrac{2}{3}\right)^3 = \dfrac{2^3}{3^3} = \dfrac{8}{27}$

f. $(x^5)^3 = x^{5 \cdot 3} = x^{15}$

g. $3^{-2} = \dfrac{1}{3^2} = \dfrac{1}{9}$

Multiplication of Monomials [5.3]

To multiply two monomials, multiply coefficients and add exponents.

2. $(5x^2)(3x^4) = 15x^6$

Division of Monomials [5.3]

To divide two monomials, divide coefficients and subtract exponents.

3. $\dfrac{12x^9}{4x^5} = 3x^4$

Scientific Notation [5.1, 5.2, 5.3]

A number is in scientific notation when it is written as the product of a number between 1 and 10 and an integer power of 10.

4. $768{,}000 = 7.68 \times 10^5$
$0.00039 = 3.9 \times 10^{-4}$

Addition of Polynomials [5.4]

To add two polynomials, add coefficients of similar terms.

5. $(3x^2 - 2x + 1) + (2x^2 + 7x - 3)$
$= 5x^2 + 5x - 2$

Subtraction of Polynomials [5.4]

To subtract one polynomial from another, add the opposite of the second to the first.

6. $(3x + 5) - (4x - 3)$
$= 3x + 5 - 4x + 3$
$= -x + 8$

Multiplication of Polynomials [5.5]

To multiply a polynomial by a monomial, we apply the distributive property. To multiply two binomials we use the FOIL method. In other situations we use the COLUMN method. Each method achieves the same result: To multiply any two polynomials, we multiply each term in the first polynomial by each term in the second polynomial.

7. a. $2a^2(5a^2 + 3a - 2) = 10a^4 + 6a^3 - 4a^2$

b. $(x + 2)(3x - 1) = 3x^2 - x + 6x - 2 = 3x^2 + 5x - 2$

c.
$$\begin{array}{r} 2x^2 - 3x + 4 \\ 3x - 2 \\ \hline 6x^3 - 9x^2 + 12x \\ -4x^2 + 6x - 8 \\ \hline 6x^3 - 13x^2 + 18x - 8 \end{array}$$

Special Products [5.6]

$$(a + b)^2 = a^2 + 2ab + b^2$$
$$(a - b)^2 = a^2 - 2ab + b^2$$
Binomial squares

$$(a + b)(a - b) = a^2 - b^2$$
Difference of two squares

8. $(x + 3)^2 = x^2 + 6x + 9$
$(x - 3)^2 = x^2 - 6x + 9$
$(x + 3)(x - 3) = x^2 - 9$

Dividing a Polynomial by a Monomial [5.7]

To divide a polynomial by a monomial, divide each term in the polynomial by the monomial.

9. $\dfrac{12x^3 - 18x^2}{6x} = 2x^2 - 3x$

Long Division with Polynomials [5.8]

Division with polynomials is similar to long division with whole numbers. The steps in the process are estimate, multiply, subtract, and bring down the next term. The divisors in all the long-division problems in this chapter were binomials.

10.
$$\begin{array}{r} x - 2 \\ x - 3 \overline{) \; x^2 - 5x + 8} \\ \not{+}\, x^2 \not{-}\, 3x \quad \\ \hline -2x + 8 \\ \not{-}\, 2x \not{+}\, 6 \\ \hline 2 \end{array}$$

$x^2 - 5x + 8 \div x - 3$
$= x - 2 + \dfrac{2}{x - 3}$

COMMON MISTAKES

1. If a term contains a variable that is raised to a power, then the exponent on the variable is associated only with that variable, unless there are parentheses; that is, the expression $3x^2$ means $3 \cdot x \cdot x$, not $3x \cdot 3x$. It is a mistake to write $3x^2$ as $9x^2$. The only way to end up with $9x^2$ is to start with $(3x)^2$.
2. It is a mistake to add nonsimilar terms. For example, $2x$ and $3x^2$ are nonsimilar terms and therefore cannot be combined; that is, $2x + 3x^2 \neq 5x^3$. If you were to substitute 10 for x in the preceding expression, you would see that the two sides are not equal.
3. It is a mistake to distribute exponents over sums and differences; that is, $(a + b)^2 \neq a^2 + b^2$. Convince yourself of this by letting $a = 2$ and $b = 3$ and then simplifying both sides.
4. Another common mistake can occur when dividing a polynomial by a monomial. Here is an example:

$$\frac{x + \cancel{2}}{\cancel{2}} = x + 1 \qquad \text{Mistake}$$

The mistake here is in not dividing both terms in the numerator by 2. The correct way to divide $x + 2$ by 2 looks like this:

$$\frac{x + 2}{2} = \frac{x}{2} + \frac{2}{2} \qquad \text{Correct}$$

$$= \frac{x}{2} + 1$$

Chapter 5 Review

Simplify. [5.1]

1. $(-1)^3$
2. -8^2
3. $\left(\frac{3}{7}\right)^2$
4. $y^3 \cdot y^9$
5. $x^{15} \cdot x^7 \cdot x^5 \cdot x^3$
6. $(x^7)^5$
7. $(2^6)^4$
8. $(3y)^3$
9. $(-2xyz)^3$

Simplify each expression. Any answers that contain exponents should contain positive exponents only. [5.2]

10. 7^{-2}
11. $4x^{-5}$
12. $(3y)^{-3}$
13. $\frac{a^9}{a^3}$
14. $\left(\frac{x^3}{x^5}\right)^2$
15. $\frac{x^9}{x^{-6}}$
16. $\frac{x^{-7}}{x^{-2}}$
17. $(-3xy)^0$
18. $3^0 - 5^1 + 5^0$

Simplify. Any answers that contain exponents should contain positive exponents only. [5.1, 5.2]

19. $(3x^3y^2)^2$
20. $(2a^3b^2)^4(2a^5b^6)^2$
21. $(-3xy^2)^{-3}$
22. $\frac{(b^3)^4(b^2)^5}{(b^7)^3}$
23. $\frac{(x^{-3})^3(x^6)^{-1}}{(x^{-5})^{-4}}$

Simplify. Write all answers with positive exponents only. [5.3]

24. $\frac{(2x^4)(15x^9)}{(6x^6)}$
25. $\frac{(10x^3y^5)(21x^2y^6)}{(7xy^3)(5x^9y)}$
26. $\frac{21a^{10}}{3a^4} - \frac{18a^{17}}{6a^{11}}$
27. $\frac{8x^8y^3}{2x^3y} - \frac{10x^6y^9}{5xy^7}$

Simplify, and write all answers in scientific notation. [5.3]

28. $(3.2 \times 10^3)(2 \times 10^4)$
29. $\frac{4.6 \times 10^5}{2 \times 10^{-3}}$
30. $\frac{(4 \times 10^6)(6 \times 10^5)}{3 \times 10^8}$

Perform the following additions and subtractions. [5.4]

31. $(3a^2 - 5a + 5) + (5a^2 - 7a - 8)$
32. $(-7x^2 + 3x - 6) - (8x^2 - 4x + 7) + (3x^2 - 2x - 1)$
33. Subtract $8x^2 + 3x - 2$ from $4x^2 - 3x - 2$.
34. Find the value of $2x^2 - 3x + 5$ when $x = 3$.

Multiply. [5.5]

35. $3x(4x - 7)$
36. $8x^3y(3x^2y - 5xy^2 + 4y^3)$
37. $(a + 1)(a^2 + 5a - 4)$
38. $(x + 5)(x^2 - 5x + 25)$
39. $(3x - 7)(2x - 5)$
40. $\left(5y + \frac{1}{5}\right)\left(5y - \frac{1}{5}\right)$
41. $(a^2 - 3)(a^2 + 3)$

Perform the indicated operations. [5.6]

42. $(a - 5)^2$

43. $(3x + 4)^2$

44. $(y^2 + 3)^2$

45. Divide $10ab + 20a^2$ by $-5a$. [5.7]

46. Divide $40x^5y^4 - 32x^3y^3 - 16x^2y$ by $-8xy$. [5.7]

Divide using long division. [5.8]

47. $\dfrac{x^2 + 15x + 54}{x + 6}$

48. $\dfrac{6x^2 + 13x - 5}{3x - 1}$

49. $\dfrac{x^3 + 64}{x + 4}$

50. $\dfrac{3x^2 - 7x + 10}{3x + 2}$

51. $\dfrac{2x^3 - 7x^2 + 6x + 10}{2x + 1}$

Chapter 5 Cumulative Review

Simplify.

1. $-\left(-\frac{3}{4}\right)$
2. $-|-9|$
3. $2 \cdot 7 + 10$
4. $6 \cdot 7 + 7 \cdot 9$
5. $10 + (-15)$
6. $-15 - (-3)$
7. $-7\left(-\frac{1}{7}\right)$
8. $(-9)(-7)$
9. $6\left(-\frac{1}{2}x + \frac{1}{6}y\right)$
10. $6(4a + 2) - 3(5a - 1)$
11. $y^{10} \cdot y^6$
12. $(2y)^4$
13. $(3a^2b^5)^3(2a^6b^7)^2$
14. $\dfrac{(12xy^5)^3(16x^2y^2)}{(8x^3y^3)(3x^5y)}$

Simplify and write your answer in scientific notation.

15. $(3.5 \times 10^3)(8 \times 10^{-6})$
16. $\dfrac{3.5 \times 10^{-7}}{7 \times 10^{-3}}$

Perform the indicated operations.

17. $10a^2b - 5a^2b + a^2b$
18. $(-4x^2 - 5x + 2) + (3x^2 - 6x + 1) - (-x^2 + 2x - 7)$
19. $(5x - 1)^2$
20. $(x - 1)(x^2 + x + 1)$

Solve each equation.

21. $8 - 2y + 3y = 12$
22. $6a - 5 = 4a$
23. $18 = 3(2x - 2)$
24. $2(3x + 5) + 8 = 2x + 10$

Solve the inequality, and graph the solution set.

25. $-\frac{a}{6} > 4$
26. $3(2t - 5) - 7 \le 5(3t + 1) + 5$
27. $-3 < 2x + 3 < 5$
28. $-5 \le 4x + 3 \le 11$

Divide.

29. $\dfrac{4x^2 + 8x - 10}{2x - 3}$
30. $\dfrac{15x^5 - 10x^2 + 20x}{5x^5}$

Graph on a rectangular coordinate system.

31. $x + y = 3$
32. $x = 2$

Solve each system by graphing.

33. $x + 2y = 1$
 $x + 2y = 2$
34. $3x + 2y = 4$
 $3x + 6y = 12$

Solve each system.

35. $x + 2y = 5$
 $x - y = 2$
36. $x + 2y = 5$
 $3x + 6y = 14$
37. $2x + y = -3$
 $x - 3y = -5$
38. $\frac{1}{6}x + \frac{1}{4}y = 1$
 $\frac{6}{5}x - y = \frac{8}{5}$
39. $x + y = 9$
 $y = x + 1$
40. $4x + 5y = 25$
 $2y = x - 3$
41. Find the value of $8x - 3$ when $x = 3$.
42. Find x when y is 8 in the equation $y = 3x - 1$.
43. Solve $A = \frac{1}{2}bh$ for h.
44. Solve $\frac{y - 1}{x} = -2$ for y.

For the set $\{-3, -\sqrt{2}, 0, \frac{3}{4}, 1.5, \pi\}$ list all of the

45. Whole numbers
46. Irrational numbers

Give the property that justifies the statement.

47. $5 + (-5) = 0$
48. $a + 2 = 2 + a$
49. Find the x- and y-intercepts for the equation $y = 2x + 4$.
50. Find the slope of the line $2x - 5y = 7$.

Use the illustration to answer the following questions.

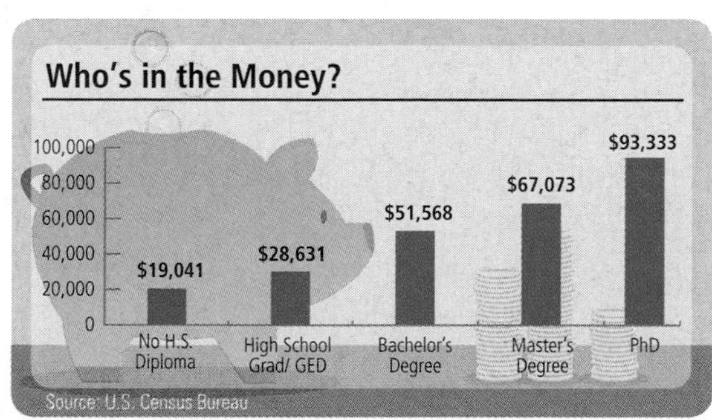

51. How much more is the average income of a worker with a master's degree than that of a worker with a bachelor's degree?
52. On average, how much less does a high school graduate earn than a worker with a bachelor's degree?

Chapter 5 Test

Simplify each of the following expressions. [5.1]

1. $(-3)^4$

2. $\left(\frac{3}{4}\right)^2$

3. $(3x^3)^2(2x^4)^3$

Simplify each expression. Write all answers with positive exponents only. [5.2]

4. 3^{-2}

5. $(3a^4b^2)^0$

6. $\frac{a^{-3}}{a^{-5}}$

7. $\frac{(x^{-2})^3(x^{-3})^{-5}}{(x^{-4})^{-2}}$

8. Write 0.0278 in scientific notation. [5.2]

9. Write 2.43×10^5 in expanded form. [5.1]

Simplify. Write all answers with positive exponents only. [5.3]

10. $\frac{35x^2y^4z}{70x^6y^2z}$

11. $\frac{(6a^2b)(9a^3b^2)}{18a^4b^3}$

12. $\frac{24x^7}{3x^2} + \frac{14x^9}{7x^4}$

13. $\frac{(2.4 \times 10^5)(4.5 \times 10^{-2})}{1.2 \times 10^{-6}}$

Add or subtract as indicated. [5.4]

14. $8x^2 - 4x + 6x + 2$

15. $(5x^2 - 3x + 4) - (2x^2 - 7x - 2)$

16. Subtract $3x - 4$ from $6x - 8$. [5.4]

17. Find the value of $2y^2 - 3y - 4$ when y is -2. [5.4]

Multiply. [5.5]

18. $2a^2(3a^2 - 5a + 4)$

19. $\left(x + \frac{1}{2}\right)\left(x + \frac{1}{3}\right)$

20. $(4x - 5)(2x + 3)$

21. $(x - 3)(x^2 + 3x + 9)$

Multiply. [5.6]

22. $(x + 5)^2$

23. $(3a - 2b)^2$

24. $(3x - 4y)(3x + 4y)$

25. $(a^2 - 3)(a^2 + 3)$

26. Divide $10x^3 + 15x^2 - 5x$ by $5x$. [5.7]

Divide. [5.8]

27. $\frac{8x^2 - 6x - 5}{2x - 3}$

28. $\frac{3x^3 - 2x + 1}{x - 3}$

29. Volume Find the volume of a cube if the length of a side is 2.5 centimeters. [5.1]

30. Volume Find the volume of a rectangular solid if the length is five times the width, and the height is one fifth the width. [5.3]

The illustration below shows the number of cattle (in millions) residing in various nations.

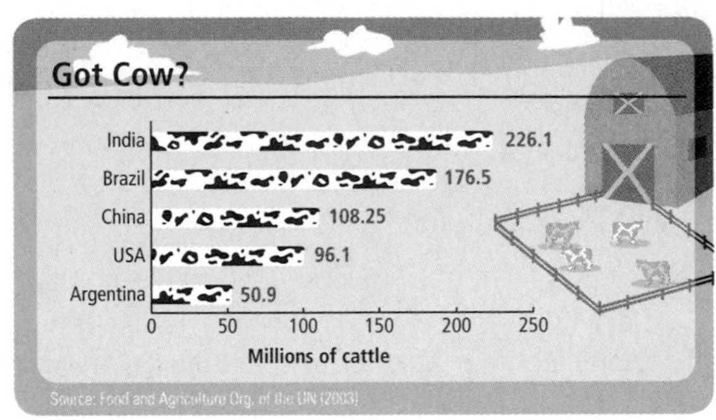

Write in scientific notation the number of cows residing in the given country. [5.1]

31. Brazil

32. USA

The illustration below shows the top five margins of victory for Nascar.

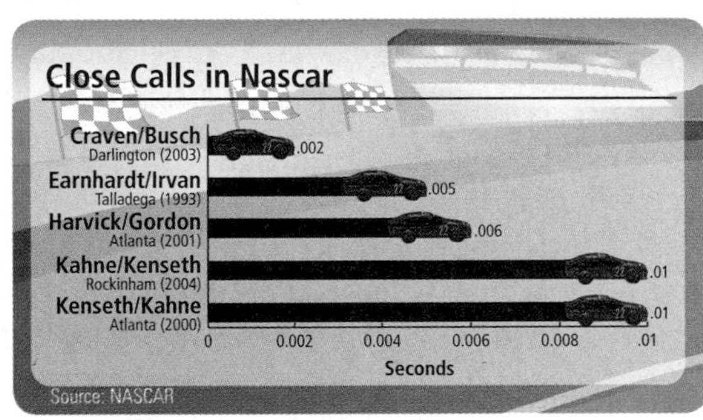

Write in scientific notation the margin of victory for the given race. [5.2]

33. Earnhardt/Irvan

34. Kahne/Kenseth

Chapter 5 Projects

EXPONENTS AND POLYNOMIALS

GROUP PROJECT

Discovering Pascal's Triangle

Number of People 3

Time Needed 20 minutes

Equipment Paper and pencils

Background The triangular array of numbers shown here is known as Pascal's triangle, after the French philosopher Blaise Pascal (1623–1662).

1

1 1

1 2 1

1 3 3 1

1 4 6 4 1

1 5 10 10 5 1

Procedure Look at Pascal's triangle and discover how the numbers in each row of the triangle are obtained from the numbers in the row above it.

1. Once you have discovered how to extend the triangle, write the next two rows.
2. Pascal's triangle can be linked to the Fibonacci sequence by rewriting Pascal's triangle so that the 1's on the left side of the triangle line up under one another, and the other columns are equally spaced to the right of the first column. Rewrite Pascal's triangle as indicated and then look along the diagonals of the new array until you discover how the Fibonacci sequence can be obtained from it.
3. The diagram below shows Pascal's triangle as written in Japan in 1781. Use your knowledge of Pascal's triangle to translate the numbers written in Japanese into our number system. Then write down the Japanese numbers from 1 to 20.

Pascal's triangle in Japanese (1781)

RESEARCH PROJECT

Working Mathematicians

It may seem at times as if all the mathematicians of note lived 100 or more years ago. However, that is not the case. There are mathematicians doing research today who are discovering new mathematical ideas and extending what is known about mathematics.

Use the Internet to find a mathematician working in the field today. Find out what drew them to mathematics in the first place, what it took for them to be successful, and what they like about their career in mathematics. Then summarize your results into an essay that gives anyone reading it a profile of a working mathematician.

Factoring

Chapter Outline

Introduction

The image above is of the Sydney Opera House as it appears in Google Earth. Although cross sections of the shells that are so prominent are actually arcs on a circle, they were originally designed using parabolas, then later using ellipses. Circles, parabolas, and ellipses are from a general category of geometric figures called conic sections. Conic sections are formed by the intersection of a plane and a cone, as shown in the diagram below.

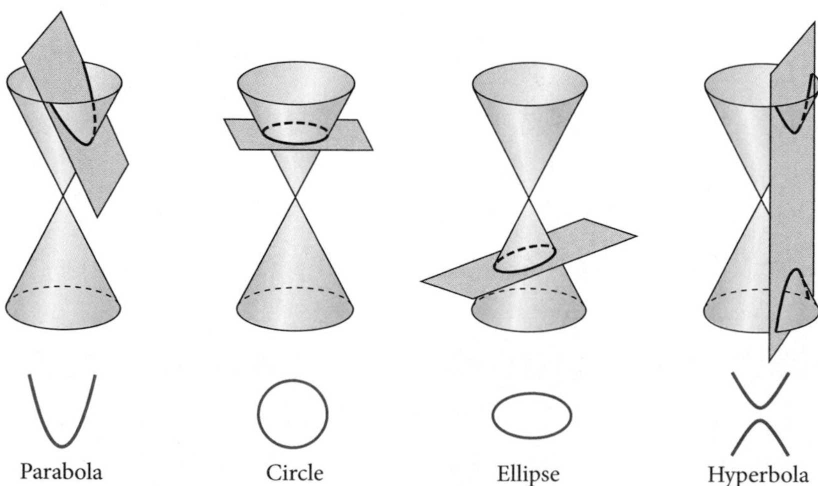

Conic sections have been studied for over 2,000 years. Later in the book we will graph some parabolas. But, in order to do that, we need to be able to solve quadratic equations, and before that we need a good working knowledge of factoring. Factoring is a tool that we use in many different places in mathematics. It is the main topic of this chapter.

Chapter Pretest

Factor out the greatest common factor. [6.1]

1. $6x + 18$
2. $12a^2b - 24ab + 8ab^2$

Factor by grouping. [6.1]

3. $x^2 + 3ax - 2bx - 6ab$
4. $15y - 5xy - 12 + 4x$

Factor the following completely. [6.2–6.6]

5. $x^2 + x - 12$
6. $x^2 - 4x - 21$
7. $x^2 - 25$
8. $x^4 - 16$
9. $x^2 + 36$
10. $18x^2 - 32y^2$
11. $x^3 + 4x^2 - 3x - 12$
12. $x^2 + bx - 3x - 3b$
13. $4x^2 - 6x - 10$
14. $4n^2 + 13n - 12$
15. $12c^2 + c - 6$
16. $12x^3 + 12x^2 - 9x$
17. $x^3 + 125y^3$
18. $54b^3 - 128$

Solve the following equations. [6.7]

19. $x^2 - 2x - 15 = 0$
20. $x^2 - 7x + 12 = 0$
21. $x^2 - 25 = 0$
22. $x^2 = 5x + 14$
23. $x^2 + x = 30$
24. $y^3 = 9y$
25. $2x^2 = -5x + 12$
26. $15x^3 - 65x^2 - 150x = 0$

Getting Ready for Chapter 6

Divide.

1. $\dfrac{y^3 - 16y^2 + 64y}{y}$
2. $\dfrac{5x^3 + 35x^2 + 60x}{5x}$
3. $\dfrac{-12x^4 + 48x^3 + 144x^2}{-12x^2}$
4. $\dfrac{16x^5 + 20x^4 + 60x^3}{4x^3}$

Multiply

5. $(x - 4)(x + 4)$
6. $(x - 6)(x + 6)$
7. $(x - 8)(x + 8)$
8. $(x^2 - 4)(x^2 + 4)$
9. $x(x^2 + 2x + 4)$
10. $2x(4x^2 - 10x + 25)$
11. $-2(x^2 + 2x + 4)$
12. $5(4x^2 - 10x + 25)$
13. $(x - 2)(x^2 + 2x + 4)$
14. $(2x + 5)(4x^2 - 10x + 25)$
15. $3x^2(x + 3)(x - 3)$
16. $2x^3(x + 5)^2$
17. $y^3(y^2 + 36)$
18. $(3a - 4)(2a - 1)$
19. $6x(x - 4)(x + 2)$
20. $2a^3b(a^2 + 3a + 1)$

6.1 The Greatest Common Factor and Factoring by Grouping

Objectives

A Factor the greatest common factor from a polynomial.

B Factor by grouping.

Examples now playing at MathTV.com/books

In Chapter 1 we used the following diagram to illustrate the relationship between multiplication and factoring.

Multiplication

$$\text{Factors} \rightarrow 3 \cdot 5 = 15 \leftarrow \text{Product}$$

Factoring

A similar relationship holds for multiplication of polynomials. Reading the following diagram from left to right, we say the product of the binomials $x + 2$ and $x + 3$ is the trinomial $x^2 + 5x + 6$. However, if we read in the other direction, we can say that $x^2 + 5x + 6$ factors into the product of $x + 2$ and $x + 3$.

Multiplication

$$\text{Factors} \rightarrow (x + 2)(x + 3) = x^2 + 5x + 6 \leftarrow \text{Product}$$

Factoring

In this chapter we develop a systematic method of factoring polynomials.

A Factoring Out the Greatest Common Factor

In this section we will apply the distributive property to polynomials to factor from them what is called the greatest common factor.

> **Definition**
> The **greatest common factor** for a polynomial is the largest monomial that divides (is a factor of) each term of the polynomial.

We use the term *largest monomial* to mean the monomial with the greatest coefficient and highest power of the variable.

EXAMPLE 1 Find the greatest common factor for the polynomial:

$$3x^5 + 12x^2$$

SOLUTION The terms of the polynomial are $3x^5$ and $12x^2$. The largest number that divides the coefficients is 3, and the highest power of x that is a factor of x^5 and x^2 is x^2. Therefore, the greatest common factor for $3x^5 + 12x^2$ is $3x^2$; that is, $3x^2$ is the largest monomial that divides each term of $3x^5 + 12x^2$.

EXAMPLE 2 Find the greatest common factor for:

$$8a^3b^2 + 16a^2b^3 + 20a^3b^3$$

SOLUTION The largest number that divides each of the coefficients is 4. The highest power of the variable that is a factor of a^3b^2, a^2b^3, and a^3b^3 is a^2b^2. The greatest common factor for $8a^3b^2 + 16a^2b^3 + 20a^3b^3$ is $4a^2b^2$. It is the largest monomial that is a factor of each term.

Once we have recognized the greatest common factor of a polynomial, we can apply the distributive property and factor it out of each term. We rewrite the polynomial as the product of its greatest common factor with the polynomial that remains after the greatest common factor has been factored from each term in the original polynomial.

PRACTICE PROBLEMS

1. Find the greatest common factor:
$$10x^6 + 15x^4$$

2. Find the greatest common factor:
$$27a^4b^5 + 18a^3b^3 + 9a^3b^2$$

Answers

1. $5x^4$ **2.** $9a^3b^2$

3. Factor the greatest common factor from:
$5x + 15$

EXAMPLE 3 Factor the greatest common factor from $3x - 15$.

SOLUTION The greatest common factor for the terms $3x$ and 15 is 3. We can rewrite both $3x$ and 15 so that the greatest common factor 3 is showing in each term. It is important to realize that $3x$ means $3 \cdot x$. The 3 and the x are not "stuck" together:

$$3x - 15 = 3 \cdot x - 3 \cdot 5$$

Now, applying the distributive property, we have:

$$3 \cdot x - 3 \cdot 5 = 3(x - 5)$$

To check a factoring problem like this, we can multiply 3 and $x - 5$ to get $3x - 15$, which is what we started with. Factoring is simply a procedure by which we change sums and differences into products. In this case we changed the difference $3x - 15$ into the product $3(x - 5)$. Note, however, that we have not changed the meaning or value of the expression. The expression we end up with is equivalent to the expression we started with.

4. Factor the greatest common factor from:
$25x^4 - 35x^3$

EXAMPLE 4 Factor the greatest common factor from:

$$5x^3 - 15x^2$$

SOLUTION The greatest common factor is $5x^2$. We rewrite the polynomial as:

$$5x^3 - 15x^2 = 5x^2 \cdot x - 5x^2 \cdot 3$$

Then we apply the distributive property to get:

$$5x^2 \cdot x - 5x^2 \cdot 3 = 5x^2(x - 3)$$

To check our work, we simply multiply $5x^2$ and $(x - 3)$ to get $5x^3 - 15x^2$, which is our original polynomial.

5. Factor the greatest common factor from:
$20x^8 - 12x^7 + 16x^6$

EXAMPLE 5 Factor the greatest common factor from:

$$16x^5 - 20x^4 + 8x^3$$

SOLUTION The greatest common factor is $4x^3$. We rewrite the polynomial so we can see the greatest common factor $4x^3$ in each term; then we apply the distributive property to factor it out.

$$16x^5 - 20x^4 + 8x^3 = 4x^3 \cdot 4x^2 - 4x^3 \cdot 5x + 4x^3 \cdot 2$$
$$= 4x^3(4x^2 - 5x + 2)$$

6. Factor the greatest common factor from:
$8xy^3 - 16x^2y^2 + 8x^3y$

EXAMPLE 6 Factor the greatest common factor from:

$$6x^3y - 18x^2y^2 + 12xy^3$$

SOLUTION The greatest common factor is $6xy$. We rewrite the polynomial in terms of $6xy$ and then apply the distributive property as follows:

$$6x^3y - 18x^2y^2 + 12xy^3 = 6xy \cdot x^2 - 6xy \cdot 3xy + 6xy \cdot 2y^2$$
$$= 6xy(x^2 - 3xy + 2y^2)$$

Answers
3. $5(x + 3)$ **4.** $5x^3(5x - 7)$
5. $4x^6(5x^2 - 3x + 4)$
6. $8xy(y^2 - 2xy + x^2)$

EXAMPLE 7 Factor the greatest common factor from:

$$3a^2b - 6a^3b^2 + 9a^3b^3$$

SOLUTION The greatest common factor is $3a^2b$:

$$3a^2b - 6a^3b^2 + 9a^3b^3 = 3a^2b(1) - 3a^2b(2ab) + 3a^2b(3ab^2)$$

$$= 3a^2b(1 - 2ab + 3ab^2)$$

7. Factor the greatest common factor from:
$2ab^2 - 6a^2b^2 + 8a^3b^2$

B Factoring by Grouping

To develop our next method of factoring, called *factoring by grouping,* we start by examining the polynomial $xc + yc$. The greatest common factor for the two terms is c. Factoring c from each term we have:

$$xc + yc = c(x + y)$$

But suppose that c itself was a more complicated expression, such as $a + b$, so that the expression we were trying to factor was $x(a + b) + y(a + b)$, instead of $xc + yc$. The greatest common factor for $x(a + b) + y(a + b)$ is $(a + b)$. Factoring this common factor from each term looks like this:

$$x(a + b) + y(a + b) = (a + b)(x + y)$$

To see how all of this applies to factoring polynomials, consider the polynomial

$$xy + 3x + 2y + 6$$

There is no greatest common factor other than the number 1. However, if we group the terms together two at a time, we can factor an x from the first two terms and a 2 from the last two terms:

$$xy + 3x + 2y + 6 = x(y + 3) + 2(y + 3)$$

The expression on the right can be thought of as having two terms: $x(y + 3)$ and $2(y + 3)$. Each of these expressions contains the common factor $y + 3$, which can be factored out using the distributive property:

$$x(y + 3) + 2(y + 3) = (y + 3)(x + 2)$$

This last expression is in factored form. The process we used to obtain it is called factoring by grouping. Here are some additional examples.

EXAMPLE 8 Factor $ax + bx + ay + by$.

SOLUTION We begin by factoring x from the first two terms and y from the last two terms:

$$ax + bx + ay + by = x(a + b) + y(a + b)$$

$$= (a + b)(x + y)$$

To convince yourself that this is factored correctly, multiply the two factors $(a + b)$ and $(x + y)$.

8. Factor $ax - bx + ay - by$.

Answers

7. $2ab^2(1 - 3a + 4a^2)$
8. $(a - b)(x + y)$

9. Factor by grouping:
$5ax - 3ay + 10x - 6y$

EXAMPLE 9 Factor by grouping: $3ax - 2a + 15x - 10$.

SOLUTION First, we factor a from the first two terms and 5 from the last two terms. Then, we factor $3x - 2$ from the remaining two expressions:

$$3ax - 2a + 15x - 10 = a(3x - 2) + 5(3x - 2)$$
$$= (3x - 2)(a + 5)$$

Again, multiplying $(3x - 2)$ and $(a + 5)$ will convince you that these are the correct factors.

10. Factor completely:
$8x^2 - 12x + 10x - 15$

EXAMPLE 10 Factor $6x^2 - 3x - 4x + 2$ by grouping.

SOLUTION The first two terms have $3x$ in common, and the last two terms have either a 2 or a -2 in common. Suppose we factor $3x$ from the first two terms and 2 from the last two terms. We get:

$$6x^2 - 3x - 4x + 2 = 3x(2x - 1) + 2(-2x + 1)$$

We can't go any further because there is no common factor that will allow us to factor further. However, if we factor -2, instead of 2, from the last two terms, our problem is solved:

$$6x^2 - 3x - 4x + 2 = 3x(2x - 1) - 2(2x - 1)$$
$$= (2x - 1)(3x - 2)$$

In this case, factoring -2 from the last two terms gives us an expression that can be factored further.

11. Factor:
$3x^2 + 7bx - 3xy - 7by$

EXAMPLE 11 Factor $2x^2 + 5ax - 2xy - 5ay$.

SOLUTION From the first two terms we factor x. From the second two terms we must factor $-y$ so that the binomial that remains after we do so matches the binomial produced by the first two terms:

$$2x^2 + 5ax - 2xy - 5ay = x(2x + 5a) - y(2x + 5a)$$
$$= (2x + 5a)(x - y)$$

Another way to accomplish the same result is to use the commutative property to interchange the middle two terms, and then factor by grouping:

$$2x^2 + 5ax - 2xy - 5ay = 2x^2 - 2xy + 5ax - 5ay \quad \text{Commutative property}$$
$$= 2x(x - y) + 5a(x - y)$$
$$= (x - y)(2x + 5a)$$

This is the same result we obtained previously.

Getting Ready for Class

After reading through the preceding section, respond in your own words and in complete sentences.

1. What is the greatest common factor for a polynomial?
2. After factoring a polynomial, how can you check your result?
3. When would you try to factor by grouping?
4. What is the relationship between multiplication and factoring?

Answers
9. $(5x - 3y)(a + 2)$
10. $(2x - 3)(4x + 5)$
11. $(3x + 7b)(x - y)$

Problem Set 6.1

A Factor the following by taking out the greatest common factor. **[Examples 1–7]**

1. $15x + 25$

2. $14x + 21$

3. $6a + 9$

4. $8a + 10$

5. $4x - 8y$

6. $9x - 12y$

7. $3x^2 - 6x - 9$

8. $2x^2 + 6x + 4$

9. $3a^2 - 3a - 60$

10. $2a^2 - 18a + 28$

11. $24y^2 - 52y + 24$

12. $18y^2 + 48y + 32$

13. $9x^2 - 8x^3$

14. $7x^3 - 4x^2$

15. $13a^2 - 26a^3$

16. $5a^2 - 10a^3$

17. $21x^2y - 28xy^2$

18. $30xy^2 - 25x^2y$

19. $22a^2b^2 - 11ab^2$

20. $15x^3 - 25x^2 + 30x$

21. $7x^3 + 21x^2 - 28x$

22. $16x^4 - 20x^2 - 16x$

23. $121y^4 - 11x^4$

24. $25a^4 - 5b^4$

25. $100x^4 - 50x^3 + 25x^2$

26. $36x^5 + 72x^3 - 81x^2$

27. $8a^2 + 16b^2 + 32c^2$

28. $9a^2 - 18b^2 - 27c^2$

29. $4a^2b - 16ab^2 + 32a^2b^2$

30. $5ab^2 + 10a^2b^2 + 15a^2b$

A Factor the following by taking out the greatest common factor.

31. $121a^3b^2 - 22a^2b^3 + 33a^3b^3$

32. $20a^4b^3 - 18a^3b^4 + 22a^4b^4$

33. $12x^2y^3 - 72x^5y^3 - 36x^4y^4$

34. $49xy - 21x^2y^2 + 35x^3y^3$

B Factor by grouping. [Example 8–11]

35. $xy + 5x + 3y + 15$

36. $xy + 2x + 4y + 8$

37. $xy + 6x + 2y + 12$

38. $xy + 2y + 6x + 12$

39. $ab + 7a - 3b - 21$

40. $ab + 3b - 7a - 21$

41. $ax - bx + ay - by$

42. $ax - ay + bx - by$

43. $2ax + 6x - 5a - 15$

44. $3ax + 21x - a - 7$

45. $3xb - 4b - 6x + 8$

46. $3xb - 4b - 15x + 20$

47. $x^2 + ax + 2x + 2a$

48. $x^2 + ax + 3x + 3a$

49. $x^2 - ax - bx + ab$

50. $x^2 + ax - bx - ab$

B Factor by grouping. You can group the terms together two at a time or three at a time. Either way will produce the same result.

51. $ax + ay + bx + by + cx + cy$

52. $ax + bx + cx + ay + by + cy$

B Factor the following polynomials by grouping the terms together two at a time. [Examples 9–11]

53. $6x^2 + 9x + 4x + 6$

54. $6x^2 - 9x - 4x + 6$

55. $20x^2 - 2x + 50x - 5$

56. $20x^2 + 25x + 4x + 5$

57. $20x^2 + 4x + 25x + 5$

58. $20x^2 + 4x - 25x - 5$

59. $x^3 + 2x^2 + 3x + 6$

60. $x^3 - 5x^2 - 4x + 20$

61. $6x^3 - 4x^2 + 15x - 10$

62. $8x^3 - 12x^2 + 14x - 21$

63. The greatest common factor of the binomial $3x + 6$ is 3. The greatest common factor of the binomial $2x + 4$ is 2. What is the greatest common factor of their product $(3x + 6)(2x + 4)$ when it has been multiplied out?

64. The greatest common factors of the binomials $4x + 2$ and $5x + 10$ are 2 and 5, respectively. What is the greatest common factor of their product $(4x + 2)(5x + 10)$ when it has been multiplied out?

65. The following factorization is incorrect. Find the mistake, and correct the right-hand side:

$$12x^2 + 6x + 3 = 3(4x^2 + 2x)$$

66. Find the mistake in the following factorization, and then rewrite the right-hand side correctly:

$$10x^2 + 2x + 6 = 2(5x^2 + 3)$$

Getting Ready for the Next Section

Multiply each of the following.

67. $(x - 7)(x + 2)$
68. $(x - 7)(x - 2)$
69. $(x - 3)(x + 2)$
70. $(x + 3)(x - 2)$

71. $(x + 3)(x^2 - 3x + 9)$
72. $(x - 2)(x^2 + 2x + 4)$
73. $(2x + 1)(x^2 + 4x - 3)$
74. $(3x + 2)(x^2 - 2x - 4)$

75. $3x^4(6x^3 - 4x^2 + 2x)$
76. $2x^4(5x^3 + 4x^2 - 3x)$
77. $\left(x + \frac{1}{3}\right)\left(x + \frac{2}{3}\right)$
78. $\left(x + \frac{1}{4}\right)\left(x + \frac{3}{4}\right)$

79. $(6x + 4y)(2x - 3y)$
80. $(8a - 3b)(4a - 5b)$
81. $(9a + 1)(9a - 1)$
82. $(7b + 1)(7b + 1)$

83. $(x - 9)(x - 9)$
84. $(x - 8)(x - 8)$
85. $(x + 2)(x^2 - 2x + 4)$
86. $(x - 3)(x^2 + 3x + 9)$

Maintaining Your Skills

Divide.

87. $\dfrac{y^3 - 16y^2 + 64y}{y}$
88. $\dfrac{5x^3 + 35x^2 + 60x}{5x}$
89. $\dfrac{-12x^4 + 48x^3 + 144x^2}{-12x^2}$

90. $\dfrac{16x^5 + 20x^4 + 60x^3}{4x^3}$
91. $\dfrac{-18y^5 + 63y^4 - 108y^3}{-9y^3}$
92. $\dfrac{36y^6 - 66y^5 + 54y^4}{6y^4}$

Subtract.

93. $(5x^2 + 5x - 4) - (3x^2 - 2x + 7)$
94. $(7x^4 - 4x^2 - 5) - (2x^4 - 4x^2 + 5)$
95. Subtract $4x - 5$ from $7x + 3$.

96. Subtract $3x + 2$ from $-6x + 1$.
97. Subtract $2x^2 - 4x$ from $5x^2 - 5$.
98. Subtract $6x^2 + 3$ from $2x^2 - 4x$.

In this section we will factor trinomials in which the coefficient of the squared term is 1. The more familiar we are with multiplication of binomials, the easier factoring trinomials will be.

Recall multiplication of binomials from Chapter 5:

$$(x + 3)(x + 4) = x^2 + 7x + 12$$
$$(x - 5)(x + 2) = x^2 - 3x - 10$$

The first term in the answer is the product of the first terms in each binomial. The last term in the answer is the product of the last terms in each binomial. The middle term in the answer comes from adding the product of the outside terms to the product of the inside terms.

Let's have a and b represent real numbers and look at the product of $(x + a)$ and $(x + b)$:

$$\begin{aligned}(x + a)(x + b) &= x^2 + ax + bx + ab \\ &= x^2 + (a + b)x + ab\end{aligned}$$

The coefficient of the middle term is the sum of a and b. The last term is the product of a and b. Writing this as a factoring problem, we have:

$$x^2 + \underbrace{(a + b)}_{\text{Sum}}x + \underbrace{ab}_{\text{Product}} = (x + a)(x + b)$$

A Factoring with a Leading Coefficient of 1

To factor a trinomial in which the coefficient of x^2 is 1, we need only find the numbers a and b whose sum is the coefficient of the middle term and whose product is the constant term (last term).

EXAMPLE 1 Factor $x^2 + 8x + 12$.

SOLUTION The coefficient of x^2 is 1. We need two numbers whose sum is 8 and whose product is 12. The numbers are 6 and 2:

$$x^2 + 8x + 12 = (x + 6)(x + 2)$$

We can easily check our work by multiplying $(x + 6)$ and $(x + 2)$

$$\begin{aligned}\text{Check:}\quad (x + 6)(x + 2) &= x^2 + 6x + 2x + 12 \\ &= x^2 + 8x + 12\end{aligned}$$

EXAMPLE 2 Factor $x^2 - 2x - 15$.

SOLUTION The coefficient of x^2 is again 1. We need to find a pair of numbers whose sum is -2 and whose product is -15. Here are all the possibilities for products that are -15.

Products	Sums
$-1(15) = -15$	$-1 + 15 = 14$
$1(-15) = -15$	$1 + (-15) = -14$
$-5(3) = -15$	$-5 + 3 = -2$
$5(-3) = -15$	$5 + (-3) = 2$

The third line gives us what we want. The factors of $x^2 - 2x - 15$ are $(x - 5)$ and $(x + 3)$:

$$x^2 - 2x - 15 = (x - 5)(x + 3)$$

Objectives

A Factor a trinomial whose leading coefficient is the number 1.

B Factor a polynomial by first factoring out the greatest common factor and then factoring the polynomial that remains.

Examples now playing at MathTV.com/books

Note As you will see as we progress through the book, factoring is a tool that is used in solving a number of problems. Before seeing how it is used, however, we first must learn how to do it. So, in this section and the two sections that follow, we will be developing our factoring skills.

PRACTICE PROBLEMS

1. Factor $x^2 + 5x + 6$.

2. Factor $x^2 + 2x - 15$.

Note Again, we can check our results by multiplying our factors to see if their product is the original polynomial.

Answers
1. $(x + 2)(x + 3)$ 2. $(x + 5)(x - 3)$

3. Factor $2x^2 - 2x - 24$.

In Example 3 we began by factoring out the greatest common factor. The first step in factoring any trinomial is to look for the greatest common factor. If the trinomial in question has a greatest common factor other than 1, we factor it out first and then try to factor the trinomial that remains.

4. Factor $3x^3 + 18x^2 + 15x$.

5. Factor $x^2 + 7xy + 12y^2$.

Trinomials in which the coefficient of the second-degree term is 1 are the easiest to factor. Success in factoring any type of polynomial is directly related to the amount of time spent working the problems. The more we practice, the more accomplished we become at factoring.

Answers

3. $2(x - 4)(x + 3)$
4. $3x(x + 5)(x + 1)$
5. $(x + 3y)(x + 4y)$

B Factoring Out the Greatest Common Factor

EXAMPLE 3 Factor $2x^2 + 10x - 28$.

SOLUTION The coefficient of x^2 is 2. We begin by factoring out the greatest common factor, which is 2:

$$2x^2 + 10x - 28 = 2(x^2 + 5x - 14)$$

Now, we factor the remaining trinomial by finding a pair of numbers whose sum is 5 and whose product is -14. Here are the possibilities:

Products	Sums
$-1(14) = -14$	$-1 + 14 = 13$
$1(-14) = -14$	$1 + (-14) = -13$
$-7(2) = -14$	$-7 + 2 = -5$
$7(-2) = -14$	$7 + (-2) = 5$

From the last line we see that the factors of $x^2 + 5x - 14$ are $(x + 7)$ and $(x - 2)$. Here is the complete solution:

$$2x^2 + 10x - 28 = 2(x^2 + 5x - 14)$$
$$= 2(x + 7)(x - 2)$$

EXAMPLE 4 Factor $3x^3 - 3x^2 - 18x$.

SOLUTION We begin by factoring out the greatest common factor, which is $3x$. Then we factor the remaining trinomial. Without showing the table of products and sums as we did in Examples 2 and 3, here is the complete solution:

$$3x^3 - 3x^2 - 18x = 3x(x^2 - x - 6)$$
$$= 3x(x - 3)(x + 2)$$

EXAMPLE 5 Factor $x^2 + 8xy + 12y^2$.

SOLUTION This time we need two expressions whose product is $12y^2$ and whose sum is $8y$. The two expressions are $6y$ and $2y$ (see Example 1 in this section):

$$x^2 + 8xy + 12y^2 = (x + 6y)(x + 2y)$$

You should convince yourself that these factors are correct by finding their product.

Getting Ready for Class

After reading through the preceding section, respond in your own words and in complete sentences.

1. When the leading coefficient of a trinomial is 1, what is the relationship between the other two coefficients and the factors of the trinomial?
2. When factoring polynomials, what should you look for first?
3. How can you check to see that you have factored a trinomial correctly?
4. Describe how you would find the factors of $x^2 + 8x + 12$.

Problem Set 6.2

A Factor the following trinomials. [Examples 1–2]

1. $x^2 + 7x + 12$

2. $x^2 + 7x + 10$

3. $x^2 + 3x + 2$

4. $x^2 + 7x + 6$

5. $a^2 + 10a + 21$

6. $a^2 - 7a + 12$

7. $x^2 - 7x + 10$

8. $x^2 - 3x + 2$

9. $y^2 - 10y + 21$

10. $y^2 - 7y + 6$

11. $x^2 - x - 12$

12. $x^2 - 4x - 5$

13. $y^2 + y - 12$

14. $y^2 + 3y - 18$

15. $x^2 + 5x - 14$

16. $x^2 - 5x - 24$

17. $r^2 - 8r - 9$

18. $r^2 - r - 2$

19. $x^2 - x - 30$

20. $x^2 + 8x + 12$

21. $a^2 + 15a + 56$

22. $a^2 - 9a + 20$

23. $y^2 - y - 42$

24. $y^2 + y - 42$

25. $x^2 + 13x + 42$

26. $x^2 - 13x + 42$

B Factor the following problems completely. First, factor out the greatest common factor, and then factor the remaining trinomial. [Examples 3–4]

27. $2x^2 + 6x + 4$

28. $3x^2 - 6x - 9$

29. $3a^2 - 3a - 60$

30. $2a^2 - 18a + 28$

31. $100x^2 - 500x + 600$

32. $100x^2 - 900x + 2{,}000$

33. $100p^2 - 1{,}300p + 4{,}000$

34. $100p^2 - 1{,}200p + 3{,}200$

35. $x^4 - x^3 - 12x^2$

36. $x^4 - 11x^3 + 24x^2$

37. $2r^3 + 4r^2 - 30r$

38. $5r^3 + 45r^2 + 100r$

39. $2y^4 - 6y^3 - 8y^2$

40. $3r^3 - 3r^2 - 6r$

41. $x^5 + 4x^4 + 4x^3$

42. $x^5 + 13x^4 + 42x^3$

43. $3y^4 - 12y^3 - 15y^2$

44. $5y^4 - 10y^3 + 5y^2$

45. $4x^4 - 52x^3 + 144x^2$

46. $3x^3 - 3x^2 - 18x$

A Factor the following trinomials. [Example 5]

47. $x^2 + 5xy + 6y^2$

48. $x^2 - 5xy + 6y^2$

49. $x^2 - 9xy + 20y^2$

50. $x^2 + 9xy + 20y^2$

51. $a^2 + 2ab - 8b^2$

52. $a^2 - 2ab - 8b^2$

53. $a^2 - 10ab + 25b^2$

54. $a^2 + 6ab + 9b^2$

55. $a^2 + 10ab + 25b^2$

56. $a^2 - 6ab + 9b^2$

57. $x^2 + 2xa - 48a^2$

58. $x^2 - 3xa - 10a^2$

59. $x^2 - 5xb - 36b^2$

60. $x^2 - 13xb + 36b^2$

61. $x^4 - 5x^2 + 6$

62. $x^6 - 2x^3 - 15$

63. $x^2 - 80x - 2{,}000$

64. $x^2 - 190x - 2{,}000$

65. $x^2 - x + \frac{1}{4}$

66. $x^2 - \frac{2}{3}x + \frac{1}{9}$

67. $x^2 + 0.6x + 0.08$

68. $x^2 + 0.8x + 0.15$

69. If one of the factors of $x^2 + 24x + 128$ is $x + 8$, what is the other factor?

70. If one factor of $x^2 + 260x + 2{,}500$ is $x + 10$, what is the other factor?

71. What polynomial, when factored, gives $(4x + 3)(x - 1)$?

72. What polynomial factors to $(4x - 3)(x + 1)$?

Getting Ready for the Next Section

Multiply using the FOIL method.

73. $(6a + 1)(a + 2)$

74. $(6a - 1)(a - 2)$

75. $(3a + 2)(2a + 1)$

76. $(3a - 2)(2a - 1)$

77. $(6a + 2)(a + 1)$

78. $(3a + 1)(2a + 2)$

Maintaining Your Skills

Simplify each expression. Write using only positive exponents.

79. $\left(-\frac{2}{5}\right)^2$

80. $\left(-\frac{3}{8}\right)^2$

81. $(3a^3)^2(2a^2)^3$

82. $(-4x^4)^2(2x^5)^4$

83. $\frac{(4x)^{-7}}{(4x)^{-5}}$

84. $\frac{(2x)^{-3}}{(2x)^{-5}}$

85. $\frac{12a^5b^3}{72a^2b^5}$

86. $\frac{25x^5y^3}{50x^2y^7}$

87. $\frac{15x^{-5}y^3}{45x^2y^5}$

88. $\frac{25a^2b^7}{75a^5b^3}$

89. $(-7x^3y)(3xy^4)$

90. $(9a^6b^4)(6a^4b^3)$

91. $(-5a^3b^{-1})(4a^{-2}b^4)$

92. $(-3a^2b^{-4})(6a^5b^{-2})$

93. $(9a^2b^3)(-3a^3b^5)$

94. $(-7a^5b^8)(6a^7b^4)$

More Trinomials to Factor

6.3

Objectives

A Factor a trinomial whose leading coefficient is other than 1.

B Factor a polynomial by first factoring out the greatest common factor and then factoring the polynomial that remains.

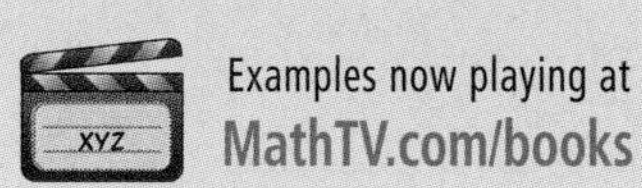

Examples now playing at
MathTV.com/books

We now will consider trinomials whose greatest common factor is 1 and whose leading coefficient (the coefficient of the squared term) is a number other than 1. We present two methods for factoring trinomials of this type. The first method involves listing possible factors until the correct pair of factors is found. This requires a certain amount of trial and error. The second method is based on the factoring by grouping process that we covered previously. Either method can be used to factor trinomials whose leading coefficient is a number other than 1.

A Method 1: Factoring $ax^2 + bx + c$ by trial and error

Suppose we want to factor the trinomial $2x^2 - 5x - 3$. We know that the factors (if they exist) will be a pair of binomials. The product of their first terms is $2x^2$ and the product of their last term is -3. Let's list all the possible factors along with the trinomial that would result if we were to multiply them together. Remember, the middle term comes from the product of the inside terms plus the product of the outside terms.

Binomial Factors	First Term	Middle Term	Last Term
$(2x - 3)(x + 1)$	$2x^2$	$-x$	-3
$(2x + 3)(x - 1)$	$2x^2$	$+x$	-3
$(2x - 1)(x + 3)$	$2x^2$	$+5x$	-3
$(2x + 1)(x - 3)$	$2x^2$	$-5x$	-3

We can see from the last line that the factors of $2x^2 - 5x - 3$ are $(2x + 1)$ and $(x - 3)$. There is no straightforward way, as there was in the previous section, to find the factors, other than by trial and error or by simply listing all the possibilities. We look for possible factors that, when multiplied, will give the correct first and last terms, and then we see if we can adjust them to give the correct middle term.

EXAMPLE 1 Factor $6a^2 + 7a + 2$.

SOLUTION We list all the possible pairs of factors that, when multiplied together, give a trinomial whose first term is $6a^2$ and whose last term is $+2$.

Binomial Factors	First Term	Middle Term	Last Term
$(6a + 1)(a + 2)$	$6a^2$	$+13a$	$+2$
$(6a - 1)(a - 2)$	$6a^2$	$-13a$	$+2$
$(3a + 2)(2a + 1)$	$6a^2$	$+7a$	$+2$
$(3a - 2)(2a - 1)$	$6a^2$	$-7a$	$+2$

The factors of $6a^2 + 7a + 2$ are $(3a + 2)$ and $(2a + 1)$.

Check: $(3a + 2)(2a + 1) = 6a^2 + 7a + 2$

Notice that in the preceding list we did not include the factors $(6a + 2)$ and $(a + 1)$. We do not need to try these since the first factor has a 2 common to each term and so could be factored again, giving $2(3a + 1)(a + 1)$. Since our original trinomial, $6a^2 + 7a + 2$, did *not* have a greatest common factor of 2, neither of its factors will.

PRACTICE PROBLEMS

1. Factor $2a^2 + 7a + 6$.

Note

Remember, we can always check our results by multiplying the factors we have and comparing that product with our original polynomial.

Answer

1. $(2a + 3)(a + 2)$

2. Factor $4x^2 - x - 5$.

EXAMPLE 2 Factor $4x^2 - x - 3$.

SOLUTION We list all the possible factors that, when multiplied, give a trinomial whose first term is $4x^2$ and whose last term is -3.

Binomial Factors	First Term	Middle Term	Last Term
$(4x + 1)(x - 3)$	$4x^2$	$-11x$	-3
$(4x - 1)(x + 3)$	$4x^2$	$+11x$	-3
$(4x + 3)(x - 1)$	$4x^2$	$-x$	-3
$(4x - 3)(x + 1)$	$4x^2$	$+x$	-3
$(2x + 1)(2x - 3)$	$4x^2$	$-4x$	-3
$(2x - 1)(2x + 3)$	$4x^2$	$+4x$	-3

The third line shows that the factors are $(4x + 3)$ and $(x - 1)$.

$$\text{Check:}\quad (4x + 3)(x - 1) = 4x^2 - x - 3$$

You will find that the more practice you have at factoring this type of trinomial, the faster you will get the correct factors. You will pick up some shortcuts along the way, or you may come across a system of eliminating some factors as possibilities. Whatever works best for you is the method you should use. Factoring is a very important tool, and you must be good at it.

B Factoring Out the Greatest Common Factor

3. Factor $30y^2 + 5y - 5$.

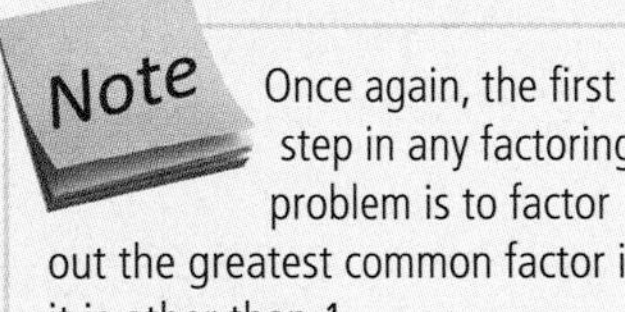

Once again, the first step in any factoring problem is to factor out the greatest common factor if it is other than 1.

EXAMPLE 3 Factor $12y^3 + 10y^2 - 12y$.

SOLUTION We begin by factoring out the greatest common factor, $2y$:

$$12y^3 + 10y^2 - 12y = 2y(6y^2 + 5y - 6)$$

We now list all possible factors of a trinomial with the first term $6y^2$ and last term -6, along with the associated middle terms.

Possible Factors	Middle Term When Multiplied
$(3y + 2)(2y - 3)$	$-5y$
$(3y - 2)(2y + 3)$	$+5y$
$(6y + 1)(y - 6)$	$-35y$
$(6y - 1)(y + 6)$	$+35y$

The second line gives the correct factors. The complete problem is:

$$\begin{aligned} 12y^3 + 10y^2 - 12y &= 2y(6y^2 + 5y - 6) \\ &= 2y(3y - 2)(2y + 3) \end{aligned}$$

4. Factor $10x^2y^2 + 5xy^3 - 30y^4$.

EXAMPLE 4 Factor $30x^2y - 5xy^2 - 10y^3$.

SOLUTION The greatest common factor is $5y$:

$$\begin{aligned} 30x^2y - 5xy^2 - 10y^3 &= 5y(6x^2 - xy - 2y^2) \\ &= 5y(2x + y)(3x - 2y) \end{aligned}$$

Method 2: Factoring $ax^2 + bx + c$ by grouping

Recall that previously that we used factoring by grouping to factor the polynomial $6x^2 - 3x - 4x + 2$. We began by factoring $3x$ from the first two terms and -2 from the last two terms. For review, here is the complete problem:

Answers
2. $(4x - 5)(x + 1)$
3. $5(3y - 1)(2y + 1)$
4. $5y^2(2x - 3y)(x + 2y)$

$$6x^2 - 3x - 4x + 2 = 3x(2x - 1) - 2(2x - 1)$$
$$= (2x - 1)(3x - 2)$$

Now, let's back up a little and notice that our original polynomial $6x^2 - 3x - 4x + 2$ can be simplified to $6x^2 - 7x + 2$ by adding $-3x$ and $-4x$. This means that $6x^2 - 7x + 2$ can be factored to $(2x - 1)(3x - 2)$ by the grouping method shown previously. The key to using this process is to rewrite the middle term $-7x$ as $-3x - 4x$.

To generalize this discussion, here are the steps we use to factor trinomials by grouping.

Strategy Factoring $ax^2 + bx + c$ by Grouping

Step 1: Form the product ac.

Step 2: Find a pair of numbers whose product is ac and whose sum is b.

Step 3: Rewrite the polynomial to be factored so that the middle term bx is written as the sum of two terms whose coefficients are the two numbers found in step 2.

Step 4: Factor by grouping.

EXAMPLE 5 Factor $3x^2 - 10x - 8$ using these steps.

SOLUTION The trinomial $3x^2 - 10x - 8$ has the form $ax^2 + bx + c$, where $a = 3$, $b = -10$, and $c = -8$.

Step 1: The product ac is $3(-8) = -24$.

Step 2: We need to find two numbers whose product is -24 and whose sum is -10. Let's systematically begin to list all the pairs of numbers whose product is -24 to find the pair whose sum is -10.

Product	*Sum*
$-24(1) = -24$	$-24 + 1 = -23$
$-12(2) = -24$	$-12 + 2 = -10$

We stop here because we have found the pair of numbers whose product is -24 and whose sum is -10. The numbers are -12 and 2.

Step 3: We now rewrite our original trinomial so the middle term $-10x$ is written as the sum of $-12x$ and $2x$:

$$3x^2 - 10x - 8 = 3x^2 - 12x + 2x - 8$$

Step 4: Factoring by grouping, we have:

$$3x^2 - 12x + 2x - 8 = 3x(x - 4) + 2(x - 4)$$
$$= (x - 4)(3x + 2)$$

We can check our work by multiplying $x - 4$ and $3x + 2$ to get $3x^2 - 10x - 8$. ■

5. Factor $2x^2 + 7x - 15$

EXAMPLE 6 Factor $4x^2 - x - 3$.

SOLUTION In this case, $a = 4$, $b = -1$, and $c = -3$. The product ac is $4(-3) = -12$. We need a pair of numbers whose product is -12 and whose sum is -1. We begin listing pairs of numbers whose product is -12 and whose sum is -1.

6. Factor $6x^2 + x - 5$

Answer
5. $(x + 5)(2x - 3)$

Product	*Sum*
$-12(1) = -12$	$-12 + 1 = -11$
$-6(2) = -12$	$-6 + 2 = -4$
$-4(3) = -12$	$-4 + 3 = -1$

We stop here because we have found the pair of numbers for which we are looking. They are -4 and 3. Next, we rewrite the middle term $-x$ as the sum $-4x + 3x$ and proceed to factor by grouping.

$$\begin{aligned} 4x^2 - x - 3 &= 4x^2 - 4x + 3x - 3 \\ &= 4x(x - 1) + 3(x - 1) \\ &= (x - 1)(4x + 3) \end{aligned}$$

Compare this procedure and the result with those shown in Example 2 of this section.

7. Factor $15x^2 - 2x - 8$

EXAMPLE 7 Factor $8x^2 - 2x - 15$.

SOLUTION The product ac is $8(-15) = -120$. There are many pairs of numbers whose product is -120. We are looking for the pair whose sum is also -2. The numbers are -12 and 10. Writing $-2x$ as $-12x + 10x$ and then factoring by grouping, we have:

$$\begin{aligned} 8x^2 - 2x - 15 &= 8x^2 - 12x + 10x - 15 \\ &= 4x(2x - 3) + 5(2x - 3) \\ &= (2x - 3)(4x + 5) \end{aligned}$$

8. Assume the ball in Example 8 is tossed from a building that is 48 feet tall and has an upward velocity of 32 feet per second. The equation that gives the height of the ball above the ground at any time t is $h = 48 + 32t - 16t^2$. Factor the right side of this equation and then find the h when t is 3.

EXAMPLE 8 A ball is tossed into the air with an upward velocity of 16 feet per second from the top of a building 32 feet high. The equation that gives the height of the ball above the ground at any time t is

$$h = 32 + 16t - 16t^2$$

Factor the right side of this equation and then find h when t is 2.

SOLUTION We begin by factoring out the greatest common factor, 16. Then, we factor the trinomial that remains:

$$\begin{aligned} h &= 32 + 16t - 16t^2 \\ h &= 16(2 + t - t^2) \\ h &= 16(2 - t)(1 + t) \end{aligned}$$

Letting $t = 2$ in the equation, we have

$$h = 16(0)(3) = 0$$

When t is 2, h is 0.

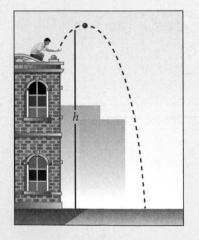

Getting Ready for Class

After reading through the preceding section, respond in your own words and in complete sentences.

1. What is the first step in factoring a trinomial?
2. Describe the criteria you would use to set up a table of possible factors of a trinomial.
3. What does it mean if you factor a trinomial and one of your factors has a greatest common factor of 3?
4. Describe how you would look for possible factors of $6a^2 + 7a + 2$.

Answers

6. $(x + 1)(6x - 5)$
7. $(3x + 2)(5x - 4)$
8. $h = 16(3 - t)(1 + t)$; When t is 3, h is 0.

Problem Set 6.3

A Factor the following trinomials. [Examples 1–2, 5–7]

1. $2x^2 + 7x + 3$

2. $2x^2 + 5x + 3$

3. $2a^2 - a - 3$

4. $2a^2 + a - 3$

5. $3x^2 + 2x - 5$

6. $3x^2 - 2x - 5$

7. $3y^2 - 14y - 5$

8. $3y^2 + 14y - 5$

9. $6x^2 + 13x + 6$

10. $6x^2 - 13x + 6$

11. $4x^2 - 12xy + 9y^2$

12. $4x^2 + 12xy + 9y^2$

13. $4y^2 - 11y - 3$

14. $4y^2 + y - 3$

15. $20x^2 - 41x + 20$

16. $20x^2 + 9x - 20$

17. $20a^2 + 48ab - 5b^2$

18. $20a^2 + 29ab + 5b^2$

19. $20x^2 - 21x - 5$

20. $20x^2 - 48x - 5$

21. $12m^2 + 16m - 3$

22. $12m^2 + 20m + 3$

23. $20x^2 + 37x + 15$

24. $20x^2 + 13x - 15$

25. $12a^2 - 25ab + 12b^2$

26. $12a^2 + 7ab - 12b^2$

27. $3x^2 - xy - 14y^2$

28. $3x^2 + 19xy - 14y^2$

29. $14x^2 + 29x - 15$

30. $14x^2 + 11x - 15$

31. $6x^2 - 43x + 55$

32. $6x^2 - 7x - 55$

33. $15t^2 - 67t + 38$

34. $15t^2 - 79t - 34$

B Factor each of the following completely. Look first for the greatest common factor. [Examples 3–7]

35. $4x^2 + 2x - 6$

36. $6x^2 - 51x + 63$

37. $24a^2 - 50a + 24$

38. $18a^2 + 48a + 32$

39. $10x^3 - 23x^2 + 12x$

40. $10x^4 + 7x^3 - 12x^2$

41. $6x^4 - 11x^3 - 10x^2$

42. $6x^3 + 19x^2 + 10x$

43. $10a^3 - 6a^2 - 4a$

44. $6a^3 + 15a^2 + 9a$

45. $15x^3 - 102x^2 - 21x$

46. $2x^4 - 24x^3 + 64x^2$

47. $35y^3 - 60y^2 - 20y$

48. $14y^4 - 32y^3 + 8y^2$

49. $15a^4 - 2a^3 - a^2$

50. $10a^5 - 17a^4 + 3a^3$

51. $12x^2y - 34xy^2 + 14y^3$

52. $12x^2y - 46xy^2 + 14y^3$

53. Evaluate the expression $2x^2 + 7x + 3$ and the expression $(2x + 1)(x + 3)$ for $x = 2$.

54. Evaluate the expression $2a^2 - a - 3$ and the expression $(2a - 3)(a + 1)$ for $a = 5$.

55. What polynomial factors to $(2x + 3)(2x - 3)$?

56. What polynomial factors to $(5x + 4)(5x - 4)$?

57. What polynomial factors to $(x + 3)(x - 3)(x^2 + 9)$?

58. What polynomial factors to $(x + 2)(x - 2)(x^2 + 4)$?

59. One factor of $12x^2 - 71x + 105$ is $x - 3$. Find the other factor.

60. One factor of $18x^2 + 121x - 35$ is $x + 7$. Find the other factor.

61. One factor of $54x^2 + 111x + 56$ is $6x + 7$. Find the other factor.

62. One factor of $63x^2 + 110x + 48$ is $7x + 6$. Find the other factor.

63. One factor of $16t^2 - 64t + 48$ is $t - 1$. Find the other factors, then write the polynomial in factored form.

64. One factor of $16t^2 - 32t + 12$ is $2t - 1$. Find the other factors, then write the polynomial in factored form.

Applying the Concepts [Example 8]

Skyscrapers The chart shows the heights of the three tallest buildings in the world. Use the chart to answer Problems 65 and 66.

65. If you drop an object off the top of the Sears Tower, the height that the object is from the ground after t seconds is given by the equation $h = 1{,}450 - 16t^2$. Factor the right hand side of this equation. How far from the ground is the object after 6 seconds?

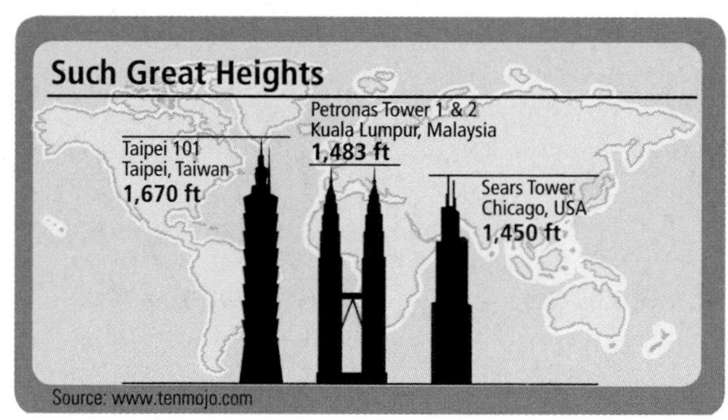

66. If you throw an object off the top of the Taipei 101 tower with an initial upward velocity of 6 feet per second, the height that the object is from the ground after t seconds is given by the equation $h = 1{,}670 + 6t - 16t^2$. Factor this equation.

67. Archery Margaret shoots an arrow into the air. The equation for the height (in feet) of the tip of the arrow is:

$$h = 8 + 62t - 16t^2$$

a. Factor the right side of this equation.

b. Fill in the table for various heights of the arrow, using the factored form of the equation.

Time t (seconds)	Height h (feet)
0	
1	
2	
3	
4	

68. Coin Toss At the beginning of every football game, the referee flips a coin to see who will kick off. The equation that gives the height (in feet) of the coin tossed in the air is:

$$h = 6 + 29t - 16t^2$$

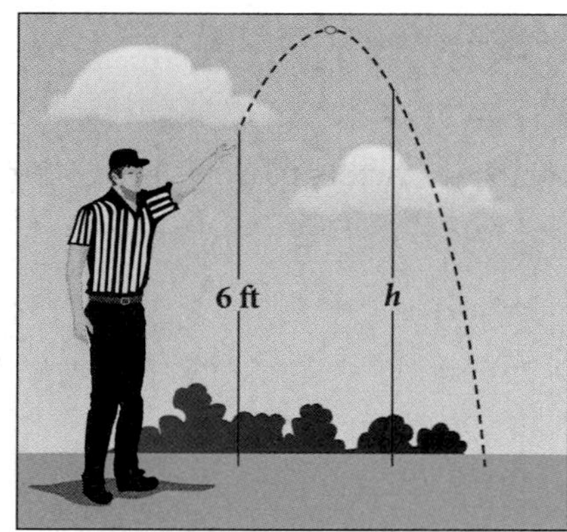

a. Factor the right side of this equation.

b. Use the factored form of the equation to find the height of the quarter after 0 seconds, 1 second, and 2 seconds.

Getting Ready for the Next Section

Multiply each of the following.

69. $(x + 3)(x - 3)$

70. $(x - 4)(x + 4)$

71. $(x + 5)(x - 5)$

72. $(x - 6)(x + 6)$

73. $(x + 7)(x - 7)$

74. $(x - 8)(x + 8)$

75. $(x + 9)(x - 9)$

76. $(x - 10)(x + 10)$

77. $(2x - 3y)(2x + 3y)$

78. $(5x - 6y)(5x + 6y)$

79. $(x^2 + 4)(x + 2)(x - 2)$

80. $(x^2 + 9)(x + 3)(x - 3)$

81. $(x + 3)^2$

82. $(x - 4)^2$

83. $(x + 5)^2$

84. $(x - 6)^2$

85. $(x + 7)^2$

86. $(x - 8)^2$

87. $(x + 9)^2$

88. $(x - 10)^2$

89. $(2x + 3)^2$

90. $(3x - y)^2$

91. $(4x - 2y)^2$

92. $(5x - 6y)^2$

Maintaining Your Skills

Perform the following additions and subtractions.

93. $(6x^3 - 4x^2 + 2x) + (9x^2 - 6x + 3)$

94. $(6x^3 - 4x^2 + 2x) - (9x^2 - 6x + 3)$

95. $(-7x^4 + 4x^3 - 6x) + (8x^4 + 7x^3 - 9)$

96. $(-7x^4 + 4x^3 - 6x) - (8x^4 + 7x^3 - 9)$

97. $(2x^5 + 3x^3 + 4x) + (5x^3 - 6x - 7)$

98. $(2x^5 + 3x^3 + 4x) - (5x^3 - 6x - 7)$

99. $(-8x^5 - 5x^4 + 7) + (7x^4 + 2x^2 + 5)$

100. $(-8x^5 - 5x^4 + 7) - (7x^4 + 2x^2 + 5)$

101. $\frac{24x^3y^7}{6x^{-2}y^4} + \frac{27x^{-2}y^{10}}{9x^{-7}y^7}$

102. $\frac{15x^8y^4}{5x^2y^2} - \frac{4x^7y^5}{2xy^3}$

103. $\frac{18a^5b^9}{3a^3b^6} - \frac{48a^{-3}b^{-1}}{16a^{-5}b^{-4}}$

104. $\frac{54a^{-3}b^5}{6a^{-7}b^{-2}} - \frac{32a^6b^5}{8a^2b^{-2}}$

The Difference of Two Squares

6.4

Objectives

A Factor the difference of two squares.

B Factor a perfect square trinomial.

In Chapter 5 we listed the following three special products:

$$(a + b)^2 = (a + b)(a + b) = a^2 + 2ab + b^2$$
$$(a - b)^2 = (a - b)(a - b) = a^2 - 2ab + b^2$$
$$(a + b)(a - b) = a^2 - b^2$$

Since factoring is the reverse of multiplication, we can also consider the three special products as three special factorings:

$$a^2 + 2ab + b^2 = (a + b)^2$$
$$a^2 - 2ab + b^2 = (a - b)^2$$
$$a^2 - b^2 = (a + b)(a - b)$$

Any trinomial of the form $a^2 + 2ab + b^2$ or $a^2 - 2ab + b^2$ can be factored by the methods of Section 6.3. The last line is the factoring to obtain the difference of two squares. The difference of two squares always factors in this way. Again, these are patterns you must be able to recognize on sight.

A The Difference of Two Squares

EXAMPLE 1 Factor $16x^2 - 25$.

SOLUTION We can see that the first term is a perfect square, and the last term is also. This fact becomes even more obvious if we rewrite the problem as:

$$16x^2 - 25 = (4x)^2 - (5)^2$$

The first term is the square of the quantity $4x$, and the last term is the square of 5. The completed problem looks like this:

$$16x^2 - 25 = (4x)^2 - (5)^2$$
$$= (4x + 5)(4x - 5)$$

To check our results, we multiply:

$$(4x + 5)(4x - 5) = 16x^2 + 20x - 20x - 25$$
$$= 16x^2 - 25$$

EXAMPLE 2 Factor $36a^2 - 1$.

SOLUTION We rewrite the two terms to show they are perfect squares and then factor. Remember, 1 is its own square, $1^2 = 1$.

$$36a^2 - 1 = (6a)^2 - (1)^2$$
$$= (6a + 1)(6a - 1)$$

To check our results, we multiply:

$$(6a + 1)(6a - 1) = 36a^2 + 6a - 6a - 1$$
$$= 36a^2 - 1$$

EXAMPLE 3 Factor $x^4 - y^4$.

SOLUTION x^4 is the perfect square $(x^2)^2$, and y^4 is $(y^2)^2$:

$$x^4 - y^4 = (x^2)^2 - (y^2)^2$$
$$= (x^2 - y^2)(x^2 + y^2)$$

PRACTICE PROBLEMS

1. Factor $9x^2 - 25$.

2. Factor $49a^2 - 1$.

3. Factor $a^4 - b^4$.

Answers

1. $(3x + 5)(3x - 5)$

2. $(7a + 1)(7a - 1)$

The factor $(x^2 - y^2)$ is itself the difference of two squares and therefore can be factored again. The factor $(x^2 + y^2)$ is the *sum* of two squares and cannot be factored again. The complete solution is this:

$$\begin{aligned} x^4 - y^4 &= (x^2)^2 - (y^2)^2 \\ &= (x^2 - y^2)(x^2 + y^2) \\ &= (x + y)(x - y)(x^2 + y^2) \end{aligned}$$

Note If you think the sum of two squares $x^2 + y^2$ factors, you should try it. Write down the factors you think it has, and then multiply them using the foil method. You won't get $x^2 + y^2$.

B Perfect Square Trinomials

4. Factor $4x^2 + 12x + 9$.

EXAMPLE 4 Factor $25x^2 - 60x + 36$.

SOLUTION Although this trinomial can be factored by the method we used in Section 6.3, we notice that the first and last terms are the perfect squares $(5x)^2$ and $(6)^2$. Before going through the method for factoring trinomials by listing all possible factors, we can check to see if $25x^2 - 60x + 36$ factors to $(5x - 6)^2$. We need only multiply to check:

$$\begin{aligned} (5x - 6)^2 &= (5x - 6)(5x - 6) \\ &= 25x^2 - 30x - 30x + 36 \\ &= 25x^2 - 60x + 36 \end{aligned}$$

The trinomial $25x^2 - 60x + 36$ factors to $(5x - 6)(5x - 6) = (5x - 6)^2$.

Note As we have indicated before, perfect square trinomials like the ones in Examples 4 and 5 can be factored by the methods developed in previous sections. Recognizing that they factor to binomial squares simply saves time in factoring.

5. Factor $6x^2 + 24x + 24$.

EXAMPLE 5 Factor $5x^2 + 30x + 45$.

SOLUTION We begin by factoring out the greatest common factor, which is 5. Then we notice that the trinomial that remains is a perfect square trinomial:

$$\begin{aligned} 5x^2 + 30x + 45 &= 5(x^2 + 6x + 9) \\ &= 5(x + 3)^2 \end{aligned}$$

6. Factor $(x - 4)^2 - 36$.

EXAMPLE 6 Factor $(x - 3)^2 - 25$.

SOLUTION This example has the form $a^2 - b^2$, where a is $x - 3$ and b is 5. We factor it according to the formula for the difference of two squares:

$$\begin{aligned} (x - 3)^2 - 25 &= (x - 3)^2 - 5^2 && \text{Write 25 as } 5^2 \\ &= [(x - 3) - 5][(x - 3) + 5] && \text{Factor} \\ &= (x - 8)(x + 2) && \text{Simplify} \end{aligned}$$

Notice in this example we could have expanded $(x - 3)^2$, subtracted 25, and then factored to obtain the same result:

$$\begin{aligned} (x - 3)^2 - 25 &= x^2 - 6x + 9 - 25 && \text{Expand } (x - 3)^2 \\ &= x^2 - 6x - 16 && \text{Simplify} \\ &= (x - 8)(x + 2) && \text{Factor} \end{aligned}$$

Getting Ready for Class

After reading through the preceding section, respond in your own words and in complete sentences.

1. Describe how you factor the difference of two squares.
2. What is a perfect square trinomial?
3. How do you know when you've factored completely?
4. Describe how you would factor $25x^2 - 60x + 36$.

Answers

3. $(a^2 + b^2)(a + b)(a - b)$
4. $(2x + 3)^2$ 5. $6(x + 2)^2$
6. $(x - 10)(x + 2)$

Problem Set 6.4

A Factor the following. **[Examples 1–3]**

1. $x^2 - 9$
2. $x^2 - 25$
3. $a^2 - 36$
4. $a^2 - 64$
5. $x^2 - 49$
6. $x^2 - 121$
7. $4a^2 - 16$
8. $4a^2 + 16$
9. $9x^2 + 25$
10. $16x^2 - 36$
11. $25x^2 - 169$
12. $x^2 - y^2$
13. $9a^2 - 16b^2$
14. $49a^2 - 25b^2$
15. $9 - m^2$
16. $16 - m^2$
17. $25 - 4x^2$
18. $36 - 49y^2$
19. $2x^2 - 18$
20. $3x^2 - 27$
21. $32a^2 - 128$
22. $3a^3 - 48a$
23. $8x^2y - 18y$
24. $50a^2b - 72b$
25. $a^4 - b^4$
26. $a^4 - 16$
27. $16m^4 - 81$
28. $81 - m^4$
29. $3x^3y - 75xy^3$
30. $2xy^3 - 8x^3y$

B Factor the following. [Examples 4–6]

31. $x^2 - 2x + 1$

32. $x^2 - 6x + 9$

33. $x^2 + 2x + 1$

34. $x^2 + 6x + 9$

35. $a^2 - 10a + 25$

36. $a^2 + 10a + 25$

37. $y^2 + 4y + 4$

38. $y^2 - 8y + 16$

39. $x^2 - 4x + 4$

40. $x^2 + 8x + 16$

41. $m^2 - 12m + 36$

42. $m^2 + 12m + 36$

43. $4a^2 + 12a + 9$

44. $9a^2 - 12a + 4$

45. $49x^2 - 14x + 1$

46. $64x^2 - 16x + 1$

47. $9y^2 - 30y + 25$

48. $25y^2 + 30y + 9$

49. $x^2 + 10xy + 25y^2$

50. $25x^2 + 10xy + y^2$

51. $9a^2 + 6ab + b^2$

52. $9a^2 - 6ab + b^2$

53. $y^2 - 3y + \frac{9}{4}$

54. $y^2 + 3y + \frac{9}{4}$

55. $a^2 + a + \frac{1}{4}$

56. $a^2 - 5a + \frac{25}{4}$

57. $x^2 - 7x + \frac{49}{4}$

58. $x^2 + 9x + \frac{81}{4}$

59. $x^2 - \frac{3}{4}x + \frac{9}{64}$

60. $x^2 - \frac{3}{2}x + \frac{9}{16}$

Factor the following by first factoring out the greatest common factor.

61. $3a^2 + 18a + 27$

62. $4a^2 - 16a + 16$

63. $2x^2 + 20xy + 50y^2$

64. $3x^2 + 30xy + 75y^2$

65. $x^3 + 4x^2 + 4x$

66. $a^3 - 10a^2 + 25a$

67. $y^4 - 8y^3 + 16y^2$

68. $x^4 + 12x^3 + 36x^2$

69. $5x^3 + 30x^2y + 45xy^2$

70. $12x^2y - 36xy^2 + 27y^3$

71. $12y^4 - 60y^3 + 75y^2$

72. $18a^4 - 12a^3b + 2a^2b^2$

Factor by grouping the first three terms together.

73. $x^2 + 6x + 9 - y^2$

74. $x^2 + 10x + 25 - y^2$

75. $x^2 + 2xy + y^2 - 9$

76. $a^2 + 2ab + b^2 - 25$

77. Find a value for b so that the polynomial $x^2 + bx + 49$ factors to $(x + 7)^2$.

78. Find a value of b so that the polynomial $x^2 + bx + 81$ factors to $(x + 9)^2$.

79. Find the value of c for which the polynomial $x^2 + 10x + c$ factors to $(x + 5)^2$.

80. Find the value of a for which the polynomial $ax^2 + 12x + 9$ factors to $(2x + 3)^2$.

Getting Ready for the Next Section

Multiply each of the following.

81. **a.** 1^3 **b.** 2^3 **c.** 3^3 **d.** 4^3 **e.** 5^3

82. **a.** $(-1)^3$ **b.** $(-2)^3$ **c.** $(-3)^3$ **d.** $(-4)^3$ **e.** $(-5)^3$

83. **a.** $x(x^2 - x + 1)$ **b.** $1(x^2 - x + 1)$ **c.** $(x + 1)(x^2 - x + 1)$

84. **a.** $x(x^2 + x + 1)$ **b.** $-1(x^2 + x + 1)$ **c.** $(x - 1)(x^2 + x + 1)$

85. **a.** $x(x^2 - 2x + 4)$ **b.** $2(x^2 - 2x + 4)$ **c.** $(x + 2)(x^2 - 2x + 4)$

86. **a.** $x(x^2 + 2x + 4)$ **b.** $-2(x^2 + 2x + 4)$ **c.** $(x - 2)(x^2 + 2x + 4)$

87. a. $x(x^2 - 3x + 9)$ b. $3(x^2 - 3x + 9)$ c. $(x + 3)(x^2 - 3x + 9)$

88. a. $x(x^2 + 3x + 9)$ b. $-3(x^2 + 3x + 9)$ c. $(x - 3)(x^2 + 3x + 9)$

89. a. $x(x^2 - 4x + 16)$ b. $4(x^2 - 4x + 16)$ c. $(x + 4)(x^2 - 4x + 16)$

90. a. $x(x^2 + 4x + 16)$ b. $-4(x^2 + 4x + 16)$ c. $(x - 4)(x^2 + 4x + 16)$

91. a. $x(x^2 - 5x + 25)$ b. $5(x^2 - 5x + 25)$ c. $(x + 5)(x^2 - 5x + 25)$

92. a. $x(x^2 + 5x + 25)$ b. $-5(x^2 + 5x + 25)$ c. $(x - 5)(x^2 + 5x + 25)$

Maintaining Your Skills

93. **Picture Messaging** The graph shows the number of text messages Mike sent each month in 2008. According to the chart, 32 picture messages were sent during August. If Verizon Wireless charged $6.05 for the month, how much were they charging Mike for picture messages? Round to the nearest cent.

94. **Triathlon** The chart shows the number of participants in the top five triathlons in the United States. What was the average number of participants in the five races?

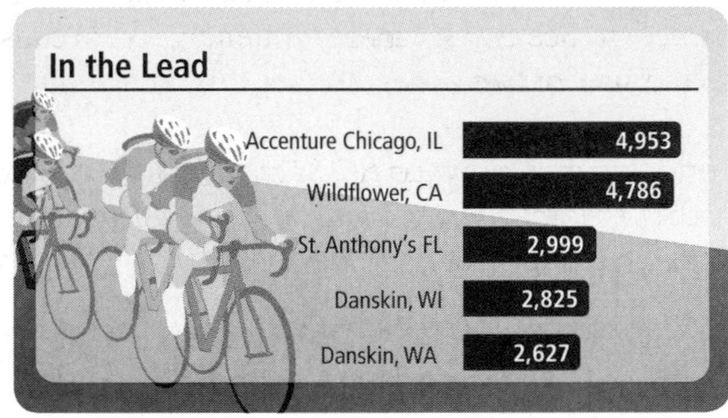

Divide.

95. $\dfrac{24y^3 - 36y^2 - 18y}{6y}$

96. $\dfrac{77y^3 + 35y^2 + 14y}{-7y}$

97. $\dfrac{48x^7 - 36x^5 + 12x^2}{4x^2}$

98. $\dfrac{-50x^5 + 15x^4 + 10x^2}{5x^2}$

99. $\dfrac{18x^7 + 12x^6 - 6x^5}{-3x^4}$

100. $\dfrac{-64x^5 - 18x^4 - 56x^3}{2x^3}$

101. $\dfrac{-42x^5 + 24x^4 - 66x^2}{6x^2}$

102. $\dfrac{63x^7 - 27x^6 - 99x^5}{-9x^4}$

Use long division to divide.

103. $\dfrac{x^2 - 5x + 8}{x - 3}$

104. $\dfrac{x^2 + 7x + 12}{x + 4}$

105. $\dfrac{6x^2 + 5x + 3}{2x + 3}$

106. $\dfrac{x^3 + 27}{x + 3}$

Extending the Concepts

Factor.

107. $t^2 - \dfrac{2}{5}t + \dfrac{1}{25}$

108. $t^2 + \dfrac{6}{5}t + \dfrac{9}{25}$

The Sum and Difference of Two Cubes

6.5

Objectives

A Factor the sum or difference of two cubes.

Previously, we factored a variety of polynomials. Among them were polynomials that were the difference of two squares. The formula we used to factor the difference of two squares looks like this:

$$a^2 - b^2 = (a + b)(a - b)$$

If we ran across a binomial that had the form of the difference of two squares, we factored it by applying this formula. For example, to factor $x^2 - 25$, we simply notice that it can be written in the form $x^2 - 5^2$, which looks like the difference of two squares. According to the formula above, this binomial factors into $(x + 5)(x - 5)$.

A Factoring the Sum of Difference of Two Cubes

In this section we want to use two new formulas that will allow us to factor the sum and difference of two cubes. For example, we want to factor the binomial $x^3 - 8$, which is the difference of two cubes. To see that it is the difference of two cubes, notice that it can be written $x^3 - 2^3$. We also want to factor $y^3 + 27$, which is the sum of two cubes. (To see this, notice that $y^3 + 27$ can be written as $y^3 + 3^3$.)

The formulas that allow us to factor the sum of two cubes and the difference of two cubes are not as simple as the formula for factoring the difference of two squares. Here is what they look like:

$$a^3 + b^3 = (a + b)(a^2 - ab + b^2)$$
$$a^3 - b^3 = (a - b)(a^2 + ab + b^2)$$

Let's begin our work with these two formulas by showing that they are true. To do so, we multiply out the right side of each formula.

EXAMPLE 1 Verify the two formulas.

SOLUTION We verify the formulas by multiplying the right sides and comparing the results with the left sides:

$$\begin{array}{rrrr} a^2 & - ab & + b^2 & \\ & a & + b & \\ \hline a^3 & - a^2b & + ab^2 & \\ & a^2b & - ab^2 & + b^3 \\ \hline a^3 & & & + b^3 \end{array} \qquad \begin{array}{rrrr} a^2 & + ab & + b^2 & \\ & a & - b & \\ \hline a^3 & + a^2b & + ab^2 & \\ & - a^2b & - ab^2 & - b^3 \\ \hline a^3 & & & - b^3 \end{array}$$

The first formula is correct. The second formula is correct.

PRACTICE PROBLEMS

1. Show that $(x - 2)(x^2 + 2x + 4) = x^3 - 8$

Here are some examples that use the formulas for factoring the sum and difference of two cubes.

EXAMPLE 2 Factor $x^3 - 8$.

SOLUTION Since the two terms are perfect cubes, we write them as such and apply the formula:

$$\begin{aligned} x^3 - 8 &= x^3 - 2^3 \\ &= (x - 2)(x^2 + 2x + 2^2) \\ &= (x - 2)(x^2 + 2x + 4) \end{aligned}$$

2. Factor $x^3 + 8$.

Answers

1. See solutions section.
2. $(x + 2)(x^2 - 2x + 4)$

3. Factor $y^3 - 27$.

EXAMPLE 3

Factor $y^3 + 27$.

SOLUTION Proceeding as we did in Example 2, we first write 27 as 3^3. Then, we apply the formula for factoring the sum of two cubes, which is $a^3 + b^3 = (a + b)(a^2 - ab + b^2)$:

$$\begin{aligned} y^3 + 27 &= y^3 + 3^3 \\ &= (y + 3)(y^2 - 3y + 3^2) \\ &= (y + 3)(y^2 - 3y + 9) \end{aligned}$$

Factor.

4. $27 + x^3$

EXAMPLE 4

Factor $64 + t^3$.

SOLUTION The first term is the cube of 4 and the second term is the cube of t. Therefore,

$$\begin{aligned} 64 + t^3 &= 4^3 + t^3 \\ &= (4 + t)(16 - 4t + t^2) \end{aligned}$$

5. $8x^3 + y^3$

EXAMPLE 5

Factor $27x^3 + 125y^3$.

SOLUTION Writing both terms as perfect cubes, we have

$$\begin{aligned} 27x^3 + 125y^3 &= (3x)^3 + (5y)^3 \\ &= (3x + 5y)(9x^2 - 15xy + 25y^2) \end{aligned}$$

6. $a^3 - \frac{1}{27}$

EXAMPLE 6

Factor $a^3 - \frac{1}{8}$.

SOLUTION The first term is the cube of a, whereas the second term is the cube of $\frac{1}{2}$:

$$\begin{aligned} a^3 - \frac{1}{8} &= a^3 - \left(\frac{1}{2}\right)^3 \\ &= \left(a - \frac{1}{2}\right)\left(a^2 + \frac{1}{2}a + \frac{1}{4}\right) \end{aligned}$$

7. $x^6 - 1$

EXAMPLE 7

Factor $x^6 - y^6$.

SOLUTION We have a choice of how we want to write the two terms to begin. We can write the expression as the difference of two squares, $(x^3)^2 - (y^3)^2$, or as the difference of two cubes, $(x^2)^3 - (y^2)^3$. It is better to use the difference of two squares if we have a choice:

$$\begin{aligned} x^6 - y^6 &= (x^3)^2 - (y^3)^2 \\ &= (x^3 - y^3)(x^3 + y^3) \\ &= (x - y)(x^2 + xy + y^2)(x + y)(x^2 - xy + y^2) \end{aligned}$$

Getting Ready for Class

After reading through the preceding section, respond in your own words and in complete sentences.

1. How can you check your work when factoring?
2. Why are the numbers 8, 27, 64, and 125 used so frequently in this section?
3. List the cubes of the numbers 1 through 10.
4. How are you going to remember that the sum of two cubes factors, while the sum of two squares is prime?

Answers

3. $(y - 3)(y^2 + 3y + 9)$
4. $(3 + x)(9 - 3x + x^2)$
5. $(2x + y)(4x^2 - 2xy + y^2)$
6. $\left(a - \frac{1}{3}\right)\left(a^2 + \frac{1}{3}a + \frac{1}{9}\right)$
7. $(x + 1)(x^2 - x + 1)(x - 1)(x^2 + x + 1)$

Problem Set 6.5

A Factor each of the following as the sum or difference of two cubes. [Examples 1–7]

1. $x^3 - y^3$

2. $x^3 + y^3$

3. $a^3 + 8$

4. $a^3 - 8$

5. $27 + x^3$

6. $27 - x^3$

7. $y^3 - 1$

8. $y^3 + 1$

9. $y^3 - 64$

10. $y^3 + 64$

11. $125h^3 - t^3$

12. $t^3 + 125h^3$

13. $x^3 - 216$

14. $216 + x^3$

15. $2y^3 - 54$

16. $81 + 3y^3$

17. $2a^3 - 128b^3$

18. $128a^3 + 2b^3$

19. $2x^3 + 432y^3$

20. $432x^3 - 2y^3$

21. $10a^3 - 640b^3$

22. $640a^3 + 10b^3$

23. $10r^3 - 1{,}250$

24. $10r^3 + 1{,}250$

25. $64 + 27a^3$

26. $27 - 64a^3$

27. $8x^3 - 27y^3$

28. $27x^3 - 8y^3$

29. $t^3 + \frac{1}{27}$

30. $t^3 - \frac{1}{27}$

31. $27x^3 - \frac{1}{27}$

32. $8x^3 + \frac{1}{8}$

33. $64a^3 + 125b^3$

34. $125a^3 - 27b^3$

35. $\frac{1}{8}x^3 - \frac{1}{27}y^3$

36. $\frac{1}{27}x^3 + \frac{1}{8}y^3$

37. $a^6 - b^6$

38. $x^6 - 64y^6$

39. $64x^6 - y^6$

40. $x^6 - (3y)^6$

41. $x^6 - (5y)^6$

42. $(4x)^6 - (7y)^6$

Getting Ready for the Next Section

Multiply each of the following.

43. $2x^3(x + 2)(x - 2)$

44. $3x^2(x + 3)(x - 3)$

45. $3x^2(x - 3)^2$

46. $2x^3(x + 5)^2$

47. $y(y^2 + 25)$

48. $y^3(y^2 + 36)$

49. $(5a - 2)(3a + 1)$

50. $(3a - 4)(2a - 1)$

51. $4x^2(x - 5)(x + 2)$

52. $6x(x - 4)(x + 2)$

53. $2ab^3(b^2 - 4b + 1)$

54. $2a^3b(a^2 + 3a + 1)$

Maintaining Your Skills

55. U.S. Energy The pie chart shows where Americans get their energy. In 2006, Americans used 101.6 quadrillion Btu of energy. How many Btu did Americans use from Petroleum?

56. Gas Prices The chart shows the breakdown in gas prices. If gasoline costs \$4.05, how much is paid for the actual oil? Round to the nearest cent.

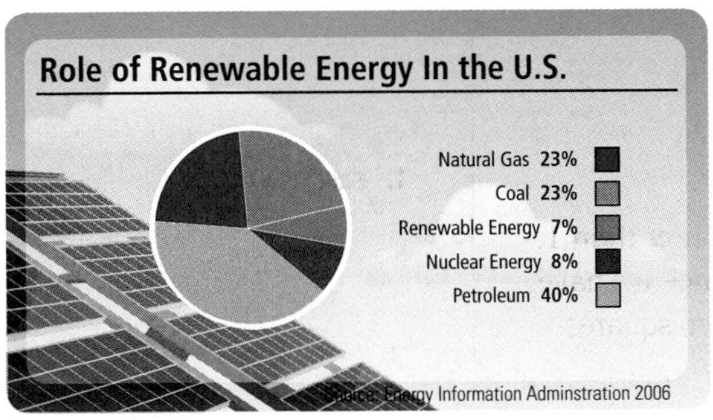

What we pay for in a gallon of gasoline

Gasoline: 10%, 5%, 10%, 75%

Diesel: 10%, 5%, 21%, 64%

Taxes
Distributing & Marketing
Refining
Crude Oil

Source: Energy Information Administration 2008

Solve each equation for x.

57. $2x - 6y = 8$

58. $-3x + 9y = 12$

59. $4x - 6y = 8$

60. $-20x + 15y = -10$

Solve each equation for y.

61. $3x - 6y = -18$

62. $-3x + 9y = -18$

63. $4x - 6y = 24$

64. $-20x + 5y = -10$

6.6 Factoring: A General Review

In this section we will review the different methods of factoring that we presented previously. Prior to this section, the polynomials you worked with were grouped together according to the method used to factor them; for example, in Section 6.4 all the polynomials you factored were either the difference of two squares or perfect square trinomials. What usually happens in a situation like this is that you become proficient at factoring the kind of polynomial you are working with at the time but have trouble when given a variety of polynomials to factor.

We begin this section with a checklist that can be used in factoring polynomials of any type. When you have finished this section and the problem set that follows, you want to be proficient enough at factoring that the checklist is second nature.

Objectives

A Factor a variety of polynomials.

Strategy Factoring a Polynomial

Step 1: If the polynomial has a greatest common factor other than 1, then factor out the greatest common factor.

Step 2: If the polynomial has two terms (it is a binomial), then see if it is the difference of two squares or the sum or difference of two cubes, and then factor accordingly. Remember, if it is the sum of two squares, it will not factor.

Step 3: If the polynomial has three terms (a trinomial), then either it is a perfect square trinomial, which will factor into the square of a binomial, or it is not a perfect square trinomial, in which case you use the trial and error method developed in Section 6.3.

Step 4: If the polynomial has more than three terms, try to factor it by grouping.

Step 5: As a final check, see if any of the factors you have written can be factored further. If you have overlooked a common factor, you can catch it here.

A Factoring a Variety of Polynomials

Here are some examples illustrating how we use the checklist.

EXAMPLE 1 Factor $2x^5 - 8x^3$.

SOLUTION First, we check to see if the greatest common factor is other than 1. Since the greatest common factor is $2x^3$, we begin by factoring it out. Once we have done so, we notice that the binomial that remains is the difference of two squares:

$$2x^5 - 8x^3 = 2x^3(x^2 - 4) \quad \text{Factor out the greatest common factor, } 2x^3$$
$$= 2x^3(x + 2)(x - 2) \quad \text{Factor the difference of two squares}$$

Note that the greatest common factor $2x^3$ that we factored from each term in the first step of Example 1 remains as part of the answer to the problem; that is because it is one of the factors of the original binomial. Remember, the expression we end up with when factoring must be equal to the expression we start with. We can't just drop a factor and expect the resulting expression to equal the original expression.

PRACTICE PROBLEMS

1. Factor $3x^4 - 27x^2$.

Answer

1. $3x^2(x + 3)(x - 3)$

EXAMPLE 2 Factor $3x^4 - 18x^3 + 27x^2$.

SOLUTION Step 1 is to factor out the greatest common factor, $3x^2$. After we have done so, we notice that the trinomial that remains is a perfect square trinomial, which will factor as the square of a binomial:

$$3x^4 - 18x^3 + 27x^2 = 3x^2(x^2 - 6x + 9) \quad \text{Factor out } 3x^2$$
$$= 3x^2(x - 3)^2 \quad x^2 - 6x + 9 \text{ is the square of } x - 3$$

EXAMPLE 3 Factor $y^3 + 25y$.

SOLUTION We begin by factoring out the y that is common to both terms. The binomial that remains after we have done so is the sum of two squares, which does not factor, so after the first step we are finished:

$$y^3 + 25y = y(y^2 + 25) \quad \text{Factor out the greatest common factor, } y\text{; then notice that } y^2 + 25 \text{ cannot be factored further}$$

EXAMPLE 4 Factor $6a^2 - 11a + 4$.

SOLUTION Here we have a trinomial that does not have a greatest common factor other than 1. Since it is not a perfect square trinomial, we factor it by trial and error; that is, we look for binomial factors the product of whose first terms is $6a^2$ and the product of whose last terms is 4. Then we look for the combination of these types of binomials whose product gives us a middle term of $-11a$. Without showing all the different possibilities, here is the answer:

$$6a^2 - 11a + 4 = (3a - 4)(2a - 1)$$

EXAMPLE 5 Factor $6x^3 - 12x^2 - 48x$.

SOLUTION This trinomial has a greatest common factor of $6x$. The trinomial that remains after the $6x$ has been factored from each term must be factored by trial and error:

$$6x^3 - 12x^2 - 48x = 6x(x^2 - 2x - 8)$$
$$= 6x(x - 4)(x + 2)$$

EXAMPLE 6 Factor $2ab^5 + 8ab^4 + 2ab^3$.

SOLUTION The greatest common factor is $2ab^3$. We begin by factoring it from each term. After that we find the trinomial that remains cannot be factored further:

$$2ab^5 + 8ab^4 + 2ab^3 = 2ab^3(b^2 + 4b + 1)$$

EXAMPLE 7 Factor $xy + 8x + 3y + 24$.

SOLUTION Since our polynomial has four terms, we try factoring by grouping:

$$xy + 8x + 3y + 24 = x(y + 8) + 3(y + 8)$$
$$= (y + 8)(x + 3)$$

Getting Ready for Class

After reading through the preceding section, respond in your own words and in complete sentences.

1. What is the first step in factoring any polynomial?
2. If a polynomial has four terms, what method of factoring should you try?
3. If a polynomial has two terms, what method of factoring should you try?
4. What is the last step in factoring any polynomial?

2. Factor $2x^5 + 20x^4 + 50x^3$.

3. Factor $y^5 + 36y^3$.

4. Factor $15a^2 - a - 2$.

5. Factor $4x^4 - 12x^3 - 40x^2$.

6. Factor $2a^5b + 6a^4b + 2a^3b$.

7. Factor $3ab + 9a + 2b + 6$.

Answers
2. $2x^3(x + 5)^2$ **3.** $y^3(y^2 + 36)$
4. $(5a - 2)(3a + 1)$
5. $4x^2(x - 5)(x + 2)$
6. $2a^3b(a^2 + 3a + 1)$
7. $(b + 3)(3a + 2)$

Problem Set 6.6

A Factor each of the following polynomials completely; that is, once you are finished factoring, none of the factors you obtain should be factorable. Also, note that the even-numbered problems are not necessarily similar to the odd-numbered problems that precede them in this problem set. **[Examples 1–7]**

1. $x^2 - 81$

2. $x^2 - 18x + 81$

3. $x^2 + 2x - 15$

4. $15x^2 + 11x - 6$

5. $x^2 + 6x + 9$

6. $12x^2 - 11x + 2$

7. $y^2 - 10y + 25$

8. $21y^2 - 25y - 4$

9. $2a^3b + 6a^2b + 2ab$

10. $6a^2 - ab - 15b^2$

11. $x^2 + x + 1$

12. $2x^2 - 4x + 2$

13. $12a^2 - 75$

14. $18a^2 - 50$

15. $9x^2 - 12xy + 4y^2$

16. $x^3 - x^2$

17. $4x^3 + 16xy^2$

18. $16x^2 + 49y^2$

19. $2y^3 + 20y^2 + 50y$

20. $3y^2 - 9y - 30$

21. $a^6 + 4a^4b^2$

22. $5a^2 - 45b^2$

23. $xy + 3x + 4y + 12$

24. $xy + 7x + 6y + 42$

25. $x^3 - 27$

26. $x^4 - 81$

27. $xy - 5x + 2y - 10$

28. $xy - 7x + 3y - 21$

29. $5a^2 + 10ab + 5b^2$

30. $3a^3b^2 + 15a^2b^2 + 3ab^2$

31. $x^2 + 49$

32. $16 - x^4$

33. $3x^2 + 15xy + 18y^2$

34. $3x^2 + 27xy + 54y^2$

35. $2x^2 + 15x - 38$

36. $2x^2 + 7x - 85$

37. $100x^2 - 300x + 200$

38. $100x^2 - 400x + 300$

39. $x^2 - 64$

40. $9x^2 - 4$

41. $x^2 + 3x + ax + 3a$

42. $x^2 + 4x + bx + 4b$

43. $49a^7 - 9a^5$

44. $8a^3 + 1$

45. $49x^2 + 9y^2$

46. $12x^4 - 62x^3 + 70x^2$

47. $25a^3 + 20a^2 + 3a$

48. $36a^4 - 100a^2$

49. $xa - xb + ay - by$

50. $xy - bx + ay - ab$

51. $48a^4b - 3a^2b$

52. $18a^4b^2 - 12a^3b^3 + 8a^2b^4$

53. $20x^4 - 45x^2$

54. $16x^3 + 16x^2 + 3x$

55. $3x^2 + 35xy - 82y^2$

56. $3x^2 + 37xy - 86y^2$

57. $16x^5 - 44x^4 + 30x^3$

58. $16x^2 + 16x - 1$

59. $2x^2 + 2ax + 3x + 3a$

60. $2x^2 + 2ax + 5x + 5a$

61. $y^4 - 1$

62. $a^7 + 8a^4b^3$

63. $12x^4y^2 + 36x^3y^3 + 27x^2y^4$

64. $16x^3y^2 - 4xy^2$

65. $16t^2 - 64t + 48$

66. $16t^2 - 32t + 12$

67. $54x^2 + 111x + 56$

68. $63x^2 + 110x + 48$

Getting Ready for the Next Section

Solve each equation.

69. $3x - 6 = 9$

70. $5x - 1 = 14$

71. $2x + 3 = 0$

72. $4x - 5 = 0$

73. $4x + 3 = 0$

74. $3x - 1 = 0$

Maintaining Your Skills

Solve each equation.

75. $-2(x + 4) = -10$

76. $\frac{3}{4}(-4x - 8) = 21$

77. $\frac{3}{5}x + 4 = 22$

78. $-10 = 4 - \frac{7}{4}x$

79. $6x - 4(9 - x) = -96$

80. $-2(x - 5) + 5x = 4 - 9$

81. $2x - 3(4x - 7) = -3x$

82. $\frac{3}{4}(8 + x) = \frac{1}{5}(5x - 15)$

83. $\frac{1}{2}x - \frac{5}{12} = \frac{1}{12}x + \frac{5}{12}$

84. $\frac{3}{10}x + \frac{5}{2} = \frac{3}{5}x - \frac{1}{2}$

6.7 Solving Equations by Factoring

Objectives

A Solve an equation by writing it in standard form and then factoring.

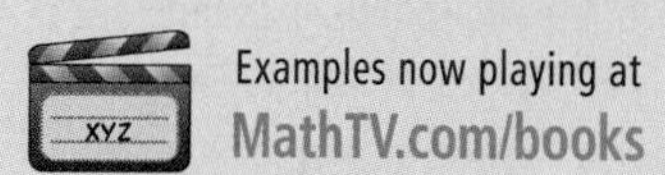

In this section we will use the methods of factoring developed in previous sections, along with a special property of 0, to solve quadratic equations.

Definition

Any equation that can be put in the form $ax^2 + bx + c = 0$, where a, b, and c are real numbers ($a \neq 0$), is called a **quadratic equation.** The equation $ax^2 + bx + c = 0$ is called **standard form** for a quadratic equation:

$$\underset{\text{an } x^2 \text{ term}}{a(\text{variable})^2} + \underset{\text{an } x \text{ term}}{b(\text{variable})} + \underset{\text{and a constant term}}{c(\text{absence of the variable})} = 0$$

The number 0 has a special property. If we multiply two numbers and the product is 0, then one or both of the original two numbers must be 0. In symbols, this property looks like this.

Zero-Factor Property

Let a and b represent real numbers. If $a \cdot b = 0$, then $a = 0$ or $b = 0$.

Suppose we want to solve the quadratic equation $x^2 + 5x + 6 = 0$. We can factor the left side into $(x + 2)(x + 3)$. Then we have:

$$x^2 + 5x + 6 = 0$$
$$(x + 2)(x + 3) = 0$$

Now, $(x + 2)$ and $(x + 3)$ both represent real numbers. Their product is 0; therefore, either $(x + 3)$ is 0 or $(x + 2)$ is 0. Either way we have a solution to our equation. We use the property of 0 stated to finish the problem:

$$x^2 + 5x + 6 = 0$$
$$(x + 2)(x + 3) = 0$$
$$x + 2 = 0 \quad \text{or} \quad x + 3 = 0$$
$$x = -2 \quad \text{or} \quad x = -3$$

Our solution set is $\{-2, -3\}$. Our equation has two solutions. To check our solutions we have to check each one separately to see that they both produce a true statement when used in place of the variable:

When $x = -3$

the equation $x^2 + 5x + 6 = 0$

becomes $(-3)^2 + 5(-3) + 6 \stackrel{?}{=} 0$

$9 + (-15) + 6 = 0$

$0 = 0$

When $x = -2$

the equation $x^2 + 5x + 6 = 0$

becomes $(-2)^2 + 5(-2) + 6 \stackrel{?}{=} 0$

$4 + (-10) + 6 = 0$

$0 = 0$

We have solved a quadratic equation by replacing it with two linear equations in one variable.

Notice that to solve a quadratic equation by this method, it must be possible to factor it. If we can't factor it, we can't solve it by this method. We will learn how to solve quadratic equations that do not factor when we get to Chapter 9.

A Solving a Quadratic Equation by Factoring

Strategy **Solving a Quadratic Equation by Factoring**

Step 1: Put the equation in standard form; that is, 0 on one side and decreasing powers of the variable on the other.

Step 2: Factor completely.

Step 3: Use the zero-factor property to set each variable factor from step 2 to 0.

Step 4: Solve each equation produced in step 3.

Step 5: Check each solution, if necessary.

PRACTICE PROBLEMS

1. Solve the equation:
$$2x^2 + 5x = 3$$

EXAMPLE 1 Solve the equation $2x^2 - 5x = 12$.

SOLUTION

Step 1: Begin by adding -12 to both sides, so the equation is in standard form:

$$2x^2 - 5x = 12$$
$$2x^2 - 5x - 12 = 0$$

Step 2: Factor the left side completely:

$$(2x + 3)(x - 4) = 0$$

Step 3: Set each factor to 0:

$$2x + 3 = 0 \quad \text{or} \quad x - 4 = 0$$

Step 4: Solve each of the equations from step 3:

$$2x + 3 = 0 \qquad x - 4 = 0$$
$$2x = -3 \qquad x = 4$$
$$x = -\frac{3}{2}$$

Step 5: Substitute each solution into $2x^2 - 5x = 12$ to check:

Check: $-\frac{3}{2}$	Check: 4
$2\left(-\frac{3}{2}\right)^2 - 5\left(-\frac{3}{2}\right) \stackrel{?}{=} 12$	$2(4)^2 - 5(4) \stackrel{?}{=} 12$
$2\left(\frac{9}{4}\right) + 5\left(\frac{3}{2}\right) = 12$	$2(16) - 20 = 12$
$\frac{9}{2} + \frac{15}{2} = 12$	$32 - 20 = 12$
$\frac{24}{2} = 12$	$12 = 12$
$12 = 12$	

Answer

1. $\frac{1}{2}, -3$

EXAMPLE 2 Solve for a: $16a^2 - 25 = 0$.

SOLUTION The equation is already in standard form:

$$16a^2 - 25 = 0$$
$$(4a - 5)(4a + 5) = 0 \quad \text{Factor left side}$$
$$4a - 5 = 0 \quad \text{or} \quad 4a + 5 = 0 \quad \text{Set each factor to 0}$$
$$4a = 5 \qquad 4a = -5 \quad \text{Solve the resulting equations}$$
$$a = \frac{5}{4} \qquad a = -\frac{5}{4}$$

2. Solve for a: $49a^2 - 16 = 0$.

EXAMPLE 3 Solve $4x^2 = 8x$.

SOLUTION We begin by adding $-8x$ to each side of the equation to put it in standard form. Then we factor the left side of the equation by factoring out the greatest common factor.

$$4x^2 = 8x$$
$$4x^2 - 8x = 0 \quad \text{Add } -8x \text{ to each side}$$
$$4x(x - 2) = 0 \quad \text{Factor the left side}$$
$$4x = 0 \quad \text{or} \quad x - 2 = 0 \quad \text{Set each factor to 0}$$
$$x = 0 \quad \text{or} \quad x = 2 \quad \text{Solve the resulting equations}$$

The solutions are 0 and 2.

3. Solve $5x^2 = -2x$.

EXAMPLE 4 Solve $x(2x + 3) = 44$.

SOLUTION We must multiply out the left side first and then put the equation in standard form:

$$x(2x + 3) = 44$$
$$2x^2 + 3x = 44 \quad \text{Multiply out the left side}$$
$$2x^2 + 3x - 44 = 0 \quad \text{Add } -44 \text{ to each side}$$
$$(2x + 11)(x - 4) = 0 \quad \text{Factor the left side}$$
$$2x + 11 = 0 \quad \text{or} \quad x - 4 = 0 \quad \text{Set each factor to 0}$$
$$2x = -11 \quad \text{or} \quad x = 4 \quad \text{Solve the resulting equations}$$
$$x = -\frac{11}{2}$$

The two solutions are $-\frac{11}{2}$ and 4.

4. Solve $x(13 - x) = 40$.

EXAMPLE 5 Solve for x: $5^2 = x^2 + (x + 1)^2$.

SOLUTION Before we can put this equation in standard form we must square the binomial. Remember, to square a binomial, we use the formula $(a + b)^2 = a^2 + 2ab + b^2$:

$$5^2 = x^2 + (x + 1)^2$$
$$25 = x^2 + x^2 + 2x + 1 \quad \text{Expand } 5^2 \text{ and } (x + 1)^2$$
$$25 = 2x^2 + 2x + 1 \quad \text{Simplify the right side}$$
$$0 = 2x^2 + 2x - 24 \quad \text{Add } -25 \text{ to each side}$$
$$0 = 2(x^2 + x - 12) \quad \text{Factor out 2}$$
$$0 = 2(x + 4)(x - 3) \quad \text{Factor completely}$$
$$x + 4 = 0 \quad \text{or} \quad x - 3 = 0 \quad \text{Set each variable factor to 0}$$
$$x = -4 \quad \text{or} \quad x = 3$$

5. Solve $(x + 2)^2 = (x + 1)^2 + x^2$.

Answers

2. $-\frac{4}{7}, \frac{4}{7}$ **3.** $0, -\frac{2}{5}$ **4.** 5, 8

6. Solve $8x^3 = 2x^2 + 10x$.

Note, in the second to last line, that we do not set 2 equal to 0. That is because 2 can never be 0. It is always 2. We only use the zero-factor property to set variable factors to 0 because they are the only factors that can possibly be 0.

Also notice that it makes no difference which side of the equation is 0 when we write the equation in standard form.

Although the equation in the next example is not a quadratic equation, it can be solved by the method shown in the first five examples.

EXAMPLE 6 Solve $24x^3 = -10x^2 + 6x$ for x.

SOLUTION First, we write the equation in standard form:

$$24x^3 + 10x^2 - 6x = 0 \quad \text{Standard form}$$

$$2x(12x^2 + 5x - 3) = 0 \quad \text{Factor out } 2x$$

$$2x(3x - 1)(4x + 3) = 0 \quad \text{Factor remaining trinomial}$$

$$2x = 0 \quad \text{or} \quad 3x - 1 = 0 \quad \text{or} \quad 4x + 3 = 0 \quad \text{Set factors to 0}$$

$$x = 0 \quad \text{or} \quad x = \frac{1}{3} \quad \text{or} \quad x = -\frac{3}{4} \quad \text{Solutions}$$

Getting Ready for Class

After reading through the preceding section, respond in your own words and in complete sentences.

1. When is an equation in standard form?
2. What is the first step in solving an equation by factoring?
3. Describe the zero-factor property in your own words.
4. Describe how you would solve the equation $2x^2 - 5x = 12$.

Answers

5. $-1, 3$ **6.** $0, \frac{5}{4}, -1$

Problem Set 6.7

A The following equations are already in factored form. Use the zero factor property to set the factors to 0 and solve.

1. $(x + 2)(x - 1) = 0$

2. $(x + 3)(x + 2) = 0$

3. $(a - 4)(a - 5) = 0$

4. $(a + 6)(a - 1) = 0$

5. $x(x + 1)(x - 3) = 0$

6. $x(2x + 1)(x - 5) = 0$

7. $(3x + 2)(2x + 3) = 0$

8. $(4x - 5)(x - 6) = 0$

9. $m(3m + 4)(3m - 4) = 0$

10. $m(2m - 5)(3m - 1) = 0$

11. $2y(3y + 1)(5y + 3) = 0$

12. $3y(2y - 3)(3y - 4) = 0$

Solve the following equations. [Examples 1–6]

13. $x^2 + 3x + 2 = 0$

14. $x^2 - x - 6 = 0$

15. $x^2 - 9x + 20 = 0$

16. $x^2 + 2x - 3 = 0$

17. $a^2 - 2a - 24 = 0$

18. $a^2 - 11a + 30 = 0$

19. $100x^2 - 500x + 600 = 0$

20. $100x^2 - 300x + 200 = 0$

21. $x^2 = -6x - 9$

22. $x^2 = 10x - 25$

23. $a^2 - 16 = 0$

24. $a^2 - 36 = 0$

25. $2x^2 + 5x - 12 = 0$

26. $3x^2 + 14x - 5 = 0$

27. $9x^2 + 12x + 4 = 0$

28. $12x^2 - 24x + 9 = 0$

29. $a^2 + 25 = 10a$

30. $a^2 + 16 = 8a$

31. $2x^2 = 3x + 20$

32. $6x^2 = x + 2$

33. $3m^2 = 20 - 7m$

34. $2m^2 = -18 + 15m$

35. $4x^2 - 49 = 0$

36. $16x^2 - 25 = 0$

37. $x^2 + 6x = 0$

38. $x^2 - 8x = 0$

39. $x^2 - 3x = 0$

40. $x^2 + 5x = 0$

41. $2x^2 = 8x$

42. $2x^2 = 10x$

43. $3x^2 = 15x$

44. $5x^2 = 15x$

45. $1{,}400 = 400 + 700x - 100x^2$

46. $2{,}700 = 700 + 900x - 100x^2$

47. $6x^2 = -5x + 4$

48. $9x^2 = 12x - 4$

49. $x(2x - 3) = 20$

50. $x(3x - 5) = 12$

51. $t(t + 2) = 80$

52. $t(t + 2) = 99$

53. $4{,}000 = (1{,}300 - 100p)p$

54. $3{,}200 = (1{,}200 - 100p)p$

55. $x(14 - x) = 48$

56. $x(12 - x) = 32$

57. $(x + 5)^2 = 2x + 9$

58. $(x + 7)^2 = 2x + 13$

59. $(y - 6)^2 = y - 4$

60. $(y + 4)^2 = y + 6$

61. $10^2 = (x + 2)^2 + x^2$

62. $15^2 = (x + 3)^2 + x^2$

63. $2x^3 + 11x^2 + 12x = 0$

64. $3x^3 + 17x^2 + 10x = 0$

65. $4y^3 - 2y^2 - 30y = 0$

66. $9y^3 + 6y^2 - 24y = 0$

67. $8x^3 + 16x^2 = 10x$

68. $24x^3 - 22x^2 = -4x$

69. $20a^3 = -18a^2 + 18a$

70. $12a^3 = -2a^2 + 10a$

71. $16t^2 - 32t + 12 = 0$

72. $16t^2 - 64t + 48 = 0$

Simplify each side as much as possible, then solve the equation.

73. $(a - 5)(a + 4) = -2a$

74. $(a + 2)(a - 3) = -2a$

75. $3x(x + 1) - 2x(x - 5) = -42$

76. $4x(x - 2) - 3x(x - 4) = -3$

77. $2x(x + 3) = x(x + 2) - 3$

78. $3x(x - 3) = 2x(x - 4) + 6$

79. $a(a - 3) + 6 = 2a$

80. $a(a - 4) + 8 = 2a$

81. $15(x + 20) + 15x = 2x(x + 20)$

82. $15(x + 8) + 15x = 2x(x + 8)$

83. $15 = a(a + 2)$

84. $6 = a(a - 5)$

Use factoring by grouping to solve the following equations.

85. $x^3 + 3x^2 - 4x - 12 = 0$

86. $x^3 + 5x^2 - 9x - 45 = 0$

87. $x^3 + x^2 - 16x - 16 = 0$

88. $4x^3 + 12x^2 - 9x - 27 = 0$

89. Find a quadratic equation that has two solutions; $x = 3$ and $x = 5$. Write your answer in standard form.

90. Find a quadratic equation that has two solutions; $x = 9$ and $x = 1$. Write your answer in standard form.

91. Find a quadratic equation that has the two given solutions.

a. $x = 3$ and $x = 2$.

b. $x = 1$ and $x = 6$.

c. $x = 3$ and $x = -2$.

92. Find a quadratic equation that has the two given solutions.

a. $x = 4$ and $x = 5$.

b. $x = 2$ and $x = 10$.

c. $x = -4$ and $x = 5$.

Getting Ready for the Next Section

Write each sentence as an algebraic equation.

93. The product of two consecutive integers is 72.

94. The product of two consecutive even integers is 80.

95. The product of two consecutive odd integers is 99.

96. The product of two consecutive odd integers is 63.

97. The product of two consecutive even integers is 10 less than 5 times their sum.

98. The product of two consecutive odd integers is 1 less than 4 times their sum.

The following word problems are taken from the book *Academic Algebra,* written by William J. Milne and published by the American Book Company in 1901. Solve each problem.

99. Cost of a Bicycle and a Suit A bicycle and a suit cost \$90. How much did each cost, if the bicycle cost 5 times as much as the suit?

100. Cost of a Cow and a Calf A man bought a cow and a calf for \$36, paying 8 times as much for the cow as for the calf. What was the cost of each?

101. Cost of a House and a Lot A house and a lot cost \$3,000. If the house cost 4 times as much as the lot, what was the cost of each?

102. Daily Wages A plumber and two helpers together earned \$7.50 per day. How much did each earn per day, if the plumber earned 4 times as much as each helper?

Maintaining Your Skills

Use the properties of exponents to simplify each expression.

103. 2^{-3}

104. 5^{-2}

105. $\frac{x^5}{x^{-3}}$

106. $\frac{x^{-2}}{x^{-5}}$

107. $\frac{(x^2)^3}{(x^{-3})^4}$

108. $\frac{(x^2)^{-4}(x^{-2})^3}{(x^{-3})^{-5}}$

109. Write the number 0.0056 in scientific notation.

110. Write the number 2.34×10^{-4} in expanded form.

111. Write the number 5,670,000,000 in scientific notation.

112. Write the number 0.00000567 in scientific notation.

Applications

6.8

Objectives

A Solve number problems by factoring.

B Solve geometry problems by factoring.

C Use the Pythagorean Theorem to solve problems.

In this section we will look at some application problems, the solutions to which require solving a quadratic equation. We will also introduce the Pythagorean theorem, one of the oldest theorems in the history of mathematics. The person whose name we associate with the theorem, Pythagoras (of Samos), was a Greek philosopher and mathematician who lived from about 560 B.C. to 480 B.C. According to the British philosopher Bertrand Russell, Pythagoras was "intellectually one of the most important men that ever lived."

Also in this section, the solutions to the examples show only the essential steps from our Blueprint for Problem Solving. Recall that step 1 is done mentally; we read the problem and mentally list the items that are known and the items that are unknown. This is an essential part of problem solving. However, now that you have had experience with application problems, you are doing step 1 automatically.

A Number Problems

EXAMPLE 1 The product of two consecutive odd integers is 63. Find the integers.

SOLUTION Let x = the first odd integer; then $x + 2$ = the second odd integer. An equation that describes the situation is:

$$x(x + 2) = 63 \quad \text{Their product is 63}$$

We solve the equation:

$$x(x + 2) = 63$$
$$x^2 + 2x = 63$$
$$x^2 + 2x - 63 = 0$$
$$(x - 7)(x + 9) = 0$$
$$x - 7 = 0 \quad \text{or} \quad x + 9 = 0$$
$$x = 7 \quad \text{or} \quad x = -9$$

If the first odd integer is 7, the next odd integer is $7 + 2 = 9$. If the first odd integer is -9, the next consecutive odd integer is $-9 + 2 = -7$. We have two pairs of consecutive odd integers that are solutions. They are 7, 9 and $-9, -7$.

We check to see that their products are 63:

$$7(9) = 63$$
$$-7(-9) = 63$$

PRACTICE PROBLEMS

1. The product of two consecutive even integers is 48. Find the integers. (*Hint:* If x is one of the even integers, then $x + 2$ is the next consecutive even integer.)

Suppose we know that the sum of two numbers is 50. We want to find a way to represent each number using only one variable. If we let x represent one of the two numbers, how can we represent the other? Let's suppose for a moment that x turns out to be 30. Then the other number will be 20, because their sum is 50; that is, if two numbers add up to 50 and one of them is 30, then the other must be $50 - 30 = 20$. Generalizing this to any number x, we see that if two numbers have

Answer

1. $-8, -6$ and $6, 8$

a sum of 50 and one of the numbers is x, then the other must be $50 - x$. The table that follows shows some additional examples.

If two numbers have a sum of	and one of them is	then the other must be
50	x	$50 - x$
100	x	$100 - x$
10	y	$10 - y$
12	n	$12 - n$

Now, let's look at an example that uses this idea.

2. The sum of two numbers is 14. Their product is 48. Find the two numbers.

EXAMPLE 2 The sum of two numbers is 13. Their product is 40. Find the numbers.

SOLUTION If we let x represent one of the numbers, then $13 - x$ must be the other number because their sum is 13. Since their product is 40, we can write:

$$x(13 - x) = 40 \quad \text{The product of the two numbers is 40}$$

$$13x - x^2 = 40 \quad \text{Multiply the left side}$$

$$x^2 - 13x = -40 \quad \text{Multiply both sides by } -1 \text{ and reverse the order of the terms on the left side}$$

$$x^2 - 13x + 40 = 0 \quad \text{Add 40 to each side}$$

$$(x - 8)(x - 5) = 0 \quad \text{Factor the left side}$$

$$x - 8 = 0 \quad \text{or} \quad x - 5 = 0$$

$$x = 8 \qquad\qquad x = 5$$

The two solutions are 8 and 5. If x is 8, then the other number is $13 - x = 13 - 8 = 5$. Likewise, if x is 5, the other number is $13 - x = 13 - 5 = 8$. Therefore, the two numbers we are looking for are 8 and 5. Their sum is 13 and their product is 40.

B Geometry Problems

Many word problems dealing with area can best be described algebraically by quadratic equations.

3. The length of a rectangle is 2 more than twice the width. The area is 60 square inches. Find the dimensions.

EXAMPLE 3 The length of a rectangle is 3 more than twice the width. The area is 44 square inches. Find the dimensions (find the length and width).

SOLUTION As shown in Figure 1, let x = the width of the rectangle. Then $2x + 3$ = the length of the rectangle because the length is three more than twice the width.

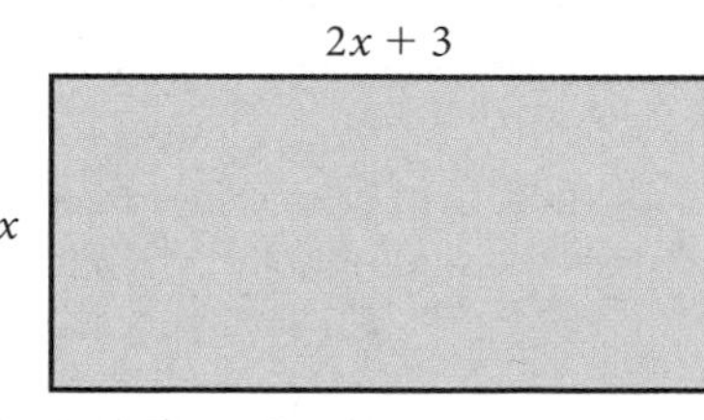

FIGURE 1

Answer
2. 6, 8

Since the area is 44 square inches, an equation that describes the situation is

$$x(2x + 3) = 44 \qquad \text{Length} \cdot \text{width} = \text{area}$$

We now solve the equation:

$$x(2x + 3) = 44$$

$$2x^2 + 3x = 44$$

$$2x^2 + 3x - 44 = 0$$

$$(2x + 11)(x - 4) = 0$$

$$2x + 11 = 0 \quad \text{or} \quad x - 4 = 0$$

$$x = -\frac{11}{2} \quad \text{or} \quad x = 4$$

The solution $x = -\frac{11}{2}$ cannot be used since length and width are always given in positive units. The width is 4. The length is 3 more than twice the width or $2(4) + 3 = 11$.

Width = 4 inches

Length = 11 inches

The solutions check in the original problem since $4(11) = 44$.

EXAMPLE 4 The numerical value of the area of a square is twice its perimeter. What is the length of its side?

SOLUTION As shown in Figure 2, let x = the length of its side. Then x^2 = the area of the square and $4x$ = the perimeter of the square:

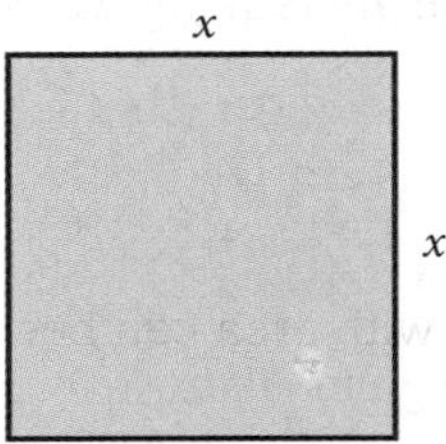

FIGURE 2

An equation that describes the situation is

$$x^2 = 2(4x) \qquad \text{The area is 2 times the perimeter}$$

$$x^2 = 8x$$

$$x^2 - 8x = 0$$

$$x(x - 8) = 0$$

$$x(x - 8) = 0$$

$$x = 0 \quad \text{or} \quad x = 8$$

Since $x = 0$ does not make sense in our original problem, we use $x = 8$. If the side has length 8, then the perimeter is $4(8) = 32$ and the area is $8^2 = 64$. Since 64 is twice 32, our solution is correct.

4. The height of a triangle is 3 more than the base. The area is 20 square inches. Find the base.

Answers

3. Width is 5 inches; length is 12 inches
4. Base is 5 inches, height is 8 inches

C The Pythagorean Theorem

FACTS FROM GEOMETRY: The Pythagorean Theorem

Next, we will work some problems involving the Pythagorean theorem. It may interest you to know that Pythagoras formed a secret society around the year 540 B.C. Known as the Pythagoreans, members kept no written record of their work; everything was handed down by spoken word. They influenced not only mathematics, but religion, science, medicine, and music as well. Among other things, they discovered the correlation between musical notes and the reciprocals of counting numbers, $\frac{1}{2}, \frac{1}{3}, \frac{1}{4}$, and so on. In their daily lives, they followed strict dietary and moral rules to achieve a higher rank in future lives.

Pythagorean Theorem

In any right triangle (Figure 3), the square of the longer side (called the hypotenuse) is equal to the sum of the squares of the other two sides (called legs).

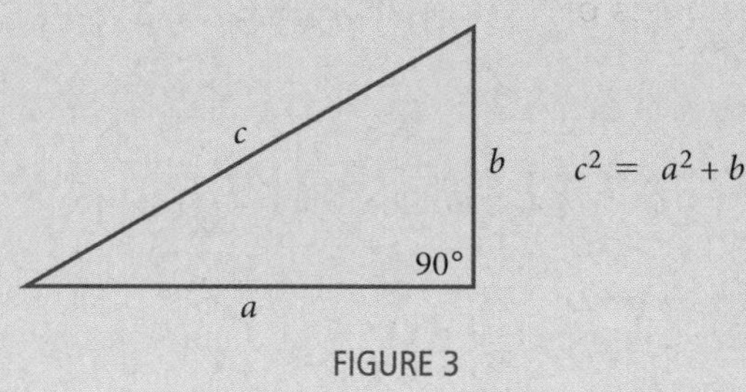

FIGURE 3

5. The longer leg of a right triangle is 2 more than twice the shorter leg. The hypotenuse is 3 more than twice the shorter leg. Find the length of each side. (*Hint:* Let x = the length of the shorter leg. Then the longer leg is $2x + 2$. Now write an expression for the hypotenuse, and you will be on your way.)

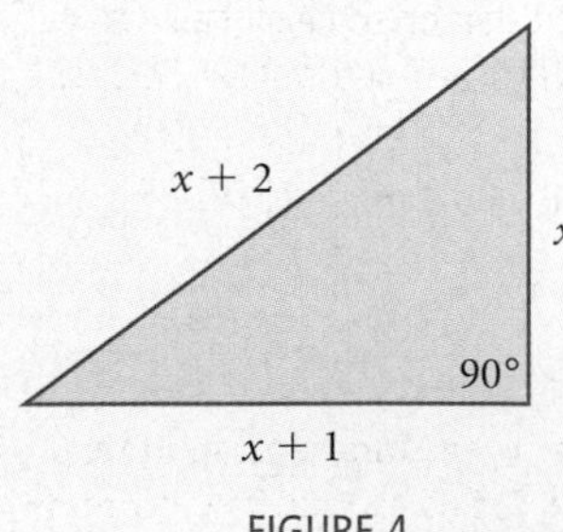

FIGURE 4

EXAMPLE 5 The three sides of a right triangle are three consecutive integers. Find the lengths of the three sides.

SOLUTION

$$\begin{aligned} \text{Let } x &= \text{the first integer (shortest side)} \\ \text{then } x + 1 &= \text{the next consecutive integer} \\ \text{and } x + 2 &= \text{the consecutive integer (longest side)} \end{aligned}$$

A diagram of the triangle is shown in Figure 4.

The Pythagorean theorem tells us that the square of the longest side $(x + 2)^2$ is equal to the sum of the squares of the two shorter sides, $(x + 1)^2 + x^2$. Here is the equation:

$$(x + 2)^2 = (x + 1)^2 + x^2$$

$$x^2 + 4x + 4 = x^2 + 2x + 1 + x^2 \quad \text{Expand squares}$$

$$x^2 - 2x - 3 = 0 \quad \text{Standard form}$$

$$(x - 3)(x + 1) = 0 \quad \text{Factor}$$

$$x - 3 = 0 \quad \text{or} \quad x + 1 = 0 \quad \text{Set factors to 0}$$

$$x = 3 \quad \text{or} \quad x = -1$$

Since a triangle cannot have a side with a negative number for its length, we must not use -1 for a solution to our original problem; therefore, the shortest side is 3. The other two sides are the next two consecutive integers, 4 and 5.

Answer

5. 5, 12, 13

EXAMPLE 6 The hypotenuse of a right triangle is 5 inches, and the lengths of the two legs (the other two sides) are given by two consecutive integers. Find the lengths of the two legs.

SOLUTION If we let x = the length of the shortest side, then the other side must be $x + 1$. A diagram of the triangle is shown in Figure 5.

6. The hypotenuse of a right triangle is 10 inches, and the lengths of the two sides are given by two consecutive even integers. Find the two sides.

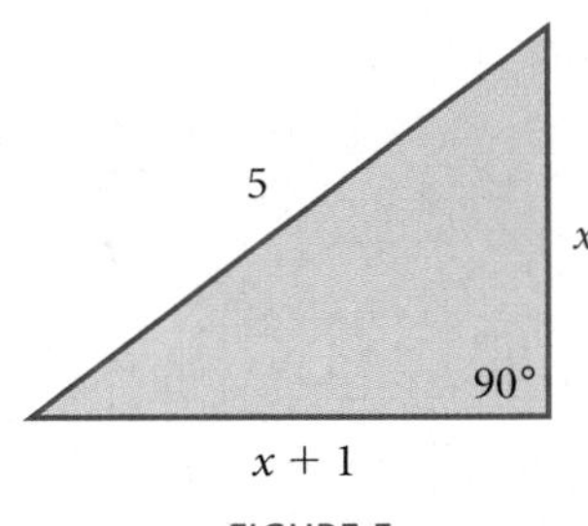

FIGURE 5

The Pythagorean theorem tells us that the square of the longest side, 5^2, is equal to the sum of the squares of the two shorter sides, $x^2 + (x + 1)^2$. Here is the equation:

$5^2 = x^2 + (x + 1)^2$	Pythagorean theorem
$25 = x^2 + x^2 + 2x + 1$	Expand 5^2 and $(x + 1)^2$
$25 = 2x^2 + 2x + 1$	Simplify the right side
$0 = 2x^2 + 2x - 24$	Add -25 to each side
$0 = 2(x^2 + x - 12)$	Factor out 2
$0 = 2(x + 4)(x - 3)$	Factor completely
$x + 4 = 0$ or $x - 3 = 0$	Set variable factors to 0
$x = -4$ or $x = 3$	

Since a triangle cannot have a side with a negative number for its length, we cannot use -4; therefore, the shortest side must be 3 inches. The next side is $x + 1 = 3 + 1 = 4$ inches. Since the hypotenuse is 5, we can check our solutions with the Pythagorean theorem as shown in Figure 6.

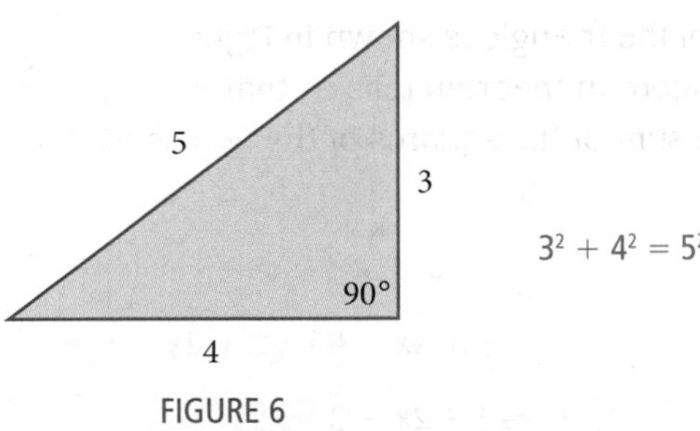

FIGURE 6

If you are planning on taking finite mathematics, statistics, or business calculus in the future, Examples 7 and 8 will give you a head start on some of the problems you will see in those classes.

EXAMPLE 7 A company can manufacture x hundred items for a total cost of $C = 300 + 500x - 100x^2$. How many items were manufactured if the total cost is $900?

SOLUTION We are looking for x when C is 900. We begin by substituting 900 for C in the cost equation. Then we solve for x:

When	$C = 900$
the equation	$C = 300 + 500x - 100x^2$
becomes	$900 = 300 + 500x - 100x^2$

7. Using the information in Example 7, find the number of items that should be made if the total cost is to be $700.

Answer

6. The lengths of the sides are 6 inches and 8 inches.

We can write this equation in standard form by adding -300, $-500x$, and $100x^2$ to each side. The result looks like this:

$$\begin{aligned} 100x^2 - 500x + 600 &= 0 \\ 100(x^2 - 5x + 6) &= 0 && \text{Factor out 100} \\ 100(x - 2)(x - 3) &= 0 && \text{Factor completely} \\ x - 2 = 0 \quad \text{or} \quad x - 3 &= 0 && \text{Set variable factors to 0} \\ x = 2 \quad \text{or} \quad x &= 3 \end{aligned}$$

Our solutions are two and three, which means that the company can manufacture 200 items or 300 items for a total cost of $900.

8. Use the information in Example 8 to find the selling price that should be charged if the weekly revenue is to be $4,200.

EXAMPLE 8 A manufacturer of small portable radios knows that the number of radios she can sell each week is related to the price of the radios by the equation $x = 1{,}300 - 100p$ (x is the number of radios and p is the price per radio). What price should she charge for the radios to have a weekly revenue of $4,000?

SOLUTION First, we must find the revenue equation. The equation for total revenue is $R = xp$, where x is the number of units sold and p is the price per unit. Since we want R in terms of p, we substitute $1{,}300 - 100p$ for x in the equation $R = xp$:

$$\begin{aligned} &\text{If} && R = xp \\ &\text{and} && x = 1{,}300 - 100p \\ &\text{then} && R = (1{,}300 - 100p)p \end{aligned}$$

We want to find p when R is 4,000. Substituting 4,000 for R in the equation gives us:

$$4{,}000 = (1{,}300 - 100p)p$$

If we multiply out the right side, we have:

$$4{,}000 = 1{,}300p - 100p^2$$

To write this equation in standard form, we add $100p^2$ and $-1{,}300p$ to each side:

$$\begin{aligned} 100p^2 - 1{,}300p + 4{,}000 &= 0 && \text{Add } 100p^2 \text{ and } -1{,}300p \text{ to each side} \\ 100(p^2 - 13p + 40) &= 0 && \text{Factor out 100} \\ 100(p - 5)(p - 8) &= 0 && \text{Factor completely} \\ p - 5 = 0 \quad \text{or} \quad p - 8 &= 0 && \text{Set variable factors to 0} \\ p = 5 \quad \text{or} \quad p &= 8 \end{aligned}$$

If she sells the radios for $5 each or for $8 each, she will have a weekly revenue of $4,000.

Getting Ready for Class

After reading through the preceding section, respond in your own words and in complete sentences.

1. What are consecutive integers?
2. Explain the Pythagorean theorem in words.
3. Write an application problem for which the solution depends on solving the equation $x(x + 1) = 12$.
4. Write an application problem for which the solution depends on solving the equation $x(2x - 3) = 40$.

Answers

7. 100 items or 400 items
8. $6.00 or $7.00

Problem Set 6.8

A Number Problems [Examples 1–2]

Solve the following word problems. Be sure to show the equation used.

1. The product of two consecutive even integers is 80. Find the two integers.

2. The product of two consecutive integers is 72. Find the two integers.

3. The product of two consecutive odd integers is 99. Find the two integers.

4. The product of two consecutive integers is 132. Find the two integers.

5. The product of two consecutive even integers is 10 less than 5 times their sum. Find the two integers.

6. The product of two consecutive odd integers is 1 less than 4 times their sum. Find the two integers.

7. The sum of two numbers is 14. Their product is 48. Find the numbers.

8. The sum of two numbers is 12. Their product is 32. Find the numbers.

9. One number is 2 more than 5 times another. Their product is 24. Find the numbers.

10. One number is 1 more than twice another. Their product is 55. Find the numbers.

11. One number is 4 times another. Their product is 4 times their sum. Find the numbers.

12. One number is 2 more than twice another. Their product is 2 more than twice their sum. Find the numbers.

B Geometry Problems [Examples 3–4]

13. The length of a rectangle is 1 more than the width. The area is 12 square inches. Find the dimensions.

14. The length of a rectangle is 3 more than twice the width. The area is 44 square inches. Find the dimensions.

15. The height of a triangle is twice the base. The area is 9 square inches. Find the base.

16. The height of a triangle is 2 more than twice the base. The area is 20 square feet. Find the base.

17. The hypotenuse of a right triangle is 10 inches. The lengths of the two legs are given by two consecutive even integers. Find the lengths of the two legs.

18. The hypotenuse of a right triangle is 15 inches. One of the legs is 3 inches more than the other. Find the lengths of the two legs.

19. The shorter leg of a right triangle is 5 meters. The hypotenuse is 1 meter longer than the longer leg. Find the length of the longer leg.

20. The shorter leg of a right triangle is 12 yards. If the hypotenuse is 20 yards, how long is the other leg?

Business Problems [Examples 7–8]

21. A company can manufacture x hundred items for a total cost of $C = 400 + 700x - 100x^2$. Find x if the total cost is \$1,400.

22. If the total cost C of manufacturing x hundred items is given by the equation $C = 700 + 900x - 100x^2$, find x when C is \$2,700.

23. The total cost C of manufacturing x hundred DVD's is given by the equation

$$C = 600 + 1{,}000x - 100x^2$$

Find x if the total cost is \$2,200.

24. The total cost C of manufacturing x hundred pen and pencil sets is given by the equation

$$C = 500 + 800x - 100x^2$$

Find x when C is \$1,700.

25. A company that manufactures typewriter ribbons knows that the number of ribbons it can sell each week, x, is related to the price p per ribbon by the equation $x = 1{,}200 - 100p$. At what price should the company sell the ribbons if it wants the weekly revenue to be \$3,200? (*Remember:* The equation for revenue is $R = xp$.)

26. A company manufactures small pen drives for home computers. It knows from experience that the number of pen drives it can sell each day, x, is related to the price p per pen drive by the equation $x = 800 - 100p$. At what price should the company sell the pen drives if it wants the daily revenue to be \$1,200?

27. The relationship between the number of calculators a company sells per week, x, and the price p of each calculator is given by the equation $x = 1{,}700 - 100p$. At what price should the calculators be sold if the weekly revenue is to be \$7,000?

28. The relationship between the number of pencil sharpeners a company can sell each week, x, and the price p of each sharpener is given by the equation $x = 1{,}800 - 100p$. At what price should the sharpeners be sold if the weekly revenue is to be \$7,200?

[Examples 5–6]

29. **Pythagorean Theorem** A 13-foot ladder is placed so that it reaches to a point on the wall that is 2 feet higher than twice the distance from the base of the wall to the base of the ladder.
   a. How far from the wall is the base of the ladder?
   b. How high does the ladder reach?

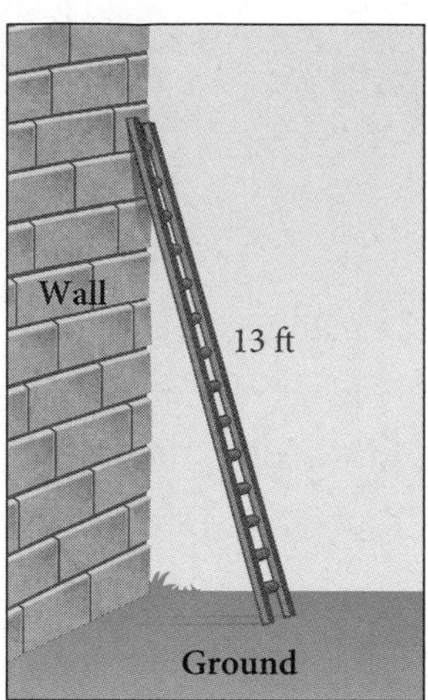

30. **Height of a Projectile** If a rocket is fired vertically into the air with a speed of 240 feet per second, its height at time t seconds is given by $h(t) = -16t^2 + 240t$. Here is a graph of its height at various times, with the details left out:

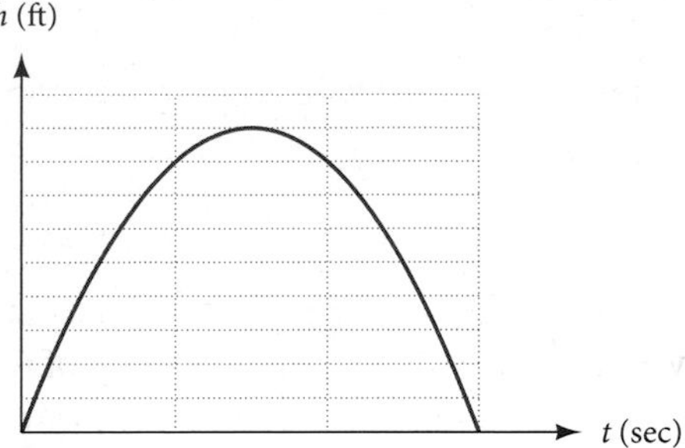

At what time(s) will the rocket be the following number of feet above the ground?
   a. 704 feet
   b. 896 feet
   c. Why do parts a and b each have two answers?
   d. How long will the rocket be in the air? (*Hint:* How high is it when it hits the ground?)
   e. When the equation for part d is solved, one of the answers is $t = 0$ second. What does this represent?

31. **Projectile Motion** A gun fires a bullet almost straight up from the edge of a 100-foot cliff. If the bullet leaves the gun with a speed of 396 feet per second, its height at time t is given by $h(t) = -16t^2 + 396t + 100$, measured from the ground below the cliff.
   a. When will the bullet land on the ground below the cliff? (*Hint:* What is its height when it lands? Remember that we are measuring from the ground below, not from the cliff.)
   b. Make a table showing the bullet's height every five seconds, from the time it is fired ($t = 0$) to the time it lands. (*Note:* It is faster to substitute into the factored form.)

t (sec)	h (feet)
0	
5	
10	
15	
20	
25	

32. **Constructing a Box** I have a piece of cardboard that is twice as long as it is wide. If I cut a 2-inch by 2-inch square from each corner and fold up the resulting flaps, I get a box with a volume of 32 cubic inches. What are the dimensions of the cardboard?

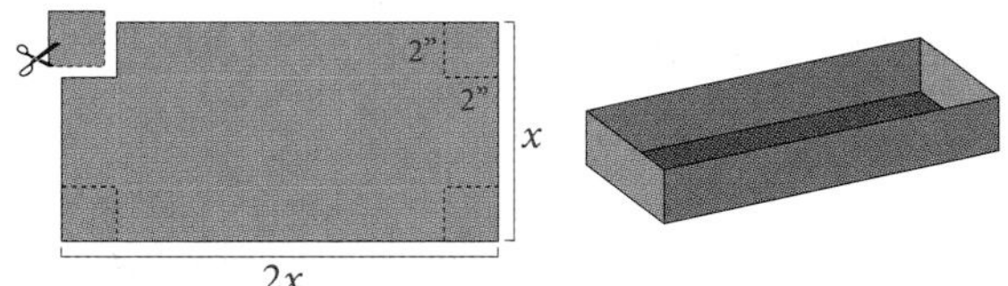

Skyscrapers The chart shows the heights of the three tallest buildings in the world. Use the chart to answer Problems 33 and 34.

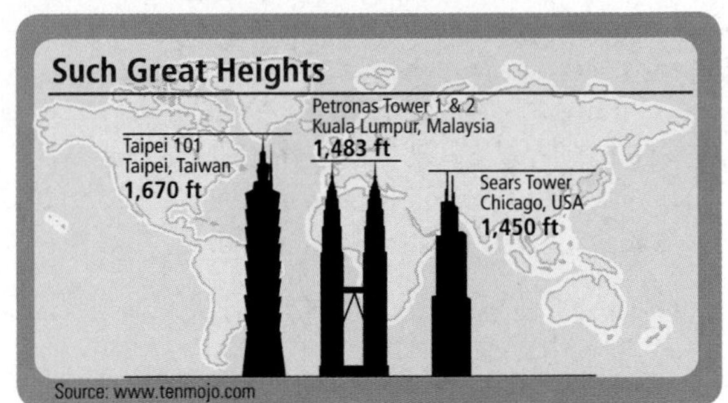

33. If you drop an object off the top of the Sears Tower, the height that the object is from the ground after t seconds is given by the equation $h = 1{,}450 - 16t^2$.

a. When is the object 1,386 feet above the ground?

b. How far has the object fallen after 5 seconds?

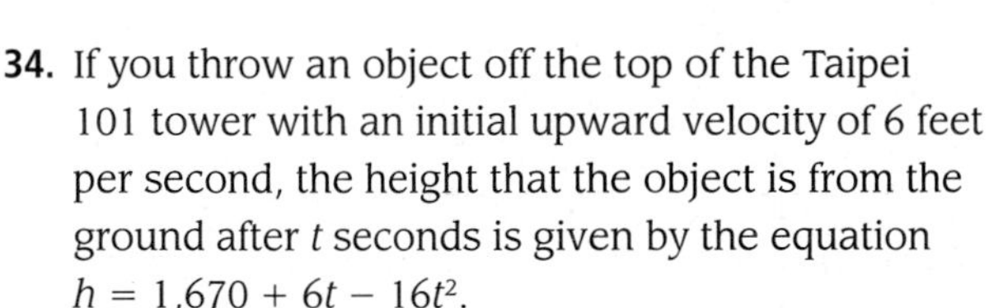

34. If you throw an object off the top of the Taipei 101 tower with an initial upward velocity of 6 feet per second, the height that the object is from the ground after t seconds is given by the equation $h = 1{,}670 + 6t - 16t^2$.

a. When is the object 1,660 feet above the ground?

b. When is the object 928 feet above the ground?

Maintaining Your Skills

Simplify each expression. (Write all answers with positive exponents only.)

35. $(5x^3)^2(2x^6)^3$

36. 2^{-3}

37. $\dfrac{x^4}{x^{-3}}$

38. $\dfrac{(20x^2y^3)(5x^4y)}{(2xy^5)(10x^2y^3)}$

39. $(2 \times 10^{-4})(4 \times 10^5)$

40. $\dfrac{9 \times 10^{-3}}{3 \times 10^{-2}}$

41. $20ab^2 - 16ab^2 + 6ab^2$

42. Subtract $6x^2 - 5x - 7$ from $9x^2 + 3x - 2$.

Multiply.

43. $2x^2(3x^2 + 3x - 1)$

44. $(2x + 3)(5x - 2)$

45. $(3y - 5)^2$

46. $(a - 4)(a^2 + 4a + 16)$

47. $(2a^2 + 7)(2a^2 - 7)$

48. Divide $15x^{10} - 10x^8 + 25x^6$ by $5x^6$.

Chapter 6 Summary

Greatest Common Factor [6.1]

The largest monomial that divides each term of a polynomial is called the greatest common factor for that polynomial. We begin all factoring by factoring out the greatest common factor.

EXAMPLES

1. $8x^4 - 10x^3 + 6x^2$
$= 2x^2 \cdot 4x^2 - 2x^2 \cdot 5x + 2x^2 \cdot 3$
$= 2x^2(4x^2 - 5x + 3)$

Factoring Trinomials [6.2, 6.3]

One method of factoring a trinomial is to list all pairs of binomials the product of whose first terms gives the first term of the trinomial and the product of whose last terms gives the last term of the trinomial. We then choose the pair that gives the correct middle term for the original trinomial.

2. $x^2 + 5x + 6 = (x + 2)(x + 3)$
$x^2 - 5x + 6 = (x - 2)(x - 3)$
$6x^2 - x - 2 = (2x + 1)(3x - 2)$
$6x^2 + 7x + 2 = (2x + 1)(3x + 2)$

Special Factorings [6.4]

$$a^2 + 2ab + b^2 = (a + b)^2$$
$$a^2 - 2ab + b^2 = (a - b)^2$$
$$a^2 - b^2 = (a + b)(a - b)$$

3. $x^2 + 10x + 25 = (x + 5)^2$
$x^2 - 10x + 25 = (x - 5)^2$
$x^2 - 25 = (x + 5)(x - 5)$

Sum and Difference of Two Cubes [6.5]

$a^3 - b^3 = (a - b)(a^2 + ab + b^2)$ Difference of two cubes

$a^3 + b^3 = (a + b)(a^2 - ab + b^2)$ Sum of two cubes

4. $x^3 - 27 = (x - 3)(x^2 + 3x + 9)$
$x^3 + 27 = (x + 3)(x^2 - 3x + 9)$

Strategy for Factoring a Polynomial [6.6]

Step 1: If the polynomial has a greatest common factor other than 1, then factor out the greatest common factor.

Step 2: If the polynomial has two terms (it is a binomial), then see if it is the difference of two squares or the sum or difference of two cubes, and then factor accordingly. Remember, if it is the sum of two squares, it will not factor.

Step 3: If the polynomial has three terms (a trinomial), then it is either a perfect square trinomial that will factor into the square of a binomial, or it is not a perfect square trinomial, in which case you use the trial and error method developed in Section 6.3.

Step 4: If the polynomial has more than three terms, then try to factor it by grouping.

Step 5: As a final check, see if any of the factors you have written can be factored further. If you have overlooked a common factor, you can catch it here.

5. a. $2x^5 - 8x^3 = 2x^3(x^2 - 4)$
$= 2x^3(x + 2)(x - 2)$

b. $3x^4 - 18x^3 + 27x^2$
$= 3x^2(x^2 - 6x + 9)$
$= 3x^2(x - 3)^2$

c. $6x^3 - 12x^2 - 48x$
$= 6x(x^2 - 2x - 8)$
$= 6x(x - 4)(x + 2)$

d. $x^2 + ax + bx + ab$
$= x(x + a) + b(x + a)$
$= (x + a)(x + b)$

Strategy for Solving a Quadratic Equation [6.7]

6. Solve $x^2 - 6x = -8$.

$$x^2 - 6x + 8 = 0$$
$$(x - 4)(x - 2) = 0$$
$$x - 4 = 0 \quad \text{or} \quad x - 2 = 0$$
$$x = 4 \quad \text{or} \quad x = 2$$

Both solutions check.

Step 1: Write the equation in standard form:

$$ax^2 + bx + c = 0$$

Step 2: Factor completely.

Step 3: Set each variable factor equal to 0.

Step 4: Solve the equations found in step 3.

Step 5: Check solutions, if necessary.

The Pythagorean Theorem [6.8]

7. The hypotenuse of a right triangle is 5 inches, and the lengths of the two legs (the other two sides) are given by two consecutive integers. Find the lengths of the two legs.

If we let x = the length of the shortest side, then the other side must be $x + 1$. The Pythagorean theorem tells us that

$$5^2 = x^2 + (x + 1)^2$$
$$25 = x^2 + x^2 + 2x + 1$$
$$25 = 2x^2 + 2x + 1$$
$$0 = 2x^2 + 2x - 24$$
$$0 = 2(x^2 + x - 12)$$
$$0 = 2(x + 4)(x - 3)$$
$$x + 4 = 0 \quad \text{or} \quad x - 3 = 0$$
$$x = -4 \quad \text{or} \quad x = 3$$

Since a triangle cannot have a side with a negative number for its length, we cannot use -4. One leg is $x = 3$ and the other leg is $x + 1 = 3 + 1 = 4$.

In any right triangle, the square of the longest side (called the hypotenuse) is equal to the sum of the squares of the other two sides (called legs).

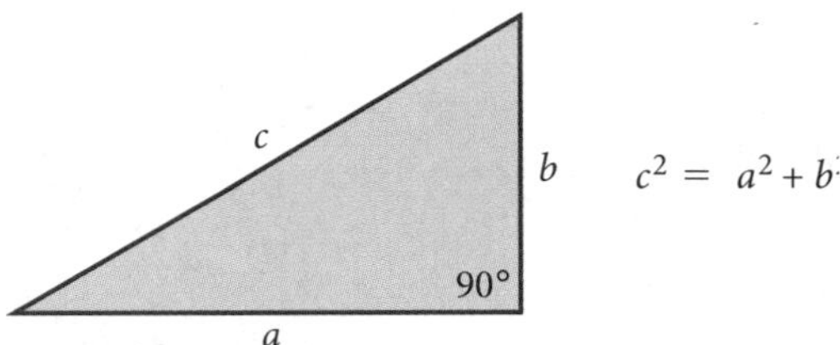

COMMON MISTAKES

It is a mistake to apply the zero-factor property to numbers other than zero. For example, consider the equation $(x - 3)(x + 4) = 18$. A fairly common mistake is to attempt to solve it with the following steps:

$$(x - 3)(x + 4) = 18$$
$$x - 3 = 18 \quad \text{or} \quad x + 4 = 18 \leftarrow \text{Mistake}$$
$$x = 21 \quad \text{or} \quad x = 14$$

These are obviously not solutions, as a quick check will verify:

Check: $x = 21$		Check: $x = 14$
$(21 - 3)(21 + 4) \stackrel{?}{=} 18$		$(14 - 3)(14 + 4) \stackrel{?}{=} 18$
$18 \cdot 25 = 18$		$11 \cdot 18 = 18$
$450 = 18$	$\xleftrightarrow{\text{false statements}}$	$198 = 18$

The mistake is in setting each factor equal to 18. It is not necessarily true that when the product of two numbers is 18, either one of them is itself 18. The correct solution looks like this:

$$(x - 3)(x + 4) = 18$$
$$x^2 + x - 12 = 18$$
$$x^2 + x - 30 = 0$$
$$(x + 6)(x - 5) = 0$$
$$x + 6 = 0 \quad \text{or} \quad x - 5 = 0$$
$$x = -6 \quad \text{or} \quad x = 5$$

To avoid this mistake, remember that before you factor a quadratic equation, you must write it in standard form. It is in standard form only when 0 is on one side and decreasing powers of the variable are on the other.

Chapter 6 Review

Factor the following by factoring out the greatest common factor. [6.1]

1. $10x - 20$
2. $4x^3 - 9x^2$
3. $5x - 5y$
4. $7x^3 + 2x$
5. $8x + 4$
6. $2x^2 + 14x + 6$
7. $24y^2 - 40y + 48$
8. $30xy^3 - 45x^3y^2$
9. $49a^3 - 14b^3$
10. $6ab^2 + 18a^3b^3 - 24a^2b$

Factor by grouping. [6.1]

11. $xy + bx + ay + ab$
12. $xy + 4x - 5y - 20$
13. $2xy + 10x - 3y - 15$
14. $5x^2 - 4ax - 10bx + 8ab$

Factor the following trinomials. [6.2]

15. $y^2 + 9y + 14$
16. $w^2 + 15w + 50$
17. $a^2 - 14a + 48$
18. $r^2 - 18r + 72$
19. $y^2 + 20y + 99$
20. $y^2 + 8y + 12$

Factor the following trinomials. [6.3]

21. $2x^2 + 13x + 15$
22. $4y^2 - 12y + 5$
23. $5y^2 + 11y + 6$
24. $20a^2 - 27a + 9$
25. $6r^2 + 5rt - 6t^2$
26. $10x^2 - 29x - 21$

Factor the following if possible. [6.4]

27. $n^2 - 81$
28. $4y^2 - 9$
29. $x^2 + 49$
30. $36y^2 - 121x^2$
31. $64a^2 - 121b^2$
32. $64 - 9m^2$

Factor the following. [6.4]

33. $y^2 + 20y + 100$
34. $m^2 - 16m + 64$
35. $64t^2 + 16t + 1$
36. $16n^2 - 24n + 9$
37. $4r^2 - 12rt + 9t^2$
38. $9m^2 + 30mn + 25n^2$

Factor the following. [6.2]

39. $2x^2 + 20x + 48$
40. $a^3 - 10a^2 + 21a$
41. $3m^3 - 18m^2 - 21m$
42. $5y^4 + 10y^3 - 40y^2$

Factor the following trinomials. [6.3]

43. $8x^2 + 16x + 6$

44. $3a^3 - 14a^2 - 5a$

45. $20m^3 - 34m^2 + 6m$

46. $30x^2y - 55xy^2 + 15y^3$

Factor the following. [6.4]

47. $4x^2 + 40x + 100$

48. $4x^3 + 12x^2 + 9x$

49. $5x^2 - 45$

50. $12x^3 - 27xy^2$

Factor the following. [6.5]

51. $8a^3 - b^3$

52. $27x^3 + 8y^3$

53. $125x^3 - 64y^3$

Factor the following polynomials completely. [6.6]

54. $6a^3b + 33a^2b^2 + 15ab^3$

55. $x^5 - x^3$

56. $4y^6 + 9y^4$

57. $12x^5 + 20x^4y - 8x^3y^2$

58. $30a^4b + 35a^3b^2 - 15a^2b^3$

59. $18a^3b^2 + 3a^2b^3 - 6ab^4$

Solve. [6.7]

60. $(x - 5)(x + 2) = 0$

61. $3(2y + 5)(2y - 5) = 0$

62. $m^2 + 3m = 10$

63. $a^2 - 49 = 0$

64. $m^2 - 9m = 0$

65. $6y^2 = -13y - 6$

66. $9x^4 + 9x^3 = 10x^2$

Solve the following word problems. [6.8]

67. Number Problem The product of two consecutive even integers is 120. Find the two integers.

68. Number Problem The product of two consecutive integers is 110. Find the two integers.

69. Number Problem The product of two consecutive odd integers is 1 less than 3 times their sum. Find the integers.

70. Number Problem The sum of two numbers is 20. Their product is 75. Find the numbers.

71. Number Problem One number is 1 less than twice another. Their product is 66. Find the numbers.

72. Geometry The height of a triangle is 8 times the base. The area is 16 square inches. Find the base.

73. Babies Use the chart to round the number of babies born to women between 30 and 39 to the nearest hundred thousand and then write your answer in scientific notation.

74. Golf The chart shows the lengths of the longest golf courses used for the U.S. open. Write the length of the Medinah Golf Course in scientific notation.

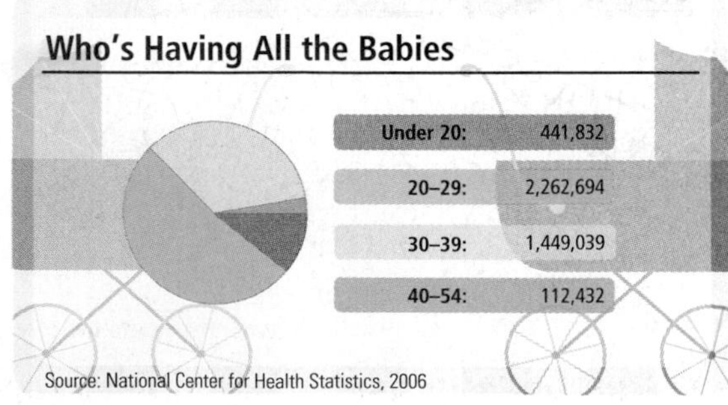

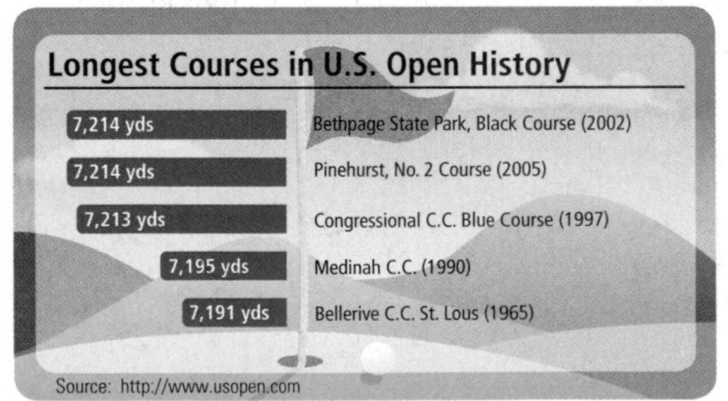

Chapter 6 Cumulative Review

Simplify.

1. $9 + (-7) + (-8)$
2. $20 - (-9)$
3. $\dfrac{-63}{-7}$
4. $\dfrac{9(-2)}{-2}$
5. $(-4)^3$
6. 9^{-2}
7. $\dfrac{-3(4-7) - 5(7-2)}{-5-2-1}$
8. $\dfrac{4^2 - 8^2}{(4-8)^2}$
9. $-a + 3 + 6a - 8$
10. $6 - 2(4a + 2) - 5$
11. $(x^4)^{10}$
12. $(9xy)^0$
13. $(5x - 2)(3x + 4)$
14. $(a^2 + 7)(a^2 - 7)$

Solve each equation.

15. $3x = -18$
16. $\dfrac{x}{2} = 5$
17. $-\dfrac{x}{3} = 7$
18. $\dfrac{1}{2}(4t - 1) + \dfrac{1}{3} = -\dfrac{25}{6}$
19. $4m(m - 7)(2m - 7) = 0$
20. $16x^2 - 81 = 0$

Solve each inequality.

21. $-2x > -8$
22. $-3x \geq -6$ or$2x - 7 \geq 7$

Graph on a rectangular coordinate system.

23. $y = -3x$
24. $y = 2$
25. $2x + 3y \geq 6$
26. $x < -2$
27. Which of the ordered pairs (0, 3), (4, 0), and $\left(\frac{16}{3}, 1\right)$ are solutions to the equation $3x - 4y = 12$?
28. Find the x- and y-intercepts for the equation $10x - 3y = 30$.
29. Find the slope of the line that passes through the points (7, 3) and (−2, −4).
30. Find the equation for the line with slope $\frac{2}{3}$ that passes through (−6, 4).
31. Find the equation for the line with slope $-\frac{2}{5}$ and y-intercept $-\frac{2}{3}$.
32. Find the slope and y-intercept for the equation $3x - 4y = -16$.

Solve each system by graphing.

33. $y = x + 3$; $x + y = -1$
34. $2x - 2y = 3$; $2x - 2y = 2$

Solve each system.

35. $-x + y = 3$; $x + y = 7$
36. $5x + 7y = -18$; $8x + 3y = 4$
37. $2x + y = 4$; $x = y - 1$
38. $x + y = 5{,}000$; $0.04x + 0.06y = 270$

Factor completely.

39. $n^2 - 5n - 36$
40. $14x^2 + 31xy - 10y^2$
41. $16 - a^2$
42. $49x^2 - 14x + 1$
43. $45x^2y - 30xy^2 + 5y^3$
44. $18x^3 - 3x^2y - 3xy^2$
45. $3xy + 15x - 2y - 10$
46. $a^3 + 64$

Give the opposite, reciprocal, and absolute value of the given number.

47. -2
48. $\dfrac{1}{5}$

Divide using long division.

49. $\dfrac{4x^2 - 7x - 13}{x - 3}$
50. $\dfrac{x^3 - 27}{x - 3}$

51. **Cutting a Board into Two Pieces** A 72-inch board is to be cut into two pieces. One piece is to be 4 inches longer than the other. How long is each piece?
52. **Hamburgers and Fries** Sheila bought burgers and fries for her children and some friends. The burgers cost \$2.05 each, and the fries are \$0.85 each. She bought a total of 14 items, for a total cost of \$19.10. How many of each did she buy?

Use the pie chart to answer the following questions.

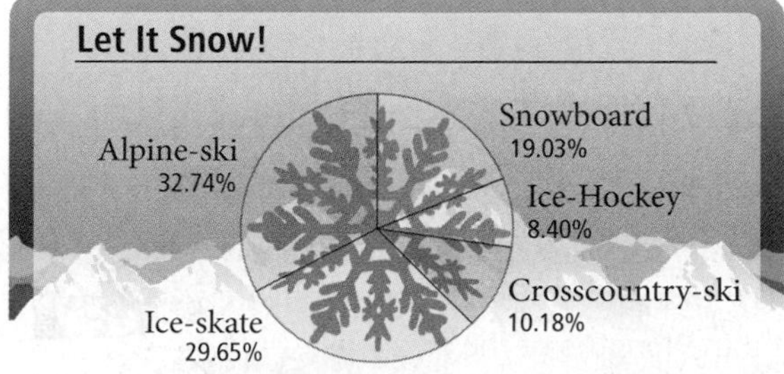

Source: U.S. Census Bureau

Of the 22.6 million Americans who participate in winter sports, how many millions would choose the following as their favorite? Round your answers to the nearest tenth.

53. Snowboarding
54. Ice-skating

Chapter 6 Test

Factor out the greatest common factor. [6.1]

1. $5x - 10$

2. $18x^2y - 9xy - 36xy^2$

Factor by grouping. [6.1]

3. $x^2 + 2ax - 3bx - 6ab$

4. $xy + 4x - 7y - 28$

Factor the following completely. [6.2–6.6]

5. $x^2 - 5x + 6$

6. $x^2 - x - 6$

7. $a^2 - 16$

8. $x^2 + 25$

9. $x^4 - 81$

10. $27x^2 - 75y^2$

11. $x^3 + 5x^2 - 9x - 45$

12. $x^2 - bx + 5x - 5b$

13. $4a^2 + 22a + 10$

14. $3m^2 - 3m - 18$

15. $6y^2 + 7y - 5$

16. $12x^3 - 14x^2 - 10x$

17. $a^3 + 64b^3$

18. $54x^3 - 16$

Solve the following equations. [6.7]

19. $x^2 + 7x + 12 = 0$

20. $x^2 - 4x + 4 = 0$

21. $x^2 - 36 = 0$

22. $x^2 = x + 20$

23. $x^2 - 11x = -30$

24. $y^3 = 16y$

25. $2a^2 = a + 15$

26. $30x^3 - 20x^2 = 10x$

Solve the following word problems. Be sure to show the equation used. [6.8]

27. Number Problem Two numbers have a sum of 20. Their product is 64. Find the numbers.

28. Consecutive Integers The product of two consecutive odd integers is 7 more than their sum. Find the integers.

29. Geometry The length of a rectangle is 5 feet more than 3 times the width. The area is 42 square feet. Find the dimensions.

30. Geometry One leg of a right triangle is 2 meters more than twice the other. The hypotenuse is 13 meters. Find the lengths of the two legs.

31. Production Cost A company can manufacture x hundred items for a total cost C, given the equation $C = 200 + 500x - 100x^2$. How many items can be manufactured if the total cost is to be \$800?

32. Price and Revenue A manufacturer knows that the number of items he can sell each week, x, is related to the price p of each item by the equation $x = 900 - 100p$. What price should he charge for each item to have a weekly revenue of \$1,800? (*Remember:* $R = xp$.)

33. Pythagorean Theorem A 17 foot ladder is placed so that it leans against a wall at a point that is one foot less than twice the distance from the wall to the base of the ladder.

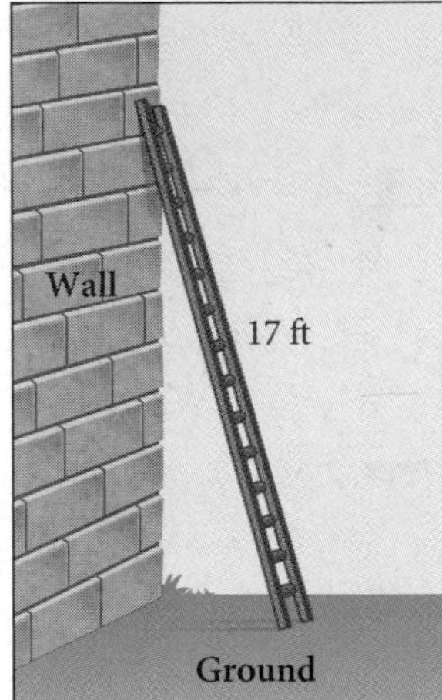

a. How far from the wall is the ladder?

b. How high does the ladder reach?

34. Projectile Motion A ball is thrown into the air with an upward velocity of 40 feet per second from a building that is 24 feet high. The equation that gives the height of the ball above the ground at time t is

$h = 24 + 40t - 16t^2$

a. At what time(s) will the ball be 40 feet above the ground?

b. When will the ball hit the ground?

Chapter 6 Projects

FACTORING

GROUP PROJECT

Visual Factoring

Number of People 2 or 3

Time Needed 10–15 minutes

Equipment Pencil, graph paper, and scissors

Background When a geometric figure is divided into smaller figures, the area of the original figure and the area of any rearrangement of the smaller figures must be the same. We can use this fact to help visualize some factoring problems.

Procedure Use the diagram below to work the following problems.

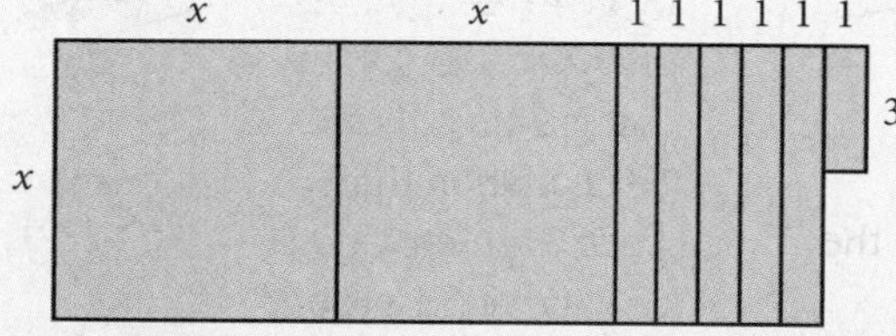

1. Write a polynomial involving x that gives the area of the diagram.
2. Factor the polynomial found in Part 1.
3. Copy the figure onto graph paper, then cut along the lines so that you end up with 2 squares and 6 rectangles.
4. Rearrange the pieces from Part 3 to show that the factorization you did in Part 2 is correct.

RESEARCH PROJECT

Factoring and Internet Security

The security of the information on computers is directly related to factoring of whole numbers. The key lies in the fact that multiplying whole numbers is a straightforward, simple task, whereas factoring can be very time-consuming. For example, multiplying the numbers 1,234 and 3,433 to obtain 4,236,322 takes very little time, even if done by hand. But given the number 4,236,322, finding its factors, even with a calculator or computer, is more than you want to try. The discipline that studies how to make and break codes is cryptography. The current Web browsers, such as Internet Explorer and Netscape, use a system called RSA public-key cryptosystem invented by Adi Shamir of Israel's Weizmann Institute of Science. In 1999 Shamir announced that he had found a method of factoring large numbers quickly that will put the current Internet security system at risk.

Research the connection between computer security and factoring, the RSA cryptosystem, and the current state of security on the Internet, and then write an essay summarizing your results.

Rational Expressions

7

Chapter Outline

Introduction

The Google Earth photograph shows the New Library in Alexandria, Egypt, which opened in 2003. Within the new library are rare artifacts from the ancient library in Alexandria, which existed from approximately 300 BC to 415 AD. One of the most famous books kept in the ancient library was *Euclid's Elements*, written in approximately 300 BC. The photograph below is from a later printing of *Euclid's Elements*.

SPL/Photo Researchers, Inc.

Hypatia of Alexandria, the first woman mentioned in the history of mathematics, was the last librarian at the ancient library. She studied and revised the *Elements* to make it more understandable for the other students at the library. Proportions, one of the topics we will study in this chapter, is also a topic in *Euclid's Elements*.

Chapter Pretest

Reduce to lowest terms. [7.1]

1. $\dfrac{x^2 - 9}{x^2 - 6x + 9}$

2. $\dfrac{15a + 30}{5a^2 - 10a - 40}$

3. $\dfrac{xy - 2x + 3y - 6}{y^3 - 2y^2 - y + 2}$

Multiply or divide as indicated. [7.2]

4. $\dfrac{2x - 6}{3} \cdot \dfrac{9}{4x - 12}$

5. $\dfrac{x^2 - 9}{x - 4} \div \dfrac{x + 3}{x^2 - 16}$

6. $\dfrac{x^2 + x - 6}{x^2 + 4x + 3} \div \dfrac{x^2 + 2x - 8}{2x^2 - x - 3}$

7. $(x^2 - 16)\left(\dfrac{x - 1}{x - 4}\right)$

Add or subtract as indicated. [7.3]

8. $\dfrac{7}{x - 8} - \dfrac{9}{x - 8}$

9. $\dfrac{x}{x^2 - 16} + \dfrac{3}{3x - 12}$

10. $\dfrac{3}{(x - 3)(x + 3)} - \dfrac{1}{(x - 3)(x - 1)}$

Solve the following equations. [7.4]

11. $\dfrac{3}{5} = \dfrac{x + 3}{7}$

12. $\dfrac{25}{x - 3} = \dfrac{7}{x}$

13. $\dfrac{6}{x - 3} - \dfrac{5}{x + 1} = \dfrac{7}{x^2 - 2x - 3}$

Simplify each complex fraction. [7.6]

14. $\dfrac{1 + \dfrac{2}{x}}{1 - \dfrac{2}{x}}$

15. $\dfrac{1 - \dfrac{9}{x^2}}{1 - \dfrac{1}{x} - \dfrac{6}{x^2}}$

16. **Direct Variation** Suppose y varies directly with the cube of x. If y is 16 when x is 2, find y when x is 3. [7.8]

17. **Inverse Variation** If y varies inversely with the square of x, and y is 8 when x is 3, find y when x is 6. [7.8]

Getting Ready for Chapter 7

Simplify

1. $\dfrac{2}{9} \cdot \dfrac{15}{22}$

2. $\dfrac{3}{5} \div \dfrac{15}{7}$

3. $1 - \dfrac{5}{3}$

4. $1 + \dfrac{3}{5}$

5. $\dfrac{5}{0}$

6. $\dfrac{1}{10} + \dfrac{3}{14}$

7. $\dfrac{1}{21} + \dfrac{4}{15}$

Multiply.

8. $x(x + 2)$

9. $(x + 3)(x - 4)$

Factor completely.

10. $x^2 - 25$

11. $2x - 4$

12. $x^2 + 7x + 12$

13. $a^2 - 4a$

14. $xy^2 + y$

Solve.

15. $72 = 2x$

16. $15 = 3x - 3$

17. $a^2 - a - 20 = -2a$

18. $x^2 + 2x = 8$

19. Use the formula $y = \dfrac{20}{x}$ to find y when $x = 10$.

20. Use the formula $y = 2x^2$ to find x when $y = 72$.

Reducing Rational Expressions to Lowest Terms

7.1

Objectives

A Find the restrictions on the variable in a rational expression.

B Reduce a rational expression to lowest terms.

C Work problems involving ratios.

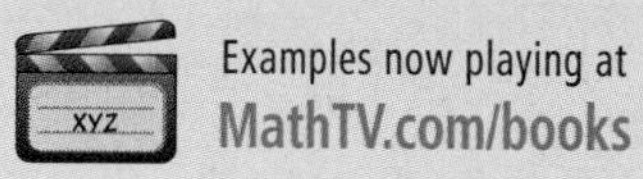

In Chapter 1 we defined the set of rational numbers to be the set of all numbers that could be put in the form $\frac{a}{b}$, where a and b are integers ($b \neq 0$):

$$\text{Rational numbers} = \left\{ \frac{a}{b} \,\middle|\, a \text{ and } b \text{ are integers, } b \neq 0 \right\}$$

A *rational expression* is any expression that can be put in the form $\frac{P}{Q}$, where P and Q are polynomials and $Q \neq 0$:

$$\text{Rational expressions} = \left\{ \frac{P}{Q} \,\middle|\, P \text{ and } Q \text{ are polynomials, } Q \neq 0 \right\}$$

Each of the following is an example of a rational expression:

$$\frac{2x + 3}{x} \qquad \frac{x^2 - 6x + 9}{x^2 - 4} \qquad \frac{5}{x^2 + 6} \qquad \frac{2x^2 + 3x + 4}{2}$$

For the rational expression

$$\frac{x^2 - 6x + 9}{x^2 - 4}$$

the polynomial on top, $x^2 - 6x + 9$, is called the numerator, and the polynomial on the bottom, $x^2 - 4$, is called the denominator. The same is true of the other rational expressions.

A Restrictions on Variables

When working with rational expressions, we must be careful that we do not use a value of the variable that will give us a denominator of zero. Remember, division by zero is not defined.

EXAMPLES State the restrictions on the variable in the following rational expressions:

1. $\frac{x + 2}{x - 3}$

SOLUTION The variable x can be any real number except $x = 3$ since, when $x = 3$, the denominator is $3 - 3 = 0$. We state this restriction by writing $x \neq 3$.

2. $\frac{5}{x^2 - x - 6}$

SOLUTION If we factor the denominator, we have $x^2 - x - 6 = (x - 3)(x + 2)$. If either of the factors is zero, the whole denominator is zero. Our restrictions are $x \neq 3$ and $x \neq -2$ since either one makes $x^2 - x - 6 = 0$.

We will not always list each restriction on a rational expression, but we should be aware of them and keep in mind that no rational expression can have a denominator of zero.

PRACTICE PROBLEMS

State the restrictions on the variable:

1. $\frac{x + 5}{x - 4}$

2. $\frac{6}{x^2 + x - 6}$

Answers

1. $x \neq 4$ **2.** $x \neq -3, x \neq 2$

The two fundamental properties of rational expressions are listed next. We will use these two properties many times in this chapter.

Properties of Rational Expressions

Property 1
Multiplying the numerator and denominator of a rational expression by the same nonzero quantity will not change the value of the rational expression.

Property 2
Dividing the numerator and denominator of a rational expression by the same nonzero quantity will not change the value of the rational expression.

B Reducing Rational Expressions

We can use Property 2 to reduce rational expressions to lowest terms. Since this process is almost identical to the process of reducing fractions to lowest terms, let's recall how the fraction $\frac{6}{15}$ is reduced to lowest terms:

$$\frac{6}{15} = \frac{2 \cdot 3}{5 \cdot 3} \quad \text{Factor numerator and denominator}$$

$$= \frac{2 \cdot \cancel{3}}{5 \cdot \cancel{3}} \quad \text{Divide out the common factor, 3}$$

$$= \frac{2}{5} \quad \text{Reduce to lowest terms}$$

The same procedure applies to reducing rational expressions to lowest terms. The process is summarized in the following rule.

Rule
To reduce a rational expression to lowest terms, first factor the numerator and denominator completely and then divide both the numerator and denominator by any factors they have in common.

EXAMPLE 3 Reduce $\frac{x^2 - 9}{x^2 + 5x + 6}$ to lowest terms.

SOLUTION We begin by factoring:

$$\frac{x^2 - 9}{x^2 + 5x + 6} = \frac{(x - 3)(x + 3)}{(x + 2)(x + 3)}$$

Notice that both polynomials contain the factor $(x + 3)$. If we divide the numerator by $(x + 3)$, we are left with $(x - 3)$. If we divide the denominator by $(x + 3)$, we are left with $(x + 2)$. The complete solution looks like this:

$$\frac{x^2 - 9}{x^2 + 5x + 6} = \frac{(x - 3)\cancel{(x + 3)}}{(x + 2)\cancel{(x + 3)}} \quad \text{Factor the numerator and denominator completely}$$

$$= \frac{x - 3}{x + 2} \quad \text{Divide out the common factor, } x + 3$$

3. Reduce to lowest terms:

$$\frac{x^2 - 4}{x^2 - 2x - 8}$$

Answer

3. $\frac{x - 2}{x - 4}$

It is convenient to draw a line through the factors as we divide them out. It is especially helpful when the problems become longer.

EXAMPLE 4 Reduce to lowest terms $\frac{10a + 20}{5a^2 - 20}$.

SOLUTION We begin by factoring out the greatest common factor from the numerator and denominator:

$$\frac{10a + 20}{5a^2 - 20} = \frac{10(a + 2)}{5(a^2 - 4)}$$ Factor out the greatest common factor from the numerator and denominator

$$= \frac{10\cancel{(a + 2)}}{5\cancel{(a + 2)}(a - 2)}$$ Factor the denominator as the difference of two squares

$$= \frac{2}{a - 2}$$ Divide out the common factors 5 and $a + 2$

EXAMPLE 5 Reduce $\frac{2x^3 + 2x^2 - 24x}{x^3 + 2x^2 - 8x}$ to lowest terms.

SOLUTION We begin by factoring the numerator and denominator completely. Then we divide out all factors common to the numerator and denominator. Here is what it looks like:

$$\frac{2x^3 + 2x^2 - 24x}{x^3 + 2x^2 - 8x} = \frac{2x(x^2 + x - 12)}{x(x^2 + 2x - 8)}$$ Factor out the greatest common factor first

$$= \frac{2\cancel{x}(x - 3)\cancel{(x + 4)}}{\cancel{x}(x - 2)\cancel{(x + 4)}}$$ Factor the remaining trinomials

$$= \frac{2(x - 3)}{x - 2}$$ Divide out the factors common to the numerator and denominator

EXAMPLE 6 Reduce $\frac{x - 5}{x^2 - 25}$ to lowest terms.

SOLUTION

$$\frac{x - 5}{x^2 - 25} = \frac{\cancel{x - 5}}{\cancel{(x - 5)}(x + 5)}$$ Factor numerator and denominator completely

$$= \frac{1}{x + 5}$$ Divide out the common factor, $x - 5$

EXAMPLE 7 Reduce $\frac{x^3 + y^3}{x^2 - y^2}$ to lowest terms.

SOLUTION We begin by factoring the numerator and denominator completely. (Remember, we can only reduce to lowest terms when the numerator and denominator are in factored form. Trying to reduce before factoring will only lead to mistakes.)

$$\frac{x^3 + y^3}{x^2 - y^2} = \frac{\cancel{(x + y)}(x^2 - xy + y^2)}{\cancel{(x + y)}(x - y)}$$ Factor

$$= \frac{x^2 - xy + y^2}{x - y}$$ Divide out the common factor

4. Reduce to lowest terms: $\frac{5a + 15}{10a^2 - 90}$

5. Reduce to lowest terms: $\frac{x^3 - x^2 - 2x}{x^3 + 4x^2 + 3x}$

6. Reduce to lowest terms: $\frac{x + 3}{x^2 - 9}$

7. Reduce to lowest terms: $\frac{x^3 - 8}{x^2 - 4}$

Answers

4. $\frac{1}{2(a - 3)}$ 5. $\frac{x - 2}{x + 3}$ 6. $\frac{1}{x - 3}$

7. $\frac{x^2 + 2x + 4}{x + 2}$

C Ratios

For the rest of this section we will concern ourselves with ratios, a topic closely related to reducing fractions and rational expressions to lowest terms. Let's start with a definition.

Definition

If a and b are any two numbers, $b \neq 0$, then the **ratio** of a and b is

$$\frac{a}{b}$$

As you can see, ratios are another name for fractions or rational numbers. They are a way of comparing quantities. Since we also can think of $\frac{a}{b}$ as the quotient of a and b, ratios are also quotients. The following table gives some ratios in words and as fractions.

With ratios it is common to leave the 1 in the denominator.

Ratio	As a Fraction	In Lowest Terms
25 to 75	$\frac{25}{75}$	$\frac{1}{3}$
8 to 2	$\frac{8}{2}$	$\frac{4}{1}$
20 to 16	$\frac{20}{16}$	$\frac{5}{4}$

EXAMPLE 8 A solution of hydrochloric acid (HCl) and water contains 49 milliliters of water and 21 milliliters of HCl. Find the ratio of HCl to water and of HCl to the total volume of the solution.

SOLUTION The ratio of HCl to water is 21 to 49, or

$$\frac{21}{49} = \frac{3}{7}$$

The amount of total solution volume is $49 + 21 = 70$ milliliters. Therefore, the ratio of HCl to total solution is 21 to 70, or

$$\frac{21}{70} = \frac{3}{10}$$

8. A solution of alcohol and water contains 27 milliliters of alcohol and 54 milliliters of water. Find the ratio of alcohol to water and the ratio of alcohol to total volume.

Rate Equation

Many of the problems in this chapter will use what is called the *rate equation*. You use this equation on an intuitive level when you are estimating how long it will take you to drive long distances. For example, if you drive at 50 miles per hour for 2 hours, you will travel 100 miles. Here is the rate equation:

$$\text{Distance} = \text{rate} \cdot \text{time}$$

$$d = r \cdot t$$

The rate equation has two equivalent forms, the most common of which is obtained by solving for r. Here it is:

$$r = \frac{d}{t}$$

The rate r in the rate equation is the ratio of distance to time and also is referred to as *average speed*. The units for rate are miles per hour, feet per second, kilometers per hour, and so on.

Answer

8. $\frac{1}{2}$ and $\frac{1}{3}$

EXAMPLE 9 The Forest chair lift at the Northstar ski resort in Lake Tahoe is 5,603 feet long. If a ride on this chair lift takes 11 minutes, what is the average speed of the lift in feet per minute?

SOLUTION To find the speed of the lift, we find the ratio of distance covered to time. (Our answer is rounded to the nearest whole number.)

$$\text{Rate} = \frac{\text{distance}}{\text{time}} = \frac{5{,}603 \text{ feet}}{11 \text{ minutes}} = \frac{5{,}603}{11} \text{ feet/minute} = 509 \text{ feet/minute}$$

Note how we separate the numerical part of the problem from the units. In the next section, we will convert this rate to miles per hour.

EXAMPLE 10 A Ferris wheel was built in St. Louis in 1986. It is named *Colossus*. The circumference of the wheel is 518 feet. It has 40 cars, each of which holds 6 passengers. A trip around the wheel takes 40 seconds. Find the average speed of a rider on *Colossus*.

SOLUTION To find the average speed, we divide the distance traveled, which in this case is the circumference, by the time it takes to travel once around the wheel.

$$r = \frac{d}{t} = \frac{518 \text{ feet}}{40 \text{ seconds}} = 13.0 \text{ (rounded)}$$

The average speed of a rider on the *Colossus* is approximately 13.0 feet per second.

In the next section, you will convert the ratio into an equivalent ratio that gives the speed of the rider in miles per hour.

Getting Ready for Class

After reading through the preceding section, respond in your own words and in complete sentences.

1. How do you reduce a rational expression to lowest terms?
2. What are the properties we use to manipulate rational expressions?
3. For what values of the variable is a rational expression undefined?
4. What is a ratio?

9. The Bear Paw chair lift at the Northstar ski resort in Lake Tahoe is 772 feet long. If a ride on this chair lift takes 2.2 minutes, what is the average speed of the lift in feet per minute? (Round to the nearest whole number.)

The Forest Chair Lift
Northstar Ski Resort, Lake Tahoe

10. A Ferris wheel has a circumference of 840 feet. A trip around the wheel takes 80 seconds. Find the average speed of a rider on this Ferris wheel.

Answers

9. 351 feet per minute
10. 10.5 feet per second

Problem Set 7.1

1. Simplify each expression.
   a. $\frac{5 + 1}{25 - 1}$
   b. $\frac{x + 1}{x^2 - 1}$
   c. $\frac{x^2 - x}{x^2 - 1}$
   d. $\frac{x^3 - 1}{x^2 - 1}$
   e. $\frac{x^3 - 1}{x^3 - x^2}$

2. Simplify each expression.
   a. $\frac{25 - 30 + 9}{25 - 9}$
   b. $\frac{x^2 - 6x + 9}{x^2 - 9}$
   c. $\frac{x^2 - 10x + 9}{x^2 - 9x}$
   d. $\frac{x^2 + 3x + ax + 3a}{x^2 - 9}$
   e. $\frac{x^3 + 27}{x^3 - 9x}$

A B Reduce the following rational expressions to lowest terms, if possible. Also, specify any restrictions on the variable in Problems 3 through 12. **[Examples 1–7]**

3. $\frac{5}{5x - 10}$
4. $\frac{-4}{2x - 8}$
5. $\frac{a - 3}{a^2 - 9}$
6. $\frac{a + 4}{a^2 - 16}$
7. $\frac{x + 5}{x^2 - 25}$
8. $\frac{x - 2}{x^2 - 4}$
9. $\frac{2x^2 - 8}{4}$
10. $\frac{5x - 10}{x - 2}$
11. $\frac{2x - 10}{3x - 6}$
12. $\frac{4x - 8}{x - 2}$
13. $\frac{10a + 20}{5a + 10}$
14. $\frac{11a + 33}{6a + 18}$
15. $\frac{5x^2 - 5}{4x + 4}$
16. $\frac{7x^2 - 28}{2x + 4}$
17. $\frac{x - 3}{x^2 - 6x + 9}$
18. $\frac{x^2 - 10x + 25}{x - 5}$
19. $\frac{3x + 15}{3x^2 + 24x + 45}$
20. $\frac{5x + 15}{5x^2 + 40x + 75}$
21. $\frac{a^2 - 3a}{a^3 - 8a^2 + 15a}$
22. $\frac{a^2 + 3a}{a^3 - 2a^2 - 15a}$

23. $\dfrac{3x - 2}{9x^2 - 4}$

24. $\dfrac{2x - 3}{4x^2 - 9}$

25. $\dfrac{x^2 + 8x + 15}{x^2 + 5x + 6}$

26. $\dfrac{x^2 - 8x + 15}{x^2 - x - 6}$

27. $\dfrac{2m^3 - 2m^2 - 12m}{m^2 - 5m + 6}$

28. $\dfrac{2m^3 + 4m^2 - 6m}{m^2 - m - 12}$

29. $\dfrac{x^3 + 3x^2 - 4x}{x^3 - 16x}$

30. $\dfrac{3a^2 - 8a + 4}{9a^3 - 4a}$

31. $\dfrac{4x^3 - 10x^2 + 6x}{2x^3 + x^2 - 3x}$

32. $\dfrac{3a^3 - 8a^2 + 5a}{4a^3 - 5a^2 + 1a}$

33. $\dfrac{4x^2 - 12x + 9}{4x^2 - 9}$

34. $\dfrac{5x^2 + 18x - 8}{5x^2 + 13x - 6}$

35. $\dfrac{x + 3}{x^4 - 81}$

36. $\dfrac{x^2 + 9}{x^4 - 81}$

37. $\dfrac{3x^2 + x - 10}{x^4 - 16}$

38. $\dfrac{5x^2 - 16x + 12}{x^4 - 16}$

39. $\dfrac{42x^3 - 20x^2 - 48x}{6x^2 - 5x - 4}$

40. $\dfrac{36x^3 + 132x^2 - 135x}{6x^2 + 25x - 9}$

41. $\dfrac{x^3 - y^3}{x^2 - y^2}$

42. $\dfrac{x^3 + y^3}{x^2 - y^2}$

43. $\dfrac{x^3 + 8}{x^2 - 4}$

44. $\dfrac{x^3 - 125}{x^2 - 25}$

45. $\dfrac{x^3 + 8}{x^2 + x - 2}$

46. $\dfrac{x^2 - 2x - 3}{x^3 - 27}$

B To reduce each of the following rational expressions to lowest terms, you will have to use factoring by grouping. Be sure to factor each numerator and denominator completely before dividing out any common factors. (Remember, factoring by grouping takes two steps.)

47. $\dfrac{xy + 3x + 2y + 6}{xy + 3x + 5y + 15}$

48. $\dfrac{xy + 7x + 4y + 28}{xy + 3x + 4y + 12}$

49. $\dfrac{x^2 - 3x + ax - 3a}{x^2 - 3x + bx - 3b}$

50. $\dfrac{x^2 - 6x + ax - 6a}{x^2 - 7x + ax - 7a}$

51. $\dfrac{xy + bx + ay + ab}{xy + bx + 3y + 3b}$

52. $\dfrac{x^2 + 5x + ax + 5a}{x^2 + 5x + bx + 5b}$

The next two problems are intended to give you practice reading, and paying attention to, the instructions that accompany the problems you are working. Working these problems is an excellent way to get ready for a test or quiz

53. Work each problem according to the instructions given.

a. Add: $(x^2 - 4x) + (4x - 16)$

b. Subtract: $(x^2 - 4x) - (4x - 16)$

c. Multiply: $(x^2 - 4x)(4x - 16)$

d. Reduce: $\dfrac{x^2 - 4x}{4x - 16}$

54. Work each problem according to the instructions given.

a. Add: $(9x^2 - 3x) + (6x - 2)$

b. Subtract: $(9x^2 - 3x) - (6x - 2)$

c. Multiply: $(9x^2 - 3x)(6x - 2)$

d. Reduce: $\dfrac{9x^2 - 3x}{6x - 2}$

C Write each ratio as a fraction in lowest terms. [Example 8]

55. 8 to 6

56. 6 to 8

57. 200 to 250

58. 250 to 200

59. 32 to 4

60. 4 to 32

Applying the Concepts [Examples 9–10]

61. Cars The chart shows the fastest cars in America. Which car or cars could travel 390 miles in one-and-one-half hours?

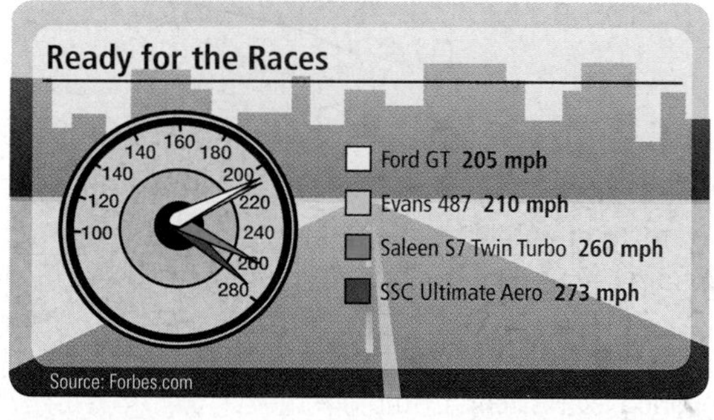

62. Fifth Avenue Mile The chart shows the times of the five fastest runners for this year's Continental Airlines Fifth Avenue Mile. Insert $<$ or $>$ to make each statement regarding the runners' speeds true (based on the times given below).

a. Silva Webb

b. Angwenyi Famiglietti

c. Mottram Webb

Fastest on Fifth

Continental Airlines Fifth Avenue Mile

Craig Mottram, AUS	3:49.90
Alan Webb, USA	3:51.40
Elkanah Angwenyi, KEN	3:54.30
Anthony Famiglietti, USA	3:57.10
Rui Silva, POR	3:57.40

Source: www.coolrunning.com, 2005

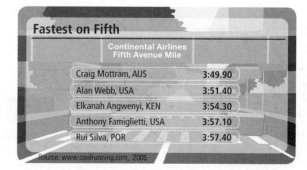

63. Speed of a Car A car travels 122 miles in 3 hours. Find the average speed of the car in miles per hour. Round to the nearest tenth.

64. Speed of a Bullet A bullet fired from a gun travels a distance of 4,500 feet in 3 seconds. Find the average speed of the bullet in feet per second.

65. **Ferris Wheel** The first Ferris wheel was designed and built by George Ferris in 1893. It was a large wheel with a circumference of 785 feet. If one trip around the circumference of the wheel took 20 minutes, find the average speed of a rider in feet per minute.

66. **Ferris Wheel** In 1897 a large Ferris wheel was built in Vienna; it is still in operation today. Known as *The Great Wheel,* it has a circumference of 618 feet. If one trip around the wheel takes 15 minutes, find the average speed of a rider on this wheel in feet per minute.

67. **Ferris Wheel** A person riding a Ferris Wheel travels once around the wheel, a distance of 188 feet, in 30 seconds. What is the average speed of the rider in feet per second? Round to the nearest tenth.

68. **Average Speed** Tina is training for a biathlon. As part of her training, she runs an 8-mile course, 2 miles of which is on level ground and 6 miles of which is downhill. It takes her 20 minutes to run the level part of the course and 40 minutes to run the downhill part of the course. Find her average speed in minutes per mile and in miles per minute for each part of the course. Round to the nearest hundredth, if rounding is necessary.

69. **Fuel Consumption** An economy car travels 168 miles on 3.5 gallons of gas. Give the average fuel consumption of the car in miles per gallon.

70. **Fuel Consumption** A luxury car travels 100 miles on 8 gallons of gas. Give the average fuel consumption of the car in miles per gallon.

71. **Comparing Expressions** Replace x with 5 and y with 4 in the expression

$$\frac{x^2 - y^2}{x - y}$$

and simplify the result. Is the result equal to $5 - 4$ or $5 + 4$?

72. **Comparing Expressions** Replace x with 2 in the expression

$$\frac{x^3 - 1}{x - 1}$$

and simplify the result. Your answer should be equal to what you would get if you replaced x with 2 in $x^2 + x + 1$.

73. **Comparing Expressions** Complete the following table; then show why the table turns out as it does.

x	$\frac{x-3}{3-x}$
−2	
−1	
0	
1	
2	

74. **Comparing Expressions** Complete the following table; then show why the table turns out as it does.

x	$\frac{25-x^2}{x^2-25}$
−4	
−2	
0	
2	
4	

75. Comparing Expressions You know from reading through Example 6 in this section that $\frac{x-5}{x^2-25} = \frac{1}{x+5}$. Compare these expressions by completing the following table. (Be careful—not all the rows have equal entries.)

x	$\frac{x-5}{x^2-25}$	$\frac{1}{x+5}$
0		
2		
−2		
5		
−5		

76. Comparing Expressions You know from your work in this section that $\frac{x^2-6x+9}{x^2-9} = \frac{x-3}{x+3}$. Compare these expressions by completing the following table. (Be careful—not all the rows have equal entries.)

x	$\frac{x^2-6x+9}{x^2-9}$	$\frac{x-3}{x+3}$
−3		
−2		
−1		
0		
1		
2		
3		

Getting Ready for the Next Section

Perform the indicated operation.

77. $\frac{3}{4} \cdot \frac{10}{21}$

78. $\frac{2}{90} \cdot \frac{15}{22}$

79. $\frac{4}{5} \div \frac{8}{9}$

80. $\frac{3}{5} \div \frac{15}{7}$

Factor completely.

81. $x^2 - 9$

82. $x^2 - 25$

83. $3x - 9$

84. $2x - 4$

85. $x^2 - x - 20$

86. $x^2 + 7x + 12$

87. $a^2 + 5a$

88. $a^2 - 4a$

Reduce to lowest terms.

89. $\frac{a(a+5)(a-5)(a+4)}{a^2+5a}$

90. $\frac{a(a+2)(a-4)(a+5)}{a^2-4a}$

Multiply. Give the answers as decimals rounded to the nearest tenth.

91. $\frac{5{,}603}{11} \cdot \frac{1}{5{,}280} \cdot \frac{60}{1}$

92. $\frac{772}{2.2} \cdot \frac{1}{5{,}280} \cdot \frac{60}{1}$

Maintaining Your Skills

Simplify.

93. $\frac{27x^5}{9x^2} - \frac{45x^8}{15x^5}$

94. $\frac{36x^9}{4x} - \frac{45x^3}{5x^{-5}}$

95. $\frac{72a^3b^7}{9ab^5} + \frac{64a^5b^3}{8a^3b}$

96. $\frac{80a^5b^{11}}{10a^2b} + \frac{33a^6b^{12}}{11a^3b^2}$

Divide.

97. $\frac{38x^7 + 42x^5 - 84x^3}{2x^3}$

98. $\frac{49x^6 - 63x^4 - 35x^2}{7x^2}$

99. $\frac{28a^5b^5 + 36ab^4 - 44a^4b}{4ab}$

100. $\frac{30a^3b - 12a^2b^2 + 6ab^3}{6ab}$

101. **Stock Market** One method of comparing stocks on the stock market is the price to earnings ratio, or P/E.

$$\text{P/E} = \frac{\text{Current Stock Price}}{\text{Earnings per Share}}$$

Most stocks have a P/E between 25 and 40. A stock with a P/E of less than 25 may be undervalued, while a stock with a P/E greater than 40 may be overvalued. Fill in the P/E for each stock listed in the table below. Based on your results, are any of the stocks undervalued?

Stock	Price	Earnings per Share	P/E
Yahoo	37.80	1.07	
Google	381.24	4.51	
Disney	24.96	1.34	
Nike	85.46	4.88	
Ebay	40.96	0.73	

Multiplication and Division of Rational Expressions

7.2

Objectives

A Multiply and divide rational expressions by factoring and then dividing out common factors.

B Convert between units using unit analysis.

Examples now playing at MathTV.com/books

A Multiplying and Dividing Rational Expressions

Recall that to multiply two fractions we simply multiply numerators and multiply denominators and then reduce to lowest terms, if possible:

$$\frac{3}{4} \cdot \frac{10}{21} = \frac{30}{84} \quad \begin{array}{l}\leftarrow \text{Multiply numerators}\\ \leftarrow \text{Multiply denominators}\end{array}$$

$$= \frac{5}{14} \quad \leftarrow \text{Reduce to lowest terms}$$

Recall also that the same result can be achieved by factoring numerators and denominators first and then dividing out the factors they have in common:

$$\frac{3}{4} \cdot \frac{10}{21} = \frac{3}{2 \cdot 2} \cdot \frac{2 \cdot 5}{3 \cdot 7} \qquad \text{Factor}$$

$$= \frac{\cancel{3} \cdot \cancel{2} \cdot 5}{\cancel{2} \cdot 2 \cdot \cancel{3} \cdot 7} \quad \begin{array}{l}\leftarrow \text{Multiply numerators}\\ \leftarrow \text{Multiply denominators}\end{array}$$

$$= \frac{5}{14} \qquad \text{Divide out common factors}$$

We can apply the second process to the product of two rational expressions, as the following example illustrates.

EXAMPLE 1 Multiply $\dfrac{x-2}{x+3} \cdot \dfrac{x^2-9}{2x-4}$.

SOLUTION We begin by factoring numerators and denominators as much as possible. Then we multiply the numerators and denominators. The last step consists of dividing out all factors common to the numerator and denominator:

$$\frac{x-2}{x+3} \cdot \frac{x^2-9}{2x-4} = \frac{x-2}{x+3} \cdot \frac{(x-3)(x+3)}{2(x-2)} \qquad \text{Factor completely}$$

$$= \frac{\cancel{(x-2)}(x-3)\cancel{(x+3)}}{\cancel{(x+3)}(2)\cancel{(x-2)}} \qquad \text{Multiply numerators and denominators}$$

$$= \frac{x-3}{2} \qquad \text{Divide out common factors}$$

In Chapter 1 we defined division as the equivalent of multiplication by the reciprocal. This is how it looks with fractions:

$$\frac{4}{5} \div \frac{8}{9} = \frac{4}{5} \cdot \frac{9}{8} \qquad \text{Division as multiplication by the reciprocal}$$

$$= \frac{\cancel{2} \cdot \cancel{2} \cdot 3 \cdot 3}{5 \cdot \cancel{2} \cdot \cancel{2} \cdot 2}$$

$$= \frac{9}{10} \qquad \text{Factor and divide out common factors}$$

The same idea holds for division with rational expressions. The rational expression that follows the division symbol is called the *divisor;* to divide, we multiply by the reciprocal of the divisor.

PRACTICE PROBLEMS

1. Multiply $\dfrac{x-5}{x+2} \cdot \dfrac{x^2-4}{3x-15}$.

Answer

1. $\frac{x-2}{3}$

2. Divide:
$\frac{4x + 8}{x^2 - x - 6} \div \frac{x^2 + 7x + 12}{x^2 - 9}$

EXAMPLE 2 Divide $\frac{3x - 9}{x^2 - x - 20} \div \frac{x^2 + 2x - 15}{x^2 - 25}$.

SOLUTION We begin by taking the reciprocal of the divisor and writing the problem again in terms of multiplication. We then factor, multiply, and, finally, divide out all factors common to the numerator and denominator of the resulting expression. The complete solution looks like this:

$$\frac{3x - 9}{x^2 - x - 20} \div \frac{x^2 + 2x - 15}{x^2 - 25}$$

$$= \frac{3x - 9}{x^2 - x - 20} \cdot \frac{x^2 - 25}{x^2 + 2x - 15} \quad \text{Multiply by the reciprocal of the divisor}$$

$$= \frac{3(x - 3)}{(x + 4)(x - 5)} \cdot \frac{(x - 5)(x + 5)}{(x + 5)(x - 3)} \quad \text{Factor}$$

$$= \frac{3(x - 3)(x - 5)(x + 5)}{(x + 4)(x - 5)(x + 5)(x - 3)} \quad \text{Multiply}$$

$$= \frac{3}{x + 4} \quad \text{Divide out common factors}$$

As you can see, factoring is the single most important tool we use in working with rational expressions. Most of the work we have done or will do with rational expressions is accomplished most easily if the rational expressions are in factored form. Here are some more examples of multiplication and division with rational expressions.

3. Multiply $\frac{2a + 10}{a^3} \cdot \frac{a^2}{3a + 15}$.

EXAMPLE 3 Multiply $\frac{3a + 6}{a^2} \cdot \frac{a}{2a + 4}$.

SOLUTION

$$\frac{3a + 6}{a^2} \cdot \frac{a}{2a + 4}$$

$$= \frac{3(a + 2)}{a^2} \cdot \frac{a}{2(a + 2)} \quad \text{Factor completely}$$

$$= \frac{3(a + 2)a}{a^2(2)(a + 2)} \quad \text{Multiply}$$

$$= \frac{3}{2a} \quad \text{Divide numerator and denominator by common factors } a(a + 2)$$

4. Divide:
$\frac{x^2 - x - 12}{x^2 - 16} \div \frac{x^2 + 6x + 9}{2x + 8}$

EXAMPLE 4 Divide $\frac{x^2 + 7x + 12}{x^2 - 16} \div \frac{x^2 + 6x + 9}{2x - 8}$.

SOLUTION

$$\frac{x^2 + 7x + 12}{x^2 - 16} \div \frac{x^2 + 6x + 9}{2x - 8}$$

$$= \frac{x^2 + 7x + 12}{x^2 - 16} \cdot \frac{2x - 8}{x^2 + 6x + 9} \quad \text{Division is multiplication by the reciprocal}$$

$$= \frac{(x + 3)(x + 4)(2)(x - 4)}{(x - 4)(x + 4)(x + 3)(x + 3)} \quad \text{Factor and multiply}$$

$$= \frac{2}{x + 3} \quad \text{Divide out common factors}$$

In this example we factored and multiplied the two expressions in a single step. This saves writing the problem one extra time.

Answers

2. $\frac{4}{x + 4}$ 3. $\frac{2}{3a}$

EXAMPLE 5 Multiply $(x^2 - 49)\left(\frac{x+4}{x+7}\right)$.

SOLUTION We can think of the polynomial $x^2 - 49$ as having a denominator of 1. Thinking of $x^2 - 49$ in this way allows us to proceed as we did in previous examples:

$$(x^2 - 49)\left(\frac{x+4}{x+7}\right) = \frac{x^2 - 49}{1} \cdot \frac{x+4}{x+7} \quad \text{Write } x^2 - 49 \text{ with denominator 1}$$

$$= \frac{\cancel{(x+7)}(x-7)(x+4)}{\cancel{x+7}} \quad \text{Factor and multiply}$$

$$= (x-7)(x+4) \quad \text{Divide out common factors}$$

We can leave the answer in this form or multiply to get $x^2 - 3x - 28$. In this section let's agree to leave our answers in factored form.

EXAMPLE 6 Multiply $a(a+5)(a-5)\left(\frac{a+4}{a^2+5a}\right)$.

SOLUTION We can think of the expression $a(a+5)(a-5)$ as having a denominator of 1:

$$a(a+5)(a-5)\left(\frac{a+4}{a^2+5a}\right)$$

$$= \frac{a(a+5)(a-5)}{1} \cdot \frac{a+4}{a^2+5a}$$

$$= \frac{\cancel{a}\cancel{(a+5)}(a-5)(a+4)}{\cancel{a}\cancel{(a+5)}} \quad \text{Factor and multiply}$$

$$= (a-5)(a+4) \quad \text{Divide out common factors}$$

B Unit Analysis

Unit analysis is a method of converting between units of measure by multiplying by the number 1. Here is our first illustration: Suppose you are flying in a commercial airliner and the pilot tells you the plane has reached its cruising altitude of 35,000 feet. How many miles is the plane above the ground?

If you know that 1 mile is 5,280 feet, then it is simply a matter of deciding what to do with the two numbers, 5,280 and 35,000. By using unit analysis, this decision is unnecessary:

$$35{,}000 \text{ feet} = \frac{35{,}000 \text{ feet}}{1} \cdot \frac{1 \text{ mile}}{5{,}280 \text{ feet}}$$

We treat the units common to the numerator and denominator in the same way we treat factors common to the numerator and denominator; common units can be divided out, just as common factors are. In the previous expression, we have feet common to the numerator and denominator. Dividing them out leaves us with miles only. Here is the complete solution:

$$35{,}000 \text{ feet} = \frac{35{,}000 \cancel{\text{feet}}}{1} \cdot \frac{1 \text{ mile}}{5{,}280 \cancel{\text{feet}}}$$

$$= \frac{35{,}000}{5{,}280} \text{ miles}$$

$$= 6.6 \text{ miles to the nearest tenth of a mile}$$

5. Multiply $(x^2 - 9)\left(\frac{x+2}{x+3}\right)$.

6. Multiply:

$a(a+2)(a-4)\left(\frac{a+5}{a^2-4a}\right)$

Answers

4. $\frac{2}{x+3}$ **5.** $(x-3)(x+2)$

6. $(a+2)(a+5)$

The expression $\frac{1 \text{ mile}}{5{,}280 \text{ feet}}$ is called a *conversion factor*. It is simply the number 1 written in a convenient form. Because it is the number 1, we can multiply any other number by it and always be sure we have not changed that number. The key to unit analysis is choosing the right conversion factors.

EXAMPLE 7 The Mall of America in the Twin Cities covers 78 acres of land. If 1 square mile = 640 acres, how many square miles does the Mall of America cover? Round your answer to the nearest hundredth of a square mile.

SOLUTION We are starting with acres and want to end up with square miles. We need to multiply by a conversion factor that will allow acres to divide out and leave us with square miles:

$$78 \text{ acres} = \frac{78 \cancel{\text{ acres}}}{1} \cdot \frac{1 \text{ square mile}}{640 \cancel{\text{ acres}}}$$

$$= \frac{78}{640} \text{ square miles}$$

$$= 0.12 \text{ square miles to the nearest hundredth}$$

EXAMPLE 8 The Forest chair lift at the Northstar ski resort in Lake Tahoe is 5,603 feet long. If a ride on this chair lift takes 11 minutes, what is the average speed of the lift in miles per hour?

SOLUTION First, we find the speed of the lift in feet per second, as we did in Example 9 of Section 7.1, by taking the ratio of distance to time.

$$\text{Rate} = \frac{\text{distance}}{\text{time}} = \frac{5{,}603 \text{ feet}}{11 \text{ minutes}} = \frac{5{,}603}{11} \text{ feet per minute}$$

$$= 509 \text{ feet per minute}$$

Next, we convert feet per minute to miles per hour. To do this, we need to know that

$$1 \text{ mile} = 5{,}280 \text{ feet}$$
$$1 \text{ hour} = 60 \text{ minutes}$$

$$\text{Speed} = 509 \text{ feet per minute} = \frac{509 \cancel{\text{ feet}}}{1 \cancel{\text{ minute}}} \cdot \frac{1 \text{ mile}}{5{,}280 \cancel{\text{ feet}}} \cdot \frac{60 \cancel{\text{ minutes}}}{1 \text{ hour}}$$

$$= \frac{509 \cdot 60}{5{,}280} \text{ miles per hour}$$

$$= 5.8 \text{ miles per hour to the nearest tenth}$$

Getting Ready for Class

After reading through the preceding section, respond in your own words and in complete sentences.

1. How do we multiply rational expressions?
2. Explain the steps used to divide rational expressions.
3. What part does factoring play in multiplying and dividing rational expressions?
4. Why are all conversion factors the same as the number 1?

7. The South Coast Plaza shopping mall in Costa Mesa, California, covers an area of 2,224,750 square feet. If 1 acre = 43,560 square feet, how many acres does the South Coast Plaza shopping mall cover? Round your answer to the nearest tenth of an acre.

8. The Bear Paw chair lift at the Northstar ski resort in Lake Tahoe is 772 feet long. If a ride on this chair lift takes 2.2 minutes, what is the average speed of the lift in miles per hour?

The Forest Chair Lift
Northstar Ski Resort, Lake Tahoe

Answers
7. 51.1 acres
8. 4.0 miles per hour

Problem Set 7.2

A Multiply or divide as indicated. Be sure to reduce all answers to lowest terms. (The numerator and denominator of the answer should not have any factors in common.) **[Examples 1–4]**

1. $\frac{x+y}{3} \cdot \frac{6}{x+y}$

2. $\frac{x-1}{x+1} \cdot \frac{5}{x-1}$

3. $\frac{2x+10}{x^2} \cdot \frac{x^3}{4x+20}$

4. $\frac{3x^4}{3x-6} \cdot \frac{x-2}{x^2}$

5. $\frac{9}{2a-8} \div \frac{3}{a-4}$

6. $\frac{8}{a^2-25} \div \frac{16}{a+5}$

7. $\frac{x+1}{x^2-9} \div \frac{2x+2}{x+3}$

8. $\frac{11}{x-2} \div \frac{22}{2x^2-8}$

9. $\frac{a^2+5a}{7a} \cdot \frac{4a^2}{a^2+4a}$

10. $\frac{4a^2+4a}{a^2-25} \cdot \frac{a^2-5a}{8a}$

11. $\frac{y^2-5y+6}{2y+4} \div \frac{2y-6}{y+2}$

12. $\frac{y^2-7y}{3y^2-48} \div \frac{y^2-9}{y^2-7y+12}$

13. $\dfrac{2x - 8}{x^2 - 4} \cdot \dfrac{x^2 + 6x + 8}{x - 4}$

14. $\dfrac{x^2 + 5x + 1}{7x - 7} \cdot \dfrac{x - 1}{x^2 + 5x + 1}$

15. $\dfrac{x - 1}{x^2 - x - 6} \cdot \dfrac{x^2 + 5x + 6}{x^2 - 1}$

16. $\dfrac{x^2 - 3x - 10}{x^2 - 4x + 3} \cdot \dfrac{x^2 - 5x + 6}{x^2 - 3x - 10}$

17. $\dfrac{a^2 + 10a + 25}{a + 5} \div \dfrac{a^2 - 25}{a - 5}$

18. $\dfrac{a^2 + a - 2}{a^2 + 5a + 6} \div \dfrac{a - 1}{a}$

19. $\dfrac{y^3 - 5y^2}{y^4 + 3y^3 + 2y^2} \div \dfrac{y^2 - 5y + 6}{y^2 - 2y - 3}$

20. $\dfrac{y^2 - 5y}{y^2 + 7y + 12} \div \dfrac{y^3 - 7y^2 + 10y}{y^2 + 9y + 18}$

21. $\dfrac{2x^2 + 17x + 21}{x^2 + 2x - 35} \cdot \dfrac{x^3 - 125}{2x^2 - 7x - 15}$

22. $\dfrac{x^2 + x - 42}{4x^2 + 31x + 21} \cdot \dfrac{4x^2 - 5x - 6}{x^3 - 8}$

23. $\dfrac{2x^2 + 10x + 12}{4x^2 + 24x + 32} \cdot \dfrac{2x^2 + 18x + 40}{x^2 + 8x + 15}$

24. $\dfrac{3x^2 - 3}{6x^2 + 18x + 12} \cdot \dfrac{2x^2 - 8}{x^2 - 3x + 2}$

25. $\dfrac{2a^2 + 7a + 3}{a^2 - 16} \div \dfrac{4a^2 + 8a + 3}{2a^2 - 5a - 12}$

26. $\dfrac{3a^2 + 7a - 20}{a^2 + 3a - 4} \div \dfrac{3a^2 - 2a - 5}{a^2 - 2a + 1}$

27. $\dfrac{4y^2 - 12y + 9}{y^2 - 36} \div \dfrac{2y^2 - 5y + 3}{y^2 + 5y - 6}$

28. $\dfrac{5y^2 - 6y + 1}{y^2 - 1} \div \dfrac{16y^2 - 9}{4y^2 + 7y + 3}$

29. $\dfrac{x^2 - 1}{6x^2 + 42x + 60} \cdot \dfrac{7x^2 + 17x + 6}{x^3 + 1} \cdot \dfrac{6x + 30}{7x^2 - 11x - 6}$

30. $\dfrac{4x^2 - 1}{3x - 15} \cdot \dfrac{4x^2 - 17x - 15}{4x^2 - 9x - 9} \cdot \dfrac{3x - 9}{8x^3 - 1}$

31. $\dfrac{18x^3 + 21x^2 - 60x}{21x^2 - 25x - 4} \cdot \dfrac{28x^2 - 17x - 3}{16x^3 + 28x^2 - 30x}$

32. $\dfrac{56x^3 + 54x^2 - 20x}{8x^2 - 2x - 15} \cdot \dfrac{6x^2 + 5x - 21}{63x^3 + 129x^2 - 42x}$

The next two problems are intended to give you practice reading, and paying attention to, the instructions that accompany the problems you are working. Working these problems is an excellent way to get ready for a test or quiz

33. Work each problem according to the instructions given.

a. Simplify: $\dfrac{9 - 1}{27 - 1}$

b. Reduce: $\dfrac{x^2 - 1}{x^3 - 1}$

c. Multiply: $\dfrac{x^2 - 1}{x^3 - 1} \cdot \dfrac{x - 1}{x + 1}$

d. Divide: $\dfrac{x^2 - 1}{x^3 - 1} \div \dfrac{x - 1}{x^2 + x + 1}$

34. Work each problem according to the instructions given.

a. Simplify: $\dfrac{16 - 9}{16 + 24 + 9}$

b. Reduce: $\dfrac{4x^2 - 9}{4x^2 + 12x + 9}$

c. Multiply: $\dfrac{4x^2 - 9}{4x^2 + 12x + 9} \cdot \dfrac{2x + 3}{2x - 3}$

d. Divide: $\dfrac{4x^2 - 9}{4x^2 + 12x + 9} \div \dfrac{2x + 3}{2x - 3}$

A Multiply the following expressions using the method shown in Examples 5 and 6 in this section. [Examples 5–6]

35. $(x^2 - 9)\left(\frac{2}{x + 3}\right)$

36. $(x^2 - 9)\left(\frac{-3}{x - 3}\right)$

37. $a(a + 5)(a - 5)\left(\frac{2}{a^2 - 25}\right)$

38. $a(a^2 - 4)\left(\frac{a}{a + 2}\right)$

39. $(x^2 - x - 6)\left(\frac{x + 1}{x - 3}\right)$

40. $(x^2 - 2x - 8)\left(\frac{x + 3}{x - 4}\right)$

41. $(x^2 - 4x - 5)\left(\frac{-2x}{x + 1}\right)$

42. $(x^2 - 6x + 8)\left(\frac{4x}{x - 2}\right)$

A Each of the following problems involves some factoring by grouping. Remember, before you can divide out factors common to the numerators and denominators of a product, you must factor completely.

43. $\frac{x^2 - 9}{x^2 - 3x} \cdot \frac{2x + 10}{xy + 5x + 3y + 15}$

44. $\frac{x^2 - 16}{x^2 - 4x} \cdot \frac{3x + 18}{xy + 6x + 4y + 24}$

45. $\dfrac{2x^2 + 4x}{x^2 - y^2} \cdot \dfrac{x^2 + 3x + xy + 3y}{x^2 + 5x + 6}$

46. $\dfrac{x^2 - 25}{3x^2 + 3xy} \cdot \dfrac{x^2 + 4x + xy + 4y}{x^2 + 9x + 20}$

47. $\dfrac{x^3 - 3x^2 + 4x - 12}{x^4 - 16} \cdot \dfrac{3x^2 + 5x - 2}{3x^2 - 10x + 3}$

48. $\dfrac{x^3 - 5x^2 + 9x - 45}{x^4 - 81} \cdot \dfrac{5x^2 + 18x + 9}{5x^2 - 22x - 15}$

Simplify each expression. Work inside parentheses first, and then divide out common factors.

49. $\left(1 - \frac{1}{2}\right)\left(1 - \frac{1}{3}\right)\left(1 - \frac{1}{4}\right)\left(1 - \frac{1}{5}\right)$

50. $\left(1 + \frac{1}{2}\right)\left(1 + \frac{1}{3}\right)\left(1 + \frac{1}{4}\right)\left(1 + \frac{1}{5}\right)$

The dots in the following problems represent factors not written that are in the same pattern as the surrounding factors. Simplify.

51. $\left(1 - \frac{1}{2}\right)\left(1 - \frac{1}{3}\right)\left(1 - \frac{1}{4}\right) \ldots \left(1 - \frac{1}{99}\right)\left(1 - \frac{1}{100}\right)$

52. $\left(1 - \frac{1}{3}\right)\left(1 - \frac{1}{4}\right)\left(1 - \frac{1}{5}\right) \ldots \left(1 - \frac{1}{98}\right)\left(1 - \frac{1}{99}\right)$

B Applying the Concepts [Examples 7–8]

Cars The chart shows the fastest cars in America. Use the chart to answer Problems 53 and 54.

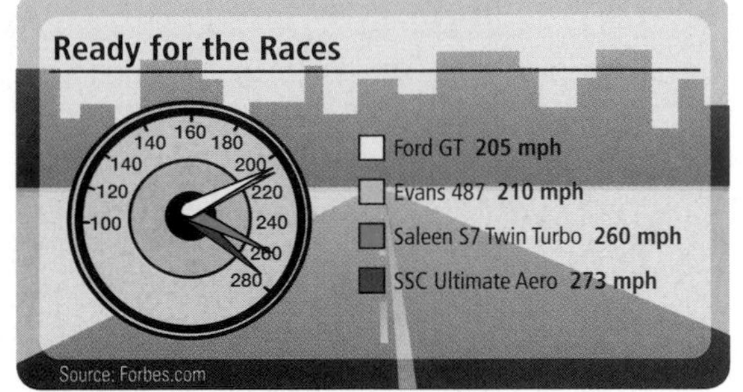

53. Convert the speed of the Ford GT to feet per second. Round to the nearest tenth.

54. Convert the speed of the Saleen S7 Twin Turbo to feet per second. Round to the nearest tenth.

55. Mount Whitney The top of Mount Whitney, the highest point in California, is 14,494 feet above sea level. Give this height in miles to the nearest tenth of a mile.

56. Motor Displacement The relationship between liters and cubic inches, both of which are measures of volume, is 0.0164 liters = 1 cubic inch. If a Ford Mustang has a motor with a displacement of 4.9 liters, what is the displacement in cubic inches? Round your answer to the nearest cubic inch.

57. Speed of Sound The speed of sound is 1,088 feet per second. Convert the speed of sound to miles per hour. Round your answer to the nearest whole number.

58. Average Speed A car travels 122 miles in 3 hours. Find the average speed of the car in feet per second. Round to the nearest whole number.

59. Ferris Wheel As we mentioned in Problem Set 7.1, the first Ferris wheel was built in 1893. It was a large wheel with a circumference of 785 feet. If one trip around the circumference of the wheel took 20 minutes, find the average speed of a rider in miles per hour. Round to the nearest hundredth.

60. Unit Analysis The photograph shows the Cone Nebula as seen by the Hubble telescope in April 2002. The distance across the photograph is about 2.5 light-years. If we assume light travels 186,000 miles in 1 second, we can find the number of miles in 1 light-year by converting 186,000 miles/second to miles/year. Find the number of miles in 1 light-year. Write your answer in expanded form and in scientific notation.

NASA

61. Ferris Wheel A Ferris wheel called *Colossus* has a circumference of 518 feet. If a trip around the circumference of *Colossus* takes 40 seconds, find the average speed of a rider in miles per hour. Round to the nearest tenth.

62. Ferris Wheel A person riding a Ferris wheel travels once around the wheel, a distance of 188 feet, in 30 seconds. What is the average speed of the rider in miles per hour? Round to the nearest tenth.

Getting Ready for the Next Section

Perform the indicated operation.

63. $\frac{1}{5} + \frac{3}{5}$

64. $\frac{1}{7} + \frac{5}{7}$

65. $\frac{1}{10} + \frac{3}{14}$

66. $\frac{1}{21} + \frac{4}{15}$

67. $\frac{1}{10} - \frac{3}{14}$

68. $\frac{1}{21} - \frac{4}{15}$

Multiply.

69. $2(x - 3)$

70. $x(x + 2)$

71. $(x + 4)(x - 5)$

72. $(x + 3)(x - 4)$

Reduce to lowest terms.

73. $\frac{x + 3}{x^2 - 9}$

74. $\frac{x + 7}{x^2 - 49}$

75. $\frac{x^2 - x - 30}{2(x + 5)(x - 5)}$

76. $\frac{x^2 - x - 20}{2(x + 4)(x - 4)}$

Simplify.

77. $(x + 4)(x - 5) - 10$

78. $(x + 3)(x - 4) - 8$

Maintaining Your Skills

Add the following fractions.

79. $\frac{1}{2} + \frac{5}{2}$

80. $\frac{2}{3} + \frac{8}{3}$

81. $2 + \frac{3}{4}$

82. $1 + \frac{4}{7}$

Simplify each term, then add.

83. $\frac{10x^4}{2x^2} + \frac{12x^6}{3x^4}$

84. $\frac{32x^8}{8x^3} + \frac{27x^7}{3x^2}$

85. $\frac{12a^2b^5}{3ab^3} + \frac{14a^4b^7}{7a^3b^5}$

86. $\frac{16a^3b^2}{4ab} + \frac{25a^6b^5}{5a^4b^4}$

Addition and Subtraction of Rational Expressions

7.3

A Adding and Subtracting with the Same Denominator

In Chapter 1 we combined fractions having the same denominator by combining their numerators and putting the result over the common denominator. We use the same process to add two rational expressions with the same denominator.

EXAMPLE 1 Add $\frac{5}{x} + \frac{3}{x}$.

SOLUTION Adding numerators, we have:

$$\frac{5}{x} + \frac{3}{x} = \frac{8}{x}$$

EXAMPLE 2 Add $\frac{x}{x^2 - 9} + \frac{3}{x^2 - 9}$.

SOLUTION Since both expressions have the same denominator, we add numerators and reduce to lowest terms:

$$\frac{x}{x^2 - 9} + \frac{3}{x^2 - 9} = \frac{x + 3}{x^2 - 9}$$

$$= \frac{\cancel{x+3}}{(\cancel{x+3})(x - 3)}$$

$$= \frac{1}{x - 3}$$

Reduce to lowest terms by factoring the denominator and then dividing out the common factor $x + 3$

B Adding and Subtracting with Different Denominators

Remember, that the distributive property allows us to add rational expressions by simply adding numerators. Because of this, we must begin all addition problems involving rational expressions by first making sure all the expressions have the same denominator.

> **Definition**
> The **least common denominator** (LCD) for a set of denominators is the simplest quantity that is exactly divisible by all the denominators.

EXAMPLE 3 Add $\frac{1}{10} + \frac{3}{14}$.

SOLUTION

Step 1: Find the LCD for 10 and 14. To do so, we factor each denominator and build the LCD from the factors:

$$\left.\begin{aligned} 10 &= 2 \cdot 5 \\ 14 &= 2 \cdot 7 \end{aligned}\right\} \quad \text{LCD} = 2 \cdot 5 \cdot 7 = 70$$

We know the LCD is divisible by 10 because it contains the factors 2 and 5. It is also divisible by 14 because it contains the factors 2 and 7.

Objectives

A Add and subtract rational expressions that have the same denominators.

B Add and subtract rational expressions that have different denominators.

Examples now playing at MathTV.com/books

PRACTICE PROBLEMS

1. Add $\frac{2}{x} + \frac{7}{x}$.

2. Add $\frac{x}{x^2 - 49} + \frac{7}{x^2 - 49}$.

3. Add $\frac{1}{21} + \frac{4}{15}$.

If you have had a difficult time in the past with addition and subtraction of fractions with different denominators, this is the time to get it straightened out. Go over Example 3 as many times as is necessary for you to understand the process.

Answers

1. $\frac{9}{x}$ 2. $\frac{1}{x - 7}$

Step 2: Change to equivalent fractions that each have denominator 70. To accomplish this task, we multiply the numerator and denominator of each fraction by the factor of the LCD that is not also a factor of its denominator:

Original Fractions		*Denominators in Factored Form*		*Multiply by Factor Needed to Obtain LCD*		*These Have the Same Value as the Original Fractions*
$\frac{1}{10}$	=	$\frac{1}{2 \cdot 5}$	=	$\frac{1}{2 \cdot 5} \cdot \frac{\mathbf{7}}{\mathbf{7}}$	=	$\frac{7}{70}$
$\frac{3}{14}$	=	$\frac{3}{2 \cdot 7}$	=	$\frac{3}{2 \cdot 7} \cdot \frac{\mathbf{5}}{\mathbf{5}}$	=	$\frac{15}{70}$

The fraction $\frac{7}{70}$ has the same value as the fraction $\frac{1}{10}$. Likewise, the fractions $\frac{15}{70}$ and $\frac{3}{14}$ are equivalent; they have the same value.

Step 3: Add numerators and put the result over the LCD:

$$\frac{7}{70} + \frac{15}{70} = \frac{7 + 15}{70} = \frac{22}{70}$$

Step 4: Reduce to lowest terms:

$$\frac{22}{70} = \frac{11}{35}$$ Divide numerator and denominator by 2

The main idea in adding fractions is to write each fraction again with the LCD for a denominator. Once we have done that, we simply add numerators. The same process can be used to add rational expressions, as the next example illustrates.

4. Subtract $\frac{2}{x} - \frac{1}{3}$.

EXAMPLE 4 Subtract $\frac{3}{x} - \frac{1}{2}$.

SOLUTION

Step 1: The LCD for x and 2 is $2x$. It is the smallest expression divisible by x and by 2.

Step 2: To change to equivalent expressions with the denominator $2x$, we multiply the first fraction by $\frac{2}{2}$ and the second by $\frac{x}{x}$:

$$\frac{3}{x} \cdot \frac{\mathbf{2}}{\mathbf{2}} = \frac{6}{2x}$$

$$\frac{1}{2} \cdot \frac{\boldsymbol{x}}{\boldsymbol{x}} = \frac{x}{2x}$$

Step 3: Subtracting numerators of the rational expressions in step 2, we have:

$$\frac{6}{2x} - \frac{x}{2x} = \frac{6 - x}{2x}$$

Step 4: Since $6 - x$ and $2x$ do not have any factors in common, we cannot reduce any further. Here is the complete solution:

$$\begin{aligned} \frac{3}{x} - \frac{1}{2} &= \frac{3}{x} \cdot \frac{\mathbf{2}}{\mathbf{2}} - \frac{1}{2} \cdot \frac{\boldsymbol{x}}{\boldsymbol{x}} \\ &= \frac{6}{2x} - \frac{x}{2x} \\ &= \frac{6 - x}{2x} \end{aligned}$$

Answers

3. $\frac{11}{35}$ **4.** $\frac{6 - x}{3x}$

EXAMPLE 5 Add $\frac{5}{2x-6} + \frac{x}{x-3}$.

SOLUTION If we factor $2x - 6$, we have $2x - 6 = 2(x - 3)$. We need only multiply the second rational expression in our problem by $\frac{2}{2}$ to have two expressions with the same denominator:

$$\frac{5}{2x-6} + \frac{x}{x-3} = \frac{5}{2(x-3)} + \frac{x}{x-3}$$
$$= \frac{5}{2(x-3)} + \frac{\mathbf{2}}{\mathbf{2}}\left(\frac{x}{x-3}\right)$$
$$= \frac{5}{2(x-3)} + \frac{2x}{2(x-3)}$$
$$= \frac{2x+5}{2(x-3)}$$

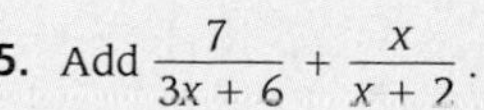

5. Add $\frac{7}{3x+6} + \frac{x}{x+2}$.

EXAMPLE 6 Add $\frac{1}{x+4} + \frac{8}{x^2-16}$.

SOLUTION After writing each denominator in factored form, we find that the least common denominator is $(x + 4)(x - 4)$. To change the first rational expression to an equivalent rational expression with the common denominator, we multiply its numerator and denominator by $x - 4$:

$$\frac{1}{x+4} + \frac{8}{x^2-16}$$
$$= \frac{1}{x+4} + \frac{8}{(x+4)(x-4)} \qquad \text{Factor each denominator}$$
$$= \frac{1}{x+4} \cdot \frac{\mathbf{x-4}}{\mathbf{x-4}} + \frac{8}{(x+4)(x-4)} \qquad \text{Change to equivalent rational expressions}$$
$$= \frac{x-4}{(x+4)(x-4)} + \frac{8}{(x+4)(x-4)} \qquad \text{Multiply}$$
$$= \frac{x+4}{(x+4)(x-4)} \qquad \text{Add numerators}$$
$$= \frac{1}{x-4} \qquad \text{Divide out common factor } x+4$$

6. Add $\frac{1}{x-3} + \frac{-6}{x^2-9}$.

In the last step we reduced the rational expression to lowest terms by dividing out the common factor of $x + 4$.

EXAMPLE 7 Add $\frac{2}{x^2+5x+6} + \frac{x}{x^2-9}$.

SOLUTION

Step 1: We factor each denominator and build the LCD from the factors:

$$\left.\begin{aligned} x^2 + 5x + 6 &= (x+2)(x+3) \\ x^2 - 9 &= (x+3)(x-3) \end{aligned}\right\} \quad \text{LCD} = (x+2)(x+3)(x-3)$$

Step 2: Change to equivalent rational expressions:

$$\frac{2}{x^2+5x+6} = \frac{2}{(x+2)(x+3)} \cdot \frac{\mathbf{(x-3)}}{\mathbf{(x-3)}} = \frac{2x-6}{(x+2)(x+3)(x-3)}$$

$$\frac{x}{x^2-9} = \frac{x}{(x+3)(x-3)} \cdot \frac{\mathbf{(x+2)}}{\mathbf{(x+2)}} = \frac{x^2+2x}{(x+2)(x+3)(x-3)}$$

Step 3: Add numerators of the rational expressions produced in step 2:

$$\frac{2x-6}{(x+2)(x+3)(x-3)} + \frac{x^2+2x}{(x+2)(x+3)(x-3)}$$
$$= \frac{x^2+4x-6}{(x+2)(x+3)(x-3)}$$

The numerator and denominator do not have any factors in common.

7. Add:

$$\frac{5}{x^2-7x+12} + \frac{1}{x^2-9}$$

Answers

5. $\frac{7+3x}{3(x+2)}$ 6. $\frac{1}{x+3}$

7. $\frac{6x+11}{(x-4)(x-3)(x+3)}$

8. Subtract:
$$\frac{x+3}{2x+8} - \frac{4}{x^2-16}$$

Note: In the second step we replaced subtraction by addition of the opposite. There seems to be less chance for error when this is done on longer problems.

9. Write an expression for the difference of twice a number and its reciprocal. Then simplify.

EXAMPLE 8 Subtract $\frac{x+4}{2x+10} - \frac{5}{x^2-25}$.

SOLUTION We begin by factoring each denominator:

$$\frac{x+4}{2x+10} - \frac{5}{x^2-25} = \frac{x+4}{2(x+5)} - \frac{5}{(x+5)(x-5)}$$

The LCD is $2(x+5)(x-5)$. Completing the problem, we have:

$$= \frac{x+4}{2(x+5)} \cdot \frac{\mathbf{(x-5)}}{\mathbf{(x-5)}} + \frac{-5}{(x+5)(x-5)} \cdot \frac{\mathbf{2}}{\mathbf{2}}$$

$$= \frac{x^2-x-20}{2(x+5)(x-5)} + \frac{-10}{2(x+5)(x-5)}$$

$$= \frac{x^2-x-30}{2(x+5)(x-5)}$$

To see if this expression will reduce, we factor the numerator into $(x-6)(x+5)$:

$$= \frac{(x-6)\cancel{(x+5)}}{2\cancel{(x+5)}(x-5)}$$

$$= \frac{x-6}{2(x-5)}$$

EXAMPLE 9 Write an expression for the sum of a number and its reciprocal, and then simplify that expression.

SOLUTION If we let x = the number, then its reciprocal is $\frac{1}{x}$. To find the sum of the number and its reciprocal, we add them:

$$x + \frac{1}{x}$$

The first term x can be thought of as having a denominator of 1. Since the denominators are 1 and x, the least common denominator is x.

$$x + \frac{1}{x} = \frac{x}{1} + \frac{1}{x} \qquad \text{Write } x \text{ as } \frac{x}{1}$$

$$= \frac{x}{1} \cdot \frac{\mathbf{x}}{\mathbf{x}} + \frac{1}{x} \qquad \text{The LCD is } x$$

$$= \frac{x^2}{x} + \frac{1}{x}$$

$$= \frac{x^2+1}{x} \qquad \text{Add numerators}$$

Getting Ready for Class

After reading through the preceding section, respond in your own words and in complete sentences.

1. How do we add two rational expressions that have the same denominator?
2. What is the least common denominator for two fractions?
3. What role does factoring play in finding a least common denominator?
4. Explain how to find a common denominator for two rational expressions.

Answers

8. $\frac{x-5}{2(x-4)}$ 9. $\frac{2x^2-1}{x}$

Problem Set 7.3

A Find the following sums and differences. [Examples 1–2]

1. $\frac{3}{x} + \frac{4}{x}$
2. $\frac{5}{x} + \frac{3}{x}$
3. $\frac{9}{a} - \frac{5}{a}$
4. $\frac{8}{a} - \frac{7}{a}$
5. $\frac{1}{x+1} + \frac{x}{x+1}$
6. $\frac{x}{x-3} - \frac{3}{x-3}$
7. $\frac{y^2}{y-1} - \frac{1}{y-1}$
8. $\frac{y^2}{y+3} - \frac{9}{y+3}$
9. $\frac{x^2}{x+2} + \frac{4x+4}{x+2}$
10. $\frac{x^2-6x}{x-3} + \frac{9}{x-3}$
11. $\frac{x^2}{x-2} - \frac{4x-4}{x-2}$
12. $\frac{x^2}{x-5} - \frac{10x-25}{x-5}$
13. $\frac{x+2}{x+6} - \frac{x-4}{x+6}$
14. $\frac{x+5}{x+2} - \frac{x+3}{x+2}$

B [Examples 4–8]

15. $\frac{y}{2} - \frac{2}{y}$
16. $\frac{3}{y} + \frac{y}{3}$
17. $\frac{1}{2} + \frac{a}{3}$
18. $\frac{2}{3} + \frac{2a}{5}$
19. $\frac{x}{x+1} + \frac{3}{4}$
20. $\frac{x}{x-3} + \frac{1}{3}$

21. $\frac{x+1}{x-2} - \frac{4x+7}{5x-10}$

22. $\frac{3x+1}{2x-6} - \frac{x+2}{x-3}$

23. $\frac{4x-2}{3x+12} - \frac{x-2}{x+4}$

24. $\frac{6x+5}{5x-25} - \frac{x+2}{x-5}$

25. $\frac{6}{x(x-2)} + \frac{3}{x}$

26. $\frac{10}{x(x+5)} - \frac{2}{x}$

27. $\frac{4}{a} - \frac{12}{a^2+3a}$

28. $\frac{5}{a} + \frac{20}{a^2-4a}$

29. $\frac{2}{x+5} - \frac{10}{x^2-25}$

30. $\frac{6}{x^2-1} + \frac{3}{x+1}$

31. $\frac{x-4}{x-3} + \frac{6}{x^2-9}$

32. $\frac{x+1}{x-1} - \frac{4}{x^2-1}$

33. $\frac{a-4}{a-3} + \frac{5}{a^2-a-6}$

34. $\frac{a+2}{a+1} + \frac{7}{a^2-5a-6}$

35. $\frac{8}{x^2-16} - \frac{7}{x^2-x-12}$

36. $\frac{6}{x^2-9} - \frac{5}{x^2-x-6}$

37. $\frac{4y}{y^2+6y+5} - \frac{3y}{y^2+5y+4}$

38. $\frac{3y}{y^2+7y+10} - \frac{2y}{y^2+6y+8}$

39. $\dfrac{4x+1}{x^2+5x+4} - \dfrac{x+3}{x^2+4x+3}$

40. $\dfrac{2x-1}{x^2+x-6} - \dfrac{x+2}{x^2+5x+6}$

41. $\dfrac{1}{x} + \dfrac{x}{3x+9} - \dfrac{3}{x^2+3x}$

42. $\dfrac{1}{x} + \dfrac{x}{2x+4} - \dfrac{2}{x^2+2x}$

43. Work each problem according to the instructions given.

a. Multiply: $\dfrac{4}{9} \cdot \dfrac{1}{6}$

b. Divide: $\dfrac{4}{9} \div \dfrac{1}{6}$

c. Add: $\dfrac{4}{9} + \dfrac{1}{6}$

d. Multiply: $\dfrac{x+2}{x-2} \cdot \dfrac{3x+10}{x^2-4}$

e. Divide: $\dfrac{x+2}{x-2} \div \dfrac{3x+10}{x^2-4}$

f. Subtract: $\dfrac{x+2}{x-2} - \dfrac{3x+10}{x^2-4}$

44. Work each problem according to the instructions given.

a. Multiply: $\dfrac{9}{25} \cdot \dfrac{1}{15}$

b. Divide: $\dfrac{9}{25} \div \dfrac{1}{15}$

c. Subtract: $\dfrac{9}{25} - \dfrac{1}{15}$

d. Multiply: $\dfrac{3x-2}{3x+2} \cdot \dfrac{15x+6}{9x^2-4}$

e. Divide: $\dfrac{3x-2}{3x+2} \div \dfrac{15x+6}{9x^2-4}$

f. Subtract: $\dfrac{3x+2}{3x-2} - \dfrac{15x+6}{9x^2-4}$

Complete the following tables.

45.

Number x	Reciprocal $\frac{1}{x}$	Sum $1+\frac{1}{x}$	Sum $\frac{x+1}{x}$
1			
2			
3			
4			

46.

Number x	Reciprocal $\frac{1}{x}$	Difference $1-\frac{1}{x}$	Difference $\frac{x-1}{x}$
1			
2			
3			
4			

47.

x	$x+\frac{4}{x}$	$\frac{x^2+4}{x}$	$x+4$
1			
2			
3			
4			

48.

x	$2x+\frac{6}{x}$	$\frac{2x^2+6}{x}$	$2x+6$
1			
2			
3			
4			

B Add or subtract as indicated. [Example 9]

49. $1 + \dfrac{1}{x+2}$

50. $1 - \dfrac{1}{x+2}$

51. $1 - \dfrac{1}{x+3}$

52. $1 + \dfrac{1}{x+3}$

Getting Ready for the Next Section

Simplify.

53. $6\left(\frac{1}{2}\right)$ **54.** $10\left(\frac{1}{5}\right)$ **55.** $\frac{0}{5}$ **56.** $\frac{0}{2}$

57. $\frac{5}{0}$ **58.** $\frac{2}{0}$ **59.** $1 - \frac{5}{2}$ **60.** $1 - \frac{5}{3}$

Use the distributive property to simplify.

61. $6\left(\frac{x}{3} + \frac{5}{2}\right)$ **62.** $10\left(\frac{x}{2} + \frac{3}{5}\right)$ **63.** $x^2\left(1 - \frac{5}{x}\right)$ **64.** $x^2\left(1 - \frac{3}{x}\right)$

Solve.

65. $2x + 15 = 3$ **66.** $15 = 3x - 3$ **67.** $-2x - 9 = x - 3$ **68.** $a^2 - a - 20 = -2a$

Maintaining Your Skills

Solve each equation.

69. $2x + 3(x - 3) = 6$ **70.** $4x - 2(x - 5) = 6$ **71.** $x - 3(x + 3) = x - 3$

72. $x - 4(x + 4) = x - 4$ **73.** $7 - 2(3x + 1) = 4x + 3$ **74.** $8 - 5(2x - 1) = 2x + 4$

Solve each quadratic equation.

75. $x^2 + 5x + 6 = 0$ **76.** $x^2 - 5x + 6 = 0$ **77.** $x^2 - x = 6$

78. $x^2 + x = 6$ **79.** $x^2 - 5x = 0$ **80.** $x^2 - 6x = 0$

81. Temperature The chart shows the temperatures for some of the world's hottest places. We can find the temperature in Celsius using the equation, $C = \left(\frac{5}{9}\right)(F - 32)$, where C is the temperature in Celsius and F is the temperature in Fahrenheit. Find the temperature of Al'Aziziyah, Libya, in Celsius.

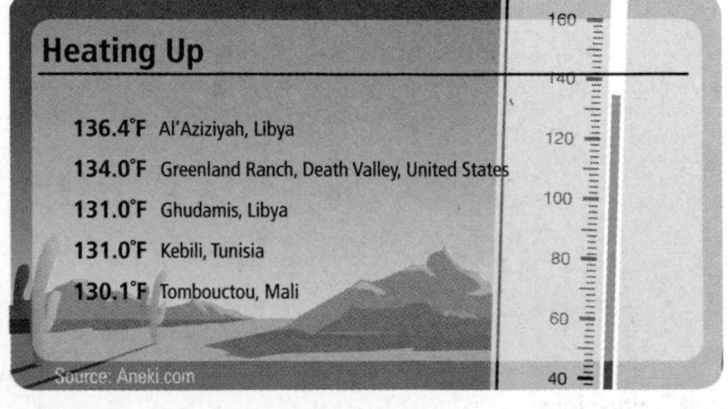

82. Population The map shows the average number of days spent commuting per year in the United States' largest cities. We can find the average number of hours per day spent in the car with the equation, $y = \left(\frac{24}{365}\right)x$, where y is hours per day and x is days per year. Find the hours per day spent commuting in Phoenix, Arizona. Round to the nearest hundredth.

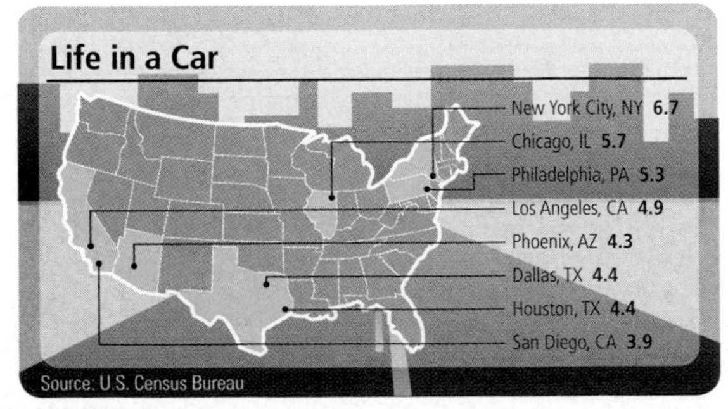

Equations Involving Rational Expressions

7.4

Objectives

A Solve equations that contain rational expressions.

Examples now playing at MathTV.com/books

A Solving Rational Equations

The first step in solving an equation that contains one or more rational expressions is to find the LCD for all denominators in the equation. Once the LCD has been found, we multiply both sides of the equation by it. The resulting equation should be equivalent to the original one (unless we inadvertently multiplied by zero) and free from any denominators except the number 1.

EXAMPLE 1 Solve $\frac{x}{3} + \frac{5}{2} = \frac{1}{2}$ for x.

SOLUTION The LCD for 3 and 2 is 6. If we multiply both sides by 6, we have:

$$6\left(\frac{x}{3} + \frac{5}{2}\right) = 6\left(\frac{1}{2}\right) \quad \text{Multiply both sides by 6}$$

$$6\left(\frac{x}{3}\right) + 6\left(\frac{5}{2}\right) = 6\left(\frac{1}{2}\right) \quad \text{Distributive property}$$

$$2x + 15 = 3$$

$$2x = -12$$

$$x = -6$$

We can check our solution by replacing x with -6 in the original equation:

$$-\frac{6}{3} + \frac{5}{2} \stackrel{?}{=} \frac{1}{2}$$

$$\frac{1}{2} = \frac{1}{2}$$

Multiplying both sides of an equation containing fractions by the LCD clears the equation of all denominators, because the LCD has the property that all denominators will divide it evenly.

EXAMPLE 2 Solve for x: $\frac{3}{x-1} = \frac{3}{5}$.

SOLUTION The LCD for $(x - 1)$ and 5 is $5(x - 1)$. Multiplying both sides by $5(x - 1)$, we have:

$$\frac{5\cancel{(x-1)} \cdot 3}{\cancel{x-1}} = \frac{\cancel{5}(x-1) \cdot 3}{\cancel{5}}$$

$$5 \cdot 3 = (x - 1) \cdot 3$$

$$15 = 3x - 3$$

$$18 = 3x$$

$$6 = x$$

If we substitute $x = 6$ into the original equation, we have:

$$\frac{3}{6-1} \stackrel{?}{=} \frac{3}{5}$$

$$\frac{3}{5} = \frac{3}{5}$$

The solution set is $\{6\}$.

PRACTICE PROBLEMS

1. Solve $\frac{x}{2} + \frac{3}{5} = \frac{1}{5}$ for x.

2. Solve for x: $\frac{4}{x-5} = \frac{4}{3}$.

Answers

1. $-\frac{4}{5}$ 2. 8

3. Solve $1 - \frac{3}{x} = -\frac{2}{x^2}$.

EXAMPLE 3 Solve $1 - \frac{5}{x} = \frac{-6}{x^2}$.

SOLUTION The LCD is x^2. Multiplying both sides by x^2, we have

$$x^2\left(1 - \frac{5}{x}\right) = x^2\left(\frac{-6}{x^2}\right) \quad \text{Multiply both sides by } x^2$$

$$x^2(1) - x^2\left(\frac{5}{x}\right) = x^2\left(\frac{-6}{x^2}\right) \quad \text{Apply distributive property to the left side}$$

$$x^2 - 5x = -6 \quad \text{Simplify each side}$$

We have a quadratic equation, which we write in standard form, factor, and solve as we did in Chapter 6.

$$x^2 - 5x + 6 = 0 \quad \text{Standard form}$$

$$(x - 2)(x - 3) = 0 \quad \text{Factor}$$

$$x - 2 = 0 \quad \text{or} \quad x - 3 = 0 \quad \text{Set factors equal to 0}$$

$$x = 2 \quad \text{or} \quad x = 3$$

The two possible solutions are 2 and 3. Checking each in the original equation, we find they both give true statements. They are both solutions to the original equation:

Check $x = 2$ | Check $x = 3$

$$1 - \frac{5}{2} \stackrel{?}{=} \frac{-6}{4} \qquad 1 - \frac{5}{3} \stackrel{?}{=} \frac{-6}{9}$$

$$\frac{2}{2} - \frac{5}{2} = -\frac{3}{2} \qquad \frac{3}{3} - \frac{5}{3} = -\frac{2}{3}$$

$$-\frac{3}{2} = -\frac{3}{2} \qquad -\frac{2}{3} = -\frac{2}{3}$$

4. Solve:
$\frac{x}{x^2 - 4} - \frac{2}{x - 2} = \frac{1}{x + 2}$

EXAMPLE 4 Solve $\frac{x}{x^2 - 9} - \frac{3}{x - 3} = \frac{1}{x + 3}$.

SOLUTION The factors of $x^2 - 9$ are $(x + 3)(x - 3)$. The LCD, then, is $(x + 3)(x - 3)$:

$$\cancel{(x+3)}\cancel{(x-3)} \cdot \frac{x}{\cancel{(x+3)}\cancel{(x-3)}} + (x+3)\cancel{(x-3)} \cdot \frac{-3}{\cancel{x-3}}$$

$$= \cancel{(x+3)}(x-3) \cdot \frac{1}{\cancel{x+3}}$$

$$x + (x + 3)(-3) = (x - 3) \cdot 1$$

$$x + (-3x) + (-9) = x - 3$$

$$-2x - 9 = x - 3$$

$$-3x = 6$$

$$x = -2$$

The solution is $x = -2$. It checks when substituted for x in the original equation.

Answers
3. 1 or 2 4. −1

EXAMPLE 5 Solve $\frac{x}{x-3} + \frac{3}{2} = \frac{3}{x-3}$.

SOLUTION We begin by multiplying each term on both sides of the equation by the LCD, $2(x - 3)$:

$$2(x-3) \cdot \frac{x}{x-3} + 2(x-3) \cdot \frac{3}{2} = 2(x-3) \cdot \frac{3}{x-3}$$

$$2x + (x - 3) \cdot 3 = 2 \cdot 3$$

$$2x + 3x - 9 = 6$$

$$5x - 9 = 6$$

$$5x = 15$$

$$x = 3$$

Our only possible solution is $x = 3$. If we substitute $x = 3$ into our original equation, we get:

$$\frac{3}{3-3} + \frac{3}{2} \stackrel{?}{=} \frac{3}{3-3}$$

$$\frac{3}{0} + \frac{3}{2} = \frac{3}{0}$$

Two of the terms are undefined, so the equation is meaningless. What has happened is that we have multiplied both sides of the original equation by zero. The equation produced by doing this is not equivalent to our original equation. We always must check our solution when we multiply both sides of an equation by an expression containing the variable to make sure we have not multiplied both sides by zero.

Our original equation has no solution; that is, there is no real number x such that:

$$\frac{x}{x-3} + \frac{3}{2} = \frac{3}{x-3}$$

The solution set is $\varnothing$.

EXAMPLE 6 Solve $\frac{a+4}{a^2+5a} = \frac{-2}{a^2-25}$ for a.

SOLUTION Factoring each denominator, we have:

$$a^2 + 5a = a(a + 5)$$

$$a^2 - 25 = (a + 5)(a - 5)$$

The LCD is $a(a + 5)(a - 5)$. Multiplying both sides of the equation by the LCD gives us:

$$a(a+5)(a-5) \cdot \frac{a+4}{a(a+5)} = \frac{-2}{(a+5)(a-5)} \cdot a(a+5)(a-5)$$

$$(a - 5)(a + 4) = -2a$$

$$a^2 - a - 20 = -2a$$

5. Solve:

$$\frac{x}{x-5} + \frac{5}{2} = \frac{5}{x-5}$$

6. Solve for a:

$$\frac{a+2}{a^2+3a} = \frac{-2}{a^2-9}$$

Answer

5. No solution

The result is a quadratic equation, which we write in standard form, factor, and solve:

$$a^2 + a - 20 = 0 \quad \text{Add } 2a \text{ to both sides}$$

$$(a + 5)(a - 4) = 0 \quad \text{Factor}$$

$$a + 5 = 0 \quad \text{or} \quad a - 4 = 0 \quad \text{Set each factor to 0}$$

$$a = -5 \quad \text{or} \quad a = 4$$

The two possible solutions are -5 and 4. There is no problem with the 4. It checks when substituted for a in the original equation. However, -5 is not a solution. Substituting -5 into the original equation gives:

$$\frac{-5 + 4}{(-5)^2 + 5(-5)} \stackrel{?}{=} \frac{-2}{(-5)^2 - 25}$$

$$\frac{-1}{0} = \frac{-2}{0}$$

This indicates -5 is not a solution. The solution is 4.

Getting Ready for Class

After reading through the preceding section, respond in your own words and in complete sentences.

1. What is the first step in solving an equation that contains rational expressions?
2. Explain how to find the LCD used to clear an equation of fractions.
3. When will a solution to an equation containing rational expressions be extraneous?
4. When do we check for extraneous solutions to an equation containing rational expressions?

Answer
6. 2

Problem Set 7.4

A Solve the following equations. Be sure to check each answer in the original equation if you multiply both sides by an expression that contains the variable. [Examples 1–6]

1. $\frac{x}{3} + \frac{1}{2} = -\frac{1}{2}$

2. $\frac{x}{2} + \frac{4}{3} = -\frac{2}{3}$

3. $\frac{4}{a} = \frac{1}{5}$

4. $\frac{2}{3} = \frac{6}{a}$

5. $\frac{3}{x} + 1 = \frac{2}{x}$

6. $\frac{4}{x} + 3 = \frac{1}{x}$

7. $\frac{3}{a} - \frac{2}{a} = \frac{1}{5}$

8. $\frac{7}{a} + \frac{1}{a} = 2$

9. $\frac{3}{x} + 2 = \frac{1}{2}$

10. $\frac{5}{x} + 3 = \frac{4}{3}$

11. $\frac{1}{y} - \frac{1}{2} = -\frac{1}{4}$

12. $\frac{3}{y} - \frac{4}{5} = -\frac{1}{5}$

13. $1 - \frac{8}{x} = \frac{-15}{x^2}$

14. $1 - \frac{3}{x} = \frac{-2}{x^2}$

15. $\frac{x}{2} - \frac{4}{x} = -\frac{7}{2}$

16. $\frac{x}{2} - \frac{5}{x} = -\frac{3}{2}$

17. $\frac{x-3}{2} + \frac{2x}{3} = \frac{5}{6}$

18. $\frac{x-2}{3} + \frac{5x}{2} = 5$

19. $\frac{x+1}{3} + \frac{x-3}{4} = \frac{1}{6}$

20. $\frac{x+2}{3} + \frac{x-1}{5} = -\frac{3}{5}$

21. $\frac{6}{x+2} = \frac{3}{5}$

22. $\frac{4}{x+3} = \frac{1}{2}$

23. $\frac{3}{y-2} = \frac{2}{y-3}$

24. $\frac{5}{y+1} = \frac{4}{y+2}$

25. $\frac{x}{x-2} + \frac{2}{3} = \frac{2}{x-2}$

26. $\frac{x}{x-5} + \frac{1}{5} = \frac{5}{x-5}$

27. $\frac{x}{x-2} + \frac{3}{2} = \frac{9}{2(x-2)}$

28. $\frac{x}{x+1} + \frac{4}{5} = \frac{-14}{5(x+1)}$

29. $\frac{5}{x+2} + \frac{1}{x+3} = \frac{-1}{x^2+5x+6}$

30. $\frac{3}{x-1} + \frac{2}{x+3} = \frac{-3}{x^2+2x-3}$

31. $\frac{8}{x^2 - 4} + \frac{3}{x + 2} = \frac{1}{x - 2}$

32. $\frac{10}{x^2 - 25} - \frac{1}{x - 5} = \frac{3}{x + 5}$

33. $\frac{a}{2} + \frac{3}{a - 3} = \frac{a}{a - 3}$

34. $\frac{a}{2} + \frac{4}{a - 4} = \frac{a}{a - 4}$

35. $\frac{6}{y^2 - 4} = \frac{4}{y^2 + 2y}$

36. $\frac{2}{y^2 - 9} = \frac{5}{y^2 - 3y}$

37. $\frac{2}{a^2 - 9} = \frac{3}{a^2 + a - 12}$

38. $\frac{2}{a^2 - 1} = \frac{6}{a^2 - 2a - 3}$

39. $\frac{3x}{x - 5} - \frac{2x}{x + 1} = \frac{-42}{x^2 - 4x - 5}$

40. $\frac{4x}{x - 4} - \frac{3x}{x - 2} = \frac{-3}{x^2 - 6x + 8}$

41. $\frac{2x}{x + 2} = \frac{x}{x + 3} - \frac{3}{x^2 + 5x + 6}$

42. $\frac{3x}{x - 4} = \frac{2x}{x - 3} + \frac{6}{x^2 - 7x + 12}$

43. Solve each equation.

a. $5x - 1 = 0$

b. $\frac{5}{x} - 1 = 0$

c. $\frac{x}{5} - 1 = \frac{2}{3}$

d. $\frac{5}{x} - 1 = \frac{2}{3}$

e. $\frac{5}{x^2} + 5 = \frac{26}{x}$

44. Solve each equation.

a. $2x - 3 = 0$

b. $2 - \frac{3}{x} = 0$

c. $\frac{x}{3} - 2 = \frac{1}{2}$

d. $\frac{3}{x} - 2 = \frac{1}{2}$

e. $\frac{1}{x} + \frac{3}{x^2} = 2$

45. Work each problem according to the instructions given.

a. Divide: $\frac{7}{a^2 - 5a - 6} \div \frac{a + 2}{a + 1}$

b. Add: $\frac{7}{a^2 - 5a - 6} + \frac{a + 2}{a + 1}$

c. Solve: $\frac{7}{a^2 - 5a - 6} + \frac{a + 2}{a + 1} = 2$

46. Work each problem according to the instructions given.

a. Divide: $\frac{6}{x^2 - 9} \div \frac{x - 4}{x - 3}$

b. Add: $\frac{6}{x^2 - 9} + \frac{x - 4}{x - 3}$

c. Solve: $\frac{6}{x^2 - 9} + \frac{x - 4}{x - 3} = \frac{3}{4}$

Getting Ready for the Next Section

Solve.

47. $\frac{1}{x} + \frac{1}{2x} = \frac{9}{2}$

48. $\frac{50}{x+5} = \frac{30}{x-5}$

49. $\frac{1}{10} - \frac{1}{15} = \frac{1}{x}$

50. $\frac{15}{x} + \frac{15}{x+20} = 2$

Find the value of $y = \frac{-6}{x}$ for the given value of x.

51. $x = -6$

52. $x = -3$

53. $x = 2$

54. $x = 1$

Maintaining Your Skills

55. Google Earth This Google Earth image is of the London Eye. The distance around the wheel is 420 meters. If it takes a person 27 minutes to complete a full revolution, what is the speed of the London Eye in centimeters per second? Round to the nearest whole number.

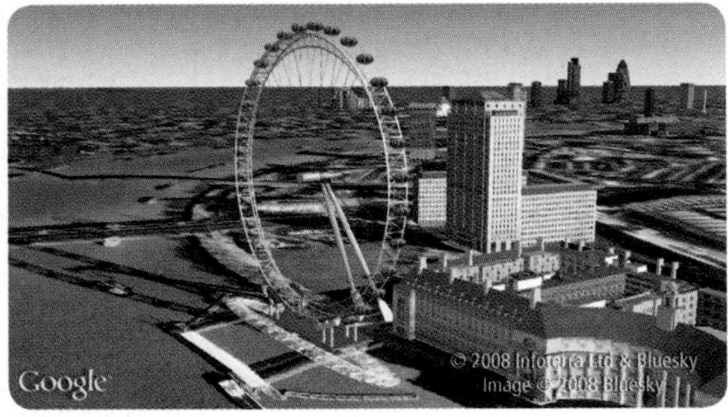

56. Google Earth The Google Earth image shows three cities in Colorado. If the distance between North Washington and Edgewater is 4.7 miles, and the distance from Edgewater to Denver is 4 miles, what is the distance from Denver to North Washington? Round to the nearest tenth.

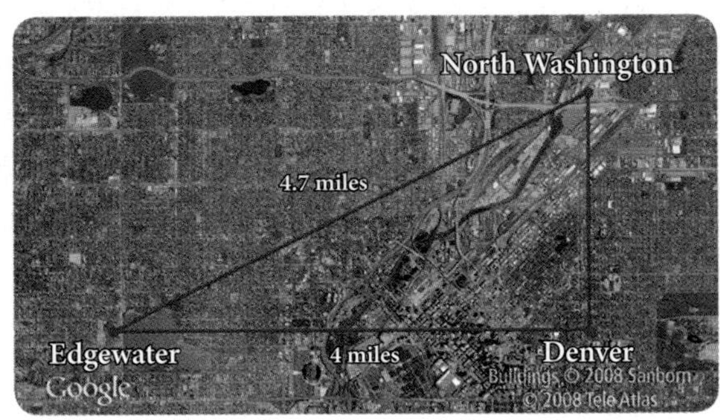

57. Number Problem If twice the difference of a number and 3 were decreased by 5, the result would be 3. Find the number.

58. Number Problem If 3 times the sum of a number and 2 were increased by 6, the result would be 27. Find the number.

59. Geometry The length of a rectangle is 5 inches more than twice the width. The perimeter is 34 inches. Find the length and width.

60. Geometry The length of a rectangle is 2 feet more than 3 times the width. The perimeter is 44 feet. Find the length and width.

61. Number Problem The product of two consecutive even integers is 48. Find the two integers.

62. Number Problem The product of two consecutive odd integers is 35. Find the two integers.

63. Geometry The hypotenuse (the longest side) of a right triangle is 10 inches, and the lengths of the two legs (the other two sides) are given by two consecutive even integers. Find the lengths of the two legs.

64. Geometry One leg of a right triangle is 2 feet more than twice the other. If the hypotenuse is 13 feet, find the lengths of the two legs.

A Applications Involving Rational Expressions

In this section we will solve some word problems whose equations involve rational expressions. Like the other word problems we have encountered, the more you work with them, the easier they become.

EXAMPLE 1 One number is twice another. The sum of their reciprocals is $\frac{9}{2}$. Find the two numbers.

SOLUTION Let x = the smaller number. The larger then must be $2x$. Their reciprocals are $\frac{1}{x}$ and $\frac{1}{2x}$, respectively. An equation that describes the situation is:

$$\frac{1}{x} + \frac{1}{2x} = \frac{9}{2}$$

We can multiply both sides by the LCD, $2x$, and then solve the resulting equation:

$$2x\left(\frac{1}{x}\right) + 2x\left(\frac{1}{2x}\right) = 2x\left(\frac{9}{2}\right)$$

$$2 + 1 = 9x$$

$$3 = 9x$$

$$x = \frac{3}{9} = \frac{1}{3}$$

The smaller number is $\frac{1}{3}$. The other number is twice as large, or $\frac{2}{3}$. If we add their reciprocals, we have:

$$\frac{3}{1} + \frac{3}{2} = \frac{6}{2} + \frac{3}{2} = \frac{9}{2}$$

The solutions check with the original problem.

EXAMPLE 2 A boat travels 30 miles up a river in the same amount of time it takes to travel 50 miles down the same river. If the current is 5 miles per hour, what is the speed of the boat in still water?

SOLUTION The easiest way to work a problem like this is with a table. The top row of the table is labeled with d for distance, r for rate, and t for time. The left column of the table is labeled with the two trips: upstream and downstream. Here is what the table looks like:

	d	r	t
Upstream			
Downstream			

The next step is to read the problem over again and fill in as much of the table as we can with the information in the problem. The distance the boat travels upstream is 30 miles and the distance downstream is 50 miles. Since we are asked for the speed of the boat in still water, we will let that be x. If the speed of the boat in still water is x, then its speed upstream (against the current) must be

Objectives

A Solve applications whose solutions depend on solving an equation containing rational expressions.

B Graph an equation involving a rational expression.

Examples now playing at MathTV.com/books

PRACTICE PROBLEMS

1. One number is three times another. The sum of their reciprocals is $\frac{4}{9}$. Find the numbers.

2. A boat travels 26 miles up a river in the same amount of time it takes to travel 34 miles down the same river. If the current is 2 miles per hour, what is the speed of the boat in still water? (*Hint:* Begin by filling in the table that follows.)

	d	r	t
Upstream			
Downstream			

Answers

1. 3, 9

Note There are two things to note about this problem. The first is that to solve the equation $d = r \cdot t$ for t, we divide each side by r, like this:

$$\frac{d}{r} = \frac{r \cdot t}{r}$$

$$\frac{d}{r} = t$$

The second thing is this: The speed of the boat in still water is the rate at which it would be traveling if there were no current; that is, it is the speed of the boat through the water. Since the water itself is moving at 5 miles per hour, the boat is going 5 miles per hour slower when it travels against the current and 5 miles per hour faster when it travels with the current.

$x - 5$, and its speed downstream (with the current) must be $x + 5$. Putting these four quantities into the appropriate positions in the table, we have

	d	r	t
Upstream	30	$x - 5$	
Downstream	50	$x + 5$	

The last positions in the table are filled in by using the equation $t = \frac{d}{r}$.

	d	r	t
Upstream	30	$x - 5$	$\frac{30}{x - 5}$
Downstream	50	$x + 5$	$\frac{50}{x + 5}$

Reading the problem again, we find that the time for the trip upstream is equal to the time for the trip downstream. Setting these two quantities equal to each other, we have our equation:

$$\text{Time (downstream)} = \text{time (upstream)}$$

$$\frac{50}{x + 5} = \frac{30}{x - 5}$$

The LCD is $(x + 5)(x - 5)$. We multiply both sides of the equation by the LCD to clear it of all denominators. Here is the solution:

$$\cancel{(x + 5)}(x - 5) \cdot \frac{50}{\cancel{x + 5}} = (x + 5)\cancel{(x - 5)} \cdot \frac{30}{\cancel{x - 5}}$$

$$50x - 250 = 30x + 150$$

$$20x = 400$$

$$x = 20$$

The speed of the boat in still water is 20 miles per hour.

EXAMPLE 3 Tina is training for a biathlon. To train for the bicycle portion, she rides her bike 15 miles up a hill and then 15 miles back down the same hill. The complete trip takes her 2 hours. If her downhill speed is 20 miles per hour faster than her uphill speed, how fast does she ride uphill?

3. Repeat Example 3 if Tina's downhill speed is 8 miles per hour faster than her uphill speed.

SOLUTION Again, we make a table. As in the previous example, we label the top row with distance, rate, and time. We label the left column with the two trips, uphill and downhill.

	d	r	t
Uphill			
Downhill			

Next, we fill in the table with as much information as we can from the problem. We know the distance traveled is 15 miles uphill and 15 miles downhill, which allows us to fill in the distance column. To fill in the rate column, we first note that she rides 20 miles per hour faster downhill than uphill. Therefore, if we let x equal her rate uphill, then her rate downhill is $x + 20$. Filling in the table with this information gives us

Answer
2. 15 miles per hour

	d	r	t
Uphill	15	x	
Downhill	15	$x + 20$	

Since time is distance divided by rate, $t = d/r$, we can fill in the last column in the table.

	d	r	t
Uphill	15	x	$\frac{15}{x}$
Downhill	15	$x + 20$	$\frac{15}{x+20}$

Rereading the problem, we find that the total time (the time riding uphill plus the time riding downhill) is two hours. We write our equation as follows:

$$\text{Time (uphill)} + \text{time (downhill)} = 2$$

$$\frac{15}{x} + \frac{15}{x+20} = 2$$

We solve this equation for x by first finding the LCD and then multiplying each term in the equation by it to clear the equation of all denominators. Our LCD is $x(x + 20)$. Here is our solution:

$$x(x+20)\frac{15}{x} + x(x+20)\frac{15}{x+20} = 2 \cdot [x(x+20)]$$

$$15(x + 20) + 15x = 2x(x + 20)$$

$$15x + 300 + 15x = 2x^2 + 40x$$

$$0 = 2x^2 + 10x - 300$$

$$0 = x^2 + 5x - 150 \qquad \text{Divide both sides by 2}$$

$$0 = (x + 15)(x - 10)$$

$$x = -15 \quad \text{or} \quad x = 10$$

Since we cannot have a negative speed, our only solution is $x = 10$. Tina rides her bike at a rate of 10 miles per hour when going uphill. (Her downhill speed is $x + 20 = 30$ miles per hour.)

EXAMPLE 4 An inlet pipe can fill a water tank in 10 hours, while an outlet pipe can empty the same tank in 15 hours. By mistake, both pipes are left open. How long will it take to fill the water tank with both pipes open?

SOLUTION Let x = amount of time to fill the tank with both pipes open.

One method of solving this type of problem is to think in terms of how much of the job is done by a pipe in 1 hour.

1. If the inlet pipe fills the tank in 10 hours, then in 1 hour the inlet pipe fills $\frac{1}{10}$ of the tank.
2. If the outlet pipe empties the tank in 15 hours, then in 1 hour the outlet pipe empties $\frac{1}{15}$ of the tank.
3. If it takes x hours to fill the tank with both pipes open, then in 1 hour the tank is $\frac{1}{x}$ full.

4. A tub can be filled by the cold water faucet in 8 minutes. The drain empties the tub in 12 minutes. How long will it take to fill the tub if both the faucet and the drain are opened?

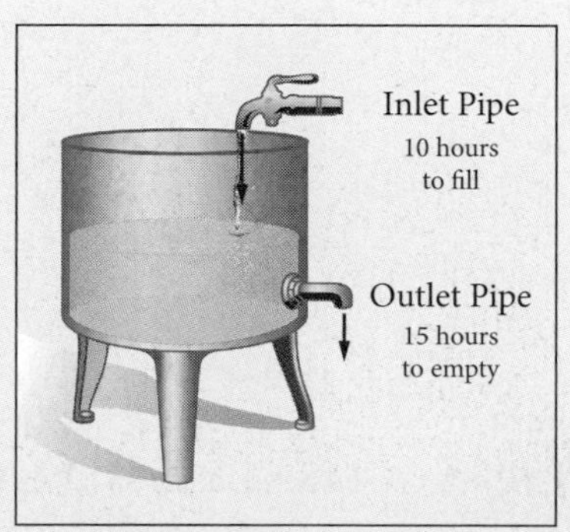

Answer
3. 12 miles/hour

Note In solving a problem of this type, we have to assume that the thing doing the work (whether it is a pipe, a person, or a machine) is working at a constant rate; that is, as much work gets done in the first hour as is done in the last hour and any other hour in between.

Here is how we set up the equation. *In 1 hour,*

$\frac{1}{10}$	$-$	$\frac{1}{15}$	$=$	$\frac{1}{x}$
Amount of water let in by inlet pipe		Amount of water let out by outlet pipe		Total amount of water into tank

The LCD for our equation is $30x$. We multiply both sides by the LCD and solve:

$$30x\left(\frac{1}{10}\right) - 30x\left(\frac{1}{15}\right) = 30x\left(\frac{1}{x}\right)$$

$$3x - 2x = 30$$

$$x = 30$$

It takes 30 hours with both pipes open to fill the tank.

B Graphing Rational Expressions

EXAMPLE 5 Graph the equation $y = \frac{1}{x}$.

SOLUTION Since this is the first time we have graphed an equation of this form, we will make a table of values for x and y that satisfy the equation. Before we do, let's make some generalizations about the graph (Figure 1).

First, notice that since y is equal to 1 divided by x, y will be positive when x is positive. (The quotient of two positive numbers is a positive number.) Likewise, when x is negative, y will be negative. In other words, x and y always will have the same sign. Thus, our graph will appear in quadrants I and III only because in those quadrants x and y have the same sign.

Next, notice that the expression $\frac{1}{x}$ will be undefined when x is 0, meaning that there is no value of y corresponding to $x = 0$. Because of this, the graph will not cross the y-axis. Further, the graph will not cross the x-axis either. If we try to find the x-intercept by letting $y = 0$, we have

$$0 = \frac{1}{x}$$

But there is no value of x to divide into 1 to obtain 0. Therefore, since there is no solution to this equation, our graph will not cross the x-axis.

To summarize, we can expect to find the graph in quadrants I and III only, and the graph will cross neither axis.

x	y
-3	$-\frac{1}{3}$
-2	$-\frac{1}{2}$
-1	-1
$-\frac{1}{2}$	-2
$-\frac{1}{3}$	-3
0	Undefined
$\frac{1}{3}$	3
$\frac{1}{2}$	2
1	1
2	$\frac{1}{2}$
3	$\frac{1}{3}$

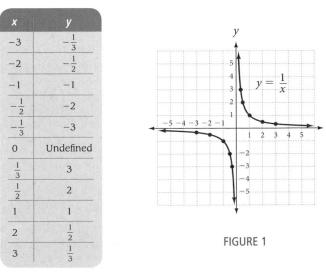

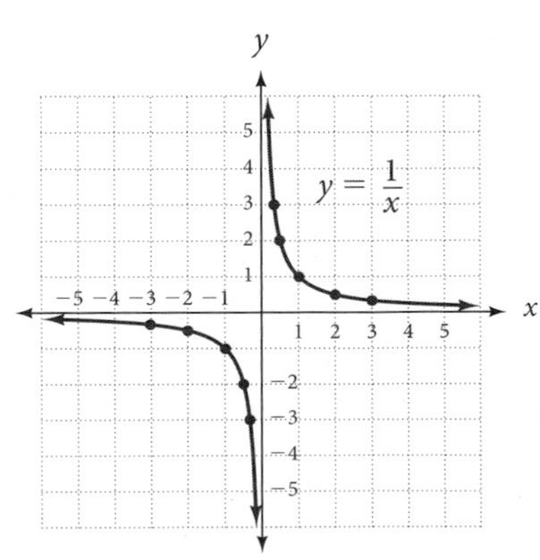

FIGURE 1

5. Graph the equation $y = \frac{2}{x}$.

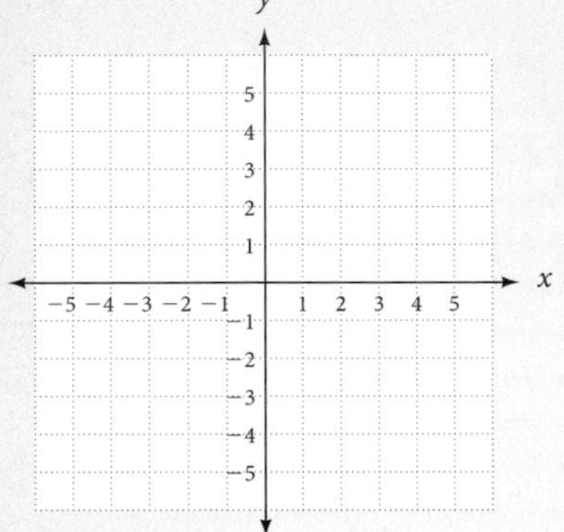

Answers

4. 24 minutes
5. See Solutions to Selected Practice Problems

EXAMPLE 6

Graph the equation $y = \frac{-6}{x}$.

SOLUTION Since y is -6 divided by x, when x is positive, y will be negative (a negative divided by a positive is negative), and when x is negative, y will be positive (a negative divided by a negative). Thus, the graph (Figure 2) will appear in quadrants II and IV only. As was the case in Example 5, the graph will not cross either axis.

x	y
−6	1
−3	2
−2	3
−1	6
0	Undefined
1	−6
2	−3
3	−2
6	−1

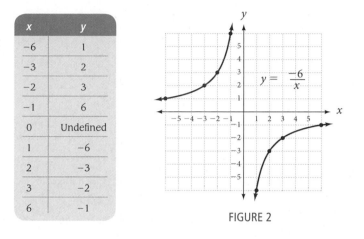

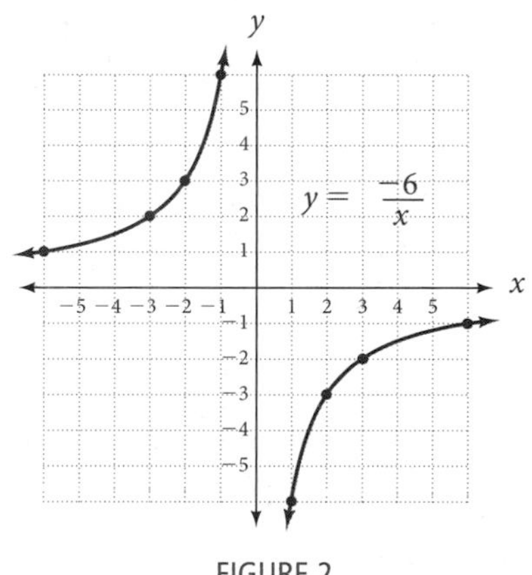

FIGURE 2

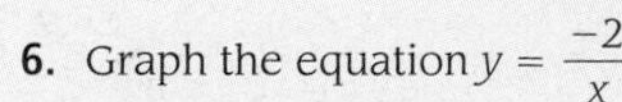

6. Graph the equation $y = \frac{-2}{x}$.

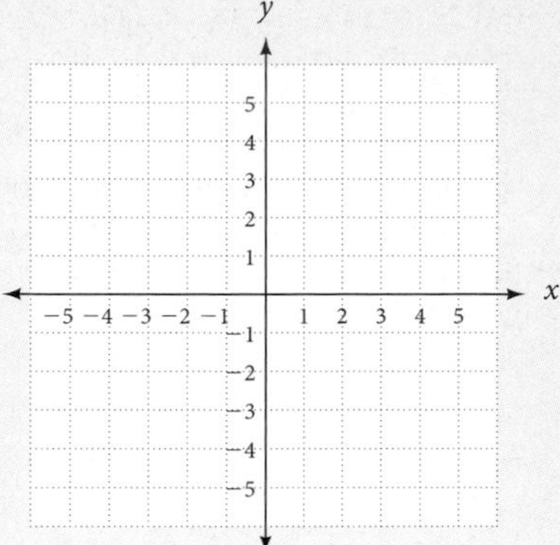

Getting Ready for Class

After reading through the preceding section, respond in your own words and in complete sentences.

1. Write an application problem for which the solution depends on solving the equation $\frac{1}{2} + \frac{1}{3} = \frac{1}{x}$.
2. How does the current of a river affect the speed of a motor boat traveling against the current?
3. How does the current of a river affect the speed of a motor boat traveling in the same direction as the current?
4. What is the relationship between the total number of minutes it takes for a drain to empty a sink and the amount of water that drains out of the sink in 1 minute?

Answer

6. See Solutions to Selected Practice Problems

Problem Set 7.5

A Number Problems [Example 1]

1. One number is 3 times as large as another. The sum of their reciprocals is $\frac{16}{3}$. Find the two numbers.

2. If $\frac{3}{5}$ is added to twice the reciprocal of a number, the result is 1. Find the number.

3. The sum of a number and its reciprocal is $\frac{13}{6}$. Find the number.

4. The sum of a number and 10 times its reciprocal is 7. Find the number.

5. If a certain number is added to both the numerator and denominator of the fraction $\frac{7}{9}$, the result is $\frac{5}{7}$. Find the number.

6. The numerator of a certain fraction is 2 more than the denominator. If $\frac{1}{3}$ is added to the fraction, the result is 2. Find the fraction.

7. The sum of the reciprocals of two consecutive even integers is $\frac{5}{12}$. Find the integers.

8. The sum of the reciprocals of two consecutive integers is $\frac{7}{12}$. Find the two integers.

A Motion Problems [Examples 2–3]

9. A boat travels 26 miles up the river in the same amount of time it takes to travel 38 miles down the same river. If the current is 3 miles per hour, what is the speed of the boat in still water?

	d	r	t
Upstream			
Downstream			

10. A boat can travel 9 miles up a river in the same amount of time it takes to travel 11 miles down the same river. If the current is 2 miles per hour, what is the speed of the boat in still water?

	d	r	t
Upstream			
Downstream			

11. An airplane flying against the wind travels 140 miles in the same amount of time it would take the same plane to travel 160 miles with the wind. If the wind speed is a constant 20 miles per hour, how fast would the plane travel in still air?

12. An airplane flying against the wind travels 500 miles in the same amount of time that it would take to travel 600 miles with the wind. If the speed of the wind is 50 miles per hour, what is the speed of the plane in still air?

13. One plane can travel 20 miles per hour faster than another. One of them goes 285 miles in the same time it takes the other to go 255 miles. What are their speeds?

14. One car travels 300 miles in the same amount of time it takes a second car traveling 5 miles per hour slower than the first to go 275 miles. What are the speeds of the cars?

15. Tina, whom we mentioned in Example 3 of this section, is training for a biathlon. To train for the running portion of the race, she runs 8 miles each day, over the same course. The first 2 miles of the course is on level ground, while the last 6 miles is downhill. She runs 3 miles per hour slower on level ground than she runs downhill. If the complete course takes 1 hour, how fast does she run on the downhill part of the course?

16. Jerri is training for the same biathlon as Tina (Example 3 and Problem 15). To train for the bicycle portion of the race, she rides 24 miles out a straight road, then turns around and rides 24 miles back. The trip out is against the wind, whereas the trip back is with the wind. If she rides 10 miles per hour faster with the wind then she does against the wind, and the complete trip out and back takes 2 hours, how fast does she ride when she rides against the wind?

17. To train for the running of a triathlon, Jerri jogs 1 hour each day over the same 9-mile course. Five miles of the course is downhill, whereas the other 4 miles is on level ground. Jerri figures that she runs 2 miles per hour faster downhill than she runs on level ground. Find the rate at which Jerri runs on level ground.

18. Travis paddles his kayak in the harbor at Morro Bay, California, where the incoming tide has caused a current in the water. From the point where he enters the water, he paddles 1 mile against the current, then turns around and paddles 1 mile back to where he started. His average speed when paddling with the current is 4 miles per hour faster than his speed against the current. If the complete trip (out and back) takes him 1.2 hours, find his average speed when he paddles against the current.

A Work Problems [Example 4]

19. An inlet pipe can fill a pool in 12 hours, while an outlet pipe can empty it in 15 hours. If both pipes are left open, how long will it take to fill the pool.

20. A water tank can be filled in 20 hours by an inlet pipe and emptied in 25 hours by an outlet pipe. How long will it take to fill the tank if both pipes are left open?

21. A bathtub can be filled by the cold water faucet in 10 minutes and by the hot water faucet in 12 minutes. How long does it take to fill the tub if both faucets are open?

22. A water faucet can fill a sink in 6 minutes, whereas the drain can empty it in 4 minutes. If the sink is full, how long will it take to empty if both the faucet and the drain are open?

23. A sink can be filled by the cold water faucet in 3 minutes. The drain can empty a full sink in 4 minutes. If the sink is empty and both the cold water faucet and the drain are open, how long will it take the sink to overflow?

24. A bathtub can be filled by the cold water faucet in 9 minutes and by the hot water faucet in 10 minutes. The drain can empty the tub in 5 minutes. Can the tub be filled if both faucets and the drain are open?

B Graph each of the following equations. [Examples 5–6]

25. $y = \frac{-4}{x}$

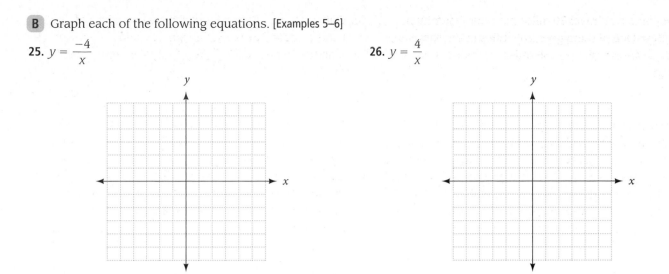

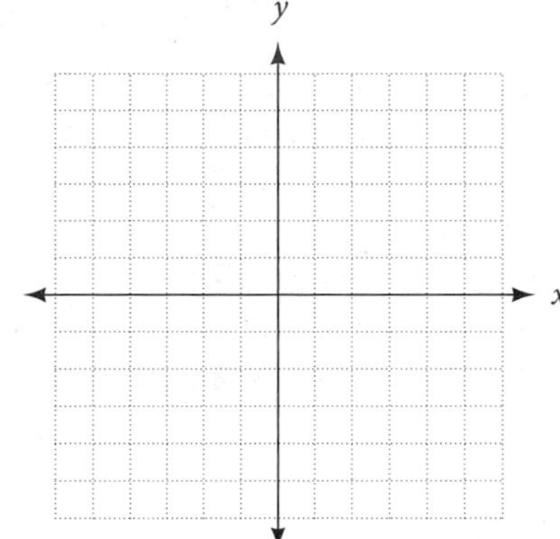

26. $y = \frac{4}{x}$

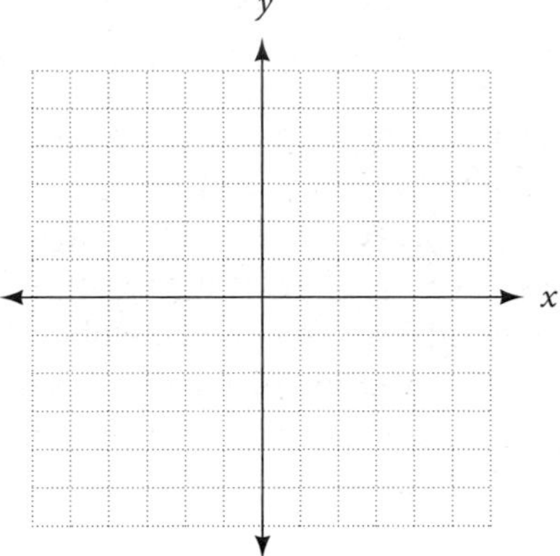

27. $y = \frac{8}{x}$

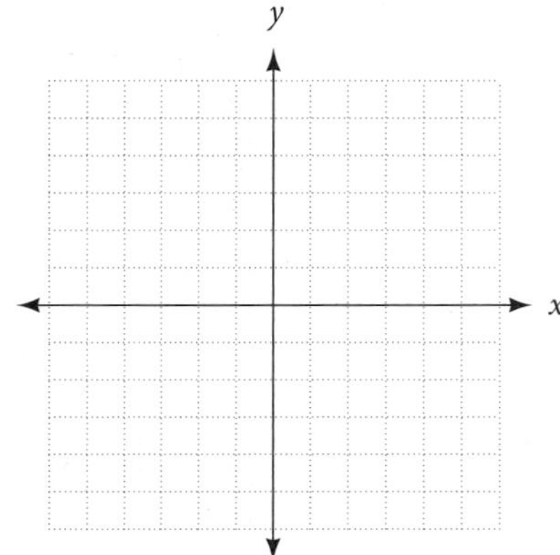

28. $y = \frac{-8}{x}$

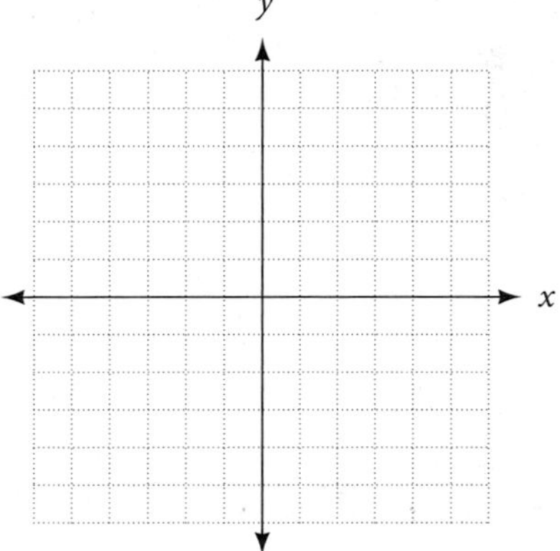

29. Graph $y = \frac{3}{x}$ and $x + y = 4$ on the same coordinate system. At what points do the two graphs intersect?

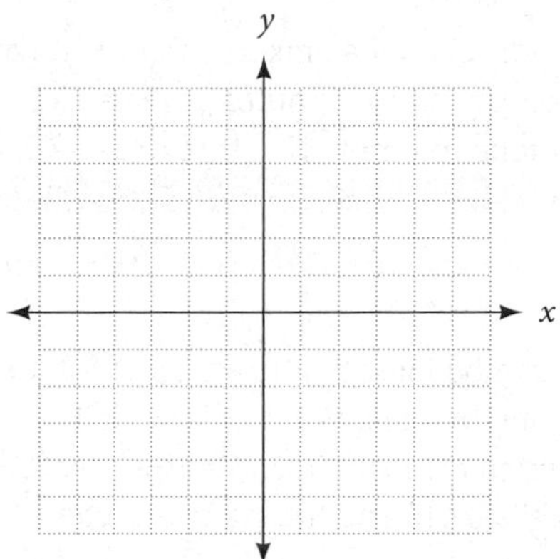

30. Graph $y = \frac{4}{x}$ and $x - y = 3$ on the same coordinate system. At what points do the two graphs intersect?

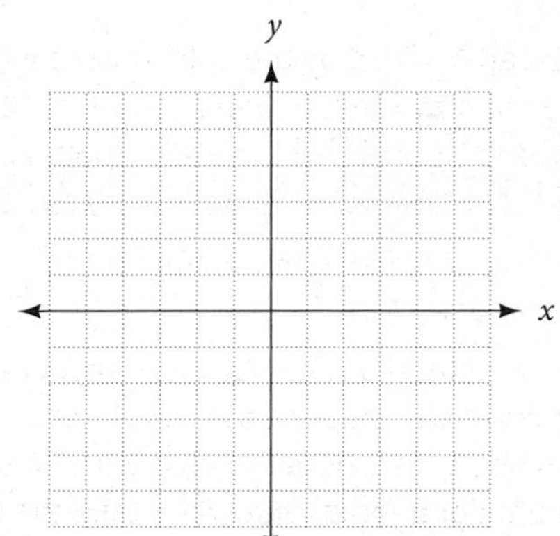

Getting Ready for the Next Section

Simplify.

31. $\frac{1}{2} \div \frac{2}{3}$

32. $\frac{1}{3} \div \frac{3}{4}$

33. $1 + \frac{1}{2}$

34. $1 + \frac{2}{3}$

35. $y^5 \cdot \frac{2x^3}{y^2}$

36. $y^7 \cdot \frac{3x^5}{y^4}$

37. $\frac{2x^3}{y^2} \cdot \frac{y^5}{4x}$

38. $\frac{3x^5}{y^4} \cdot \frac{y^7}{6x^2}$

Factor.

39. $x^2y + x$

40. $xy^2 + y$

Reduce.

41. $\frac{2x^3y^2}{4x}$

42. $\frac{3x^5y^3}{6x^2}$

43. $\frac{x^2 - 4}{x^2 - x - 6}$

44. $\frac{x^2 - 9}{x^2 - 5x + 6}$

Maintaining Your Skills

45. Population The map shows the most populated cities in the United States. If the population of New York City is about 42% of the state's population, what is the approximate population of the state? Round to the nearest tenth.

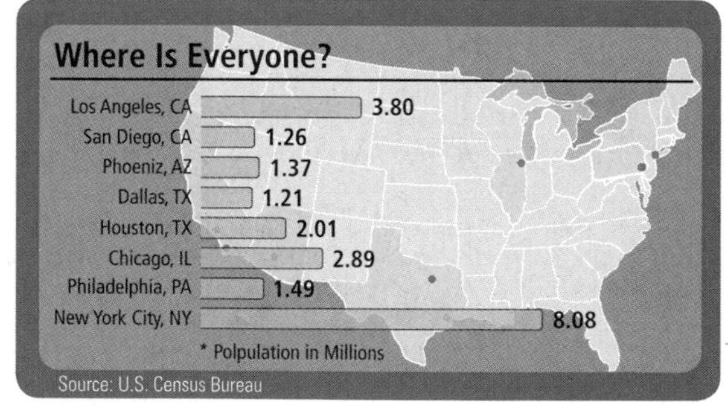

46. Horse Racing The graph shows the total amount of money wagered on the Kentucky Derby. What was the percent increase in wagers from 1995 to 2005? Round to the nearest tenth of a percent.

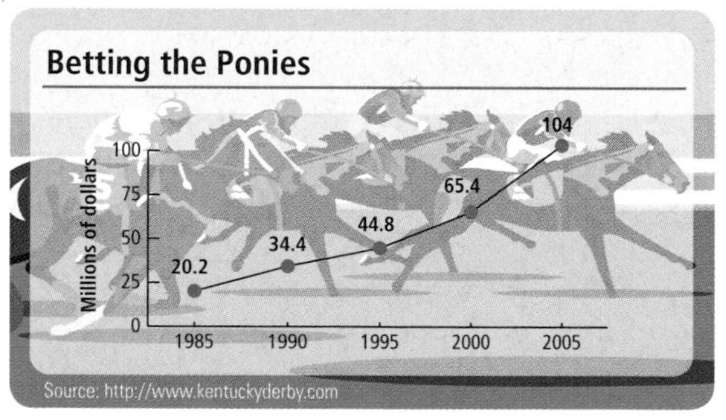

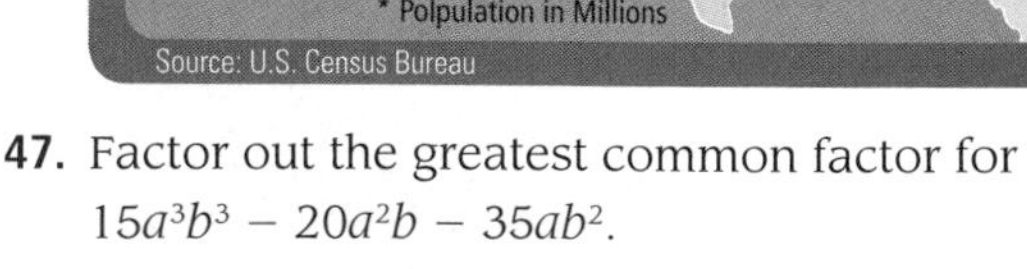

47. Factor out the greatest common factor for $15a^3b^3 - 20a^2b - 35ab^2$.

48. Factor by grouping $3ax - 2a + 15x - 10$.

Factor completely.

49. $x^2 - 4x - 12$

50. $4x^2 - 20xy + 25y^2$

51. $x^4 - 16$

52. $2x^2 + xy - 21y^2$

53. $5x^3 - 25x^2 - 30x$

Objectives

A Simplify a complex fraction.

Examples now playing at MathTV.com/books

A complex fraction is a fraction or rational expression that contains other fractions in its numerator or denominator. Each of the following is a complex fraction:

$$\frac{\frac{1}{2}}{\frac{2}{3}} \qquad \frac{x + \frac{1}{y}}{y + \frac{1}{x}} \qquad \frac{\frac{a+1}{a^2-9}}{\frac{2}{a+3}}$$

A Simplying Complex Fractions

We will begin this section by simplifying the first of these complex fractions. Before we do, though, let's agree on some vocabulary. So that we won't have to use phrases such as the "numerator of the denominator," let's call the numerator of a complex fraction the *top* and the denominator of a complex fraction the *bottom*.

EXAMPLE 1 Simplify $\frac{\frac{1}{2}}{\frac{2}{3}}$.

SOLUTION There are two methods we can use to solve this problem.

METHOD 1 We can multiply the top and bottom of this complex fraction by the LCD for both fractions. In this case the LCD is 6:

$$\frac{\frac{1}{2}}{\frac{2}{3}} = \frac{\mathbf{6} \cdot \frac{1}{2}}{\mathbf{6} \cdot \frac{2}{3}} = \frac{3}{4}$$

METHOD 2 We can treat this as a division problem. To divide by $\frac{2}{3}$, we multiply by its reciprocal $\frac{3}{2}$:

$$\frac{\frac{1}{2}}{\frac{2}{3}} = \frac{1}{2} \cdot \frac{3}{2} = \frac{3}{4}$$

Using either method, we obtain the same result.

EXAMPLE 2 Simplify: $\frac{\frac{2x^3}{y^2}}{\frac{4x}{y^5}}$

SOLUTION

METHOD 1 The LCD for each rational expression is y^5. Multiplying the top and bottom of the complex fraction by y^5, we have:

$$\frac{\frac{2x^3}{y^2}}{\frac{4x}{y^5}} = \frac{\mathbf{y^5} \cdot \frac{2x^3}{y^2}}{\mathbf{y^5} \cdot \frac{4x}{y^5}} = \frac{2x^3y^3}{4x} = \frac{x^2y^3}{2}$$

METHOD 2 To divide by $\frac{4x}{y^5}$ we multiply by its reciprocal, $\frac{y^5}{4x}$:

$$\frac{\frac{2x^3}{y^2}}{\frac{4x}{y^5}} = \frac{2x^3}{y^2} \cdot \frac{y^5}{4x} = \frac{x^2y^3}{2}$$

Again the result is the same, whether we use Method 1 or Method 2.

PRACTICE PROBLEMS

1. Simplify $\frac{\frac{1}{3}}{\frac{3}{4}}$.

2. Simplify $\frac{\frac{3x^5}{y^4}}{\frac{6x^2}{y^7}}$.

Answers

1. $\frac{4}{9}$ **2.** $\frac{x^3y^3}{2}$

3. Simplify $\dfrac{y - \dfrac{1}{x}}{x - \dfrac{1}{y}}$.

EXAMPLE 3 Simplify: $\dfrac{x + \dfrac{1}{y}}{y + \dfrac{1}{x}}$

SOLUTION To apply Method 2 as we did in the first two examples, we would have to simplify the top and bottom separately to obtain a single rational expression for both before we could multiply by the reciprocal. It is much easier, in this case, to multiply the top and bottom by the LCD xy:

$$\frac{x + \frac{1}{y}}{y + \frac{1}{x}} = \frac{\mathbf{xy}\left(x + \frac{1}{y}\right)}{\mathbf{xy}\left(y + \frac{1}{x}\right)} \quad \text{Multiply top and bottom by } xy$$

$$= \frac{xy \cdot x + xy \cdot \frac{1}{y}}{xy \cdot y + xy \cdot \frac{1}{x}} \quad \text{Distributive property}$$

$$= \frac{x^2y + x}{xy^2 + y} \quad \text{Simplify}$$

We can factor an x from $x^2y + x$ and a y from $xy^2 + y$ and then reduce to lowest terms:

$$= \frac{x(xy + 1)}{y(xy + 1)}$$

$$= \frac{x}{y}$$

4. Simplify $\dfrac{1 - \dfrac{9}{x^2}}{1 - \dfrac{5}{x} + \dfrac{6}{x^2}}$.

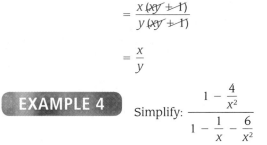

EXAMPLE 4 Simplify: $\dfrac{1 - \dfrac{4}{x^2}}{1 - \dfrac{1}{x} - \dfrac{6}{x^2}}$

SOLUTION The easiest way to simplify this complex fraction is to multiply the top and bottom by the LCD, x^2:

$$\frac{1 - \frac{4}{x^2}}{1 - \frac{1}{x} - \frac{6}{x^2}} = \frac{\mathbf{x^2}\left(1 - \frac{4}{x^2}\right)}{\mathbf{x^2}\left(1 - \frac{1}{x} - \frac{6}{x^2}\right)} \quad \text{Multiply top and bottom by } x^2$$

$$= \frac{x^2 \cdot 1 - x^2 \cdot \frac{4}{x^2}}{x^2 \cdot 1 - x^2 \cdot \frac{1}{x} - x^2 \cdot \frac{6}{x^2}} \quad \text{Distributive property}$$

$$= \frac{x^2 - 4}{x^2 - x - 6} \quad \text{Simplify}$$

$$= \frac{(x - 2)(x + 2)}{(x - 3)(x + 2)} \quad \text{Factor}$$

$$= \frac{x - 2}{x - 3} \quad \text{Reduce}$$

In our next example, we find the relationship between a sequence of complex fractions and the numbers in the Fibonacci sequence.

Answers

3. $\frac{y}{x}$ **4.** $\frac{x + 3}{x - 2}$

EXAMPLE 5 Simplify each term in the following sequence, and then explain how this sequence is related to the Fibonacci sequence:

$$1 + \frac{1}{1+1},\ 1 + \frac{1}{1 + \frac{1}{1+1}},\ 1 + \frac{1}{1 + \frac{1}{1 + \frac{1}{1+1}}},\ \ldots$$

SOLUTION We can simplify our work somewhat if we notice that the first term $1 + \frac{1}{1+1}$ is the larger denominator in the second term and that the second term is the largest denominator in the third term:

$$\text{First term:}\quad 1 + \frac{1}{1+1} = 1 + \frac{1}{2} = \frac{2}{2} + \frac{1}{2} = \frac{3}{2}$$

$$\text{Second term:}\quad 1 + \frac{1}{1 + \frac{1}{1+1}} = 1 + \frac{1}{\frac{3}{2}} = 1 + \frac{2}{3} = \frac{3}{3} + \frac{2}{3} = \frac{5}{3}$$

$$\text{Third term:}\quad 1 + \frac{1}{1 + \frac{1}{1 + \frac{1}{1+1}}} = 1 + \frac{1}{\frac{5}{3}} = 1 + \frac{3}{5} = \frac{5}{5} + \frac{3}{5} = \frac{8}{5}$$

Here are the simplified numbers for the first three terms in our sequence:

$$\frac{3}{2}, \frac{5}{3}, \frac{8}{5}, \ldots$$

Recall the Fibonacci sequence:

$$1, 1, 2, 3, 5, 8, 13, 21, \ldots$$

As you can see, each term in the sequence we have simplified is the ratio of two consecutive numbers in the Fibonacci sequence. If the pattern continues in this manner, the next number in our sequence will be $\frac{13}{8}$.

Getting Ready for Class

After reading through the preceding section, respond in your own words and in complete sentences.

1. What is a complex fraction?
2. Explain one method of simplifying complex fractions.
3. How is a least common denominator used to simplify a complex fraction?
4. What types of complex fractions can be rewritten as division problems?

5. Simplify each term in the following sequence:

$$1 + \frac{2}{1+2},\quad 1 + \frac{2}{1 + \frac{2}{1+2}}, \ldots$$

$$1 + \frac{2}{1 + \frac{2}{1 + \frac{2}{1+2}}},$$

Answer

5. $\frac{5}{3}, \frac{11}{5}, \frac{21}{11}$

Problem Set 7.6

A Simplify each complex fraction. [Examples 1–4]

1. $\dfrac{\frac{3}{4}}{\frac{1}{8}}$

2. $\dfrac{\frac{1}{3}}{\frac{5}{6}}$

3. $\dfrac{\frac{2}{3}}{4}$

4. $\dfrac{5}{\frac{1}{2}}$

5. $\dfrac{\frac{x^2}{y}}{\frac{x}{y^3}}$

6. $\dfrac{\frac{x^5}{y^3}}{\frac{x^2}{y^8}}$

7. $\dfrac{\frac{4x^3}{y^6}}{\frac{8x^2}{y^7}}$

8. $\dfrac{\frac{6x^4}{y}}{\frac{2x}{y^5}}$

9. $\dfrac{y + \frac{1}{x}}{x + \frac{1}{y}}$

10. $\dfrac{y - \frac{1}{x}}{x - \frac{1}{y}}$

11. $\dfrac{1 + \frac{1}{a}}{1 - \frac{1}{a}}$

12. $\dfrac{\frac{1}{a} - 1}{\frac{1}{a} + 1}$

13. $\dfrac{\frac{x + 1}{x^2 - 9}}{\frac{2}{x + 3}}$

14. $\dfrac{\frac{3}{x - 5}}{\frac{x + 1}{x^2 - 25}}$

15. $\dfrac{\frac{1}{a + 2}}{\frac{1}{a^2 - a - 6}}$

16. $\dfrac{\frac{1}{a^2 + 5a + 6}}{\frac{1}{a + 3}}$

17. $\dfrac{1 - \dfrac{9}{y^2}}{1 - \dfrac{1}{y} - \dfrac{6}{y^2}}$

18. $\dfrac{1 - \dfrac{4}{y^2}}{1 - \dfrac{2}{y} - \dfrac{8}{y^2}}$

19. $\dfrac{\dfrac{1}{y} + \dfrac{1}{x}}{\dfrac{1}{xy}}$

20. $\dfrac{\dfrac{1}{xy}}{\dfrac{1}{y} - \dfrac{1}{x}}$

21. $\dfrac{1 - \dfrac{1}{a^2}}{1 - \dfrac{1}{a}}$

22. $\dfrac{1 + \dfrac{1}{a}}{1 - \dfrac{1}{a^2}}$

23. $\dfrac{\dfrac{1}{10x} - \dfrac{y}{10x^2}}{\dfrac{1}{10} - \dfrac{y}{10x}}$

24. $\dfrac{\dfrac{1}{2x} + \dfrac{y}{2x^2}}{\dfrac{1}{4} + \dfrac{y}{4x}}$

25. $\dfrac{\dfrac{1}{a+1} + 2}{\dfrac{1}{a+1} + 3}$

26. $\dfrac{\dfrac{2}{a+1} + 3}{\dfrac{3}{a+1} + 4}$

Although the following problems do not contain complex fractions, they do involve more than one operation. Simplify inside the parentheses first, then multiply. [Example 5]

27. $\left(1 - \dfrac{1}{x}\right)\left(1 - \dfrac{1}{x+1}\right)\left(1 - \dfrac{1}{x+2}\right)$

28. $\left(1 + \dfrac{1}{x}\right)\left(1 + \dfrac{1}{x+1}\right)\left(1 + \dfrac{1}{x+2}\right)$

29. $\left(1 + \frac{1}{x+3}\right)\left(1 + \frac{1}{x+2}\right)\left(1 + \frac{1}{x+1}\right)$

30. $\left(1 - \frac{1}{x+3}\right)\left(1 - \frac{1}{x+2}\right)\left(1 - \frac{1}{x+1}\right)$

31. Simplify each term in the following sequence.

$$2 + \frac{1}{2+1},\ 2 + \frac{1}{2 + \frac{1}{2+1}},\ 2 + \frac{1}{2 + \frac{1}{2 + \frac{1}{2+1}}},\ \ldots$$

32. Simplify each term in the following sequence.

$$2 + \frac{3}{2+3},\ 2 + \frac{3}{2 + \frac{3}{2+3}},\ 2 + \frac{3}{2 + \frac{3}{2 + \frac{3}{2+3}}},\ \ldots$$

Complete the following tables.

33.

Number x	Reciprocal $\frac{1}{x}$	Quotient $\frac{x}{\frac{1}{x}}$	Square x^2
1			
2			
3			
4			

34.

Number x	Reciprocal $\frac{1}{x}$	Quotient $\frac{\frac{1}{x}}{x}$	Square x^2
1			
2			
3			
4			

35.

Number x	Reciprocal $\frac{1}{x}$	Sum $1 + \frac{1}{x}$	Quotient $\frac{1 + \frac{1}{x}}{\frac{1}{x}}$
1			
2			
3			
4			

36.

Number x	Reciprocal $\frac{1}{x}$	difference $1 - \frac{1}{x}$	Quotient $\frac{1 - \frac{1}{x}}{\frac{1}{x}}$
1			
2			
3			
4			

Applying the Concepts

Cars The chart shows the fastest cars in America. Use the chart to answer Problems 37 and 38.

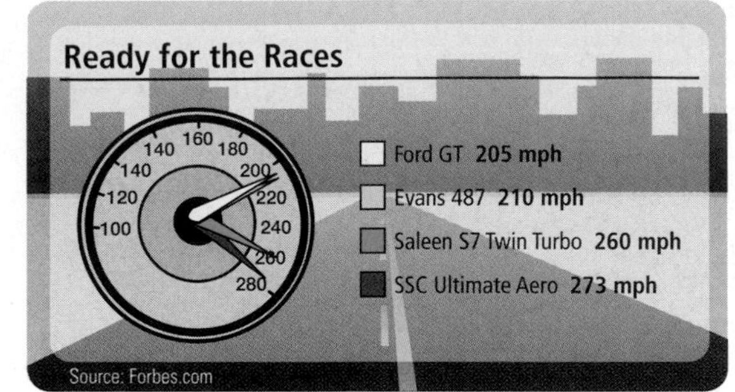

37. To convert miles per hour to kilometers per hour, we can use the complex fraction $\frac{x}{\left(\frac{100}{161}\right)}$ where x is the speed in miles per hour. What is the top speed of the Ford GT in kilometer per hour? Round to the nearest whole number.

38. To convert miles per hour to meters per second we use the complex fraction, $\frac{\left(\frac{x}{3,600}\right)}{\left(\frac{1}{1,609}\right)}$, where x is the speed in miles per hour. What is the top speed of the Saleen S7 Twin Turbo in meters per second? Round to the nearest whole number.

Getting Ready for the Next Section

Solve.

39. $21 = 6x$

40. $72 = 2x$

41. $x^2 + x = 6$

42. $x^2 + 2x = 8$

Maintaining Your Skills

Solve each inequality.

43. $2x + 3 < 5$

44. $3x - 2 > 7$

45. $-3x \le 21$

46. $-5x \ge -10$

47. $-2x + 8 > -4$

48. $-4x - 1 < 11$

49. $4 - 2(x + 1) \ge -2$

50. $6 - 2(x + 3) \le -8$

Objectives

A Solve a proportion.

B Solve application problems involving proportions.

XYZ
Examples now playing at
MathTV.com/books

A proportion is two equal ratios; that is, if $\frac{a}{b}$ and $\frac{c}{d}$ are ratios, then:

$$\frac{a}{b} = \frac{c}{d}$$

is a proportion.

Each of the four numbers in a proportion is called a *term* of the proportion. We number the terms as follows:

$$\begin{array}{r} \text{First term} \rightarrow a \\ \text{Second term} \rightarrow b \end{array} = \begin{array}{l} c \leftarrow \text{Third term} \\ d \leftarrow \text{Fourth term} \end{array}$$

The first and fourth terms are called the *extremes,* and the second and third terms are called the *means:*

$$\text{Means} \quad \frac{a}{b} = \frac{c}{d} \quad \text{Extremes}$$

For example, in the proportion:

$$\frac{3}{8} = \frac{12}{32}$$

the extremes are 3 and 32, and the means are 8 and 12.

Means-Extremes Property

If a, b, c, and d are real numbers with $b \neq 0$ and $d \neq 0$, then

if $\quad \frac{a}{b} = \frac{c}{d}$

then $\quad ad = bc$

In words: In any proportion, the product of the extremes is equal to the product of the means.

A Solving Proportions

This property of proportions comes from the multiplication property of equality. We can use it to solve for a missing term in a proportion.

EXAMPLE 1 Solve the proportion $\frac{3}{x} = \frac{6}{7}$ for x.

SOLUTION We could solve for x by using the method developed in Section 7.4; that is, multiplying both sides by the LCD $7x$. Instead, let's use our new means-extremes property:

$\frac{3}{x} = \frac{6}{7}$ Extremes are 3 and 7; means are x and 6

$21 = 6x$ Product of extremes = product of means

$\frac{21}{6} = x$ Divide both sides by 6

$x = \frac{7}{2}$ Reduce to lowest terms

PRACTICE PROBLEMS

1. Solve for x in the proportion:

$$\frac{2}{9} = \frac{8}{x}$$

Answer

1. 36

2. Solve for x: $\frac{x+2}{4} = \frac{2}{x}$.

EXAMPLE 2 Solve for x : $\frac{x+1}{2} = \frac{3}{x}$.

SOLUTION Again, we want to point out that we could solve for x by using the method we used in Section 7.4. Using the means-extremes property is simply an alternative to the method developed in Section 7.4:

$$\frac{x+1}{2} = \frac{3}{x}$$ Extremes are $x + 1$ and x; means are 2 and 3

$$x^2 + x = 6$$ Product of extremes = product of means

$$x^2 + x - 6 = 0$$ Standard form for a quadratic equation

$$(x + 3)(x - 2) = 0$$ Factor

$$x + 3 = 0 \quad \text{or} \quad x - 2 = 0$$ Set factors equal to 0

$$x = -3 \quad \text{or} \quad x = 2$$

This time we have two solutions: −3 and 2.

B Applications with Proportions

3. Using the information in Example 3, determine how many out of 1,650 parts can be expected to be defective.

EXAMPLE 3 A manufacturer knows that during a production run, 8 out of every 100 parts produced by a certain machine will be defective. If the machine produces 1,450 parts, how many can be expected to be defective?

SOLUTION The ratio of defective parts to total parts produced is $\frac{8}{100}$. If we let x represent the numbers of defective parts out of the total of 1,450 parts, then we can write this ratio again as $\frac{x}{1,450}$. This gives us a proportion to solve:

Defective parts in numerator → x; Total parts in denominator → 1,450

$$\frac{x}{1,450} = \frac{8}{100}$$ Extremes are x and 100; means are 1,450 and 8

$$100x = 11,600$$ Product of extremes = product of means

$$x = 116$$

The manufacturer can expect 116 defective parts out of the total of 1,450 parts if the machine usually produces 8 defective parts for every 100 parts it produces.

4. What is the distance between two cities 2.0 inches apart on the map in Example 4?

EXAMPLE 4 The scale on a map indicates that 1 inch on the map corresponds to an actual distance of 85 miles. Two cities are 3.5 inches apart on the map. What is the actual distance between the two cities?

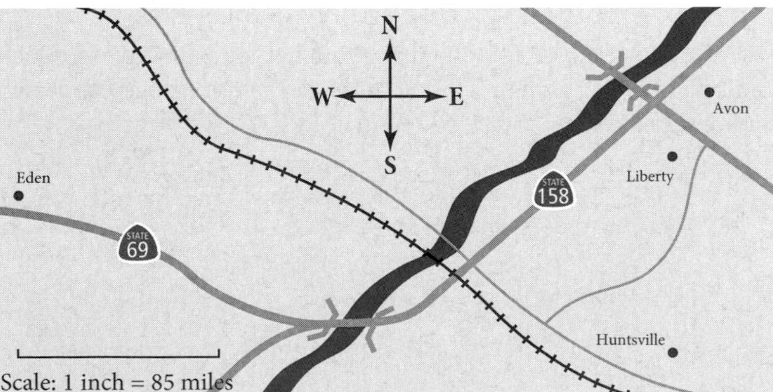

Answers
2. −4, 2 3. 132 parts

SOLUTION We let x represent the actual distance between the two cities. The proportion is

$$\begin{array}{r} \text{Miles} \rightarrow \\ \text{Inches} \rightarrow \end{array} \frac{x}{3.5} = \frac{85}{1} \begin{array}{l} \leftarrow \text{Miles} \\ \leftarrow \text{Inches} \end{array}$$

$$x \cdot 1 = 3.5(85)$$

$$x = 297.5 \text{ miles}$$

EXAMPLE 5 A woman drives her car 270 miles in 6 hours. If she continues at the same rate, how far will she travel in 10 hours?

SOLUTION We let x represent the distance traveled in 10 hours. Using x, we translate the problem into the following proportion:

$$\begin{array}{r} \text{Miles} \rightarrow \\ \text{Hours} \rightarrow \end{array} \frac{x}{10} = \frac{270}{6} \begin{array}{l} \leftarrow \text{Miles} \\ \leftarrow \text{Hours} \end{array}$$

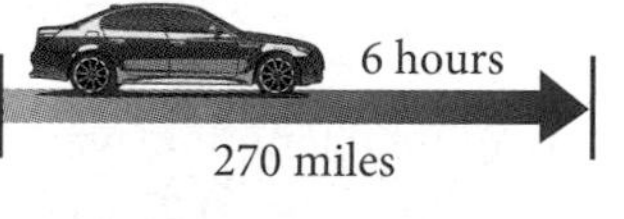

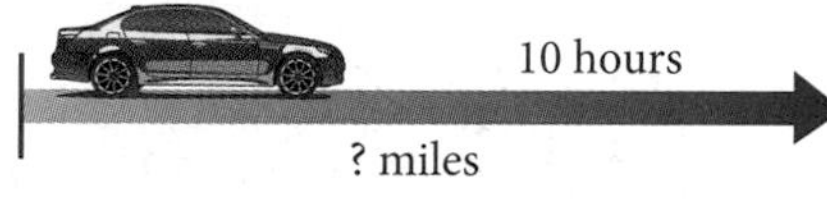

Notice that the two ratios in the proportion compare the same quantities. That is, both ratios compare miles to hours. In words this proportion says:

x miles is to 10 hours as 270 miles is to 6 hours

$$\frac{x}{10} = \frac{270}{6}$$

Next, we solve the proportion.

$$x \cdot 6 = 10 \cdot 270$$

$$x \cdot 6 = 2{,}700$$

$$\frac{x \cdot \cancel{6}}{\cancel{6}} = \frac{2{,}700}{6}$$

$$x = 450 \text{ miles}$$

If the woman continues at the same rate, she will travel 450 miles in 10 hours.

5. If a man travels 250 miles in 5 hours, how far will he travel in 9 hours if he continues at the same rate?

Similar Triangles

Two triangles that have the same shape are similar when their corresponding sides are proportional, or have the same ratio. The triangles below are similar.

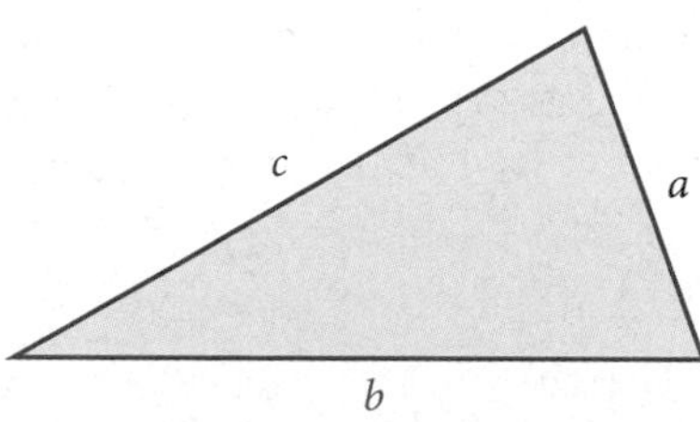

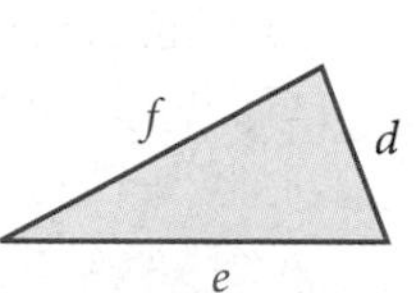

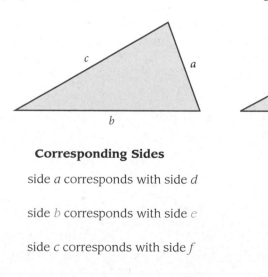

Corresponding Sides	**Ratio**
side a corresponds with side d	$\frac{a}{d}$
side b corresponds with side e	$\frac{b}{e}$
side c corresponds with side f	$\frac{c}{f}$

LABELING TRIANGLES

Recall from Chapter 2 that one way to label the important parts of a triangle is to label the vertices with capital letters and the sides with lower-case letters.

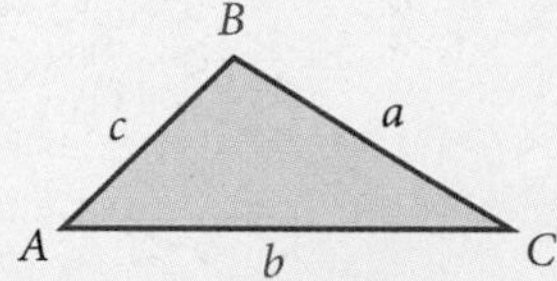

Notice that side a is opposite vertex A, side b is opposite vertex B, and side c is opposite vertex C. Also, because each vertex is the vertex of one of the angles of the triangle, we refer to the three interior angles as A, B, and C.

Answers

4. 170 miles **5.** 450 miles

Because their corresponding sides are proportional, we write

$$\frac{a}{d} = \frac{b}{e} = \frac{c}{f}$$

The two triangles below are similar. Find side x.

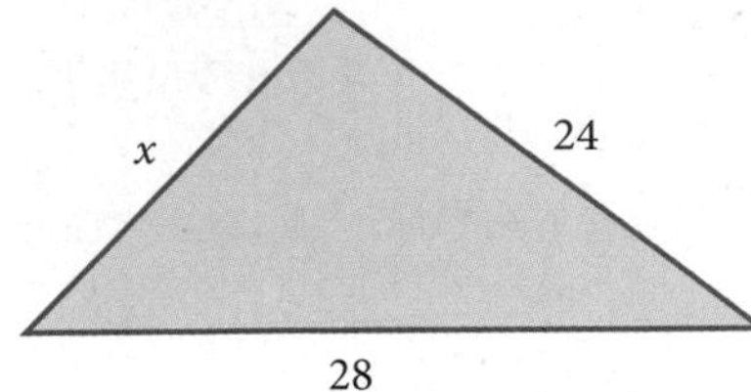

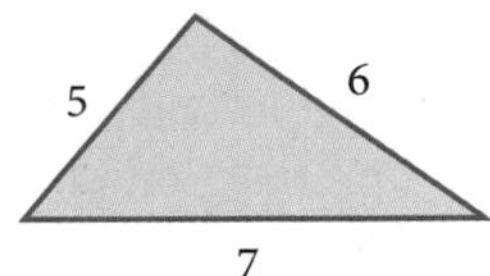

SOLUTION To find the length x, we set up a proportion of equal ratios. The ratio of x to 5 is equal to the ratio of 24 to 6 and to the ratio of 28 to 7. Algebraically we have

$$\frac{x}{5} = \frac{24}{6} \quad \text{and} \quad \frac{x}{5} = \frac{28}{7}$$

We can solve either proportion to get our answer. The first gives us

$$\frac{x}{5} = 4 \qquad \frac{24}{6} = 4$$

$$x = 4 \cdot 5 \qquad \text{Multiply both sides by 5}$$

$$x = 20 \qquad \text{Simplify}$$

6. The two triangles below are similar. Find the missing side, x.

(Triangles: 10, 14; 25, x)

Getting Ready for Class

After reading through the preceding section, respond in your own words and in complete sentences.

1. What is a proportion?
2. What are the means and extremes of a proportion?
3. What is the relationship between the means and the extremes in a proportion? (It is called the means-extremes property of proportions.)
4. How are ratios and proportions related?

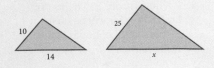

Answer
6. 35

Problem Set 7.7

A Solve each of the following proportions. [Examples 1–2]

1. $\frac{x}{2} = \frac{6}{12}$

2. $\frac{x}{4} = \frac{6}{8}$

3. $\frac{2}{5} = \frac{4}{x}$

4. $\frac{3}{8} = \frac{9}{x}$

5. $\frac{10}{20} = \frac{20}{x}$

6. $\frac{15}{60} = \frac{60}{x}$

7. $\frac{a}{3} = \frac{5}{12}$

8. $\frac{a}{2} = \frac{7}{20}$

9. $\frac{2}{x} = \frac{6}{7}$

10. $\frac{4}{x} = \frac{6}{7}$

11. $\frac{x+1}{3} = \frac{4}{x}$

12. $\frac{x+1}{6} = \frac{7}{x}$

13. $\frac{x}{2} = \frac{8}{x}$

14. $\frac{x}{9} = \frac{4}{x}$

15. $\frac{4}{a+2} = \frac{a}{2}$

16. $\frac{3}{a+2} = \frac{a}{5}$

17. $\frac{1}{x} = \frac{x-5}{6}$

18. $\frac{1}{x} = \frac{x-6}{7}$

B Applying the Concepts [Examples 3–5]

19. **Google Earth** The Google Earth image shows the energy consumption for parts of Europe. In 2006, the ratio of the oil consumption of Switzerland to the oil consumption of the Czech Republic was 9 to 7. If the oil consumption of the Czech Republic was 9.8 million metric tons, what was the oil consumption of Switzerland?

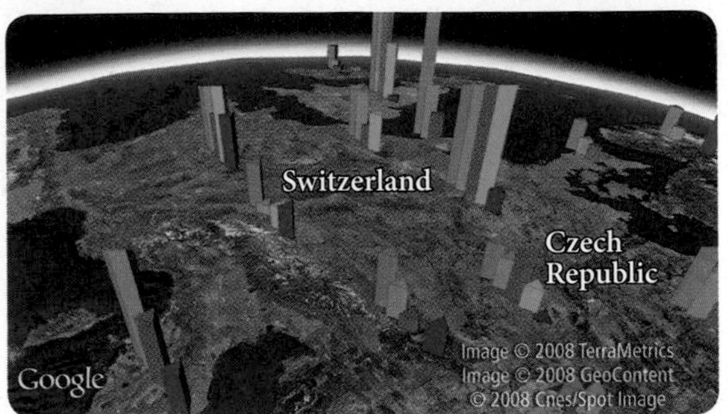

20. **Eiffel Tower** The Eiffel Tower at the Paris Las Vegas Hotel is a replica of the Eiffel Tower in France. The heights of the Eiffel Tower in Las Vegas and the one in France are 460 feet and 1,063 feet respectively. The base of the Eiffel Tower in France is 410 feet wide. What is the width of the base of the Eiffel Tower in Las Vegas? Round to the nearest foot.

21. **Baseball** A baseball player gets 6 hits in the first 18 games of the season. If he continues hitting at the same rate, how many hits will he get in the first 45 games?

22. **Basketball** A basketball player makes 8 of 12 free throws in the first game of the season. If she shoots with the same accuracy in the second game, how many of the 15 free throws she attempts will she make?

23. **Mixture Problem** A solution contains 12 milliliters of alcohol and 16 milliliters of water. If another solution is to have the same concentration of alcohol in water but is to contain 28 milliliters of water, how much alcohol must it contain?

24. **Mixture Problem** A solution contains 15 milliliters of HCl and 42 milliliters of water. If another solution is to have the same concentration of HCl in water but is to contain 140 milliliters of water, how much HCl must it contain?

25. **Nutrition** If 100 grams of ice cream contains 13 grams of fat, how much fat is in 350 grams of ice cream?

26. **Nutrition** A 6-ounce serving of grapefruit juice contains 159 grams of water. How many grams of water are in 20 ounces of grapefruit juice?

27. **Map Reading** A map is drawn so that every 3.5 inches on the map corresponds to an actual distance of 100 miles. If the actual distance between the two cities is 420 miles, how far apart are they on the map?

28. **Map Reading** The scale on a map indicates that 1 inch on the map corresponds to an actual distance of 105 miles. Two cities are 4.5 inches apart on the map. What is the actual distance between the two cities?

On the map shown here, 0.5 inches on the map is equal to 5 miles. Use the information from the map to work Problems 29 through 32.

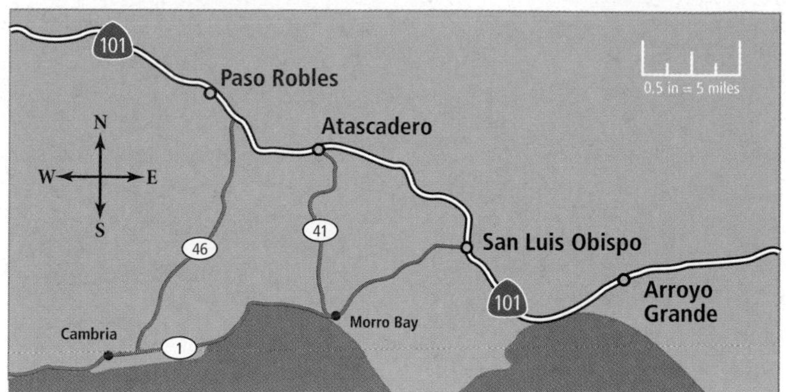

29. **Map Reading** Suppose San Luis Obispo is 1.25 inches from Arroyo Grande on the map. How far apart are the two cities?

30. **Map Reading** Suppose San Luis Obispo is 3.4 inches from Paso Robles on the map. How far apart are the two cities?

31. **Driving Time** If Ava drives from Paso Robles to San Luis Obispo in 46 minutes, how long will it take her to drive from San Luis Obispo to Arroyo Grande, if she drives at the same speed? Round to the nearest minute.

32. **Driving Time** If Brooke drives from Arroyo Grande to San Luis Obispo in 15 minutes, how long will it take her to drive from San Luis Obispo to Paso Robles, if she drives at the same speed? Round to the nearest minute.

33. **Distance** A man drives his car 245 miles in 5 hours. At this rate, how far will he travel in 7 hours?

34. **Distance** An airplane flies 1,380 miles in 3 hours. How far will it fly in 5 hours?

For each pair of similar triangles, set up a proportion in order to find the unknown. **[Example 6]**

35.

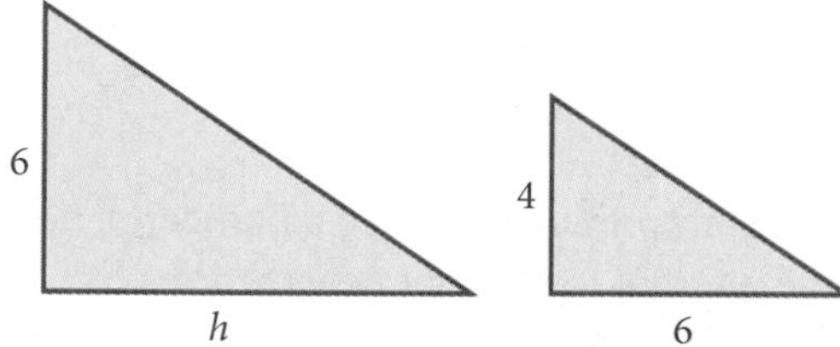

36.

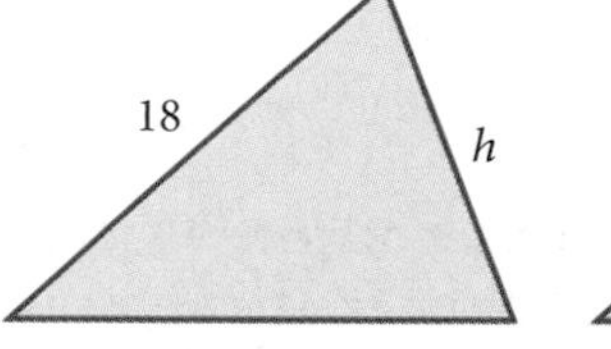

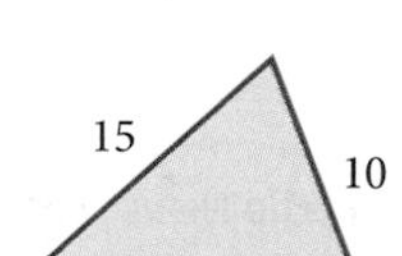

37.

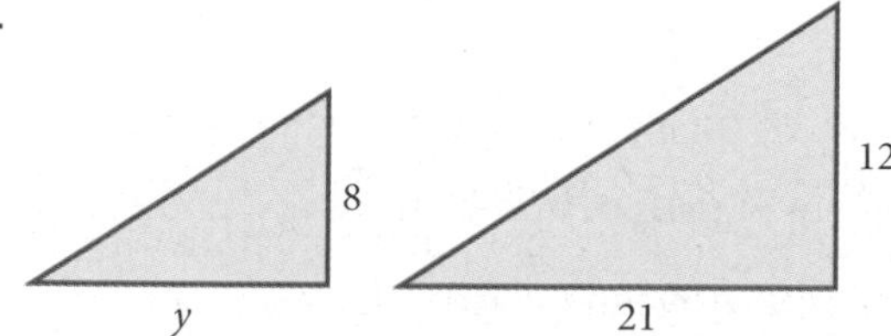

38.

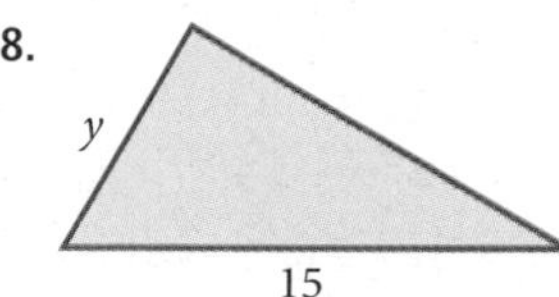

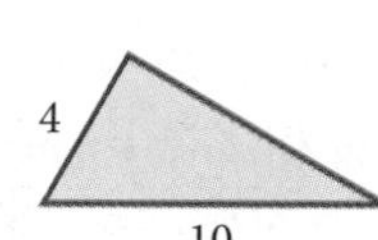

Getting Ready for the Next Section

Use the formula $y = 2x^2$ to find x when

39. $y = 50$

40. $y = 72$

Use the formula $y = Kx$ to find K when

41. $y = 15$ and $x = 3$

42. $y = 72$ and $x = 4$

Use the formula $y = Kx^2$ to find K when

43. $y = 32$ and $x = 4$

44. $y = 45$ and $x = 3$

Maintaining Your Skills

Reduce to lowest terms.

45. $\dfrac{x^2 - x - 6}{x^2 - 9}$

46. $\dfrac{xy + 5x + 3y + 15}{x^2 + ax + 3x + 3a}$

Multiply or divide, as indicated.

47. $\dfrac{x^2 - 25}{x + 4} \cdot \dfrac{2x + 8}{x^2 - 9x + 20}$

48. $\dfrac{3x + 6}{x^2 + 4x + 3} \div \dfrac{x^2 + x - 2}{x^2 + 2x - 3}$

Add or subtract, as indicated.

49. $\dfrac{x}{x^2 - 16} + \dfrac{4}{x^2 - 16}$

50. $\dfrac{2}{x^2 - 1} - \dfrac{5}{x^2 + 3x - 4}$

51. Proportions in the News Searching Google News for news articles that involve proportions produce a number of items involving proportions, the top three of which are shown below. Search the Internet for current news articles by searching on the word *proportion*. Then find an article that the material in this section helps you understand.

Callahan Decries Commotion Over Gesture
. . . "I don't use that type of demeanor, and I never have. This is way blown out of **proportion.** I don't know where we get all these . . .
Rapid City Journal, SD, November 2, 2005

Council to Look Again at Inn Plans
. . . meeting months ago. "On paper the size of the inn seemed to be all out of **proportion** with the rest of the park," he said. "It was . . .
Wairarapa Times Age, New Zealand, November 2, 2005

The Dilemma of Halting Clinical Trials Early
. . . What's more, the authors found, a large **proportion** of the trials are funded by the drugs and devices industry, and involve drugs with large market potential in . . .
MedPage Today, NJ, November 1, 2005

Objectives

A Solve problems involving direct and inverse variation.

Examples now playing at
MathTV.com/books

A Variation

Two variables are said to *vary directly* if one is a constant multiple of the other. For instance, y varies directly as x if $y = Kx$, where K is a constant. The constant K is called the constant of variation. The following table gives the relation between direct variation statements and their equivalent algebraic equations.

Statement	Equation (K = constant of variation)
y varies directly as x	$y = Kx$
y varies directly as the square of x	$y = Kx^2$
s varies directly as the square root of t	$s = K\sqrt{t}$
r varies directly as the cube of s	$r = Ks^3$

Any time we run across a statement similar to those in the table, we immediately can write an equivalent equation involving variables and a constant of variation K.

EXAMPLE 1 Suppose y varies directly as x. When y is 15, x is 3. Find y when x is 4.

SOLUTION From the first sentence we can write the relationship between x and y as:

$$y = Kx$$

We now use the second sentence to find the value of K. Since y is 15 when x is 3, we have:

$$15 = K(3) \quad \text{or} \quad K = 5$$

Now we can rewrite the relationship between x and y more specifically as:

$$y = 5x$$

To find the value of y when x is 4 we simply substitute $x = 4$ into our last equation. Substituting:

$$x = 4$$

into:

$$y = 5x$$

we have:

$$y = 5(4)$$

$$y = 20$$

PRACTICE PROBLEMS

1. y varies directly as x. When y is 12, x is 4. Find y when x is 9.

EXAMPLE 2 Suppose y varies directly as the square of x. When x is 4, y is 32. Find x when y is 50.

SOLUTION The first sentence gives us:

$$y = Kx^2$$

Since y is 32 when x is 4, we have:

$$32 = K(4)^2$$

$$32 = 16K$$

$$K = 2$$

2. Suppose y varies directly as the square of x. When y is 45, x is 3. Find x when y is 80.

Answer
1. 27

The equation now becomes:

$$y = 2x^2$$

When y is 50, we have:

$$50 = 2x^2$$

$$25 = x^2$$

$$x = \pm 5$$

There are two possible solutions, $x = 5$ or $x = -5$.

3. The amount of money Bob makes varies directly with the number of hours he works. If he earns $12.30 for 3 hours' work, how much will he earn if he works 8 hours?

EXAMPLE 3 The cost of a certain kind of candy varies directly with the weight of the candy. If 12 ounces of the candy cost $1.68, how much will 16 ounces cost?

SOLUTION Let x = the number of ounces of candy and y = the cost of the candy. Then $y = Kx$. Since y is 1.68 when x is 12, we have:

$$1.68 = K \cdot 12$$

$$K = \frac{1.68}{12}$$

$$= 0.14$$

The equation must be:

$$y = 0.14x$$

When x is 16, we have:

$$y = 0.14(16)$$

$$= 2.24$$

The cost of 16 ounces of candy is $2.24.

Inverse Variation

Two variables are said to *vary inversely* if one is a constant multiple of the reciprocal of the other. For example, y varies inversely as x if $y = \frac{K}{x}$, where K is a real number constant. Again, K is called the constant of variation. The table that follows gives some examples of inverse variation statements and their associated algebraic equations.

Statement	Equation (K = constant of variation)
y varies inversely as x	$y = \frac{K}{x}$
y varies inversely as the square of x	$y = \frac{K}{x^2}$
F varies inversely as the square root of t	$F = \frac{K}{\sqrt{t}}$
r varies inversely as the cube of s	$r = \frac{K}{s^3}$

Every inverse variation statement has an associated inverse variation equation.

Answers
2. ± 4 3. $32.80

EXAMPLE 4 Suppose y varies inversely as x. When y is 4, x is 5. Find y when x is 10.

SOLUTION The first sentence gives us the relationship between x and y:

$$y = \frac{K}{x}$$

We use the second sentence to find the value of the constant K:

$$4 = \frac{K}{5}$$

or:

$$K = 20$$

We can now write the relationship between x and y more specifically as:

$$y = \frac{20}{x}$$

We use this equation to find the value of y when x is 10. Substituting:

$$x = 10$$

into:

$$y = \frac{20}{x}$$

we have:

$$y = \frac{20}{10}$$

$$y = 2$$

4. Suppose y varies inversely with x. When y is 15, x is 5. Find y when x is 9.

EXAMPLE 5 The intensity (I) of light from a source varies inversely as the square of the distance (d) from the source. Ten feet away from the source the intensity is 200 candlepower. What is the intensity 5 feet from the source?

SOLUTION

$$I = \frac{K}{d^2}$$

Since $I = 200$ when $d = 10$, we have:

$$200 = \frac{K}{10^2}$$

$$200 = \frac{K}{100}$$

$$K = 20{,}000$$

The equation becomes:

$$I = \frac{20{,}000}{d^2}$$

5. Using the information in Example 5, find the intensity at 20 feet from the light source.

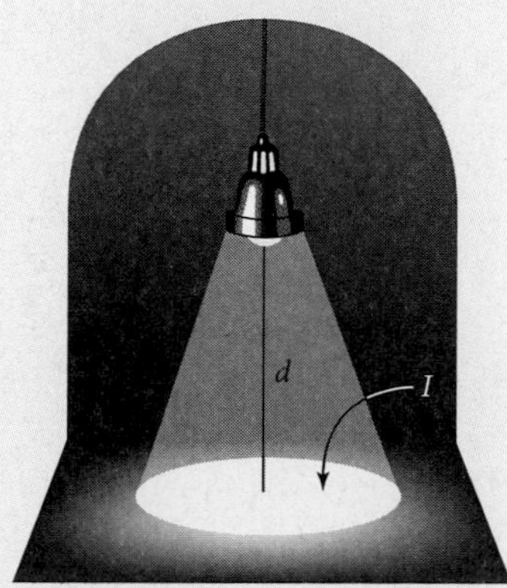

Answer

4. $\frac{25}{3}$

When $d = 5$, we have

$$I = \frac{20,000}{5^2}$$

$$= \frac{20,000}{25}$$

$$= 800 \text{ candlepower}$$

Getting Ready for Class

After reading through the preceding section, respond in your own words and in complete sentences.

1. What does it mean when we say "y varies directly with x"?

2. Give an example of a sentence that is a direct variation statement.

3. Translate the equation $y = \frac{K}{x}$ into words.

4. Give an example of an everyday situation where one quantity varies inversely with another.

Answer

5. 50 candlepower

Problem Set 7.8

A For each of the following problems, y varies directly as x. [Example 1]

1. If $y = 10$ when $x = 5$, find y when x is 4.

2. If $y = 20$ when $x = 4$, find y when x is 11.

3. If $y = 39$ when $x = 3$, find y when x is 10.

4. If $y = -18$ when $x = 6$, find y when x is 3.

5. If $y = -24$ when $x = 4$, find x when y is -30.

6. If $y = 30$ when $x = -15$, find x when y is 8.

7. If $y = -7$ when $x = -1$, find x when y is -21.

8. If $y = 30$ when $x = 4$, find y when x is 7.

For each of the following problems, y varies directly as the square of x. [Example 2]

9. If $y = 75$ when $x = 5$, find y when x is 1.

10. If $y = -72$ when $x = 6$, find y when x is 3.

11. If $y = 48$ when $x = 4$, find y when x is 9.

12. If $y = 27$ when $x = 3$, find x when y is 75.

For each of the following problems, y varies inversely with x. [Example 4]

13. If $y = 5$ when $x = 2$, find y when x is 5.

14. If $y = 2$ when $x = 10$, find y when x is 4.

15. If $y = 2$ when $x = 1$, find y when x is 4.

16. If $y = 4$ when $x = 3$, find y when x is 6.

17. If $y = 5$ when $x = 3$, find x when y is 15.

18. If $y = 12$ when $x = 10$, find x when y is 60.

19. If $y = 10$ when $x = 10$, find x when y is 20.

20. If $y = 15$ when $x = 2$, find x when y is 6.

A For each of the following problems, y varies inversely as the square of x.

21. If $y = 4$ when $x = 5$, find y when x is 2.

22. If $y = 5$ when $x = 2$, y when x is 6.

23. If $y = 4$ when $x = 3$, find y when x is 2.

24. If $y = 9$ when $x = 4$, find y when x is 3.

A Applying the Concepts [Examples 3, 5]

25. **Cars** The chart shows the fastest cars in America. The time that it takes a car to travel a given distance varies inversely with the speed of the car. If an Evans 487 traveled a certain distance in 2 hours, how long would it take the Saleen S7 Twin Turbo to travel the same distance?

Ready for the Races

Ford GT 205 mph
Evans 487 210 mph
Saleen S7 Twin Turbo 260 mph
SSC Ultimate Aero 273 mph

Source: Forbes.com

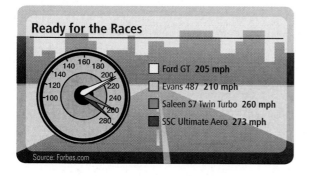

26. **Mountains** The map shows the heights of the tallest mountains in the world. The height of a mountain in feet is directly proportional to its height in miles. If Mount Everest is 5.5 miles high, what is the height of Kangchenjunga? Round to the nearest tenth of a mile.

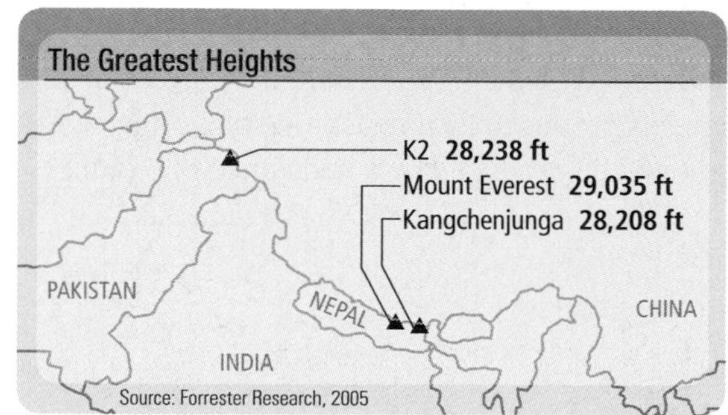

Source: Forrester Research, 2005

27. **Tension in a Spring** The tension t in a spring varies directly with the distance d the spring is stretched. If the tension is 42 pounds when the spring is stretched 2 inches, find the tension when the spring is stretched twice as far.

28. **Fill Time** The time t it takes to fill a bucket varies directly with the volume g of the bucket. If it takes 1 minute to fill a 4-gallon bucket, how long will it take to fill a 6-gallon bucket?

29. **Electricity** The power P in an electric circuit varies directly with the square of the current I. If $P = 30$ when $I = 2$, find P when $I = 7$.

30. **Electricity** The resistance R in an electric circuit varies directly with the voltage V. If $R = 20$ when $V = 120$, find R when $V = 240$.

31. **Wages** The amount of money M a woman makes per week varies directly with the number of hours h she works per week. If she works 20 hours and earns \$157, how much does she make if she works 30 hours?

32. **Volume** The volume V of a gas varies directly as the temperature T. If $V = 3$ when $T = 150$, find V when T is 200.

33. **Weight** The weight F of a body varies inversely with the square of the distance d between the body and the center of the Earth. If a man weighs 150 pounds 4,000 miles from the center of the Earth, how much will he weigh at a distance of 5,000 miles from the center of the Earth?

34. **Light Intensity** The intensity I of a light source varies inversely with the square of the distance d from the source. Four feet from the source, the intensity is 9 foot-candles. What is the intensity 3 feet from the source?

35. **Electricity** The current I in an electric circuit varies inversely with the resistance R. If a current of 30 amperes is produced by a resistance of 2 ohms, what current will be produced by a resistance of 5 ohms?

36. **Pressure** The pressure exerted by a gas on the container in which it is held varies inversely with the volume of the container. A pressure of 40 pounds per square inch is exerted on a container of volume 2 cubic feet. What is the pressure on a container whose volume is 8 cubic feet?

Maintaining Your Skills

Solve each system of equations by the elimination method.

37. $2x + y = 3$
    $3x - y = 7$

38. $3x - y = -6$
    $4x + y = -8$

39. $4x - 5y = 1$
    $x - 2y = -2$

40. $6x - 4y = 2$
    $2x + y = 10$

Solve by the substitution method.

41. $5x + 2y = 7$
    $y = 3x - 2$

42. $-7x - 5y = -1$
    $y = x + 5$

43. $2x - 3y = 4$
    $x = 2y + 1$

44. $4x - 5y = 2$
    $x = 2y - 1$

Chapter 7 Summary

Rational Numbers [7.1]

Any number that can be put in the form $\frac{a}{b}$, where a and b are integers ($b \neq 0$), is called a rational number.

Multiplying or dividing the numerator and denominator of a rational number by the same nonzero number never changes the value of the rational number.

EXAMPLES

1. We can reduce $\frac{6}{8}$ to lowest terms by dividing the numerator and denominator by their greatest common factor 2:

$$\frac{6}{8} = \frac{\cancel{2} \cdot 3}{\cancel{2} \cdot 4} = \frac{3}{4}$$

Rational Expressions [7.1]

Any expression of the form $\frac{P}{Q}$, where P and Q are polynomials ($Q \neq 0$), is a rational expression.

Multiplying or dividing the numerator and denominator of a rational expression by the same nonzero quantity always produces a rational expression equivalent to the original one.

2. We reduce rational expressions to lowest terms by factoring the numerator and denominator and then dividing out any factors they have in common:

$$\frac{x-3}{x^2-9} = \frac{\cancel{x-3}}{(\cancel{x-3})(x+3)} = \frac{1}{x+3}$$

Multiplication [7.2]

To multiply two rational numbers or two rational expressions, multiply numerators, multiply denominators, and divide out any factors common to the numerator and denominator:

For rational numbers $\frac{a}{b}$ and $\frac{c}{d}$, $\quad \frac{a}{b} \cdot \frac{c}{d} = \frac{ac}{bd}$

For rational expressions $\frac{P}{Q}$ and $\frac{R}{S}$, $\quad \frac{P}{Q} \cdot \frac{R}{S} = \frac{PR}{QS}$

3. $$\frac{x-1}{x^2+2x-3} \cdot \frac{x^2-9}{x-2}$$

$$= \frac{\cancel{x-1}}{(\cancel{x+3})(\cancel{x-1})} \cdot \frac{(x-3)(\cancel{x+3})}{x-2}$$

$$= \frac{x-3}{x-2}$$

Division [7.2]

To divide by a rational number or rational expression, simply multiply by its reciprocal:

For rational numbers $\frac{a}{b}$ and $\frac{c}{d}$, $\quad \frac{a}{b} \div \frac{c}{d} = \frac{a}{b} \cdot \frac{d}{c}$

For rational expressions $\frac{P}{Q}$ and $\frac{R}{S}$, $\quad \frac{P}{Q} \div \frac{R}{S} = \frac{P}{Q} \cdot \frac{S}{R}$

4. $$\frac{2x}{x^2-25} \div \frac{4}{x-5}$$

$$= \frac{2x}{(\cancel{x-5})(x+5)} \cdot \frac{(\cancel{x-5})}{4}$$

$$= \frac{x}{2(x+5)}$$

Addition [7.3]

To add two rational numbers or rational expressions, find a common denominator, change each expression to an equivalent expression having the common denominator, and then add numerators and reduce if possible:

For rational numbers $\frac{a}{c}$ and $\frac{b}{c}$, $\quad \frac{a}{c} + \frac{b}{c} = \frac{a+b}{c}$

For rational expressions $\frac{P}{S}$ and $\frac{Q}{S}$, $\quad \frac{P}{S} + \frac{Q}{S} = \frac{P+Q}{S}$

5. $$\frac{3}{x-1} + \frac{x}{2}$$

$$= \frac{3}{x-1} \cdot \frac{\mathbf{2}}{\mathbf{2}} + \frac{x}{2} \cdot \frac{\mathbf{x-1}}{\mathbf{x-1}}$$

$$= \frac{6}{2(x-1)} + \frac{x^2-x}{2(x-1)}$$

$$= \frac{x^2-x+6}{2(x-1)}$$

Subtraction [7.3]

6. $\frac{x}{x^2 - 4} - \frac{2}{x^2 - 4}$

$= \frac{x - 2}{x^2 - 4}$

$= \frac{\cancel{x-2}}{(\cancel{x-2})(x + 2)}$

$= \frac{1}{x + 2}$

To subtract a rational number or rational expression, simply add its opposite:

For rational numbers $\frac{a}{c}$ and $\frac{b}{c}$, $\frac{a}{c} - \frac{b}{c} = \frac{a}{c} + \left(\frac{-b}{c}\right)$

For rational expressions $\frac{P}{S}$ and $\frac{Q}{S}$, $\frac{P}{S} - \frac{Q}{S} = \frac{P}{S} + \left(\frac{-Q}{S}\right)$

Equations [7.4]

7. Solve $\frac{1}{2} + \frac{3}{x} = 5$.

$2x\left(\frac{1}{2}\right) + 2x\left(\frac{3}{x}\right) = 2x(5)$

$x + 6 = 10x$

$6 = 9x$

$x = \frac{2}{3}$

To solve equations involving rational expressions, first find the least common denominator (LCD) for all denominators. Then multiply both sides by the LCD and solve as usual. Check all solutions in the original equation to be sure there are no undefined terms.

Complex Fractions [7.6]

8. $\frac{1 - \frac{4}{x}}{x - \frac{16}{x}} = \frac{x\left(1 - \frac{4}{x}\right)}{x\left(x - \frac{16}{x}\right)}$

$= \frac{x - 4}{x^2 - 16}$

$= \frac{\cancel{x-4}}{(\cancel{x-4})(x + 4)}$

$= \frac{1}{x + 4}$

A rational expression that contains a fraction in its numerator or denominator is called a complex fraction. The most common method of simplifying a complex fraction is to multiply the top and bottom by the LCD for all denominators.

Ratio and Proportion [7.1, 7.7]

9. Solve for x: $\frac{3}{x} = \frac{5}{20}$.

$3 \cdot 20 = 5 \cdot x$

$60 = 5x$

$x = 12$

The ratio of a to b is:

$$\frac{a}{b}$$

Two equal ratios form a proportion. In the proportion

$$\frac{a}{b} = \frac{c}{d}$$

a and d are the *extremes,* and b and c are the *means.* In any proportion the product of the extremes is equal to the product of the means.

Direct Variation [7.8]

10. If y varies directly with the square of x, then:

$y = Kx^2$

The variable y is said to vary directly with the variable x if $y = Kx$, where K is a real number.

Inverse Variation [7.8]

11. If y varies inversely with the cube of x, then

$y = \frac{K}{x^3}$

The variable y is said to vary inversely with the variable x if $y = \frac{K}{x}$, where K is a real number.

Chapter 7 Review

The numbers in brackets refer to the sections of the text in which similar problems can be found. Reduce to lowest terms. Also specify any restriction on the variable. [7.1]

1. $\dfrac{7}{14x - 28}$

2. $\dfrac{a + 6}{a^2 - 36}$

3. $\dfrac{8x - 4}{4x + 12}$

4. $\dfrac{x + 4}{x^2 + 8x + 16}$

5. $\dfrac{3x^3 + 16x^2 - 12x}{2x^3 + 9x^2 - 18x}$

6. $\dfrac{x + 2}{x^4 - 16}$

7. $\dfrac{x^2 + 5x - 14}{x + 7}$

8. $\dfrac{a^2 + 16a + 64}{a + 8}$

9. $\dfrac{xy + bx + ay + ab}{xy + 5x + ay + 5a}$

Multiply or divide as indicated. [7.2]

10. $\dfrac{3x + 9}{x^2} \cdot \dfrac{x^3}{6x + 18}$

11. $\dfrac{x^2 + 8x + 16}{x^2 + x - 12} \div \dfrac{x^2 - 16}{x^2 - x - 6}$

12. $(a^2 - 4a - 12)\left(\dfrac{a - 6}{a + 2}\right)$

13. $\dfrac{3x^2 - 2x - 1}{x^2 + 6x + 8} \div \dfrac{3x^2 + 13x + 4}{x^2 + 8x + 16}$

Find the following sums and differences. [7.3]

14. $\dfrac{2x}{2x + 3} + \dfrac{3}{2x + 3}$

15. $\dfrac{x^2}{x - 9} - \dfrac{18x - 81}{x - 9}$

16. $\dfrac{a + 4}{a + 8} - \dfrac{a - 9}{a + 8}$

17. $\dfrac{x}{x + 9} + \dfrac{5}{x}$

18. $\dfrac{5}{4x + 20} + \dfrac{x}{x + 5}$

19. $\dfrac{3}{x^2 - 36} - \dfrac{2}{x^2 - 4x - 12}$

20. $\dfrac{3a}{a^2 + 8a + 15} - \dfrac{2}{a + 5}$

Solve each equation. [7.4]

21. $\dfrac{3}{x} + \dfrac{1}{2} = \dfrac{5}{x}$

22. $\dfrac{a}{a - 3} = \dfrac{3}{2}$

23. $1 - \dfrac{7}{x} = \dfrac{-6}{x^2}$

24. $\dfrac{3}{x + 6} - \dfrac{1}{x - 2} = \dfrac{-8}{x^2 + 4x - 12}$

25. $\dfrac{2}{y^2 - 16} = \dfrac{10}{y^2 + 4y}$

26. Number Problem The sum of a number and 7 times its reciprocal is $\frac{16}{3}$. Find the number. [7.5]

27. Distance, Rate, and Time A boat travels 48 miles up a river in the same amount of time it takes to travel 72 miles down the same river. If the current is 3 miles per hour, what is the speed of the boat in still water? [7.5]

28. Filling a Pool An inlet pipe can fill a pool in 21 hours, whereas an outlet pipe can empty it in 28 hours. If both pipes are left open, how long will it take to fill the pool? [7.5]

Simplify each complex fraction. [7.6]

29. $\dfrac{\dfrac{x+4}{x^2-16}}{\dfrac{2}{x-4}}$

30. $\dfrac{1-\dfrac{9}{y^2}}{1+\dfrac{4}{y}-\dfrac{21}{y^2}}$

31. $\dfrac{\dfrac{1}{a-2}+4}{\dfrac{1}{a-2}+1}$

32. Write the ratio of 40 to 100 as a fraction in lowest terms. [7.7]

33. If there are 60 seconds in 1 minute, what is the ratio of 40 seconds to 3 minutes? [7.7]

Solve each proportion. [7.7]

34. $\dfrac{x}{9}=\dfrac{4}{3}$

35. $\dfrac{a}{3}=\dfrac{12}{a}$

36. $\dfrac{8}{x-2}=\dfrac{x}{6}$

Work the following problems involving variation. [7.8]

37. y varies directly as x. If $y=-20$ when $x=4$, find y when $x=7$.

38. y varies inversely with x. If $y=3$ when $x=2$, find y when $x=12$.

Chapter 7 Cumulative Review

Simplify.

1. $8 - 11$
2. $-20 + 14$
3. $\frac{-48}{12}$
4. $\frac{1}{6}(-18)$
5. $5x - 4 - 9x$
6. $8 - x - 4$
7. $\frac{x^{-9}}{x^{-13}}$
8. $\left(\frac{x^5}{x^3}\right)^{-2}$
9. $4^1 + 9^0 + (-7)^0$
10. $\frac{(x^{-4})^{-3}(x^{-3})^4}{x^0}$
11. $4x - 7x$
12. $(4a^3 - 10a^2 + 6) - (6a^3 + 5a - 7)$
13. $\frac{x^2}{x - 7} - \frac{14x - 49}{x - 7}$
14. $\frac{6}{10x + 30} + \frac{x}{x + 3}$

Solve each equation.

15. $4x - 3 = 8x + 5$
16. $\frac{3}{4}(8x - 12) = \frac{1}{2}(4x + 4)$
17. $98r^2 - 18 = 0$
18. $6x^4 = 33x^3 - 42x^2$
19. $\frac{5}{x} - \frac{1}{3} = \frac{3}{x}$
20. $\frac{4}{x - 5} - \frac{3}{x + 2} = \frac{28}{x^2 - 3x - 10}$

Solve each system.

21. $x + 2y = 1$
    $x - y = 4$
22. $9x + 14y = -4$
    $15x - 8y = 9$
23. $x = 4y - 3$
    $2x - 5y = 9$
24. $\frac{1}{2}x + \frac{1}{3}y = -1$
    $\frac{1}{3}x = \frac{1}{4}y + 5$

Graph each equation on a rectangular coordinate system.

25. $y = -3x + 2$
26. $y = \frac{1}{3}x$

Solve each inequality

27. $-\frac{a}{3} \le -2$
28. $-3x < 9$

Factor completely.

29. $xy + 5x + ay + 5a$
30. $a^2 + 2a - 35$
31. $20y^2 - 27y + 9$
32. $4r^2 - 9t^2$
33. $3x^2 + 12y^2$
34. $16x^2 + 72xy + 81y^2$

Solve the following systems by graphing.

35. $2x + y = 4$
    $3x - 2y = 6$
36. $4x + 7y = -1$
    $3x - 2y = -8$

37. For the equation $2x + 5y = 10$, find the x- and y-intercepts.
38. For the equation $3x - y = 4$, find the slope and y-intercept.
39. Find the equation for the line with slope -5 and y-intercept -1.
40. Find the equation for the line with slope $-\frac{2}{5}$ that passes through $(-2, -1)$.
41. Subtract -2 from 6.
42. Add -2 to the product of -3 and 4.

Give the property that justifies the statement.

43. $(2 + x) + 3 = 2 + (x + 3)$
44. $(2 + x) + 3 = (x + 2) + 3$
45. Divide $\frac{x^2 - 3x - 28}{x + 4}$.
46. Multiply $\frac{6x - 12}{6x + 12} \cdot \frac{3x + 3}{12x - 24}$.

Reduce to lowest terms

47. $\frac{5a + 10}{10a + 20}$
48. $\frac{2xy + 10x + 3y + 15}{3xy + 15x + 2y + 10}$

49. **Direct Variation** y varies directly as the cube of x. If $y = -32$ when $x = 2$, find y when $x = 3$.
50. **Inverse Variation** y varies inversely with x. If $y = 12$ when $x = 2$, find y when $x = 6$.

The illustration shows the volume of soda, in gallons, consumed annually per capita. (Note: 1 gallon ≈ 3.79 liters.)

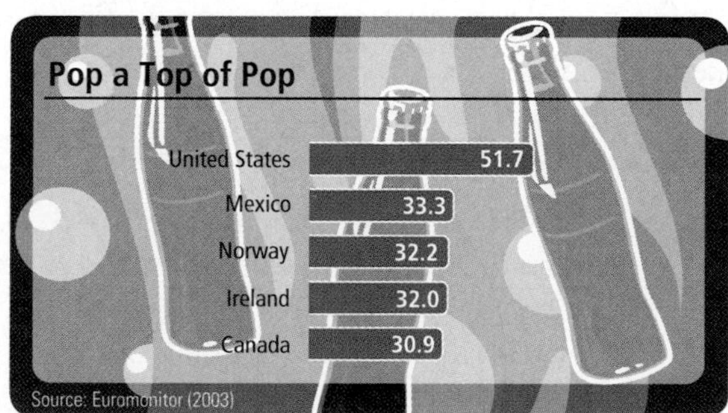

51. How many liters of soda does the U.S. consume annually per capita?
52. How many liters of soda does Ireland consume anually per capita?

Chapter 7 Test

Reduce to lowest terms. [7.1]

1. $\dfrac{x^2 - 16}{x^2 - 8x + 16}$

2. $\dfrac{10a + 20}{5a^2 + 20a + 20}$

3. $\dfrac{xy + 7x + 5y + 35}{x^2 + ax + 5x + 5a}$

Multiply or divide as indicated. [7.2]

4. $\dfrac{3x - 12}{4} \cdot \dfrac{8}{2x - 8}$

5. $\dfrac{x^2 - 49}{x + 1} \div \dfrac{x + 7}{x^2 - 1}$

6. $\dfrac{x^2 - 3x - 10}{x^2 - 8x + 15} \div \dfrac{3x^2 + 2x - 8}{x^2 + x - 12}$

7. $(x^2 - 9)\left(\dfrac{x + 2}{x + 3}\right)$

Add or subtract as indicated. [7.3]

8. $\dfrac{3}{x - 2} - \dfrac{6}{x - 2}$

9. $\dfrac{x}{x^2 - 9} + \dfrac{4}{4x - 12}$

10. $\dfrac{2x}{x^2 - 1} + \dfrac{x}{x^2 - 3x + 2}$

Solve the following equations. [7.4]

11. $\dfrac{7}{5} = \dfrac{x + 2}{3}$

12. $\dfrac{10}{x + 4} = \dfrac{6}{x} - \dfrac{4}{x}$

13. $\dfrac{3}{x - 2} - \dfrac{4}{x + 1} = \dfrac{5}{x^2 - x - 2}$

Solve the following problems. [7.5]

14. Speed of a Boat A boat travels 26 miles up a river in the same amount of time it takes to travel 34 miles down the same river. If the current is 2 miles per hour, what is the speed of the boat in still water?

15. Emptying a Pool An inlet pipe can fill a pool in 15 hours, whereas an outlet pipe can empty it in 12 hours. If the pool is full and both pipes are open, how long will it take to empty?

Solve the following problems involving ratio and proportion. [7.7]

16. Ratio A solution of alcohol and water contains 27 milliliters of alcohol and 54 milliliters of water. What is the ratio of alcohol to water and the ratio of alcohol to total volume?

17. Ratio A manufacturer knows that during a production run 8 out of every 100 parts produced by a certain machine will be defective. If the machine produces 1,650 parts, how many can be expected to be defective?

Simplify each complex fraction. [7.6]

18. $\dfrac{1 + \dfrac{1}{x}}{1 - \dfrac{1}{x}}$

19. $\dfrac{1 - \dfrac{16}{x^2}}{1 - \dfrac{2}{x} - \dfrac{8}{x^2}}$

20. Direct Variation Suppose y varies directly with the square of x. If y is 36 when x is 3, find y when x is 5. [7.8]

21. Inverse Variation If y varies inversely with x, and y is 6 when x is 3, find y when x is 9. [7.8]

22. Ratio Use the illustration below to find the ratio of the sound emitted by a blue whale to the sound of normal conversation. Round to the nearest tenth. [7.1]

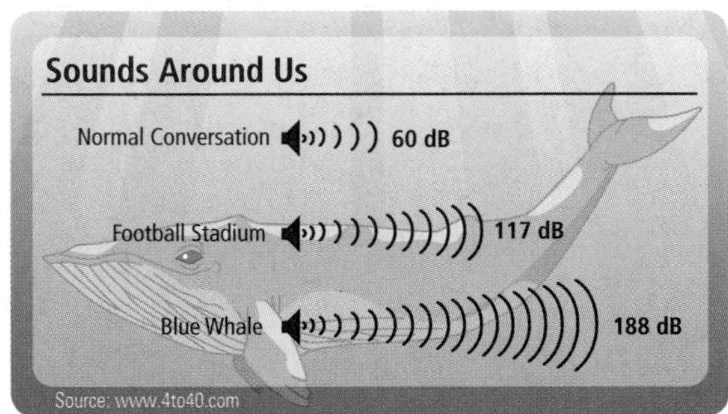

23. Unit Analysis The illustration below shows the speeds of the four fastest American cars. Convert each speed to feet per second to the nearest tenth. [7.2]

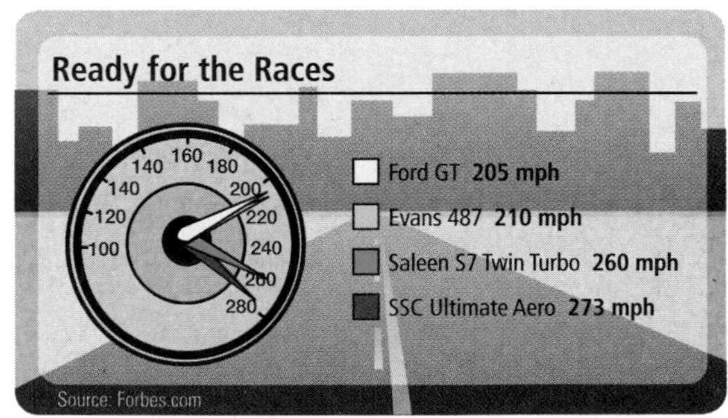

Chapter 7 Projects

RATIONAL EXPRESSIONS

GROUP PROJECT

Kayak Race

Number of People 2-3

Time Needed 20 minutes

Equipment Paper and pencil

Background In a kayak race, the participants must paddle a kayak 450 meters down a river and then return 450 meters up the river to the starting point (see figure). Susan has deduced correctly that the total time t (in seconds) depends on the speed c (in meters per second) of the water according to the following expression:

$$t = \frac{450}{v + c} + \frac{450}{v - c}$$

where v is the speed of the kayak relative to the water (the speed of the kayak in still water).

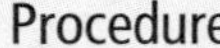

Procedure 1. Fill in the following table.

Time t (seconds)	Speed of Kayak Relative to the Water v (meters/second)	Current of the River c (meters/second)
240		1
300		2
	4	3
	3	1
540	3	
	3	3

2. If the kayak race were conducted in the still waters of a lake, do you think that the total time of a given participant would be greater than, equal to, or smaller than the time in the river? Justify your answer.

3. Suppose Peter can paddle his kayak at 4.1 meters per second and that the speed of the current is 4.1 meters per second. What will happen when Peter makes the turn and tries to come back up the river? How does this situation show up in the equation for total time?

RESEARCH PROJECT

Bertrand Russell

Here is a quote taken from the beginning of the first sentence in the book *Principles of Mathematics* by the British philosopher and mathematician, Bertrand Russell.

> Pure Mathematics is the class of all propositions of the form "p implies q," where p and q are propositions containing one or more variables . . .

He is using the phrase "p implies q" in the same way mathematicians use the phrase "If A, then B." Conditional statements are an introduction to the foundations on which all of mathematics is built.

Write an essay on the life of Bertrand Russell. In the essay, indicate what purpose he had for writing and publishing his book *Principles of Mathematics.* Write in complete sentences and organize your work just as you would if you were writing a paper for an English class.

Bettmann/Corbis

Bertrand Russell

Roots and Radicals

8

Chapter Outline

Introduction

The Petronas Towers, shown in the Google Earth photograph, are the tallest twin buildings in the world. They are in Kuala Lumpur, Malaysia. The towers themselves are very complex structures, yet they are built on a simple design involving two overlapping squares, as shown below.

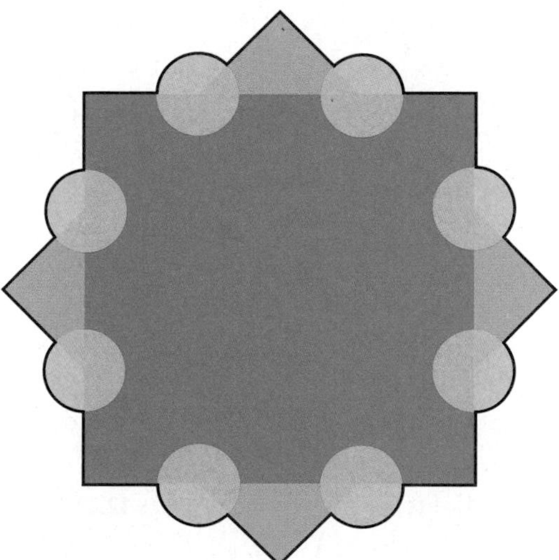

Complex structures often arise out of very simple geometric concepts. In this chapter we discover an interesting geometric figure, the golden rectangle, that is constructed from a square. Although the golden rectangle starts from a simple square, it becomes more interesting and complex the more closely we examine it.

Chapter Pretest

Find the following roots. [8.1]

1. $\sqrt{36}$
2. $-\sqrt{81}$
3. The square root of 121
4. $\sqrt[3]{64}$
5. $\sqrt[3]{27}$
6. $-\sqrt[4]{625}$

Put the following expressions into simplified form. [8.2, 8.3]

7. $\sqrt{72}$
8. $\sqrt{96}$
9. $\sqrt{\frac{3}{5}}$
10. $\frac{6}{\sqrt[3]{9}}$
11. $4\sqrt{12y^2}$
12. $\sqrt{\frac{18x^3y^5}{3}}$

Combine. [8.4]

13. $3\sqrt{20} - \sqrt{45}$
14. $3m\sqrt{12} + 2\sqrt{3m^2}$

Multiply. [8.5]

15. $\sqrt{2}(\sqrt{3} - 3)$
16. $(\sqrt{3} - 5)(\sqrt{3} + 2)$
17. $(\sqrt{x} - 9)(\sqrt{x} + 9)$
18. $(\sqrt{7} - \sqrt{2})^2$

Divide. (Rationalize the denominator.) [8.5]

19. $\frac{\sqrt{5} - \sqrt{7}}{\sqrt{5} + \sqrt{7}}$
20. $\frac{\sqrt{x}}{\sqrt{x} - 3}$

Solve the following equations. [8.6]

21. $\sqrt{4x - 3} - 3 = 4$
22. $\sqrt{2x + 8} = x + 4$
23. $\sqrt{4x - 3} + 8 = 3$

For each equation, complete the given table and then use the table.

24. $y = \sqrt{x} + 2$

x	y
0	
1	
4	
9	

25. $y = \sqrt{x + 2}$

x	y
−2	
−1	
2	
7	

26. $y = \sqrt[3]{x} - 2$

x	y
−8	
−1	
0	
1	
8	

27. $y = \sqrt[3]{x + 2}$

x	y
−10	
−3	
−2	
−1	
6	

Getting Ready for Chapter 8

Expand and simplify the following.

1. 1^2
2. 2^2
3. 3^2
4. 4^2
5. 5^2
6. 6^2
7. 7^2
8. 8^2
9. 9^2
10. 10^2
11. 11^2
12. 12^2
13. 15^2
14. 16^2
15. 2^3
16. 3^3
17. 4^3
18. 5^3

Definitions and Common Roots

8.1

Objectives

A Find a root of a number.

B Find the root of an expression containing a variable.

C Solve an application problem involving roots.

XYZ
Examples now playing at
MathTV.com/books

In Chapter 5 we developed notation (exponents) that would take us from a number to its square. If we wanted the square of 5, we wrote $5^2 = 25$. In this section we will use another type of notation that will take us in the reverse direction, from the square of a number back to the number itself.

In general we are interested in going from a number, say, 49, back to the number we squared to get 49. Since the square of 7 is 49, we say 7 is a square root of 49. The notation we use looks like this:

$$\sqrt{49}$$

NOTATION In the expression $\sqrt{49}$, 49 is called the *radicand;* $\sqrt{}$ is the *radical sign;* and the complete expression, $\sqrt{49}$, is called the *radical.*

> **Definition**
> If x represents any positive real number, then the expression $\sqrt{x}$ is the **positive square root** of x. It is the *positive* number we square to get x.
>
> The expression $-\sqrt{x}$ is the **negative square root** of x. It is the negative number we square to get x.

A Square Roots of Positive Numbers

Every positive number has two square roots, one positive and the other negative. Some books refer to the positive square root of a number as the principal root.

EXAMPLE 1 The positive square root of 25 is 5 and can be written $\sqrt{25} = 5$. The negative square root of 25 is -5 and can be written $-\sqrt{25} = -5$.

If we want to consider the negative square root of a number, we must put a negative sign in front of the radical. It is a common mistake to think of $\sqrt{25}$ as meaning either 5 or -5. The expression $\sqrt{25}$ means the *positive* square root of 25, which is 5. If we want the negative square root, we write $-\sqrt{25}$ to begin with.

EXAMPLE 2 Find the following root: $\sqrt{49}$

SOLUTION $\sqrt{49} = 7$ 7 is the positive number we square to get 49

EXAMPLE 3 Find the following root: $-\sqrt{49}$

SOLUTION $-\sqrt{49} = -7$ -7 is the negative number we square to get 49

EXAMPLE 4 Find the following root: $\sqrt{121}$

SOLUTION $\sqrt{121} = 11$ 11 is the positive number we square to get 121

EXAMPLE 5 The positive square root of 17 is written $\sqrt{17}$. The negative square root of 17 is written $-\sqrt{17}$.

PRACTICE PROBLEMS

1. Name the two square roots of 36.

Find the following roots.

2. $\sqrt{81}$
3. $-\sqrt{81}$
4. $\sqrt{144}$
5. Write the two square roots of 23.

Answers

1. 6, -6 **2.** 9 **3.** -9
4. 12 **5.** $\sqrt{23}$, $-\sqrt{23}$

We have no other exact representation for the two roots in Example 5. Since 17 itself is not a perfect square (the square of an integer), its two square roots, $\sqrt{17}$ and $-\sqrt{17}$, are irrational numbers. They have a place on the real number line but cannot be written as the ratio of two integers. The square roots of any number that is not itself a perfect square are irrational numbers.

6. Fill in the following table, as in Example 6.

Number	Positive Square Root	Negative Square Root
16		
4		
11		
23		
400		

EXAMPLE 6

Number	Positive Square Root	Negative Square Root	Roots Are
9	3	-3	Rational numbers
36	6	-6	Rational numbers
7	$\sqrt{7}$	$-\sqrt{7}$	Irrational numbers
22	$\sqrt{22}$	$-\sqrt{22}$	Irrational numbers
100	10	-10	Rational numbers

Square Root of Zero

The number 0 is the only real number with one square root. It is also its own square root:

$$\sqrt{0} = 0$$

Square Roots of Negative Numbers

Negative numbers have square roots, but their square roots are not real numbers. They do not have a place on the real number line. We will consider square roots of negative numbers later in the book.

7. Is there a real number that is a square root of -25?

EXAMPLE 7 The expression $\sqrt{-4}$ does not represent a real number since there is no real number we can square and end up with -4. The same is true of square roots of any negative number.

Other Roots

There are many other roots of numbers besides square roots, although square roots seem to be the most commonly used. The cube root of a number is the number we cube (raise to the third power) to get the original number. The cube root of 8 is 2 since $2^3 = 8$. The cube root of 27 is 3 since $3^3 = 27$. The notation for cube roots looks like this:

The 3 is called the *index* $\rightarrow \sqrt[3]{8} = 2$

$$\rightarrow \sqrt[3]{27} = 3$$

We can go as high as we want with roots. The fourth root of 16 is 2 because $2^4 = 16$. We can write this in symbols as $\sqrt[4]{16} = 2$.

Here is a list of the most common roots. They are the roots that will come up most often in the remainder of the book, and they should be memorized.

With even roots—square roots, fourth roots, sixth roots, and so on—we cannot have negative numbers *under* the radical sign. With odd roots, negative numbers under the radical sign do not cause problems.

Answers

6. 16: 4, -4; 4: 2, -2; 11: $\sqrt{11}$, $-\sqrt{11}$; 23: $\sqrt{23}$, $-\sqrt{23}$; 400: 20, -20 **7.** No

Square Roots		Cube Roots	Fourth Roots
$\sqrt{1} = 1$	$\sqrt{49} = 7$	$\sqrt[3]{1} = 1$	$\sqrt[4]{1} = 1$
$\sqrt{4} = 2$	$\sqrt{64} = 8$	$\sqrt[3]{8} = 2$	$\sqrt[4]{16} = 2$
$\sqrt{9} = 3$	$\sqrt{81} = 9$	$\sqrt[3]{27} = 3$	$\sqrt[4]{81} = 3$
$\sqrt{16} = 4$	$\sqrt{100} = 10$	$\sqrt[3]{64} = 4$	$\sqrt[4]{256} = 4$
$\sqrt{25} = 5$	$\sqrt{121} = 11$	$\sqrt[3]{125} = 5$	$\sqrt[4]{625} = 5$
$\sqrt{36} = 6$	$\sqrt{144} = 12$		

EXAMPLE 8 Find the following root, if possible: $\sqrt[3]{-8}$

SOLUTION $\sqrt[3]{-8} = -2$ Because $(-2)^3 = -8$

EXAMPLE 9 Find the following root, if possible: $\sqrt[3]{-27}$

SOLUTION $\sqrt[3]{-27} = -3$ Because $(-3)^3 = -27$

EXAMPLE 10 Find the following root, if possible: $\sqrt{-4}$

SOLUTION $\sqrt{-4}$ Not a real number since there is no real number whose square is -4

EXAMPLE 11 Find the following root, if possible: $-\sqrt{4}$

SOLUTION $-\sqrt{4} = -2$ Because -2 is the negative number we square to get 4

EXAMPLE 12 Find the following root, if possible: $\sqrt[4]{-16}$

SOLUTION $\sqrt[4]{-16}$ Not a real number since there is no real number that can be raised to the fourth power to obtain -16

EXAMPLE 13 Find the following root, if possible: $-\sqrt[4]{16}$

SOLUTION $-\sqrt[4]{16} = -2$ Because -2 is the negative number we raise to the fourth power to get 16

B Variables Under the Radical Sign

In this chapter, unless we say otherwise, we will assume that all variables that appear under a radical sign represent positive numbers. That way we can simplify expressions involving radicals that contain variables. Here are some examples.

EXAMPLE 14 Simplify $\sqrt{49x^2}$.

SOLUTION We are looking for the expression we square to get $49x^2$. Since the square of 7 is 49 and the square of x is x^2, we can square $7x$ and get $49x^2$:

$$\sqrt{49x^2} = 7x \quad \text{Because } (7x)^2 = 49x^2$$

Find the following roots, if possible.

8. $\sqrt[3]{-1}$

9. $\sqrt[3]{-125}$

10. $\sqrt{-9}$

11. $-\sqrt{9}$

12. $\sqrt[4]{-81}$

13. $-\sqrt[4]{81}$

At first it may be difficult to see the difference in some of these examples. Generally, we have to be careful with even roots of negative numbers; that is, if the index on the radical is an even number, then we cannot have a negative number under the radical sign. That is why $\sqrt{-4}$ and $\sqrt[4]{-16}$ are not real numbers.

14. Simplify $\sqrt{9a^2}$.

Answers
8. -1 **9.** -5
10. Not a real number **11.** -3
12. Not a real number **13.** -3
14. $3a$

15. Simplify $\sqrt{25x^2y^2}$.

EXAMPLE 15 Simplify $\sqrt{16a^2b^2}$.

SOLUTION We want an expression whose square is $16a^2b^2$. That expression is $4ab$:

$$\sqrt{16a^2b^2} = 4ab \qquad \text{Because } (4ab)^2 = 16a^2b^2$$

16. Simplify $\sqrt[3]{8x^3}$.

EXAMPLE 16 Simplify $\sqrt[3]{125a^3}$.

SOLUTION We are looking for the expression we cube to get $125a^3$. That expression is $5a$:

$$\sqrt[3]{125a^3} = 5a \qquad \text{Because } (5a)^3 = 125a^3$$

17. Simplify $\sqrt{x^8}$.

EXAMPLE 17 Simplify $\sqrt{x^6}$.

SOLUTION The number we square to obtain x^6 is x^3.

$$\sqrt{x^6} = x^3 \qquad \text{Because } (x^3)^2 = x^6$$

18. Simplify $\sqrt[4]{81a^8b^{12}}$.

EXAMPLE 18 Simplify $\sqrt[4]{16a^8b^4}$.

SOLUTION The number we raise to the fourth power to obtain $16a^8b^4$ is $2a^2b$.

$$\sqrt[4]{16a^8b^4} = 2a^2b \qquad \text{Because } (2a^2b)^4 = 16a^8b^4$$

C Applications Involving Roots

There are many application problems involving radicals that require decimal approximations. When we need a decimal approximation to a square root, we can use a calculator.

19. In 1988 you purchased a baseball card for \$56. Two years later you sold the same card for \$91. Find the annual rate of return on your investment to the nearest tenth of a percent.

EXAMPLE 19 If you invest P dollars in an account and after 2 years the account has A dollars in it, then the annual rate of return r on the money you originally invested is given by the formula

$$r = \frac{\sqrt{A} - \sqrt{P}}{\sqrt{P}}$$

Suppose you pay \$65 for a coin collection and find that the same coins sell for \$84 two years later. Find the annual rate of return on your investment.

SOLUTION Substituting $A = 84$ and $P = 65$ in the formula, we have

$$r = \frac{\sqrt{84} - \sqrt{65}}{\sqrt{65}}$$

From either the table of square roots or a calculator, we find that $\sqrt{84} \approx 9.165$ and $\sqrt{65} \approx 8.062$. Using these numbers in our formula gives us

$$r \approx \frac{9.165 - 8.062}{8.062}$$

$$= \frac{1.103}{8.062}$$

$$\approx 0.137 \text{ or } 13.7\%$$

To earn as much as this in a savings account that compounds interest once a year, you have to find an account that pays 13.7% in annual interest.

Answers
15. $5xy$ **16.** $2x$ **17.** x^4 **18.** $3a^2b^3$
19. approximately 0.2748 or 27.5%

The Pythagorean Theorem Again

Now that we have some experience working with square roots, we can rewrite the Pythagorean theorem using a square root. In Figure 1, if triangle *ABC* is a right triangle with $C = 90°$, then the length of the longest side is the *square root* of the sum of the squares of the other two sides.

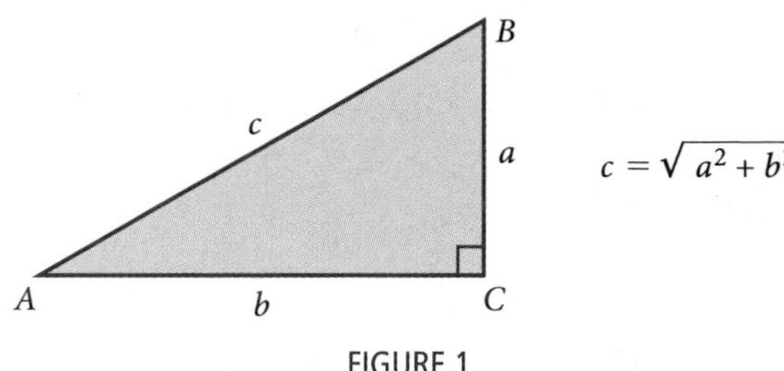

$$c = \sqrt{a^2 + b^2}$$

FIGURE 1

EXAMPLE 20 A tent pole is 8 feet in length and makes an angle of 90° with the ground. One end of a rope is attached to the top of the pole, and the other end of the rope is anchored to the ground 6 feet from the bottom of the pole. Find the length of the rope.

SOLUTION The diagram in Figure 2 is a visual representation of the situation. To find the length of the pole, we apply the Pythagorean theorem.

$$x = \sqrt{6^2 + 8^2}$$
$$x = \sqrt{36 + 64}$$
$$x = \sqrt{100}$$
$$x = 10 \text{ feet}$$

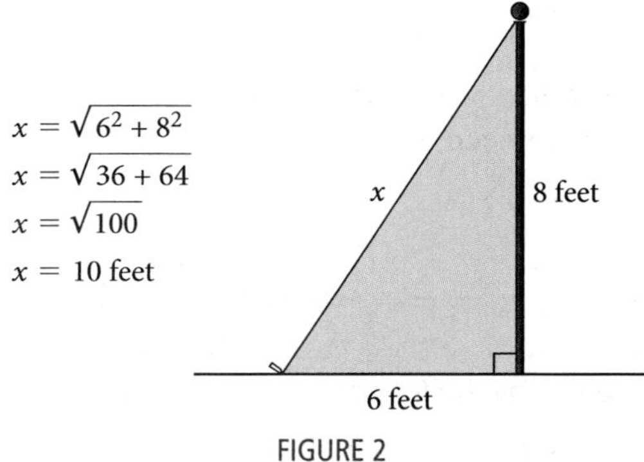

FIGURE 2

20. Find x in the right triangle shown here.

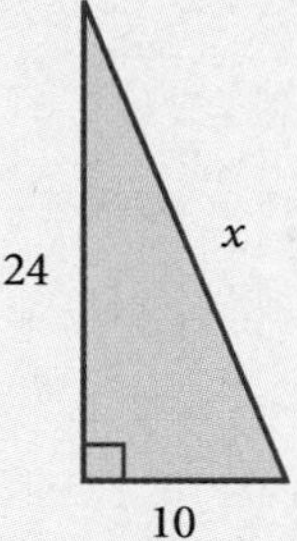

Getting Ready for Class

After reading through the preceding section, respond in your own words and in complete sentences.

1. Every real number has two square roots. Explain the notation we use to tell them apart. Use the square roots of 3 for examples.
2. Explain why a square root of -4 is not a real number.
3. We use the notation $\sqrt{17}$ to represent the positive square root of 17. Explain why there isn't a simpler way to express the positive square root of 17.
4. Write a list of the notation and terms associated with radicals.

Answer
20. 26

Problem Set 8.1

A Find the following roots. If the root does not exist as a real number, write "not a real number." [Examples 1–13]

1. $\sqrt{9}$
2. $\sqrt{16}$
3. $-\sqrt{9}$
4. $-\sqrt{16}$
5. $\sqrt{-25}$
6. $\sqrt{-36}$
7. $-\sqrt{144}$
8. $\sqrt{256}$
9. $\sqrt{625}$
10. $-\sqrt{625}$
11. $\sqrt{-49}$
12. $\sqrt{-169}$
13. $-\sqrt{64}$
14. $-\sqrt{25}$
15. $-\sqrt{100}$
16. $\sqrt{121}$
17. $\sqrt{1{,}225}$
18. $-\sqrt{1{,}681}$
19. $\sqrt[4]{1}$
20. $-\sqrt[4]{81}$
21. $\sqrt[3]{-8}$
22. $\sqrt[3]{125}$
23. $-\sqrt[3]{125}$
24. $-\sqrt[3]{-8}$
25. $\sqrt[3]{-1}$
26. $-\sqrt[3]{-1}$
27. $\sqrt[3]{-27}$
28. $-\sqrt[3]{27}$
29. $-\sqrt[4]{16}$
30. $\sqrt[4]{-16}$

B Assume all variables are positive, and find the following roots. [Examples 14–18]

31. $\sqrt{x^2}$
32. $\sqrt{a^2}$
33. $\sqrt{9x^2}$
34. $\sqrt{25x^2}$
35. $\sqrt{x^2y^2}$
36. $\sqrt{a^2b^2}$
37. $\sqrt{(a+b)^2}$
38. $\sqrt{(x+y)^2}$
39. $\sqrt{49x^2y^2}$
40. $\sqrt{81x^2y^2}$
41. $\sqrt[3]{x^3}$
42. $\sqrt[3]{a^3}$
43. $\sqrt[3]{8x^3}$
44. $\sqrt[3]{27x^3}$
45. $\sqrt{x^4}$
46. $\sqrt{x^6}$
47. $\sqrt{36a^6}$
48. $\sqrt{64a^4}$
49. $\sqrt{25a^8b^4}$
50. $\sqrt{16a^4b^8}$
51. $\sqrt[3]{x^6}$
52. $\sqrt[3]{x^9}$
53. $\sqrt[3]{27a^{12}}$
54. $\sqrt[3]{8a^6}$
55. $\sqrt[4]{x^8}$
56. $\sqrt[4]{x^{12}}$

57. Simplify each expression. Assume all variables are positive.

a. $\sqrt[4]{16}$ **b.** $\sqrt[4]{x^4}$

c. $\sqrt[4]{x^8}$ **d.** $\sqrt[4]{16x^8y^{12}}$

58. Simplify each expression. Assume all variables are positive.

a. $\sqrt[5]{32}$ **b.** $\sqrt[5]{y^5}$

c. $\sqrt[5]{y^{10}}$ **d.** $\sqrt[5]{32x^{10}y^{20}}$

Simplify each expression.

59. $\sqrt{9 \cdot 16}$

60. $\sqrt{64}\sqrt{36}$

61. $\sqrt{25}\sqrt{16}$

62. $\sqrt{144}\sqrt{25}$

63. $\sqrt{9} + \sqrt{16}$

64. $\sqrt{64} + \sqrt{36}$

65. $\sqrt{9 + 16}$

66. $\sqrt{64 + 36}$

67. $\sqrt{144} + \sqrt{25}$

68. $\sqrt{25} - \sqrt{16}$

69. $\sqrt{144 + 25}$

70. $\sqrt{25 - 16}$

71. Use the approximation $\sqrt{5} \approx 2.236$ and find approximations for each expression.

a. $\frac{1 + \sqrt{5}}{2}$

b. $\frac{1 - \sqrt{5}}{2}$

c. $\frac{1 + \sqrt{5}}{2} + \frac{1 - \sqrt{5}}{2}$

72. Use the approximation $\sqrt{3} \approx 1.732$ and find approximations for each expression.

a. $\frac{1 + \sqrt{3}}{2}$

b. $\frac{1 - \sqrt{3}}{2}$

c. $\frac{1 + \sqrt{3}}{2} + \frac{1 - \sqrt{3}}{2}$

73. Evaluate each root.

a. $\sqrt{9}$

b. $\sqrt{900}$

c. $\sqrt{0.09}$

74. Evaluate each root.

a. $\sqrt[3]{27}$

b. $\sqrt[3]{0.027}$

c. $\sqrt[3]{27{,}000}$

Simplify each of the following pairs of expressions.

75. $\frac{5 + \sqrt{49}}{2}$ and $\frac{5 - \sqrt{49}}{2}$

76. $\frac{3 + \sqrt{25}}{4}$ and $\frac{3 - \sqrt{25}}{4}$

77. $\frac{2 + \sqrt{16}}{2}$ and $\frac{2 - \sqrt{16}}{2}$

78. $\frac{3 + \sqrt{9}}{2}$ and $\frac{3 - \sqrt{9}}{2}$

79. We know that the trinomial $x^2 + 6x + 9$ is the square of the binomial $x + 3$; that is,

$$x^2 + 6x + 9 = (x + 3)^2$$

Use this fact to find $\sqrt{x^2 + 6x + 9}$.

80. Use the fact that $x^2 + 10x + 25 = (x + 5)^2$ to find $\sqrt{x^2 + 10x + 25}$.

C Applying the Concepts [Examples 19–20]

81. Google Earth The Google Earth image shows a right triangle between three cities in Southern California. If the distance between Pomona and Ontario is 5.7 miles, and the distance between Ontario and Upland is 3.6 miles, what is the distance between Pomona and Upland? Round to the nearest tenth of a mile.

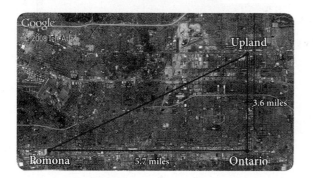

82. Building Problem The higher you are above the ground, the farther you can see. The distance in miles that you can see from h feet above the ground is given by the formula $d = \sqrt{\frac{3h}{2}}$. Use this formula to calculate the distance you can see from the top of each of the buildings shown in the image. Round to the nearest mile.

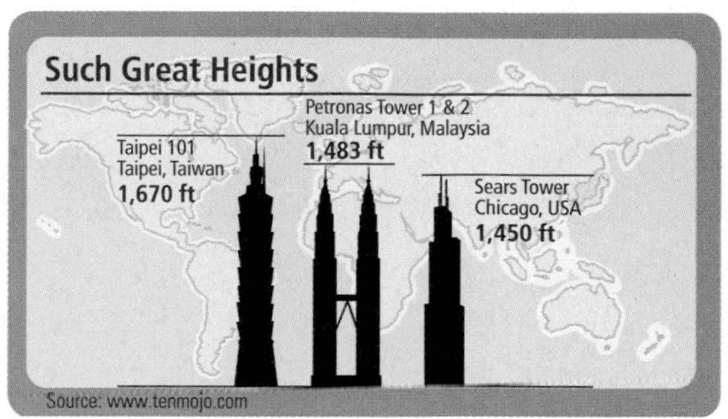

Find x in each of the following right triangles.

83.

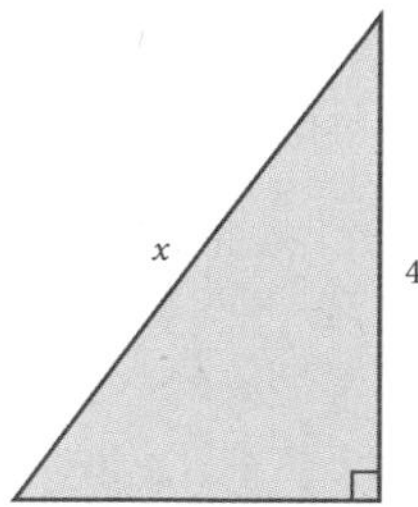

84.

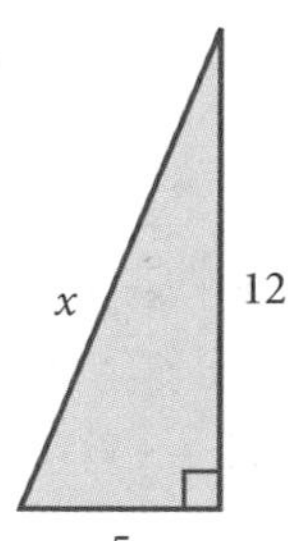

85.

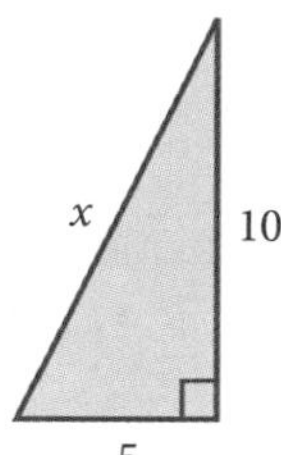

86.

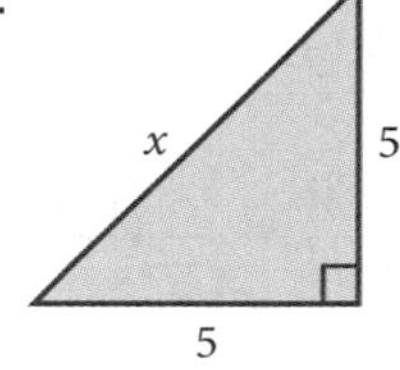

87. Geometry One end of a wire is attached to the top of a 24-foot pole; the other end of the wire is anchored to the ground 18 feet from the bottom of the pole. If the pole makes an angle of 90° with the ground, find the length of the wire.

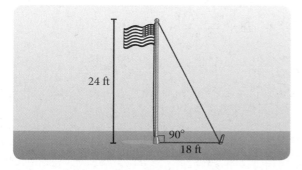

88. Geometry The screen on a television set is in the shape of a rectangle. If the length is 20 inches and the width is 12 inches, how many inches is it from one corner of the screen to the opposite corner? Round to the nearest tenth.

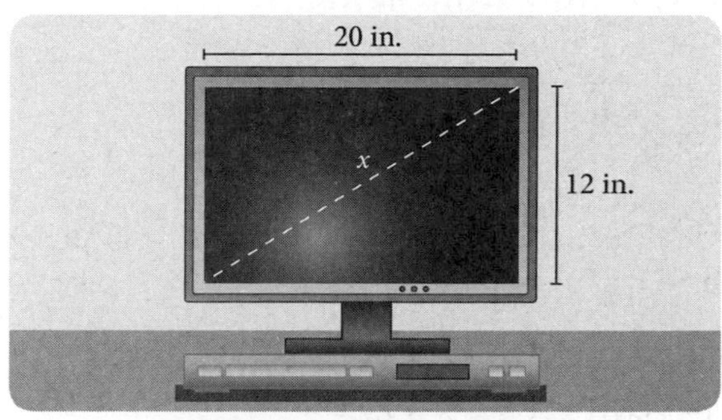

89. Geometry Two children are trying to cross a stream. They want to use a log that goes from one bank to the other. If the left bank is 5 feet higher than the right bank and the stream is 12 feet wide, how long must a log be to just barely reach?

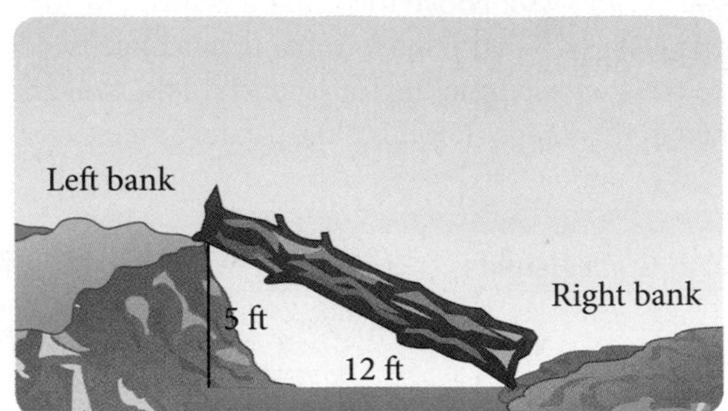

90. Geometry, Surveying A surveying team wants to calculate how long a straight tunnel through a mountain will be. They find a place where the lines to the two ends of the proposed tunnel form a right angle. The distances are shown. How long will the tunnel be? Give the exact answer, and, if you have a calculator, provide an approximate answer to two decimal places.

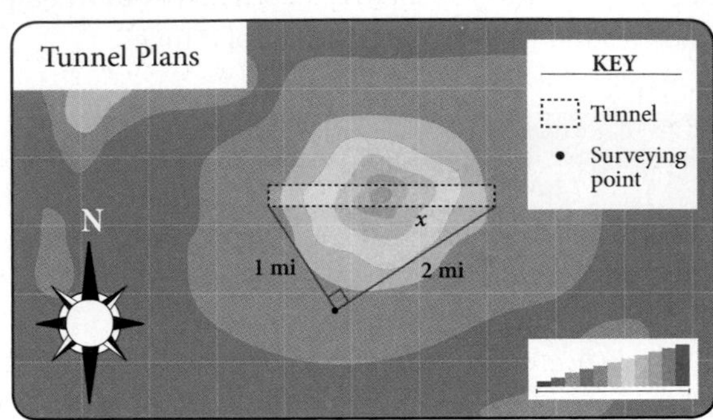

Getting Ready for the Next Section

Simplify.

91. $3 \cdot \sqrt{16}$

92. $6 \cdot \sqrt{4}$

Factor each of the following numbers into the product of two numbers, one of which is a perfect square. (Remember from Chapter 1, a perfect square is 1, 4, 9, 16, 25, 36, . . ., etc.)

93. 75

94. 12

95. 50

96. 20

97. 40

98. 18

99. Factor x^4 from x^5.

100. Factor x^2 from x^3.

101. Factor $4x^2$ from $12x^2$.

102. Factor $4x^2$ from $20x^2$.

103. Factor $25x^2y^2$ from $50x^3y^2$.

104. Factor $25x^2y^2$ from $75x^2y^3$.

Maintaining Your Skills

Reduce each rational expression to lowest terms.

105. $\dfrac{x^2 - 16}{x + 4}$

106. $\dfrac{x - 5}{x^2 - 25}$

107. $\dfrac{10a + 20}{5a^2 - 20}$

108. $\dfrac{8a - 16}{4a^2 - 16}$

109. $\dfrac{2x^2 - 5x - 3}{x^2 - 3x}$

110. $\dfrac{x^2 - 5x}{3x^2 - 13x - 10}$

111. $\dfrac{xy + 3x + 2y + 6}{xy + 3x + ay + 3a}$

112. $\dfrac{xy + 5x + 4y + 20}{x^2 + bx + 4x + 4b}$

Properties of Radicals

8.2

Objectives

A Use Property 1 for radicals to simplify a radical expression.

B Use Property 2 for radicals to simplify a radical expression.

C Use a combination of Properties 1 and 2 to simplify a radical expression.

Examples now playing at
MathTV.com/books

In this section we will consider the first part of what is called *simplified form* for radical expressions. A radical expression is any expression containing a radical, whether it is a square root, a cube root, or a higher root. Simplified form for a radical expression is the form that is easiest to work with. The first step in putting a radical expression in simplified form is to take as much out from under the radical sign as possible. To do this, we first must develop two properties of radicals in general.

A Property 1 for Radicals

Consider the following two problems:

$$\sqrt{9 \cdot 16} = \sqrt{144} = 12$$

$$\sqrt{9} \cdot \sqrt{16} = 3 \cdot 4 = 12$$

Since the answers to both are equal, the original problems also must be equal; that is, $\sqrt{9 \cdot 16} = \sqrt{9} \cdot \sqrt{16}$. We can generalize this property as follows.

Property 1 for Radicals

If x and x represent nonnegative real numbers, then it is always true that

$$\sqrt{xy} = \sqrt{x}\sqrt{y}$$

In words: The square root of a product is the product of the square roots.

We can use this property to simplify radical expressions.

EXAMPLE 1 Simplify $\sqrt{20}$.

SOLUTION To simplify $\sqrt{20}$, we want to take as much out from under the radical sign as possible. We begin by looking for the largest perfect square that is a factor of 20. The largest perfect square that divides 20 is 4, so we write 20 as $4 \cdot 5$:

$$\sqrt{20} = \sqrt{4 \cdot 5}$$

Next, we apply the first property of radicals and write

$$\sqrt{4 \cdot 5} = \sqrt{4}\sqrt{5}$$

And since $\sqrt{4} = 2$, we have

$$\sqrt{4}\sqrt{5} = 2\sqrt{5}$$

The expression $2\sqrt{5}$ is the simplified form of $\sqrt{20}$ since we have taken as much out from under the radical sign as possible.

PRACTICE PROBLEMS

1. Simplify $\sqrt{12}$.

Working a problem like the one in Example 1 depends on recognizing the largest perfect square that divides (is a factor of) the number under the radical sign. The set of perfect squares is the set

$$\{1, 4, 9, 16, 25, 36, \dots\}$$

To simplify an expression like $\sqrt{20}$, we first must find the largest number in this set that is a factor of the number under the radical sign.

EXAMPLE 2 Simplify $\sqrt{75}$.

SOLUTION Since 25 is the largest perfect square that divides 75, we have

$$\begin{aligned} \sqrt{75} &= \sqrt{25 \cdot 3} && \text{Factor 75 into } 25 \cdot 3 \\ &= \sqrt{25}\sqrt{3} && \text{Property 1 for radicals} \\ &= 5\sqrt{3} && \sqrt{25} = 5 \end{aligned}$$

The expression $5\sqrt{3}$ is the simplified form for $\sqrt{75}$ since we have taken as much out from under the radical sign as possible.

2. Simplify $\sqrt{50}$.

Answers

1. $2\sqrt{3}$ **2.** $5\sqrt{2}$

The next two examples involve square roots of expressions that contain variables. Remember, we are assuming that all variables that appear under a radical sign represent positive numbers.

3. Simplify $\sqrt{36x^3}$.

EXAMPLE 3 Simplify $\sqrt{25x^3}$.

SOLUTION The largest perfect square that is a factor of $25x^3$ is $25x^2$. We write $25x^3$ as $25x^2 \cdot x$ and apply Property 1:

$$\begin{aligned} \sqrt{25x^3} &= \sqrt{25x^2 \cdot x} && \text{Factor } 25x^3 \text{ into } 25x^2 \cdot x \\ &= \sqrt{25x^2}\,\sqrt{x} && \text{Property 1 for radicals} \\ &= 5x\sqrt{x} && \sqrt{25x^2} = 5x \end{aligned}$$

4. Simplify $\sqrt{50y^4}$.

EXAMPLE 4 Simplify $\sqrt{18y^4}$.

SOLUTION The largest perfect square that is a factor of $18y^4$ is $9y^4$. We write $18y^4$ as $9y^4 \cdot 2$ and apply Property 1:

$$\begin{aligned} \sqrt{18y^4} &= \sqrt{9y^4 \cdot 2} && \text{Factor } 18y^4 \text{ into } 9y^4 \cdot 2 \\ &= \sqrt{9y^4}\,\sqrt{2} && \text{Property 1 for radicals} \\ &= 3y^2\sqrt{2} && \sqrt{9y^4} = 3y^2 \end{aligned}$$

5. Simplify $6\sqrt{20}$.

EXAMPLE 5 Simplify $3\sqrt{32}$.

SOLUTION We want to get as much out from under $\sqrt{32}$ as possible. Since 16 is the largest perfect square that divides 32, we have:

$$\begin{aligned} 3\sqrt{32} &= 3\sqrt{16 \cdot 2} && \text{Factor 32 into } 16 \cdot 2 \\ &= 3\sqrt{16}\,\sqrt{2} && \text{Property 1 for radicals} \\ &= 3 \cdot 4\sqrt{2} && \sqrt{16} = 4 \\ &= 12\sqrt{2} && 3 \cdot 4 = 12 \end{aligned}$$

Although we have stated Property 1 for radicals in terms of square roots only, it holds for higher roots as well. If we were to state Property 1 again for cube roots, it would look like this:

$$\sqrt[3]{xy} = \sqrt[3]{x}\,\sqrt[3]{y}$$

6. Simplify $\sqrt[3]{40x^3}$.

EXAMPLE 6 Simplify $\sqrt[3]{24x^3}$.

SOLUTION Since we are simplifying a cube root, we look for the largest perfect cube that is a factor of $24x^3$. Since 8 is a perfect cube, the largest perfect cube that is a factor of $24x^3$ is $8x^3$.

$$\begin{aligned} \sqrt[3]{24x^3} &= \sqrt[3]{8x^3 \cdot 3} && \text{Factor } 24x^3 \text{ into } 8x^3 \cdot 3 \\ &= \sqrt[3]{8x^3}\,\sqrt[3]{3} && \text{Property 1 for radicals} \\ &= 2x\sqrt[3]{3} && \sqrt[3]{8x^3} = 2x \end{aligned}$$

B Property 2 for Radicals

The second property of radicals has to do with division. The property becomes apparent when we consider the following two problems:

Answers
3. $6x\sqrt{x}$ 4. $5y^2\sqrt{2}$ 5. $12\sqrt{5}$
6. $2x\sqrt[3]{5}$

$$\sqrt{\frac{64}{16}} = \sqrt{4} = 2$$

$$\frac{\sqrt{64}}{\sqrt{16}} = \frac{8}{4} = 2$$

Since the answers in each case are equal, the original problems must be also:

$$\sqrt{\frac{64}{16}} = \frac{\sqrt{64}}{\sqrt{16}}$$

Here is the property in general.

Property 2 for Radicals

If x and y both represent nonnegative real numbers and $y \neq 0$, then it is always true that

$$\sqrt{\frac{x}{y}} = \frac{\sqrt{x}}{\sqrt{y}}$$

In words: The square root of a quotient is the quotient of the square roots.

Although we have stated Property 2 for square roots only, it holds for higher roots as well.

We can use Property 2 for radicals in much the same way as we used Property 1 to simplify radical expressions.

EXAMPLE 7 Simplify $\sqrt{\frac{49}{81}}$.

7. Simplify $\sqrt{\frac{16}{25}}$.

SOLUTION We begin by applying Property 2 for radicals to separate the fraction into two separate radicals. Then we simplify each radical separately:

$$\sqrt{\frac{49}{81}} = \frac{\sqrt{49}}{\sqrt{81}} \quad \text{Property 2 for radicals}$$

$$= \frac{7}{9} \quad \sqrt{49} = 7 \text{ and } \sqrt{81} = 9$$

EXAMPLE 8 Simplify $\sqrt[4]{\frac{81}{16}}$.

8. Simplify $\sqrt[3]{\frac{125}{64}}$.

SOLUTION Remember, although Property 2 has been stated in terms of square roots, it holds for higher roots as well. Proceeding as we did in Example 7, we have

$$\sqrt[4]{\frac{81}{16}} = \frac{\sqrt[4]{81}}{\sqrt[4]{16}} \quad \text{Property 2}$$

$$= \frac{3}{2} \quad \sqrt[4]{81} = 3 \text{ and } \sqrt[4]{16} = 2$$

C Combining Properties 1 and 2

EXAMPLE 9 Simplify $\sqrt{\frac{50}{49}}$.

9. Simplify $\sqrt{\frac{18}{25}}$.

SOLUTION Applying Property 2 for radicals and then simplifying each resulting radical separately, we have

$$\sqrt{\frac{50}{49}} = \frac{\sqrt{50}}{\sqrt{49}} \quad \text{Property 2 for radicals}$$

$$= \frac{\sqrt{25 \cdot 2}}{7} \quad \text{Factor } 50 = 25 \cdot 2,\ \sqrt{49} = 7$$

Answers

7. $\frac{4}{5}$ 8. $\frac{5}{4}$ 9. $\frac{3\sqrt{2}}{5}$

$$= \frac{\sqrt{25}\,\sqrt{2}}{7} \qquad \text{Property 1 for radicals}$$

$$= \frac{5\sqrt{2}}{7} \qquad \sqrt{25} = 5$$

10. Simplify $\sqrt{\frac{20x^2}{9}}$.

EXAMPLE 10 Simplify $\sqrt{\frac{12x^2}{25}}$.

SOLUTION Proceeding as we have in the previous three examples, we use Property 2 for radicals to separate the numerator and denominator into two separate radicals. Then, we simplify each radical separately:

$$\sqrt{\frac{12x^2}{25}} = \frac{\sqrt{12x^2}}{\sqrt{25}} \qquad \text{Property 2 for radicals}$$

$$= \frac{\sqrt{4x^2 \cdot 3}}{5} \qquad \text{Factor } 12x^2 = 4x^2 \cdot 3,\ \sqrt{25} = 5$$

$$= \frac{\sqrt{4x^2}\,\sqrt{3}}{5} \qquad \text{Property 1 for radicals}$$

$$= \frac{2x\sqrt{3}}{5} \qquad \sqrt{4x^2} = 2x$$

11. Simplify $\sqrt{\frac{75x^2y^3}{36}}$.

EXAMPLE 11 Simplify $\sqrt{\frac{50x^3y^2}{49}}$.

SOLUTION We begin by taking the square roots of $50x^3y^2$ and 49 separately and then writing $\sqrt{49}$ as 7:

$$\sqrt{\frac{50x^3y^2}{49}} = \frac{\sqrt{50x^3y^2}}{\sqrt{49}} \qquad \text{Property 2}$$

$$= \frac{\sqrt{50x^3y^2}}{7} \qquad \sqrt{49} = 7$$

To simplify the numerator of this last expression, we determine that the largest perfect square that is a factor of $50x^3y^2$ is $25x^2y^2$. Continuing, we have

$$= \frac{\sqrt{25x^2y^2 \cdot 2x}}{7} \qquad \text{Factor } 50x^3y^2 \text{ into } 25x^2y^2 \cdot 2x$$

$$= \frac{\sqrt{25x^2y^2}\,\sqrt{2x}}{7} \qquad \text{Property 1}$$

$$= \frac{5xy\sqrt{2x}}{7} \qquad \sqrt{25x^2y^2} = 5xy$$

Getting Ready for Class

After reading through the preceding section, respond in your own words and in complete sentences.

1. Describe Property 1 for radicals.
2. Describe Property 2 for radicals.
3. Explain why this statement is false: "The square root of a sum is the sum of the square roots."
4. Describe how you would apply Property 1 for radicals to $\sqrt{20}$.

Answers

10. $\frac{2x\sqrt{5}}{3}$ 11. $\frac{5xy\sqrt{3y}}{6}$

Problem Set 8.2

A Use Property 1 for radicals to simplify the following radical expressions as much as possible. Assume all variables represent positive numbers. **[Examples 1–6]**

1. $\sqrt{8}$

2. $\sqrt{18}$

3. $\sqrt{12}$

4. $\sqrt{27}$

5. $\sqrt[3]{24}$

6. $\sqrt[3]{54}$

7. $\sqrt{50x^2}$

8. $\sqrt{32x^2}$

9. $\sqrt{45a^2b^2}$

10. $\sqrt{128a^2b^2}$

11. $\sqrt[3]{54x^3}$

12. $\sqrt[3]{128x^3}$

13. $\sqrt{32x^4}$

14. $\sqrt{48x^4}$

15. $5\sqrt{80}$

16. $3\sqrt{125}$

17. $\frac{1}{2}\sqrt{28x^3}$

18. $\frac{2}{3}\sqrt{54x^3}$

19. $x\sqrt[3]{8x^4}$

20. $x\sqrt[3]{8x^5}$

21. $2a\sqrt[3]{27a^5}$

22. $3a\sqrt[3]{27a^4}$

23. $\frac{4}{3}\sqrt{45a^3}$

24. $\frac{3}{5}\sqrt{300a^3}$

25. $3\sqrt{50xy^2}$

26. $4\sqrt{18xy^2}$

27. $7\sqrt{12x^2y}$

28. $6\sqrt{20x^2y}$

B Use Property 2 for radicals to simplify each of the following. Assume all variables represent positive numbers. [Examples 7, 8]

29. $\sqrt{\dfrac{16}{25}}$

30. $\sqrt{\dfrac{81}{64}}$

31. $\sqrt{\dfrac{4}{9}}$

32. $\sqrt{\dfrac{49}{16}}$

33. $\sqrt[3]{\dfrac{8}{27}}$

34. $\sqrt[3]{\dfrac{64}{27}}$

35. $\sqrt[4]{\dfrac{16}{81}}$

36. $\sqrt[4]{\dfrac{81}{16}}$

37. $\sqrt{\dfrac{100x^2}{25}}$

38. $\sqrt{\dfrac{100x^2}{4}}$

39. $\sqrt{\dfrac{81a^2b^2}{9}}$

40. $\sqrt{\dfrac{64a^2b^2}{16}}$

41. $\sqrt[3]{\dfrac{27x^3}{8y^3}}$

42. $\sqrt[3]{\dfrac{125x^3}{64y^3}}$

C Use combinations of Properties 1 and 2 for radicals to simplify the following problems as much as possible. Assume all variables represent positive numbers. [Examples 9–11]

43. $\sqrt{\dfrac{50}{9}}$

44. $\sqrt{\dfrac{32}{49}}$

45. $\sqrt{\dfrac{75}{25}}$

46. $\sqrt{\dfrac{300}{4}}$

47. $\sqrt{\dfrac{128}{49}}$

48. $\sqrt{\dfrac{32}{64}}$

49. $\sqrt{\dfrac{288x}{25}}$

50. $\sqrt{\dfrac{28y}{81}}$

51. $\sqrt{\dfrac{54a^2}{25}}$

52. $\sqrt{\dfrac{243a^2}{49}}$

53. $\dfrac{3\sqrt{50}}{2}$

54. $\dfrac{5\sqrt{48}}{3}$

55. $\dfrac{7\sqrt{28y^2}}{3}$

56. $\dfrac{9\sqrt{243x^2}}{2}$

57. $\dfrac{5\sqrt{72a^2b^2}}{\sqrt{36}}$

58. $\dfrac{2\sqrt{27a^2b^2}}{\sqrt{9}}$

59. $\dfrac{6\sqrt{8x^2y}}{\sqrt{4}}$

60. $\dfrac{5\sqrt{32xy^2}}{\sqrt{25}}$

61. Simplify $\sqrt{b^2 - 4ac}$ if

a. $a = 2, b = 4, c = -3$

b. $a = 1, b = 1, c = -6$

c. $a = 1, b = 1, c = -11$

d. $a = 3, b = 6, c = 2$

62. Simplify $\sqrt{b^2 - 4ac}$ if

a. $a = -3, b = -4, c = 2$

b. $a = 1, b = -3, c = 2$

c. $a = 4, b = 8, c = 1$

d. $a = 1, b = -4, c = 1$

Simplify.

63. a. $\sqrt{32x^{10}y^5}$

b. $\sqrt[3]{32x^{10}y^5}$

c. $\sqrt[4]{32x^{10}y^5}$

d. $\sqrt[5]{32x^{10}y^5}$

64. a. $\sqrt{16x^8y^4}$

b. $\sqrt{16x^4y^8}$

c. $\sqrt[3]{16x^8y^4}$

d. $\sqrt[4]{16x^8y^4}$

65. a. $\sqrt{4}$

b. $\sqrt{0.04}$

c. $\sqrt{400}$

d. $\sqrt{0.0004}$

66. a. $\sqrt[3]{8}$

b. $\sqrt[3]{0.008}$

c. $\sqrt[3]{80}$

d. $\sqrt[3]{8{,}000}$

Use a calculator to help complete the following tables. If an answer needs rounding, round to the nearest thousandth.

67.

x	$\sqrt{x}$	$2\sqrt{x}$	$\sqrt{4x}$
1			
2			
3			
4			

68.

x	$\sqrt{x}$	$2\sqrt{x}$	$\sqrt{4x}$
1			
4			
9			
16			

69.

x	$\sqrt{x}$	$3\sqrt{x}$	$\sqrt{9x}$
1			
2			
3			
4			

70.

x	$\sqrt{x}$	$3\sqrt{x}$	$\sqrt{9x}$
1			
4			
9			
16			

Getting Ready for the Next Section

Simplify.

71. $\sqrt{4x^3y^2}$

72. $\sqrt{9x^2y^3}$

73. $\frac{6}{2}\sqrt{16}$

74. $\frac{8}{4}\sqrt{9}$

75. $\frac{\sqrt{2}}{\sqrt{4}}$

76. $\frac{\sqrt{6}}{\sqrt{9}}$

77. $\frac{\sqrt[3]{18}}{\sqrt[3]{27}}$

78. $\frac{\sqrt[3]{12}}{\sqrt[3]{8}}$

Multiply.

79. $\frac{\sqrt{2}}{\sqrt{3}} \cdot \frac{\sqrt{3}}{\sqrt{3}}$

80. $\frac{\sqrt{y}}{\sqrt{2}} \cdot \frac{\sqrt{2}}{\sqrt{2}}$

81. $\sqrt[3]{3} \cdot \sqrt[3]{9}$

82. $\sqrt[3]{4} \cdot \sqrt[3]{2}$

Maintaining Your Skills

Multiply or divide as indicated.

83. $\frac{8x}{x^2 - 5x} \cdot \frac{x^2 - 25}{4x^2 + 4x}$

84. $\frac{x^2 + 4x}{4x^2} \cdot \frac{7x}{x^2 + 5x}$

85. $\frac{x^2 + 3x - 4}{3x^2 + 7x - 20} \div \frac{x^2 - 2x + 1}{3x^2 - 2x - 5}$

86. $\frac{x^2 - 16}{2x^2 + 7x + 3} \div \frac{2x^2 - 5x - 12}{4x^2 + 8x + 3}$

87. $(x^2 - 36)\left(\frac{x + 3}{x - 6}\right)$

88. $(x^2 - 49)\left(\frac{x + 5}{x + 7}\right)$

Simplified Form for Radicals

Objectives

A Use both properties of radicals to write a radical expression in simplified form.

B Rationalize the denominator in a radical expression that contains only one term in the denominator.

Examples now playing at MathTV.com/books

A Simplified Form for Radicals

Radical expressions that are in simplified form are generally easier to work with. A radical expression is in simplified form if it has three special characteristics.

Definition

A radical expression is in **simplified form** if

1. There are no perfect squares that are factors of the quantity under the square root sign, no perfect cubes that are factors of the quantity under the cube root sign, and so on. We want as little as possible under the radical sign.
2. There are no fractions under the radical sign.
3. There are no radicals in the denominator.

A radical expression that has these three characteristics is said to be in simplified form. As we will see, simplified form is not always the least complicated expression. In many cases, the simplified expression looks more complicated than the original expression. The important thing about simplified form for radicals is that simplified expressions are easier to work with.

The tools we will use to put radical expressions into simplified form are the properties of radicals. We list the properties again for clarity.

Note Simplified form for radicals is the form that we work toward when simplifying radicals. The properties of radicals are the tools we use to get us to simplified form.

Properties of Radicals

If a and b represent any two nonnegative real numbers, then it is always true that

1. $\sqrt{a}\sqrt{b} = \sqrt{a \cdot b}$
2. $\dfrac{\sqrt{a}}{\sqrt{b}} = \sqrt{\dfrac{a}{b}}$ $\quad b \neq 0$
3. $\sqrt{a}\sqrt{a} = (\sqrt{a})^2 = a$ $\quad$ This property comes directly from the definition of radicals

The following examples illustrate how we put a radical expression into simplified form using the three properties of radicals. Although the properties are stated for square roots only, they hold for all roots. [Property 3 written for cube roots would be $\sqrt[3]{a}\,\sqrt[3]{a}\,\sqrt[3]{a} = (\sqrt[3]{a})^3 = a$.]

B Rationalizing the Denominator

EXAMPLE 1 Put $\sqrt{\dfrac{1}{2}}$ into simplified form.

SOLUTION The expression $\sqrt{\frac{1}{2}}$ is not in simplified form because there is a fraction under the radical sign. We can change this by applying Property 2 for radicals:

$$\sqrt{\frac{1}{2}} = \frac{\sqrt{1}}{\sqrt{2}} \qquad \text{Property 2 for radicals}$$

$$= \frac{1}{\sqrt{2}} \qquad \sqrt{1} = 1$$

PRACTICE PROBLEMS

1. Put $\sqrt{\dfrac{1}{3}}$ in simplified form.

The expression $\frac{1}{\sqrt{2}}$ is not in simplified form because there is a radical sign in the denominator. If we multiply the numerator and denominator of $\frac{1}{\sqrt{2}}$ by $\sqrt{2}$, the denominator becomes $\sqrt{2} \cdot \sqrt{2} = 2$:

$$\frac{1}{\sqrt{2}} = \frac{1}{\sqrt{2}} \cdot \frac{\sqrt{\mathbf{2}}}{\sqrt{\mathbf{2}}} \qquad \text{Multiply numerator and denominator by } \sqrt{2}$$

$$= \frac{\sqrt{2}}{2} \qquad 1 \cdot \sqrt{2} = \sqrt{2}, \quad \sqrt{2} \cdot \sqrt{2} = \sqrt{4} = 2$$

If we check the expression $\frac{\sqrt{2}}{2}$ against our definition of simplified form for radicals, we find that all three rules hold. There are no perfect squares that are factors of 2. There are no fractions under the radical sign. No radicals appear in the denominator. The expression $\frac{\sqrt{2}}{2}$, therefore, must be in simplified form.

2. Write $\sqrt{\frac{3}{5}}$ in simplified form.

EXAMPLE 2 Write $\sqrt{\frac{2}{3}}$ in simplified form.

SOLUTION We proceed as we did in Example 1:

$$\sqrt{\frac{2}{3}} = \frac{\sqrt{2}}{\sqrt{3}} \qquad \text{Use Property 2 to separate radicals}$$

$$= \frac{\sqrt{2}}{\sqrt{3}} \cdot \frac{\sqrt{\mathbf{3}}}{\sqrt{\mathbf{3}}} \qquad \text{Multiply by } \frac{\sqrt{3}}{\sqrt{3}} \text{ to remove the radical from the denominator}$$

$$= \frac{\sqrt{6}}{3} \qquad \sqrt{2} \cdot \sqrt{3} = \sqrt{6}, \quad \sqrt{3} \cdot \sqrt{3} = \sqrt{9} = 3$$

3. Put the expression $\frac{8\sqrt{18}}{4\sqrt{2}}$ in simplified form.

EXAMPLE 3 Put the expression $\frac{6\sqrt{20}}{2\sqrt{5}}$ into simplified form.

SOLUTION Although there are many ways to begin this problem, we notice that 20 is divisible by 5. Using Property 2 for radicals as the first step, we can quickly put the expression into simplified form:

$$\frac{6\sqrt{20}}{2\sqrt{5}} = \frac{6}{2}\sqrt{\frac{20}{5}} \qquad \text{Property 2 for radicals}$$

$$= 3\sqrt{4} \qquad \frac{20}{5} = 4$$

$$= 3 \cdot 2 \qquad \sqrt{4} = 2$$

$$= 6$$

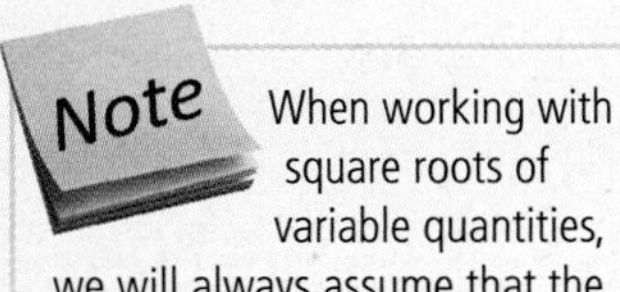

When working with square roots of variable quantities, we will always assume that the variables represent positive numbers. That way we can say that $\sqrt{x^2} = x$.

4. Simplify $\sqrt{\frac{9x^2y^3}{2}}$.

EXAMPLE 4 Simplify $\sqrt{\frac{4x^3y^2}{3}}$.

SOLUTION We begin by separating the numerator and denominator and then taking the perfect squares out of the numerator:

$$\sqrt{\frac{4x^3y^2}{3}} = \frac{\sqrt{4x^3y^2}}{\sqrt{3}} \qquad \text{Property 2 for radicals}$$

$$= \frac{\sqrt{4x^2y^2}\,\sqrt{x}}{\sqrt{3}} \qquad \text{Property 1 for radicals}$$

$$= \frac{2xy\sqrt{x}}{\sqrt{3}} \qquad \sqrt{4x^2y^2} = 2xy$$

The only thing keeping our expression from being in simplified form is the $\sqrt{3}$ in the denominator. We can take care of this by multiplying the numerator and denominator by $\sqrt{3}$:

Answers

1. $\frac{\sqrt{3}}{3}$ 2. $\frac{\sqrt{15}}{5}$ 3. 6

$$\frac{2xy\sqrt{x}}{\sqrt{3}} = \frac{2xy\sqrt{x}}{\sqrt{3}} \cdot \frac{\sqrt{3}}{\sqrt{3}}$$ Multiply numerator and denominator by $\sqrt{3}$

$$= \frac{2xy\sqrt{3x}}{3}$$ $\sqrt{3} \cdot \sqrt{3} = \sqrt{9} = 3$

Although the final expression in Example 4 may look more complicated than the original expression, it is in simplified form. The last step is called *rationalizing the denominator.* We have taken the radical out of the denominator and replaced it with a rational number.

EXAMPLE 5 Simplify $\sqrt[3]{\frac{2}{3}}$.

SOLUTION We can apply Property 2 first to separate the cube roots:

$$\sqrt[3]{\frac{2}{3}} = \frac{\sqrt[3]{2}}{\sqrt[3]{3}}$$

To write this expression in simplified form, we must remove the radical from the denominator. Since the radical is a cube root, we will need to multiply it by an expression that will give us a perfect cube under that cube root. We can accomplish this by multiplying the numerator and denominator by $\sqrt[3]{9}$. Here is what it looks like:

$$\frac{\sqrt[3]{2}}{\sqrt[3]{3}} = \frac{\sqrt[3]{2}}{\sqrt[3]{3}} \cdot \frac{\sqrt[3]{9}}{\sqrt[3]{9}}$$

$$= \frac{\sqrt[3]{18}}{\sqrt[3]{27}}$$ $\sqrt[3]{2} \cdot \sqrt[3]{9} = \sqrt[3]{18}$
$\sqrt[3]{3} \cdot \sqrt[3]{9} = \sqrt[3]{27}$

$$= \frac{\sqrt[3]{18}}{3}$$ $\sqrt[3]{27} = 3$

To see why multiplying the numerator and denominator by $\sqrt[3]{9}$ works in this example, you first must convince yourself that multiplying the numerator and denominator by $\sqrt[3]{3}$ would not have worked.

EXAMPLE 6 Simplify $\sqrt[3]{\frac{1}{4}}$.

SOLUTION We begin by separating the numerator and denominator:

$$\sqrt[3]{\frac{1}{4}} = \frac{\sqrt[3]{1}}{\sqrt[3]{4}}$$ Property 2 for radicals

$$= \frac{1}{\sqrt[3]{4}}$$ $\sqrt[3]{1} = 1$

To rationalize the denominator, we need to have a perfect cube under the cube root sign. If we multiply the numerator and denominator by $\sqrt[3]{2}$, we will have $\sqrt[3]{4} \cdot \sqrt[3]{2} = \sqrt[3]{8}$ in the denominator:

$$\frac{1}{\sqrt[3]{4}} = \frac{1}{\sqrt[3]{4}} \cdot \frac{\sqrt[3]{2}}{\sqrt[3]{2}}$$ Multiply numerator and denominator by $\sqrt[3]{2}$

$$= \frac{\sqrt[3]{2}}{\sqrt[3]{8}}$$ $\sqrt[3]{4} \cdot \sqrt[3]{2} = \sqrt[3]{8}$

$$= \frac{\sqrt[3]{2}}{2}$$ $\sqrt[3]{8} = 2$

The final expression has no radical sign in the denominator and therefore is in simplified form.

5. Simplify $\sqrt[3]{\frac{3}{2}}$.

6. Simplify $\sqrt[3]{\frac{1}{9}}$.

Answers

4. $\frac{3xy\sqrt{2y}}{2}$ 5. $\frac{\sqrt[3]{12}}{2}$ 6. $\frac{\sqrt[3]{3}}{3}$

The Spiral of Roots

To visualize the square roots of the positive integers, we can construct the spiral of roots, which we mentioned in the introduction to this chapter. To begin, we draw two line segments, each of length 1, at right angles to each other. Then we use the Pythagorean theorem to find the length of the diagonal. Figure 1 illustrates.

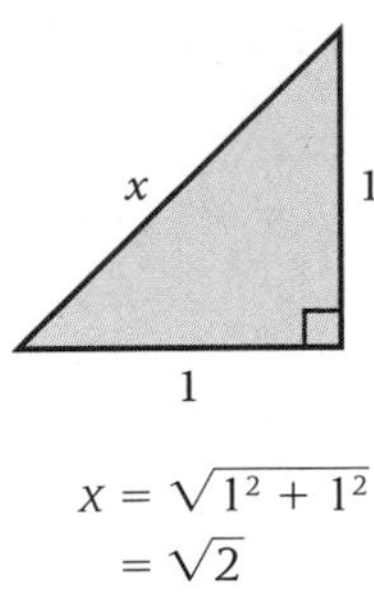

$$x = \sqrt{1^2 + 1^2}$$
$$= \sqrt{2}$$

FIGURE 1

Next, we construct a second triangle by connecting a line segment of length 1 to the end of the first diagonal so that the angle formed is a right angle. We find the length of the second diagonal using the Pythagorean theorem. Figure 2 illustrates this procedure. Continuing to draw new triangles by connecting line segments of length 1 to the end of each new diagonal, so that the angle formed is a right angle, the spiral of roots begins to appear (see Figure 3).

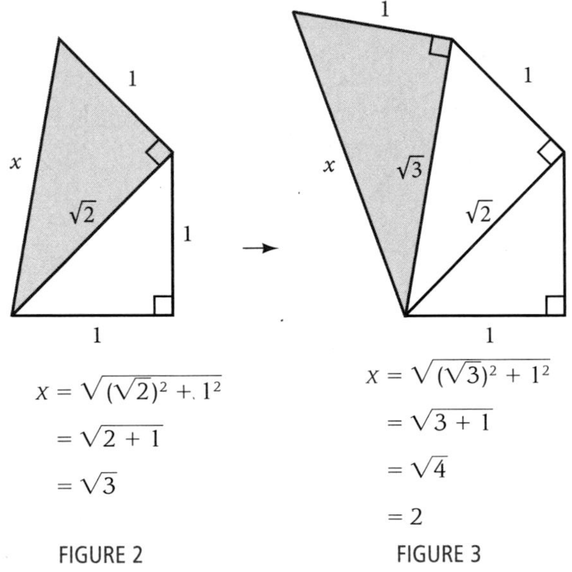

$$x = \sqrt{(\sqrt{2})^2 + 1^2}$$
$$= \sqrt{2 + 1}$$
$$= \sqrt{3}$$

FIGURE 2

$$x = \sqrt{(\sqrt{3})^2 + 1^2}$$
$$= \sqrt{3 + 1}$$
$$= \sqrt{4}$$
$$= 2$$

FIGURE 3

Getting Ready for Class

After reading through the preceding section, respond in your own words and in complete sentences.

1. Describe how you would put $\sqrt{\frac{1}{2}}$ in simplified form.

2. What is simplified form for an expression that contains a square root?

3. What does it mean to rationalize the denominator in an expression?

4. What is useful about the spiral of roots?

Problem Set 8.3

A B Put each of the following radical expressions into simplified form. Assume all variables represent positive numbers. [Examples 1–6]

1. $\sqrt{\frac{1}{2}}$
2. $\sqrt{\frac{1}{5}}$
3. $\sqrt{\frac{1}{3}}$
4. $\sqrt{\frac{1}{6}}$
5. $\sqrt{\frac{2}{5}}$
6. $\sqrt{\frac{3}{7}}$
7. $\sqrt{\frac{3}{2}}$
8. $\sqrt{\frac{5}{3}}$
9. $\sqrt{\frac{20}{3}}$
10. $\sqrt{\frac{32}{5}}$
11. $\sqrt{\frac{45}{6}}$
12. $\sqrt{\frac{48}{7}}$
13. $\sqrt{\frac{20}{5}}$
14. $\sqrt{\frac{12}{3}}$
15. $\frac{\sqrt{21}}{\sqrt{3}}$
16. $\frac{\sqrt{21}}{\sqrt{7}}$
17. $\frac{\sqrt{35}}{\sqrt{7}}$
18. $\frac{\sqrt{35}}{\sqrt{5}}$
19. $\frac{10\sqrt{15}}{5\sqrt{3}}$
20. $\frac{4\sqrt{12}}{8\sqrt{3}}$
21. $\frac{6\sqrt{21}}{3\sqrt{7}}$
22. $\frac{8\sqrt{50}}{16\sqrt{2}}$
23. $\frac{6\sqrt{35}}{12\sqrt{5}}$
24. $\frac{8\sqrt{35}}{16\sqrt{7}}$

25. $\sqrt{\frac{4x^2y^2}{2}}$ **26.** $\sqrt{\frac{9x^2y^2}{3}}$ **27.** $\sqrt{\frac{5x^2y}{3}}$ **28.** $\sqrt{\frac{7x^2y}{5}}$

29. $\sqrt{\frac{16a^4}{5}}$ **30.** $\sqrt{\frac{25a^4}{7}}$ **31.** $\sqrt{\frac{72a^5}{5}}$ **32.** $\sqrt{\frac{12a^5}{5}}$

33. $\sqrt{\frac{20x^2y^3}{3}}$ **34.** $\sqrt{\frac{27x^2y^3}{2}}$ **35.** $\frac{2\sqrt{20x^2y^3}}{3}$ **36.** $\frac{5\sqrt{27x^3y^2}}{2}$

37. $\frac{6\sqrt{54a^2b^3}}{5}$ **38.** $\frac{7\sqrt{75a^3b^2}}{6}$ **39.** $\frac{3\sqrt{72x^4}}{\sqrt{2x}}$ **40.** $\frac{2\sqrt{45x^4}}{\sqrt{5x}}$

41. $\sqrt[3]{\frac{1}{2}}$ **42.** $\sqrt[3]{\frac{1}{4}}$ **43.** $\sqrt[3]{\frac{1}{9}}$ **44.** $\sqrt[3]{\frac{1}{3}}$ **45.** $\sqrt[3]{\frac{3}{2}}$ **46.** $\sqrt[3]{\frac{7}{9}}$

47. Rationalize the denominator

a. $\frac{6}{\sqrt{\pi}}$

b. $\sqrt{\frac{A}{\pi}}$

c. $\sqrt[3]{\frac{3V}{4\pi}}$

d. $\frac{2}{\sqrt[3]{2\pi}}$

48. Rationalize the denominator

a. $\frac{3}{\sqrt{2}}$

b. $\sqrt{\frac{3}{2}}$

c. $\sqrt[3]{\frac{3}{2}}$

d. $\sqrt[3]{\frac{3}{4}}$

Use a calculator to help complete the following tables. If an answer needs rounding, round to the nearest thousandth.

49.

x	$\sqrt{x}$	$\frac{1}{\sqrt{x}}$	$\frac{\sqrt{x}}{x}$
1			
2			
3			
4			
5			
6			

50.

x	$\sqrt{x}$	$\frac{1}{\sqrt{x}}$	$\frac{\sqrt{x}}{x}$
1			
4			
9			
16			
25			
36			

51.

x	$\sqrt{x^2}$	$\sqrt{x^3}$	$x\sqrt{x}$
1			
2			
3			
4			
5			
6			

52.

x	$\sqrt{x^2}$	$\sqrt{x^3}$	$x\sqrt{x}$
1			
4			
9			
16			
25			
36			

Applying the Concepts

53. Google Earth The Google Earth image shows three cities in southern California that form a right triangle. If the distance from Upland to Ontario is 3.6 miles, and the distance from Pomona to Upland is 6.7 miles, what is the area of the triangle formed by these cities? Round to the nearest tenth.

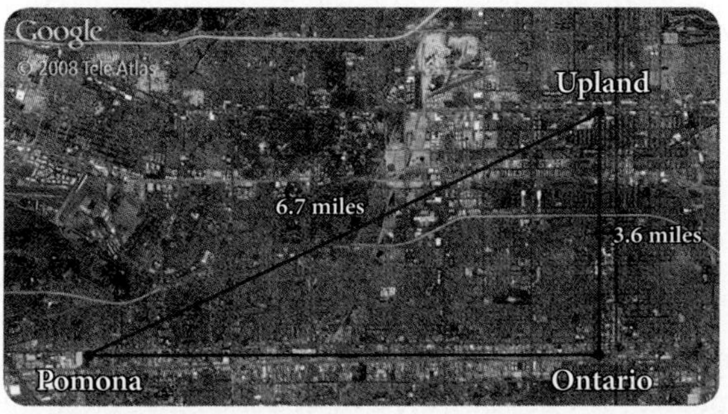

54. Google Earth The Google Earth image shows three cities in Colorado. If the distance between Denver to North Washington is 2.5 miles, and the distance from Edgewater to Denver is 4 miles, what is the distance from North Washington and Edgewater? Round to the nearest tenth.

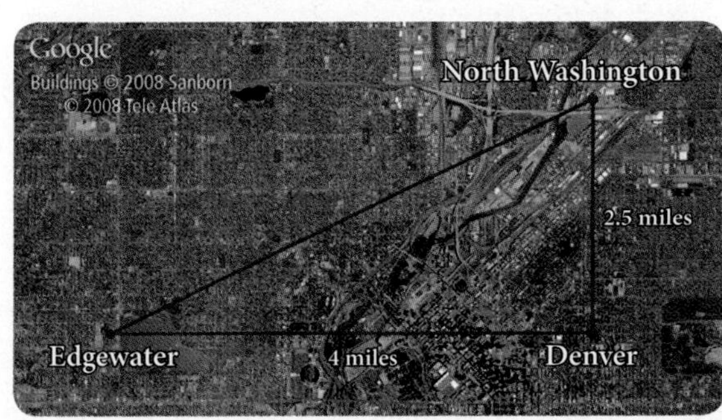

Getting Ready for the Next Section

Combine like terms.

55. $15x + 8x$

56. $6x + 20x$

57. $25y + 3y - y$

58. $12y + 4y - y$

59. $2ab + 5ab$

60. $3ab + 7ab$

61. $2xy - 9xy + 50x$

62. $2xy - 18xy + 3x$

Maintaining Your Skills

Use the distributive property to combine the following.

63. $3x + 7x$

64. $3x - 7x$

65. $15x + 8x$

66. $15x - 8x$

67. $7a - 3a + 6a$

68. $25a + 3a - a$

Add or subtract as indicated.

69. $\dfrac{x^2}{x+5} + \dfrac{10x+25}{x+5}$

70. $\dfrac{x^2}{x-3} - \dfrac{9}{x-3}$

71. $\dfrac{a}{3} + \dfrac{2}{5}$

72. $\dfrac{4}{a} + \dfrac{2}{3}$

73. $\dfrac{6}{a^2-9} - \dfrac{5}{a^2-a-6}$

74. $\dfrac{4a}{a^2+6a+5} - \dfrac{3a}{a^2+5a+4}$

Extending the Concepts

75. Spiral of Roots Construct your own spiral of roots by using a ruler. Draw the first triangle by using two 1-inch lines. The first diagonal will have a length of $\sqrt{2}$ inches. Each new triangle will be formed by drawing a 1-inch line segment at the end of the previous diagonal so that the angle formed is 90°.

76. Spiral of Roots Construct a spiral of roots by using the line segments of length 2 inches. The length of the first diagonal will be $2\sqrt{2}$ inches. The length of the second diagonal will be $2\sqrt{3}$ inches.

77. Number Sequence Simplify the terms in the following sequence. The result will be a sequence that gives the lengths of the diagonals in the spiral of roots you constructed in Problem 75.

$$\sqrt{1^2+1}, \sqrt{(\sqrt{2})^2+1}, \sqrt{(\sqrt{3})^2+1}, \ldots$$

78. Number Sequence Simplify the terms in the following sequence. The result will be a sequence that gives the lengths of the diagonals in the spiral of roots you constructed in Problem 76.

$$\sqrt{2^2+4}, \sqrt{(2\sqrt{2})^2+4}, \sqrt{(2\sqrt{3})^2+4}$$

Addition and Subtraction of Radical Expressions

8.4

Objectives

A Add and subtract similar radical expressions.

Examples now playing at
MathTV.com/books

A Addition and Subtraction of Radical Expressions

To add two or more radical expressions, we apply the distributive property. Adding radical expressions is similar to adding similar terms of polynomials.

EXAMPLE 1 Combine terms in the expression $3\sqrt{5} - 7\sqrt{5}$.

SOLUTION The two terms $3\sqrt{5}$ and $7\sqrt{5}$ each have $\sqrt{5}$ in common. Since $3\sqrt{5}$ means 3 times $\sqrt{5}$, or $3 \cdot \sqrt{5}$, we apply the distributive property:

$$\begin{aligned} 3\sqrt{5} - 7\sqrt{5} &= (3 - 7)\sqrt{5} && \text{Distributive property} \\ &= -4\sqrt{5} && 3 - 7 = -4 \end{aligned}$$

Since we use the distributive property to add radical expressions, each expression must contain exactly the same radical.

EXAMPLE 2 Combine terms in the expression $7\sqrt{2} - 3\sqrt{2} + 6\sqrt{2}$.

SOLUTION

$$\begin{aligned} 7\sqrt{2} - 3\sqrt{2} + 6\sqrt{2} &= (7 - 3 + 6)\sqrt{2} && \text{Distributive property} \\ &= 10\sqrt{2} && \text{Addition} \end{aligned}$$

In Examples 1 and 2, each term was a radical expression in simplified form. If one or more terms are not in simplified form, we must put them into simplified form and then combine terms, if possible.

> **Rule**
> To combine two or more radical expressions, put each expression in simplified form, and then apply the distributive property, if possible.

EXAMPLE 3 Combine terms in the expression $3\sqrt{50} + 2\sqrt{32}$.

SOLUTION We begin by putting each term into simplified form:

$$\begin{aligned} 3\sqrt{50} + 2\sqrt{32} &= 3\sqrt{25}\,\sqrt{2} + 2\sqrt{16}\,\sqrt{2} && \text{Property 1 for radicals} \\ &= 3 \cdot 5\sqrt{2} + 2 \cdot 4\sqrt{2} && \sqrt{25} = 5 \text{ and } \sqrt{16} = 4 \\ &= 15\sqrt{2} + 8\sqrt{2} && \text{Multiplication} \end{aligned}$$

Applying the distributive property to the last line, we have

$$\begin{aligned} 15\sqrt{2} + 8\sqrt{2} &= (15 + 8)\sqrt{2} && \text{Distributive property} \\ &= 23\sqrt{2} && 15 + 8 = 23 \end{aligned}$$

EXAMPLE 4 Combine terms in the expression $5\sqrt{75} + \sqrt{27} - \sqrt{3}$.

SOLUTION

$$\begin{aligned} 5\sqrt{75} + \sqrt{27} - \sqrt{3} &= 5\sqrt{25}\,\sqrt{3} + \sqrt{9}\,\sqrt{3} - \sqrt{3} && \text{Property 1 for radicals} \\ &= 5 \cdot 5\sqrt{3} + 3\sqrt{3} - \sqrt{3} && \sqrt{25} = 5 \text{ and } \sqrt{9} = 3 \\ &= 25\sqrt{3} + 3\sqrt{3} - \sqrt{3} && 5 \cdot 5 = 25 \\ &= (25 + 3 - 1)\sqrt{3} && \text{Distributive property} \\ &= 27\sqrt{3} && \text{Addition} \end{aligned}$$

PRACTICE PROBLEMS

1. Combine terms in the expression: $5\sqrt{2} - 8\sqrt{2}$
2. Combine terms in the expression: $9\sqrt{3} - 4\sqrt{3} + \sqrt{3}$
3. Combine terms in the expression: $3\sqrt{12} + 5\sqrt{48}$
4. Combine terms in the expression: $4\sqrt{18} + \sqrt{32} - \sqrt{2}$

Answers
1. $-3\sqrt{2}$ **2.** $6\sqrt{3}$ **3.** $26\sqrt{3}$
4. $15\sqrt{2}$

The most time-consuming part of combining most radical expressions is simplifying each term in the expression. Once this has been done, applying the distributive property is simple and fast.

5. Simplify $a\sqrt{18} + 7\sqrt{2a^2}$.

EXAMPLE 5 Simplify $a\sqrt{12} + 5\sqrt{3a^2}$.

SOLUTION We must assume that a represents a positive number. Then we simplify each term in the expression by putting it in simplified form for radicals:

$a\sqrt{12} + 5\sqrt{3a^2} = a\sqrt{4}\,\sqrt{3} + 5\sqrt{a^2}\,\sqrt{3}$	Property 1 for radicals
$= a \cdot 2\sqrt{3} + 5 \cdot a\sqrt{3}$	$\sqrt{4} = 2$ and $\sqrt{a^2} = a$
$= 2a\sqrt{3} + 5a\sqrt{3}$	Commutative property
$= (2a + 5a)\sqrt{3}$	Distributive property
$= 7a\sqrt{3}$	Addition

6. Combine terms in the expression: $\sqrt{12x^3} - 6x\sqrt{27x} + \sqrt{9x^2}$

EXAMPLE 6 Combine terms in the expression:

$$\sqrt{20x^3} - 3x\sqrt{45x} + 10\sqrt{25x^2}$$

(Assume x is a positive real number.)

SOLUTION

$$\begin{aligned}\sqrt{20x^3} - 3x\sqrt{45x} + 10\sqrt{25x^2} &= \sqrt{4x^2}\,\sqrt{5x} - 3x\sqrt{9}\,\sqrt{5x} + 10\sqrt{25x^2}\\ &= 2x\sqrt{5x} - 3x \cdot 3\sqrt{5x} + 10 \cdot 5x\\ &= 2x\sqrt{5x} - 9x\sqrt{5x} + 50x\end{aligned}$$

Each term is now in simplified form. The best we can do next is to combine the first two terms. The last term does not have the common radical $\sqrt{5x}$.

$$\begin{aligned}2x\sqrt{5x} - 9x\sqrt{5x} + 50x &= (2x - 9x)\sqrt{5x} + 50x\\ &= -7x\sqrt{5x} + 50x\end{aligned}$$

We have, in any case, succeeded in reducing the number of terms in our original problem.

Our next example involves an expression that is similar to many of the expressions we will find when we solve quadratic equations in Chapter 9.

7. Simplify $\dfrac{6 - \sqrt{8}}{2}$.

EXAMPLE 7 Simplify $\dfrac{6 + \sqrt{12}}{4}$.

SOLUTION We begin by writing $\sqrt{12}$ as $2\sqrt{3}$:

$$\frac{6 + \sqrt{12}}{4} = \frac{6 + 2\sqrt{3}}{4}$$

Factor 2 from the numerator and denominator and then reduce to lowest terms.

$$\begin{aligned}\frac{6 + 2\sqrt{3}}{4} &= \frac{\cancel{2}(3 + \sqrt{3})}{\cancel{2} \cdot 2}\\ &= \frac{3 + \sqrt{3}}{2}\end{aligned}$$

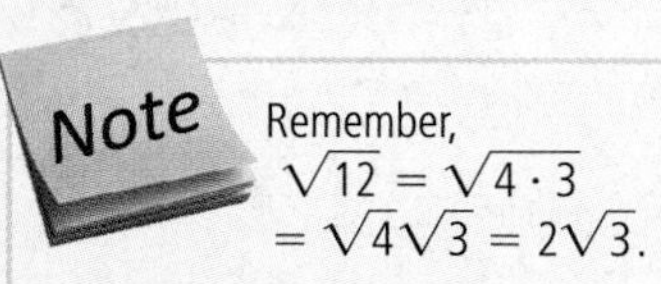

Remember, $\sqrt{12} = \sqrt{4 \cdot 3} = \sqrt{4}\sqrt{3} = 2\sqrt{3}$.

Getting Ready for Class

After reading through the preceding section, respond in your own words and in complete sentences.

1. What are similar radicals?
2. When can we add two radical expressions?
3. What is the first step when adding or subtracting expressions containing radicals?
4. It is not possible to add $\sqrt{2}$ and $\sqrt{3}$. Explain why.

Answers

5. $10a\sqrt{2}$ **6.** $-16x\sqrt{3x} + 3x$
7. $3 - \sqrt{2}$

Problem Set 8.4

A In each of the following problems, simplify each term, if necessary, and then use the distributive property to combine terms, if possible. [Examples 1–4]

1. a. $3x + 4x$
   b. $3y + 4y$
   c. $3\sqrt{2} + 4\sqrt{2}$
   d. $3\sqrt{5} + 4\sqrt{5}$

2. a. $7x + 2x$
   b. $7t + 2t$
   c. $7\sqrt{3} + 2\sqrt{3}$
   d. $7\sqrt{x} + 2\sqrt{x}$

3. a. $x + 6x$
   b. $t + 6t$
   c. $\sqrt{3} + 6\sqrt{3}$
   d. $\sqrt{x} + 6\sqrt{x}$

4. a. $x + 10x$
   b. $y + 10y$
   c. $\sqrt{2} + 10\sqrt{2}$
   d. $\sqrt{7} + 10\sqrt{7}$

5. $3\sqrt{2} + 4\sqrt{2}$
6. $7\sqrt{3} + 2\sqrt{3}$
7. $9\sqrt{5} - 7\sqrt{5}$
8. $6\sqrt{7} - 10\sqrt{7}$
9. $\sqrt{3} + 6\sqrt{3}$
10. $\sqrt{2} + 10\sqrt{2}$
11. $\frac{5}{8}\sqrt{5} - \frac{3}{7}\sqrt{5}$
12. $\frac{5}{6}\sqrt{11} - \frac{7}{9}\sqrt{11}$
13. $14\sqrt{13} - \sqrt{13}$
14. $-2\sqrt{6} - 9\sqrt{6}$
15. $-3\sqrt{10} + 9\sqrt{10}$
16. $11\sqrt{11} + \sqrt{11}$
17. $5\sqrt{5} + \sqrt{5}$
18. $\sqrt{6} - 10\sqrt{6}$
19. $\sqrt{8} + 2\sqrt{2}$
20. $\sqrt{20} + 3\sqrt{5}$
21. $3\sqrt{3} - \sqrt{27}$
22. $4\sqrt{5} - \sqrt{80}$
23. $5\sqrt{12} - 10\sqrt{48}$
24. $3\sqrt{300} - 5\sqrt{27}$
25. $-\sqrt{75} - \sqrt{3}$
26. $5\sqrt{20} + 8\sqrt{80}$
27. $\frac{1}{5}\sqrt{75} - \frac{1}{2}\sqrt{12}$
28. $\frac{1}{2}\sqrt{24} + \frac{1}{5}\sqrt{150}$

29. $\frac{3}{4}\sqrt{8} + \frac{3}{10}\sqrt{75}$

30. $\frac{5}{6}\sqrt{54} - \frac{3}{4}\sqrt{24}$

31. $\sqrt{27} - 2\sqrt{12} + \sqrt{3}$

32. $\sqrt{20} + 3\sqrt{45} - \sqrt{5}$

33. $\frac{5}{6}\sqrt{72} - \frac{3}{8}\sqrt{8} + \frac{3}{10}\sqrt{50}$

34. $\frac{3}{4}\sqrt{24} - \frac{5}{6}\sqrt{54} - \frac{7}{10}\sqrt{150}$

35. $5\sqrt{7} + 2\sqrt{28} - 4\sqrt{63}$

36. $3\sqrt{3} - 5\sqrt{27} + 8\sqrt{75}$

37. $6\sqrt{48} - 2\sqrt{12} + 5\sqrt{27}$

38. $5\sqrt{50} + 8\sqrt{12} - \sqrt{32}$

39. $6\sqrt{48} - \sqrt{72} - 3\sqrt{300}$

40. $7\sqrt{44} - 8\sqrt{99} + \sqrt{176}$

41. $9 + 6\sqrt{2} + 2 - 18 - 6\sqrt{2}$

42. $4 + 4\sqrt{3} + 3 - 8 - 4\sqrt{3}$

43. $4 + 4\sqrt{5} + 5 - 8 - 4\sqrt{5} - 1$

44. $1 - 2\sqrt{2} + 2 - 2 + 2\sqrt{2} - 1$

A All variables in the following problems represent positive real numbers. Simplify each term, and combine, if possible. [Examples 5–6]

45. $\sqrt{x^3} + x\sqrt{x}$

46. $2\sqrt{x} - 2\sqrt{4x}$

47. $5\sqrt{3a^2} - a\sqrt{3}$

48. $6a\sqrt{a} + 7\sqrt{a^3}$

49. $5\sqrt{8x^3} + x\sqrt{50x}$

50. $2\sqrt{27x^2} - x\sqrt{48}$

51. $3\sqrt{75x^3y} - 2x\sqrt{3xy}$

52. $9\sqrt{24x^3y^2} - 5x\sqrt{54xy^2}$

53. $\sqrt{20ab^2} - b\sqrt{45a}$

54. $4\sqrt{a^3b^2} - 5a\sqrt{ab^2}$

55. $9\sqrt{18x^3} - 2x\sqrt{48x}$

56. $8\sqrt{72x^2} - x\sqrt{8}$

57. $7\sqrt{50x^2y} + 8x\sqrt{8y} - 7\sqrt{32x^2y}$

58. $6\sqrt{44x^3y^3} - 8x\sqrt{99xy^3} - 6y\sqrt{176x^3y}$

A Simplify each expression. [Example 7]

59. $\dfrac{6 + 2\sqrt{2}}{2}$

60. $\dfrac{6 - 2\sqrt{3}}{2}$

61. $\dfrac{9 - 3\sqrt{3}}{3}$

62. $\dfrac{-8 + 4\sqrt{2}}{2}$

63. $\dfrac{8 - \sqrt{24}}{6}$

64. $\dfrac{8 + \sqrt{48}}{8}$

65. $\dfrac{6 + \sqrt{8}}{2}$

66. $\dfrac{4 - \sqrt{12}}{2}$

67. $\dfrac{-10 + \sqrt{50}}{10}$

68. $\dfrac{-12 + \sqrt{20}}{6}$

69. Use the approximation $\sqrt{3} \approx 1.732$ to simplify the following.

a. $\dfrac{3 + \sqrt{12}}{2}$

b. $\dfrac{3 - \sqrt{12}}{2}$

c. $\dfrac{3 + \sqrt{12}}{2} + \dfrac{3 - \sqrt{12}}{2}$

70. Use the approximation $\sqrt{5} \approx 2.236$ to simplify the following.

a. $\dfrac{1 + \sqrt{20}}{2}$

b. $\dfrac{1 - \sqrt{20}}{2}$

c. $\dfrac{1 + \sqrt{20}}{2} + \dfrac{1 - \sqrt{20}}{2}$

Use a calculator to help complete the following tables. If an answer needs rounding, round to the nearest thousandth.

71.

x	$\sqrt{x^2+9}$	$x+3$
1		
2		
3		
4		
5		
6		

72.

x	$\sqrt{x^2+16}$	$x+4$
1		
2		
3		
4		
5		
6		

73.

x	$\sqrt{x+3}$	$\sqrt{x}+\sqrt{3}$
1		
2		
3		
4		
5		
6		

74.

x	$\sqrt{x+4}$	$\sqrt{x}+2$
1		
2		
3		
4		
5		
6		

75. Comparing Expressions The following statement is false. Correct the right side to make the statement true.

$$4\sqrt{3} + 5\sqrt{3} = 9\sqrt{6}$$

76. Comparing Expressions The following statement is false. Correct the right side to make the statement true.

$$7\sqrt{5} - 3\sqrt{5} = 4\sqrt{25}$$

Getting Ready for the Next Section

Multiply. Assume any variables represent positive numbers.

77. $\sqrt{5}\sqrt{2}$

78. $\sqrt{3}\sqrt{2}$

79. $\sqrt{5}\sqrt{5}$

80. $\sqrt{3}\sqrt{3}$

81. $\sqrt{x}\sqrt{x}$

82. $\sqrt{y}\sqrt{y}$

83. $\sqrt{5}\sqrt{7}$

84. $\sqrt{5}\sqrt{3}$

Combine like terms.

85. $5 + 7\sqrt{5} + 2\sqrt{5} + 14$

86. $3 + 5\sqrt{3} + 2\sqrt{3} + 10$

87. $x - 7\sqrt{x} + 3\sqrt{x} - 21$

88. $x - 6\sqrt{x} + 8\sqrt{x} - 48$

Maintaining Your Skills

Multiply.

89. $(3x + y)^2$

90. $(2x - 3y)^2$

91. $(3x - 4y)(3x + 4y)$

92. $(7x + 2y)(7x - 2y)$

Solve each equation.

93. $\frac{x}{3} - \frac{1}{2} = \frac{5}{2}$

94. $\frac{3}{x} + \frac{1}{5} = \frac{4}{5}$

95. $1 - \frac{5}{x} = -\frac{6}{x^2}$

96. $1 - \frac{1}{x} = \frac{6}{x^2}$

97. $\frac{a}{a-4} - \frac{a}{2} = \frac{4}{a-4}$

98. $\frac{a}{a-3} - \frac{a}{2} = \frac{3}{a-3}$

Multiplication and Division of Radicals

8.5

Objectives

A Multiply radical expressions.

B Rationalize the denominator in a radical expression that contains two terms in the denominator.

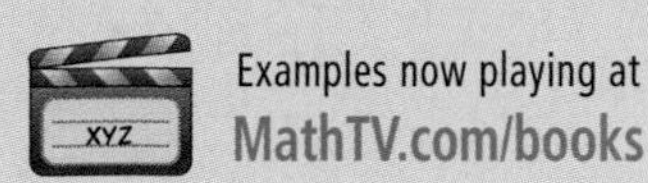
XYZ
Examples now playing at
MathTV.com/books

In this section we will look at multiplication and division of expressions that contain radicals. As you will see, multiplication of expressions that contain radicals is very similar to multiplication of polynomials. The division problems in this section are just an extension of the work we did previously when we rationalized denominators.

A Multiplying Radical Expressions

EXAMPLE 1 Multiply $(3\sqrt{5})(2\sqrt{7})$.

SOLUTION We can rearrange the order and grouping of the numbers in this product by applying the commutative and associative properties. Following that, we apply Property 1 for radicals and multiply:

$$(3\sqrt{5})(2\sqrt{7}) = (3 \cdot 2)(\sqrt{5}\,\sqrt{7}) \quad \text{Commutative and associative properties}$$
$$= (3 \cdot 2)(\sqrt{5 \cdot 7}) \quad \text{Property 1 for radicals}$$
$$= 6\sqrt{35} \quad \text{Multiplication}$$

In actual practice, it is not necessary to show either of the first two steps, although you may want to show them on the first few problems you work, just to be sure you understand them.

EXAMPLE 2 Multiply $\sqrt{5}(\sqrt{2} + \sqrt{5})$.

SOLUTION

$$\sqrt{5}(\sqrt{2} + \sqrt{5}) = \sqrt{5} \cdot \sqrt{2} + \sqrt{5} \cdot \sqrt{5} \quad \text{Distributive property}$$
$$= \sqrt{10} + 5 \quad \text{Multiplication}$$

EXAMPLE 3 Multiply $3\sqrt{2}(2\sqrt{5} + 5\sqrt{3})$.

SOLUTION

$$3\sqrt{2}(2\sqrt{5} + 5\sqrt{3}) = 3\sqrt{2} \cdot 2\sqrt{5} + 3\sqrt{2} \cdot 5\sqrt{3} \quad \text{Distributive property}$$
$$= 3 \cdot 2 \cdot \sqrt{2}\,\sqrt{5} + 3 \cdot 5\sqrt{2}\,\sqrt{3} \quad \text{Commutative property}$$
$$= 6\sqrt{10} + 15\sqrt{6}$$

Each item in the last line is in simplified form, so the problem is complete.

EXAMPLE 4 Multiply $(\sqrt{5} + 2)(\sqrt{5} + 7)$.

SOLUTION We multiply using the FOIL method that we used to multiply binomials:

$$(\sqrt{5} + 2)(\sqrt{5} + 7) = \underset{F}{\sqrt{5} \cdot \sqrt{5}} + \underset{O}{7\sqrt{5}} + \underset{I}{2\sqrt{5}} + \underset{L}{14}$$
$$= 5 + 9\sqrt{5} + 14$$
$$= 19 + 9\sqrt{5}$$

We must be careful not to try to simplify further by adding 19 and 9. We can add only radical expressions that have a common radical part; 19 and $9\sqrt{5}$ are not similar.

PRACTICE PROBLEMS

1. Multiply $(4\sqrt{3})(7\sqrt{5})$.
2. Multiply $\sqrt{3}(\sqrt{2} + \sqrt{3})$.
3. Multiply $5\sqrt{2}(2\sqrt{3} + 4\sqrt{5})$.
4. Multiply $(\sqrt{3} + 5)(\sqrt{3} + 2)$.

Answers

1. $28\sqrt{15}$ **2.** $\sqrt{6} + 3$
3. $10\sqrt{6} + 20\sqrt{10}$ **4.** $13 + 7\sqrt{3}$

5. Multiply $(\sqrt{x} + 8)(\sqrt{x} - 6)$.

EXAMPLE 5 Multiply $(\sqrt{x} + 3)(\sqrt{x} - 7)$.

SOLUTION Remember, we are assuming that any variables that appear under a radical represent positive numbers.

$$(\sqrt{x} + 3)(\sqrt{x} - 7) = \underset{F}{\sqrt{x}\sqrt{x}} - \underset{O}{7\sqrt{x}} + \underset{I}{3\sqrt{x}} - \underset{L}{21}$$
$$= x - 4\sqrt{x} - 21$$

6. Expand and simplify $(\sqrt{5} - 4)^2$.

EXAMPLE 6 Expand and simplify $(\sqrt{3} - 2)^2$.

SOLUTION Multiplying $\sqrt{3} - 2$ times itself, we have

$$\begin{aligned}(\sqrt{3} - 2)^2 &= (\sqrt{3} - 2)(\sqrt{3} - 2) \\ &= \sqrt{3}\sqrt{3} - 2\sqrt{3} - 2\sqrt{3} + 4 \\ &= 3 - 4\sqrt{3} + 4 \\ &= 7 - 4\sqrt{3}\end{aligned}$$

7. Multiply $(\sqrt{7} - \sqrt{3})(\sqrt{7} + \sqrt{3})$.

EXAMPLE 7 Multiply $(\sqrt{5} + \sqrt{2})(\sqrt{5} - \sqrt{2})$.

SOLUTION We can apply the formula $(x + y)(x - y) = x^2 - y^2$ to obtain

$$\begin{aligned}(\sqrt{5} + \sqrt{2})(\sqrt{5} - \sqrt{2}) &= (\sqrt{5})^2 - (\sqrt{2})^2 \\ &= 5 - 2 \\ &= 3\end{aligned}$$

We also could have multiplied the two expressions using the FOIL method. If we were to do so, the work would look like this:

$$\begin{aligned}(\sqrt{5} + \sqrt{2})(\sqrt{5} - \sqrt{2}) &= \underset{F}{\sqrt{5}\sqrt{5}} - \underset{O}{\sqrt{2}\sqrt{5}} + \underset{I}{\sqrt{2}\sqrt{5}} - \underset{L}{\sqrt{2}\sqrt{2}} \\ &= 5 - \sqrt{10} + \sqrt{10} - 2 \\ &= 5 - 2 \\ &= 3\end{aligned}$$

In either case, the product is 3. Also, the expressions $\sqrt{5} + \sqrt{2}$ and $\sqrt{5} - \sqrt{2}$ are called *conjugates* of each other.

8. Multiply $(\sqrt{x} + \sqrt{y})(\sqrt{x} - \sqrt{y})$.

EXAMPLE 8 Multiply $(\sqrt{a} + \sqrt{b})(\sqrt{a} - \sqrt{b})$.

SOLUTION We can apply the formula $(x + y)(x - y) = x^2 - y^2$ to obtain

$$\begin{aligned}(\sqrt{a} + \sqrt{b})(\sqrt{a} - \sqrt{b}) &= (\sqrt{a})^2 - (\sqrt{b})^2 \\ &= a - b\end{aligned}$$

B Rationalizing the Denominator

9. Rationalize the denominator in the expression

$$\frac{\sqrt{5}}{\sqrt{5} - \sqrt{3}}$$

EXAMPLE 9 Rationalize the denominator in the expression

$$\frac{\sqrt{3}}{\sqrt{3} - \sqrt{2}}$$

SOLUTION To remove the two radicals in the denominator, we must multiply both the numerator and denominator by $\sqrt{3} + \sqrt{2}$. That way, when we multiply $\sqrt{3} - \sqrt{2}$ and $\sqrt{3} + \sqrt{2}$, we will obtain the difference of two rational numbers in the denominator:

Answers
5. $x + 2\sqrt{x} - 48$ **6.** $21 - 8\sqrt{5}$
7. 4 **8.** $x - y$

$$\frac{\sqrt{3}}{\sqrt{3}-\sqrt{2}} = \frac{\sqrt{3}}{(\sqrt{3}-\sqrt{2})} \cdot \frac{(\sqrt{\mathbf{3}}+\sqrt{\mathbf{2}})}{(\sqrt{\mathbf{3}}+\sqrt{\mathbf{2}})}$$

$$= \frac{\sqrt{3}\sqrt{3}+\sqrt{3}\sqrt{2}}{(\sqrt{3})^2-(\sqrt{2})^2}$$

$$= \frac{3+\sqrt{6}}{3-2}$$

$$= \frac{3+\sqrt{6}}{1}$$

$$= 3+\sqrt{6}$$

EXAMPLE 10 Rationalize the denominator in the expression $\frac{2}{5-\sqrt{3}}$.

SOLUTION We use the same procedure as in Example 9. Multiply the numerator and denominator by the conjugate of the denominator, which is $5+\sqrt{3}$:

$$\left(\frac{2}{5-\sqrt{3}}\right)\left(\frac{\mathbf{5}+\sqrt{\mathbf{3}}}{\mathbf{5}+\sqrt{\mathbf{3}}}\right) = \frac{10+2\sqrt{3}}{5^2-(\sqrt{3})^2}$$

$$= \frac{10+2\sqrt{3}}{25-3}$$

$$= \frac{10+2\sqrt{3}}{22}$$

The numerator and denominator of this last expression have a factor of 2 in common. We can reduce to lowest terms by dividing out the common factor 2. Continuing, we have

$$= \frac{\cancel{2}(5+\sqrt{3})}{\cancel{2}\cdot 11}$$

$$= \frac{5+\sqrt{3}}{11}$$

The final expression is in simplified form.

EXAMPLE 11 Rationalize the denominator in the expression

$$\frac{\sqrt{2}+\sqrt{3}}{\sqrt{2}-\sqrt{3}}$$

SOLUTION We remove the two radicals in the denominator by multiplying both the numerator and denominator by the conjugate of $\sqrt{2}-\sqrt{3}$, which is $\sqrt{2}+\sqrt{3}$:

$$\frac{\sqrt{2}+\sqrt{3}}{\sqrt{2}-\sqrt{3}} = \left(\frac{\sqrt{2}+\sqrt{3}}{\sqrt{2}-\sqrt{3}}\right)\frac{(\sqrt{\mathbf{2}}+\sqrt{\mathbf{3}})}{(\sqrt{\mathbf{2}}+\sqrt{\mathbf{3}})}$$

$$= \frac{\sqrt{2}\sqrt{2}+\sqrt{2}\sqrt{3}+\sqrt{3}\sqrt{2}+\sqrt{3}\sqrt{3}}{(\sqrt{2})^2-(\sqrt{3})^2}$$

$$= \frac{2+\sqrt{6}+\sqrt{6}+3}{2-3}$$

$$= \frac{5+2\sqrt{6}}{-1}$$

$$= -(5+2\sqrt{6}) \text{ or } -5-2\sqrt{6}$$

10. Rationalize the denominator in the expression

$$\frac{2}{3-\sqrt{7}}$$

11. Rationalize the denominator:

$$\frac{\sqrt{3}+\sqrt{2}}{\sqrt{3}-\sqrt{2}}$$

Answers

9. $\frac{5+\sqrt{15}}{2}$ **10.** $3+\sqrt{7}$

11. $5+2\sqrt{6}$

Getting Ready for Class

After reading through the preceding section, respond in your own words and in complete sentences.

1. Describe how you would use the commutative and associative properties to multiply $3\sqrt{5}$ and $2\sqrt{7}$.
2. Explain why $(\sqrt{5} + 7)^2$ is not the same as $5 + 49$.
3. Explain in words how you would rationalize the denominator in the expression $\frac{\sqrt{3}}{\sqrt{3} - \sqrt{2}}$.
4. What are conjugates?

Problem Set 8.5

A Perform the following multiplications. All answers should be in simplified form for radical expressions. **[Examples 1–8]**

1. $\sqrt{3}\sqrt{2}$

2. $\sqrt{5}\sqrt{6}$

3. $\sqrt{6}\sqrt{2}$

4. $\sqrt{6}\sqrt{3}$

5. $(2\sqrt{3})(5\sqrt{7})$

6. $(3\sqrt{2})(4\sqrt{5})$

7. $(4\sqrt{3})(2\sqrt{6})$

8. $(7\sqrt{6})(3\sqrt{2})$

9. $(2\sqrt{2})^2$

10. $(-2\sqrt{3})^2$

11. $(-2\sqrt{6})^2$

12. $(5\sqrt{2})^2$

13. $[(-1 + 5\sqrt{2}) + 1]^2$

14. $(-8 + 2\sqrt{6} + 8)^2$

15. $[2(-3 + \sqrt{2}) + 6]^2$

16. $[2(3 - \sqrt{3}) - 6]^2$

17. $\sqrt{2}(\sqrt{3} - 1)$

18. $\sqrt{3}(\sqrt{5} + 2)$

19. $\sqrt{2}(\sqrt{3} + \sqrt{2})$

20. $\sqrt{5}(\sqrt{7} - \sqrt{5})$

21. $\sqrt{3}(2\sqrt{2} + \sqrt{3})$

22. $\sqrt{11}(3\sqrt{2} - \sqrt{11})$

23. $\sqrt{3}(2\sqrt{3} - \sqrt{5})$

24. $\sqrt{7}(\sqrt{14} - \sqrt{7})$

25. $2\sqrt{3}(\sqrt{2} + \sqrt{5})$

26. $3\sqrt{2}(\sqrt{3} + \sqrt{2})$

27. $(\sqrt{2} + 1)^2$

28. $(\sqrt{5} - 4)^2$

29. $(\sqrt{x} + 3)^2$

30. $(\sqrt{x} - 4)^2$

31. $(5 - \sqrt{2})^2$

32. $(2 + \sqrt{5})^2$

33. $\left(\sqrt{a} - \frac{1}{2}\right)^2$

34. $\left(\sqrt{a} + \frac{1}{2}\right)^2$

35. $(3 + \sqrt{7})^2$

36. $(3 - \sqrt{2})^2$

37. $(\sqrt{5} + 3)(\sqrt{5} + 2)$

38. $(\sqrt{7} + 4)(\sqrt{7} - 5)$

39. $(\sqrt{2} - 5)(\sqrt{2} + 6)$

40. $(\sqrt{3} + 8)(\sqrt{3} - 2)$

41. $\left(\sqrt{3} + \frac{1}{2}\right)\left(\sqrt{2} + \frac{1}{3}\right)$

42. $\left(\sqrt{5} - \frac{1}{4}\right)\left(\sqrt{3} + \frac{1}{5}\right)$

43. $(\sqrt{x} + 6)(\sqrt{x} - 6)$

44. $(\sqrt{x} + 7)(\sqrt{x} - 7)$

45. $\left(\sqrt{a} + \frac{1}{3}\right)\left(\sqrt{a} + \frac{2}{3}\right)$

46. $\left(\sqrt{a} + \frac{1}{4}\right)\left(\sqrt{a} + \frac{3}{4}\right)$

47. $(\sqrt{5} - 2)(\sqrt{5} + 2)$

48. $(\sqrt{6} - 3)(\sqrt{6} + 3)$

49. $(2\sqrt{7} + 3)(3\sqrt{7} - 4)$

50. $(3\sqrt{5} + 1)(4\sqrt{5} + 3)$

51. $(2\sqrt{x} + 4)(3\sqrt{x} + 2)$

52. $(3\sqrt{x} + 5)(4\sqrt{x} + 2)$

53. $(7\sqrt{a} + 2\sqrt{b})(7\sqrt{a} - 2\sqrt{b})$

54. $(3\sqrt{a} - 2\sqrt{b})(3\sqrt{a} + 2\sqrt{b})$

55. $(3 + \sqrt{2})^2 - 6(3 + \sqrt{2})$

56. $(3 - \sqrt{2})^2 - 6(3 - \sqrt{2})$

57. $(2 + \sqrt{5})^2 - 4(2 + \sqrt{5})$

58. $(2 + \sqrt{2})^2 - 4(2 + \sqrt{2})$

59. $(7 - \sqrt{5})^2 - 14(7 - \sqrt{5}) + 44$

60. $(5 - \sqrt{7})^2 - 10(5 - \sqrt{7}) - 24$

B Rationalize the denominator. All answers should be expressed in simplified form. [Examples 9–11]

61. $\dfrac{\sqrt{3}}{\sqrt{5} - \sqrt{2}}$

62. $\dfrac{\sqrt{2}}{\sqrt{6} + \sqrt{3}}$

63. $\dfrac{\sqrt{5}}{\sqrt{5} + \sqrt{2}}$

64. $\dfrac{\sqrt{7}}{\sqrt{7} - \sqrt{2}}$

65. $\dfrac{8}{3 - \sqrt{5}}$

66. $\dfrac{10}{5 + \sqrt{5}}$

67. $\dfrac{\sqrt{3} + \sqrt{2}}{\sqrt{3} - \sqrt{2}}$

68. $\dfrac{\sqrt{5} - \sqrt{2}}{\sqrt{5} + \sqrt{2}}$

69. $\dfrac{\sqrt{7} - \sqrt{3}}{\sqrt{7} + \sqrt{3}}$

70. $\dfrac{\sqrt{11} - \sqrt{6}}{\sqrt{11} + \sqrt{6}}$

71. $\dfrac{\sqrt{x} + 2}{\sqrt{x} - 2}$

72. $\dfrac{\sqrt{x} - 3}{\sqrt{x} + 3}$

73. $\dfrac{\sqrt{5} - \sqrt{2}}{\sqrt{5} + \sqrt{3}}$

74. $\dfrac{\sqrt{7} - \sqrt{3}}{\sqrt{5} + \sqrt{2}}$

75. Work each problem according to the instructions given.

a. Subtract: $(\sqrt{7} + \sqrt{3}) - (\sqrt{7} - \sqrt{3})$

b. Multiply: $(\sqrt{7} + \sqrt{3})(\sqrt{7} - \sqrt{3})$

c. Square: $(\sqrt{7} - \sqrt{3})^2$

d. Divide: $\dfrac{\sqrt{7} - \sqrt{3}}{\sqrt{7} + \sqrt{3}}$

76. Work each problem according to the instructions given.

a. Subtract: $(\sqrt{11} - \sqrt{6}) - (\sqrt{11} + \sqrt{6})$

b. Multiply: $(\sqrt{11} - \sqrt{6})(\sqrt{11} + \sqrt{6})$

c. Square: $(\sqrt{11} - \sqrt{6})^2$

d. Divide: $\dfrac{\sqrt{11} - \sqrt{6}}{\sqrt{11} + \sqrt{6}}$

77. Work each problem according to the instructions given.

a. Add: $(\sqrt{x} + 2) + (\sqrt{x} - 2)$

b. Multiply: $(\sqrt{x} + 2)(\sqrt{x} - 2)$

c. Square: $(\sqrt{x} + 2)^2$

d. Divide: $\dfrac{\sqrt{x} + 2}{\sqrt{x} - 2}$

78. Work each problem according to the instructions given.

a. Add: $(\sqrt{x} - 3) + (\sqrt{x} + 3)$

b. Multiply: $(\sqrt{x} - 3)(\sqrt{x} + 3)$

c. Square: $(\sqrt{x} + 3)^2$

d. Divide: $\dfrac{\sqrt{x} + 3}{\sqrt{x} - 3}$

79. Work each problem according to the instructions given.

a. Add: $(5 + \sqrt{2}) + (5 - \sqrt{2})$

b. Multiply: $(5 + \sqrt{2})(5 - \sqrt{2})$

c. Square: $(5 + \sqrt{2})^2$

d. Divide: $\dfrac{5 + \sqrt{2}}{5 - \sqrt{2}}$

80. Work each problem according to the instructions given.

a. Add: $(2 + \sqrt{3}) + (2 - \sqrt{3})$

b. Multiply: $(2 + \sqrt{3})(2 - \sqrt{3})$

c. Square: $(2 + \sqrt{3})^2$

d. Divide: $\dfrac{2 + \sqrt{3}}{2 - \sqrt{3}}$

81. Work each problem according to the instructions given.

a. Add: $\sqrt{2} + (\sqrt{6} + \sqrt{2})$

b. Multiply: $\sqrt{2}(\sqrt{6} + \sqrt{2})$

c. Divide: $\dfrac{\sqrt{6} + \sqrt{2}}{\sqrt{2}}$

d. Divide: $\dfrac{\sqrt{2}}{\sqrt{6} + \sqrt{2}}$

82. Work each problem according to the instructions given.

a. Add: $\sqrt{5} + (\sqrt{5} + \sqrt{10})$

b. Multiply: $\sqrt{5}(\sqrt{5} + \sqrt{10})$

c. Divide: $\dfrac{\sqrt{5} + \sqrt{10}}{\sqrt{5}}$

d. Divide: $\dfrac{\sqrt{5}}{\sqrt{5} + \sqrt{10}}$

83. Work each problem according to the instructions given.

a. Add: $\left(\dfrac{1 + \sqrt{5}}{2}\right) + \left(\dfrac{1 - \sqrt{5}}{2}\right)$

b. Multiply: $\left(\dfrac{1 + \sqrt{5}}{2}\right)\left(\dfrac{1 - \sqrt{5}}{2}\right)$

84. Work each problem according to the instructions given.

a. Add: $\left(\dfrac{1 + \sqrt{3}}{2}\right) + \left(\dfrac{1 - \sqrt{3}}{2}\right)$

b. Multiply: $\left(\dfrac{1 + \sqrt{3}}{2}\right)\left(\dfrac{1 - \sqrt{3}}{2}\right)$

Applying the Concepts

85. Comparing Expressions The following statement is false. Correct the right side to make the statement true.

$$2(3\sqrt{5}) = 6\sqrt{15}$$

86. Comparing Expressions The following statement is false. Correct the right side to make the statement true.

$$5(2\sqrt{6}) = 10\sqrt{30}$$

87. Comparing Expressions The following statement is false. Correct the right side to make the statement true.

$$(\sqrt{3} + 7)^2 = 3 + 49$$

88. Comparing Expressions The following statement is false. Correct the right side to make the statement true.

$$(\sqrt{5} + \sqrt{2})^2 = 5 + 2$$

Getting Ready for the Next Section

Simplify.

89. 7^2

90. 5^2

91. $(-9)^2$

92. $(-4)^2$

93. $(\sqrt{x+1})^2$

94. $(\sqrt{x+2})^2$

95. $(\sqrt{2x-3})^2$

96. $(\sqrt{3x-1})^2$

97. $(x+3)^2$

98. $(x+2)^2$

Solve.

99. $3a - 2 = 4$

100. $2a - 3 = 25$

101. $x + 15 = x^2 + 6x + 9$

102. $x + 8 = x^2 + 4x + 4$

Determine whether the given numbers are solutions to the equation.

103. $\sqrt{x+15} = x + 3$

a. $x = -6$ **b.** $x = 1$

104. $\sqrt{x+8} = x + 2$

a. $x = -4$ **b.** $x = 1$

105. Evaluate $y = 3\sqrt{x}$ for $x = -4, -1, 0, 1, 4, 9, 16$.

106. Evaluate $y = \sqrt[3]{x}$ for $x = -27, -8, -1, 0, 1, 8, 27$.

Maintaining Your Skills

Solve each equation.

107. $x^2 + 5x - 6 = 0$

108. $x^2 + 5x + 6 = 0$

109. $x^2 - 3x = 0$

110. $x^2 + 5x = 0$

Solve each proportion.

111. $\frac{x}{3} = \frac{27}{x}$

112. $\frac{x}{2} = \frac{8}{x}$

113. $\frac{x}{5} = \frac{3}{x+2}$

114. $\frac{x}{2} = \frac{4}{x+2}$

Equations Involving Radicals

8.6

Objectives

A Solve equations that contain radicals.

B Graph equations that contain radicals.

XYZ
Examples now playing at
MathTV.com/books

A Solving Equations with Radicals

To solve equations that contain one or more radical expressions, we need an additional property. From our work with exponents we know that if two quantities are equal, then so are the squares of those quantities; that is, for real numbers a and b,

if $\quad a = b$

then $\quad a^2 = b^2$

The only problem with squaring both sides of an equation is that occasionally we will change a false statement into a true statement. Let's take the false statement $3 = -3$ as an example.

$3 = -3$ A false statement

$(3)^2 = (-3)^2$ Square both sides

$9 = 9$ A true statement

We can avoid this problem by always checking our solutions if, at any time during the process of solving an equation, we have squared both sides of the equation. Here is how the property is stated.

Squaring Property of Equality
We can square both sides of an equation any time it is convenient to do so, as long as we check all solutions in the original equation.

EXAMPLE 1 Solve for x : $\sqrt{x + 1} = 7$.

SOLUTION To solve this equation by our usual methods, we must first eliminate the radical sign. We can accomplish this by squaring both sides of the equation:

$\sqrt{x + 1} = 7$

$(\sqrt{x + 1})^2 = 7^2$ Square both sides

$x + 1 = 49$

$x = 48$

To check our solution, we substitute $x = 48$ into the original equation:

$\sqrt{48 + 1} \stackrel{?}{=} 7$

$\sqrt{49} = 7$

$7 = 7$ A true statement

The solution checks.

PRACTICE PROBLEMS

1. Solve for x: $\sqrt{x + 2} = 5$.

Answers

1. 23

2. Solve for x: $\sqrt{3x-1} = -4$.

EXAMPLE 2

Solve for x : $\sqrt{2x-3} = -9$.

SOLUTION We square both sides and proceed as in Example 1:

$$\begin{aligned} \sqrt{2x-3} &= -9 \\ (\sqrt{2x-3})^2 &= (-9)^2 && \text{Square both sides} \\ 2x - 3 &= 81 \\ 2x &= 84 \\ x &= 42 \end{aligned}$$

Checking our solution in the original equation, we have

$$\begin{aligned} \sqrt{2(42)-3} &\stackrel{?}{=} -9 \\ \sqrt{84-3} &= -9 \\ \sqrt{81} &= -9 \\ 9 &= -9 && \text{A false statement} \end{aligned}$$

Our solution does not check because we end up with a false statement.

Note As you can see, when we check $x = 42$ in the original equation, we find that it is not a solution to the equation. Actually, it was apparent from the beginning that the equation had no solution; that is, no matter what x is, the equation

$$\sqrt{2x-3} = -9$$

never can be true because the left side is a positive number (or zero) for any value of x, and the right side is always negative.

Squaring both sides of the equation in Example 2 produced what is called an *extraneous solution*. This happens occasionally when we use the squaring property of equality. We can always eliminate extraneous solutions by checking each solution in the original equation.

3. Solve for a: $\sqrt{2a-3} + 4 = 9$.

EXAMPLE 3

Solve for a: $\sqrt{3a-2} + 3 = 5$.

SOLUTION Before we can square both sides to eliminate the radical, we must isolate the radical on the left side of the equation. To do so, we add -3 to both sides:

$$\begin{aligned} \sqrt{3a-2} + 3 &= 5 \\ \sqrt{3a-2} &= 2 && \text{Add } -3 \text{ to both sides} \\ (\sqrt{3a-2})^2 &= 2^2 && \text{Square both sides} \\ 3a - 2 &= 4 \\ 3a &= 6 \\ a &= 2 \end{aligned}$$

Checking $a = 2$ in the original equation, we have

$$\begin{aligned} \sqrt{3 \cdot 2 - 2} + 3 &\stackrel{?}{=} 5 \\ \sqrt{4} + 3 &= 5 \\ 5 &= 5 && \text{A true statement} \end{aligned}$$

4. Solve for x: $\sqrt{x+8} = x + 2$.

EXAMPLE 4

Solve for x : $\sqrt{x+15} = x + 3$.

SOLUTION We begin by squaring both sides:

$$\begin{aligned} (\sqrt{x+15})^2 &= (x+3)^2 && \text{Square both sides} \\ x + 15 &= x^2 + 6x + 9 \end{aligned}$$

We have a quadratic equation. We put it into standard form by adding $-x$ and -15 to both sides. Then we factor and solve as usual:

$$\begin{aligned} 0 &= x^2 + 5x - 6 && \text{Standard form} \\ 0 &= (x+6)(x-1) && \text{Factor} \\ x + 6 = 0 \quad &\text{or} \quad x - 1 = 0 && \text{Set factors equal to 0} \\ x = -6 \quad &\text{or} \quad x = 1 \end{aligned}$$

Answers
2. No solution 3. 14

We check each solution in the original equation:

Check −6	*Check 1*
$\sqrt{-6+15} \stackrel{?}{=} -6+3$	$\sqrt{1+15} \stackrel{?}{=} 1+3$
$\sqrt{9} = -3$	$\sqrt{16} = 4$
$3 = -3$	$4 = 4$
A false statement	A true statement

Since $x = -6$ does not check in the original equation, it cannot be a solution. The only solution is $x = 1$.

B Graphing Equations with Radicals

EXAMPLE 5 Graph $y = \sqrt{x}$ and $y = \sqrt[3]{x}$.

SOLUTION The graphs are shown in Figures 1 and 2. Notice that the graph of $y = \sqrt{x}$ appears in the first quadrant only because in the equation $y = \sqrt{x}$, x and y cannot be negative.

The graph of $y = \sqrt[3]{x}$ appears in quadrants I and III, since the cube root of a positive number is also a positive number and the cube root of a negative number is a negative number; that is, when x is positive, y will be positive, and when x is negative, y will be negative.

The graphs of both equations will contain the origin since $y = 0$ when $x = 0$ in both equations.

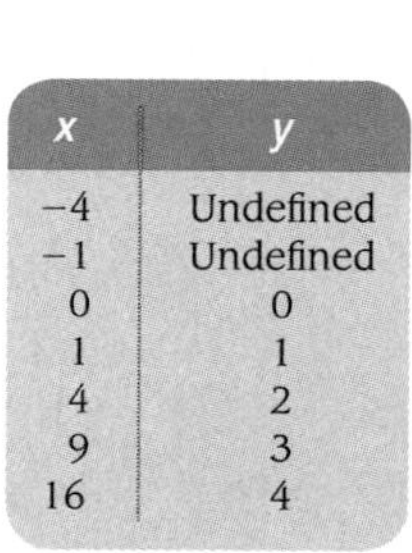

x	y
−4	Undefined
−1	Undefined
0	0
1	1
4	2
9	3
16	4

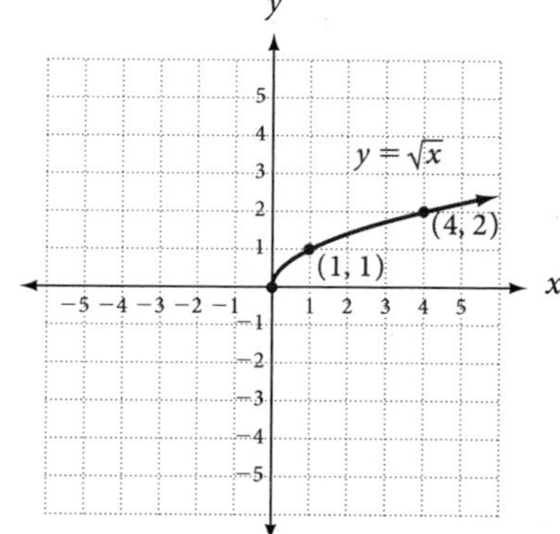

FIGURE 1

x	y
−27	−3
−8	−2
−1	−1
0	0
1	1
8	2
27	3

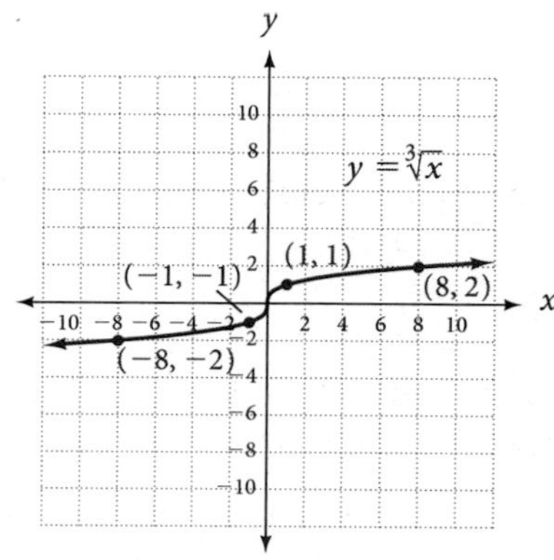

FIGURE 2

5. Graph $y = 3\sqrt{x}$.

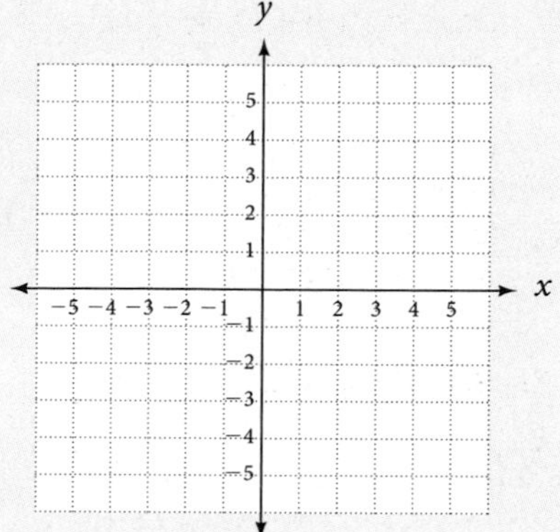

Answers

4. 1

5. See Solutions to Selected Practice Problems

Getting Ready for Class

After reading through the preceding section, respond in your own words and in complete sentences.

1. What is the squaring property of equality?
2. Under what conditions do we obtain extraneous solutions to equations that contain radical expressions?
3. Why do we check our solutions to equations when we have used the squaring property of equality?
4. Why are there no solutions to the equation $\sqrt{2x - 9} = -9$?

Problem Set 8.6

A Solve each equation by applying the squaring property of equality. Be sure to check all solutions in the original equation. [Examples 1–4]

1. $\sqrt{x+1} = 2$

2. $\sqrt{x-3} = 4$

3. $\sqrt{x+5} = 7$

4. $\sqrt{x+8} = 5$

5. $\sqrt{x-9} = -6$

6. $\sqrt{x+10} = -3$

7. $\sqrt{x-5} = -4$

8. $\sqrt{x+7} = -5$

9. $\sqrt{x-8} = 0$

10. $\sqrt{x-9} = 0$

11. $\sqrt{2x+1} = 3$

12. $\sqrt{2x-5} = 7$

13. $\sqrt{2x-3} = -5$

14. $\sqrt{3x-8} = -4$

15. $\sqrt{3x+6} = 2$

16. $\sqrt{5x-1} = 5$

17. $2\sqrt{x} = 10$

18. $3\sqrt{x} = 9$

19. $3\sqrt{a} = 6$

20. $2\sqrt{a} = 12$

21. $\sqrt{3x+4} - 3 = 2$

22. $\sqrt{2x-1} + 2 = 5$

23. $\sqrt{5y-4} - 2 = 4$

24. $\sqrt{3y+1} + 7 = 2$

25. $\sqrt{2x+1} + 5 = 2$

26. $\sqrt{6x-8} - 1 = 3$

27. $\sqrt{x+3} = x - 3$

28. $\sqrt{x-3} = x - 3$

29. $\sqrt{a+2} = a + 2$

30. $\sqrt{a+10} = a - 2$

31. $\sqrt{2x+9} = x + 5$

32. $\sqrt{x+6} = x + 4$

33. $\sqrt{y-4} = y - 6$

34. $\sqrt{2y+13} = y + 7$

35. Solve each equation.

a. $\sqrt{y} - 4 = 6$

b. $\sqrt{y-4} = 6$

c. $\sqrt{y-4} = -6$

d. $\sqrt{y-4} = y - 6$

36. Solve each equation.

a. $\sqrt{2y} + 15 = 7$

b. $\sqrt{2y+15} = 7$

c. $\sqrt{2y+15} = y$

d. $\sqrt{2y+15} = y + 6$

37. Solve each equation.

a. $x - 3 = 0$

b. $\sqrt{x} - 3 = 0$

c. $\sqrt{x-3} = 0$

d. $\sqrt{x} + 3 = 0$

e. $\sqrt{x} + 3 = 5$

f. $\sqrt{x} + 3 = -5$

g. $x - 3 = \sqrt{5-x}$

38. Solve each equation.

a. $x - 2 = 0$

b. $\sqrt{x} - 2 = 0$

c. $\sqrt{x} + 2 = 0$

d. $\sqrt{x+2} = 0$

e. $\sqrt{x} + 2 = 7$

f. $x - 2 = \sqrt{2x-1}$

39. Mountains The map shows the heights of the tallest mountains in the world. The time t in seconds it takes an object to fall d feet is given by the equation

$$t = \frac{1}{4}\sqrt{d}$$

If it were possible to drop a penny off the top of Mount Everest and have it free fall all the way to the bottom, how long would the penny fall for? Round to the nearest hundredth.

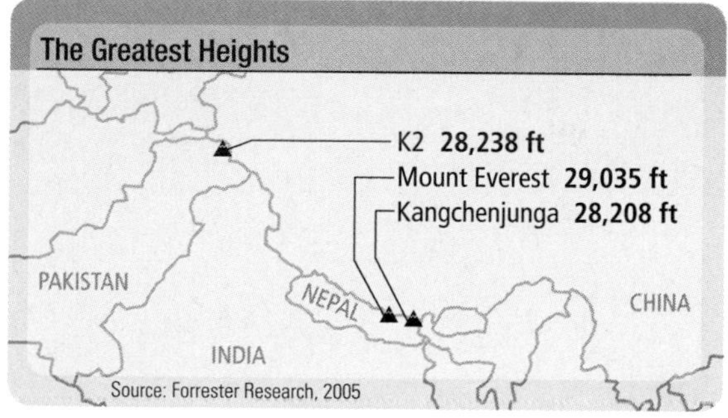

40. Gravity If an object is dropped of the top of a building x feet tall, the amount of time t (in seconds) that it takes for the object to be h feet from the ground is given by the formula

$$t = \frac{1}{4}\sqrt{x - h}$$

If an object is dropped off the top of one of the Petronas Towers, how long will it take for the object to be 187 feet from the ground?

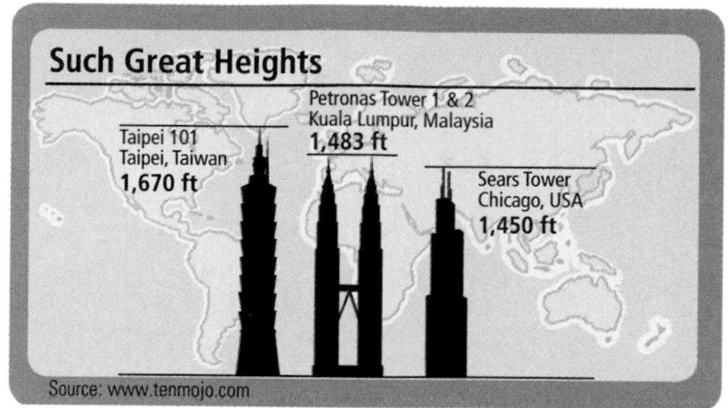

41. Pendulum Problem The time (in seconds) it takes for the pendulum on a clock to swing through one complete cycle is given by the formula

$$T = \frac{11}{7}\sqrt{\frac{L}{2}}$$

where L is the length of the pendulum, in feet. The following table was constructed using this formula. Use the template to draw a line graph of the information in the table.

1 sec

Length L (feet)	Time T (seconds)
1	1.11
2	1.57
3	1.92
4	2.22
5	2.48
6	2.72

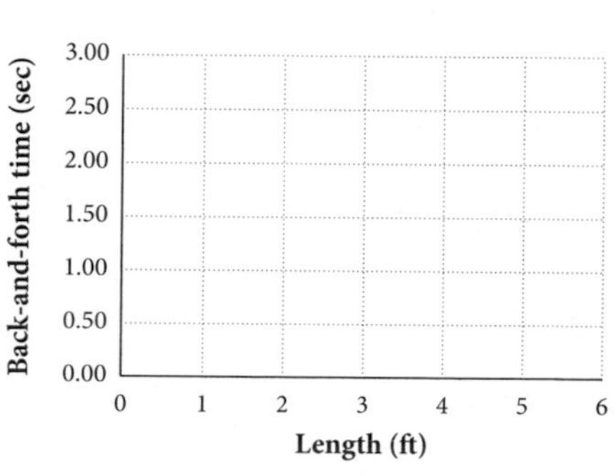

42. Lighthouse Problem The higher you are above the ground, the farther you can see. If your view is unobstructed, then the distance in miles that you can see from h feet above the ground is given by the formula

$$d = \sqrt{\frac{3h}{2}}$$

The following table was constructed using this formula. Use the template to draw a line graph of the information in the table.

Height h (feet)	Distance d (miles)
10	3.9
50	8.7
90	11.6
130	14.0
170	16.0
190	16.9

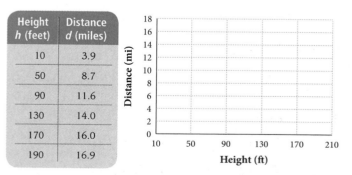

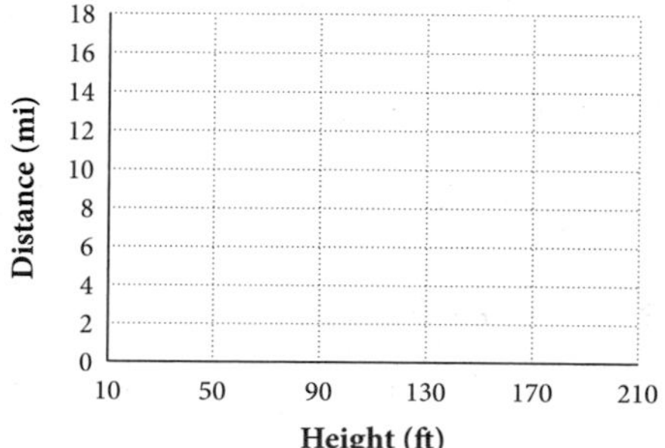

B From each equation, complete the given table and sketch each graph. [Example 5]

43. $y = \sqrt{x}$

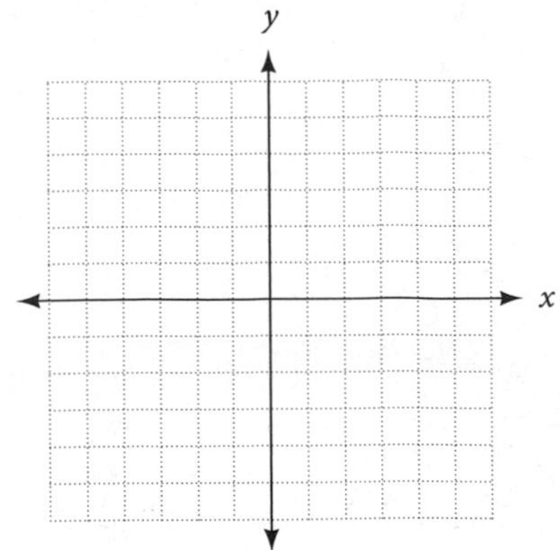

x	y
0	
1	
2	
3	
4	

44. $y = \sqrt[3]{x}$

x	y
−8	
−4	
−1	
0	
4	
8	

45. $y = 2\sqrt{x}$

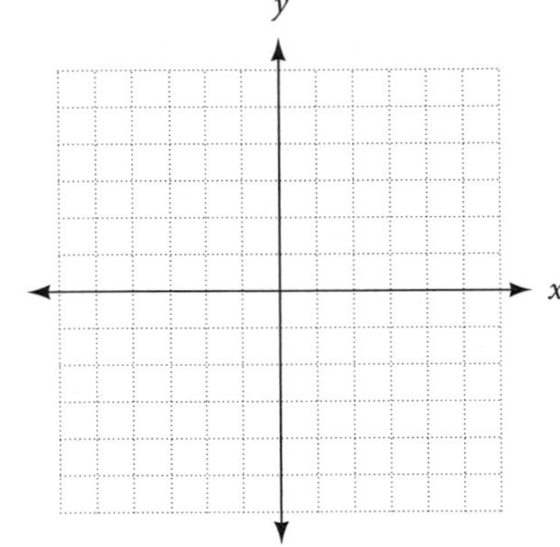

x	y
0	
1	
4	
9	

46. $y = 3\sqrt[3]{x}$

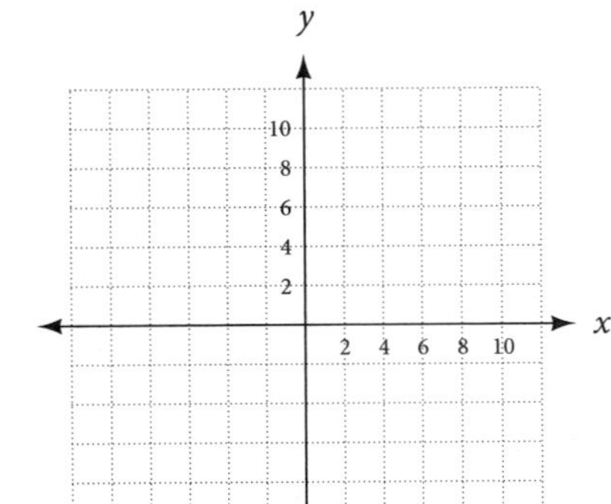

x	y
−8	
−1	
0	
1	
8	

47. $y = \sqrt{x} + 2$

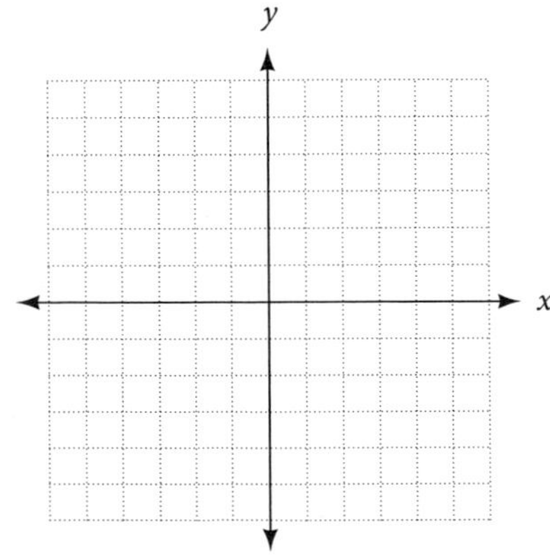

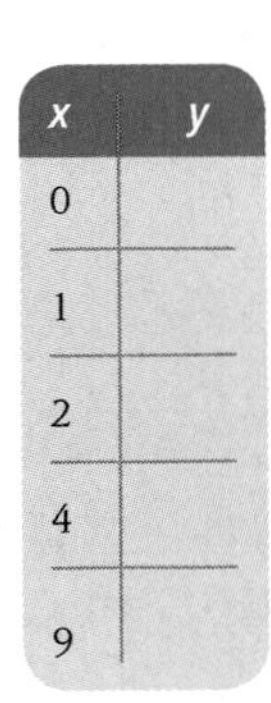

x	y
0	
1	
2	
4	
9	

48. $y = \sqrt[3]{x} + 3$

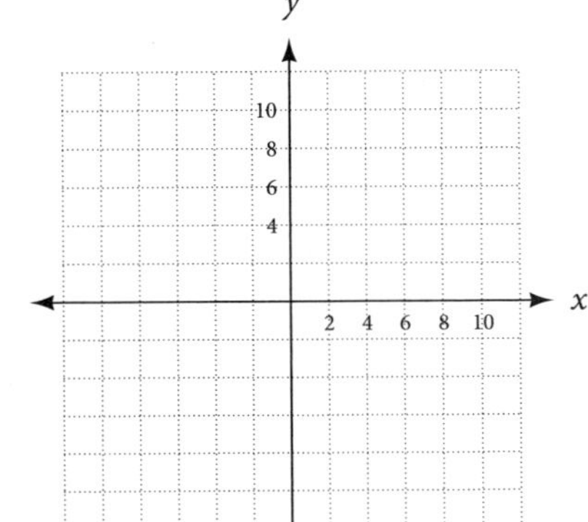

x	y
−8	
−1	
0	
1	
8	

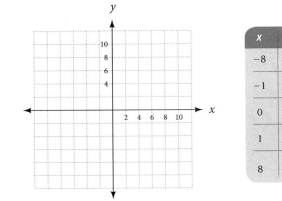

49. Number Problem The sum of a number and 2 is equal to the positive square root of 8 times the number. Find the number.

50. Number Problem The sum of twice a number and 1 is equal to 3 times the positive square root of the number. Find the number.

51. Number Problem The difference of a number and 3 is equal to twice the positive square root of the number. Find the number.

52. Number Problem The difference of a number and 2 is equal to the positive square root of the number. Find the number.

53. Pendulum Problem The number of seconds T it takes the pendulum of a grandfather clock to swing through one complete cycle is given by the formula

$$T = \frac{11}{7}\sqrt{\frac{L}{2}}$$

where L is the length, in feet, of the pendulum. Find how long the pendulum must be for one complete cycle to take 2 seconds by substituting 2 for T in the formula and then solving for L.

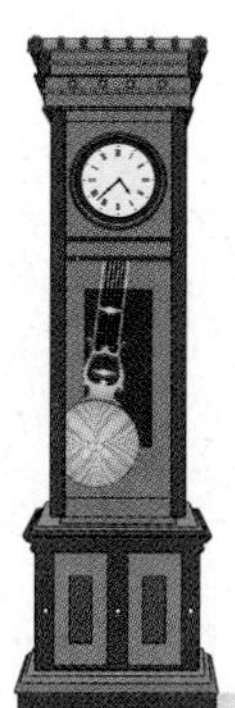

54. Pendulum Problem How long must the pendulum on a grandfather clock be if one complete cycle is to take 1 second?

Maintaining Your Skills

55. Reduce to lowest terms $\frac{x^2 - x - 6}{x^2 - 9}$.

56. Divide using long division $\frac{x^2 - 2x + 6}{x - 4}$.

Perform the indicated operations.

57. $\frac{x^2 - 25}{x + 4} \cdot \frac{2x + 8}{x^2 - 9x + 20}$

58. $\frac{3x + 6}{x^2 + 4x + 3} \div \frac{x^2 + x - 2}{x^2 + 2x - 3}$

59. $\frac{x}{x^2 - 16} + \frac{4}{x^2 - 16}$

60. $\frac{2}{x^2 - 1} - \frac{5}{x^2 + 3x - 4}$

61. $\dfrac{1 - \frac{25}{x^2}}{1 - \frac{8}{x} + \frac{15}{x^2}}$

Solve each equation.

62. $\frac{x}{2} - \frac{5}{x} = -\frac{3}{2}$

63. $\frac{x}{x^2 - 9} - \frac{3}{x - 3} = \frac{1}{x + 3}$

64. Speed of a Boat A boat travels 30 miles up a river in the same amount of time it takes to travel 50 miles down the same river. If the current is 5 miles per hour, what is the speed of the boat in still water?

65. Filling a Pool A pool can be filled by an inlet pipe in 8 hours. The drain will empty the pool in 12 hours. How long will it take to fill the pool if both the inlet pipe and the drain are open?

66. Mixture Problem If 30 liters of a certain solution contains 2 liters of alcohol, how much alcohol is in 45 liters of the same solution?

67. y varies directly with x. If $y = 8$ when x is 12, find y when x is 36.

Chapter 8 Summary

Roots [8.1]

Every positive real number x has two square roots, one positive and one negative. The positive square root is written $\sqrt{x}$. The negative square root of x is written $-\sqrt{x}$. In both cases the square root of x is a number we square to get x.

The cube root of x is written $\sqrt[3]{x}$ and is the number we cube to get x.

EXAMPLES

1. The two square roots of 9 are 3 and −3:

$\sqrt{9} = 3$ and $-\sqrt{9} = -3$

Notation [8.1]

In the expression $\sqrt[3]{8}$, 8 is called the *radicand,* 3 is the *index,* $\sqrt{}$ is called the *radical sign,* and the whole expression $\sqrt[3]{8}$ is called the *radical.*

2. Index → $\sqrt[3]{24}$ ← Radical sign; 24 ← Radicand; $\sqrt[3]{24}$: Radical

Properties of Radicals [8.2]

If a and b represent nonnegative real numbers, then

1. $\sqrt{a}\sqrt{b} = \sqrt{ab}$ — The product of the square roots is the square root of the product
2. $\dfrac{\sqrt{a}}{\sqrt{b}} = \sqrt{\dfrac{a}{b}} \quad (b \neq 0)$ — The quotient of the square roots is the square root of the quotient
3. $\sqrt{a} \cdot \sqrt{a} = (\sqrt{a})^2 = a$ — This property shows that squaring and square roots are inverse operations

3. a. $\sqrt{3} \cdot \sqrt{2} = \sqrt{3 \cdot 2} = \sqrt{6}$

b. $\dfrac{\sqrt{12}}{\sqrt{3}} = \sqrt{\dfrac{12}{3}} = \sqrt{4} = 2$

c. $\sqrt{5} \cdot \sqrt{5} = (\sqrt{5})^2 = 5$

Simplified Form for Radicals [8.3]

A radical expression is in simplified form if:

1. There are no perfect squares that are factors of the quantity under the square root sign, no perfect cubes that are factors of the quantity under the cube root sign, and so on. We want as little as possible under the radical sign.
2. There are no fractions under the radical sign.
3. There are no radicals in the denominator.

4. Simplify $\sqrt{20}$ and $\sqrt{\dfrac{2}{3}}$.

$\sqrt{20} = \sqrt{4 \cdot 5} = \sqrt{4}\sqrt{5} = 2\sqrt{5}$

$\sqrt{\dfrac{2}{3}} = \dfrac{\sqrt{2}}{\sqrt{3}} = \dfrac{\sqrt{2}}{\sqrt{3}} \cdot \dfrac{\sqrt{\mathbf{3}}}{\sqrt{\mathbf{3}}} = \dfrac{\sqrt{6}}{3}$

Addition and Subtraction of Radical Expressions [8.4]

We add and subtract radical expressions by using the distributive property to combine terms that have the same radical parts. If the radicals are not in simplified form, we begin by writing them in simplified form and then combining similar terms, if possible.

5. a. $5\sqrt{7} + 3\sqrt{7} = 8\sqrt{7}$

b. $2\sqrt{18} - 3\sqrt{50}$
$= 2 \cdot 3\sqrt{2} - 3 \cdot 5\sqrt{2}$
$= 6\sqrt{2} - 15\sqrt{2}$
$= -9\sqrt{2}$

Multiplication of Radical Expressions [8.5]

6. **a.** $\sqrt{3}(\sqrt{5} - \sqrt{3}) = \sqrt{15} - 3$
   b. $(\sqrt{7} + 3)(\sqrt{7} - 5)$
   $= 7 - 5\sqrt{7} + 3\sqrt{7} - 15$
   $= -8 - 2\sqrt{7}$

We multiply radical expressions by applying the distributive property or the FOIL method.

Division of Radical Expressions [8.5]

7. $\dfrac{7}{\sqrt{5} - \sqrt{3}}$

$= \dfrac{7}{\sqrt{5} - \sqrt{3}} \cdot \dfrac{\sqrt{5} + \sqrt{3}}{\sqrt{5} + \sqrt{3}}$

$= \dfrac{7\sqrt{5} + 7\sqrt{3}}{2}$

To divide by an expression like $\sqrt{5} - \sqrt{3}$, we multiply the numerator and denominator by its conjugate, $\sqrt{5} + \sqrt{3}$. This process also is called rationalizing the denominator.

Squaring Property of Equality [8.6]

8. Solve $\sqrt{x - 3} = 2$

$$(\sqrt{x - 3})^2 = 2^2$$
$$x - 3 = 4$$
$$x = 7$$

The solution checks in the original equation.

We are free to square both sides of an equation whenever it is convenient, as long as we check all solutions in the original equation. We must check solutions because squaring both sides of an equation occasionally produces extraneous solutions.

COMMON MISTAKES

1. A very common mistake with radicals is to think of $\sqrt{25}$ as representing both the positive and negative square roots of 25. The notation $\sqrt{25}$ stands for the *positive* square root of 25. If we want the negative square root of 25, we write $-\sqrt{25}$.
2. The most common mistake when working with radicals is to try to apply a property similar to Property 1 for radicals involving addition instead of multiplication. Here is an example:

$$\sqrt{16 + 9} = \sqrt{16} + \sqrt{9} \qquad \text{Mistake}$$

Although this example looks like it may be true, it isn't. If we carry it out further, the mistake becomes obvious:

$$\sqrt{16 + 9} \stackrel{?}{=} \sqrt{16} + \sqrt{9}$$
$$\sqrt{25} = 4 + 3$$
$$5 = 7 \qquad \text{False}$$

3. It is a mistake to try to simplify expressions like $2 + 3\sqrt{7}$. The 2 and 3 cannot be combined because the terms they appear in are not similar. Therefore, $2 + 3\sqrt{7} \neq 5\sqrt{7}$. The expression $2 + 3\sqrt{7}$ cannot be simplified further.

Chapter 8 Review

Find the following roots. Assume all variables are positive. [8.1]

1. $\sqrt{25}$ **2.** $\sqrt{169}$ **3.** $\sqrt[3]{-1}$ **4.** $\sqrt[4]{625}$ **5.** $\sqrt{100x^2y^4}$ **6.** $\sqrt[3]{8a^3}$

Simplify. Assume all variables represent positive numbers. [8.2]

7. $\sqrt{24}$ **8.** $\sqrt{60x^2}$ **9.** $\sqrt{90x^3y^4}$ **10.** $-\sqrt{32}$ **11.** $3\sqrt{20x^3y}$

Simplify. Assume all variables represent positive numbers. [8.2]

12. $\sqrt{\dfrac{3}{49}}$ **13.** $\sqrt{\dfrac{8}{81}}$ **14.** $\sqrt{\dfrac{49}{64}}$ **15.** $\sqrt{\dfrac{49a^2b^2}{16}}$

Simplify. Assume all variables represent positive numbers. [8.2]

16. $\sqrt{\dfrac{80}{49}}$ **17.** $\sqrt{\dfrac{40a^2}{121}}$ **18.** $\dfrac{5\sqrt{84}}{7}$ **19.** $\dfrac{3\sqrt{120a^2b^2}}{\sqrt{25}}$ **20.** $\dfrac{-5\sqrt{20x^3y^2}}{\sqrt{144}}$

Write in simplest form. Assume all variables represent positive numbers. [8.3]

21. $\dfrac{2}{\sqrt{7}}$ **22.** $\sqrt{\dfrac{32}{5}}$ **23.** $\sqrt{\dfrac{5}{48}}$ **24.** $\dfrac{-3\sqrt{60}}{\sqrt{5}}$ **25.** $\sqrt{\dfrac{32ab^2}{3}}$ **26.** $\sqrt[3]{\dfrac{3}{4}}$

Write in simplest form. Assume all variables represent positive numbers. [8.5]

27. $\dfrac{3}{\sqrt{3}-4}$ **28.** $\dfrac{2}{3+\sqrt{7}}$ **29.** $\dfrac{3}{\sqrt{5}-\sqrt{2}}$ **30.** $\dfrac{\sqrt{5}}{\sqrt{3}-\sqrt{5}}$ **31.** $\dfrac{\sqrt{5}-\sqrt{2}}{\sqrt{5}+\sqrt{2}}$ **32.** $\dfrac{\sqrt{x}+3}{\sqrt{x}-3}$

Combine the following expressions. [8.4]

33. $3\sqrt{5}-7\sqrt{5}$ **34.** $3\sqrt{27}-5\sqrt{48}$ **35.** $-2\sqrt{45}-5\sqrt{80}+2\sqrt{20}$

36. $3\sqrt{50x^2}-x\sqrt{200}$ **37.** $\sqrt{40a^3b^2}-a\sqrt{90ab^2}$

Multiply. Write all answers in simplest form. [8.5]

38. $\sqrt{3}(\sqrt{3} + 3)$

39. $4\sqrt{2}(\sqrt{3} + \sqrt{5})$

40. $(\sqrt{x} + 7)(\sqrt{x} - 7)$

41. $(2\sqrt{5} - 4)(\sqrt{5} + 3)$

42. $(\sqrt{x} + 5)^2$

Solve each equation. [8.6]

43. $\sqrt{x - 3} = 3$

44. $\sqrt{3x - 5} = 4$

45. $5\sqrt{a} = 20$

46. $\sqrt{3x - 7} + 6 = 2$

47. $\sqrt{2x + 1} + 10 = 8$

48. $\sqrt{7x + 1} = x + 1$

Find x in each of the following right triangles.

49.

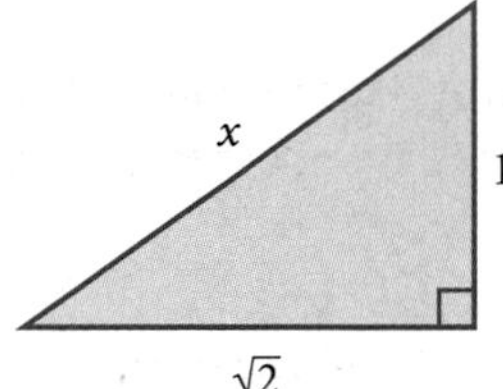

50.

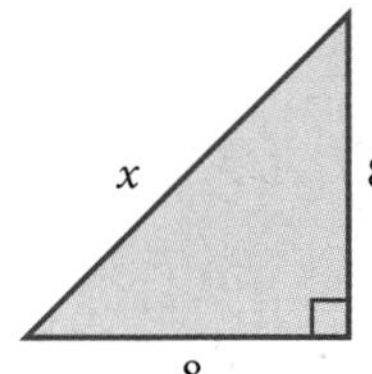

Chapter 8 Cumulative Review

Simplify.

1. $\left(\frac{4}{5}\right)\left(\frac{5}{4}\right)$
2. $-(-3)$
3. $-\left|-\frac{1}{2}\right|$
4. $-40 + (-5)$
5. $10 - 8 - 11$
6. $-3(4)(-2)$
7. $\frac{a^{10}}{a^2}$
8. $\frac{(b^6)^2(b^3)^4}{(b^{10})^3}$
9. $\frac{24a^{12}}{6a^3} + \frac{30a^{24}}{10a^{15}}$
10. $\frac{50x^8y^8}{25x^4y^2} + \frac{28x^7y^7}{14x^3y}$
11. $\left(\frac{1}{2}y + 2\right)\left(\frac{1}{2}y - 2\right)$
12. $5x(8x + 3)$
13. $\dfrac{\frac{1}{a+6} + 3}{\frac{1}{a+6} + 2}$
14. $\dfrac{\frac{x-2}{x^2+6x+8}}{\frac{4}{x+4}}$
15. $\frac{7a}{a^2 - 3a - 54} + \frac{5}{a - 9}$
16. $\frac{3x}{3x - 4} - \frac{4}{3x - 4}$

Simplify. Assume all variables represent positive numbers.

17. $\sqrt{120x^4y^3}$
18. $\sqrt{\frac{90a^2}{169}}$

Solve.

19. $3(5x - 1) = 6(2x + 3) - 21$
20. $5x - 1 = 9x + 9$
21. $x(3x + 2)(x - 4) = 0$
22. $1 + \frac{12}{x} = \frac{-35}{x^2}$
23. $\sqrt{x + 4} = 5$
24. $\sqrt{x - 3} + 5 = 0$

Solve each system.

25. $x + y = 1$
    $x - y = 3$
26. $2x + 5y = 4$
    $3x - 2y = -13$
27. $x - y = 5$
    $y = 3x - 1$
28. $x - y = 5$
    $y = -3x - 1$

Graph on a rectangular coordinate system.

29. $3x + 2y = 6$
30. $x = 1$
31. $y > 3x - 4$
32. $y < 2$
33. For the equation $y = -x + 7$, find the x- and y-intercepts.
34. For the equation $y = -3x + 2$, find the slope and y-intercept.
35. Find the equation for the line with slope -1 that passes through $(3, -3)$.
36. Find y if the line through $(-2, -3)$ and $(2, y)$ has slope -2.

Factor completely.

37. $r^2 + r - 20$
38. $y^4 - 81$
39. $x^5 - x^4 - 30x^3$
40. $80x^3 - 5xy^2$

Simplify and write your answer in scientific notation.

41. $(5 \times 10^5)(2.1 \times 10^3)$
42. $\frac{(6 \times 10^5)(6 \times 10^{-3})}{9 \times 10^{-4}}$

Rationalize the denominator.

43. $\frac{5}{\sqrt{3}}$
44. $\frac{\sqrt{7} + \sqrt{3}}{\sqrt{7} - \sqrt{3}}$
45. Find the value of the expression $2x + 9$ when $x = -2$.
46. Solve $P = 2l + 2w$ for l.
47. Multiply $\frac{x^2 + 3x}{x^2 + 4x + 4} \cdot \frac{x^2 - 5x - 14}{x^2 + 6x + 9}$.
48. Divide $\frac{x^2 + 5x + 6}{x^2 - x - 6} \div \frac{x^2 + 6x + 9}{2x^2 - 5x - 3}$.
49. **Geometry** A rectangle has a perimeter of 48 feet. If the length is twice the width, find the dimensions.
50. **Number Problem** One number is 5 more than twice another number. The sum of the numbers is 35. Find the numbers.

Use the illustration to answer the following questions.

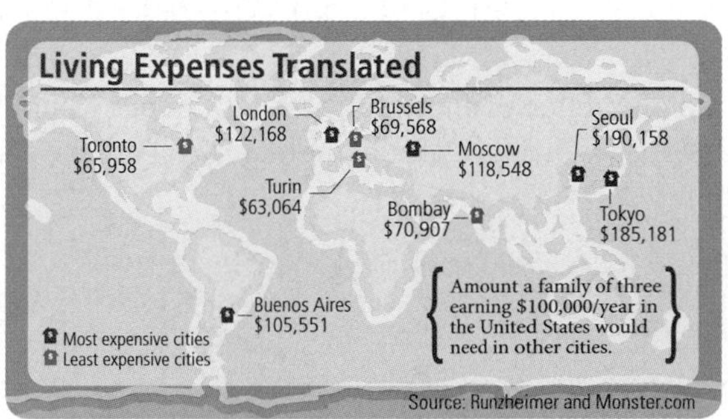

51. How much more does it cost for a family to live in Moscow than in Toronto?
52. What is the ratio of the cost to live in Seoul to the cost to live in Bombay? Round your answer to the nearest hundredth.

Chapter 8 Test

Find the following roots. [8.1]

1. $\sqrt{16}$
2. $-\sqrt{36}$
3. The square root of 49
4. $\sqrt[3]{27}$
5. $\sqrt[3]{-8}$
6. $-\sqrt[4]{81}$

Put the following expressions into simplified form. [8.2, 8.3]

7. $\sqrt{75}$
8. $\sqrt{32}$
9. $\sqrt{\dfrac{2}{3}}$
10. $\dfrac{1}{\sqrt[3]{4}}$
11. $3\sqrt{50x^2}$
12. $\sqrt{\dfrac{12x^2y^3}{5}}$

Combine. [8.4]

13. $5\sqrt{12} - 2\sqrt{27}$
14. $2x\sqrt{18} + 5\sqrt{2x^2}$

Multiply. [8.5]

15. $\sqrt{3}(\sqrt{5} - 2)$
16. $(\sqrt{5} + 7)(\sqrt{5} - 8)$
17. $(\sqrt{x} + 6)(\sqrt{x} - 6)$
18. $(\sqrt{5} - \sqrt{3})^2$

Divide. (Rationalize the denominator.) [8.5]

19. $\dfrac{\sqrt{7} - \sqrt{3}}{\sqrt{7} + \sqrt{3}}$
20. $\dfrac{\sqrt{x}}{\sqrt{x} + 5}$

Solve the following equations. [8.6]

21. $\sqrt{2x + 1} + 2 = 7$
22. $\sqrt{3x + 1} + 6 = 2$
23. $\sqrt{2x - 3} = x - 3$
24. The difference of a number and 4 is equal to 3 times the positive square root of the number. Find the number.
25. Find x in the following right triangle. [8.1]

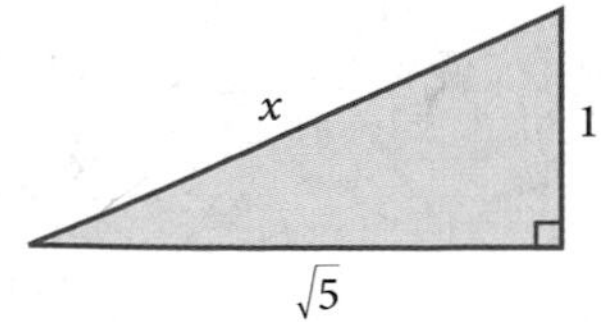

For each equation, complete the given table and then use the table values to match the equation with its graph. [8.6]

26. $y = \sqrt{x} - 1$

x	y
0	
1	
4	
9	

27. $y = \sqrt{x + 1}$

x	y
−1	
0	
3	
8	

28. $y = \sqrt[3]{x} - 1$

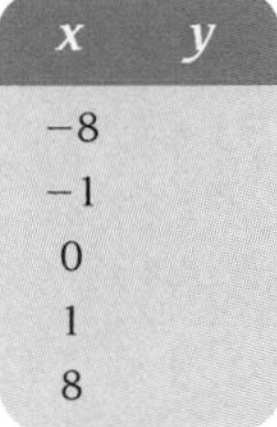

x	y
−8	
−1	
0	
1	
8	

29. $y = \sqrt[3]{x + 1}$

x	y
−9	
−2	
−1	
0	
7	

A.

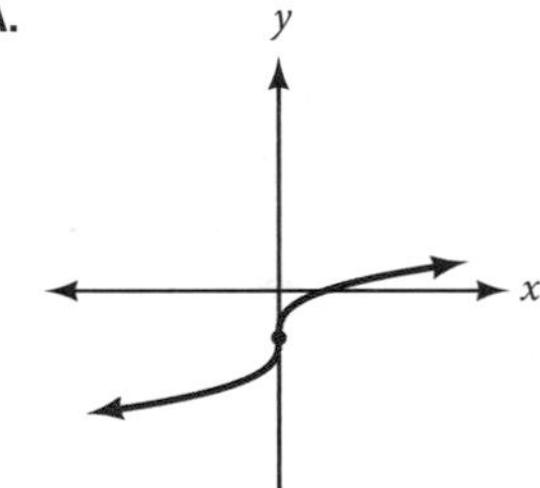

B.

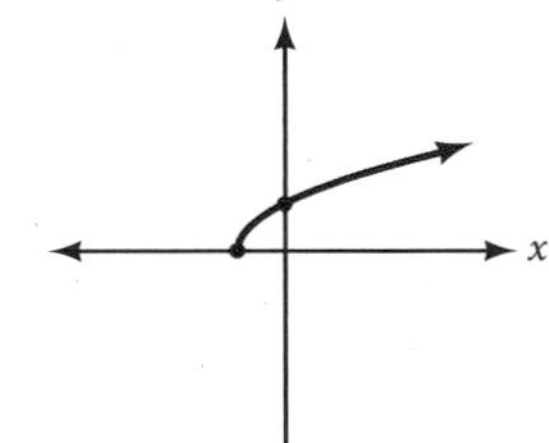

C.

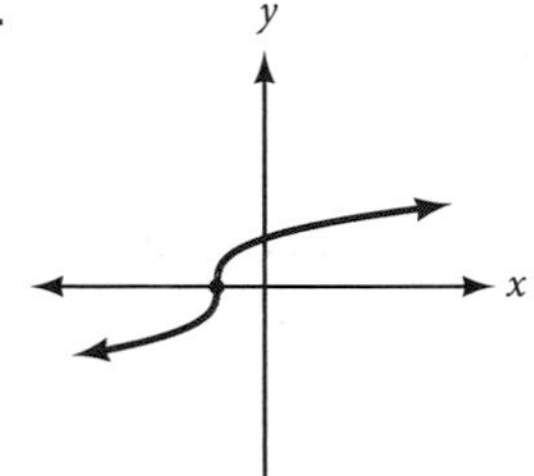

D.

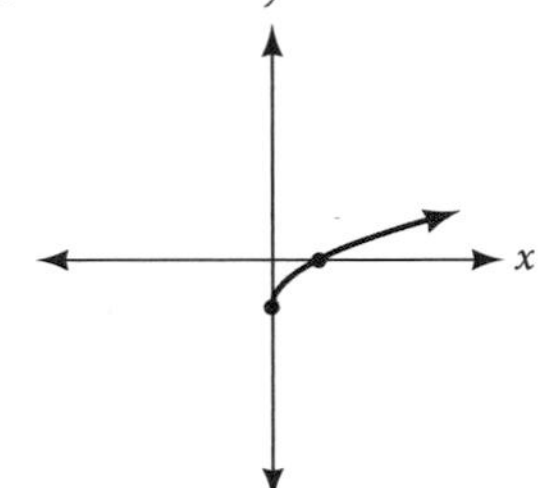

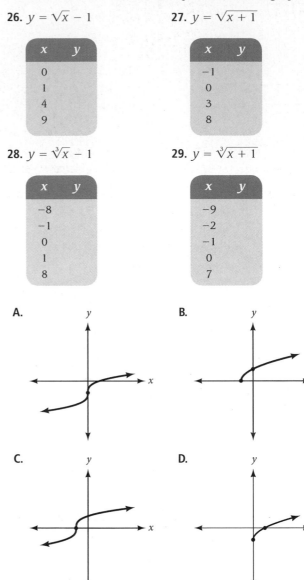

Chapter 8 Projects

ROOTS AND RADICALS

GROUP PROJECT

Unwinding the Spiral of Roots

Number of People 2–3

Time Needed 8–12 minutes

Equipment Pencil, ruler, graph paper, scissors, and tape

Background In this chapter, we used the Spiral of Roots to visualize square roots of positive integers. If we "unwind" the spiral of roots, we can produce the graph of a simple equation on a rectangular coordinate system.

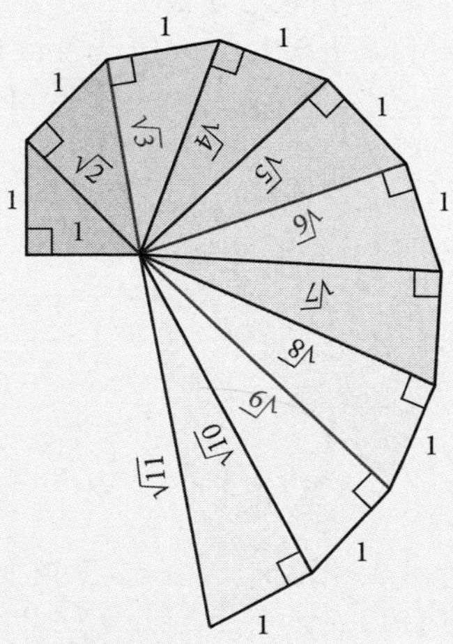

Procedure

1. Carefully cut out each triangle from the Spiral of Roots above.
2. Line up the triangles horizontally on the coordinate system shown here so that the side of length 1 is on the x-axis and the hypotenuse is on the left. Note that the first triangle is shown in place, and the outline of the second triangle is next to it. The 1-unit side of each triangle should fit in each of the 1-unit spaces on the x-axis.
3. On the coordinate system, plot a point at the tip of each triangle. Then, connect these points with a smooth curve.
4. What is the equation of the curve you have just drawn?

RESEARCH PROJECT

Connections

Although it may not look like it, the three items shown here are related very closely to one another. Your job is to find the connection.

A Continued Fraction

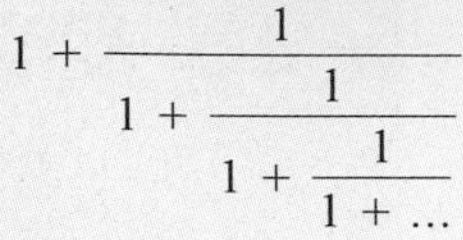

$$1 + \cfrac{1}{1 + \cfrac{1}{1 + \cfrac{1}{1 + \dots}}}$$

The Fibonacci Sequence

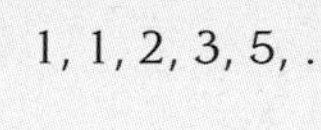

1, 1, 2, 3, 5, . . .

The Golden Rectangle

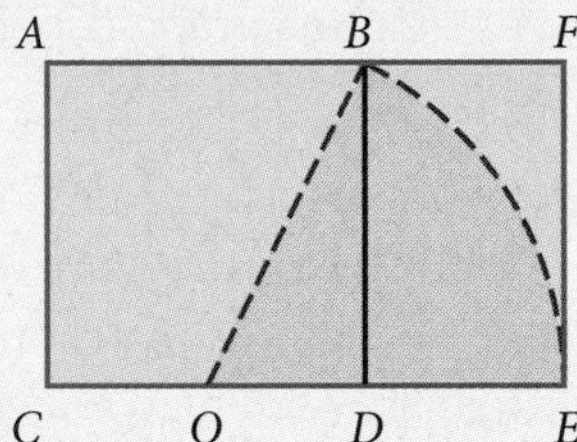

Step 1: The dots in the continued fraction indicate that the pattern shown continues indefinitely. This means that there is no way for us to simplify this expression, as we have simplified the expressions in this chapter. However, we can begin to understand the continued fraction, and what it simplifies to, by working with the following sequence of expressions. Simplify each expression. Write each answer as a fraction, in lowest terms.

$$1 + \frac{1}{1+1} \qquad 1 + \cfrac{1}{1 + \cfrac{1}{1+1}} \qquad 1 + \cfrac{1}{1 + \cfrac{1}{1 + \cfrac{1}{1+1}}} \qquad 1 + \cfrac{1}{1 + \cfrac{1}{1 + \cfrac{1}{1 + \cfrac{1}{1+1}}}}$$

Step 2: Compare the fractional answers to step 1 with the numbers in the Fibonacci sequence. Based on your observation, give the answer to the following problem, without actually doing any arithmetic.

$$\cfrac{1}{1 + \cfrac{1}{1 + \cfrac{1}{1 + \cfrac{1}{1 + \cfrac{1}{1+1}}}}}$$

Step 3: Continue the sequence of simplified fractions you have written in steps 1 and 2, until you have nine numbers in the sequence. Convert each of these numbers to a decimal, accurate to four places past the decimal point.

Step 4: Find a decimal approximation to the golden ratio $\frac{1+\sqrt{5}}{2}$, accurate to four places past the decimal point.

Step 5: Compare the results in steps 3 and 4, and then make a conjecture about what number the continued fraction would simplify to, if it were actually possible to simplify it.

Quadratic Equations

9

Chapter Outline

Introduction

The city of Khiva, in Uzbekistan, is the birthplace of the mathematician, al-Khwarizmi. Although he was born in Khiva, he spent most of his life in Baghdad, Iraq. The Islamic mathematician wrote on Hindu-Arabic numerals and is credited with being the first to use zero as a place holder. His book *Hisab al-jabr w'al-muqabala* is considered as the first book to be written on algebra. Below is a postage stamp issued by the USSR in 1983, to commemorate the 1200th anniversary of al-Khwarizmi's birth.

"That fondness for science, ... that affability and condescension which God shows to the learned, that promptitude with which he protects and supports them in the elucidation of obscurities and in the removal of difficulties, has encouraged me to compose a short work on calculating by *al-jabr* and *al-muqabala*, confining it to what is easiest and most useful in arithmetic."

--al-Khwarizmi

In his book, al-Khwarizmi solves quadratic equations by a method called completing the square, one of the topics we will study in this chapter.

Chapter Pretest

Solve the following quadratic equations. [9.1, 9.3, 9.5]

1. $x^2 - 8x - 9 = 0$

2. $(x - 2)^2 = 18$

3. $\left(x - \frac{7}{3}\right)^2 = -\frac{50}{9}$

4. $\frac{1}{3}x^2 = \frac{5}{6}x - \frac{1}{2}$

5. $4x^2 + 7x - 2$

6. $(x - 4)(x + 1) = 6$

7. $4x^2 - 20x + 25 = 0$

8. Solve $x^2 - 4x - 9 = 0$ by completing the square. [9.2]

Write as a complex number. [9.5]

9. $\sqrt{-36}$

10. $\sqrt{-169}$

11. $\sqrt{-45}$

12. $\sqrt{-12}$

Work the following problems involving complex numbers. [9.4]

13. $(4i + 3) + (4 + 6i)$

14. $(5 - 3i) - (2 - 7i)$

15. $(5 - i)(5 + i)$

16. $(2 + 4i)(3 - i)$

17. $\frac{i}{3 + i}$

18. $\frac{4 - i}{4 + i}$

Graph the following equations. [9.6]

19. $y = x^2 + 3$

20. $y = (x + 2)^2$

21. $y = (x - 4)^2 + 2$

22. $y = x^2 - 6x + 8$

23. Find the height of the isosceles triangle. [9.1]

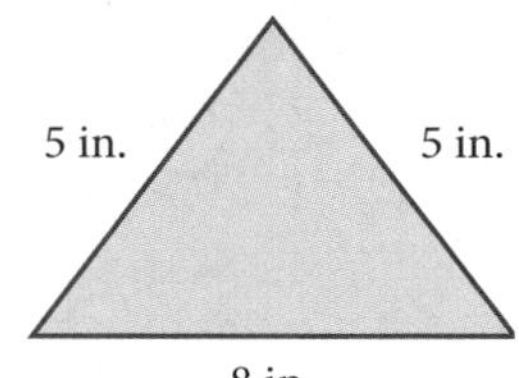

Getting Ready for Chapter 9

Simplify.

1. $\left[\frac{1}{2}(-10)\right]^2$

2. $\left(\frac{1}{2} \cdot 5\right)^2$

3. $\frac{5 - \sqrt{49}}{2}$

4. $\frac{-4 + \sqrt{40}}{4}$

5. $(2 + 7\sqrt{2}) + (3 - 5\sqrt{2})$

6. $(3 - 8\sqrt{2}) - (2 + 4\sqrt{2}) + (3 - \sqrt{2})$

7. $\frac{6x + 5}{5x - 25} - \frac{x + 2}{x - 5}$

8. $\frac{x + 1}{2x - 2} - \frac{2}{x^2 - 1}$

9. $\sqrt{49}$

10. $\sqrt{12}$

Factor.

11. $y^2 - 12y + 36$

12. $x^2 + 12x + 36$

13. $10x^2 + 19x - 15$

14. $6x^2 + x - 2$

15. $y^2 + 3y + \frac{9}{4}$

16. $y^2 - 5y + \frac{25}{4}$

Use the equation $y = (x - 5)^2 + 2$ to find the value of y when:

17. $x = 3$

18. $x = 2$

19. $x = 1$

20. $x = 4$

More Quadratic Equations

9.1

Objectives

A Solve a quadratic equation by taking the square root of both sides of the equation.

B Solve an application problem involving a quadratic equation.

Examples now playing at
MathTV.com/books

A Square Root Property

Consider the equation $x^2 = 9$. Inspection shows that there are two solutions: $x = 3$ and $x = -3$, the two square roots of 9. Since every positive real number has two square roots, we can write the following property.

> **Square Root Property for Equations**
> For all positive real numbers b:
>
> If $a^2 = b$, then $a = \sqrt{b}$ or $a = -\sqrt{b}$

NOTATION A shorthand notation for

$$a = \sqrt{b} \text{ or } a = -\sqrt{b}$$

is:

$$a = \pm\sqrt{b}$$

which is read "a is plus or minus the square root of b."

We can use the square root property any time we feel it is helpful. We must make sure, however, that we include both the positive and the negative square roots.

EXAMPLE 1 Solve for x : $x^2 = 7$.

SOLUTION

$$x^2 = 7$$

$$x = \pm\sqrt{7} \quad \text{Square root property}$$

The two solutions are $\sqrt{7}$ and $-\sqrt{7}$.

PRACTICE PROBLEMS

1. Solve for x: $x^2 = 5$.

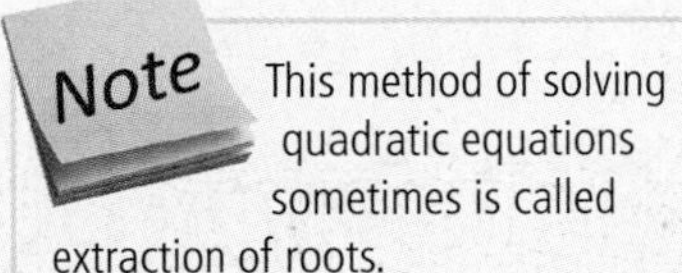

This method of solving quadratic equations sometimes is called extraction of roots.

EXAMPLE 2 Solve for y : $3y^2 = 60$.

SOLUTION We begin by dividing both sides by 3 (which is the same as multiplying both sides by $\frac{1}{3}$):

$$3y^2 = 60$$

$$y^2 = 20 \quad \text{Divide each side by 3}$$

$$y = \pm\sqrt{20} \quad \text{Square root property}$$

$$y = \pm 2\sqrt{5} \quad \sqrt{20} = \sqrt{4 \cdot 5} = \sqrt{4}\sqrt{5} = 2\sqrt{5}$$

Our two solutions are $2\sqrt{5}$ and $-2\sqrt{5}$. Each of them will yield a true statement when used in place of the variable in the original equation, $3y^2 = 60$.

2. Solve for y: $5y^2 = 60$.

EXAMPLE 3 Solve for a : $(a + 3)^2 = 16$.

SOLUTION We begin by applying the square root property for equations:

$$(a + 3)^2 = 16$$

$$a + 3 = \pm 4$$

At this point we add -3 to both sides to get:

$$a = -3 \pm 4$$

3. Solve for a: $(a + 2)^2 = 25$.

Answers
1. $\pm\sqrt{5}$ **2.** $\pm 2\sqrt{3}$

which we can write as:

$$a = -3 + 4 \quad \text{or} \quad a = -3 - 4$$
$$a = 1 \quad \text{or} \quad a = -7$$

Our solutions are 1 and −7.

4. Solve for x: $(2x - 5)^2 = 49$.

EXAMPLE 4 Solve for x : $(3x - 2)^2 = 25$.

SOLUTION

$$(3x - 2)^2 = 25$$
$$3x - 2 = \pm 5$$

Adding 2 to both sides, we have:

$$3x = 2 \pm 5$$

Dividing both sides by 3 gives us:

$$x = \frac{2 \pm 5}{3}$$

We separate the preceding equation into two separate statements:

$$x = \frac{2 + 5}{3} \quad \text{or} \quad x = \frac{2 - 5}{3}$$
$$x = \frac{7}{3} \quad \text{or} \quad x = \frac{-3}{3} = -1$$

Note We can solve the equation in Example 4 by factoring (as we did in Section 6.7) if we first expand $(3x - 2)^2$.

$$(3x - 2)^2 = 25$$
$$9x^2 - 12x + 4 = 25$$
$$9x^2 - 12x - 21 = 0$$
$$3(3x^2 - 4x - 7) = 0$$
$$3(3x - 7)(x + 1) = 0$$
$$x = \frac{7}{3} \quad \text{or} \quad x = -1$$

5. Solve for x: $(2x - 3)^2 = 7$.

EXAMPLE 5 Solve for y : $(4y - 5)^2 = 6$.

SOLUTION

$$(4y - 5)^2 = 6$$
$$4y - 5 = \pm\sqrt{6}$$
$$4y = 5 \pm \sqrt{6} \qquad \text{Add 5 to both sides}$$
$$y = \frac{5 \pm \sqrt{6}}{4} \qquad \text{Divide both sides by 4}$$

Since $\sqrt{6}$ is irrational, we cannot simplify the expression further. The solution set is $\left\{\frac{5 + \sqrt{6}}{4}, \frac{5 - \sqrt{6}}{4}\right\}$.

6. Solve for x: $(2x - 6)^2 = 12$.

EXAMPLE 6 Solve for x : $(2x + 6)^2 = 8$.

SOLUTION

$$(2x + 6)^2 = 8$$
$$2x + 6 = \pm\sqrt{8}$$
$$2x + 6 = \pm 2\sqrt{2} \qquad \sqrt{8} = \sqrt{4 \cdot 2} = 2\sqrt{2}$$
$$2x = -6 \pm 2\sqrt{2} \qquad \text{Add } -6 \text{ to both sides}$$
$$x = \frac{-6 \pm 2\sqrt{2}}{2} \qquad \text{Divide each side by 2}$$

Answers

3. 3, −7 **4.** 6, −1 **5.** $\frac{3 \pm \sqrt{7}}{2}$

We can reduce the previous expression to lowest terms by factoring a 2 from each term in the numerator and then dividing that 2 by the 2 in the denominator. This is equivalent to dividing each term in the numerator by the 2 in the denominator. Here is what it looks like:

$$x = \frac{2(-3 \pm \sqrt{2})}{2} \quad \text{Factor a 2 from each term in numerator}$$

$$x = -3 \pm \sqrt{2} \quad \text{Divide numerator and denominator by 2}$$

The two solutions are $-3 + \sqrt{2}$ and $-3 - \sqrt{2}$.

We can check our two solutions in the original equation. Let's check our first solution, $-3 + \sqrt{2}$.

When $x = -3 + \sqrt{2}$

the equation $(2x + 6)^2 = 8$

becomes $[2(-3 + \sqrt{2}) + 6]^2 \stackrel{?}{=} 8$

$$(-6 + 2\sqrt{2} + 6)^2 = 8$$

$$(2\sqrt{2})^2 = 8$$

$$4 \cdot 2 = 8$$

$$8 = 8 \quad \text{A true statement}$$

The second solution, $-3 - \sqrt{2}$, checks also.

We are showing the check here so you can see that the irrational number $-3 + \sqrt{2}$ is a solution to $(2x + 6)^2 = 8$. Some people don't believe it at first.

B Applications Involving Quadratic Equations

EXAMPLE 7 If an object is dropped from a height of h feet, the amount of time in seconds it will take for the object to reach the ground (ignoring the resistance of air) is given by the formula

$$h = 16t^2$$

Solve this formula for t.

SOLUTION To solve for t, we apply the square root property:

$$h = 16t^2 \quad \text{Original formula}$$

$$\pm\sqrt{h} = 4t \quad \text{Square root property}$$

$$\pm\frac{\sqrt{h}}{4} = t \quad \text{Divide each side by 4}$$

Since t represents the time it takes for the object to fall h feet, t will never be negative. Therefore, the formula that gives t in terms of h is:

$$t = \frac{\sqrt{h}}{4}$$

Whenever we are solving an application problem like this one and we obtain a result that includes the ± sign, we must ask ourselves if the result actually can be negative. If it cannot be, we delete the negative result and use only the positive result.

7. The formula for the area of a circle is given by the equation
$A = \pi r^2$
Solve this formula for r.

Answers

6. $3 \pm \sqrt{3}$ **7.** $\sqrt{\frac{A}{\pi}}$

Facts From Geometry: Special Triangles

An *equilateral triangle* (Figure 1) is a triangle with three sides of equal length. If all three sides in a triangle have the same length, then the three interior angles in the triangle must also be equal. Since the sum of the interior angles in a triangle is always 180°, each of the three interior angles in any equilateral triangle must be 60°.

An *isosceles triangle* (Figure 2) is a triangle with two sides of equal length. Angles A and B in the isosceles triangle in Figure 2 are called the *base angles*; they are the angles opposite the two equal sides. In every isosceles triangle, the base angles are equal.

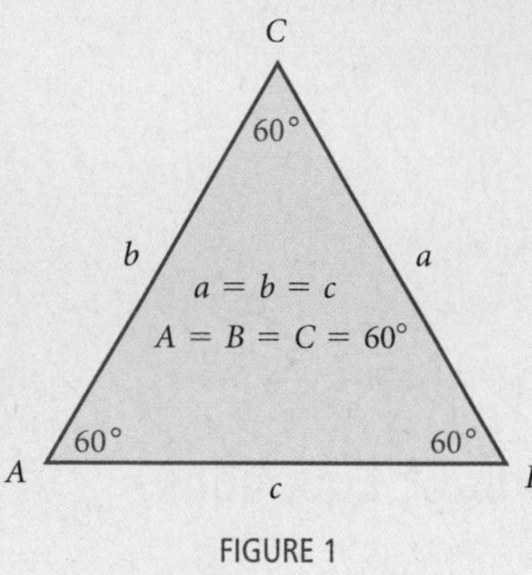

FIGURE 1

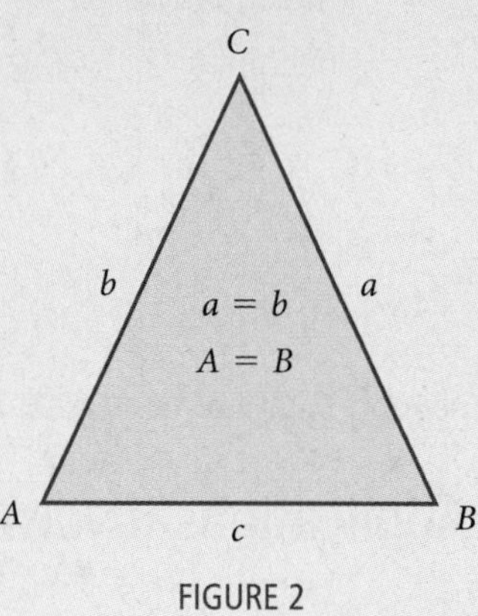

FIGURE 2

8. The lengths of all three sides of an equilateral triangle are 6 inches. Find the height of the triangle.

EXAMPLE 8 The lengths of all three sides of an equilateral triangle are 8 centimeters. Find the height of the triangle.

SOLUTION Because the three sides of an equilateral triangle are equal, the height always divides the base into two equal line segments. Figure 3 illustrates this fact. We find the height by applying the Pythagorean theorem.

$$8^2 = x^2 + 4^2$$
$$64 = x^2 + 16$$
$$x^2 = 64 - 16$$
$$x^2 = 48$$
$$x = \sqrt{48}$$
$$x = \sqrt{16 \cdot 3}$$
$$x = 4\sqrt{3} \text{ cm}$$

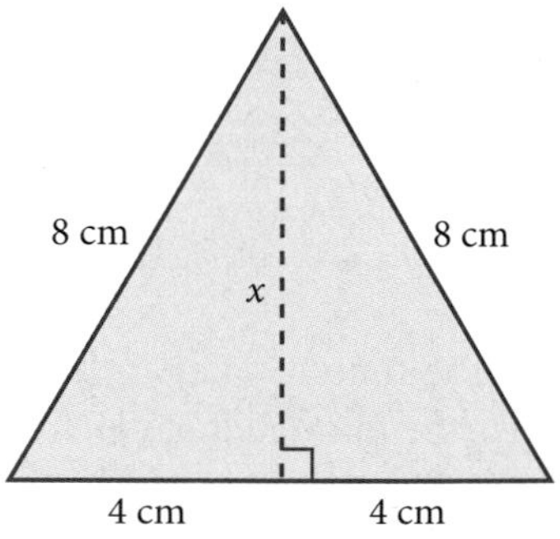

8 cm 8 cm x 4 cm 4 cm

FIGURE 3

Note A calculator gives a decimal approximation of 6.9 centimeters for the height of the triangle in Example 8.

Getting Ready for Class

After reading through the preceding section, respond in your own words and in complete sentences.

1. What is the square root property for equations?
2. Describe how you would solve the equation $x^2 = 7$.
3. What does the symbol $\pm$ stand for?
4. What is an isosceles triangle?

Answer
8. $3\sqrt{3}$ inches

Problem Set 9.1

A Solve each of the following equations using the methods learned in this section. [Examples 1–6]

1. $x^2 = 9$

2. $x^2 = 16$

3. $a^2 = 25$

4. $a^2 = 36$

5. $y^2 = 8$

6. $y^2 = 75$

7. $2x^2 = 100$

8. $2x^2 = 54$

9. $3a^2 = 54$

10. $2a^2 = 64$

11. $(x + 2)^2 = 4$

12. $(x - 3)^2 = 16$

13. $(x + 1)^2 = 25$

14. $(x + 3)^2 = 64$

15. $(a - 5)^2 = 75$

16. $(a - 4)^2 = 32$

17. $(y + 1)^2 = 50$

18. $(y - 5)^2 = 27$

19. $(2x + 1)^2 = 25$

20. $(3x - 2)^2 = 16$

21. $(4a - 5)^2 = 36$

22. $(2a + 6)^2 = 64$

23. $(3y - 1)^2 = 12$

24. $(5y - 4)^2 = 12$

25. $(6x + 2)^2 = 27$

26. $(8x - 1)^2 = 20$

27. $(3x - 9)^2 = 27$

28. $(2x + 8)^2 = 32$

29. $(3x + 6)^2 = 45$

30. $(5x - 10)^2 = 75$

31. $(2y - 4)^2 = 8$

32. $(4y - 6)^2 = 48$

33. $\left(x - \frac{2}{3}\right)^2 = \frac{25}{9}$ **34.** $\left(x - \frac{3}{4}\right)^2 = \frac{49}{16}$ **35.** $\left(x + \frac{1}{2}\right)^2 = \frac{7}{4}$ **36.** $\left(x + \frac{1}{3}\right)^2 = \frac{5}{9}$

37. $\left(a - \frac{4}{5}\right)^2 = \frac{12}{25}$ **38.** $\left(a - \frac{3}{7}\right)^2 = \frac{18}{49}$

Since $a^2 + 2ab + b^2$ can be written as $(a + b)^2$, each of the following equations can be solved using our square root method. The first step is to write the trinomial on the left side of the equal sign as the square of a binomial. Solve each equation.

39. $x^2 + 10x + 25 = 7$ **40.** $x^2 + 6x + 9 = 11$ **41.** $x^2 - 2x + 1 = 9$

42. $x^2 + 8x + 16 = 25$ **43.** $x^2 + 12x + 36 = 8$ **44.** $x^2 - 4x + 4 = 12$

45. For the equation $x^2 = 3$.
a. Can you solve it by factoring?
b. Solve it

46. Consider the equation $x^2 - 5 = 0$.
a. Can you solve it by factoring?
b. Solve it

47. Consider the equation $(x - 3)^2 = 4$.
a. Can it be solved by factoring?
b. Solve it

48. For the equation $x^2 - 10x + 25 = 1$.
a. Can it be solved by factoring?
b. Solve it

49. Is $3 + \sqrt{3}$ a solution to $(2x - 6)^2 = 12$?

50. Is $2 + \sqrt{2}$ a solution to $(3x + 6)^2 = 18$?

The next two problems will give you practice with a variety of equations.

51. Solve each equation.
a. $2x - 1 = 0$
b. $2x - 1 = 4$
c. $(2x - 1)^2 = 4$
d. $\sqrt{2x} - 1 = 0$
e. $\frac{1}{2x} - 1 = \frac{1}{4}$

52. Solve each equation.
a. $x + 5 = 0$
b. $x + 5 = 8$
c. $(x + 5)^2 = 8$
d. $\sqrt{x} + 5 = 8$
e. $\frac{1}{x} + 5 = \frac{1}{2}$

B Applying the Concepts [Examples 7–9]

53. **Checking Solutions** Check the solution

$$x = -1 + 5\sqrt{2}$$

in the equation $(x + 1)^2 = 50$.

54. **Checking Solutions** Check the solution

$$x = -8 + 2\sqrt{6}$$

in the equation $(x + 8)^2 = 24$.

55. The equation $x^2 - 10x + 22 = 0$ has solutions $5 + \sqrt{3}$ and $5 - \sqrt{3}$. Find:
   a. The sum of the solutions: $(5 + \sqrt{3}) + (5 - \sqrt{3})$
   b. The product of the solutions: $(5 + \sqrt{3})(5 - \sqrt{3})$

56. The equation $x^2 + 14x + 44 = 0$ has solutions $-7 + \sqrt{5}$ and $-7 - \sqrt{5}$. Find:
   a. The sum of the solutions
   b. The product of the solutions

57. **Geometry** The area of a circle with radius r is $A = \pi r^2$. Find the radius if the area is 36π square feet.

58. **Geometry** Solve the formula $A = \pi r^2$ for r.

59. **Falling Object** A baseball is dropped from the top of a 100-ft building. It will take t seconds to hit the ground, according to the formula $100 = 16t^2$. Find t.

60. **Ball Toss** A ball thrown straight up into the air takes t seconds to reach a height of 25 feet, according to the formula $25 = 16t^2$. Find t.

61. **Number Problem** The square of the sum of a number and 3 is 16. Find the number. (There are two solutions.)

62. **Number Problem** The square of the sum of twice a number and 3 is 25. Find the number. (There are two solutions.)

63. **Investing** If you invest \$100 in an account with interest rate r compounded annually, the amount of money A in the account after 2 years is given by the formula

$$A = 100(1 + r)^2$$

Solve this formula for r.

64. **Investing** If you invest P dollars in an account with interest rate r compounded annually, the amount of money A in the account after 2 years is given by the formula

$$A = P(1 + r)^2$$

Solve this formula for r.

65. Geometry The lengths of all three sides of an equilateral triangle are 10 feet. Find the height of the triangle.

66. Geometry The lengths of all three sides of an equilateral triangle are 12 meters. Find the height of the triangle.

67. Geometry The front of a tent forms an equilateral triangle with sides of 6 feet. Can a person 5 feet 8 inches tall stand up inside the tent?

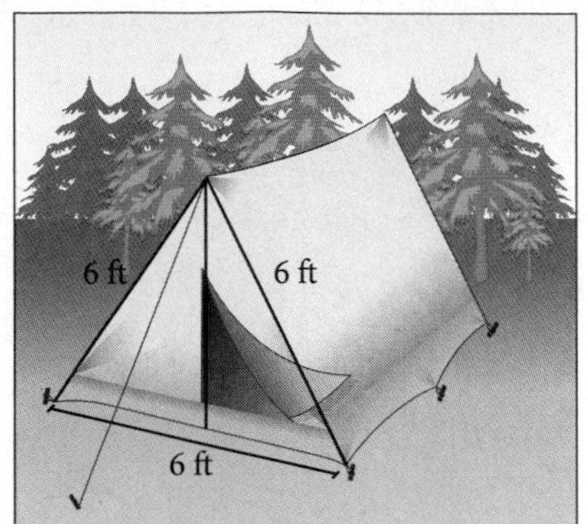

68. Geometry The front of a tent forms an equilateral triangle. The tent must be constructed so that a person 6 feet tall can stand up inside. Find the length of the three sides of the front of the tent.

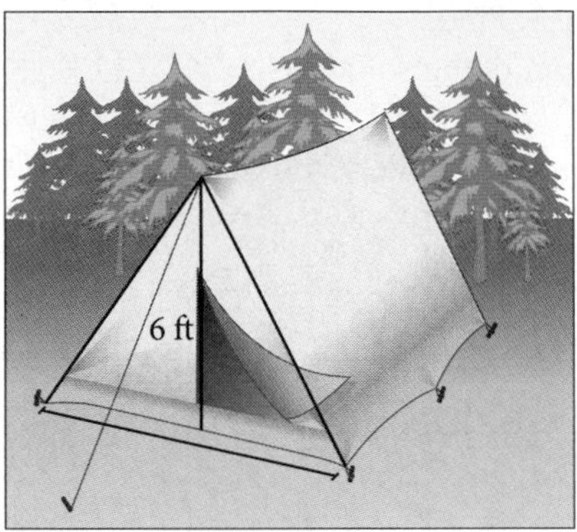

69. Geometry The base of an isosceles triangle is 8 feet, and the length of the two equal sides is 5 feet. Find the height of the triangle.

70. Geometry An isosceles triangle has a base 8 feet long. If the length of the two equal sides is 6 feet, find the height of the triangle.

Displacement The displacement, in cubic inches, of a car engine is given by the formula

$$d = \pi \cdot s \cdot c \cdot \left(\frac{1}{2} \cdot b\right)^2$$

where s is the stroke and b is the bore, as shown here, and c is the number of cylinders. Calculate the bore for each of the following cars. Use 3.14 to approximate π. Round answers to the nearest tenth.

71. BMW M Coupe Six cylinders, 3.53 inches of stroke, and 192 cubic inches of displacement

72. Chevrolet Corvette Eight cylinders, 3.62 inches of stroke, and 346 cubic inches of displacement

Getting Ready for the Next Section

Simplify.

73. $\left(\frac{1}{2} \cdot 18\right)^2$

74. $\left(\frac{1}{2} \cdot 8\right)^2$

75. $\left[\frac{1}{2}(-2)\right]^2$

76. $\left[\frac{1}{2}(-10)\right]^2$

77. $\left(\frac{1}{2} \cdot 3\right)^2$

78. $\left(\frac{1}{2} \cdot 5\right)^2$

79. $\frac{2x^2 + 16}{2}$

80. $\frac{3x^2 - 3x}{3}$

Factor.

81. $x^2 + 6x + 9$

82. $x^2 + 12x + 36$

83. $y^2 - 3y + \frac{9}{4}$

84. $y^2 - 5y + \frac{25}{4}$

Maintaining Your Skills

Multiply.

85. $(x - 5)^2$

86. $(x + 5)^2$

Factor.

87. $x^2 - 12x + 36$

88. $x^2 + 12x + 36$

89. $x^2 + 4x + 4$

90. $x^2 - 4x + 4$

Find the following roots.

91. $\sqrt[3]{8}$

92. $\sqrt[3]{27}$

93. $\sqrt[4]{16}$

94. $\sqrt[4]{81}$

95. **Pitchers** The chart shows the number of strikeouts for active starting pitchers with the most strikeouts per game. Using the chart,and knowing he threw 2,734 innings, how many strikeouts did Pedro Martinez throw per inning? Round to the nearest hundredth.

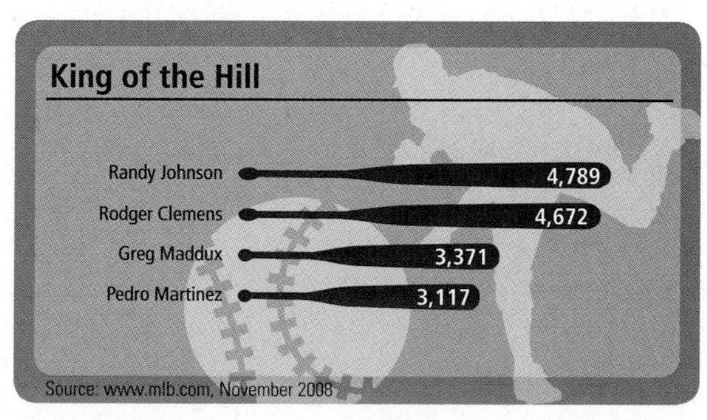

96. **100 Meters** The chart shows the fastest times for the women's 100 meters in the Olympics. Use the chart to find Merlene Ottey's speed in miles per hour. Round to the nearest tenth. [Hint: 1 mile = 1609.34 meters.]

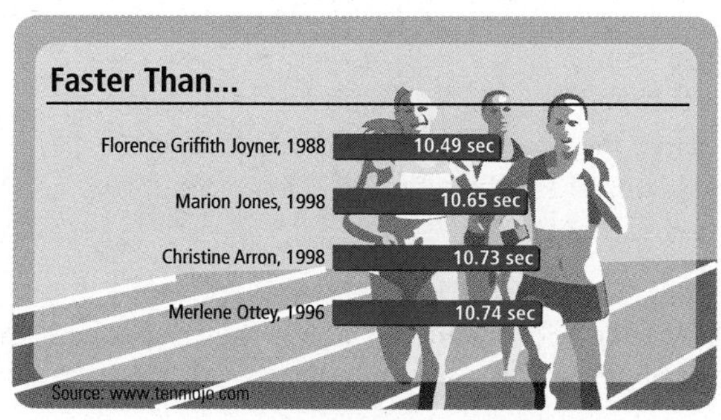

Objectives

A Solve a quadratic equation by completing the square.

Examples now playing at MathTV.com/books

A Completing the Square

In this section we will develop a method of solving quadratic equations that works whether or not the equation can be factored. Since we will be working with the individual terms of trinomials, we need some new definitions so that we can keep our vocabulary straight.

Definition

In the trinomial $2x^2 + 3x + 4$, the first term, $2x^2$, is called the **quadratic term**; the middle term, $3x$, is called the **linear term**; and the last term, 4, is called the **constant term**.

Now consider the following list of perfect square trinomials and their corresponding binomial squares:

$$x^2 + 6x + 9 = (x + 3)^2$$
$$x^2 - 8x + 16 = (x - 4)^2$$
$$x^2 - 10x + 25 = (x - 5)^2$$
$$x^2 + 12x + 36 = (x + 6)^2$$

In each case the coefficient of x^2 is 1. A more important observation, however, stems from the relationship between the linear terms (middle terms) and the constant terms (last terms). Notice that the constant in each case is the square of half the coefficient of x in the linear term; that is,

1. For the first trinomial, $x^2 + 6x + 9$, the last term, 9, is the square of half the coefficient of the middle term: $9 = \left(\frac{6}{2}\right)^2$.
2. For the second trinomial, $x^2 - 8x + 16$, we have $16 = \left[\frac{1}{2}(-8)\right]^2$.
3. For the trinomial $x^2 - 10x + 25$, it also holds: $25 = \left[\frac{1}{2}(-10)\right]^2$.

Check and see that it also works for the final trinomial.

In summary, then, for every perfect square trinomial in which the coefficient of x^2 is 1, the final term is always the square of half the coefficient of the linear term. We can use this fact to build our own perfect square trinomials.

EXAMPLES Write the correct final term to each of the following expressions so that each becomes a perfect square trinomial and then factor.

1. $x^2 - 2x$

SOLUTION The coefficient of the linear term is -2. If we take half of -2, we get -1, the square of which is 1. Adding the 1 as the final term, we have the perfect square trinomial:

$$x^2 - 2x + 1 = (x - 1)^2$$

2. $x^2 + 18x$

SOLUTION Half of 18 is 9, the square of which is 81. If we add 81 at the end, we have:

$$x^2 + 18x + 81 = (x + 9)^2$$

3. $x^2 + 3x$

SOLUTION Half of 3 is $\frac{3}{2}$, the square of which is $\frac{9}{4}$:

$$x^2 + 3x + \frac{9}{4} = \left(x + \frac{3}{2}\right)^2$$

PRACTICE PROBLEMS

Add the correct term to each binomial so that it becomes a perfect square trinomial and then factor.

1. $x^2 + 2x$
2. $x^2 - 10x$
3. $x^2 + 5x$

Answers

1. $x^2 + 2x + 1 = (x + 1)^2$
2. $x^2 - 10x + 25 = (x - 5)^2$
3. $x^2 + 5x + \frac{25}{4} = \left(x + \frac{5}{2}\right)^2$

We can use this procedure, along with the method developed in Section 9.1, to solve some quadratic equations.

4. Solve $x^2 - 6x - 7 = 0$ by completing the square.

EXAMPLE 4

Solve $x^2 - 6x + 5 = 0$ by completing the square.

SOLUTION We begin by adding -5 to both sides of the equation. We want just $x^2 - 6x$ on the left side so that we can add on our own final term to get a perfect square trinomial:

$$x^2 - 6x + 5 = 0$$

$$x^2 - 6x = -5 \quad \text{Add } -5 \text{ to both sides}$$

Now we can add 9 to both sides and the left side will be a perfect square:

$$x^2 - 6x + \mathbf{9} = -5 + \mathbf{9}$$

$$(x - 3)^2 = 4$$

The final line is in the form of the equations we solved in Section 9.1:

$$x - 3 = \pm 2 \quad \text{Square root property}$$

$$x = 3 \pm 2 \quad \text{Add 3 to both sides}$$

$$x = 3 + 2 \quad \text{or} \quad x = 3 - 2$$

$$x = 5 \quad \text{or} \quad x = 1$$

The two solutions are 5 and 1.

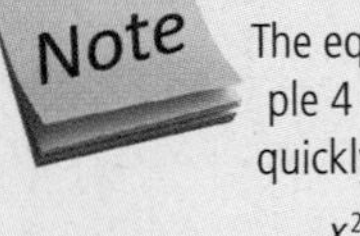

The equation in Example 4 can be solved quickly by factoring:

$$x^2 - 6x + 5 = 0$$

$$(x - 5)(x - 1) = 0$$

$$x - 5 = 0 \quad \text{or} \quad x - 1 = 0$$

$$x = 5 \quad \text{or} \quad x = 1$$

The reason we didn't solve it by factoring is that we want to practice completing the square on some simple equations.

The preceding method of solution is called *completing the square.*

5. Solve $3x^2 - 3x - 18 = 0$ by completing the square.

EXAMPLE 5

Solve by completing the square $2x^2 + 16x - 18 = 0$.

SOLUTION We begin by moving the constant term to the other side:

$$2x^2 + 16x - 18 = 0$$

$$2x^2 + 16x = 18 \quad \text{Add 18 to both sides}$$

To complete the square, we must be sure the coefficient of x^2 is 1. To accomplish this, we divide both sides by 2:

$$\frac{2x^2}{2} + \frac{16x}{2} = \frac{18}{2}$$

$$x^2 + 8x = 9$$

We now complete the square by adding the square of half the coefficient of the linear term to both sides:

$$x^2 + 8x + \mathbf{16} = 9 + \mathbf{16} \quad \text{Add 16 to both sides}$$

$$(x + 4)^2 = 25$$

$$x + 4 = \pm 5 \quad \text{Square root property}$$

$$x = -4 \pm 5 \quad \text{Add } -4 \text{ to both sides}$$

$$x = -4 + 5 \quad \text{or} \quad x = -4 - 5$$

$$x = 1 \quad \text{or} \quad x = -9$$

The solution set arrived at by completing the square is $\{1, -9\}$.

Answers
4. $-1, 7$ 5. $-2, 3$

We will now summarize the preceding examples by listing the steps involved in solving quadratic equations by completing the square.

Strategy Solving a Quadratic Equation by Completing the Square

Step 1: Put the equation in the form $ax^2 + bx = c$. This usually involves moving only the constant term to the opposite side.

Step 2: Make sure the coefficient of the squared term is 1. If it is not 1, simply divide both sides by whatever it is.

Step 3: Add the square of half the coefficient of the linear term to both sides of the equation.

Step 4: Write the left hand side of the equation as a binomial square, apply the square root property, and solve.

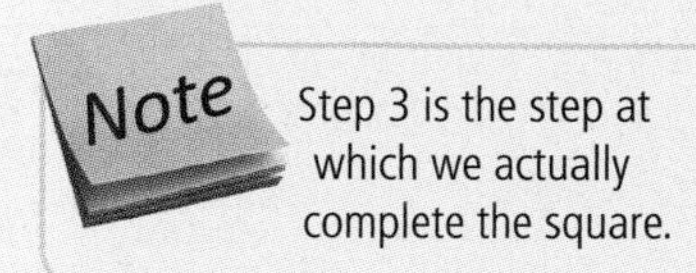

Note
Step 3 is the step at which we actually complete the square.

Here is one final example.

EXAMPLE 6 Solve for y: $3y^2 - 9y + 3 = 0$.

6. Solve for y: $2y^2 - 8y + 2 = 0$.

SOLUTION

$$3y^2 - 9y + 3 = 0$$

$$3y^2 - 9y = -3 \qquad \text{Add } -3 \text{ to both sides}$$

$$y^2 - 3y = -1 \qquad \text{Divide by 3}$$

$$y^2 - 3y + \frac{9}{4} = -1 + \frac{9}{4} \qquad \text{Complete the square}$$

$$\left(y - \frac{3}{2}\right)^2 = \frac{5}{4} \qquad -1 + \frac{9}{4} = -\frac{4}{4} + \frac{9}{4} = \frac{5}{4}$$

$$y - \frac{3}{2} = \pm\frac{\sqrt{5}}{2} \qquad \text{Square root property}$$

$$y = \frac{3}{2} \pm \frac{\sqrt{5}}{2} \qquad \text{Add } \frac{3}{2} \text{ to both sides}$$

$$y = \frac{3}{2} + \frac{\sqrt{5}}{2} \quad \text{or} \quad y = \frac{3}{2} - \frac{\sqrt{5}}{2}$$

$$y = \frac{3 + \sqrt{5}}{2} \quad \text{or} \quad y = \frac{3 - \sqrt{5}}{2}$$

The solutions are $\frac{3 + \sqrt{5}}{2}$ and $\frac{3 - \sqrt{5}}{2}$, which can be written in a shorter form as:

$$\frac{3 \pm \sqrt{5}}{2}$$

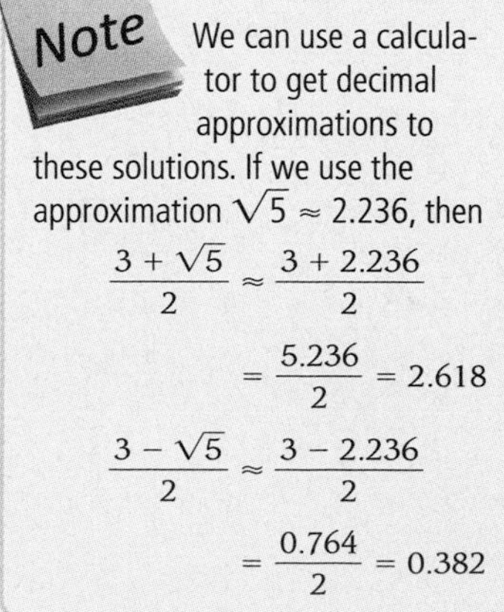

Note
We can use a calculator to get decimal approximations to these solutions. If we use the approximation $\sqrt{5} \approx 2.236$, then
$\frac{3 + \sqrt{5}}{2} \approx \frac{3 + 2.236}{2} = \frac{5.236}{2} = 2.618$
$\frac{3 - \sqrt{5}}{2} \approx \frac{3 - 2.236}{2} = \frac{0.764}{2} = 0.382$

Answer

6. $2 \pm \sqrt{3}$

Getting Ready for Class

After reading through the preceding section, respond in your own words and in complete sentences.

1. What kind of equation do we solve using the method of completing the square?

2. Explain in words how you would complete the square on $x^2 - 6x = 5$.

3. What is the first step in completing the square?

4. What is the linear term in a trinomial?

Problem Set 9.2

A Give the correct final term for each of the following expressions to ensure that the resulting trinomial is a perfect square trinomial. [Examples 1–3]

1. $x^2 + 6x$

2. $x^2 - 10x$

3. $x^2 + 2x$

4. $x^2 + 14x$

5. $y^2 - 8y$

6. $y^2 + 12y$

7. $y^2 - 2y$

8. $y^2 - 6y$

9. $x^2 + 16x$

10. $x^2 - 4x$

11. $a^2 - 3a$

12. $a^2 + 5a$

13. $x^2 - 7x$

14. $x^2 - 9x$

15. $y^2 + y$

16. $y^2 - y$

17. $x^2 - \frac{3}{2}x$

18. $x^2 + \frac{2}{3}x$

A Solve each of the following equations by completing the square. Follow the steps in the Strategy box given at the end of this section. [Examples 4–6]

19. $x^2 + 4x = 12$

20. $x^2 - 2x = 8$

21. $x^2 - 6x = 16$

22. $x^2 + 12x = -27$

23. $a^2 + 2a = 3$

24. $a^2 - 8a = -7$

25. $x^2 - 10x = 0$

26. $x^2 + 4x = 0$

27. $y^2 + 2y - 15 = 0$

28. $y^2 - 10y - 11 = 0$

29. $x^2 + 4x - 3 = 0$

30. $x^2 + 6x + 5 = 0$

31. $x^2 - 4x = 4$

32. $x^2 + 4x = -1$

33. $a^2 = 7a + 8$

34. $a^2 = 3a + 1$

35. $4x^2 + 8x - 4 = 0$

36. $3x^2 + 12x + 6 = 0$

37. $2x^2 + 2x - 4 = 0$

38. $4x^2 + 4x - 3 = 0$

39. $4x^2 + 8x + 1 = 0$

40. $3x^2 + 6x + 2 = 0$

41. $2x^2 - 2x = 1$

42. $3x^2 - 3x = 1$

43. $4a^2 - 4a + 1 = 0$

44. $2a^2 + 4a + 1 = 0$

45. $3y^2 - 9y = 2$

46. $5y^2 - 10y = 4$

47. Solve each equation $x^2 - 2x = 0$.

a. By factoring

b. By completing the square

48. Solve each equation $x^2 + 3x = 40$.

a. By factoring

b. By completing the square

49. Is $x = 3 - \sqrt{2}$ a solution to $x^2 + 6x = -7$.

50. Is $x = 2 + \sqrt{2}$ a solution to $x^2 - 4x = -1$.

51. The equation $x^2 - 6x = 0$ can be solved by completing the square. Is there a faster way to solve it?

52. The equation $(11a - 24)^2 = 144$ can be solved by expanding the left side then writng it in standard form and factoring. Is there an easier way?

53. Suppose you solve the equation $x^2 - 6x = 7$ by completing the square, and then find the solutions are -1 and 7. Was there another way to solve the equation?

54. Suppose you solve the equation $x^2 - 5x = 24$ by completing the square, and then find the solutions are -3 and 8. Was there another way to solve the equation?

Applying the Concepts

55. Computer Screen An advertisement for a laptop computer indicates it has a 14-inch viewing screen. This means that the diagonal of the screen measures 14 inches. If the ratio of the length to the width of the screen is 3 to 4, we can represent the length with $4x$ and the width with $3x$, and then use the Pythagorean theorem to solve for x. Once we have x, the length will be $4x$ and the width will be $3x$. Find the length and width of this computer screen to the nearest tenth of an inch.

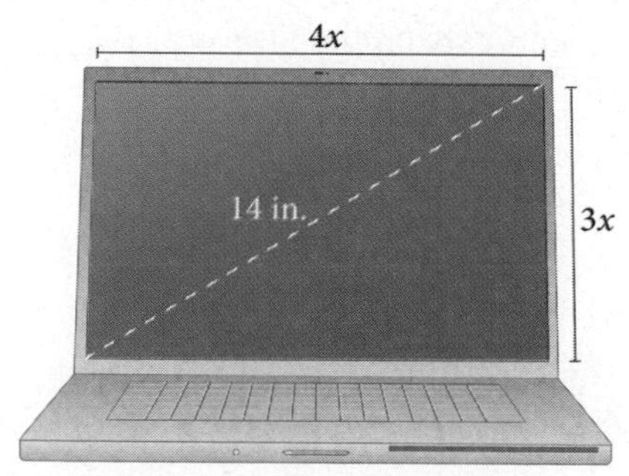

56. Television Screen A 25-inch television has a rectangular screen on which the diagonal measures 25 inches. The ratio of the length to the width of the screen is 3 to 4. Let the length equal $4x$ and the width equal $3x$, and then use the Pythagorean theorem to solve for x. Then find the length and width of this television screen.

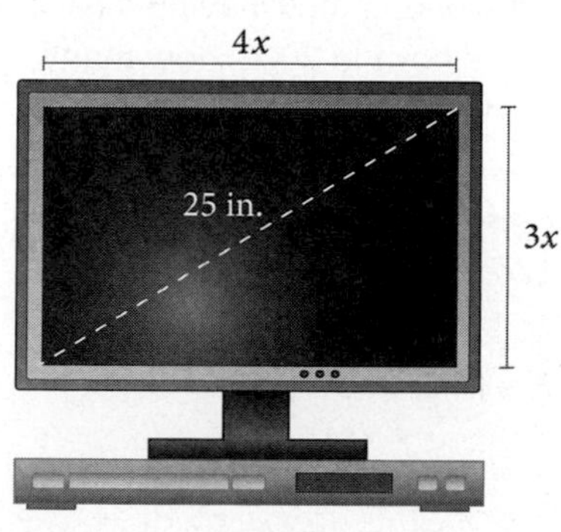

Getting Ready for the Next Section

Write the quadratic equation in standard form ($ax^2 + bx + c = 0$).

57. $2x^2 = -4x + 3$ **58.** $3x^2 = -4x + 2$ **59.** $(x - 2)(x + 3) = 5$ **60.** $(x - 1)(x + 2) = 4$

Identify the coefficient of x^2, the coefficient of x, and the constant term.

61. $x^2 - 5x - 6$ **62.** $x^2 - 6x + 7$ **63.** $2x^2 + 4x - 3$ **64.** $3x^2 + 4x - 2$

Find the value of $b^2 - 4ac$ for the given values of a, b, and c.

65. $a = 1, b = -5, c = -6$ **66.** $a = 1, b = -6, c = 7$ **67.** $a = 2, b = 4, c = -3$ **68.** $a = 3, b = 4, c = -2$

Simplify.

69. $\dfrac{5 + \sqrt{49}}{2}$ **70.** $\dfrac{5 - \sqrt{49}}{2}$ **71.** $\dfrac{-4 - \sqrt{40}}{4}$ **72** $\dfrac{-4 + \sqrt{40}}{4}$

Maintaining Your Skills

Find the value of each expression if $a = 2$, $b = 4$, and $c = -3$.

73. $2a$ **74.** b^2 **75.** $4ac$ **76.** $b^2 - 4ac$

77. $\sqrt{b^2 - 4ac}$ **78.** $-b + \sqrt{b^2 - 4ac}$

Put each expression in simplified form for radicals.

79. $\sqrt{12}$ **80.** $\sqrt{50x^2}$ **81.** $\sqrt{20x^2y^3}$ **82.** $3\sqrt{48x^4}$ **83.** $\sqrt{\dfrac{81}{25}}$ **84.** $\dfrac{6\sqrt{8x^2y}}{\sqrt{9}}$

85. Mothers The chart shows the percentage of women who continue working after having a baby. If 4 million babies were born in 2002, how many mothers remained at home with them?

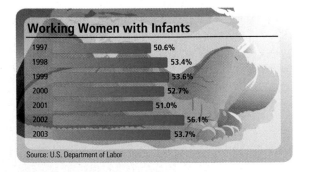

Working Women with Infants

Year	Percent
1997	50.6%
1998	53.4%
1999	53.6%
2000	52.7%
2001	51.0%
2002	56.1%
2003	53.7%

Source: U.S. Department of Labor

86. Health Care The graph shows the rising cost of health care. What is the estimated percent increase between 2011 and 2014? Round to the nearest tenth of a percent.

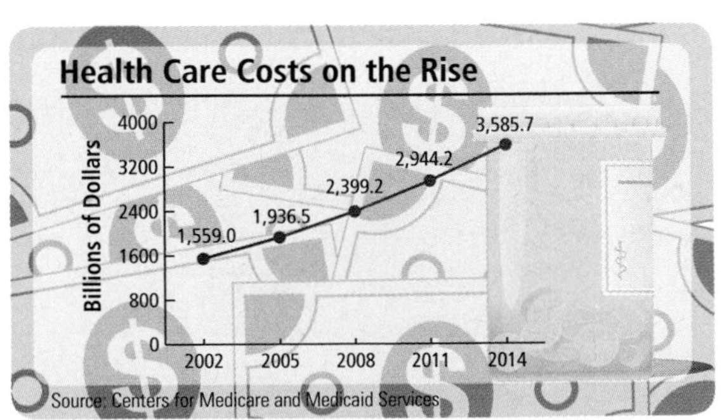

The Quadratic Formula

9.3

Objectives

A Solve a quadratic equation by using the quadratic formula.

Examples now playing at MathTV.com/books

In this section we will derive the quadratic formula. It is one formula that you will use in almost all types of mathematics. We will first state the formula as a theorem and then prove it. The proof is based on the method of completing the square developed in the preceding section.

A The Quadratic Formula

The Quadratic Theorem

For any quadratic equation in the form $ax^2 + bx + c = 0$, where a, b, and c are real numbers and $a \neq 0$, the two solutions are:

$$x = \frac{-b + \sqrt{b^2 - 4ac}}{2a} \quad \text{and} \quad x = \frac{-b - \sqrt{b^2 - 4ac}}{2a}$$

Proof We will prove the theorem by completing the square on:

$$ax^2 + bx + c = 0$$

Adding $-c$ to both sides, we have:

$$ax^2 + bx = -c$$

To make the coefficient of x^2 one, we divide both sides by a:

$$\frac{ax^2}{a} + \frac{bx}{a} = -\frac{c}{a}$$

$$x^2 + \frac{b}{a}x = -\frac{c}{a}$$

Now, to complete the square, we add the square of half of $\frac{b}{a}$ to both sides:

$$x^2 + \frac{b}{a}x + \left(\frac{b}{2a}\right)^2 = -\frac{c}{a} + \left(\frac{b}{2a}\right)^2 \qquad \frac{1}{2} \text{ of } \frac{b}{a} \text{ is } \frac{b}{2a}$$

Let's simplify the right side separately:

$$-\frac{c}{a} + \left(\frac{b}{2a}\right)^2 = -\frac{c}{a} + \frac{b^2}{4a^2}$$

The least common denominator is $4a^2$. We multiply the numerator and denominator of $-\frac{c}{a}$ by $4a$ to give it the common denominator. Then we combine numerators:

$$\frac{\mathbf{4a}}{\mathbf{4a}}\left(-\frac{c}{a}\right) + \frac{b^2}{4a^2} = -\frac{4ac}{4a^2} + \frac{b^2}{4a^2}$$

$$= \frac{-4ac + b^2}{4a^2}$$

$$= \frac{b^2 - 4ac}{4a^2}$$

Now, back to the equation. We use our simplified expression for the right side:

$$x^2 + \frac{b}{a}x + \left(\frac{b}{2a}\right)^2 = \frac{b^2 - 4ac}{4a^2}$$

$$\left(x + \frac{b}{2a}\right)^2 = \frac{b^2 - 4ac}{4a^2}$$

This is one of the few times in the course where we actually get to show a proof. The proof shown here is the reason the quadratic formula looks the way it does. We will use the quadratic formula in every example we do in this section. As you read through the examples, you may find yourself wondering why some parts of the formula are the way they are. If that happens, come back to this proof and see for yourself.

Applying the square root property, we have:

$$x + \frac{b}{2a} = \pm\frac{\sqrt{b^2 - 4ac}}{2a}$$

$$x = \frac{-b}{2a} \pm \frac{\sqrt{b^2 - 4ac}}{2a} \qquad \text{Add } \frac{-b}{2a} \text{ to both sides}$$

$$x = \frac{-b \pm \sqrt{b^2 - 4ac}}{2a}$$

Note This formula is called the quadratic formula. You will see it many times if you continue taking math classes. By the time you are finished with this section and the problems in the problem set, you should have it memorized.

Our proof is now complete. What we have is this: If our equation is in the form $ax^2 + bx + c = 0$ (standard form), then the solution can always be found by using the quadratic formula:

$$x = \frac{-b \pm \sqrt{b^2 - 4ac}}{2a}$$

PRACTICE PROBLEMS

1. Use the quadratic formula to solve $x^2 - 3x + 2 = 0$.

EXAMPLE 1 Solve $x^2 - 5x - 6 = 0$ by using the quadratic formula.

SOLUTION To use the quadratic formula, we must make sure the equation is in standard form; identify a, b, and c; substitute them into the formula; and work out the arithmetic.

For the equation $x^2 - 5x - 6 = 0$, $a = 1$, $b = -5$, and $c = -6$:

$$x = \frac{-b \pm \sqrt{b^2 - 4ac}}{2a} = \frac{-(-5) \pm \sqrt{(-5)^2 - 4(1)(-6)}}{2(1)}$$

$$= \frac{5 \pm \sqrt{49}}{2}$$

$$= \frac{5 \pm 7}{2}$$

$$x = \frac{5 + 7}{2} \quad \text{or} \quad x = \frac{5 - 7}{2}$$

$$x = \frac{12}{2} \qquad x = -\frac{2}{2}$$

$$x = 6 \qquad x = -1$$

The two solutions are 6 and -1.

Note Whenever the solutions to our quadratic equations turn out to be rational numbers, as in Example 1, it means the original equation could have been solved by factoring. (We didn't solve the equation in Example 1 by factoring because we were trying to get some practice with the quadratic formula.)

2. Solve for x: $3x^2 = -4x + 2$.

EXAMPLE 2 Solve for x : $2x^2 = -4x + 3$.

SOLUTION Before we can identify a, b, and c, we must write the equation in standard form. To do so, we add $4x$ and -3 to each side of the equation:

$$2x^2 = -4x + 3$$

$$2x^2 + 4x - 3 = 0 \qquad \text{Add } 4x \text{ and } -3 \text{ to each side}$$

Now that the equation is in standard form, we see that $a = 2$, $b = 4$, and $c = -3$. Using the quadratic formula we have:

$$x = \frac{-b \pm \sqrt{b^2 - 4ac}}{2a}$$

$$= \frac{-4 \pm \sqrt{4^2 - 4(2)(-3)}}{2(2)}$$

$$= \frac{-4 \pm \sqrt{40}}{4}$$

$$= \frac{-4 \pm 2\sqrt{10}}{4}$$

Answer

1. 2, 1

We can reduce the final expression in the preceding equation to lowest terms by factoring 2 from the numerator and denominator and then dividing it out:

$$x = \frac{\not{2}(-2 \pm \sqrt{10})}{\not{2} \cdot 2}$$

$$= \frac{-2 \pm \sqrt{10}}{2}$$

Our two solutions are $\frac{-2 + \sqrt{10}}{2}$ and $\frac{-2 - \sqrt{10}}{2}$.

EXAMPLE 3 Solve for x: $(x - 2)(x + 3) = 5$.

3. Solve for x: $(x - 1)(x + 2) = 4$.

SOLUTION We must put the equation into standard form before we can use the quadratic formula:

$$(x - 2)(x + 3) = 5$$

$$x^2 + x - 6 = 5 \quad \text{Multiply out the left side}$$

$$x^2 + x - 11 = 0 \quad \text{Add } -5 \text{ to each side}$$

Now, $a = 1$, $b = 1$, and $c = -11$; therefore:

$$x = \frac{-1 \pm \sqrt{1^2 - 4(1)(-11)}}{2(1)}$$

$$= \frac{-1 \pm \sqrt{45}}{2}$$

$$= \frac{-1 \pm 3\sqrt{5}}{2}$$

The solution set is $\left\{\frac{-1 + 3\sqrt{5}}{2}, \frac{-1 - 3\sqrt{5}}{2}\right\}$.

EXAMPLE 4 Solve $x^2 - 6x = -7$.

4. Solve $x^2 - 4x = -1$.

SOLUTION We begin by writing the equation in standard form:

$$x^2 - 6x = -7$$

$$x^2 - 6x + 7 = 0 \quad \text{Add 7 to each side}$$

Using $a = 1$, $b = -6$, and $c = 7$ in the quadratic formula

$$x = \frac{-b \pm \sqrt{b^2 - 4ac}}{2a}$$

we have:

$$x = \frac{-(-6) \pm \sqrt{(-6)^2 - 4(1)(7)}}{2(1)}$$

$$= \frac{6 \pm \sqrt{36 - 28}}{2}$$

$$= \frac{6 \pm \sqrt{8}}{2}$$

$$= \frac{6 \pm 2\sqrt{2}}{2}$$

Answers

2. $\frac{-2 \pm \sqrt{10}}{3}$ 3. 2, −3

The two terms in the numerator have a 2 in common. We reduce to lowest terms by factoring the 2 from the numerator and then dividing numerator and denominator by 2:

$$= \frac{\cancel{2}(3 \pm \sqrt{2})}{\cancel{2}}$$

$$= 3 \pm \sqrt{2}$$

The two solutions are $3 + \sqrt{2}$ and $3 - \sqrt{2}$. This time, let's check our solutions in the original equation $x^2 - 6x = -7$.

Checking $x = 3 + \sqrt{2}$, we have:

$$(3 + \sqrt{2})^2 - 6(3 + \sqrt{2}) \stackrel{?}{=} -7$$

$$9 + 6\sqrt{2} + 2 - 18 - 6\sqrt{2} = -7 \quad \text{Multiply}$$

$$11 - 18 + 6\sqrt{2} - 6\sqrt{2} = -7 \quad \text{Add 9 and 2}$$

$$-7 + 0 = -7 \quad \text{Subtraction}$$

$$-7 = -7 \quad \text{A true statement}$$

Checking $x = 3 - \sqrt{2}$, we have:

$$(3 - \sqrt{2})^2 - 6(3 - \sqrt{2}) \stackrel{?}{=} -7$$

$$9 - 6\sqrt{2} + 2 - 18 + 6\sqrt{2} = -7 \quad \text{Multiply}$$

$$11 - 18 - 6\sqrt{2} + 6\sqrt{2} = -7 \quad \text{Add 9 and 2}$$

$$-7 + 0 = -7 \quad \text{Subtraction}$$

$$-7 = -7 \quad \text{A true statement}$$

As you can see, both solutions yield true statements when used in place of the variable in the original equation.

Getting Ready for Class

After reading through the preceding section, respond in your own words and in complete sentences.

1. What is the quadratic formula?
2. Under what circumstances should the quadratic formula be applied?
3. What is standard form for a quadratic equation?
4. What is the first step in solving a quadratic equation using the quadratic formula?

Answer
4. $2 \pm \sqrt{3}$

Problem Set 9.3

A Solve the following equations by using the quadratic formula. [Examples 1–4]

1. $x^2 + 3x + 2 = 0$
2. $x^2 - 5x + 6 = 0$
3. $x^2 + 5x + 6 = 0$
4. $x^2 - 7x - 8 = 0$
5. $x^2 + 6x + 9 = 0$
6. $x^2 - 10x + 25 = 0$
7. $x^2 + 6x + 7 = 0$
8. $x^2 - 4x - 1 = 0$
9. $2x^2 + 5x + 3 = 0$
10. $2x^2 + 3x - 20 = 0$
11. $4x^2 + 8x + 1 = 0$
12. $3x^2 + 6x + 2 = 0$

13. $x^2 - 2x + 1 = 0$

14. $x^2 + 2x - 3 = 0$

15. $x^2 - 5x = 7$

16. $2x^2 - 6x = 8$

17. $6x^2 - x - 2 = 0$

18. $6x^2 + 5x - 4 = 0$

19. $(x - 2)(x + 1) = 3$

20. $(x - 8)(x + 7) = 5$

21. $(2x - 3)(x + 2) = 1$

22. $(4x - 5)(x - 3) = 6$

23. $2x^2 - 3x = 5$

24. $3x^2 - 4x = 5$

25. $2x^2 = -6x + 7$

26. $5x^2 = -6x + 3$

27. $3x^2 = -4x + 2$

28. $3x^2 = 4x + 2$

29. $2x^2 - 5 = 2x$

30. $5x^2 + 1 = -10x$

31. Solve the equation $2x^3 + 3x^2 - 4x = 0$ by first factoring out the common factor x and then using the quadratic formula. There are three solutions to this equation.

32. Solve the equation $5y^3 - 10y^2 + 4y = 0$ by first factoring out the common factor y and then using the quadratic formula.

33. To apply the quadratic formula to the equation $3x^2 - 4x = 0$, you have to notice that $c = 0$. Solve the equation using the quadratic formula.

34. Solve the equation $9x^2 - 16 = 0$ using the quadratic formula. (Notice $b = 0$.)

Applying the Concepts

Skyscrapers The chart shows the heights of the three tallest buildings in the world. Use the chart to answer Problems 35 and 36.

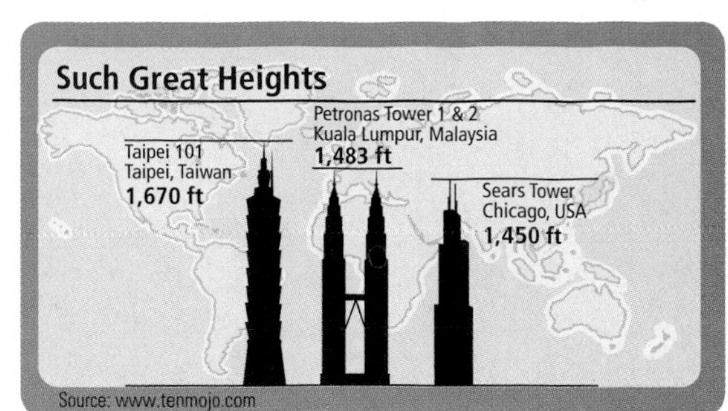

35. If you drop an object off the top of the Sears Tower, the height that the object is from the ground after t seconds is given by the equation $h = 1{,}450 - 16t^2$. When does the object reach a height of zero feet?

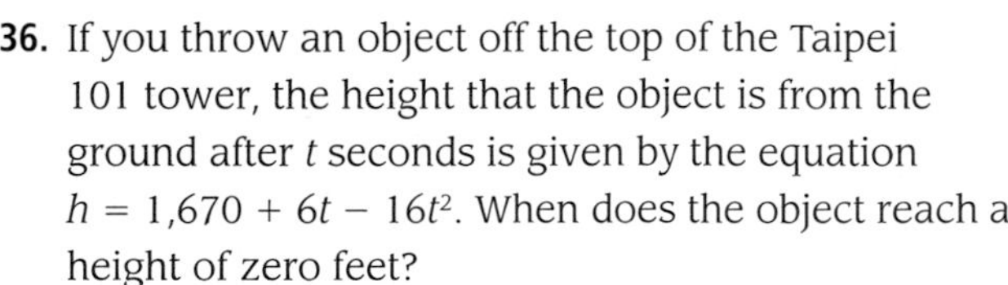

36. If you throw an object off the top of the Taipei 101 tower, the height that the object is from the ground after t seconds is given by the equation $h = 1{,}670 + 6t - 16t^2$. When does the object reach a height of zero feet?

37. Archery Margaret shoots an arrow into the air. The equation for the height (in feet) of the tip of the arrow is $h = 8 + 64t - 16t^2$. To find the time at which the arrow is 56 feet above the ground, we replace h with 56 to obtain

$$56 = 8 + 64t - 16t^2$$

Solve this equation for t to find the times at which the arrow is 56 feet above the ground.

38. Coin Toss At the beginning of every football game, the referee flips a coin to see who will kick off. The equation that gives the height (in feet) of the coin tossed in the air is $h = 6 + 32t - 16t^2$. To find the times at which the coin is 18 feet above the ground we substitute 18 for h in the equation, giving us

$$18 = 6 + 32t - 16t^2$$

Solve this equation for t to find the times at which the coin is 18 feet above the ground.

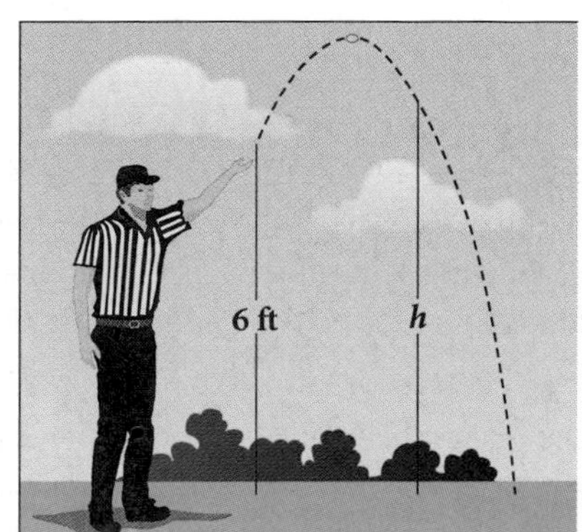

Getting Ready for the Next Section

Combine.

39. $(3 + 4\sqrt{2}) + (2 - 6\sqrt{2})$

40. $(2 + 7\sqrt{2}) + (3 - 5\sqrt{2})$

41. $(2 - 5\sqrt{2}) - (3 + 7\sqrt{2}) + (2 - \sqrt{2})$

42. $(3 - 8\sqrt{2}) - (2 + 4\sqrt{2}) + (3 - \sqrt{2})$

Multiply.

43. $4\sqrt{2}(3 + 5\sqrt{2})$

44. $5\sqrt{2}(2 + 3\sqrt{2})$

45. $(3 + 2\sqrt{2})(4 - 3\sqrt{2})$

46. $(2 - 3\sqrt{2})(5 - \sqrt{2})$

Rationalize the denominator.

47. $\dfrac{2}{3 + \sqrt{5}}$

48. $\dfrac{2}{5 + \sqrt{3}}$

49. $\dfrac{2 + \sqrt{3}}{2 - \sqrt{3}}$

50. $\dfrac{3 + \sqrt{5}}{3 - \sqrt{5}}$

Maintaining Your Skills

Add or subtract as indicated.

51. $\dfrac{3}{x - 2} + \dfrac{6}{2x - 4}$

52. $\dfrac{6x + 5}{5x - 25} - \dfrac{x + 2}{x - 5}$

53. $\dfrac{x}{x^2 - 9} + \dfrac{4}{4x - 12}$

54. $\dfrac{x + 1}{2x - 2} - \dfrac{2}{x^2 - 1}$

55. $\dfrac{2x}{x^2 - 1} + \dfrac{x}{x^2 - 3x + 2}$

56. $\dfrac{4x}{x^2 + 6x + 5} - \dfrac{3x}{x^2 + 5x + 4}$

Extending the Concepts

57. Solve the following equation by first multiplying both sides by the LCD and then applying the quadratic formula to the result.

$$\frac{1}{2}x^2 - \frac{1}{2}x - \frac{1}{6} = 0$$

58. Solve the following equation by first multiplying both sides by the LCD and then apply the quadratic formula to the result.

$$\frac{1}{2}y^2 - y - \frac{3}{2} = 0$$

Complex Numbers

9.4

Objectives

A Add and subtract complex numbers.

B Multiply two complex numbers.

C Divide two complex numbers.

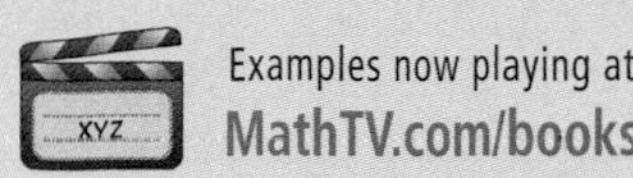
Examples now playing at
MathTV.com/books

To solve quadratic equations such as $x^2 = -4$, we have to introduce a new set of numbers. If we try to solve $x^2 = -4$ using real numbers, we always get no solution. There is no real number whose square is -4.

The new set of numbers is called the *complex numbers* and is based on the following definition.

Definition
The **number** i is a number such that $i = \sqrt{-1}$.

The first thing we notice about this definition is that i is not a real number. There are no real numbers that represent the square root of -1. The other observation we make about i is $i^2 = -1$. If $i = \sqrt{-1}$, then, squaring both sides, we must have $i^2 = -1$. The most common power of i is i^2. Whenever we see i^2, we can write it as -1. We are now ready for a definition of complex numbers.

Definition
A **complex number** is any number that can be put in the form $a + bi$, where a and b are real numbers and $i = \sqrt{-1}$.

Note The form $a + bi$ is called standard form for complex numbers. The definition in the box indicates that if a number can be written in the form $a + bi$, then it is a complex number.

Example The following are complex numbers:

$$3 + 4i \qquad \frac{1}{2} - 6i \qquad 8 + i\sqrt{2} \qquad \frac{3}{4} - 2i\sqrt{5}$$

Example The number $4i$ is a complex number because $4i = 0 + 4i$.

Example The number 8 is a complex number because $8 = 8 + 0i$.

From the previous example we can see that the real numbers are a subset of the complex numbers because any real number x can be written as $x + 0i$.

A Addition and Subtraction of Complex Numbers

We add and subtract complex numbers according to the same procedure we used to add and subtract polynomials: We combine similar terms.

EXAMPLE 1 Combine $(3 + 4i) + (2 - 6i)$.

SOLUTION

$$\begin{aligned}(3 + 4i) + (2 - 6i) &= (3 + 2) + (4i - 6i) && \text{Commutative and associative properties}\\ &= 5 + (-2i) && \text{Combine similar terms}\\ &= 5 - 2i\end{aligned}$$

EXAMPLE 2 Combine $(2 - 5i) - (3 + 7i) + (2 - i)$.

SOLUTION

$$\begin{aligned}(2 - 5i) - (3 + 7i) + (2 - i) &= 2 - 5i - 3 - 7i + 2 - i\\ &= (2 - 3 + 2) + (-5i - 7i - i)\\ &= 1 - 13i\end{aligned}$$

PRACTICE PROBLEMS

1. Combine $(2 + 7i) + (3 - 5i)$.

2. Combine:
$(3 - 8i) - (2 + 4i) + (3 - i)$

Answers

1. $5 + 2i$ **2.** $4 - 13i$

B Multiplication of Complex Numbers

Multiplication of complex numbers is very similar to multiplication of polynomials. We can simplify many answers by using the fact that $i^2 = -1$.

3. Multiply $5i(2 + 3i)$.

EXAMPLE 3 Multiply $4i(3 + 5i)$.

SOLUTION

$$\begin{aligned} 4i(3 + 5i) &= 4i(3) + 4i(5i) && \text{Distributive property} \\ &= 12i + 20i^2 && \text{Multiplication} \\ &= 12i + 20(-1) && i^2 = -1 \\ &= -20 + 12i \end{aligned}$$

4. Multiply $(2 - 3i)(5 - i)$.

EXAMPLE 4 Multiply $(3 + 2i)(4 - 3i)$.

SOLUTION

$$\begin{aligned} (3 + 2i)(4 - 3i) &= 3 \cdot 4 + 3(-3i) + 2i(4) + 2i(-3i) && \text{FOIL method} \\ &= 12 - 9i + 8i - 6i^2 \\ &= 12 - 9i + 8i - 6(-1) && i^2 = -1 \\ &= (12 + 6) + (-9i + 8i) \\ &= 18 - i \end{aligned}$$

C Division of Complex Numbers

We divide complex numbers by applying the same process we used to rationalize denominators.

5. Divide $\dfrac{3 - i}{4 + 2i}$.

EXAMPLE 5 Divide $\dfrac{2 + i}{5 + 2i}$.

SOLUTION The conjugate of the denominator is $5 - 2i$:

$$\begin{aligned} \left(\frac{2 + i}{5 + 2i}\right)\left(\frac{5 - 2i}{5 - 2i}\right) &= \frac{10 - 4i + 5i - 2i^2}{25 - 4i^2} \\ &= \frac{10 - 4i + 5i - 2(-1)}{25 - 4(-1)} && i^2 = -1 \\ &= \frac{12 + i}{29} \end{aligned}$$

When we write our answer in standard form for complex numbers, we get:

$$\frac{12}{29} + \frac{1}{29}i$$

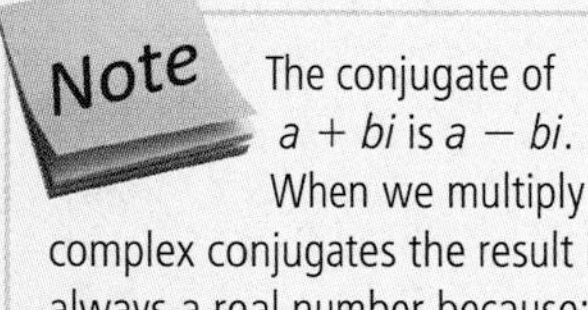

The conjugate of $a + bi$ is $a - bi$. When we multiply complex conjugates the result is always a real number because:

$$\begin{aligned} (a + bi)(a - bi) &= a^2 - (bi)^2 \\ &= a^2 - b^2i^2 \\ &= a^2 - b^2(-1) \\ &= a^2 + b^2 \end{aligned}$$

which is a real number.

Getting Ready for Class

After reading through the preceding section, respond in your own words and in complete sentences.

1. What is the number i?
2. What is a complex number?
3. What kind of number will always result when we multiply complex conjugates?
4. Explain how to divide complex numbers.

Answers

3. $-15 + 10i$ **4.** $7 - 17i$ **5.** $\frac{1 - i}{2}$

Problem Set 9.4

A Combine the following complex numbers. **[Examples 1–2]**

1. $(3 - 2i) + 3i$

2. $(5 - 4i) - 8i$

3. $(6 + 2i) - 10i$

4. $(8 - 10i) + 7i$

5. $(11 + 9i) - 9i$

6. $(12 + 2i) + 6i$

7. $(3 + 2i) + (6 - i)$

8. $(4 + 8i) - (7 + i)$

9. $(5 + 7i) - (6 + 8i)$

10. $(11 + 6i) - (3 + 6i)$

11. $(9 - i) + (2 - i)$

12. $(8 + 3i) - (8 - 3i)$

13. $(6 + i) - 4i - (2 - i)$

14. $(3 + 2i) - 5i - (5 + 4i)$

15. $(6 - 11i) + 3i + (2 + i)$

16. $(3 + 4i) - (5 + 7i) - (6 - i)$

17. $(2 + 3i) - (6 - 2i) + (3 - i)$

18. $(8 + 9i) + (5 - 6i) - (4 - 3i)$

B Multiply the following complex numbers. [Examples 3–4]

19. $3(2 - i)$

20. $4(5 + 3i)$

21. $2i(8 - 7i)$

22. $-3i(2 + 5i)$

23. $(2 + i)(4 - i)$

24. $(6 + 3i)(4 + 3i)$

25. $(2 + i)(3 - 5i)$

26. $(4 - i)(2 - i)$

27. $(3 + 5i)(3 - 5i)$

28. $(8 + 6i)(8 - 6i)$

29. $(2 + i)(2 - i)$

30. $(3 + i)(3 - i)$

C Divide the following complex numbers. [Example 5]

31. $\dfrac{2}{3 - 2i}$

32. $\dfrac{3}{5 + 6i}$

33. $\dfrac{-3i}{2 + 3i}$

34. $\dfrac{4i}{3 + i}$

35. $\dfrac{6i}{3 - i}$

36. $\dfrac{-7i}{5 - 4i}$

37. $\dfrac{2 + i}{2 - i}$

38. $\dfrac{3 + 2i}{3 - 2i}$

39. $\dfrac{4 + 5i}{3 - 6i}$

40. $\dfrac{-2 + i}{5 + 6i}$

Applying the Concepts

41. **Google Earth** This is a Google Earth photo of Lake Clark National Park in Alaska. Lake Clark has an average temperature of 40 degrees Fahrenheit. Use the formula $C = \frac{5}{9}(F - 32)$ to find its average temperature in Celsius. Round to the nearest degree.

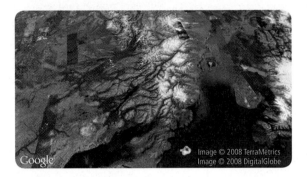

42. **Temperature** The chart shows the temperatures for some of the world's hottest places. Convert the temperature in Tombouctou to Kelvin. Round to the nearest tenth of a degree. [Hint: $C = \frac{5}{9}(F - 32)$ converts a temperature from Fahrenheit to Celsius. To convert Celsius to Kelvin, add 273 to the temperature.

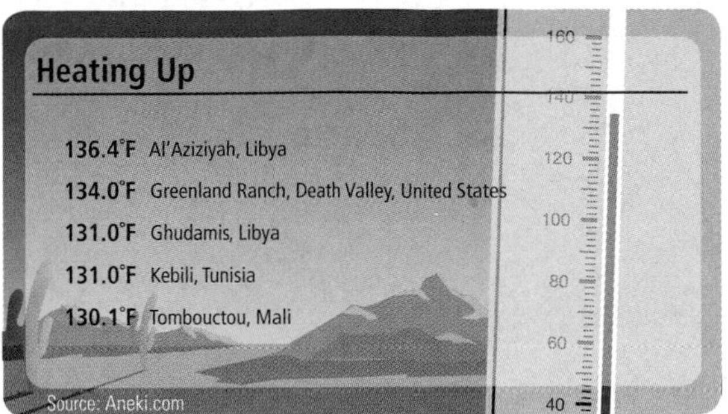

Heating Up
136.4°F Al'Aziziyah, Libya
134.0°F Greenland Ranch, Death Valley, United States
131.0°F Ghudamis, Libya
131.0°F Kebili, Tunisia
130.1°F Tombouctou, Mali
Source: Aneki.com
160
140
120
100
80
60
40

Getting Ready for the Next Section

Simplify.

43. $\sqrt{36}$

44. $\sqrt{49}$

45. $\sqrt{75}$

46. 12

Solve for x.

47. $(x + 2)^2 = 9$

48. $(x - 3)^2 = 9$

49. $\frac{1}{10}x^2 - \frac{1}{5}x = \frac{1}{2}$

50. $\frac{1}{18}x^2 - \frac{2}{9}x = \frac{13}{18}$

51. $(2x - 3)(2x - 1) = 4$

52. $(x - 1)(3x - 3) = 10$

Maintaining Your Skills

Solve each equation by taking the square root of each side.

53. $(x - 3)^2 = 25$

54. $(x - 2)^2 = 9$

55. $(2x - 6)^2 = 16$

56. $(2x + 1)^2 = 49$

57. $(x + 3)^2 = 12$

58. $(x + 3)^2 = 8$

Put each expression into simplified form for radicals.

59. $\sqrt{\frac{1}{2}}$

60. $\frac{5}{6}$

61. $\sqrt{\frac{8x^2y^3}{3}}$

62. $\sqrt{\frac{45xy^4}{7}}$

63. $\sqrt[3]{\frac{1}{4}}$

64. $\sqrt[3]{\frac{2}{3}}$

Extending the Concepts

65. Use the FOIL method to multiply $(x + 3i)(x - 3i)$.

66. Use the FOIL method to multiply $(x + 5i)(x - 5i)$.

67. The opposite of i is $-i$. The reciprocal of i is $\frac{1}{i}$. Multiply the numerator and denominator of $\frac{1}{i}$ by i and simplify the result to see that the opposite of i and the reciprocal of i are the same number.

68. If $i^2 = -1$, what are i^3 and i^4? (*Hint:* $i^3 = i^2 \cdot i$.)

Complex Solutions to Quadratic Equations

9.5

Objectives

A Write square roots of negative numbers as complex numbers.

B Solve quadratic equations with complex solutions.

Examples now playing at MathTV.com/books

A Complex Numbers

The quadratic formula tells us that the solutions to equations of the form $ax^2 + bx + c = 0$ are always:

$$x = \frac{-b \pm \sqrt{b^2 - 4ac}}{2a}$$

The part of the quadratic formula under the radical sign is called the *discriminant:*

$$\text{Discriminant} = b^2 - 4ac$$

When the discriminant is negative, we have to deal with the square root of a negative number. We handle square roots of negative numbers by using the definition $i = \sqrt{-1}$. To illustrate, suppose we want to simplify an expression that contains $\sqrt{-9}$, which is not a real number. We begin by writing $\sqrt{-9}$ as $\sqrt{9(-1)}$. Then, we write this expression as the product of two separate radicals: $\sqrt{9}\sqrt{-1}$. Applying the definition $i = \sqrt{-1}$ to this last expression, we have

$$\sqrt{9}\sqrt{-1} = 3i$$

As you may recall from the previous section, the number $3i$ is called a complex number. Here are some further examples.

EXAMPLE 1 Write the following radicals as complex numbers:

a. $\sqrt{-4} = \sqrt{4(-1)} = \sqrt{4}\sqrt{-1} = 2i$
b. $\sqrt{-36} = \sqrt{36(-1)} = \sqrt{36}\sqrt{-1} = 6i$
c. $\sqrt{-7} = \sqrt{7(-1)} = \sqrt{7}\sqrt{-1} = i\sqrt{7}$
d. $\sqrt{-75} = \sqrt{75(-1)} = \sqrt{75}\sqrt{-1} = 5i\sqrt{3}$

In parts (c) and (d) of Example 1, we wrote i before the radical because it is less confusing that way. If we put i after the radical, it is sometimes mistaken for being under the radical.

B Quadratic Equations with Complex Solutions

Let's see how complex numbers relate to quadratic equations by looking at some examples of quadratic equations whose solutions are complex numbers.

EXAMPLE 2 Solve for x : $(x + 2)^2 = -9$.

SOLUTION We can solve this equation by expanding the left side, putting the result into standard form, and then applying the quadratic formula. It is faster, however, simply to apply the square root property:

$(x + 2)^2 = -9$	
$x + 2 = \pm\sqrt{-9}$	Square root property
$x + 2 = \pm 3i$	$\sqrt{-9} = \sqrt{9}\sqrt{-1} = 3i$
$x = -2 \pm 3i$	Add -2 to both sides

The solution set contains two complex solutions. Notice that the two solutions are conjugates.

The solution set is $\{-2 + 3i, -2 - 3i\}$.

PRACTICE PROBLEMS

1. Write as complex numbers:

a. $\sqrt{-9}$
b. $\sqrt{-49}$
c. $\sqrt{-6}$
d. $\sqrt{-12}$

2. Solve for x: $(x - 3)^2 = -4$.

Answers

1. a. $3i$ b. $7i$ c. $i\sqrt{6}$ d. $2i\sqrt{3}$
2. $3 \pm 2i$

EXAMPLE 3 Solve for x: $\frac{1}{10}x^2 - \frac{1}{5}x = -\frac{1}{2}$.

SOLUTION It will be easier to apply the quadratic formula if we clear the equation of fractions. Multiplying both sides of the equation by the LCD 10 gives us:

$$x^2 - 2x = -5$$

Next, we add 5 to both sides to put the equation into standard form:

$$x^2 - 2x + 5 = 0 \quad \text{Add 5 to both sides}$$

Applying the quadratic formula with $a = 1$, $b = -2$, and $c = 5$, we have:

$$x = \frac{-(-2) \pm \sqrt{(-2)^2 - 4(1)(5)}}{2(1)} = \frac{2 \pm \sqrt{-16}}{2} = \frac{2 \pm 4i}{2}$$

Dividing the numerator and denominator by 2, we have the two solutions:

$$x = 1 \pm 2i$$

The two solutions are $1 + 2i$ and $1 - 2i$.

EXAMPLE 4 Solve $(2x - 3)(2x - 1) = -4$.

SOLUTION We multiply the binomials on the left side and then add 4 to each side to write the equation in standard form. From there we identify a, b, and c and apply the quadratic formula:

$$(2x - 3)(2x - 1) = -4$$
$$4x^2 - 8x + 3 = -4 \quad \text{Multiply binomials on left side}$$
$$4x^2 - 8x + 7 = 0 \quad \text{Add 4 to each side}$$

Placing $a = 4$, $b = -8$, and $c = 7$ in the quadratic formula we have:

$$x = \frac{-(-8) \pm \sqrt{(-8)^2 - 4(4)(7)}}{2(4)}$$
$$= \frac{8 \pm \sqrt{64 - 112}}{8}$$
$$= \frac{8 \pm \sqrt{-48}}{8}$$
$$= \frac{8 \pm 4i\sqrt{3}}{8} \quad \sqrt{-48} = i\sqrt{48} = i\sqrt{16}\sqrt{3} = 4i\sqrt{3}$$

To reduce this final expression to lowest terms, we factor a 4 from the numerator and then divide the numerator and denominator by 4:

$$= \frac{4(2 \pm i\sqrt{3})}{4 \cdot 2}$$
$$= \frac{2 \pm i\sqrt{3}}{2}$$

3. Solve for x:

$\frac{1}{18}x^2 - \frac{2}{9}x = -\frac{13}{18}$

4. Solve $(x - 1)(3x - 3) = 10$.

Note It would be a mistake to try to reduce this final expression further. Sometimes first-year algebra students will try to divide the 2 in the denominator into the 2 in the numerator, which is a mistake. Remember, when we reduce to lowest terms, we do so by dividing the numerator and denominator by any factors they have in common. In this case 2 is not a factor of the numerator. This expression is in lowest terms.

Getting Ready for Class

After reading through the preceding section, respond in your own words and in complete sentences.

1. Describe how you would write $\sqrt{-4}$ in terms of the number i.
2. When would the quadratic formula result in complex solutions?
3. How can we avoid working with fractions when using the quadratic formula?
4. Describe how you would simplify the expression $\frac{8 \pm 4i\sqrt{3}}{8}$.

Answers

3. $x = 2 \pm 3i$ 4. $\frac{3 \pm \sqrt{30}}{3}$

Problem Set 9.5

A Write the following radicals as complex numbers. **[Example 1]**

1. $\sqrt{-16}$ **2.** $\sqrt{-25}$ **3.** $\sqrt{-49}$ **4.** $\sqrt{-81}$ **5.** $\sqrt{-6}$ **6.** $\sqrt{-10}$

7. $\sqrt{-11}$ **8.** $\sqrt{-19}$ **9.** $\sqrt{-32}$ **10.** $\sqrt{-288}$ **11.** $\sqrt{-50}$ **12.** $\sqrt{-45}$

13. $\sqrt{-8}$ **14.** $\sqrt{-24}$ **15.** $\sqrt{-48}$ **16.** $\sqrt{-27}$

B Solve the following quadratic equations. Use whatever method seems to fit the situation or is convenient for you. [Examples 2–4]

17. $x^2 = 2x - 2$

18. $x^2 = 4x - 5$

19. $x^2 - 4x = -4$

20. $x^2 - 4x = 4$

21. $2x^2 + 5x = 12$

22. $2x^2 + 30 = 16x$

23. $(x - 2)^2 = -4$

24. $(x - 5)^2 = -25$

25. $\left(x + \frac{1}{2}\right)^2 = -\frac{9}{4}$

26. $\left(x - \frac{1}{4}\right)^2 = -\frac{1}{2}$

27. $\left(x - \frac{1}{2}\right)^2 = -\frac{27}{36}$

28. $\left(x + \frac{1}{2}\right)^2 = -\frac{32}{64}$

29. $x^2 + x + 1 = 0$

30. $x^2 - 3x + 4 = 0$

31. $x^2 - 5x + 6 = 0$

32. $x^2 + 2x + 2 = 0$

33. $\frac{1}{2}x^2 + \frac{1}{3}x + \frac{1}{6} = 0$

34. $\frac{1}{5}x^2 + \frac{1}{20}x + \frac{1}{4} = 0$

35. $\frac{1}{3}x^2 = -\frac{1}{2}x + \frac{1}{3}$

36. $\frac{1}{2}x^2 = -\frac{1}{3}x + \frac{1}{6}$

37. $(x + 2)(x - 3) = 5$

38. $(x - 1)(x + 1) = 6$

39. $(x - 5)(x - 3) = -10$

40. $(x - 2)(x - 4) = -5$

41. $(2x - 2)(x - 3) = 9$

42. $(x - 1)(2x + 6) = 9$

43. Intercepts The graph of $y = x^2 - 4x + 5$ is shown in the following figure. As you can see, the graph does not cross the x-axis. If the graph did cross the x-axis, the x-intercepts would be solutions to the equation

$$0 = x^2 - 4x + 5$$

Solve this equation. The solutions will confirm the fact that the graph cannot cross the x-axis.

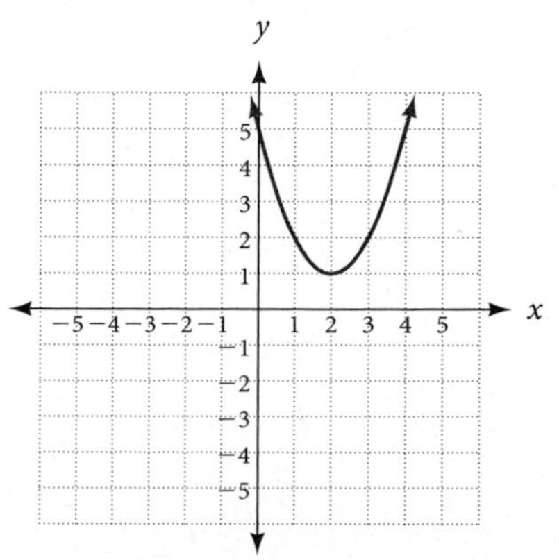

44. Intercepts The graph of $y = x^2 + 2x + 5$ is shown in the following figure. As you can see, the graph does not cross the x-axis. If the graph did cross the x-axis, the x-intercepts would be solutions to the equation

$$0 = x^2 + 2x + 5$$

Solve this equation.

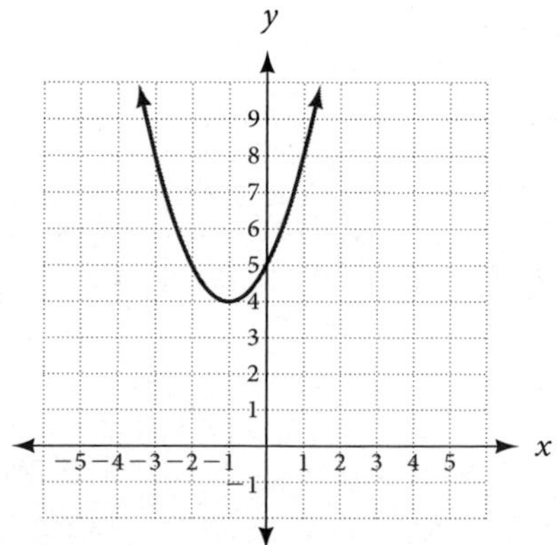

45. Is $x = 2 + 2i$ a solution to the equation $x^2 - 4x + 8 = 0$?

46. Is $x = 5 + 3i$ a solution to the equation $x^2 - 10x + 34 = 0$?

47. If one solution to a quadratic equation is $3 + 7i$, what do you think the other solution is?

48. If one solution to a quadratic equation is $4 - 2i$, what do you think the other solution is?

Applying the Concepts

49. Google Earth The Google Earth map shows Denali National Park, home to Mount McKinley, the tallest mountain in North America. It is also home to 14 different wolf packs. If Denali National Park covers 7,370 square miles, what is the average size of each pack's territory? Round to the nearest tenth.

50. Google Earth This Google Earth image is of Mount Rainier National Park in Washington. Every year, the park receives over 2 million visitors during months of operation, May to October. What is average number of visitors every month?

Getting Ready for the Next Section

Use the equation $y = (x + 1)^2 - 3$ to find the value of y when

51. $x = -4$

52. $x = -3$

53. $x = -2$

54. $x = -1$

55. $x = 1$

56. $x = 2$

Maintaining Your Skills

Write each term in simplified form for radicals. Then combine similar terms.

57. $3\sqrt{50} + 2\sqrt{32}$

58. $4\sqrt{18} + \sqrt{32} - \sqrt{2}$

59. $\sqrt{24} - \sqrt{54} - \sqrt{150}$

60. $\sqrt{72} - \sqrt{8} + \sqrt{50}$

61. $2\sqrt{27x^2} - x\sqrt{48}$

62. $5\sqrt{8x^3} + x\sqrt{50x}$

Objectives

A Graph a parabola.

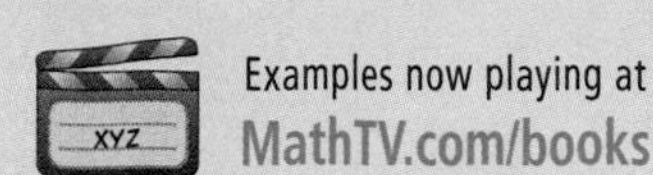

A Graphing Parabolas

In this section we will graph equations of the form $y = ax^2 + bx + c$ and equations that can be put into this form. The graphs of these types of equations all have similar shapes.

We will begin this section by graphing the simplest quadratic equation, $y = x^2$. To get the idea of the shape of this graph, we find some ordered pairs that are solutions. We can do this by setting up the following table:

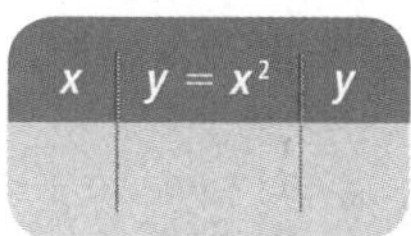

x	$y = x^2$	y

We can choose any convenient numbers for x and then use the equation $y = x^2$ to find the corresponding values for y. Let's use the values -3, -2, -1, 0, 1, 2, and 3 for x and find corresponding values for y. Here is how the table looks when we let x have these values:

x	$y = x^2$	y
-3	$y = (-3)^2 = 9$	9
-2	$y = (-2)^2 = 4$	4
-1	$y = (-1)^2 = 1$	1
0	$y = 0^2 = 0$	0
1	$y = 1^2 = 1$	1
2	$y = 2^2 = 4$	4
3	$y = 3^2 = 9$	9

The table gives us the solutions $(-3, 9)$, $(-2, 4)$, $(-1, 1)$, $(0, 0)$, $(1, 1)$, $(2, 4)$, and $(3, 9)$ for the equation $y = x^2$. We plot each of the points on a rectangular coordinate system and draw a smooth curve through them, as shown in Figure 1.

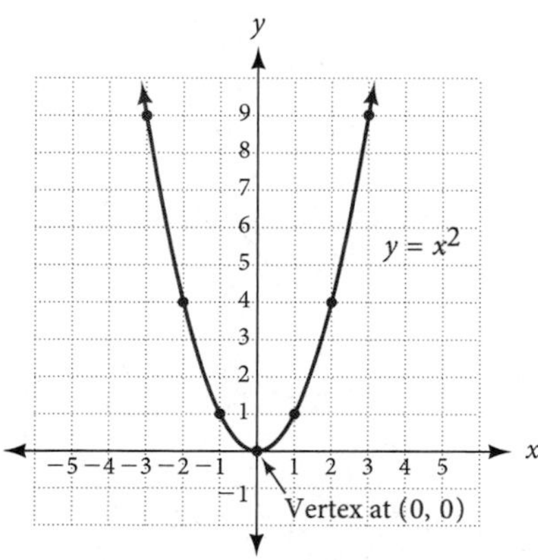

FIGURE 1

This graph is called a *parabola.* All equations of the form $y = ax^2 + bx + c$ $(a \neq 0)$ produce parabolas when graphed.

Note that the point $(0, 0)$ is called the *vertex* of the parabola in Figure 1. It is the lowest point on this graph. There are some parabolas, though, that open downward. For those parabolas, the vertex is the highest point on the graph.

PRACTICE PROBLEMS

1. Graph the equation:
$y = x^2 + 2$

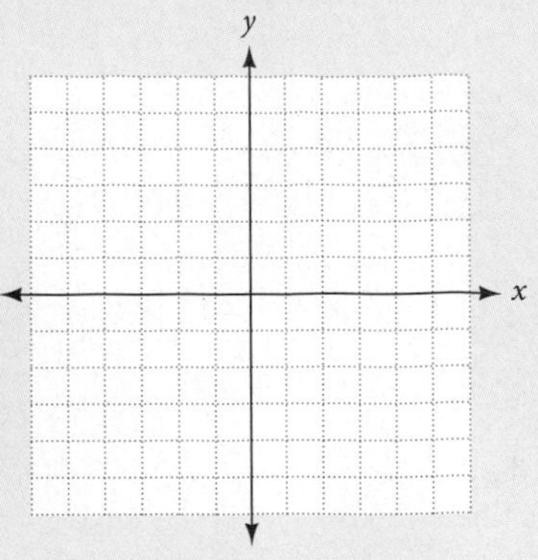

2. Graph $y = (x + 3)^2$.

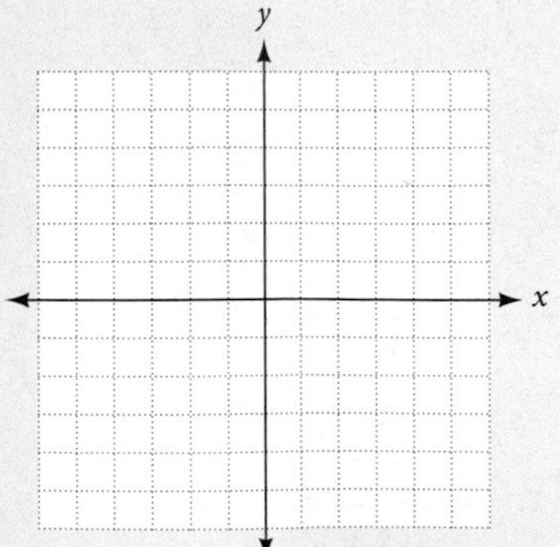

Answers

For answers to problems in this section, see the Solutions to Selected Practice Problems.

EXAMPLE 1 Graph the equation $y = x^2 - 3$.

SOLUTION We begin by making a table using convenient values for x:

x	$y = x^2 - 3$	y
−2	$y = (-2)^2 - 3 = 4 - 3 = 1$	1
−1	$y = (-1)^2 - 3 = 1 - 3 = -2$	−2
0	$y = 0^2 - 3 = -3$	−3
1	$y = 1^2 - 3 = 1 - 3 = -2$	−2
2	$y = 2^2 - 3 = 4 - 3 = 1$	1

The table gives us the ordered pairs $(-2, 1)$, $(-1, -2)$, $(0, -3)$, $(1, -2)$, and $(2, 1)$ as solutions to $y = x^2 - 3$. The graph is shown in Figure 2.

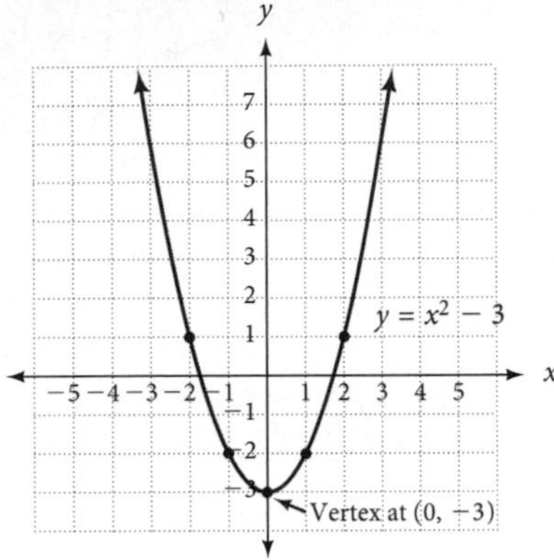

FIGURE 2

EXAMPLE 2 Graph $y = (x - 2)^2$.

SOLUTION

x	$y = (x - 2)^2$	y
−1	$y = (-1 - 2)^2 = (-3)^2 = 9$	9
0	$y = (0 - 2)^2 = (-2)^2 = 4$	4
1	$y = (1 - 2)^2 = (-1)^2 = 1$	1
2	$y = (2 - 2)^2 = 0^2 = 0$	0

We can continue the table if we feel more solutions will make the graph clearer.

3	$y = (3 - 2)^2 = 1^2 = 1$	1
4	$y = (4 - 2)^2 = 2^2 = 4$	4
5	$y = (5 - 2)^2 = 3^2 = 9$	9

Putting the results of the table onto a coordinate system, we have the graph in Figure 3.

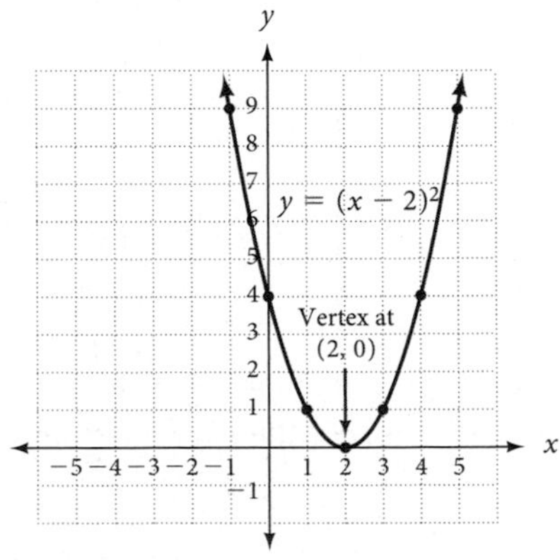

FIGURE 3

When graphing parabolas, it is sometimes easier to obtain the correct graph if you have some idea what the graph will look like before you start making your table.

If you look over the first two examples closely, you will see that the graphs in Figures 2 and 3 have the same shape as the graph of $y = x^2$. The difference is in the position of the vertex. For example, the graph of $y = x^2 - 3$, as shown in Figure 2, looks like the graph of $y = x^2$ with its vertex moved down three units vertically. Similarly, the graph of $y = (x - 2)^2$, shown in Figure 3, looks like the graph of $y = x^2$ with its vertex moved two units to the right.

Without showing the tables necessary to graph them, Figures 4–7 are four more parabolas and their corresponding equations. Can you see the correlation between the numbers in the equations and the position of the graphs on the coordinate systems?

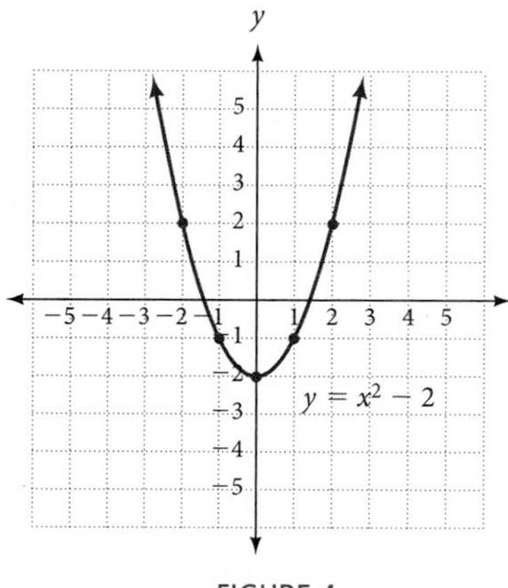

FIGURE 4

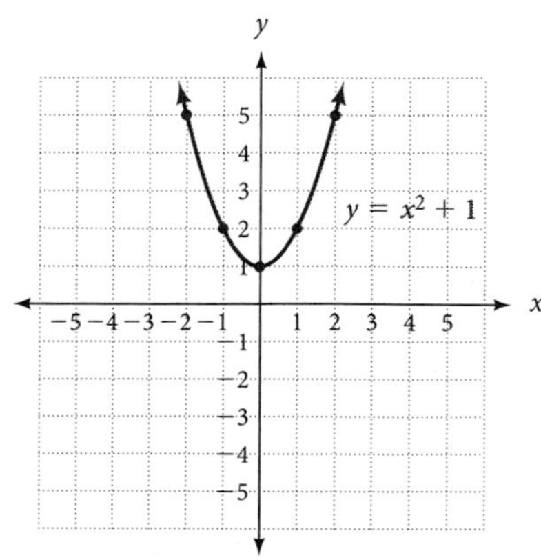

FIGURE 5

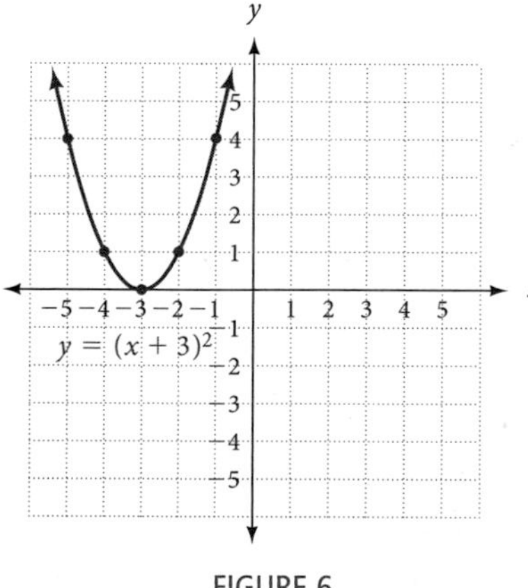

FIGURE 6

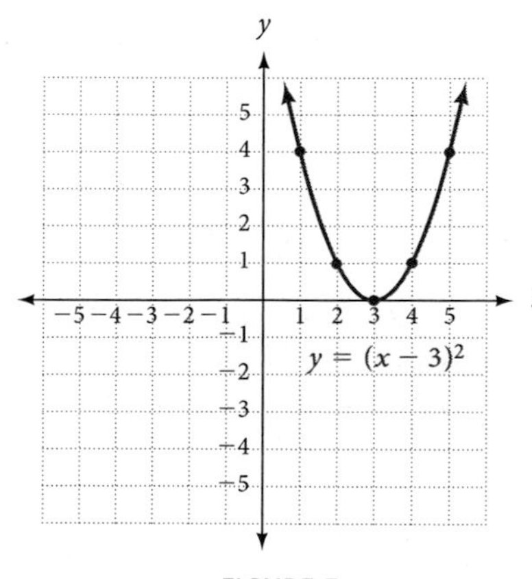

FIGURE 7

For our next example, we will graph a parabola in which the vertex has been moved both vertically and horizontally from the origin.

EXAMPLE 3 Graph $y = (x + 1)^2 - 3$.

SOLUTION

x	$y = (x + 1)^2 - 3$	y
−4	$y = (-4 + 1)^2 - 3 = 9 - 3$	6
−3	$y = (-3 + 1)^2 - 3 = 4 - 3$	1
−1	$y = (-1 + 1)^2 - 3 = 0 - 3$	−3
1	$y = (1 + 1)^2 - 3 = 4 - 3$	1
2	$y = (2 + 1)^2 - 3 = 9 - 3$	6

3. Graph $y = (x - 1)^2 - 3$.

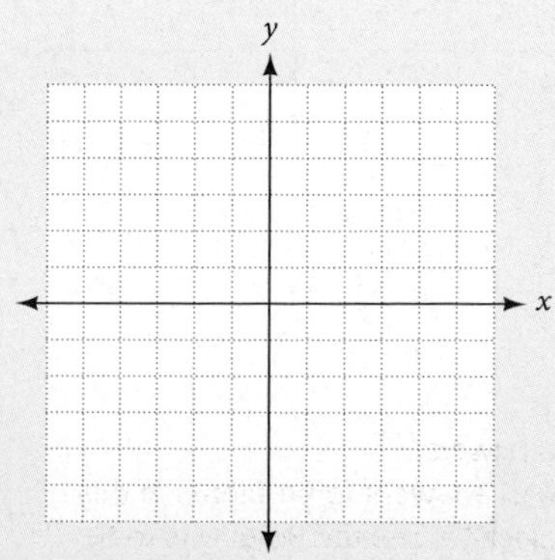

Graphing the results of the table, we have Figure 8.

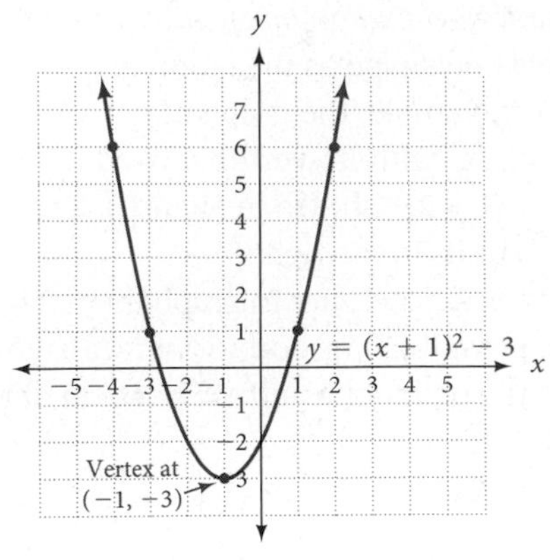

FIGURE 8

We summarize the information shown in the preceding examples and graphs with the following statement.

Graphing Parabolas

The graph of each of the following equations is a parabola with the same basic shape as the graph of $y = x^2$.

Equation	Vertex	To obtain the graph, move the graph of $y = x^2$
$y = x^2 + k$	$(0, k)$	k units vertically
$y = (x - h)^2$	$(h, 0)$	h units horizontally
$y = (x - h)^2 + k$	(h, k)	h units horizontally and k units vertically

Our last example for this section shows what we do if the equation of the parabola we want to graph is not in the form $y = (x - h)^2 + k$.

4. Graph $y = x^2 - 8x + 11$.

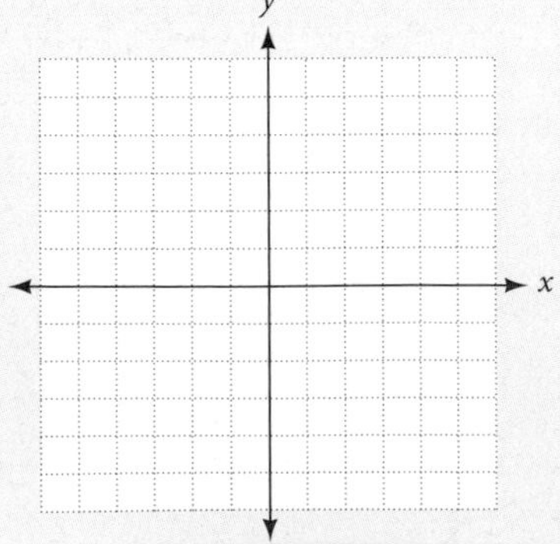

EXAMPLE 4 Graph $y = x^2 - 6x + 5$.

SOLUTION We could graph this equation by making a table of values of x and y as we have previously. However, there is an easier way. If we can write the equation in the form

$$y = (x - h)^2 + k$$

then, to graph it, we simply place the vertex at (h, k) and draw a graph from there that has the same shape as $y = x^2$.

To write our equation in the form given in the preceding equation, we simply complete the square on the first two terms on the right side of the equation, $x^2 - 6x$. We do so by adding 9 to and subtracting 9 from the right side of the equation. This amounts to adding 0 to the equation, so we know we haven't changed its solutions. This is what it looks like:

$$y = (x^2 - 6x) + 5$$

$$y = (x^2 - 6x + \mathbf{9}) + 5 - \mathbf{9}$$

$$y = (x - 3)^2 - 4$$

This final equation has the form we are looking for. Our graph will have the same shape as the graph of $y = x^2$. The vertex of our graph is at $(3, -4)$. The graph is shown in Figure 9.

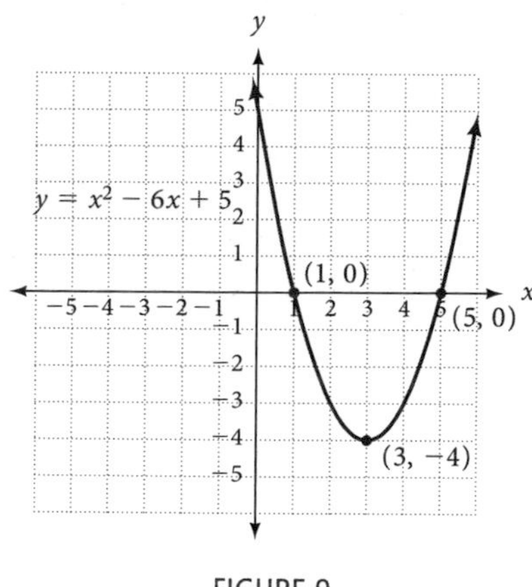

FIGURE 9

One final note: We can check to see that the x-intercepts in Figure 9 are correct by finding the x-intercepts from our original equation. Remember, the x-intercepts occur when y is 0; that is, to find the x-intercepts, we let $y = 0$ and solve for x:

$$0 = x^2 - 6x + 5$$

$$0 = (x - 5)(x - 1)$$

$$x = 5 \quad \text{or} \quad x = 1$$

Getting Ready for Class

After reading through the preceding section, respond in your own words and in complete sentences.

1. What is a parabola?
2. What is the vertex of a parabola?
3. How do you find the x-intercepts of a parabola?
4. Describe the relationship between the graph of $y = x^2$ and the graph of $y = x^2 - 2$.

Problem Set 9.6

A Graph each of the following equations. [Examples 1–3]

1. $y = x^2 - 4$

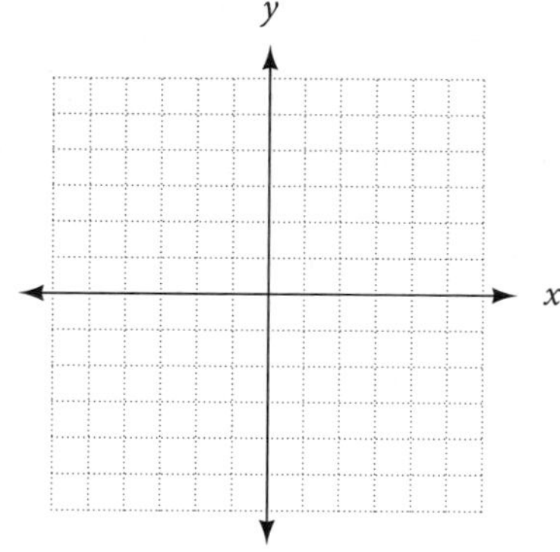

2. $y = x^2 + 2$

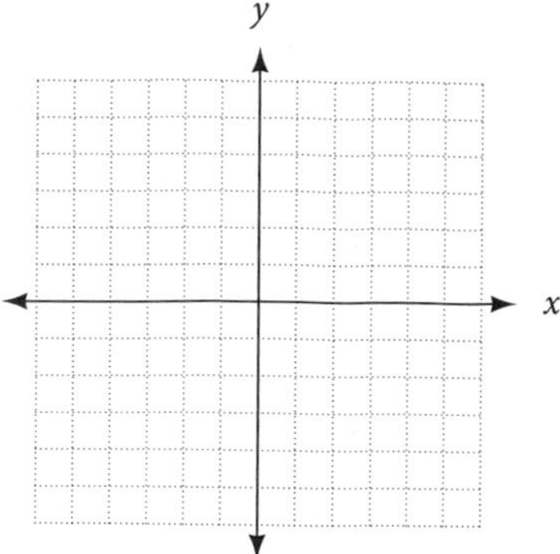

3. $y = x^2 + 5$

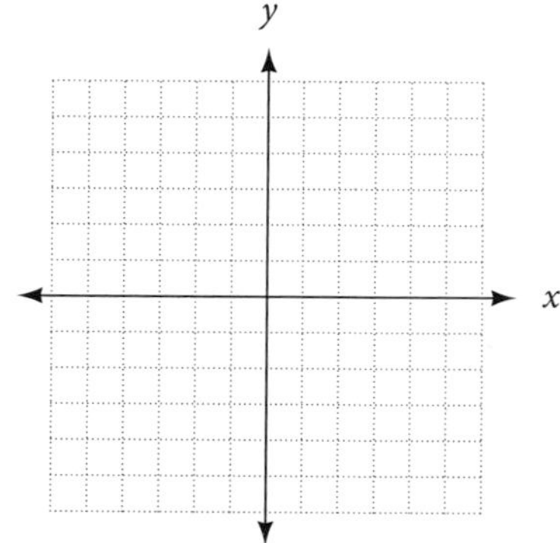

4. $y = x^2 - 2$

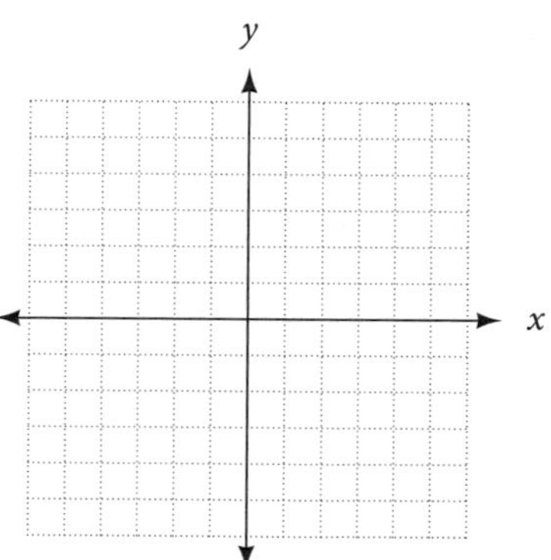

5. $y = (x + 2)^2$

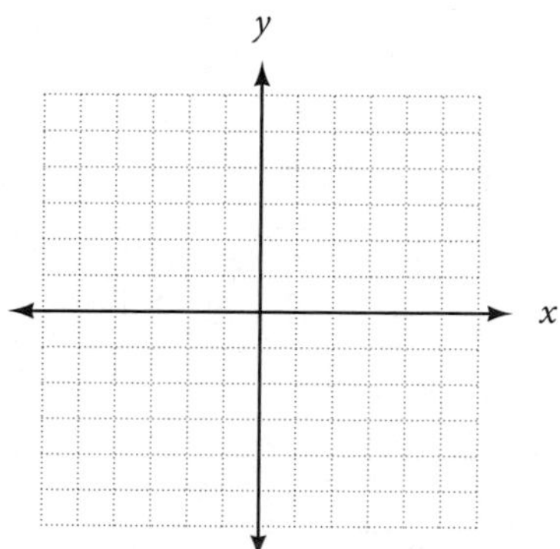

6. $y = (x + 5)^2$

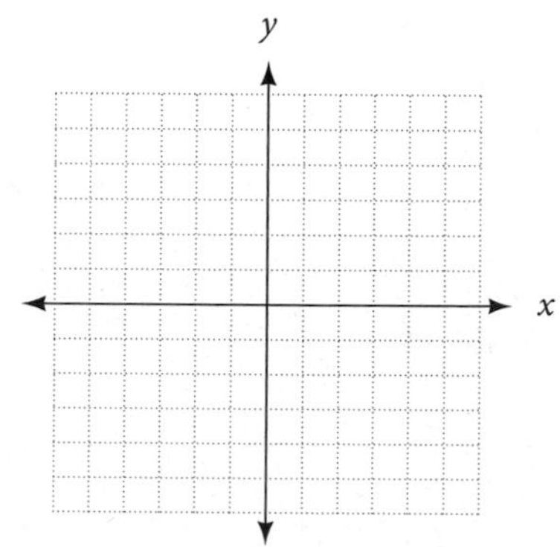

7. $y = (x - 3)^2$

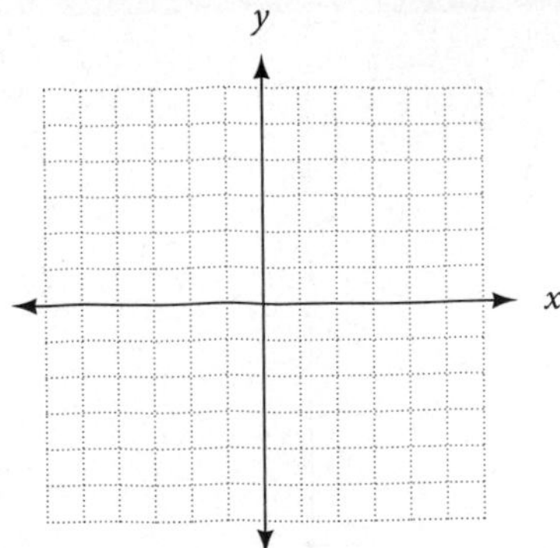

8. $y = (x - 2)^2$

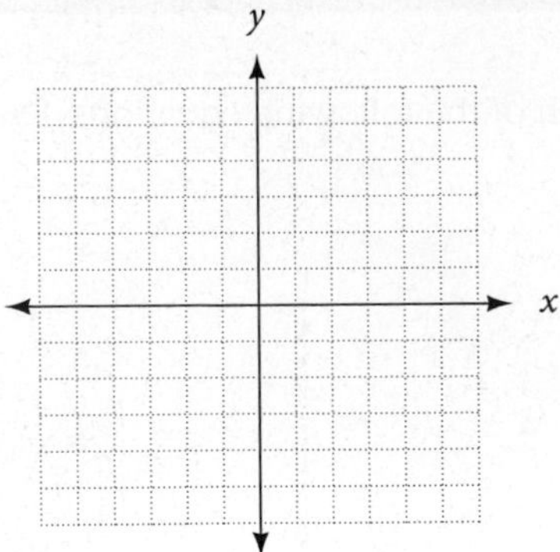

9. $y = (x - 5)^2$

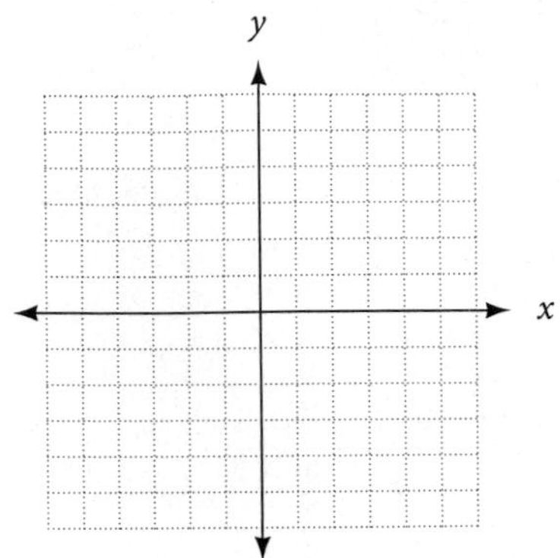

10. $y = (x + 3)^2$

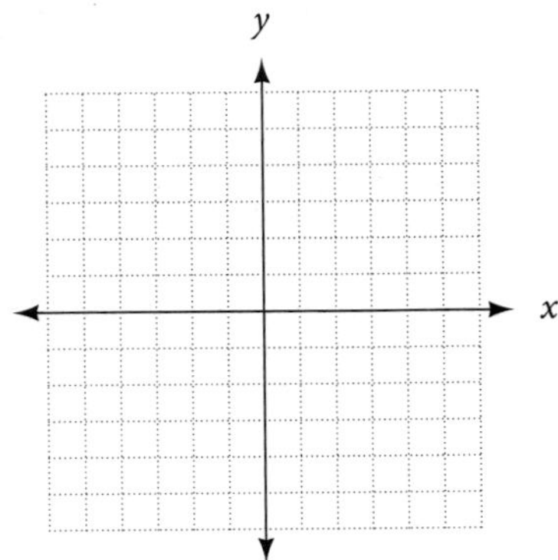

11. $y = (x + 1)^2 - 2$

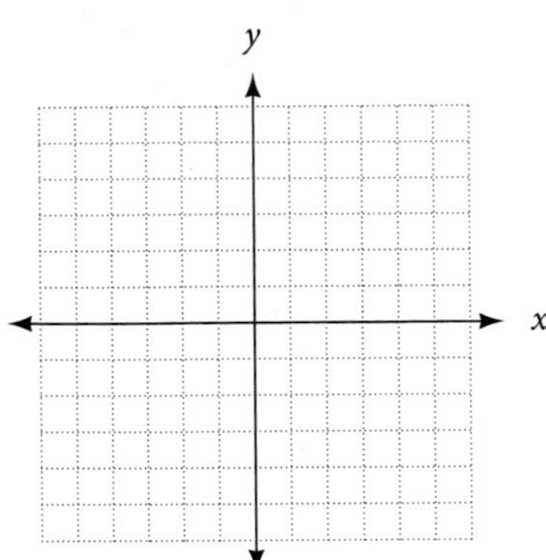

12. $y = (x - 1)^2 + 2$

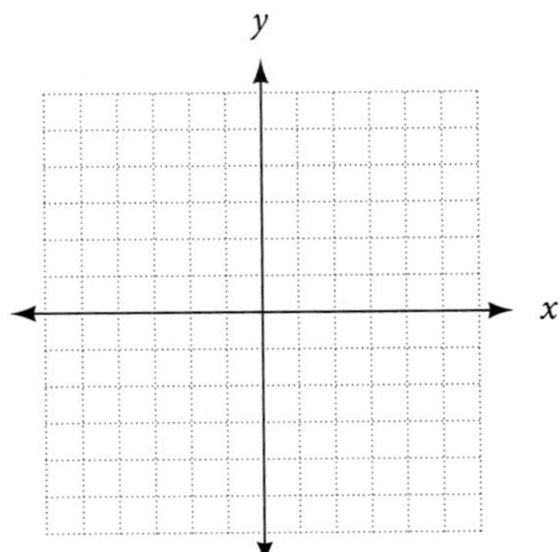

13. $y = (x + 2)^2 - 3$

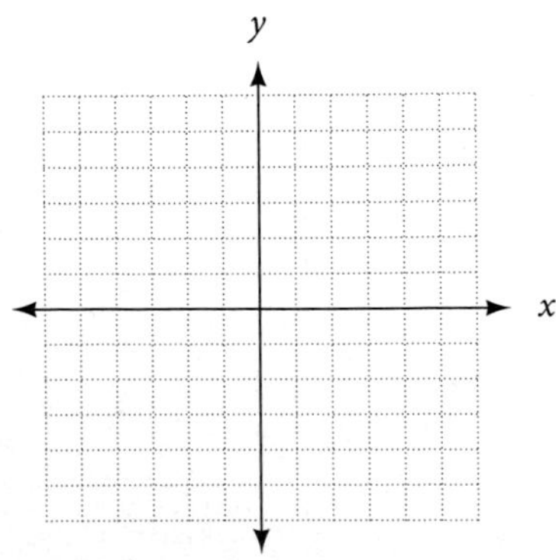

14. $y = (x - 2)^2 + 3$

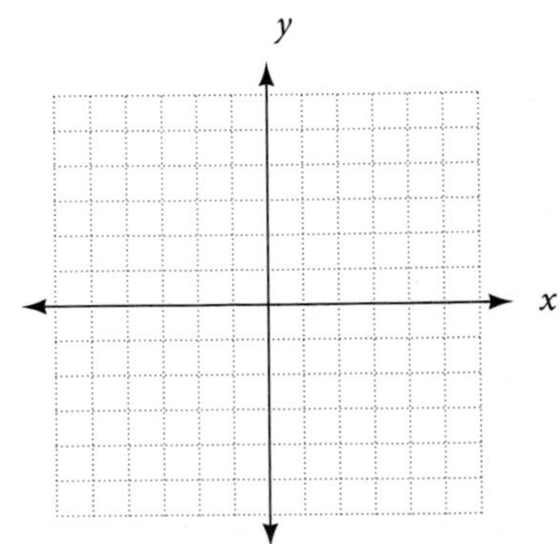

15. $y = (x - 3)^2 + 2$

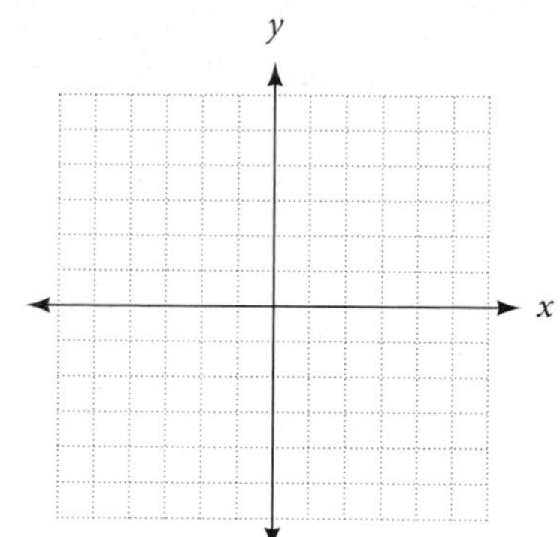

16. $y = (x + 4)^2 + 1$

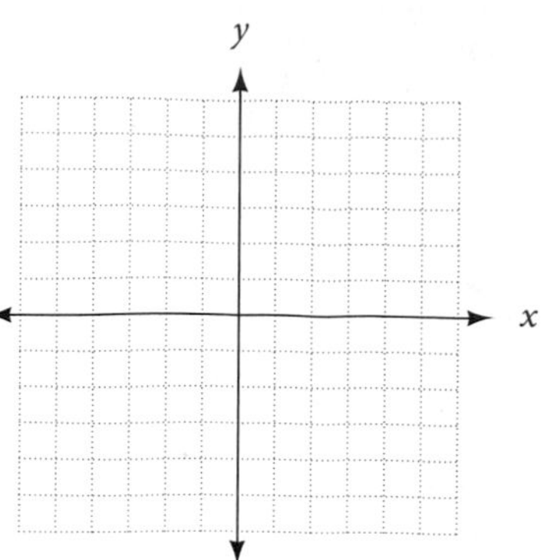

A Graph each of the following equations. Begin by completing the square on the first two terms. **[Example 4]**

17. $y = x^2 + 6x + 5$

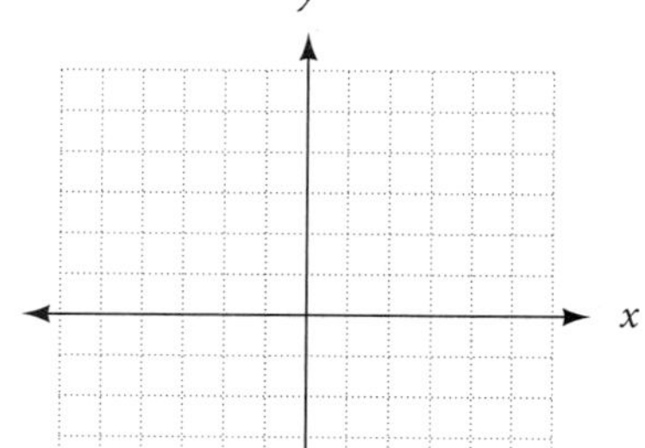

18. $y = x^2 - 8x + 12$

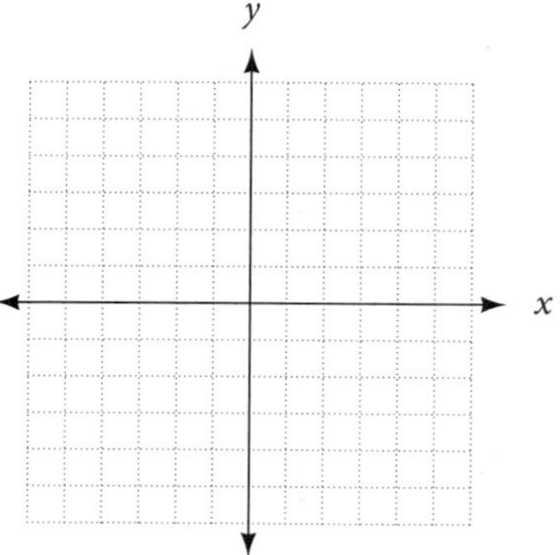

19. $y = x^2 - 2x - 3$

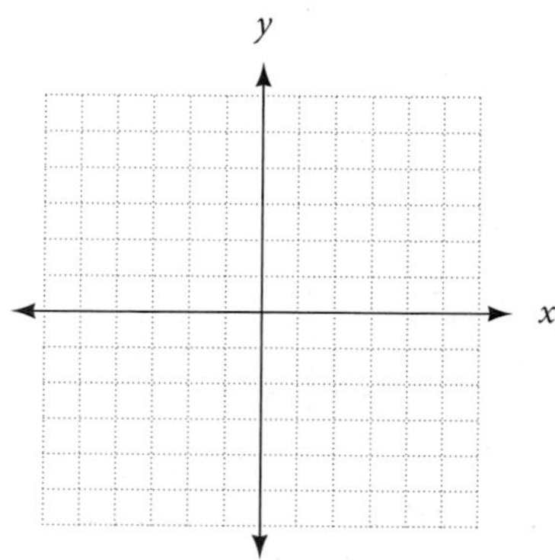

20. $y = x^2 + 2x - 3$

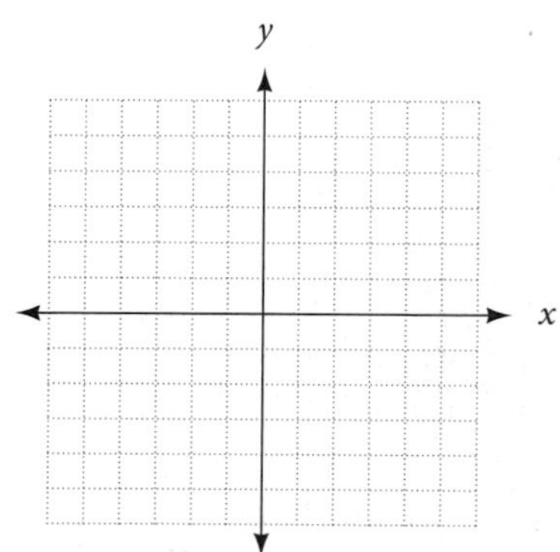

The following equations have graphs that are also parabolas. However, the graphs of these equations will open downward; the vertex of each will be the highest point on the graph. Graph each equation by first making a table of ordered pairs using the given values of x.

21. $y = 4 - x^2 \quad x = -3, -2, -1, 0, 1, 2, 3$

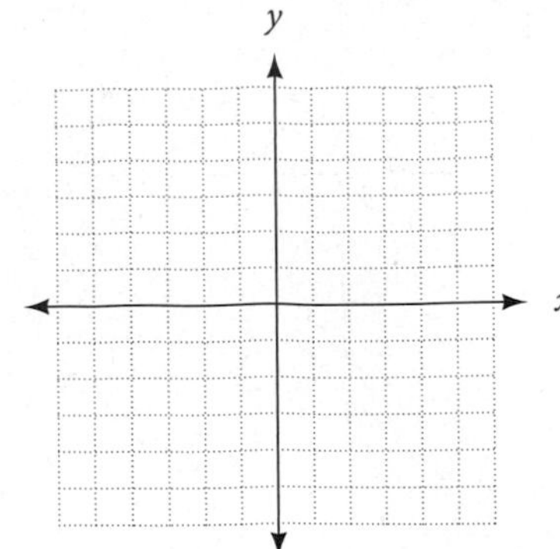

22. $y = 3 - x^2 \quad x = -3, -2, -1, 0, 1, 2, 3$

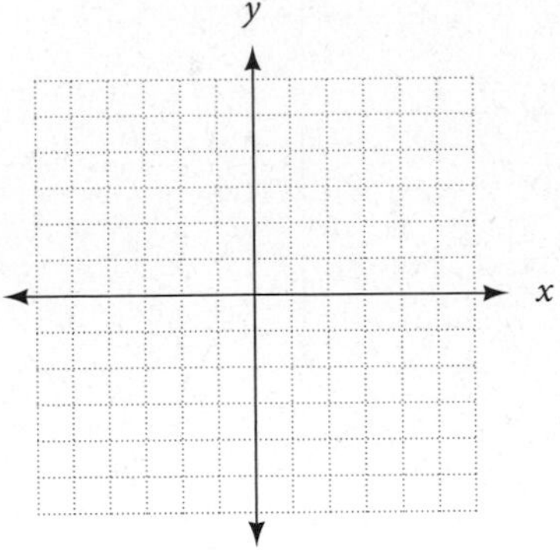

23. $y = -1 - x^2 \quad x = -2, -1, 0, 1, 2$

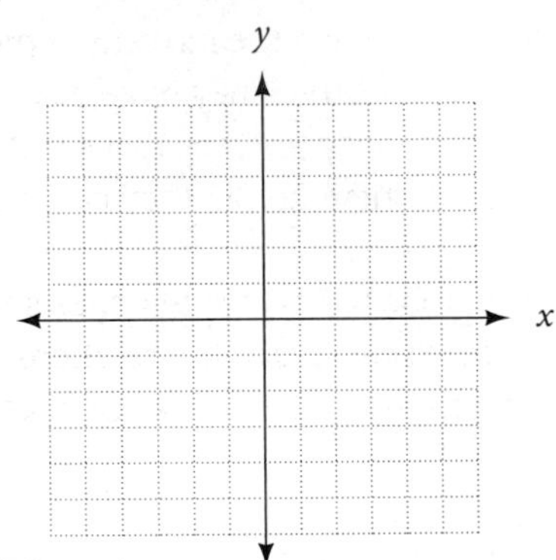

24. $y = -2 - x^2 \quad x = -2, -1, 0, 1, 2$

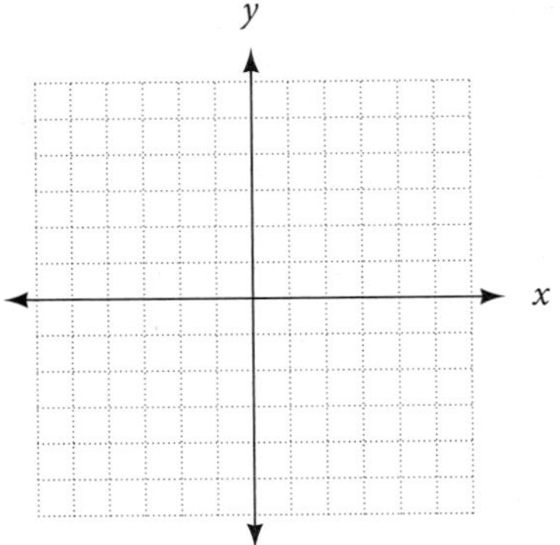

25. Graph the line $y = x + 2$ and the parabola $y = x^2$ on the same coordinate system. Name the points where the two graphs intersect.

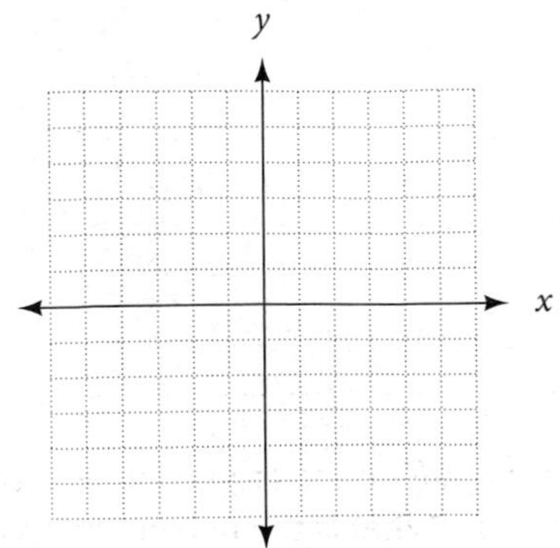

26. Graph the line $y = x$ and the parabola $y = x^2 - 2$ on the same coordinate system. Name the points where the two graphs intersect.

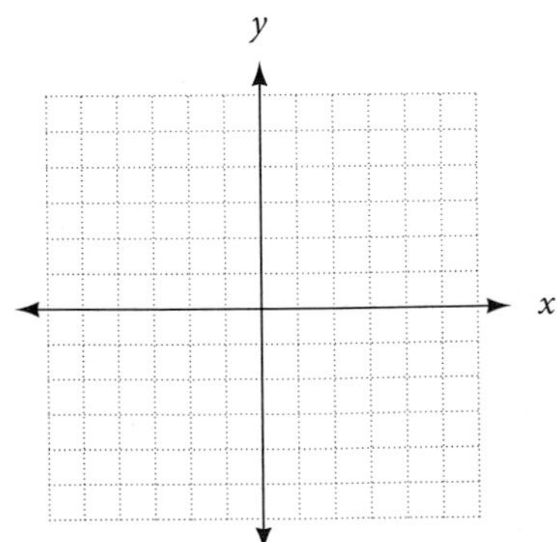

27. Graph the parabola $y = 2x^2$ and the parabola $y = \frac{1}{2}x^2$ on the same coordinate system.

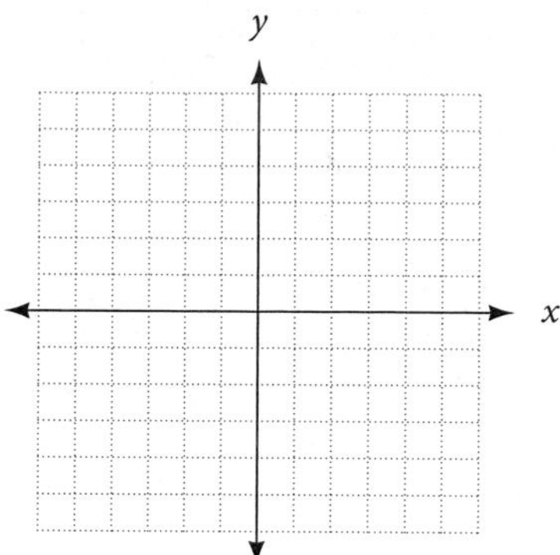

28. Graph the parabola $y = 3x^2$ and the parabola $y = \frac{1}{3}x^2$ on the same coordinate system.

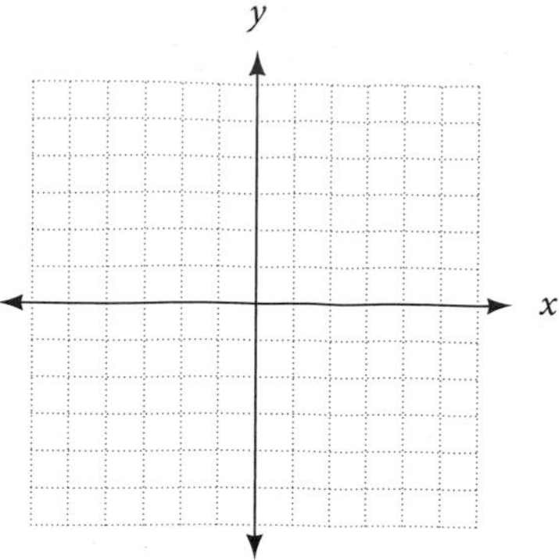

Applying the Concepts

29. **Comet Orbit** If a comet approaches the sun with *exactly* the right velocity and direction, its orbit around the sun will be a parabola. (Such a comet only will be seen once and then will leave the solar system forever.) A certain comet's orbit is given by $y = x^2 - 0.25$, where both x and y are measured in millions of miles, and the sun's position is at the origin (0, 0).

   a. Graph the comet's orbit between $x = -2$ and $x = 2$.

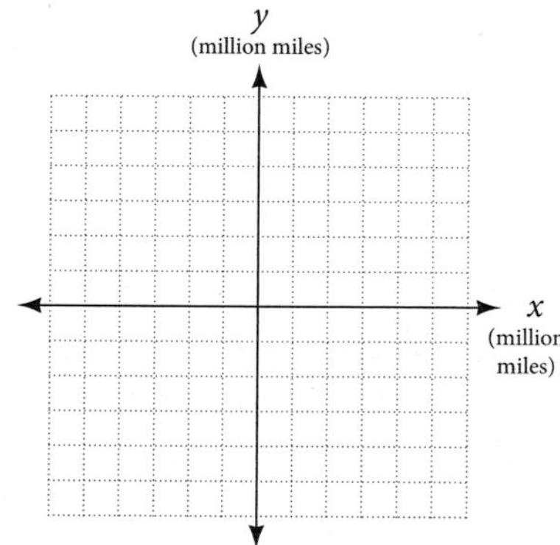

   b. How far is the comet from the sun when the comet is at the vertex of its path?

   c. How far is the comet from the sun when the comet crosses the x-axis?

30. **Maximum Profit** A company wants to know what price they should charge to make the maximum profit on their inexpensive digital watches. They have discovered that if they sell the watches for \$4, they can sell 2 million of them, but if they sell them for \$5, they can only sell 1 million.

   a. Let x be the number of *million* watches sold at p dollars per watch. Assume that the relationship between x and p is linear and find an equation that gives that relationship.

   b. Remember that revenue is price times number of items sold: $R = xp$. Solve the equation from part (a) for x and substitute the result into the equation $R = xp$ in place of x.

   c. Graph the equation from part (b) on a coordinate system where the horizontal axis is the p-axis and the vertical axis is the R-axis.

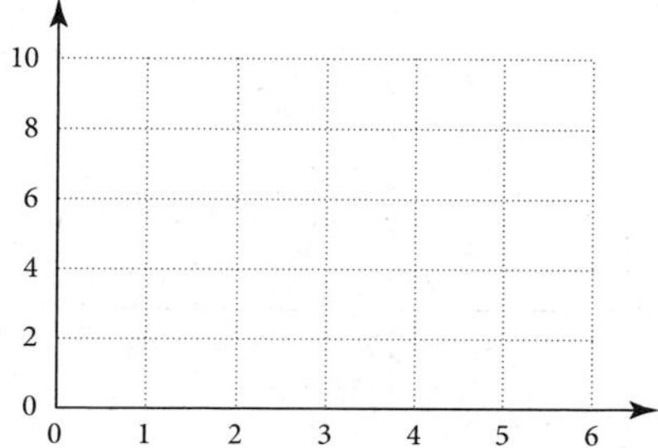

   d. The highest point on the graph will give you the maximum revenue. What is the maximum revenue?

Maintaining Your Skills

Find each root.

31. $\sqrt{49}$

32. $\sqrt[3]{-8}$

Write in simplified form for radicals.

33. $\sqrt{50}$

34. $2\sqrt{18x^2y^3}$

35. $\sqrt{\dfrac{2}{5}}$

36. $\sqrt[3]{\dfrac{1}{2}}$

Perform the indicated operations.

37. $3\sqrt{12} + 5\sqrt{27}$

38. $\sqrt{3}(\sqrt{3} - 4)$

39. $(\sqrt{6} + 2)(\sqrt{6} - 5)$

40. $(\sqrt{x} + 3)^2$

Rationalize the denominator.

41. $\dfrac{8}{\sqrt{5} - \sqrt{3}}$

42. $\dfrac{\sqrt{5} - \sqrt{2}}{\sqrt{5} + \sqrt{2}}$

Solve for x.

43. $\sqrt{2x - 5} = 3$

44. $\sqrt{x + 15} = x + 3$

Chapter 9 Summary

Solving Quadratic Equations of the Form $(ax + b)^2 = c$ [9.1]

We can solve equations of the form $(ax + b)^2 = c$ by applying the square root property for equations to write:

$$ax + b = \pm\sqrt{c}$$

EXAMPLES

1. $(x - 3)^2 = 25$
$x - 3 = \pm 5$
$x = 3 \pm 5$
$x = -2$ or $x = 8$

Solving by Completing the Square [9.2]

To complete the square on a quadratic equation as a method of solution, we use the following steps:

Step 1: Move the constant term to one side and the variable terms to the other. Then, divide each side by the coefficient of x^2 if it is other than 1.

Step 2: Take the square of half the coefficient of the linear term and add it to both sides of the equation.

Step 3: Write the left side as a binomial square and then apply the square root property.

Step 4: Solve the resulting equation.

2. $x^2 - 6x + 2 = 0$
$x^2 - 6x = -2$
$x^2 - 6x + \mathbf{9} = -2 + \mathbf{9}$
$(x - 3)^2 = 7$
$x - 3 = \pm\sqrt{7}$
$x = 3 \pm \sqrt{7}$

The Quadratic Formula [9.3, 9.5]

Any equation that is in the form $ax^2 + bx + c = 0$, where $a \neq 0$, has as its solutions:

$$x = \frac{-b \pm \sqrt{b^2 - 4ac}}{2a}$$

The expression under the square root sign, $b^2 - 4ac$, is known as the discriminant. When the discriminant is negative, the solutions are complex numbers.

3. If $2x^2 + 3x - 4 = 0$

then $x = \dfrac{-3 \pm \sqrt{9 - 4(2)(-4)}}{2(2)}$

$= \dfrac{-3 \pm \sqrt{41}}{4}$

Complex Numbers [9.4]

Any number that can be put in the form $a + bi$, where $i = \sqrt{-1}$, is called a complex number.

4. The numbers 5, $3i$, $2 + 4i$, and $7 - i$ are all complex numbers.

Addition and Subtraction of Complex Numbers [9.4]

We add (or subtract) complex numbers by using the same procedure we used to add (or subtract) polynomials: We combine similar terms.

5. $(3 + 4i) + (6 - 7i)$
$= (3 + 6) + (4i - 7i)$
$= 9 - 3i$

Multiplication of Complex Numbers [9.4]

6. $(2 + 3i)(3 - i)$
$= 6 - 2i + 9i - 3i^2$
$= 6 + 7i + 3$
$= 9 + 7i$

We multiply complex numbers in the same way we multiply binomials. The result, however, can be simplified further by substituting -1 for i^2 whenever it appears.

Division of Complex Numbers [9.4]

7. $\frac{3}{2 + 5i} = \frac{3}{2 + 5i} \cdot \frac{\mathbf{2 - 5i}}{\mathbf{2 - 5i}}$
$= \frac{6 - 15i}{29}$

Division with complex numbers is accomplished with the method for rationalizing the denominator that we developed while working with radical expressions. If the denominator has the form $a + bi$, we multiply both the numerator and the denominator by its conjugate, $a - bi$.

Graphing Parabolas [9.6]

8. Graph $y = x^2 - 2$.

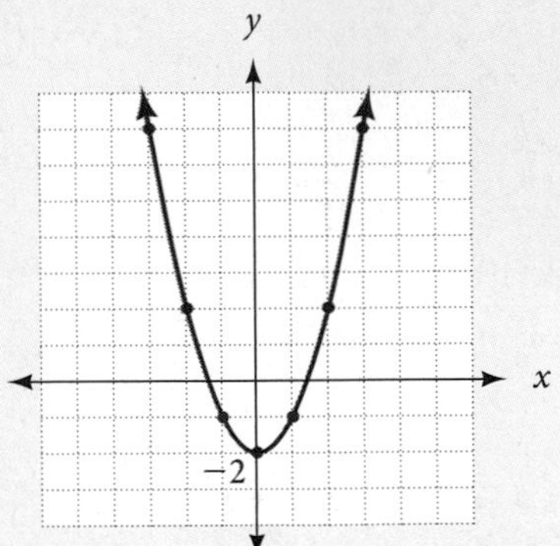

The graph of each of the following equations is a parabola with the same basic shape as the graph of $y = x^2$.

Equation	Vertex	To obtain the graph, move the graph of $y = x^2$
$y = x^2 + k$	$(0, k)$	k units vertically
$y = (x - h)^2$	$(h, 0)$	h units horizontally
$y = (x - h)^2 + k$	(h, k)	h units horizontally and k units vertically

COMMON MISTAKES

1. The most common mistake when working with complex numbers is to say $i = -1$. It does not; i is the *square root* of -1, not -1 itself.
2. The most common mistake when working with the quadratic formula is to try to identify the constants a, b, and c before putting the equation into standard form.

Chapter 9 Review

Solve each quadratic equation. [9.1, 9.5]

1. $a^2 = 32$
2. $a^2 = 60$
3. $2x^2 = 32$
4. $(x + 3)^2 = 36$
5. $(x - 2)^2 = 81$
6. $(3x + 2)^2 = 16$
7. $(2x + 5)^2 = 32$
8. $(3x - 4)^2 = 27$
9. $\left(x - \frac{2}{3}\right)^2 = -\frac{25}{9}$

Solve by completing the square. [9.2]

10. $x^2 + 8x = 4$
11. $x^2 - 8x = 4$
12. $x^2 - 4x - 7 = 0$
13. $x^2 + 4x + 3 = 0$
14. $a^2 = 9a + 3$
15. $a^2 = 5a + 6$
16. $2x^2 + 4x - 6 = 0$
17. $3x^2 - 6x - 2 = 0$

Solve by using the quadratic formula. [9.3, 9.5]

18. $x^2 + 7x + 12 = 0$
19. $x^2 - 8x + 16 = 0$
20. $x^2 + 5x + 7 = 0$
21. $2x^2 = -8x + 5$
22. $3x^2 = -3x - 4$
23. $\frac{1}{5}x^2 - \frac{1}{2}x = \frac{3}{10}$
24. $(2x + 1)(2x - 3) = -6$

Add and subtract the following complex numbers. [9.4]

25. $(4 - 3i) + 5i$
26. $(2 - 5i) - 3i$
27. $(5 + 6i) + (5 - i)$
28. $(2 + 5i) + (3 - 7i)$
29. $(3 - 2i) - (3 - i)$
30. $(5 - 7i) - (6 - 2i)$
31. $(3 + i) - 5i - (4 - i)$
32. $(2 - 3i) - (5 - 2i) + (5 - i)$

Multiply the following complex numbers. [9.4]

33. $2(3 - i)$
34. $-6(4 + 2i)$
35. $4i(6 - 5i)$
36. $(2 - i)(3 + i)$
37. $(3 - 4i)(5 + i)$
38. $(3 - i)^2$
39. $(4 + i)(4 - i)$
40. $(3 + 2i)(3 - 2i)$

Divide the following complex numbers. [9.4]

41. $\frac{i}{3 + i}$
42. $\frac{i}{2 - i}$
43. $\frac{5}{2 + 5i}$
44. $\frac{2i}{4 - i}$
45. $\frac{-3i}{3 - 2i}$
46. $\frac{3 + i}{3 - i}$
47. $\frac{4 - 5i}{4 + 5i}$
48. $\frac{2 + 3i}{3 - 5i}$

Write the following radicals as complex numbers. [9.5]

49. $\sqrt{-36}$
50. $\sqrt{-144}$
51. $\sqrt{-17}$
52. $\sqrt{-31}$
53. $\sqrt{-40}$
54. $\sqrt{-72}$
55. $\sqrt{-200}$
56. $\sqrt{-242}$

Graph each of the following equations. [9.6]

57. $y = x^2 + 2$
58. $y = (x - 2)^2$
59. $y = (x + 2)^2$
60. $y = (x + 3)^2 - 2$
61. $y = x^2 + 4x + 7$
62. $y = x^2 - 4x + 7$

Chapter 9 Cumulative Review

Simplify.

1. $\dfrac{9(-6) - 10}{2(-8)}$

2. $-\dfrac{4}{5} \div \dfrac{8}{15}$

3. $\dfrac{x^4}{x^{-8}}$

4. $\dfrac{(3x^5)(20x^3)}{15x^{10}}$

5. $\dfrac{\dfrac{3x^4}{y^5}}{\dfrac{9x^3}{y}}$

6. $\dfrac{1 - \dfrac{16}{y^2}}{1 - \dfrac{4}{y} - \dfrac{32}{y^2}}$

7. $\sqrt{81}$

8. $-\sqrt[3]{81}$

9. $\dfrac{10}{\sqrt{2}}$

10. $\dfrac{5}{\sqrt{7} + \sqrt{3}}$

11. $(3 + 3i) - 7i - (2 + 2i)$

12. $(3 + 5i) + (4 - 3i) - (6 - i)$

13. $(a - 2)(a^2 - 6a + 7)$

14. $\dfrac{x^2 + 5x - 24}{x + 8}$

Simplify. Assume all variables represent positive numbers.

15. $\sqrt{121x^4y^2}$

16. $\sqrt{\dfrac{200b^2}{81}}$

Solve each equation.

17. $x - \dfrac{3}{4} = \dfrac{5}{6}$

18. $\dfrac{2}{3}a = \dfrac{1}{6}a + 1$

19. $7 - 4(3x + 4) = -9x$

20. $3x^2 = 7x + 20$

21. $5x^2 = -15x$

22. $(2x - 3)^2 = 49$

23. $\dfrac{a}{a + 4} = \dfrac{7}{3}$

24. $\dfrac{x}{3} = \dfrac{6}{x - 3}$

Graph the solution set on the number line.

25. $-5 \le 2x - 1 \le 7$

26. $x < -5$ or $x > 2$

Graph on a rectangular coordinate system.

27. $x - y = 5$

28. $y = -3x + 4$

29. $y = x^2 - 2$

30. $y = (x - 3)^2 - 2$

Solve each system.

31. $x = y - 3$
$2x + 3y = 4$

32. $2x + 3y = 2$
$3x - y = 3$

Factor completely.

33. $x^2 - 5x - 24$

34. $12x^2 - 11xy - 15y^2$

35. $25 - y^2$

36. $25x^2 - 30xy + 9y^2$

37. Find the slope of the line that passes through the points $(3, -2)$ and $(1, 8)$.

38. Find the slope and y-intercept for the equation $8x - 4y = -12$.

39. Divide $15a^3b - 10a^2b^2 - 20ab^3$ by $5ab$.

40. Divide $\dfrac{3x^3 - 7x^2 + 14x - 10}{3x - 1}$.

Subtract.

41. $\dfrac{a - 3}{a - 7} - \dfrac{a + 10}{a - 7}$

42. $\dfrac{-1}{x^2 - 4} - \dfrac{-2}{x^2 - 4x - 12}$

Combine the expressions. Assume all variables represent positive numbers.

43. $5\sqrt{200} + 9\sqrt{50}$

44. $5\sqrt{63x^2} - x\sqrt{28}$

Divide

45. $\dfrac{8 + i}{i}$

46. $\dfrac{3 - 2i}{5 - i}$

47. **Investing** Randy invested $20,000 in two accounts. One paid 5% simple interest, and the other paid 7%. He earned $800 more in interest on the 7% account than he did on the 5% account. How much did he invest in each account?

48. **Investing** Ms. Jones invested $18,000 in two accounts. One account pays 6% simple interest, and the other pays 8%. Her total interest for the year was $1,290. How much did she have in each account?

49. **Direct Variation** y varies directly as the square root of x. If $y = 10$ when $x = 4$, find y when $x = 9$.

50. **Inverse Variation** y varies inversely as the square of x. If $y = 32$ when $x = 1$, find y when $x = 4$.

Use the illustration to answer the following questions.

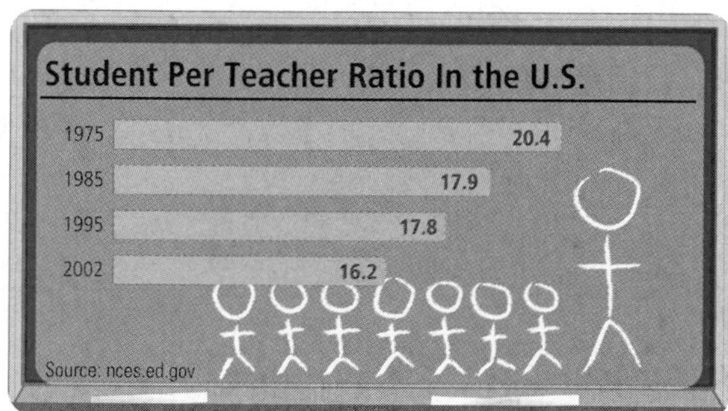

51. If a 1975 school had 11,016 students, how many teachers would it have?

52. If a 2002 school had 11,016 students, how many teachers would it have?

Chapter 9 Test

Solve the following quadratic equations. [9.1, 9.3, 9.5]

1. $x^2 - 7x - 8 = 0$
2. $(x - 3)^2 = 12$
3. $\left(x - \frac{5}{2}\right)^2 = -\frac{75}{4}$
4. $\frac{1}{3}x^2 = \frac{1}{2}x - \frac{5}{6}$
5. $3x^2 = -2x + 1$
6. $(x + 2)(x - 1) = 6$
7. $9x^2 + 12x + 4 = 0$
8. Solve $x^2 - 6x - 6 = 0$ by completing the square. [9.2]

Write as a complex number. [9.5]

9. $\sqrt{-9}$
10. $\sqrt{-121}$
11. $\sqrt{-72}$
12. $\sqrt{-18}$

Work the following problems involving complex numbers. [9.4]

13. $(3i + 1) + (2 + 5i)$
14. $(6 - 2i) - (7 - 4i)$
15. $(2 + i)(2 - i)$
16. $(3 + 2i)(1 + i)$
17. $\frac{i}{3 - i}$
18. $\frac{2 + i}{2 - i}$

Graph the following equations. [9.6]

19. $y = x^2 - 4$
20. $y = (x - 4)^2$
21. $y = (x + 3)^2 - 4$
22. $y = x^2 - 6x + 11$
23. Find the height of the isosceles triangle. [9.1]

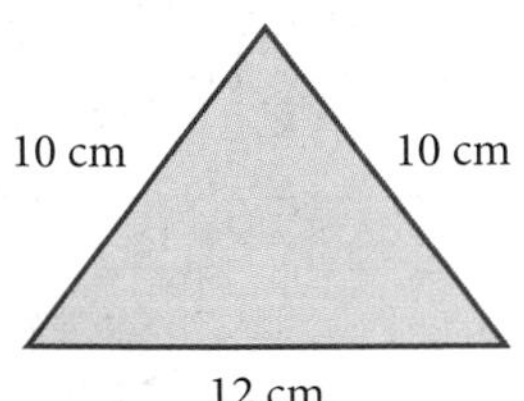

Match each equation with its graph. [9.6]

24. $y = x^2 + 2$
25. $y = (x + 2)^2$
26. $y = x^2 - 2$
27. $y = (x - 2)^2$

A

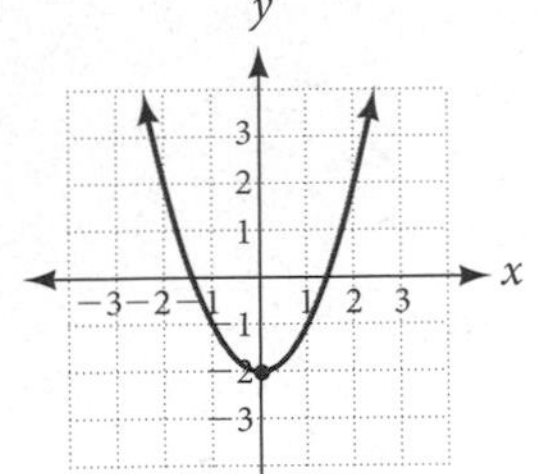

B

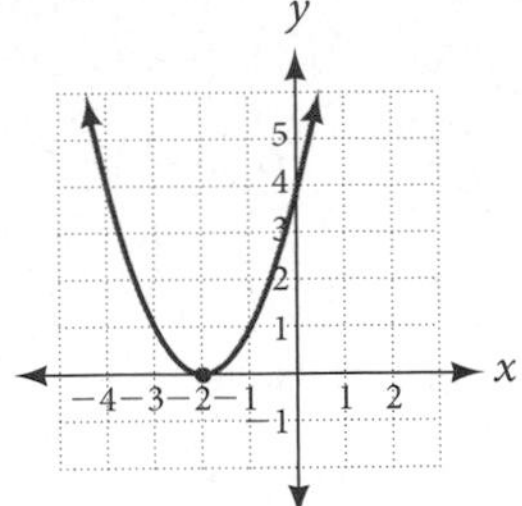

C

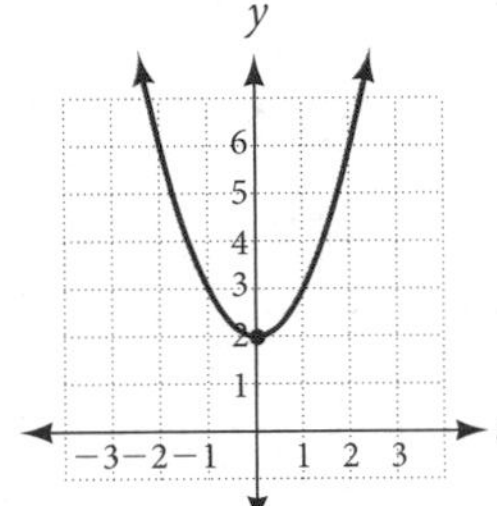

D

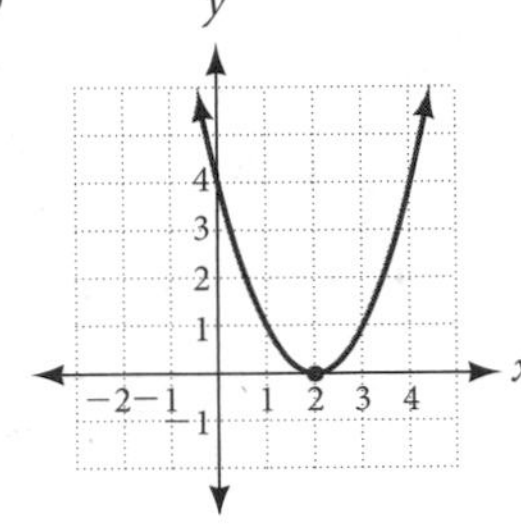

Match each equation with its graph. [9.6]

28. $y = (x + 3)^2 - 1$
29. $y = (x - 3)^2 - 1$
30. $y = (x + 1)^2 + 3$
31. $y = (x - 1)^2 + 3$

A.

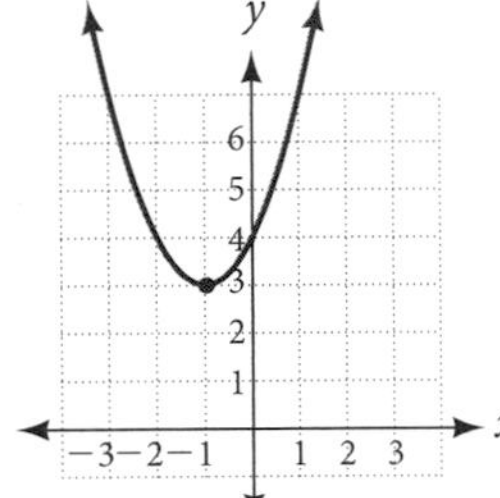

B.

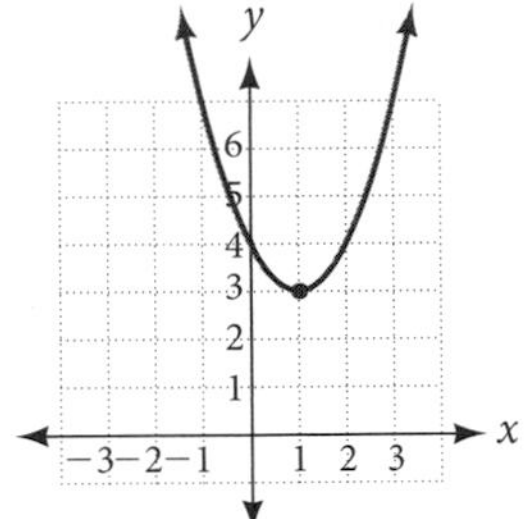

C.

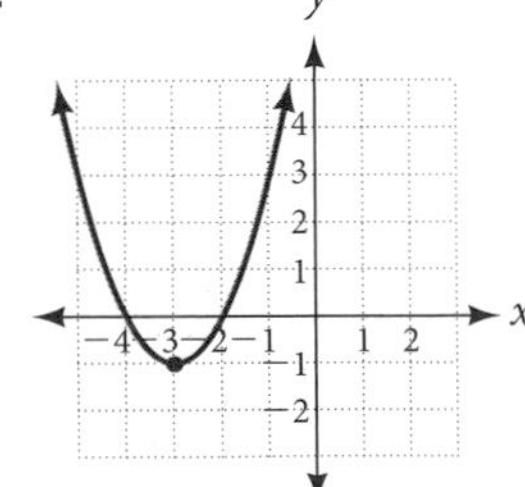

D.

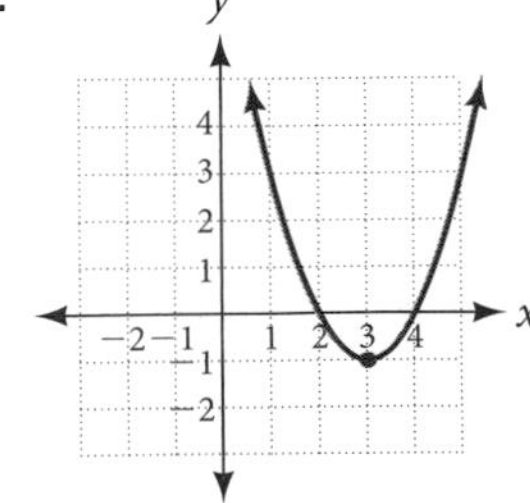

Chapter 9 Projects

QUADRATIC EQUATIONS

GROUP PROJECT

Proving the Pythagorean Theorem

Number of People 1–2

Time Needed 5–8 minutes

Equipment pencil, graph paper, and scissors

Background The postage stamp shown here was issued by San Marino, a small state in Europe, in 1982. It depicts the mathematician Pythagoras. As you know, the Pythagorean theorem is used extensively throughout mathematics, and we have used it many times in this book. We can use the diagram shown here to give a visual "proof" of this theorem, similar to the original proof given by the Pythagoreans.

Procedure The shaded triangle shown here is a right triangle.

1. Notice the square representing c^2 has dotted lines making up four additional triangles and a square.
2. Cut out the square labeled c^2, and cut along the dotted lines.
3. Use all the pieces you have cut out to cover up the squares representing a^2 and b^2. Doing so is a visual "proof" that $c^2 = a^2 + b^2$.

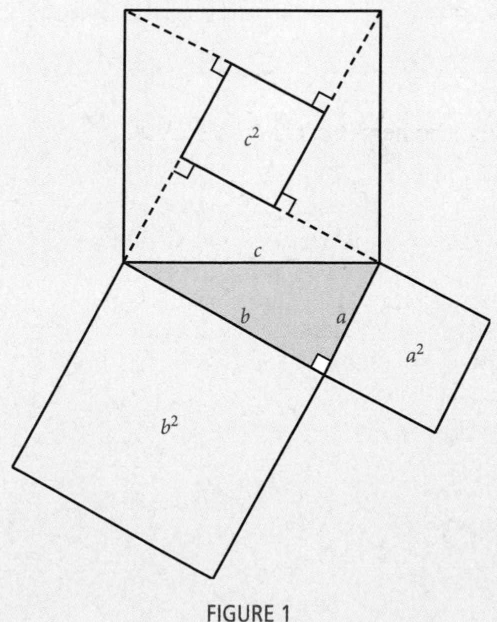

FIGURE 1

RESEARCH PROJECT

A Continued Fraction and the Golden Ratio

You may recall that the continued fraction

$$1 + \cfrac{1}{1 + \cfrac{1}{1 + \cfrac{1}{1 + \ldots}}}$$

is equal to the golden ratio.

The same conclusion can be found with the quadratic formula by using the following conditional statement

$$\text{If } x = 1 + \cfrac{1}{1 + \cfrac{1}{1 + \cfrac{1}{1 + \ldots}}}, \text{ then } x = 1 + \frac{1}{x}$$

Work with this conditional statement until you see that it is true. Then solve the equation $x = 1 + \frac{1}{x}$. Write an essay in which you explain in your own words why the conditional statement is true. Then show the details of the solution to the equation $x = 1 + \frac{1}{x}$. The message that you want to get across in your essay is that the continued fraction shown here is actually the same as the golden ratio.

Appendix A

RESOURCES

By gathering resources early in the term, before you need help, the information about these resources will be available to you when they are needed.

Instructor

Knowing the contact information for your instructor is very important. You may already have this information from the course syllabus. It is a good idea to write it down again.

Name ______________________ Office Location ______________________
Available Hours: M __________ T __________ W __________ TH __________ F __________
Phone Number ______________ ext. __________ E-mail Address ______________ @ __________

Tutoring Center

Many schools offer tutoring, free of charge to their students. If this is the case at your school, find out when and where tutoring is offered.

Tutoring Location ______________________ Phone Number ______________ ext. __________
Available Hours: M __________ T __________ W __________ TH __________ F __________

Computer Lab

Many schools offer a computer lab where students can use the online resources and software available with their textbook. Other students using the same software and websites as you can be very helpful. Find out where the computer lab at your school is located.

Computer Lab Location ______________________ Phone Number ______________ ext. __________
Available Hours: M __________ T __________ W __________ TH __________ F __________

Videos

A complete set of videos is available to your school. These videos feature the author of your textbook presenting full-length, 15- to 20-minute lessons from every section of your textbook. If you miss class, or find yourself behind, these tapes will prove very useful.

Videotape Location ______________________ Phone Number ______________ ext. __________
Available Hours: M __________ T __________ W __________ TH __________ F __________

Classmates

Form a study group and meet on a regular basis. When you meet try to speak to each other using proper mathematical language. That is, use the words that you see in the definition and property boxes in your textbook.

Name ______________________ Phone ______________ E-mail ______________________
Name ______________________ Phone ______________ E-mail ______________________
Name ______________________ Phone ______________ E-mail ______________________

Introduction to Functions B

The ad shown in the margin appeared in the "Help Wanted" section of the local newspaper the day I was writing this section of the book. We can use the information in the ad to start an informal discussion of functions.

HELP WANTED

YARD PERSON

Full time 40 hrs. with weekend work required. Cleaning & loading trucks. \$7.50/hr.

Valid CDL with clean record & drug screen required.

Submit current MVR to KCI, 225 Suburban Rd., SLO 93405 (805) 555-3304.

An Informal Look at Functions

To begin with, suppose you have a job that pays \$7.50 per hour and that you work anywhere from 0 to 40 hours per week. The amount of money you make in one week depends on the number of hours you work that week. In mathematics, we say that your weekly earnings are a *function* of the number of hours you work. If we let the variable x represent hours and the variable y represent the money you make, then the relationship between x and y can be written as

$$y = 7.5x \qquad \text{for} \qquad 0 \le x \le 40$$

EXAMPLE 1 Construct a table and graph for the function

$$y = 7.5x \qquad \text{for} \qquad 0 \le x \le 40$$

SOLUTION Table 1 gives some of the paired data that satisfy the equation $y = 7.5x$. Figure 1 is the graph of the equation with the restriction $0 \le x \le 40$.

TABLE 1 Weekly Wages

Hours Worked x	Rule $y = 7.5x$	Pay y
0	$y = 7.5(0)$	0
10	$y = 7.5(10)$	75
20	$y = 7.5(20)$	150
30	$y = 7.5(30)$	225
40	$y = 7.5(40)$	300

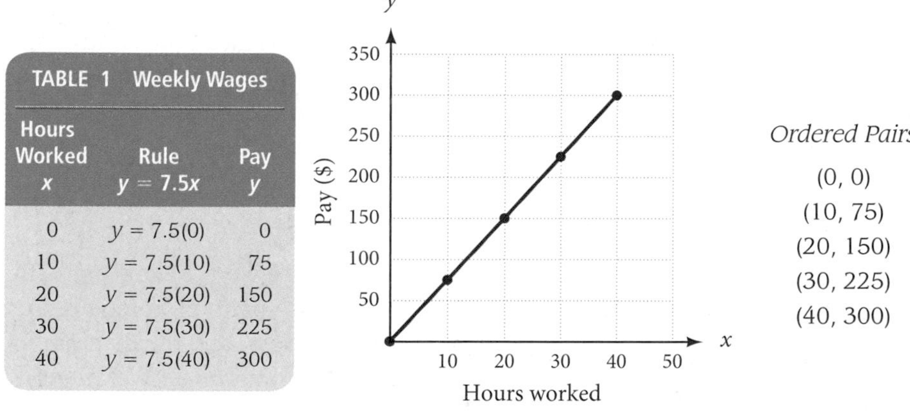

FIGURE 1 Weekly wages at \$7.50 per hour

The equation $y = 7.5x$ with the restriction $0 \le x \le 40$, Table 1, and Figure 1 are three ways to describe the same relationship between the number of hours you work in one week and your gross pay for that week. In all three, we input values of x, and then use the function rule to output values of y.

Domain and Range of a Function

We began this discussion by saying that the number of hours worked during the week was from 0 to 40, so these are the values that x can assume. From the line graph in Figure 1, we see that the values of y range from 0 to 300. We call the complete set of values that x can assume the *domain* of the function. The values that are assigned to y are called the *range* of the function.

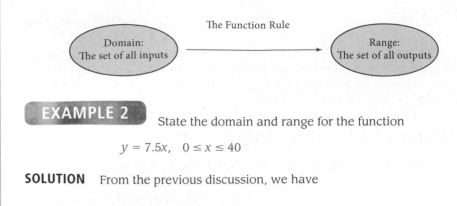

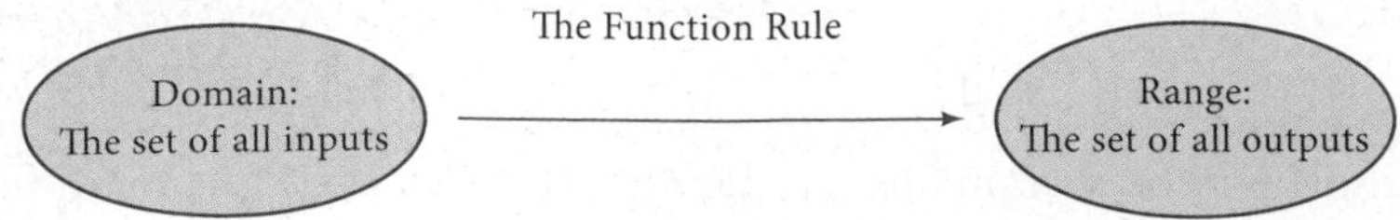

EXAMPLE 2 State the domain and range for the function

$$y = 7.5x, \quad 0 \le x \le 40$$

SOLUTION From the previous discussion, we have

$$\text{Domain} = \{x \mid 0 \le x \le 40\}$$
$$\text{Range} = \{y \mid 0 \le y \le 300\}$$

Function Maps

Another way to visualize the relationship between x and y is with the diagram in Figure 2, which we call a *function map*.

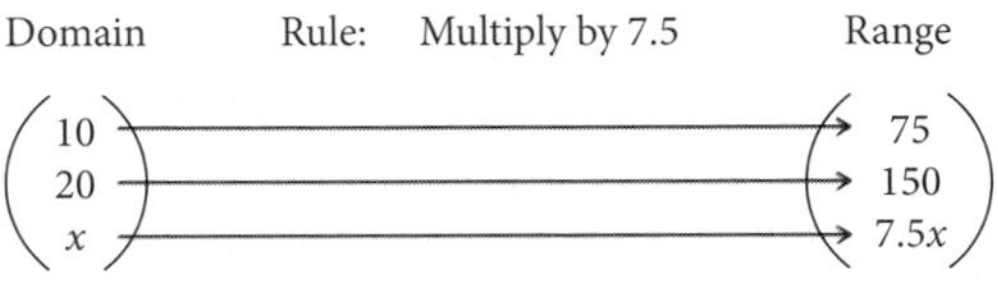

FIGURE 2 A function map

Although the diagram in Figure 2 does not show all the values that x and y can assume, it does give us a visual description of how x and y are related. It shows that values of y in the range come from values of x in the domain according to a specific rule (multiply by 7.5 each time).

A Formal Look at Functions

What is apparent from the preceding discussion is that we are working with paired data. The solutions to the equation $y = 7.5x$ are pairs of numbers; the points on the line graph in Figure 1 come from paired data; and the diagram in Figure 2 pairs numbers in the domain with numbers in the range. We are now ready for the formal definition of a function.

Definition
A **function** is a rule that pairs each element in one set, called the **domain**, with exactly one element from a second set, called the **range**.

In other words, a function is a rule for which each input is paired with exactly one output.

Functions as Ordered Pairs

The function rule $y = 7.5x$ from Example 1 produces ordered pairs of numbers (x, y). The same thing happens with all functions: the function rule produces ordered pairs of numbers. We use this result to write an alternative definition for a function.

Alternate Definition

A **function** is a set of ordered pairs in which no two different ordered pairs have the same first coordinate. The set of all first coordinates is called the **domain** of the function. The set of all second coordinates is called the **range** of the function.

The restriction on first coordinates in the alternate definition keeps us from assigning a number in the domain to more than one number in the range.

A Relationship That is Not a Function

You may be wondering if any sets of paired data fail to qualify as functions. The answer is yes, as the next example reveals.

EXAMPLE 3 Table 2 shows the prices of used Ford Mustangs that were listed in the local newspaper. The diagram in Figure 3 gives a visual representation of the data in Table 2. Why is this data not a function?

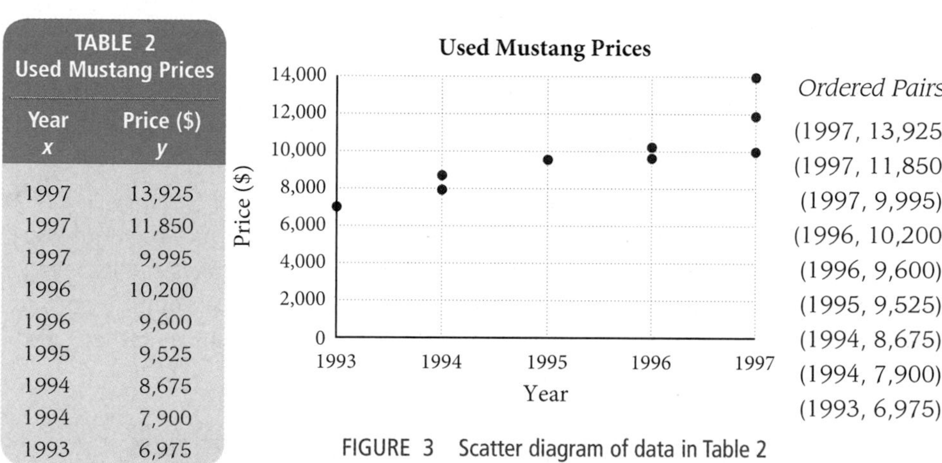

TABLE 2 Used Mustang Prices

Year x	Price ($) y
1997	13,925
1997	11,850
1997	9,995
1996	10,200
1996	9,600
1995	9,525
1994	8,675
1994	7,900
1993	6,975

FIGURE 3 Scatter diagram of data in Table 2

SOLUTION In Table 2, the year 1997 is paired with three different prices: \$13,925, \$11,850, and \$9,995. That is enough to disqualify the data from belonging to a function. For a set of paired data to be considered a function, each number in the domain must be paired with exactly one number in the range.

Still, there is a relationship between the first coordinates and second coordinates in the used-car data. It is not a function relationship, but it is a relationship. To classify all relationships specified by ordered pairs, whether they are functions or not, we include the following two definitions.

Definition

A **relation** is a rule that pairs each element in one set, called the **domain**, with *one or more elements* from a second set, called the **range**.

Alternate Definition

A **relation** is a set of ordered pairs. The set of all first coordinates is the **domain** of the relation. The set of all second coordinates is the **range** of the function.

Here are some facts that will help clarify the distinction between relations and functions:

1. Any rule that assigns numbers from one set to numbers in another set is a relation. If that rule makes the assignment such that no input has more than one output, then it is also a function.
2. Any set of ordered pairs is a relation. If none of the first coordinates of those ordered pairs is repeated, the set of ordered pairs is also a function.
3. Every function is a relation.
4. Not every relation is a function.

Graphing Relations and Functions

To give ourselves a wider perspective on functions and relations, we consider some equations whose graphs are not straight lines.

EXAMPLE 4 Kendra tosses a softball into the air with an underhand motion. The distance of the ball above her hand at any time is given by the function

$$h = 32t - 16t^2 \qquad \text{for} \qquad 0 \le t \le 2$$

where h is the height of the ball in feet and t is the time in seconds. Construct a table that gives the height of the ball at quarter-second intervals, starting with $t = 0$ and ending with $t = 2$. Construct a line graph from the table.

SOLUTION We construct Table 3 using the following values of t: 0, $\frac{1}{4}$, $\frac{1}{2}$, $\frac{3}{4}$, 1, $\frac{5}{4}$, $\frac{3}{2}$, $\frac{7}{4}$, 2. The values of h come from substituting these values of t into the equation $h = 32t - 16t^2$. (This equation comes from physics. If you take a physics class, you will learn how to derive this equation.) Then we construct the graph in Figure 4 from the table. The graph appears only in the first quadrant because neither t nor h can be negative.

TABLE 3 Tossing a Softball into the Air

Time (sec) t	Function Rule h $32t$ $16t^2$	Distance (ft) h
0	$h = 32(0) - 16(0)^2 = 0 - 0 = 0$	0
$\frac{1}{4}$	$h = 32\left(\frac{1}{4}\right) - 16\left(\frac{1}{4}\right)^2 = 8 - 1 = 7$	7
$\frac{1}{2}$	$h = 32\left(\frac{1}{2}\right) - 16\left(\frac{1}{2}\right)^2 = 16 - 4 = 12$	12
$\frac{3}{4}$	$h = 32\left(\frac{3}{4}\right) - 16\left(\frac{3}{4}\right)^2 = 24 - 9 = 15$	15
1	$h = 32(1) - 16(1)^2 = 32 - 16 = 16$	16
$\frac{5}{4}$	$h = 32\left(\frac{5}{4}\right) - 16\left(\frac{5}{4}\right)^2 = 40 - 25 = 15$	15
$\frac{3}{2}$	$h = 32\left(\frac{3}{2}\right) - 16\left(\frac{3}{2}\right)^2 = 48 - 36 = 12$	12
$\frac{7}{4}$	$h = 32\left(\frac{7}{4}\right) - 16\left(\frac{7}{4}\right)^2 = 56 - 49 = 7$	7
2	$h = 32(2) - 16(2)^2 = 64 - 64 = 0$	0

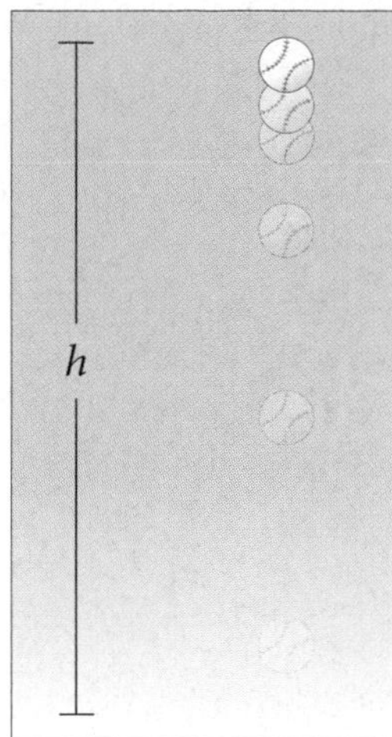

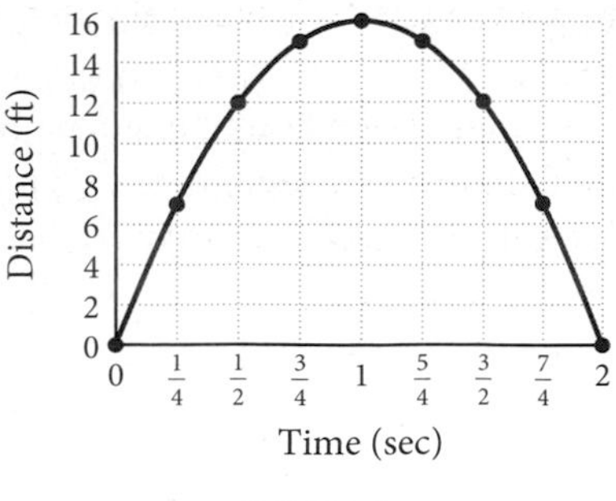

FIGURE 4

The domain is given by the inequality that follows the equation; it is

$$\text{Domain} = \{t \mid 0 \le t \le 2\}$$

The range is the set of all outputs that are possible by substituting the values of *t* from the domain into the equation. From our table and graph, it seems that the range is

$$Range = \{h \mid 0 \le h \le 16\}$$

EXAMPLE 5 Sketch the graph of $x = y^2$.

SOLUTION Without going into much detail, we graph the equation $x = y^2$ by finding a number of ordered pairs that satisfy the equation, plotting these points, then drawing a smooth curve that connects them. A table of values for x and y that satisfy the equation follows, along with the graph of $x = y^2$ shown in Figure 5.

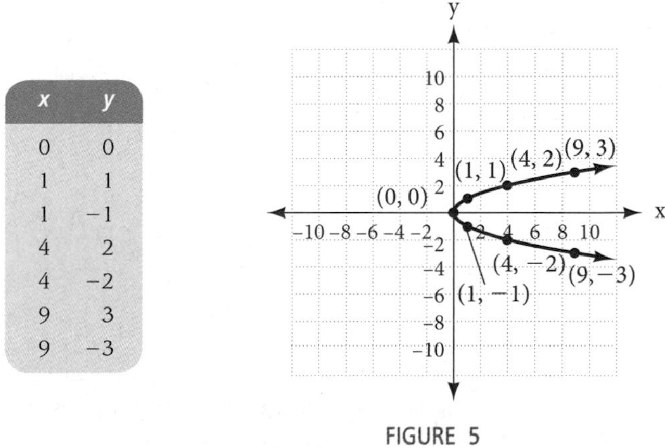

x	y
0	0
1	1
1	−1
4	2
4	−2
9	3
9	−3

FIGURE 5

As you can see from looking at the table and the graph in Figure 5, several ordered pairs whose graphs lie on the curve have repeated first coordinates, for instance (1, 1) and (1, −1), (4, 2) and (4, −2), as well as (9, 3) and (9, −3). The graph is therefore not the graph of a function.

Vertical Line Test

Look back at the scatter diagram for used Mustang prices shown in Figure 3. Notice that some of the points on the diagram lie above and below each other along vertical lines. This is an indication that the data does not constitute a function.

Two data points that lie on the same vertical line must have come from two ordered pairs with the same first coordinates.

Now, look at the graph shown in Figure 5. The reason this graph is the graph of a relation, but not of a function, is that some points on the graph have the same first coordinates, for example, the points (4, 2) and (4, −2). Furthermore, any time two points on a graph have the same first coordinates, those points must lie on a vertical line. [To convince yourself, connect the points (4, 2) and (4, −2) with a straight line. You will see that it must be a vertical line.] This allows us to write the following test that uses the graph to determine whether a relation is also a function.

Vertical Line Test

If a vertical line crosses the graph of a relation in more than one place, the relation cannot be a function. If no vertical line can be found that crosses a graph in more than one place, then the graph represents a function.

If we look back to the graph of $h = 32t - 16t^2$ as shown in Figure 4, we see that no vertical line can be found that crosses this graph in more than one place. The graph shown in Figure 4 is therefore the graph of a function.

EXAMPLE 6 Graph $y = |x|$. Use the graph to determine if this is a function. State the domain and range.

SOLUTION We let x take on values of −4, −3, −2, −1, 0, 1, 2, 3, and 4. The corresponding values of y are shown in the table. The graph is shown in Figure 6.

x	y
−4	4
−3	3
−2	2
−1	1
0	0
1	1
2	2
3	3
4	4

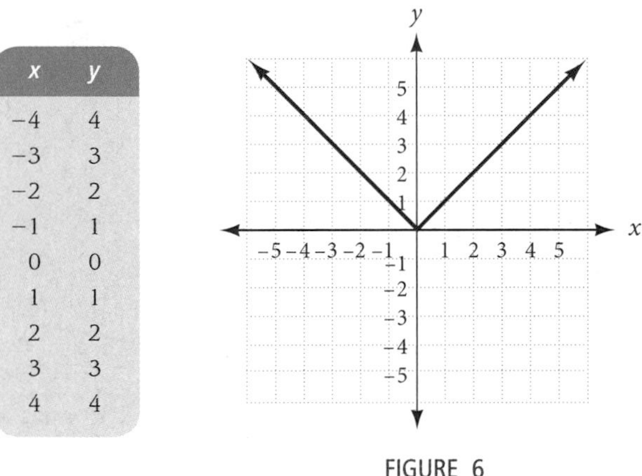

FIGURE 6

Since no vertical line can be found that crosses the graph in more than one place, $y = |x|$ is a function. The domain is all real numbers. The range is $\{y \mid y \geq 0\}$.

Getting Ready for Class

After reading through the preceding section, respond in your own words and in complete sentences.

1. What is a function?
2. What is the vertical line test?
3. Is every line the graph of a function? Explain.
4. Which variable is usually associated with the domain of a function?

Problem Set B

[Example 1]

1. Suppose you have a job that pays $8.50 per hour and you work anywhere from 10 to 40 hours per week.
   a. Write an equation, with a restriction on the variable x, that gives the amount of money, y, you will earn for working x hours in one week.
   b. Use the function rule you have written in part **a** to complete Table 4.

TABLE 4 Weekly Wages

Hours Worked x	Function Rule	Gross Pay ($) y
10		
20		
30		
40		

   c. Construct a line graph from the information in Table 4.
   d. State the domain and range of this function.
   e. What is the minimum amount you can earn in a week with this job? What is the maximum amount?

2. The ad shown here was in the local newspaper. Suppose you are hired for the job described in the ad.

   HELP WANTED
   ESPRESSO BAR OPERATOR
   Must be dependable, honest, service-oriented. Coffee exp. desired. 15-30 hrs per week. $5.25/hr. Start 5/31. Apply in person.
   Espresso Yourself
   Central Coast Mall
   Deadline May 23rd

   a. If x is the number of hours you work per week and y is your weekly gross pay, write the equation for y. (Be sure to include any restrictions on the variable x that are given in the ad.)
   b. Use the function rule you have written in part **a** to complete Table 5.

TABLE 5 Weekly Wages

Hours Worked x	Function Rule	Gross Pay ($) y
15		
20		
25		
30		

   c. Construct a line graph from the information in Table 5.
   d. State the domain and range of this function.
   e. What is the minimum amount you can earn in a week with this job? What is the maximum amount?

For each of the following relations, give the domain and range, and indicate which are also functions. [Example 2]

3. $\{(1, 3), (2, 5), (4, 1)\}$
4. $\{(3, 1), (5, 7), (2, 3)\}$
5. $\{(-1, 3), (1, 3), (2, -5)\}$
6. $\{(3, -4), (-1, 5), (3, 2)\}$
7. $\{(7, -1), (3, -1), (7, 4)\}$
8. $\{(5, -2), (3, -2), (5, -1)\}$

State whether each of the following graphs represents a function.

9. 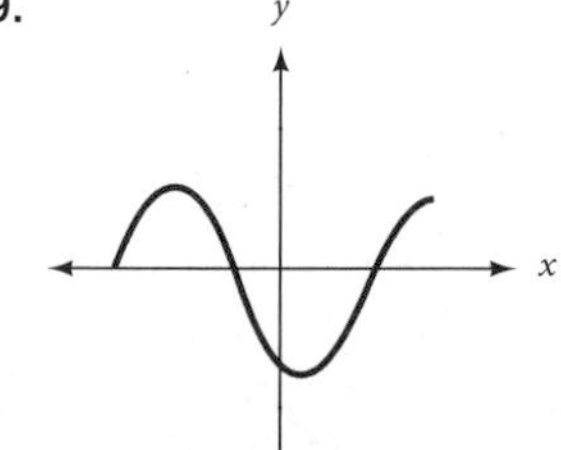

10. 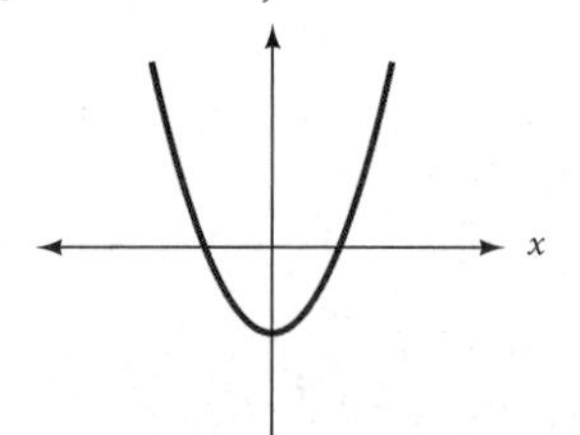

11. 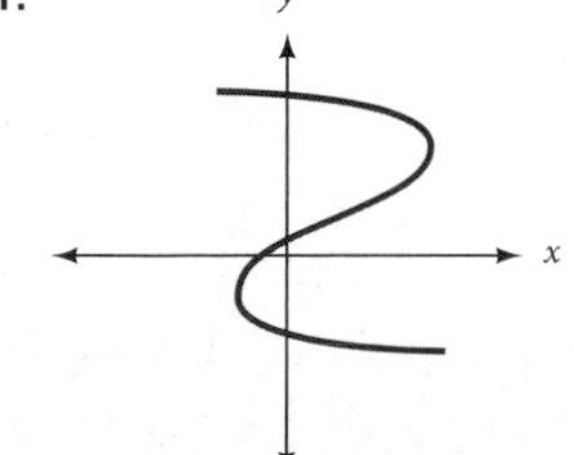

12. 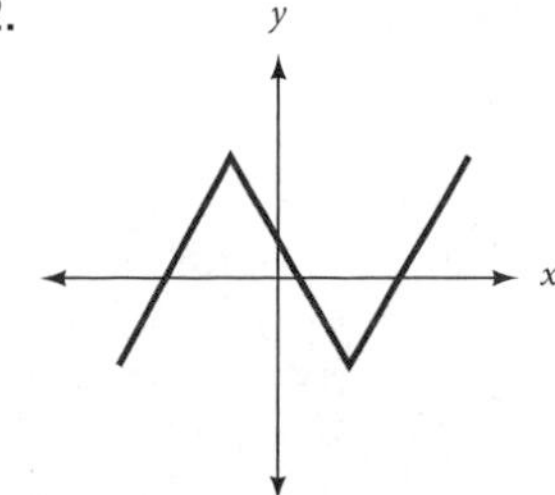

13.

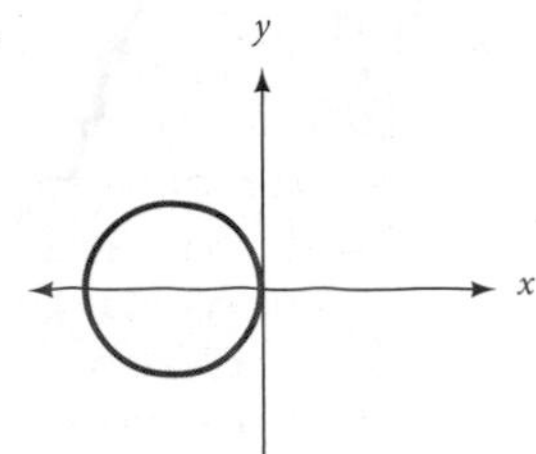

14.

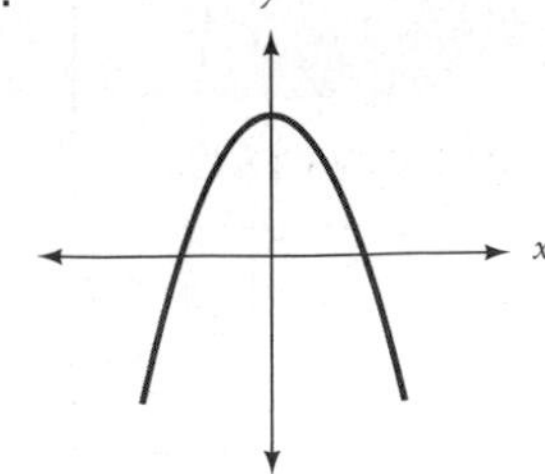

15.

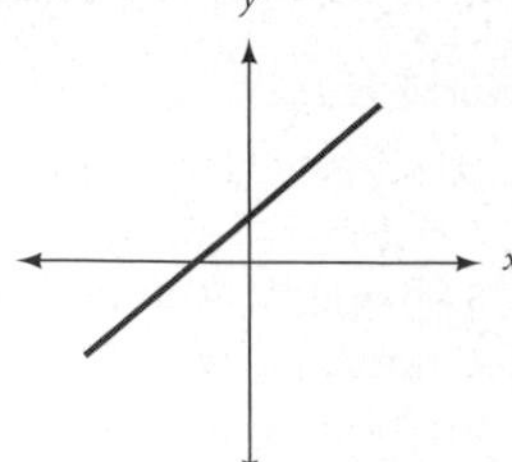

16.

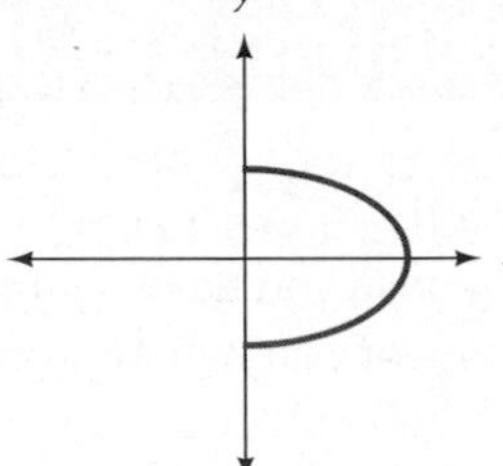

17.

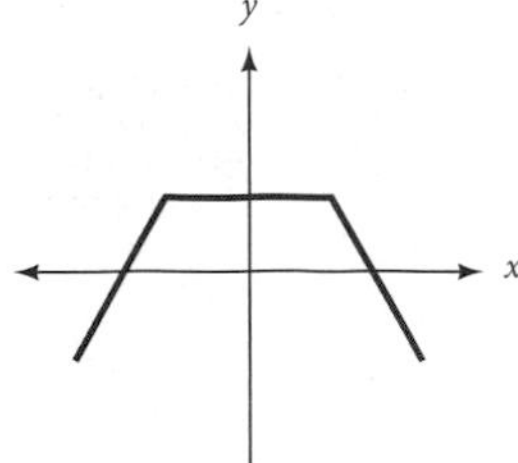

18.

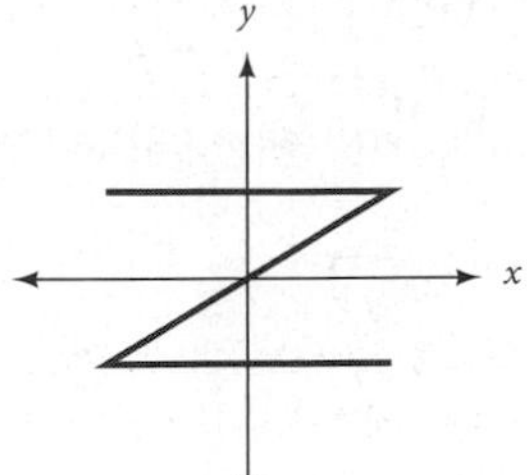

[Example 4]

19. **Tossing a Coin** Hali tosses a quarter into the air with an underhand motion. The distance the quarter is above her hand at any time is given by the function

$$h = 16t - 16t^2 \quad \text{for } 0 \le t \le 1$$

where h is the height of the quarter in feet, and t is the time in seconds.

a. Fill in the table.

Time (sec)	Function Rule	Distance (ft)
t	$h = 16t - 16t^2$	h
0		
0.1		
0.2		
0.3		
0.4		
0.5		
0.6		
0.7		
0.8		
0.9		
1		

b. State the domain and range of this function.

c. Use the data from the table to graph the function.

20. **Intensity of Light** The following formula gives the intensity of light that falls on a surface at various distances from a 100-watt light bulb:

$$I = \frac{120}{d^2} \quad \text{for} \quad d > 0$$

where I is the intensity of light (in lumens per square foot) and d is the distance (in feet) from the light bulb to the surface.

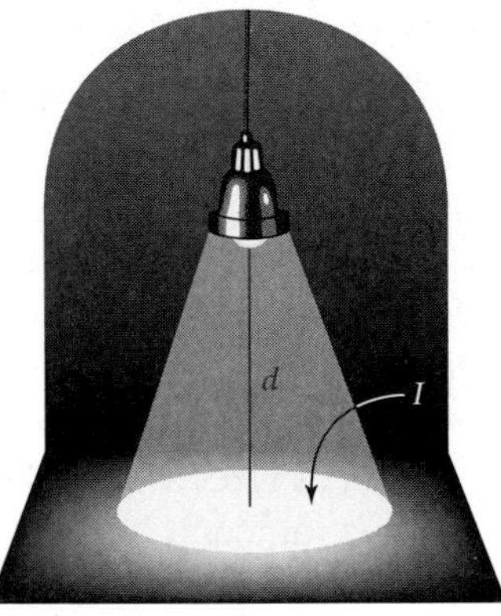

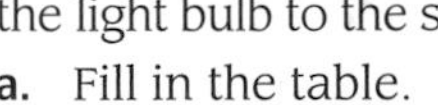

a. Fill in the table.

Distance (ft)	Function Rule	Intensity
d		I
1		
2		
3		
4		
5		
6		

b. Use the data from the table in part **a** to construct a line graph of the function.

Determine the domain and range of the following functions. Assume the *entire* function is shown.

21.

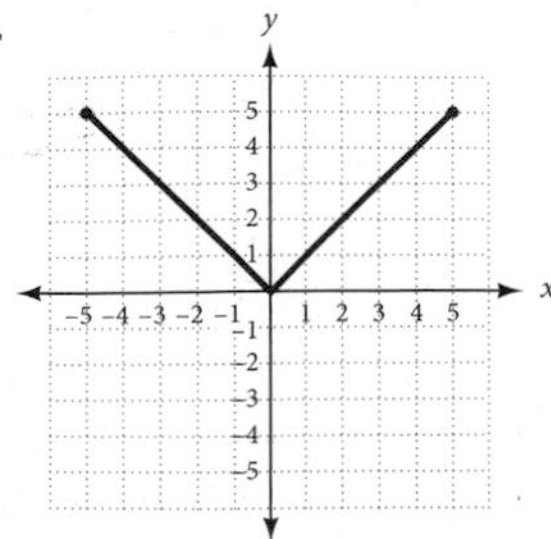

22.

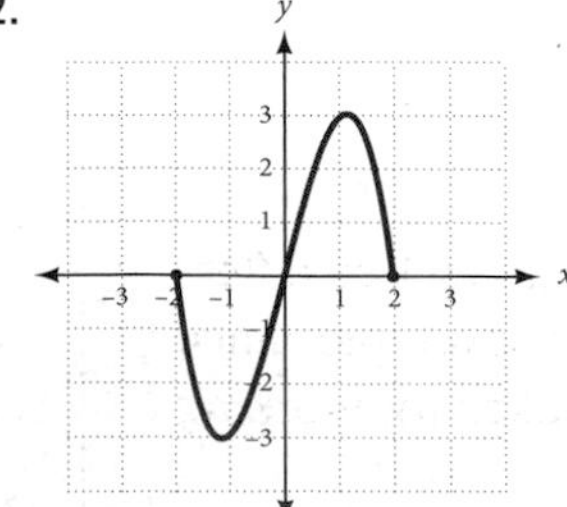

23.

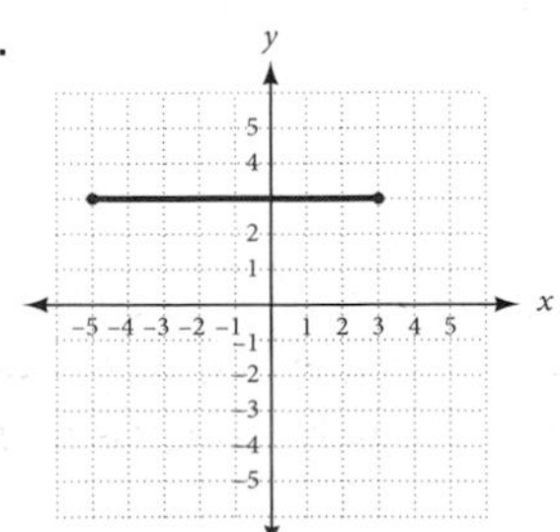

24.

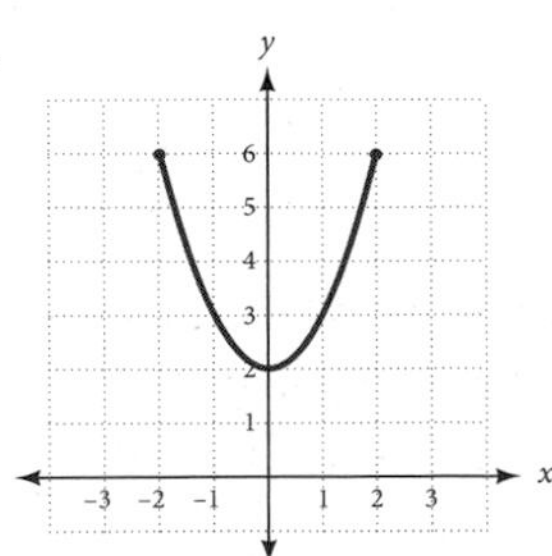

Function Notation

Let's return to the discussion that introduced us to functions. If a job pays \$7.50 per hour for working from 0 to 40 hours a week, then the amount of money y earned in one week is a function of the number of hours worked x. The exact relationship between x and y is written

$$y = 7.5x \quad \text{for} \quad 0 \le x \le 40$$

Since the amount of money earned y depends on the number of hours worked x, we call y the *dependent variable* and x the *independent variable*. If we let f represent all the ordered pairs produced by the equation, then we can write this:

$$f = \{(x, y) \mid y = 7.5x \text{ and } 0 \le x \le 40\}$$

Once we have named a function with a letter, we can use a new notation to represent the dependent variable y. The new notation for y is $f(x)$. It is read "f of x" and can be used instead of the variable y when working with functions. The notation y and the notation $f(x)$ are equivalent. That is,

$$y = 7.5x \Leftrightarrow f(x) = 7.5x$$

When we use the notation $f(x)$ we are using *function notation*. The benefit of using function notation is that we can write more information with fewer symbols than we can by using just the variable y. For example, asking how much money a person will make for working 20 hours is simply a matter of asking for $f(20)$. Without function notation, we would have to say, "Find the value of y that corresponds to a value of $x = 20$." To illustrate further, using the variable y, we can say "y is 150 when x is 20." Using the notation $f(x)$, we simply say "$f(20) = 150$." Each expression indicates that you will earn \$150 for working 20 hours.

HELP WANTED

ESPRESSO BAR OPERATOR

Must be dependable, honest, service-oriented. Coffee exp. desired. 15-30 hrs per week. \$5.25/hr. Start 5/31. Apply in person.

Espresso Yourself
Central Coast Mall
Deadline May 23rd

EXAMPLE 1 If $f(x) = 7.5x$, find $f(0)$, $f(10)$, and $f(20)$.

SOLUTION To find $f(0)$, we substitute 0 for x in the expression $7.5x$ and simplify. We find $f(10)$ and $f(20)$ in a similar manner—by substitution.

$$\begin{aligned} \text{If} \quad f(x) &= 7.5x \\ \text{then} \quad f(0) &= 7.5(0) = 0 \\ f(10) &= 7.5(10) = 75 \\ f(20) &= 7.5(20) = 150 \end{aligned}$$

Note Some students like to think of functions as machines. Values of x are put into the machine, which transforms them into values of $f(x)$, which are then output by the machine.

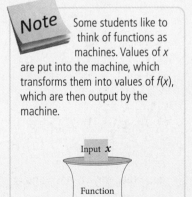

If we changed the example in the discussion that opened this section so the hourly wage was \$6.50 per hour, we would have a new equation to work with:

$$y = 6.5x \quad \text{for} \quad 0 \le x \le 40$$

Suppose we name this new function with the letter g. Then

$$g = \{(x, y) \mid y = 6.5x \text{ and } 0 \le x \le 40\}$$

and

$$g(x) = 6.5x$$

If we want to talk about both functions in the same discussion, having two different letters, f and g, makes it easy to distinguish between them. For example, since $f(x) = 7.5x$ and $g(x) = 6.5x$, asking how much money a person makes for working

20 hours is simply a matter of asking for $f(20)$ or $g(20)$, avoiding any confusion over which hourly wage we are talking about.

The diagrams shown in Figure 1 further illustrate the similarities and differences between the two functions we have been discussing.

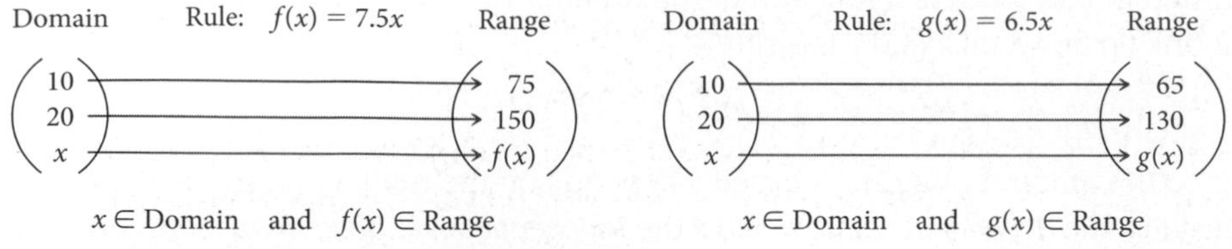

FIGURE 1 Function Maps

Function Notation and Graphs

We can visualize the relationship between x and $f(x)$ or $g(x)$ on the graphs of the two functions. Figure 2 shows the graph of $f(x) = 7.5x$ along with two additional line segments. The horizontal line segment corresponds to $x = 20$ and the vertical line segment corresponds to $f(20)$. Figure 3 shows the graph of $g(x) = 6.5x$ along with the horizontal line segment that corresponds to $x = 20$, and the vertical line segment that corresponds to $g(20)$. (Note that the domain in each case is restricted to $0 \le x \le 40$.)

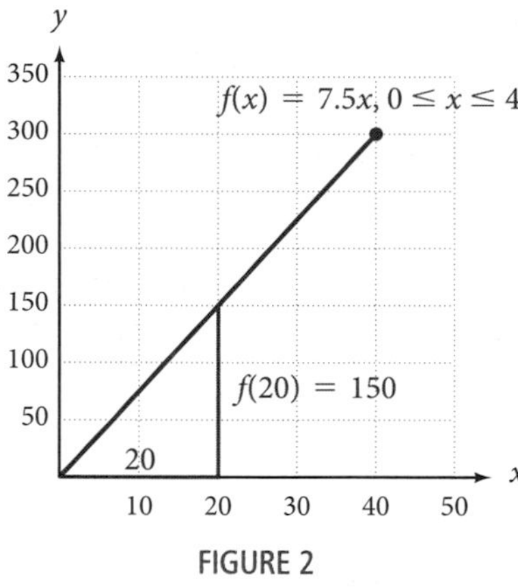

FIGURE 2

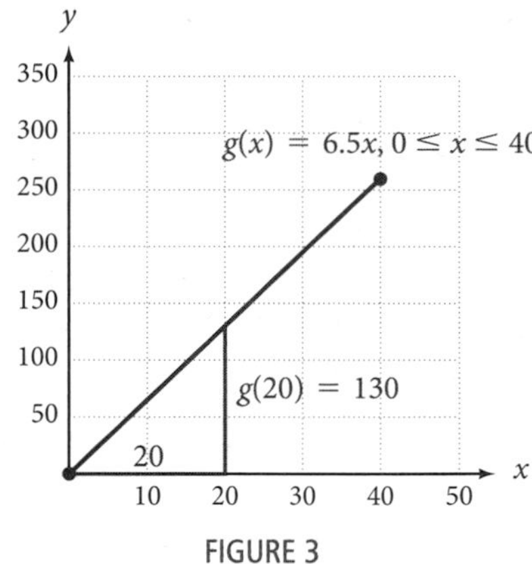

FIGURE 3

Using Function Notation

The remaining examples in this section show a variety of ways to use and interpret function notation.

EXAMPLE 2 If it takes Lorena t minutes to run a mile, then her average speed $s(t)$ in miles per hour is given by the formula

$$s(t) = \frac{60}{t} \quad \text{for} \quad t > 0$$

Find $s(10)$ and $s(8)$, and then explain what they mean.

SOLUTION To find $s(10)$, we substitute 10 for t in the equation and simplify:

$$s(\mathbf{10}) = \frac{60}{\mathbf{10}} = 6$$

In words: When Lorena runs a mile in 10 minutes, her average speed is 6 miles per hour.

We calculate s(8) by substituting 8 for t in the equation. Doing so gives us

$$s(\mathbf{8}) = \frac{60}{\mathbf{8}} = 7.5$$

In words: Running a mile in 8 minutes is running at a rate of 7.5 miles per hour.

EXAMPLE 3 A painting is purchased as an investment for \$125. If its value increases continuously so that it doubles every 5 years, then its value is given by the function

$$V(t) = 125 \cdot 2^{t/5} \quad \text{for} \quad t \geq 0$$

where t is the number of years since the painting was purchased, and $V(t)$ is its value (in dollars) at time t. Find $V(5)$ and $V(10)$, and explain what they mean.

SOLUTION The expression $V(5)$ is the value of the painting when $t = 5$ (5 years after it is purchased). We calculate $V(5)$ by substituting 5 for t in the equation $V(t) = 125 \cdot 2^{t/5}$. Here is our work:

$$V(\mathbf{5}) = 125 \cdot 2^{\mathbf{5}/5} = 125 \cdot 2^1 = 125 \cdot 2 = 250$$

In words: After 5 years, the painting is worth \$250.

The expression $V(10)$ is the value of the painting after 10 years. To find this number, we substitute 10 for t in the equation:

$$V(\mathbf{10}) = 125 \cdot 2^{\mathbf{10}/5} = 125 \cdot 2^2 = 125 \cdot 4 = 500$$

In words: The value of the painting 10 years after it is purchased is \$500.

EXAMPLE 4 A balloon has the shape of a sphere with a radius of 3 inches. Use the following formulas to find the volume and surface area of the balloon.

$$V(r) = \frac{4}{3}\pi r^3 \qquad S(r) = 4\pi r^2$$

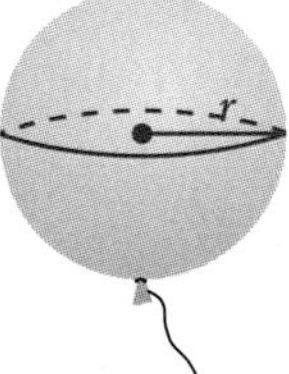

SOLUTION As you can see, we have used function notation to write the two formulas for volume and surface area and each quantity is a function of the radius. To find these quantities when the radius is 3 inches, we evaluate $V(3)$ and $S(3)$:

$$V(\mathbf{3}) = \frac{4}{3}\pi \mathbf{3}^3 = \frac{4}{3}\pi 27$$

$= 36\pi$ cubic inches, or 113 cubic inches — To the nearest whole number

$$S(\mathbf{3}) = 4\pi \mathbf{3}^2$$

$= 36\pi$ square inches, or 113 square inches — To the nearest whole number

The fact that $V(3) = 36\pi$ means that the ordered pair $(3, 36\pi)$ belongs to the function V. Likewise, the fact that $S(3) = 36\pi$ tells us that the ordered pair $(3, 36\pi)$ is a member of function S.

We can generalize the discussion at the end of Example 4 this way:

$$(a, b) \in f \quad \text{if and only if} \quad f(a) = b$$

EXAMPLE 5 If $f(x) = 3x^2 + 2x - 1$, find $f(0)$, $f(3)$, and $f(-2)$.

SOLUTION Since $f(x) = 3x^2 + 2x - 1$, we have

$$f(\mathbf{0}) = 3(\mathbf{0})^2 + 2(\mathbf{0}) - 1 = 0 + 0 - 1 = -1$$

$$f(\mathbf{3}) = 3(\mathbf{3})^2 + 2(\mathbf{3}) - 1 = 27 + 6 - 1 = 32$$

$$f(\mathbf{-2}) = 3(\mathbf{-2})^2 + 2(\mathbf{-2}) - 1 = 12 - 4 - 1 = 7$$

In Example 5, the function f is defined by the equation $f(x) = 3x^2 + 2x - 1$. We could just as easily have said $y = 3x^2 + 2x - 1$. That is, $y = f(x)$. Saying "$f(-2) = 7$" is exactly the same as saying "y is 7 when x is -2."

EXAMPLE 6 If $f(x) = 4x - 1$ and $g(x) = x^2 + 2$, then

$$f(\mathbf{5}) = 4(\mathbf{5}) - 1 = 19 \quad \text{and} \quad g(\mathbf{5}) = \mathbf{5}^2 + 2 = 27$$

$$f(\mathbf{-2}) = 4(\mathbf{-2}) - 1 = -9 \quad \text{and} \quad g(\mathbf{-2}) = (\mathbf{-2})^2 + 2 = 6$$

$$f(\mathbf{0}) = 4(\mathbf{0}) - 1 = -1 \quad \text{and} \quad g(\mathbf{0}) = \mathbf{0}^2 + 2 = 2$$

$$f(\mathbf{z}) = 4\mathbf{z} - 1 \quad \text{and} \quad g(\mathbf{z}) = \mathbf{z}^2 + 2$$

$$f(\mathbf{a}) = 4\mathbf{a} - 1 \quad \text{and} \quad g(\mathbf{a}) = \mathbf{a}^2 + 2$$

EXAMPLE 7 If the function f is given by

$$f = \{(-2, 0), (3, -1), (2, 4), (7, 5)\}$$

then

$$f(-2) = 0, f(3) = -1, f(2) = 4, \text{ and } f(7) = 5$$

EXAMPLE 8 If $f(x) = 2x^2$ and $g(x) = 3x - 1$, find

a. $f[g(2)]$ **b.** $g[f(2)]$

SOLUTION The expression $f[g(2)]$ is read "f of g of 2."

a. Because $g(2) = 3(2) - 1 = 5$,

$$f[g(2)] = f(5) = 2(5)^2 = 50$$

b. Because $f(2) = 2(2)^2 = 8$,

$$g[f(2)] = g(8) = 3(8) - 1 = 23$$

Getting Ready for Class

After reading through the preceding section, respond in your own words and in complete sentences.

1. Explain what you are calculating when you find $f(2)$ for a given function f.
2. If $s(t) = \frac{60}{t}$ how do you find $s(10)$?
3. If $f(2) = 3$ for a function f, what is the relationship between the numbers 2 and 3 and the graph of f?
4. If $f(6) = 0$ for a particular function f, then you can immediately graph one of the intercepts. Explain.

Problem Set C

Let $f(x) = 2x - 5$ and $g(x) = x^2 + 3x + 4$. Evaluate the following. **[Examples 1, 5–6]**

1. $f(2)$ **2.** $f(3)$ **3.** $f(-3)$ **4.** $g(-2)$ **5.** $g(-1)$ **6.** $f(-4)$

7. $g(-3)$ **8.** $g(2)$ **9.** $g(4) + f(4)$ **10.** $f(2) - g(3)$ **11.** $f(3) - g(2)$ **12.** $g(-1) + f(-1)$

Let $f(x) = 3x^2 - 4x + 1$ and $g(x) = 2x - 1$. Evaluate the following.

13. $f(0)$ **14.** $g(0)$ **15.** $g(-4)$ **16.** $f(1)$ **17.** $f(-1)$ **18.** $g(-1)$ **19.** $g(10)$

20. $f(10)$ **21.** $f(3)$ **22.** $g(3)$ **23.** $g\left(\frac{1}{2}\right)$ **24.** $g\left(\frac{1}{4}\right)$ **25.** $f(a)$ **26.** $g(b)$

If $f = \{(1, 4), (-2, 0), \left(3, \frac{1}{2}\right), (\pi, 0)\}$ and $g = \{(1, 1), (-2, 2), \left(\frac{1}{2}, 0\right)\}$, find each of the following values of f and g. **[Example 7]**

27. $f(1)$ **28.** $g(1)$ **29.** $g\left(\frac{1}{2}\right)$ **30.** $f(3)$ **31.** $g(-2)$ **32.** $f(\pi)$

Let $f(x) = 2x^2 - 8$ and $g(x) = \frac{1}{2}x + 1$. Evaluate each of the following. **[Example 8]**

33. $f(0)$ **34.** $g(0)$ **35.** $g(-4)$ **36.** $f(1)$ **37.** $f(a)$ **38.** $g(z)$ **39.** $f(b)$

40. $g(t)$ **41.** $f[g(2)]$ **42.** $g[f(2)]$ **43.** $g[f(-1)]$ **44.** $f[g(-2)]$ **45.** $g[f(0)]$ **46.** $f[g(0)]$

47. Graph the function $f(x) = \frac{1}{2}x + 2$. Then draw and label the line segments that represent $x = 4$ and $f(4)$.

48. Graph the function $f(x) = -\frac{1}{2}x + 6$. Then draw and label the line segments that represent $x = 4$ and $f(4)$.

49. For the function $f(x) = \frac{1}{2}x + 2$, find the value of x for which $f(x) = x$.

50. For the function $f(x) = -\frac{1}{2}x + 6$, find the value of x for which $f(x) = x$.

51. Graph the function $f(x) = x^2$. Then draw and label the line segments that represent $x = 1$ and $f(1)$, $x = 2$ and $f(2)$ and, finally, $x = 3$ and $f(3)$.

52. Graph the function $f(x) = x^2 - 2$. Then draw and label the line segments that represent $x = 2$ and $f(2)$ and the line segments corresponding to $x = 3$ and $f(3)$.

Applying the Concepts [Examples 2–4]

53. Investing in Art A painting is purchased as an investment for \$150. If its value increases continuously so that it doubles every 3 years, then its value is given by the function

$$V(t) = 150 \cdot 2^{t/3} \quad \text{for} \quad t \geq 0$$

where t is the number of years since the painting was purchased and $V(t)$ is its value (in dollars) at time t. Find $V(3)$ and $V(6)$, and then explain what they mean.

54. Average Speed If it takes Minke t minutes to run a mile, then her average speed $s(t)$, in miles per hour, is given by the formula

$$s(t) = \frac{60}{t} \quad \text{for} \quad t > 0$$

Find $s(4)$ and $s(5)$, and then explain what they mean.

55. **Dimensions of a Rectangle** The length of a rectangle is 3 inches more than twice the width. Let x represent the width of the rectangle and $P(x)$ represent the perimeter of the rectangle. Use function notation to write the relationship between x and $P(x)$, noting any restrictions on the variable x.

56. **Dimensions of a Rectangle** The length of a rectangle is 3 inches more than twice the width. Let x represent the width of the rectangle and $A(x)$ represent the area of the rectangle. Use function notation to write the relationship between x and $A(x)$, noting any restrictions on the variable x.

Area of a Circle The formula for the area A of a circle with radius r can be written with function notation as $A(r) = \pi r^2$.

57. Find $A(2)$, $A(5)$, and $A(10)$. (Use $\pi \approx 3.14$).

58. Why doesn't it make sense to ask for $A(-10)$?

59. **Cost of a Phone Call** Suppose a phone company charges 33¢ for the first minute and 24¢ for each additional minute to place a long-distance call between 5 p.m. and 11 p.m. If x is the number of additional minutes and $f(x)$ is the cost of the call, then $f(x) = 0.24x + 0.33$.

   a. How much does it cost to talk for 10 minutes?

   b. What does f(5) represent in this problem?

   c. If a call costs \$1.29, how long was it?

60. **Cost of a Phone Call** The same phone company mentioned in Problem 59 charges 52¢ for the first minute and 36¢ for each additional minute to place a long-distance call between 8 a.m. and 5 p.m.

   a. Let $g(x)$ be the total cost of a long-distance call between 8 a.m. and 5 p.m., and write an equation for $g(x)$.

   b. Find $g(5)$.

   c. Find the difference in price between a 10-minute call made between 8 a.m. and 5 p.m. and the same call made between 5 p.m. and 11 p.m.

Straight-Line Depreciation Straight-line depreciation is an accounting method used to help spread the cost of new equipment over a number of years. It takes into account both the cost when new and the salvage value, which is the value of the equipment at the time it gets replaced.

61. **Value of a Copy Machine** The function $V(t) = -3{,}300t + 18{,}000$, where V is value and t is time in years, can be used to find the value of a large copy machine during the first 5 years of use.

   a. What is the value of the copier after 3 years and 9 months?

   b. What is the salvage value of this copier if it is replaced after 5 years?

   c. State the domain of this function.

   d. Sketch the graph of this function.

   e. What is the range of this function?

   f. After how many years will the copier be worth only \$10,000?

62. **Value of a Forklift** The function

$$V(t) = -16{,}500t + 125{,}000$$

where V is value and t is time in years, can be used to find the value of an electric forklift during the first 6 years of use.

   a. What is the value of the forklift after 2 years and 3 months?

   b. What is the salvage value of this forklift?

   c. State the domain of this function.

   d. Sketch the graph of this function.

   e. What is the range of this function?

   f. After how many years will the forklift be worth only \$45,000? Let $f(x) = \frac{1}{x+3}$ and $g(x) = \frac{1}{x} + 1$.

Solutions to Selected Practice Problems

Solutions to all practice problems that require more than one step are shown here. Before you look at these solutions to see where you have made a mistake, you should try the problem you are working on twice. If you do not get the correct answer the second time you work the problem, then the solution shown here should show you where you have gone wrong.

Chapter 1

Section 1.1

6. $7^2 = 7 \cdot 7 = 49$

7. $3^4 = 3 \cdot 3 \cdot 3 \cdot 3 = 81$

8. $10^5 = 10 \cdot 10 \cdot 10 \cdot 10 \cdot 10 = 100{,}000$

9. $4 + 6 \cdot 7 = 4 + 42$
$= 46$

10. $18 \div 6 \cdot 2 = 3 \cdot 2$
$= 6$

11. $5[4 + 3(7 + 2 \cdot 4)] = 5[4 + 3(7 + 8)]$
$= 5[4 + 3(15)]$
$= 5[4 + 45]$
$= 5[49]$
$= 245$

12. $12 + 8 \div 2 + 4 \cdot 5 = 12 + 4 + 20$
$= 16 + 20$
$= 36$

13. $3^4 + 2^5 \div 8 - 5^2 = 81 + 32 \div 8 - 25$
$= 81 + 4 - 25$
$= 85 - 25$
$= 60$

14. $2(70) + 5 + 3(50) = 140 + 5 + 150$
$= 145 + 150$
$= 295$

15. **a.** For the first sequence, each number is 4 more than the number before it; therefore, the next number will be $15 + 4 = 19$.
b. For the sequence in part b, each number is 3 times the number before it; therefore, the next number in the sequence will be $3 \cdot 27 = 81$.
c. For the sequence in part c, a pattern develops when we look at the differences between the numbers:
Proceeding in the same manner, we add 6 to the last term, giving us $14 + 6 = 20$.

2 5 9 14 . . .
3 4 5

16. Each term in this sequence is the sum of the two previous terms. To extend this sequence, we add 10 and 16 to obtain 26. Continuing in this manner, we find the first 10 terms as follows:

2, 2, 4, 6, 10, 16, 26, 42, 68, 110

Section 1.2

1.
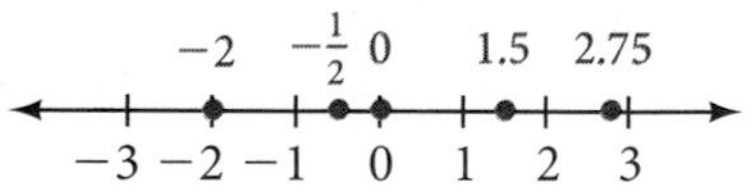

2. $\frac{5}{8} = \frac{5 \cdot \mathbf{6}}{8 \cdot \mathbf{6}} = \frac{30}{48}$

6. $|7 - 2| = |5| = 5$

7. $|2 \cdot 3^2 + 5 \cdot 2^2| = |2 \cdot 9 + 5 \cdot 4|$
$= |18 + 20|$
$= |38|$
$= 38$

8. $|10 - 4| - |11 - 9| = |6| - |2|$
$= 6 - 2$
$= 4$

13. $\frac{2}{3} \cdot \frac{7}{9} = \frac{2 \cdot 7}{3 \cdot 9} = \frac{14}{27}$

14. $8\left(\frac{1}{5}\right) = \frac{8}{1} \cdot \frac{1}{5} = \frac{8}{5}$

15. $\left(\frac{11}{12}\right)^2 = \frac{11}{12} \cdot \frac{11}{12} = \frac{121}{144}$

16. The reciprocal of 6 is $\frac{1}{6}$ because: $6\left(\frac{1}{6}\right) = \frac{6}{1}\left(\frac{1}{6}\right) = \frac{6}{6} = 1$

17. The reciprocal of 3 is $\frac{1}{3}$ because: $3\left(\frac{1}{3}\right) = \frac{3}{1}\left(\frac{1}{3}\right) = \frac{3}{3} = 1$

18. The reciprocal of $\frac{1}{2}$ is 2 because: $\frac{1}{2}(2) = \frac{1}{2}\left(\frac{2}{1}\right) = \frac{2}{2} = 1$

19. The reciprocal of $\frac{2}{3}$ is $\frac{3}{2}$ because: $\frac{2}{3}\left(\frac{3}{2}\right) = \frac{6}{6} = 1$

20. **a.** Perimeter $= 4s = 4(4) = 16$ feet
Area $= s^2 = 4^2 = 16$ feet^2
b. Perimeter $= 2l + 2w = 2(7) + 2(5) = 24$ inches
Area $= lw = (7)(5) = 35$ inches^2
c. Perimeter $= a + b + c = 40 + 50 + 30 = 120$ meters
Area $= \frac{1}{2}bh = \frac{1}{2}(50)(24) = 600$ meters^2

Section 1.3

1. Starting at the origin, move 2 units in the positive direction and then 5 units in the negative direction.

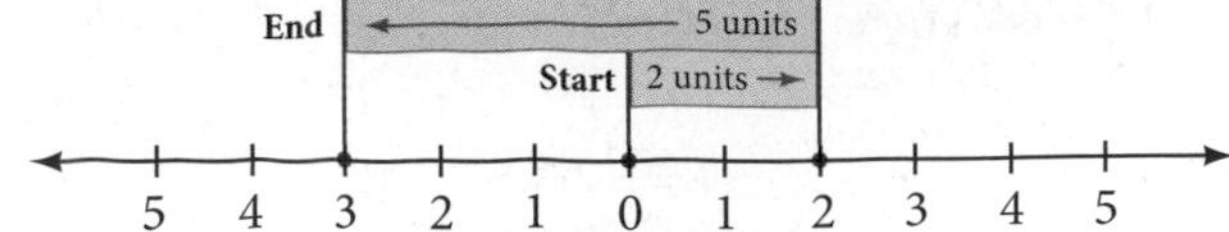

We end up at −3; therefore, 2 + (−5) = −3.

2. Starting at the origin, move 2 units in the negative direction and then 5 units in the positive direction.

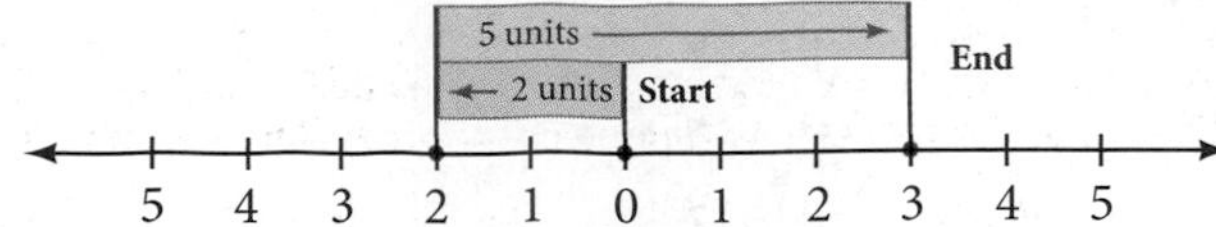

We end up at +3; therefore, −2 + 5 = 3.

3. Starting at the origin, move 2 units in the negative direction and then 5 units in the negative direction.

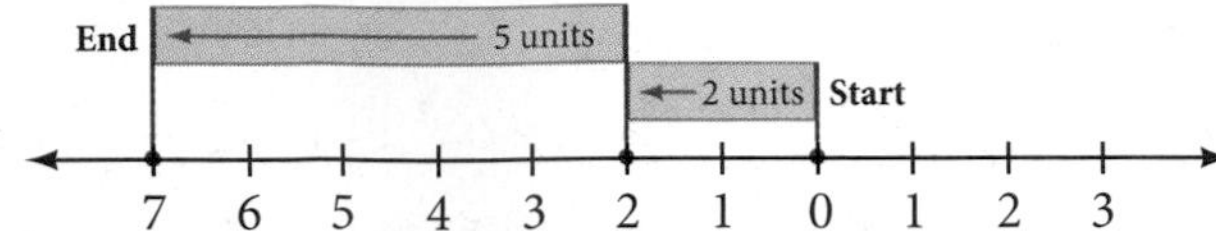

We end up at −7; therefore, −2 + (−5) = −7.

4. Starting at the origin, move 9 units in the positive direction and then 4 units in the negative direction.

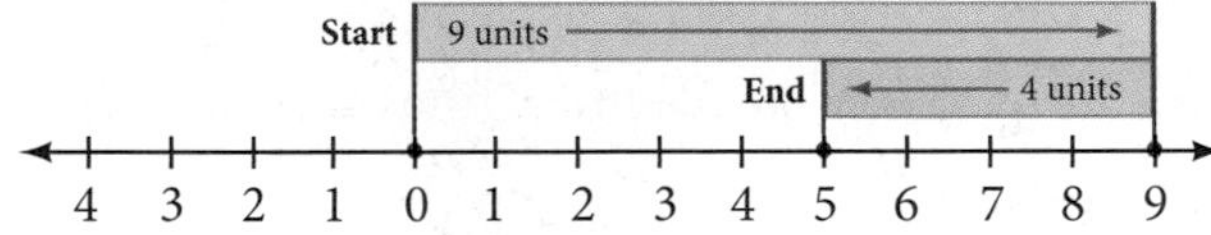

We end up at +5; therefore, 9 + (−4) = 5.

5. Starting at the origin, move 7 units in the positive direction and then 3 units in the negative direction.

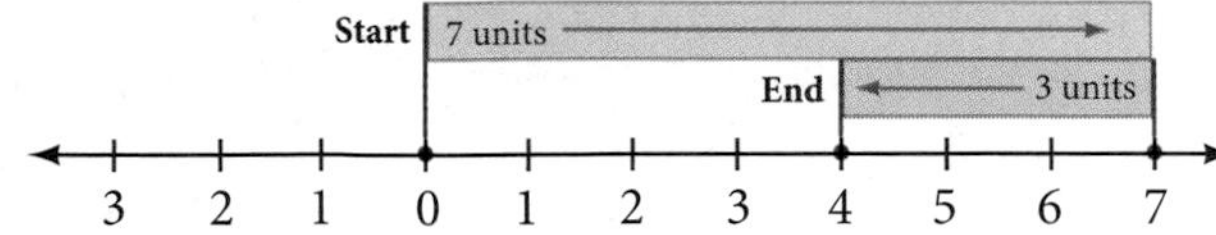

We end up at +4; therefore, 7 + (−3) = 4.

6. Starting at the origin, move 10 units in the negative direction and then 12 units in the positive direction.

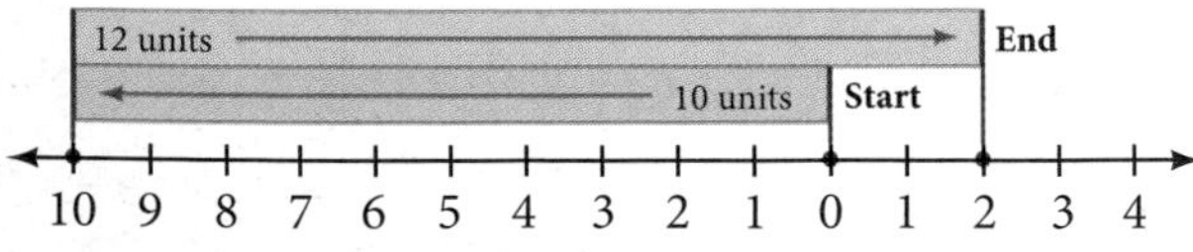

We end up at +2; therefore, −10 + 12 = 2.

7. Starting at the origin, move 4 units in the negative direction and then 6 units in the negative direction.

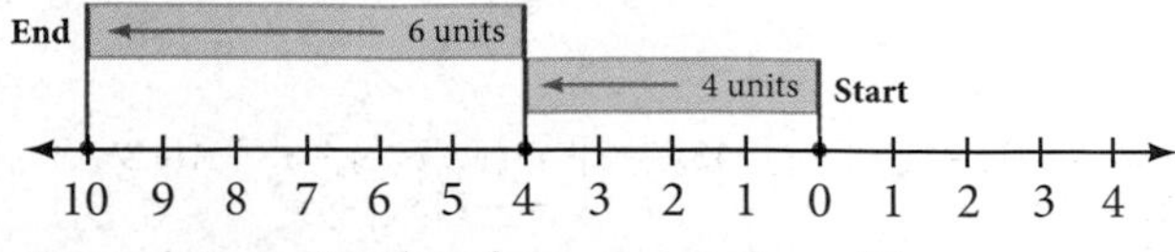

We end up at −10; therefore, −4 + (−6) = −10.

8.
$$\begin{aligned} 9 + 12 &= 21 \\ 9 + (-12) &= -3 \\ -9 + 12 &= 3 \\ -9 + (-12) &= -21 \end{aligned}$$

9.
$$\begin{aligned} 6 + 10 &= 16 \\ 6 + (-10) &= -4 \\ -6 + 10 &= 4 \\ -6 + (-10) &= -16 \end{aligned}$$

10.
$$\begin{aligned} -5 + 2 + (-7) &= -3 + (-7) \\ &= -10 \end{aligned}$$

11.
$$\begin{aligned} -4 + [2 + (-3)] + (-1) &= -4 + (-1) + (-1) \\ &= -5 + (-1) \\ &= -6 \end{aligned}$$

12.
$$\begin{aligned} -5 + 3(-4 + 7) + (-10) &= -5 + 3(3) + (-10) \\ &= -5 + 9 + (-10) \\ &= 4 + (-10) \\ &= -6 \end{aligned}$$

13. **a.** In this sequence each term is found by adding 4 to the term preceding it. Therefore, the next two terms in the sequence will be 17 and 21.
b. Each term in this sequence comes from adding 0.5 to the term preceding it; therefore, the next two terms will be 7.5 and 8.
c. Each term in this sequence comes from adding -4 to the preceding term. The next two terms in the sequence will be -8 and -12.

Section 1.4

1.
$$\begin{aligned} 7 - 4 &= 7 + (-4) = 3 \\ -7 - 4 &= -7 + (-4) = -11 \\ 7 - (-4) &= 7 + 4 = 11 \\ -7 - (-4) &= -7 + 4 = -3 \end{aligned}$$

2.
$$\begin{aligned} 6 - 4 &= 6 + (-4) = 2 \\ -6 - 4 &= -6 + (-4) = -10 \\ 6 - (-4) &= 6 + 4 = 10 \\ -6 - (-4) &= -6 + 4 = -2 \end{aligned}$$

3.
$$\begin{aligned} 4 + (-2) - 3 &= 4 + (-2) + (-3) \\ &= 2 + (-3) \\ &= -1 \end{aligned}$$

4.
$$\begin{aligned} 9 - 4 - 5 &= 9 + (-4) + (-5) \\ &= 5 + (-5) \\ &= 0 \end{aligned}$$

5.
$$\begin{aligned} -3 - (-5 + 1) - 6 &= -3 - (-4) - 6 \\ &= -3 + 4 + (-6) \\ &= 1 + (-6) \\ &= -5 \end{aligned}$$

6.
$$\begin{aligned} 3 \cdot 9 - 4 \cdot 10 - 5 \cdot 11 &= 27 - 40 - 55 \\ &= 27 + (-40) + (-55) \\ &= -13 + (-55) \\ &= -68 \end{aligned}$$

7.
$$\begin{aligned} 4 \cdot 2^5 - 3 \cdot 5^2 &= 4 \cdot 32 - 3 \cdot 25 \\ &= 128 - 75 \\ &= 53 \end{aligned}$$

8.
$$\begin{aligned} -5 - 8 &= -5 + (-8) \\ &= -13 \end{aligned}$$

9.
$$\begin{aligned} 8 - (-3) &= 8 + 3 \\ &= 11 \end{aligned}$$

10.
$$\begin{aligned} 8 - 6 &= 8 + (-6) \\ &= 2 \end{aligned}$$

11.
$$\begin{aligned} 4 - (-7) &= 4 + 7 \\ &= 11 \end{aligned}$$

12. **a.** The two angles are complementary angles, so we can find x by subtracting 45° from 90°:
$$x = 90° - 45° = 45°$$
b. The two angles are supplementary. To find x, we subtract 60° from 180°:
$$x = 180° - 60° = 120°$$

Section 1.5

3.
$$\begin{aligned} 4 + x + 7 &= x + 4 + 7 \\ &= x + 11 \end{aligned}$$

4.
$$\begin{aligned} 3 + (7 + x) &= (3 + 7) + x \\ &= 10 + x \end{aligned}$$

5.
$$\begin{aligned} 3(4x) &= (3 \cdot 4)x \\ &= 12x \end{aligned}$$

6.
$$\begin{aligned} \frac{1}{3}(3x) &= \left(\frac{1}{3} \cdot 3\right)x \\ &= 1 \cdot x \\ &= x \end{aligned}$$

7.
$$\begin{aligned} 4\left(\frac{1}{4}x\right) &= \left(\frac{4}{1} \cdot \frac{1}{4}\right)x \\ &= x \end{aligned}$$

8.
$$\begin{aligned} 15\left(\frac{3}{5}x\right) &= \left(\frac{15}{1} \cdot \frac{3}{5}\right)x \\ &= \frac{45}{5}x \\ &= 9x \end{aligned}$$

9.
$$\begin{aligned} 3(x + 2) &= 3 \cdot x + 3 \cdot 2 \\ &= 3x + 6 \end{aligned}$$

10.
$$\begin{aligned} 4(2x - 5) &= 4 \cdot 2x - 4 \cdot 5 \\ &= 8x - 20 \end{aligned}$$

11.
$$5(x + y) = 5x + 5y$$

12.
$$\begin{aligned} 3(7x - 6y) &= 3 \cdot 7x - 3 \cdot 6y \\ &= 21x - 18y \end{aligned}$$

13.
$$\begin{aligned} \frac{1}{3}(3x + 6) &= \frac{1}{3} \cdot 3x + \frac{1}{3} \cdot 6 \\ &= x + 2 \end{aligned}$$

14.
$$\begin{aligned} 5(7a - 3) + 2 &= 5 \cdot 7a - 5 \cdot 3 + 2 \\ &= 35a - 15 + 2 \\ &= 35a - 13 \end{aligned}$$

15.
$$\begin{aligned} x\left(2 + \frac{2}{x}\right) &= 2x + \frac{2x}{x} \\ &= 2x + 2 \end{aligned}$$

16.
$$\begin{aligned} 4\left(\frac{1}{4}x - 3\right) &= \left(\frac{4}{1} \cdot \frac{1}{4}\right)x - 4 \cdot 3 \\ &= x - 12 \end{aligned}$$

17.
$$\begin{aligned} 8\left(\frac{3}{4}x + \frac{1}{2}y\right) &= \left(\frac{8}{1} \cdot \frac{3}{4}\right)x + \left(\frac{8}{1} \cdot \frac{1}{2}\right)y \\ &= \frac{24}{4}x + \frac{8}{2}y \\ &= 6x + 4y \end{aligned}$$

Section 1.6

11.
$$\begin{aligned} -7(-6)(-1) &= 42(-1) \\ &= -42 \end{aligned}$$

12.
$$\begin{aligned} 2(-8) + 3(-7) - 4 &= -16 + (-21) - 4 \\ &= -37 - 4 \\ &= -41 \end{aligned}$$

13.
$$\begin{aligned} (-3)^5 &= (-3)(-3)(-3)(-3)(-3) \\ &= -243 \end{aligned}$$

14. $-5(-2)^3 - 7(-3)^2 = -5(-8) - 7(9)$
$= 40 - 63$
$= -23$

15. $7 - 3(4 - 9) = 7 - 3(-5)$
$= 7 - (-15)$
$= 22$

16. $-\frac{2}{3}\left(\frac{5}{9}\right) = -\frac{2 \cdot 5}{3 \cdot 9}$
$= -\frac{10}{27}$

17. $-8\left(\frac{1}{2}\right) = -\frac{8 \cdot 1}{1 \cdot 2}$
$= -\frac{8}{2}$
$= -4$

18. $-\frac{3}{4}\left(-\frac{4}{3}\right) = \frac{3 \cdot 4}{4 \cdot 3}$
$= \frac{12}{12}$
$= 1$

19. $-2(680) + 510 = -1{,}360 + 510$
$= -850$ calories

20. $-5(3x) = (-5 \cdot 3)x$
$= -15x$

21. $4(-8y) = [4(-8)]y$
$= -32y$

22. $-3\left(-\frac{1}{3}x\right) = \left[-3\left(-\frac{1}{3}\right)x\right]$
$= 1x$
$= x$

23. $-3(a + 5) = -3a + (-3)(5)$
$= -3a + (-15)$
$= -3a - 15$

24. $-5(2x + 1) = -5(2x) + (-5)(1)$
$= -10x + (-5)$
$= -10x - 5$

25. $-\frac{1}{2}(4x - 12) = -\frac{1}{2}(4x) - \left(-\frac{1}{2}\right)(12)$
$= -2x - (-6)$
$= -2x + 6$

26. $-3(2x - 6) - 7 = -3(2x) - (-3)(6) - 7$
$= -6x - (-18) - 7$
$= -6x + 18 - 7$
$= -6x + 11$

27. a. $-\frac{3}{4}(0) + 2 = 0 + 2$
$= 2$

b. $-\frac{3}{4}(4) + 2 = -3 + 2$
$= -1$

c. $-\frac{3}{4}\left(-\frac{8}{3}\right) + 2 = \frac{24}{12} + 2$
$= 2 + 2$
$= 4$

28. a. $3(4) - 5(0) = 12 - 0$
$= 12$

b. $3(0) - 5(-2) = 0 + 10$
$= 10$

c. $3(-2) - 5(3) = -6 - 15$
$= -21$

29. a. Starting with 4, each term in the sequence is 2 times the previous term. The next number in the sequence will be $2 \cdot 16 = 32$.
b. Starting with 2, each term in the sequence is multiplied by -3 to get the next term in the sequence. Multiplying 18 by -3, we get $18(-3) = -54$.
c. Starting with $\frac{1}{9}$, each term in the sequence is multiplied by 3 to get the next term. Multiplying 1 by 3, we get $3(1) = 3$.

Section 1.7

1. $\frac{12}{4} = 12\left(\frac{1}{4}\right) = 3$

2. $\frac{12}{-4} = 12\left(-\frac{1}{4}\right) = -3$

3. $\frac{-12}{4} = -12\left(\frac{1}{4}\right) = -3$

4. $\frac{-12}{-4} = -12\left(-\frac{1}{4}\right) = 3$

5. $\frac{18}{3} = 18\left(\frac{1}{3}\right) = 6$

6. $\frac{18}{-3} = 18\left(-\frac{1}{3}\right) = -6$

7. $\frac{-18}{3} = -18\left(\frac{1}{3}\right) = -6$

8. $\frac{-18}{-3} = -18\left(-\frac{1}{3}\right) = 6$

9. $\frac{30}{-10} = 30\left(-\frac{1}{10}\right) = -3$

10. $\frac{-50}{-25} = -50\left(-\frac{1}{25}\right) = 2$

11. $\frac{-21}{3} = -21\left(\frac{1}{3}\right) = -7$

12. $\frac{3}{4} \div \frac{5}{7} = \frac{3}{4} \cdot \frac{7}{5}$
$= \frac{21}{20}$

13. $-\frac{7}{8} \div \frac{3}{5} = -\frac{7}{8} \cdot \frac{5}{3}$
$= -\frac{35}{24}$

14. $10 \div \left(-\frac{5}{6}\right) = 10 \cdot \left(-\frac{6}{5}\right)$
$= \frac{10}{1} \cdot \left(-\frac{6}{5}\right)$
$= -\frac{60}{5}$
$= -12$

15. $\frac{-6(5)}{9} = \frac{-30}{9}$
$= -\frac{10}{3}$

16. $\frac{10}{-7 - 1} = \frac{10}{-8}$
$= -\frac{5}{4}$

17. $\frac{-6 - 6}{-2 - 4} = \frac{-12}{-6}$
$= 2$

18. $\frac{3(-4) + 9}{6} = \frac{-12 + 9}{6}$
$= \frac{-3}{6}$
$= -\frac{1}{2}$

19. $\frac{6(-2) + 5(-3)}{5(4) - 11} = \frac{-12 + (-15)}{20 - 11}$
$= \frac{-27}{9}$
$= -3$

20. $\frac{4^2 - 2^2}{-4 + 2} = \frac{16 - 4}{-2}$
$= \frac{12}{-2}$
$= -6$

21. $\frac{(4 + 3)^2}{-4^2 - 3^2} = \frac{7^2}{-16 - 9}$
$= \frac{49}{-25}$
$= -\frac{49}{25}$

22. $8\left(\frac{x}{4}\right) = \left(\frac{8}{1} \cdot \frac{1}{4}\right)x$
$= 2x$

23. $x\left(\frac{4}{x} - 5\right) = \frac{x}{1} \cdot \frac{4}{x} - 5x$
$= 4 - 5x$

Section 1.8

4. $90 = 9 \cdot 10$
$= 3 \cdot 3 \cdot 2 \cdot 5$
$= 2 \cdot 3^2 \cdot 5$

5. $420 = 42 \cdot 10$
$= 6 \cdot 7 \cdot 2 \cdot 5$
$= 2 \cdot 3 \cdot 7 \cdot 2 \cdot 5$
$= 2^2 \cdot 3 \cdot 5 \cdot 7$

6. $\frac{154}{1{,}155} = \frac{2 \cdot \cancel{7} \cdot \cancel{11}}{3 \cdot 5 \cdot \cancel{7} \cdot \cancel{11}}$
$= \frac{2}{3 \cdot 5}$
$= \frac{2}{15}$

Section 1.9

1. $\frac{3}{10} + \frac{1}{10} = \frac{3+1}{10}$
$= \frac{4}{10}$
$= \frac{2}{5}$

2. $\frac{a-5}{12} + \frac{3}{12} = \frac{a-5+3}{12}$
$= \frac{a-2}{12}$

3. $\frac{8}{x} - \frac{5}{x} = \frac{8-5}{x}$
$= \frac{3}{x}$

4. $\frac{5}{9} - \frac{8}{9} + \frac{5}{9} = \frac{5-8+5}{9}$
$= \frac{2}{9}$

5. $\left.\begin{array}{l} 18 = 2 \cdot 3 \cdot 3 \\ 14 = 2 \cdot 7 \end{array}\right\}$ LCD $= 2 \cdot 3 \cdot 3 \cdot 7$
$= 126$

6. $\frac{5}{18} + \frac{3}{14} = \frac{5 \cdot \mathbf{7}}{18 \cdot \mathbf{7}} + \frac{3 \cdot \mathbf{9}}{14 \cdot \mathbf{9}}$
$= \frac{35}{126} + \frac{27}{126}$
$= \frac{62}{126}$
$= \frac{31}{63}$

7. $\left.\begin{array}{l} 9 = 3 \cdot 3 \\ 15 = 3 \cdot 5 \end{array}\right\}$ LCD $= 3 \cdot 3 \cdot 5$
$= 45$

8. $\frac{2}{9} + \frac{4}{15} = \frac{2 \cdot \mathbf{5}}{9 \cdot \mathbf{5}} + \frac{4 \cdot \mathbf{3}}{15 \cdot \mathbf{3}}$
$= \frac{10}{45} + \frac{12}{45}$
$= \frac{22}{45}$

9. LCD = 100; $\frac{8}{25} - \frac{3}{20} = \frac{8 \cdot \mathbf{4}}{25 \cdot \mathbf{4}} - \frac{3 \cdot \mathbf{5}}{20 \cdot \mathbf{5}}$
$= \frac{32}{100} - \frac{15}{100}$
$= \frac{17}{100}$

10. LCD = 36; $\frac{1}{9} + \frac{1}{4} + \frac{1}{6} = \frac{1 \cdot \mathbf{4}}{9 \cdot \mathbf{4}} + \frac{1 \cdot \mathbf{9}}{4 \cdot \mathbf{9}} + \frac{1 \cdot \mathbf{6}}{6 \cdot \mathbf{6}}$
$= \frac{4}{36} + \frac{9}{36} + \frac{6}{36}$
$= \frac{19}{36}$

11. $2 - \frac{3}{4} = \frac{2}{1} - \frac{3}{4}$
$= \frac{2 \cdot \mathbf{4}}{1 \cdot \mathbf{4}} - \frac{3}{4}$
$= \frac{8}{4} - \frac{3}{4}$
$= \frac{5}{4}$

12. **a.** Adding $-\frac{1}{3}$ to each term produces the next term. The next term in this sequence will come from adding $-\frac{1}{3} + \left(-\frac{1}{3}\right) = -\frac{2}{3}$. This is an arithmetic sequence.
b. Each term in this sequence comes from adding $\frac{1}{3}$ to the term preceding it. The fourth term will be $1 + \frac{1}{3} = \frac{4}{3}$. This is also an arithmetic sequence.
c. This is a geometric sequence in which each term is found by multiplying the term preceding it by $\frac{1}{3}$. The fourth term will be $\frac{1}{9} \cdot \frac{1}{3} = \frac{1}{27}$.

13. LCD = 20; $\frac{x}{4} - \frac{1}{5} = \frac{x \cdot \mathbf{5}}{4 \cdot \mathbf{5}} - \frac{1 \cdot \mathbf{4}}{5 \cdot \mathbf{4}}$
$= \frac{5x}{20} - \frac{4}{20}$
$= \frac{5x-4}{20}$

14. $\frac{5}{x} + \frac{2}{3} = \frac{5 \cdot \mathbf{3}}{x \cdot \mathbf{3}} + \frac{2 \cdot \mathbf{x}}{3 \cdot \mathbf{x}}$
$= \frac{15}{3x} + \frac{2x}{3x}$
$= \frac{15+2x}{3x}$

15. Method1: $\frac{1}{2}x + \frac{3}{4}x = \frac{\mathbf{2} \cdot x}{\mathbf{2} \cdot 2} + \frac{3x}{4}$ LCD
$= \frac{2x+3x}{4}$ Add numerators
$= \frac{5x}{4}$

Method2: $\frac{1}{2}x + \frac{3}{4}x = \left(\frac{1}{2} + \frac{3}{4}\right)x$ Distributive property
$= \left(\frac{2 \cdot 1}{2 \cdot 2} + \frac{3}{4}\right)x$ LCD
$= \left(2 + \frac{3}{4}\right)x$
$= \frac{5}{4}x$

Chapter 2

Section 2.1

1. $5x + 2x = (5 + 2)x$
$= 7x$

2. $12a - 7a = (12 - 7)a$
$= 5a$

3. $6y - 8y + 5y = (6 - 8 + 5)y$
$= 3y$

4. $2x + 6 + 3x - 5 = (2x + 3x) + (6 - 5)$
$= 5x + 1$

5. $8a - 2 - 4a + 5 = (8a - 4a) + (-2 + 5)$
$= 4a + 3$

6. $9x + 1 - x - 6 = (9x - x) + (1 - 6)$
$= 8x - 5$

7. $4(3x - 5) - 2 = 12x - 20 - 2$
$= 12x - 22$

8. $5 - 2(4y + 1) = 5 - 8y - 2$
$= -8y + 3$

9. $6(x - 3) - (7x + 2) = 6x - 18 - 7x - 2$
$= -x - 20$

10. When $x = 3$
then $4x - 7$
becomes $4(3) - 7 = 12 - 7$
$= 5$

11. When $a = -5$
then $2a + 4$
becomes $2(-5) + 4 = -10 + 4$
$= -6$

12. When $x = -2$
then $2x - 5 + 6x$
becomes $2(-2) - 5 + 6(-2)$
$= -4 - 5 - 12$
$= -21$

13. When $x = 10$
then $7x - 3 - 4x$
becomes $7(10) - 3 - 4(10)$
$= 70 - 3 - 40$
$= 27$

14. When $y = -2$
then $y^2 - 10y + 25$
becomes $(-2)^2 - 10(-2) + 25$
$= 4 + 20 + 25$
$= 49$

15. When $a = -3$ and $b = 4$
then $5a - 3b - 2$
becomes $5(-3) - 3(4) - 2$
$= -15 - 12 - 2$
$= -29$

16. When $x = 5$ and $y = 3$
then $x^2 - 4xy + 4y^2$
becomes $5^2 - 4(5)(3) + 4 \cdot 3^2$
$= 25 - 60 + 4 \cdot 9$
$= 25 - 60 + 36$
$= 1$

17. Substituting the given values we have
When $n = 1, 2n + 1 = 3$
When $n = 2, 2n + 1 = 5$
When $n = 3, 2n + 1 = 7$
When $n = 4, 2n + 1 = 9$

Section 2.2

1. $2(4) + 3 \stackrel{?}{=} 7$
$8 + 3 \stackrel{?}{=} 7$
$11 \neq 7$
No

2. $8 \stackrel{?}{=} 3\left(\frac{4}{3}\right) + 4$
$8 \stackrel{?}{=} 4 + 4$
$8 = 8$
Yes

3. **a.** Yes, because $x = 6$ makes both of them true statements.
b. No, $a - 5 = 3$ is equivalent to $a = 8$, not $a = 6$.
c. Yes, they all have 3 for a solution.

4. $x - 3 = 10$
$x - 3 + \mathbf{3} = 10 + \mathbf{3}$
$x + 0 = 13$
$x = 13$

5. $a + \frac{2}{3} = -\frac{1}{6}$
$a + \frac{2}{3} + \left(-\mathbf{\frac{2}{3}}\right) = -\frac{1}{6} + \left(-\mathbf{\frac{2}{3}}\right)$
$a + 0 = -\frac{5}{6}$
$a = -\frac{5}{6}$

6. $-2.7 + x = 8.1$
$-2.7 + \mathbf{2.7} + x = 8.1 + \mathbf{2.7}$
$x = 10.8$

7. $-4x + 2 = -3x$
$-4x + \mathbf{4x} + 2 = -3x + \mathbf{4x}$
$2 = x$

8. $5(3a - 4) - 14a = 25$
$15a - 20 - 14a = 25$
$a - 20 = 25$
$a - 20 + \mathbf{20} = 25 + \mathbf{20}$
$a = 45$

9. $5x - 4 = 4x + 6$
$5x + (\mathbf{-4x}) - 4 = 4x + (\mathbf{-4x}) + 6$
$x - 4 = 6$
$x - 4 + \mathbf{4} = 6 + \mathbf{4}$
$x = 10$

Section 2.3

1. $3a = -27$
$\mathbf{\frac{1}{3}}(3a) = \mathbf{\frac{1}{3}}(-27)$
$\frac{1}{3} \cdot 3\,a = \frac{1}{3}(-27)$
$a = -9$

2. $\frac{t}{4} = 6$
$\mathbf{4}\left(\frac{t}{4}\right) = \mathbf{4}(6)$
$\left(4 \cdot \frac{1}{4}\right) t = 4(6)$
$t = 24$

3. $\frac{2}{5}y = 8$
$\mathbf{\frac{5}{2}}\left(\frac{2}{5}\right)y = \mathbf{\frac{5}{2}}(8)$
$\frac{5}{2} \cdot \frac{2}{5}y = \frac{5}{2} \cdot \frac{8}{1}$
$y = 20$

4. $3 + 7 = 6x + 8x - 9x$
$10 = 5x$
$\mathbf{\frac{1}{5}}\ 10) = \mathbf{\frac{1}{5}}(5x)$
$2 = x$

5.
$$\begin{aligned} 4x + 3 &= -13 \\ 4x + 3 + (\mathbf{-3}) &= -13 + (\mathbf{-3}) \\ 4x &= -16 \\ \frac{\mathbf{1}}{\mathbf{4}}(4x) &= \frac{\mathbf{1}}{\mathbf{4}}(-16) \\ x &= -4 \end{aligned}$$

6.
$$\begin{aligned} 7x &= 4x + 15 \\ 7x + (\mathbf{-4x}) &= 4x + (\mathbf{-4x}) + 15 \\ 3x &= 15 \\ \frac{\mathbf{1}}{\mathbf{3}}(3x) &= \frac{\mathbf{1}}{\mathbf{3}}(15) \\ x &= 5 \end{aligned}$$

7.
$$\begin{aligned} 2x - 5 &= -3x + 10 \\ 2x + \mathbf{3x} - 5 &= -3x + \mathbf{3x} + 10 \\ 5x - 5 &= 10 \\ 5x - 5 + \mathbf{5} &= 10 + \mathbf{5} \\ 5x &= 15 \\ \frac{\mathbf{1}}{\mathbf{5}}(5x) &= \frac{\mathbf{1}}{\mathbf{5}}(15) \\ x &= 3 \end{aligned}$$

8. Multiply each side by the LCD 10:
$$\begin{aligned} \mathbf{10}\left(\frac{3}{5}x + \frac{1}{2}\right) &= \mathbf{10}\left(-\frac{7}{10}\right) \\ \mathbf{10}\left(\frac{3}{5}x\right) + \mathbf{10}\left(\frac{1}{2}\right) &= \mathbf{10}\left(-\frac{7}{10}\right) \\ 6x + 5 &= -7 \\ 6x &= -12 \\ x &= -2 \end{aligned}$$

Section 2.4

1.
$$\begin{aligned} 3(x + 4) &= 6 \\ 3x + 12 &= 6 \\ 3x + 12 + (\mathbf{-12}) &= 6 + (\mathbf{-12}) \\ 3x &= -6 \\ \frac{\mathbf{1}}{\mathbf{3}}(3x) &= \frac{\mathbf{1}}{\mathbf{3}}(-6) \\ x &= -2 \end{aligned}$$

2.
$$\begin{aligned} 2(x - 3) + 3 &= 9 \\ 2x - 6 + 3 &= 9 \\ 2x - 3 &= 9 \\ 2x - 3 + \mathbf{3} &= 9 + \mathbf{3} \\ 2x &= 12 \\ \frac{\mathbf{1}}{\mathbf{2}}(2x) &= \frac{\mathbf{1}}{\mathbf{2}}(12) \\ x &= 6 \end{aligned}$$

3.
$$\begin{aligned} 7(x - 3) + 5 &= 4(3x - 2) - 8 \\ 7x - 21 + 5 &= 12x - 8 - 8 \\ 7x - 16 &= 12x - 16 \\ 7x + (\mathbf{-12x}) - 16 &= 12x + (\mathbf{-12x}) - 16 \\ -5x - 16 &= -16 \\ -5x - 16 + \mathbf{16} &= -16 + \mathbf{16} \\ -\frac{\mathbf{1}}{\mathbf{5}}(-5x) &= -\frac{\mathbf{1}}{\mathbf{5}}(0) \\ x &= 0 \end{aligned}$$

4.
$$\begin{aligned} 0.06x + 0.04(x + 7{,}000) &= 680 \\ 0.06x + 0.04x + 0.04(7{,}000) &= 680 \\ 0.10x + 280 &= 680 \\ 0.10x &= 400 \\ x &= 4{,}000 \end{aligned}$$

5.
$$\begin{aligned} 4 - 2(3y + 1) &= -16 \\ 4 - 6y - 2 &= -16 \\ -6y + 2 &= -16 \\ -6y + 2 + (\mathbf{-2}) &= -16 + (\mathbf{-2}) \\ -6y &= -18 \\ -\frac{\mathbf{1}}{\mathbf{6}}(-6y) &= -\frac{\mathbf{1}}{\mathbf{6}}(-18) \\ y &= 3 \end{aligned}$$

6.
$$\begin{aligned} 4(3x - 2) - (6x - 5) &= 6 - (3x + 1) \\ 12x - 8 - 6x + 5 &= 6 - 3x - 1 \\ 6x - 3 &= -3x + 5 \\ 9x - 3 &= 5 \\ 9x &= 8 \\ x &= \frac{8}{9} \end{aligned}$$

Section 2.5

1. When $P = 80$ and $w = 6$ the formula $P = 2l + 2w$ becomes
$$\begin{aligned} 80 &= 2l + 2(6) \\ 80 &= 2l + 12 \\ 68 &= 2l \\ 34 &= l \end{aligned}$$

2. When $y = 9$ then $3x + 2y = 6$ becomes
$$\begin{aligned} 3x + 2(9) &= 6 \\ 3x + 18 &= 6 \\ 3x &= -12 \\ x &= -4 \end{aligned}$$

3.
$$\begin{aligned} 5x - 4y &= 20 \\ -4y &= -5x + 20 \\ -\frac{\mathbf{1}}{\mathbf{4}}(-4y) &= -\frac{\mathbf{1}}{\mathbf{4}}(-5x + 20) \\ y &= \frac{5}{4}x - 5 \end{aligned}$$

4.
$$\begin{aligned} P &= 2w + 2l \\ P - 2w &= 2l \\ \frac{P - 2w}{2} &= l \end{aligned}$$

5.
$$\begin{aligned} \frac{y - 2}{x} &= \frac{4}{3} \\ x \cdot \frac{y - 2}{x} &= \frac{4}{3} \cdot x \\ y - 2 &= \frac{4}{3}x \\ y &= \frac{4}{3}x + 2 \end{aligned}$$

6. With $x = 35°$ we use the formulas for finding the complement and the supplement of an angle:

The complement of 35° is 90° − 35° = 55°

The supplement of 35° is 180° − 35° = 145°

7. $x = 0.25 \cdot 74$
$x = 18.5$

8. $84x = 21$
$$x = \frac{21}{84} = 0.25 = 25\%$$

9. $35 = 0.40x$
$$x = \frac{35}{0.40}$$
$x = 87.5$

10. $x = 0.28 \cdot 150$
$x = 42$

Section 2.6

1. Let x = the number:
$$\begin{aligned} 2x + 5 &= 11 \\ 2x &= 6 \\ x &= 3 \end{aligned}$$

2. Let x = the number
$$\begin{aligned} 3(x - 2) - 6 &= 3 \\ 3x - 6 - 6 &= 3 \\ 3x - 12 &= 3 \\ 3x &= 15 \\ x &= 5 \end{aligned}$$

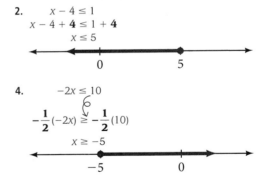

3. Let x = Mary's age now.

	Five Years Ago	Now
Pete	x	$x + 5$
Mary	$x - 5$	x

The equation is:

$$x = 2(x - 5)$$
$$x = 2x - 10$$
$$-x = -10$$
$$x = 10$$

Mary is 10; Pete is 15.

4. Let x = the width.

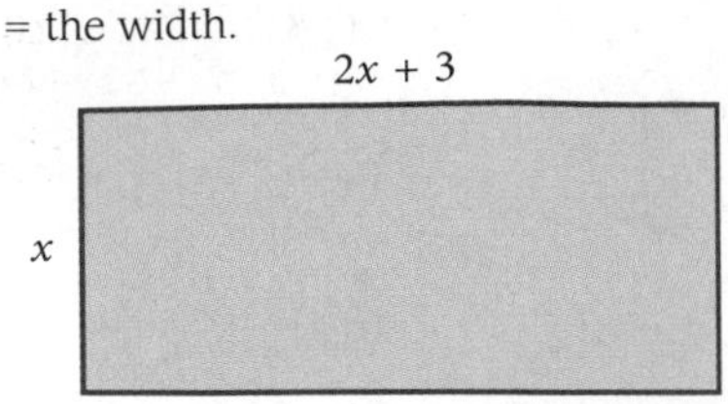

The equation is

$$2(2x + 3) + 2x = 36$$
$$4x + 6 + 2x = 36$$
$$6x + 6 = 36$$
$$6x = 30$$
$$x = 5$$

The width is 5; the length is $2(5) + 3 = 13$.

5. Let x = the number of quarters.

	Quarters	Dimes
Number of	x	$x + 7$
Value of	$25x$	$10(x + 7)$

$$25x + 10(x + 7) = 175$$
$$25x + 10x + 70 = 175$$
$$35x + 70 = 175$$
$$35x = 105$$
$$x = 3$$

She has 3 quarters and 10 dimes.

Section 2.7

1. Let x = the number

$$x + (x + 1) = 27$$
$$2x + 1 = 27$$
$$2x = 26$$
$$x = 13 \quad \text{and} \quad x + 1 = 14$$

2. Interest earned at 4% + interest earned at 5% = total interest earned

$$0.04x + 0.05(x + 3{,}000) = 510$$
$$0.04x + 0.05x + 150 = 510$$
$$0.09x + 150 = 510$$
$$0.09x = 360$$
$$x = 4{,}000$$

Alfredo put \$4,000 into his savings account on July 1, 1992.

3. Let x = the smallest angle.

Then $x + 3x + 5x = 180°$

$$9x = 180°$$
$$x = 20°$$
$$3x = 60°$$
$$5x = 100°$$

Section 2.8

1.

$$x + 3 < 5$$
$$x + 3 + (\mathbf{-3}) < 5 + (\mathbf{-3})$$
$$x < 2$$

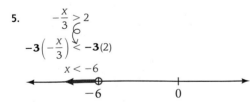

2.

$$x - 4 \le 1$$
$$x - 4 + \mathbf{4} \le 1 + \mathbf{4}$$
$$x \le 5$$

3.

$$4a < 12$$
$$\frac{\mathbf{1}}{\mathbf{4}}(4a) < \frac{\mathbf{1}}{\mathbf{4}}(12)$$
$$a < 3$$

4.

$$-2x \le 10$$
$$-\frac{\mathbf{1}}{\mathbf{2}}(-2x) \ge -\frac{\mathbf{1}}{\mathbf{2}}(10)$$
$$x \ge -5$$

5.

$$-\frac{x}{3} > 2$$
$$\mathbf{-3}\left(-\frac{x}{3}\right) < \mathbf{-3}(2)$$
$$x < -6$$

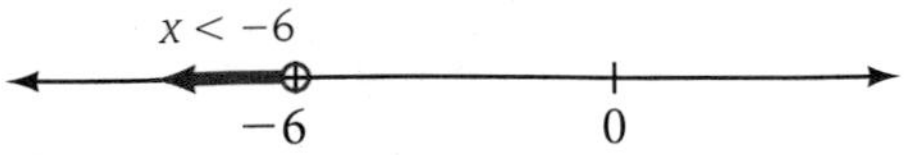

6. Method 1

$$4.5x + 2.31 > 6.3x - 4.89$$
$$4.5x + 2.31 + (\mathbf{-2.31}) > 6.3x - 4.89 + (\mathbf{-2.31})$$
$$4.5x > 6.3x - 7.2$$
$$4.5x + (\mathbf{-6.3x}) > 6.3x + (\mathbf{-6.3x}) - 7.2$$
$$-1.8x > -7.2$$
$$\frac{-1.8x}{-1.8} < \frac{-7.2}{-1.8}$$
$$x < 4$$

Method 2

$$4.5x + 2.31 > 6.3x - 4.89$$
$$100(4.5x + 2.31) > 100(6.3x - 4.89)$$
$$450x + 231 > 630x - 489$$
$$450x + 231 + (\mathbf{-231}) > 630x - 489 + (\mathbf{-231})$$
$$450x > 630x - 720$$
$$450x + (\mathbf{-630x}) > 630x + (\mathbf{-630x}) - 720$$
$$-180x > -720$$
$$\frac{-180x}{-180} < \frac{-720}{-180}$$
$$x < 4$$

7. $$2(x-5) \geq -6$$
$$2x - 10 \geq -6$$
$$2x - 10 + \mathbf{10} \geq -6 + \mathbf{10}$$
$$2x \geq 4$$
$$\frac{\mathbf{1}}{\mathbf{2}}(2x) \geq \frac{\mathbf{1}}{\mathbf{2}}(4)$$
$$x \geq 2$$

0 2

8. $$3(1-2x) + 5 < 3x - 1$$
$$3 - 6x + 5 < 3x - 1$$
$$-6x + 8 < 3x - 1$$
$$-6x + (\mathbf{-3x}) + 8 < 3x + (\mathbf{-3x}) - 1$$
$$-9x + 8 < -1$$
$$-9x + 8 + (\mathbf{-8}) < -1 + (\mathbf{-8})$$
$$-9x < -9$$
$$-\frac{\mathbf{1}}{\mathbf{9}}(-9x) > -\frac{\mathbf{1}}{\mathbf{9}}(-9)$$
$$x > 1$$

0 1

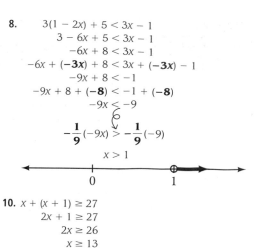

9. $$2x - 3y < 6$$
$$2x - 3y + \mathbf{3y} < 6 + \mathbf{3y}$$
$$2x < 6 + 3y$$
$$\frac{2x}{2} < \frac{6}{2} + \frac{3y}{2}$$
$$x < \frac{3}{2}y + 3$$

10. $$x + (x+1) \geq 27$$
$$2x + 1 \geq 27$$
$$2x \geq 26$$
$$x \geq 13$$

Section 2.9

1.

−3 2

2.

3. $3x + 1 \geq 7$ and $-2x > -8$

$3x \geq 6$

$\frac{\mathbf{1}}{\mathbf{3}}(3x) \geq \frac{\mathbf{1}}{\mathbf{3}}(6)$ and $-\frac{\mathbf{1}}{\mathbf{2}}(-2x) < -\frac{\mathbf{1}}{\mathbf{2}}(-8)$

$x \geq 2$ and $x < 4$

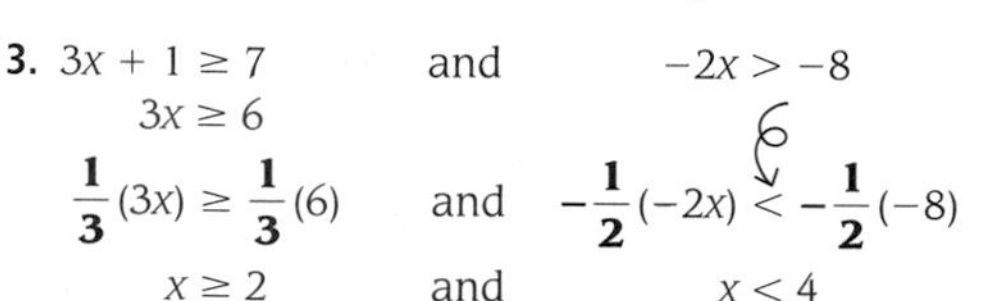

4. $$-3 \leq 4x + 1 \leq 9$$
$$-4 \leq 4x \leq 8$$
$$-1 \leq x \leq 2$$

−1 2

Chapter 3

Section 3.1

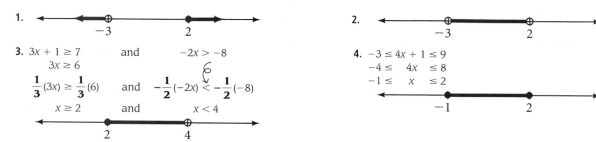

1.

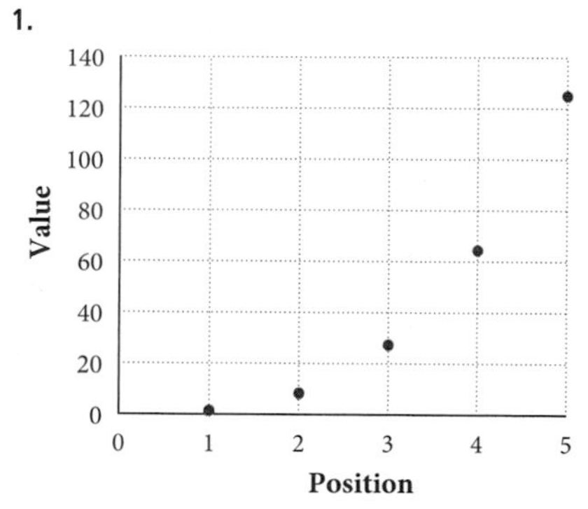

2.

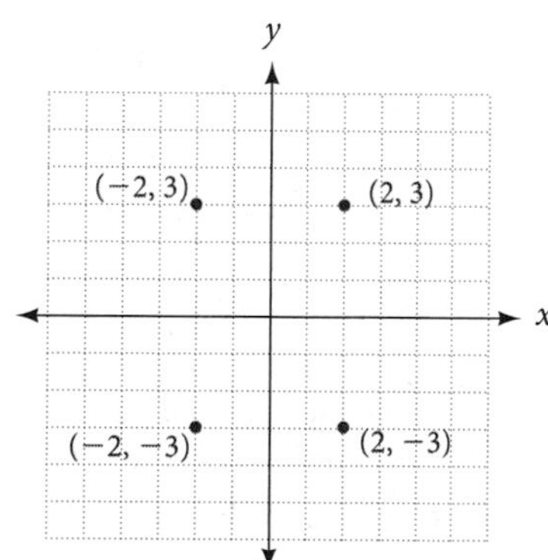

3.

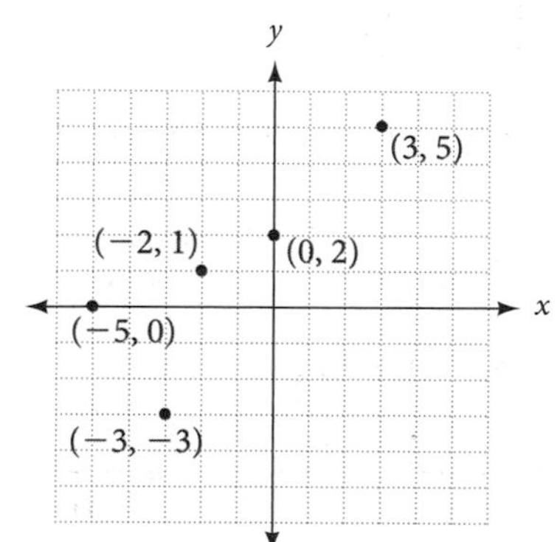

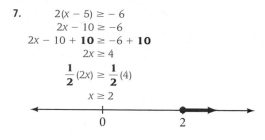

Section 3.2

1. To complete (0,),
we let $x = 0$:
$2(0) + 5y = 10$
$5y = 10$
$y = 2$
The ordered pair is (0, 2).
To complete (, 1),
we let $y = 1$:
$2x + 5(1) = 10$
$2x + 5 = 10$
$2x = 5$
$x = \frac{5}{2}$
The ordered pair is $(\frac{5}{2}, 1)$.
To complete (5,),
we let $x = 5$:
$2(5) + 5y = 10$
$10 + 5y = 10$
$5y = 0$
$y = 0$
The ordered pair is (5, 0).

2. When $x = 0$, we have:
$3(0) - 2y = 12$
$-2y = 12$
$y = -6$
When $y = 3$, we have:
$3x - 2(3) = 12$
$3x - 6 = 12$
$3x = 18$
$x = 6$
When $y = 0$, we have:
$3x - 2(0) = 12$
$3x = 12$
$x = 4$
When $x = -3$, we have:
$3(-3) - 2y = 12$
$-9 - 2y = 12$
$-2y = 21$
$y = -\frac{21}{2}$
The completed table is:

x	y
0	−6
6	3
4	0
−3	$-\frac{21}{2}$

3. When $x = 0$, we have:
$y = 3(0) - 2$
$y = -2$
When $x = 2$, we have:
$y = 3(2) - 2$
$y = 6 - 2$
$y = 4$
When $y = 7$, we have:
$7 = 3x - 2$
$9 = 3x$
$3 = x$
When $y = 0$, we have:
$0 = 3x - 2$
$2 = 3x$
$\frac{2}{3} = x$
The completed table is:

x	y
0	−2
2	4
3	7
$\frac{2}{3}$	0

4. Try (0, 1) in $y = 4x + 1$
$1 \stackrel{?}{=} 4(0) + 1$
$1 = 0 + 1$
$1 = 1$ A true statement
Try (3, 11) in $y = 4x + 1$
$11 \stackrel{?}{=} 4(3) + 1$
$11 = 12 + 1$
$11 = 13$ A false statement
Try (2, 9) in $y = 4x + 1$
$9 \stackrel{?}{=} 4(2) + 1$
$9 = 8 + 1$
$9 = 9$ A true statement
The ordered pairs (0, 1) and (2, 9) are solutions to $y = 4x + 1$, but (3, 11) is not.

Section 3.3

1. To graph $x + y = 3$, we must first find some ordered pairs that are solutions. Here are a few convenient ones: (0, 3), (3, 0) (1, 2), (2, 1), (−1, 4), and (4, −1). There are many others.

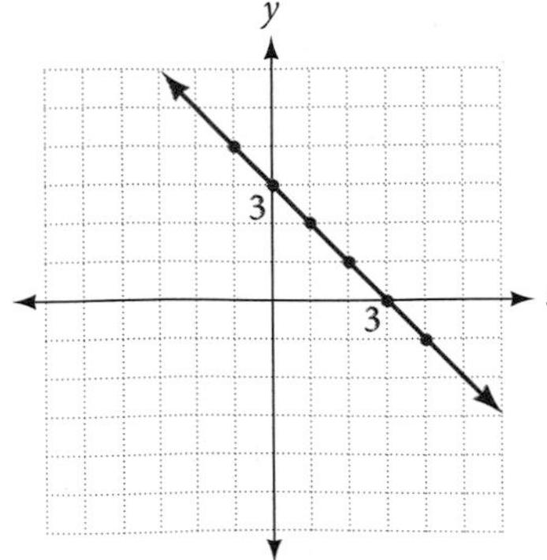

2. To find some ordered pairs that are solutions to $y = 2x + 3$, we can let x take on values of −2, 0, and 1 and then find the values of y that go with these values of x.
When $x = -2, y = 2(-2) + 3$
$= -4 + 3$
$= -1$
So (−2, −1) is one solution.
When $x = 0, y = 2(0) + 3$
$= 0 + 3$
$= 3$
So (0, 3) is another solution.
When $x = 1, y = 2(1) + 3$
$= 2 + 3$
$= 5$
So (1, 5) is third solution.
There are many other solutions as well because for any number you substitute for x, there is a y value to go with it.

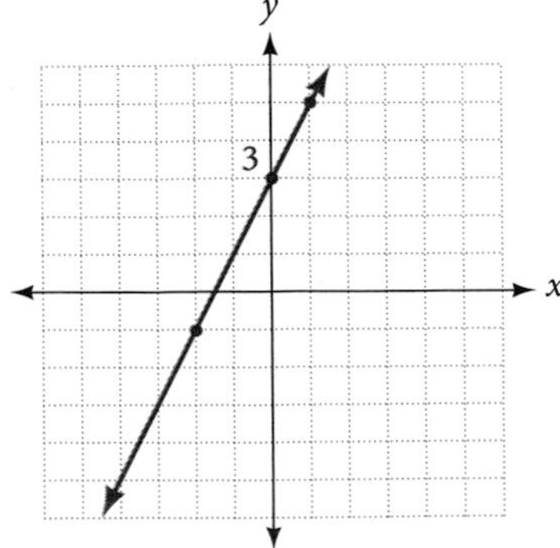

3. To graph $y = \frac{3}{2}x - 3$, we can first find some ordered pairs that are solutions, as follows:
Let $x = 0$, then
$y = \frac{3}{2}(0) - 3$
$= -3$
So (0, −3) is one solution.
Let $x = 2$, then
$y = \frac{3}{2}(2) - 3$
$= 3 - 3$
$= 0$
So (2, 0) is another solution. Now let x (or y) be some other number, and then solve for y (or x).

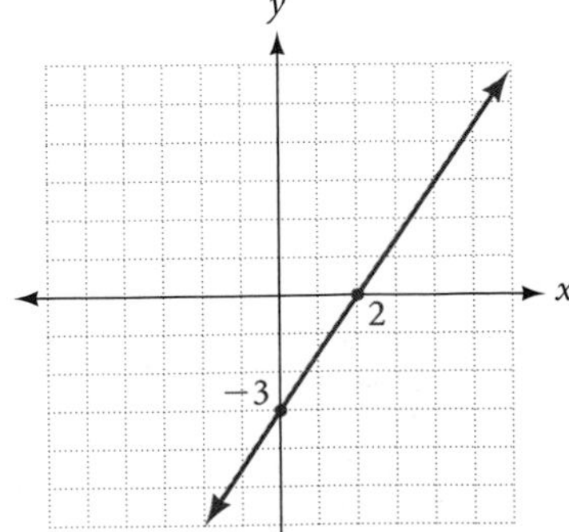

4. $2x - 4y = 8$

$$-4y = -2x + 8$$

$$y = \frac{1}{2}x - 2$$

When $x = 0$: $y = -2$

$x = 2$: $y = -1$

$x = 4$: $y = 0$

Graphing the ordered pairs $(0, -2)$, $(2, -1)$, $(4, 0)$:

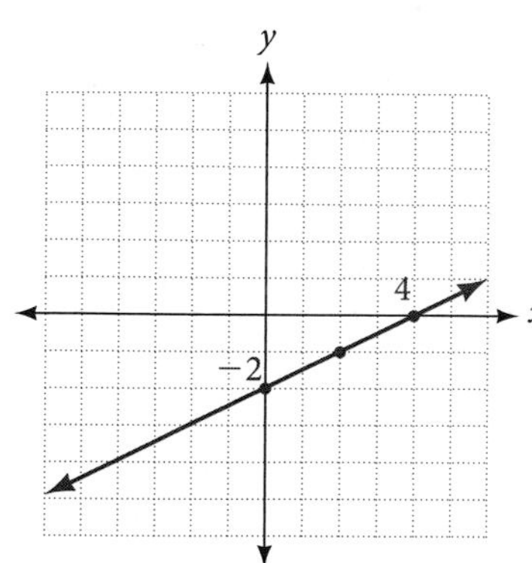

5.

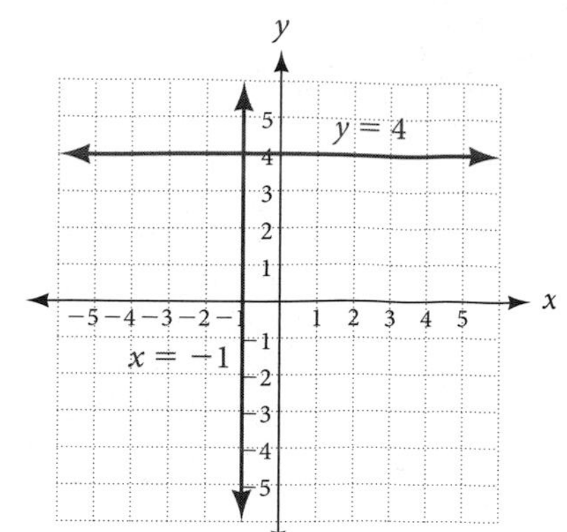

Section 3.4

1. x-intercept:
When $y = 0$
the equation becomes

$$2x - 5y = 10$$
$$2x - 5(0) = 10$$
$$2x = 10$$
$$x = 5$$

The graph crosses the x-axis at $(5, 0)$, which means the x-intercept is 5.

y-intercept:
When $x = 0$
the equation becomes

$$2x - 5y = 10$$
$$2(0) - 5y = 10$$
$$-5y = 10$$
$$y = -2$$

The graph crosses the y-axis at $(0, -2)$, which means the y-intercept is -2.

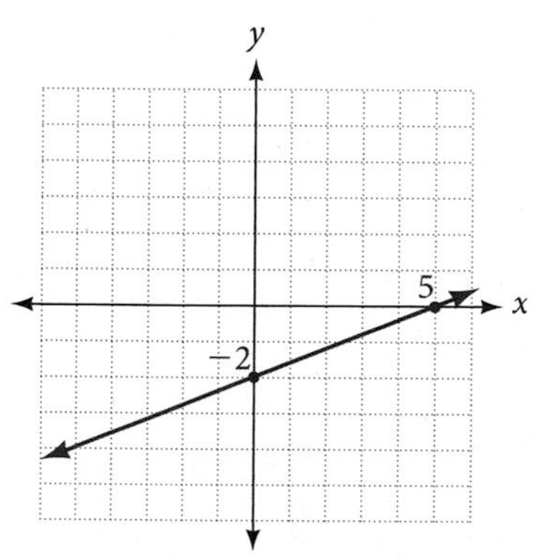

2. x-intercept:
When $y = 0$
the equation becomes

$$3x - y = 6$$
$$3x - 0 = 6$$
$$3x = 6$$
$$x = 2$$

The x-intercept is 2.

y-intercept:
When $x = 0$
the equation becomes

$$3x - y = 6$$
$$3(0) - y = 6$$
$$-y = 6$$
$$y = -6$$

The y-intercept is -6.

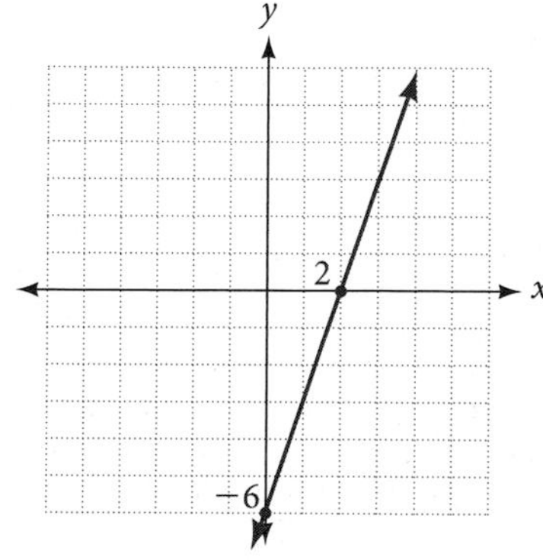

3. x-intercept:
When $y = 0$, we have

$$0 = \frac{3}{2}x - 3$$
$$3 = \frac{3}{2}x$$
$$\frac{2}{3}(3) = x$$
$$2 = x$$

The x-intercept is 2.

y-intercept:
When $x = 0$, we have

$$y = \frac{3}{2}(0) - 3$$
$$y = -3$$

The y-intercept is -3.

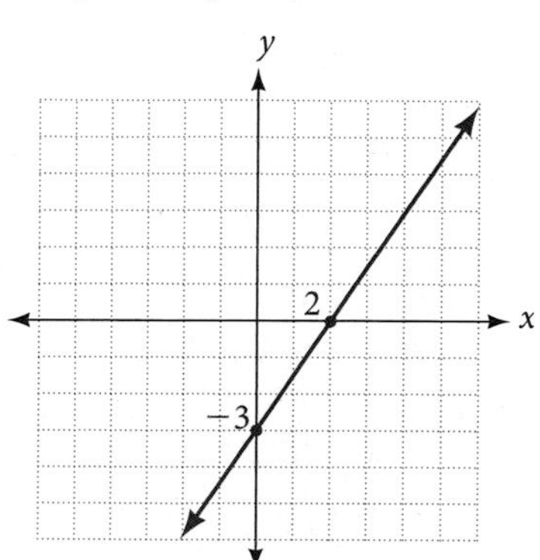

Section 3.5

1. $\frac{6-2}{3-5} = \frac{4}{-2} = -2$

2. $\frac{1-(-6)}{-5-4} = \frac{1+6}{-9} = -\frac{7}{9}$

3.

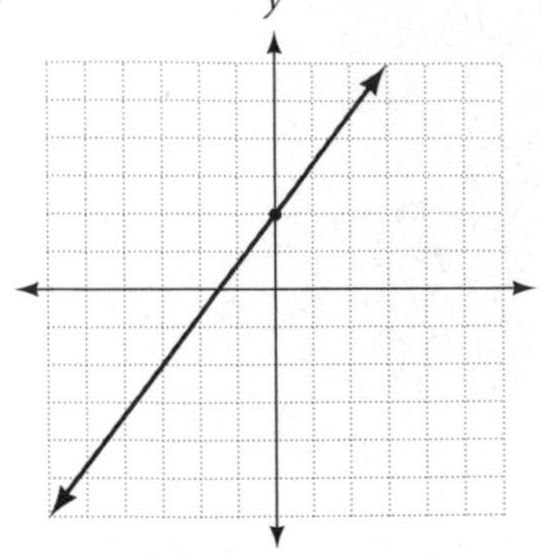

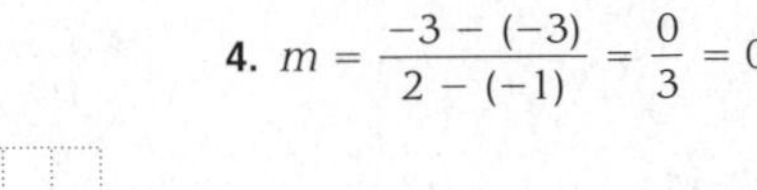

4. $m = \frac{-3-(-3)}{2-(-1)} = \frac{0}{3} = 0$

Section 3.6

1.

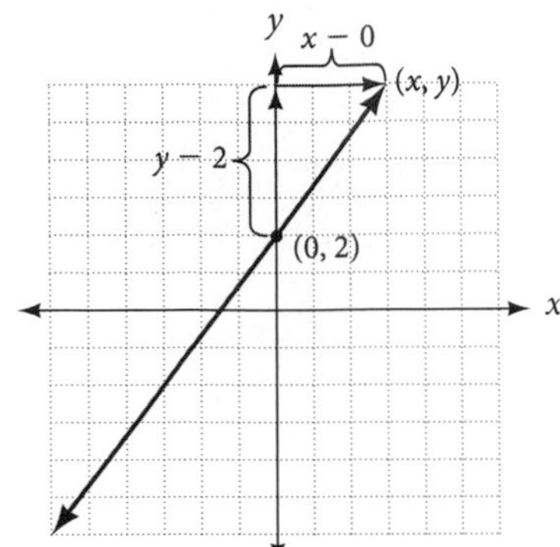

$$\text{Slope} = \frac{\text{vertical change}}{\text{horizontal change}}$$

$$\frac{y-2}{x-0} = \frac{4}{3}$$

$$\frac{y-2}{x} = \frac{4}{3}$$

$$y - 2 = \frac{4}{3}x$$

$$y = \frac{4}{3}x + 2$$

2. Substituting $m = -\frac{2}{3}$ and $b = 3$ into the equation $y = mx + b$, we have $y = -\frac{2}{3}x + 3$

3. $-4x + y = -2$
$y = 4x - 2$
$m = 4;\ b = -2$

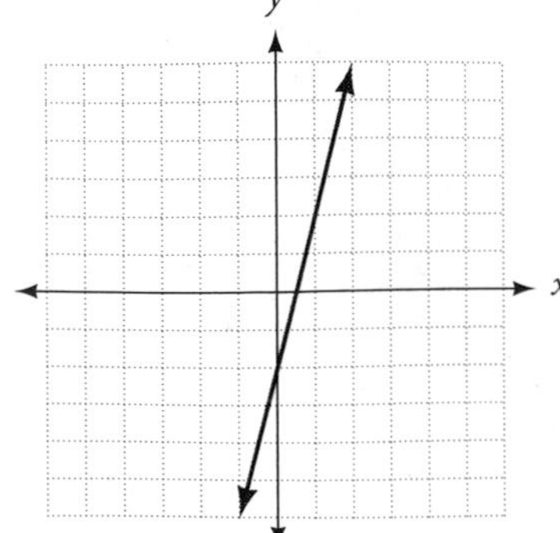

4. $4x - 3y = 9$
$-3y = -4x + 9$
$y = \frac{4}{3}x - 3$
$m = \frac{4}{3};\ b = -3$

5. $y - 5 = 4(x - 3)$
$y - 5 = 4x - 12$
$y = 4x - 7$

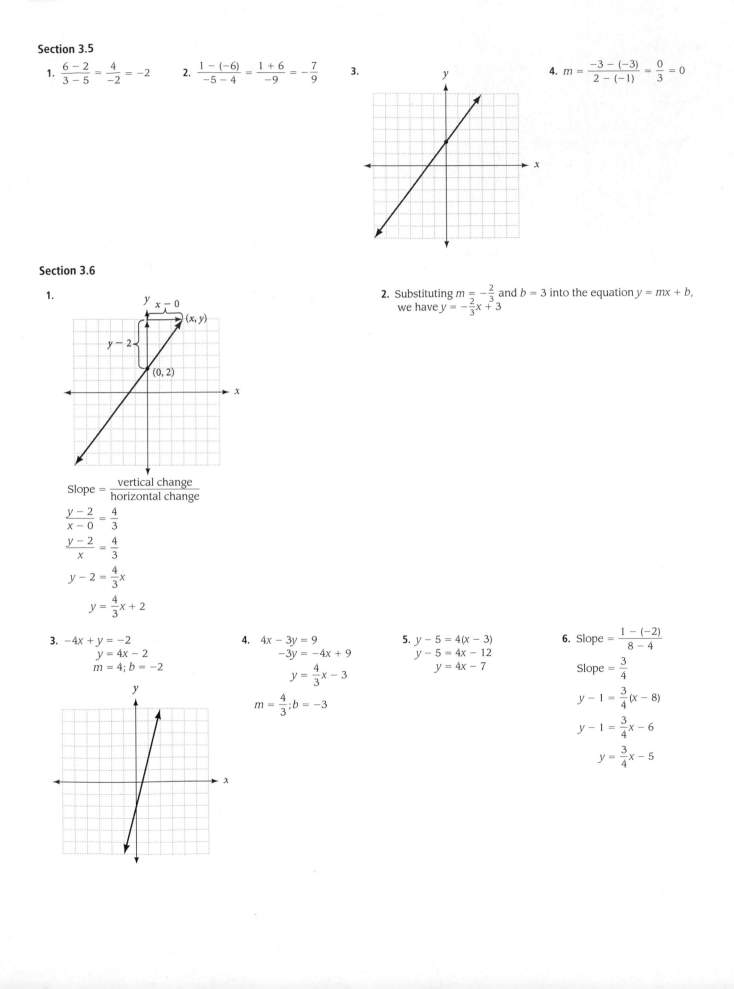

6. $\text{Slope} = \frac{1-(-2)}{8-4}$
$\text{Slope} = \frac{3}{4}$
$y - 1 = \frac{3}{4}(x - 8)$
$y - 1 = \frac{3}{4}x - 6$
$y = \frac{3}{4}x - 5$

Section 3.7

1.

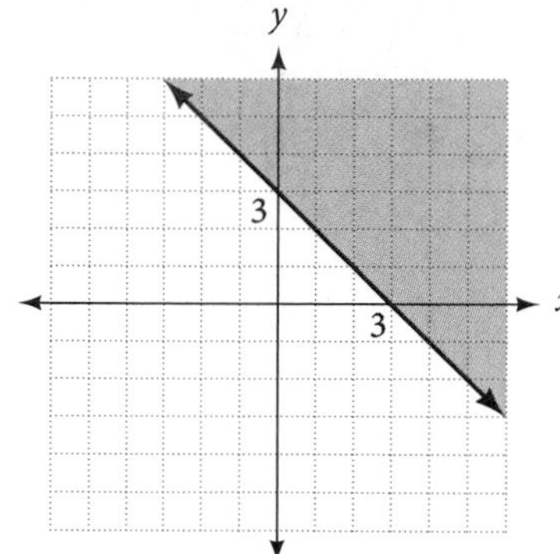

2.

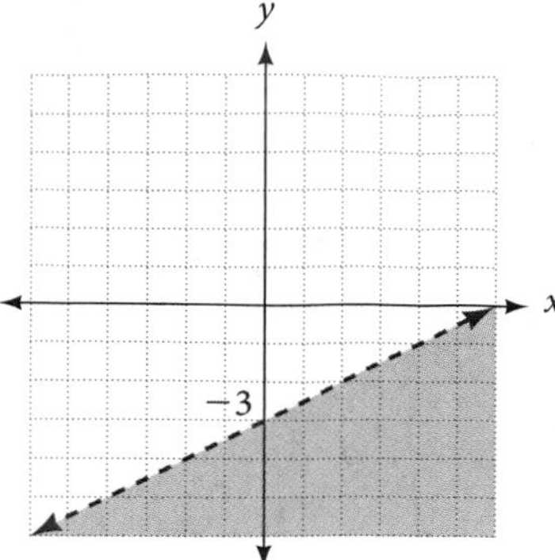

3.

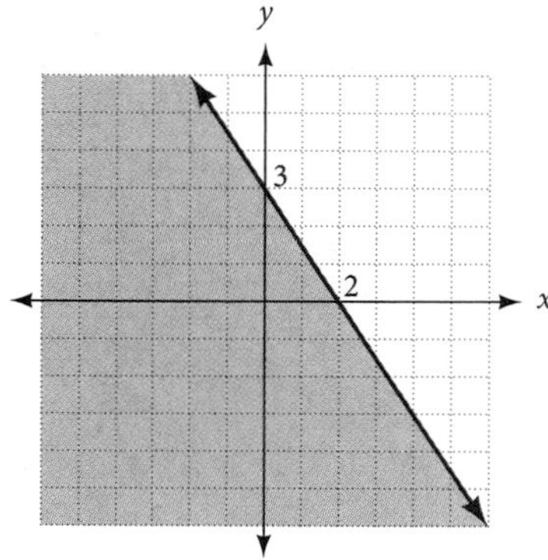

4.

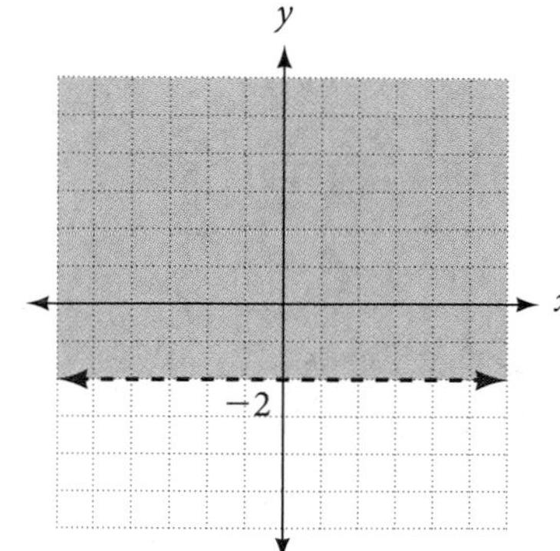

Chapter 4

Section 4.1

1.

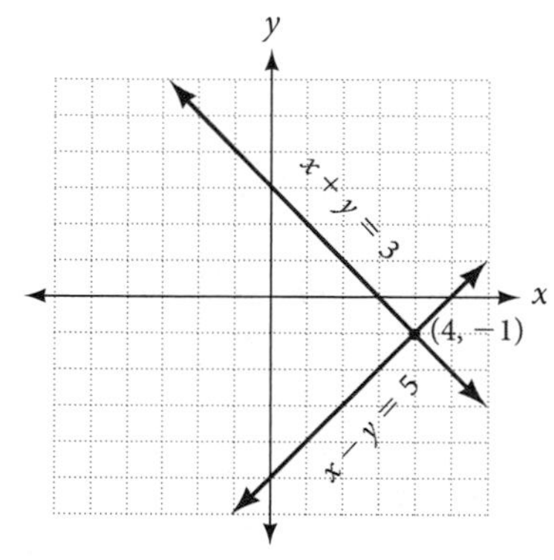

2.

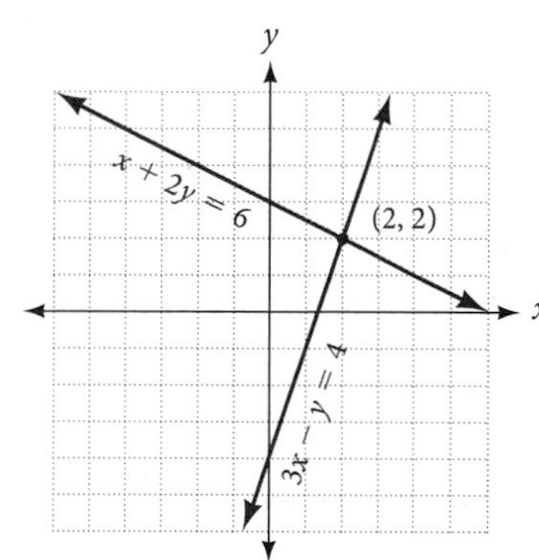

3.

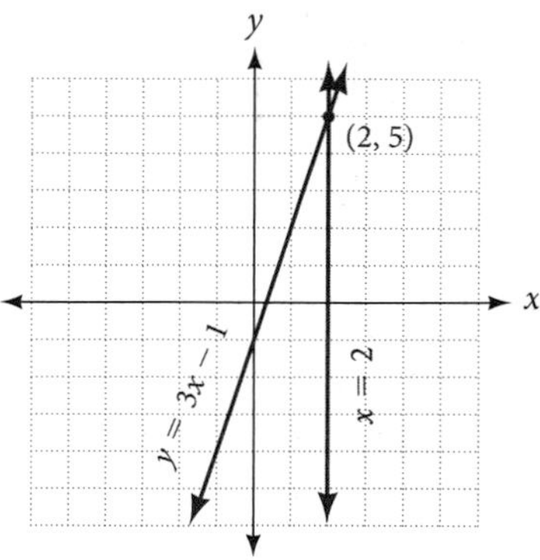

4.

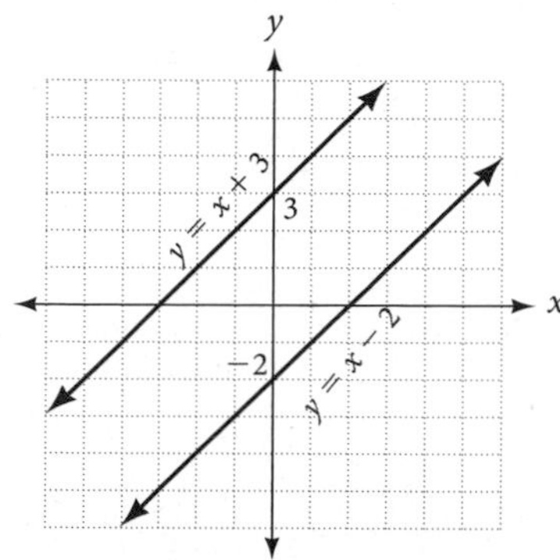

5.

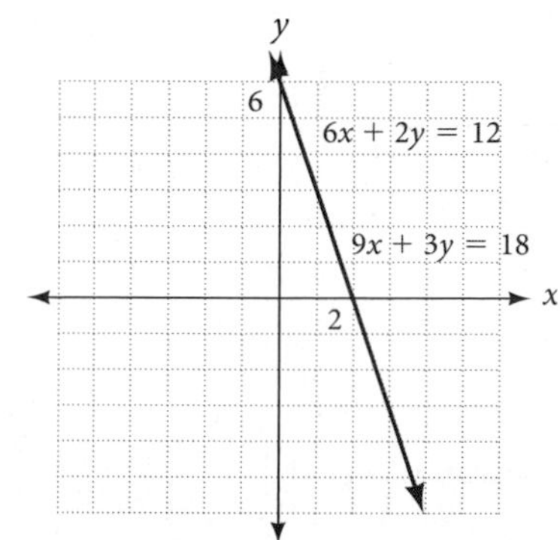

Section 4.2

1.
$$\begin{array}{r} x + y = 3 \\ x - y = 5 \\ \hline 2x + 0 = 8 \\ 2x = 8 \\ x = 4 \end{array}$$
Substituting $x = 4$ into $x + y = 3$ gives us $y = -1$. The solution is $(4, -1)$.

2.
$$\begin{array}{l} x + 3y = 5 \xrightarrow{\text{No change}} \\ x - y = 1 \xrightarrow[\text{Multiply by} -1]{} \end{array} \begin{array}{r} x + 3y = 5 \\ -x + y = -1 \\ \hline 0 + 4y = 4 \\ 4y = 4 \\ y = 1 \end{array}$$
Substituting $y = 1$ into $x + 3y = 5$ gives us $x = 2$. The solution is $(2, 1)$.

3.
$$\begin{array}{l} 3x - y = 7 \xrightarrow{\text{2 times both sides}} \\ x + 2y = 7 \xrightarrow[\text{No change}]{} \end{array} \begin{array}{r} 6x - 2y = 14 \\ x + 2y = 7 \\ \hline 7x + 0 = 21 \\ 7x = 21 \\ x = 3 \end{array}$$
When $x = 3$, the equation $x + 2y = 7$ gives us $y = 2$. The solution is $(3, 2)$.

4.
$$\begin{array}{l} 3x + 2y = 3 \xrightarrow{-2 \text{ times each side}} \\ 2x + 5y = 13 \xrightarrow[\text{3 times each side}]{} \end{array} \begin{array}{r} -6x - 4y = -6 \\ 6x + 15y = 39 \\ \hline 0 + 11y = 33 \\ 11y = 33 \\ y = 3 \end{array}$$
Substituting $y = 3$ into any of the previous equations gives $x = -1$. The solution is $(-1, 3)$.

5.
$$\begin{array}{l} 5x + 4y = -6 \xrightarrow{\text{2 times each side}} \\ 2x + 3y = -8 \xrightarrow[-5 \text{ times each side}]{} \end{array} \begin{array}{r} 10x + 8y = -12 \\ -10x - 15y = 40 \\ \hline 0 - 7y = 28 \\ -7y = 28 \\ y = -4 \end{array}$$
Using $y = -4$ in either of the original two equations gives $x = 2$. The solution is $(2, -4)$.

6.
$$\frac{1}{3}x + \frac{1}{2}y = 1 \xrightarrow{\text{Multiply by 6}} 2x + 3y = 6$$
$$x + \frac{3}{4}y = 0 \xrightarrow{\text{Multiply by 4}} 4x + 3y = 0$$
Now we can eliminate y by multiplying the top equation by -1 and leaving the bottom equation unchanged:
$$\begin{array}{l} 2x + 3y = 6 \xrightarrow{\text{Multiply by} -1} \\ 4x + 3y = 0 \xrightarrow[\text{No change}]{} \end{array} \begin{array}{r} -2x - 3y = -6 \\ 4x + 3y = 0 \\ \hline 2x = -6 \\ x = -3 \end{array}$$
Substituting $x = -3$ into $4x + 3y = 0$ gives us $y = 4$. The solution is $(-3, 4)$.

7.
$$\begin{array}{l} x - 3y = 2 \xrightarrow{\text{Multiply by 3}} \\ -3x + 9y = 2 \xrightarrow[\text{No change}]{} \end{array} \begin{array}{r} 3x - 9y = 6 \\ -3x + 9y = 2 \\ \hline 0 = 8 \end{array}$$
Both variables are eliminated and the resulting statement $0 = 8$ is false. The lines are parallel and there is no solution to the system.

8.
$$\begin{array}{l} 5x - y = 1 \xrightarrow{-2 \text{ times each side}} \\ 10x - 2y = 2 \xrightarrow[\text{No change}]{} \end{array} \begin{array}{r} -10x + 2y = -2 \\ 10x - 2y = 2 \\ \hline 0 + 0 = 0 \\ 0 = 0 \end{array}$$
We have eliminated both variables and are left with a true statement indicating that the two lines coincide. Any ordered pair that is a solution to one of the equations is a solution to the other one as well.

Section 4.3

1. $x + y = 3$
$y = x + 5$
Substituting $x + 5$ for y in the first equation gives us
$$\begin{aligned} x + (x + 5) &= 3 \\ 2x + 5 &= 3 \\ 2x &= -2 \\ x &= -1 \end{aligned}$$
When $x = -1$ in either of the two original equations, y is 4. The solution is $(-1, 4)$.

2. Substituting $y = 2x + 2$ into $5x - 4y = -2$ gives us
$$\begin{aligned} 5x - 4(2x + 2) &= -2 \\ 5x - 8x - 8 &= -2 \\ -3x - 8 &= -2 \\ -3x &= 6 \\ x &= -2 \end{aligned}$$
Putting $x = -2$ into $y = 2x + 2$ gives us $y = -2$. The solution is $(-2, -2)$.

3. Solving $x - 4y = -5$ for x gives us $x = 4y - 5$. Substituting this expression for x in the equation $3x - 2y = 5$ gives us
$$\begin{aligned} 3(4y - 5) - 2y &= 5 \\ 12y - 15 - 2y &= 5 \\ 10y - 15 &= 5 \\ 10y &= 20 \\ y &= 2 \end{aligned}$$
Putting $y = 2$ into $x = 4y - 5$ gives us $x = 3$. The solution is $(3, 2)$.

4. Solving $5x - y = 1$ for y gives us $y = 5x - 1$. Substituting this expression into the equation $-2x + 3y = 10$ in place of y gives us
$$\begin{aligned} -2x + 3(5x - 1) &= 10 \\ -2x + 15x - 3 &= 10 \\ 13x - 3 &= 10 \\ 13x &= 13 \\ x &= 1 \end{aligned}$$
When x is 1, y is 4. The solution is $(1, 4)$.

5. Substituting $-2x - 5$ for y in the equation $6x + 3y = 1$ gives us
$$\begin{aligned} 6x + 3(-2x - 5) &= 1 \\ 6x - 6x - 15 &- 1 \\ -15 &= 1 \quad \text{A false statement} \end{aligned}$$
We have eliminated both variables and are left with a false statement indicating that the lines are parallel. There is no solution to the system.

6. If we let x = the number of bags collected each month, then the equations for each company are:

$$\text{Company 1: } y = 1.5x + 13$$
$$\text{Company 2: } y = x + 21.85$$

We replace y in the equation for Company 2 with $1.5x + 13$ from the equation for Company 1.

$$1.5x + 13 = x + 21.85$$
$$1.5x - x = 21.85 - 13$$
$$0.5x = 8.85$$
$$x = 17.70$$

The monthly bill is based on the number of bags used for garbage. In this scenario, x must be a whole number. Therefore, the bills from the two companies will never be for the same amount.

$$\text{Company 1: } y = 1.5x + 13$$
$$\text{When } x = 17, y = \$38.50$$
$$\text{When } x = 18, y = \$40.00$$
$$\text{Company 2: } y = x + 21.85$$
$$\text{When } x = 17, y = \$38.85$$
$$\text{When } x = 18, y = \$39.85$$

Section 4.4

1. Let x = one number and y = the other number:
$x + y = 9$
$y = 2x + 3$
Substituting $2x + 3$ from the second equation into the first equation we have:
$x + (2x + 3) = 9$
$3x + 3 = 9$
$3x = 6$
$x = 2$
One number is 2, the other is $2(2) + 3 = 7$.

2. Let x = the amount invested at 6% and y = the amount invested at 7%:

$$\begin{aligned} x + y &= 10{,}000 \xrightarrow{-6 \text{ times each side}} & -6x - 6y &= -60{,}000 \\ 0.06x + 0.07y &= 630 \xrightarrow{100 \text{ times each side}} & 6x + 7y &= 63{,}000 \\ & & y &= 3{,}000 \end{aligned}$$

The amount invested at 7% is \$3,000. The amount invested at 6% is \$7,000.

3. Let x = the number of dimes and y = the number of quarters:

$$\begin{aligned} x + y &= 14 \xrightarrow{-10 \text{ times each side}} & -10x - 10y &= -140 \\ 0.10x + 0.25y &= 1.85 \xrightarrow{100 \text{ times each side}} & 10x + 25y &= 185 \\ & & 15y &= 45 \\ & & y &= 3 \end{aligned}$$

He has 3 quarters and 11 dimes.

4. Let x = the number of gallons of 30% solution and y = the number of gallons of 60% solution:
The system of equations is:
$x + y = 25$
$0.30x + 0.60y = 0.48(25)$
Solving the first equation for y gives $y = 25 - x$, which we use in place of y in the second equation:
$0.30x + 0.60(25 - x) = 0.48(25)$
$30x + 60(25 - x) = 48(25)$ Multiply each side by 100.
$30x + 1{,}500 - 60x = 1{,}200$
$-30x + 1{,}500 = 1{,}200$
$-30x = -300$
$x = 10$
We need 10 gallons of 30% solution and 15 gallons of 60% solution to get 25 gallons of 48% solution.

Chapter 5

Section 5.1

1. $5^3 = 5 \cdot 5 \cdot 5 = 125$

2. $-2^4 = -2 \cdot 2 \cdot 2 \cdot 2 = -16$

3. $(-3)^4 = (-3)(-3)(-3)(-3) = 81$

4. $\left(-\frac{2}{3}\right)^2 = \left(-\frac{2}{3}\right)\left(-\frac{2}{3}\right) = \frac{4}{9}$

5. $5^4 \cdot 5^5 = 5^{4+5} = 5^9$

6. $x^3 \cdot x^7 = x^{3+7} = x^{10}$

7. $4^5 \cdot 4^2 \cdot 4^6 = 4^{5+2+6} = 4^{13}$

8. $(3^4)^5 = 3^{4 \cdot 5} = 3^{20}$

9. $(x^7)^2 = x^{7 \cdot 2} = x^{14}$

10. $(7x)^2 = 7^2 \cdot x^2 = 49x^2$

11. $(3xy)^4 = 3^4 \cdot x^4 \cdot y^4 = 81x^4y^4$

12. $(4x^3)^2 = 4^2(x^3)^2 = 16x^6$

13. $\left(-\frac{1}{3}x^3y^2\right)^2 = \left(-\frac{1}{3}\right)^2(x^3)^2(y^2)^2 = \frac{1}{9}x^6y^4$

14. $(x^5)^2(x^4)^3 = x^{10} \cdot x^{12} = x^{22}$

15. $(3y)^2(2y^3) = 9y^2 \cdot 2y^3 = 18y^5$

16. $(3x^3y^2)^2(2xy^4)^3 = 3^2(x^3)^2(y^2)^2(2)^3x^3(y^4)^3$
$= 9x^6y^4 \cdot 8x^3y^{12}$
$= 72x^9y^{16}$

17. Move the decimal point 6 places to the left and multiply by 10^6: 4.81×10^6.

18. $3.05 \times 10^5 = 3.05 \times 100{,}000 = 305{,}000$

Section 5.2

1. $3^{-2} = \frac{1}{3^2} = \frac{1}{9}$

2. $4^{-3} = \frac{1}{4^3} = \frac{1}{64}$

3. $5x^{-4} = 5 \cdot \frac{1}{x^4} = \frac{5}{x^4}$

4. $\frac{x^{10}}{x^4} = x^{10-4} = x^6$

5. $\frac{x^5}{x^7} = x^{5-7} = x^{-2} = \frac{1}{x^2}$

6. $\dfrac{2^{21}}{2^{25}} = 2^{21-25} = 2^{-4} = \dfrac{1}{2^4} = \dfrac{1}{16}$ 7. $\left(\dfrac{x}{5}\right)^2 = \dfrac{x^2}{5^2} = \dfrac{x^2}{25}$ 8. $\left(\dfrac{2}{a}\right)^3 = \dfrac{2^3}{a^3} = \dfrac{8}{a^3}$ 9. $\left(\dfrac{3}{4}\right)^3 = \dfrac{3^3}{4^3} = \dfrac{27}{64}$ 10. $10^0 = 1$

11. $10^1 = 10$ 12. $6^1 + 6^0 = 6 + 1 = 7$ 13. $(3x^5y^2)^0 = 1$ 14. $\dfrac{(2x^3)^2}{x^4} = \dfrac{4x^6}{x^4} = 4x^2$ 15. $\dfrac{x^{-6}}{(x^3)^4} = \dfrac{x^{-6}}{x^{12}} = x^{-6-12} = x^{-18} = \dfrac{1}{x^{18}}$

16. $\left(\dfrac{y^8}{y^3}\right)^2 = (y^5)^2 = y^{10}$ 17. $(2x^4)^{-2} = \dfrac{1}{(2x^4)^2} = \dfrac{1}{4x^8}$ 18. $x^{-6} \cdot x^2 = x^{-6+2} = x^{-4} = \dfrac{1}{x^4}$ 19. $\dfrac{a^5(a^{-2})^4}{(a^{-3})^2} = \dfrac{a^5 \cdot a^{-8}}{a^{-6}} = \dfrac{a^{-3}}{a^{-6}} = a^{-3-(-6)} = a^3$

20. Area of square 1 $= x^2$
Area of square 2 $= (4x)^2 = 16x^2$
To find out how many smaller squares it will take to cover the larger square, we divide the area of the larger square by the area of the smaller square.
$$\frac{16x^2}{x^2} = 16$$

21. Volume of box 1 $= x^3$
Volume of box 2 $= (4x)^3 = 64x^3$
To find out how many smaller boxes will fit inside the larger box, we divide the volume of the larger box by the volume of the smaller box.
$$\frac{64x^3}{x^3} = 64$$

22. a. Move decimal point left 3 places; multiply by 10^3; 4.73×10^3
b. $4.73 \times 10^1 = 4.73 \times 10 = 47.3$
c. Move decimal point right 1 place; multiply by 10^{-1}; 4.73×10^{-1}
d. $4.73 \times 10^{-3} = 4.73 \times \dfrac{1}{10^3} = 4.73 \times \dfrac{1}{1{,}000} = \dfrac{4.73}{1{,}000}$
$= 0.00473$

Section 5.3

1. $(-2x^3)(5x^2) = (-2 \cdot 5)(x^3 \cdot x^2)$
$= -10x^5$

2. $(4x^6y^3)(2xy^2) = (4 \cdot 2)(x^6 \cdot x)(y^3 \cdot y^2)$
$= 8x^7y^5$

3. $\dfrac{25x^4}{5x} = \dfrac{25}{5} \cdot \dfrac{x^4}{x}$
$= 5x^3$

4. $\dfrac{27x^4y^3}{9xy^2} = \dfrac{27}{9} \cdot \dfrac{x^4}{x} \cdot \dfrac{y^3}{y^2}$
$= 3x^3y$

5. $\dfrac{13a^6b^2}{39a^4b^7} = \dfrac{13}{39} \cdot \dfrac{a^6}{a^4} \cdot \dfrac{b^2}{b^7}$
$= \dfrac{1}{3} \cdot a^2 \cdot \dfrac{1}{b^5}$
$= \dfrac{a^2}{3b^5}$

6. $(3 \times 10^6)(3 \times 10^{-8}) = (3 \times 3)(10^6 \times 10^{-8})$
$= 9 \times 10^{-2}$

7. $\dfrac{4.8 \times 10^{20}}{2 \times 10^{12}} = \dfrac{4.8}{2} \times \dfrac{10^{20}}{10^{12}}$
$= 2.4 \times 10^8$

8. $-4x^2 + 9x^2 = (-4 + 9)x^2$
$= 5x^2$

9. $12x^2y - 15x^2y = (12 - 15)x^2y$
$= -3x^2y$

10. $6x^2 + 9x^3$ cannot be simplified further.

11. $\dfrac{(10x^3y^2)(6xy^4)}{12x^2y^3} = \dfrac{60x^4y^6}{12x^2y^3}$
$= 5x^2y^3$

12. $\dfrac{(1.2 \times 10^6)(6.3 \times 10^{-5})}{6 \times 10^{-10}} = \dfrac{(1.2)(6.3)}{6} \times \dfrac{(10^6)(10^{-5})}{10^{-10}}$
$= 1.26 \times 10^{11}$

13. $\dfrac{24x^7}{3x^2} + \dfrac{14x^9}{7x^4} = 8x^5 + 2x^5$
$= 10x^5$

14. $x^2\left(1 - \dfrac{4}{x}\right) = x^2 \cdot 1 - x^2 \cdot \dfrac{4}{x}$
$= x^2 - \dfrac{4x^2}{x}$
$= x^2 - 4x$

15. $3a\left(\dfrac{1}{a} - \dfrac{1}{3}\right) = 3a \cdot \dfrac{1}{a} - 3a \cdot \dfrac{1}{3}$
$= \dfrac{3a}{a} - \dfrac{3a}{3}$
$= 3 - a$

16. We make a diagram of the object with the dimensions labeled as given in the problem.

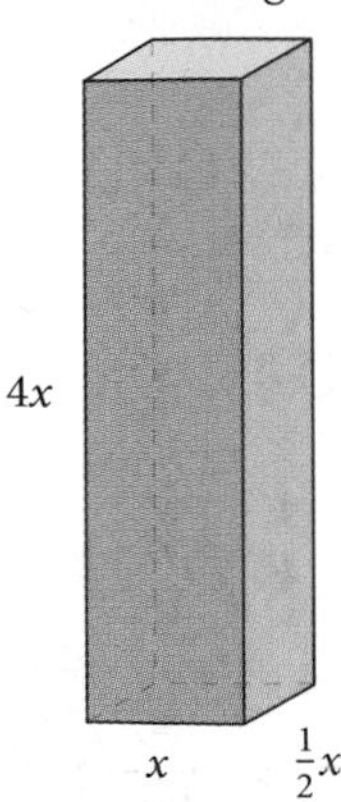

The volume is the product of the three dimensions.
$$V = 4x \cdot x \cdot \frac{1}{2}x$$
$$V = 2x^3$$

Section 5.4

1. $(2x^2 + 3x + 4) + (5x^2 - 6x - 3)$
$= (2x^2 + 5x^2) + (3x - 6x) + (4 - 3)$
$= 7x^2 - 3x + 1$

2. $x^2 + 7x + 4x + 28 = x^2 + 11x + 28$

3. $(5x^2 + x + 2) - (x^2 + 3x + 7)$
$= 5x^2 + x + 2 - x^2 - 3x - 7$
$= (5x^2 - x^2) + (x - 3x) + (2 - 7)$
$= 4x^2 - 2x - 5$

4. $(x^2 + x + 1) - (-4x^2 - 3x + 8) = x^2 + x + 1 + 4x^2 + 3x - 8$
$= (x^2 + 4x^2) + (x + 3x) + (1 - 8)$
$= 5x^2 + 4x - 7$

5. $2(-3)^2 - (-3) + 3 = 2(9) + 3 + 3$
$= 18 + 3 + 3$
$= 24$

Section 5.5

1. $2x^3(4x^2 - 5x + 3)$
$= 2x^3(4x^2) - 2x^3(5x) + 2x^3(3)$
$= 8x^5 - 10x^4 + 6x^3$

2. $(2x - 4)(3x + 2)$
$= 2x(3x + 2) - 4(3x + 2)$
$= 2x(3x) + 2x(2) - 4(3x) - 4(2)$
$= 6x^2 + 4x - 12x - 8$
$= 6x^2 - 8x - 8$

3. $(3x + 5)(2x - 1)$
F O I L
$= 3x(2x) + 3x(-1) + 5(2x) + 5(-1)$
$= 6x^2 - 3x + 10x - 5$
$= 6x^2 + 7x - 5$

4. $$\begin{array}{r} 2x^2 - 3x + 4 \\ 3x - 2 \\ \hline 6x^3 - 9x^2 + 12x \\ -4x^2 + 6x - 8 \\ \hline 6x^3 - 13x^2 + 18x - 8 \end{array}$$

5. $5a^2(3a^2 - 2a + 4)$
$= 5a^2(3a^2) - 5a^2(2a) + 5a^2(4)$
$= 15a^4 - 10a^3 + 20a^2$

6. $(x + 3)(y - 5) = xy - 5x + 3y - 15$
F O I L

7. $(x + 2y)(a + b) = ax + bx + 2ay + 2by$
F O I L

8. $(4x + 1)(3x + 5)$
F O I L
$= 4x(3x) + 4x(5) + 1(3x) + 1(5)$
$= 12x^2 + 20x + 3x + 5$
$= 12x^2 + 23x + 5$

9. Let x = the width. The length is $3x + 5$. To find the area, multiply length and width: $A = x(3x + 5) = 3x^2 + 5x$

10. Since $x = 700 - 110p$, we can substitute $700 - 110p$ for x in the revenue equation to obtain
$$R = (700 - 110p)p$$
$$R = 700p - 110p^2$$

Section 5.6

1. $(4x - 5)^2 = (4x - 5)(4x - 5)$
$= 16x^2 - 20x - 20x + 25$
$= 16x^2 - 40x + 25$

2. $(x + y)^2 = (x + y)(x + y)$
$= x^2 + 2xy + y^2$

3. $(x - y)^2 = (x - y)(x - y)$
$= x^2 - 2xy + y^2$

4. $(x - 3)^2 = x^2 + 2(x)(-3) + 9$
$= x^2 - 6x + 9$

5. $(x + 5)^2 = x^2 + 2(x)(5) + 25$
$= x^2 + 10x + 25$

6. $(3x - 2)^2 = 9x^2 + 2(3x)(-2) + 4$
$= 9x^2 - 12x + 4$

7. $(4x + 5)^2 = 16x^2 + 2(4x)(5) + 25$
$= 16x^2 + 40x + 25$

8. $(3x - 5)(3x + 5) = 9x^2 + 15x - 15x - 25$
$= 9x^2 - 25$

9. $(x - 4)(x + 4) = x^2 + 4x - 4x - 16$
$= x^2 - 16$

10. $(4x - 1)(4x + 1) = 16x^2 + 4x - 4x - 1$
$= 16x^2 - 1$

11. $(x + 2)(x - 2) = x^2 - 4$

12. $(a + 7)(a - 7) = a^2 - 49$

13. $(6a + 1)(6a - 1) = 36a^2 - 1$

14. $(8x - 2y)(8x + 2y) = 64x^2 - 4y^2$

15. $(4a - 3b)(4a + 3b) = 16a^2 - 9b^2$

16. $x^2 + (x - 1)^2 + (x + 1)^2 = x^2 + (x^2 - 2x + 1) + (x^2 + 2x + 1)$
$= 3x^2 + 2$

Section 5.7

1. $\dfrac{8x^3 - 12x^2}{4x} = \dfrac{8x^3}{4x} - \dfrac{12x^2}{4x}$
$= 2x^2 - 3x$

2. $\dfrac{5x^2 - 10}{5} = \dfrac{5x^2}{5} - \dfrac{10}{5}$
$= x^2 - 2$

3. $\dfrac{8x^2 - 2}{2} = \dfrac{8x^2}{2} - \dfrac{2}{2}$
$= 4x^2 - 1$

4. $\dfrac{15x^4 - 10x^3}{5x} = \dfrac{15x^4}{5x} - \dfrac{10x^3}{5x}$
$= 3x^3 - 2x^2$

5. $\dfrac{9x^3y^2 - 18x^2y^3}{-9xy} = \dfrac{9x^3y^2}{-9xy} - \dfrac{18x^2y^3}{-9xy}$
$= -x^2y - (-2xy^2)$
$= -x^2y + 2xy^2$

6. $\dfrac{21x^3y^2 + 14x^2y^2 - 7x^2y^3}{7x^2y} = \dfrac{21x^3y^2}{7x^2y} + \dfrac{14x^2y^2}{7x^2y} - \dfrac{7x^2y^3}{7x^2y}$
$= 3xy + 2y - y^2$

Section 5.8

1. $$\begin{array}{r} 122 \\ 35\overline{)4281} \\ 35\downarrow \\ \hline 78 \\ 70\downarrow \\ \hline 81 \\ 70 \\ \hline 11 \end{array}$$
$\dfrac{4,281}{35} = 122 + \dfrac{11}{35}$

2. $$\begin{array}{r} x - 3 \\ x - 2\overline{)\ x^2 - 5x + 8} \\ \not{-}x^2 \not{+} 2x \quad \downarrow \\ \hline -3x + 8 \\ \not{+}3x \not{-} 6 \\ \hline 2 \end{array}$$
$\dfrac{x^2 - 5x + 8}{x - 2} = x - 3 + \dfrac{2}{x - 2}$

3. $$\begin{array}{r} 4x + 3 \\ 2x - 3 \overline{)\ 8x^2 - 6x - 5} \\ \underline{-8x^2 + 12x} \\ 6x - 5 \\ \underline{-6x + 9} \\ 4 \end{array}$$

$$\frac{8x^2 - 6x - 5}{2x - 3} = 4x + 3 + \frac{4}{2x - 3}$$

4. $$\begin{array}{r} 3x^2 + 9x + 25 \\ x - 3 \overline{)\ 3x^3 + 0x^2 - 2x + 1} \\ \underline{-3x^3 + 9x^2} \\ 9x^2 - 2x \\ \underline{-9x^2 + 27x} \\ 25x + 1 \\ \underline{-25x + 75} \\ 76 \end{array}$$

$$\frac{3x^3 - 2x + 1}{x - 3} = 3x^2 + 9x + 25 + \frac{76}{x - 3}$$

Chapter 6

Section 6.1

1. The largest number that divides 10 and 15 is 5, and the highest power of x that is a factor of x^6 and x^4 is x^4. Therefore, the greatest common factor for the polynomial $10x^6 + 15x^4$ is $5x^4$.

2. The largest number that divides 27, 18, and 9 is 9, and the highest power of ab that is a factor of a^4b^5, a^3b^3, and a^3b^2 is a^3b^2. Therefore, the greatest common factor for the polynomial $27a^4b^5 + 18a^3b^3 + 9a^3b^2$ is $9a^3b^2$.

3. $5x + 15 = 5 \cdot x + 5 \cdot 3$
   $= 5(x + 3)$

4. $25x^4 - 35x^3 = 5x^3 \cdot 5x - 5x^3 \cdot 7$
   $= 5x^3(5x - 7)$

5. $20x^8 - 12x^7 + 16x^6 = 4x^6 \cdot 5x^2 - 4x^6 \cdot 3x + 4x^6 \cdot 4$
   $= 4x^6(5x^2 - 3x + 4)$

6. $8xy^3 - 16x^2y^2 + 8x^3y = 8xy \cdot y^2 - 8xy \cdot 2xy + 8xy \cdot x^2$
   $= 8xy(y^2 - 2xy + x^2)$

7. $2ab^2 - 6a^2b^2 + 8a^3b^2 = 2ab^2 \cdot 1 - 2ab^2 \cdot 3a + 2ab^2 \cdot 4a^2$
   $= 2ab^2(1 - 3a + 4a^2)$

8. $ax - bx + ay - by = x(a - b) + y(a - b)$
   $= (a - b)(x + y)$

9. $5ax - 3ay + 10x - 6y = a(5x - 3y) + 2(5x - 3y)$
   $= (5x - 3y)(a + 2)$

10. $8x^2 - 12x + 10x - 15 = 4x(2x - 3) + 5(2x - 3)$
    $= (4x + 5)(2x - 3)$

11. $3x^2 + 7bx - 3xy - 7by = x(3x + 7b) - y(3x + 7b)$
    $= (3x + 7b)(x - y)$

Section 6.2

1. We want two numbers whose product is 6 and whose sum is 5. The numbers are 2 and 3.
   $x^2 + 5x + 6 = (x + 2)(x + 3)$

2. The two numbers whose product is -15 and whose sum is 2 are 5 and -3.
   $x^2 + 2x - 15 = (x + 5)(x - 3)$

3. $2x^2 - 2x - 24 = 2(x^2 - x - 12)$
   $= 2(x - 4)(x + 3)$

4. $3x^3 + 18x^2 + 15x = 3x(x^2 + 6x + 5)$
   $= 3x(x + 5)(x + 1)$

5. Two expressions whose product is $12y^2$ and whose sum is $7y$ are $3y$ and $4y$.
   $x^2 + 7xy + 12y^2 = (x + 3y)(x + 4y)$

Section 6.3

1. We list all the possible pairs of factors that, when multiplied together, give a trinomial whose first term is $2a^2$ and whose last term is $+6$.

Binomial Factors	First Term	Middle Term	Last Term
$(2a + 1)(a + 6)$	$2a^2$	$+13a$	$+6$
$(2a - 1)(a - 6)$	$2a^2$	$-13a$	$+6$
$(2a + 3)(a + 2)$	$2a^2$	$+7a$	$+6$
$(2a - 3)(a - 2)$	$2a^2$	$-7a$	$+6$

The factors of $2a^2 + 7a + 6$ are $(2a + 3)$ and $(a + 2)$. Therefore,
$2a^2 + 7a + 6 = (2a + 3)(a + 2)$

2. We list all the possible pairs of factors that, when multiplied together, give a trinomial whose first term is $4x^2$ and whose last term is -5.

Binomial Factors	First Term	Middle Term	Last Term
$(4x + 1)(x - 5)$	$4x^2$	$-19x$	-5
$(4x - 1)(x + 5)$	$4x^2$	$+19x$	-5
$(4x + 5)(x - 1)$	$4x^2$	$+x$	-5
$(4x - 5)(x + 1)$	$4x^2$	$-x$	-5
$(2x + 1)(2x - 5)$	$4x^2$	$-8x$	-5
$(2x - 1)(2x + 5)$	$4x^2$	$+8x$	-5

The factors of $4x^2 - x - 5$ are $(4x - 5)$ and $(x + 1)$. Therefore,
$4x^2 - x - 5 = (4x - 5)(x + 1)$

3. $30y^2 + 5y - 5 = 5(6y^2 + y - 1)$
   $= 5(3y - 1)(2y + 1)$

4. $10x^2y^2 + 5xy^3 - 30y^4 = 5y^2(2x^2 + xy - 6y^2)$
   $= 5y^2(2x - 3y)(x + 2y)$

5. The trinomial $2x^2 + 7x - 15$ has the form $ax^2 + bx + c$, where $a = 2$, $b = 7$, and $c = -15$.
Step 1 The product ac is $2(-15) = -30$.
Step 2 We need to find two numbers whose product is -30 and whose sum is 7. We list all the pairs of numbers whose product is -30.

Product	*Sum*
$30(-1) = -30$	$30 + (-1) = 29$
$15(-2) = -30$	$15 + (-2) = 13$
$10(-3) = -30$	$10 + (-3) = 7$

Step 3 Now we rewrite our original trinomial so the middle term, $7x$, is written as the sum of $10x$ and $-3x$.
$$2x^2 + 7x - 15 = 2x^2 + 10x - 3x - 15$$
Step 4 Factoring by grouping, we have:
$$\begin{aligned} 2x^2 + 10x - 3x - 15 &= 2x(x + 5) - 3(x + 5) \\ &= (x + 5)(2x - 3) \end{aligned}$$

6. In the trinomial $6x^2 + x - 5$, $a = 6$, $b = 1$, and $c = -5$. The product of ac is $6(-5) = -30$. We need two numbers whose product is -30 and whose sum is 1.

Product	*Sum*
$30(-1) = -30$	$30 + (-1) = 29$
$15(-2) = -30$	$15 + (-2) = 13$
$10(-3) = -30$	$10 + (-3) = 7$
$6(-5) = -30$	$6 + (-5) = 1$

Now we write the middle term, x, as the sum of $6x$ and $-5x$ and proceed to factor by grouping.
$$\begin{aligned} 6x^2 + x - 5 &= 6x^2 + 6x - 5x - 5 \\ &= 6x(x + 1) - 5(x + 1) \\ &= (x + 1)(6x - 5) \end{aligned}$$

7. In the trinomial $15x^2 - 2x - 8$, the product of the a and c terms is $15(-8) = -120$. We are looking for a pair of numbers whose product is -120 and whose sum is -2. The two numbers are -12 and 10. Writing $-2x$ as $-12x + 10x$, we have
$$\begin{aligned} 15x^2 - 2x - 8 &= 15x^2 - 12x + 10x - 8 \\ &= 3x(5x - 4) + 2(5x - 4) \\ &= (3x + 2)(5x - 4) \end{aligned}$$

8. We begin by factoring out the greatest common factor, 16. The we factor the trinomial that remains.
$$\begin{aligned} h &= 48 + 32t - 16t^2 \\ h &= 16(3 + 2t - t^2) \\ h &= 16(3 - t)(1 + t) \end{aligned}$$
Letting $t = 3$ in this equation, we have
$$\begin{aligned} h &= 16(3 - 3)(1 + 3) \\ &= 16(0)(4) \\ &= 0 \end{aligned}$$
When $t = 3$, $h = 0$.

Section 6.4

1. $$\begin{aligned} 9x^2 - 25 &= (3x)^2 - 5^2 \\ &= (3x + 5)(3x - 5) \end{aligned}$$

2. $$\begin{aligned} 49a^2 - 1 &= (7a)^2 - 1^2 \\ &= (7a + 1)(7a - 1) \end{aligned}$$

3. $$\begin{aligned} a^4 - b^4 &= (a^2)^2 - (b^2)^2 \\ &= (a^2 + b^2)(a^2 - b^2) \\ &= (a^2 + b^2)(a + b)(a - b) \end{aligned}$$

4. $$\begin{aligned} 4x^2 + 12x + 9 &= 2^2x^2 + 2(2x)(3) + 3^2 \\ &= (2x + 3)^2 \end{aligned}$$

5. $$\begin{aligned} 6x^2 + 24x + 24 &= 6(x^2 + 4x + 4) \\ &= 6(x + 2)^2 \end{aligned}$$

6. $$\begin{aligned} (x - 4)^2 - 36 &= (x - 4)^2 - 6^2 \\ &= [(x - 4) - 6][(x - 4) + 6] \\ &= (x - 10)(x + 2) \end{aligned}$$

Section 6.5

1. We use the distributive property to expand the product:
$$\begin{aligned} (x - 2)(x^2 + 2x + 4) &= x^3 + 2x^2 + 4x \\ &\quad\ \ - 2x^2 - 4x - 8 \\ &= x^3 \qquad\qquad\qquad - 8 \end{aligned}$$

2. The first term is the cube of x and the second term is the cube of 2. Therefore,
$$\begin{aligned} x^3 + 8 = x^3 + 2^3 &= (x + 2)(x^2 - 2x + 2^2) \\ &= (x + 2)(x^2 - 2x + 4) \end{aligned}$$

3. The first term is the cube of y and the second term is the cube of 3. Therefore,
$$\begin{aligned} y^3 - 27 = y^3 - 3^3 &= (y - 3)(y^2 + 3y + 3^2) \\ &= (y - 3)(y^2 + 3y + 9) \end{aligned}$$

4. The first term is the cube of 3 and the second term is the cube of x. Therefore,
$$\begin{aligned} 27 + x^3 = 3^3 + x^3 &= (3 + x)(3^2 - 3x + x^2) \\ &= (3 + x)(9 - 3x + x^2) \end{aligned}$$

5. $$\begin{aligned} 8x^3 + y^3 &= (2x)^3 + y^3 \\ &= (2x + y)[(2x)^2 - (2x)(y) + y^2] \\ &= (2x + y)[(4x^2 - 2xy + y^2) \end{aligned}$$

6. $$\begin{aligned} a^3 - \frac{1}{27} &= a^3 - \left(\frac{1}{3}\right)^3 = \left(a - \frac{1}{3}\right)\left[a^2 + a\left(\frac{1}{3}\right) + \left(\frac{1}{3}\right)^2\right] \\ &= \left(a - \frac{1}{3}\right)\left(a^2 + \frac{1}{3}a + \frac{1}{9}\right) \end{aligned}$$

7. $$\begin{aligned} x^6 - 1 = (x^3)^2 - 1^2 &= (x^3 + 1)(x^3 - 1) \\ &= (x + 1)(x^2 - x + 1)(x - 1)(x^2 + x + 1) \end{aligned}$$

Section 6.6

1. $$\begin{aligned} 3x^4 - 27x^2 &= 3x^2(x^2 - 9) \\ &= 3x^2(x + 3)(x - 3) \end{aligned}$$

2. $$\begin{aligned} 2x^5 + 20x^4 + 50x^3 &= 2x^3(x^2 + 10x + 25) \\ &= 2x^3(x + 5)^2 \end{aligned}$$

3. $y^5 + 36y^3 = y^3(y^2 + 36)$

4. $15a^2 - a - 2 = (5a - 2)(3a + 1)$

5. $$\begin{aligned} 4x^4 - 12x^3 - 40x^2 &= 4x^2(x^2 - 3x - 10) \\ &= 4x^2(x - 5)(x + 2) \end{aligned}$$

6. $2a^5b + 6a^4b + 2a^3b = 2a^3b(a^2 + 3a + 1)$

7. $$\begin{aligned} 3ab + 9a + 2b + 6 &= 3a(b + 3) + 2(b + 3) \\ &= (b + 3)(3a + 2) \end{aligned}$$

Section 6.7

1. $$\begin{aligned} 2x^2 + 5x &= 3 \\ 2x^2 + 5x - 3 &= 0 \\ (2x - 1)(x + 3) &= 0 \\ 2x - 1 = 0 \quad &\text{or} \quad x + 3 = 0 \\ 2x = 1 \quad &\text{or} \quad x = -3 \\ x = \frac{1}{2} \end{aligned}$$

2. $$\begin{aligned} 49a^2 - 16 &= 0 \\ (7a + 4)(7a - 4) &= 0 \\ 7a + 4 = 0 \quad &\text{or} \quad 7a - 4 = 0 \\ 7a = -4 \quad &\text{or} \quad 7a = 4 \\ a = -\frac{4}{7} \quad &\text{or} \quad a = \frac{4}{7} \end{aligned}$$

3. $$\begin{aligned} 5x^2 &= -2x \\ 5x^2 + 2x &= 0 \\ x(5x + 2) &= 0 \\ x = 0 \quad &\text{or} \quad 5x + 2 = 0 \\ & x = -\frac{2}{5} \end{aligned}$$

4. $$\begin{aligned} x(13 - x) &= 40 \\ 13x - x^2 &= 40 \\ 0 &= x^2 - 13x + 40 \\ 0 &= (x - 5)(x - 8) \\ x - 5 = 0 \quad &\text{or} \quad x - 8 = 0 \\ x = 5 \quad &\text{or} \quad x = 8 \end{aligned}$$

5. $$\begin{aligned} (x + 2)^2 &= (x + 1)^2 + x^2 \\ x^2 + 4x + 4 &= x^2 + 2x + 1 + x^2 \\ x^2 - 2x - 3 &= 0 \\ (x - 3)(x + 1) &= 0 \\ x - 3 = 0 \quad &\text{or} \quad x + 1 = 0 \\ x = 3 \quad &\text{or} \quad x = -1 \end{aligned}$$

6. $$\begin{aligned} 8x^3 &= 2x^2 + 10x \\ 8x^3 - 2x^2 - 10x &= 0 \\ 2x(4x^2 - x - 5) &= 0 \\ 2x(4x - 5)(x + 1) &= 0 \\ 2x = 0 \quad \text{or} \quad 4x - 5 &= 0 \quad \text{or} \quad x + 1 = 0 \\ x = 0 \qquad x &= \frac{5}{4} \qquad x = -1 \end{aligned}$$

Section 6.8

1. Let x and $x + 2$ be the two even integers.

$$\begin{aligned} x(x + 2) &= 48 \\ x^2 + 2x &= 48 \\ x^2 + 2x - 48 &= 0 \\ (x + 8)(x - 6) &= 0 \\ x + 8 = 0 \quad &\text{or} \quad x - 6 = 0 \\ x = -8 \quad & \quad x = 6 \end{aligned}$$

If $x = -8$, then $x + 2 = -6$.
If $x = 6$, then $x + 2 = 8$.

2. If x = one of the numbers, then $14 - x$ is the other.

$$\begin{aligned} x(14 - x) &= 48 \\ 14x - x^2 &= 48 \\ x^2 - 14x &= -48 \quad \text{Multiply each side by } -1 \\ x^2 - 14x + 48 &= 0 \\ (x - 6)(x - 8) &= 0 \\ x - 6 = 0 \quad &\text{or} \quad x - 8 = 0 \\ x = 6 \quad &\text{or} \quad x = 8 \end{aligned}$$

If $x = 6$, then $14 - x = 8$.
If $x = 8$, then $14 - x = 6$.

3. Let x = the width.

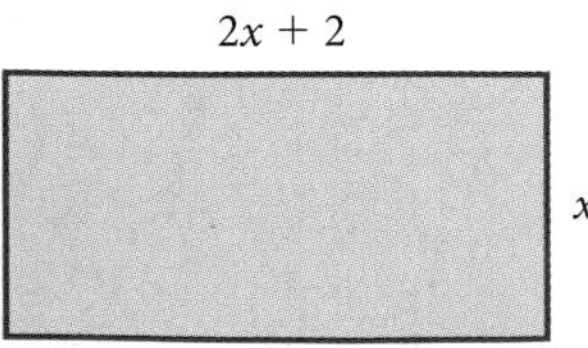

$$\begin{aligned} x(2x + 2) &= 60 \\ 2x^2 + 2x &= 60 \\ 2x^2 + 2x - 60 &= 0 \\ x^2 + x - 30 &= 0 \quad \text{Divide both sides by 2} \\ (x + 6)(x - 5) &= 0 \\ x + 6 = 0 \quad &\text{or} \quad x - 5 = 0 \\ x = -6 \quad &\text{or} \quad x = 5 \end{aligned}$$

Since the width cannot be negative, it is 5. The length is $2(5) + 2 = 12$.

4. Let x = the base.

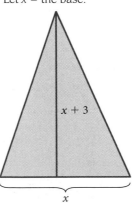

$x + 3$

x

$$\begin{aligned} \frac{1}{2}x(x + 3) &= 20 \\ x(x + 3) &= 40 \quad \text{Multiply each side by 2.} \\ x^2 + 3x &= 40 \\ x^2 + 3x - 40 &= 0 \\ (x + 8)(x - 5) &= 0 \\ x + 8 = 0 \quad &\text{or} \quad x - 5 = 0 \\ x = -8 \quad &\text{or} \quad x = 5 \end{aligned}$$

The base is 5 and the height is $5 + 3 = 8$.

5. Let x = the length of the shorter leg.

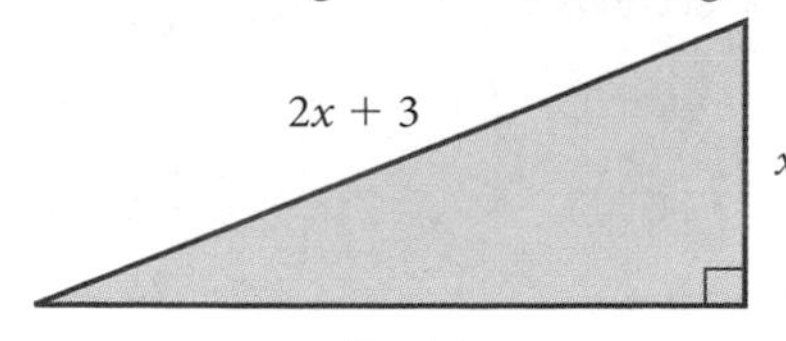

$2x + 2$

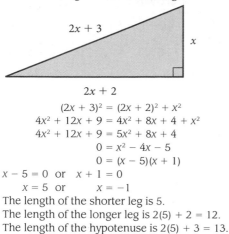

$$\begin{aligned} (2x+3)^2 &= (2x+2)^2 + x^2 \\ 4x^2 + 12x + 9 &= 4x^2 + 8x + 4 + x^2 \\ 4x^2 + 12x + 9 &= 5x^2 + 8x + 4 \\ 0 &= x^2 - 4x - 5 \\ 0 &= (x-5)(x+1) \end{aligned}$$

$$x - 5 = 0 \quad \text{or} \quad x + 1 = 0$$
$$x = 5 \quad \text{or} \quad x = -1$$

The length of the shorter leg is 5.
The length of the longer leg is $2(5) + 2 = 12$.
The length of the hypotenuse is $2(5) + 3 = 13$.

6. Let x = the shorter leg.
Let $x + 2$ = the longer leg.

$$\begin{aligned} x^2 + (x+2)^2 &= 10^2 \\ x^2 + x^2 + 4x + 4 &= 100 \\ 2x^2 + 4x - 96 &= 0 \\ 2(x^2 + 2x - 48) &= 0 \\ 2(x+8)(x-6) &= 0 \\ x = 6 \quad \text{or} \quad x &= -8 \end{aligned}$$

The length of the shorter side is 6 inches.
The length of the longer side is 8 inches.

7. We are looking for x when C is 700. We begin by substituting 700 for C in the cost equation. Then we solve for x:

When	$C = 700$
the equation	$C = 300 + 500x - 100x^2$
becomes	$700 = 300 + 500x - 100x^2$

We write the equation in standard form by adding -300, $-500x$, and $100x^2$ to each side. The result of doing so looks like this:

$$\begin{aligned} 100x^2 - 500x + 400 &= 0 \\ 100(x^2 - 5x + 4) &= 0 \\ 100(x-4)(x-1) &= 0 \\ x - 4 = 0 \quad \text{or} \quad x - 1 &= 0 \\ x = 4 \quad \text{or} \quad x &= 1 \end{aligned}$$

Our solutions are 1 and 4, which means a company can manufacture one hundred items or four hundred items for a total cost of \$700.

8. First, we must find the revenue equation. The equation for total revenue is $R = xp$, where x is the number of units sold and p is the price per unit. Since we want R in terms of p, we substitute $1{,}300 - 100p$ for x in the equation $R = xp$:

If	$R = xp$
and	$x = 1{,}300 - 100p$
then	$R = (1{,}300 - 100p)p$

We want to find p when R is 4,200. Substituting 4,200 for R in the equation gives us

$$4{,}200 = (1{,}300 - 100p)p$$

If we multiply out the right side, we have

$$4{,}200 = 1{,}300p - 100p^2$$

To write this equation in standard form, we add $100p^2$ and $-1{,}300p$ to each side.

$$\begin{aligned} 100p^2 - 1{,}300p + 4{,}200 &= 0 \\ 100(p^2 - 13p + 42) &= 0 \\ 100(p-7)(p-6) &= 0 \\ p - 7 = 0 \quad \text{or} \quad p - 6 &= 0 \\ p = 7 \quad \text{or} \quad p &= 6 \end{aligned}$$

If she sells the radios for \$6 each or \$7 each, she will have a weekly revenue of \$4,200.

Chapter 7

Section 7.1

1. The variable x can be any real number except $x = 4$, since, when $x = 4$, the denominator is $4 - 4 = 0$. The restriction is $x \neq 4$.

2. $\dfrac{6}{x^2 + x - 6} = \dfrac{6}{(x+3)(x-2)}$

If either of the factors is zero, the whole denominator is zero. The restrictions are $x \neq -3$ and $x \neq 2$, since either one makes $x^2 + x - 6 = 0$.

3. $\dfrac{x^2 - 4}{x^2 - 2x - 8} = \dfrac{\cancel{(x+2)}(x-2)}{(x-4)\cancel{(x+2)}} = \dfrac{x-2}{x-4}$

4. $\dfrac{5a + 15}{10a^2 - 90} = \dfrac{5(a+3)}{10(a^2 - 9)} = \dfrac{\overset{1}{\cancel{5}}(a+3)}{\underset{2}{\cancel{10}}(a+3)(a-3)} = \dfrac{1}{2(a-3)}$

5. $\dfrac{x^3 - x^2 - 2x}{x^3 + 4x^2 + 3x} = \dfrac{x(x^2 - x - 2)}{x(x^2 + 4x + 3)} = \dfrac{\cancel{x}(x-2)\cancel{(x+1)}}{\cancel{x}(x+3)\cancel{(x+1)}} = \dfrac{x-2}{x+3}$

6. $\dfrac{x+3}{x^2 - 9} = \dfrac{\cancel{x+3}}{\cancel{(x+3)}(x-3)} = \dfrac{1}{x-3}$

7. $\dfrac{x^3 - 8}{x^2 - 4} = \dfrac{\cancel{(x-2)}(x^2 + 2x + 4)}{\cancel{(x-2)}(x+2)} = \dfrac{x^2 + 2x + 4}{x+2}$

8. $\dfrac{27}{54} = \dfrac{1}{2}$; $27 + 54 = 81$, so $\dfrac{27}{81} = \dfrac{1}{3}$

9. Rate $= \dfrac{\text{distance}}{\text{time}} = \dfrac{772 \text{ feet}}{2.2 \text{ minutes}} = 351$ feet per minute to the nearest foot

10. $r = \dfrac{d}{t} = \dfrac{840 \text{ feet}}{80 \text{ seconds}} = 10.5$ feet per second

Section 7.2

1.
$$\frac{x-5}{x+2}\cdot\frac{x^2-4}{3x-15} = \frac{x-5}{x+2}\cdot\frac{(x+2)(x-2)}{3(x-5)}$$
$$= \frac{\cancel{(x-5)}\cancel{(x+2)}(x-2)}{\cancel{(x+2)}(3)\cancel{(x-5)}}$$
$$= \frac{x-2}{3}$$

2.
$$\frac{4x+8}{x^2-x-6} \div \frac{x^2+7x+12}{x^2-9}$$
$$= \frac{4x+8}{x^2-x-6}\cdot\frac{x^2-9}{x^2+7x+12}$$
$$= \frac{4\cancel{(x+2)}\cancel{(x+3)}\cancel{(x-3)}}{\cancel{(x-3)}\cancel{(x+2)}\cancel{(x+3)}(x+4)}$$
$$= \frac{4}{x+4}$$

3.
$$\frac{2a+10}{a^3}\cdot\frac{a^2}{3a+15} = \frac{2\cancel{(a+5)}\cancel{a^2}}{\cancel{a^3}(3)\cancel{(a+5)}}$$
$$= \frac{2}{3a}$$

4.
$$\frac{x^2-x-12}{x^2-16} \div \frac{x^2+6x+9}{2x+8}$$
$$= \frac{x^2-x-12}{x^2-16}\cdot\frac{2x+8}{x^2+6x+9}$$
$$= \frac{\cancel{(x-4)}\cancel{(x+3)}(2)\cancel{(x+4)}}{\cancel{(x-4)}\cancel{(x+4)}\cancel{(x+3)}(x+3)}$$
$$= \frac{2}{x+3}$$

5.
$$(x^2-9)\left(\frac{x+2}{x+3}\right) = \frac{x^2-9}{1}\cdot\frac{x+2}{x+3}$$
$$= \frac{\cancel{(x+3)}(x-3)(x+2)}{1\cancel{(x+3)}}$$
$$= (x-3)(x+2)$$

6.
$$a(a+2)(a-4)\left(\frac{a+5}{a^2-4a}\right)$$
$$= \frac{a(a+2)(a-4)}{1}\cdot\left(\frac{a+5}{a^2-4a}\right)$$
$$= \frac{\cancel{a}(a+2)\cancel{(a-4)}(a+5)}{1\cancel{(a)}\cancel{(a-4)}}$$
$$= (a+2)(a+5)$$

7. Acres $= \dfrac{2{,}224{,}750 \text{ feet}^2}{1}\cdot\dfrac{1 \text{ acre}}{43{,}560 \text{ feet}^2} = 51.1$ acres

8.
$$\frac{772 \text{ feet}}{2.2 \text{ minutes}}\cdot\frac{1 \text{ mile}}{5{,}280 \text{ feet}}\cdot\frac{60 \text{ minutes}}{1 \text{ hour}} = \frac{772(60)}{2.2(5{,}280)} \text{ miles per hour}$$
$$= 4.0 \text{ miles per hour}$$

Section 7.3

1.
$$\frac{2}{x}+\frac{7}{x} = \frac{2+7}{x} = \frac{9}{x}$$

2.
$$\frac{x}{x^2-49}+\frac{7}{x^2-49} = \frac{x+7}{x^2-49}$$
$$= \frac{\cancel{x+7}}{\cancel{(x+7)}(x-7)}$$
$$= \frac{1}{x-7}$$

3. To find the LCD for 21 and 15, we factor each and build the LCD from the factors.

$$\left.\begin{aligned}21 &= 3\cdot 7\\ 15 &= 3\cdot 5\end{aligned}\right\}\text{LCD} = 3\cdot 5\cdot 7 = 105$$
$$\frac{1}{21} = \frac{1}{3\cdot 7} = \frac{1}{3\cdot 7}\cdot\frac{\mathbf{5}}{\mathbf{5}} = \frac{5}{105}$$
$$\frac{4}{15} = \frac{4}{3\cdot 5} = \frac{4}{3\cdot 5}\cdot\frac{\mathbf{7}}{\mathbf{7}} = \frac{28}{105}$$
$$\frac{5}{105}+\frac{28}{105} = \frac{33}{105} = \frac{11}{35}$$

4.
$$\frac{2}{x}-\frac{1}{3} = \frac{2}{x}\cdot\frac{\mathbf{3}}{\mathbf{3}}-\frac{1}{3}\cdot\frac{\mathbf{x}}{\mathbf{x}}$$
$$= \frac{6-x}{3x}$$

5.
$$\frac{7}{3x+6}+\frac{x}{x+2} = \frac{7}{3(x+2)}+\frac{x}{x+2}$$
$$= \frac{7}{3(x+2)}+\frac{\mathbf{3}}{\mathbf{3}}\cdot\frac{x}{x+2}$$
$$= \frac{7+3x}{3(x+2)}$$

6.
$$\frac{1}{x-3}+\frac{-6}{x^2-9} = \frac{1}{x-3}+\frac{-6}{(x-3)(x+3)}$$
$$= \frac{1}{x-3}\cdot\frac{\mathbf{x+3}}{\mathbf{x+3}}+\frac{-6}{(x-3)(x+3)}$$
$$= \frac{x+3+(-6)}{(x-3)(x+3)}$$
$$= \frac{\cancel{x-3}}{\cancel{(x-3)}(x+3)}$$
$$= \frac{1}{x+3}$$

7.
$$\frac{5}{x^2-7x+12}+\frac{1}{x^2-9}$$
$$= \frac{5}{(x-4)(x-3)}+\frac{1}{(x-3)(x+3)}$$
$$= \frac{5}{(x-4)(x-3)}\cdot\frac{\mathbf{x+3}}{\mathbf{x+3}}+\frac{1}{(x-3)(x+3)}\cdot\frac{\mathbf{x-4}}{\mathbf{x-4}}$$
$$= \frac{5(x+3)+(x-4)}{(x-4)(x-3)(x+3)}$$
$$= \frac{6x+11}{(x-4)(x-3)(x+3)}$$

8.
$$\frac{x+3}{2x+8}-\frac{4}{x^2-16} = \frac{x+3}{2(x+4)}+\frac{-4}{(x-4)(x+4)}$$
$$= \frac{\mathbf{(x-4)}}{\mathbf{(x-4)}}\cdot\frac{x+3}{2(x+4)}+\frac{-4}{(x-4)(x+4)}\cdot\frac{\mathbf{2}}{\mathbf{2}}$$
$$= \frac{x^2-x-12}{2(x-4)(x+4)}+\frac{-8}{2(x-4)(x+4)}$$
$$= \frac{x^2-x-20}{2(x-4)(x+4)}$$
$$= \frac{(x-5)(x+4)}{2(x-4)(x+4)}$$
$$= \frac{x-5}{2(x-4)}$$

9.
$$2x-\frac{1}{x} = \frac{2x}{1}\cdot\frac{x}{x}-\frac{1}{x}$$
$$= \frac{2x^2-1}{x}$$

Section 7.4

1.
$$\frac{x}{2} + \frac{3}{5} = \frac{1}{5} \quad \text{LCD} = 10$$
$$10\left(\frac{x}{2} + \frac{3}{5}\right) = 10\left(\frac{1}{5}\right)$$
$$\overset{5}{\cancel{10}}\left(\frac{x}{\cancel{2}}\right) + \overset{2}{\cancel{10}}\left(\frac{3}{\cancel{5}}\right) = \overset{2}{\cancel{10}}\left(\frac{1}{\cancel{5}}\right)$$
$$5x + 6 = 2$$
$$5x = -4$$
$$x = -\frac{4}{5}$$

2.
$$\frac{4}{x-5} = \frac{4}{3} \quad \text{LCD} = 3(x-5)$$
$$\frac{3\cancel{(x-5)} \cdot 4}{\cancel{x-5}} = \cancel{3}(x-5) \cdot \frac{4}{\cancel{3}}$$
$$12 = (x-5)4$$
$$12 = 4x - 20$$
$$32 = 4x$$
$$8 = x$$

3.
$$1 - \frac{3}{x} = \frac{-2}{x^2} \quad \text{LCD} = x^2$$
$$x^2\left(1 - \frac{3}{x}\right) = x^2\left(\frac{-2}{x^2}\right)$$
$$x^2(1) - x^2\left(\frac{3}{x}\right) = x^2\left(\frac{-2}{x^2}\right)$$
$$x^2 - 3x = -2$$
$$x^2 - 3x + 2 = 0$$
$$(x-2)(x-1) = 0$$
$$x - 2 = 0 \quad \text{or} \quad x - 1 = 0$$
$$x = 2 \quad \text{or} \quad x = 1$$

4.
$$\frac{x}{x^2-4} - \frac{2}{x-2} = \frac{1}{x+2} \quad \text{LCD} = (x+2)(x-2)$$
$$\cancel{(x+2)}\cancel{(x-2)} \cdot \frac{x}{\cancel{(x+2)}\cancel{(x-2)}} - (x+2)\cancel{(x-2)} \cdot \frac{2}{\cancel{x-2}} = \cancel{(x+2)}(x-2) \cdot \frac{1}{\cancel{x+2}}$$
$$x - (x+2)2 = (x-2)1$$
$$x - 2x - 4 = x - 2$$
$$-x - 4 = x - 2$$
$$-2x - 4 = -2$$
$$-2x = 2$$
$$x = -1$$

5.
$$\frac{x}{x-5} + \frac{5}{2} = \frac{5}{x-5} \quad \text{LCD} = 2(x-5)$$
$$2\cancel{(x-5)} \cdot \frac{x}{\cancel{x-5}} + \cancel{2}(x-5) \cdot \frac{5}{\cancel{2}} = 2\cancel{(x-5)} \cdot \frac{5}{\cancel{x-5}}$$
$$2x + (x-5)5 = 2 \cdot 5$$
$$2x + 5x - 25 = 10$$
$$7x - 25 = 10$$
$$7x = 35$$
$$x = 5$$

When $x = 5$, the original equation contains undefined terms. Therefore, there is no solution.

6.
$$\frac{a+2}{a^2+3a} = \frac{-2}{a^2-9}$$
$$\frac{a+2}{a(a+3)} = \frac{-2}{(a+3)(a-3)} \quad \text{LCD} = a(a+3)(a-3)$$
$$\cancel{a}\cancel{(a+3)}(a-3) \cdot \frac{a+2}{\cancel{a}\cancel{(a+3)}} = \frac{-2}{\cancel{(a+3)}\cancel{(a-3)}} \cdot a\cancel{(a+3)}\cancel{(a-3)}$$
$$(a-3)(a+2) = -2a$$
$$a^2 - a - 6 = -2a$$
$$a^2 + a - 6 = 0$$
$$(a+3)(a-2) = 0$$
$$a + 3 = 0 \quad \text{or} \quad a - 2 = 0$$
$$a = -3 \quad \text{or} \quad a = 2$$

The two possible solutions are -3 and 2, but the only one that checks when used in place of a in the original equation is 2.

Section 7.5

1. Let x = one of the numbers; then the other is $3x$.
$$\frac{1}{x} + \frac{1}{3x} = \frac{4}{9} \quad \text{LCD} = 9x$$
$$9x \cdot \frac{1}{x} + 9x \cdot \frac{1}{3x} = 9x \cdot \frac{4}{9}$$
$$9 + 3 = 4x$$
$$12 = 4x$$

One number is 3 and the other is 9.

2. Let x = the speed of the boat in still water.

	d	r	t
Upstream	26	$x-2$	$\frac{26}{x-2}$
Downstream	34	$x+2$	$\frac{34}{x+2}$

$$\frac{26}{x-2} = \frac{34}{x+2} \quad \text{LCD} = (x-2)(x+2)$$
$$\cancel{(x-2)}(x+2) \cdot \frac{26}{\cancel{x-2}} = \frac{34}{\cancel{x+2}} \cdot (x-2)\cancel{(x+2)}$$
$$(x+2)26 = 34(x-2)$$
$$26x + 52 = 34x - 68$$
$$-8x + 52 = -68$$
$$-8x = -120$$
$$x = 15 \text{ miles per hour}$$

3. We make a table and label the top row with distance, rate, and time. We label the left column with the two trips, uphill and downhill.

	d	r	t
Uphill			
Downhill			

Now, we fill in the table with the information from the problem. Both the uphill and downhill distances are 15 miles. Since we don't know her uphill rate, we will call it x; her downhill rate is $x + 8$. And, finally, since $t = \frac{d}{r}$, we can fill in the last column of the table.

	d	r	t
Uphill	15	x	$\frac{15}{x}$
Downhill	15	$x + 8$	$\frac{15}{x + 8}$

From the problem, her total time is 2 hours. We write an equation as follows:

$$\frac{15}{x} + \frac{15}{x + 8} = 2$$

To solve this problem, we multiply each term by the LCD, which is $x(x + 8)$. Here is our solution:

$$\begin{aligned} \frac{15}{x} + \frac{15}{x + 8} &= 2 \\ x(x + 8)\frac{15}{x} + x(x + 8)\frac{15}{x + 8} &= x(x + 8)2 \\ 15(x + 8) + 15x &= 2x(x + 8) \\ 15x + 120 + 15x &= 2x^2 + 16x \\ 0 &= 2x^2 - 14x - 120 \\ 0 &= x^2 - 7x - 60 \\ 0 &= (x - 12)(x + 5) \\ x = -5 \quad \text{or} \quad x &= 12 \end{aligned}$$

Since we cannot have a negative speed, our only solution is $x = 12$. Tina rides her bike at a rate of 12 miles per hour uphill. Her downhill rate is $x + 8 = 20$ miles per hour.

4. Let x = amount of time to fill tub with faucet and drain opened. In 1 minute, the faucet fills $\frac{1}{8}$ of the tub, the drain empties $\frac{1}{12}$ of the tub, and the tub is $\frac{1}{x}$ full. Thus, in 1 minute,

$\frac{1}{8}$	$-$	$\frac{1}{12}$	$=$	$\frac{1}{x}$
Amount of water let in by faucet	$-$	Amount of water let out by drain	$=$	Total amount of water into tub

Multiply both sides by the LCD, which is 24x, and solve.

$$\begin{aligned} 24x\left(\frac{1}{8}\right) - 24x\left(\frac{1}{12}\right) &= 24x\left(\frac{1}{x}\right) \\ 3x - 2x &= 24 \\ x &= 24 \end{aligned}$$

The tub will fill in 24 minutes.

5. $y = \frac{2}{x}$

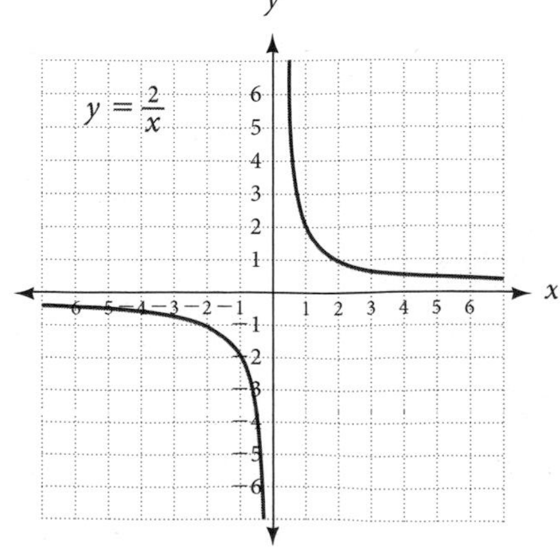

6. $y = -\frac{2}{x}$

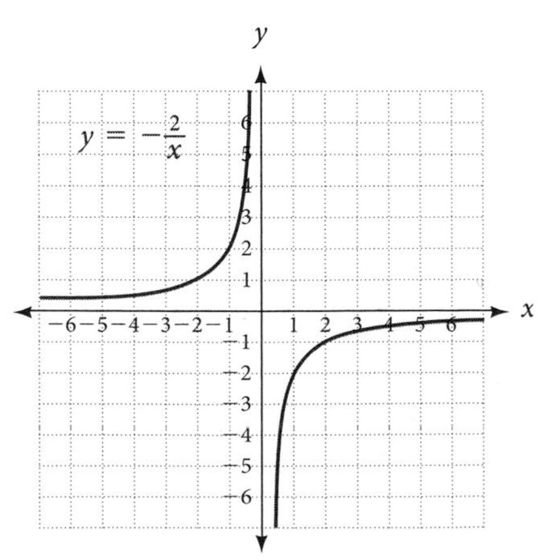

Section 7.6

1. $\dfrac{\frac{1}{3}}{\frac{3}{4}} = \dfrac{1}{3} \cdot \dfrac{4}{3} = \dfrac{4}{9}$

2. $\dfrac{\frac{3x^5}{y^4}}{\frac{6x^2}{y^7}} = \dfrac{3x^5}{y^4} \cdot \dfrac{y^7}{6x^2}$

$= \dfrac{x^3y^3}{2}$

3. $\dfrac{y - \frac{1}{x}}{x - \frac{1}{y}} = \dfrac{\left(y - \frac{1}{x}\right) \cdot xy}{\left(x - \frac{1}{y}\right) \cdot xy}$

$= \dfrac{y \cdot xy - \frac{1}{x} \cdot xy}{x \cdot xy - \frac{1}{y} \cdot xy}$

$= \dfrac{xy^2 - y}{x^2y - x} = \dfrac{y(xy - 1)}{x(xy - 1)} = \dfrac{y}{x}$

4. $\dfrac{1 - \frac{9}{x^2}}{1 - \frac{5}{x} + \frac{6}{x^2}} = \dfrac{\left(1 - \frac{9}{x^2}\right)x^2}{\left(1 - \frac{5}{x} + \frac{6}{x^2}\right)x^2}$

$= \dfrac{1 \cdot x^2 - \frac{9}{x^2} \cdot x^2}{1 \cdot x^2 - \frac{5}{x} \cdot x^2 + \frac{6}{x^2} \cdot x^2}$

$= \dfrac{x^2 - 9}{x^2 - 5x + 6}$

$= \dfrac{(x + 3)(x - 3)}{(x - 3)(x - 2)}$

$= \dfrac{x + 3}{x - 2}$

5. First, we simplify each term.

First term: $1 + \dfrac{2}{1 + 2} = \dfrac{3}{3} + \dfrac{2}{3} = \dfrac{5}{3}$

Second term: $1 + \dfrac{2}{1 + \frac{2}{1 + 2}} = 1 + \dfrac{2}{\frac{5}{3}} = 1 + \dfrac{6}{5} = \dfrac{11}{5}$

Third term: $1 + \dfrac{2}{1 + \frac{2}{1 + \frac{2}{1 + 2}}} = 1 + \dfrac{10}{11} = \dfrac{21}{11}$

Section 7.7

1. $\dfrac{2}{9} = \dfrac{8}{x}$

$2x = 9 \cdot 8$

$2x = 72$

$x = 36$

2. $\dfrac{x + 2}{4} = \dfrac{2}{x}$

$x(x + 2) = 4 \cdot 2$

$x^2 + 2x = 8$

$x^2 + 2x - 8 = 0$

$(x + 4)(x - 2) = 0$

$x + 4 = 0$ or $x - 2 = 0$

$x = -4$ or $x = 2$

3. $\dfrac{x}{1{,}650} = \dfrac{8}{100}$

$100x = 13{,}200$

$x = 132$

4. $\dfrac{x}{2} = \dfrac{85}{1}$

$1 \cdot x = 85(2)$

$x = 170$ miles

5. $\dfrac{x}{9} = \dfrac{250}{5}$

$5x = (9)250$

$5x = 2250$

$x = 450$ miles

6. $\dfrac{x}{14} = \dfrac{25}{10}$

$10x = 350$

$x = 35$

Section 7.8

1. $y = kx \longrightarrow y = 3x$

$12 = k \cdot 4 \qquad y = 3(9) = 27$

$k = 3$

2. $y = kx^2 \longrightarrow y = 5x^2$

$45 = k(3)^2 \qquad 80 = 5x^2$

$45 = k \cdot 9 \qquad 16 = x^2$

$k = 5 \qquad x = \pm 4$

3. Let y = the amount he earns and x = the hours he works.

$y = kx \longrightarrow y = 4.1x$

$12.30 = k \cdot 3 \qquad y = 4.1(8)$

$k = 4.1 \qquad y = \$32.80$

4. $y = \dfrac{k}{x} \longrightarrow y = \dfrac{75}{x}$

$15 = \dfrac{k}{5} \qquad y = \dfrac{75}{9}$

$k = 75 \qquad y = \dfrac{25}{3}$

5. $I = \dfrac{20{,}000}{d^2}$

$I = \dfrac{20{,}000}{20^2}$

$I = \dfrac{20{,}000}{400}$

$I = 50$ candlepower

Chapter 8

Section 8.1

1. The positive square root of 36 is 6 and can be written $\sqrt{36} = 6$. The negative square root of 36 is -6 and can be written $-\sqrt{36} = -6$.

2. $\sqrt{81} = 9$ because 9 is the positive number we square to get 81.

3. $-\sqrt{81} = -9$ because -9 is the negative number we square to get 81.

4. $\sqrt{144} = 12$ because 12 is the positive number we square to get 144.

5. The positive square root of 23 is written $\sqrt{23}$. The negative square root of 23 is written $-\sqrt{23}$.

6.

Number	Positive Square Root	Negative Square Root
16	4	-4
4	2	-2
11	$\sqrt{11}$	$-\sqrt{11}$
23	$\sqrt{23}$	$-\sqrt{23}$
400	20	-20

7. No, the expression $\sqrt{-25}$ does not represent a real number since there is no real number whose square is -25.

8. $\sqrt[3]{-1} = -1$ because $(-1)^3 = -1$.

9. $\sqrt[3]{-125} = -5$ because $(-5)^3 = -125$.

10. $\sqrt{-9}$, not a real number since there is no real number whose square is -9.

11. $-\sqrt{9} = -3$ because -3 is the negative number whose square is 9.

12. $\sqrt[4]{-81}$, not a real number since there is no real number that can be raised to the fourth power to obtain -81.

13. $-\sqrt[4]{81} = -3$ because -3 is the negative number we raise to the fourth power to get 81.

14. Since the square of 3 is 9 and the square of a is a^2, we can square $3a$ and get $9a^2$. Therefore,
$$\sqrt{9a^2} = 3a \qquad \text{because } (3a)^2 = 9a^2$$

15. Since the square of 5 is 25 and the square of xy is x^2y^2, we can square $5xy$ and get $25x^2y^2$. Therefore,
$$\sqrt{25x^2y^2} = 5xy \qquad \text{because } (5xy)^2 = 25x^2y^2$$

16. Since the cube of 2 is 8 and the cube of x is x^3, we can cube $2x$ and get $8x^3$. Therefore,
$$\sqrt[3]{8x^3} = 2x \qquad \text{because } (2x)^3 = 8x^3$$

17. The number we square to obtain x^8 is x^4.
$$\sqrt{x^8} = x^4 \qquad \text{because } (x^4)^2 = x^8$$

18. The number we raise to the fourth power to obtain $81a^8b^{12}$ is $3a^2b^3$.
$$\sqrt[4]{81a^8b^{12}} = 3a^2b^3 \qquad \text{because } (3a^2b^3)^4 = 81a^8b^{12}$$

19.
$$\begin{aligned} r &= \frac{\sqrt{91} - \sqrt{56}}{\sqrt{56}} \\ &\approx \frac{9.5394 - 7.4833}{7.4833} \\ &= \frac{2.0561}{7.4833} \\ &\approx 0.2748 \text{ or } 27.5\% \end{aligned}$$

20.
$$\begin{aligned} x &= \sqrt{(10)^2 + (24)^2} \\ &= \sqrt{100 + 576} \\ &= \sqrt{676} \\ &= 26 \end{aligned}$$

Section 8.2

1. $\sqrt{12} = \sqrt{4 \cdot 3} = \sqrt{4}\,\sqrt{3} = 2\sqrt{3}$

2. $\sqrt{50} = \sqrt{25 \cdot 2} = \sqrt{25}\,\sqrt{2} = 5\sqrt{2}$

3. $\sqrt{36x^3} = \sqrt{36x^2 \cdot x} = \sqrt{36x^2}\sqrt{x} = 6x\sqrt{x}$

4. $\sqrt{50y^4} = \sqrt{25y^4 \cdot 2} = \sqrt{25y^4}\,\sqrt{2} = 5y^2\sqrt{2}$

5. $6\sqrt{20} = 6\sqrt{4 \cdot 5} = 6\sqrt{4}\,\sqrt{5} = 6 \cdot 2\sqrt{5} = 12\sqrt{5}$

6. $\sqrt[3]{40x^3} = \sqrt[3]{8x^3 \cdot 5} = \sqrt[3]{8x^3} \cdot \sqrt[3]{5} = 2x\,\sqrt[3]{5}$

7. $\sqrt{\dfrac{16}{25}} = \dfrac{\sqrt{16}}{\sqrt{25}} = \dfrac{4}{5}$

8. $\sqrt[3]{\dfrac{125}{64}} = \dfrac{\sqrt[3]{125}}{\sqrt[3]{64}} = \dfrac{5}{4}$

9. $\sqrt{\dfrac{18}{25}} = \dfrac{\sqrt{18}}{\sqrt{25}} = \dfrac{\sqrt{9 \cdot 2}}{\sqrt{25}} = \dfrac{\sqrt{9}\,\sqrt{2}}{\sqrt{25}} = \dfrac{3\sqrt{2}}{5}$

10. $\sqrt{\dfrac{20x^2}{9}} = \dfrac{\sqrt{4x^2 \cdot 5}}{\sqrt{9}} = \dfrac{\sqrt{4x^2}\,\sqrt{5}}{\sqrt{9}} = \dfrac{2x\sqrt{5}}{3}$

11. $\sqrt{\dfrac{75x^2y^3}{36}} = \dfrac{\sqrt{35x^2y^2 \cdot 3y}}{\sqrt{36}} = \dfrac{\sqrt{25x^2y^2}\,\sqrt{3y}}{\sqrt{36}} = \dfrac{5xy\sqrt{3y}}{6}$

Section 8.3

1. $\sqrt{\frac{1}{3}} = \frac{\sqrt{1}}{\sqrt{3}} = \frac{1}{\sqrt{3}} \cdot \frac{\sqrt{3}}{\sqrt{3}} = \frac{\sqrt{3}}{3}$

2. $\sqrt{\frac{3}{5}} = \frac{\sqrt{3}}{\sqrt{5}} = \frac{\sqrt{3}}{\sqrt{5}} \cdot \frac{\sqrt{5}}{\sqrt{5}} = \frac{\sqrt{15}}{5}$

3. $\frac{8\sqrt{18}}{4\sqrt{2}} = \frac{8}{4}\sqrt{\frac{18}{2}} = 2\sqrt{9} = 2 \cdot 3 = 6$

4. $\sqrt{\frac{9x^2y^3}{2}} = \frac{\sqrt{9x^2y^2 \cdot y}}{\sqrt{2}}$
$= \frac{\sqrt{9x^2y^2}\sqrt{y}}{\sqrt{2}}$
$= \frac{3xy\sqrt{y}}{\sqrt{2}} \cdot \frac{\sqrt{2}}{\sqrt{2}}$
$= \frac{3xy\sqrt{2y}}{2}$

5. $\sqrt[3]{\frac{3}{2}} = \frac{\sqrt[3]{3}}{\sqrt[3]{2}}$
$= \frac{\sqrt[3]{3}}{\sqrt[3]{2}} \cdot \frac{\sqrt[3]{4}}{\sqrt[3]{4}}$
$= \frac{\sqrt[3]{12}}{\sqrt[3]{8}}$
$= \frac{\sqrt[3]{12}}{2}$

6. $\sqrt[3]{\frac{1}{9}} = \frac{\sqrt[3]{1}}{\sqrt[3]{9}}$
$= \frac{1}{\sqrt[3]{9}} \cdot \frac{\sqrt[3]{3}}{\sqrt[3]{3}}$
$= \frac{\sqrt[3]{3}}{\sqrt[3]{27}}$
$= \frac{\sqrt[3]{3}}{3}$

Section 8.4

1. $5\sqrt{2} - 8\sqrt{2} = (5 - 8)\sqrt{2}$
$= -3\sqrt{2}$

2. $9\sqrt{3} - 4\sqrt{3} + \sqrt{3} = (9 - 4 + 1)\sqrt{3}$
$= 6\sqrt{3}$

3. $3\sqrt{12} + 5\sqrt{48} = 3\sqrt{4}\sqrt{3} + 5\sqrt{16}\sqrt{3}$
$= 3 \cdot 2\sqrt{3} + 5 \cdot 4\sqrt{3}$
$= 6\sqrt{3} + 20\sqrt{3}$
$= (6 + 20)\sqrt{3}$
$= 26\sqrt{3}$

4. $4\sqrt{18} + \sqrt{32} - \sqrt{2}$
$= 4\sqrt{9}\sqrt{2} + \sqrt{16}\sqrt{2} - \sqrt{2}$
$= 4 \cdot 3\sqrt{2} + 4\sqrt{2} - \sqrt{2}$
$= 12\sqrt{2} + 4\sqrt{2} - \sqrt{2}$
$= (12 + 4 - 1)\sqrt{2}$
$= 15\sqrt{2}$

5. $a\sqrt{18} + 7\sqrt{2a^2}$
$= a\sqrt{9}\sqrt{2} + 7\sqrt{a^2}\sqrt{2}$
$= 3a\sqrt{2} + 7a\sqrt{2}$
$= 10a\sqrt{2}$

6. $\sqrt{12x^3} - 6x\sqrt{27x} + \sqrt{9x^2}$
$= \sqrt{4x^2}\sqrt{3x} - 6x\sqrt{9}\sqrt{3x} + 3x$
$= 2x\sqrt{3x} - 6x \cdot 3\sqrt{3x} + 3x$
$= 2x\sqrt{3x} - 18x\sqrt{3x} + 3x$
$= (2x - 18x)\sqrt{3x} + 3x$
$= -16x\sqrt{3x} + 3x$

7. $\frac{6 - \sqrt{8}}{2} = \frac{6 - 2\sqrt{2}}{2}$
$= \frac{\cancel{2}(3 - \sqrt{2})}{\cancel{2}}$
$= 3 - \sqrt{2}$

Section 8.5

1. $(4\sqrt{3})(7\sqrt{5}) = (4 \cdot 7)(\sqrt{3} \cdot \sqrt{5})$
$= 28\sqrt{15}$

2. $\sqrt{3}(\sqrt{2} + \sqrt{3}) = \sqrt{3}\sqrt{2} + \sqrt{3}\sqrt{3}$
$= \sqrt{6} + 3$

3. $5\sqrt{2}(2\sqrt{3} + 4\sqrt{5}) = 5\sqrt{2} \cdot 2\sqrt{3} + 5\sqrt{2} \cdot 4\sqrt{5}$
$= 10\sqrt{6} + 20\sqrt{10}$

4. $(\sqrt{3} + 5)(\sqrt{3} + 2) = \sqrt{3}\sqrt{3} + 2\sqrt{3} + 5\sqrt{3} + 10$
$= 3 + 7\sqrt{3} + 10$
$= 13 + 7\sqrt{3}$

5. $(\sqrt{x} + 8)(\sqrt{x} - 6) = \sqrt{x}\sqrt{x} - 6\sqrt{x} + 8\sqrt{x} - 48$
$= x + 2\sqrt{x} - 48$

6. $(\sqrt{5} - 4)^2 = (\sqrt{5} - 4)(\sqrt{5} - 4)$
$= \sqrt{5}\sqrt{5} - 4\sqrt{5} - 4\sqrt{5} + 16$
$= 5 - 8\sqrt{5} + 16$
$= 21 - 8\sqrt{5}$

7. $(\sqrt{7} - \sqrt{3})(\sqrt{7} + \sqrt{3}) = (\sqrt{7})^2 - (\sqrt{3})^2$
$= 7 - 3$
$= 4$

8. $(\sqrt{x} + \sqrt{y})(\sqrt{x} - \sqrt{y}) = (\sqrt{x})^2 - (\sqrt{y})^2$
$= x - y$

9. $\frac{\sqrt{5}}{\sqrt{5} - \sqrt{3}} = \frac{\sqrt{5}}{\sqrt{5} - \sqrt{3}} \cdot \frac{\sqrt{5} + \sqrt{3}}{\sqrt{5} + \sqrt{3}}$
$= \frac{\sqrt{5}\sqrt{5} + \sqrt{5}\sqrt{3}}{(\sqrt{5})^2 - (\sqrt{3})^2}$
$= \frac{5 + \sqrt{15}}{5 - 3}$
$= \frac{5 + \sqrt{15}}{2}$

10. $\frac{2}{3 - \sqrt{7}} = \frac{2}{3 - \sqrt{7}} \cdot \frac{3 + \sqrt{7}}{3 + \sqrt{7}}$
$= \frac{6 + 2\sqrt{7}}{9 - 7}$
$= \frac{6 + 2\sqrt{7}}{2}$
$= \frac{\cancel{2}(3 + \sqrt{7})}{\cancel{2}}$
$= 3 + \sqrt{7}$

11. $\frac{\sqrt{3} + \sqrt{2}}{\sqrt{3} - \sqrt{2}} = \frac{\sqrt{3} + \sqrt{2}}{\sqrt{3} - \sqrt{2}} \cdot \frac{\sqrt{3} + \sqrt{2}}{\sqrt{3} + \sqrt{2}}$
$= \frac{3 + \sqrt{6} + \sqrt{6} + 2}{(\sqrt{3})^2 - (\sqrt{2})^2}$
$= \frac{5 + 2\sqrt{6}}{3 - 2}$
$= 5 + 2\sqrt{6}$

Section 8.6

1. $\sqrt{x + 2} = 5$
$(\sqrt{x + 2})^2 = 5^2$
$x + 2 = 25$
$x = 23$
The solution checks.

2. $\sqrt{3x - 1} = -4$ has no solution because the left side is always positive or 0, but the right side is negative.

3. $\sqrt{2a - 3} + 4 = 9$
$\sqrt{2a - 3} = 5$
$(\sqrt{2a - 3})^2 = 5^2$
$2a - 3 = 25$
$2a = 28$
$a = 14$
The solution checks.

4. $\sqrt{x + 8} = x + 2$
$(\sqrt{x + 8})^2 = (x + 2)^2$
$x + 8 = x^2 + 4x + 4$
$0 = x^2 + 3x - 4$
$0 = (x + 4)(x - 1)$
$x + 4 = 0$ or $x - 1 = 0$
$x = -4$ or $x = 1$
Only $x = 1$ checks when substituted for x in the original equation.

5. $y = 3\sqrt{x}$

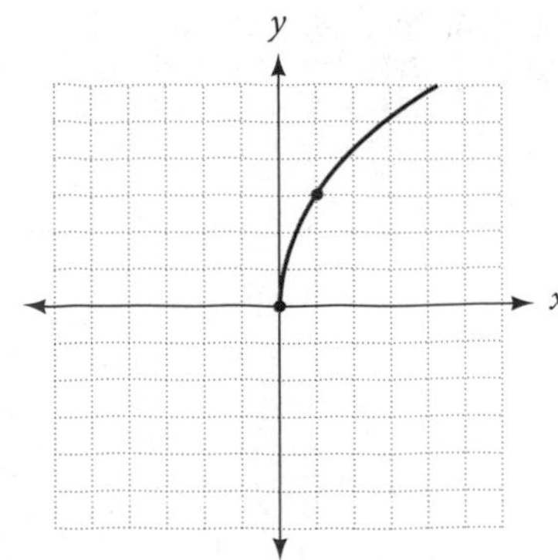

Chapter 9

Section 9.1

1. $x^2 = 5$
$x = \pm\sqrt{5}$

2. $5y^2 = 60$
$y^2 = 12$
$y = \pm\sqrt{12}$
$y = \pm 2\sqrt{3}$

3. $(a + 2)^2 = 25$
$a + 2 = \pm 5$
$a = -2 \pm 5$
$a = -2 + 5$ or $a = -2 - 5$
$= 3$ $\quad = -7$

4. $(2x - 5)^2 = 49$
$2x - 5 = \pm 7$
$2x = 5 \pm 7$
$x = \dfrac{5 \pm 7}{2}$
$x = \dfrac{5 + 7}{2}$ or $x = \dfrac{5 - 7}{2}$
$= 6$ $\quad = -1$

5. $(2x - 3)^2 = 7$
$2x - 3 = \pm\sqrt{7}$
$2x = 3 \pm \sqrt{7}$
$x = \dfrac{3 \pm \sqrt{7}}{2}$

6. $(2x - 6)^2 = 12$
$2x - 6 = \pm\sqrt{12}$
$2x - 6 = \pm 2\sqrt{3}$
$2x = 6 \pm 2\sqrt{3}$
$x = \dfrac{6 \pm 2\sqrt{3}}{2}$
$= \dfrac{2(3 \pm \sqrt{3})}{2}$
$= 3 \pm \sqrt{3}$

7. $A = \pi r^2$
$\dfrac{A}{\pi} = r^2$
$r = \pm\sqrt{\dfrac{A}{\pi}}$
The radius of a circle can never be negative; therefore, the formula which gives r in terms of A and π is
$r = \sqrt{\dfrac{A}{\pi}}$

8. $6^2 = x^2 + 3^2$
$36 = x^2 + 9$
$27 = x^2$
$x = 3\sqrt{3}$ inches

Section 9.2

1. The coefficient of the linear term is +2. If we take the square of half of +2, $[\frac{1}{2}(+2)]^2$, we get 1. Adding the 1 as the last term, we have the perfect square trinomial:

$$x^2 + 2x + 1 = (x + 1)^2$$

2. The coefficient of the linear term is −10. If we take the square of half of −10, $[\frac{1}{2}(-10)]^2$, we get 25. Adding the 25 as the last term, we have the perfect square trinomial:

$$x^2 - 10x + 25 = (x - 5)^2$$

3. The coefficient of the linear term is +5. If we take the square of half of +5, $[\frac{1}{2}(+5)]^2$, we get $\frac{25}{4}$. Adding the $\frac{25}{4}$ as the last term, we have the perfect square trinomial:

$$x^2 + 5x + \frac{25}{4} = \left(x + \frac{5}{2}\right)^2$$

4.
$$\begin{aligned} x^2 - 6x - 7 &= 0 \\ x^2 - 6x &= 7 \\ x^2 - 6x + 9 &= 7 + 9 \\ (x-3)^2 &= 16 \\ x - 3 &= \pm 4 \\ x &= 3 \pm 4 \end{aligned}$$
$$x = 3 + 4 = 7 \quad \text{or} \quad x = 3 - 4 = -1$$

5.
$$\begin{aligned} 3x^2 - 3x - 18 &= 0 \\ 3x^2 - 3x &= 18 \\ x^2 - x &= 6 \\ x^2 - x + \frac{1}{4} &= 6 + \frac{1}{4} \\ \left(x - \frac{1}{2}\right)^2 &= \frac{25}{4} \\ x - \frac{1}{2} &= \pm\frac{5}{2} \\ x &= \frac{1}{2} \pm \frac{5}{2} \end{aligned}$$
$$x = \frac{1}{2} + \frac{5}{2} = 3 \quad \text{or} \quad x = \frac{1}{2} - \frac{5}{2} = -2$$

6.
$$\begin{aligned} 2y^2 - 8y + 2 &= 0 \\ 2y^2 - 8y &= -2 \\ y^2 - 4y &= -1 \\ y^2 - 4y + 4 &= -1 + 4 \\ (y-2)^2 &= 3 \\ y - 2 &= \pm\sqrt{3} \\ y &= 2 \pm \sqrt{3} \end{aligned}$$

Section 9.3

1.
$$x^2 - 3x + 2 = 0$$
$$a = 1, b = -3, c = 2$$
$$x = \frac{-(-3) \pm \sqrt{9 - 4(1)(2)}}{2(1)} = \frac{3 \pm \sqrt{1}}{2} = \frac{3 \pm 1}{2}$$
$$x = \frac{3+1}{2} = 2 \quad \text{or} \quad x = \frac{3-1}{2} = 1$$

2.
$$3x^2 = -4x + 2$$
$$3x^2 + 4x - 2 = 0$$
$$a = 3, b = 4, c = -2$$
$$x = \frac{-4 \pm \sqrt{16 - 4(3)(-2)}}{2(3)} = \frac{-4 \pm \sqrt{16 + 24}}{6} = \frac{-4 \pm \sqrt{40}}{6} = \frac{-4 \pm 2\sqrt{10}}{6} = \frac{\cancel{2}(-2 \pm \sqrt{10})}{\cancel{2} \cdot 3} = \frac{-2 \pm \sqrt{10}}{3}$$

3.
$$(x-1)(x+2) = 4$$
$$x^2 + x - 2 = 4$$
$$x^2 + x - 6 = 0$$
$$a = 1, b = 1, c = -6$$
$$x = \frac{-1 \pm \sqrt{1 - 4(1)(-6)}}{2(1)} = \frac{-1 \pm \sqrt{1 + 24}}{2} = \frac{-1 \pm \sqrt{25}}{2} = \frac{-1 \pm 5}{2}$$
$$x = \frac{-1+5}{2} = 2 \quad \text{or} \quad x = \frac{-1-5}{2} = -3$$

4.
$$x^2 - 4x = -1$$
$$x^2 - 4x + 1 = 0$$
$$a = 1, b = -4, c = 1$$
$$x = \frac{4 \pm \sqrt{16 - 4(1)(1)}}{2(1)} = \frac{4 \pm \sqrt{12}}{2} = \frac{4 \pm 2\sqrt{3}}{2} = \frac{\cancel{2}(2 \pm \sqrt{3})}{\cancel{2}} = 2 \pm \sqrt{3}$$

Section 9.4

1.
$$(2 + 7i) + (3 - 5i) = (2 + 3) + (7i - 5i) = 5 + 2i$$

2.
$$(3 - 8i) - (2 + 4i) + (3 - i) = (3 - 2 + 3) + (-8i - 4i - i) = 4 - 13i$$

3.
$$5i(2 + 3i) = 5i(2) + 5i(3i) = 10i + 15i^2 = 10i + 15(-1) = -15 + 10i$$

4.
$$(2 - 3i)(5 - i) = 2 \cdot 5 - 2i - 3i \cdot 5 - 3i(-i) = 10 - 2i - 15i + 3i^2 = 10 - 17i + 3(-1) = 10 - 17i - 3 = 7 - 17i$$

5.
$$\frac{3-i}{4+2i} = \frac{3-i}{4+2i} \cdot \frac{4-2i}{4-2i} = \frac{12 - 6i - 4i + 2i^2}{16 - 4i^2} = \frac{12 - 10i + 2(-1)}{16 - 4(-1)} = \frac{12 - 10i - 2}{16 + 4} = \frac{10 - 10i}{20} = \frac{\cancel{10}(1-i)}{\cancel{10} \cdot 2} = \frac{1-i}{2}$$

Section 9.5

1. a. $\sqrt{-9} = \sqrt{9}\sqrt{-1}$
$= 3i$
b. $\sqrt{-49} = \sqrt{49}\sqrt{-1}$
$= 7i$
c. $\sqrt{-6} = \sqrt{6}\sqrt{-1}$
$= \sqrt{6} \cdot i$
$= i\sqrt{6}$
d. $\sqrt{-12} = \sqrt{12}\sqrt{-1}$
$= \sqrt{4 \cdot 3}\sqrt{-1}$
$= 2\sqrt{3} \cdot i$
$= 2i\sqrt{3}$

2. $(x - 3)^2 = -4$
$x - 3 = \pm\sqrt{-4}$
$x - 3 = \pm 2i$
$x = 3 \pm 2i$

3. $\frac{1}{18}x^2 - \frac{2}{9}x = -\frac{13}{18}$
$x^2 - 4x = -13$
$x^2 - 4x + 13 = 0$
$x = \frac{-(-4) \pm \sqrt{(-4)^2 - 4(1)(13)}}{2(1)}$
$x = \frac{4 \pm \sqrt{16 - 52}}{2}$
$x = \frac{4 \pm \sqrt{-36}}{2}$
$x = \frac{4 \pm 6i}{2}$
$x = 2 \pm 3i$

4. $(x - 1)(3x - 3) = 10$
$3x^2 - 6x + 3 = 10$
$3x^2 - 6x - 7 = 0$
$a = 3, b = -6, c = -7$
$x = \frac{6 \pm \sqrt{36 - 4(3)(-7)}}{2(3)}$
$x = \frac{6 \pm \sqrt{36 + 84}}{6}$
$x = \frac{6 \pm \sqrt{120}}{6}$
$x = \frac{6 \pm 2\sqrt{30}}{6}$
$x = \frac{\cancel{2}(3 \pm \sqrt{30})}{\cancel{2} \cdot 3}$
$x = \frac{3 \pm \sqrt{30}}{3}$

Section 9.6

1.

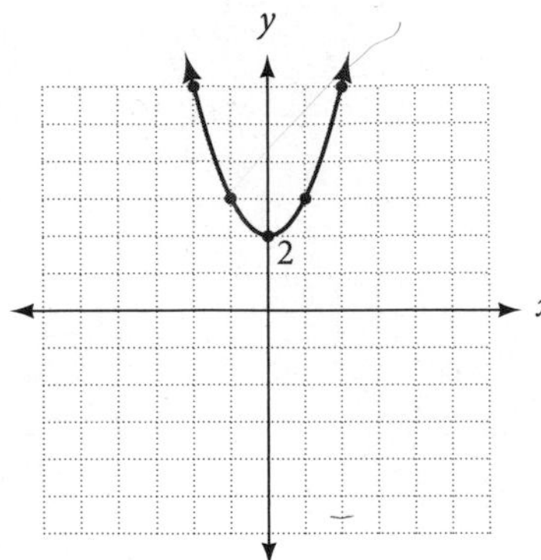

2.

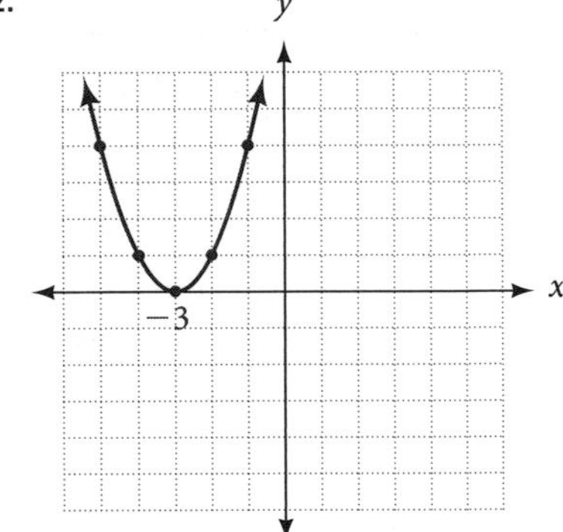

3.

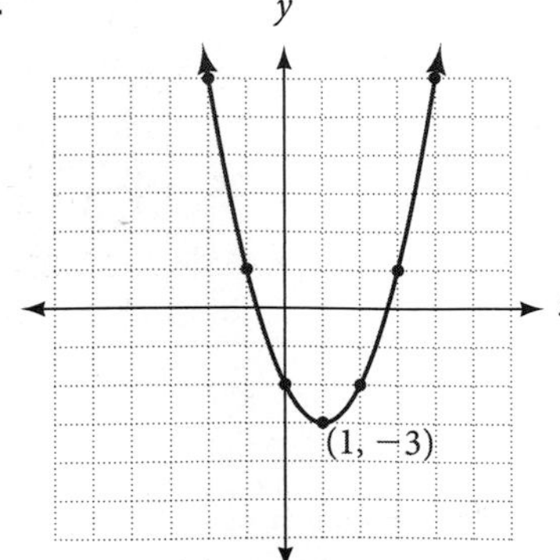

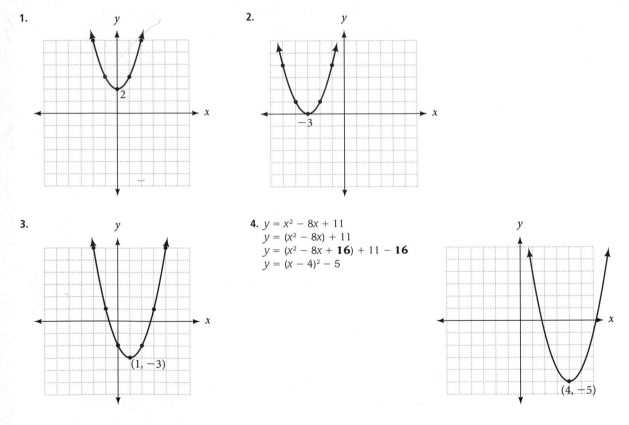

4. $y = x^2 - 8x + 11$
$y = (x^2 - 8x) + 11$
$y = (x^2 - 8x + \mathbf{16}) + 11 - \mathbf{16}$
$y = (x - 4)^2 - 5$

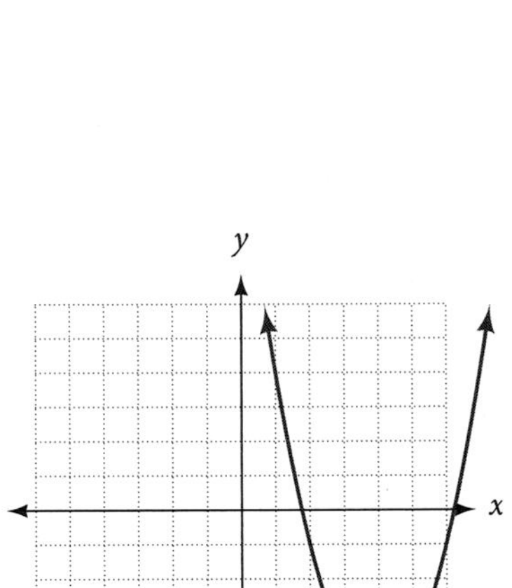

Answers to Odd-Numbered Problems

Chapter 1

Pretest for Chapter 1

1. $15 - x = 12$ **2.** $6a = 30$ **3.** 2 **4.** 10 **5.** $6, -\frac{1}{6}, 6$ **6.** $-\frac{5}{3}, \frac{3}{5}, \frac{5}{3}$ **7.** -3 **8.** 12 **9.** -11 **10.** -7 **11.** -28 **12.** 24 **13.** -3 **14.** $-\frac{27}{8}$ **15.** -13 **16.** -10 **17.** 2 **18.** -6 **19.** $12 + 3x$ **20.** $-15y$ **21.** $-10x + 15$ **22.** $2x + 4$ **23.** $-3, 2$ **24.** $-3, -\frac{1}{2}, 2$ **25.** $\sqrt{5}, \pi$ **26.** $-3, -\frac{1}{2}, 2, \sqrt{5}, \pi$

Getting Ready for Chapter 1

1. 125 **2.** 3 **3.** 0.5 **4.** 0.5 **5.** 1.4 **6.** 1.8 **7.** 5 **8.** 3 **9.** 45 **10.** 32 **11.** 29 **12.** 4 **13.** 13 **14.** 44 **15.** 10 **16.** 21 **17.** 210 **18.** 231 **19.** 630 **20.** 3

Problem Set 1.1

1. $x + 5 = 14$ **3.** $5y < 30$ **5.** $3y \le y + 6$ **7.** $\frac{x}{3} = x + 2$ **9.** 9 **11.** 49 **13.** 8 **15.** 64 **17.** 16 **19.** 100 **21.** 121 **23.** 11 **25.** 16 **27.** 17 **29.** 42 **31.** 30 **33.** 30 **35.** 24 **37.** 80 **39.** 27 **41.** 35 **43.** 13 **45.** 4 **47.** 37 **49.** 37 **51.** 16 **53.** 16 **55.** 81 **57.** 41 **59.** 345 **61.** 2,345 **63.** 2 **65.** 148 **67.** 36 **69.** 36 **71.** 58 **73.** 62 **75.** 0.17 **77.** 0.22 **79.** 2.3 **81.** 5.55 **83.** 240 **85.** 10 **87.** 3,000 **89.** 2,000 **91.** 35.33 **93.** 0.75 **95.** 0.33 **97.** 272.73 **99.** 100 **101.** 9 **103.** 18 **105.** 8 **107.** 12 **109.** 18 **111.** 42 **113.** 53 **115.** Two million, eighty-two thousand, four hundred ninety-seven **117.** 10 **119.** 420 **121.** $410,814 **123. a.** 600 mg **b.** 231 mg **125.**

Activity	Calories Burned in 1 Hour by a 150-Pound Person
Bicycling	374
Bowling	265
Handball	680
Jogging	680
Skiing	544

127. 5 **129.** 10 **131.** 25 **133.** 10

Problem Set 1.2

1–7.

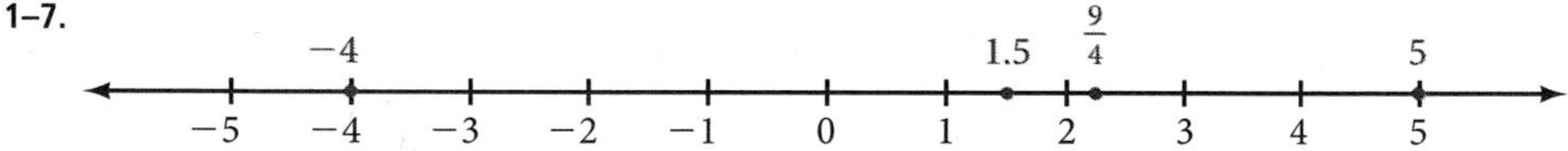

9. $\frac{18}{24}$ **11.** $\frac{12}{24}$ **13.** $\frac{15}{24}$ **15.** $\frac{36}{60}$ **17.** $\frac{22}{60}$ **19.** 2 **21.** 25 **23.** 6 **25.** $-10, \frac{1}{10}, 10$ **27.** $-\frac{3}{4}, \frac{4}{3}, \frac{3}{4}$ **29.** $-\frac{11}{2}, \frac{2}{11}, \frac{11}{2}$ **31.** $3, -\frac{1}{3}, 3$ **33.** $\frac{2}{5}, -\frac{5}{2}, \frac{2}{5}$ **35.** $-x, \frac{1}{x}, |x|$ **37.** $<$ **39.** $>$ **41.** $>$ **43.** $>$ **45.** $<$ **47.** $<$ **49.** 6 **51.** 22 **53.** 3 **55.** 7 **57.** 3 **59.** $\frac{8}{15}$ **61.** $\frac{3}{2}$ **63.** $\frac{5}{4}$ **65.** 1 **67.** 1 **69.** 1 **71. a.** 2 **b.** 4 **c.** 8 **d.** 0.03 **73. a.** 6 **b.** 12 **c.** 24 **d.** 0.09 **75.** $\frac{9}{16}$ **77.** $\frac{8}{27}$ **79.** $\frac{1}{10,000}$ **81. a.** 4 **b.** 14 **c.** 24 **d.** 34 **83. a.** 12 **b.** 20 **c.** 100 **d.** 1,024 **85. a.** 17 **b.** 25 **c.** 25 **87.** $\frac{1}{9}$ **89.** $\frac{1}{25}$ **91.** 4 inches; 1 inch^2 **93.** 4.5 inches; 1.125 inches^2 **95.** 10.25 centimeters; 5 centimeters^2 **97.** $-8, -2$ **99.** $-64°$ F; $-54°$ F **101.** 1150 ft **103.** $-15°$F **105.** $-25°$F **107.** -100 feet; -105 feet **109.** 1,387 calories **111.** 654 more calories

Problem Set 1.3

1. $3 + 5 = 8$; $3 + (-5) = -2$, $-3 + 5 = 2$; $-3 + (-5) = -8$ **3.** $15 + 20 = 35$; $15 + (-20) = -5$, $-15 + 20 = 5$; $-15 + (-20) = -35$ **5.** 3 **7.** -7 **9.** -14 **11.** -3 **13.** -25 **15.** -12 **17.** -19 **19.** -25 **21.** -8 **23.** -4 **25.** 6 **27.** 6 **29.** 8 **31.** -4 **33.** -14 **35.** -17 **37.** 4 **39.** 3 **41.** 15 **43.** -8 **45.** 12 **47.** 11 **49.** 10.8 **51.** 980 **53.** 1.70 **55.** 23, 28 **57.** 30, 35 **59.** $0, -5$ **61.** $-12, -18$ **63.** $-4, -8$ **65.** Yes **67.** $5 + 9 = 14$ **69.** $[-7 + (-5)] + 4 = -8$ **71.** $[-2 + (-3)] + 10 = 5$ **73.** 3 **75.** -3 **77.** $71.3 million **79.** $-12 + 4$ **81.** $10 + (-6) + (-8) = -\$4$ **83.** $-30 + 40 = 10$

Problem Set 1.4

1. -3 **3.** -6 **5.** 0 **7.** -10 **9.** -16 **11.** -12 **13.** -7 **15.** 35 **17.** 0 **19.** -4 **21.** 4 **23.** -24 **25.** -28 **27.** 25 **29.** 4 **31.** 7 **33.** 17 **35.** 8 **37.** 4 **39.** 18 **41.** 10 **43.** 17 **45.** 1 **47.** 1 **49.** 27 **51.** -26

53. −2 **55.** 68 **57.** −11.3 **59.** −3.6 **61. a.** −9 **b.** −10 **c.** −3 **63.** $-7-4=-11$ **65.** $12-(-8)=20$ **67.** $-5-(-7)=2$ **69.** $[4+(-5)]-17=-18$ **71.** $8-5=3$ **73.** $-8-5=-13$ **75.** $8-(-5)=13$ **77.** 10 **79.** −2 **81.** 1,418 strikeouts **83.** 1,500 − 730 **85.** −35 + 15 − 20 = −\$40 **87.** 98 − 65 − 53 = −\$20 **89.** \$4,500, \$3,950, \$3,400, \$2,850, \$2,300; yes **91.** 769 feet **93.** 439 feet **95.** 2 seconds **97.** 35° **99.** 60° **101. a.**

Day	Plant Height (inches)	Day	Plant Height (inches)
0	0	6	6
1	0.5	7	9
2	1	8	13
3	1.5	9	18
4	3	10	23
5	4		

b. 11.5 inches

Problem Set 1.5

1. Commutative **3.** Multiplicative inverse **5.** Commutative **7.** Distributive **9.** Commutative, associative **11.** Commutative, associative **13.** Commutative **15.** Commutative, associative **17.** Commutative **19.** Additive inverse **21.** $3x+6$ **23.** $9a+9b$ **25.** 0 **27.** 0 **29.** 10 **31.** $(4+2)+x=6+x$ **33.** $x+(2+7)=x+9$ **35.** $(3\cdot5)x=15x$ **37.** $(9\cdot6)y=54y$ **39.** $\left(\frac{1}{2}\cdot3\right)a=\frac{3}{2}a$ **41.** $\left(\frac{1}{3}\cdot3\right)x=x$ **43.** $\left(\frac{1}{2}\cdot2\right)y=y$ **45.** $\left(\frac{3}{4}\cdot\frac{4}{3}\right)x=x$ **47.** $\left(\frac{6}{5}\cdot\frac{5}{6}\right)a=a$ **49.** $8x+16$ **51.** $8x-16$ **53.** $4y+4$ **55.** $18x+15$ **57.** $6a+14$ **59.** $54y-72$ **61.** $\frac{3}{2}x-3$ **63.** $x+2$ **65.** $3x+3y$ **67.** $8a-8b$ **69.** $12x+18y$ **71.** $12a-8b$ **73.** $3x+2y$ **75.** $4a+25$ **77.** $6x+12$ **79.** $14x+38$ **81.** $0.09x+180$ **83.** $0.15x+75$ **85.** $3x-2y$ **87.** $4x+3y$ **89.** $2x+1$ **91.** $6x-3$ **93.** $5x+10$ **95.** $6x+5$ **97.** $5x+6$ **99.** $6m-5$ **101.** $7+3x$ **103.** $3x-2y$ **105.** $0.09x+180$ **107.** $0.12x+60$ **109.** $a+1$ **111.** $1-a$ **113.** No **115.** No, not commutative **117.** 3(6,200 − 100) = 18,600 − 300 = \$18,300 **119.** $8\div4\neq4\div8$ **121.** 12(2,400 − 480) = \$23,040; 12(2,400) − 12(480) = \$23,040

Problem Set 1.6

1. −42 **3.** −16 **5.** 3 **7.** 121 **9.** 6 **11.** −60 **13.** 24 **15.** 49 **17.** −27 **19.** 6 **21.** 10 **23.** 9 **25.** 45 **27.** 14 **29.** −2 **31.** 216 **33.** −2 **35.** −18 **37.** 29 **39.** 38 **41.** −5 **43.** 37 **45. a.** 60 **b.** −23 **c.** 17 **d.** −2 **47.** $-\frac{10}{21}$ **49.** −4 **51.** $\frac{9}{16}$ **53. a.** 27 **b.** 3 **c.** −27 **d.** −27 **55.** 9 **57.** $\frac{25}{4}$ **59.** 4 **61.** $\frac{9}{4}$ **63.** $-8x$ **65.** $42x$ **67.** x **69.** $-4a-8$ **71.** $-\frac{3}{2}x+3$ **73.** $-6x+8$ **75.** $-15x-30$ **77.** $-12x-20y$ **79.** $-6x-10y$ **81.** $-\frac{3}{2}x+3$ **83.** $-\frac{2}{3}x+2$ **85.** $\frac{2}{3}x-2$ **87.** $-2x+y$ **89. a.** 2 **b.** 1 **c.** 3 **91. a.** 3 **b.** 3 **c.** −4 **93. a.** 12 **b.** 12 **95. a.** 44 **b.** 44 **97.** −25 **99.** $2(-4x)=-8x$ **101.** −26 **103.** 8 **105.** −80 **107.** $\frac{1}{8}$ **109.** −24 **111.** $3x-11$ **113.** 14 **115.** 81 **117.** $\frac{9}{4}$ **119.** \$2.50 **121.** \$60 **123.** 1° F **125.** 100 mg **127.** 465 calories

Problem Set 1.7

1. −2 **3.** −3 **5.** $-\frac{1}{3}$ **7.** 3 **9.** $\frac{1}{7}$ **11.** 0 **13.** 9 **15.** −15 **17.** −36 **19.** $-\frac{1}{4}$ **21.** $\frac{16}{15}$ **23.** $\frac{4}{3}$ **25.** $-\frac{8}{13}$ **27.** −1 **29.** 1 **31.** $\frac{3}{5}$ **33.** $-\frac{5}{3}$ **35.** −2 **37.** −3 **39.** Undefined **41.** Undefined **43.** 5 **45.** $-\frac{7}{3}$ **47.** −1 **49.** −7 **51.** $\frac{15}{17}$ **53.** $-\frac{32}{17}$ **55.** $\frac{1}{3}$ **57.** 1 **59.** 1 **61.** −2 **63.** $\frac{9}{7}$ **65.** $\frac{16}{11}$ **67.** −1 **69.** $\frac{7}{20}=0.35$ **71.** −1 **73. a.** $\frac{3}{2}$ **b.** $\frac{3}{2}$ **75. a.** $-\frac{5}{7}$ **b.** $-\frac{5}{7}$ **77. a.** 2.618 **b.** 0.382 **c.** 3 **79. a.** 25 **b.** −25 **c.** −25 **d.** −25 **e.** 25 **81. a.** 10 **b.** 0 **c.** −100 **d.** −20 **83.** $5x+6$ **85.** $3x+20$ **87.** $3+x$ **89.** $3x-7y$ **91.** 1 **93.** $2-y$ **95.** 3 **97.** −10 **99.** −3 **101.** −8 **103.** 401 yards **105.** \$350 **107.** Drops 3.5° F each hour **109.** 3 capsules **111.** 0.3 mg **113. a.** \$20,000 **b.** \$50,000 **c.** Yes

Problem Set 1.8

1. 0, 1 **3.** $-3, -2.5, 0, 1, \frac{3}{2}$ **5.** All **7.** −10, −8, −2, 9 **9.** π **11.** T **13.** F **15.** F **17.** T **19.** Composite, $2^4\cdot3$ **21.** Prime **23.** Composite, $3\cdot11\cdot31$ **25.** $2^4\cdot3^2$ **27.** $2\cdot19$ **29.** $3\cdot5\cdot7$ **31.** $2^2\cdot3^2\cdot5$ **33.** $5\cdot7\cdot11$ **35.** 11^2 **37.** $2^2\cdot3\cdot5\cdot7$ **39.** $2^2\cdot5\cdot31$ **41.** $\frac{7}{11}$ **43.** $\frac{5}{7}$ **45.** $\frac{11}{13}$ **47.** $\frac{14}{15}$ **49.** $\frac{5}{9}$ **51.** $\frac{5}{8}$ **53. a.** −30 **b.** 130 **c.** −4,000 **d.** $-\frac{5}{8}$ **55.** $\frac{2}{9}$ **57.** $\frac{3}{2}$ **59.** 64 **61.** 509 **63. a.** 0.35 **b.** 0.28 **c.** 0.25 **65.** $6^3=(2\cdot3)^3=2^3\cdot3^3$ **67.** $9^4\cdot16^2=(3^2)^4(2^4)^2=2^8\cdot3^8$ **69.** $3\cdot8+3\cdot7+3\cdot5=24+21+15=60=2^2\cdot3\cdot5$ **71.** $2^2\cdot5\cdot13$ **73.** Irrational numbers **75.** 8, 21, 34

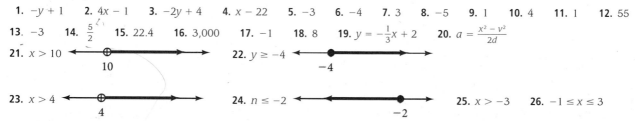

Problem Set 1.9

1. $\frac{2}{3}$ **3.** $-\frac{1}{4}$ **5.** $\frac{1}{2}$ **7.** $\frac{x-1}{3}$ **9.** $\frac{3}{2}$ **11.** $\frac{x+6}{2}$ **13.** $-\frac{3}{5}$ **15.** $\frac{10}{a}$ **17.** $\frac{7}{8}$ **19.** $\frac{1}{10}$ **21.** $\frac{7}{9}$ **23.** $\frac{7}{3}$ **25.** $\frac{1}{4}$ **27.** $\frac{7}{6}$ **29.** $\frac{19}{24}$ **31.** $\frac{13}{60}$ **33.** $\frac{29}{35}$ **35.** $\frac{949}{1{,}260}$ **37.** $\frac{13}{420}$ **39.** $\frac{41}{24}$ **41.** $\frac{5}{4}$ **43.** $-\frac{3}{2}$ **45.** $\frac{3}{2}$ **47.** $-\frac{2}{3}$ **49.** $\frac{7}{3}$ **51.** $\frac{1}{125}$ **53.** 4 **55.** −2 **57.** −18 **59.** −16 **61.** −2 **63.** 2 **65.** −144 **67.** −162 **69.** $\frac{5x+4}{20}$ **71.** $\frac{a+4}{12}$ **73.** $\frac{9x+4}{12}$ **75.** $\frac{10+3x}{5x}$ **77.** $\frac{3y+28}{7y}$ **79.** $\frac{60+19a}{20a}$ **81.** $\frac{2}{3}x$ **83.** $-\frac{1}{4}x$ **85.** $\frac{14}{15}x$ **87.** $\frac{11}{12}x$ **89.** $\frac{41}{40}x$ **91.** $\frac{x-1}{x}$ **93. a.** $\frac{1}{4}$ **b.** $\frac{5}{4}$ **c.** $-\frac{3}{8}$ **d.** $-\frac{3}{2}$ **95.** $\frac{1}{3}$ **97.** $\frac{3}{4}$ **99. a.** $\frac{3}{2}$ **b.** $\frac{4}{3}$ **c.** $\frac{5}{4}$ **101. a.** 8 **b.** 7 **c.** 8 **103.** $\frac{47}{60}$ inch

Chapter 1 Review

1. $-7 + (-10) = -17$ **2.** $(-7 + 4) + 5 = 2$ **3.** $(-3 + 12) + 5 = 14$ **4.** $4 - 9 = -5$ **5.** $9 - (-3) = 12$ **6.** $-7 - (-9) = 2$ **7.** $(-3)(-7) - 6 = 15$ **8.** $5(-6) + 10 = -20$ **9.** $2[(-8)(3x)] = -48x$ **10.** $\frac{-25}{-5} = 5$ **11.** 1.8 **12.** −10 **13.** $-6, \frac{1}{6}$ **14.** $\frac{12}{5}, -\frac{5}{12}$ **15.** −5 **16.** $-\frac{5}{4}$ **17.** 3 **18.** −38 **19.** −25 **20.** 1 **21.** −3 **22.** 22 **23.** −4 **24.** −20 **25.** −30 **26.** −12 **27.** −24 **28.** 12 **29.** −4 **30.** $-\frac{2}{3}$ **31.** 23 **32.** 47 **33.** −3 **34.** −35 **35.** 32 **36.** 2 **37.** 20 **38.** −3 **39.** −98 **40.** 70 **41.** 2 **42.** $\frac{17}{2}$ **43.** Undefined **44.** 3 **45.** Associative **46.** Multiplicative identity **47.** Commutative **48.** Additive inverse **49.** Commutative, associative **50.** Distributive **51.** $12 + x$ **52.** $28a$ **53.** x **54.** y **55.** $14x + 21$ **56.** $6a - 12$ **57.** $\frac{5}{2}x - 3$ **58.** $-\frac{3}{2}x + 3$ **59.** $-\frac{1}{3}, 0, 5, -4.5, \frac{2}{5}, -3$ **60.** 0, 5 **61.** $\sqrt{7}, \pi$ **62.** 0, 5, −3 **63.** $2 \cdot 3^2 \cdot 5$ **64.** $2^3 \cdot 3 \cdot 5 \cdot 7$ **65.** $\frac{173}{210}$ **66.** $\frac{2x+7}{12}$ **67.** −2 **68.** 810 **69.** 8 **70.** 12 **71.** −1 **72.** $\frac{1}{16}$

Chapter 1 Test

1. $x + 3 = 8$ **2.** $5y = 15$ **3.** 40 **4.** 16 **5.** $4, -\frac{1}{4}, 4$ **6.** $-\frac{3}{4}, \frac{4}{3}, \frac{3}{4}$ **7.** −4 **8.** 17 **9.** −12 **10.** 0 **11.** c **12.** e **13.** d **14.** a **15.** −21 **16.** 64 **17.** −2 **18.** $-\frac{8}{27}$ **19.** 4 **20.** 204 **21.** 25 **22.** 52 **23.** 2 **24.** 2 **25.** $8 + 2x$ **26.** $10x$ **27.** $6x + 10$ **28.** $-2x + 1$ **29.** −8, 1 **30.** $-8, \frac{3}{4}, 1, 1.5$ **31.** $\sqrt{2}$ **32.** $-8, \frac{3}{4}, 1, \sqrt{2}, 1.5$ **33.** $2^4 \cdot 37$ **34.** $2^2 \cdot 5 \cdot 67$ **35.** $\frac{25}{42}$ **36.** $\frac{8}{x}$ **37.** $8 + (-3) = 5$ **38.** $-24 - 2 = -26$ **39.** $(-5)(-4) = 20$ **40.** $\frac{-24}{-2} = 12$ **41.** 12 **42.** $\frac{1}{2}$ **43.** 425 million **44.** 1.35 billion **45.** 18 million

Chapter 2

Pretest for Chapter 2

1. $-y + 1$ **2.** $4x - 1$ **3.** $-2y + 4$ **4.** $x - 22$ **5.** −3 **6.** −4 **7.** 3 **8.** −5 **9.** 1 **10.** 4 **11.** 1 **12.** 55 **13.** −3 **14.** $\frac{5}{2}$ **15.** 22.4 **16.** 3,000 **17.** −1 **18.** 8 **19.** $y = -\frac{1}{3}x + 2$ **20.** $a = \frac{x^2 - v^2}{2d}$

21. $x > 10$ 

22. $y \ge -4$ 

23. $x > 4$ 

24. $n \le -2$ 

25. $x > -3$ **26.** $-1 \le x \le 3$

Getting Ready for Chapter 2

1. 4 **2.** −14 **3.** 33 **4.** −35 **5.** $-6y - 3$ **6.** 10.8 **7.** $-\frac{5}{4}$ **8.** $-\frac{6}{5}$ **9.** x **10.** x **11.** x **12.** x **13.** $-3x - 4$ **14.** −9.7 **15.** $0.04x + 280$ **16.** $0.09x + 180$ **17.** $3x - 11$ **18.** $5x - 13$ **19.** $10x - 43$ **20.** $12x - 22$

Problem Set 2.1

1. $-3x$ **3.** $-a$ **5.** $12x$ **7.** $6a$ **9.** $6x - 3$ **11.** $7a + 5$ **13.** $5x - 5$ **15.** $4a + 2$ **17.** $-9x - 2$ **19.** $12a + 3$ **21.** $10x - 1$ **23.** $21y + 6$ **25.** $-6x + 8$ **27.** $-2a + 3$ **29.** $-4x + 26$ **31.** $4y - 16$ **33.** $-6x - 1$ **35.** $2x - 12$ **37.** $10a + 33$ **39.** $4x - 9$ **41.** $7y - 39$ **43.** $-19x - 14$ **45.** 5 **47.** −9 **49.** 4 **51.** 4 **53.** −37 **55.** −41 **57.** 64 **59.** 64 **61.** 144 **63.** 144 **65.** 3 **67.** 0 **69.** 15 **71.** 6

73. a.

n	1	2	3	4
$3n$	3	6	9	12

b.

n	1	2	3	4
n^3	1	8	27	64

75. 1, 4, 7, 10, . . . an arithmetic sequence **77.** 0, 1, 4, 9, . . . a sequence of squares **79.** $-6y + 4$ **81.** $0.17x$ **83.** $2x$ **85.** $5x - 4$ **87.** $7x - 5$ **89.** $-2x - 9$ **91.** $7x + 2$ **93.** $-7x + 6$ **95.** $7x$ **97.** $-y$ **99.** $10y$ **101.** $0.17x + 180$ **103.** $0.22x + 60$ **105.** $a = 4$; $13x$ **107.** 49 **109.** 40 **111. a.** 42° F **b.** 28° F **c.** −14° F **113. a.** $37.50 **b.** $40.00 **c.** $42.50 **115.** $0.71G$; $887.50 **117.** 74 mph **119.** 12 **121.** −3 **123.** −9.7 **125.** $-\frac{5}{4}$ **127.** −3 **129.** $a - 12$ **131.** $-\frac{7}{2}$ **133.** $\frac{51}{40}$ **135.** 7 **137.** $-\frac{1}{24}$ **139.** $-\frac{7}{9}$ **141.** $-\frac{23}{420}$

Problem Set 2.2

1. 11 **3.** 4 **5.** $-\frac{3}{4}$ **7.** −5.8 **9.** −17 **11.** $-\frac{1}{8}$ **13.** −6 **15.** −3.6 **17.** −7 **19.** $-\frac{7}{45}$ **21.** 3 **23.** $\frac{11}{8}$ **25.** 21 **27.** 7 **29.** 3.5 **31.** 22 **33.** −2 **35.** −16 **37.** −3 **39.** 10 **41.** −12 **43.** 4 **45.** 2 **47.** −5 **49.** −1 **51.** −3 **53.** 8 **55.** −8 **57.** 2 **59.** 11 **61. a.** $\frac{3}{2}$ **b.** 1 **c.** $-\frac{3}{2}$ **d.** −4 **e.** $\frac{8}{5}$ **63. a.** 6% **b.** 5% **c.** 2% **d.** 75% **65.** $x + 55 + 55 = 180$; 70° **67.** 100 dB **69.** y **71.** x **73.** 6 **75.** 6 **77.** −9 **79.** $-\frac{15}{8}$ **81.** −18 **83.** $-\frac{5}{4}$ **85.** $3x$ **87.** $18x$ **89.** x **91.** y **93.** x **95.** a

Problem Set 2.3

1. 2 **3.** 4 **5.** $-\frac{1}{2}$ **7.** −2 **9.** 3 **11.** 4 **13.** 0 **15.** 0 **17.** 6 **19.** −50 **21.** $\frac{3}{2}$ **23.** 12 **25.** −3 **27.** 32 **29.** −8 **31.** $\frac{1}{2}$ **33.** 4 **35.** 8 **37.** −4 **39.** 4 **41.** −15 **43.** $-\frac{1}{2}$ **45.** 3 **47.** 1 **49.** $\frac{1}{4}$ **51.** −3 **53.** 3 **55.** 2 **57.** $-\frac{3}{2}$ **59.** $-\frac{3}{2}$ **61.** 1 **63.** 1 **65.** −2 **67.** −2 **69.** $-\frac{4}{5}$ **71.** 1 **73.** $\frac{7}{2}$ **75.** $\frac{5}{4}$ **77.** 40 **79.** $\frac{7}{3}$ **81.** 200 tickets **83.** $1,391 per month **85.** 725 ft. **87.** 2 **89.** 6 **91.** 3,000 **93.** $3x - 11$ **95.** $0.09x + 180$ **97.** $-6y + 4$ **99.** $4x - 11$ **101.** $5x$ **103.** $0.17x$ **105.** $6x - 10$ **107.** $\frac{3}{2}x + 3$ **109.** $-x + 2$ **111.** $10x - 43$ **113.** $-3x - 13$ **115.** $-6y + 4$

Problem Set 2.4

1. 3 **3.** −2 **5.** −1 **7.** 2 **9.** −4 **11.** −2 **13.** 0 **15.** 1 **17.** $\frac{1}{2}$ **19.** 7 **21.** 8 **23.** $-\frac{1}{3}$ **25.** $\frac{3}{4}$ **27.** 75 **29.** 2 **31.** 6 **33.** 8 **35.** 0 **37.** $\frac{3}{7}$ **39.** 1 **41.** 1 **43.** −1 **45.** 6 **47.** $\frac{3}{4}$ **49.** 3 **51.** $\frac{3}{4}$ **53.** 8 **55.** 6 **57.** −2 **59.** −2 **61.** 2 **63.** −6 **65.** 2 **67.** 20 **69.** 4,000 **71.** 700 **73.** 11 **75.** 7 **77. a.** $\frac{5}{4}$ **b.** $\frac{15}{2}$ **c.** $6x + 20$ **d.** 15 **e.** $4x - 20$ **f.** $\frac{45}{2}$ **79.** 2004 **81.** 14 **83.** −3 **85.** $\frac{1}{4}$ **87.** $\frac{1}{3}$ **89.** $-\frac{3}{2}x + 3$ **91.** $\frac{3}{2}$ **93.** 4 **95.** 1

97.

x	$3(x + 2)$	$3x + 2$	$3x + 6$
0	6	2	6
1	9	5	9
2	12	8	12
3	15	11	15

99.

a	$(2a + 1)^2$	$4a^2 + 4a + 1$
1	9	9
2	25	25
3	49	49

Problem Set 2.5

1. 100 feet **3. a.** 2 **b.** $\frac{3}{2}$ **c.** 0 **5. a.** 2 **b.** −3 **c.** 1 **7.** 0 **9.** 2 **11.** 15 **13.** 10 **15.** −2 **17.** 1 **19. a.** 2 **b.** 4 **21. a.** 5 **b.** 18 **23. a.** 20 **b.** 75 **25.** $l = \frac{A}{w}$ **27.** $h = \frac{V}{lw}$ **29.** $a = P - b - c$ **31.** $x = 3y - 1$ **33.** $y = 3x + 6$ **35.** $y = -\frac{2}{3}x + 2$ **37.** $y = -2x - 5$ **39.** $y = -\frac{2}{3}x + 1$ **41.** $w = \frac{P - 2l}{2}$ **43.** $v = \frac{h - 16t^2}{t}$ **45.** $h = \frac{A - \pi r^2}{2\pi r}$ **47. a.** $y = -2x - 5$ **b.** $y = 4x - 7$ **49. a.** $y = \frac{3}{4}x + \frac{1}{4}$ **b.** $y = \frac{3}{4}x - 5$ **51. a.** $y = \frac{3}{5}x + 1$ **b.** $y = \frac{1}{2}x + 2$ **53.** $y = \frac{3}{7}x - 3$ **55.** $y = 2x + 8$ **57. a.** $\frac{15}{4}$ **b.** 17 **c.** $y = -\frac{4}{5}x + 4$ **d.** $x = -\frac{5}{4}x + 5$ **59.** 60°; 150° **61.** 45°; 135° **63.** 10 **65.** 240 **67.** 25% **69.** 35% **71.** 64 **73.** 2,000 **75.** 100° C; yes **77.** 20° C; yes **79.** $C = \frac{5}{9}(F - 32)$ **81.** 13.9% **83.** 60% **85.** 26.5% **87.** 7 meters **89.** $\frac{3}{2}$ or 1.5 inches **91.** 132 feet **93.** $\frac{2}{9}$ centimeters **95.** The sum of 4 and 1 is 5.

97. The difference of 6 and 2 is 4. **99.** The difference of a number and 5 is -12.
101. The sum of a number and 3 is four times the difference of that number and 3. **103.** $2(6 + 3) = 18$ **105.** $2(5) + 3 = 13$
107. $x + 5 = 13$ **109.** $5(x + 7) = 30$ **111. a.** 95 **b.** -41 **c.** -95 **d.** 41 **113. a.** 9 **b.** -73 **c.** 9 **d.** -73

Problem Set 2.6

Along with the answers to the odd-numbered problems in this problem set and the next, we are including some of the equations used to solve the problems. Be sure that you try the problems on your own before looking to see the correct equations.

1. 8 **3.** $2x + 4 = 14$; 5 **5.** -1 **7.** The two numbers are x and $x + 2$; $x + (x + 2) = 8$; 3 and 5 **9.** 6 and 14
11. Shelley is 39, and Michele is 36 **13.** Evan is 11, and Cody is 22 **15.** Barney is 27; Fred is 31 **17.** Lacy is 16; Jack is 32
19. Patrick is 18; Pat is 38 **21.** $s = 9$ inches **23.** $s = 15$ feet **25.** 11 feet, 18 feet, 33 feet **27.** 26 feet, 13 feet, 14 feet
29. Length = 11 inches; width = 6 inches **31.** Length = 25 inches; width = 9 inches **33.** Length = 15 feet; width = 3 feet
35. 9 dimes; 14 quarters **37.** 12 quarters; 27 nickels **39.** 8 nickels; 17 dimes **41.** 7 nickels; 10 dimes; 12 quarters
43. 3 nickels; 9 dimes; 6 quarters **45.** $5x$ **47.** $1.075x$ **49.** $0.09x + 180$ **51.** 6,000 **53.** 30 **55.** 4 is less than 10.
57. 9 is greater than or equal to -5 **59.** $<$ **61.** $<$ **63.** 2 **65.** 12
67. a. e **b.** i **c.** x **d.** The larger the area of space used to store the letter in a typecase, the more often the letter is printed.

Problem Set 2.7

1. 5 and 6 **3.** -4 and -5 **5.** 13 and 15 **7.** 52 and 54 **9.** -14 and -16 **11.** 17, 19, and 21 **13.** 42, 44, and 46
15. \$4,000 invested at 8%, \$6,000 invested at 9% **17.** \$700 invested at 10%, \$1,200 invested at 12%
19. \$500 at 8%, \$1,000 at 9%, \$1,500 at 10% **21.** 45°, 45°, 90° **23.** 22.5°, 45°, 112.5° **25.** 53°, 90° **27.** 80°, 60°, 40°
29. 16 adult and 22 children's tickets **31.** 16 minutes **33.** 39 hours **35.** They are in offices 7329 and 7331.
37. Kendra is 8 years old and Marissa is 10 years old. **39.** Jeff **41.** \$10.38 **43.** Length = 12 meters; width = 10 meters
45. 59°, 60°, 61° **47.** \$54.00 **49.** Yes **51. a.** 9 **b.** 3 **c.** -9 **d.** -3 **53. a.** -8 **b.** 8 **c.** 8 **d.** -8 **55.** -2.3125
57. $\frac{10}{3}$ **59.** -3 **61.** -8 **63.** 0 **65.** 8 **67.** -24 **69.** 28 **71.** $-\frac{1}{2}$ **73.** -16 **75.** -25 **77.** 24

Problem Set 2.8

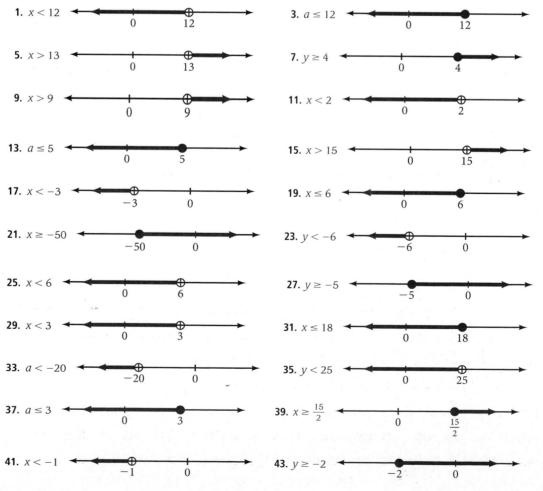

1. $x < 12$ **3.** $a \le 12$

5. $x > 13$ (0, 13) **7.** $y \ge 4$ (0, 4)

9. $x > 9$ **11.** $x < 2$

13. $a \le 5$ **15.** $x > 15$

17. $x < -3$ (-3, 0) **19.** $x \le 6$ (0, 6)

21. $x \ge -50$ (-50, 0) **23.** $y < -6$ (-6, 0)

25. $x < 6$ **27.** $y \ge -5$

29. $x < 3$ (0, 3) **31.** $x \le 18$ (0, 18)

33. $a < -20$ (-20, 0) **35.** $y < 25$ (0, 25)

37. $a \le 3$ (0, 3) **39.** $x \ge \frac{15}{2}$ (0, $\frac{15}{2}$)

41. $x < -1$ (-1, 0) **43.** $y \ge -2$ (-2, 0)

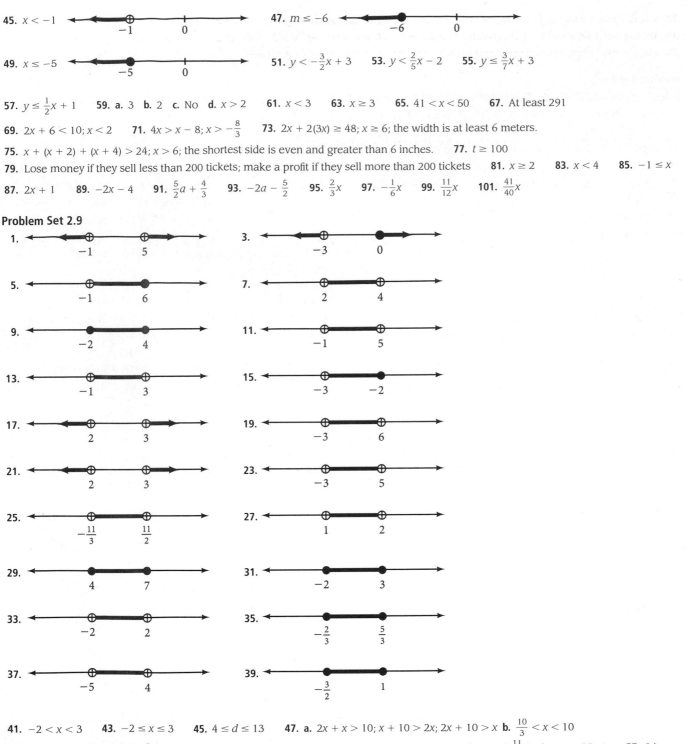

45. $x < -1$ (−1, 0) **47.** $m \leq -6$ (−6, 0)

49. $x \leq -5$ (−5, 0) **51.** $y < -\frac{3}{2}x + 3$ **53.** $y < \frac{2}{5}x - 2$ **55.** $y \leq \frac{3}{7}x + 3$

57. $y \leq \frac{1}{2}x + 1$ **59. a.** 3 **b.** 2 **c.** No **d.** $x > 2$ **61.** $x < 3$ **63.** $x \geq 3$ **65.** $41 < x < 50$ **67.** At least 291

69. $2x + 6 < 10; x < 2$ **71.** $4x > x - 8; x > -\frac{8}{3}$ **73.** $2x + 2(3x) \geq 48; x \geq 6$; the width is at least 6 meters.

75. $x + (x + 2) + (x + 4) > 24; x > 6$; the shortest side is even and greater than 6 inches. **77.** $t \geq 100$

79. Lose money if they sell less than 200 tickets; make a profit if they sell more than 200 tickets **81.** $x \geq 2$ **83.** $x < 4$ **85.** $-1 \leq x$

87. $2x + 1$ **89.** $-2x - 4$ **91.** $\frac{5}{2}a + \frac{4}{3}$ **93.** $-2a - \frac{5}{2}$ **95.** $\frac{2}{3}x$ **97.** $-\frac{1}{6}x$ **99.** $\frac{11}{12}x$ **101.** $\frac{41}{40}x$

Problem Set 2.9

1. (−1, 5) **3.** (−3, 0)

5. (−1, 6) **7.** (2, 4)

9. (−2, 4) **11.** (−1, 5)

13. (−1, 3) **15.** (−3, −2)

17. (2, 3) **19.** (−3, 6)

21. (2, 3) **23.** (−3, 5)

25. **27.**

29. **31.**

33. **35.**

37. (−5, 4) **39.** ($-\frac{3}{2}$, 1)

41. $-2 < x < 3$ **43.** $-2 \leq x \leq 3$ **45.** $4 \leq d \leq 13$ **47. a.** $2x + x > 10; x + 10 > 2x; 2x + 10 > x$ **b.** $\frac{10}{3} < x < 10$

49. (50, 266) **51.** $4 < x < 5$ **53.** The width is between 3 inches and $\frac{11}{2}$ inches. **55.** 8 **57.** 24

59. 25% **61.** 10% **63.** 80 **65.** 400 **67.** −5 **69.** 5 **71.** 7 **73.** 9 **75.** 6 **77.** $2x - 3$ **79.** −3, 0, 2

Chapter 2 Review

1. $-3x$ **2.** $-2x - 3$ **3.** $4a - 7$ **4.** $-2a + 3$ **5.** $-6y$ **6.** $-6x + 1$ **7.** 19 **8.** −11 **9.** −18 **10.** −13 **11.** 8 **12.** 6 **13.** −8 **14.** $\frac{15}{14}$ **15.** 2 **16.** −5 **17.** −5 **18.** 0 **19.** 12 **20.** −8 **21.** −1 **22.** 1 **23.** 5 **24.** 2 **25.** 0 **26.** 0 **27.** 10 **28.** 2 **29.** 2 **30.** $-\frac{8}{3}$ **31.** 0 **32.** −5 **33.** 0 **34.** −4 **35.** −8 **36.** 4 **37.** $y = \frac{2}{5}x - 2$

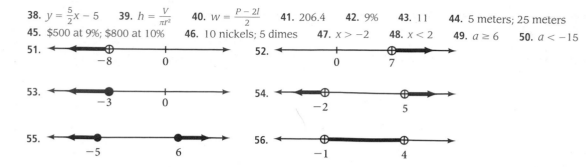

38. $y = \frac{5}{2}x - 5$ **39.** $h = \frac{V}{\pi r^2}$ **40.** $w = \frac{P - 2l}{2}$ **41.** 206.4 **42.** 9% **43.** 11 **44.** 5 meters; 25 meters
45. \$500 at 9%; \$800 at 10% **46.** 10 nickels; 5 dimes **47.** $x > -2$ **48.** $x < 2$ **49.** $a \geq 6$ **50.** $a < -15$

51. −8 0 **52.** 0 7

53. −3 0 **54.** −2 5

55. −5 6 **56.** −1 4

Chapter 2 Cumulative Review

1. −14 **2.** −7 **3.** −12 **4.** 100 **5.** 30 **6.** 8 **7.** $-\frac{8}{27}$ **8.** 16 **9.** $-\frac{1}{4}$ **10.** $\frac{3}{4}$ **11.** $-\frac{4}{5}$ **12.** $\frac{8}{35}$ **13.** $-\frac{8}{3}$
14. 0 **15.** $\frac{8}{9}$ **16.** $-\frac{1}{15}$ **17.** $2x$ **18.** $-72x$ **19.** $2x - 1$ **20.** $11x + 7$ **21.** 4 **22.** −16 **23.** −50 **24.** $-\frac{3}{2}$ **25.** 5
26. $\frac{3}{2}$ **27.** 1 **28.** $\frac{1}{3}$ **29.** 12 **30.** 48 **31.** $c = P - a - b$ **32.** $y = -\frac{3}{4}x + 3$
33. $x \leq 6$ 6 **34.** $x > 3$ 3
35. $x < 2$ or $x > 3$ 2 3 **36.** $-3 < x < 4$ −3 4 **37.** $x - 5 = 12$ **38.** $x + 7 = 4$
39. −5, −3 **40.** −5, −3, −1.7, 2.3, $\frac{12}{7}$ **41.** 11 **42.** 25 **43.** 15 **44.** 75% **45.** 14 **46.** 12 **47.** 48°, 90° **48.** 65°
49. 18 hours **50.** 3 oz **51.** \$837.3 million **52.** \$180.8 million

Chapter 2 Test

1. $-4x + 5$ **2.** $3a - 4$ **3.** $-3y - 12$ **4.** $11x + 28$ **5.** 22 **6.** 25

7. a.

n	$(n+1)^2$
1	4
2	9
3	16
4	25

b.

n	n^2+1
1	2
2	5
3	10
4	17

8. 6 **9.** $\frac{4}{3}$ **10.** 2 **11.** $-\frac{1}{2}$ **12.** −3 **13.** 70 **14.** −3 **15.** −1

16. 5.7 **17.** 2,000 **18.** 3 **19.** −8 **20.** $y = -\frac{2}{5}x + 4$ **21.** $v = \frac{h - x - 16t^2}{t}$ **22.** Rick is 20, Dave is 40
23. ℓ = 20 in., w = 10 in. **24.** 8 quarters, 15 dimes **25.** \$800 at 7% and \$1,400 at 9%
26. $x < 1$ 1 **27.** $a < -4$ −4

28. $x \leq -3$ −3 **29.** $m \geq -2$ −2

30. $x \leq 2$ or $x \geq 5$ 2 5 **31.** $-3 < x < 5$ −3 5 **32.** $x > -4$ **33.** $-3 \leq x \leq 1$

34. $x < -4$ or $x > 1$ **35.** 11.3% **36.** 13.3%

Chapter 3

Pretest for Chapter 3

1-4.

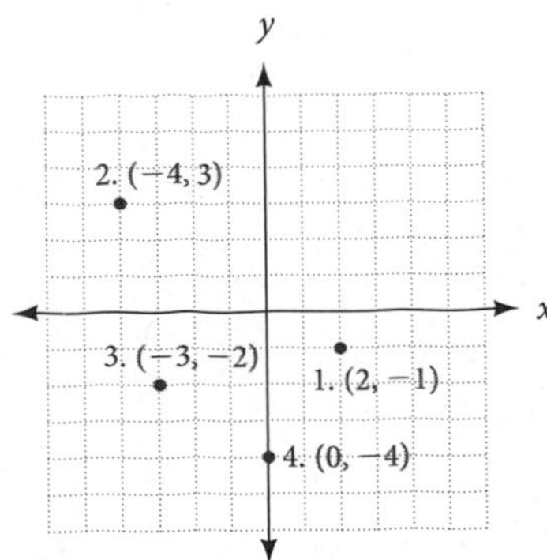

5. (0, −3), (2, 0), (4, 3), (−2, −6) **6.** (0, 7), (4, −5)

7. x-intercept = (2, 0); y-intercept = (0, −4) **8.** x-intercept = (−4, 0), y-intercept = (0, 6) **9.** $-\frac{1}{2}$ **10.** $-\frac{8}{7}$ **11.** −3

12. No slope **13.** $y = -\frac{1}{2}x + 3$ **14.** $y = 3x - 5$ **15.** $y = -\frac{2}{3}x - 2$ **16.** $y = -\frac{6}{5}x + \frac{18}{5}$

Getting Ready for Chapter 3

1. 3 **2.** 9 **3.** 1 **4.** 0 **5.** $\frac{3}{2}x - 3$ **6.** $-\frac{2}{3}x + 2$ **7.** 0 **8.** 15 **9.** 4 **10.** 2 **11.** −2 **12.** 9 **13.** 1 **14.** −2

15. 1 **16.** 0 **17.** $y = \frac{7}{2}x - 7$ **18.** $y = -\frac{2}{5}x - \frac{4}{5}$ **19.** $y = -2x - 5$ **20.** $y = -\frac{2}{3}x + 1$

Problem Set 3.1

1–17 .

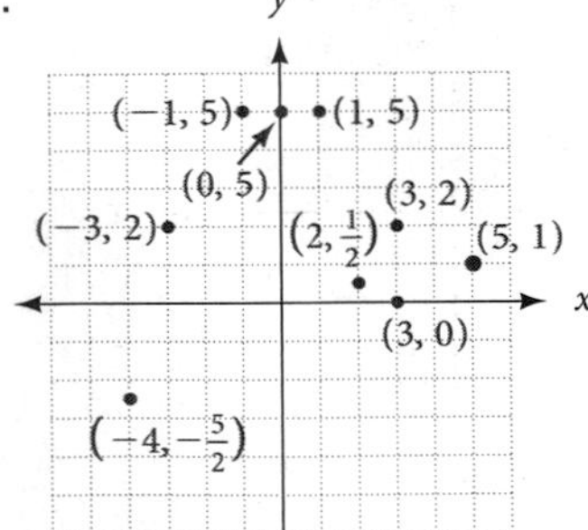

19. (−4, 4) **21.** (−4, 2) **23.** (−3, 0) **25.** (2, −2) **27.** (−5, −5)

29. Yes **31.** No

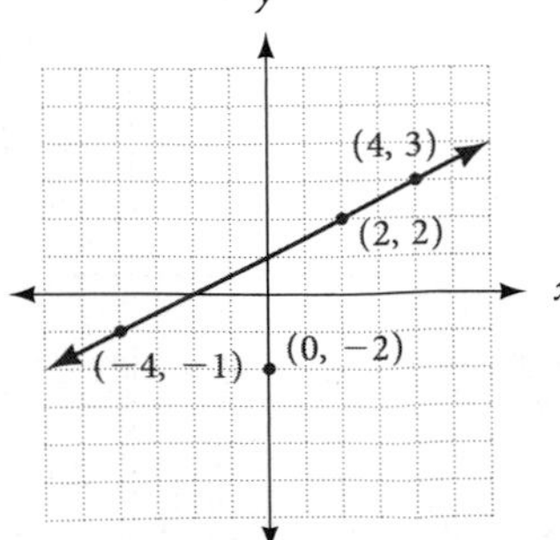

33. Yes **35.** No

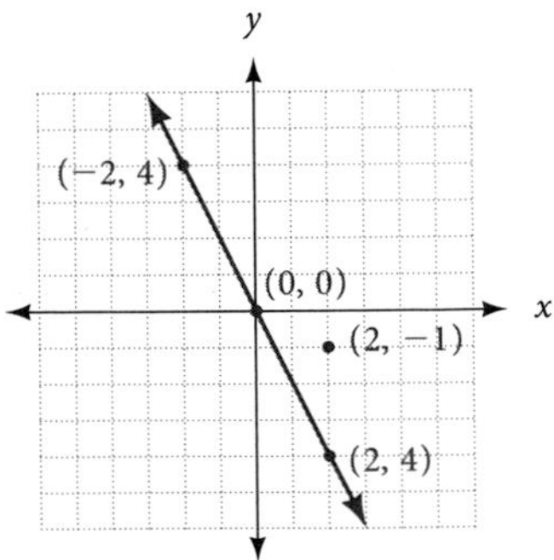

37. Yes **39.** No

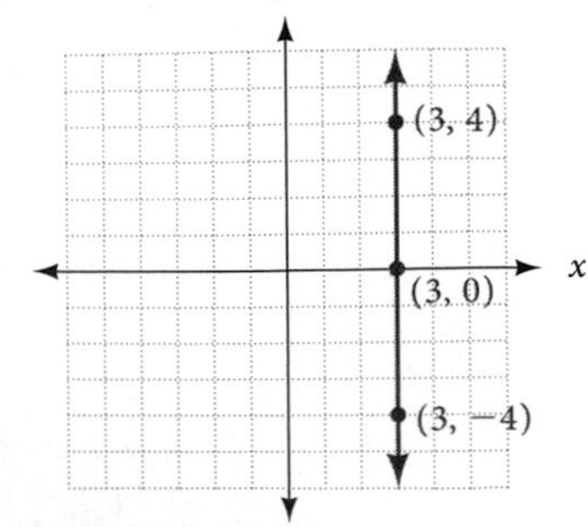

41. No **43.** No

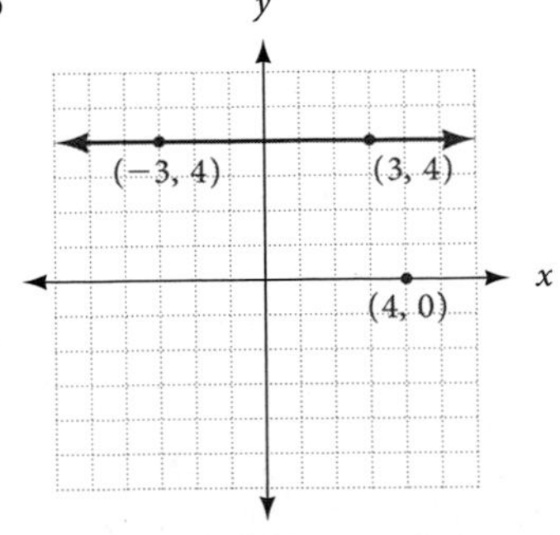

45. **a.** 60 watts **b.** 20 watts **c.** 100 watts

47. **a.** (5, 40), (10, 80), (15, 120), answers may vary **b.** \$320 **c.** 30 hours **d.** No, she should make \$280 for working 35 hours.

49. $A = (1, 2),\ B = (6, 7)$ **51.** $A = (2, 2),\ B = (2, 5),\ C = (7, 5)$ **53.** **a.** -3 **b.** 6 **c.** 0 **d.** -4 **55.** **a.** 4 **b.** 2 **c.** -1 **d.** 9

57. $\frac{x+3}{5}$ **59.** $\frac{2-a}{7}$ **61.** $\frac{1-2y}{14}$ **63.** $\frac{x+6}{2x}$ **65.** $\frac{x}{6}$ **67.** $\frac{8+x}{2x}$

Problem Set 3.2

1. (0, 6), (3, 0), (6, -6) **3.** (0, 3), (4, 0), (-4, 6) **5.** (1, 1), $\left(\frac{3}{4}, 0\right)$, (5, 17) **7.** (2, 13), (1, 6), (0, -1) **9.** (-5, 4), (-5, -3), (-5, 0)

11.

x	y
1	3
−3	−9
4	12
6	18

13.

x	y
0	0
$-\frac{1}{2}$	−2
−3	−12
3	12

15.

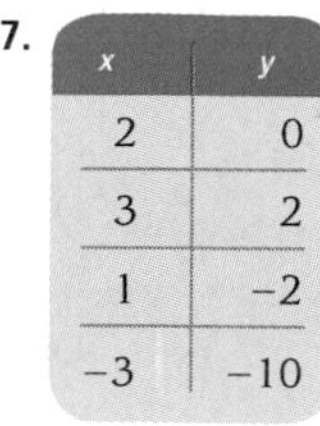

x	y
2	3
3	2
5	0
9	−4

17.

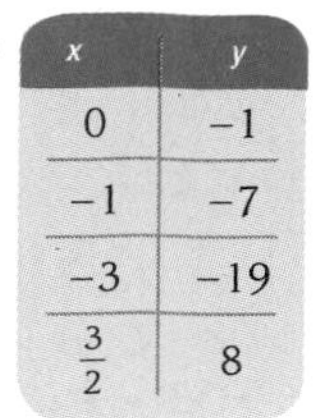

x	y
2	0
3	2
1	−2
−3	−10

19.

x	y
0	−1
−1	−7
−3	−19
$\frac{3}{2}$	8

21. (0, -2) **23.** (1, 5), (0, -2), (-2, -16) **25.** (1, 6), (-2, -12), (0, 0) **27.** (2, -2) **29.** (3, 0), (3, -3)

31. (2000, 20,000), (2003, 41,000), (2004, 48,000) **33.** 12 inches

35. **a.** Yes **b.** No, she should earn \$108 for working 9 hours. **c.** No, she should earn \$84 for working 7 hours. **d.** Yes

37. **a.** \$375,000 **b.** At the end of six years **c.** No, the crane will be worth \$195,000 after nine years. **d.** \$600,000

39. -3 **41.** 2 **43.** 0 **45.** $y = -5x + 4$ **47.** $y = \frac{3}{2}x - 3$ **49.** 6 **51.** 3 **53.** 6 **55.** 4

Problem Set 3.3

1.

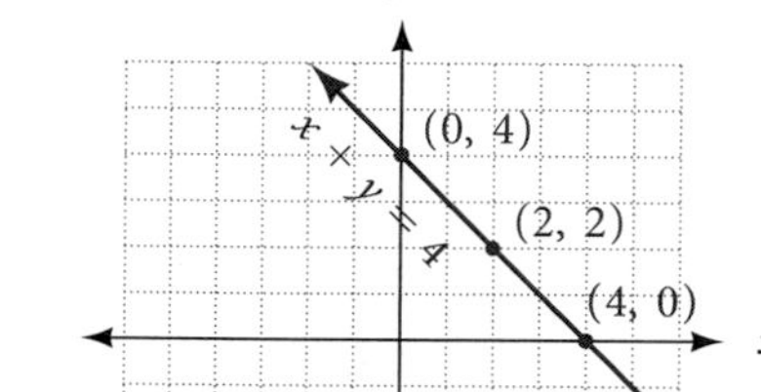

3.

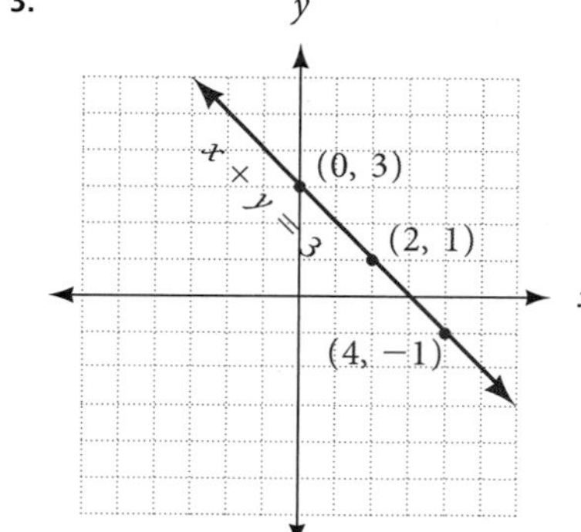

5.

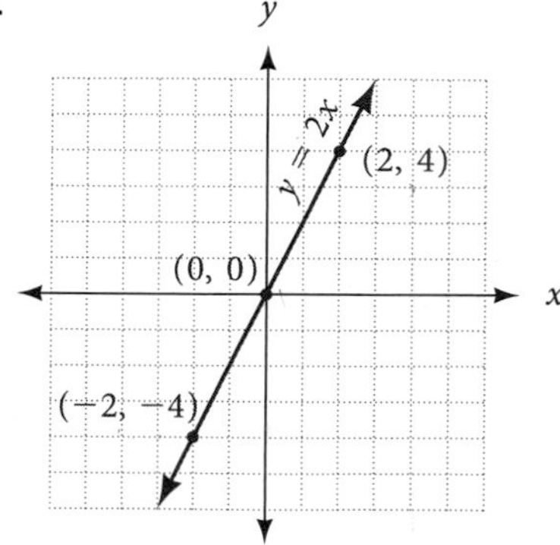

7.

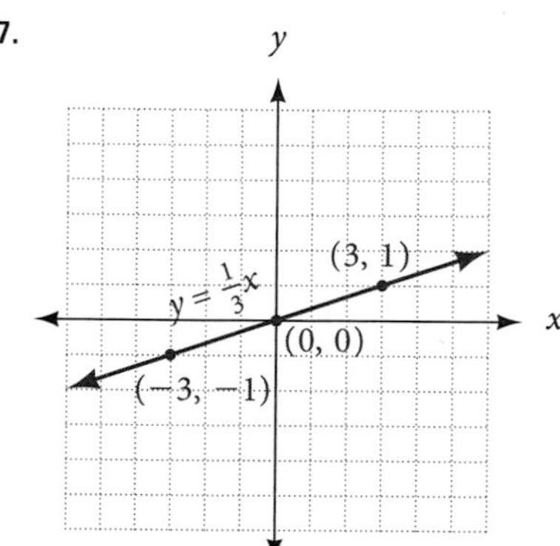

9.

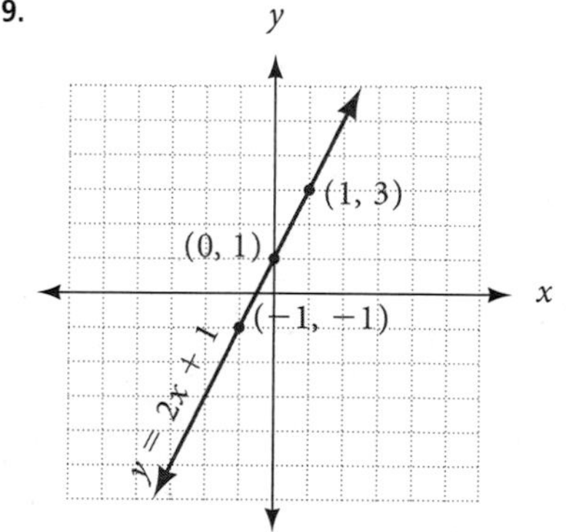

11.

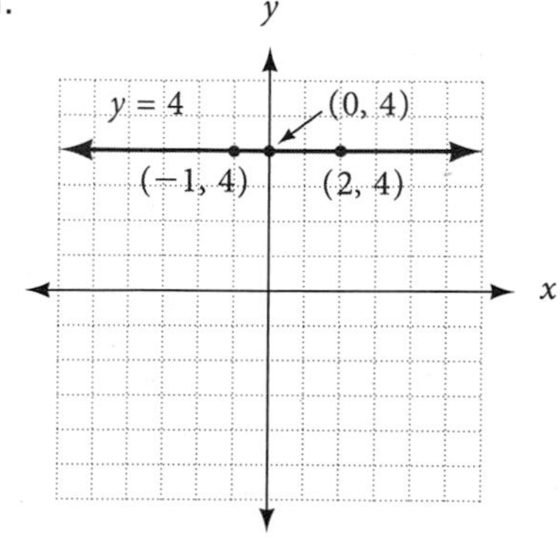

13.

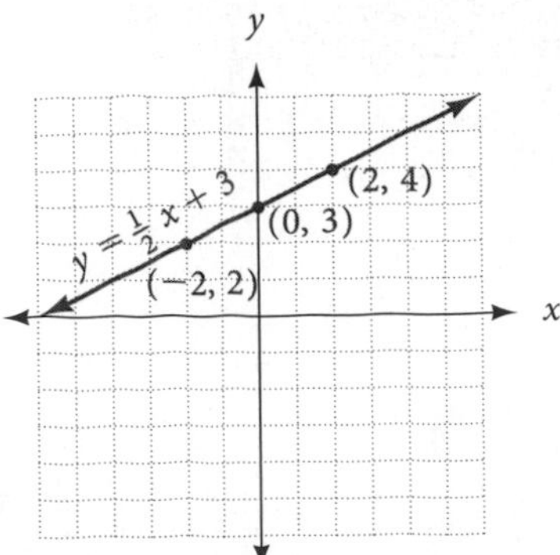

15.

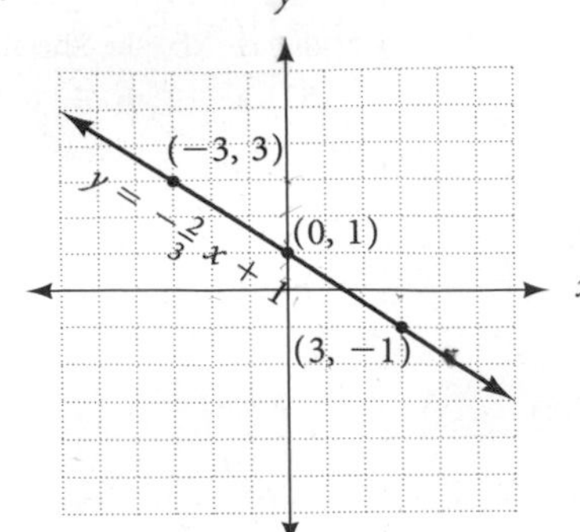

17.

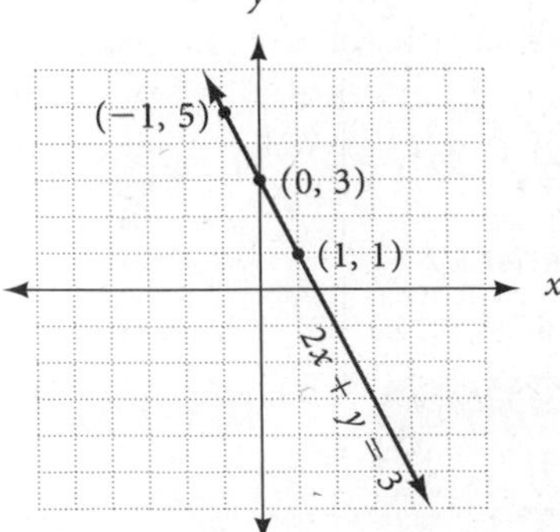

19.

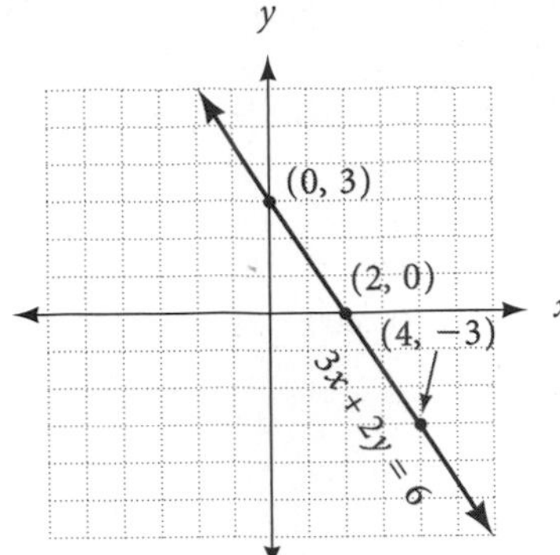

21.

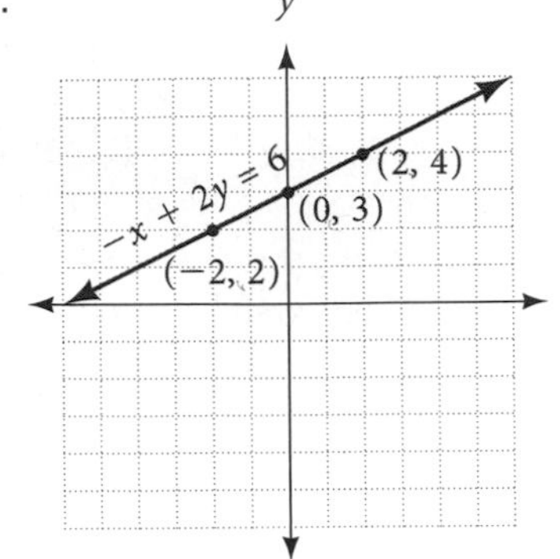

23.

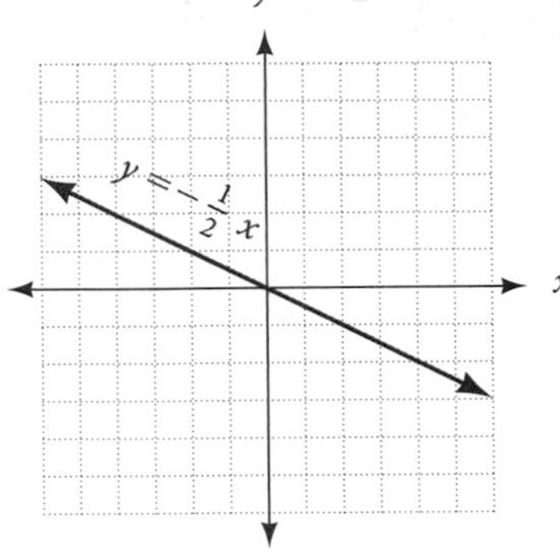

25.

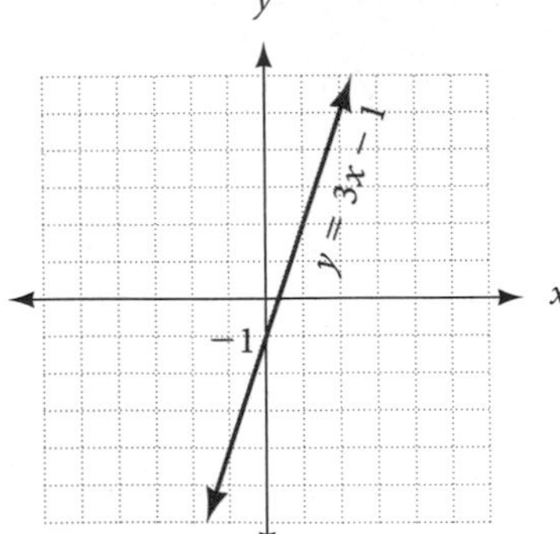

27.

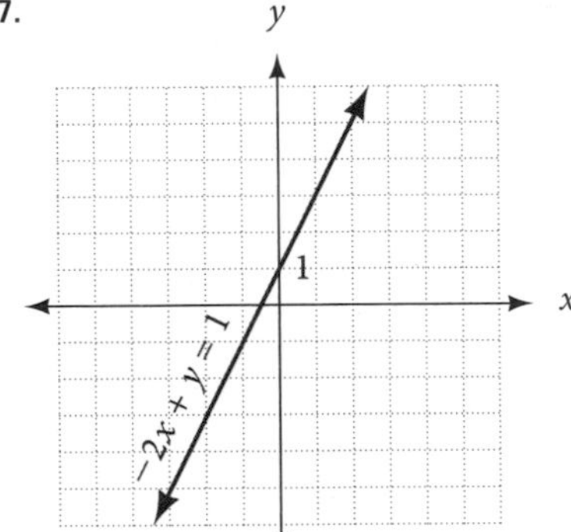

29.

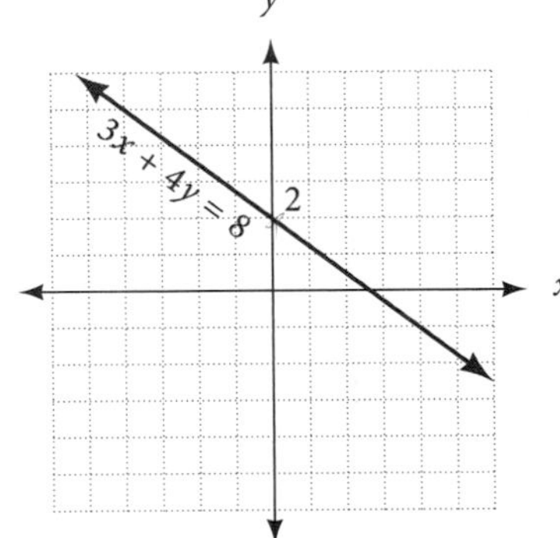

31.

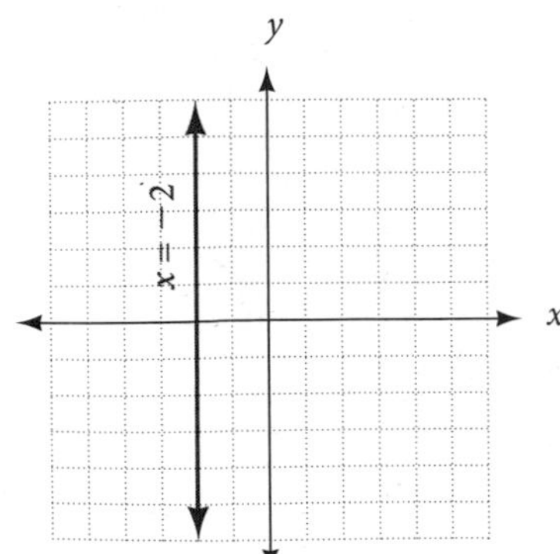

33.

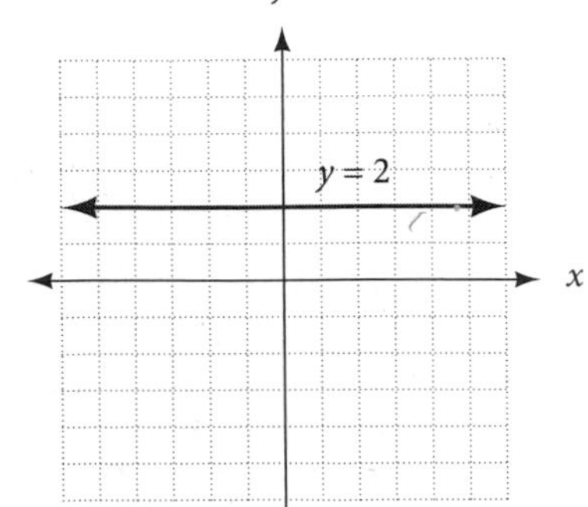

35.

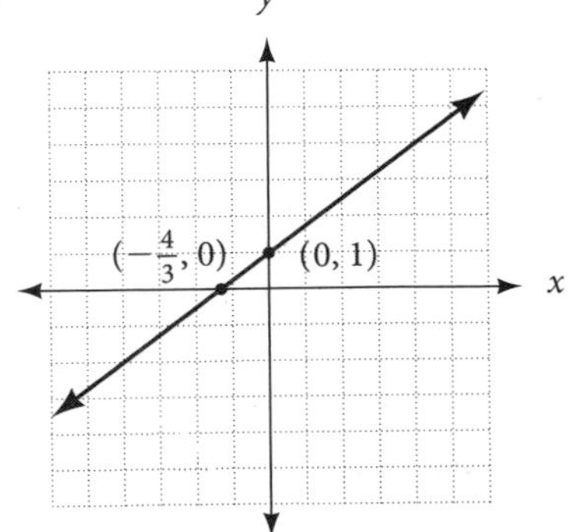

37.

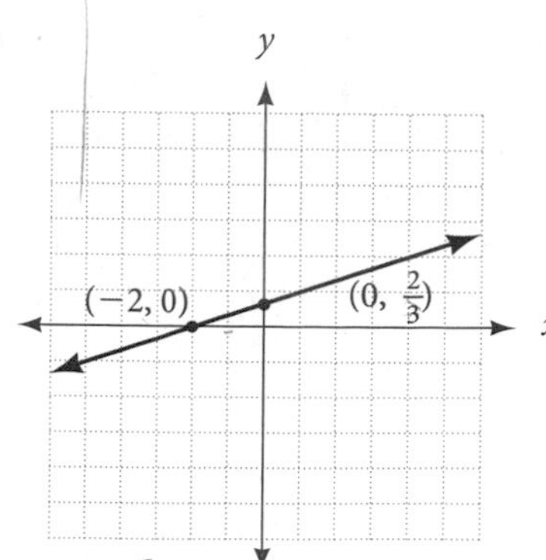

39.

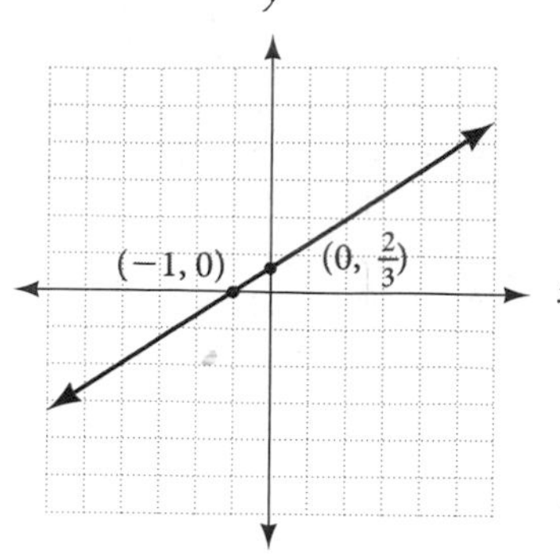

41.

Equation	H, V, and/or O
$x = 3$	V
$y = 3$	H
$y = 3x$	O
$y = 0$	O, H

43.

Equation	H, V, and/or O
$x = -\frac{3}{5}$	V
$y = -\frac{3}{5}$	H
$y = -\frac{3}{5}x$	O
$x = 0$	O, V

45. a. $\frac{5}{2}$ b. 5 c. 2 d.

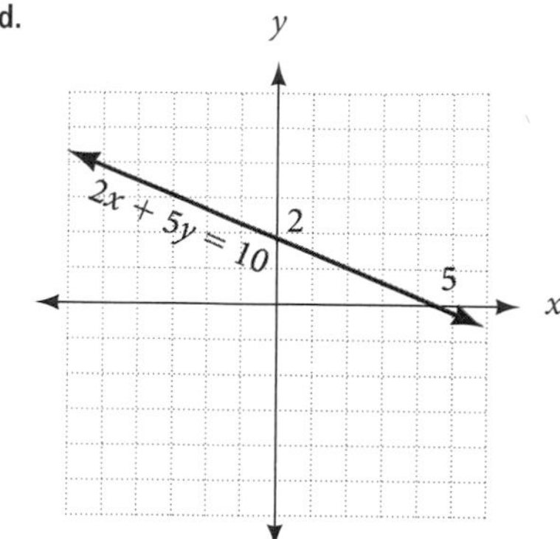

e. $y = -\frac{2}{5}x + 2$

47. $y = 21$ 49. a. 2 b. −3 51. a. −4 b. 2 53. a. 6 b. 2 55. $2x + 5$ 57. $2x - 6$ 59. $3x + \frac{15}{2}$ 61. $6x + 9$ 63. $2x + 50$ 65. $6x - 28$ 67. $\frac{3}{2}x + 9y$ 69. $\frac{5}{2}x - 5y + 6$

Problem Set 3.4

1.

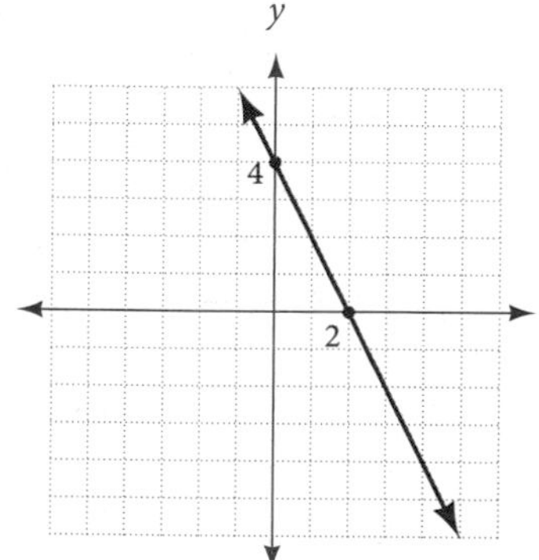

3.

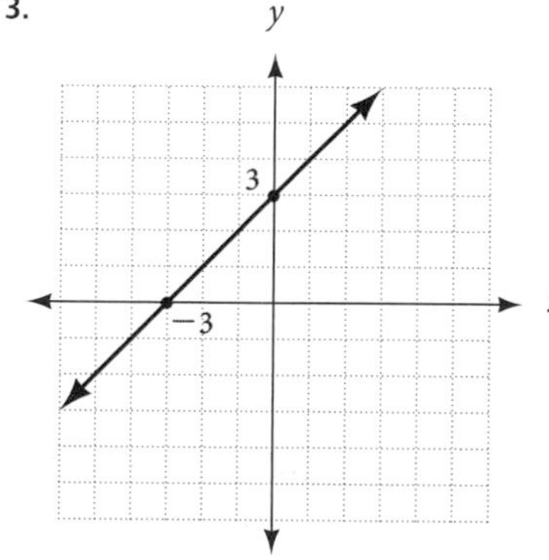

5.

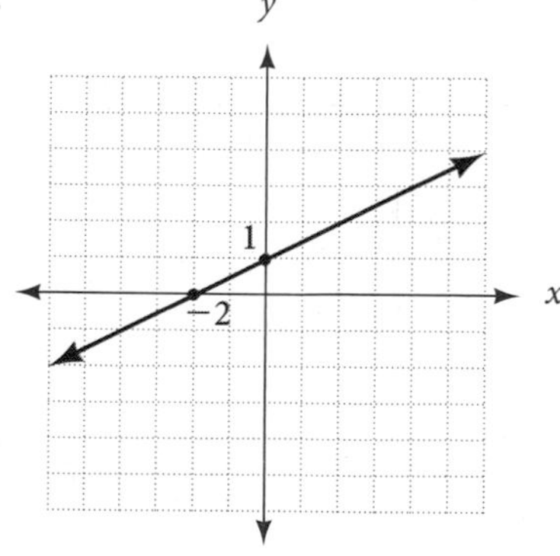

7.

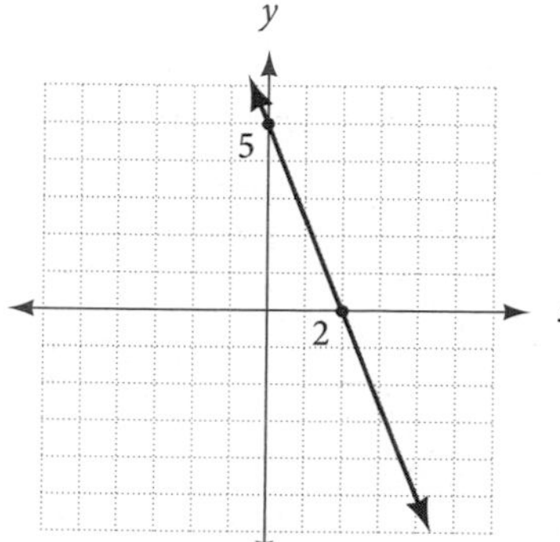

9.

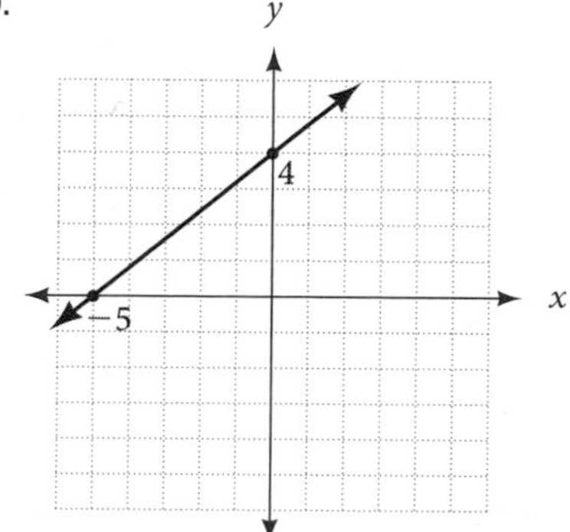

11.

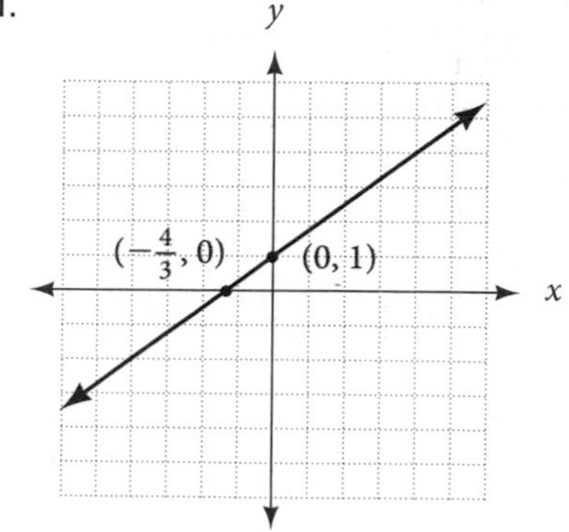

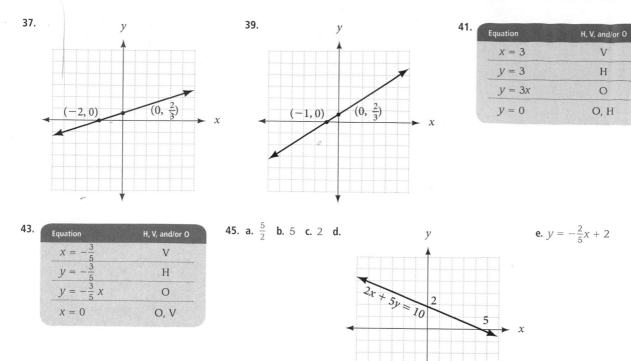

13.

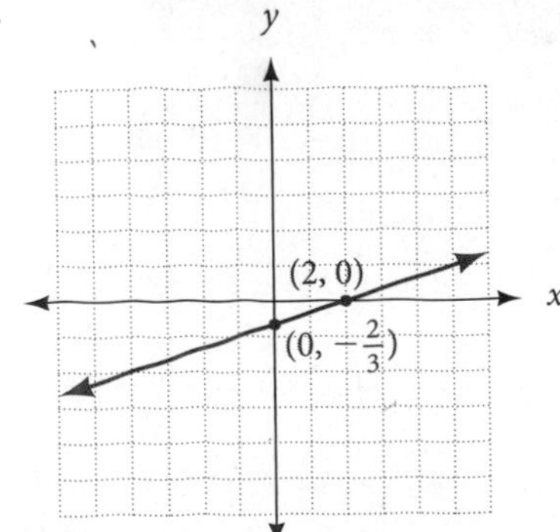

15.

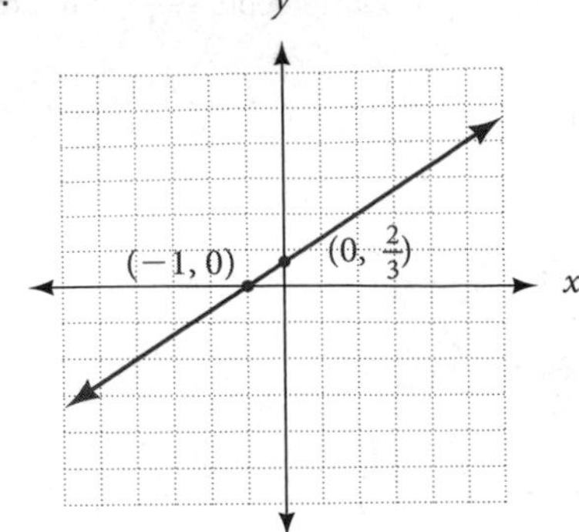

17.

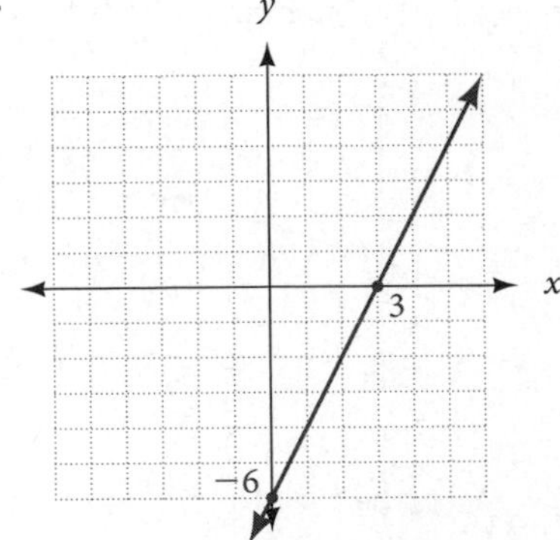

19.

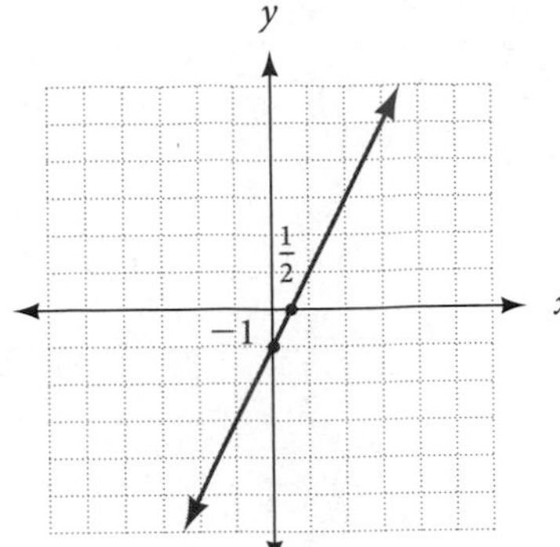

21.

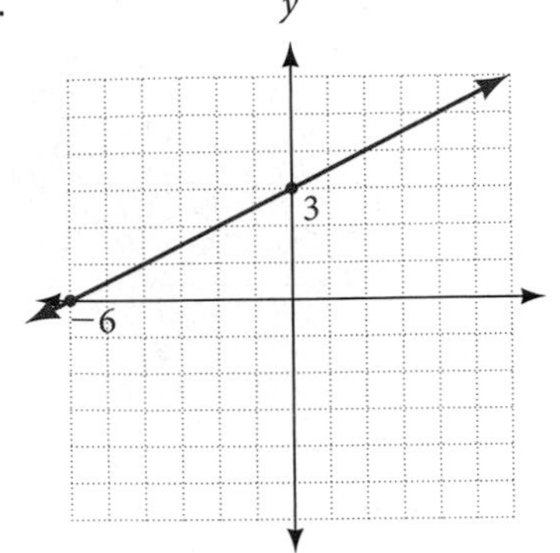

23.

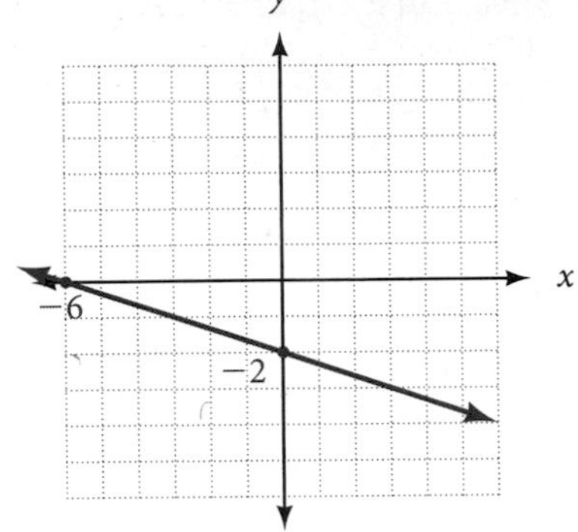

25.

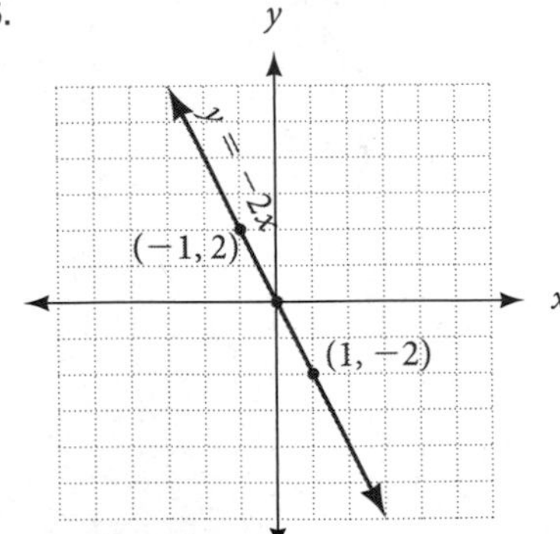

27.

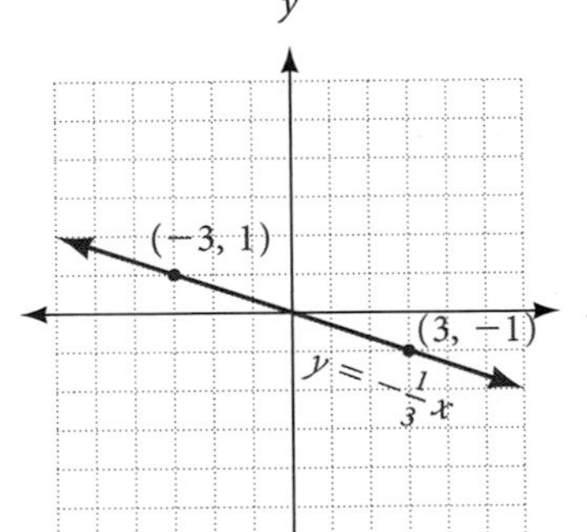

29.

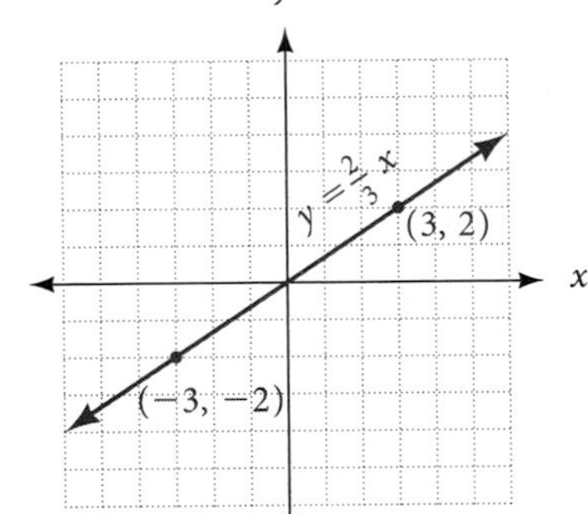

31.

Equation	x-intercept	y-intercept
$3x + 4y = 12$	4	3
$3x + 4y = 4$	$\frac{4}{3}$	1
$3x + 4y = 3$	1	$\frac{3}{4}$
$3x + 4y = 2$	$\frac{2}{3}$	$\frac{1}{2}$

33.

Equation	x-intercept	y-intercept
$x - 3y = 2$	2	$-\frac{2}{3}$
$y = \frac{1}{3}x - \frac{2}{3}$	2	$-\frac{2}{3}$
$x - 3y = 0$	0	0
$y = \frac{1}{3}x$	0	0

35. **a.** 0 **b.** $-\frac{3}{2}$ **c.** 1 **d.**

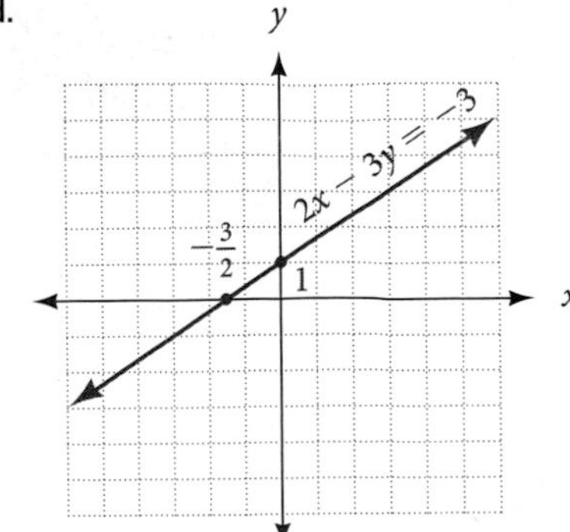

e. $y = \frac{2}{3}x + 1$

37.

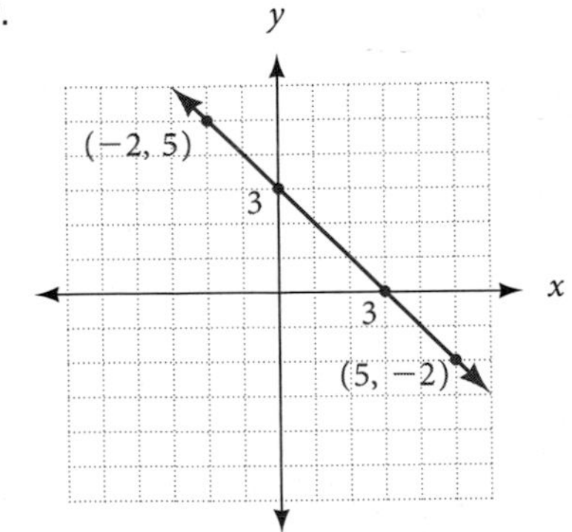

39. x-intercept = −1; y-intercept = −3 **41.**

x	y
−2	1
0	−1
−1	0
1	−2

43. x-intercept: $-\frac{400}{3}$; y-intercept: $\frac{16}{7}$

45. a. 10x + 12y = 480 **b.** x-intercept = 48; y-intercept = 40 **c.**

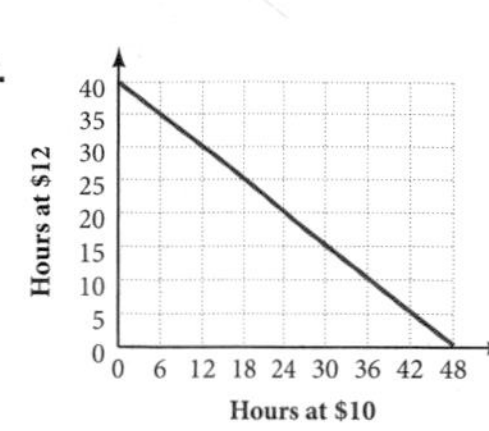

d. 10 hours **e.** 18 hours

47. a. $\frac{3}{2}$ **b.** $\frac{3}{2}$ **49. a.** $\frac{3}{2}$ **b.** $\frac{3}{2}$ **51.** 12 **53.** −11 **55.** −6 **57.** 7 **59.** −2 **61.** −1 **63.** 14 **65.** −6

Problem Set 3.5

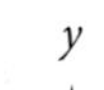

1.

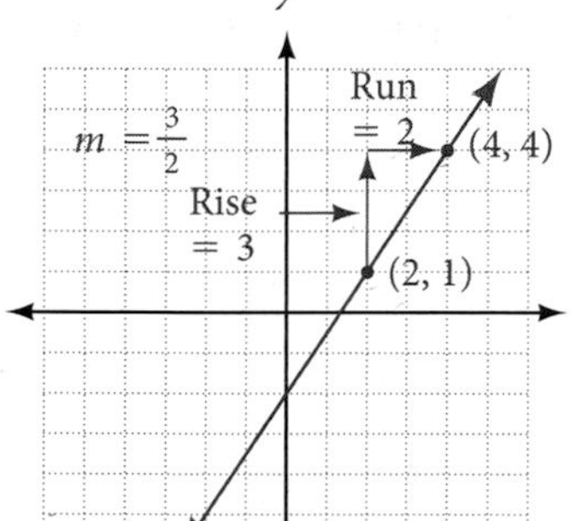

3.

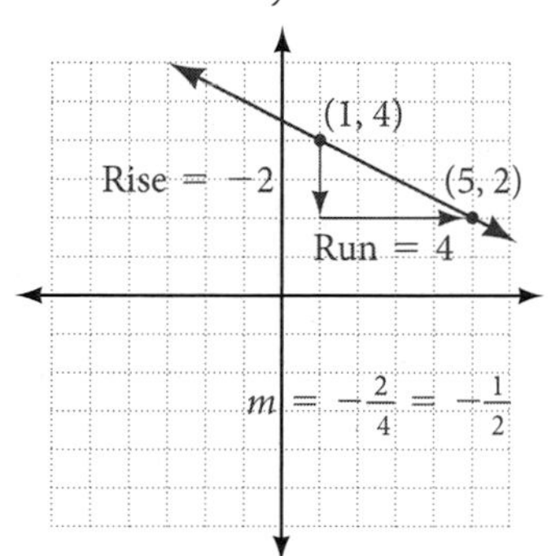

5.

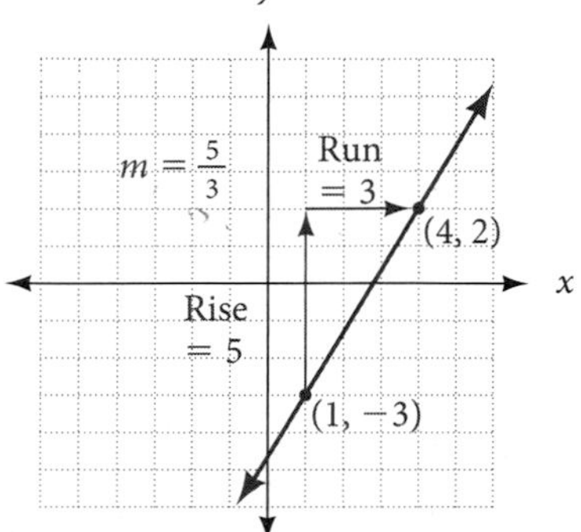

7.

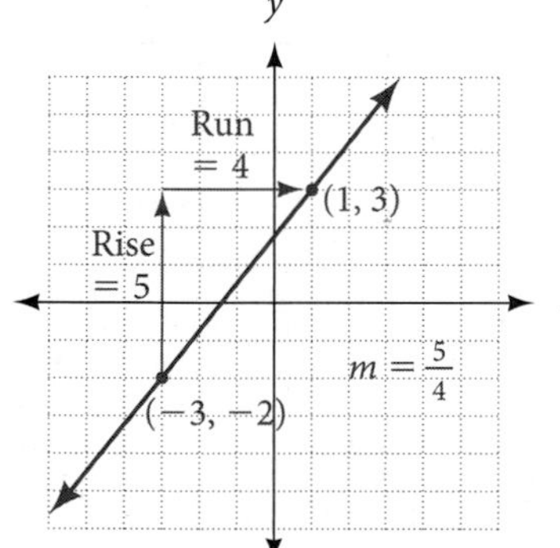

9.

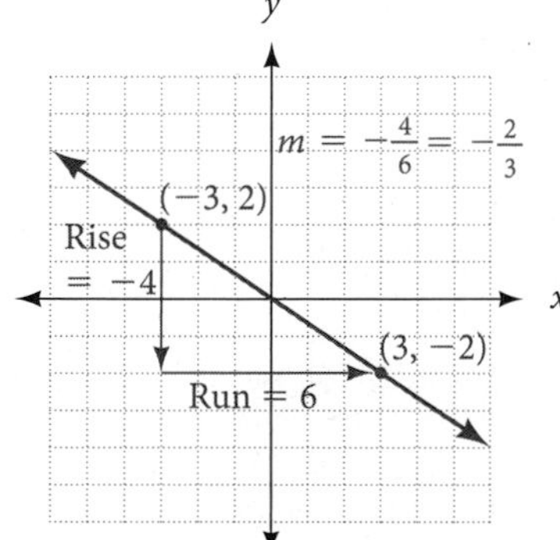

11.

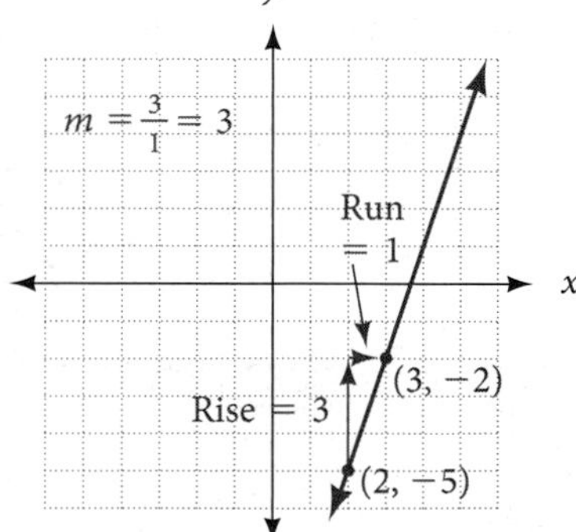

13.

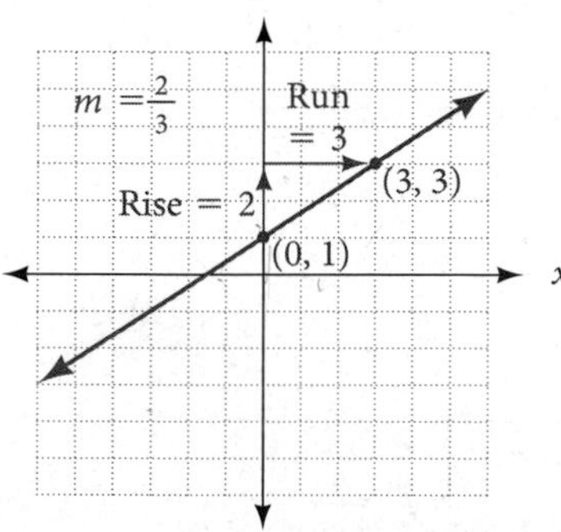

15.

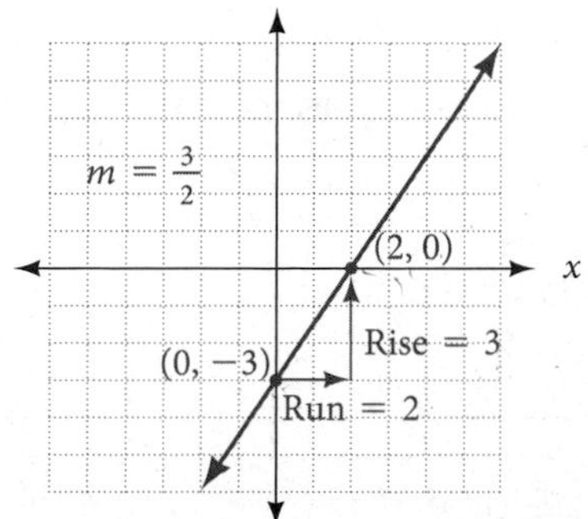

17.

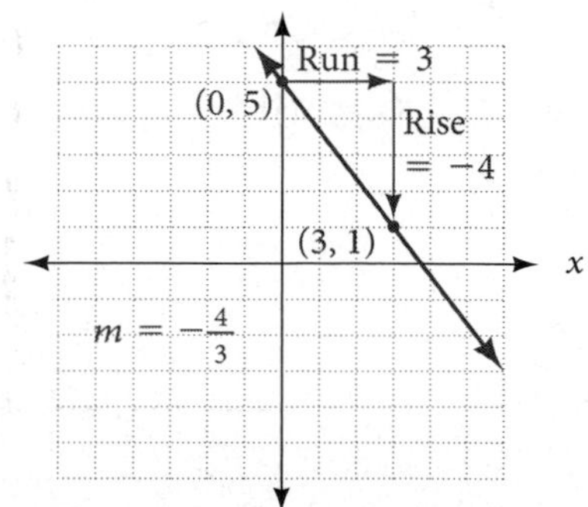

19.

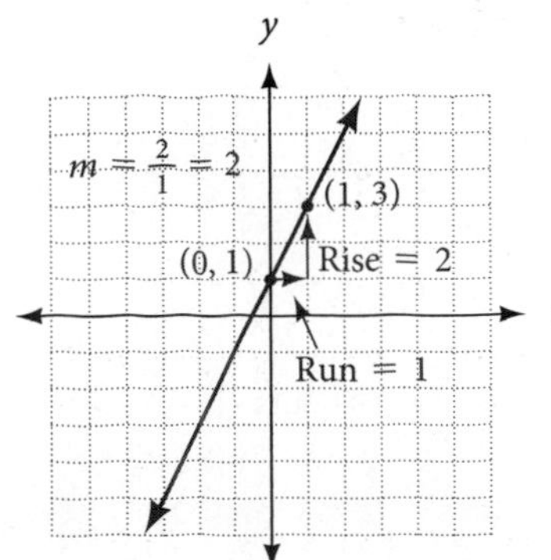

21.

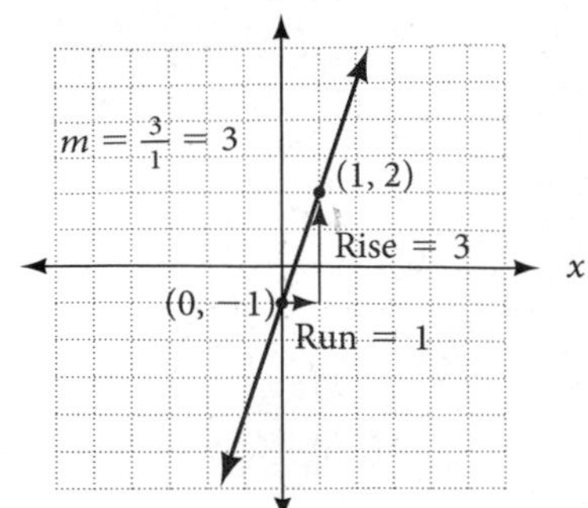

23. Slope = 3, y-intercept = 2

25. Slope = 2, y-intercept = −2

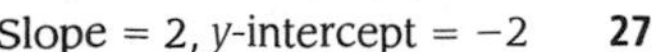

27.

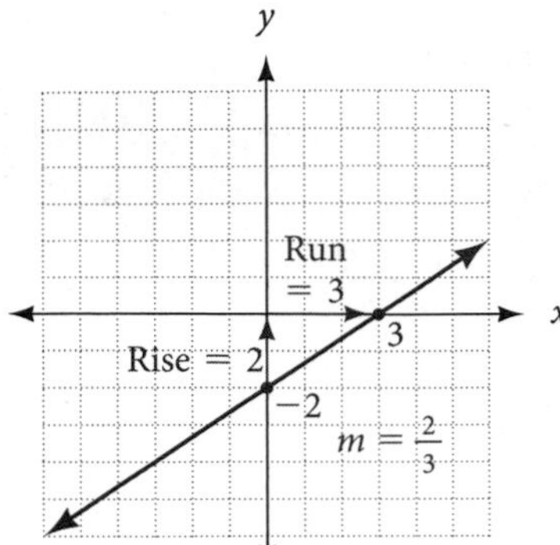

29.

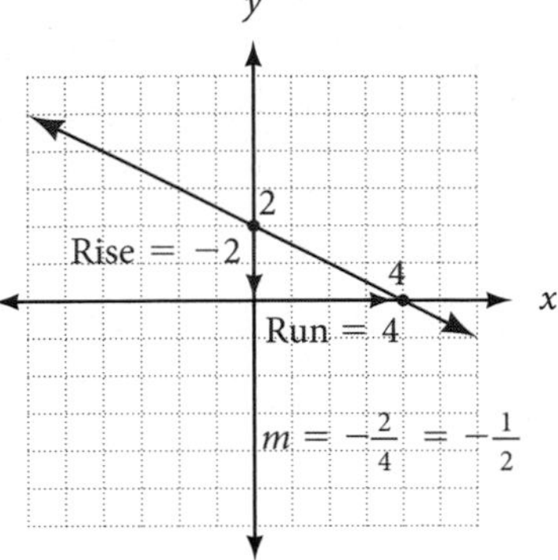

31.

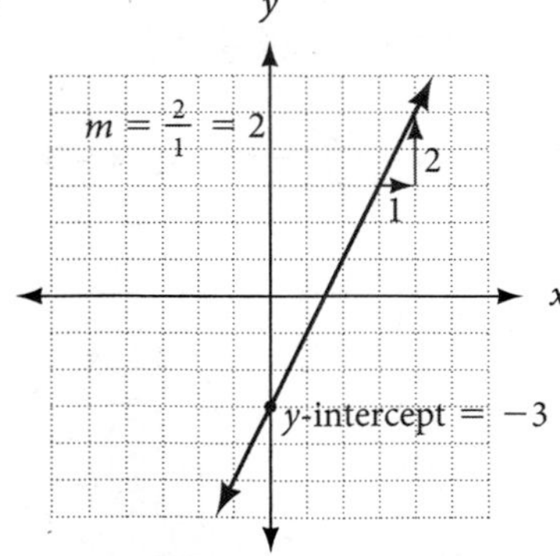

33.

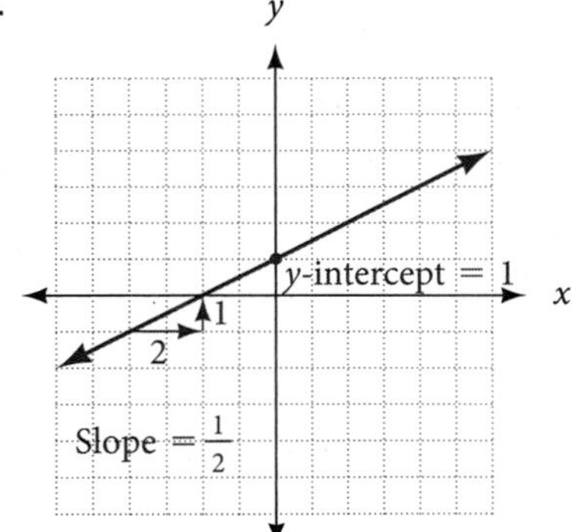

35. 6

37.

Equation	Slope
$x = 3$	Undefined
$y = 3$	0
$y = 3x$	3

39.

Equation	Slope
$x = -\frac{2}{3}$	0
$y = -\frac{2}{3}$	Undefined
$y = -\frac{2}{3}x$	$-\frac{2}{3}$

41. 154.23 **43.** Slopes: A, 3.3; B, 3.1; C, 5.3; D, 1.9 **45.** $y = 2x + 4$ **47.** $y = -3x + 3$ **49.** $y = \frac{4}{5}x - 4$ **51.** $y = -2x - 5$ **53.** $y = -\frac{2}{3}x + 1$ **55.** $y = \frac{3}{2}x + 1$ **57.** 3 **59.** 15 **61.** −6 **63.** 6 **65.** −39 **67.** 4 **69.** 10 **71.** −20

Problem Set 3.6

1. $y = \frac{2}{3}x + 1$ **3.** $y = \frac{3}{2}x - 1$ **5.** $y = -\frac{2}{5}x + 3$ **7.** $y = 2x - 4$

9. $m = 2, b = 4$

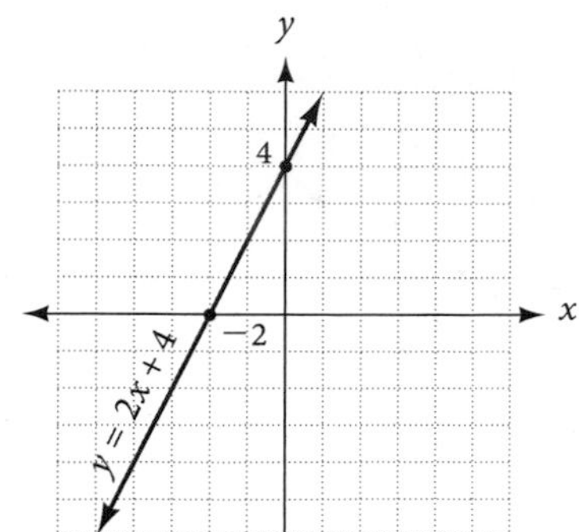

11. $m = -3, b = 3$

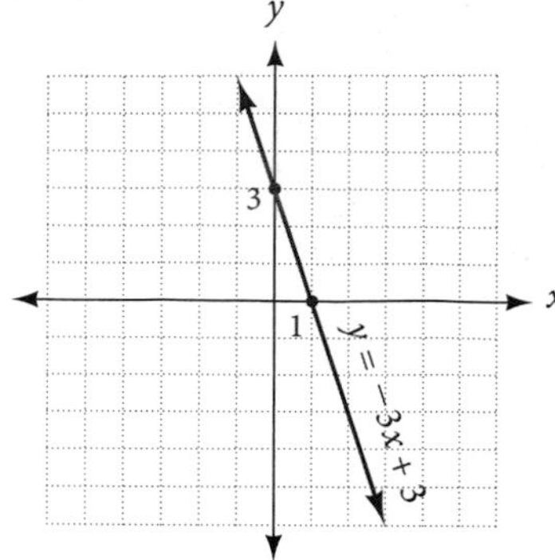

13. $m = -\frac{3}{2}, b = 3$

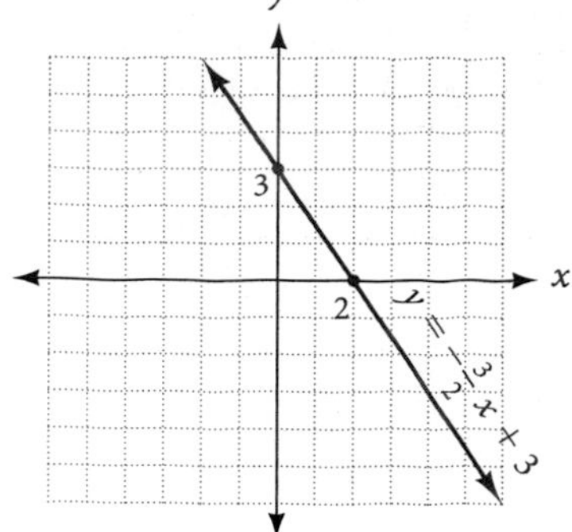

15. $m = \frac{4}{5}, b = -4$

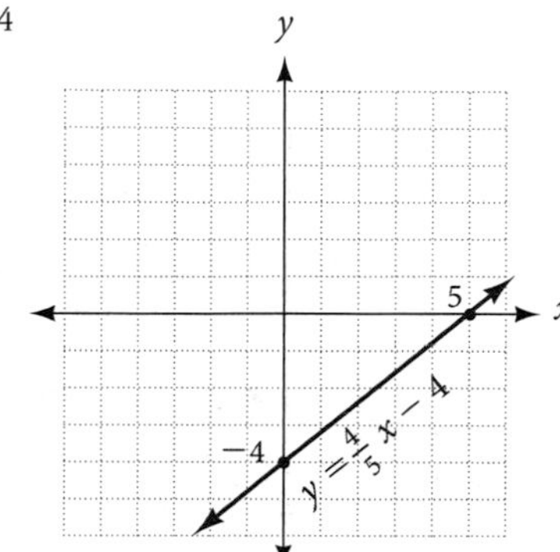

17. $m = -\frac{2}{5}, b = -2$

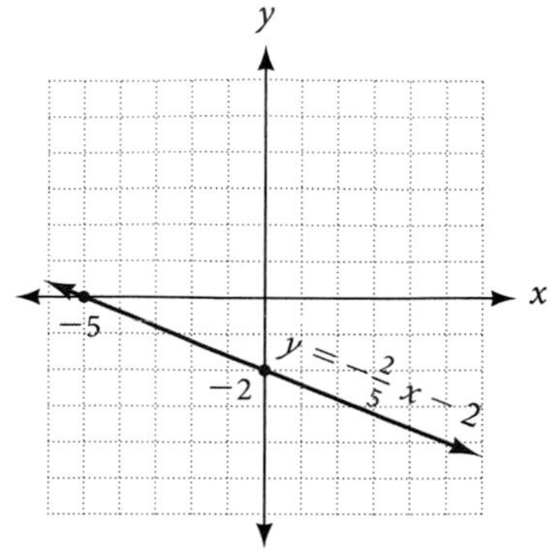

19. $y = 2x - 1$ **21.** $y = -\frac{1}{2}x - 1$ **23.** $y = \frac{3}{2}x - 6$ **25.** $y = -3x + 1$ **27.** $y = x - 2$ **29.** $y = 2x - 3$ **31.** $y = \frac{4}{3}x + 2$
33. $y = -\frac{2}{3}x - 3$ **35.** $m = 3, b = 3, y = 3x + 3$ **37.** $m = \frac{1}{4}, b = -1, y = \frac{1}{4}x - 1$
39. **a.**  $-\frac{5}{2}$ **b.** $y = 2x + 6$ **c.** 6 **d.** 2 **e.**

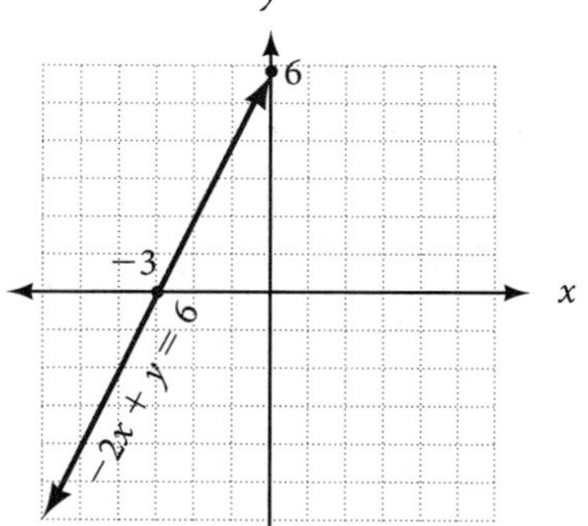

41. $y = -\frac{2}{3}x + 2$ **43.** $y = -\frac{5}{2}x - 5$ **45.** $x = 3$

47. $y = 7.72x - 15{,}374.6$ **49.** **a.** \$6,000 **b.** 3 years **c.** Slope = −3,000 **d.** \$3,000 **e.** $V = -3000t + 21{,}000$

51.

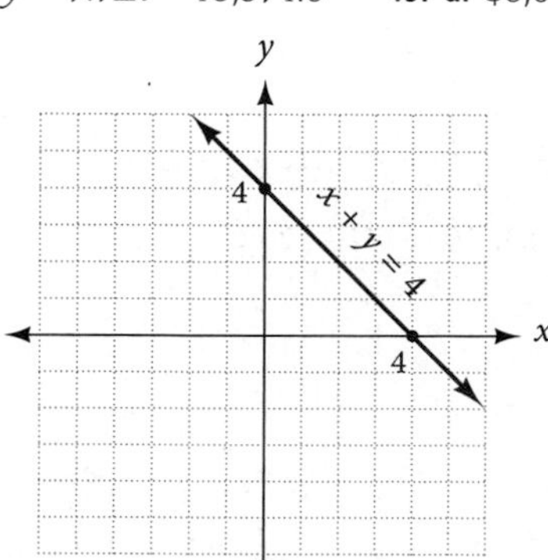

53.

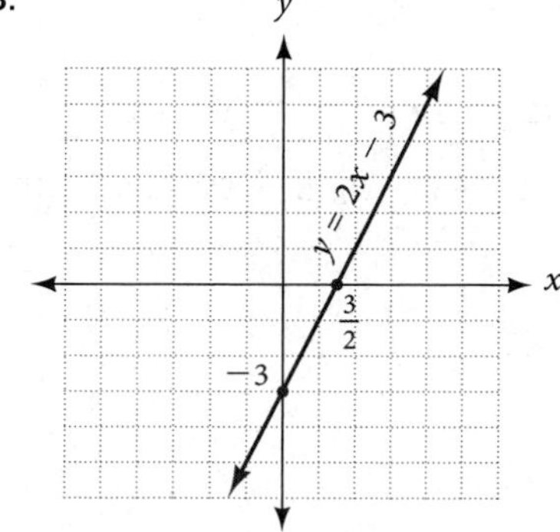

55.

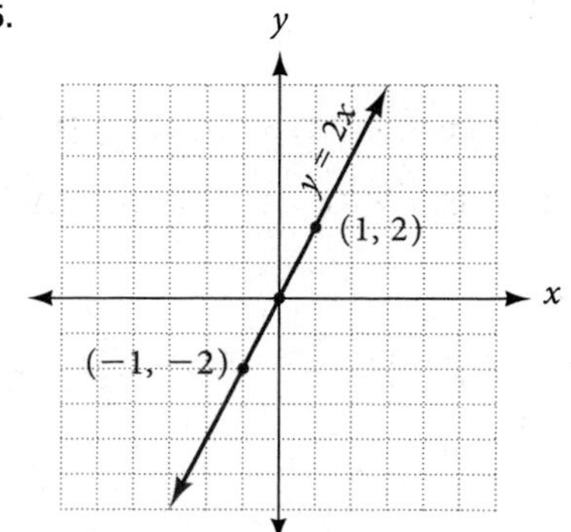

57. 6 **59.** −2 **61.** 4 **63.** −11 **65.** 10 **67.** −38 **69.** −12 **71.** −18

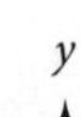

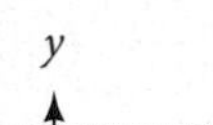

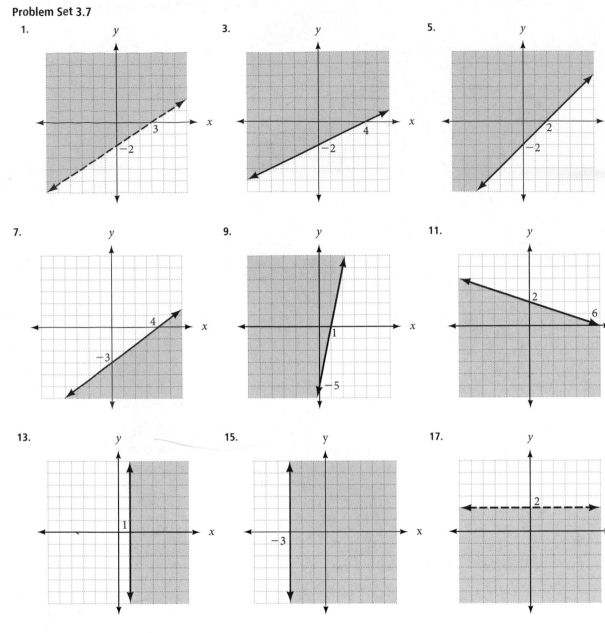

Problem Set 3.7

1.

3.

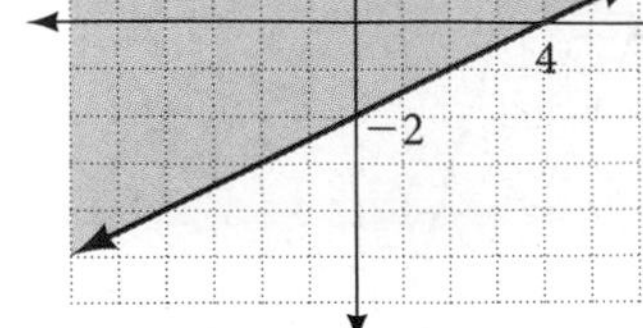

5.

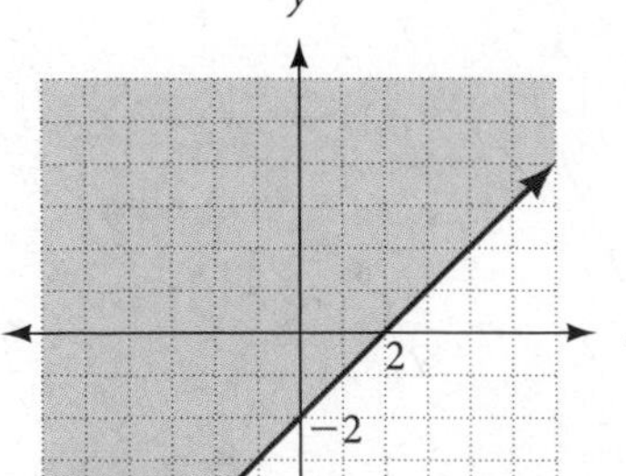

7.

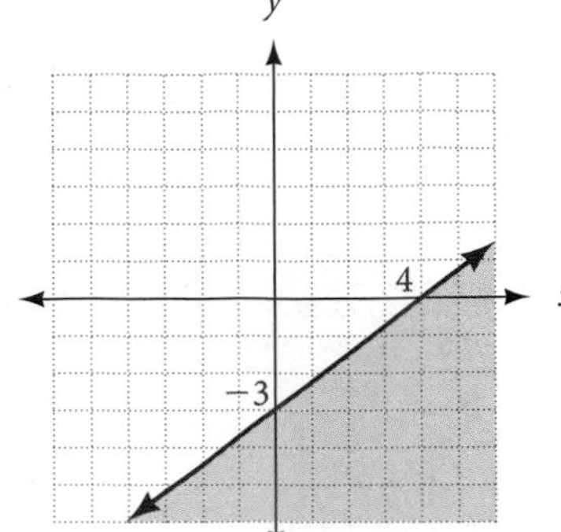

9.

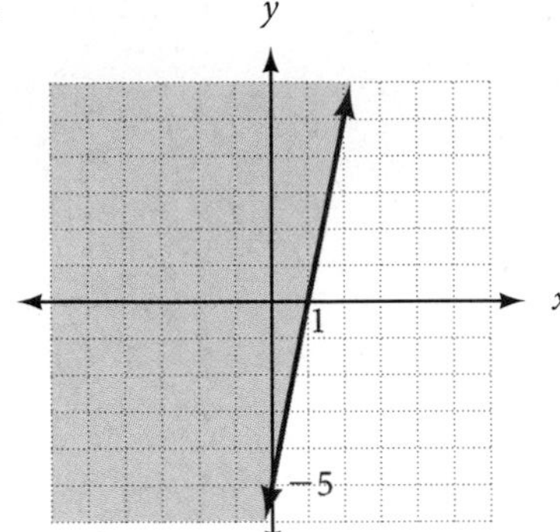

11.

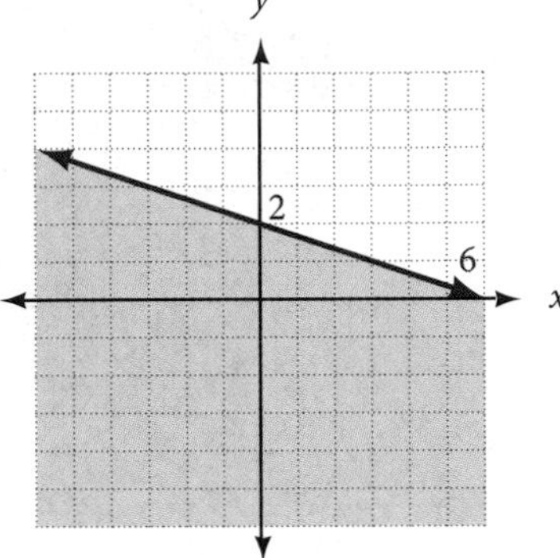

13.

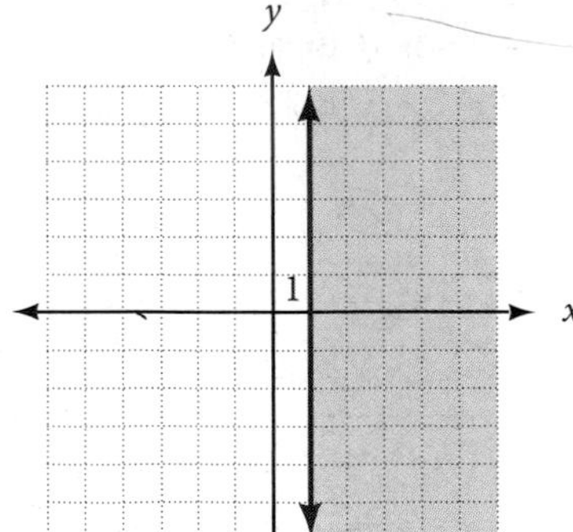

15.

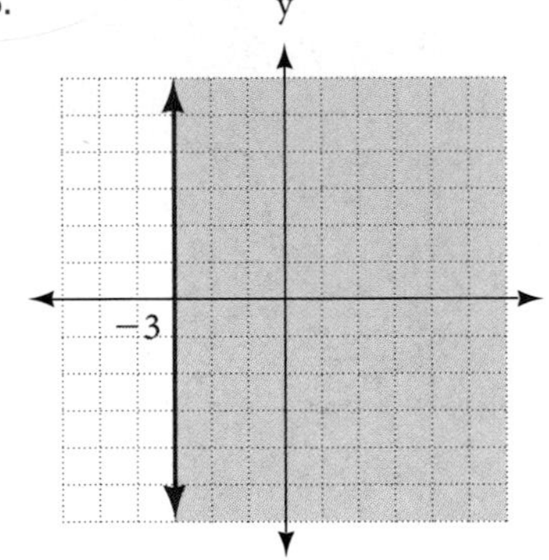

17.

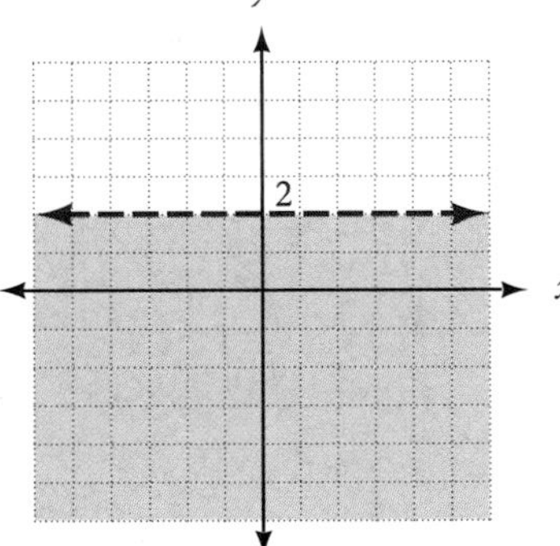

19.

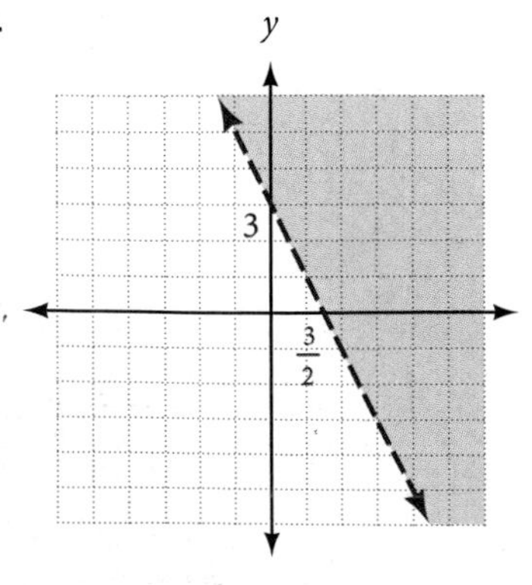

21.

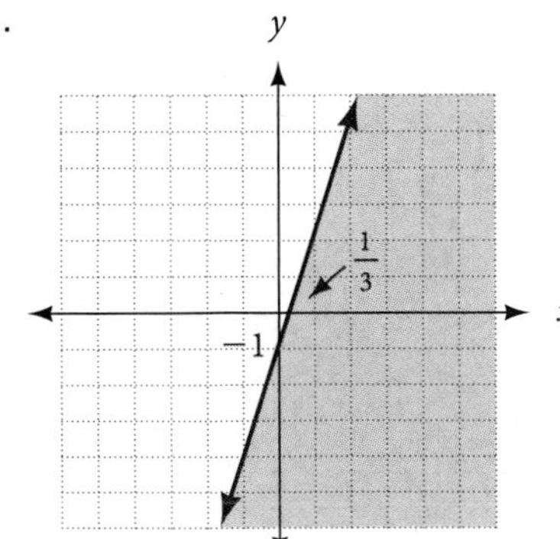

23.

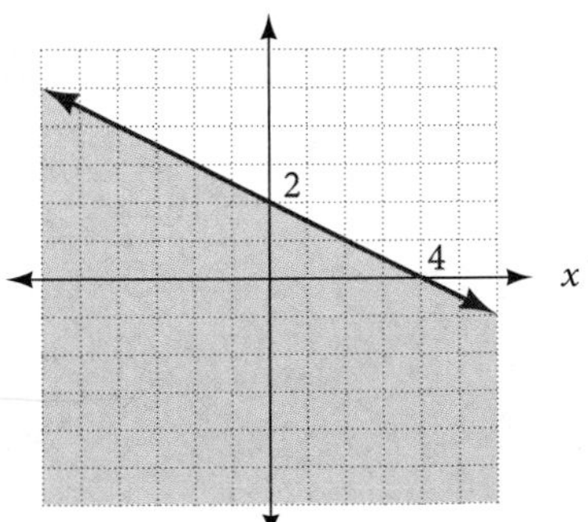

25. a. $y < \frac{8}{3}$ **b.** $y > -\frac{8}{3}$ **c.** $y = -\frac{4}{3}x + 4$ **d.**

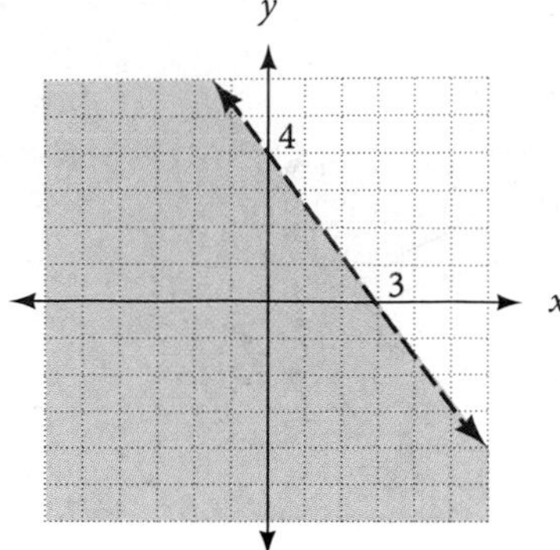

27. a. $y = \frac{2}{5}x + 2$ **b.** $y < \frac{2}{5}x + 2$ **c.** $y > \frac{2}{5}x + 2$

29.

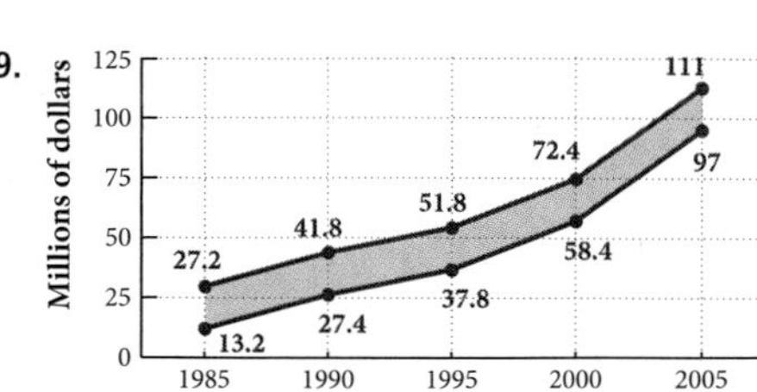

31. $-6x + 11$ **33.** -8 **35.** -4

37. $w = \frac{P - 2l}{2}$ **39.** (number line) -1 0 **41.** $y \geq \frac{3}{2}x - 6$ **43.** Width 2 inches, length 11 inches

Chapter 3 Review

1. $(4, -6), (0, 6), (1, 3), (2, 0)$ **2.** $(5, -2), (0, -4), (15, 2), (10, 0)$ **3.** $(4, 2), (2, -2), \left(\frac{9}{2}, 3\right)$ **4.** $(2, 13), \left(-\frac{3}{5}, 0\right), \left(-\frac{6}{5}, -3\right)$

5. $(2, -3), (-1, -3), (-3, -3)$ **6.** $(6, 5), (6, 0), (6, -1)$ **7.** $\left(2, -\frac{3}{2}\right)$ **8.** $\left(-\frac{8}{3}, -1\right), (-3, -2)$

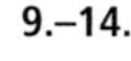

9.–14.

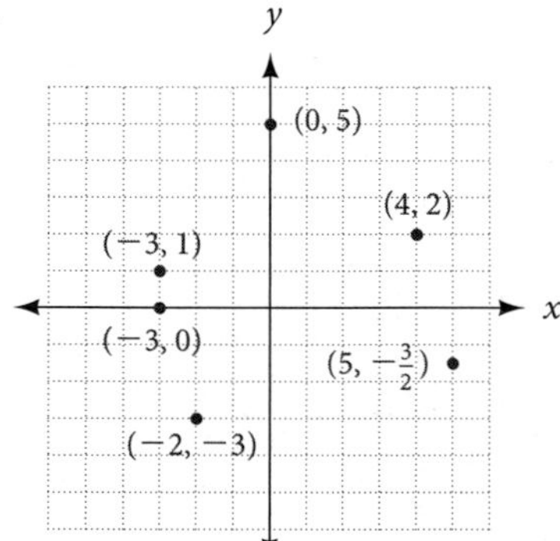

15. $(-2, 0), (0, -2), (1, -3)$

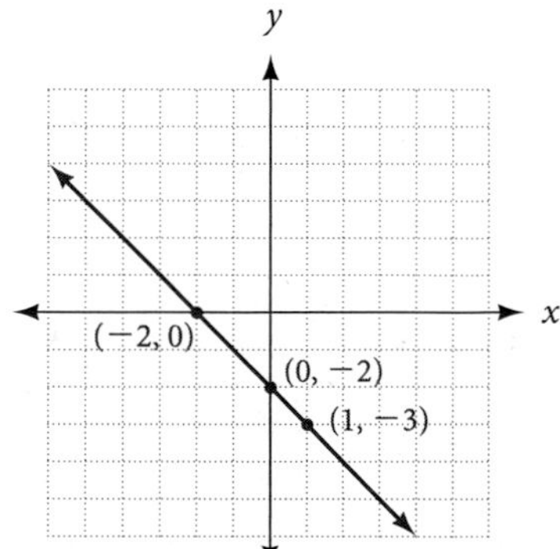

16. $(-1, -3), (1, 3), (0, 0)$

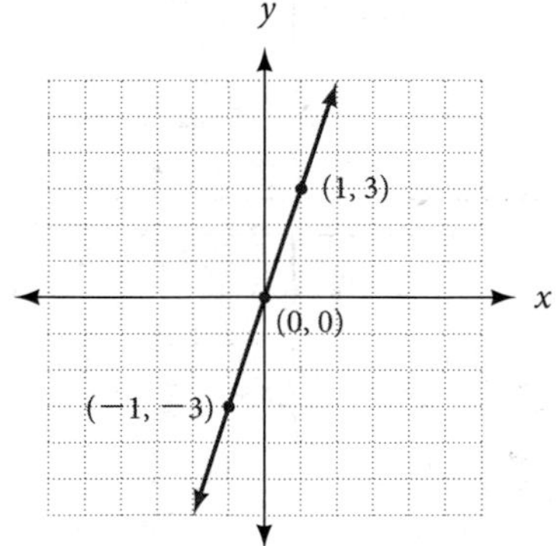

17. $(1, 1), (0, -1), (-1, -3)$

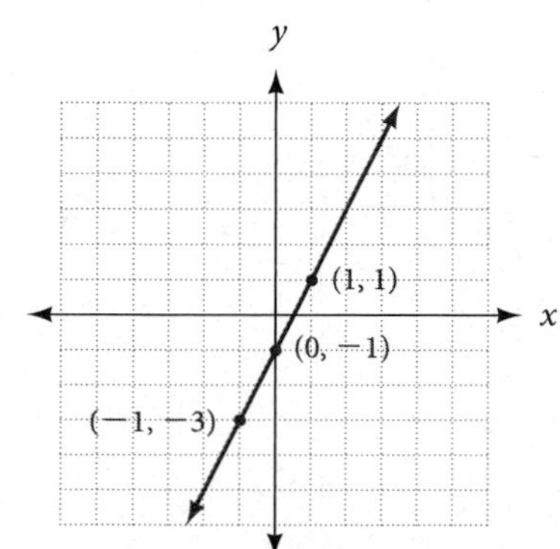

18. $(-3, 0), (-3, 5), (-3, -5)$

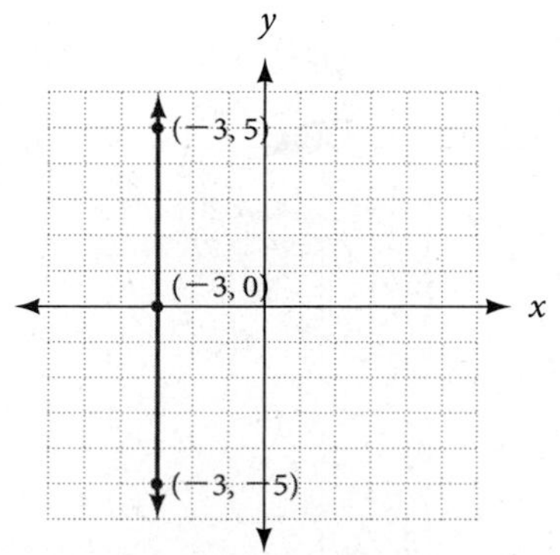

19.

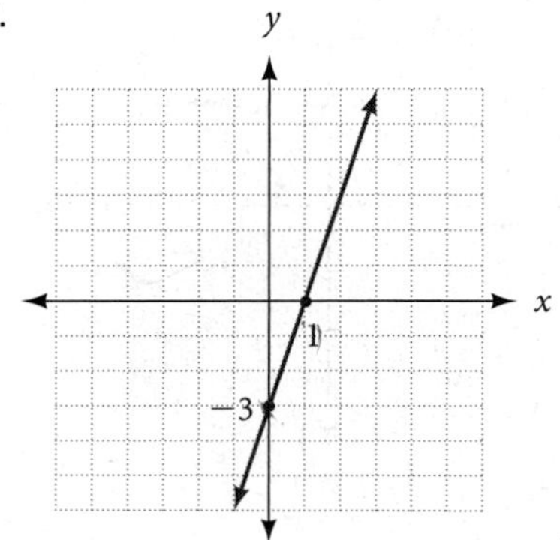

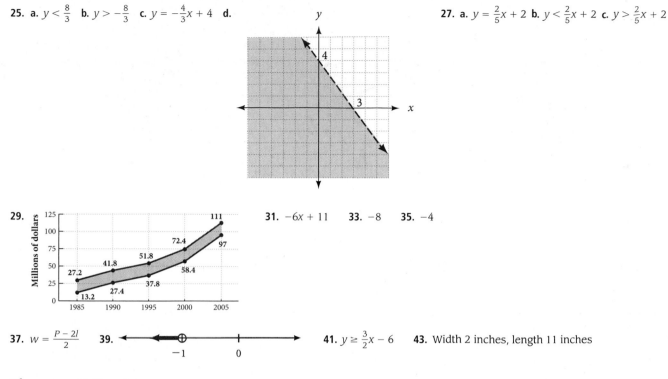

20.

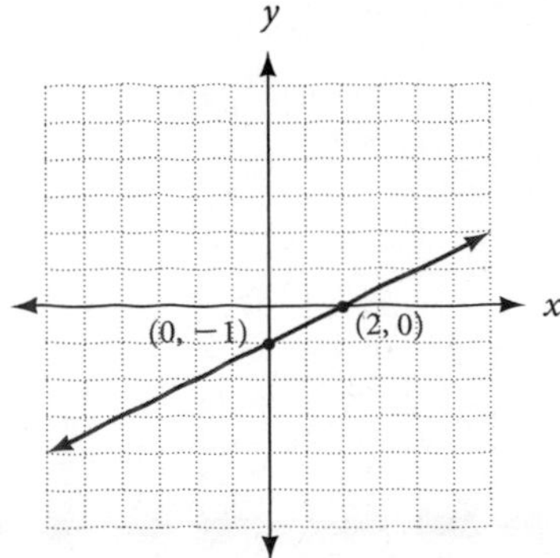

21.

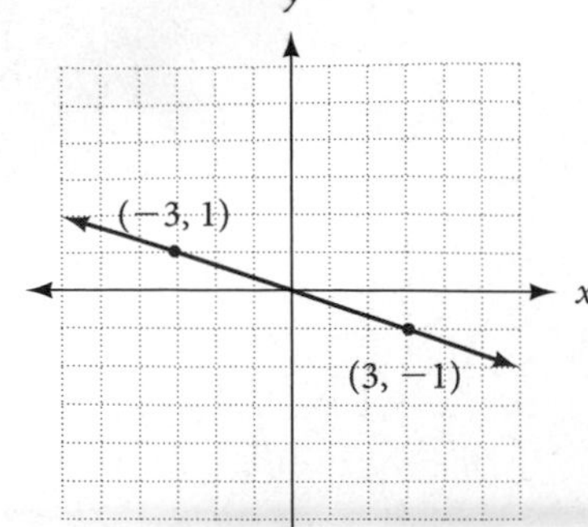

22.

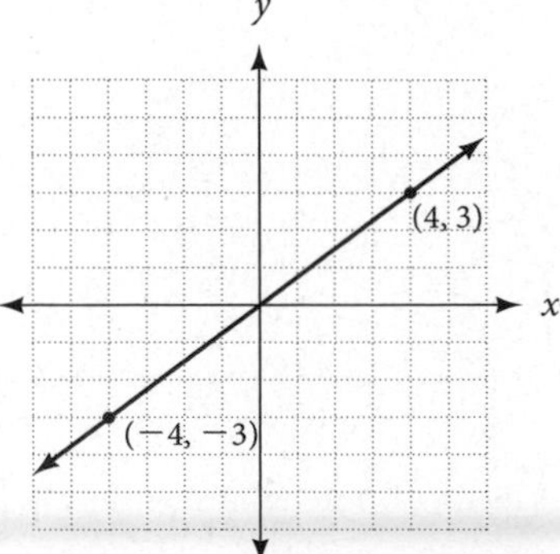

23.

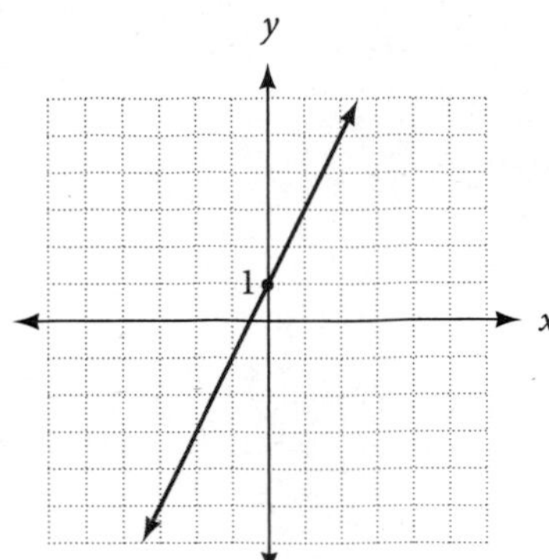

24.

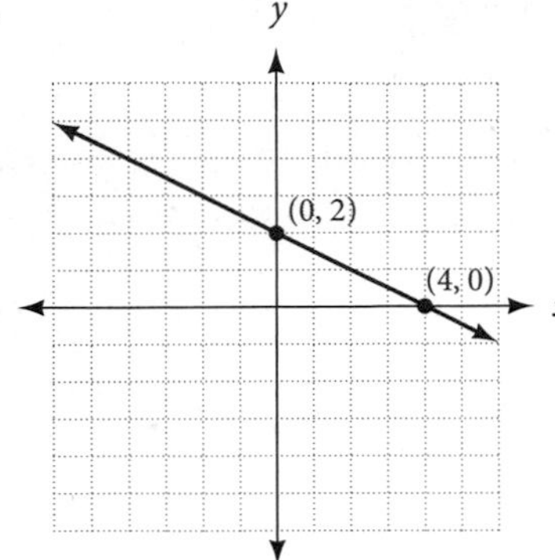

25.

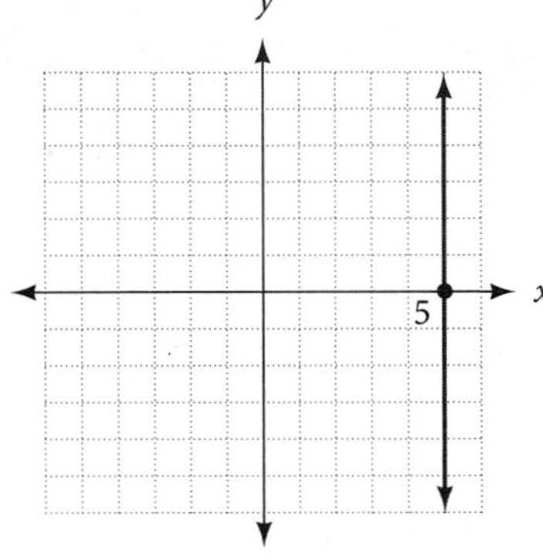

26.

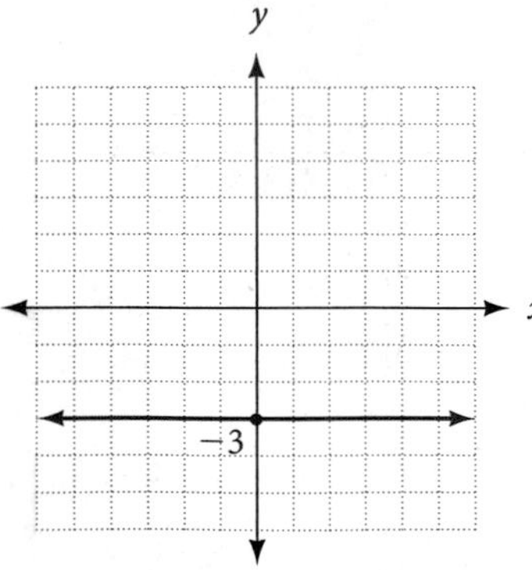

27.

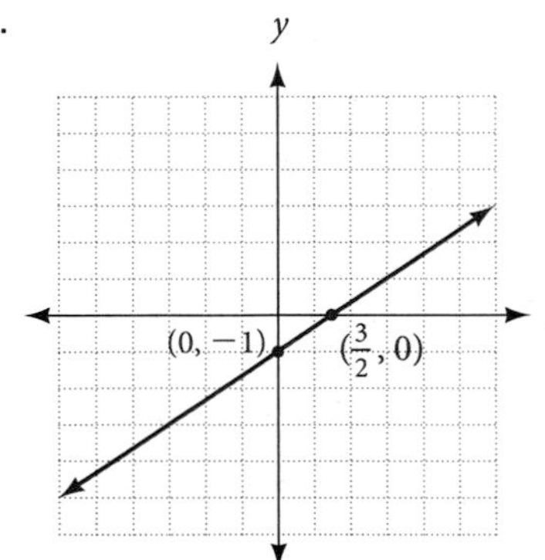

28.

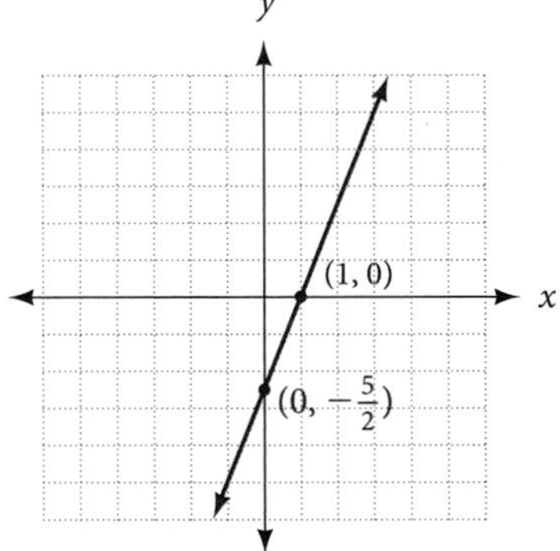

29. x-intercept = 2; y-intercept = −6 **30.** x-intercept = 12; y-intercept = −4 **31.** x-intercept = 3; y-intercept = −3 **32.** x-intercept = 2; y-intercept = −6 **33.** no x-intercept; y-intercept = −5 **34.** x-intercept = 4; no y-intercept **35.** $m = 2$ **36.** $m = -1$ **37.** $m = 2$ **38.** $m = 2$ **39.** $x = 6$ **40.** $y = -25$ **41.** $y = -2x + 2$ **42.** $y = \frac{1}{2}x + 1$ **43.** $y = -\frac{3}{4}x + \frac{1}{4}$ **44.** $y = 5$ **45.** $y = 3x + 2$ **46.** $y = -x + 6$ **47.** $y = -\frac{1}{3}x + \frac{3}{4}$ **48.** $y = 0$ **49.** $m = 4, b = -1$ **50.** $m = -2, b = -5$ **51.** $m = -2, b = 3$ **52.** $m = -\frac{5}{2}, b = 4$

53.

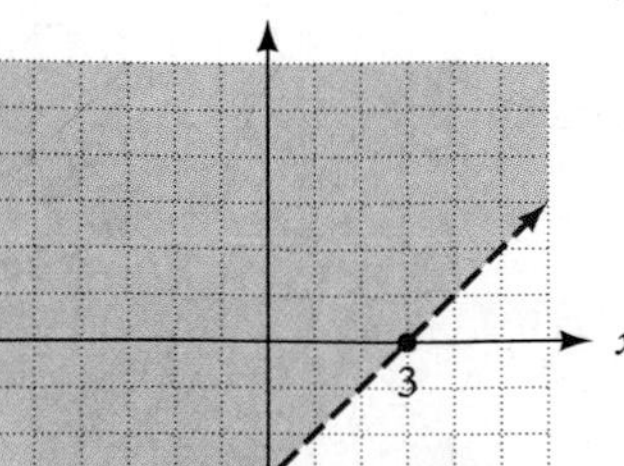

54.

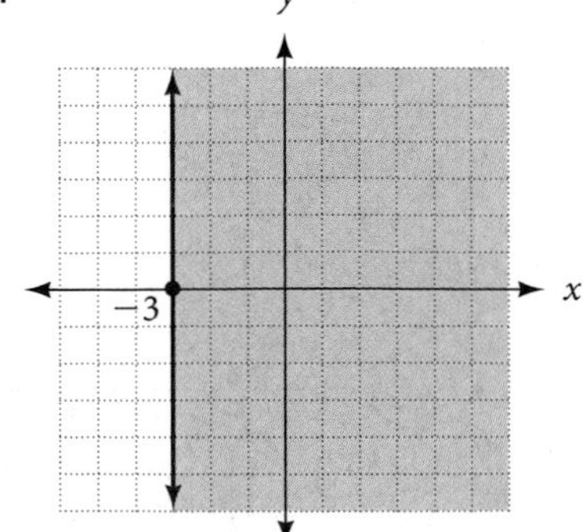

55.

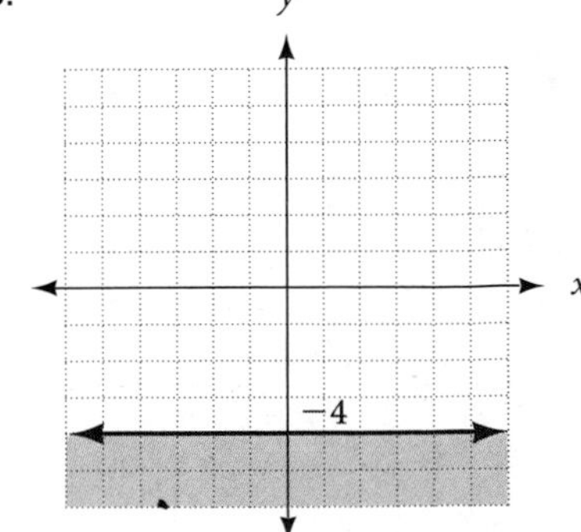

56.

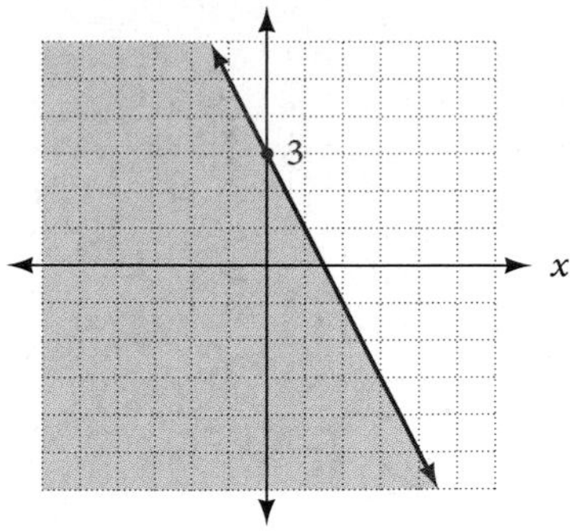

Chapter 3 Cumulative Review

1. -5 **2.** 22 **3.** 0 **4.** 18 **5.** 55 **6.** -70 **7.** Undefined **8.** $\frac{1}{4}$ **9.** $\frac{5}{9}$ **10.** 0 **11.** $40x$ **12.** $a - 3$
13. 4 **14.** 2 **15.** 2 **16.** -4 **17.** $-\frac{14}{5}$ **18.** 150 **19.** $x < -7$

20. $x \leq -2$

21.

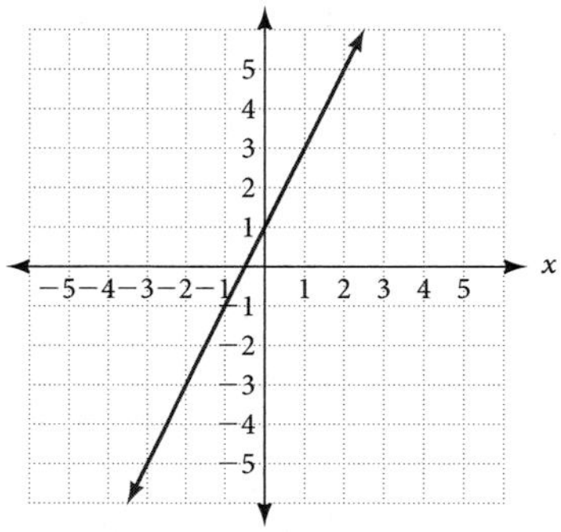

22.

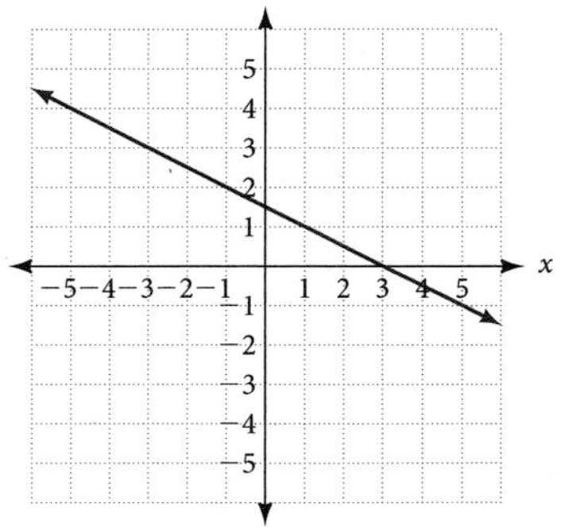

23.

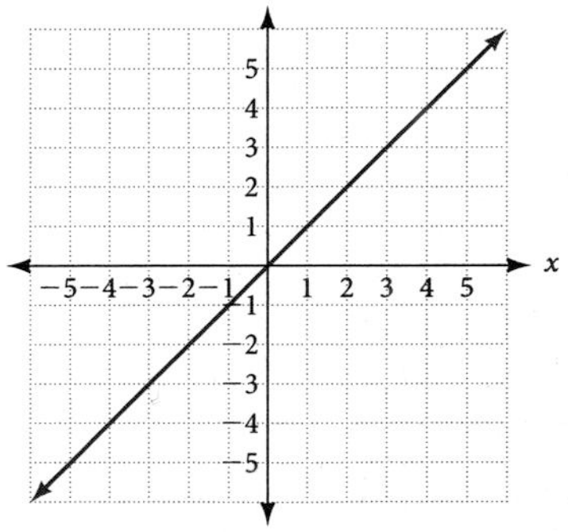

24.

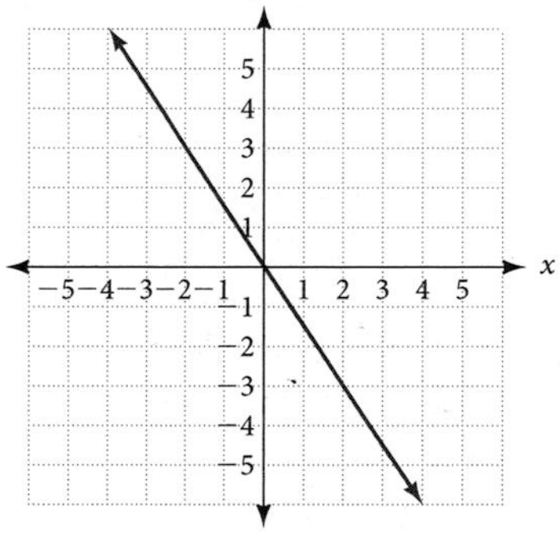

25.

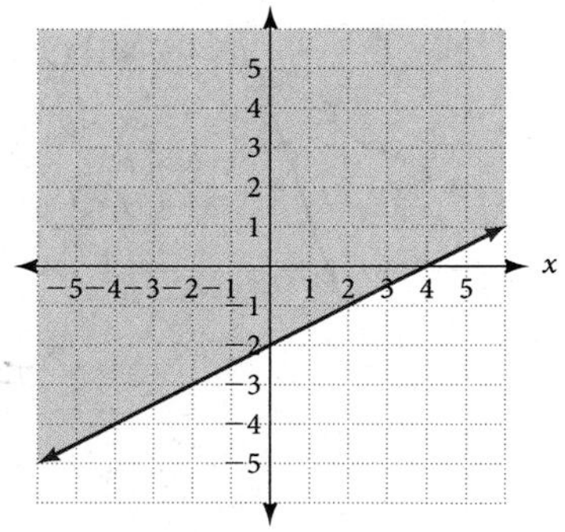

26.

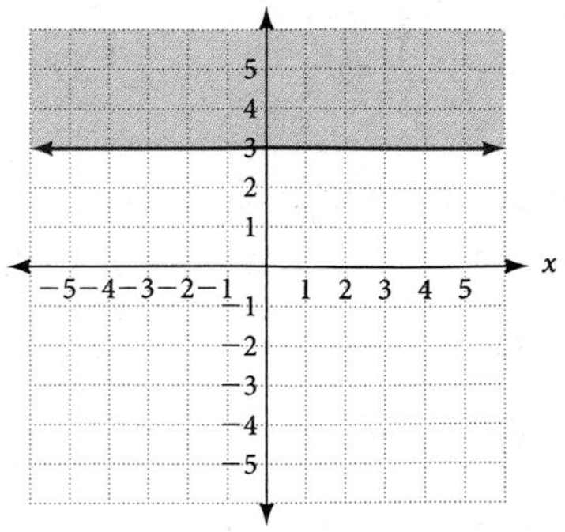

27.

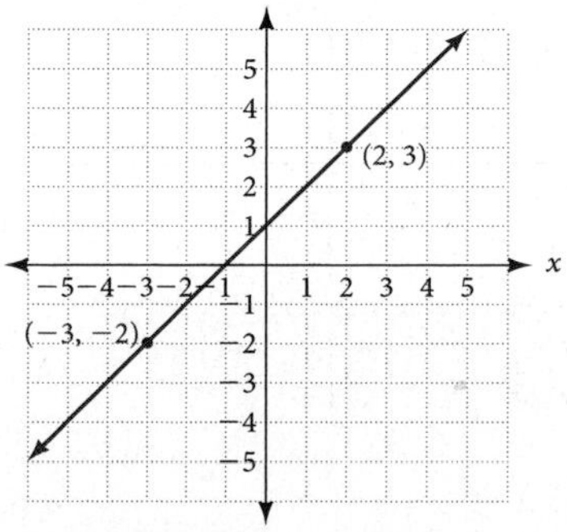

no, yes

28.

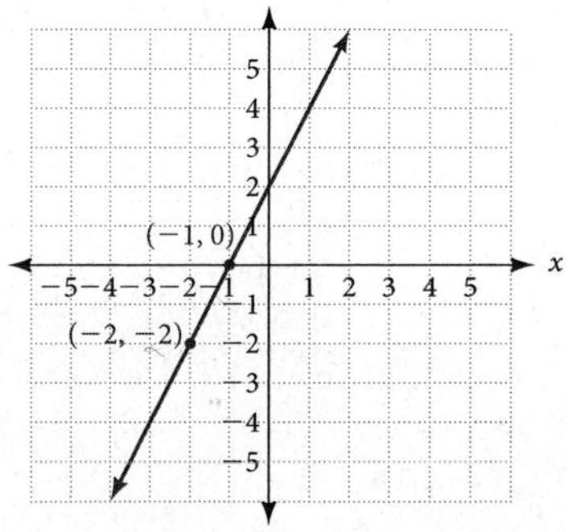

y-intercept = (0, 2)

29. x-intercept = (5, 0); y-intercept = (0, 2)

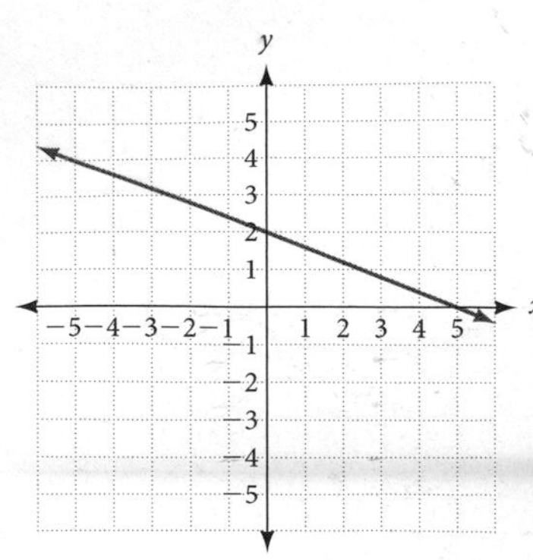

30.

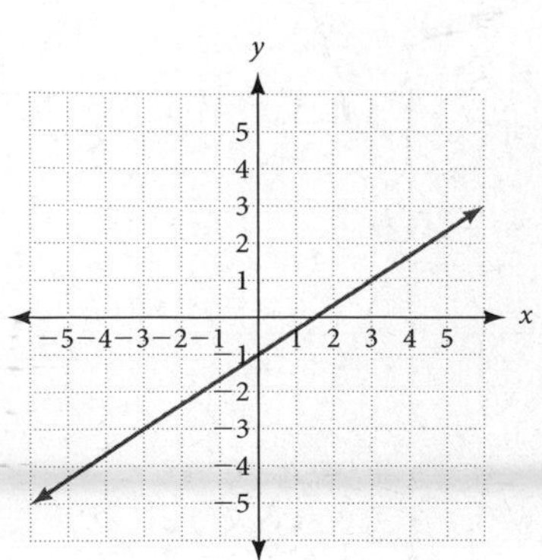

31. $m = \frac{4}{3}$ **32.** $y = 3x + 1$ **33.** $m = -\frac{2}{3}$ **34.** $y = \frac{1}{2}x + 4$ **35.** $y = 2x - 1$ **36.** $y = \frac{4}{3}x - 4$ **37.** (4, 3)(0, −3)

38. $(0, 2)(4, \frac{2}{5})$ **39.** $15 - 2(11) = -7$ **40.** −17 **41.** $\frac{2}{3}, -\frac{3}{2}, \frac{2}{3}$ **42.** $-2, \frac{1}{2}, 2$ **43.** $\frac{1}{36}$ **44.** $\frac{13}{5}$ **45.** 9 **46.** −3

47. length = 9 in., width = 5 in. **48.** 4 quarters, 12 dimes **49.** $-\frac{2}{5}$; The poverty rate is decreasing 0.4 percent per year

50. $\frac{3}{10}$; The poverty rate is increasing 0.3 percent per year

Chapter 3 Test

1-4.

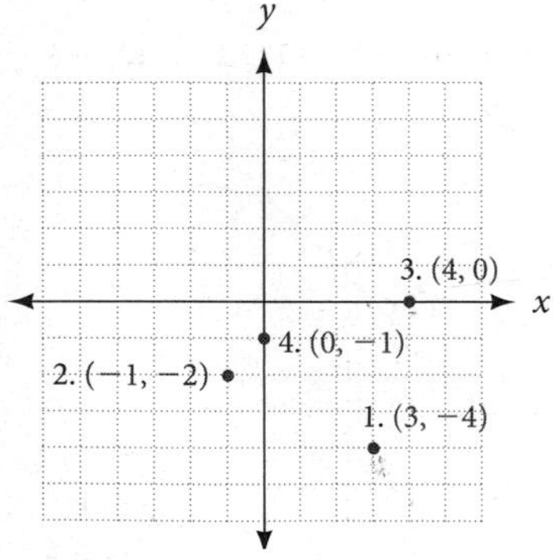

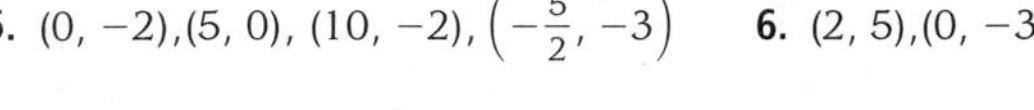

5. $(0, -2), (5, 0), (10, -2), \left(-\frac{5}{2}, -3\right)$ **6.** (2, 5), (0, −3)

7.

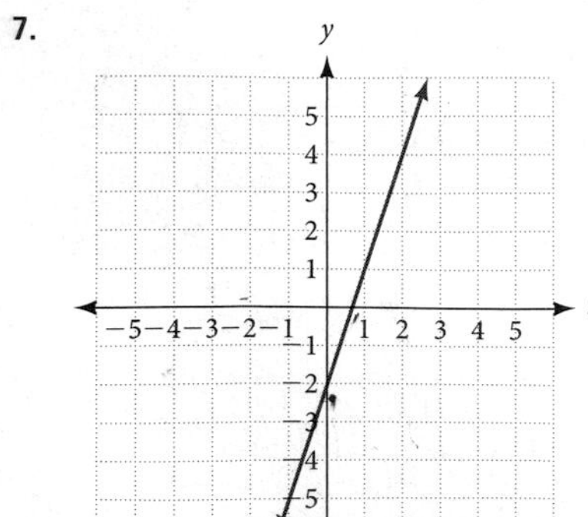

8.

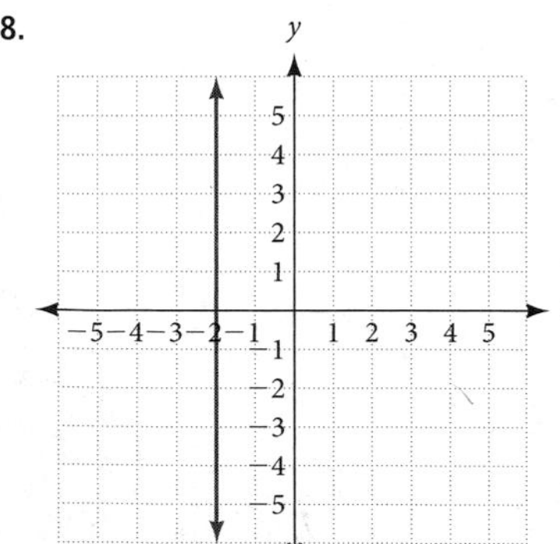

9. x-intercept = 5; y-intercept = −3

10. x-intercept = $-\frac{2}{3}$; y-intercept = 1 **11.** x-intercept = 3; y-intercept = −2 **12.** −2 **13.** $-\frac{13}{5}$ **14.** $-\frac{1}{4}$ **15.** 0

16. No slope **17.** $y = 3x + 20$ **18.** $y = 4x + 8$ **19.** $y = 3x + 10$ **20.** $y = 2x - 2$ **21.** $y = \frac{4}{3}x + 4$

22.

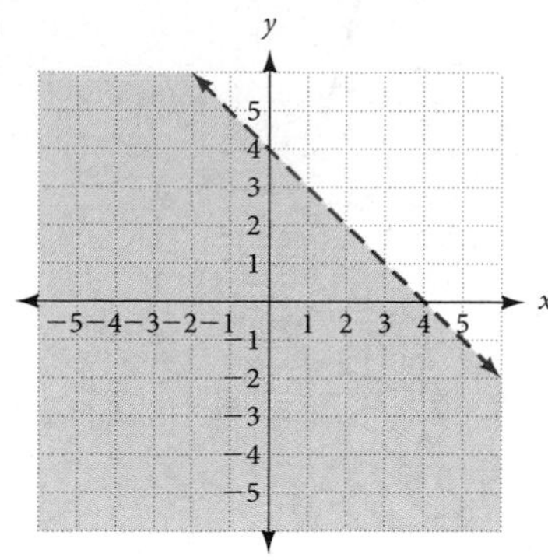

23.

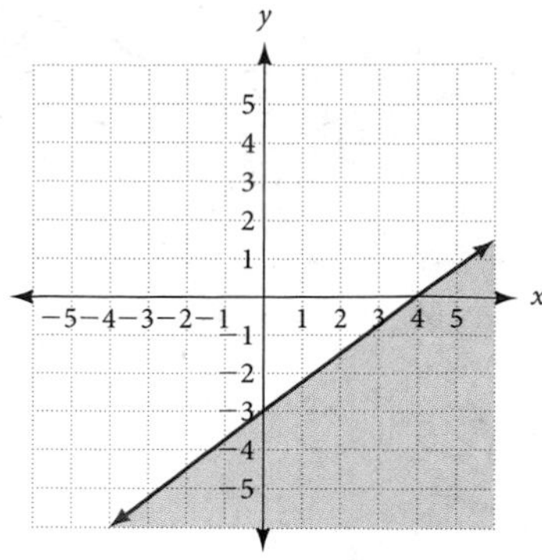

24. (1986, 37), (1989, 26), (1996, 24), (1998, 27), (2000, 24), (2004, 33)

Chapter 4

Pretest for Chapter 4

1. (0, 6) **2.** (3, 2) **3.** (7, 4) **4.** (−2, 4) **5.** (−4, 5) **6.** (1, 2) **7.** (−6, 3) **8.** Lines coincide **9.** (4, 0) **10.** (−5, 3) **11.** (−3, 4) **12.** (11, 3)

Getting Ready for Chapter 4

1. $3y$ **2.** $7x$ **3.** $-3y$ **4.** $-6x - 8y$ **5.** $12x + 16y$ **6.** $6x - \frac{9}{4}y$ **7.** 2 **8.** 4 **9.** −4 **10.** −2 **11.** 2 **12.** −4 **13.** 0 **14.** 5

Problem Set 4.1

1.

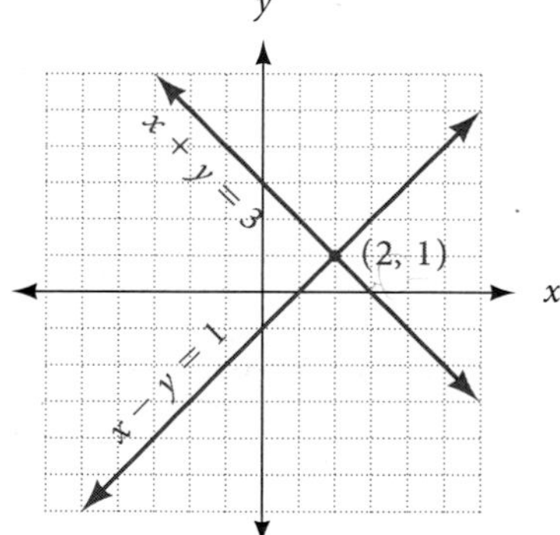

3.

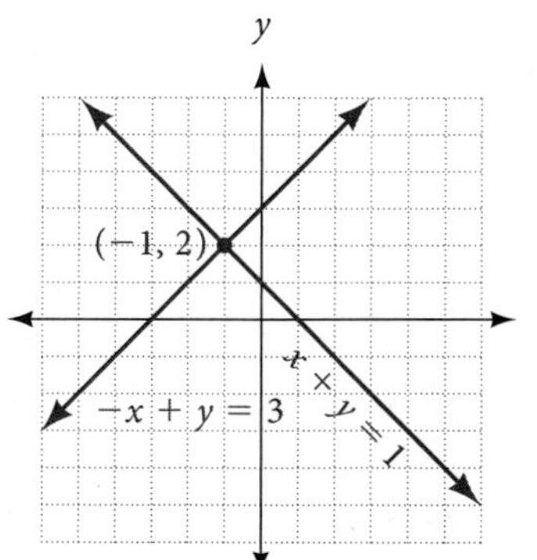

5.

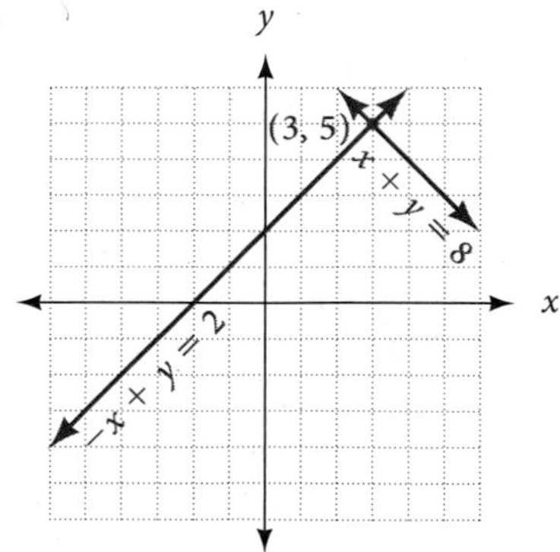

7.

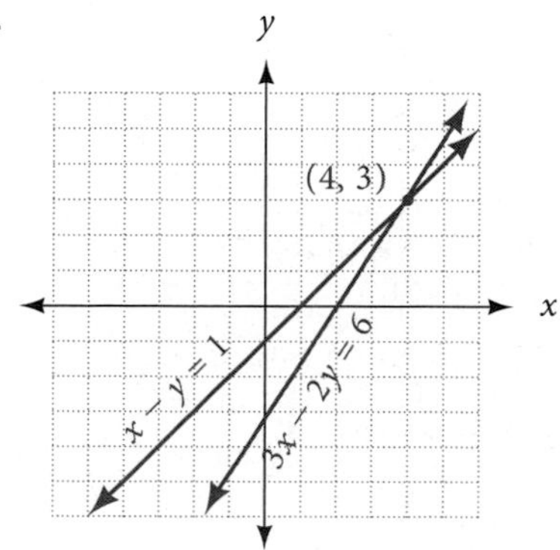

9.

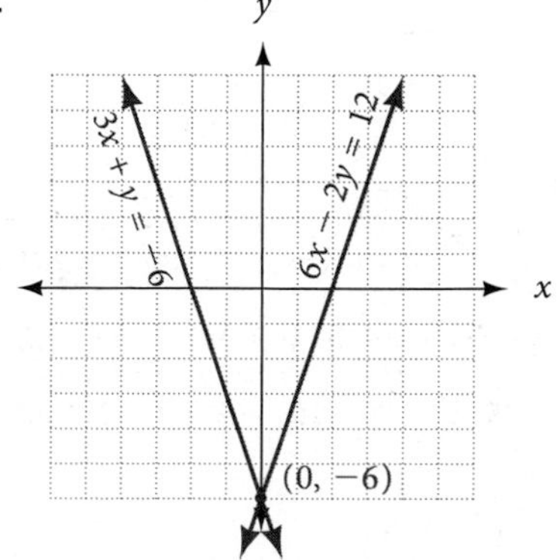

11.

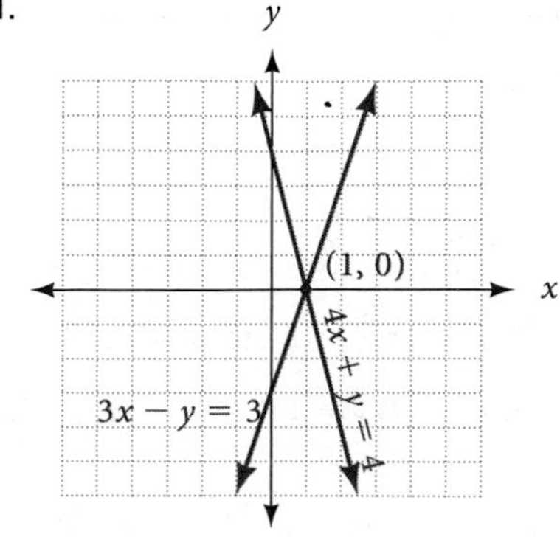

13.

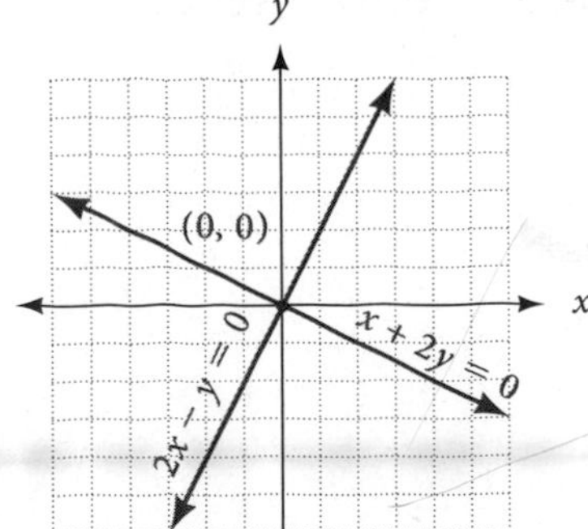

15.

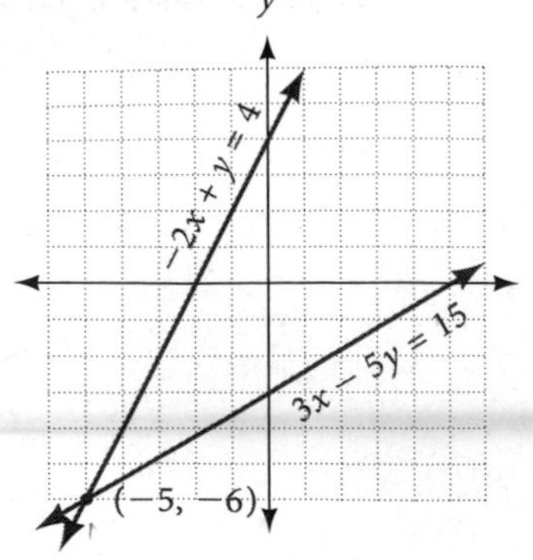

17.

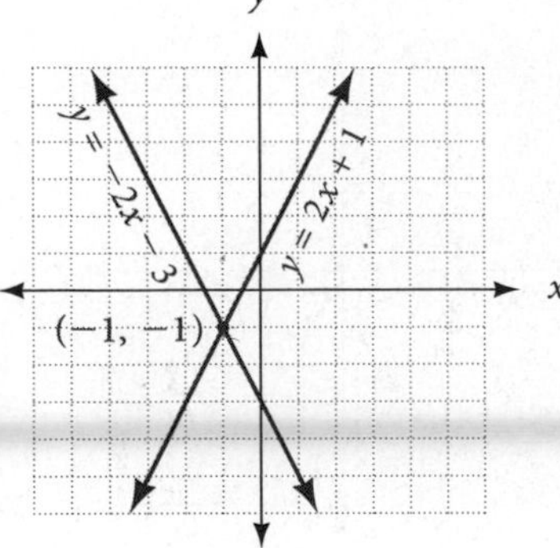

19.

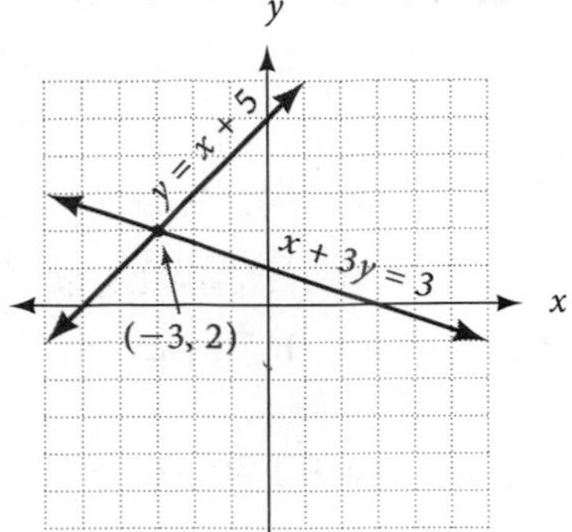

21.

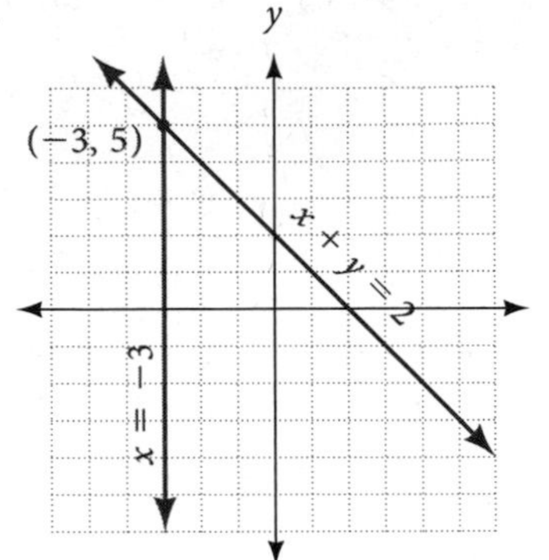

23.

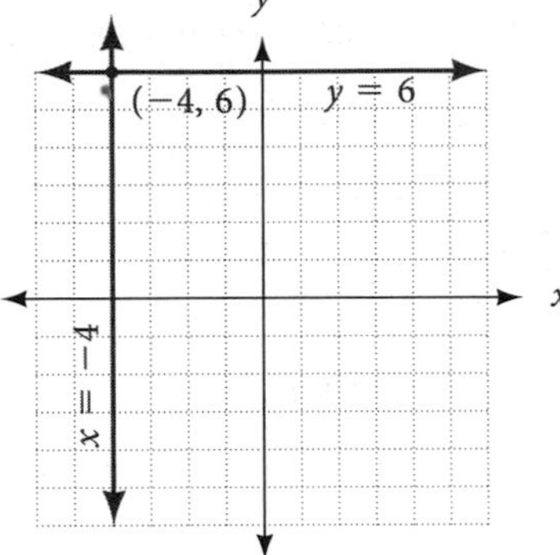

25. ∅

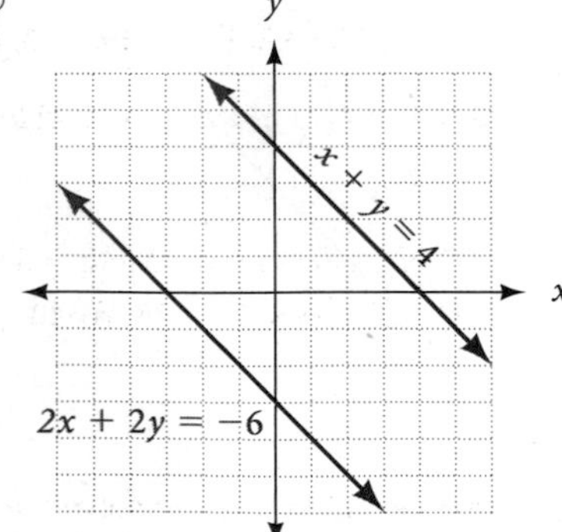

27. Any point on the line

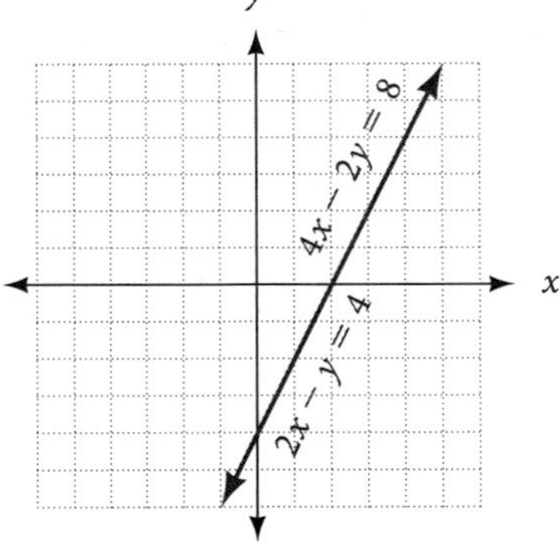

29. **a.** $4x - 5y$ **b.** 1 **c.** −2 **d.**

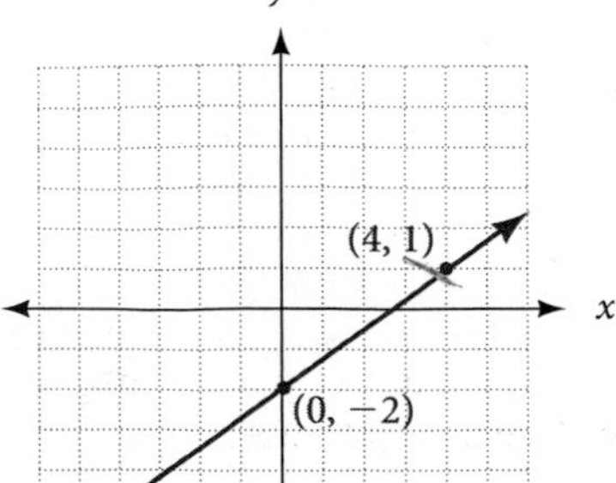

e. (0, −2)

31.

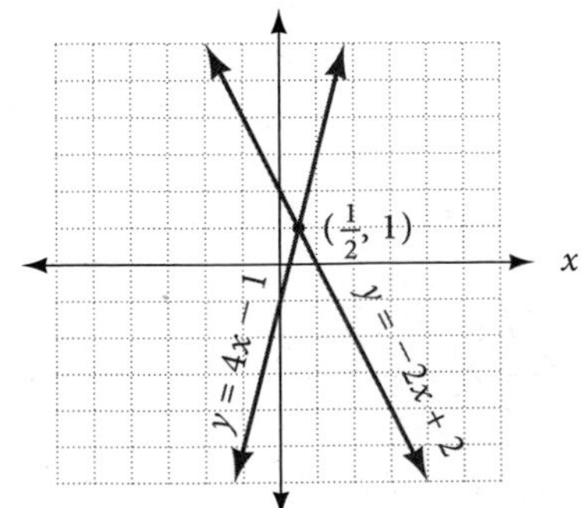

33.

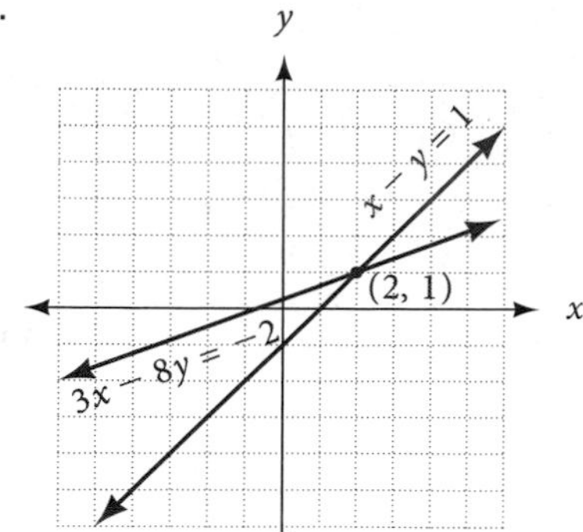

35. Yes **37. a.** 25 hours **b.** Gigi's **c.** Marcy's **39.** $2x$ **41.** $7x$ **43.** $13x$

45. $-12x - 20y$ **47.** $3x + 8y$ **49.** $-4x + 2y$ **51.** 1 **53.** 0 **55.** −5 **57.** $10x + 75$ **59.** $11x - 35$ **61.** $-x + 58$ **63.** $-0.02x + 6$ **65.** $-0.003x - 28$ **67.** $1.31x - 125$

Problem Set 4.2

1. (2, 1) **3.** (3, 7) **5.** (2, −5) **7.** (−1, 0) **9.** Lines coincide **11.** (4, 8) **13.** $\left(\frac{1}{5}, 1\right)$ **15.** (1, 0) **17.** (−1, −2) **19.** $\left(-5, \frac{3}{4}\right)$ **21.** (−4, 5) **23.** (−3, −10) **25.** (3, 2) **27.** $\left(5, \frac{1}{3}\right)$ **29.** $\left(-2, \frac{2}{3}\right)$ **31.** (2, 2) **33.** Lines are parallel; ∅ **35.** (1, 1) **37.** Lines are parallel; ∅ **39.** (10, 12) **41.** (6, 8) **43.** (7,000, 8,000) **45.** (3,000, 8,000) **47.** (11, 12) **49.** (10, 12) **51.** 2001 **53.** 1 **55.** 2 **57.** 1 **59.** $x = 3y - 1$ **61.** 1 **63.** 5 **65.** 34.5 **67.** 33.95 **69.** slope = 3; y-int = −3 **71.** slope = $\frac{2}{5}$; y-int = −5 **73.** slope = −1 **75.** slope = 2 **77.** 11 **79.** 9 **81.** $y = 3x$ **83.** $y = x - 2$

Problem Set 4.3

1. (4, 7) **3.** (3, 17) **5.** $\left(\frac{3}{2}, 2\right)$ **7.** (2, 4) **9.** (0, 4) **11.** (−1, 3) **13.** (1, 1) **15.** (2, −3) **17.** $\left(-2, \frac{3}{5}\right)$ **19.** (−3, 5) **21.** Lines are parallel; ∅ **23.** (3, 1) **25.** $\left(\frac{1}{2}, \frac{3}{4}\right)$ **27.** (2, 6) **29.** (4, 4) **31.** (5, −2) **33.** (18, 10) **35.** Lines coincide **37.** (8, 4) **39.** (6, 12) **41.** (16, 32) **43.** (10, 12) **45. a.** $\frac{25}{4}$ **b.** $y = \frac{4}{5}x - 4$ **c.** $x = y + 5$ **d.** (5, 0) **47.** 2001 **49. a.** 1,000 miles **b.** Car **c.** Truck **d.** Miles ≥ 0 **51.** 3 and 23 **53.** 15 and 24 **55.** Length = 23 in.; Width = 6 in. **57.** 14 nickels. and 10 dimes **59.** (3, 17) **61.** (7,000, 8,000) **63.** 47 **65.** 14 **67.** 70 **69.** 35 **71.** 5 **73.** 6,540 **75.** 1,760 **77.** 20 **79.** 63 **81.** 53

Problem Set 4.4

As you can see, in addition to the answers to the problems we sometimes have included the system of equations used to solve the problems. Remember, you should attempt the problem on your own before checking your answers or equations.

1. $x + y = 25$, $y = x + 5$ The two numbers are 10 and 15. **3.** 3 and 12 **5.** $x - y = 5$, $x = 2y + 1$ The two numbers are 9 and 4. **7.** 6 and 29

9. Let x = the amount invested at 6% and y = the amount invested at 8%.
$x + y = 20{,}000$
$0.06x + 0.08y = 1{,}380$
He has $9,000 at 8% and $11,000 at 6%.

11. $2,000 at 6%, $8,000 at 5% **13.** 6 nickels, 8 quarters

15. Let x = the number of dimes and y = the number of quarters.
$x + y = 21$
$0.10x + 0.25y = 3.45$
He has 12 dimes and 9 quarters.

17. Let x = the number of liters of 50% solution and y = the number of liters of 20% solution.
$x + y = 18$
$0.50x + 0.20y = 0.30(18)$
6 liters of 50% solution
12 liters of 20% solution

19. 10 gallons of 10% solution, 20 gallons of 7% solution **21.** 20 adults, 50 kids **23.** 16 feet wide, 32 feet long **25.** 33 $5 chips, 12 $25 chips **27.** 50 at $11, 100 at $20 **29.** 2 modules and 6 fixed racks **31.** (−2, 2), (0, 3), (2, 4)

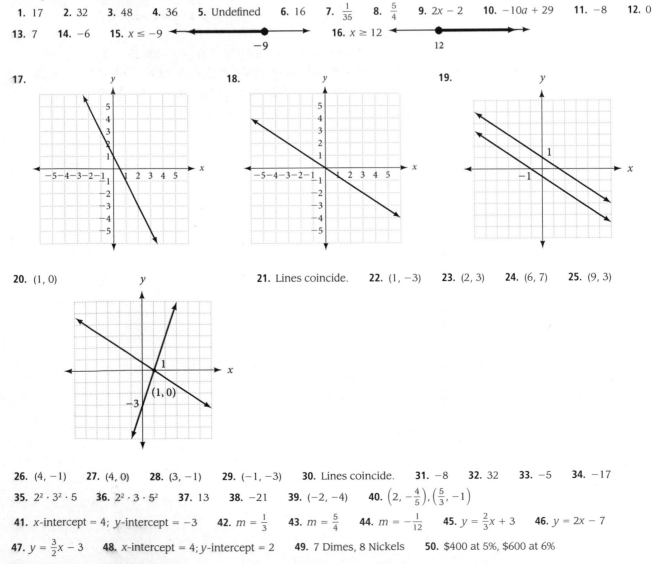

33.

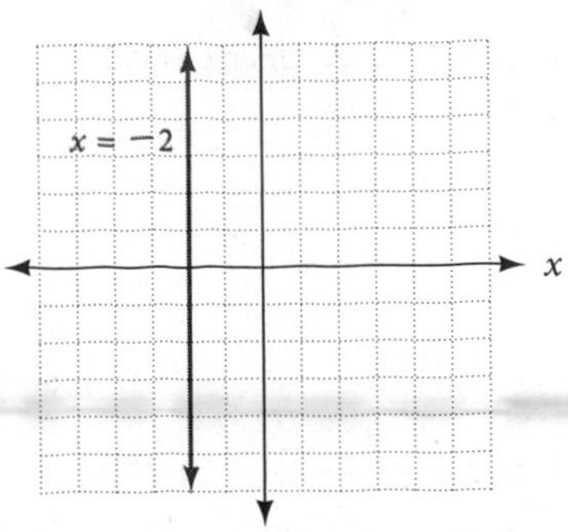

35. 2 **37.** $y = \frac{1}{2}x + 2$ **39.** $y = 2x + 1$

Chapter 4 Review

1. (4, −2) **2.** (−3, 2) **3.** (3, −2) **4.** (2, 0) **5.** (2, 1) **6.** (−1, −2) **7.** (1, −3) **8.** (−2, 5) **9.** Lines coincide **10.** (2, −2) **11.** (1, 1) **12.** Lines are parallel; ∅ **13.** (−2, −3) **14.** (5, −1) **15.** (−2, 7) **16.** (−4, −2) **17.** (−1, 4) **18.** (2, −6) **19.** Lines are parallel; ∅ **20.** (−3, −10) **21.** (1, −2) **22.** Lines coincide **23.** 10, 8 **24.** 24, 8 **25.** \$4,000 at 4%, \$8,000 at 5% **26.** \$3,000 at 6%, \$11,000 at 8% **27.** 10 dimes, 7 nickels **28.** 9 dimes, 6 quarters **29.** 40 liters of 10% solution, 10 liters of 20% solution **30.** 20 liters of 25% solution, 20 liters of 15% solution

Chapter 4 Cumulative Review

1. 17 **2.** 32 **3.** 48 **4.** 36 **5.** Undefined **6.** 16 **7.** $\frac{1}{35}$ **8.** $\frac{5}{4}$ **9.** $2x - 2$ **10.** $-10a + 29$ **11.** −8 **12.** 0 **13.** 7 **14.** −6 **15.** $x \le -9$

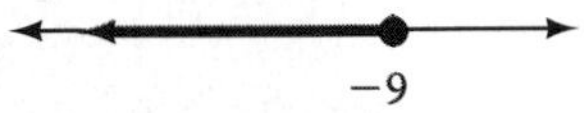

16. $x \ge 12$

17.

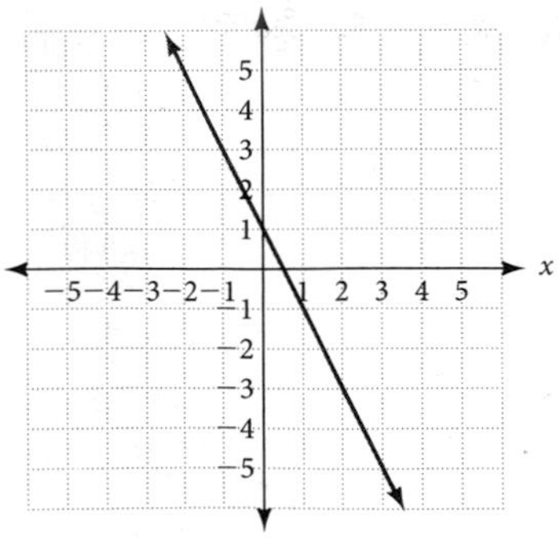

18.

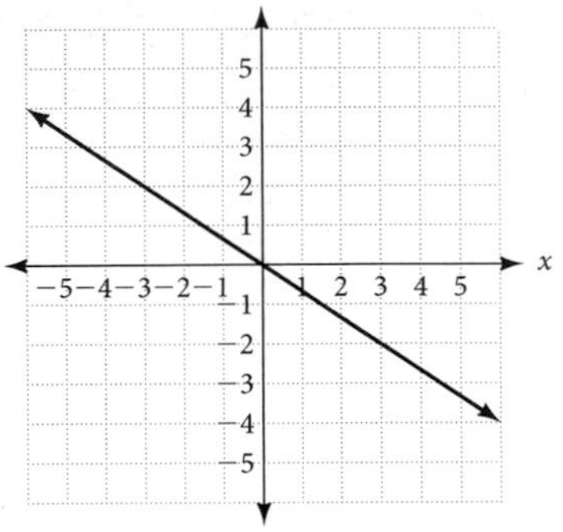

19.

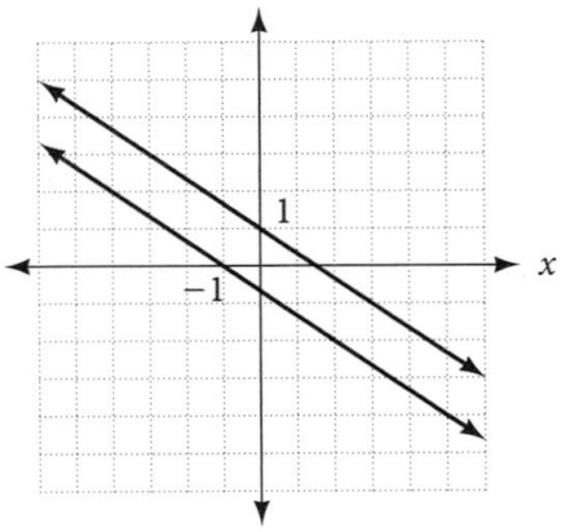

20. (1, 0)

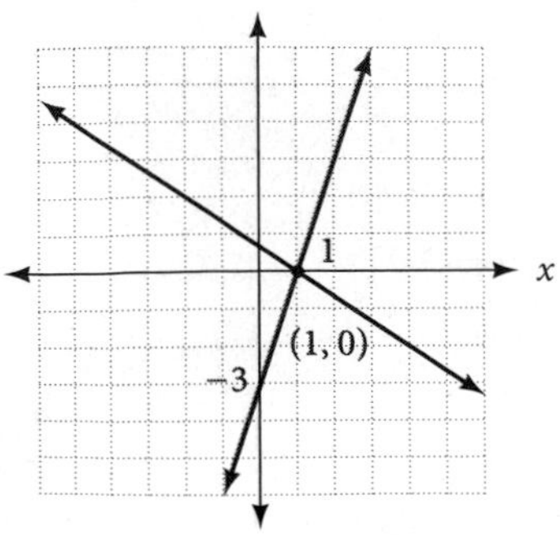

21. Lines coincide. **22.** (1, −3) **23.** (2, 3) **24.** (6, 7) **25.** (9, 3)

26. (4, −1) **27.** (4, 0) **28.** (3, −1) **29.** (−1, −3) **30.** Lines coincide. **31.** −8 **32.** 32 **33.** −5 **34.** −17 **35.** $2^2 \cdot 3^2 \cdot 5$ **36.** $2^2 \cdot 3 \cdot 5^2$ **37.** 13 **38.** −21 **39.** (−2, −4) **40.** $\left(2, -\frac{4}{5}\right), \left(\frac{5}{3}, -1\right)$ **41.** x-intercept = 4; y-intercept = −3 **42.** $m = \frac{1}{3}$ **43.** $m = \frac{5}{4}$ **44.** $m = -\frac{1}{12}$ **45.** $y = \frac{2}{3}x + 3$ **46.** $y = 2x - 7$ **47.** $y = \frac{3}{2}x - 3$ **48.** x-intercept = 4; y-intercept = 2 **49.** 7 Dimes, 8 Nickels **50.** \$400 at 5%, \$600 at 6% **51.** 156 people **52.** 250 people

Chapter 4 Test

1. $(-4, 2)$ **2.** $(1, 2)$ **3.** $(3, -2)$ **4.** Lines are parallel; $\varnothing$ **5.** $(-3, -4)$ **6.** $(5, -3)$ **7.** $(2, -2)$ **8.** Lines coincide. **9.** $(2, 7)$ **10.** $(4, 3)$ **11.** $(2, 9)$ **12.** $(-5, -1)$ **13.** 5, 7 **14.** 3, 12 **15.** \$6,000 at 9%, \$4,000 at 11% **16.** 7 nickles, 5 quarters **17.** 25 ft by 60 ft **18. a.** 150 miles **b.** Company 1 **c.** $x > 150$

Chapter 5

Pretest for Chapter 5

1. -32 **2.** $\frac{8}{27}$ **3.** $128x^{13}$ **4.** $\frac{1}{16}$ **5.** 1 **6.** x^3 **7.** $\frac{1}{x^3}$ **8.** 4.307×10^{-2} **9.** 7,630,000 **10.** $\frac{y^3z^2}{3x^2}$ **11.** $\frac{a^3b^2}{2}$ **12.** $12x^3$ **13.** 7.5×10^7 **14.** $9x^2 + 5x + 4$ **15.** $2x^2 + 6x - 2$ **16.** $5x - 4$ **17.** 21 **18.** $15x^4 - 6x^3 + 12x^2$ **19.** $x^2 - \frac{1}{12}x - \frac{1}{12}$ **20.** $10x^2 - 3x - 18$ **21.** $x^3 + 64$ **22.** $x^2 - 12x + 36$ **23.** $4a^2 + 16ab + 16b^2$ **24.** $9x^2 - 36$ **25.** $x^4 - 16$ **26.** $3x^2 - 6x + 1$ **27.** $3x - 1 - \frac{5}{3x - 1}$ **28.** $4x + 11 + \frac{50}{x - 4}$

Getting Ready for Chapter 5

1. $\frac{5}{14}$ **2.** 3.2 **3.** -11 **4.** -7 **5.** -14 **6.** 14 **7.** 16 **8.** 26 **9.** $-4x^2 + 2x + 6$ **10.** $9x^2$ **11.** $-12x$ **12.** $6x^2$ **13.** $-13x$ **14.** $-x$ **15.** $10x^2$ **16.** -4 **17.** $-4x^2 - 20x$ **18.** $-4x^2 - 14x$ **19.** $2x^3 - 10x^2$ **20.** $10x^2 - 50x$

Problem Set 5.1

1. Base 4; exponent 2; 16 **3.** Base 0.3; exponent 2; 0.09 **5.** Base 4; exponent 3; 64 **7.** Base -5; exponent 2; 25 **9.** Base 2; exponent 3; -8 **11.** Base 3; exponent 4; 81 **13.** Base $\frac{2}{3}$; exponent 2; $\frac{4}{9}$ **15.** Base $\frac{1}{2}$; exponent 4; $\frac{1}{16}$

17. a.

Number x	Square x^2
1	1
2	4
3	9
4	16
5	25
6	36
7	49

b. Either *larger* or *greater* will work. **19.** x^9 **21.** y^{30} **23.** 2^{12} **25.** x^{28} **27.** x^{10} **29.** 5^{12} **31.** y^9 **33.** 2^{50} **35.** a^{3x} **37.** b^{xy} **39.** $16x^2$ **41.** $32y^5$ **43.** $81x^4$ **45.** $0.25a^2b^2$ **47.** $64x^3y^3z^3$ **49.** $8x^{12}$ **51.** $16a^6$ **53.** x^{14} **55.** a^{11} **57.** $128x^7$ **59.** $432x^{10}$ **61.** $16x^4y^6$ **63.** $\frac{8}{27}a^{12}b^{15}$ **65.** x^2 **67.** $4x$ **69.** 2 **71.** $4x$ **73. a.** 32 **b.** 64 **c.** 32 **d.** 64

75.

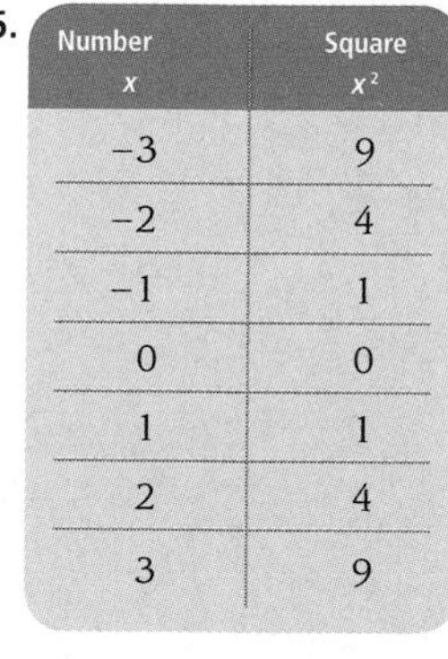

Number x	Square x^2
−3	9
−2	4
−1	1
0	0
1	1
2	4
3	9

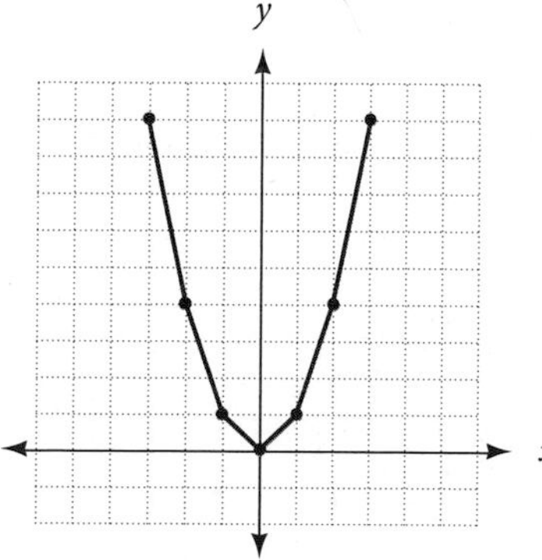

77.

Number x	Square x^2
-2.5	6.25
-1.5	2.25
-0.5	0.25
0	0
0.5	0.25
1.5	2.25
2.5	6.25

79. 4.32×10^4 **81.** 5.7×10^2 **83.** 2.38×10^5 **85.** 2,490 **87.** 352 **89.** 28,000

91. 275,625 sq ft **93.** 27 inches^3 **95.** 15.6 inches^3 **97.** 36 inches^3 **99.** 6.5×10^8 seconds **101.** \$740,000 **103.** \$180,000 **105.** 219 inches^3 **107.** 182 inches^3 **109.** -3 **111.** 11 **113.** -5 **115.** 5 **117.** 2 **119.** 6 **121.** 4 **123.** 3 **125.** 2^7 **127.** $2 \cdot 5^3$ **129.** $2^4 \cdot 3^2 \cdot 5$ **131.** $2^2 \cdot 5 \cdot 41$ **133.** $2^3 \cdot 3^3$ **135.** $2^3 \cdot 3^3 \cdot 5^3$ **137.** 5^6 **139.** $2^6 \cdot 3^3$

Problem Set 5.2

1. $\frac{1}{9}$ **3.** $\frac{1}{36}$ **5.** $\frac{1}{64}$ **7.** $\frac{1}{125}$ **9.** $\frac{2}{x^3}$ **11.** $\frac{1}{8x^3}$ **13.** $\frac{1}{25y^2}$ **15.** $\frac{1}{100}$

17.

Number x	Square x^2	Power of 2 2^x
-3	9	$\frac{1}{8}$
-2	4	$\frac{1}{4}$
-1	1	$\frac{1}{2}$
0	0	1
1	1	2
2	4	4
3	9	8

19. $\frac{1}{25}$ **21.** x^6 **23.** 64 **25.** $8x^3$ **27.** 6^{10} **29.** $\frac{1}{6^{10}}$ **31.** $\frac{1}{2^8}$ **33.** 2^8

35. $27x^3$ **37.** $81x^4y^4$ **39.** 1 **41.** $2a^2b$ **43.** $\frac{1}{49y^6}$ **45.** $\frac{1}{x^8}$ **47.** $\frac{1}{y^3}$ **49.** x^2 **51.** a^6 **53.** $\frac{1}{y^9}$ **55.** y^{40} **57.** $\frac{1}{x}$ **59.** x^9 **61.** a^{16} **63.** $\frac{1}{a^4}$ **65. a.** 32 **b.** 32 **c.** $\frac{1}{32}$ **d.** $\frac{1}{32}$ **67. a.** $\frac{1}{5}$ **b.** $\frac{1}{8}$ **c.** $\frac{1}{x}$ **d.** $\frac{1}{x^2}$

69.

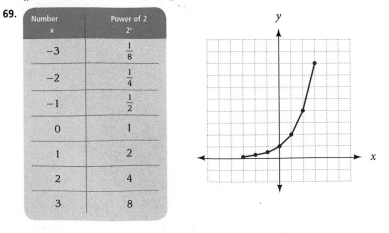

Number x	Power of 2 2^x
-3	$\frac{1}{8}$
-2	$\frac{1}{4}$
-1	$\frac{1}{2}$
0	1
1	2
2	4
3	8

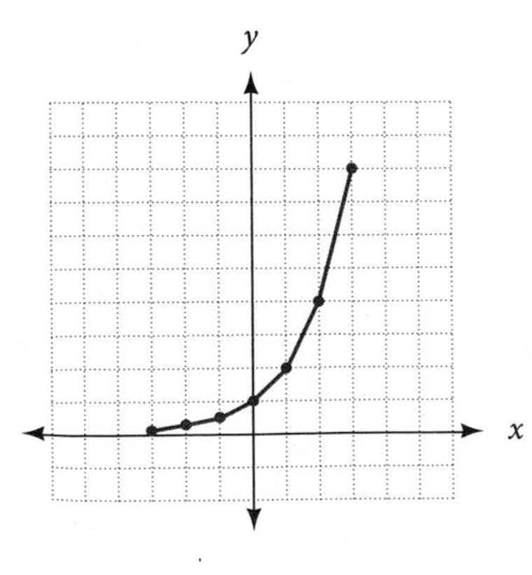

71. 4.8×10^{-3} **73.** 2.5×10^1 **75.** 9×10^{-6}

77.

Expanded Form	Scientific Notation $n \times 10^r$
0.000357	3.57×10^{-4}
0.00357	3.57×10^{-3}
0.0357	3.57×10^{-2}
0.357	3.57×10^{-1}
3.57	3.57×10^{0}
35.7	3.57×10^{1}
357	3.57×10^{2}
3,570	3.57×10^{3}
35,700	3.57×10^{4}

79. 0.00423 **81.** 0.00008 **83.** 4.2 **85.** $\$4.82 \times 10^7$ **87.** 0.002

89. 2.5×10^4 **91.** 2.35×10^5 **93.** 8.2×10^{-4} **95.** 100 inches^2; 400 inches^2; 4 **97.** x^2; $4x^2$; 4 **99.** 216 inches^3; 1,728 inches^3; 8 **101.** x^3; $8x^3$; 8 **103.** 13.5 **105.** 8 **107.** 26.52 **109.** 12 **111.** x^8 **113.** x **115.** $\frac{1}{y^2}$ **117.** 340 **119.** $7x$ **121.** $2a$ **123.** $10y$

Problem Set 5.3

1. $12x^7$ **3.** $-16y^{11}$ **5.** $32x^2$ **7.** $200a^6$ **9.** $-24a^3b^3$ **11.** $24x^6y^8$ **13.** $3x$ **15.** $\frac{6}{y^3}$ **17.** $\frac{1}{2a}$ **19.** $-\frac{3a}{b^2}$ **21.** $\frac{x^2}{9z^2}$

23.

a	b	ab	$\frac{a}{b}$	$\frac{b}{a}$
10	$5x$	$50x$	$\frac{2}{x}$	$\frac{x}{2}$
$20x^3$	$6x^2$	$120x^5$	$\frac{10x}{3}$	$\frac{3}{10x}$
$25x^5$	$5x^4$	$125x^9$	$5x$	$\frac{1}{5x}$
$3x^{-2}$	$3x^2$	9	$\frac{1}{x^4}$	x^4
$-2y^4$	$8y^7$	$-16y^{11}$	$-\frac{1}{4y^3}$	$-4y^3$

25. 6×10^8 **27.** 1.75×10^{-1} **29.** 1.21×10^{-6} **31.** 4.2×10^3

33. 3×10^{10} **35.** 5×10^{-3} **37.** $8x^2$ **39.** $-11x^5$ **41.** 0 **43.** $4x^3$ **45.** $31ab^2$ **47.**

a	b	ab	$a + b$
$5x$	$3x$	$15x^2$	$8x$
$4x^2$	$2x^2$	$8x^4$	$6x^2$
$3x^3$	$6x^3$	$18x^6$	$9x^3$
$2x^4$	$-3x^4$	$-6x^8$	$-x^4$
x^5	$7x^5$	$7x^{10}$	$8x^5$

49. $4x^3$ **51.** $\frac{1}{b^2}$ **53.** $\frac{6y^{10}}{x^4}$

55. 2×10^6 **57.** 1×10^1 **59.** 4.2×10^{-6} **61.** $9x^3$ **63.** $-20a^2$ **65.** $6x^5y^2$ **67.** $x^2y + x$ **69.** $x + y$ **71.** $x^2 - 4$ **73.** $x^2 - x - 6$ **75.** $x^2 - 5x$ **77.** $x^2 - 8x$ **79. a.** 2 **b.** $-3b$ **c.** $5b^2$ **81. a.** $3xy$ **b.** $2y$ **c.** $-\frac{y^2}{2}$ **83.** -5 **85.** 6 **87.** 76 **89.** $6x^2$ **91.** $2x$ **93.** $-2x-9$ **95.** 11 **97.** -8 **99.** 9 **101.** 0 **103.**

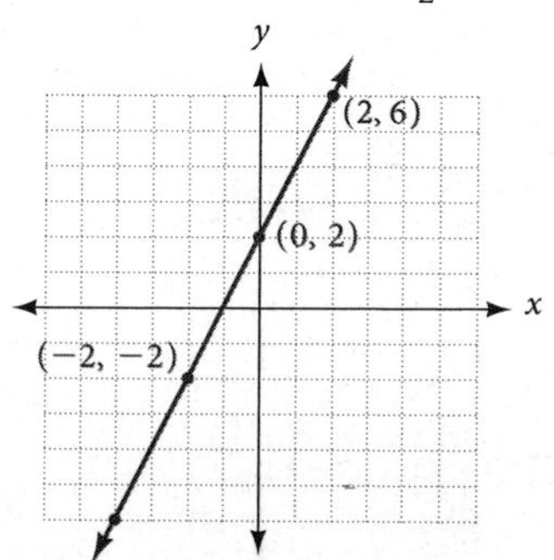

105.

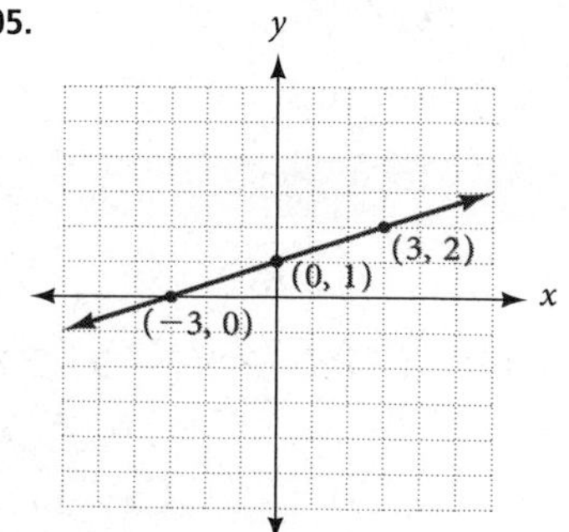

Problem Set 5.4

1. Trinomial, 3 **3.** Trinomial, 3 **5.** Binomial, 1 **7.** Binomial, 2 **9.** Monomial, 2 **11.** Monomial, 0 **13.** $5x^2 + 5x + 9$ **15.** $5a^2 - 9a + 7$ **17.** $x^2 + 6x + 8$ **19.** $6x^2 - 13x + 5$ **21.** $x^2 - 9$ **23.** $3y^2 - 11y + 10$ **25.** $6x^3 + 5x^2 - 4x + 3$ **27.** $2x^2 - x + 1$ **29.** $2a^2 - 2a - 2$ **31.** $-\frac{1}{9}x^3 - \frac{2}{3}x^2 - \frac{5}{2}x + \frac{7}{4}$ **33.** $-4y^2 + 15y - 22$ **35.** $-2x$ **37.** $4x$ **39.** $x^3 - 27$ **41.** $x^3 + 6x^2 + 12x + 8$ **43.** $4x - 8$ **45.** $x^2 - 33x + 63$ **47.** $8y^2 + 4y + 26$ **49.** $75x^2 - 150x - 75$ **51.** $12x + 2$ **53.** 4 **55. a.** 0 **b.** 0 **57. a.** 2,200 **b.** −1,800 **59.** 1.44×10^{12} in² **61.** 18π inches³ **63.** First year $51{,}568x + 67{,}073y$; Second year $51{,}568(2x) + 67{,}073y$; $154{,}704x + 134{,}146y$ **65.** 5 **67.** −6 **69.** $-20x^2$ **71.** $-21x$ **73.** $2x$ **75.** $6x - 18$ **77.** $-15x^2$ **79.** $6x^3$ **81.** $6x^4$

Problem Set 5.5

1. $6x^2 + 2x$ **3.** $6x^4 - 4x^3 + 2x^2$ **5.** $2a^3b - 2a^2b^2 + 2ab$ **7.** $3y^4 + 9y^3 + 12y^2$ **9.** $8x^5y^2 + 12x^4y^3 + 32x^2y^4$ **11.** $x^2 + 7x + 12$ **13.** $x^2 + 7x + 6$ **15.** $x^2 + 2x + \frac{3}{4}$ **17.** $a^2 + 2a - 15$ **19.** $xy + bx - ay - ab$ **21.** $x^2 - 36$ **23.** $y^2 - \frac{25}{36}$ **25.** $2x^2 - 11x + 12$ **27.** $2a^2 + 3a - 2$ **29.** $6x^2 - 19x + 10$ **31.** $2ax + 8x + 3a + 12$ **33.** $25x^2 - 16$ **35.** $2x^2 + \frac{5}{2}x - \frac{3}{4}$ **37.** $3 - 10a + 8a^2$

39.

	x	3
x	x^2	$3x$
2	$2x$	6

$(x + 2)(x + 3) = x^2 + 2x + 3x + 6$
$= x^2 + 5x + 6$

41.

	x	x	2
x	x^2	x^2	$2x$
1	x	x	2

$(x + 1)(2x + 2) = 2x^2 + 4x + 2$

43. $a^3 - 6a^2 + 11a - 6$ **45.** $x^3 + 8$ **47.** $2x^3 + 17x^2 + 26x + 9$ **49.** $5x^4 - 13x^3 + 20x^2 + 7x + 5$ **51.** $2x^4 + x^2 - 15$ **53.** $6a^6 + 15a^4 + 4a^2 + 10$ **55.** $x^3 + 12x^2 + 47x + 60$ **57.** $x^2 - 5x + 8$ **59.** $8x^2 - 6x - 5$ **61.** $x^2 - x - 30$ **63.** $x^2 + 4x - 6$ **65.** $x^2 + 13x$ **67.** $x^2 + 2x - 3$ **69.** $a^2 - 3a + 6$ **71. a.** $x^2 - 1$ **b.** $x^2 + 2x + 1$ **c.** $x^3 + 3x^2 + 3x + 1$ **d.** $x^4 + 4x^3 + 6x^2 + 4x + 1$ **73. a.** $x^2 - 1$ **b.** $x^2 - x - 2$ **c.** $x^2 - 2x - 3$ **d.** $x^2 - 3x - 4$ **75.** 0 **77.** −5 **79.** −2, 3 **81.** Never 0 **83.** $x(6{,}200 - 25x) = 6{,}200x - 25x^2$; \$18,375 **85.** $A = x(2x + 5) = 2x^2 + 5x$ **87.** $A = x(x + 1) = x^2 + x$ **89.** $R = (1{,}200 - 100p)p = 1{,}200p - 100p^2$ **91.** 169 **93.** $-10x$ **95.** 0 **97.** 0 **99.** $-12x + 16$ **101.** $x^2 + x - 2$ **103.** $x^2 + 6x + 9$ **105.**

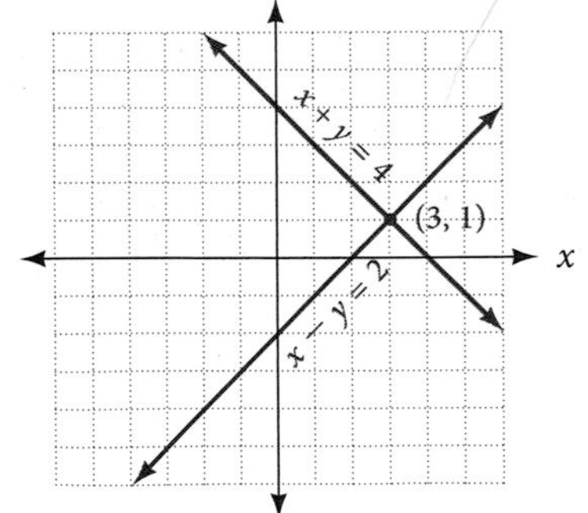

107. (1, −1) **109.** $\left(-\frac{1}{2}, -\frac{1}{2}\right)$ **111.** 6 dimes, 5 quarters

Problem Set 5.6

1. $x^2 - 4x + 4$ **3.** $a^2 + 6a + 9$ **5.** $x^2 - 10x + 25$ **7.** $a^2 - a + \frac{1}{4}$ **9.** $x^2 + 20x + 100$ **11.** $a^2 + 1.6a + 0.64$ **13.** $4x^2 - 4x + 1$ **15.** $16a^2 + 40a + 25$ **17.** $9x^2 - 12x + 4$ **19.** $9a^2 + 30ab + 25b^2$ **21.** $16x^2 - 40xy + 25y^2$ **23.** $49m^2 + 28mn + 4n^2$ **25.** $36x^2 - 120xy + 100y^2$ **27.** $x^4 + 10x^2 + 25$ **29.** $a^4 + 2a^2 + 1$ **31.** $y^2 + 3y + \frac{9}{4}$ **33.** $a^2 + a + \frac{1}{4}$ **35.** $x^2 + \frac{3}{2}x + \frac{9}{16}$ **37.** $t^2 + \frac{2}{5}t + \frac{1}{25}$

39.

x	$(x + 3)^2$	$x^2 + 9$	$x^2 + 6x + 9$
1	16	10	16
2	25	13	25
3	36	18	36
4	49	25	49

41.

a	b	$(a + b)^2$	$a^2 + b^2$	$a^2 + ab + b^2$	$a^2 + 2ab + b^2$
1	1	4	2	3	4
3	5	64	34	49	64
3	4	49	25	37	49
4	5	81	41	61	81

43. $a^2 - 25$ **45.** $y^2 - 1$ **47.** $81 - x^2$ **49.** $4x^2 - 25$ **51.** $16x^2 - \frac{1}{9}$ **53.** $4a^2 - 49$ **55.** $36 - 49x^2$ **57.** $x^4 - 9$ **59.** $a^4 - 16$

61. $25y^8 - 64$ **63.** $2x^2 - 34$ **65.** $-12x^2 + 20x + 8$ **67.** $a^2 + 4a + 6$ **69.** $8x^3 + 36x^2 + 54x + 27$ **71. a.** 11 **b.** 1 **c.** 11 **73. a.** 1 **b.** 13 **c.** 1 **75.** Both equal 25. **77.** $x^2 + (x + 1)^2 = 2x^2 + 2x + 1$ **79.** $x^2 + (x + 1)^2 + (x + 2)^2 = 3x^2 + 6x + 5$ **81.** $a^2 + ab + ba + b^2 = a^2 + 2ab + b^2$ **83.** $2x^2$ **85.** x^2 **87.** $3x$ **89.** $3xy$ **91.** $5x$ **93.** $\frac{a^4}{2b^2}$ **95.** $\frac{11}{14}$

97.

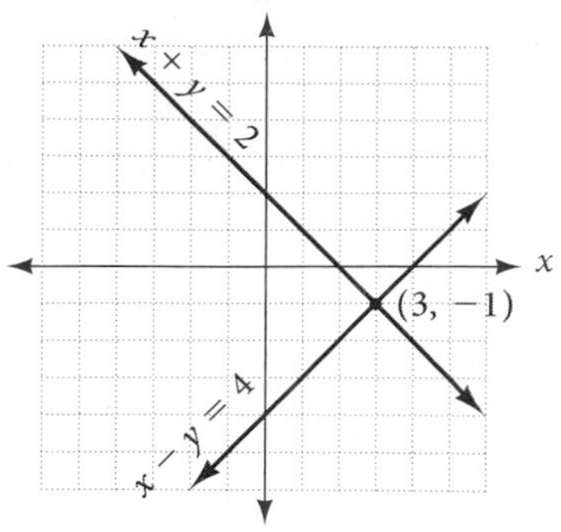

99.

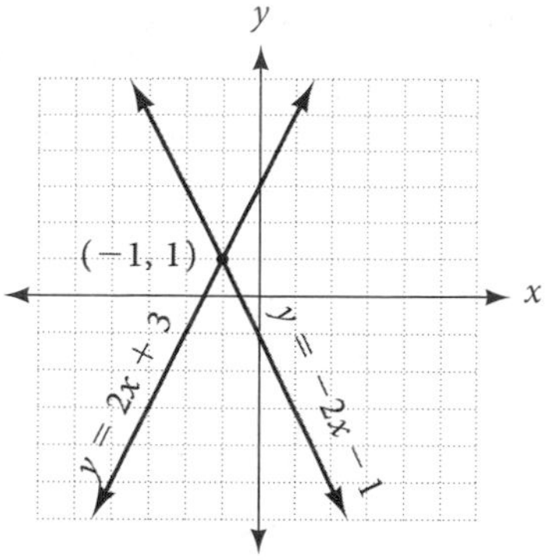

Problem Set 5.7

1. $x - 2$ **3.** $3 - 2x^2$ **5.** $5xy - 2y$ **7.** $7x^4 - 6x^3 + 5x^2$ **9.** $10x^4 - 5x^2 + 1$ **11.** $-4a + 2$ **13.** $-8a^4 - 12a^3$ **15.** $-4b - 5a$ **17.** $-6a^2b + 3ab^2 - 7b^3$ **19.** $-\frac{a}{2} - b - \frac{b^2}{2a}$ **21.** $3x + 4y$ **23.** $-y + 3$ **25.** $x^2 + 8x - 9$ **27.** $y^2 - 3y + 1$ **29.** $5y - 4$ **31.** $xy - x^2y^2$ **33.** $-1 + xy$ **35.** $-a + 1$ **37.** $x^2 - 3xy + y^2$ **39.** $2 - 3b + 5b^2$ **41.** $-2xy + 1$ **43.** $xy - \frac{1}{2}$ **45.** $\frac{1}{4x} - \frac{1}{2a} + \frac{3}{4}$ **47.** $\frac{4x^2}{3} + \frac{2}{3x} + \frac{1}{x^2}$ **49.** $3a^{3m} - 9a^m$ **51.** $2x^{4m} - 5x^{2m} + 7$ **53.** $3x^2 - x + 6$ **55.** 4 **57.** $x + 5$ **59. a.** 7 **b.** 7 **c.** 23 **61. a.** 19 **b.** 34 **c.** 19 **63. a.** 2 **b.** 2 **65.** Both equal 7. **67.** $\frac{3(10) + 8}{2} = 19; 3(10) + 4 = 34$ **69.** $146\frac{20}{27}$ **71.** $2x + 5$ **73.** $x^2 - 3x$ **75.** $2x^3 - 10x^2$ **77.** $-2x$ **79.** 2 **81.** 107 dB **83.** $(7, -1)$ **85.** $(2, 3)$ **87.** $(1, 1)$ **89.** Lines coincide.

Problem Set 5.8

1. $x - 2$ **3.** $a + 4$ **5.** $x - 3$ **7.** $x + 3$ **9.** $a - 5$ **11.** $x + 2 + \frac{2}{x + 3}$ **13.** $a - 2 + \frac{12}{a + 5}$ **15.** $x + 4 + \frac{9}{x - 2}$ **17.** $x + 4 + \frac{-10}{x + 1}$ **19.** $a + 1 + \frac{-1}{a + 2}$ **21.** $x - 3 + \frac{17}{2x + 4}$ **23.** $3a - 2 + \frac{7}{2a + 3}$ **25.** $2a^2 - a - 3$ **27.** $x^2 - x + 5$ **29.** $x^2 + x + 1$ **31.** $x^2 + 2x + 4$ **33. a.** 19 **b.** 19 **c.** 5 **35. a.** 7 **b.** $\frac{25}{7}$ **c.** $\frac{25}{7}$ **37.** 401 yards **39.** 5, 20 **41.** \$800 at 8%, \$400 at 9% **43.** Eight \$5 bills, twelve \$10 bills **45.** 10 gallons of 20%, 6 gallons of 60%

Chapter 5 Review

1. -1 **2.** -64 **3.** $\frac{9}{49}$ **4.** y^{12} **5.** x^{30} **6.** x^{35} **7.** 2^{24} **8.** $27y^3$ **9.** $-8x^3y^3z^3$ **10.** $\frac{1}{49}$ **11.** $\frac{4}{x^5}$ **12.** $\frac{1}{27y^3}$ **13.** a^6 **14.** $\frac{1}{x^4}$ **15.** x^{15} **16.** $\frac{1}{x^5}$ **17.** 1 **18.** -3 **19.** $9x^6y^4$ **20.** $64a^{22}b^{20}$ **21.** $-\frac{1}{27x^3y^6}$ **22.** b **23.** $\frac{1}{x^{35}}$ **24.** $5x^7$ **25.** $\frac{6y^7}{x^5}$ **26.** $4a^6$ **27.** $2x^5y^2$ **28.** 6.4×10^7 **29.** 2.3×10^8 **30.** 8×10^3 **31.** $8a^2 - 12a - 3$ **32.** $-12x^2 + 5x - 14$ **33.** $-4x^2 - 6x$ **34.** 14 **35.** $12x^2 - 21x$ **36.** $24x^5y^2 - 40x^4y^3 + 32x^3y^4$ **37.** $a^3 + 6a^2 + a - 4$ **38.** $x^3 + 125$ **39.** $6x^2 - 29x + 35$ **40.** $25y^2 - \frac{1}{25}$ **41.** $a^4 - 9$ **42.** $a^2 - 10a + 25$ **43.** $9x^2 + 24x + 16$ **44.** $y^4 + 6y^2 + 9$ **45.** $-2b - 4a$ **46.** $-5x^4y^3 + 4x^2y^2 + 2x$ **47.** $x + 9$ **48.** $2x + 5$ **49.** $x^2 - 4x + 16$ **50.** $x - 3 + \frac{16}{3x + 2}$ **51.** $x^2 - 4x + 5 + \frac{5}{2x + 1}$

Chapter 5 Cumulative Review

1. $\frac{3}{4}$ **2.** -9 **3.** 24 **4.** 105 **5.** -5 **6.** -12 **7.** 1 **8.** 63 **9.** $-3x + y$ **10.** $9a + 15$ **11.** y^{16} **12.** $16y^4$ **13.** $108a^{18}b^{29}$ **14.** $\frac{1152y^{13}}{x^3}$ **15.** 2.8×10^{-2} **16.** 5×10^{-5} **17.** $6a^2b$ **18.** $-13x + 10$ **19.** $25x^2 - 10x + 1$ **20.** $x^3 - 1$ **21.** 4 **22.** $\frac{5}{2}$ **23.** 4 **24.** -2 **25.** $a < -24$

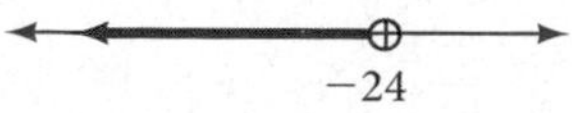

26. $t \geq -\frac{32}{9}$

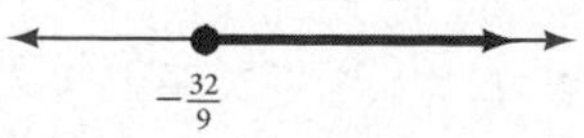

27. $-3 < x < 1$

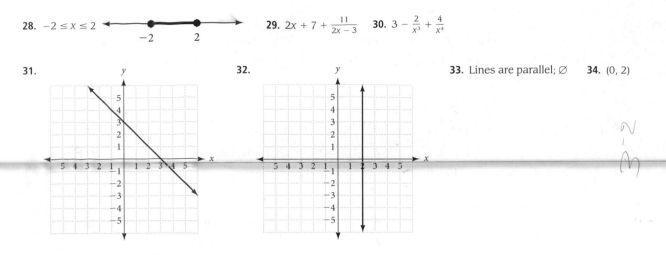

28. $-2 \le x \le 2$ 

29. $2x + 7 + \frac{11}{2x - 3}$ **30.** $3 - \frac{2}{x^3} + \frac{4}{x^4}$

31. 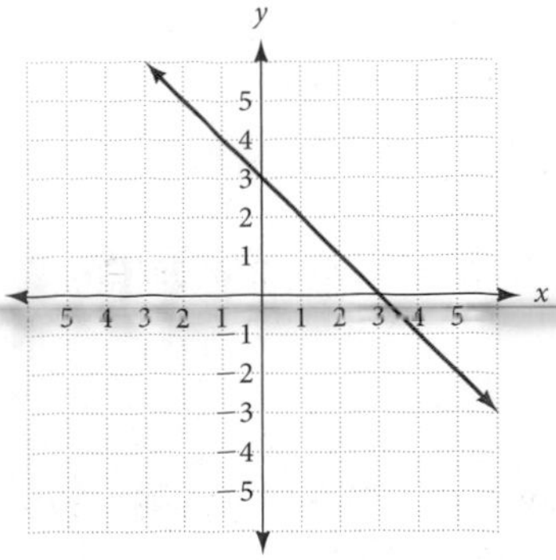

32. 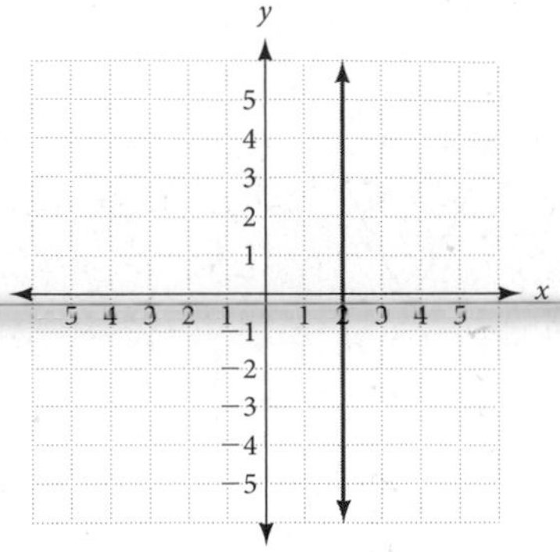

33. Lines are parallel; ∅ **34.** (0, 2)

35. (3, 1) **36.** Lines are parallel; ∅ **37.** (−2, 1) **38.** (3, 2) **39.** (4, 5) **40.** (5, 1) **41.** 21 **42.** 3 **43.** $h = \frac{2A}{b}$
44. $y = -2x + 1$ **45.** 0 **46.** $-\sqrt{2}, \pi$ **47.** Additive inverse property **48.** Commutative property of addition
49. x-intercept = −2; y-intercept = 4 **50.** $m = \frac{2}{5}$ **51.** $15,505 more **52.** $22,937 less

Chapter 5 Test

1. 81 **2.** $\frac{9}{16}$ **3.** $72x^{18}$ **4.** $\frac{1}{9}$ **5.** 1 **6.** a^2 **7.** x **8.** 2.78×10^{-2} **9.** 243,000 **10.** $\frac{y^2}{2x^4}$ **11.** $3a$ **12.** $10x^5$
13. 9×10^9 **14.** $8x^2 + 2x + 2$ **15.** $3x^2 + 4x + 6$ **16.** $3x - 4$ **17.** 10 **18.** $6a^4 - 10a^3 + 8a^2$ **19.** $x^2 + \frac{5}{6}x + \frac{1}{6}$
20. $8x^2 + 2x - 15$ **21.** $x^3 - 27$ **22.** $x^2 + 10x + 25$ **23.** $9a^2 - 12ab + 4b^2$ **24.** $9x^2 - 16y^2$ **25.** $a^4 - 9$ **26.** $2x^2 + 3x - 1$
27. $4x + 3 + \frac{4}{2x - 3}$ **28.** $3x^2 + 9x + 25 + \frac{76}{x - 3}$ **29.** 15.625cm³ **30.** $V = w^3$ **31.** 1.765×10^8 **32.** 9.61×10^7
33. 5×10^{-3} **34.** 1×10^{-2}

Chapter 6

Pretest for Chapter 6

1. $6(x + 3)$ **2.** $4ab(3a - 6 + 2b)$ **3.** $(x - 2b)(x + 3a)$ **4.** $(3 - x)(5y - 4)$ **5.** $(x - 3)(x + 4)$ **6.** $(x - 7)(x + 3)$
7. $(x - 5)(x + 5)$ **8.** $(x - 2)(x + 2)(x^2 + 4)$ **9.** Does not factor **10.** $2(3x - 4y)(3x + 4y)$ **11.** $(x^2 - 3)(x + 4)$
12. $(x - 3)(x + b)$ **13.** $2(2x - 5)(x + 1)$ **14.** $(4n - 3)(n + 4)$ **15.** $(3c - 2)(4c + 3)$ **16.** $3x(2x - 1)(2x + 3)$
17. $(x + 5y)(x^2 - 5xy + 25y^2)$ **18.** $2(3b - 4)(9b^2 + 12b + 16)$ **19.** −3, 5 **20.** 3, 4 **21.** −5, 5 **22.** −2, 7 **23.** −6, 5
24. −3, 0, 3 **25.** $-4, \frac{3}{2}$ **26.** $-\frac{5}{3}, 6, 0$

Getting Ready for Chapter 6

1. $y^2 - 16y + 64$ **2.** $x^2 + 7x + 12$ **3.** $x^2 - 4x - 12$ **4.** $4x^2 + 5x + 15$ **5.** $x^2 - 16$ **6.** $x^2 - 36$ **7.** $x^2 - 64$ **8.** $x^4 - 16$
9. $x^3 + 2x^2 + 4x$ **10.** $8x^3 - 20x^2 + 50x$ **11.** $-2x^2 - 4x - 8$ **12.** $20x^2 - 50x + 125$ **13.** $x^3 - 8$ **14.** $8x^3 + 125$
15. $3x^4 - 27x^2$ **16.** $2x^5 + 20x^4 + 50x^3$ **17.** $y^5 + 36y^3$ **18.** $6a^2 - 11a + 4$ **19.** $6x^3 - 12x^2 - 48x$ **20.** $2a^5b + 6a^4b + 2a^3b$

Problem Set 6.1

1. $5(3x + 5)$ **3.** $3(2a + 3)$ **5.** $4(x - 2y)$ **7.** $3(x^2 - 2x - 3)$ **9.** $3(a^2 - a - 20)$ **11.** $4(6y^2 - 13y + 6)$ **13.** $x^2(9 - 8x)$
15. $13a^2(1 - 2a)$ **17.** $7xy(3x - 4y)$ **19.** $11ab^2(2a - 1)$ **21.** $7x(x^2 + 3x - 4)$ **23.** $11(11y^4 - x^4)$ **25.** $25x^2(4x^2 - 2x + 1)$
27. $8(a^2 + 2b^2 + 4c^2)$ **29.** $4ab(a - 4b + 8ab)$ **31.** $11a^2b^2(11a - 2b + 3ab)$ **33.** $12x^2y^3(1 - 6x^3 - 3x^2y)$ **35.** $(x + 3)(y + 5)$
37. $(x + 2)(y + 6)$ **39.** $(a - 3)(b + 7)$ **41.** $(a - b)(x + y)$ **43.** $(2x - 5)(a + 3)$ **45.** $(b - 2)(3x - 4)$ **47.** $(x + 2)(x + a)$
49. $(x - b)(x - a)$ **51.** $(x + y)(a + b + c)$ **53.** $(3x + 2)(2x + 3)$ **55.** $(10x - 1)(2x + 5)$ **57.** $(4x + 5)(5x + 1)$ **59.** $(x + 2)(x^2 + 3)$
61. $(3x - 2)(2x^2 + 5)$ **63.** 6 **65.** $3(4x^2 + 2x + 1)$ **67.** $x^2 - 5x - 14$ **69.** $x^2 - x - 6$ **71.** $x^3 + 27$ **73.** $2x^3 + 9x^2 - 2x - 3$
75. $18x^7 - 12x^6 + 6x^5$ **77.** $x^2 + x + \frac{2}{9}$ **79.** $12x^2 - 10xy - 12y^2$ **81.** $81a^2 - 1$ **83.** $x^2 - 18x + 81$ **85.** $x^3 + 8$
87. $y^2 - 16y + 64$ **89.** $x^2 - 4x - 12$ **91.** $2y^2 - 7y + 12$ **93.** $2x^2 + 7x - 11$ **95.** $3x + 8$ **97.** $3x^2 + 4x - 5$

Problem Set 6.2

1. $(x + 3)(x + 4)$ **3.** $(x + 1)(x + 2)$ **5.** $(a + 3)(a + 7)$ **7.** $(x - 2)(x - 5)$ **9.** $(y - 3)(y - 7)$ **11.** $(x - 4)(x + 3)$ **13.** $(y + 4)(y - 3)$ **15.** $(x + 7)(x - 2)$ **17.** $(r - 9)(r + 1)$ **19.** $(x - 6)(x + 5)$ **21.** $(a + 7)(a + 8)$ **23.** $(y + 6)(y - 7)$ **25.** $(x + 6)(x + 7)$ **27.** $2(x + 1)(x + 2)$ **29.** $3(a + 4)(a - 5)$ **31.** $100(x - 2)(x - 3)$ **33.** $100(p - 5)(p - 8)$ **35.** $x^2(x + 3)(x - 4)$ **37.** $2r(r + 5)(r - 3)$ **39.** $2y^2(y + 1)(y - 4)$ **41.** $x^3(x + 2)(x + 2)$ **43.** $3y^2(y + 1)(y - 5)$ **45.** $4x^2(x - 4)(x - 9)$ **47.** $(x + 2y)(x + 3y)$ **49.** $(x - 4y)(x - 5y)$ **51.** $(a + 4b)(a - 2b)$ **53.** $(a - 5b)(a - 5b)$ **55.** $(a + 5b)(a + 5b)$ **57.** $(x - 6a)(x + 8a)$ **59.** $(x + 4b)(x - 9b)$ **61.** $(x^2 - 3)(x^2 - 2)$ **63.** $(x - 100)(x + 20)$ **65.** $\left(x - \frac{1}{2}\right)\left(x - \frac{1}{2}\right)$ **67.** $(x + 0.2)(x + 0.4)$ **69.** $x + 16$ **71.** $4x^2 - x - 3$ **73.** $6a^2 + 13a + 2$ **75.** $6a^2 + 7a + 2$ **77.** $6a^2 + 8a + 2$ **79.** $\frac{4}{25}$ **81.** $72a^{12}$ **83.** $\frac{1}{16x^2}$ **85.** $\frac{a^3}{6b^2}$ **87.** $\frac{1}{3x^7y^2}$ **89.** $-21x^4y^5$ **91.** $-20ab^3$ **93.** $-27a^5b^8$

Problem Set 6.3

1. $(2x + 1)(x + 3)$ **3.** $(2a - 3)(a + 1)$ **5.** $(3x + 5)(x - 1)$ **7.** $(3y + 1)(y - 5)$ **9.** $(2x + 3)(3x + 2)$ **11.** $(2x - 3y)(2x - 3y)$ **13.** $(4y + 1)(y - 3)$ **15.** $(4x - 5)(5x - 4)$ **17.** $(10a - b)(2a + 5b)$ **19.** $(4x - 5)(5x + 1)$ **21.** $(6m - 1)(2m + 3)$ **23.** $(4x + 5)(5x + 3)$ **25.** $(3a - 4b)(4a - 3b)$ **27.** $(3x - 7y)(x + 2y)$ **29.** $(2x + 5)(7x - 3)$ **31.** $(3x - 5)(2x - 11)$ **33.** $(5t - 19)(3t - 2)$ **35.** $2(2x + 3)(x - 1)$ **37.** $2(4a - 3)(3a - 4)$ **39.** $x(5x - 4)(2x - 3)$ **41.** $x^2(3x + 2)(2x - 5)$ **43.** $2a(5a + 2)(a - 1)$ **45.** $3x(5x + 1)(x - 7)$ **47.** $5y(7y + 2)(y - 2)$ **49.** $a^2(5a + 1)(3a - 1)$ **51.** $2y(2x - y)(3x - 7y)$ **53.** Both equal 25. **55.** $4x^2 - 9$ **57.** $x^4 - 81$ **59.** $12x - 35$ **61.** $9x + 8$ **63.** $16(t - 1)(t - 3)$ **65.** $h = 2(725 - 8t^2)$, 874 feet **67. a.** $h = 2(4 - t)(1 + 8t)$ **b.**

Time t (seconds)	Height h (feet)
0	8
1	54
2	68
3	50
4	0

69. $x^2 - 9$ **71.** $x^2 - 25$ **73.** $x^2 - 49$ **75.** $x^2 - 81$ **77.** $4x^2 - 9y^2$ **79.** $x^4 - 16$ **81.** $x^2 + 6x + 9$ **83.** $x^2 + 10x + 25$ **85.** $x^2 + 14x + 49$ **87.** $x^2 + 18x + 81$ **89.** $4x^2 + 12x + 9$ **91.** $16x^2 - 16xy + 4y^2$ **93.** $6x^3 + 5x^2 - 4x + 3$ **95.** $x^4 + 11x^3 - 6x - 9$ **97.** $2x^5 + 8x^3 - 2x - 7$ **99.** $-8x^5 + 2x^4 + 2x^2 + 12$ **101.** $7x^5y^3$ **103.** $3a^2b^3$

Problem Set 6.4

1. $(x + 3)(x - 3)$ **3.** $(a + 6)(a - 6)$ **5.** $(x + 7)(x - 7)$ **7.** $4(a + 2)(a - 2)$ **9.** Cannot be factored **11.** $(5x + 13)(5x - 13)$ **13.** $(3a + 4b)(3a - 4b)$ **15.** $(3 + m)(3 - m)$ **17.** $(5 + 2x)(5 - 2x)$ **19.** $2(x + 3)(x - 3)$ **21.** $32(a + 2)(a - 2)$ **23.** $2y(2x + 3)(2x - 3)$ **25.** $(a^2 + b^2)(a + b)(a - b)$ **27.** $(4m^2 + 9)(2m + 3)(2m - 3)$ **29.** $3xy(x + 5y)(x - 5y)$ **31.** $(x - 1)^2$ **33.** $(x + 1)^2$ **35.** $(a - 5)^2$ **37.** $(y + 2)^2$ **39.** $(x - 2)^2$ **41.** $(m - 6)^2$ **43.** $(2a + 3)^2$ **45.** $(7x - 1)^2$ **47.** $(3y - 5)^2$ **49.** $(x + 5y)^2$ **51.** $(3a + b)^2$ **53.** $\left(y - \frac{3}{2}\right)^2$ **55.** $\left(a + \frac{1}{2}\right)^2$ **57.** $\left(x - \frac{7}{2}\right)^2$ **59.** $\left(x - \frac{3}{8}\right)^2$ **61.** $3(a + 3)^2$ **63.** $2(x + 5y)^2$ **65.** $x(x + 2)^2$ **67.** $y^2(y - 4)^2$ **69.** $5x(x + 3y)^2$ **71.** $3y^2(2y - 5)^2$ **73.** $(x + 3 + y)(x + 3 - y)$ **75.** $(x + y + 3)(x + y - 3)$ **77.** 14 **79.** 25 **81. a.** 1 **b.** 8 **c.** 27 **d.** 64 **e.** 125 **83. a.** $x^3 - x^2 + x$ **b.** $x^2 - x + 1$ **c.** $x^3 + 1$ **85. a.** $x^3 - 2x^2 + 4x$ **b.** $2x^2 - 4x + 8$ **c.** $x^3 + 8$ **87. a.** $x^3 - 3x^2 + 9x$ **b.** $3x^2 - 9x + 27$ **c.** $x^3 + 27$ **89. a.** $x^3 - 4x^2 + 16x$ **b.** $4x^2 - 16x + 64$ **c.** $x^3 + 64$ **91. a.** $x^3 - 5x^2 + 25x$ **b.** $5x^2 - 25x + 125$ **c.** $x^3 + 125$ **93.** \$0.19 **95.** $4y^2 - 6y - 3$ **97.** $12x^5 - 9x^3 + 3$ **99.** $-6x^3 - 4x^2 + 2x$ **101.** $-7x^3 + 4x^2 - 11$ **103.** $x - 2 + \frac{2}{x - 3}$ **105.** $3x - 2 + \frac{9}{2x + 3}$ **107.** $\left(t - \frac{1}{5}\right)^2$

Problem Set 6.5

1. $(x - y)(x^2 + xy + y^2)$ **3.** $(a + 2)(a^2 - 2a + 4)$ **5.** $(3 + x)(9 - 3x + x^2)$ **7.** $(y - 1)(y^2 + y + 1)$ **9.** $(y - 4)(y^2 + 4y + 16)$ **11.** $(5h - t)(25h^2 + 5ht + t^2)$ **13.** $(x - 6)(x^2 + 6x + 36)$ **15.** $2(y - 3)(y^2 + 3y + 9)$ **17.** $2(a - 4b)(a^2 + 4ab + 16b^2)$ **19.** $2(x + 6y)(x^2 - 6xy + 36y^2)$ **21.** $10(a - 4b)(a^2 + 4ab + 16b^2)$ **23.** $10(r - 5)(r^2 + 5r + 25)$ **25.** $(4 + 3a)(16 - 12a + 9a^2)$ **27.** $(2x - 3y)(4x^2 + 6xy + 9y^2)$ **29.** $\left(t + \frac{1}{3}\right)\left(t^2 - \frac{1}{3}t + \frac{1}{9}\right)$ **31.** $\left(3x - \frac{1}{3}\right)\left(9x^2 + x + \frac{1}{9}\right)$ **33.** $(4a + 5b)(16a^2 - 20ab + 25b^2)$ **35.** $\left(\frac{1}{2}x - \frac{1}{3}y\right)\left(\frac{1}{4}x^2 + \frac{1}{6}xy + \frac{1}{9}y^2\right)$ **37.** $(a - b)(a^2 + ab + b^2)(a + b)(a^2 - ab + b^2)$ **39.** $(2x - y)(4x^2 + 2xy + y^2)(2x + y)(4x^2 - 2xy + y^2)$

41. $(x - 5y)(x^2 + 5xy + 25y^2)(x + 5y)(x^2 - 5xy + 25y^2)$ **43.** $2x^5 - 8x^3$ **45.** $3x^4 - 18x^3 + 27x^2$ **47.** $y^3 + 25y$ **49.** $15a^2 - a - 2$ **51.** $4x^4 - 12x^3 - 40x^2$ **53.** $2ab^5 - 8ab^4 + 2ab^3$ **55.** 40.64 quadrillion Btu **57.** $x = 3y + 4$ **59.** $x = \frac{3}{2}y + 2$ **61.** $y = \frac{1}{2}x + 3$ **63.** $y = \frac{2}{3}x - 4$

Problem Set 6.6

1. $(x + 9)(x - 9)$ **3.** $(x + 5)(x - 3)$ **5.** $(x + 3)^2$ **7.** $(y - 5)^2$ **9.** $2ab(a^2 + 3a + 1)$ **11.** Cannot be factored **13.** $3(2a + 5)(2a - 5)$ **15.** $(3x - 2y)^2$ **17.** $4x(x^2 + 4y^2)$ **19.** $2y(y + 5)^2$ **21.** $a^4(a^2 + 4b^2)$ **23.** $(x + 4)(y + 3)$ **25.** $(x - 3)(x^2 + 3x + 9)$ **27.** $(x + 2)(y - 5)$ **29.** $5(a + b)^2$ **31.** Cannot be factored **33.** $3(x + 2y)(x + 3y)$ **35.** $(2x + 19)(x - 2)$ **37.** $100(x - 2)(x - 1)$ **39.** $(x + 8)(x - 8)$ **41.** $(x + a)(x + 3)$ **43.** $a^5(7a + 3)(7a - 3)$ **45.** Cannot be factored **47.** $a(5a + 1)(5a + 3)$ **49.** $(x + y)(a - b)$ **51.** $3a^2b(4a + 1)(4a - 1)$ **53.** $5x^2(2x + 3)(2x - 3)$ **55.** $(3x + 41y)(x - 2y)$ **57.** $2x^3(2x - 3)(4x - 5)$ **59.** $(2x + 3)(x + a)$ **61.** $(y^2 + 1)(y + 1)(y - 1)$ **63.** $3x^2y^2(2x + 3y)^2$ **65.** $16(t - 1)(t - 3)$ **67.** $(9x + 8)(6x + 7)$ **69.** 5 **71.** $-\frac{3}{2}$ **73.** $-\frac{3}{4}$ **75.** 1 **77.** 30 **79.** -6 **81.** 3 **83.** 2

Problem Set 6.7

1. $-2, 1$ **3.** $4, 5$ **5.** $0, -1, 3$ **7.** $-\frac{2}{3}, -\frac{3}{2}$ **9.** $0, -\frac{4}{3}, \frac{4}{3}$ **11.** $0, -\frac{1}{3}, -\frac{3}{5}$ **13.** $-1, -2$ **15.** $4, 5$ **17.** $6, -4$ **19.** $2, 3$ **21.** -3 **23.** $4, -4$ **25.** $\frac{3}{2}, -4$ **27.** $-\frac{2}{3}$ **29.** 5 **31.** $4, -\frac{5}{2}$ **33.** $\frac{5}{3}, -4$ **35.** $\frac{7}{2}, -\frac{7}{2}$ **37.** $0, -6$ **39.** $0, 3$ **41.** $0, 4$ **43.** $0, 5$ **45.** $2, 5$ **47.** $\frac{1}{2}, -\frac{4}{3}$ **49.** $4, -\frac{5}{2}$ **51.** $8, -10$ **53.** $5, 8$ **55.** $6, 8$ **57.** -4 **59.** $5, 8$ **61.** $6, -8$ **63.** $0, -\frac{3}{2}, -4$ **65.** $0, 3, -\frac{5}{2}$ **67.** $0, \frac{1}{2}, -\frac{5}{2}$ **69.** $0, \frac{3}{5}, -\frac{3}{2}$ **71.** $\frac{1}{2}, \frac{3}{2}$ **73.** $-5, 4$ **75.** $-7, -6$ **77.** $-3, -1$ **79.** $2, 3$ **81.** $-15, 10$ **83.** $-5, 3$ **85.** $-3, -2, 2$ **87.** $-4, -1, 4$ **89.** $x^2 - 8x + 15 = 0$ **91. a.** $x^2 - 5x + 6 = 0$ **b.** $x^2 - 7x + 6 = 0$ **c.** $x^2 - x - 6 = 0$ **93.** $x(x + 1) = 72$ **95.** $x(x + 2) = 99$ **97.** $x(x + 2) = 5[x + (x + 2)] - 10$ **99.** Bicycle \$75, suit \$15 **101.** House \$2,400, lot \$600 **103.** $\frac{1}{8}$ **105.** x^8 **107.** x^{18} **109.** 5.6×10^{-3} **111.** 5.67×10^9

Problem Set 6.8

1. Two consecutive even integers are x and $x + 2$; $x(x + 2) = 80$; 8, 10 and $-10, -8$ **3.** 9, 11 and $-11, -9$ **5.** $x(x + 2) = 5(x + x + 2) - 10$; 8, 10 and 0, 2 **7.** 8, 6 **9.** The numbers are x and $5x + 2$; $x(5x + 2) = 24$; 2, 12 and $-\frac{12}{5}, -10$ **11.** 5, 20 and 0, 0 **13.** Let x = the width; $x(x + 1) = 12$; width 3 inches, length 4 inches **15.** Let x = the base; $\frac{1}{2}(x)(2x) = 9$; base 3 inches **17.** $x^2 + (x + 2)^2 = 10^2$; 6 inches and 8 inches **19.** 12 meters **21.** $1{,}400 = 400 + 700x - 100x^2$; 200 or 500 items **23.** 200 DVD's or 800 DVD's **25.** $R = xp = (1{,}200 - 100p)p = 3{,}200$; \$4 or \$8 **27.** \$7 or \$10 **29. a.** 5 feet **b.** 12 feet **31. a.** 25 seconds later **b.**

t (sec)	h (feet)
0	100
5	1,680
10	2,460
15	2,440
20	1,620
25	0

33. a. 2 seconds **b.** 400 feet **35.** $200x^{24}$ **37.** x^7 **39.** 8×10^1 **41.** $10ab^2$ **43.** $6x^4 + 6x^3 - 2x^2$ **45.** $9y^2 - 30y + 25$ **47.** $4a^4 - 49$

Chapter 6 Review

1. $10(x - 2)$ **2.** $x^2(4x - 9)$ **3.** $5(x - y)$ **4.** $x(7x^2 + 2)$ **5.** $4(2x + 1)$ **6.** $2(x^2 + 7x + 3)$ **7.** $8(3y^2 - 5y + 6)$ **8.** $15xy^2(2y - 3x^2)$ **9.** $7(7a^3 - 2b^3)$ **10.** $6ab(b + 3a^2b^2 - 4a)$ **11.** $(x + a)(y + b)$ **12.** $(x - 5)(y + 4)$ **13.** $(2x - 3)(y + 5)$ **14.** $(5x - 4a)(x - 2b)$ **15.** $(y + 7)(y + 2)$ **16.** $(w + 5)(w + 10)$ **17.** $(a - 6)(a - 8)$ **18.** $(r - 6)(r - 12)$ **19.** $(y + 9)(y + 11)$ **20.** $(y + 6)(y + 2)$ **21.** $(2x + 3)(x + 5)$ **22.** $(2y - 5)(2y - 1)$ **23.** $(5y + 6)(y + 1)$ **24.** $(5a - 3)(4a - 3)$ **25.** $(2r + 3t)(3r - 2t)$ **26.** $(2x - 7)(5x + 3)$ **27.** $(n + 9)(n - 9)$ **28.** $(2y + 3)(2y - 3)$ **29.** Cannot be factored. **30.** $(6y + 11x)(6y - 11x)$ **31.** $(8a + 11b)(8a - 11b)$ **32.** $(8 + 3m)(8 - 3m)$ **33.** $(y + 10)^2$ **34.** $(m - 8)^2$ **35.** $(8t + 1)^2$ **36.** $(4n - 3)^2$ **37.** $(2r - 3t)^2$ **38.** $(3m + 5n)^2$ **39.** $2(x + 4)(x + 6)$ **40.** $a(a - 3)(a - 7)$ **41.** $3m(m + 1)(m - 7)$ **42.** $5y^2(y - 2)(y + 4)$ **43.** $2(2x + 1)(2x + 3)$

44. $a(3a + 1)(a - 5)$ **45.** $2m(2m - 3)(5m - 1)$ **46.** $5y(2x - 3y)(3x - y)$ **47.** $4(x + 5)^2$ **48.** $x(2x + 3)^2$ **49.** $5(x + 3)(x - 3)$ **50.** $3x(2x + 3y)(2x - 3y)$ **51.** $(2a - b)(4a^2 + 2ab + b^2)$ **52.** $(3x + 2y)(9x^2 - 6xy + 4y^2)$ **53.** $(5x - 4y)(25x^2 + 20xy + 16y^2)$ **54.** $3ab(2a + b)(a + 5b)$ **55.** $x^3(x + 1)(x - 1)$ **56.** $y^4(4y^2 + 9)$ **57.** $4x^3(x + 2y)(3x - y)$ **58.** $5a^2b(3a - b)(2a + 3b)$ **59.** $3ab^2(2a - b)(3a + 2b)$ **60.** $-2, 5$ **61.** $-\frac{5}{2}, \frac{5}{2}$ **62.** $2, -5$ **63.** $7, -7$ **64.** $0, 9$ **65.** $-\frac{3}{2}, -\frac{2}{3}$ **66.** $0, -\frac{5}{3}, \frac{2}{3}$ **67.** 10, 12 and $-12, -10$ **68.** 10, 11 and $-11, -10$ **69.** 5, 7, and $-1, 1$ **70.** 5, 15 **71.** 6, 11 and $-\frac{11}{2}, -12$ **72.** 2 inches **73.** 1.4×10^6 babies **74.** 7.195×10^3 yds

Chapter 6 Cumulative Review

1. -6 **2.** 29 **3.** 9 **4.** 9 **5.** -64 **6.** $\frac{1}{81}$ **7.** 2 **8.** -3 **9.** $5a - 5$ **10.** $-8a - 3$ **11.** x^{40} **12.** 1 **13.** $15x^2 + 14x - 8$ **14.** $a^4 - 49$ **15.** -6 **16.** 10 **17.** -21 **18.** -2 **19.** $0, \frac{7}{2}, 7$ **20.** $-\frac{9}{4}, \frac{9}{4}$ **21.** $x < 4$ **22.** $x \le 2$ or $x \ge 7$

23.

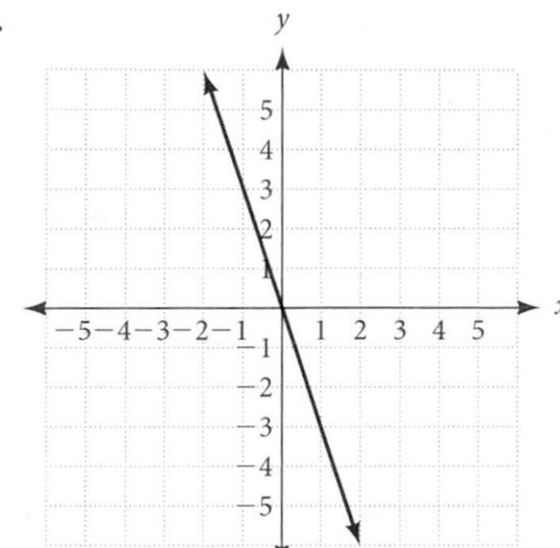

24.

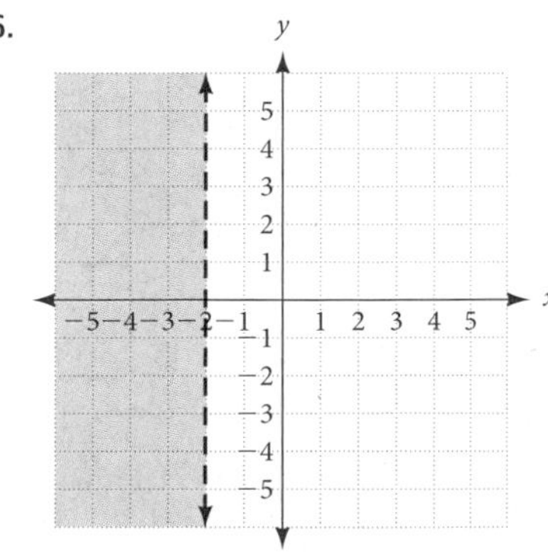

25. **26.** **27.** $(4, 0)$ and $\left(\frac{16}{3}, 1\right)$

28. x-intercept = $(3, 0)$; y-intercept = $(0, -10)$ **29.** $m = \frac{7}{9}$ **30.** $\frac{2}{3}x + 8$ **31.** $y = -\frac{2}{5}x - \frac{2}{3}$ **32.** slope = $\frac{3}{4}$; y-intercept = $(0, 4)$

33. $(-2, 1)$

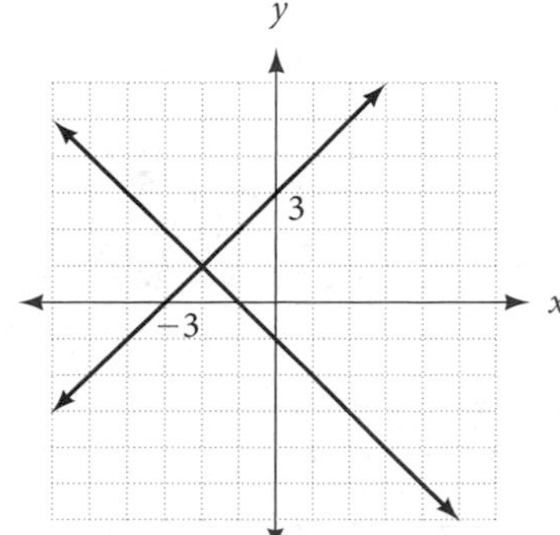

34. Lines are parallel; $\varnothing$

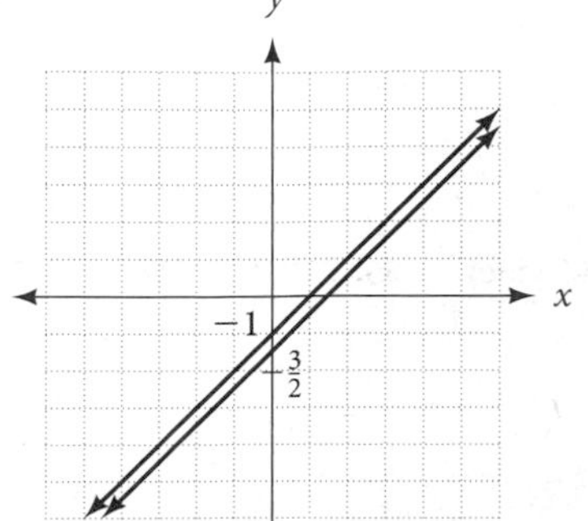

35. $(2, 5)$

36. $(2, -4)$ **37.** $(1, 2)$ **38.** $(1500, 3500)$ **39.** $(n - 9)(n + 4)$ **40.** $(2x + 5y)(7x - 2y)$ **41.** $(4 - a)(4 + a)$ **42.** $(7x - 1)^2$ **43.** $5y(3x - y)^2$ **44.** $3x(3x + y)(2x - y)$ **45.** $(3x - 2)(y + 5)$ **46.** $(a + 4)(a^2 - 4a + 16)$ **47.** $2, -\frac{1}{2}, 2$ **48.** $-\frac{1}{5}, 5, \frac{1}{5}$ **49.** $4x + 5 + \frac{2}{x - 3}$ **50.** $x^2 + 3x + 9$ **51.** 34 inches, 38 inches **52.** 6 burgers, 8 fries **53.** 4.3 million **54.** 6.7 million

Chapter 6 Test

1. $5(x-2)$ **2.** $9xy(2x-1-4y)$ **3.** $(x-3b)(x+2a)$ **4.** $(y+4)(x-7)$ **5.** $(x-3)(x-2)$ **6.** $(x-3)(x+2)$ **7.** $(a-4)(a+4)$ **8.** Does not factor **9.** $(x^2+9)(x-3)(x+3)$ **10.** $3(3x-5y)(3x+5y)$ **11.** $(x-3)(x+3)(x+5)$ **12.** $(x+5)(x-b)$ **13.** $2(2a+1)(a+5)$ **14.** $3(m+2)(m-3)$ **15.** $(2y-1)(3y+5)$ **16.** $2x(2x+1)(3x-5)$ **17.** $(a+4b)(a^2-4ab+16b^2)$ **18.** $2(3x-2)(9x^2+6x+4)$ **19.** $-3, -4$ **20.** 2 **21.** $-6, 6$ **22.** $-4, 5$ **23.** 5, 6 **24.** $-4, 0, 4$ **25.** $-\frac{5}{2}, 3$ **26.** $-\frac{1}{3}, 0, 1$ **27.** 4 and 16 **28.** $-3, -1$ and 3, 5 **29.** 3 ft by 14 ft **30.** 5 m, 12 m **31.** 200 or 300 **32.** \$3 or \$6 **33. a.** 8 ft **b.** 15 ft **34. a.** $t = \frac{1}{2}$, 2 sec **b.** $t = 3$ sec

Chapter 7

Pretest for Chapter 7

1. $\frac{x+3}{x-3}$ **2.** $\frac{3}{a-4}$ **3.** $\frac{x+3}{(y-1)(y+1)}$ **4.** $\frac{3}{2}$ **5.** $(x-3)(x+4)$ **6.** $\frac{2x-3}{x+4}$ **7.** $(x+4)(x-1)$ **8.** $-\frac{2}{x-8}$ **9.** $\frac{2x+4}{(x-4)(x+4)}$ **10.** $\frac{2}{(x+3)(x-1)}$ **11.** $\frac{6}{5}$ **12.** $-\frac{7}{6}$ **13.** -14 **14.** $\frac{x+2}{x-2}$ **15.** $\frac{x+3}{x+2}$ **16.** 54 **17.** 2

Getting Ready for Chapter 7

1. $\frac{5}{33}$ **2.** $\frac{7}{25}$ **3.** $-\frac{2}{3}$ **4.** $\frac{8}{5}$ **5.** Undefined **6.** $\frac{11}{35}$ **7.** $\frac{11}{35}$ **8.** x^2+2x **9.** x^2-x-12 **10.** $(x+5)(x-5)$ **11.** $2(x-2)$ **12.** $(x+3)(x+4)$ **13.** $a(a-4)$ **14.** $y(xy+1)$ **15.** 36 **16.** 6 **17.** $-5, 4$ **18.** $-4, 2$ **19.** 2 **20.** $-6, 6$

Problem Set 7.1

1. a. $\frac{1}{4}$ **b.** $\frac{1}{x-1}$; $x \neq \pm 1$ **c.** $\frac{x}{x+1}$; $x \neq \pm 1$ **d.** $\frac{x^2+x+1}{x+1}$; $x \neq \pm 1$ **e.** $\frac{x^2+x+1}{x^2}$; $x \neq 0, 1$ **3.** $\frac{1}{x-2}$, $x \neq 2$ **5.** $\frac{1}{a+3}$, $a \neq -3, 3$ **7.** $\frac{1}{x-5}$, $x \neq -5, 5$ **9.** $\frac{(x+2)(x-2)}{2}$ **11.** $\frac{2(x-5)}{3(x-2)}$, $x \neq 2$ **13.** 2 **15.** $\frac{5(x-1)}{4}$ **17.** $\frac{1}{x-3}$ **19.** $\frac{1}{x+3}$ **21.** $\frac{1}{a-5}$ **23.** $\frac{1}{3x+2}$ **25.** $\frac{x+5}{x+2}$ **27.** $\frac{2m(m+2)}{m-2}$ **29.** $\frac{x-1}{x-4}$ **31.** $\frac{2(2x-3)}{2x+3}$ **33.** $\frac{2x-3}{2x+3}$ **35.** $\frac{1}{(x^2+9)(x-3)}$ **37.** $\frac{3x-5}{(x^2+4)(x-2)}$ **39.** $\frac{2x(7x+6)}{2x+1}$ **41.** $\frac{x^2+xy+y^2}{x+y}$ **43.** $\frac{x^2-2x+4}{x-2}$ **45.** $\frac{x^2-2x+4}{x-1}$ **47.** $\frac{x+2}{x+5}$ **49.** $\frac{x+a}{x+b}$ **51.** $\frac{x+a}{x+3}$ **53. a.** x^2-16 **b.** $x^2-8x+16$ **c.** $4x^3-32x^2+64$ **d.** $\frac{x}{4}$ **55.** $\frac{4}{3}$ **57.** $\frac{4}{5}$ **59.** $\frac{8}{1}$ **61.** Saleen S7 Twin Turbo and the SSC Ultimate Aero **63.** 40.7 miles/hour **65.** 39.25 feet/minute **67.** 6.3 feet/second **69.** 48 miles/gallon **71.** $5+4=9$ **73.**

x	$\frac{x-3}{3-x}$
−2	−1
−1	−1
0	−1
1	−1
2	−1

The entries are all −1 because the numerator and denominator are opposites. Or, $\frac{x-3}{3-x} = \frac{-1(x-3)}{3-x} = -1$.

75.

x	$\frac{x-5}{x^2-25}$	$\frac{1}{x+5}$
0	$\frac{1}{5}$	$\frac{1}{5}$
2	$\frac{1}{7}$	$\frac{1}{7}$
−2	$\frac{1}{3}$	$\frac{1}{3}$
5	Undefined	$\frac{1}{10}$
−5	Undefined	Undefined

77. $\frac{5}{14}$ **79.** $\frac{9}{10}$ **81.** $(x+3)(x-3)$ **83.** $3(x-3)$ **85.** $(x-5)(x+4)$ **87.** $a(a+5)$ **89.** $(a-5)(a+4)$ **91.** 5.8 **93.** 0 **95.** $16a^2b^2$ **97.** $19x^4+21x^2-42$ **99.** $7a^4b^4+9b^3-11a^3$ **101.** Disney, Nike

company	p/e
Yahoo	35.33
Google	84.53
Disney	18.63
Nike	17.51
Ebay	56.11

Problem Set 7.2

1. 2 **3.** $\frac{x}{2}$ **5.** $\frac{3}{2}$ **7.** $\frac{1}{2(x-3)}$ **9.** $\frac{4a(a+5)}{7(a+4)}$ **11.** $\frac{y-2}{4}$ **13.** $\frac{2(x+4)}{x-2}$ **15.** $\frac{x+3}{(x-3)(x+1)}$ **17.** 1 **19.** $\frac{y-5}{(y+2)(y-2)}$ **21.** $\frac{x^2+5x+25}{x-5}$ **23.** 1 **25.** $\frac{a+3}{a+4}$ **27.** $\frac{2y-3}{y-6}$ **29.** $\frac{x-1}{(x-2)(x^2-x+1)}$ **31.** $\frac{3}{2}$ **33. a.** $\frac{4}{13}$ **b.** $\frac{x+1}{x^2+x+1}$ **c.** $\frac{x-1}{x^2+x+1}$ **d.** $\frac{x+1}{x-1}$ **35.** $2(x-3)$ **37.** $2a$ **39.** $(x+2)(x+1)$ **41.** $-2x(x-5)$ **43.** $\frac{2(x+5)}{x(y+5)}$ **45.** $\frac{2x}{x-y}$ **47.** $\frac{1}{x-2}$ **49.** $\frac{1}{5}$ **51.** $\frac{1}{100}$ **53.** 300.7 ft/s **55.** 2.7 miles **57.** 742 miles/hour **59.** 0.45 mile/hour **61.** 8.8 miles/hour **63.** $\frac{4}{5}$ **65.** $\frac{11}{35}$ **67.** $-\frac{4}{35}$ **69.** $2x-6$ **71.** x^2-x-20 **73.** $\frac{1}{x-3}$ **75.** $\frac{x-6}{2(x-5)}$ **77.** x^2-x-30 **79.** 3 **81.** $\frac{11}{4}$ **83.** $9x^2$ **85.** $6ab^2$

Problem Set 7.3

1. $\frac{7}{x}$ **3.** $\frac{4}{a}$ **5.** 1 **7.** $y+1$ **9.** $x+2$ **11.** $x-2$ **13.** $\frac{6}{x+6}$ **15.** $\frac{(y+2)(y-2)}{2y}$ **17.** $\frac{3+2a}{6}$ **19.** $\frac{7x+3}{4(x+1)}$ **21.** $\frac{1}{5}$ **23.** $\frac{1}{3}$ **25.** $\frac{3}{x-2}$ **27.** $\frac{4}{a+3}$ **29.** $\frac{2(x-10)}{(x+5)(x-5)}$ **31.** $\frac{x+2}{x+3}$ **33.** $\frac{a+1}{a+2}$ **35.** $\frac{1}{(x+3)(x+4)}$ **37.** $\frac{y}{(y+5)(y+4)}$ **39.** $\frac{3(x-1)}{(x+4)(x+1)}$ **41.** $\frac{1}{3}$ **43. a.** $\frac{2}{27}$ **b.** $\frac{8}{3}$ **c.** $\frac{11}{18}$ **d.** $\frac{3x+10}{(x-2)^2}$ **e.** $\frac{(x+2)^2}{3x+10}$ **f.** $\frac{x+3}{x+2}$

45.

Number x	Reciprocal $\frac{1}{x}$	Sum $1+\frac{1}{x}$	Sum $\frac{x+1}{x}$
1	1	2	2
2	$\frac{1}{2}$	$\frac{3}{2}$	$\frac{3}{2}$
3	$\frac{1}{3}$	$\frac{4}{3}$	$\frac{4}{3}$
4	$\frac{1}{4}$	$\frac{5}{4}$	$\frac{5}{4}$

47.

x	$x+\frac{4}{x}$	$\frac{x^2+4}{x}$	$x+4$
1	5	5	5
2	4	4	6
3	$\frac{13}{3}$	$\frac{13}{3}$	7
4	5	5	8

49. $\frac{x+3}{x+2}$ **51.** $\frac{x+2}{x+3}$ **53.** 3 **55.** 0 **57.** Undefined **59.** $-\frac{3}{2}$ **61.** $2x+15$ **63.** x^2-5x **65.** -6 **67.** -2 **69.** 3 **71.** -2 **73.** $\frac{1}{5}$ **75.** $-2, -3$ **77.** $3, -2$ **79.** $0, 5$ **81.** 58° C

Problem Set 7.4

1. -3 **3.** 20 **5.** -1 **7.** 5 **9.** -2 **11.** 4 **13.** 3, 5 **15.** -8, 1 **17.** 2 **19.** 1 **21.** 8 **23.** 5 **25.** Possible solution 2, which does not check; ∅ **27.** 3 **29.** Possible solution -3, which does not check; ∅ **31.** 0 **33.** Possible solutions 2 and 3, but only 2 checks; 2 **35.** -4 **37.** -1 **39.** $-6, -7$ **41.** Possible solutions -3 and -1, but only -1 checks; -1 **43. a.** $\frac{1}{5}$ **b.** 5 **c.** $\frac{25}{3}$ **d.** 3 **e.** $\frac{1}{5}$, 5 **45. a.** $\frac{7}{(a-6)(a+2)}$ **b.** $\frac{a-5}{a-6}$ **c.** Possible 7, -1; only 7 **47.** $\frac{1}{3}$ **49.** 30 **51.** 1 **53.** -3 **55.** 26 centimeters per second **57.** 7 **59.** Length 13 inches, width 4 inches **61.** 6, 8, or -8, -6 **63.** 6 inches, 8 inches

Problem Set 7.5

1. $\frac{1}{4}, \frac{3}{4}$ **3.** $x+\frac{1}{x}=\frac{13}{6}$; $\frac{2}{3}$ and $\frac{3}{2}$ **5.** -2 **7.** $\frac{1}{x}+\frac{1}{x+2}=\frac{5}{12}$; 4 and 6

9. Let x = the speed of the boat in still water.

	d	r	t
Upstream	26	$x-3$	$\frac{26}{x-3}$
Downstream	38	$x+3$	$\frac{38}{x+3}$

The equation is $\frac{26}{x-3}=\frac{38}{x+3}$; x = 16 miles per hour.

11. 300 miles/hour **13.** 170 miles/hour, 190 miles/hour **15.** 9 miles/hour **17.** 8 miles/hour **19.** $\frac{1}{12}-\frac{1}{15}=\frac{1}{x}$; 60 hours **21.** $5\frac{5}{11}$ minutes **23.** 12 minutes

25.

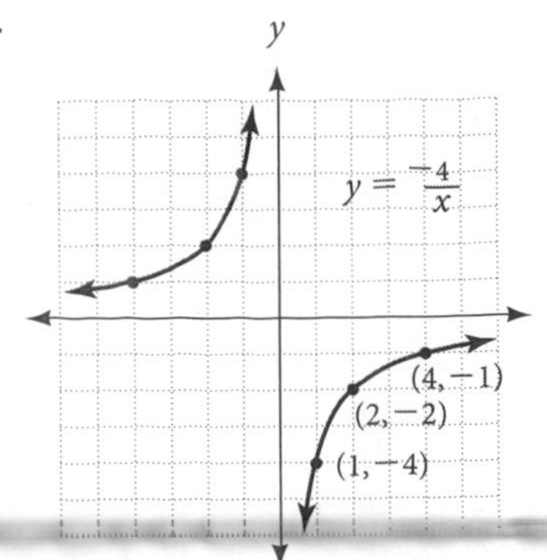

27.

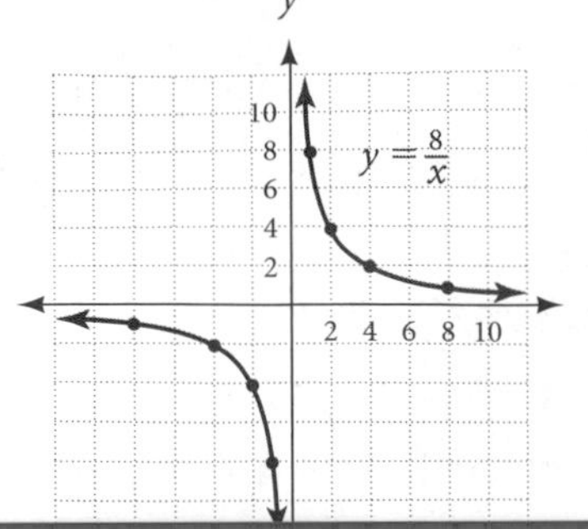

29.

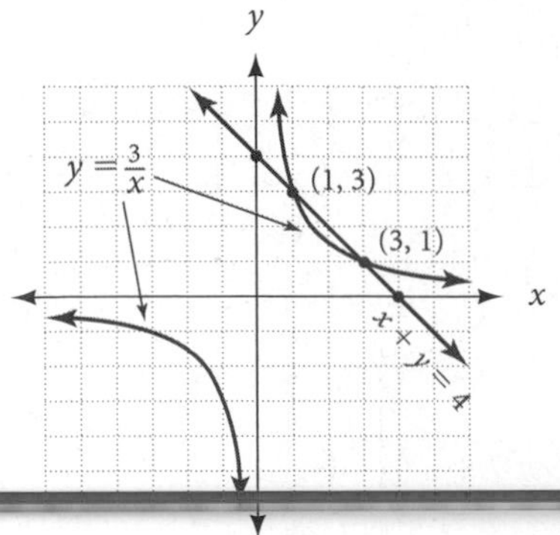

31. $\frac{3}{4}$ **33.** $\frac{3}{2}$ **35.** $2x^3y^3$ **37.** $\frac{x^2y^3}{2}$ **39.** $x(xy + 1)$ **41.** $\frac{x^2y^2}{2}$ **43.** $\frac{x-2}{x-3}$ **45.** About 19.2 million
47. $5ab(3a^2b^2 - 4a - 7b)$ **49.** $(x - 6)(x + 2)$ **51.** $(x^2 + 4)(x + 2)(x - 2)$ **53.** $5x(x - 6)(x + 1)$

Problem Set 7.6

1. 6 **3.** $\frac{1}{6}$ **5.** xy^2 **7.** $\frac{xy}{2}$ **9.** $\frac{y}{x}$ **11.** $\frac{a+1}{a-1}$ **13.** $\frac{x+1}{2(x-3)}$ **15.** $a - 3$ **17.** $\frac{y+3}{y+2}$ **19.** $x + y$ **21.** $\frac{a+1}{a}$ **23.** $\frac{1}{x}$
25. $\frac{2a+3}{3a+4}$ **27.** $\frac{x-1}{x+2}$ **29.** $\frac{x+4}{x+1}$ **31.** $\frac{7}{3}, \frac{17}{7}, \frac{41}{17}$

33.

Number x	Reciprocal $\frac{1}{x}$	Quotient $\frac{x}{\frac{1}{x}}$	Square x^2
1	1	1	1
2	$\frac{1}{2}$	4	4
3	$\frac{1}{3}$	9	9
4	$\frac{1}{4}$	16	16

35.

Number x	Reciprocal $\frac{1}{x}$	Sum $1 + \frac{1}{x}$	Quotient $\frac{1 + \frac{1}{x}}{\frac{1}{x}}$
1	1	2	2
2	$\frac{1}{2}$	$\frac{3}{2}$	3
3	$\frac{1}{3}$	$\frac{4}{3}$	4
4	$\frac{1}{4}$	$\frac{5}{4}$	5

37. 330 kilometers per hour **39.** $\frac{7}{2}$ **41.** −3, 2 **43.** $x < 1$ **45.** $x \geq -7$ **47.** $x < 6$ **49.** $x \leq 2$

Problem Set 7.7

1. 1 **3.** 10 **5.** 40 **7.** $\frac{5}{4}$ **9.** $\frac{7}{3}$ **11.** 3, −4 **13.** 4, −4 **15.** 2, −4 **17.** 6, −1 **19.** 12.6 million metric tons
21. 15 hits **23.** 21 milliliters **25.** 45.5 grams **27.** 14.7 inches **29.** 12.5 miles **31.** 17 minutes **33.** 343 miles **35.** $h = 9$
37. $y = 14$ **39.** −5, 5 **41.** 5 **43.** 2 **45.** $\frac{x+2}{x+3}$ **47.** $\frac{2(x+5)}{x-4}$ **49.** $\frac{1}{x-4}$

Problem Set 7.8

1. 8 **3.** 130 **5.** 5 **7.** −3 **9.** 3 **11.** 243 **13.** 2 **15.** $\frac{1}{2}$ **17.** 1 **19.** 5 **21.** 25 **23.** 9 **25.** $\frac{103}{65} \approx 1.6$ hours
27. 84 pounds **29.** $\frac{735}{2}$ or 367.5 **31.** $235.50 **33.** 96 pounds **35.** 12 amperes **37.** (2, −1) **39.** (4, 3) **41.** (1, 1)
43. (5, 2)

Chapter 7 Review

1. $\frac{1}{2(x-2)}, x \neq 2$ **2.** $\frac{1}{a-6}, a \neq -6, 6$ **3.** $\frac{2x-1}{x+3}, x \neq -3$ **4.** $\frac{1}{x+4}, x \neq -4$ **5.** $\frac{3x-2}{2x-3}, x \neq \frac{3}{2}, -6, 0$ **6.** $\frac{1}{(x-2)(x^2+4)}, x \neq \pm 2$
7. $x - 2, x \neq -7$ **8.** $a + 8, a \neq -8$ **9.** $\frac{y+b}{y+5}, x \neq -a, y \neq -5$ **10.** $\frac{x}{2}$ **11.** $\frac{x+2}{x-4}$ **12.** $(a - 6)^2$ **13.** $\frac{x-1}{x+2}$ **14.** 1 **15.** $x - 9$
16. $\frac{13}{a+8}$ **17.** $\frac{x^2+5x+45}{x(x+9)}$ **18.** $\frac{4x+5}{4(x+5)}$ **19.** $\frac{1}{(x+6)(x+2)}$ **20.** $\frac{a-6}{(a+5)(a+3)}$ **21.** 4 **22.** 9 **23.** 1, 6
24. Possible solution 2, which does not check; ∅ **25.** 5 **26.** 3 and $\frac{7}{3}$ **27.** 15 miles/hour **28.** 84 hours **29.** $\frac{1}{2}$
30. $\frac{y+3}{y+7}$ **31.** $\frac{4a-7}{a-1}$ **32.** $\frac{2}{5}$ **33.** $\frac{2}{9}$ **34.** 12 **35.** −6, 6 **36.** −6, 8 **37.** −35 **38.** $\frac{1}{2}$

Chapter 7 Cumulative Review

1. -3 **2.** -6 **3.** -4 **4.** -3 **5.** $-4x - 4$ **6.** $4 - x$ **7.** x^4 **8.** $\frac{1}{x^4}$ **9.** 6 **10.** 1 **11.** $-3x$

12. $-2a^3 - 10a^2 - 5a + 13$ **13.** $x - 7$ **14.** $\frac{5x + 3}{5(x + 3)}$ **15.** -2 **16.** $\frac{11}{4}$ **17.** $-\frac{3}{7}, \frac{3}{7}$ **18.** $0, 2, \frac{7}{2}$ **19.** 6 **20.** No solution

21. $(3, -1)$ **22.** $\left(\frac{1}{3}, -\frac{1}{2}\right)$ **23.** $(17, 5)$ **24.** $(6, -12)$

25.

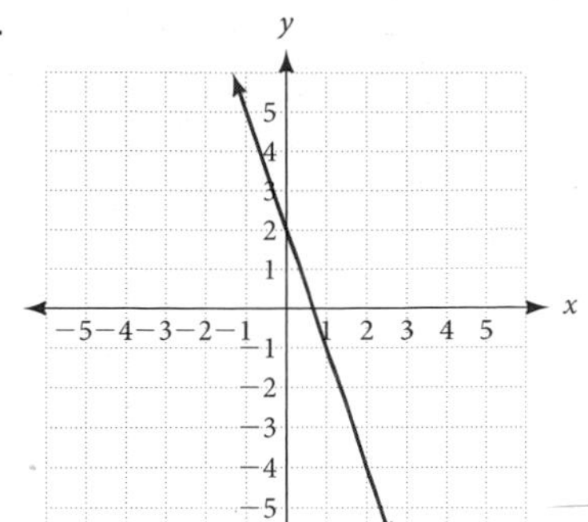

26.

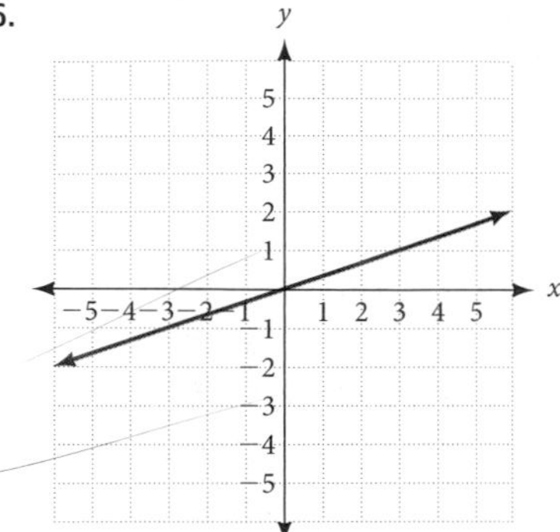

27. $a \geq 6$ **28.** $x > -3$

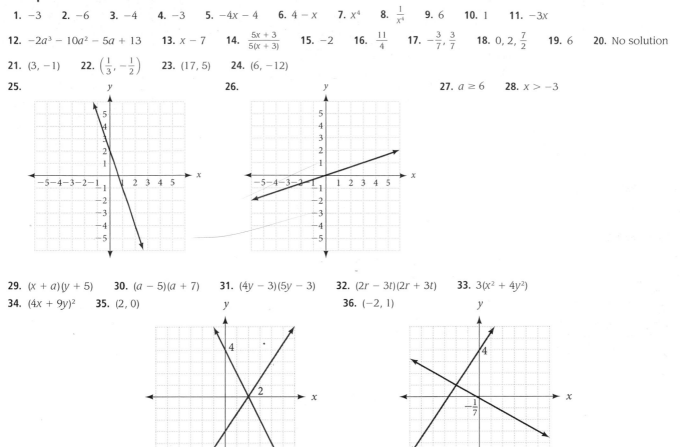

29. $(x + a)(y + 5)$ **30.** $(a - 5)(a + 7)$ **31.** $(4y - 3)(5y - 3)$ **32.** $(2r - 3t)(2r + 3t)$ **33.** $3(x^2 + 4y^2)$

34. $(4x + 9y)^2$ **35.** $(2, 0)$

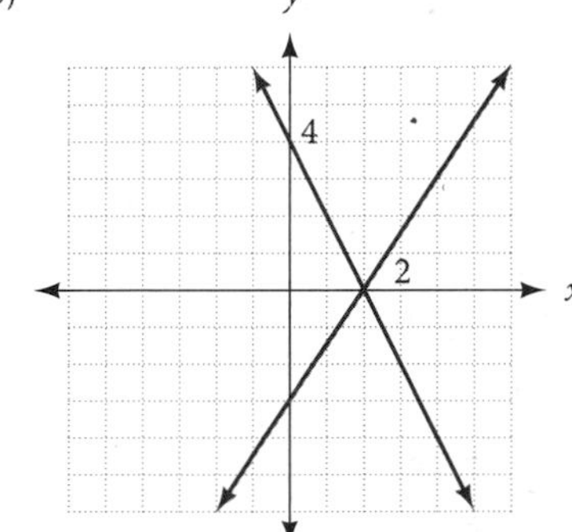

36. $(-2, 1)$

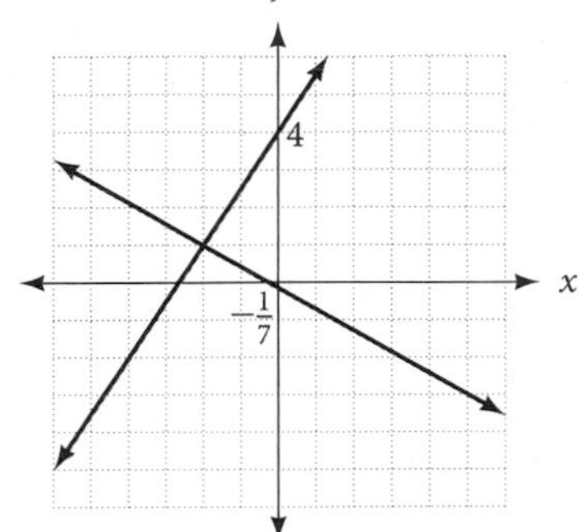

37. x-intercept = $(5, 0)$; y-intercept = $(0, 2)$ **38.** $m = 3$; y-intercept = $(0, -4)$ **39.** $y = -5x - 1$ **40.** $y = -\frac{2}{5}x - \frac{9}{5}$

41. 8 **42.** -14 **43.** Associative property of addition **44.** Commutative property of addition **45.** $x - 7$

46. $\frac{x + 1}{4(x + 2)}$ **47.** $\frac{1}{2}$ **48.** $\frac{2x + 3}{3x + 2}$ **49.** -108 **50.** 4 **51.** 195.9 liters **52.** 121.3 liters

Chapter 7 Test

1. $\frac{x + 4}{x - 4}$ **2.** $\frac{2}{a + 2}$ **3.** $\frac{y + 7}{x + a}$ **4.** 3 **5.** $(x - 7)(x - 1)$ **6.** $\frac{x + 4}{3x - 4}$ **7.** $(x - 3)(x + 2)$ **8.** $-\frac{3}{x - 2}$ **9.** $\frac{2x + 3}{(x - 3)(x + 3)}$

10. $\frac{3x}{(x - 2)(x + 1)}$ **11.** $\frac{11}{5}$ **12.** 1 **13.** 6 **14.** 15 miles/hour **15.** 60 hours **16.** $\frac{1}{2}, \frac{1}{3}$ **17.** 132 parts **18.** $\frac{x + 1}{x - 1}$

19. $\frac{x + 4}{x + 2}$ **20.** 100 **21.** 2 **22.** 3.1

23. Ford GT: 300.7 ft/sec.; Evans 487: 302.1 ft/sec.; S7 Twin Turbo: 381.3 ft/sec.; SSC Ultimate Aero: 400.4 ft/sec

Chapter 8

Pretest for Chapter 8

1. 6 **2.** -9 **3.** 11 **4.** 4 **5.** 3 **6.** -5 **7.** $6\sqrt{2}$ **8.** $4\sqrt{6}$ **9.** $\frac{\sqrt{15}}{5}$ **10.** $2\sqrt[3]{3}$ **11.** $8y\sqrt{3}$ **12.** $xy^2\sqrt{2xy}$

13. $3\sqrt{5}$ **14.** $8m\sqrt{3}$ **15.** $\sqrt{6} - 3\sqrt{2}$ **16.** $-7 - 3\sqrt{3}$ **17.** $x - 81$ **18.** $9 - 2\sqrt{14}$ **19.** $-6 + \sqrt{35}$ **20.** $\frac{x + 3\sqrt{x}}{x - 9}$

21. 13 **22.** $-2, -4$ **23.** $\varnothing$

24.

x	y
0	2
1	3
4	4
9	5

25.

x	y
−2	0
−1	1
2	2
7	3

26.

x	y
−8	−4
−1	−3
0	−2
1	−1
8	0

27.

x	y
−10	−2
−3	−1
−2	0
−1	1
6	2

Getting Ready for Chapter8

1. 1 **2.** 4 **3.** 9 **4.** 16 **5.** 25 **6.** 36 **7.** 49 **8.** 64 **9.** 81 **10.** 100 **11.** 121 **12.** 144 **13.** 225 **14.** 256 **15.** 8 **16.** 27 **17.** 64 **18.** 125

Problem Set 8.1

1. 3 **3.** −3 **5.** Not a real number **7.** −12 **9.** 25 **11.** Not a real number **13.** −8 **15.** −10 **17.** 35 **19.** 1 **21.** −2 **23.** −5 **25.** −1 **27.** −3 **29.** −2 **31.** x **33.** $3x$ **35.** xy **37.** $a + b$ **39.** $7xy$ **41.** x **43.** $2x$ **45.** x^2 **47.** $6a^3$ **49.** $5a^4b^2$ **51.** x^2 **53.** $3a^4$ **55.** x^2 **57. a.** 2 **b.** x **c.** x^2 **d.** $2x^2y^3$ **59.** 12 **61.** 20 **63.** 7 **65.** 5 **67.** 17 **69.** 13 **71. a.** 1.618 **b.** −0.618 **c.** 1 **73. a.** 3 **b.** 30 **c.** 0.3 **75.** 6, −1 **77.** 3, −1 **79.** $x + 3$ **81.** 6.7 miles **83.** 5 **85.** $\sqrt{125} \approx 11.2$ **87.** 30 feet **89.** 13 feet **91.** 12 **93.** $25 \cdot 3$ **95.** $25 \cdot 2$ **97.** $4 \cdot 10$ **99.** $x^4 \cdot x$ **101.** $4x^2 \cdot 3$ **103.** $25x^2y^2 \cdot 2x$ **105.** $x - 4$ **107.** $\frac{2}{a-2}$ **109.** $\frac{2x+1}{x}$ **111.** $\frac{x+2}{x+a}$

Problem Set 8.2

1. $2\sqrt{2}$ **3.** $2\sqrt{3}$ **5.** $2\sqrt[3]{3}$ **7.** $5x\sqrt{2}$ **9.** $3ab\sqrt{5}$ **11.** $3x\sqrt[3]{2}$ **13.** $4x^2\sqrt{2}$ **15.** $20\sqrt{5}$ **17.** $x\sqrt{7x}$ **19.** $2x^2\sqrt[3]{x}$ **21.** $6a^2\sqrt[3]{a^2}$ **23.** $4a\sqrt{5a}$ **25.** $15y\sqrt{2x}$ **27.** $14x\sqrt{3y}$ **29.** $\frac{4}{5}$ **31.** $\frac{2}{3}$ **33.** $\frac{2}{3}$ **35.** $\frac{2}{3}$ **37.** $2x$ **39.** $3ab$ **41.** $\frac{3x}{2y}$ **43.** $\frac{5\sqrt{2}}{3}$ **45.** $\sqrt{3}$ **47.** $\frac{8\sqrt{2}}{7}$ **49.** $\frac{12\sqrt{2x}}{5}$ **51.** $\frac{3a\sqrt{6}}{5}$ **53.** $\frac{15\sqrt{2}}{2}$ **55.** $\frac{14y\sqrt{7}}{3}$ **57.** $5ab\sqrt{2}$ **59.** $6x\sqrt{2y}$ **61. a.** $2\sqrt{10}$ **b.** 5 **c.** $3\sqrt{5}$ **d.** $2\sqrt{3}$ **63. a.** $4x^5y^2\sqrt{2y}$ **b.** $2x^3y\sqrt[3]{4xy^2}$ **c.** $2x^2y\sqrt[4]{2x^2y}$ **d.** $2x^2y$ **65. a.** 2 **b.** 0.2 **c.** 20 **d.** 0.02

67.

x	$\sqrt{x}$	$2\sqrt{x}$	$\sqrt{4x}$
1	1	2	2
2	1.414	2.828	2.828
3	1.732	3.464	3.464
4	2	4	4

69.

x	$\sqrt{x}$	$3\sqrt{x}$	$\sqrt{9x}$
1	1	3	3
2	1.414	4.243	4.243
3	1.732	5.196	5.196
4	2	6	6

71. $2xy\sqrt{x}$ **73.** 12 **75.** $\frac{\sqrt{2}}{2}$ **77.** $\frac{\sqrt[3]{18}}{3}$ **79.** $\frac{\sqrt{6}}{3}$ **81.** 3 **83.** $\frac{2(x+5)}{x(x+1)}$ **85.** $\frac{x+1}{x-1}$ **87.** $(x + 6)(x + 3)$

Problem Set 8.3

1. $\frac{\sqrt{2}}{2}$ **3.** $\frac{\sqrt{3}}{3}$ **5.** $\frac{\sqrt{10}}{5}$ **7.** $\frac{\sqrt{6}}{2}$ **9.** $\frac{2\sqrt{15}}{3}$ **11.** $\frac{\sqrt{30}}{2}$ **13.** 2 **15.** $\sqrt{7}$ **17.** $\sqrt{5}$ **19.** $2\sqrt{5}$ **21.** $2\sqrt{3}$ **23.** $\frac{\sqrt{7}}{2}$ **25.** $xy\sqrt{2}$ **27.** $\frac{x\sqrt{15y}}{3}$ **29.** $\frac{4a^2\sqrt{5}}{5}$ **31.** $\frac{6a^2\sqrt{10a}}{5}$ **33.** $\frac{2xy\sqrt{15y}}{3}$ **35.** $\frac{4xy\sqrt{5y}}{3}$ **37.** $\frac{18ab\sqrt{6b}}{5}$ **39.** $18x\sqrt{x}$ **41.** $\frac{\sqrt[3]{4}}{2}$ **43.** $\frac{\sqrt[3]{3}}{3}$ **45.** $\frac{\sqrt[3]{12}}{2}$ **47. a.** $\frac{6\sqrt{\pi}}{\pi}$ **b.** $\frac{\sqrt{A\pi}}{\pi}$ **c.** $\frac{\sqrt[3]{6V\pi^2}}{\pi}$ **d.** $\frac{\sqrt[3]{4\pi^2}}{\pi}$

49.

x	$\sqrt{x}$	$\frac{1}{\sqrt{x}}$	$\frac{\sqrt{x}}{x}$
1	1	1	1
2	1.414	.707	.707
3	1.732	.577	.577
4	2	.5	.5
5	2.236	.447	.447
6	2.449	.408	.408

51.

x	$\sqrt{x^2}$	$\sqrt{x^3}$	$x\sqrt{x}$
1	1	1	1
2	2	2.828	2.828
3	3	5.196	5.196
4	4	8	8
5	5	11.18	11.18
6	6	14.697	14.697

53. 10.2 mi² **55.** $23x$ **57.** $27y$ **59.** $7ab$

61. $-7xy + 50x$ **63.** $10x$ **65.** $23x$ **67.** $10a$ **69.** $x + 5$ **71.** $\frac{5a+6}{15}$ **73.** $\frac{1}{(a+2)(a+3)}$ **75.**

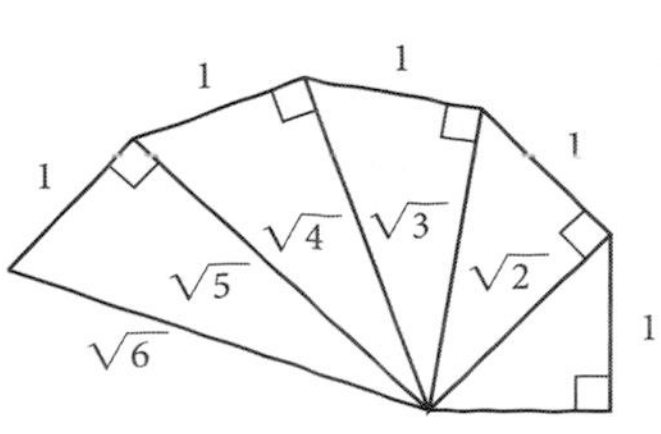

77. $\sqrt{2}, \sqrt{3}, \sqrt{4} = 2$

Problem Set 8.4

1. a. $7x$ **b.** $7y$ **c.** $7\sqrt{2}$ **d.** $7\sqrt{5}$ **3. a.** $7x$ **b.** $7t$ **c.** $7\sqrt{3}$ **d.** $7\sqrt{x}$ **5.** $7\sqrt{2}$ **7.** $2\sqrt{5}$ **9.** $7\sqrt{3}$ **11.** $\frac{11}{56}\sqrt{5}$
13. $13\sqrt{13}$ **15.** $6\sqrt{10}$ **17.** $6\sqrt{5}$ **19.** $4\sqrt{2}$ **21.** 0 **23.** $-30\sqrt{3}$ **25.** $-6\sqrt{3}$ **27.** 0 **29.** $\frac{3}{2}\sqrt{2} + \frac{3}{2}\sqrt{3}$
31. 0 **33.** $\frac{23}{4}\sqrt{2}$ **35.** $-3\sqrt{7}$ **37.** $35\sqrt{3}$ **39.** $-6\sqrt{2} - 6\sqrt{3}$ **41.** -7 **43.** 0 **45.** $2x\sqrt{x}$ **47.** $4a\sqrt{3}$
49. $15x\sqrt{2x}$ **51.** $13x\sqrt{3xy}$ **53.** $-b\sqrt{5a}$ **55.** $27x\sqrt{2x} - 8x\sqrt{3x}$ **57.** $23x\sqrt{2y}$ **59.** $3 + \sqrt{2}$ **61.** $3 - \sqrt{3}$
63. $\frac{4-\sqrt{6}}{3}$ **65.** $3 + \sqrt{2}$ **67.** $\frac{-2+\sqrt{2}}{2}$ **69. a.** 3.232 **b.** -0.232 **c.** 3

71.

x	$\sqrt{x^2+9}$	$x+3$
1	3.162	4
2	3.606	5
3	4.243	6
4	5	7
5	5.831	8
6	6.708	9

73.

x	$\sqrt{x+3}$	$\sqrt{x}+\sqrt{3}$
1	2	2.732
2	2.236	3.146
3	2.449	3.464
4	2.646	3.732
5	2.828	3.968
6	3	4.182

75. $9\sqrt{3}$ **77.** $\sqrt{10}$ **79.** 5 **81.** x **83.** $\sqrt{35}$ **85.** $19 + 9\sqrt{5}$

87. $x - 4\sqrt{x} - 21$ **89.** $9x^2 + 6xy + y^2$ **91.** $9x^2 - 16y^2$ **93.** 9 **95.** 2, 3 **97.** Possible solutions are 2 and 4; only 2 checks.

Problem Set 8.5

1. $\sqrt{6}$ **3.** $2\sqrt{3}$ **5.** $10\sqrt{21}$ **7.** $24\sqrt{2}$ **9.** 8 **11.** 24 **13.** 50 **15.** 8 **17.** $\sqrt{6} - \sqrt{2}$ **19.** $\sqrt{6} + 2$
21. $2\sqrt{6} + 3$ **23.** $6 - \sqrt{15}$ **25.** $2\sqrt{6} + 2\sqrt{15}$ **27.** $3 + 2\sqrt{2}$ **29.** $x + 6\sqrt{x} + 9$ **31.** $27 - 10\sqrt{2}$ **33.** $a - \sqrt{a} + \frac{1}{4}$
35. $16 + 6\sqrt{7}$ **37.** $11 + 5\sqrt{5}$ **39.** $-28 + \sqrt{2}$ **41.** $\sqrt{6} + \frac{1}{3}\sqrt{3} + \frac{1}{2}\sqrt{2} + \frac{1}{6}$ **43.** $x - 36$ **45.** $a + \sqrt{a} + \frac{2}{9}$
47. 1 **49.** $30 + \sqrt{7}$ **51.** $6x + 16\sqrt{x} + 8$ **53.** $49a - 4b$ **55.** -7 **57.** 1 **59.** 0 **61.** $\frac{\sqrt{15}+\sqrt{6}}{3}$ **63.** $\frac{5-\sqrt{10}}{3}$
65. $6 + 2\sqrt{5}$ **67.** $5 + 2\sqrt{6}$ **69.** $\frac{5-\sqrt{21}}{2}$ **71.** $\frac{x+4\sqrt{x}+4}{x-4}$ **73.** $\frac{5-\sqrt{15}-\sqrt{10}+\sqrt{6}}{2}$
75. a. $2\sqrt{3}$ **b.** 4 **c.** $10 - 2\sqrt{21}$ **d.** $\frac{5-\sqrt{21}}{2}$ **77. a.** $2\sqrt{x}$ **b.** $x - 4$ **c.** $x + 4\sqrt{x} + 4$ **d.** $\frac{x+4\sqrt{x}+4}{x-4}$
79. a. 10 **b.** 23 **c.** $27 + 10\sqrt{2}$ **d.** $\frac{27+10\sqrt{2}}{23}$ **81. a.** $\sqrt{6} + 2\sqrt{2}$ **b.** $2 + 2\sqrt{3}$ **c.** $1 + \sqrt{3}$ **d.** $\frac{\sqrt{3}-1}{2}$ **83. a.** 1 **b.** -1
85. $6\sqrt{5}$ **87.** $52 + 14\sqrt{3}$ **89.** 49 **91.** 81 **93.** $x + 1$ **95.** $2x - 3$ **97.** $x^2 + 6x + 9$ **99.** 2 **101.** $-6, 1$
103. a. no **b.** yes **105.** Undefined, Undefined, 0, 3, 6, 9, 12 **107.** $-6, 1$ **109.** 0, 3 **111.** $-9, 9$ **113.** $3, -5$

Problem Set 8.6

1. 3 **3.** 44 **5.** $\varnothing$ **7.** $\varnothing$ **9.** 8 **11.** 4 **13.** $\varnothing$ **15.** $-\frac{2}{3}$ **17.** 25 **19.** 4 **21.** 7 **23.** 8 **25.** $\varnothing$
27. Possible solutions are 1 and 6; only 6 checks. **29.** $-1, -2$ **31.** -4 **33.** Possible solutions are 5 and 8; only 8 checks.
35. a. 100 **b.** 40 **c.** $\varnothing$ **d.** possible 5, 8; only 8 **37. a.** 3 **b.** 9 **c.** 3 **d.** $\varnothing$ **e.** 4 **f.** $\varnothing$ **g.** possible 4, 1; only 4 **39.** 42.60 seconds

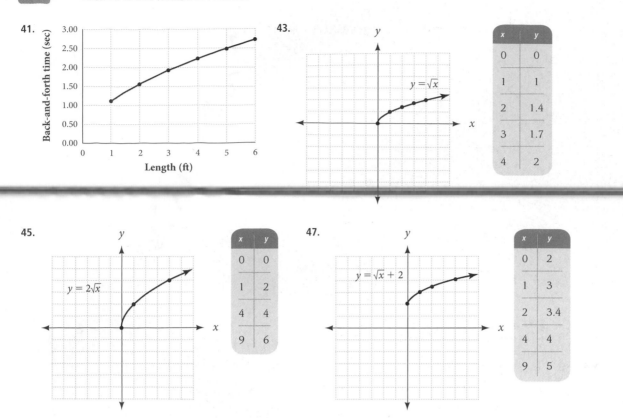

41.

43. 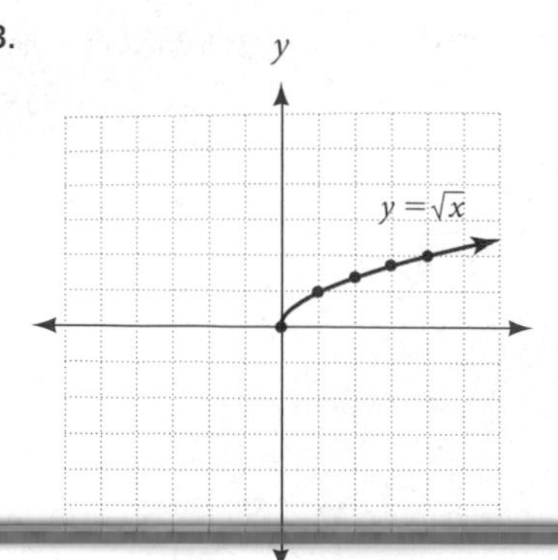

x	y
0	0
1	1
2	1.4
3	1.7
4	2

45. 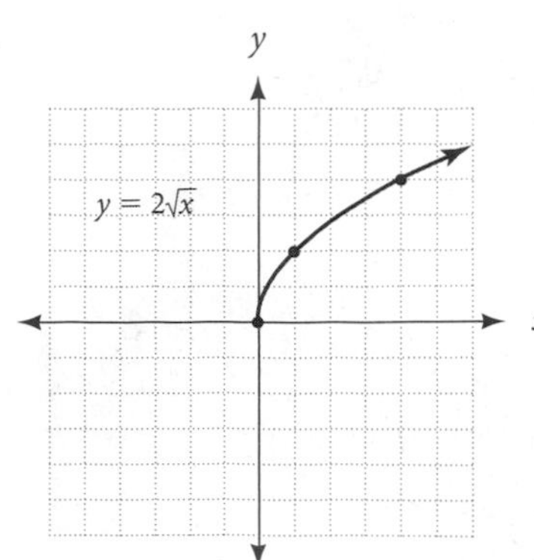

x	y
0	0
1	2
4	4
9	6

47. 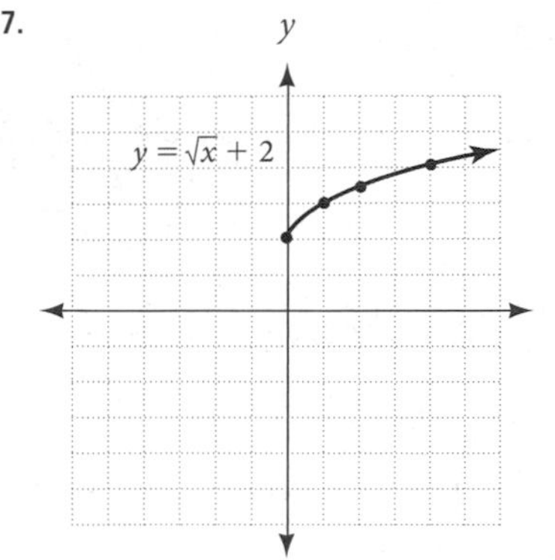

x	y
0	2
1	3
2	3.4
4	4
9	5

49. $x + 2 = \sqrt{8x}$; $x = 2$ **51.** $x - 3 = 2\sqrt{x}$; possible solutions are 1 and 9; only 9 checks. **53.** $\frac{392}{121} \approx 3.2$ feet **55.** $\frac{x+2}{x+3}$

57. $\frac{2(x+5)}{x-4}$ **59.** $\frac{1}{x-4}$ **61.** $\frac{x+5}{x-3}$ **63.** -2 **65.** 24 hours **67.** 24

Chapter 8 Review

1. 5 **2.** 13 **3.** -1 **4.** 5 **5.** $10xy^2$ **6.** $2a$ **7.** $2\sqrt{6}$ **8.** $2x\sqrt{15}$ **9.** $3xy^2\sqrt{10x}$ **10.** $-4\sqrt{2}$ **11.** $6x\sqrt{5xy}$
12. $\frac{\sqrt{3}}{7}$ **13.** $\frac{2\sqrt{2}}{9}$ **14.** $\frac{7}{8}$ **15.** $\frac{7ab}{4}$ **16.** $\frac{4\sqrt{5}}{7}$ **17.** $\frac{2a\sqrt{10}}{11}$ **18.** $\frac{10\sqrt{21}}{7}$ **19.** $\frac{6ab\sqrt{30}}{5}$ **20.** $\frac{-5xy\sqrt{5x}}{6}$ **21.** $\frac{2\sqrt{7}}{7}$
22. $\frac{4\sqrt{10}}{5}$ **23.** $\frac{\sqrt{15}}{12}$ **24.** $-6\sqrt{3}$ **25.** $\frac{4b\sqrt{6a}}{3}$ **26.** $\frac{\sqrt[3]{6}}{2}$ **27.** $\frac{-3\sqrt{3}-12}{13}$ **28.** $3-\sqrt{7}$ **29.** $\sqrt{5}+\sqrt{2}$ **30.** $\frac{-\sqrt{15}-5}{2}$
31. $\frac{7-2\sqrt{10}}{3}$ **32.** $\frac{x+6\sqrt{x}+9}{x-9}$ **33.** $-4\sqrt{5}$ **34.** $-11\sqrt{3}$ **35.** $-22\sqrt{5}$ **36.** $5x\sqrt{2}$ **37.** $-ab\sqrt{10a}$ **38.** $3+3\sqrt{3}$
39. $4\sqrt{6}+4\sqrt{10}$ **40.** $x-49$ **41.** $2\sqrt{5}-2$ **42.** $x+10\sqrt{x}+25$ **43.** 12 **44.** 7 **45.** 16 **46.** No solution
47. No solution **48.** 0, 5 **49.** $\sqrt{3}$ **50.** $8\sqrt{2}$

Chapter 8 Cumulative Review

1. 1 **2.** 3 **3.** $-\frac{1}{2}$ **4.** -45 **5.** -9 **6.** 24 **7.** a^8 **8.** $\frac{1}{b^6}$ **9.** $7a^9$ **10.** $4x^4y^6$ **11.** $\frac{1}{4}y^2 - 4$ **12.** $40x^2 + 15x$
13. $\frac{3a+19}{2a+13}$ **14.** $\frac{x-2}{4(x+2)}$ **15.** $\frac{6(2a+5)}{(a-9)(a+6)}$ **16.** 1 **17.** $2x^2y\sqrt{30y}$ **18.** $\frac{3a\sqrt{10}}{13}$ **19.** 0 **20.** $-\frac{5}{2}$ **21.** $-\frac{2}{3}$, 0, 4
22. $-7, -5$ **23.** 21 **24.** No solution **25.** $(2, -1)$ **26.** $(-3, 2)$ **27.** $(-2, -7)$ **28.** $(1, -4)$

29. 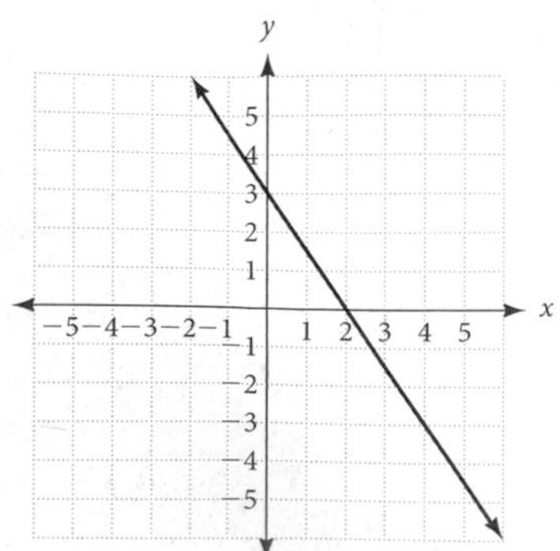

30. 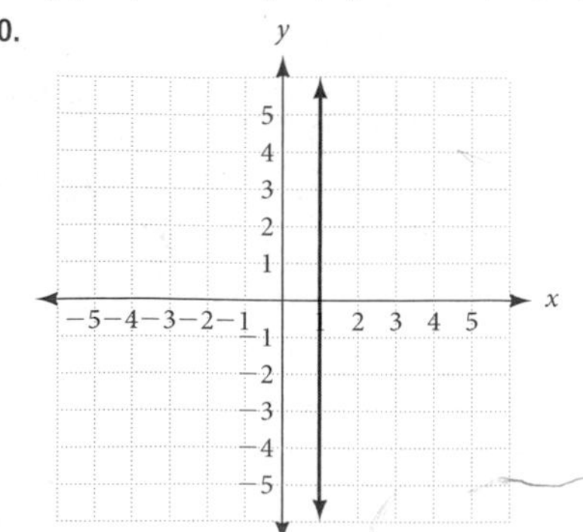

31. 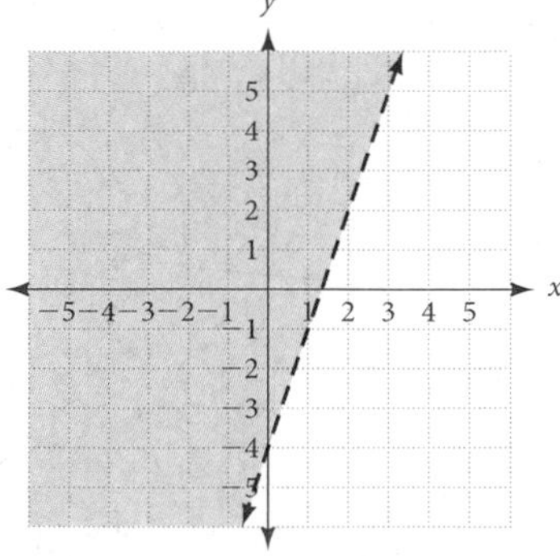

32.

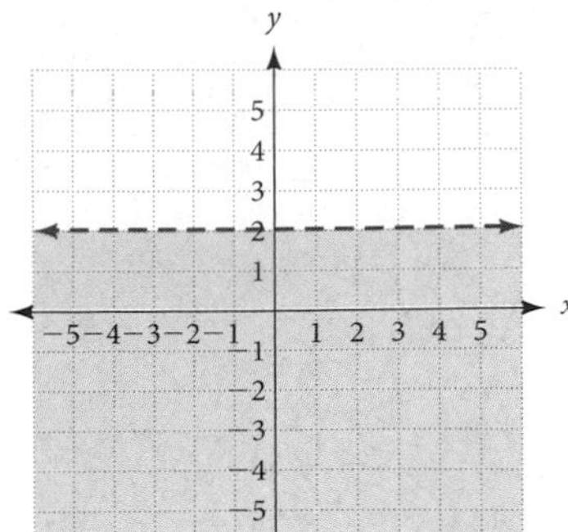

33. x-intercept = (7, 0); y-intercept = (0, 7) **34.** $m = -3$; y-intercept = (0, 2)

35. $y = -x$ **36.** −11 **37.** $(r + 5)(r - 4)$ **38.** $(y - 3)(y + 3)(y^2 + 9)$ **39.** $x^3(x + 5)(x - 6)$ **40.** $5x(4x - y)(4x + y)$
41. 1.05×10^9 **42.** 4×10^6 **43.** $\frac{5\sqrt{3}}{3}$ **44.** $\frac{5 + \sqrt{21}}{2}$ **45.** 5 **46.** $l = \frac{P - 2w}{2}$ **47.** $\frac{x(x - 7)}{(x + 2)(x + 3)}$ **48.** $\frac{2x + 1}{x + 3}$
49. width = 8 ft; length = 16 ft **50.** 10 and 25 **51.** \$52,590 more **52.** 2.68

Chapter 8 Test

1. 4 **2.** −6 **3.** 7 **4.** 3 **5.** −2 **6.** −3 **7.** $5\sqrt{3}$ **8.** $4\sqrt{2}$ **9.** $\frac{\sqrt{6}}{3}$ **10.** $\frac{\sqrt[3]{2}}{2}$ **11.** $15x\sqrt{2}$ **12.** $\frac{2xy\sqrt{15y}}{5}$
13. $4\sqrt{3}$ **14.** $11x\sqrt{2}$ **15.** $\sqrt{15} - 2\sqrt{3}$ **16.** $-51 - \sqrt{5}$ **17.** $x - 36$ **18.** $8 - 2\sqrt{15}$ **19.** $\frac{5 - \sqrt{21}}{2}$
20. $\frac{x - 5\sqrt{x}}{x - 25}$ **21.** 12 **22.** No solution **23.** 6 **24.** 16 **25.** $\sqrt{6}$

26. D

x	y
0	−1
1	0
4	1
9	2

27. B

x	y
−1	0
0	1
3	2
8	3

28. A

x	y
−8	−3
−1	−2
0	−1
1	0
8	1

29. C

x	y
−9	−2
−2	−1
−1	0
0	1
7	2

Chapter 9

Pretest for Chapter 9

1. −1, 9 **2.** $2 \pm 3\sqrt{2}$ **3.** $\frac{7}{3} \pm \frac{5i\sqrt{2}}{3}$ **4.** $\frac{3}{2}$, 1 **5.** $\frac{1}{4}$, −2 **6.** −2, 5 **7.** $\frac{5}{2}$ **8.** $2 \pm \sqrt{13}$ **9.** $6i$ **10.** $13i$ **11.** $3i\sqrt{5}$
12. $2i\sqrt{3}$ **13.** $7 + 10i$ **14.** $3 + 4i$ **15.** 26 **16.** $10 + 10i$ **17.** $\frac{1 + 3i}{10}$ **18.** $\frac{15 - 8i}{17}$

19.

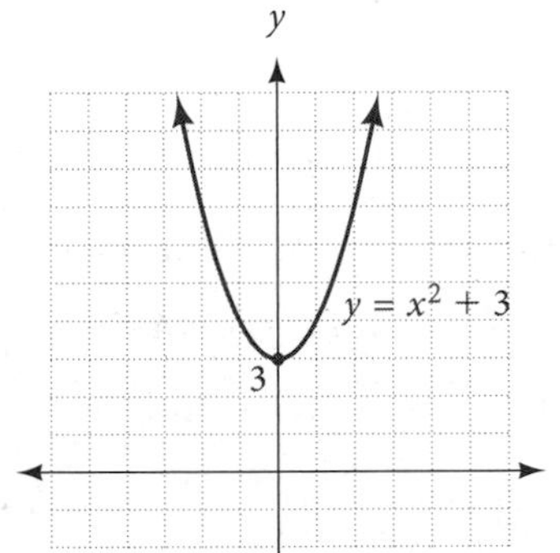

20.

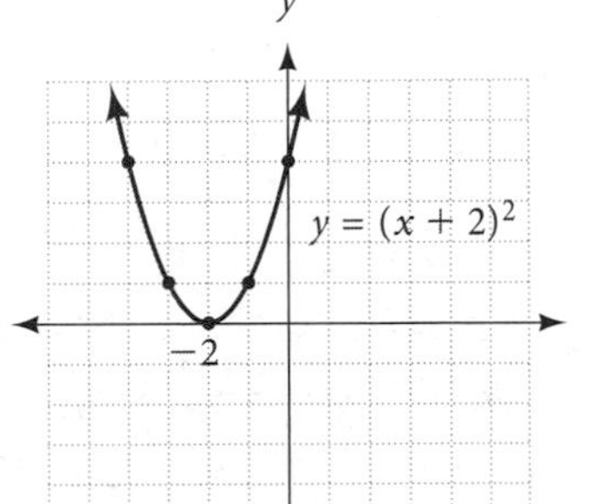

21.

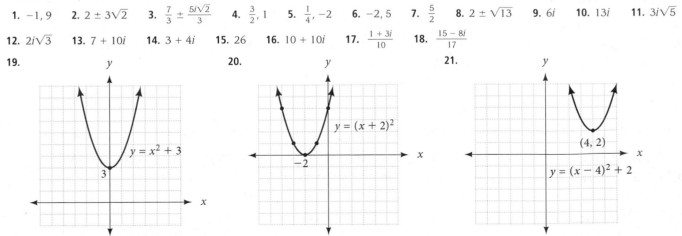

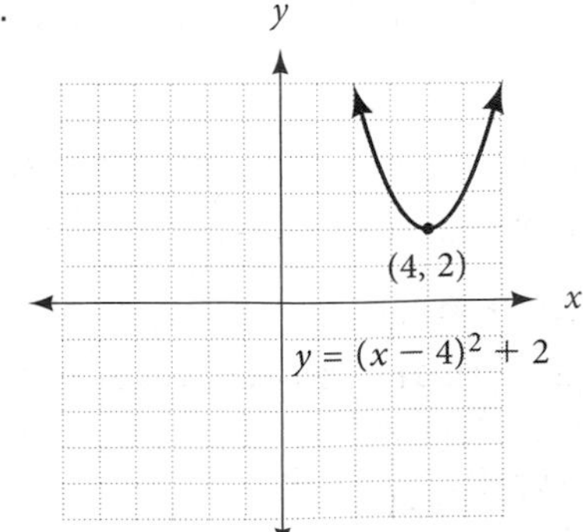

22.

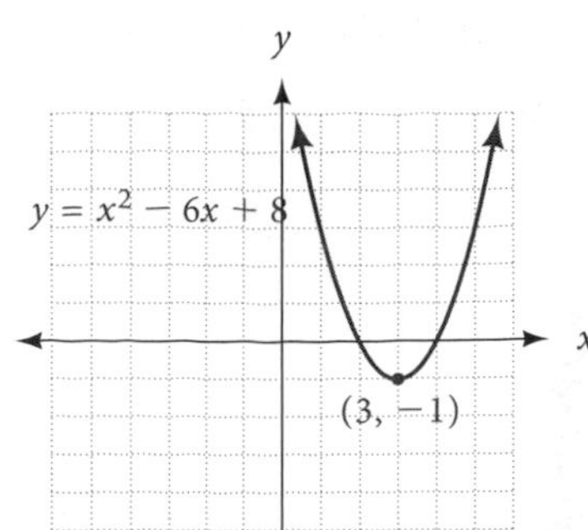

23. 3 inches

Getting Ready for Chapter 9

1. 25 **2.** $\frac{25}{4}$ **3.** -1 **4.** $\frac{-2+\sqrt{10}}{2}$ **5.** $5+2\sqrt{2}$ **6.** $4-13\sqrt{2}$ **7.** $\frac{1}{5}$ **8.** $\frac{x+3}{2(x+1)}$ **9.** 7 **10.** $2\sqrt{3}$ **11.** $(y-6)^2$ **12.** $(x+6)^2$ **13.** $(5x-3)(2x+5)$ **14.** $(2x-1)(3x+2)$ **15.** $\left(y+\frac{3}{2}\right)^2$ **16.** $\left(y-\frac{5}{2}\right)^2$ **17.** 6 **18.** 11 **19.** 18 **20.** 3

Problem Set 9.1

1. ± 3 **3.** ± 5 **5.** $\pm 2\sqrt{2}$ **7.** $\pm 5\sqrt{2}$ **9.** $\pm 3\sqrt{2}$ **11.** $0, -4$ **13.** $4, -6$ **15.** $5 \pm 5\sqrt{3}$ **17.** $-1 \pm 5\sqrt{2}$ **19.** $2, -3$ **21.** $\frac{11}{4}, -\frac{1}{4}$ **23.** $\frac{1 \pm 2\sqrt{3}}{3}$ **25.** $\frac{-2 \pm 3\sqrt{3}}{6}$ **27.** $3 \pm \sqrt{3}$ **29.** $-2 \pm \sqrt{5}$ **31.** $2 \pm \sqrt{2}$ **33.** $\frac{7}{3}, -1$ **35.** $\frac{-1 \pm \sqrt{7}}{2}$ **37.** $\frac{4 \pm 2\sqrt{3}}{5}$ **39.** $-5 \pm \sqrt{7}$ **41.** $4, -2$ **43.** $-6 \pm 2\sqrt{2}$ **45. a.** No **b.** $\pm\sqrt{3}$ **47. a.** Yes **b.** 1, 5 **49.** Yes **51. a.** $\frac{1}{2}$ **b.** $\frac{5}{2}$ **c.** $-\frac{1}{2}, \frac{3}{2}$ **d.** $\frac{1}{2}$ **e.** $\frac{2}{5}$ **53.** $(-1+5\sqrt{2}+1)^2 = (5\sqrt{2})^2 = 50$ **55. a.** 10 **b.** 22 **57.** 6 feet **59.** $\frac{5}{2}$ sec **61.** $-7, 1$ **63.** $r = \frac{-1+\sqrt{A}}{10}$ **65.** $5\sqrt{3}$ feet

67. No, tent height is $3\sqrt{3}$ feet ≈ 5.20 feet, which is less than 5 feet 8 inches $\left(5\frac{8}{12} \text{ feet} \approx 5.67 \text{ feet}\right)$. **69.** 3 feet **71.** 3.4 inches **73.** 81 **75.** 1 **77.** $\frac{9}{4}$ **79.** x^2+8 **81.** $(x+3)^2$ **83.** $\left(y-\frac{3}{2}\right)^2$ **85.** $x^2-10x+25$ **87.** $(x-6)^2$ **89.** $(x+2)^2$ **91.** 2 **93.** 2 **95.** 1.14 strikeouts per inning

Problem Set 9.2

1. 9 **3.** 1 **5.** 16 **7.** 1 **9.** 64 **11.** $\frac{9}{4}$ **13.** $\frac{49}{4}$ **15.** $\frac{1}{4}$ **17.** $\frac{9}{16}$ **19.** $-6, 2$ **21.** $-2, 8$ **23.** $-3, 1$ **25.** 0, 10 **27.** $-5, 3$ **29.** $-2 \pm \sqrt{7}$ **31.** $2 \pm 2\sqrt{2}$ **33.** $8, -1$ **35.** $-1 \pm \sqrt{2}$ **37.** $-2, 1$ **39.** $\frac{-2 \pm \sqrt{3}}{2}$ **41.** $\frac{1 \pm \sqrt{3}}{2}$ **43.** $\frac{1}{2}$ **45.** $\frac{9 \pm \sqrt{105}}{6}$ **47. a.** 0, 2 **b.** 0, 2 **49.** No **51.** Yes, by factoring **53.** Yes, by factoring **55.** Length = 11.2 inches, width = 8.4 inches **57.** $2x^2+4x-3=0$ **59.** $x^2+x-11=0$ **61.** $1, -5, -6$ **63.** $2, 4, -3$ **65.** 49 **67.** 40 **69.** 6 **71.** $\frac{-2-\sqrt{10}}{2}$ **73.** 4 **75.** -24 **77.** $2\sqrt{10}$ **79.** $2\sqrt{3}$ **81.** $2xy\sqrt{5y}$ **83.** $\frac{9}{5}$ **85.** 1,756,000 mothers

Problem Set 9.3

1. $-1, -2$ **3.** $-2, -3$ **5.** -3 **7.** $-3 \pm \sqrt{2}$ **9.** $-1, -\frac{3}{2}$ **11.** $\frac{-2 \pm \sqrt{3}}{2}$ **13.** 1 **15.** $\frac{5 \pm \sqrt{53}}{2}$ **17.** $\frac{2}{3}, -\frac{1}{2}$ **19.** $\frac{1 \pm \sqrt{21}}{2}$ **21.** $\frac{-1 \pm \sqrt{57}}{4}$ **23.** $\frac{5}{2}, -1$ **25.** $\frac{-3 \pm \sqrt{23}}{2}$ **27.** $\frac{-2 \pm \sqrt{10}}{3}$ **29.** $\frac{1 \pm \sqrt{11}}{2}$ **31.** $0, \frac{-3 \pm \sqrt{41}}{4}$ **33.** $0, \frac{4}{3}$ **35.** $\frac{5\sqrt{58}}{4}$ seconds **37.** 1 second and 3 seconds **39.** $5-2\sqrt{2}$ **41.** $1-13\sqrt{2}$ **43.** $12\sqrt{2}+40$ **45.** $-\sqrt{2}$ **47.** $\frac{3-\sqrt{5}}{2}$ **49.** $7+4\sqrt{3}$ **51.** $\frac{6}{x-2}$ **53.** $\frac{2x+3}{(x+3)(x-3)}$ **55.** $\frac{3x}{(x+1)(x-2)}$ **57.** $\frac{3 \pm \sqrt{21}}{6}$

Problem Set 9.4

1. $3+i$ **3.** $6-8i$ **5.** 11 **7.** $9+i$ **9.** $-1-i$ **11.** $11-2i$ **13.** $4-2i$ **15.** $8-7i$ **17.** $-1+4i$ **19.** $6-3i$ **21.** $14+16i$ **23.** $9+2i$ **25.** $11-7i$ **27.** 34 **29.** 5 **31.** $\frac{6+4i}{13}$ **33.** $\frac{-9-6i}{13}$ **35.** $\frac{-3+9i}{5}$ **37.** $\frac{3+4i}{5}$ **39.** $\frac{-6+13i}{15}$ **41.** 4° C **43.** 6 **45.** $5\sqrt{3}$ **47.** $-5, 1$ **49.** $1 \pm \sqrt{6}$ **51.** $\frac{2 \pm \sqrt{5}}{2}$ **53.** $8, -2$ **55.** 5, 1 **57.** $-3 \pm 2\sqrt{3}$ **59.** $\frac{\sqrt{2}}{2}$ **61.** $\frac{2xy\sqrt{6y}}{3}$ **63.** $\frac{\sqrt[3]{2}}{2}$ **65.** x^2+9 **67.** $\frac{1}{i} \cdot \frac{i}{i} = \frac{i}{i^2} = \frac{i}{-1} = -i$

Problem Set 9.5

1. $4i$ **3.** $7i$ **5.** $i\sqrt{6}$ **7.** $i\sqrt{11}$ **9.** $4i\sqrt{2}$ **11.** $5i\sqrt{2}$ **13.** $2i\sqrt{2}$ **15.** $4i\sqrt{3}$ **17.** $1 \pm i$ **19.** 2 **21.** $\frac{3}{2}, -4$

23. $2 \pm 2i$ **25.** $\frac{-1 \pm 3i}{2}$ **27.** $\frac{1 \pm i\sqrt{3}}{2}$ **29.** $\frac{-1 \pm i\sqrt{3}}{2}$ **31.** 2, 3 **33.** $\frac{-1 \pm i\sqrt{2}}{3}$ **35.** $\frac{1}{2}, -2$ **37.** $\frac{1 \pm 3\sqrt{5}}{2}$ **39.** $4 \pm 3i$

41. $\frac{4 \pm \sqrt{22}}{2}$ **43.** $2 \pm i$ **45.** yes **47.** $3 - 7i$ **49.** 526.4 square miles **51.** 6 **53.** -2 **55.** 1 **57.** $23\sqrt{2}$

59. $-6\sqrt{6}$ **61.** $2x\sqrt{3}$

Problem Set 9.6

1.

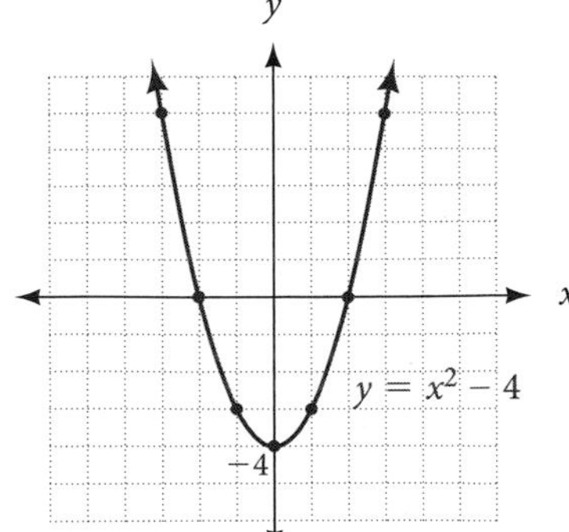

3.

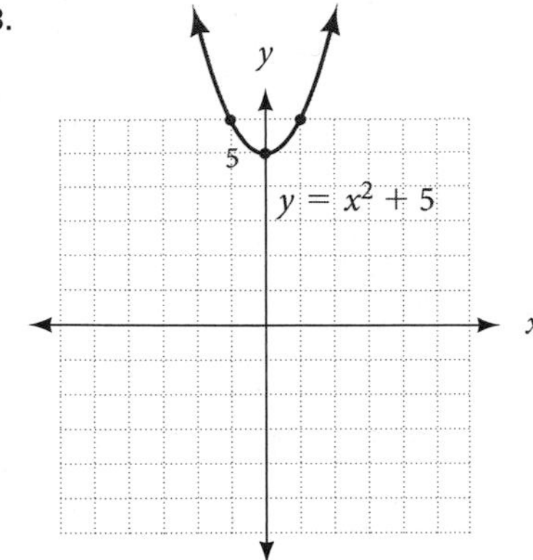

5.

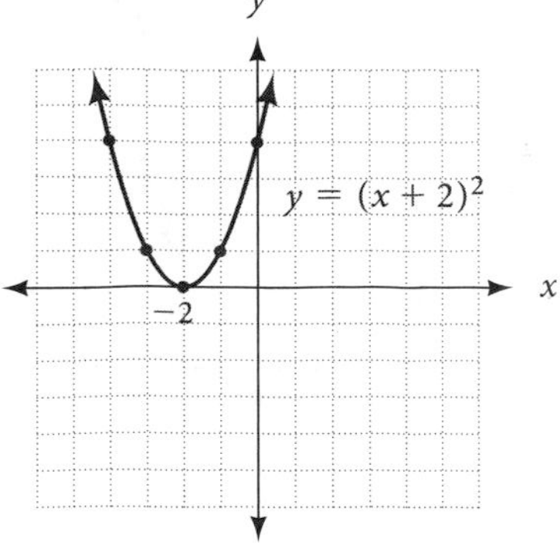

7.

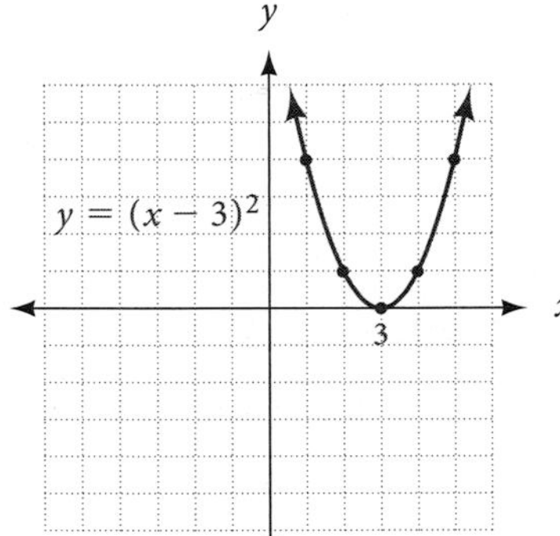

9.

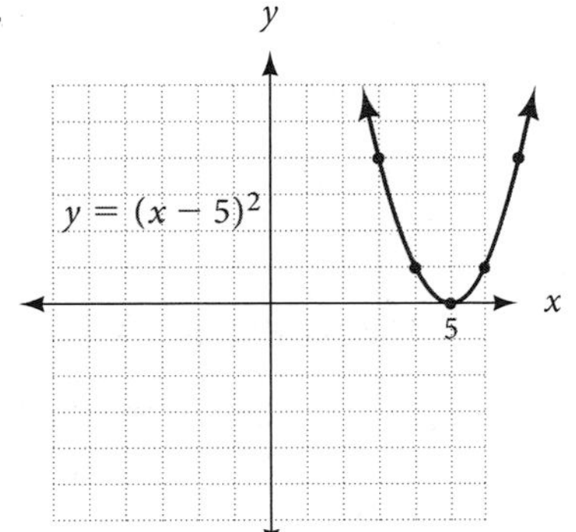

11.

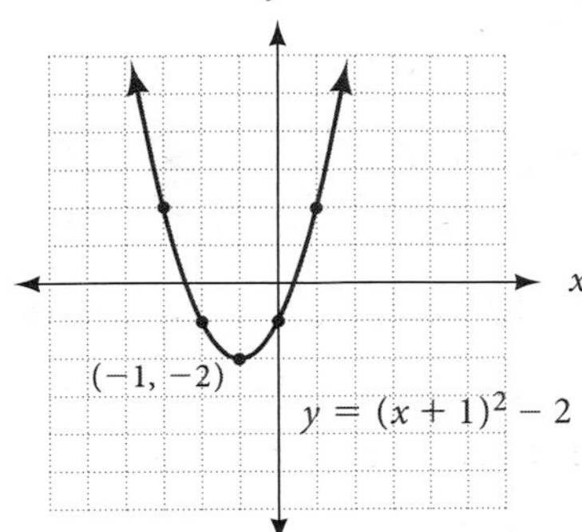

13.

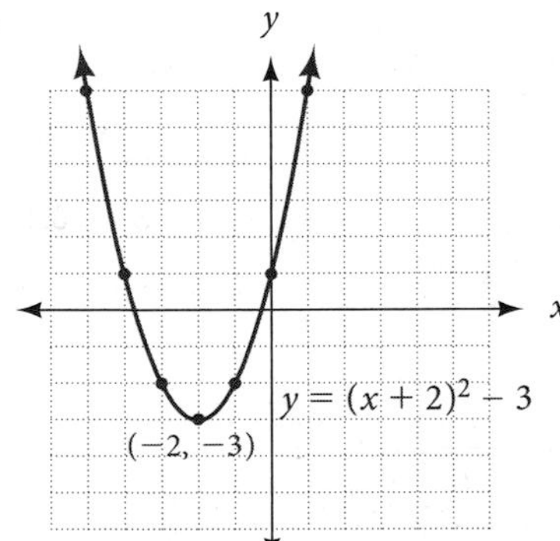

15.

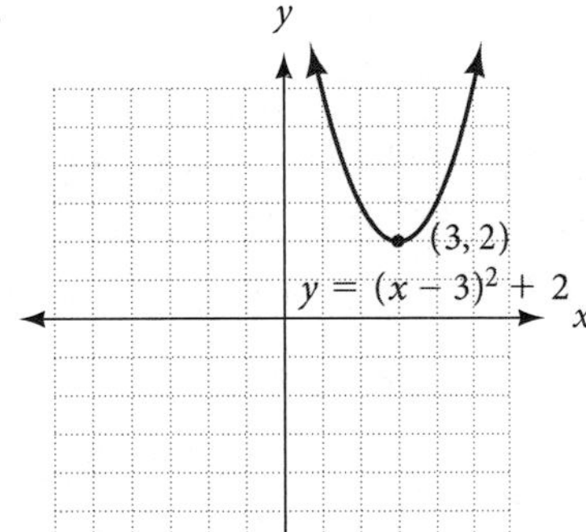

17.

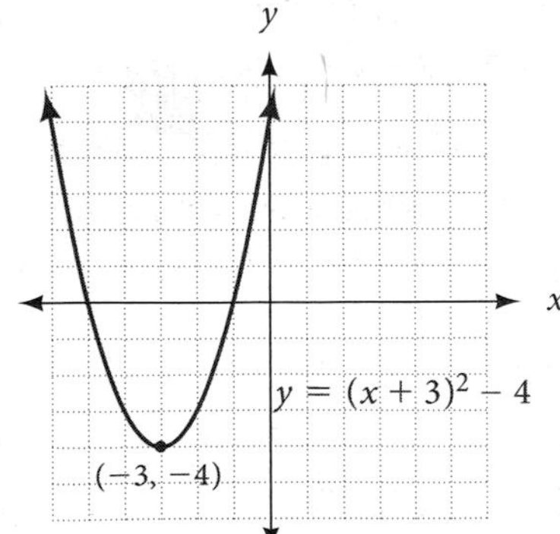

19.

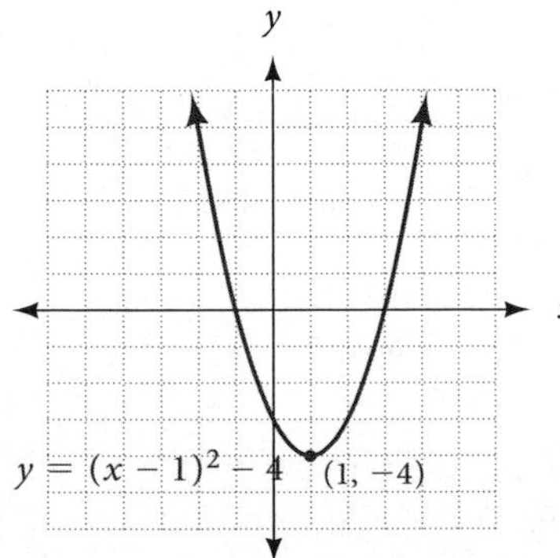

21.

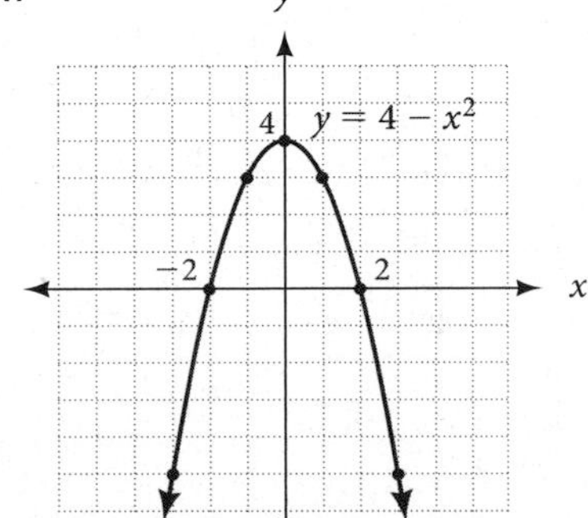

23.

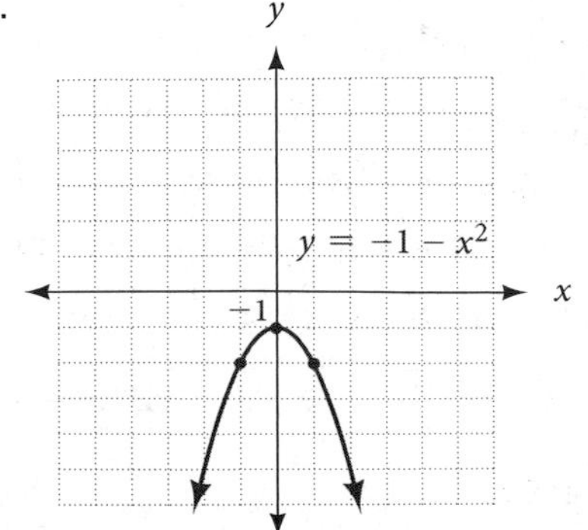

25.

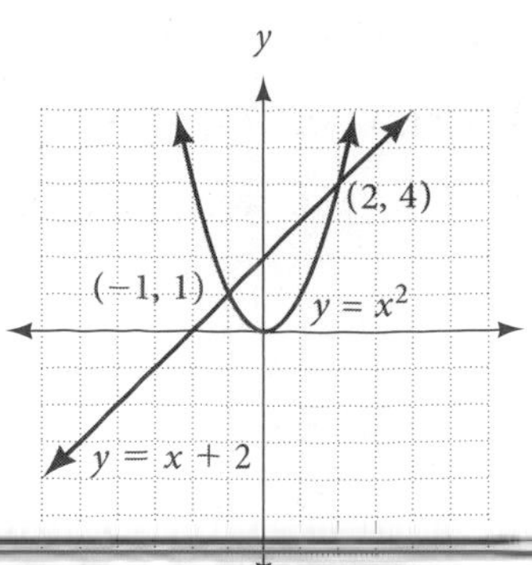

27.

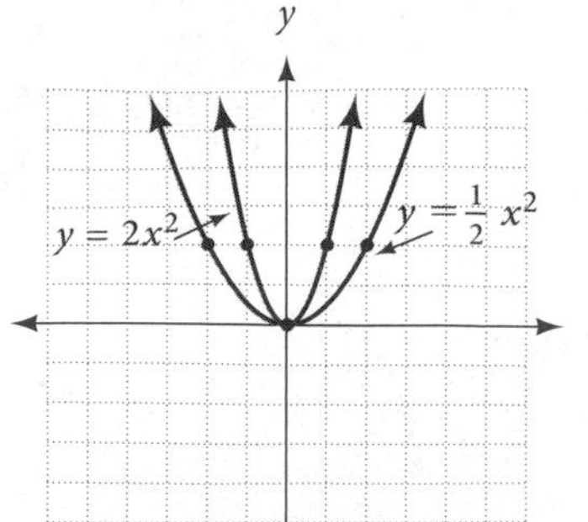

29. a.

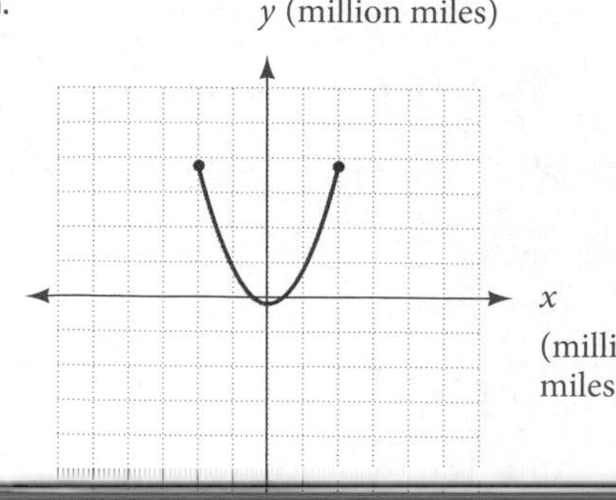

b. 250,000 miles **c.** 500,000 miles

31. 7 **33.** $5\sqrt{2}$ **35.** $\frac{\sqrt{10}}{5}$ **37.** $21\sqrt{3}$ **39.** $-4 - 3\sqrt{6}$ **41.** $4\sqrt{5} + 4\sqrt{3}$ **43.** 7

Chapter 9 Review

1. $\pm 4\sqrt{2}$ **2.** $\pm 2\sqrt{15}$ **3.** ± 4 **4.** $-9, 3$ **5.** $-7, 11$ **6.** $-2, \frac{2}{3}$ **7.** $\frac{-5 \pm 4\sqrt{2}}{2}$ **8.** $\frac{4 \pm 3\sqrt{3}}{3}$ **9.** $\frac{2}{3} \pm \frac{5}{3}i$
10. $-4 \pm 2\sqrt{5}$ **11.** $4 \pm 2\sqrt{5}$ **12.** $2 \pm \sqrt{11}$ **13.** $-3, -1$ **14.** $\frac{9 \pm \sqrt{93}}{2}$ **15.** $-1, 6$ **16.** $-3, 1$ **17.** $\frac{3 \pm \sqrt{15}}{3}$
18. $-4, -3$ **19.** 4 **20.** $\frac{-5 \pm i\sqrt{3}}{2}$ **21.** $\frac{-4 \pm \sqrt{26}}{2}$ **22.** $\frac{-3 \pm i\sqrt{39}}{6}$ **23.** $-\frac{1}{2}, 3$ **24.** $\frac{1 \pm i\sqrt{2}}{2}$ **25.** $4 + 2i$ **26.** $2 - 8i$
27. $10 + 5i$ **28.** $5 - 2i$ **29.** $-i$ **30.** $-1 - 5i$ **31.** $-1 - 3i$ **32.** $2 - 2i$ **33.** $6 - 2i$ **34.** $-24 - 12i$ **35.** $20 + 24i$
36. $7 - i$ **37.** $19 - 17i$ **38.** $8 - 6i$ **39.** 17 **40.** 13 **41.** $\frac{1 + 3i}{10}$ **42.** $\frac{-1 + 2i}{5}$ **43.** $\frac{10 - 25i}{29}$ **44.** $\frac{-2 + 8i}{17}$ **45.** $\frac{6 - 9i}{13}$
46. $\frac{4 + 3i}{5}$ **47.** $\frac{-9 - 40i}{41}$ **48.** $\frac{-9 + 19i}{34}$ **49.** $6i$ **50.** $12i$ **51.** $i\sqrt{17}$ **52.** $i\sqrt{31}$ **53.** $2i\sqrt{10}$ **54.** $6i\sqrt{2}$ **55.** $10i\sqrt{2}$
56. $11i\sqrt{2}$

57.

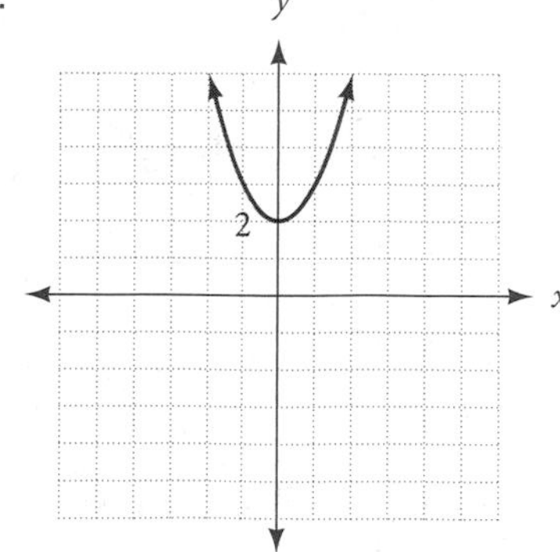

58.

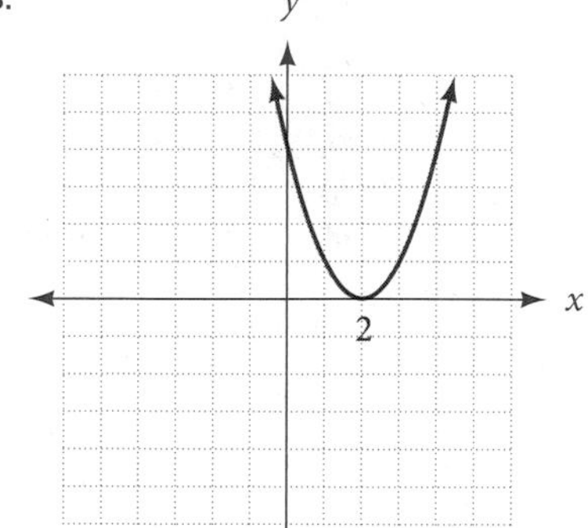

59.

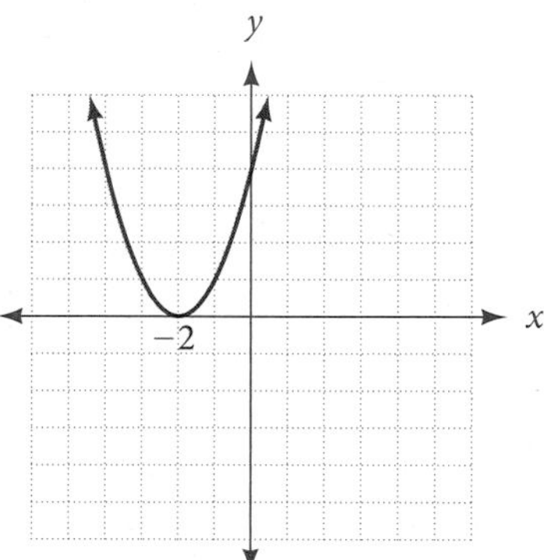

60.

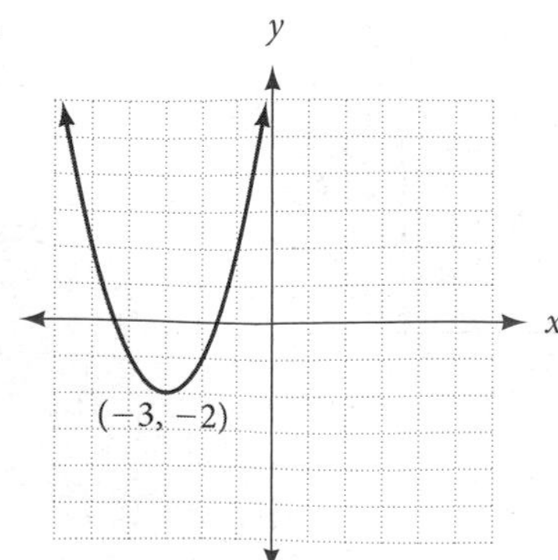

61.

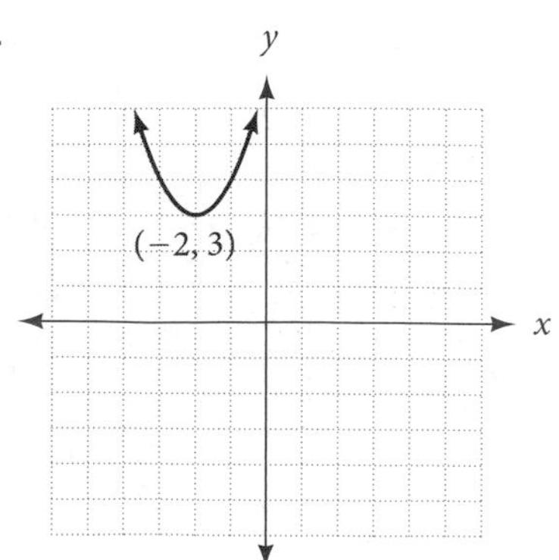

62.

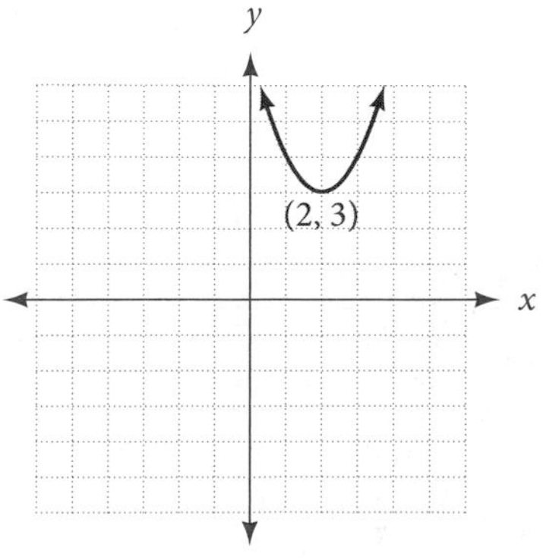

Chapter 9 Cumulative Review

1. 4 **2.** $-\frac{3}{2}$ **3.** x^{12} **4.** $\frac{4}{x^2}$ **5.** $\frac{x}{3y^4}$ **6.** $\frac{y-4}{y-8}$ **7.** 9 **8.** -3 **9.** $5\sqrt{2}$ **10.** $\frac{5\sqrt{7}-5\sqrt{3}}{4}$ **11.** $1 - 6i$ **12.** $1 + 3i$

13. $a^3 - 8a^2 + 19a - 14$ **14.** $x - 3$ **15.** $11x^2y$ **16.** $\frac{10b\sqrt{2}}{9}$ **17.** $\frac{19}{12}$ **18.** 2 **19.** -3 **20.** $-\frac{5}{3}, 4$ **21.** $-3, 0$ **22.** $-2, 5$

23. -7 **24.** $-3, 6$ **25.** (number line) -2 4 **26.** (number line) -5 2

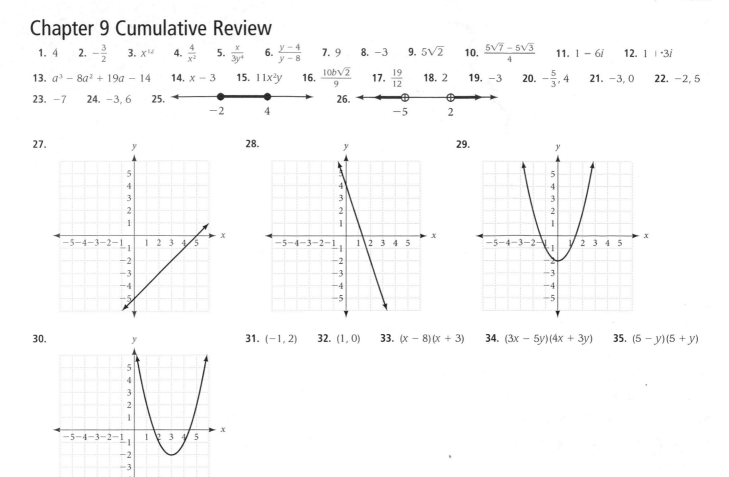

27.

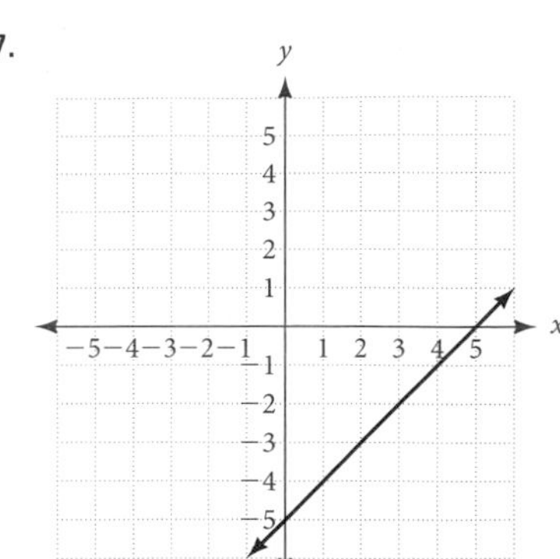

28.

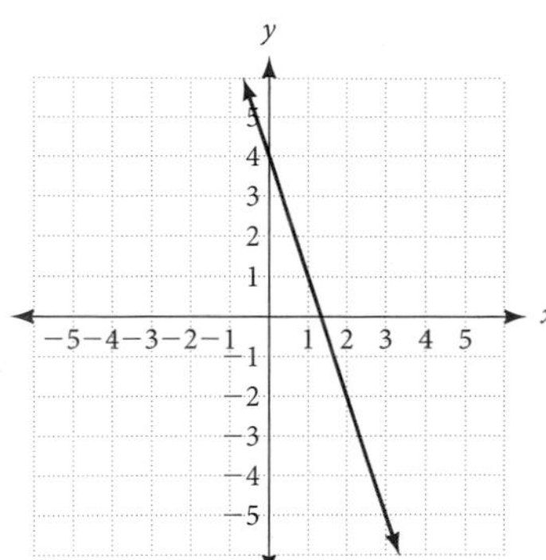

29.

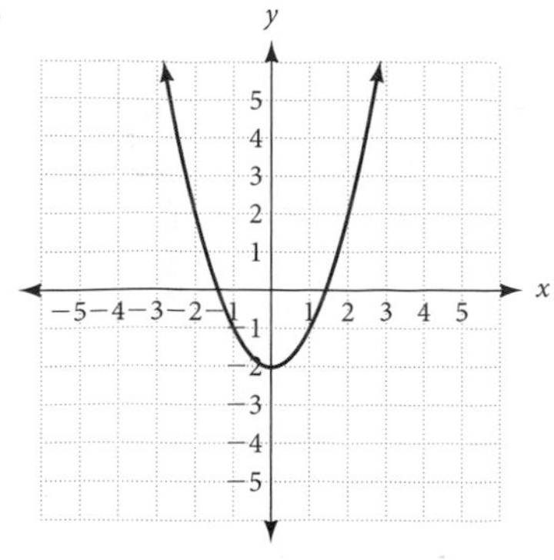

30.

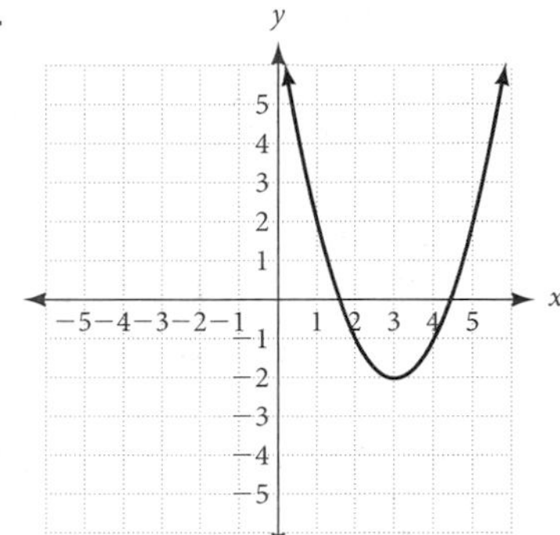

31. $(-1, 2)$ **32.** $(1, 0)$ **33.** $(x - 8)(x + 3)$ **34.** $(3x - 5y)(4x + 3y)$ **35.** $(5 - y)(5 + y)$

36. $(5x - 3y)^2$ **37.** $m = -5$ **38.** $m = 2$; y-intercept $= (0, 3)$ **39.** $3a^2 - 2ab - 4b^2$ **40.** $x^2 - 2x + 4 - \frac{6}{3x-1}$ **41.** $-\frac{13}{a-7}$

42. $\frac{1}{(x-2)(x-6)}$ **43.** $95\sqrt{2}$ **44.** $13x\sqrt{7}$ **45.** $1 - 8i$ **46.** $\frac{17-7i}{26}$ **47.** \$5,000 at 5%, \$15,000 at 7%

48. \$7,500 at 6%, \$10,500 at 8% **49.** 15 **50.** 2 **51.** 540 teachers **52.** 680 teachers

Chapter 9 Test

1. $-1, 8$ **2.** $3 \pm 2\sqrt{3}$ **3.** $\frac{5}{2} \pm \frac{5}{2}i\sqrt{3}$ **4.** $\frac{3 \pm i\sqrt{31}}{4}$ **5.** $-1, \frac{1}{3}$ **6.** $\frac{-1 \pm \sqrt{33}}{2}$ **7.** $-\frac{2}{3}$ **8.** $3 \pm \sqrt{15}$ **9.** $3i$ **10.** $11i$

11. $6i\sqrt{2}$ **12.** $3i\sqrt{2}$ **13.** $3 + 8i$ **14.** $-1 + 2i$ **15.** 5 **16.** $1 + 5i$ **17.** $\frac{-1+3i}{10}$ **18.** $\frac{3+4i}{5}$

19.

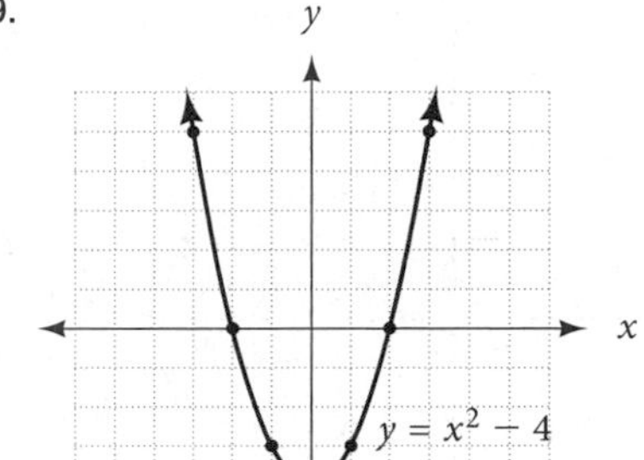

20.

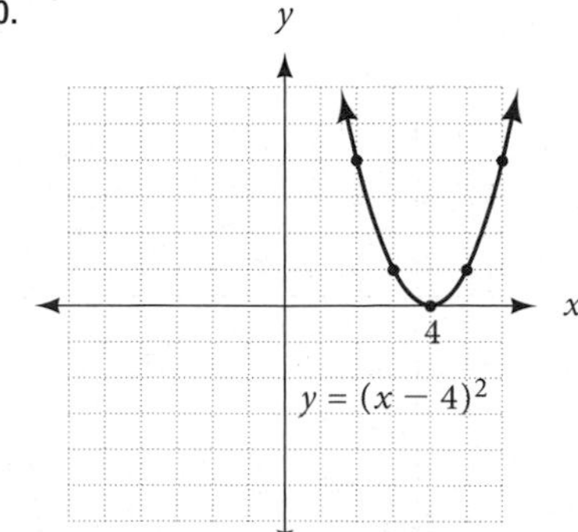

21.

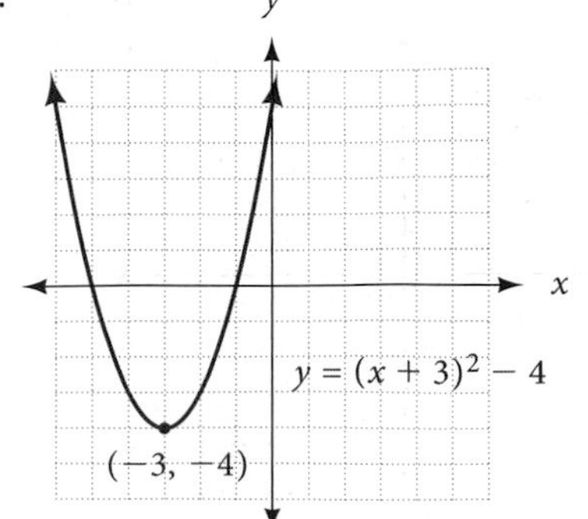

22.

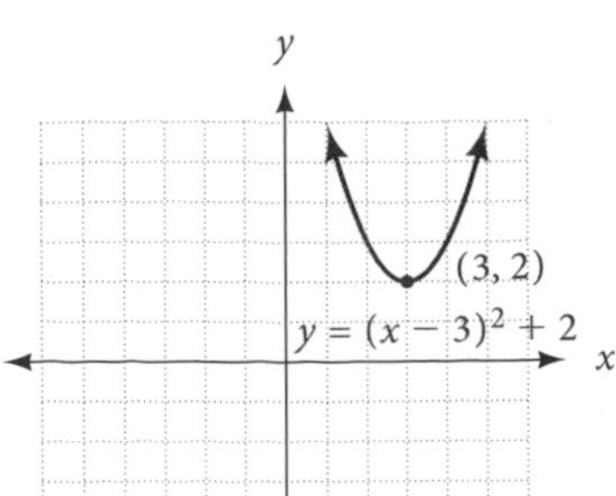

23. 8 cm **24.** C **25.** B **26.** A **27.** D **28.** C **29.** D **30.** A **31.** B

Appendix B

Problem Set A

1. a. $y = 8.5x$ for $10 \le x \le 40$ **b.**

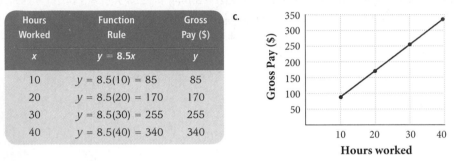

Hours Worked	Function Rule	Gross Pay ($)
x	$y = 8.5x$	y
10	$y = 8.5(10) = 85$	85
20	$y = 8.5(20) = 170$	170
30	$y = 8.5(30) = 255$	255
40	$y = 8.5(40) = 340$	340

c.

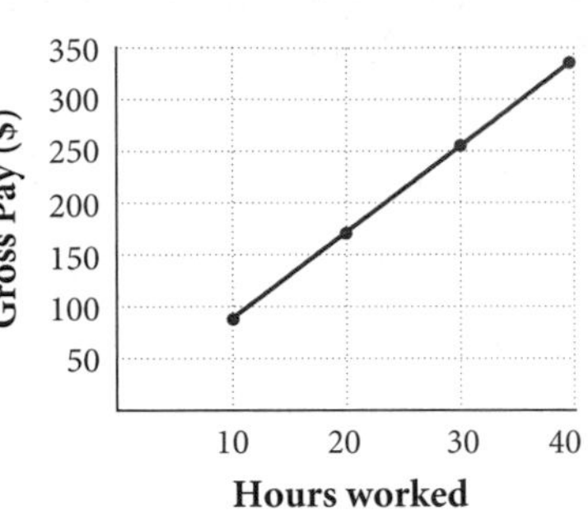

d. Domain = $\{x \mid 10 \le x \le 40\}$; Range = $\{y \mid 85 \le y \le 340\}$ **e.** Minimum = \$85; Maximum = \$340

3. Domain = $\{1, 2, 4\}$; Range = $\{3, 5, 1\}$; a function 5. Domain = $\{-1, 1, 2\}$; Range = $\{3, -5\}$; a function

7. Domain = $\{7, 3\}$; Range = $\{-1, 4\}$; not a function **9.** Yes **11.** No **13.** No **15.** Yes **17.** Yes

19. a.

Time (sec)	Function Rule	Distance (ft)
t	$h = 16t - 16t^2$	h
0	$h = 16(0) - 16(0)^2$	0
0.1	$h = 16(0.1) - 16(0.1)^2$	1.44
0.2	$h = 16(0.2) - 16(0.2)^2$	2.56
0.3	$h = 16(0.3) - 16(0.3)^2$	3.36
0.4	$h = 16(0.4) - 16(0.4)^2$	3.84
0.5	$h = 16(0.5) - 16(0.5)^2$	4
0.6	$h = 16(0.6) - 16(0.6)^2$	3.84
0.7	$h = 16(0.7) - 16(0.7)^2$	3.36
0.8	$h = 16(0.8) - 16(0.8)^2$	2.56
0.9	$h = 16(0.9) - 16(0.9)^2$	1.44
1	$h = 16(1) - 16(1)^2$	0

b. Domain = $\{t \cdot 0 \le t \le 1\}$; Range = $\{h \cdot 0 \le h \le 4\}$

c.

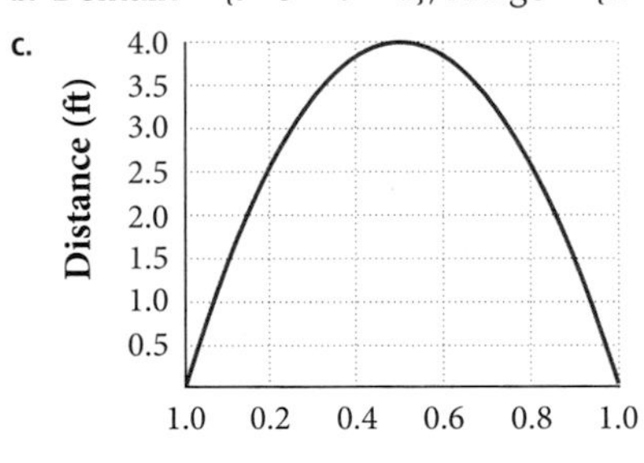

21. Domain = $\{x \mid -5 \le x \le 5\}$; Range = $\{y \mid 0 \le y \le 5\}$ **23.** Domain = $\{x \mid -5 \le x \le 3\}$; Range = $\{y \mid y = 3\}$

Appendix C

Problem Set B

1. −1 **3.** −11 **5.** 2 **7.** 4 **9.** 35 **11.** −13 **13.** 1 **15.** −9 **17.** 8 **19.** 19 **21.** 16 **23.** 0 **25.** $3a^2 - 4a + 1$ **27.** 4 **29.** 0 **31.** 2 **33.** −8 **35.** −1 **37.** $2a^2 - 8$ **39.** $2b^2 - 8$ **41.** 0 **43.** −2 **45.** −3

47.

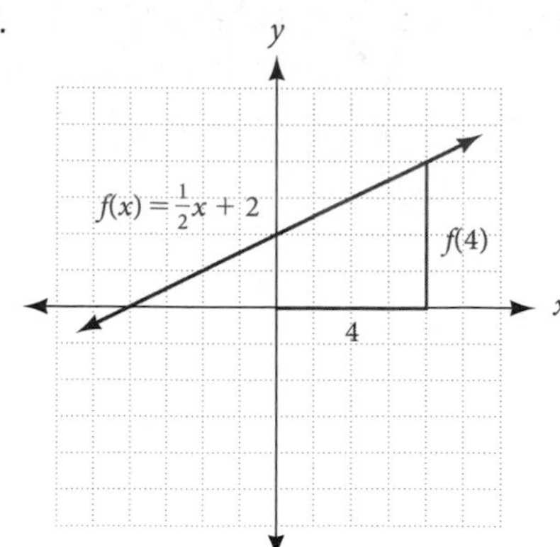

49. $x = 4$ **51.**

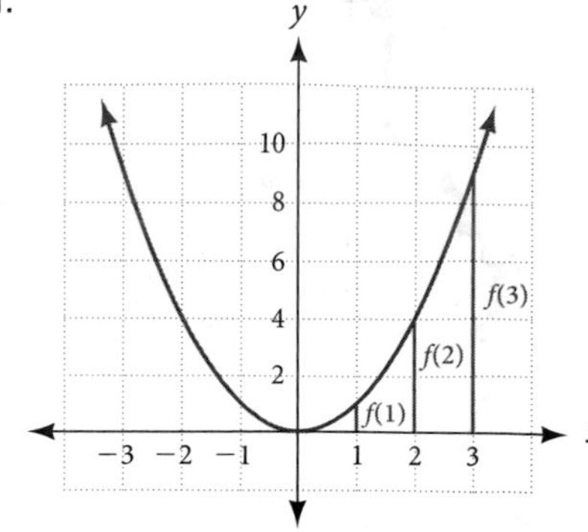

53. $V(3) = 300$, the painting is worth $300 in 3 years; $V(6) = 600$, the painting is worth $600 in 6 years

55. $P(x) = 2x + 2(2x + 3) = 6x + 6$, where $x > 0$ **57.** $A(2) = 3.14(4) = 12.56$; $A(5) = 3.14(25) = 78.5$; $A(10) = 3.14(100) = 314$

59. **a.** $2.49 **b.** $1.53 for a 6-minute call **c.** 5 minutes **61.** **a.** $5,625 **b.** $1,500 **c.** Domain = $\{t \cdot 0 \le t \le 5\}$

d.

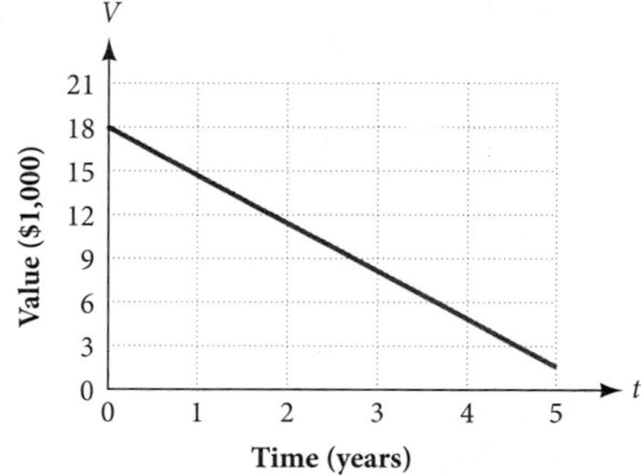

e. Range = $\{V(t) \cdot 1{,}500 \le V(t) \le 18{,}000\}$ **f.** 2.42 years

Index

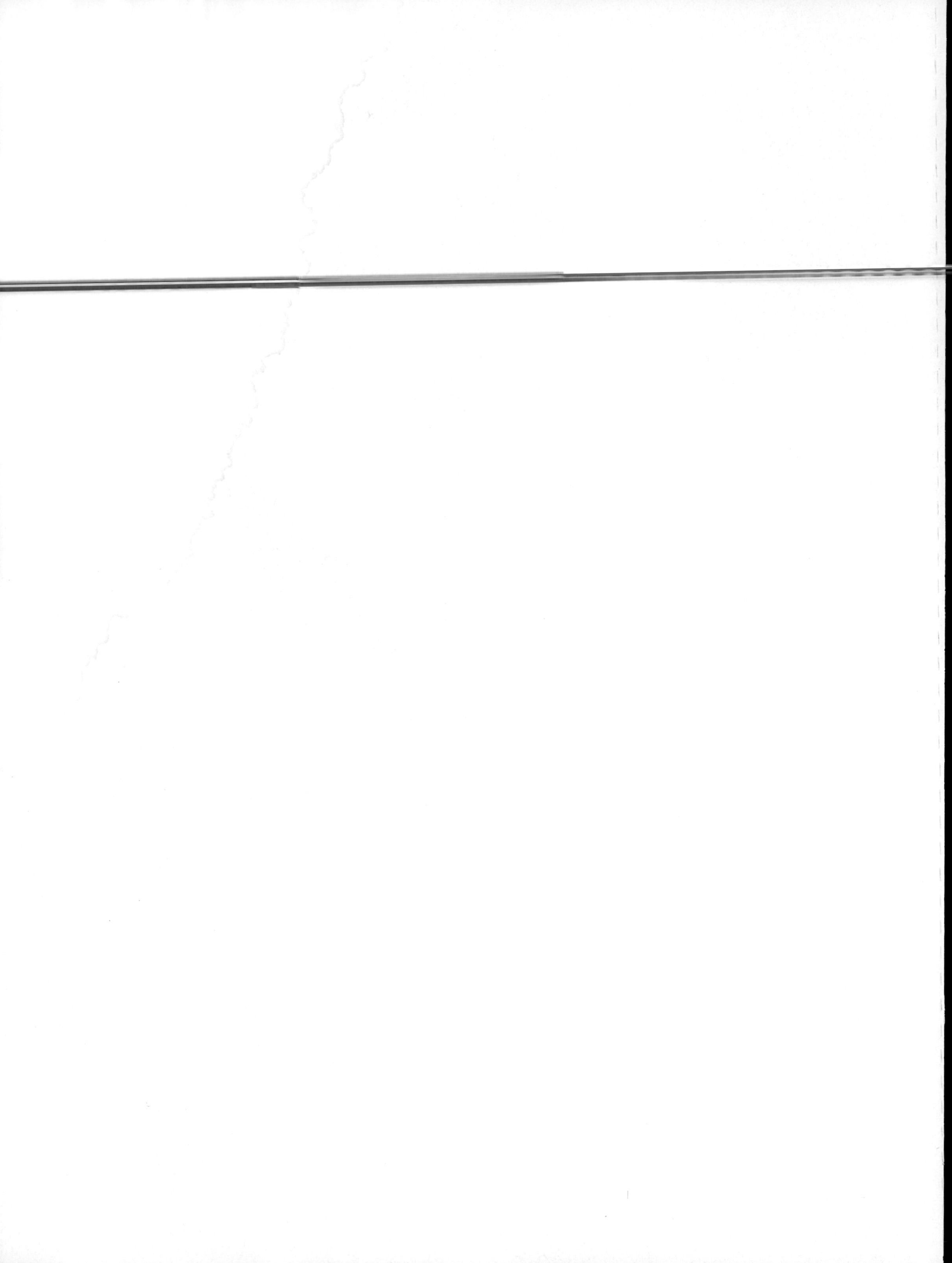